Ebook Set up and Use

Use of the ebook is subject to the single user licence at the back of this book.

Electronic access to your price book is now provided as an ebook on the VitalSource® Bookshelf platform. You can access it online or offline on your PC/Mac, smartphone or tablet. You can browse and search the content across all the books you've got, make notes and highlights and share these notes with other users.

Setting up

1. Create a VitalSource Bookshelf account at https://online.vitalsource.com/user/new (or log into your existing account if you already have one).

2. Retrieve the code by scratching off the security-protected label inside the front cover of this book. Log in to Bookshelf and click the **Redeem** menu at the top right of the screen. and Enter the code in the **Redeem code** box and press **Redeem**. Once the code has been redeemed your Spon's Price Book will download and appear in your **library**. N.B. the code in the scratch-off panel can only be used once, and has to be redeemed before end December 2021.

When you have created a Bookshelf account and redeemed the code you will be able to access the ebook online or offline on your smartphone, tablet or PC/Mac. Your notes and highlights will automatically stay in sync no matter where you make them.

Use ONLINE

1. Log in to your Bookshelf account at https://online.vitalsource.com).
2. Double-click on the title in your **library** to open the ebook.

Use OFFLINE

Download BookShelf to your PC, Mac, iOS device, Android device or Kindle Fire, and log in to your Bookshelf account to access your ebook, as follows:

On your PC.
Go to https://support.vitalsource.com/hc/en-us/ and follow the instructions to download the free VitalSource Bookshelf app to your PC or Mac. Double-click the VitalSource Bookshelf icon that appears on your desktop and log into your Bookshelf account. Select **All Titles** from the menu on the left – you should see your price book on your Bookshelf. If your Price Book does not appear, select **Update Booklist** from the **Account** menu. Double-click the price book to open it.

On your iPhone/iPod Touch/iPad
Download the free VitalSource Bookshelf App available via the iTunes App Store. Open the Bookshelf app and log into your Bookshelf account. Select **All Titles** - you should see your price book on your Bookshelf. Select the price book to open it. You can find more information at https://support.vitalsource.com/hc/en-us/categories/200134217-Bookshelf-for-iOS

On your Android™ smartphone or tablet
Download the free VitalSource Bookshelf App available via Google Play. Open the Bookshelf app and log into your Bookshelf account. You should see your price book on your Bookshelf. Select the price book to open it. You can find more information at https://support.vitalsource.com/hc/en-us/categories/200139976-Bookshelf-for-Android-and-Kindle-Fire

On your Kindle Fire
Download the free VitalSource Bookshelf App from Amazon. Open the Bookshelf app and log into your Bookshelf account. Select All Titles – you should see your price book on your Bookshelf. Select the price book to open it. You can find more information at https://support.vitalsource.com/hc/en-us/categories/200139976-Bookshelf-for-Android-and-Kindle-Fire

Support

If you have any questions about downloading Bookshelf, creating your account, or accessing and using your ebook edition, please visit http://support.vitalsource.com/

CW01467141

995128107 9

Free Updates

with three easy steps…

1. Register today on www.pricebooks.co.uk/updates

2. We'll alert you by email when new updates are posted on our website

3. Then go to www.pricebooks.co.uk/updates
    and download the update.

All four Spon Price Books – *Architects' and Builders'*, *Civil Engineering and Highway Works*, *External Works and Landscape* and *Mechanical and Electrical Services* – are supported by an updating service. Two or three updates are loaded on our website during the year, typically in November, February and May. Each gives details of changes in prices of materials, wage rates and other significant items, with regional price level adjustments for Northern Ireland, Scotland and Wales and regions of England. The updates terminate with the publication of the next annual edition.

As a purchaser of a Spon Price Book you are entitled to this updating service for this 2021 edition – free of charge. Simply register via the website www.pricebooks.co.uk/updates and we will send you an email when each update becomes available.

If you haven't got internet access or if you've some questions about the updates please write to us at Spon Price Book Updates, Spon Press Marketing Department, 3 Park Square, Milton Park, Abingdon, Oxfordshire, OX14 4RN.

Find out more about Spon books
Visit www.pricebooks.co.uk for more details.

Spon's
Mechanical and
Electrical Services
Price Book

2021

Spon's Mechanical and Electrical Services Price Book

Edited by

AECOM

2021

Fifty-second edition

CRC Press
Taylor & Francis Group

First edition 1968
Fifty-second edition published 2021
by CRC Press
2 Park Square, Milton Park, Abingdon, Oxon, OX14 4RN

and by CRC Press
Taylor & Francis, 6000 Broken Sound Parkway, NW, Suite 300, Boca Raton, FL 33487

CRC Press is an imprint of the Taylor & Francis Group, an informa business

British Library Cataloguing in Publication Data
A catalogue record for this book is available from the British Library

ISBN: 978-0-367-51405-1
Ebook: 978-1-003-05370-5
ISSN: 0305-4543

Typeset in Arial by Taylor & Francis Books

Printed and bound by CPI Group (UK) Ltd, Croydon, CR0 4YY

Contents

Preface

The Fifty-Second Edition of *Spon's Mechanical and Electrical Services Price Book* continues to cover the widest range and depth of engineering services, reflecting the many alternative systems and products that are commonly used in the industry as well as current industry trends.

During 2020, MEP output continued to outperform the general construction industry, but there was a further tempering of forecast growth levels within both the London market and the UK in general. Yet again, the inflation drivers experienced in previous years have continued with tender price inflation of 2–4% experienced.

Key influences

- Continued skills shortage
- UK Sterling exchange rate against the Euro and US Dollar
- General uncertainty associated with Brexit
- Commodity price volatility
- Brexit

At the time the book data was being collated the Coronavirus pandemic is resulting in unique and extraordinary impacts on many industry sectors. Expansive and immediate disruption to populations, economies and businesses was quickly introduced by the public health crisis, as governments across the world introduced measures to contain the Coronavirus outbreak. These mitigation measures directly address the public health risks and implications from Coronavirus and COVID–19, but an economic demand and supply shock is also a consequence of the actions.

With all this in mind we believe that UK construction supply chain firms will face an array of risks during 2020 as the fallout from the crisis evolves.

Due to the evolving and still uncertain outlook, for the purposes of forming this baseline scenario on market drivers, the following assumptions have been used:

- The full UK/RoI lockdown ended in May/June but the social distancing measures are still in place.
- Construction sites begin reopening in 2020 Q2, steadily increasing in number over 2020 Q3.
- Coronavirus infection transfer risk is mitigated by ongoing social distancing measures in force across the population. Consequently, no further full lockdowns were implemented in the second half of 2020. However, it should be acknowledged that further lockdowns could take place if the transfer rate of infection begins to increase once again in the second half of 2020
- Social distancing rules are implemented on construction sites where possible, which have negative impacts on site productivity.
- Construction output across 2020 Q2 was severely impacted because of site lockdowns, furloughs and decisions made by businesses to pause their construction activity. In the medium-term (6–12 months) output improves from restarted construction activity but is still materially impacted though to a lesser extent than 2020 Q2.
- Underlying construction demand into the medium-term remains broadly in place, although some projects are deferred or held because of funding challenges or client-side decisions.
- Risks and impacts from a 'no deal' Brexit are excluded
- Risks and impacts of the coronavirus on construction market are excluded

The wage increases for 2021 with both Mechanical and Electrical trades have yet to be agreed, with the Mechanical trade rates being reviewed in October by the HVCA and Electrical trade rates being reviewed by the JIB with a target of concluding in the summer.

Before referring to prices or other information in the book, readers are advised to study the 'Directions' which precede each section of the Materials Costs/Measured Work Prices. As before, no allowance has been made in any of the sections for Value Added Tax.

The order of the book reflects the order of the estimating process, from broad outline costs through to detailed unit rate items.

The approximate estimating section has been revised to provide up to date key data in terms of square metre rates, all-in-rates for key elements and selected specialist activities and elemental analyses on a comprehensive range of building types.

The prime purpose of the Materials Costs/Measured Work Prices part is to provide industry average prices for mechanical and electrical services, giving a reasonably accurate indication of their likely cost. Supplementary information is included which will enable readers to make adjustments to suit their own requirements. It cannot be emphasized too strongly that it is not intended that these prices are used in the preparation of an actual tender without adjustment for the circumstances of the particular project in terms of productivity, locality, project size and current market conditions. Adjustments should be made to standard rates for time, location, local conditions, site constraints and any other factor likely to affect the costs of a specific scheme. Readers are referred to the build up of the gang rates, where allowances are included for supervision, labour related insurances, and where the percentage allowances for overhead, profit and preliminaries are defined.

Readers are reminded of the service available on the Spon's website detailing significant changes to the published information.

www.pricebooks.co.uk/updates

As with previous editions the Editors invite the views of readers, critical or otherwise, which might usefully be considered when preparing future editions of this work.

Whilst every effort is made to ensure the accuracy of the information given in this publication, neither the Editors nor Publishers in any way accept liability for loss of any kind resulting from the use made by any person of such information.

In conclusion, the Editors record their appreciation of the indispensable assistance received from the many individuals and organizations in compiling this book.

AECOM Limited
Aldgate Tower
2 Leman Street
London E1 8FA

Special Acknowledgements

The Editors wish to record their appreciation of the special assistance given by the following organizations in the compilation of this edition.

HOTCHKISS AIR SUPPLY

Hampden Park Industrial Estate
Eastbourne
East Sussex
BN22 9 AX
Tel: 01323 501234
Email: info@Hotchkiss.co.uk
www.Hotchkiss.co.uk

Abbey

23–24 Riverside House
Lower Southend Road
Wickford
Essex
SS11 8BB
Tel: 01268 572116
Fax: 01268 572117
Email: general@abbeythermal.com

T.Clarke
BUILDING SERVICES GROUP

45 Moorfields
London
EC2Y 9AE
Tel: 020 7997 7400
Email: info@tclarke.co.uk
www.tclarke.co.uk

DORNAN

Dornan Engineering Ltd
114a Cromwell Road
Kensington
London
SW7 4ES
Tel: 020 7340 1030
Email: info@dornangroup.com
www.dornan.ie

DESIGN INTENT

Design Intent International Ltd
Unit 4 Eaton Court Road
Colmworth Business Park
Eaton Socon, St Neots
PE19 8ER
Email: lhayes@design-intent-international.com

TRYKA
● ● ● LED.

Tryka LED Ltd
Unit 3 Station Works
Station Road, Shepreth
Hertfordshire
SG8 6PZ
Email: info@tryka.com
Tel: 01763 260666
www.tryka.com

MJL MICHAEL J.LONSDALE
Creating the right environment

Michael J Lonsdale Limited
Unit 1 Langley Quay
Waterside Drive
Langley
Berkshire
SL3 6EY
Tel: 01753 588750
www.michaellonsdale.com

SPON'S 2021 PRICEBOOKS from AECOM

Spon's Architects' and Builders' Price Book 2021

Editor: AECOM

New items this year include:
- a London fringe office cost model
- a Higher Education Refurbishment cost model
- Pecafil® permanent formwork
and an expanded range of cast iron rainwater products

Hbk & VitalSource® ebook 868pp approx.
978-0-367-51402-0 £175
VitalSource® ebook
978-1-003-05367-5 £175
(inc. sales tax where appropriate)

Spon's Civil Engineering and Highway Works Price Book 2021

Editor: AECOM

This year gives more items on shafts, tunnelling, drainage and water proofing – covering some brand new materials and methods. Notes have been added to tunnelling, viaducts, D-walls and piling under the output section. The book partially reflects costs and new ways of working resulting from the Covid-19 pandemic.

Hbk & VitalSource® ebook 704 pp approx.
978-0-367-51403-7 £195
VitalSource® ebook
978-1-003-05368-2 £195
(inc. sales tax where appropriate)

Spon's External Works and Landscape Price Book 2021

Editor: AECOM in association with LandPro Ltd Landscape Surveyors

Now with a revised and updated street furniture section.
Plus several new items:
- Kinley systems: Metal edgings and systems for landscapes and podiums
- New cost evaluations of water features
- Stainless steel landscape channel drainage

Hbk & VitalSource® ebook 680pp
approx. 978-0-367-51404-4 £165
VitalSource® ebook
978-1-003-05369-9 £165
(inc. sales tax where appropriate)

Spon's Mechanical and Electrical Services Price Book 2021

Editor: AECOM

An overhaul of the uninterruptible power supply section, and revised costs for air source heat pumps.
Plus new items, including:
HDPE pipe for above ground drainage systems; fire protection mist systems; and electric vehicle chargers

Hbk & VitalSource® ebook 864pp approx.
978-0-367-51405-1 £175
VitalSource® ebook
978-1-003-05370-5 £175
(inc. sales tax where appropriate)

Receive our VitalSource® ebook free when you order any hard copy Spon 2021 Price Book

Visit www.pricebooks.co.uk

To order:
Tel: 01235 400524
Post: Taylor & Francis Customer Services, Bookpoint Ltd, 200 Milton Park, Abingdon, Oxon, OX14 4SB, UK
Email: book.orders@tandf.co.uk
A complete listing of all our books is on www.routledge.com

CRC Press
Taylor & Francis Group

Acknowledgements

The editors wish to record their appreciation of the assistance given by many individuals and organizations in the compilation of this edition.

Manufacturers, Distributors and Sub-Contractors who have contributed this year include:

Aquatronic Group Management Plc
A G M House
London Rd
Copford
Colchester
Essex CO6 1GT
Tel: 01206 215100
www.agm-plc.co.uk
LTHW Pressurization Units

Aquilar Limited
Unit 30, Lawson Hunt Industrial Park
Broadbridge Heath
Horsham
West Sussex RH12 3JR
Tel: 01403 216100
Email: info@aquilar.co.uk
www.aquilar.co.uk
Leak Detection

Babcock Wanson
7 Elstree Way
Borehamwood
Hertfordshire WD6 1SA
Tel: 0208 9537111
www.babcock-wanson.co.uk
Packaged Steam Generators

Balmoral Tanks
Rathbone Square Office 5B
24 Tanfield Road
Croydon
Surrey CR0 1AL
Tel: 01208 665410
www.balmoral-group.com
Sprinkler Tanks

Baltimore Aircoil Company
c/o Balticare Ltd
Church House
18–20 Church Street
Staines-on-Thames
TW18 4EP
Tel: 01784 460316
Email: info@balticare.co.uk
www.baltimoreaircoil.eu; www.balticare.eu
Heat Rejection

Bodet Ltd
4 Sovereign Park
Cleveland Way
Industrial Estate
Hemel Hempstead
Hertfordshire HP2 7DA
Tel: 01442 418800
www.bodet.co.uk
Clocks

Caice Acoustic Air Movement Ltd.
Riverside House,
3 Winnersh Fields,
Gazelle Close,
Winnersh,
Wokingham,
Berkshire RG41 5QS
Tel: 0118–918–6470
Email: enquiries@caice.co.uk
Silencers & Acoustic Treatment

Dewey Waters Limited
The Heritage Works
Winterstoke Rd
Weston Super Mare
Somerset BS24 9AN
Tel: 01934 421477
www.deweywaters.co.uk
Cold Water Storage Tanks

DMS Flow Measurement and Controls Limited
X-Cel House
Chrysalis Way
Langley Bridge
Eastwood
Nottinghamshire NG16 3RY
Tel: 01773 534555
Email: sales@dmsltd.com
www.dmsltd.com
Chilled Water Plant and Energy Meters

Dormakaba Ltd
Lower Moor Way
Tiverton
Devon EX16 6SS
Tel: 01884 256464
Email: info.gb@dormakaba.com
https: //www.dormakaba.com
Access Control Barriers, Revolving Doors

EA-RS Fire Engineering Limited
4 Swanbridge Industrial Park
Black Croft Road
Witham
Essex, CM8 3YN
Tel: 01376 503680
Email: onesolution@ea-rsgroup.com
www.ea-rsgroup.com
Fire Detection and Alarms

Fire Protection Ltd
Flamebar House
South Road
Templefields
Harlow
Essex CM20 2 AR
Tel: 01279 634 230
Email: info@fireprotection.co.uk
www.fireprotection.co.uk
Fire-Rated Ductwork

Fireworks Fire Protection Ltd
Amber House, Station Road
Attleborough
Norfolk NR17 2 AT
Tel: 01953 469237
Email: enquiries@fireworks-ltd.com
www.fireworks-ltd.com
Watermist System

FläktGroup Limited
Dolphin House
Moreton Business Park
Moreton-on-Lugg
Hereford HR4 8DS
Tel: 0845 608 4446
Email: appliedsystems.uk@flaktgroup.com
www.flaktgroup.com
Fans

Frontline Security Solutions Ltd
Reflex House
The Vale
Chalfont St Peter
Bucks SL9 9RZ
Tel: 01753 482 248
http: //www.fsslimited.com/
Access Control and Security Detection Alarm

Harlequin
Harlequin Manufacturing Ltd,
21 Clarehill Road,
Moira, County Armagh,
Northern Ireland
BT67 0PB
Tel: 028 9261 1077
Email: info@harlequin-mfg.com
www.harlequinplastics.co.uk/
Fuel Oil Storage Tanks

Hoval Limited
Northgate
Newark
Notts NG24 1JN
Tel: 01636 672711
www.hoval.co.uk
Storage Cylinders/Calorifiers/Commercial Oil Boilers/MVHR/HIU

HRS Hevac Ltd
10–12 Caxton Way
Watford Business Park
Watford
Herts WD18 8JY
Tel: 01923 232335
Email: mail@hrshevac.co.uk
www.hrshevac.co.uk
Heat Exchangers

Hudevad
Record Hall Business Centre
Rm 215 Hatton Garden
London, EC1N 7RJ
Tel: 02476 881200
https://hudevad.com/
Heat Emitters & Radiators

Hydrotec (UK) Limited
Hydrotec House
5 Mannor Courtyard
Hughenden Avenue
High Wycombe
Buckinghamshire HP13 5RE
Tel: 01494 796040
www.hydrotec.com
Cleaning and Chemical Treatment

Industrial Engineering Plastics
Passfield Mill Business Park
Passfield
Liphook
Hampshire GU30 7QU
Tel: 0121 771 2828
Email: sales@ieplastics.co.uk
www.iep-ltd.co.uk
Plastic

Kohler Uninterruptible Power Ltd
Woodgate
Bartley Wood Business Park
Hook
Hampshire RG27 9XA
Tel: 01256 386700
Email: uksales.ups@kohler.com
www.kohler-ups.co.uk
UPS

Mitsubishi Electric
Wolvers Home Farm
Ironsbottom
Sidlow
Reigate
Surrey RH2 8QG
Tel: 01737387170
https://les.mitsubishielectric.co.uk/
Local Cooling Units/FCU

Ormandy Rycroft Engineering
Duncombe Road
Bradford
West Yorkshire BD8 9TB
Tel: 01274 490911
www.ormandygroup.com
Calorifiers

Purified Air
Lyon House
Lyon Road
Romford
RM1 2BG
Tel: 0800 018 4000
Email: info@purifiedair.com
www.purifiedair.com
Air Filtration

Rock Clean Energy
Unit 5, Buckholt Business Centre,
Buckholt Drive,
Worcester, WR4 9ND
Tel: 0330 223 4566
Email: enquire@rockcleanenergy.co.uk
www.rockcleanenergy.co.uk
Battery Storage Systems

Rung Heating
Unit 4 A, Henstridge Trading Estate,
Henstridge,
Somerset BA8 0TG
Tel: 01963 364600
Email: sales@rung-heating.co.uk
www.rungheating.co.uk
Flue Systems

Safelincs Ltd
33 West Street, Alford
Lincolnshire LN13 9FX
Tel: 01507 464179
Email: support@safelincs.co.uk
www.safelincs.co.uk
Fire Protection Equipment

Schneider Electric
Cardinal Place
80 Victoria Street
London SW1E 5JL
Tel: 0870 608 8608
Email: gb-customerservices@gb.schneider-electric.com
www.se.com/uk/en
LV Switch & Distribution Boards/Breakers & Fusers

Socomec Limited
Knowl Piece
Wilbury Way
Hitchin
Hertfordshire SG4 0TY
Tel: 01462 440033
www.socomec.com
Automatic Transfer Switches

Swegon Air Management
Stourbridge Rd
Bridgnorth,
Shropshire WV15 5BB
Tel: 01746 761921
Email: sales@swegonair.co.uk
www.swegonair.co.uk
Dampers and Access Hatches

Swegon Limited
The Swegon Pavilion
St Cross Chambers
Upper Marsh Lane
Hoddesdon
Hertfordshire EN11 8LQ
Tel 0800 093 7929
Email: sales@swegon.co.uk
www.swegon.com
Chilled Beams

Transtherm Advanced Cooling Solutions
Banner Park, Wickmans Drive,
Coventry CV4 9XA
Tel: 0247 647 1120
Transtherm.co.uk
Dry Air Coolers

Utile Engineering Company Ltd
New Street,
Irthlingborough
Northamptonshire NN9 5UG
Tel: 01933 650216
Email: sherringf@utileengineering.com
www.utileengineering.com
Gas Boosters

Vent Axia/Volution Ventilation UK
Fleming Way
Crawley
West Sussex RH10 9YX
Tel: 0344 856 0591
Email: sales@vent-axia.com
www.vent-axia.com
Fans

Waterloo Air Products PLC
Mills Road
Aylesford
Kent ME20 7NB
Tel: 01622 711511
Email: sales@waterloo.co.uk
www.waterloo.co.uk
Grilles, Diffusers, Louvres, VAV Boxes

Whitecroft Lighting
Burlington Street
Ashton-under-Lyne
Lancashire OL7 0 AX
Tel: 0161 330 6811
Email: email@whitecroftlight.com
www.whitecroftlighting.com
Emergency Lighting

SPON'S 2021 PRICEBOOKS from AECOM

Spon's Architects' and Builders' Price Book 2021

Editor: AECOM

New items this year include:
- a London fringe office cost model
- a Higher Education Refurbishment cost model
- Pecafil® permanent formwork

and an expanded range of cast iron rainwater products

Hbk & VitalSource® ebook 868pp approx.
978-0-367-51402-0 £175
VitalSource® ebook
978-1-003-05367-5 £175
(inc. sales tax where appropriate)

Spon's Civil Engineering and Highway Works Price Book 2021

Editor: AECOM

This year gives more items on shafts, tunnelling, drainage and water proofing – covering some brand new materials and methods. Notes have been added to tunnelling, viaducts, D-walls and piling under the output section. The book partially reflects costs and new ways of working resulting from the Covid-19 pandemic.

Hbk & VitalSource® ebook 704 pp approx.
978-0-367-51403-7 £195
VitalSource® ebook
978-1-003-05368-2 £195
(inc. sales tax where appropriate)

Spon's External Works and Landscape Price Book 2021

Editor: AECOM in association with LandPro Ltd Landscape Surveyors

Now with a revised and updated street furniture section.
Plus several new items:
- Kinley systems: Metal edgings and systems for landscapes and podiums
- New cost evaluations of water features
- Stainless steel landscape channel drainage

Hbk & VitalSource® ebook 680pp approx. 978-0-367-51404-4 £165
VitalSource® ebook
978-1-003-05369-9 £165
(inc. sales tax where appropriate)

Spon's Mechanical and Electrical Services Price Book 2021

Editor: AECOM

An overhaul of the uninterruptible power supply section, and revised costs for air source heat pumps.
Plus new items, including:
HDPE pipe for above ground drainage systems; fire protection mist systems; and electric vehicle chargers

Hbk & VitalSource® ebook 864pp approx.
978-0-367-51405-1 £175
VitalSource® ebook
978-1-003-05370-5 £175
(inc. sales tax where appropriate)

Receive our VitalSource® ebook free when you order any hard copy Spon 2021 Price Book

Visit www.pricebooks.co.uk

To order:
Tel: 01235 400524
Post: Taylor & Francis Customer Services, Bookpoint Ltd, 200 Milton Park, Abingdon, Oxon, OX14 4SB, UK
Email: book.orders@tandf.co.uk
A complete listing of all our books is on www.routledge.com

CRC Press
Taylor & Francis Group

It is one thing to imagine a better world.
It's another to deliver it.

Designing adaptable workplaces. Putting the employees at the centre of the design. Focusing on sustainability. Our team of circa 900 cost and project managers collaborate to turn imagination into reality, providing innovative solutions to complex challenges across the globe.

Imagine it. Delivered.

BBC's new 15,000m² development in Cardiff's city centre, houses all the technology and manpower needed to enable state-of-the-art Radio and TV broadcasting and production. The new facility supports 800 staff in a mostly open plan area with breakout spaces that sit alongside Radio Studios, TV broadcast areas, Edit Suites and meeting rooms.

The style and room layout give high levels of flexibility and adaptability to meet changing needs over the life of the building.

Services provided include: building services, structural engineering, acoustics, architectural lighting, advanced modelling, BREEAM assessment, advanced security, principal designer and fire engineering.

New Aspects of Quantity Surveying Practice, 4th edition

Duncan Cartlidge

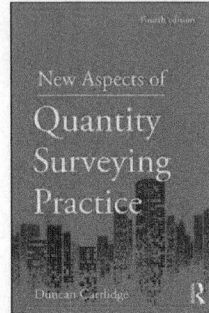

In this fourth edition of New Aspects of Quantity Surveying Practice, renowned quantity surveying author Duncan Cartlidge reviews the history of the quantity surveyor, examines and reflects on the state of current practice with a concentration on new and innovative practice, and attempts to predict the future direction of quantity surveying practice in the UK and worldwide.

The book champions the adaptability and flexibility of the quantity surveyor, whilst covering the hot topics which have emerged since the previous edition's publication, including:

- the RICS 'Futures' publication;
- Building Information Modelling (BIM);
- mergers and acquisitions;
- a more informed and critical evaluation of the NRM;
- greater discussion of ethics to reflect on the renewed industry interest;
- and a new chapter on Dispute Resolution.

As these issues create waves throughout the industry whilst it continues its global growth in emerging markets, such reflections on QS practice are now more important than ever. The book is essential reading for all Quantity Surveying students, teachers and professionals. It is particularly suited to undergraduate professional skills courses and non-cognate postgraduate students looking for an up to date understanding of the industry and the role.

December 2017: 234 x 156 mm: 282 pp
Pb: 978-1-138-67376-2: £50.99

To Order: Tel: +44 (0) 1235 400524 Fax: +44 (0) 1235 400525
or Post: Taylor and Francis Customer Services,
Bookpoint Ltd, Unit T1, 200 Milton Park, Abingdon, Oxon, OX14 4TA UK
Email: book.orders@tandf.co.uk

For a complete listing of all our titles visit:
www.tandf.co.uk

Taylor & Francis
Taylor & Francis Group

PART 1

Engineering Features

This section on Engineering Features, deals with current issues and/or technical advancements within the industry. These shall be complimented by cost models and/or itemized prices for items that form part of such.

The intention is that the book provides more than just a schedule of prices to assist the user in the preparation and evaluation of costs, but can inform the reader on current and future trends in the industry.

Contractual Procedures in the Construction Industry, 7th edition

Allan Ashworth and Srinath Perera

Contractual Procedures in the Construction Industry 7th edition aims to provide students with a comprehensive understanding of the subject, and reinforces the changes that are taking place within the construction industry. The book looks at contract law within the context of construction contracts, it examines the different procurement routes that have evolved over time and the particular aspects relating to design and construction, lean methods of construction and the advantages and disadvantages of PFI/PPP and its variants. It covers the development of partnering, supply chain management, design and build and the way that the clients and professions have adapted to change in the procurement of buildings and engineering projects.

Key features of the new edition include:

*   A revised chapter covering the concept of value for money in line with the greater emphasis on added value throughout the industry today.
*   A new chapter covering developments in information technology applications (building information modelling, blockchains, data analytics, smart contracts and others) and construction procurement.
*   Deeper coverage of the strategies that need to be considered in respect of contract selection.
*   Improved discussion of sustainability and the increasing importance of resilience in the built environment.
*   Concise descriptions of some the more important construction case laws.

March 2018: 246 x 174 mm: 458 pp
Pb: 978-1-138-69393-7: £50.99

To Order: Tel: +44 (0) 1235 400524 Fax: +44 (0) 1235 400525
or Post: Taylor and Francis Customer Services,
Bookpoint Ltd, Unit T1, 200 Milton Park, Abingdon, Oxon, OX14 4TA UK
Email: book.orders@tandf.co.uk

For a complete listing of all our titles visit:
www.tandf.co.uk

Taylor & Francis
Taylor & Francis Group

Battery Storage Technology

Introduction

The advance in battery storage technology and the gradual realization of the role it can play in developing a smarter energy system is becoming a commercial reality. The falling price of lithium-ion batteries has meant that storage has become an increasingly attractive method of reducing energy bills and dependence on the grid. Coupled with a dramatic fall in the price of PV cells, there is a promising business case to be made for the large scale employment of both technologies in both the domestic and commercial market.

In simple terms, batteries can store the energy generated by renewables and use it when required by households and businesses. This provides a reduction in electricity bills as less energy is being pulled from the grid during peak times, known as 'peak shaving'. Furthermore, as 'Time of Use Tariffs', such as Economy 7, become more widely adopted across the UK, battery storage offers consumers the opportunity to buy energy during low cost periods, avoiding the high cost peak periods. When you also consider that the battery can also provide supply security during outages, it proves to be a compelling offer.

Now that we are seeing climate change climbing higher up the political agenda, it is expected that battery storage will become an increasingly popular method for consumers to pre-empt the current market regulation, reduce their carbon footprint and not only reduce their energy bill but actually generate another revenue stream.

The current and future state of the market

Since 2010 there have been over 700,000 domestic PV installations in the UK, resulting in a PV capacity of 9 GW – representing a huge opportunity for battery storage to harness this energy. In the same period of time, lithium-ion battery storage has seen a significant fall in price from £770/kWh to £180/kWh. However, this is plateauing, with Tesla and Panasonic recently revising their prices upwards by 12% for their domestic and small business Powerwall product.

Despite this, when the relative warranties and efficiency degradation rates are taken into account, products like Tesla's still offer a cost effective solution to a consumer with sufficient load demands. While the rest of the domestic battery storage market catches up with the demand created, the market is not necessarily moving in the right direction. If battery storage is to be employed effectively in communities (which contain a mix of residential, retail and commercial space with varying loads) it needs to be done in partnership with the Distribution Network Operator (DNO).

A great example is Project SCENe's Trent Basin development, which is home to Europe's largest community battery storage system, with a capacity of 2.1 MWh. Sized to be future proof, this system aggregates demand and supply, therefore offering a far more efficient use of energy while also generating a revenue stream through a Firm Frequency Response (FFR) contract signed with National Grid.

This concept of community energy storage gets particularly exciting when potentially combined with electric vehicles (EVs). Leading researchers and the industry are now grappling with the challenge of integrating EVs into community battery storage systems with two-way charging points, which would see energy pooled between building demands and the connected vehicles (naturally stationary 95% of the time). Academics have suggested that there are potentially billions of dollars of capital investment to be saved if EVs are used in lieu of stationary storage, the resultant savings could instead accelerate the roll-out of EVs, therefore reducing the demand for the scarce supply of cobalt and other finite resources. Indeed, in the near future we could be seeing EV owners being paid when they connect to the grid and agree to controlled charging.

The near future holds a lot of promise, but what about the current market, and is it commercially viable to invest in battery storage now?

The business case

Electricity consumption can be a significant cost to a commercial or industrial consumer, and battery storage represents an opportunity to not only reduce this, but generate a return on the investment. Judging the business case for investing in battery storage systems requires an understanding of not only the technology and its associated costs but the relevant agreements in place with the grid that ensure revenue streams back to the operator.

Within the initial capital investment that is made into a battery storage system, there is not only the battery itself, but the connection costs associated with the DNO, the necessary inverter, transformer, switchgear and contract formation. While there is the sizeable initial capital expenditure to be made, the operator has to consider the coinciding operating expenditure (OPEX) over the period of planned operation. This includes the annual maintenance necessary to ensure the battery health, the physical maintenance after 10 years onwards, the additional insurance premiums and lastly the largest OPEX – the replacement cost once the lifetime of the battery has expired (every 13 years in this example). These are significant cash outflows – but outweighed when the correct inflows are planned and agreed.

The main inflows due to an operator are the savings generated via arbitrage (peak shaving), and the avoidance of both the Transmission Use of System (TNUoS) and Distribution Use of System (DUoS) charges. The TNUoS charge is for the use of the electricity transmission network and applies solely for business users and generators. This charge is based on the share of demand on the transmission network during the 'triad' periods. The triad period refers to the three half hour periods with highest system demand between November and February, which National Grid uses to determine TNUoS charges to business customers. This higher cost band can be avoided with a battery storage system, and therefore represents a cash inflow to an operator. The DUoS charge is added to all business electricity bills, is set by your DNO, and is time-banded to encourage levelled demand; this charge can be mitigated through use of a battery storage system.

These are not the only inflows due to an operator; battery storage can also be used in conjunction with a Time of Use Tariff such as Economy 7 for arbitrage and load shifting to cheaper periods. A battery storage operator can also draw revenue from the savings made in reducing their Capacity Market Levy charges – a government levy to improve supply security. The final cash inflow an operator will see in the current market conditions is revenue from providing a Firm Frequency Response (FFR) to the grid, once negotiated this can be a good source of revenue, however, this is due to be phased out over the coming years.

The cost data below shows the inflows and outflows, and calculates against an initial investment in three scenarios and the final return is then generated. It does have to be acknowledged that this analysis does not take into account the benefit of a PV generation system paired with the storage system. The storage system is assumed to be a 1.8 MW, with two hour supply time, lithium-ion battery system. The data shows there can be significant returns recognized, even without a PV system, only when all incentives are taken advantage of in the current market conditions. These are by no means precise and final numbers and it should be emphasized that these are benchmarked figures. The data should give readers an idea of what an investment in a battery storage system might look like financially, and clearly show that there is a strong commercial case to be made for battery storage systems.

Cost data: battery storage, three scenarios: TNUoS/DUoS income and FFR and Econ 7	Base position				Base position + five years FFR				Base position + five years FFR + Economy 7			
	Year 2019 1	Year 2020–28 2–10	Year 2029–38 11–20	Year 2039–48 21–30	Year 2019 1	Year 2020–28 2–10	Year 2029–38 11–20	Year 2039–48 21–38	Year 2019 1	Year 2020–28 2–10	Year 2029–38 11–20	Year 2039–48 21–38
Capital expenditure (capex)												
Estimated installation of 1.8 MW system	900,000				900,000				900,000			
DNO network connection costs	45,000				45,000				45,000			
LV inverter/ transformer/ switchgear	60,000				60,000				60,000			
Contract formation	40,000				40,000				40,000			
Operational expenditure (opex)												
Annual maintenance		27,300	44,000	65,200		27,300	44,000	65,200		28,600	46,200	68,300
Replacement cycle			1,320,900	684,900			1,320,900	684,900			1,320,900	684,900
Additional insurance premiums		86,100	110,200	127,900		86,100	110,200	127,900		109,519	110,200	127,900
Physical maintenance		500	5,800	7,400		4,000	6,200	7,900		7,500	10,500	13,500
Net cash outflows	1,045,000	113,900	1,498,900	885,400	1,045,000	117,400	1,481,300	885,900	1,045,000	122,200	1,487,800	894,600
Income												
TNUoS		618,100	1,144,700	1,955,300		618,100	1,144,700	1,955,300		618,100	1,144,700	1,955,300
DUoS		240,100	444,600	759,300		240,100	444,600	759,300		240,100	444,600	759,300
FFR						282,900	0	0		282,900	0	0
Capacity market levy		49,000	68,000	88,000		49,000	68,000	88,000		49,000	68,800	88,000
Economy 7 opportunity										145,900	271,400	463,500
		907,200	1,658,000	2,802,600		1,190,100	1,658,100	2,802,600		1,336,000	1,929,500	3,266,100
Assumed average efficiency (*fluctuates due to replacement cycle*)		89%	81%	72%		89%	81%	72%		89%	81%	72%
Net cash inflows (*after efficiency is accounted for*)	0	795,800	1,342,100	1,999,500	0	1,065,800	1,342,100	1,999,500	0	1,179,900	1,560,600	2,330,100
Net cash flows	–1,045,000	681,900	–138,800	1,114,100	–1,045,000	948,400	–139,200	1,113,600	–1,045,000	1,057,700	73,000	1,435,700
Present value of income	834,885				1,058,502				1,378,918			
Initial capital cost	–1,045,000				–1,045,000				–1,045,000			
Net present value	–210,115				13,502				333,918			
Percentage value of return on investment	–20%				1%				32%			

Quantity Surveyor's Pocket Book, 3rd Edition

D. Cartlidge

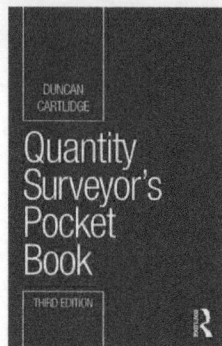

The third edition of the Quantity Surveyor's Pocket Book has been updated in line with NRM1, NRM2 and NRM3, and remains a must-have guide for students and qualified practitioners. Its focused coverage of the data, techniques and skills essential to the quantity surveying role makes it an invaluable companion for everything from initial cost advice to the final account stage.

Key features and updates included in this new edition:
- An up-to-date analysis of NRM1, 2 and 3;
- Measurement and estimating examples in NRM2 format;
- Changes in procurement practice;
- Changes in professional development, guidance notes and schemes of work;
- The increased use of NEC3 form of contract;
- The impact of BIM.

This text includes recommended formats for cost plans, developer's budgets, financial reports, financial statements and final accounts. This is the ideal concise reference for quantity surveyors, project and commercial managers, and students of any of the above.

March 2017; 186 × 123 mm, 466 pp
Pbk: 978-1-138-69836-9; £23.99

To Order: Tel: +44 (0) 1235 400524 Fax: +44 (0) 1235 400525
or Post: Taylor and Francis Customer Services,
Bookpoint Ltd, Unit T1, 200 Milton Park, Abingdon, Oxon, OX14 4TA UK
Email: book.orders@tandf.co.uk

For a complete listing of all our titles visit:
www.tandf.co.uk

Taylor & Francis
Taylor & Francis Group

Building Information Modelling (BIM)

The publication of the Government Construction Strategy in 2011, helped bring about a digital construction revolution in the UK and the rest of the work. At the centre is a package of measures with Building Information Modelling (BIM) at the heart, focussed on digital collaboration achieving best value across the project timeline. Following the UK Government's BIM mandate in 2016 and a Scottish BIM policy the following year, BIM has seen a steady rise in its use within the construction industry.

BIM has become an agent for change. A central proposition of BIM is working collaboratively in order to improve productivity efficiency and reduce waste. It also plays an important role in achieving the ambitious targets outlined in the Government Construction Strategy 2025 that aims to create more sustainably assets that are built faster and cheaper by moving towards offsite manufacturing. Working in conjunction with a number of other initiatives such as new procurement models, benchmarking and early contractor engagement, the UK Government has demonstrated considerable rewards using BIM, such as a reduction in construction costs by 20%, on public's sector projects. This is achieved by unlocking savings in capital expenditure and also the operational stages of the asset lifecycle. However, it is just not the public sector that has much to gain, the private sector has also demonstrated considerable benefits using BIM processes.

So what is BIM?

BIM is a collaborative process, enabled by technology. BIM exploits the potential in computer-based modelling technologies to provide a new way of designing buildings and managing design and construction processes. It requires new ways of working that need cultural, behavioural and technological changes in order to deliver better value and better project outcomes. Information models and structured data, are created, shared and exchanged throughout the project lifecycle in a managed way.

It requires clients to be clear about their information requirements. This means communicating to the supply chain what information they need, when they need it and who is responsible for it, in order to make key decisions at strategic points in the project. Essentially information allows the client to confirm that the asset meets its performance expectation. BIM also requires teams to come together earlier on in the process in order to bridge the gap between design, construction and operation. This ensures that the operational performance is optimized and the asset performs as intended, an area which is traditionally overlooked.

To embed and prepare the construction industry for BIM adoption, a phased approach was chosen. A chosen staging post, known as BIM Level 2, was selected for the 2016 mandate. This is a series of domain and collaborative federated models. (Federation means that models and information are linked together to create a single model of the asset.) Models, also known as information models, are a combination of geometrical and non-graphical data, prepared by different players during the project lifecycle. This allows the ability to interrogate, repurpose and verify data via graphical models and databases. Dynamic modelling software can be used to develop and manipulate these digital models to refine the design and also to test and validate its potential performance across a range of criteria, including buildability, energy performance-in-use, whole life costing, etc.

Collaborative working

A key theme of BIM is the sharing of information and data within a central online space, known as a Common Data Environment or CDE. Significant benefits are to be gained from this approach, as it ensures everyone has access to the most up-to-date data, decreasing risk and increasing accuracy. It also improves design coordination, reduces design costs and improves communications throughout the design and construction process. By working together to develop sophisticated, coordinated information models, the different design and construction disciplines can 'prototype' projects before they are built. Designs can be developed and tested 'virtually' so that the performance and cost are optimized. They can be coordinated so that many potential problems are either designed-out or avoided altogether. Ultimately, the benefits not only lie in more efficient and effective design and construction processes, but better and more certain project outcomes – better buildings that are more fit-for-purpose and meet their brief requirements and design intent.

How does BIM work?

The potential for all key project information to be stored and manipulated on a computer is what sets BIM apart from more conventional approaches, and BIM-based design solutions differ from their traditional counterparts in that they:

- Are created and developed on digital databases which enable collaboration and effective data exchange between different disciplines;
- Allow change to be managed through these databases, so that changes in one part of the database are reflected in (and coordinated through) changes in other parts; and
- Capture and preserve information for reuse by all members of the design and construction team, including facilities management (FM), and user operation and management.

Conventionally, a good deal of design and construction work is document based. Information is communicated and stored via a variety of drawings and reports that, despite being stored and distributed in digital form, are essentially 'unstructured' and thus of limited use. Not only is this information unstructured, it is also held in a variety of forms and locations that are not formally coordinated (information on individual building components, for example, are contained on drawings, specifications, bills of quantity descriptions, etc.). Such an approach has considerable potential for data conflicts and redundancy as well as risks to data integrity and security.

BIM opens up a wide range of possibilities for improvement, which includes better ways of generating, exchanging, storing and reusing project information that greatly improve communications between different design and construction disciplines through the life of the asset. As well as procuring a physical asset, the client is also procuring a 'digital asset', one which can be used for forward planning, investment decisions and operational and maintenance activities.

Of course, digitally structuring information can take many forms, from simple temporary models created for a specific purpose (i.e. a room schedule with key room dimensions in spreadsheet form) through to shared 3D whole-building models containing architectural, structural, servicing and other data all in the same place. The greater the degree of information-sharing and collaboration in the development of the models, the more accurate and complete the models will be and the greater are the benefits of using BIM. As shared models are developed, problems – such as clashes between structure and services – tend to get ironed out, and designs become more consistent and coordinated.

BIM benefits

Generally, BIM is recognized as providing a wide range of valuable benefits including:

- **Design:** Improved coordination of design and deliverables between disciplines: improved project understanding through visualization; improved design management and control, including change control; and improved understanding of design changes and implications through parametric modelling.
- **Compliance:** Ability to perform simulation and analysis for regulatory compliance; and ability to simulate and optimize energy and wider sustainability performance.
- **Costing/economics:** Ability to perform cost analysis as the design develops, and to check for adherence to budget/cost targets; ability to understand cost impacts of design changes; and improved accuracy of cost estimates.
- **Construction:** Reduction of construction risks through identification of constructability issues early in the design process; early detection and avoidance of clashes; ability to model impact of design changes on schedule and programme; and ability to integrate contractor/subcontractor design input directly to the model.

- **Operation and management:** Creation of an FM database directly from the project (as built) model; ability to perform FM costing and procurement from the model; and ability to update the model with real-time information on actual performance through the life of the building.

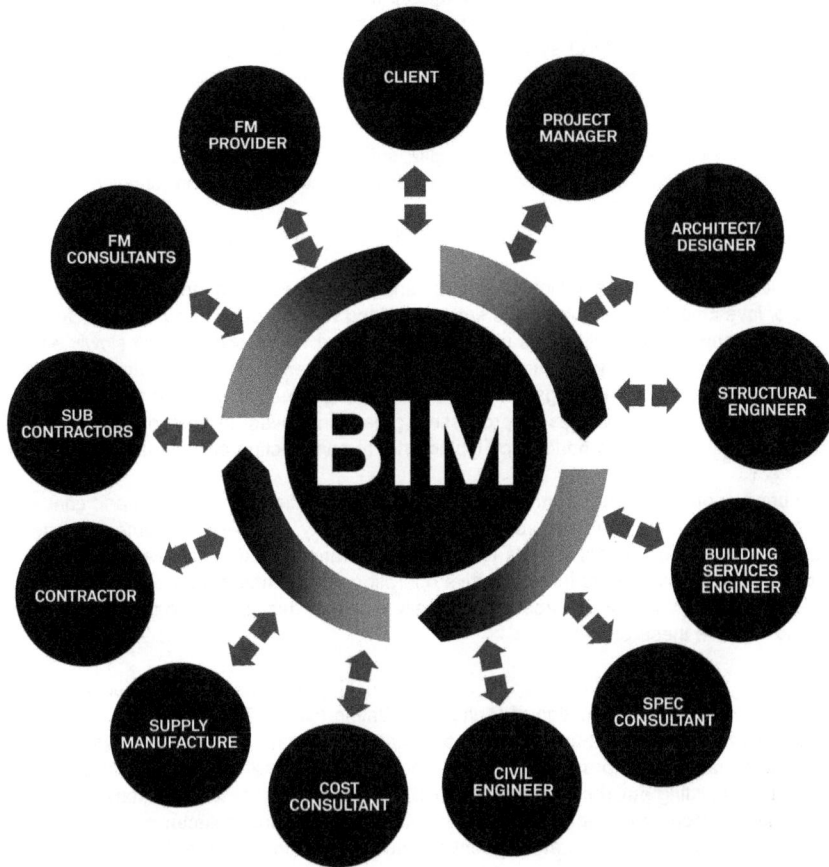

Ongoing BIM issues and risks

There are a number of uncertainties and potential risks in its adoption including:

- **Data security:** The increased use of digital data means new threats arise around cyber security. The management of sensitive projects, data and information require permissions and policies in place around who can have access and who can allow permissions. This is a key area that was recognized during the development of BIM Level 2, and a standard PAS 1192:5 has been published which addresses these issues.
- **Legal, contractual and insurance issues:** BIM Level 1 and 2 requires little change to the fundamental building blocks of copyright law, contracts or insurance. Standard forms of building contract – including design and consultant appointments often do not make specific provision for BIM, however they may cover areas such as collaborative working. A BIM Protocol can be used as a supplementary legal agreement, appended to existing building contracts. The protocol covers the ethos of collaboration by placing

obligations on parties to provide and share information at defined stages of a project. As industry moves towards the concept of a single central model, which all parties contribute directly to, a new contractual landscape will be required.

- **IT and software:** BIM Level 2 allows industry to use the 'best of breed' software. However, it does require particular information about the facilities management and operational maintenance of an asset to be supplied in a common file format, so that it can be accessed by everyone, and also in the future. The move towards greater interoperability of systems and data (where all data and information can be read by all software) will require greater use of Open Standards such as the implementation and use of Industry Foundation Classes (IFC). The variety of systems in use can impose heavy training burdens on firms who need to operate with some or all of them.

- **Awareness and expectations:** Paradoxically, BIM suffers from relatively low levels of client and sector awareness combined with very high expectations typical of new information technology developments.

A balanced view

BIM inevitably requires initial investment. New processes and plans of work require more upfront work. There will also be a degree of investment in technology and software, along with the supporting infrastructure and upskilling of the team to acquire new skills. While these new ways of working and processes are slowly embedded, there will be a significant learning curve, during which productivity and efficiency may take a dip. However, the benefits outweigh the risks to a very significant degree and the sheer transformative effect of BIM should not be under-estimated: more efficient and effective design processes; greatly improved information quality and coordination between design and construction; and better prototyping prior to construction are all made possible through BIM.

Potentially these bring enormous benefits, not only by improving the efficiency of design and construction, but by improving its effectiveness also, ultimately providing greater certainty of project outcomes and better buildings. More than that, the potential for project models to support the management of facilities in the post-construction phase is considerable and could lead to more effective operation of buildings through the whole life cycle. One of these benefits is to provide new learning about what really works in design and construction which can be fed back to inform and improve these processes.

BIM is a journey and not a destination. Already, industry is looking towards the next step and the longer term trajectory towards Level 3 with the formation of Digital Built Britain (DBB). This is built around a programme that will help us plan new infrastructure and built assets more effectively, create (or more likely assemble) it at a lower cost, and operate and maintain it more efficiently. The Level 3 programme is not just looking at BIM, but at the wider digitization of the building environment. With the BIM Level 2 standards and mandate now in place, part of the DBB programme is supporting government departments and wider public-sector clients to adopt Level 2 as 'Business As Usual'. Following this, the Level 2 Convergence programme looks at how we can start to connect construction to the smart city world of service, consumers and the citizen, using the rich data sources such as IoT. This will require the use of advanced manufacturing such as off-site manufacturing, standardization and optimization and changes to the current delivery model.

Conclusion

The most important decision clients will face is the selection and appointment of the design and construction team. Finding the right team of people with a positive approach to using BIM, to sharing information and, above all, learning from the experience is key. In this regard, clients should expect all their consultants, contractors and specialists to be familiar with BIM and its requirements; to be positively engaged in its adoption; and to be actively developing ways in which processes can be made more value-adding and effective. Many projects have seen the adoption of BIM for clash detection and coordination purposes with a BIM consultant being appointed to manage the process. Greater understanding from the design teams and demand from clients will see more traditionally based tasks being achieved through BIM in the subsequent years to come.

Building Energy Management Systems (BEMS) or Building Automation Control System (BACS)

A Building Energy Management System (BEMS) is a networked computer based control system installed in buildings that controls and monitors the building's mechanical, electrical and public health (MEP) systems and equipment as well as other ancillary plant and building systems and equipment in the most operational and energy efficient way.

Benefits

- Switches plant on and off automatically according to time, time of day, occupancy/demand and environmental conditions (e.g. senses conditions and activates devices to correct settings)
- Optimizes plant operation and services to meet building occupancy and demand requirements
- Monitors plant status and environmental conditions and thus improves standards of operation and maintenance
- Allows remote access, control and monitoring (e.g. using graphical information and alarms) via web-browser connected PC, tablet or smart phone
- Provides real-time and historical operational and energy performance data for continual building performance analysis in conjunction with Energy Management and Analytics software tools.

Types of BMS points

- Is the physical connection of a BMS point to a controller or outstation I/O terminals or 'soft' points obtained through integration by various means with 3rd party systems and equipment? The points schedule and methods of integration are critical when designing the system.
- Input – gathers data/monitor; Output – provide command/control
- Analogue points – has a variable value 0% to 100% (e.g. temperature sensor, modulating control valves)
- Digital points – binary I/O (0 or 1) (e.g. on/off control, run/off/fault status)
- Network integration points (NI) or High Level Interface (HLI) – allows to communicate or integrate to stand alone controller such as chiller, meters, boilers, generators.

BMS components – hardware and software/programming

Hardware

- **Primary network/BMS backbone**
  - o Central Virtual BEMS Server located in MER or Data Centre with distributed Web-client PCs or local stand-alone Head end equipment (desktop computer, laptop, monitors, keyboard, network printer (if required), UPS supply, etc.)
  - o Ethernet network cabling (normally Cat 5e or Cat 6 cables)
  - o Ethernet switch (located inside the control panels) or
  - o Part of converged building or client managed MEPS network
  - o Building Enterprise Level system interfaces, e.g. Energy Management Reporting, etc.
- 'Middleware' Integration layer providing systems integration, interoperability and smart building operational features

- **Secondary network**
  - o Control enclosure (CE) or Control panel (CP) or BMS outstation complete with I/O.
  - o Motor control centre (MCC) incorporating BEMS outstations in control section (Typically Form 2, 3 or 4).
  - o Local operator interfaces or touch screen display panels.
  - o Unitary controller for FCU/VAV/chilled beams.
  - o Power cables and isolators (from MCC panel to mechanical equipment) and associated containment.
  - o Control cables (from MCC/CE panel to field devices) and field network cabling (to 3rd party equipment, e.g. Inverters, packaged plant, chillers, etc.).
  - o Field devices (control valves, meters, sensors/transducers, VSD/Inverter, Soft-start, DOL starter, star-delta starter, actuators).
  - o Consider options for Wireless controllers, sensors, switches, occupancy PIRs, etc.

Software/programming

- Management level software residing on central server or head-end (typically licensed per user or per point)
- BMS communications protocol – Refers to the data communication within the various layers of the of BMS from PC/server software to the different type of outstation/controller/sensors, meters and other 'edge' devices (e.g. Modbus, M-Bus, LonWorks, BACnet (IP/MSTP), KNX or proprietary).
- Graphical User Interface – Provide true real-time dynamic representation of the installed plant and connected systems. Provide visualization of all points and necessary overrides and adjustments. Colour graphics shall incorporate automatic updating of real time field data.
- Setting up of system network security and operator access regime.
- Configure alarm/reporting strategy and historical data logging.
- Software system integration (e.g. energy management system, chillers, boiler, CHP, generator, water treatment plant, closed control units, blind control, lighting control, security and access controls).
- Application of additional Building Performance Analytics software to provide enhanced intelligent alarming and Energy Management reporting, etc.

BMS Pricing – *Quick Estimate for a typical medium size commercial office building:*

Shell and Core:

1.	Head end equipment (Workstation) *Extra cost:* • *Additional remote workstation*	£10,000–£20,000 *£5,000–£10,000*
2.	Network connection (BMS backbone – Copper)	£15,000–£30,000
3.	Main plant – cost per BMS points *Factors that affect the cost:* • *UPS and touch pad display for each control panel* • *Panels enclosure – fire-rated or GRP enclosure*	£550–£900 per pts
4.	Control enclosure (CE) for future tenants. Additional I/O modules will be part of fitting out cost *Factors that affect the cost:* • *UPS and touch pad display for each control panel* • *Panels enclosure – fire-rated or GRP enclosure*	£5,000–£10,000 per panel
5.	Landlord FCUs – depending on points per/FCU controller *Factors that affect the cost:* • *Local controller for cellular office, meeting rooms, etc.*	£450–£850 per unit
6.	Software, graphics, commissioning, engineering	Included in BMS pts rate
7.	Trade contract preliminaries	20% to 30%

Category A, B or C Office Fit-out:

1.	Tenant plants – cost per points utilizing existing MCC/CE panel but require additional I/O modules	£330–£550 per pts
2.	Energy meters a. LTHW/CHW meters b. Electricity meters (meters by Electrical TC) *Factors that affect the cost:* • *Size of heat meter or type of meter such as ultrasonic flow meter*	£850–£1,750 £250–£350
3.	Intelligent unitary controller (IUC) – depends on points per/IUC controller and the air conditioning solution such as 4-pipe FCU, 2-pipe w/electric heating, VAV, chilled beams, chilled ceiling *Factors that affect the cost:* • *Local controller for cellular office/meeting room* • *For chilled beam and chilled ceiling, controls depend on the zoning* • *Depend also on zone controls*	£650–£850 per unit
4.	Software, graphics, commissioning *Extra over cost:* • *Integration to other standalone system (e.g. lighting control, AV, fire alarm, security, central battery)*	Included in pts rate
5.	Trade contract preliminaries	20% to 30%

Capital Allowances

Introduction

Capital Allowances provide tax relief by prescribing a statutory rate of depreciation for tax purposes in place of that used for accounting purposes. They are utilized by government to provide an incentive to invest in capital equipment, including assets within commercial property, by allowing the majority of taxpayers a deduction from taxable profits for certain types of capital expenditure, thereby reducing or deferring tax liabilities.

The capital allowances most commonly applicable to real estate are those given for capital expenditure on existing commercial buildings in disadvantaged areas, and plant and machinery in all buildings other than residential dwellings, except for common areas. Relief for certain expenditure on industrial buildings and hotels was withdrawn from April 2011, although the ability to claim plant and machinery remains.

Enterprise Zone Allowances are also available for capital expenditure within designated areas only where there is a focus on high value manufacturing. Enhanced rates of allowances are also available on certain types of energy and water saving plant and machinery assets.

The Act

The primary legislation is contained in the Capital Allowances Act 2001. Major changes to the system were introduced in 2008, 2014 and 2018 affecting the treatment of tax relief to include plant and machinery allowances and the newly introduced Structures and Building Allowances (SBAs).

Plant and machinery

Various legislative changes and case law precedents in recent years have introduced major changes to the availability of Capital Allowances on property expenditure. The Capital Allowances Act 2001 excludes expenditure on the provision of a building from qualifying for plant and machinery, with prescribed exceptions.

List A in Section 21 of the 2001 Act sets out those assets treated as parts of buildings: -

- *Walls, floors, ceilings, doors, gates, shutters, windows and stairs.*
- *Mains services, and systems, for water, electricity and gas.*
- *Waste disposal systems.*
- *Sewerage and drainage systems.*
- *Shafts, or other structures, in which lifts, hoists, escalators and moving walkways are installed.*
- *Fire safety systems.*

Similarly, List B in Section 22 identifies excluded structures and other assets.

Both sections are, however, subject to Section 23. This section sets out expenditure, which, although being part of a building, may still be expenditure on the provision of Plant and Machinery.

Sections 21 and 22 do not affect the question whether expenditure on any item in List C is expenditure on the provision of Plant or Machinery.

List C in Section 23 is reproduced below:

1.   Machinery (including devices for providing motive power) not within any other item in this list.
2.   Gas and sewerage systems provided mainly –
     a.   to meet the particular requirements of the qualifying activity, or
     b.   to serve particular plant or machinery used for the purposes of the qualifying activity.
3.   Omitted
4.   Manufacturing or processing equipment; storage equipment (including cold rooms); display equipment; and counters, checkouts and similar equipment.
5.   Cookers, washing machines, dishwashers, refrigerators and similar equipment; washbasins, sinks, baths, showers, sanitary ware and similar equipment; and furniture and furnishings.
6.   Hoists.
7.   Sound insulation provided mainly to meet the particular requirements of the qualifying activity.
8.   Computer, telecommunication and surveillance systems (including their wiring or other links).
9.   Refrigeration or cooling equipment.
10. Fire alarm systems; sprinkler and other equipment for extinguishing or containing fires.
11. Burglar alarm systems.
12. Strong rooms in bank or building society premises; safes.
13. Partition walls, where moveable and intended to be moved in the course of the qualifying activity.
14. Decorative assets provided for the enjoyment of the public in hotel, restaurant or similar trades.
15. Advertising hoardings; signs, displays and similar assets.
16. Swimming pools (including diving boards, slides & structures on which such boards or slides are mounted).
17. Any glasshouse constructed so that the required environment (namely, air, heat, light, irrigation and temperature) for the growing of plants is provided automatically by means of devices forming an integral part of its structure.
18. Cold stores.
19. Caravans provided mainly for holiday lettings.
20. Buildings provided for testing aircraft engines run within the buildings.
21. Moveable buildings intended to be moved in the course of the qualifying activity.
22. The alteration of land for the purpose only of installing Plant or Machinery.
23. The provision of dry docks.
24. The provision of any jetty or similar structure provided mainly to carry Plant or Machinery.
25. The provision of pipelines or underground ducts or tunnels with a primary purpose of carrying utility conduits.
26. The provision of towers to support floodlights.
27. The provision of –
     a.   any reservoir incorporated into a water treatment works, or
     b.   any service reservoir of treated water for supply within any housing estate or other particular locality.
28. The provision of –
     a.   silos provided for temporary storage, or
     b.   storage tanks.
29. The provision of slurry pits or silage clamps.
30. The provision of fish tanks or fish ponds.
31. The provision of rails, sleepers and ballast for a railway or tramway.
32. The provision of structures and other assets for providing the setting for any ride at an amusement park or exhibition.
33. The provision of fixed zoo cages.

Main pool plant and machinery

Capital Allowances on main pool plant and machinery are currently given in the form of writing down allowances at the current rate of 18% per annum on a reducing balance basis. For every £100 of qualifying expenditure £18 is claimable in year 1, £14.76 in year 2 and so on until either all of the allowances have been claimed, or the asset is sold.

Integral features

The category of qualifying expenditure on 'integral features' was introduced with effect from April 2008. The following items are integral features:

- An electrical system (including a lighting system)
- A cold water system
- A space or water heating system, a powered system of ventilation, air cooling or air purification, and any floor or ceiling comprised in such a system
- A lift, an escalator or a moving walkway
- External solar shading

Integral Features attract a writing down allowance of 8% per annum which reduces to 6% per annum from April 2019, on a reducing balance basis.

The Integral Features legislation introduced certain assets which were not previously allowable as plant and machinery. These assets include general power and lighting, thermal insulation to existing buildings, external solar shading and cold water systems.

Thermal insulation

From April 2008, expenditure incurred on the installation of thermal insulation to existing buildings qualifies as Integral Feature plant and machinery which is available on a reducing balance basis at 8% per annum.

Long-life assets

A writing down allowance of 8% per annum is available on long-life assets.

A long-life asset is defined as plant and machinery that has an expected useful economic life of at least 25 years. The useful economic life is taken as the period from first use until it is likely to cease to be used as a fixed asset of any business. It is important to note that this is likely to be a shorter period than an asset's physical life.

Plant and machinery provided for use in a building used wholly, or mainly, as a dwelling house, showroom, hotel, office, retail shop, or similar premises, or for purposes ancillary to such use, cannot be classified as a long-life asset.

In contrast certain plant and machinery assets in buildings such as factories, cinemas, hospitals etc. could potentially be treated as long-life assets.

Case law

The fact that an item appears in List C does not automatically mean that it will qualify for capital allowances. It only means that it may potentially qualify.

Guidance about what can qualify as plant is found in case law dating back to 1887. The case of *Wimpy International Ltd and Associated Restaurants Ltd v Warland* in the late 1980s is one of the most important case law references for determining what can qualify as plant.

The Judge in that case applied three tests when considering whether or not an item is plant.

1. Is the item stock in trade? If the answer is yes, then the item is not plant.
2. Is the item used for carrying on the business? In order to pass the business use test the item must be employed in carrying on the business; it is not enough for the asset to be simply used in the business. For example, product display lighting in a retail store may be plant but general lighting in a warehouse would fail the test. *(Please note, this case law relates to the pre-Integral Feature Legislation, which introduced lighting as an eligible asset.)*

3.   Is the item the business premises, or part of the business premises? An item cannot be plant if it fails the premises test, i.e. if the business use is the premises itself, or part of the premises, or a place in which the business is conducted. The meaning of 'part of the premises' in this context should not be confused with real property law. HMRC's internal manuals suggest there are four general factors to be considered, each of which is a question of fact and degree:
- Does the item appear visually to retain a separate identity
- With what degree of permanence has it been attached to the building
- To what extent is the structure complete without it
- To what extent is it intended to be permanent, or alternatively, is it likely to be replaced within a short period.

Certain assets will qualify as plant in most cases. However, many others need to be considered on a case-by-case basis. For example, decorative assets in a hotel restaurant may be plant, but similar assets in an office reception area may be ineligible.

Refurbishment schemes

Building refurbishment projects will typically be a mixture of capital costs and revenue expenses, unless the works are so extensive that they are more appropriately classified as a redevelopment. A straightforward repair or a 'like for like' replacement of part of an asset would be a revenue expense, meaning that the entire amount can be deducted from taxable profits in the same year.

Where capital expenditure is incurred which is incidental to the installation of plant or machinery then Section 25 of the Capital Allowances Act 2001 allows it to be treated as part of the expenditure on the qualifying item. Incidental expenditure will often include parts of the building that would be otherwise disallowed, as shown in the Lists reproduced above. For example, the cost of forming a lift shaft inside an existing building would be deemed to be part of the expenditure on the provision of the new lift.

The extent of the application of Section 25 was reviewed for the first time by the Special Commissioners in December 2007 and by the First Tier Tribunal (Tax Chamber) in December 2009, in the case of JD Wetherspoon. The key areas of expenditure considered were overheads and preliminaries where it was held that such costs could be allocated on a pro-rata basis; decorative timber panelling which was found to be part of the premises and so ineligible for allowances; toilet lighting which was considered to provide an attractive ambience and qualified for allowances; and incidental building alterations, of which enclosing walls to toilets and kitchens and floor finishes did not qualify but tiled splash backs, bespoke toilet cubicles and drainage did qualify, along with the related sanitary fittings and kitchen equipment.

The Enhanced Capital Allowances scheme

The scheme is one of a series of measures introduced to ensure that the UK meets its target for reducing greenhouse gases under the Kyoto Protocol. 100% first year allowances are available on products included on the Energy Technology List published on the website at www.eca.gov.uk and other technologies supported by the scheme. All businesses will be able to claim the enhanced allowances, but only investments in new and unused Machinery and Plant can qualify.

There are currently 18 technologies with multiple sub-technologies currently covered by the scheme:

- Air-to-air energy recovery
- Automatic monitoring and targeting (AMT)
- Boiler equipment
- Combined heat and power (CHP)
- Compressed air equipment
- Heat pumps
- Heating ventilation and air conditioning (HVAC) equipment
- High speed hand air dryers
- Illuminated signs
- Lighting

- Motors and drives
- Pipework insulation
- Portable energy monitoring equipment
- Radiant and warm air heaters
- Refrigeration equipment
- Solar thermal systems and collectors
- Uninterruptible power supplies
- Waste heat to electricity conversion equipment

The Finance Act 2003 introduced a new category of environmentally beneficial plant and machinery qualifying for 100% first-year allowances. The Water Technology List includes 14 technologies:

- Cleaning in place equipment
- Efficient showers
- Efficient taps
- Efficient toilets
- Efficient washing machines
- Flow controllers
- Greywater recovery and reuse equipment
- Leakage detection equipment
- Meters and monitoring equipment
- Rainwater harvesting equipment
- Small scale slurry and sludge dewatering equipment
- Vehicle wash water reclaim units
- Water efficient industrial cleaning equipment
- Water management equipment for mechanical seals

Buildings and structures and long-life assets, as defined above, cannot qualify under the scheme. However, following the introduction of the integral features rules, lighting in any non-residential building may potentially qualify for enhanced capital allowances, if it meets the relevant criteria. The same would apply to residential common areas.

A limited payable ECA tax credit equal to 19% of the loss surrendered was also introduced for UK companies in April 2008.

From April 2012, expenditure on plant and machinery for which tariff payments are received under the renewable energy schemes introduced by the Department of Energy and Climate Change (Feed-in Tariffs or Renewable Heat Incentives), will not be entitled to enhanced capital allowances.

The 2018 Budget announced that from April 2020, this accelerated ECA benefit will no longer be available.

Structure and building allowances

On 29 October 2018 the government published a Technical Note containing details of a new Structures and Buildings Allowance (SBA) for expenditure incurred on non-residential structures and buildings, following an announcement in the October 2018 budget.

Initially, the SBA was available at a flat rate of 2% per annum over a 50-year period, but this increased to 3% per annum over a 33-year period following the March 2020 Budget. This relates to expenditure incurred on new commercial structures and buildings, including costs for new conversions, or renovations, incurred on or after 29 October 2018, but only where all the contracts for the physical construction works were entered into on, or after 29 October 2018.

The SBA extends to landlord capital contribution expenditure.

The relief will be available when the building, or structure, first comes into use and will be available for both UK and overseas assets, provided the business is within the charge to UK tax. Excluded from this relief is expenditure on dwelling houses.

The claimant must have an interest in the land on which the structure or building is constructed.

The SBA expenditure will not qualify for the annual investment allowance (AIA).

The relief will cease to be available if the building or structure is brought into residential use, or if it is demolished. However, SBA continues to be available for periods of temporary disuse.

If the building or structure is sold, the new owner takes over the remainder of the residue of the SBA over the remaining 50-year period.

Structures and buildings include offices, retail and wholesale premises, walls, bridges, tunnels, factories and warehouses. Capital expenditure on renovations or conversions of existing commercial structures or buildings will also qualify. Other qualifying uses include mines, transport undertakings, investment management businesses and professions or vocations, but UK or EEA furnished holiday lettings businesses will not qualify for SBA.

The costs of construction will include only the net direct costs related to physically constructing the building, or structure after any discounts, refunds, or other adjustments. This will include demolition costs, or any land alterations necessary for construction and direct costs required to bring the building, or structure, into existence.

SBA qualifying expenditure will also be applicable to purchases of second-hand buildings, or structures. The basis of the claim will be dependent upon the vendor's holding structure and whether the property has been used, or unused, at the time of acquisition.

Where a building, or structure, is acquired and subsequently altered, or renovated, this will trigger a new 50-year SBA qualification period, as will any further new streams of qualifying expenditure.

It is important to note that expenditure on plant and machinery does not qualify for SBA. At present, however, the draft legislation excludes the facility for a purchaser to reclassify expenditure on plant and machinery that has been included within a prior claim for SBA.

Expenditure on qualifying land remediation will also not qualify for SBA.

Annual Investment Allowance

The Annual Investment Allowance (AIAs) is available to all businesses of any size and allows a deduction for the whole AIA of qualifying expenditure on plant and machinery, including integral features and long-life assets. The AIA rates have fluctuated over the years:

- 1 April 2014 to 31 December 2015–£500,000
- 1 January 2016 to 31 December 2018–£200,000
- Currently, the AIA rate from 1 January 2019 to 31 December 2020 is £1,000,000.

For accounting periods less, or greater than, 12 months, or if claiming in periods where the rates have changed, time apportionment rules will apply to calculate hybrid rates applicable to the period of claim.

Enterprise zones

The creation of 11 Enterprise Zones was announced in the 2011 Budget. Additional zones have since been added bringing the number to 24 in total. Originally introduced in the early 1980s as a stimulus to commercial development and investment, these zones had become virtually non-existent.

Enterprise zones benefit from a number of reliefs, including a 100% first-year allowance for new and unused non-leased plant and machinery assets, where there is a focus on high-value manufacturing.

Business Premises Renovation Allowance

The Business Premises Renovation Allowance (BPRA) was first announced in December 2003. The idea behind this scheme was to bring long-term vacant properties back into productive use by providing 100% capital allowances for the cost of renovating and converting unused premises in disadvantaged areas. The legislation was included in the Finance Act 2005 and was finally implemented on 11 April 2007 following EU state aid approval.

The scheme will apply to properties within the areas specified in the Assisted Areas Order 2007 and Northern Ireland.

BPRA is available to both individuals and companies who own, or lease, business property which has been unused for 12 months or more. Allowances will be available to a person who incurs qualifying capital expenditure on the renovation of business premises.

An announcement to extend the scheme by a further five years to 2017 was made within the 2011 Budget, along with a further 11 new designated Enterprise Zones.

Legislation was introduced in the Finance Bill 2014 to clarify the scope of expenditure qualifying for relief to actual costs of construction and building work and for certain specified activities, such as architectural and surveying services. The changes came into effect for qualifying expenditure incurred on or after 1 April 2014 for businesses within the charge to corporation tax and 6 April 2014 for businesses within the charge to income tax.

Other capital allowances

Other types of allowances include those available for capital expenditure on Mineral Extraction, Research and Development, Know-How, Patents, Dredging and Assured Tenancy.

International tax depreciation

The UK is not the only tax regime offering investors, owners and occupiers valuable incentives to invest in plant and machinery and environmentally friendly equipment. Ireland, Australia, Malaysia and Singapore also have capital allowances regimes that are broadly similar to the UK and provide comparable levels of tax relief to businesses.

Many other overseas countries have tax depreciation regimes based on accounting treatment, instead of Capital Allowances. Some use a systematic basis over the useful life of the asset and others have prescribed methods spreading the cost over a statutory period, not always equating to the asset's useful life. Some regimes have prescribed statutory rates, whilst others have rates which have become acceptable to the tax authorities through practice.

Quality Auditing in Construction Projects

Abdul Razzak Rumane

This book provides construction professionals, designers, contractors and quality auditors involved in construction projects with the auditing skills and processes required to improve construction quality and make their projects more competitive and economical.

The processes within the book focus on auditing compliance to ISO, corporate quality management systems, project specific quality management systems, contract management, regulatory authorities' requirements, safety, and environmental considerations. The book is divided into seven chapters and each chapter is divided into numbered sections covering auditing-related topics that have importance or relevance for understanding quality auditing concepts for construction projects.

No other book covers construction quality auditing in such detail and with this level of practical application. It is an essential guide for construction and quality professionals, but also for students and academics interested in learning about quality auditing in construction projects.

June 2019: 234 × 156 mm: 600pp
Hb: 978-0-8153-8531-8 : £130.00

To Order: Tel: +44 (0) 1235 400524 Fax: +44 (0) 1235 400525
or Post: Taylor and Francis Customer Services,
Bookpoint Ltd, Unit T1, 200 Milton Park, Abingdon, Oxon, OX14 4TA UK
Email: book.orders@tandf.co.uk

For a complete listing of all our titles visit:
www.tandf.co.uk

Taylor & Francis
Taylor & Francis Group

Data Centre Cooling

Introduction

Cooling of data centres has evolved from cooling small clusters of servers to giant server farms. While these modern large data centres are vital components of the services economy, they consume a formidable amount of energy worldwide.It's been widely documented that the cooling systems – the chiller, humidifier and CRAC units – account for 45% of the total energy consumption of a data centre, while the IT equipment accounts for 30%. This means that 1 kWh consumed by the IT equipment requires another 1 kWh of energy to drive the cooling and auxiliary systems.

From environmental and cost-efficiency perspectives, selecting a cooling method that can reduce this energy demand is clearly beneficial. Traditionally, data centres have been air cooled with chillers, humidifiers and CRAC units, in a variety of 'cold aisle/hot aisle' approaches that aren't terribly efficient and that can result in hot spots within the data hall.

Over the years, server equipment has become more resilient and can now tolerate a greater range in temperature and humidity levels than older technology allowed. These days, legitimate alternatives to cooling are routinely considered in most data centre projects looking for greener and more efficient strategies. Currently, in the northern hemisphere, air-cooling solutions are looking more towards free cooling, which works by using air from outside combined with reclaimed heat (winter) and evaporative cooling (summer) to provide the total cooling solution throughout the year.

Water-based cooling options are a more modern approach and cool the inside of the servers by pumping cold water through pipes or plates. Water-cooled rack systems work well but have an inherent risk of leaks. Understanding the cost drivers and benefits of each are crucial to advising clients effectively.

There are many factors that drive the selection of any option, not least capital and life-cycle costs, but also the location of the data centre and the feasibility of incorporating innovations such as free cooling or aquifer thermal energy storage (ATES). Parameters that drive these decisions include the requirement for power usage efficiency (PUE) levels to hit planning stipulations, and for acoustics and total cost of ownership (TCO) levels to be optimized. PUE measures how efficiently the data centre uses input power – the larger the number the less efficient the solution. In addition, the selected method of supplying power and power resilience plays a role in the PUE calculations – and hence the overall cooling and power solution combined is what is considered against the criteria to finalize the preferred solution for the client.

In this article, we are looking solely at the merits of three cooling solutions that are currently being used on projects, to ascertain the cost drivers of each and understand the cooling-only related costs. These are:

- Air cooling by chiller and CRAC units
- Air cooling by indirect air cooling (IAC) air handling plant
- Chilled-water cooling derived from free-cooling, hybrid cooling towers with chiller assist.

Any power supply solution, associated building works and main contractor prelims are excluded.

Air cooling by chiller and CRAC units

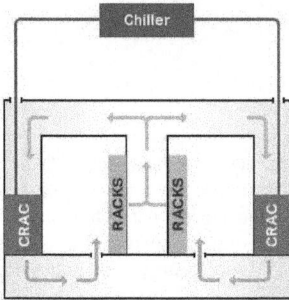

Air cooling by IAC air handling plant

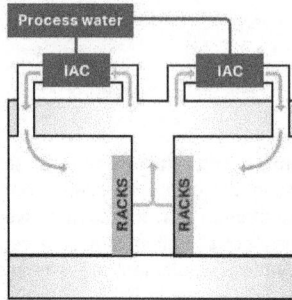

Chilled-water cooling derived from free cooling hybrid cooling towers with chiller assist

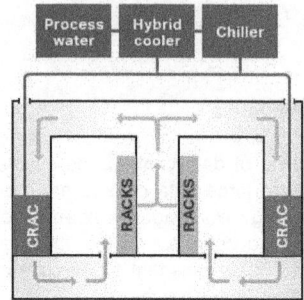

Air cooling by chiller and CRAC units

This chilled-water solution serves CRAC downflow units typically serving cold air to the data hall white space through a floor void. CRAC units normally include humidification elements to control the static within the air and all hot air is redirected back into the CRAC to remove the heat for redistribution into the white space. The source of the cooling water is via a traditional refrigeration chiller located externally, usually on the roof. There is no free cooling and chillers are sized for full peak load.

Air cooling by IAC air handling plant

This 'all air' based cooling solution incorporates an air handling plant mounted externally to the white space. Treated air is distributed to the white space via ductwork or through a plenum. Supply air is relatively low velocity to the cold aisle, giving more control than traditional floor void distribution.

The hot air is returned to the IAC via ductwork and is cooled by the outdoor ambient air at a plate heat exchanger. To assist the cooling process during warm months, the ambient air is adiabatically cooled (water evaporation), which then cools the warm air at the plate heat exchanger in the IAC unit. The water used for adiabatic cooling is bulk stored in the event of a mains supply outage. The process water is distributed from a central pump plantroom to the IAC units.

Chilled-water cooling derived from free-cooling, hybrid cooling towers with chiller assist

This chilled-water solution serves CRAC downflow units typically supplying cold air to the white space through a floor void. The source of the cooling water is via 'free cooling' cooling towers located externally, usually on the roof. Ambient air is used to cool the warm return water from the CRAC units, with adiabatic cooling added during the warmer months. At peak times, when approaching the towers cooling-load limits, refrigeration chillers are used to run in parallel with the cooling towers.

Table 1 is a summary of the pros and cons of each system. Bear in mind that numerous factors work in tandem within any given solution; for example, net-to gross area, efficiency, power load, capital expenditure cost, and TCO combine to determine the best solution. Always make sure defined parameters are set to allow measurement of any solution against these critical factors. This will ensure the best-fit solution can be determined.

Table 1: Benefits and drawbacks of three data centre cooling systems

* All based on 1500 W/m² IT Load Density requirement; Tier III certification

	1	**Traditional**	**2**	**Indirect Air Cooling**	**3**	**Cool water with Chiller Assist**
Method		Air cooling with chiller and CRAC units		Air cooling		CHW cooling with free cooling
PUE range (Cooling only)		1.4 to 2.0		1.1 to 1.3		1.1 to 1.3
Pros		• Tried and tested		• Proven technology		• Flexibility for distribution
		• Simple controls methodology		• No risk of leaks in white space		• Higher achieved white space ratio
				• Limited maintenance needed		• Simple controls methodology
				• Cost effective for low PUE		
				• Better control of supply air distribution		
Cons		• Hot spots, inefficient		• Acoustic challenges		• Large plant space required
		• Highest PUE range		• Impact on net white space for ducts		• Increased risk of water leaks
		• 24/7 operation required		• Reduced flexibility in distribution		• Higher PUE
		• No free cooling		• Space for bulk water storage required		• Great bulk water storage required
				• IAC plant to be located close to DH		• Higher maintenance costs

Most data centres use air- or water-based cooling solutions, and this is where our cost comparison has focused. The future is already in place, however, with some clients opting for immersion cooling, by which servers are immersed into a liquid coolant for direct cooling of the electronic components. Immersing servers have been shown to improve rack density, cooling capacity and other design-critical factors. Test projects where data centres are in the sea could result in some significant changes in this industry in the future. We are carrying out advisory work with Atlantis, which proposes to build a data centre on the site of its tidal-energy centre, off the coast of Scotland. It demonstrates how high-power-demand data centres could help fund the emerging tidal-power sector, thereby contributing to the future decarbonization of the data centre sector.

COST MODEL: COST COMPARISON OF COOLING METHODS

Net Technical Space (m²):	3,000	(4 nr Data Halls)
Load Density (W/m²):	1,500	
IT Load (kW):	4,500	

Cost/kW of Cooling Installation	CRAC Units and Chiller (£/kW)	Indirect Air Cooling (£/kW)	CRAC Units and Hybrid Cooler with Chiller Assist (£/kW)
CRAC Units (N+2)	240		240
Air cooled chillers (N+1) including CHW pipework distribution, pumps and other associated plant	890		
IAC Units (N+2)		760	
Process water installation to IAC units including water storage tanks, pumps, water treatment. Supply and return ductwork to IAC units including attenuation.		220	
Hybrid cooler (N+1), water cooled chillers (N+1), including CHW pipework distribution, process water storage tank and pipework distribution, pumps and others associated plant			1130
BMS Controls and power supplies to above mechanical plant	320	280	350
Total £/kW of Cooling Installation	**1450**	**1260**	**1720**

Notes:
- Hot/cold aisle containment is excluded
- Main Contractor Prelims and OHP are excluded
- Building/structural/architectural works, dedicated fresh air systems, and electrical infrastructure are excluded

Exposed Services

Background

'Traditional' Category A space has always been offered by developers within the market as designs often required minimal investment and the basic provisions have enabled them to target a wide range of potential tenants – each with varying needs. However, this norm has begun to be seen by many as inflexible and outdated.

The evolution of the office has been driven by reactions to the behavioural changes in the way people work based on the needs of a growing millennial dominated workforce and a shift towards different workplace preferences, for example, activity based working (ABW). Offices are becoming more collaborative, open and place greater emphasis on health and wellbeing. Research has shown that spaces with high ceilings positively promote the well-being of their occupants, even if there is limited natural light, as it triggers a psychological sense of freedom. The spaces within now need to be multifunctional, adaptable to changes in density and population whilst also maintaining a sense of privacy when required.

Fuelled by the growth of companies such as Google, Twitter, Amazon, all of whom champion the exposed services, prospective tenants are now placing greater emphasis on their office space. They look to use the space to attract the best talent joining the industry and to differentiate themselves from their competition. Couple this with the explosion in companies offering co-working space, such as WeWork and Fora (which represents around 4% of total office floor area within London and this is rising…). Co-working helps resolve the disconnect between a start-up's initial business plan and the 'crippling' commercial property leases which often start at five years. Start-ups and smaller businesses simply cannot justify the costs and risks associated with a lease so the need for flexible short-term space is an easy decision.

As a result, developers have had to begin to think outside the box in order to differentiate themselves in the market in order to attract new tenants. Whilst exposed services are not a new thing in CAT B it is still in its infancy in speculative CAT A developments. Prospective tenants are also looking to get as much value out of their spaces and it would be incredibly foolish to let a space where the majority if not all of the high level services (including ceilings etc.) are required to be stripped out. Developers have warmed to this and are offering 'exposed' CAT A spaces designed with flexibility and future proofing at the forefront.

Cost drivers

The largest misconception individuals have about 'exposed services' is that they are less expensive than installing a traditional CAT A utilizing a suspended ceiling as straight away you are removing the costs of a ceiling grid or tiles. It is important to realize that creating this aesthetic is not as straightforward (if done correctly) and involves a lot of additional work to achieve the perfect 'exposed' look. Traditionally, the ceiling grid and tiles were there to hide the nastiness above, ductwork, pipework and cabling, i.e. the services! As a result the following items will need to be considered, all of which represent additional cost.

Enhanced treatment to the ductwork, this could be achieved by a metal wrapping over traditional lagging to provide a clean finish. Alternatively, additional time and labour would improve the final installation.

Installation of suspended or track mounted lighting, including pendants in lieu of ceiling mounted lighting tiles and downlights – fittings are generally more expensive, take longer to install and require additional design coordination.

Painting of exposed ceiling and services in a uniform colour, which is a far more complicated job than simply painting walls. This could also later tie in with a company's overall colour scheme and branding, so in reality the 'casual' exposed look is actually the result of a lot of skilled labour.

Increased on-site design coordination and labour intensive tasks, including the careful and considered routing of electrical conduit to create a uniform and consistent finish and appearance.

Another invisible but inevitable cost consideration of exposed services is acoustics. Ceiling tiles are effective at absorbing sound and keep ambient noise levels from being disruptive. The hard surfaces associated with exposed ceilings, e.g. concrete slabs, amplify sounds and easily distract occupants who are trying to concentrate. This doesn't just include the sounds of people talking; it also includes ambient noise created by things like the HVAC system. Therefore it is important to consider the additional acoustic control options necessary to minimize noise, such as suspended baffles, acoustic panels, and spray-on solutions. The added cost for this varies depending on the options you choose and the level of design coordination required (for lighting, fire alarms, sprinklers, etc.). White noise systems can be installed to help; however, some people find the white noise itself to be disruptive and annoying.

Exposed services also come with other hidden costs besides construction costs, most notably, energy costs. Removing a suspended ceiling will mean that the HVAC system will have an increased volume and area of space to heat and cool. Research carried out by the Ceilings & Interiors Systems Association (CISCA) found that offices with suspended ceilings achieved 9–10.3% greater energy savings compared to those with exposed ceilings and services. In addition to this, the research found that exposed ceilings require more frequent cleaning, stricter cleaning regimes and periodic repainting that is not necessary with suspended ceilings.

Cost model

The cost model is based on a central London office development. The model is for the Category A fit-out to one floor occupied by a single tenant, with a net lettable area of 1,800 m². It is purely for mechanical and electrical installations and takes no account of changes to building elements. The exposed services option does not require a suspended ceiling, which would give a saving of about £6/square foot (NIA).

This is only part of the picture, however, with items required to be added back into the exposed services costs. For example, any exposed concrete surface would need dust sealing/painting at £1/square foot NIA. Acoustic panelling would be required to walls, suspended panels and shrouds above FCUs, at a cost of around £3.75-£5/square foot NIA.

Category A Works	Traditional CAT A		
	Based on a NIA 1,800 m²		
	Total (£)	£/m² on NIA	£/ft² on NIA
5.6 Space Heating and Air Treatment			
Installation of four pipe fan coil units to the perimeter zones and two pipe fan coil units to the internal zones. Including CHW and LTHW distribution pipework, ductwork distribution, plenums, diffusers, grilles and insulation. Installation of condensate drainage to fan coil units. Includes for testing and commissioning and subcontractor preliminaries.	£355,000	197.22 m²	18.32 ft²
5.8 Electrical Installations			
Installation of one tenant distribution board, LED lighting including controls based on a modular wiring system. Power to the mechanical services and general earthing and bonding. Includes for testing and commissioning and subcontractor preliminaries.	£220,000	122.22 m²	11.35 ft²

Category A Works	Traditional CAT A		
	Based on a NIA 1,800 m²		
	Total (£)	£/m² on NIA	£/ft² on NIA
5.11 Fire and Lightning Protection			
Sprinkler protection to floorplate. Includes for testing and commissioning and subcontractor preliminaries.	£45,000	25.00 m²	2.32 ft²
5.12 Communication, Security and Control Systems			
Fire detection system and BMS to fan coil units. Includes for testing and commissioning and subcontractor preliminaries.	£75,000	41.67 m²	3.87 ft²
Traditional Category A MEP Total Costs	**£695,000**	**386.11 m²**	**35.87 ft²**

Category A Works	Exposed Services CAT A		
	Based on a NIA 1,800 m²		
	Total (£)	£/m² on NIA	£/ft² on NIA
5.6 Space Heating and Air Treatment			
Installation of a four pipe fan coil units to the perimeter zones and two pipe fan coil units to the internal zones. Including CHW and LTHW distribution pipework, ductwork distribution, plenums, diffusers, grilles and insulation. Installation of condensate drainage to fan coil units. Includes for testing and commissioning and subcontractor preliminaries.			
Extra over for exposed services; including enhanced finishes to ductwork via metal cladding and enhanced grilles/diffusers.	£425,000	236.11 m²	21.94 ft²
5.8 Electrical Installations			
Installation of one tenant distribution board, LED lighting including controls based on a modular wiring system. Power to the mechanical services and general earthing and bonding. Includes for testing and commissioning and subcontractor preliminaries.			
Extra over for exposed services; including suspended fittings in lieu of ceiling mounted and increased coordination of wiring to minimize the visual impact.	£275,000	152.78 m²	14.19 ft²
5.11 Fire and Lightning Protection			
Sprinkler protection to floorplate. Includes for testing and commissioning and subcontractor preliminaries.	£45,000	25.00 m²	2.32 ft²

Category A Works	Exposed Services CAT A		
	Based on a NIA 1,800 m²		
	Total (£)	£/m² on NIA	£/ft² on NIA
5.12 Communication, Security and Control Systems			
Fire detection system and BMS to fan coil units. Includes for testing and commissioning and subcontractor preliminaries.			
Extra over for exposed services; increased coordination of wiring to minimize the visual impact.	£80,000	44.44 m²	4.13 ft²
Exposed Services Category A MEP Total Costs	**£825,000**	**458.33 m²**	**42.58 ft²**

Greywater Recycling and Rainwater Harvesting

The potential for greywater recycling and rainwater harvesting for both domestic residential and for various types of commercial building, considering the circumstances in which the systems offer benefits, both as stand-alone installations and combined.

Water usage trends

Water usage in the UK has increased dramatically over the last century or so and it is still accelerating. The current average per capita usage is estimated to be at least 150 litres per day, and the population is predicted to rise from 60 million now to 75 million by 2031, with an attendant increase in loading on water supply and drainage infrastructures.

Even at the current levels of consumption, it is clear from recent experience that long, dry summers can expose the drier regions of the UK to water shortages and restrictions. The predicted effects of climate change include reduced summer rainfall, more extreme weather patterns, and an increase in the frequency of exceptionally warm dry summers. This is likely to result in a corresponding increase in demand to satisfy more irrigation of gardens, parks, additional usage of sports facilities and other open spaces, together with additional needs for agriculture. The net effect, therefore, at least in the drier regions of the UK, is for increased demand coincident with a reduction in water resource, thereby increasing the risk of shortages.

Water applications and reuse opportunities

Average domestic water utilization can be summarized as follows, as a percentage of total usage (Source: Three Valleys Water)

- Wash hand basin 8%
- Toilet 35%
- Dishwasher 4%
- Washing machine 12%
- Shower 5%
- Kitchen sink 15%
- Bath 15%
- External use 6%

Water for drinking and cooking makes up less than 20% of the total, and more than a third of the total is used for toilet flushing. The demand for garden watering, although still relatively small, is increasing year by year and coincides with summer shortages, thereby exacerbating the problem.

In many types of building it is feasible to collect rainwater from the roof area and other surface areas and to store it, after suitable filtration, in order to meet the demand, for example, for toilet flushing, cleaning, washing machines and outdoor use – thereby saving in many cases 50% of the water demand. The other often complementary recycling approach is to collect and treat 'greywater' – the waste water from baths, showers and washbasins. Hotels, leisure centres, care homes and apartment blocks generate large volumes of waste water and therefore present a greater opportunity for recycling. With intelligent design, even offices can make worthwhile water savings by recycling greywater, not necessarily to flush all of the toilets in the building but perhaps just those in one or two primary cores, with the greywater plant and distribution pipework dimensioned accordingly.

Intuitively, rainwater harvesting and greywater recycling seem like 'the right things to do' and rainwater harvesting is already common practice in many countries in Northern continental Europe. In Germany, for example, some 60,000 to 80,000 systems are being installed every year – compared with perhaps 2,000 systems in the UK. Greywater recycling systems, which are less widespread than rainwater harvesting, have been developed over the last 20 years. The 'state of the art' system is to use biological and UV (ultra-violet) disinfection rather than chemicals, and to reduce the associated energy use through advanced technologies such as membrane filtration.

In assessing the environmental credentials of new developments, the sustainable benefits are recognized by the Building Research Establishment Environmental Assessment Method (BREEAM) whereby additional points can be gained for efficient systems – those designed to achieve enough water savings to satisfy at least 50% of the relevant demand. Furthermore, rainwater harvesting systems from several manufacturers are included in the 'Energy Technology Product List' and thereby qualify for Enhanced Capital Allowances (ECAs). Claims are allowed not only for the equipment, but also for directly associated project costs including:

- Transportation – the cost of getting equipment to the site.
- Installation – cranage (to lift heavy equipment into place), project management costs and labour, plus any necessary modifications to the site or existing equipment.
- Professional Fees – if they are directly related to the acquisition and installation of the equipment.

Rainwater harvesting

A typical rainwater harvesting system can be installed at reasonable cost if properly designed and installed at the same time as building the development/building. The collection tank can either be buried or installed in a basement area. Rainwater enters the drainage system through sealed gullies and passes through a pre-filter to remove leaves and other debris before passing into the collection tank. A submersible pump, under the control of the monitoring and sensing panel, delivers recycled rainwater on demand. The non-potable distribution pipework to the washing machine, cleaner's tap, outside tap and toilets etc. could be either a boosted system or configured for a header tank in the loft, with mains supply back-up, monitors and sensors located there instead of at the control panel.

Calculating the collection tank size brings into play the concept of system efficiency – relating the water volume saved to the annual demand. In favourable conditions – ample rainfall and large roof collection area – it would be possible in theory to achieve almost 100%. In practice, systems commonly achieve 50 to 70% efficiency, with enough storage to meet demand for typically 18 days or 2–3 weeks (BS EN 16941–1:2018), though this is subject to several variables. As well as reducing the demand for drinking quality mains supply water, rainwater harvesting tanks can if configured appropriately act as an effective storm water attenuator, thereby reducing the drainage burden and the risk of local flooding which is a benefit to the wider community. Where site rainwater attenuation is required it may be a sensible approach to consider the integration of a rainwater harvesting system to avoid duplication of some main plant. Many urban buildings are located where conditions are unfavourable for rainwater harvesting – low rainfall and small roof collection area. In these circumstances it may still be worth considering water savings through greywater recycling or communal systems, as well as considering the use of different collection surfaces.

Greywater recycling

In a greywater recycling system, waste water from baths, showers and washbasins is collected via a separate conventional fittings and pipework drainage system, to enter a pre-filtration tank which removes the larger dirt particles. This is followed by the aerobic treatment tank in which cleaning bacteria ensure that all biodegradable substances are broken down. The water then passes onto a third tank, where an ultra-filtration membrane removes all particles larger than 0.05 microns (this includes viruses and bacteria) effectively disinfecting the recycled greywater. The clean water is then stored in the fourth tank from where it is pumped on demand under the control of monitors and sensors in the control panel. Recycled greywater may then be used for toilet and urinal flushing, for laundry and general cleaning, and for outdoor use such as vehicle washing and garden irrigation – a substantial water saving for premises such as hotels. If the tank becomes depleted, the distribution is switched automatically to the mains water back-up supply. If there is insufficient plant room space, then the tanks may be buried but with adequate arrangements for maintenance access.

Combined rainwater/greywater systems

Rainwater can be integrated into a greywater scheme with very little added complication other than increased tank size with the exception of green roofs, where additional polishing (i.e. colour treatment) may be required. In situations where adequate rainwater can be readily collected and diverted to pretreatment, then heavy demands such as garden irrigation can be met more easily than with greywater alone. An additional benefit is that they reduce the risk of flooding by keeping collected storm water on site instead of passing it immediately into the drains.

Indicative system cost and payback considerations

Rainwater harvesting scenario: Office building in Leeds having a roof area of 2000 m² and accommodating 435 people over 3 floors. Local annual rainfall is 875 mm and the application is for toilet and urinal flushing. An underground collection tank of 25,000 litres has been specified to give normally 18 days storage.

Rainwater harvesting cost breakdown		Cost £
System tanks and filters and controls		21,000
Mains water back-up and distribution pump arrangement		5,000
Non-potable distribution pipework		1,000
Connections to drainage		1,000
Civil works and tank installation (assumption normal ground conditions)		8,000
System installation & commissioning		2,000
	TOTAL COST	38,000

Rainwater harvesting payback considerations

Annual water saving: 1100 m³ @ average cost £2.30/m³ = £2,530

Annual Maintenance and system energy cost = £800

Indicative payback period = 38,000/1,730 = 22 years

Greywater recycling scenario: Urban leisure hotel building with 200 bedrooms offers little opportunity for rainwater harvesting but has a greywater demand of up to 12,000 litres per day for toilet and urinal flushing, plus a laundry. The greywater recycling plant is located in a basement plant room.

Greywater recycling cost breakdown		Cost £
System tanks and controls		37,000
Mains water back-up and distribution pump arrangement		4,000
Non-potable distribution pipework		4,000
Greywater waste collection pipework		5,000
Connections to drainage		1,000
System installation & commissioning		6,000
	TOTAL COST	57,000

Greywater recycling payback considerations

Annual water saving: 4260 m³ @ average cost £2.30/m³ = £9,798

Annual Maintenance and system energy cost = £2,200

Indicative payback period = 57,000/7,598 = 7.5 years

Exclusions

- Site organization and management costs other than specialist contractor's allowances
- Contingency/design reserve
- Main contractor's overhead and profit or management fee
- Professional fees
- Tax allowances
- Value Added Tax

Conclusions

There is a justified and growing interest in saving and recycling water by way of both greywater recycling and rainwater harvesting. The financial incentive at today's water cost is not great for small or inefficient systems but water costs are predicted to rise and demand to increase – not least as a result of population growth.

The payback periods for the above scenarios are not intended to compare to potential payback periods of rainwater harvesting against greywater recycling but rather to illustrate the importance of choosing 'horses for courses'. The office building with its relatively small roof area and limited demand for toilet flushing results in a fairly inefficient system. Burying the collection tank also adds a cost so that payback exceeds 20 years. Payback periods of less than 10 years are feasible for buildings with large roofs and a large demand for toilet flushing or for other uses. Therefore sports stadia, exhibition halls, supermarkets, schools and similar structures are likely to be suitable. Where the project constraints require rainwater attenuation to be stored within tanks then it may be beneficial to provide an integrated RW attenuation and harvesting system which will optimize plant requirements.

The hotel scenario is good application for greywater recycling. Many hotels and residential developments will generate more than enough greywater to meet the demand for toilet flushing etc. and, in these circumstances, large quantities of water can be saved and recycled with attractive payback periods.

Grid Decarbonization

Introduction

Among other changes, SAP10 introduced a much lower carbon emission factor for electricity than the one currently used in Building Regulations; this is meant to reflect the grid electricity carbonization. With the GLA (Greater London Authority) encouraging developers to use SAP10 from January 2019, it is expected that developments in London will now have the opportunity to test the impacts of the decarbonizing grid on their design and servicing approach as well as on capex and opex.

In January 2019, the GLA published an updated Energy Assessment Guidance; among other changes, key has been the Mayor's approach to request developers to adopt the SAP 10 carbon factors in their Energy Assessments. This approach can have significant impacts on the selection of low and zero carbon technologies in buildings and can trigger a shift in the industry. This article examines the implications on technology selection and costs.

Background on carbon targets for London

The UN Paris Agreement aim is to strengthen the global response to the threat of climate change. To achieve this, global temperature rise this century must be kept to well below 2°C above pre-industrial levels; and efforts must be pursued to limit the temperature increase even further, to 1.5°C.

In the UK, the 2008 Climate Change Act requires at least an 80% reduction in CO_2 emissions (compared to 1990 levels). However, in May 2019, the Committee on Climate Change published 'Net Zero – the UK's contribution to stopping global warming', where it recommended that a new emissions target should be set for the UK: net-zero greenhouse gases (GHG) by 2050. A net-zero GHG target for 2050 will deliver on the commitment that the UK made by signing the Paris Agreement.

Cities have a major role to play in developing a path to a low carbon future. The Mayor's London Environment Strategy, published in May 2018, has already set an ambition for a Zero Carbon London by 2050. The ambition is London's response to the UN Paris Agreement.

The city's emissions are falling, having decreased by 25% since 1990. This is largely due to reduced gas consumption and decarbonization of the national electricity grid. To achieve the Mayor's Zero Carbon ambition by 2050, however, the rate of emissions reduction must be increased threefold over progress to date since 1990.

SAP 10 publication

SAP10 was published in July 2018. The document noted that it is not currently to be used for any official purpose, and that SAP 2012 should continue to be used for the Part L of the Building Regulations. Its intention was to provide an indication of the expected future carbon emission factors that would inform, and potentially be adopted in, any future update to the Building Regulations.

A key update in SAP 10 is that **the proposed grid electricity carbon emission factor is 0.233 kgCO$_2$ per kWh. This is less than half the figure of 0.519 kgCO$_2$ per kWh, currently used in Building Regulations**, reflecting the rapid reduction in grid electricity emission factors since the Building Regulations were last updated.

The GLA Energy Assessment Guidance (published in October 2018) encourages planning applications referable to the Mayor to use the SAP 10 emission factors from January 2019. Therefore, even though the BRE SAP 10 methodology states it is not to be used in Part L calculations, the GLA have chosen to implement the revised SAP 10 emission factors as an interim measure until a new Part L document is published.

Impacts on building services technologies

The impact of the reduced carbon intensity of grid electricity is substantial on the reported carbon savings for technologies that gain their benefit from **displacing** grid electricity. For example, carbon savings from electricity generating technologies such as PV (photovoltaics) and district energy systems served by gas CHP (Combined Heat and Power), will be reduced. On the other hand, technologies such as heat pumps, that use electricity to generate heat, will see their carbon emissions decline.

The early adoption of SAP10 emission factors by the GLA intends to enable developers to make appropriate, future-proofed technology choices, accounting for the reduced grid electricity emission factors. For example, CHP will likely no longer be the technology that can deliver the highest emission savings in heat-led developments; and alternatives such as **heat pumps** and **sources of secondary heat** will need to be considered to help deliver the zero carbon target – a target that every referable development in London is called to meet.

This shift is further supported by the policies set out in the draft New London Plan: the proposed 'heat hierarchy' lists the use of local secondary heat sources (in conjunction with heat pump, if required, and a lower temperature heating system) as the second option (after connection to an existing or planned heat network). Meanwhile, the implementation of CHP is moved further down in the hierarchy, and is supplemented by a note that attention should be paid to **impacts on local air quality.**

Shift towards heat pumps – cost considerations

With a clear shift towards heat pumps, the impacts on cost for developers and occupants must remain a key consideration.

Capex

There are two main challenges when trying to determine the impact of heat pumps on capital costs for developers:

1.   The variety of building types and scales in London; and
2.   A very wide range of costs, due to the range of heat pump systems that can be applied to each building type.

A study carried out as part of the GLA evidence base examined, among others, the cost implications.

To help tackle the challenge associated with the variety of building types, the study was narrowed down to examine cost implications for a particular type of development: a medium density residential building. This type was selected because it was considered the most challenging in terms of establishing cost implications.

Mechanical and Electrical capital costs of heat pump systems were compared against a baseline relying on a connection to a District Heating system with gas-fired CHP. Impacts were assessed in terms of:

- 'infrastructure costs', such as those associated with a ground loop in the case of a ground source heat pump;
- 'building costs', such as the central plant; and
- 'dwelling costs', such as heat interface units and hot water cylinders.

The study tackled the second challenge, the wide variety of heat pump systems, by developing a taxonomy of the possible systems and examining costs associated with each of the identified deployment solutions. The taxonomy was arranged around small, medium and large scale heat pump deployment configurations.

The study noted that a simple conclusion on capital costs is difficult to draw, given the variety of systems and the range of costs within a single system. On average, the capital costs of heat pumps were anticipated to be slightly higher than a 'business as usual' heating system with central gas boilers and a gas-fired CHP. For example, for small systems there was clearly an additional cost at the dwelling level. For medium systems, there was a range in the scale of additional costs between the three levels (infrastructure, building, dwelling). And finally, for large scale systems, costs were largely comparable. In summary, the study concluded that the potential additional capital costs compared with 'business as usual' are likely to be small in comparison to the total project costs, in the order of 0–3%.

Finally, research commissioned by the Government in 2016 examined the potential cost reductions for air source heat pumps and ground source heat pumps between the market as it was in 2014 and a mass market scenario. The conclusion was that costs could reduce by 15–20%. The reduction would be the result of a 30–50% potential cost reduction in non-equipment costs; and a 5–10% reduction in equipment costs. A larger installer base with larger companies, a consolidated supply chain, better sales channels and cheaper installation costs, were factors identified as contributing to the reduction in costs.

Opex

The main concern around operational expenses is that the move towards using electricity as the main heating fuel increases the risk of high heating bills for the user.

A GLA commissioned study examined the impacts on operating cost from the perspective of residents and concluded that, typically, heat pumps are likely to lead to a small increase in annual heating costs compared to the cheapest non-heat pump solution (communal gas boiler).

It was also noted that the impact of poor energy efficiency of a dwelling can be significant and this can be very relevant when examining existing building stock. Poor energy efficiency leads to an increase in heating demand, which negatively impacts the direct electric systems' heating costs to a greater extent than a gas-based system, as electricity has a high unit cost. As anticipated, the increased heating demand was found to have an almost proportional effect.

However, care must be taken when assessing operating costs for occupants: there are several components to the heating costs that must be considered consistently and in the unique context of every development so that the assessment can help inform the decision making process. For example, the Committee on Climate Change Report 'UK housing: Fit for the future' published in February 2019, recommended that homes should be well insulated reducing space heating demands to no more than 15–25 kWh/m^2/year (the PassivHouse standard calls for 15 kWh/m^2). Where these low heating demands can be achieved, the heating costs of an electrically-led solution could be significantly reduced and, coupled with the other benefits of lower air quality impacts and reduced carbon, can make a heat-pump-led solution advantageous over the lifecycle of a development.

Conclusion

In the following months, developers, designers and energy consultants will be called to come up with innovative solutions to address the technical and financial challenges associated with the shift these changes will cause. The goal, as set out in the Mayor's Environment Strategy, is to decarbonize London's homes and workplaces, as the city progresses towards the goal to become Zero Carbon by 2050. It is imperative, however, that the solutions developed and tested can also protect the most disadvantaged by tackling fuel poverty and ensuring that access to heating remains affordable for all.

Building Services Design for Energy Efficient Buildings, 2nd edition

Paul Tymkow *et al.*

SECONDD
EDITION

Building Services Design for Energy Efficient Buildings

Edited by PAUL TYMKOW,
SAVVAS TASSOU,
MARIA KOLOKOTRONI
AND HUSSAM JOUHARA

The role and influence of building services engineers are undergoing rapid change and are pivotal to achieving low-carbon buildings.

The essential conceptual design issues for planning the principal building services systems that influence energy efficiency are examined in detail, namely HVAC and electrical systems. In addition, the following issues are addressed:
- background issues on climate change, whole-life performance and design collaboration
- generic strategies for energy efficient, low-carbon design
- health and wellbeing and post occupancy evaluation
- building ventilation
- air conditioning and HVAC system selection
- thermal energy generation and distribution systems
- low-energy approaches for thermal control
- electrical systems, data collection, controls and monitoring
- building thermal load assessment
- building electric power load assessment
- space planning and design integration with other disciplines.

In order to deliver buildings that help mitigate climate change impacts, a new perspective is required for building services engineers, from the initial conceptual design and throughout the design collaboration with other disciplines. This book provides a contemporary introduction and guide to this new approach, for students and practitioners alike.

July 2020: 376 pp
Pb: 978-0-8153-6561-7 : £42.99

To Order: Tel: +44 (0) 1235 400524 Fax: +44 (0) 1235 400525
or Post: Taylor and Francis Customer Services,
Bookpoint Ltd, Unit T1, 200 Milton Park, Abingdon, Oxon, OX14 4TA UK
Email: book.orders@tandf.co.uk

For a complete listing of all our titles visit:
www.tandf.co.uk

Taylor & Francis
Taylor & Francis Group

Groundwater Cooling

The use of groundwater cooling systems considering the technical and cost implications of this renewable energy technology.

The application of groundwater cooling systems is quickly becoming an established technology in the UK with numerous installations having been completed for a wide range of building types, both new build and existing (refurbished).

Buildings in the UK are significant users of energy, accounting for 60% of UK carbon emissions in relation to their construction and occupation. The drivers for considering renewable technologies such as groundwater cooling are well documented and can briefly be summarized as follows:

- Government set targets – The Energy White Paper, published in 2003, setting a target of producing 10% of UK electricity from renewable sources by 2010 and the aspiration of doubling this by 2020.
- The proposed revision to the Building Regulations Part L 2006, in raising the overall energy efficiency of non-domestic buildings, through the reduction in carbon emissions, by 27%.
- Local Government policy for sustainable development. In the case of London, major new developments (i.e. City of London schemes over 30,000 m²) are required to demonstrate how they will generate a proportion of the site's delivered energy requirements from on-site renewable sources where feasible. The GLA's expectation is that, overall, large developments will contribute 10% of their energy requirement using renewables, although the actual requirement will vary from site to site. Local authorities are also likely to set lower targets for buildings which fall below the GLA's renewables threshold.
- Company policies of building developers and end users to minimize detrimental impact to the environment.

The ground as a heat source/sink

The thermal capacity of the ground can provide an efficient means of tempering the internal climate of buildings. Whereas the annual swing in mean air temperature in the UK is around 20 K, the temperature of the ground is far more stable. At the modest depth of 2 m, the swing in temperature reduces to 8 K, while at a depth of 50 m the temperature of the ground is stable at 11–13°C. This stability and ambient temperature therefore makes groundwater a useful source of renewable energy for heating and cooling systems in buildings.

Furthermore, former industrial cities like Nottingham, Birmingham, Liverpool and London have a particular problem with rising groundwater levels due to a reduction in groundwater abstraction for use in manufacturing, particularly since the 1970s. The use of groundwater for cooling is therefore encouraged by the Environment Agency in areas with rising groundwater levels as a means of combating this problem.

System types

Ground and groundwater cooling systems may be defined as either open or closed loop systems.

Open loop systems

Open loop systems generally involve the direct abstraction and use of groundwater, typically from aquifers (porous water bearing rock), although there are also systems which utilize water in former underground mineral workings. Water is abstracted via one or more boreholes and passed through a heat exchanger and then is returned to the aquifer via a separate borehole or boreholes, discharged to foul water drainage or released into a suitable available source such as a river. Typical groundwater supply temperatures are in the range 8–12°C and typical reinjection temperatures for a cooling mode operation is 12–18°C.

The properties of the aquifer are important in ensuring the necessary flow rates to achieve the energy requirements. Whilst reasonable flow rates normally are obtained from the major aquifers, such as the Chalk and the Sherwood Sandstones, there is a small risk that boreholes in these strata will not provide the required yield. In addition, reinjection of the abstracted water tends to be more difficult than abstraction and it is often necessary to install more than one recharge borehole to accept the discharge from a single abstraction borehole. This has obvious cost implications. This issue is often site-specific and an assessment of the hydrogeological conditions, in particular the depth to groundwater, should be obtained before progressing with the scheme.

All open loop abstractions from groundwater require an abstraction licence. If the water is being returned to the aquifer, it is designated as non-consumptive use. The system also requires an environmental permit to discharge the water back into the aquifer. Where it is a simple abstraction, it is regarded as a consumptive use and the abstraction charges are higher. In order to protect groundwater from the discharges from open loop GSHC systems, the Environment Agency restrict the change in temperature between the abstracted and discharged waters to no greater than +/- 8°C and the outflow water temperature must not exceed 25°C.

The efficiency of an open loop system can deteriorate if a large proportion of the reinjected water is contained within the water pumped from the abstraction borehole. This can occur if the abstraction and recharge boreholes are positioned too close to each other, allowing recirculation of the recharged water. This is not an uncommon event and in many circumstances cannot be avoided due to the size of the development. In order to minimize this effect, the abstraction and recharge boreholes should be separated by as large a distance as possible.

Open loop systems fed by groundwater at 8°C, can typically cool water to 12°C on the secondary side of the heat exchanger to serve conventional cooling systems.

Open loop systems are thermally efficient but over time can suffer from blockages caused by silt, and corrosion due to dissolved salts. As a result, additional cost may be incurred in having to provide filtration or water treatment, before the water can be used in the building.

There is also a risk that minerals can precipitate out of the groundwater, especially where the water is iron and/or manganese rich and there is either and/or a change in pressure or exposure of the water to air. This could require cleaning of the pipework and the borehole lining or some form of filtration as indicated above. Several water supply boreholes suffer from this problem, which can cause a reduction in borehole yield.

An abstraction licence and environmental permit/discharge consent need to be obtained for each installation, and this together with the maintenance and durability issues can significantly affect whole life operating costs, making this system less attractive.

A final point to consider is that access by plant will be required to the boreholes for pump replacement, maintenance and hence the boreholes should be located where access for plant is easily achievable.

Closed loop systems

Closed loop systems do not rely on the direct abstraction of water, but instead comprise a continuous pipework loop buried in the ground. Water circulates in the pipework and provides the means of heat transfer with the ground. Since groundwater is not being directly used, closed loop systems therefore suffer fewer of the operational problems of open loop systems, being designed to be virtually maintenance free, but do not contribute to the control of groundwater levels.

There are two types of closed loop system:

Vertical Boreholes – Vertical loops are inserted as U tubes into pre-drilled boreholes, typically less than 150 mm in diameter. These are backfilled with a high conductivity grout to seal the bore, prevent any cross contamination and to ensure good thermal conductivity between the pipe wall and surrounding ground. Vertical boreholes have the highest performance and means of heat rejection, but also have the highest cost due to associated drilling and excavation requirements.

Alterative systems to the U tubes are available on the market with larger diameter propriety pipework systems with flow and return pipes. The larger pipework can have better heat transfer properties, reduced resistance and therefore achieve better efficiencies or less pipework and therefore less ground area requirement than a traditional U tube system which may outweigh the additional cost per m length of the pipework.

As an alternative to having a separate borehole housing the pipe loop, it can also be integrated with the piling, where the loop is encased within the structural piles. This obviously saves on the costs of drilling and excavation since these would be carried out as part of the piling installation. The feasibility of this option would depend on marrying up the piling layout with the load requirement, and hence the number of loops, for the building.

Horizontal Loops – These are single (or pairs) of pipes laid in 2 m deep trenches, which are backfilled with fine aggregate. These obviously require a greater physical area than vertical loops but are cheaper to install. As they are located closer to the surface where ground temperatures are less stable, efficiency is lower compared to open systems. Alternatively, coiled pipework can also be used where excavation is more straightforward and a large amount of land is available. Although performance may be reduced with this system as the pipe overlaps itself, it does represent a cost effective way of maximizing the length of pipe installed and hence overall system capacity.

The case for heat pumps

Instead of using the groundwater source directly in the building, referred to as passive cooling, when coupled to a reverse cycle heat pump, substantially increased cooling loads can be achieved.

Heat is extracted from the building and transferred by the heat pump into the water circulating through the loop. As it circulates, it gives up heat to the cooler earth, with the cooler water returning to the heat pump to pick up more heat. In heating mode the cycle is reversed, with the heat being extracted from the earth and being delivered to the HVAC system.

The use of heat pumps provides greater flexibility for heating and cooling applications within the building than passive systems. Ground source heat pumps are inherently more efficient than air source heat pumps, their energy requirement is therefore lower and their associated CO_2 emissions are also reduced, so they are well suited for connection to a groundwater source.

Closed loop systems can typically achieve outputs of 50 W/m (of bore length), although this will vary with geology and borehole construction. When coupled to a reverse cycle heat pump, 1 m of vertical borehole will typically deliver 140 kWh of useful heating and 110 kWh of cooling per annum, although this will depend on hours run and length of heating and cooling seasons.

Key factors affecting cost

• The cost is obviously dependent on the type of system used. Deciding on what system is best suited to a particular project is dependent on the geological conditions, the peak cooling and heating loads of the building and its likely load profile. This in turn determines the performance required from the ground loop, in terms of area of coverage in the case of the horizontal looped system, and in the case of vertical boreholes, the depth and number or bores. The cost of the system is therefore a function of the building load.

- In the case of vertical boreholes for a closed loop system, drilling costs are a significant factor, as ground conditions can be variable, and there are potential problems in drilling through sand layers, pebble beds, gravels and clay, which may mean additional costs through having to drill additional holes or the provision of sleeving. The costs of drilling make the vertical borehole solution significantly more expensive than the equivalent horizontal loop.
- The thermal efficiency of the building is also a factor. The higher load associated with a thermally inefficient building obviously results in the requirement for a greater number of boreholes or greater area of horizontal loop coverage, however in the case of boreholes the associated cost differential between a thermally inefficient building and a thermally efficient one is substantially greater than the equivalent increase in the cost of conventional plant. Reducing the energy consumption of the building is cheaper than producing the energy from renewables and the use of renewable energy only becomes cost effective, and indeed should only be considered, when a building is energy efficient.
- With open loop systems, the principal risk in terms of operation is that the user is not in control of the quantity or quality of the water being taken out of the ground, this being dependent on the local ground and groundwater conditions. Reduced performance due to blockage (silting etc.) may lead to the system not delivering the design duties whilst bacteriological contamination may lead to the expensive water treatment or the system being taken temporarily out of operation. In order to mitigate the above risk, it may be decided to provide additional means of heat rejection and heating by mechanical means as a back-up to the borehole system, in the event of operational problems. This obviously carries a significant cost. If this additional plant were not provided, then there are space savings to be had over conventional systems due to the absence of heating, heat rejection and possibly refrigeration plant.
- With the design of open loop systems close analysis can be undertaken of existing boreholes in the area and also detailed desktop studies can be undertaken by a Hydrogeologist to determine the feasibility of the scheme and predict the yield for the proposed borehole. As noted above the yield of the borehole is subject to the local hydrogeological conditions in the location where the borehole is located and as such the delivered yield will only be validated after the borehole is completed and tested. There is a risk that you can get less (you could get more) than was predicted requiring additional boreholes (and cost) to be delivered to obtain the required yield.
- Open loop systems may lend themselves particularly well to certain applications increasing their cost effectiveness, i.e. in the case of a leisure centre, the removal of heat from the air-conditioned parts of the centre and the supply of fresh water to the swimming pool.
- In terms of the requirements for abstraction and disposal of the water for open loop systems, there are risks associated with the future availability and cost of the necessary licences, particularly in areas of high forecast energy consumption, such as the South East of England, which needs to be borne in mind when selecting a suitable system. Provided that the system is designed to be non-consumptive, a licence restriction is unlikely.
- Closed loop systems and non-consumptive open loop systems generally require balanced heating and cooling loads (if there is moving groundwater this is not as critical) in order to not cause the ground to heat up or cool down over time. If the heating and cooling loads (kWh) over the year are not balanced this can cause the ground to heat up or cool down over time causing a reduction in the efficiency of the system over time which will have a detrimental effect on the seasonal COPs and operational running costs of the system over time.
- Heat pump systems use electricity to power them and winter triad electrical costs should be considered in life cycle analysis for commercial and industrial users and appropriate controls storage systems put in place to mitigate costs.

Whilst open loop systems would suit certain applications or end user clients, for commercial buildings the risks associated with this system tend to mean that closed loop applications are the system of choice. When coupled to a reversible heat pump, the borehole acts simply as a heat sink or heat source so the problems associated with open loop systems do not arise.

Typical costs

Table 1 gives details of the typical borehole cost to an existing site in Central London, using one 140 m deep borehole working on the open loop principle, providing heat rejection for the 600 kW of cooling provided to the building. The borehole passes through rubble, river gravel terraces, clay and finally the Chalk, and is lined above the Chalk to prevent the hole collapsing. The breakdown includes all costs associated with the provision of a working borehole up to the well head, including the manhole chamber and manhole. The costs of any plant or equipment to deliver the water from the borehole are not included.

Heat is drawn out of the cooling circuit and the water is discharged into the Thames at an elevated temperature. In this instance, although the boreholes are more expensive than the dry air cooler alternative, the operating cost is significantly reduced as the system can operate at around three times the efficiency of conventional dry air coolers, so the payback period is a reasonable one. Additionally, the borehole system does not generate any noise, does not require rooftop space and does not require as much maintenance.

This is representative of a typical cost of providing a borehole for an open loop scheme within the London basin. There are obviously economies of scale to be had in drilling more than one well at the same time, with two wells saving approximately 10% of the comparative cost of two separate wells and four wells typically saving 15%.

Table 2 provides a summary of the typical range of costs that could expected for the different types of system based on current prices.

Table 1: Breakdown of the Cost of a Typical Open Loop Borehole System

Description	Cost £
General Items	
• Mobilization, insurances, demobilization on completion	21,000
• Fencing around working area for the duration of drilling and testing	2,000
• Modifications to existing LV panel and installation of new power supplies for borehole installation	15,000
Trial Hole	
• Allowance for breakout access to nearest walkway (Existing borehole on site used for trial purposes, hence no drilling costs included)	3,000
Construct Borehole	
• Drilling, using temporary casing where required, permanent casing and grouting	32,000
Borehole Cap and Chamber	
• Cap borehole with PN16 flange, construct manhole chamber in roadway, rising main, header pipework, valves, flow meter	12,000
• Permanent pump	14,000
Samples	
• Water samples	1,000
Acidization	
• Mobilization, set up and removal of equipment for acidization of borehole, carry out acidization	12,000
Development and Test Pumping	
• Mobilize pumping equipment and materials and remove on completion of testing	4,000
• Calibration test, pretest monitoring, step testing	4,000
• Constant rate testing and monitoring	20,000
• Waste removal and disposal	3,000
Reinstatement	
• Reinstatement and making good	2,000
Total	**145,000**

Table 2: Summary of the Range of Costs for Different Systems

System	Range			Notes
	Small – 4 kWth	Medium – 50 kWth	Large – 400 kWth	
Heat pump (per unit)	£3,000–5,000	£30,000–42,000	£145,000–175,000	
Slinky pipe (per installation) including excavation	£3,000–4,000	£42,000–52,000	£360,000–390,000 [1]	[1] Based on 90 nr 50 m lengths
Vertical, closed (per installation) using structural piles	N/A	£42,000–63,000	Not available	Based on 50 nr piles. Includes borehole cap and header pipework but excludes connection to pump room and heat pumps
Vertical, closed (per installation) including excavation	£2,000–3,000	£63,000–85,000	£370,000–400,000	Includes borehole cap and header pipework but excludes connection to pump room and heat pumps
Vertical, open (per installation) including excavation	£2,000–3,000	£45,000–65,000	£335,000–370,000	Excludes connection to pump room and heat exchangers

Estimator's Pocket Book, 2nd edition

Duncan Cartlidge

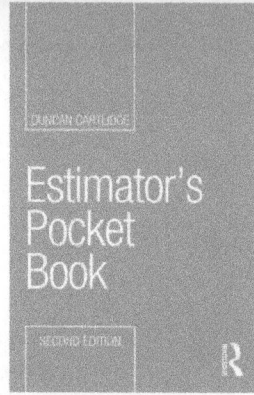

The *Estimator's Pocket Book, Second Edition* is a concise and practical reference covering the main pricing approaches, as well as useful information such as how to process sub-contractor quotations, tender settlement and adjudication. It is fully up to date with NRM2 throughout, features a look ahead to NRM3 and describes the implications of BIM for estimators.
It includes instructions on how to handle:

- the NRM order of cost estimate;
- unit-rate pricing for different trades;
- pro-rata pricing and dayworks;
- builders' quantities;
- approximate quantities.

Worked examples show how each of these techniques should be carried out in clear, easy-to-follow steps. This is the indispensable estimating reference for all quantity surveyors, cost managers, project managers and anybody else with estimating responsibilities. Particular attention is given to NRM2, but the overall focus is on the core estimating skills needed in practice. Updates to this edition include a greater reference to BIM, an update on the current state of the construction industry as well as up-to-date wage rates, legislative changes and guidance notes.

Routledge Pocket Books

February 2019: 198 × 129 mm: 292pp
Pb: 978-1-138-36670-1 : £24.99

To Order: Tel: +44 (0) 1235 400524 Fax: +44 (0) 1235 400525
or Post: Taylor and Francis Customer Services,
Bookpoint Ltd, Unit T1, 200 Milton Park, Abingdon, Oxon, OX14 4TA UK
Email: book.orders@tandf.co.uk

For a complete listing of all our titles visit:
www.tandf.co.uk

Taylor & Francis
Taylor & Francis Group

Large Scale Heat Pump

Introduction

With an increasing drive to demonstrate green credentials from developers and a demand from tenants for greener buildings the use of heat pumps to generate heating and cooling has been increasing not only as part of the heating/cooling installation but completely replacing the traditional approach of boilers and chillers.

The drive towards low-carbon fuels

There is a drive and demand from governments, developers, tenants and employees to be greener and reduce carbon. Commercial developments for offices and residential buildings are increasingly competing for tenants, sales and investors. One way to achieve the commercial edge is to be greener and demonstrate the drive to reduce carbon.

Legislation is driving the interest in the built environment to switch from traditional fuel sources to the use of lower carbon technologies which utilize renewable energy such as solar, geothermal, tidal, water and wind. Alternative heating and cooling systems are being developed to combat the ever growing issue of carbon and as a result large scale heat pumps are increasingly becoming a viable solution.

A Green grid

Whilst heat pumps demand more electricity than the traditional approach of boilers and chillers, the progressive movement of the UK electrical grid to provide 100% of its energy via renewable sources means that large heat pumps are becoming far more attractive than in the past.

Installing a heat pump that is run on electricity will see its carbon footprint reduced over its working life due to the ongoing decarbonization of the grid. The grid is going green by investing in more renewable sources of energy such as solar and wind.

This is a word of caution with demanding more and more electricity from the grid. There has been a drive in recent years to make office developments taller and increase the density of populations on floor plates to make projects economically viable. From a residential perspective, the growing population is driving the requirement for both low- and high-rise accommodation. In certain locations, electricity is becoming increasingly difficult to deliver at a cost that makes the use of heat pumps viable in terms of capex costs.

Other benefits

Despite the potential for heat pumps to lead to a higher capital cost than traditional systems, this can be offset during the lifecycle of the equipment due to higher efficiencies. The energy-saving characteristics of heat pumps make them eligible to be included within the ECA (Enhanced Capital Allowance) which means that businesses are able to set 100% of the asset costs against their taxable profits in a single tax year.

There is the potential to tap into the Renewable Heat Incentive – a government environmental programme that provides financial incentives to promote and increase the uptake of heating from renewable sources. It has two schemes (Non-Domestic and Domestic) which have separate tariffs, joining conditions, rules and application processes. Quarterly payments are given over a number of years (depending on the scheme) to eligible installations based on the amount of heat generated.

Systems comparison – Traditional vs Heat Pump

As with any alternative solution to the traditional approach there are numerous and interlinked advantages and disadvantages to be reviewed and taken account of when considering a change in strategy. The primary advantages and disadvantages are listed below.

Central Plant (Chiller/Boiler)

Advantages

- More stable heating/cooling source
- Fast heating (boiler)
- Lower initial cost
- High leaving water temperature (boiler)
- Large capacity available
- Traditional technology so known and understood by all parties
- Natural gas is still very cheap (boiler)
- Scalable

Disadvantages

- Less efficient conversion of energy
- High operating cost
- Burns fossil fuel (boiler)

Large Scale Heat Pump (ASHP)

Advantages

- Lower running cost
- Less maintenance
- Safer to use than combustion based system
- Low-carbon technology
- Efficient conversion of energy
- Longer life span
- No flue or impact on local air quality
- Industry compliant (LEED, ASHRAE)
- Provides both heating and cooling
- Does not contribute adversely to air quality

Disadvantages

- High initial cost
- Noise
- Limited heating output
- Capacity is reduced at low ambient temperature
- Relatively low flow temperature
- Requires careful design/application
- Require significant space to allow free air flow

Many of the highlighted disadvantages of the heat pump approach can be overcome. The initial high capex cost of the heat pump is often stated as the primary reason for not considering them, however with more units and manufacturers coming to market, the capital cost is reducing. Noise issues can be overcome with careful selection of plant and well-designed attenuation. The issues of limited heat output, reduced capacity at low ambient temperature, and low flow temperatures can be dealt with at early design stages. Heating issues can be dealt with via a small makeup boiler and whilst this would introduce a gas driven heat source into a scheme, the makeup boiler is considerably smaller than a traditional approach and is used intimately.

Cost model

The cost model comparison between an ASHP system and a traditional boiler and chiller central plant is based on a central London office development arranged over fourteen floors, with one basement level. The scheme has a GIFA of 15,000 m^2, with an approximate net lettable area of 11,250 m^2 – resulting in a net to gross efficiency of about 75%. The office has been designed to comply with Part L and to achieve a BREEAM Excellent rating. The occupational density (total NIA divided by the total number of workplaces in the building) of the scheme has been assumed to be 1:8.

The comparative costs between the traditional approach of boilers and chillers versus heat pump technology excludes any capital costs associated with upstream reinforcement of the electrical infrastructure as this is site and area specific and could negate any potential decision. However, it does take account of additional shell and core infrastructure to support the associated electrical load, such as additional substations, distribution boards and cabling.

	Central Plant				Heat Pump				Cost Variance
	Air Cooled Chiller + Gas Fired Boiler				Air Source Heat Pump (ASHP)				ASHP vs Cental Plant
Description	Qty	Unit	Rate	Total	Qty	Unit	Rate	Total	
Heat Source									
Boiler	1,170	kW	£47	£54,990		kW		not required	(£54,990)
Flue	100	m	£640	£64,000		m		not required	(£64,000)
Primary pump set/ Pressurization units	1	item	£58,000	£58,000	1	item	£87,000	£87,000	£29,000
Primary distribution	1	item	£87,000	£87,000	1	item	£87,000	£87,000	£0
Space Heating and Air Treatment									
Chiller	1,450	kW	£260	£377,000		kW		not required	(£377,000)
ASHP					1,450	kW	£340	£493,000	£493,000
CHW primary and secondary distribution	15,000	m²	£22	£330,000	15,000	m²	£25	£375,000	£45,000
LTHW secondary distribution	15,000	m²	£17	£255,000	15,000	m²	£21	£315,000	£60,000
Gas Installation									
Gas distribution	15,000	m²	£3	£45,000				not required	(£45,000)
Electrical Installation									
Mechanical equipment power supply	1	item	£50,000	£50,000	1	item	£50,000	£50,000	£0
Communications, Security and Control Systems									
Controls	1	item	£50,000	£50,000	1	item	£50,000	£50,000	£0
Total				£1,370,990				£1,457,000	£86,010

Notes on the above cost:

1.   Based on Shell and Core only, excludes all CAT A fit-out costs
2.   Base date of Q1 2019, and the prices used are reflective of a project procured through a competitive two-stage tender
3.   Unit rates are inclusive of MEP subcontractor testing and commissioning and MEP subcontractor preliminaries.
4.   Excludes builder's work, main contractor's preliminaries, design reserve, contingencies and inflation allowances.

The model compares a 1,450 kW cooling and 1,150 kW heating output from a boiler/chiller combination against the equivalent air source heat pump using 4 pipe water distribution systems.

The aim of the model is to present the cost differences associated with traditional plant compared to an ASHP. It should be noted that the cost variance could vary from project to project depending on but not limited to the following:

a.   Project location
b.   Project size, shape and configuration
c.   Procurement route
d.   Programme
e.   Design criteria (flow rates, material and performance specifications, etc.)

Later Living

Introduction

This cost model focuses upon a generic luxury later living development in the South of England formed of 50 number apartments comprising of 20 one bed units, 25 two bed units and 5 two bed 'close care units' that provide care and facilities for persons with dementia. The type of amenities provided can be, but are not limited to, a restaurant, hair salon, gym, spa, club lounge and parlour and the like all with 24-hour concierge and on-site nursing staff.

The approach of these style of developments is not to compromise on the level of facilities and location just because the age of the population is higher, in fact there can often be an enhancement in the type of amenities and facilities provided.

Later living developments are a lifestyle concept providing luxury developments to over 65 s. This is achieved by offering the best in design, technology, services and care, the goal is to provide on their doorstep everything the later generation need to live life to the fullest.

This article focuses on the high end later living developments located within the South of England.

Innovative technologies in the apartments

It is possible within apartments to include a number of elements that are not considered the norm in care apartments or developments. These technologies are offered incorporated within the 'close care' apartments.

The apartments are designed to ensure optional functionality incorporating innovative technologies such as 'smart' floors. The smart technology sits under the chosen floor covering within the apartment and has the ability to learn the resident's movements and set parameters so that alarms are not falsely triggered but still offers a high level of monitoring that can detect falls and alert the on-site staff to a resident in difficulties.

There has been an awful lot of research into the benefits of using lighting to assist with dementia patients and their care. Within the 'close care' apartments additional lighting may be incorporated within fixed furniture to establish zones within the apartments, low-level night-time lighting within bathrooms and kitchen and even way-finder lighting incorporated into floor finishes. The next step within lighting is the introduction of lighting control to provide circadian lighting. The control of the lighting within rooms and zones is pre-set remotely to provide the optimal level and type of lighting for individual apartments/residents.

The 'close care' units could incorporate aspirating smoke systems within apartments to detect fires at the earliest opportunity, fire suppression systems within cooker hoods that isolate the heat source and discharge a fine water mist that requires limited clean up in the event of discharge. Leak detection and flow monitor devices can be incorporated into bathrooms, kitchens and other wet areas to prevent flooding and accidental overflows from baths and sinks.

Health care

Health care is provided on site via 24-hour trained medical staff that are provided with sleeping accommodation. Within the later living developments residents can select the desired level of care that best suits their needs and medical requirements. On larger developments it is not uncommon to see a small medical facility that provides a physiotherapist, later living specialist and a doctor all on a visiting basis.

A limited number of apartments will have nurse call points and medical gases provided to them in order to meet additional care requirements of certain residents.

Concierge service and security

A development of this nature will include for in-house concierge service that is manned 24 hours a day aiming to answer the needs of all residents. Dealing with tasks such as making restaurant reservations, organizing dry cleaning or booking a taxi. Security is an important factor for residents. Access control and CCTV within later living developments are at an enhanced level and will often include CCTV within lifts, corridors and stair wells to assist in the tracking and monitoring of residents. The level of cameras and security whilst being enhanced is designed so that it is non-intrusive and discreet.

Amenity

A priority for these developments in not just to provide a place for residents to reside but to provide space to socialize and within later care developments this is critical to both the health and wellbeing of the residents.

The key operational driver is community, therefore developments that encourage interaction and engagement between residents are vital, as it helps tackle loneliness, creates social value and ensures residents can be independent, consequently, numerous amenity spaces are provided. These spaces can be quite large and numerous, thus affecting the efficiency of the development, and as a result of their incorporation within the development will come at a cost premium.

The focal point for socialization is through the Drawing rooms which are traditionally based on a library or music room.

High end luxury restaurants are provided within later life developments, furthermore a bespoke wine room is often on site enabling residents to explore and keep fine wine. The level of offering and experience is not compromised within developments such as this and can even be enhanced with feature lighting and music systems.

A fully equipped gym is specially designed for the residents with options for personal training and studios which provide personal or group Yoga and Pilates classes.

This style of development will often include a spa comprising of a swimming pool and sauna. Treatment rooms are available for spa styled massages or alternative therapist sessions such as physiotherapists.

As well as treatment rooms, beauty services are also available to residents including an in-house hair and beauty salon.

A high-quality private cinema room is available.

Private and secure gardens are available that incorporate sensory areas for residents.

Requirements of the extended family are often considered with guest suites available for overnight stays, the ability to stay with a resident within their apartment, have in-room catering and a play area for small children within the grounds.

Development details

This cost model is based on a high end later living development located in the South of England. The basis of the development is a 4-level apartment block with two cores from Ground to Level 3, totalling a GIA of 9,000 m^2 and an Apartment NIA of 5,000 m^2. There is a total of 50 apartments located from level 1 and above with the unit mix being based on 20 1-bedroom apartments at an average NIA of 80 m^2, 25 2-bedroom apartments at an average NIA of 110 m and 5 2-bed 'close care units' at an average NIA of 130 m^2.

Please note the cost model excludes;

1. Lifts
2. External works
3. Utilities
4. Inflation beyond 4th quarter 2019
5. Main contractor's oncosts
6. Fees
7. VAT

COST MODEL

Shell and Core Rates	Based on an GIA 9,000 m²		
	Total (£)	£/m² on GIA	£/ft² on GIA
5.1 Sanitaryware	£28,000	£3.11	£0.29
Sanitaryware to communal areas including changing rooms, restaurant, etc.			
Includes for subcontractor preliminaries.			
5.3 Disposal Installations	£290,000	£32.22	£2.99
Rainwater and soil and waste drainage all in HDPE pipework and insulated.			
Includes for testing and commissioning and subcontractor preliminaries.			
5.4 Water Installations	£321,000	£35.67	£3.31
Boosted cold water to apartments including plant.			
Boosted cold water distribution and hot water generation to communal areas. Includes for testing and commissioning and subcontractor preliminaries.			
5.5 Heat Source	£187,000	£20.78	£1.93
Heat generation plant and pipework, including boilers, CHP, pumps, etc.			
Flues to atmosphere. Includes for testing and commissioning and subcontractor preliminaries.			
5.6 Space Heating and Air Treatment	£653,000	£72.56	£6.74
LTHW distribution to apartments and communal areas.			
Air cooled chillers and associated plant generating chilled water.			
CHW distribution to apartments and communal areas. Includes for testing and commissioning and subcontractor preliminaries.			
5.7 Ventilation	£259,000	£28.78	£2.67
Core Smoke Extract (Colt System) with environmental boost to corridors.			
Central kitchen extract system. Includes for testing and commissioning and subcontractor preliminaries.			

5.8 Electrical Installations	£891,000	£99.00	£9.20

General LV to shell and core areas and ryefield installation to apartments.

General small power, containment and earthing and bonding.

Lighting to basement and feature lighting above ground.

Car charging points.
Includes for testing and commissioning and subcontractor preliminaries.

5.9 Lift Installation	£0	£0.00	£0.00

Excluded

5.10 Gas Installation	£32,000	£3.56	£0.33

Gas supply to plant rooms and kitchens.
Includes for testing and commissioning and subcontractor preliminaries.

5.11 Protective Installation	£156,000	£17.33	£1.61

Sprinkler installation to ground floor communal space and apartments.

Dry risers.

Lightning protection.
Includes for testing and commissioning and subcontractor preliminaries.

5.12 Communication, Security and Control Systems	£772,000	£85.78	£7.97

Fire alarm installation and disabled refuge call system.

Door entry system, IRS system, fibre installation, telephone installation and data.

Access control and closed circuit television.

General containment.

Medical alert system within shell and core areas.

BMS system including energy monitoring system.
Includes for testing and commissioning and subcontractor preliminaries.

5.13 Specialist Installation	£0	£0.00	£0.00

Not applicable.

5.14 Builders Work in Connection with Services	£108,000	£12.00	£1.11

BWIC

Utilities	£0	£0.00	£0.00

Excluded

MEP Related External Works	£0	£0.00	£0.00

Excluded

Shell and Core Total Costs	**£3,697,000**	**£410.78**	**£38.16**

Communal Areas Fit-Out Rates	Based on an NIA 1,350 m²		
	Total (£)	£/m² on NIA	£/ft² on NIA
1 Reception	£83,000	£61.48	£5.71
Includes for heating and cooling, ventilation, small power, lighting, fire detection and IT/AV.			
Excludes pendant feature lighting. Includes for testing and commissioning and subcontractor preliminaries			
2 Restaurant	£173,000	£128.15	£11.91
Includes for heating and cooling, ventilation, small power, lighting, fire detection, sound system and IT/AV.			
Excludes pendant feature lighting. Includes for testing and commissioning and subcontractor preliminaries.			
3 Restaurant Kitchen	Shell and Core		
4 Wine Room	£64,000	£47.41	£4.40
Includes for temperature control, ventilation, small power, lighting, fire detection, sound system and IT/AV.			
Excludes pendant feature lighting. Includes for testing and commissioning and subcontractor preliminaries.			
5 Gym and Studios	£83,000	£61.48	£5.71
Includes for heating and cooling, ventilation, small power, lighting, fire detection, sound system and IT/AV.			
Includes for testing and commissioning and subcontractor preliminaries.			
6 Sauna, Swimming Pool and Treatment Rooms	£314,000	£232.59	£21.61
Includes for heating and cooling, ventilation, small power, lighting, fire detection, sound system and IT/AV.			
Excludes plant for sauna and swimming pool. Includes for testing and commissioning and subcontractor preliminaries.			
7 Beauty/Hair Salon	£42,000	£31.11	£2.89
Includes for heating and cooling, ventilation, small power, lighting, fire detection, sound system and IT/AV. Includes for testing and commissioning and subcontractor preliminaries.			
8 Cinema Room	£127,000	£94.07	£8.74
Includes for heating and cooling, ventilation, small power, lighting and fire detection.			
Includes IT/AV Equipment for cinema system. Includes for testing and commissioning and subcontractor preliminaries.			
9 Drawing Rooms	£166,000	£122.96	£11.42
Includes for heating and cooling, ventilation, small power, lighting, fire detection, sound system and IT/AV.			
Excludes pendant feature lighting Includes for testing and commissioning and subcontractor preliminaries.			
Communal Areas Fit-Out Total Costs	**£1,052,000**	**£779.26**	**£72.40**

Apartment Fit-Out Rates	Based on an NIA 5,000 m²		
	Total (£)	£/m² on NIA	£/ft² on NIA
5.1 Sanitaryware	£589,000	£117.80	£10.94
1 bed comprised of an ensuite with a WC, WHB, individual bath and shower and a water closet with WC and WHB.			
2 bed comprised of an master ensuite with a WC, WHB, individual bath and shower, a secondary ensuite with a WC, WHB and shower and a water closet with WC and WHB. Includes for subcontractor preliminaries.			
5.3 Disposal Installations	£52,000	£10.40	£0.97
Waste pipework serving sanitaryware and whitegoods.			
Condensate pipework to fan coil units, interface units and mechanical ventilation heat recovery unit. Includes for testing and commissioning and subcontractor preliminaries.			
5.4 Water Installations	£234,000	£46.80	£4.35
Copper hot and cold water pipework complete with phenolic foam insulation serving sanitaryware and whitegoods.			
Includes for testing and commissioning and subcontractor preliminaries.			
5.5 Heat Source	£194,000	£38.80	£3.60
Heat interface unit with energy meter.			
Chilled water valve set with energy meter. Includes for testing and commissioning and subcontractor preliminaries.			
5.6 Space Heating and Air Treatment	£938,000	£187.60	£17.43
LTHW underfloor heating throughout providing heating.			
2 pipe fan coil units within bedrooms, living areas and kitchen to provide cooling.			
Heated towel rails within bathrooms. Includes for testing and commissioning and subcontractor preliminaries.			
5.7 Ventilation	£290,000	£58.00	£5.39
Mechanical ventilation heat recovery unit (MVHR unit) with PVC ductwork complete with attenuation and linear grilles. The fresh air and exhaust ductwork is insulated.			
MVHR boost system including override switch within the utility cupboard and kitchen, double pole light switches in bathrooms proving a boost function.			
No allowance for kitchen extract hood or ductwork. Includes for testing and commissioning and subcontractor preliminaries.			

5.8	**Electrical Installations**	£815,000	£163.00	£15.14

Consumer unit.

Small power with brushed finished sockets.

Lighting including downlights throughout, cove linear lighting to bedrooms, mirror lighting, downlighters within kitchen cabinets, low level wayfinder lighting and 5 amp sockets.
Includes for testing and commissioning and subcontractor preliminaries.

5.9	**Lift Installation**	£0	£0.00	£0.00

Not applicable

5.10	**Gas Installation**	£0	£0.00	£0.00

Not applicable

5.11	**Protective Installation**	£108,000	£21.60	£2.01

Sprinkler installation with individual flow switches.
Includes for testing and commissioning and subcontractor preliminaries.

5.12	**Communication, Security and Control Systems**	£814,000	£162.80	£15.12

LD1 Fire detection system.

TV and Data with brushed finished sockets.

Security wireways for future intruder detection installation.

Video entry panel, and smart heating/cooling and lighting control.
Includes for testing and commissioning and subcontractor preliminaries.

5.13	**Specialist Installation**	£0	£0.00	£0.00

Not applicable

5.14	**Builders Work in Connection with Services**	£81,000	£16.20	£1.51

BWIC

	Fit-Out Total Costs	**£4,115,000**	**£823.00**	**£76.46**

Extra over for close care apartments (Cost per 2-bed apartment)	£24,000	£4.80	£0.45

Extra over for uplifts to dementia care apartments including: Stand-alone flood/leak detectors to all bathrooms and kitchens, Aspirating fire detection system, Full home automation system including video entry panel, medical alert signalling/nurse call within each room, link between leak and temperature detectors with the extra care team, heating/cooling controls and Circadian lighting control.

Fit-Out Uplift Per 2-Bed Extra Care Apartment	**£24,000**	**£184.62**	**£17.15**

The Client Role in Successful Construction Projects

Jason Challender *et al.*

The Client Role in Successful Construction Projects is a practical guide for clients on how to initiate, procure and manage construction projects and developments. This book is written from the perspective of the client initiating a construction project as part of a business venture and differs from most available construction literature which can externalise the client as a risk to be managed by the design team. The book provides a practical framework for new and novice clients undertaking construction, giving them a voice and enabling them to:

- Understand the challenges that they and the project are likely to face.
- Communicate and interact effectively with key stakeholders and professionals within the industry.
- Understand in straightforward terms where they can have a positive impact on the project.
- Put in place a client-side due diligence process.
- Reduce their institutional risk and the risk of project failure.
- Discover how their standard models are able to co-exist and even transfer to a common client-side procedure for managing a construction project.

Written by clients, for clients, this book is highly recommended not only for clients, but for construction industry professionals who want to develop their own skills and enhance their working relationship with their clients. A supporting website for the book will be available, which will give practical examples of the points illustrated in the book and practical advice from specialists in the field.

May 2019: 234 × 156 mm: 318pp
Pb: 978-1-138-05821-7 : £29.99

LED Lighting

Background

LED lighting technology is now sufficiently developed to be widely accepted as the light source of choice for most lighting applications. The technology offers benefits including energy efficacy, high quality light appearance, light colour adjustability, device connectivity and controllability. This makes LED luminaires suitable for applications ranging from street lighting to office and residential lighting.

Currently available types of LED luminaries

Light emitting diodes (LEDs) are small high efficacy light sources that can be packaged into a range of products suitable to both new build and retrofit applications. In new build applications LEDs are often built in to the luminaire and form an integral part of it, which can offer some technical benefits, whereas in retrofit applications LEDs are packaged into lamp formats that can easily be inserted into existing light fixtures to replace tradition light sources. There are many types of retrofit LED lamps from opalescent and filament LED bulbs that are an alternative to halogen and compact fluorescent bulbs to linear LED tubes that are an alternative to linear fluorescent lamps.

LED efficacy

LEDs are commercially available with an efficacy of over 120 lumens per watt. This makes them far more efficient than halogen lamps, which have a typical efficacy range of 5–15 lumens per watt and are now banned in the UK, and more efficient than both fluorescent and discharge technology lamps, which have an efficacy range of 40–100 lumens per watt.

LED light colour

LEDs are available in a wide range of light colour appearances. Very warm white with a colour appearance of 1800 K to very cool white with a colour appearance of over 10,000 K is available. In addition, many saturated colour options are available that can be blended together to create millions of colour options.

Typical LED colour rendering characteristics vary from CRI 65–95+ according to a product's priority for efficacy or colour appearance fidelity. In addition, specialized LED products are available with light spectral properties tuned to specific applications including human centric circadian lighting and retail lighting requirements.

LED control

In both new build and retrofit applications a wide range of LED control options are available, both wired and wireless. This makes it possible to dim LEDs and create programmable colour change effects in both new build and retrofit applications.

Cost implications

Manufacturers and suppliers are keen to highlight potential energy saving from LED lighting technology. Although the principle of realizable savings is completely valid, detailed study of potential energy savings is essential to ensure the most beneficial and appropriate LED technologies and design approaches can be implemented.

Payback

LED lamps/luminaires are typically more expensive than traditional lighting alternatives across most types. However, the cost variance and energy saving potential for different LED lamp/luminaire types are significantly varied. Therefore, the return on investment from different LED lighting typologies and implementation approaches is not constant and requires detailed study to forecast in detail.

The following provides a return on investment overview for a sample LED Luminaire typology and excludes the cost of installation labour which is assumed to be consistent

Power cost (£)/kWh	0.12
Days in use/year	260
Hours in use/day	12

	Typical LED Downlight	Compact Fluorescent Downlight	LV Halogen (IRC) Downlight
Input Power (watts)	6	12	35
Lifetime (hours)	50,000	12,000	4,000
Replacement Lamp Cost including labour (£)	–	15.00	15.00
Annual Energy Cost per lamp (£)	2.26	4.53	13.21
Total Energy Cost for 25 luminaires per year (£)	56.50	113.25	330.28
Luminaire Unit Cost (£)	110.00	60.00	80.00
Supply only cost of 25 luminaires (£)	2,750.00	1,500.00	2,000.00
Allowance for 4 nr emergency battery packs (£)	320.00	320.00	320.00
Lamps (£)	Included	375.00	375.00
Total cost of luminaries/emergencies/ lamps (£)	3,070.00	2,195.00	2,695.00
Extra over cost of LED luminaires and emergency luminaires compared to traditional luminaires is (£)		875.00	375.00
Yearly cost of energy and yearly re-lamping allowance (compact fluorescent @ every 4 years; halogen @ every 1.5 year) (£)	56.50	207.00	580.28
The calculated yearly energy and re-lamping saving of using LED luminaires when compared to traditional luminaires is (£)		150.50	523.78
Therefore time taken in years to 'pay' for the additional cost of the LED luminaires based on the energy and re-lamping costs of 'traditional' luminaires in **years** is		**5.81**	**0.72**

Power cost (£)/kWh	0.12
Days in use/year	260
Hours in use/day	12

	Recessed LED Office Luminaire	Recessed Fluorescent Office Luminaire
Input Power (watts)	37.5	42
Typical Lifetime (hours)	50,000	12,000
Replacement Lamp Cost including labour (£)	–	20.00
Annual Energy Cost per lamp (£)	14.15	15.85
Total Energy Cost for 25 luminaires per year (£)	353.93	396.28
Luminaire Unit Cost including lamps (£)	185.00	165.00
Supply only cost of 25 luminaires including lamps (£)	4,625.00	4,125.00
Allowance for 4 nr emergency battery packs (£)	480.00	400.00
Total cost of luminaries/emergencies/lamps (£)	5,105.00	4,525.00
Extra over cost of LED luminaires and emergency luminaires compared to fluorescent luminaires is (£)		580.00
Yearly cost of energy and yearly re-lamping allowance (fluorescent @ every 4 years) (£)	353.93	521.28
The calculated yearly energy and re-lamping saving of using LED luminaires when compared to fluorescent luminaires is (£)		167.35
Therefore time taken in years to 'pay' for the additional cost of the LED luminaires based on the energy and re-lamping costs of fluorescent luminaires in **years** is		**3.47**

Notes and conclusions

1. LED luminaires do save energy.
2. LED luminaires are generally twice the price of a 'traditional' downlight but LED luminaire prices are becoming more competitive as the technology develops.
3. Some LED luminaires may not be converted to emergency so 'additional' luminaires may be required.
4. LED replacement lamps may have completely different light emitting characteristics. And, the luminaire may require modification to accept the retrofit LED lamp.

Local Energy Networks: Key to Our Low Carbon Future

For the industrial sector, long reliant on carbon-intensive gas boilers and electricity drawn from an increasingly overburdened national grid, meeting the government's legally binding net zero emissions target will require wholesale changes to the way power is bought and consumed. The race to net zero is now on — and shifting to low-carbon energy generated within local networks could prove key to meeting the state's ambitious target in time.

As well as providing an alternative energy source, eliminating or reducing reliance on centralized, third-party energy suppliers sooner rather than later could result in financial gain.

De-monopolizing the major utility companies of old means businesses can start trading in locally produced, cleaner, flexible forms of heating, cooling and power. To meet the 2050 target, carbon taxes on industry will almost certainly rise, hurting businesses which fail to adapt.

Right now, there is no single technology able to bring a building, business, or society as a whole to net zero. Each development, whether at a building, estate, town or city-wide scale, will require bespoke solutions which meet its own unique needs and features.

A raft of low-carbon components can make up a district energy network. AECOM has identified some of the most promising (refer to Figure 1 below.)

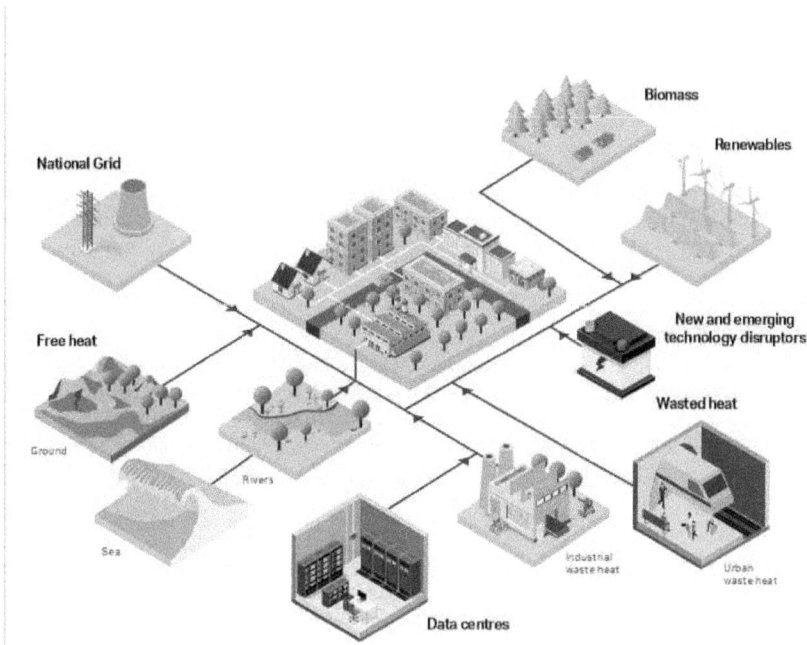

Figure 1: Decentralized Energy – linking our natural and industrial potential to create a low carbon economy.

Waste industrial heat

Industry emits a huge amount of waste heat, which is typically released into the atmosphere. This heat can be captured via heat pumps and redistributed back into a building, curbing energy consumption and carbon emissions. This recovered heat can also be integrated into a wider district heating network, providing heating and hot water across, for example, an industrial park, hospital or a town.

Natural capital

The UK may not be known for balmy temperatures, but the country is nonetheless home to rivers, lakes, coastlines and canals which hold heat and energy. Heat pumps use a similar technique to refrigeration systems to absorb and extract heat from the air, the ground or from water — used to provide both industrial and domestic heat, as well as hot water. AECOM is setting up the Natural Capital Laboratory as a testbed site near Loch Ness, Scotland, to measure and quantify natural capital. Government, NGOs, researchers and clients will carry out research and experiments on site, including the exploration of natural heat energy for local networks.

Renewables

Renewables now offer estate owners a host of low-cost energy solutions — and battery storage technologies are developing to allow businesses to store or sell the surplus power they generate. Major corporates from Google to IKEA are now procuring their own exclusive electricity supply from dedicated onshore and offshore wind farms. In the public sector, AECOM is working with Transport for London (TfL) to explore the use of on-site solar panels to help take the transport authority to net zero.

Combined heat and power systems [CHP]

In recent years CHP has stopped being a low carbon technology, due to the decarbonizing grid. However, CHP systems could still help bridge the gap between today's carbon-intensive industries and zero-carbon future; CHP systems, which tap the heat generated by thermal power generators, may provide the financial incentives that are crucial to the success of low carbon projects. It is not all bad though. In 2018 there were over 2,400 hours when the electricity grid's emissions factor was 'dirty' enough that using CHP would have saved carbon. Those periods also coincided with the colder months, when electricity is at its most expensive. We could have our cake and eat it!

Natural gas

Cleaner gases can be fed into the nation's gas systems. For example, UK energy regulator Ofgem and gas distributors Cadent and Northern Gas Networks, are trialling a blend of hydrogen in their natural gas networks to reduce carbon dioxide emissions. By switching to cleaner gases like hydrogen and biomethane, significant adaptations to the nation's existing infrastructure may be needed — raising the question of who will pay for the changes?

Carbon capture and storage (CCS)

CCS technology – whereby carbon dioxide is captured as it is emitted and then stored, typically underground – could allow carbon-emitting buildings, power plants and even vehicles to continue to operate in a net zero future. Investment and research are needed to bring CCS to commercial scale; the UK abandoned a £1 billion CCS development programme in 2015. However, the technology could enjoy a reprieve after a 2019 government report by the Committee on Climate Change branded CCS a 'necessity' if we are to reach net zero by 2050.

Building a local, low carbon future

An increasingly climate-aware public is demanding a shift towards sustainable energy – and some political parties are even lobbying for the UK to reach net zero two decades sooner – by 2030. This is not just blue sky thinking. Rapid evolutions are now the norm in the energy industry.

Take renewable energy. In 1991, the nation's first commercial wind farm was built. This year, renewable energy overtook fossil fuels as the country's primary source of power for the first time since the Industrial Revolution. The UK's pivot from fossil fuels to green energy took less than three decades.

To achieve net zero, local and central government will need to support the public and private sector's work by delivering planning laws and regulation that enable the creation of low-carbon energy networks. A net-zero society won't be created overnight; but with state support and the uptake of existing and emerging solutions, we just might hit the 2050 deadline.

AECOM works on delivering long-term cuts to large public sector estate owners. We are regularly tasked with reducing carbon emissions, energy costs, and building stronger resilience against building, system and utility failures (refer to Figure 2 below).

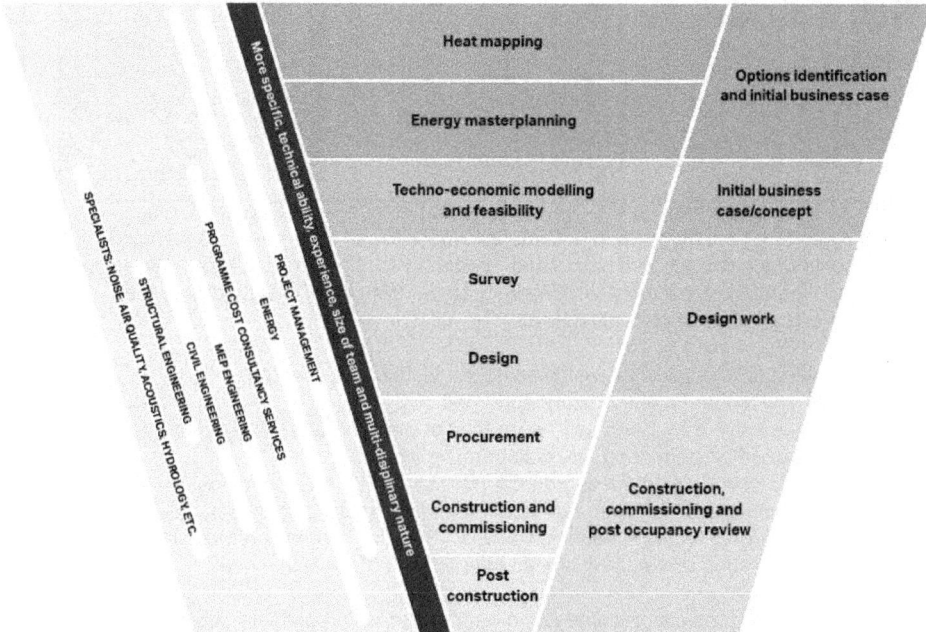

Figure 2: AECOM can offer a range of services, taking the net zero carbon projects from concept to construction

To do this, we design energy networks with electricity, heating and cooling systems all working together, tapping into the local natural potential and renewables. Feeding energy across the estate we enable load shifting and dynamic balancing to ensure a constant, clean supply of energy.

Estimating and Tendering for Construction Work, 5th edition

Martin Brook

MARTIN BROOK
Estimating and Tendering for Construction Work
FIFTH EDITION
R

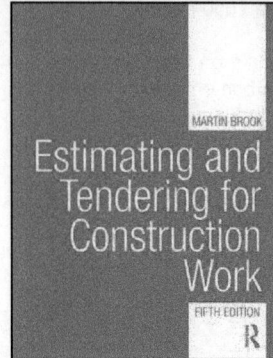

Estimators need to understand the consequences of entering into a contract, often defined by complex conditions and documents, as well as to appreciate the technical requirements of the project. Estimating and Tendering for Construction Work, 5th edition, explains the job of the estimator through every stage, from early cost studies to the creation of budgets for successful tenders.

This new edition reflects recent developments in the field and covers:

- new tendering and procurement methods
- the move from basic estimating to cost-planning and the greater emphasis placed on partnering and collaborative working
- the New Rules of Measurement (NRM1 and 2), and examines ways in which practicing estimators are implementing the guidance
- emerging technologies such as BIM (Building Information Modelling) and estimating systems which can interact with 3D design models

With the majority of projects procured using design-and-build contracts, this edition explains the contractor's role in setting costs, and design statements, to inform and control the development of a project's design.

Clearly-written and illustrated with examples, notes and technical documentation, this book is ideal for students on construction-related courses at HNC/HND and Degree levels. It is also an important source for associated professions and estimators at the outset of their careers.

July 2016; 246 × 189 mm, 334 pp
Pbk: 978-1-138-83806-2; £38.99

To Order: Tel: +44 (0) 1235 400524 Fax: +44 (0) 1235 400525
or Post: Taylor and Francis Customer Services,
Bookpoint Ltd, Unit T1, 200 Milton Park, Abingdon, Oxon, OX14 4TA UK
Email: book.orders@tandf.co.uk

For a complete listing of all our titles visit:
www.tandf.co.uk

Taylor & Francis
Taylor & Francis Group

Residential Heat Pump

Building designers are driven by legislative and market aspirations to create buildings that are more energy efficient and carbon friendly, and to work towards Grid decarbonization.

The National Grid's Electricity System Operator (ESO) is aiming for it to be zero carbon by 2025. To achieve this, the system needs to be transformed so it can be operated safely and securely at zero carbon, whenever there is sufficient renewable generation online to meet the total national load. This means reductions in the demand for high-carbon-generated electricity.

Since January 2019, the Greater London Authority (GLA) has required designers to use SAP10 emissions factors for planning permissions. These have far lower emissions factors for electricity and are forcing engineers to rethink the heating, ventilation and air conditioning (HVAC) design.

Traditionally, residential buildings include gas-fired boilers and, potentially, combined heat and power (CHP) and air or water-cooled chillers. Increasingly, these appear to fall short of the new London Plan's requirements. Also, many new residential developments are prone to overheating in risers and corridors, largely driven by internal gains and new building methods.

A heat pump energy system employing an 'ambient' or 'neutral' loop can typically offer standalone heating, cooling and hot-water systems – or a mixture of services to flats and commercial spaces. The heat pump energy system can, potentially, offer the designer more flexibility in choosing the most suitable type of plant. During the summer and winter seasons, the communal low-temperature heat loop employs plant to regulate the water temperatures circulated around the loop for heating and cooling – for example, ground and/or air-source heat pumps or a chiller. Excess thermal energy from the air source heat pumps can also be injected back into the low-temperature heat loop, giving a further reduction to plant sizes. To meet peak-time demands, thermal stores should be used to manage plant operations. Lower temperatures in the loop compared with traditional gas combined heat and power also means there is less heat loss in transmission.

There are two types of system – ambient loop, on which this cost model is based, and a central air/water loop. A system primarily consists of centralized plant, an energy loop and a selection of heat pumps. The heat pumps are connected to an energy loop, which is a water circuit maintained at 25°C flow temperature and 15°C return water temperature.

This energy loop is regulated within its operating parameters using centralized heating and cooling plant, or is connected to a wider area heat network.

The communal low-temperature heat loops are connected from the mechanical risers to a compact air source heat pump, installed in each flat and commercial mixed-use space.

Cost drivers

A heat pump ambient loop system also replaces the need for independent heat interface units (HIUs) and/or cooling interface units (CIUs) by providing hot water, space heating and/or comfort cooling within each flat or mixed-use commercial space – using radiators, underfloor heating, fan coil units or heat convectors, for example. Hot water is provided by a localized cylinder, which is charged by the heat pump.

Savings can be increased with a dual heating and cooling pump energy system, for simultaneous operation.

Compared with conventional high-pressure heating systems using HIUs in flats, a heat energy system is more efficient, because lower temperatures reduce the community energy losses from 25–30% to 2–5% and – as a result of lower distribution temperatures – overheating in risers and corridors is reduced significantly.

The units can be supplied in prefabricated, self-contained modular sections, reducing installation times and on-site labour hours, and offering design coordination and commissioning benefits.

Other potential construction costs that can be considered when designing a heat pump energy loop system include reduced sizes of distribution pipework, compared with traditional carbon steel for low-temperature hot water (LTHW) and chilled water (CHW), and a reduction in thermal insulation.

The initial costs of installing a 'heat only' pump energy system are more expensive, but savings can be achieved:

- Carbon tax payment per flat can be reduced by 40–45%
- Plantroom spaces can be decreased in size by 40–50%
- Riser spaces increased similarly by 40–50%
- Cost per kWh thermal reduced by 25–30%
- Tenant running costs reduced by 15–20%

The cost model

A traditional heating and cooling design, consisting of a 450 kW gas boiler and a 70 kWe gas CHP unit linked to HIUs, providing heating and hot water to each flat. Residents are provided with cooling via a 560 kW centralized chiller on a dedicated chilled water network linked to CIUs.

The heating-only table shows a net increase in development costs of up to £2,415 per flat by adopting a heat pump energy system for a heating-only scheme, compared with one using CHP and a gas boiler.

By using heat pump technology as the central plant, and offsetting carbon tax payments, the total net increase in development costs is reduced to £1,645 per apartment.

The heating and cooling table shows a net saving in development costs of £150 per flat by adopting a heat pump energy system for a heating and cooling scheme, compared with one using CHP and a gas boiler with chiller.

With the addition of using heat pump technology as the central plant, and offsetting carbon tax payments, the total net savings increase to £1,850 per flat.

Heating only

Description	Cost per Apartment (£)
Additional plant costs per apartment	
4kW Heat pump cylinder combination unit in apartment	3,700
Instantaneous domestic hot water (DHW) and heating production in each apartment	1,000
Sub-Total	4,700
Generated savings against traditional systems	
HIU for each apartment installation	(1,000)
Meter for HIU	(150)
Cost to fit and commission meter	(550)
Boiler size reduction	(120)
Thinner insulation on riser/corridor pipework	(50)
Savings on valve sizes and spindle lengths (four per apartment)	(70)
Reduction in pipe size on apartment spur from 28 mm to 22 mm	(20)
Change from copper to plastic	(25)
Removal of modifications to smoke extract system for corridor purge ventilation (overheating abatement)	(300)
Sub-Total	(2,285)
Net cost/saving per apartment	2,415
Changing central plant to a heat pump, removal of technologies like PV to achieve carbon compliance	(300)
Carbon tax saving due to changing to a heat pump	(470)
Sub-Total	(770)
Total net cost/savings cost per apartment (potential)	1,645

Heating and cooling

Description	Cost per Apartment (£)
Additional plant costs per apartment	
4kW Heat pump cylinder combination unit in apartment	
Additional cost of cooling in heat pump	3,700
Additional 50 l buffer tank or 7m of 100 mm pipework for Fan Coil	350
Upgrade of fan coil to work with Heat Pump	300
Instantaneous domestic hot water (DHW) and heating production in each apartment	300
Sub-Total	1,000
Generated savings against traditional systems	5,650
HIU for each apartment including installation	(1,000)
CIU for apartment including installation	(1,000)
A meter for each of the heating and cooling HIUs	(300)
Cost to fit and commission heat meters	(1,100)
Boiler size reduction	(120)
Thinner insulation specification on riser/corridor pipework	(50)
Savings on valve sizes and spindle lengths (four per apartment)	(140)
Reduction in pipe size on apartment spur from 28 mm to 22 mm	(40)
Change from copper to plastic pipework	(50)
Removal of modifications to smoke extract system for corridor purge ventilation	(300)
Removal of chilled water distribution from building risers and corridors (changing from four pipes to two pipes)	(1,500)
Removal of cooling pipework in apartment (changing from four pipes to two pipes)	(200)
Sub-Total	5,800
Net cost/saving per apartment	(150)
Changing central plant to a heat pump, removal of technologies like PV to achieve carbon compliance	(300)
Removal of chiller if adjacent to water source	(1,000)
Carbon tax saving due to changing to a heat pump	(400)
Sub-Total	(1,700)
Total net cost/savings cost per apartment (potential)	(1,850)

Net Zero Energy Building

Ming Hu

NET ZERO
ENERGY
BUILDING
PREDICTED AND
UNINTENDED
CONSEQUENCES
MING HU

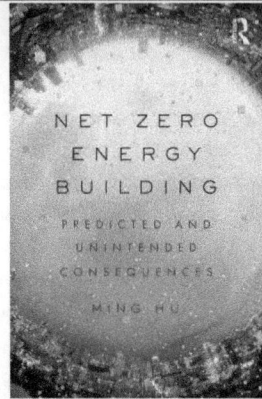

What do we mean by net zero energy? Zero operating energy? Zero energy costs? Zero emissions? There is no one answer: approaches to net zero building vary widely across the globe and are influenced by different environmental and cultural contexts.

Net Zero Energy Building: Predicted and Unintended Consequences presents a comprehensive overview of variations in 'net zero' building practices. Drawing on examples from countries such as the United States, United Kingdom, Germany, Japan, Hong Kong, and China, Ming Hu examines diverse approaches to net zero and reveals their intended and unintended consequences.

Existing approaches often focus on operating energy: how to make buildings more efficient by reducing the energy consumed by climate control, lighting, and appliances. Hu goes beyond this by analyzing overall energy consumption and environmental impact across the entire life cycle of a building—ranging from the manufacture of building materials to transportation, renovation, and demolition. Is net zero building still achievable once we look at these factors?

With clear implications for future practice, this is key reading for professionals in building design, architecture, and construction, as well as students on sustainable and green architecture courses.

April 2019: 162 pp
Pb: 978-0-8153-6780-2: £29.99

To Order: Tel: +44 (0) 1235 400524 Fax: +44 (0) 1235 400525
or Post: Taylor and Francis Customer Services,
Bookpoint Ltd, Unit T1, 200 Milton Park, Abingdon, Oxon, OX14 4TA UK
Email: book.orders@tandf.co.uk

For a complete listing of all our titles visit:
www.tandf.co.uk

Taylor & Francis
Taylor & Francis Group

RICS SKA Rating

SKA Rating is a Royal Institute of Chartered Surveyors (RICS) sustainability assessment method that helps landlords and tenants assess fit-out projects against a set of sustainability good practice criteria. Previously, although there were established tools for assessing the environmental impact of new builds, existing buildings were a blind spot in the industry. It is estimated that 11% of UK construction spending is on fit-outs and that buildings may have 30–40 fit-outs during their lifecycle. Therefore, the creation of SKA Rating in 2009 looked to address this gap in the market. Since then it has become a widely adopted assessment method with over 6,000 users and 687 certified projects. It can be used to assess the following building types:

- Offices
- Retail
- Higher Education

Fit-out projects are typically of a short duration and vary widely in scope from a quick refresh to a minor refurbishment or a comprehensive fit-out programme. SKA Rating was designed by the industry to be aligned with these relatively short timeframes for both the design and construction phases of a project. It is a flexible scheme that only assesses what's within the control of the project team and is not penalized by the limitations of the base build. This means that the project team can get rewarded for implementing a good practice fit-out, even if the building has poor transport links. Unlike other tools it provides the assessor with more ownership of what's in scope compared to other schemes that are more rigid. It is a more granular scheme that rewards incremental improvement within the fit-out and avoids complicated calculation tools or additional specialists that would not usually be appointed to a project team, which is a benefit to the client.

SKA is a freely accessible tool that is entirely online with the guidance, assessment methodology and certification generated through the tool. The aim is to promote good practice in the fit-out marketplace, so the online tool and the associated good practice guidance can be accessed by anyone (at ska-tool.rics.org). Also, anyone can train as an assessor and RICS provides a one-day Foundation course via the RICS Online Academy. Then there are separate online modules for the Offices, Retail and Higher Education Schemes.

There are currently nearly 400 accredited SKA assessors. Qualified SKA assessors can assess projects and can generate certificates once they have demonstrated that the project meets the SKA criteria.

SKA covers the following topics and issues:

- Reducing energy and water use by selecting efficient equipment and promoting metering, etc.
- Selecting materials with lower environmental impact
- Reducing construction site impacts, including waste
- Reducing pollutants such as refrigerant leakage
- Promoting health and wellbeing in the working environment, including improving internal air quality, daylight, etc.
- Rewarding more sustainable project delivery, such as registration to the Considerate Constructors Scheme and seasonal commissioning.

These issues are addressed by providing a list of good practice measures that are rewarded if the project demonstrates implementation. For example, SKA aims to reduce energy use in the completed fit-out by promoting the use of energy efficient equipment. These are assessed over three project stages:

1. **Design/Planning.** At this stage the measures in scope are identified with the project team and a route to the target rating is established. This will be dependent on the impacts of the SKA to the programme, budget and priorities of the client. Criteria will be embedded into the design specification documentation and contractor preliminaries. No certificate is generated at this stage, just confirmation that the criteria has been incorporated into the design.
2. **Delivery/Construction.** This stage involves the gathering of evidence from Operation and Maintenance manuals and other sources to prove that what has been specified has actually been delivered, and that the performance and waste benchmarks have been achieved. A Handover Certificate is produced once the SKA assessor has reviewed and approved all the available evidence and it has been uploaded to the SKA online tool.
3. **Occupancy Stage Assessment.** Finally, there is the option to review how well a fit-out has performed in use against its original brief from a year after completion. A separate certificate can be issued at this Occupancy Stage if the evidence has been discharged by the SKA Assessor.

Scoring

The below scoring is used for SKA and is based upon the as-built evidence provided at the Handover Stage by the project team:

Benefits of SKA

SKA rating is designed to be of particular use for occupiers but has benefits for other property stakeholders, including landlords, developers, consultants, fit-out contractors, and members of the supply chain.

For occupiers and tenants:

- Manage the bottom line. A decision-making tool which ranks the different aspects of a fit-out based on what will have the biggest impact on sustainability. It can be used to help the design team achieve maximum sustainability for a given budget.
- Legal/statutory compliance. Help to ensure company directors, in line with Section 172 of the Companies Act 2006, promote the success of the company and have a 'regard to the impact of the company's operations on the community and the environment'.

- Are you ISO 14001 accredited? A SKA certificate helps you ensure that a fit-out project is rated accurately and can be used to support your environmental management system.
- Customer, investor and stakeholder perception. A SKA rating certificate demonstrates that your company takes sustainability seriously and has achieved a standard with your property. This can feature in annual reports, environmental policy or tender documentation, or be displayed in a building.
- Staff engagement. Many employees increasingly place a value on a company's green credentials and wish to work in high-quality spaces. A SKA rating can be used to reinforce staff morale and complements other CSR activity.

For developers and landlords:

- Use SKA rating to set targets, use the formal assessment process to assure yourself that target performance standards are met, and finally use the certificate to report performance to stakeholders.
- Benchmark the sustainability of your fit-out across a portfolio of buildings or against your peers.
- Include the use of the RICS SKA rating in 'Green Lease' provisions to help drive up the sustainability of properties or protect a building already certified under a whole building assessment method.
- Sustainability increasingly makes good business sense. Research demonstrates an expected increase in the asset value of labelled low-carbon buildings compared with standard speculative buildings.

For consultants:

- Embed SKA rating in your standard processes to demonstrate that you follow a sustainable specification and procurement process for fit-outs.
- Use the SKA rating tool and datasheets to support the delivery of professional advice to clients on good practice in sustainable fit-outs, and to assess their designs and projects.
- Offer SKA rating certificated assessments to clients who wish to demonstrate they have achieved a more sustainable fit-out through the SKA rating accredited assessors' scheme.

For contractors:

- Use the assessment process and related guidance to make your design, specification, procurement and construction practices more environmentally sustainable.
- Demonstrate your sustainability credentials to clients and consultants by making the targeting of standards under SKA rating part of your standard tender process.

Product labelling

Product selection is a key part of SKA. This includes everything from procuring energy efficient technologies through to using furniture that have low environmental impacts during manufacture and high levels of recycled material. Grigoriou Interiors, a SKA development partner, has a label and directory for products that are proven to be SKA-compliant. This allows designers to find and specify products much more quickly and it will make the assessment process even simpler than it is now. For more information see: https: //www.rics.org/uk/about-rics/responsible-business/ska-rating/ska-rating-product-compliance-label/

Conclusion

SKA has been a great success story to date and has helped to promote sustainable good practice in fit-out and refurbishment projects. It can provide tangible and reputational benefits to occupiers and other property stakeholders and the ultimate aim of SKA is to get it widely adopted to the point that it is embedded in all projects as part of the established process.

For more information see: https://www.rics.org/uk/about-rics/responsible-business/ska-rating/

Wiring Regulations in Brief, 4th edition

Ray Tricker

**WIRING REGULATIONS
IN BRIEF**

FOURTH EDITION

RAY TRICKER

This newly updated edition of Wiring Regulations in Brief provides a user-friendly guide to the newest amendments to BS 7671 and the IET Wiring Regulations. Topic-based chapters link areas of working practice – such as earthing, cables, installations, testing and inspection, and special locations – with the specifics of the Regulations themselves. This allows quick and easy identification of the official requirements relating to the situation in front of you. The requirements of the regulations, and of related standards, are presented in an informal, easy-to-read style that strips away confusion. Packed with useful hints and tips, and highlighting the most important or mandatory requirements, this book is a concise reference on all aspects of the eighteenth edition of the IET Wiring Regulations. This handy guide provides an on-the-job reference source for Electricians, Designers, Service Engineers, Inspectors, Builders, and Students..

November 2020: 376 pp
Pb: 978-0-367-43198-3 : £39.99

To Order: Tel: +44 (0) 1235 400524 Fax: +44 (0) 1235 400525
or Post: Taylor and Francis Customer Services,
Bookpoint Ltd, Unit T1, 200 Milton Park, Abingdon, Oxon, OX14 4TA UK
Email: book.orders@tandf.co.uk

For a complete listing of all our titles visit:
www.tandf.co.uk

Taylor & Francis
Taylor & Francis Group

Smart Building Technology

Whichever smart building technology is deployed, to the users, the technology must deliver tangible benefits over and above those delivered by a traditional solution. Users shall include:

- Employees working within the building
- Visitors and guests
- Building owners, facility managers and maintenance teams.

These benefits must be able to be quantified and measured to ensure any smart building enhancements provide improvements in Comfort, Productivity and or Operational Efficiency. These benefits need to be quantified against any enhancement cost variations, both Capex and Opex.

The smart building enhancement benefits will vary across each user group, building type and industry sector.

The following discusses some of the Smart building technologies presently available to building owners and operators. Depending upon the aspirations of the client, the Smart Building system will often consist of a number of technologies integrated into a single smart building package linked to cause and effect of the building operation.

Employee and visitor apps

App and portal based technologies are being deployed within the workplace to enhance the workplace experience. Often used to support both employees and visitors to the site, a workplace App can be used to:

- Check on travel options to the building
- View weather predictions
- Pre-book visitor car parking
- Pre-book with security notifying them of your planned visit
- 1st level security check in, which can be verified against by secondary level security verification code
- In-building wayfinding
- Desk booking
- Meeting room booking
- HVAC controls interface
- Light controls
- Building amenity and facilities such as a canteen portal detailing menu options, nutrition values and allergy considerations. Location sensor linked to the App could display the canteen queue time
- Security travel updates
- General travel updates
- Facility managers can be provided with real time building performance information
- Maintenance staff can receive works orders and once completed can close down the task and await details of the next work order.

These are just some of the uses for work based Apps, delivered to a mobile device acting as the users single pane of glass.

Automatic number plate recognition and Smart parking systems

If traveling by car, at the car park entrance, Automatic Number Plate Recognition (ANPR) cameras match the registration with the vehicle listed in your profile and allow access. The same could be applied to visitors who have registered their visit via the business App or portal.

An intelligent parking system then directs you to your reserved spot.

Information from the above can be shared with Security making them aware of your arrival.

Enhanced security check-in

As you enter reception you are identified by facial recognition cameras linked to your profile. Depending upon the security procedures and present security level, before you are allowed further a second level of verification may be required. This could take the form of either:

- Security matrix barcode previously sent to your mobile device
- Smart ID card
- Finger print or retina recognition.

Vertical transportation with port destination

Knowing your meeting room or hot desk destination, vertical transport port destination would select and call the lift car. The open door to the lift awaits, poised for the journey skyward.

The vertical transport system can also detect the number of persons entering a lift car, checking numbers against the required occupancy. The system can be integrated to the CCTV monitoring system and could activate a cause and effect process commencing recording of CCTV images alerting security that tailgating has taken place.

Location sensing and wayfinding

There are many systems available today that provide location sensing. Systems can be linked to mobile phones, WiFi pendants, employee and visitor badges. In addition to providing instant location sensing they can also be used to monitor people flow through buildings and wayfinding.

Linked to the workplace App, In-building wayfinding allows users to:

- Navigate through the building finding quickest routes to meeting rooms or allocated hot desks
- Find the nearest print station, toilet or breakout area
- Identify workplace locations for individuals or workplace neighbourhoods.

Integrated Smart meeting rooms

Room booking systems not only allow the booking and scheduling of the meeting space but integrated solutions can also provision the AV connections matched to your preference, set up external video conference links and via the integrated intelligent Building Management System ensure the HVAC and lighting levels are set to the optimum level for the meeting type.

Meeting room monitoring

IoT sensors within the room can monitor the room environment for:

- Lighting levels
- Ambient noise
- CO_2
- VOC
- Power usage
- Occupancy

Linked to the iBMS the data were used to set and adjust controls to ensure users' wellbeing and comfort. Many of the systems on the market utilize WiFi enabling quick and easy deployment with limited cabled infrastructure.

The above information can also be sent back to the facilities manager where real time environment occupancy data can be stored and analyzed.

Automated and semi-automated meeting room controls

The above mentioned occupancy sensors can be used to detect when a meeting room is no longer occupied and linked to the iBMS, place the room into standby mode adjusting the heating, cooling and light levels until the next scheduled booking.

There is also the option to consider where users are incentivized to release rooms early, this could take the form of cashless vending credits.

Power over Ethernet

Within the IT sector Power over Ethernet (PoE) has been adopted for many years to power DC devices such as CCTV cameras, and WiFi Access Points.

The technology is now being adopted for office lighting. A structured cabling grid cabling is installed at high level to facilitate connection of PoE enabled devices including lighting. Lighting arrays also connect to other integrated sensors, including environmental, occupancy and footfall, enabling clients to identify where tangible business and colleague wellbeing benefits can be delivered.

Systems can use Bluetooth low energy (BLE) technology to track people's movements by facilitating wireless communication between luminaires within a lighting control system and moveable objects, e.g. security passes.

Data can then be presented in a number of ways allowing decisions to be taken about how building controls are adapted and integrated based on how the spaces are actually used.

Desk booking and utilization systems

With more and more companies moving to Agile Activity Based Working, where users are no longer allocated permanent desks and users are encouraged to select their own work place location or work group neighbourhood, desk booking systems are becoming a common offering in the modern office.

Users can easily book a desk in either their preferred location or work group neighbourhood. In addition to improving productivity it also removes the user frustration and time spent 'walking the floor(s)' looking for an available desk.

Desk monitoring and occupancy systems provide facility managers and building owners with real time floor and desk utilization data. Trend data which can be analyzed and built into predictive analysis models.

- What is our workspace utilization?
- Could we reduce our floor plate whilst still maintaining the same level of operation?
- Does the trend information show patterns when workspace utilization falls?
- On Fridays could we shut a floor?
- Are certain areas regularly underutilized? If so, why?
- Could occupancy levels be improved if the layout was changed?
- New workspace areas and layouts could be trialled with real time occupancy data provided to the facilities managers.

In-building cellular and WiFi

Some would question if In-Building Cellular and WiFi is a Smart technology but without properly designed in-building mobile communications the adoption of many Smart technologies would not be possible.

We know the deployment of WiFi will continue to increase. Traditional WiFi, based around IEEE 802.11 standards, is poised to take a significant step forward with the introduction of 802.11ax, the next standard for wireless, for a connected world where upload and download traffic will be equivalent.

There will be various speeds up to 4.8 Gbps on the 5 GHz radio and 1.15 Gbps on 2.4 GHz radio.

New and emerging technologies for In-Building Cellular (2G, 3G, 4G and the future potential for 5G) provide users with cellular coverage:

- Within buildings designed with high performing solar glass that protect users and optimize building performance but where the glass provides a shield to the transmission of cellular services
- High riser buildings where signal levels will be poor
- Basements where there is unlikely to be any coverage.

New and emerging alternative technologies including low power devices and LiFi

Devices working over lower power Bluetooth and RFID are already being deployed. Within hospitals portable equipment can be tagged allowing staff to quickly and efficiently identify equipment locations. This can include RFID tagged documents. The same technology can be applied to the office.

LiFi, a high speed bi-directional mobile networked and communications platform delivered using light to transmit data will make an impact in the Smart building arena. This is a technology seen to deliver high speed secure communications, a concern that many users have when deploying traditional WiFi.

Intelligent Building Management Systems (iBMS) integrated into a single pane of glass

An iBMS can integrate many of these technologies into a single system with a single operator interface, often referred to as a single pane of glass.

In addition to providing users with a common window into the system the iBMS systems allows services and devices to be monitored, controlled and optimized, running analytics improving building performance whilst driving down energy loads and costs.

Data on device performance can be collected and analyzed and compared against trend data. Variants can then be reviewed and actioned upon.

As an example:

A motor fitted with IoT sensors is showing a higher than average level of vibration and noise indicating a potential bearing failure. Maintenance repairs or replacement can then be scheduled in advance often providing both labour cost savings and reductions in system downtime.

As the use of Artificial Intelligence increases many of these repetitive big data number crunching processes can be undertaken using AI, with optimized alerts sent to operators and maintenance staff.

Are Smart buildings secure

Any system that connects over the internet or uses an Internet Protocol (IP) enabled interface connected to LAN switch is potentially at risk of network penetration, which everyone within the industry accepts, therefore it is important that the correct Cyber Security systems and process are put in place in advance to protect the Smart building.

Many traditional systems also fall into this category, for example:

- The majority of traditional in-building security systems solutions rely on IP enabled CCTV cameras and IP connected access controllers connected over LAN switches.
- Traditional BMS often provide remote monitoring facilities connected over the internet.
- Many meeting room booking systems can be booked remotely over the internet.

These examples, and many more, show that both traditional and Smart building solutions are both potentially at risk of network penetration/hack and therefore it is essential that a full analysis review is undertaken to minimize risk.

Risks can be minimized by:

- Undertaking a risk review process for each connected system and discipline. This needs to review the level of risk and the type of data that is associated with each discipline. For example:
  - o Security systems will contain staff personnel data and video footage which could be used to affect business operation.
  - o Cashless vending systems will contain staff personnel bank details.
  - o BMS or iBMS control critical operational plant which in the case of an illegal penetration/hack could be targeted.
- Review of device security and passwords. Devices are often sent out with default passwords which need to be changed.
- Physical network separation between in-building systems, technology disciplines and the corporate network.
- Physical security for building ingress and exit paths.
- Physical security on Main & Secondary Equipment Rooms and Building Entrance Facilities housing the external telecoms links.
- Review of firewall policy.
- Review of policy for soft and firmware updates.
- Penetration testing.
- Policy and action plan for dealing with an illegal penetration/hack. There are many documented examples where companies and organizations are aware of an illegal penetration/hack but have failed to respond or advise affected third parties.
- Ongoing review of the above.

These are just some of the activities and processes that need be undertaken to minimize penetration risk.

Summary

These are just a sample of the Smart technologies that could be deployed within the business workplace today.

The building design, construction, fit-out and handover is often over multiple years. A three to five years period is not uncommon. Yet at the same time new and emerging IoT devices and Smart building technologies are evolving at a rapid rate. IoT is being labelled as the 'Next Industrial Revolution' but it is already here.

With the growth in the IoT and Smart building arena it is very difficult to accurately predict what a Smart building will look like in 5–10 years. We know the technologies that are available today, we can evaluate new technologies as they come to market and we can analyze technology trends and make predictions.

We do know, and history has shown, that the infrastructure, be that wired, wireless or light, is the enabler for the deployment of new and emerging technologies. Without the correct infrastructure the technology cannot be deployed.

A Handbook of Sustainable Building Design and Engineering, 2nd edition

Dejan Mumovic *et al.*

A HANDBOOK OF SUSTAINABLE
BUILDING DESIGN AND ENGINEERING
An Integrated Approach to Energy,
Health and Operational Performance
EDITED BY DEJAN MUMOVIC

SECOND EDITION

The second edition of this authoritative textbook equips students with the tools they will need to tackle the challenges of sustainable building design and engineering. The book looks at how to design, engineer and monitor energy efficient buildings, how to adapt buildings to climate change, and how to make buildings healthy, comfortable and secure. New material for this edition includes sections on environmental masterplanning, renewable technologies, retrofitting, passive house design, thermal comfort and indoor air quality. With chapters and case studies from a range of international, interdisciplinary authors, the book is essential reading for students and professionals in building engineering, environmental design, construction and architecture.

October 2018: 246 × 189 mm: 604pp
Hb: 978-1-138-21547-4 : £100.00

To Order: Tel: +44 (0) 1235 400524 Fax: +44 (0) 1235 400525
or Post: Taylor and Francis Customer Services,
Bookpoint Ltd, Unit T1, 200 Milton Park, Abingdon, Oxon, OX14 4TA UK
Email: book.orders@tandf.co.uk

For a complete listing of all our titles visit:
www.tandf.co.uk

Taylor & Francis
Taylor & Francis Group

Value Added Tax

Introduction

Value Added Tax (VAT) is a tax on the consumption of goods and services. The UK introduced a domestic VAT regime when it joined the European Community in 1973. The principal source of European law in relation to VAT is Council Directive 2006/112/EC, a recast of Directive 77/388/EEC, which is currently restated and consolidated in the UK through the VAT Act 1994 and various Statutory Instruments, as amended by subsequent Finance Acts.

VAT Notice 708: Buildings and construction (August 2016) provides HMRC's interpretation of the VAT law in connection with construction works, however, the UK VAT legislation should always be referred to in conjunction with the publication. Recent VAT tribunals and court decisions since the date of this publication will affect the application of the VAT law in certain instances. The Notice is available on HM Revenue & Customs website at www.hmrc.gov.uk.

The scope of VAT

VAT is payable on:

- Supplies of goods and services made in the UK
- By a taxable person
- In the course or furtherance of business; and
- Which are not specifically exempted or zero-rated.

Rates of VAT

There are three rates of VAT:

- A standard rate, currently 20% since January 2011
- A reduced rate, currently 5%; and
- A zero rate of 0%

Additionally some supplies are exempt from VAT and others are considered outside the scope of VAT.

Recovery of VAT

When a taxpayer makes taxable supplies he must account for VAT, known as output VAT at the appropriate rate of 20%, 5% or 0%. Any VAT due then has to be declared and submitted on a VAT submission to HM Revenue & Customs and will normally be charged to the taxpayer's customers.

As a VAT registered person, the taxpayer is entitled to reclaim from HM Revenue & Customs, commonly referred to as input VAT, the VAT incurred on their purchases and expenses directly related to its business activities in respect of standard-rated, reduced-rated and zero-rated supplies. A taxable person cannot, however, reclaim VAT that relates to any non-business activities (but see below), or depending on the amount of exempt supplies they made, input VAT may be restricted or not recoverable.

At predetermined intervals the taxpayer will pay to HM Revenue & Customs the excess of VAT collected over the VAT they can reclaim. However, if the VAT reclaimed is more than the VAT collected, the taxpayer, who will be in a net repayment position, can reclaim the difference from HM Revenue & Customs.

Example

X Ltd constructs a block of flats. It sells long leases to buyers for a premium. X Ltd has constructed a new building designed as a dwelling and will have granted a long lease. This first sale of a long lease is VAT zero-rated supply. This means any VAT incurred in connection with the development which X Ltd will have paid (e.g. payments for consultants and certain preliminary services) will be recoverable. For reasons detailed below the contractor employed by X Ltd will not have charged VAT on his construction services as these should be zero-rated.

Use for business and non-business activities

Where a supply relates partly to business use and partly to non-business use then the basic rule is that it must be apportioned on a fair and reasonable basis so that only the business element is potentially recoverable. In some cases VAT on land, buildings and certain construction services purchased for both business and non-business use could be recovered in full by applying what is known as 'Lennartz' accounting to reclaim VAT relating to the non-business use and account for VAT on the non-business use over a maximum period of 10 years. Following an ECJ case restricting the scope of this approach, its application to immovable property was removed completely in January 2011 by HMRC (business brief 53/10) when UK VAT law was amended to comply with EU Directive 2009/162/EU.

Taxable persons

A taxable person is an individual, firm, company, etc. who is required to be registered for VAT. A person who makes taxable supplies above certain turnover limits is compulsorily required to be VAT registered. From 1 April 2017, the current registration limit, known as the VAT threshold, is £85,000. If the threshold is exceeded in any 12 month rolling period, or there is an expectation that the value of the taxable supplies in a single 30 day period, or you receive goods into the UK from the EU worth more than the £85,000, then you must register for UK VAT.

A person who makes taxable supplies below the limit is still entitled to be registered on a voluntary basis if they wish, for example, in order to recover input VAT incurred in relation to those taxable supplies, however output VAT will then become due on the sales and must be accounted for.

In addition, a person who is not registered for VAT in the UK but acquires goods from another EC member state, or makes distance sales in the UK above certain value limits may be required to register for VAT in the UK.

VAT exempt supplies

Where a supply is exempt from VAT this means that no output VAT is payable – but equally the person making the exempt supply cannot normally recover any of the input VAT on their own costs relating to that exempt supply.

Generally commercial property transactions such as leasing of land and buildings are exempt unless a landlord chooses to standard-rate its interest in the property by a applying for an option to tax. This means that VAT is added to rental income and also that VAT incurred on, say, an expensive refurbishment, is recoverable.

Supplies outside the scope of VAT

Supplies are outside the scope of VAT if they are:

- Made by someone who is not a taxable person
- Made outside the UK; or
- Not made in the course or furtherance of business

In course or furtherance of business

VAT must be accounted for on all taxable supplies made in the course or furtherance of business with the corresponding recovery of VAT on expenditure incurred.

If a taxpayer also carries out non-business activities then VAT incurred in relation to such supplies is generally not recoverable.

In VAT terms, business means any activity continuously performed which is mainly concerned with making supplies for a consideration. This includes:

- Anyone carrying on a trade, vocation or profession;
- The provision of membership benefits by clubs, associations and similar bodies in return for a subscription or other consideration; and
- Admission to premises for a charge.

It may also include the activities of other bodies including charities and non-profit making organizations.

Examples of non-business activities are:

- Providing free services or information;
- Maintaining some museums or particular historic sites;
- Publishing religious or political views.

Construction services

In general, the provision of construction services by a contractor will be VAT standard rated at 20%, however, there are a number of exceptions for construction services provided in relation to certain relevant residential properties and charitable buildings.

The supply of building materials is VAT standard rated at 20%, however, where these materials are supplied and installed as part of the construction services the VAT liability of those materials follows that of the construction services supplied.

Zero-rated construction services

The following construction services are VAT zero-rated, including the supply of related building materials.

The construction of new dwellings

The supply of services in the course of the construction of a new building designed for use as a dwelling or number of dwellings is zero-rated other than the services of an architect, surveyor or any other person acting as a consultant or in a supervisory capacity.

The following basic conditions must ALL be satisfied in order for the works to qualify for zero-rating:

1. A qualifying building has been, is being or will be constructed
2. Services are made 'in the course of the construction' of that building
3. Where necessary, you hold a valid certificate
4. Your services are not specifically excluded from zero-rating

The construction of a new building for 'relevant residential or charitable' use

The supply of services in the course of the construction of a building designed for use as a Relevant Residential Purpose (RRP) or Relevant Charitable Purpose (RCP), is zero-rated other than the services of an architect, surveyor or any other person acting as a consultant or in a supervisory capacity.

A 'relevant residential' use building means:

1.   A home or other institution providing residential accommodation for children;
2.   A home or other institution providing residential accommodation with personal care for persons in need of personal care by reason of old age, disablement, past or present dependence on alcohol or drugs or past or present mental disorder;
3.   A hospice;
4.   Residential accommodation for students or school pupils;
5.   Residential accommodation for members of any of the armed forces;
6.   A monastery, nunnery, or similar establishment; or
7.   An institution which is the sole or main residence of at least 90% of its residents.

A 'relevant residential' purpose building does not include use as a hospital, a prison or similar institution or as a hotel, inn or similar establishment.

A 'relevant charitable' purpose means use by a charity in either or both of the following ways:

1.   Otherwise than in the course or furtherance of a business; or
2.   As a village hall or similarly in providing social or recreational facilities for a local community.

Non-qualifying use which is not expected to exceed 10% of the time the building is normally available for use can be ignored. The calculation of business use can be based on time, floor area or head count subject to approval being acquired from HM Revenue & Customs.

The construction services can only be zero-rated if a certificate is given by the end user to the contractor carrying out the works confirming that the building is to be used for a qualifying purpose, i.e. for a 'relevant residential or charitable' purpose. It follows that such services can only be zero-rated when supplied to the end user and, unlike supplies relating to dwellings, supplies by subcontractors cannot be zero-rated.

The construction of an annex used for a 'relevant charitable' purpose

Construction services provided in the course of construction of an annexe for use entirely or partly for a 'relevant charitable' purpose can be zero-rated.

In order to qualify the annexe must:

1.   Be capable of functioning independently from the existing building;
2.   Have its own main entrance; and
3.   Be covered by a qualifying use certificate.

The conversion of a non-residential building into dwellings or the conversion of a building from non-residential use to 'relevant residential' use where the supply is to a 'relevant' housing association

The supply to a 'relevant' housing association in the course of conversion of a non-residential building or non-residential part of a building into:

1.   A new eligible dwelling designed as a dwelling or number of dwellings; or
2.   A building or part of a building for use solely for a relevant residential purpose, of any services related to the conversion other than the services of an architect, surveyor or any person acting as a consultant or in a supervisory capacity are zero-rated.

A 'relevant' housing association is defined as:

1. A private registered provider of social housing
2. A registered social landlord within the meaning of Part I of the Housing Act 1996 (Welsh registered social landlords)
3. A registered social landlord within the meaning of the Housing (Scotland) Act 2001 (Scottish registered social landlords), or
4. A registered housing association within the meaning of Part II of the Housing (Northern Ireland) Order 1992 (Northern Irish registered housing associations).

If the building is to be used for a 'relevant residential' purpose the housing association should issue a qualifying use certificate to the contractor completing the works. Subcontractors services that are not made directly to a relevant housing association are standard rated.

The development of a residential caravan park

The supply in the course of the construction of any civil engineering work 'necessary for' the development of a permanent park for residential caravans of any services related to the construction are zero-rated when a new permanent park is being developed, the civil engineering works are necessary for the development of the park and the services are not specifically excluded from zero-rating. This includes access roads, paths, drainage, sewerage and the installation of mains water, power and gas supplies.

Certain building alterations for disabled persons

Certain goods and services supplied to a disabled person, or a charity making these items and services available to disabled persons can be zero-rated. The recipient of these goods or services needs to give the supplier an appropriate written declaration that they are entitled to benefit from zero rating.

The following services (amongst others) are zero-rated:

1. The installation of specialist lifts and hoists and their repair and maintenance
2. The construction of ramps, widening doorways or passageways including any preparatory work and making good work
3. The provision, extension and adaptation of a bathroom, washroom or lavatory; and
4. Emergency alarm call systems

Approved alterations to protected buildings

The zero rate for approved alterations to protected buildings was withdrawn from 1 October 2012, other than for projects where a contract was entered into or where listed building consent (or equivalent approval for listed places of worship) had been applied for before 21 March 2012.

Provided the application was in place before 21 March 2012, zero rating will continue under the transitional rules until 30 September 2015.

All other projects will be subject to the standard rate of VAT on or after 1 October 2012.

Sale of reconstructed buildings

Since 1 October 2012 a protected building shall not be regarded as substantially reconstructed unless, when the reconstruction is completed, the reconstructed building incorporates no more of the original building than the external walls, together with other external features of architectural or historical interest. Transitional arrangements protect contracts entered into before 21 March 2012 for the first grant of a major interest in the protected building made on or before 20 March 2013.

DIY builders and converters

Private individuals who decide to construct their own home are able to reclaim VAT they pay on goods they use to construct their home by use of a special refund mechanism made by way of an application to HM Revenue & Customs. This also applies to services provided in the conversion of an existing non-residential building to form a new dwelling.

The scheme is meant to ensure that private individuals do not suffer the burden of VAT if they decide to construct their own home.

Charities may also qualify for a refund on the purchase of materials incorporated into a building used for non-business purposes where they provide their own free labour for the construction of a 'relevant charitable' use building.

Reduced-rated construction services

The following construction services are subject to the reduced rate of VAT of 5%, including the supply of related building materials.

Conversion – changing the number of dwellings

In order to qualify for the 5% rate there must be a different number of 'single household dwellings' within a building than there were before commencement of the conversion works. A 'single household dwelling' is defined as a dwelling that is designed for occupation by a single household.

These conversions can be from 'relevant residential' purpose buildings, non-residential buildings and houses in multiple occupation.

A house in multiple occupation conversion

This relates to construction services provided in the course of converting a 'single household dwelling', a number of 'single household dwellings', a non-residential building or a 'relevant residential' purpose building into a house for multiple occupation such as a bedsit accommodation.

A special residential conversion

A special residential conversion involves the conversion of a 'single household dwelling', a house in multiple occupation or a non-residential building into a 'relevant residential' purpose building such as student accommodation or a care home.

Renovation of derelict dwellings

The provision of renovation services in connection with a dwelling or 'relevant residential' purpose building that has been empty for two or more years prior to the date of commencement of construction works can be carried out at a reduced rate of VAT of 5%.

Installation of energy saving materials

A reduced rate of VAT of 5% is paid on the supply and installation of certain energy saving materials including insulation, draught stripping, central heating, hot water controls and solar panels in a residential building or a building used for a relevant charitable purpose.

Buildings that are used by charities for non-business purposes, and/or as village halls, were removed from the scope of the reduced rate for the supply of energy saving materials under legislation introduced in the Finance Bill 2013.

Grant-funded installation of heating equipment or connection of a gas supply

The grant-funded supply and installation of heating appliances, connection of a mains gas supply, supply, installation, maintenance and repair of central heating systems, and supply and installation of renewable source heating systems, to qualifying persons. A qualifying person is someone aged 60 or over or is in receipt of various specified benefits.

Grant funded installation of security goods

The grant-funded supply and installation of security goods to a qualifying person.

Housing alterations for the elderly

Certain home adaptations that support the needs of elderly people were reduced rated with effect from 1 July 2007.

Building contracts

Design and build contracts

If a contractor provides a design and build service relating to works to which the reduced or zero rate of VAT is applicable then any design costs incurred by the contractor will follow the VAT liability of the principal supply of construction services.

Management contracts

A management contractor acts as a main contractor for VAT purposes and the VAT liability of his services will follow that of the construction services provided. If the management contractor only provides advice without engaging trade contractors his services will be VAT standard rated.

Construction management and project management

The project manager or construction manager is appointed by the client to plan, manage and coordinate a construction project. This will involve establishing competitive bids for all the elements of the work and the appointment of trade contractors. The trade contractors are engaged directly by the client for their services.

The VAT liability of the trade contractors will be determined by the nature of the construction services they provide and the building being constructed.

The fees of the construction manager or project manager will be VAT standard rated. If the construction manager also provides some construction services these works may be zero or reduced rated if the works qualify.

Liquidated and ascertained damages

Liquidated damages are outside of the scope of VAT as compensation. The employer should not reduce the VAT amount due on a payment under a building contract on account of a deduction of damages. In contrast an agreed reduction in the contract price will reduce the VAT amount.

Similarly, in certain circumstances HM Revenue & Customs may agree that a claim by a contractor under a JCT or other form of contract is also compensation payment and outside the scope of VAT.

Building Performance Simulation for Design and Operation, 2nd edition

Jan Hensen *et al.*

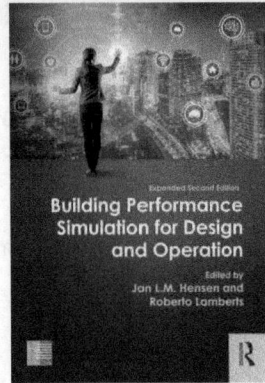

When used appropriately, building performance simulation has the potential to reduce the environmental impact of the built environment, to improve indoor quality and productivity, as well as to facilitate future innovation and technological progress in construction. Since publication of the first edition of *Building Performance Simulation for Design and Operation*, the discussion has shifted from a focus on software features to a new agenda, which centres on the effectiveness of building performance simulation in building life cycle processes.

This new edition provides a unique and comprehensive overview of building performance simulation for the complete building life cycle from conception to demolition, and from a single building to district level. It contains new chapters on building information modelling, occupant behaviour modelling, urban physics modelling, urban building energy modelling and renewable energy systems modelling. This new edition keeps the same chapter structure throughout including learning objectives, chapter summaries and assignments.

May 2019: 246 × 174 mm: 792pp
Hb: 978-1-138-39219-9 : £140.00

To Order: Tel: +44 (0) 1235 400524 Fax: +44 (0) 1235 400525
or Post: Taylor and Francis Customer Services,
Bookpoint Ltd, Unit T1, 200 Milton Park, Abingdon, Oxon, OX14 4TA UK
Email: book.orders@tandf.co.uk

For a complete listing of all our titles visit:
www.tandf.co.uk

Taylor & Francis
Taylor & Francis Group

Water Network Charges

This article gives guidance on water suppliers, new connections, infrastructure charges and network charges.

New connections and charging methodology

The Water Industry Act 1991 (the Act) (as amended by the Water Act 2014) allows the Water Services Regulation Authority (Ofwat) to set rules about the charges that a water and/or wastewater company can impose on its customers and on a water supply and/or sewerage licensee (Retailer). It also allows Ofwat to make amendments to these rules following a formal public consultation.

The methodology for charging for new connections significantly changed from April 2020, based upon the Ofwat document 'Charging Rules for New Connection Services (English Undertakers)' as published on 15 July 2019.

The principles of who can undertake the work remains as the previous rules, but the calculation of connection charges, both for water and sewerage, have been revised.

Existing appointees and New Appointments & Variations (NAVs)

Most customers in England and Wales currently receive their water and sewerage services from one of 22 appointed 'water' and 'sewerage and water only' suppliers that were in existence when the sectors were privatized. These suppliers are known as 'existing appointees' or statutory providers.

New Appointments and Variations or (NAVs) refers to small competitor water companies that are not subject to full price controls like the statutory providers. The regulator 'Ofwat' recognizes that NAVs can provide services, such as their own water infrastructure and that these services can create competition in the market place when customers are purchasing new connections.

New appointments provide challenge to existing appointees. This drives efficiencies, stimulates innovation and reveals information. They have the potential to benefit all customers, through:

- Lower prices;
- Improved service;
- Greater choice of supplier for developers and large user customers.

Providing new connections

The term 'new connection' is used to describe where a customer requires either or both:

- Access to the existing public water supply or sewerage system by means of a service pipe or lateral drain
- A new water main or public sewer.

A customer may choose their own contractor to do the work, which is then known as 'self-lay'. The statutory provider will take over responsibility for (adopt) all self-laid infrastructure that meets the terms of its agreement with the owner, developer or self-lay organization that carries out the work.

The Water Industry Act 1991 (WIA91) places a number of duties on statutory 'water only' and 'water and sewerage' companies in providing or enabling new connections for an individual property or development site.

If a property requires a new water main, sewer, service pipe or lateral drain for domestic purposes (cooking, cleaning, central heating or sanitary facilities), the owner or developer may ask the local statutory provider to install the infrastructure. For water mains and public sewers this is often referred to as 'requisitioning' the infrastructure.

Self-lay

If a development requires a new water main or sewer, you may ask the water or sewerage company to install the pipework. Alternatively, you may choose your own contractor to do the work, which is known as self-lay. The water company will take over responsibility for (adopt) self-laid pipes that meet the terms of its agreement with the developer or self-lay organization (SLO) that carries out the work.

Accreditation

Before a self-lay organization can carry out work, they must be approved by the relevant water company. Self-lay organizations can avoid having to comply with the 22 separate water companies' requirements by becoming accredited under the Water Industry Registration Scheme (WIRS) which is recognized by all the water companies and Water UK.

Under WIRS, Lloyd's Register carries out technical assessments of the service providers who elect to be accredited for contestable works associated with installing water infrastructure.

Areas covered by WIRS include:

- design
- construction
- connections
- commissioning
- project management

Lloyd's Register offers an independent assessment and registration process that includes assessing the self-lay organization's management procedures and processes.

WIRS plays an important part in supporting and expanding the self-lay market, as it promotes and ensures compliance with the highest standards of water infrastructure installation and the essential protection and maintenance of water quality.

The Water Industry Registration Scheme Advisory Panel (WIRSAP) manages the scheme, its members include representatives from water companies, self-lay organizations and the regulator Ofwat.

Other charges to consider

On 20th December 2018 Ofwat set out its final position on Wholesale Charging Rules and Charging Rules for New Connection Services (English Undertakers) in relation to network and infrastructure charging.

Example of 2020 methodology water services connection charges

This assumes a block of 90 apartments with a single point of connection via a footpath 2 m wide and the point of connection is an existing water main in highway with a distance of 3 m to edge of footpath. Connection size is 90 mm and no barrier pipe is required.

Application fee	£50
Design fee (£50 for first property and £30 for each additional property)	£2,720
Connection charge in highway	£3,740
Pipe laying (90 mm without barrier pipe, assumes 3 m in highway and 2 m in footpath)	£2,880
Inline water meters	£9,900
Total (Excl VAT)	**£19,290**

Infrastructure charges

If a supplier is providing new water supplies for domestic purposes (such as to new housing developments) it is legally entitled to levy infrastructure charges. These can be raised where premises are connected to the water company's water supply (or to the sewerage company's sewers) for the first time. These charges are intended to provide a contribution towards the costs of developing or enhancing local networks to serve new customers.

A customer will have to pay infrastructure charges when a property is connected to the water and/or wastewater networks for the first time. This is in addition to the charges for making the actual physical connection to the water main and/or public sewer. When a customer makes a connection to a network the operator is entitled to charge the customer in accordance with its published Charging Arrangements for the connection works in addition to raising infrastructure charges. Infrastructure charges apply for premises where the supply of water or provision of sewerage services is intended for domestic purposes. Water for domestic purposes refers to usage for drinking, washing, cooking, central heating and sanitary purposes for which water is supplied to premises. All other purposes (including supplies for a laundry business or for a food or drink take away business), are regarded as non-domestic purposes. *The definition is about the usage of the water and not the type of property being supplied.* Domestic sewerage purposes refers to the removal of the contents of lavatories, water which has been used for cooking or washing or surface water from the premises and associated land, with the exception of laundries and take away restaurants. *As with a potable water connection, the definition is not about the type of property.* Please note that a single development may include a combination of supplies for domestic and non-domestic purposes. For example, a development may include a number of flats (expected to be for domestic purposes), retail units (expected to be for domestic purposes), a fire-fighting supply (non-domestic purposes) and landlord supplies (non-domestic purposes). Similarly a separate supply requested for a swimming pool or a garden tap is considered to be for non-domestic purposes even if within a residential property.

The Relevant Multiplier

A Relevant Multiplier is used to calculate the amount of an infrastructure charge for the connection of premises other than dwelling houses or flats, if water is provided by a supply larger than the standard size used for new connections of houses – 25 mm or 32 mm external dia. pipe. For other properties, such as student housing, offices or care homes the regulator has introduced a pricing calculator known as a relevant multiplier. This tool is utilized to measure a cost that reflects the increased impact on a provider's network. The wastewater infrastructure charge is calculated on the same basis as the water infrastructure charge unless the customer is able to show that waste and surface water flows are not being discharged to the public sewer.

The Relevant Multiplier (RM) is a way of working out infrastructure charges for the following types of property:

- Residential properties with a single, shared supply pipe and which are subject to a 'common billing agreement'; this includes sheltered housing, student accommodation and high-rise flats.
- Non-residential properties where the supply pipe is larger than the standard size, such as office blocks.

How the Relevant Multiplier is calculated

Each water fitting (wash basin, bath, shower, etc.) is given a 'loading unit' based on the amount of water it uses. The average number of units per property is taken as 24, equal to an RM of 1.00. The operator can use this as the basis for calculating the RM for each property on a development where the RM applies. The operator will do this by adding up the loading units for all the water fittings on a development. It divides this by the number of properties to give the average loading units per property. It divides this again by 24 (the average loading units) to give the RM for each property. Details of the number of loading units assigned to each water fitting are shown below. For properties subject to a common billing agreement, the RM can be more or less than 1.00. For other properties the minimum is 1.00.

Using the Relevant Multiplier to calculate infrastructure charge

Operator will utilize the RM multiplied by the standard charge to provide the infrastructure charge for that property.

Example of a residential development

- The development consists of 90 apartments with a common billing agreement and the total loading units are 2,100.
- The total loading units (2,100) are divided by the number of properties (90) and again by the average loading unit (24). This gives an RM for each flat of 0.97 (2,100 ÷ 90 ÷ 24 = 0.97)
- The infrastructure charge for each flat is the RM of 0.97 multiplied by the standard charges.
- The infrastructure charge for the whole development is the RM multiplied by the standard charge multiplied by the number of properties.

The water infrastructure charge for the development is therefore RM 0.97 × 90 properties × £140 standard charge = £12,222.00

New to the 2020 methodology is the Income Offset. This is a credit applied to the number of Infrastructure Charges that abates the Water Infrastructure Charge and effectively provides a credit overall at current rates.

The Income Offset (Water) is therefore RM 0.97 × 90 properties × £200 Income Offset = £17,460.00 credit.

The water infrastructure charge for the development is therefore:

RM 0.97 × 90 properties × £140 standard charge = £12,222.00
RM 0.97 × 90 properties × £200 income offset = £17,460.00 credit
Total Water Infrastructure Charge = £5,238.00 credit (Excl VAT)

Wastewater connections charges

Example of 2020 methodology wastewater services connection charges

This assumes a block of 90 apartments with a single point of connection to an existing public sewer in highway less than 5 m deep. Distance from point of connection to site boundary is 6 m and pipe is 300 mm dia..

Application fee	£3,400
Connection charge	£12,600
Pipe laying (300 mm dia. × 6 m length)	£30,540
Total (Excl VAT)	**£46,540**

Example of a residential development

- The development consists of 90 apartments with a common billing agreement and the total loading units are 2,100.
- The total loading units (2,100) are divided by the number of properties (90) and again by the average loading unit (24). This gives an RM for each flat of 0.97 (2,100 ÷ 90 ÷ 24 = 0.97)
- The infrastructure charge for each flat is the RM of 0.97 multiplied by the standard charges.
- The infrastructure charge for the whole development is the RM multiplied by the standard charge multiplied by the number of properties.

The wastewater infrastructure charge for the development is therefore RM 0.97 × 90 properties × £210 standard charge = £18,333.00

New to the 2020 methodology is the Income Offset. This is a credit applied to the number of Infrastructure Charges that abates the Water Infrastructure Charge and effectively provides a credit overall at current rates.

The Income Offset (Wastewater) is therefore RM 0.97 × 90 properties × £40 Income Offset = £3,492.00 credit.

The water infrastructure charge for the development is therefore:

RM 0.97 × 90 properties × £210 standard charge = £18,333.00
RM 0.97 × 90 properties × £40 income offset = £4,382.00 credit
Total Water Infrastructure Charge = £14,841.00 (Excl VAT)

In the above worked example, the total costs of connection to water and wastewater are as follows:

Water service connection	£19,290.00
Water Infrastructure Charges (less Income Offset)	(£5,238.00)
Wastewater connection	£46,540.00
Wastewater Infrastructure Charges (less Income Offset)	£14,841.00
Total (Excl VAT)	**£75,433.00**

Choosing your water supplier

Under the current legislation, household customers are not able to change their water supplier or sewerage service provider. The water or sewerage company that supplies your property will depend on where you live.

If you are a business customer in England, you may be able to choose the retailer for your water or sewerage services. With a choice of water retailers (there may be a minimum capacity supply that applies) you may benefit from:

• better standards of service
• tailored service offerings
• advice on saving water
• lower prices
• you will still benefit from the same water quality.

Since April 2017, most non-household customers based wholly or mainly in England are now able to choose both their water and wastewater retailer. They now do not have to use a minimum amount of water as was the requirement previously. Customers will also be able to apply for a licence (WSSL) to supply themselves retail services. A licensee limited to self-supply ('self-supplier') can only supply its own premises and the premises of people or companies associated with it. Licensees limited to self-supply are not able to supply retail services to premises of third parties, such as supplying other business customers unrelated to its own business.

Regulatory information within this document has been referenced directly from the regulator Ofwat. This information relates to its current guidance on infrastructure and network charging and is correct as of December 2018.

Further Information on water distribution network areas can be found at https://www.ofwat.gov.uk/regulated-companies/ofwat-industry-overview/licences/

Building Services Handbook, 9th Edition

Fred Hall and Roger Greeno

The ninth edition of Hall and Greeno's leading textbook has been reviewed and updated in relation to the latest building and water regulations, new technology, and new legislation. For this edition, new updates includes: the reappraisal of CO_2 emissions targets, updates to sections on ventilation, fuel, A/C, refrigeration, water supply, electricity and power supply, sprinkler systems, and much more.

Building Services Handbook summarises the application of all common elements of building services practice, technique and procedure, to provide an essential information resource for students as well as practitioners working in building services, building management and the facilities administration and maintenance sectors of the construction industry. Information is presented in the highly illustrated and accessible style of the best-selling companion title *Building Construction Handbook*.

THE comprehensive reference for all construction and building services students, Building Services Handbook is ideal for a wide range of courses including NVQ and BTEC National through Higher National Certificate and Diploma to Foundation and three-year Degree level. The clear illustrations and complementary references to industry Standards combine essential guidance with a resource base for further reading and development of specific topics.

May 2017; 234 × 156 mm, 786 pp
Pbk: 978-1-138-24435-1; £36.99

To Order: Tel: +44 (0) 1235 400524 Fax: +44 (0) 1235 400525
or Post: Taylor and Francis Customer Services,
Bookpoint Ltd, Unit T1, 200 Milton Park, Abingdon, Oxon, OX14 4TA UK
Email: book.orders@tandf.co.uk

For a complete listing of all our titles visit:
www.tandf.co.uk

Taylor & Francis
Taylor & Francis Group

Digital Delivery: Transforming the Design Process

In response to increasing economic, social and environmental pressures, private and public sector developers are challenging the design and construction industry to deliver faster, greener and more efficient infrastructure solutions. Digital expert Dale Sinclair explores how a new digital approach can help the industry meet this challenge, and tackle its own long-standing productivity issues.

The UK Government has big ambitions for the country's infrastructure, with a national infrastructure programme target to deliver £650 billion worth of projects by 2025. The pipeline comprises millions of affordable homes built in multi-use communities, more efficient, sustainable transport networks and solutions, and energy infrastructure to secure low-carbon success.

Across both public and private sector developments, this vision calls for better-performing, greener buildings and infrastructure delivered faster and more cost-effectively; and, the Construction Sector Deal commits the industry to: reduce the cost of construction and the whole life costs of assets by 33%; half the time taken from inception to completion of new build; and decrease greenhouse gas emissions in the built environment by 50%, supporting the Industrial Strategy's Clean Growth Grand Challenge.

Failure to launch

Yet, despite this urgency and national focus, the UK's infrastructure ambitions continue to outpace supply – the lag exacerbated by long-standing issues in the UK design and construction industry. This includes a persistently low productivity growth rate, averaging just 0.4% per year, the ongoing prioritisation of costs over climate-change risks in some areas, and continued delays to the delivery of major infrastructure projects. Reflecting this, only 1% of housebuilders think that the target of 300,000 homes a year can be achieved by 2022.

That's, perhaps, no surprise given that the vast majority of building projects still follow the generations-old linear process of briefing, design, construction and, finally, occupation – with maintenance considerations frequently an afterthought. Typically, project teams start afresh with every project, drawing on a new client brief and site context, alongside local regulatory drivers, to shape their plan, and find it difficult with the analogue tools available to learn and apply lessons from previous successes and failures in a systematic way.

The complexities and constraints of the UK's national and local planning systems also increase project pressures as councils look to developers to help them build the major infrastructure that communities need to thrive, urgently and cost effectively. To do this, developers need to ensure a robust business plan and delivery model from the outset. Yet too many projects still run late and over budget.

To change the outcome we must change the process

For some, the answer is to standardize within the traditional design process, replicating deliverables, gateways and scopes of service across multiple projects, to save time and resources.

But the industry needs to be bolder, to go further. Digital innovation gives teams the chance to access and benefit from the latest innovations in advanced manufacturing, design and construction, and facilitate the industry's shift to a circular economy approach and creating net-zero adaptive buildings more rapidly.

Informed by our work with clients and digital expertise, AECOM has developed a new digital ecosystem to help teams realize that potential now. The platform – which brings together the latest digital tools for use across disciplines, including design for manufacture and assembly solutions – is built around a central digital library that enables knowledge capture, supports interdisciplinary workflows and applies lessons learnt to deliver smarter, greener buildings faster.

Here are three ways in which an integrated digital delivery approach like this can transform processes and improve outcomes across the entire project lifecycle and wider industry.

1. Informing decision-making from start to finish

A building is only ever as good as the quality of the spaces within it and the success of its vision to form a coherent, adaptable whole. Yet the reality is that in the earliest stages of a project, those involved don't generally have the spatial information they need to fully understand the impact of the decisions they're making on their building. Using the project detail and sector knowledge stored in digital libraries, teams can generate 3D models to visualize their design in practice and give them that insight.

It means that, well before construction, they can assess the planned use of space, products, materials, light and colour and how the operation of their building will work. Ideas and changes can be tested in the virtual world and spaces (such as those designed by sector, i.e. residential, healthcare and workplace) quickly adapted to suit new circumstances or project conditions, without losing embedded knowledge and intelligence.

With access to interdisciplinary data, teams can also see detailed information on sockets, lights, grilles and other building services from day one – limiting the potential for costly, last minute changes down the line.

2. Maximizing value across specifications

While a building's spaces determine its look and feel, its systems – ranging from structural to business services and internal wall systems – form its engine room. The building-system components of digital libraries provide fabrication and construction-ready information that enables teams to optimize these elements from an aesthetic and whole-life perspective by selecting the products, materials and solutions that can deliver the best long-term value for the project.

In addition, this better quality, more detailed design content makes it possible to repeat best-practice and well-designed spaces across programmes of projects, replacing the design standards typically used. And AECOM teams are now using digital design libraries, consisting of construction-ready spaces and systems to drive residential developments, repeatable retail plans, high-performance workplaces and adaptive healthcare and higher education buildings.

3. Blending innovation and tradition, strengthening collaboration

By having the latest tools, software and solutions in one place and ready to deploy via a digital ecosystem, including design for manufacture and assembly (DFMA) techniques such as modular construction, teams can help to eliminate waste, reduce costs and save resources across the project lifecycle. For example, the adoption of off-site modular solutions – which use advanced manufacturing technologies and processes in construction – can deliver high-performing, precision-engineered buildings equipped to lower running costs, adapt to changing needs and save energy.

In cases where modular approaches are not suitable, such as commercial offices or airport terminals, digitally integrated platforms can bring together manufacturing and construction software tools to help multidisciplinary teams and supply chain partners collaborate more efficiently, reducing downstream costs and delays and making these complex, often global, programmes simpler to manage. For example, by linking an architect's model to engineering software, you can limit the need for multiple design iterations and ensure early design decisions are as robust as possible.

Delivering for the future

Digital delivery is truly transformative, giving project teams the opportunity to not only transform the way we design and create our built environment, tackling long-standing productivity challenges in the design and construction industry, but also crucially deliver the faster, smarter, better buildings that our communities and the future demand.

How Coronavirus May Change the Future of Work

Up until very recently, the shift away from traditional office design towards more dynamic workplaces and work habits that promote wellbeing has felt relevant only to certain types of organizations, workers, demographics or personalities. The coronavirus pandemic has changed that overnight. As a consequence of self-isolation rules, many employees and organizations have been thrust into working in a completely different way. Remote working has become the norm and office buildings left empty. Organizations and their employees are having to demonstrate agility, resilience, patience and understanding to make sure that work gets done among all the disruption.

Before the crisis struck, we published an article that discussed trends that we believe would shape the 'future office'. Given that the discussion around the future of work is now entering uncharted territory, we thought it would be a good time to revisit and re-examine those trends – as much as this current crisis is going to show us what is unnecessary and obsolete, it also has the potential to create a cohesive human agreement on what it is that we value about work, which will go on to shape our future workplaces.

People and their wellbeing will matter most

Workplaces have been upping their commitment to employee wellbeing over the last decade, evident from how many times 'wellbeing' pops up as a forum or conference topic. But what will we have learned from recent events about what we value most about 'going' to work and 'being' at work? Will employees want a return to strict nine-to-five routines, or will they want more flexible working to incorporate personal wellbeing objectives? The mass stay-at-home exercise could also convince many that it's not necessary to live in a city for employment opportunities and career advancement, which up until now, has been the traditional view. We know that choice and control are the two most significant factors on a person's wellbeing – would a recalibrated option for when we work and where we live point us towards healthier living?

Workplaces will be increasingly centred on developing a community base — virtual and physical — that supports comfort, creativity, productivity and job satisfaction. But understanding and measuring the impact of workplace environmental factors on issues such as employee motivation, satisfaction, productivity and mental health is going to be even more complex moving forward. Furthermore, we are witnessing first-hand how digital technologies are rapidly reshaping what we do and how our workplaces need to be re-imagined, built and operated. Looking further ahead, the Fourth Industrial Revolution, which will be driven by automation and artificial intelligence (AI), will means that some jobs will cease to exist in the coming years while others will be created. We think that this will not change.

However, added to this mix will be an unintended assessment of job functions, criticality of different roles and operating models to come up with an optimal model of working, both for people and organizations. As more and more organizations recognize that their people are their most expensive and valuable asset, there's going to be a revaluation of the kind of work we truly value; and it may look drastically different from what we previously thought.

The prevailing logic is that as AI takes over the mundane, low-skilled tasks, people will be valued for the 'human' things we're good at (e.g. empathy, creativity, etc.): in short, people will always matter most. This probably won't be derailed by the current crisis but our perception of the spectrum of jobs that we think will require a human touch may alter. This prolonged period of non-contact may change minds about how we feel about the cashier versus the self-checkout machine conundrum.

Leveraging building and workplace data to boost wellbeing

Data is largely focused these days on understanding occupancy, movement patterns, space utilization and collaboration patterns. As we migrate back into our office buildings, we may see a shift in priorities where organizations use building and workplace data to elevate employee wellbeing and mitigate health risks instead.

Access to data may also become more democratic. As organizations deploy sensors and monitors to help reassure users that their buildings are keeping them healthy, users will also demand access to data and insights to self-regulate their own behaviours and hold their employers accountable for providing up-to-scratch workplaces.

Addressing climate challenge is still important

The climate emergency and the industry's commitment to net zero carbon by 2050 will remain a pressing issue – the ideas around design within a circular economy will continue to resonate, not just with architects and developers, but with companies and employees as well. Working within a building that is energy-efficient and built to net zero carbon will quickly become an expectation and the norm.

To achieve this, modularity will play an important part in providing adaptability, decreasing waste and redundancy of resources. By designing buildings with more modular elements that can be deconstructed, adapted and reconstructed, we can extend their life and enable resources to be salvaged and reused.

Kinder, more inclusive workplaces

When we return to the office after all these weeks of self-isolation, are we going to be kinder and more considerate to our co-workers, and focus more on connecting with each other? Will 'community' cease to be just a buzz word and become a fundamental reason and anchor for being in office buildings with fellow employees?

The intricacies of existing in a multi-generational workplace has been a much-talked about topic – even more so now that there are not four, but five generations in the workplace. In the light of our recent experiences, we think that the conversation must shift from highlighting differences to truly living out diversity and inclusion policies which will underpin how entire organizations behave, interact and collaborate.

We need to start building strong communities at work that share values but also celebrate differences: our future offices will play a critical part.

Intelligent Buildings to Boost Productivity, Efficiency and Wellbeing

The emergence of digital technology is driving the creation of more efficient, sustainable and productive workplaces. Intelligent buildings expert Tony Buckingham discusses some of the latest digital tools and trends set to shape our current and future workplaces.

You're travelling to an unfamiliar corporate office for a crucial meeting. But it's a smart-enabled building, so you're relaxed. The company app on your smartphone, containing your profile, enabled you to prepare last night and the building is expecting you.

Facial recognition cameras identify you as you enter the reception and you present a security matrix-barcode boarding pass sent to your smartphone for second-level verification. Knowing your meeting room destination, the building calls the lift.

At your floor, a location-sensing system leads you to the meeting room. As you're early the system suggests, via your smartphone, you get a coffee. You accept and take a detour to the automatic vending machine. 'The usual?' it asks, politely, with cashless vending takes care of payment.

On your return, knowing the type of device you're using, the meeting room's audio visual (AV) systems automatically activates and connects, setting up video conferencing links with the relevant offices. The intelligent building management system (iBMS) has also optimized the room's temperature and lighting for your call. You're ready.

During the meeting, without you being aware, systems all around you are collecting vital data from mechanical and electrical plant, such as Internet of Things (IoT) connected fan coil units. Armed with this real time data, Artificial Intelligence (AI) is analysing performance trends, predicting when equipment is going to need maintenance. The system can then initiate preventative alerts and even automated maintenance works orders.

Smarter, better

Smart technologies are rapidly transforming modern offices, from just places you go to do your job to interconnective hubs designed to promote creativity, collaboration and employee wellbeing.

Increasingly reliant on their smartphones as well as voice-activated solutions, such as Siri, Google and Alexa, at home and work, office users and owners are coming to expect a high-level app-based interoperability in every area of their life.

For employees, this means access to spaces and tools developed to improve their everyday work experience and wellbeing. This includes a range of environments tailored to different types of working, such as breakout areas and project 'neighbourhoods', that can be configured easily as well as, potentially in the future, the development of solutions such as 'acoustic bubbles' (see Figure 1) activated via sensors and dispatched to surround colleagues conducting impromptu meetings in an open-plan environment. Mobility and connectivity are key.

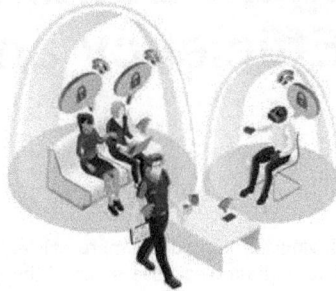

Figure 1: Acoustic bubbles in use in the workplaces of the future

Developers, occupiers and property managers want to get the most out of their space, ensuring it's used efficiently, no matter how mobile, agile and scattered around the globe their workforce and/or users might be.

To deliver on these divergent aspirations, buildings need to become smarter, facilitating systematic data collection and analysis, optimization and control, including the use of AI. Here we showcase some of the best emerging smart and intelligent building technologies to help building owners and managers better meet their users and building's needs.

Leveraging solutions from other sectors

For many years, retailers and shopping centre owners have used people-counting systems to register footfall (the location and number of people entering in a given time) and shoppers' travel paths through retail areas. Commercial offices today use similar technologies, from sensing the presence of people in meeting rooms, to analysing how they flow through a building.

Imagine you're planning to eat in the office canteen but you're in a hurry, so you check your firm's app to see how long the queue is. While you're connected, you look at the menu along with the nutritional content of the meals. Soon you'll be able to order and pay for your meal via an app, and have it delivered to your table by an autonomous robot, a technology that is already being trialled in the hospitality sector.

This new technology also makes unused meeting rooms with powered-up with lights, projectors, screens and heating a thing of the past. Linked to a company's room booking system, sensors can automatically detect somebody in the room and via iBMS, switch it to standby, saving energy and money.

With integrated, easy-to-use room controls and app-based booking systems there's also an opportunity to influence the way people manage bookings. Employees can receive rewards (such as canteen credits) if they mark a room as free after vacating it early — an action that will also switch the room to standby and enable somebody else to use it.

Furthermore, this technology provides property managers with real-time data across their real estate portfolio, letting them identify trends and understand why certain areas are well occupied while others are not. The use of sensors enables them to examine the effects of potential factors, such as ambient noise, light, temperature and ventilation.

Flexing to adapt

To enable these existing, new and emerging technologies, ICT infrastructure remains a critical part of the jigsaw, be it wired or wireless. Its design needs to be flexible and adaptive to react quickly to situations. Infrastructure is the enabler for technology.

For example, a property manager faced with regularly low occupancy levels within a selected area may wish to experiment with new floor layouts and configurations. This is already happening on a regular basis in co-working spaces, even down to the placement and use of furniture. With the correct location and workplace desk management systems in place, they can provide real-time information and occupancy trends and quickly determine if the new layouts are yielding increased occupancy. Successful models and solutions can then be rolled out to other underperforming areas.

Systems can be linked to the desk power systems, automatically turning off the power to the desk when not in use. How many times do you walk past a bank of desks as you leave to go home all with screens powered on in standby mode? Interactive desk management systems save energy and are good for the environment.

Taking control

Other trends are emerging. Studies show that a person's productivity, wellbeing and happiness increase if they work in an environment with good lighting, and comfortable temperature and air circulation. So, while the initial focus was on how we occupy space, there's a growing demand for monitoring indoor office environments, with employee wellbeing a key driver. In response, office users are now able to source cost-effective hand-held devices to measure CO_2, temperature and humidity.

By increasing the attractiveness of the building and contentment of its users, the ability to control environments and creating sustainable and inspirational spaces can also contribute to delivering better property investment returns. Landlords and tenants who embrace digital technology are both satisfying the modern-day user's needs and also differentiating themselves from competition with the potential to commanding higher returns on investment for smart workplaces.

Innovative workplaces

These innovations are just the start, with artificial intelligence and other digital technologies becoming an integral part of the way we do our jobs and live our lives.

Smart buildings are becoming intelligent buildings.

By leveraging the opportunities these solutions offer, in partnership with building users, building owners, managers and occupiers can secure improvements in productivity, efficiency and, crucially, workplace satisfaction.

The Architecture of Natural Cooling, 2nd edition

Brian Ford *et al.*

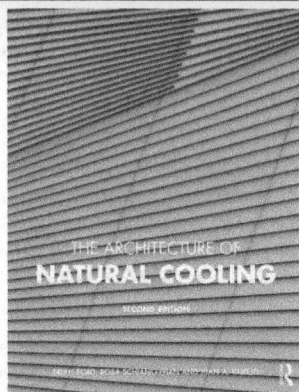

Overheating in buildings is commonplace. This book describes how we can keep cool without conventional air-conditioning: improving comfort and productivity while reducing energy costs and carbon emissions. It provides architects, engineers and policy makers with a 'how-to' guide to the application of natural cooling in new and existing buildings. It demonstrates, through reference to numerous examples, that natural cooling is viable in most climates around the world.

This completely revised and expanded second edition includes:
- An overview of natural cooling past and present.
- Guidance on the principles and strategies that can be adopted.
- A review of the applicability of different strategies.
- Explanation of simplified tools for performance assessment.
- A review of components and controls.
- A detailed evaluation of case studies from the USA, Europe, India and China.

This book is not just for the technical specialist, as it also provides a general grounding in how to avoid or minimise air-conditioning. Importantly, it demonstrates that understanding our environment, rather than fighting it, will help us to live sustainably in our rapidly warming world..

November 2019: 280 pp
Pb: 978-1-138-62907-3 : £36.99

Six Ways Data Analytics is Transforming BIM

The use of Building Information Modelling (BIM) is rapidly increasing in the design, construction, operation and maintenance of building and infrastructure assets. The reason is simple according to AECOM's James Rosenwax and Ben May: digital technology is making it easier, cheaper and more efficient to manage buildings and infrastructure projects through digital models.

Using BIM, we can make digital replicas of physical assets, processes and systems the so-called 'digital twins' of buildings. These models show three-dimensional details of construction drawings including structural, electrical, and mechanical elements, and fire safety systems. In addition to these key design elements, there are a variety of other data points that can be included such as material specifications, component costs and the sequence of the construction and maintenance requirements.

Using such digital models, which draw vast amounts of data in one place, complex calculations can be drawn. In a similar way to how analytics is used to provide insights to financial data, we can now use analytics to interpret the vast amount of data contained within building and infrastructure project models. Many organizations are creating and maintaining digital models across the full lifecycle of an asset, from early design through construction and then into ongoing operation and maintenance.

However, there is currently no 'standard' way that infrastructure owner-operators set their data requirements. They often rely on an information manager (internal or external) to manually process data, which is laborious and time-consuming.

Here we offer six reasons why building and infrastructure owners and operators should use BIM analytics tools to unleash the full potential from their building information models.

Reason 1: Data insights gathered over time help save projects time and money

The building and infrastructure sector has a poor record when it comes to productivity and efficiency. The opportunity to save time and money on projects without sacrificing the quality of outcomes is increasingly important. Project owner operators often receive models from a range of designers in a variety of formats and there are usually clashes that need to be resolved. Traditionally, this can take a couple of days for specialist BIM teams to analyse.

With access to artificial intelligence tools backed by machine learning, the days of manually extracting and interpreting data should be behind us. These digital tools allow teams to be more efficient and spend less time extracting data, increasing efficiency and accuracy, while freeing up BIM professionals to provide genuine insights and advice.

Reason 2: Easier compliance with data requirements for an audit

Audits are probably the least pleasurable and most stressful part of any project or operation. If we apply technology smartly, however, they should be a seamless part of our delivery processes.

For owner-operators, digital project models must adhere to either a global data standard (such as COBie, Uniclass or OmniClass) or an organization's custom data standards. To ensure that these requirements are met throughout the design phase, BIM consultants are often engaged to conduct an audit on the model to validate compliance. This ensures that models meet the project-information requirements specified in the contract and are compliant for audit purposes.

We can use technology to automate the compliance process, allowing more time for consultants to analyse the results and work on solutions to identified problems. Automation allows checks to be done more often, at regular 'milestones', for greater confidence that a project is on track. Automation also makes it easy to validate that the model's data complies with the required standards.

Reason 3: Cost avoidance and improved quality management through project controls

The larger and more complex projects, the bigger the chances that inadvertent and unnecessary duplication of orders are made for building materials and specialized labour. Effective supply management can lead to significant cost savings. BIM analytics makes it easier to monitor quantities of building materials to help streamline procurement and delivery schedules – all from the comfort of a site office or a mobile device. Model analytics charts and graphs make it easier to track quantities, with all suppliers and disciplines captured in a fully configurable dashboard.

In a recent project, the AECOM BIM team were able to identify that seven different strengths of concrete were being ordered. By simplifying this to three strengths of concrete, the team was able to reduce costs, including the time and cost of ordering, delivery, handling and processing. We are only just beginning to realise the benefits of having full control and visibility of all the data that is built into our projects.

Reason 4: Geometric clashes are resolved more efficiently, earlier on in a project

The advancements in digital design tools and fabrication technology has led to increasingly complex model geometry, especially in building architecture. To a lesser extent, this is also the case across roads, bridges and rail projects. There are a myriad of clash detection tools available, which thankfully makes the construction and fabrication processes more efficient. If we take this one step further and apply BIM analytics tools, this allows project owners, suppliers and designers to identify clash relationships earlier in a project's design phase, reducing the likelihood of costly on-site delays.

Reason 5: Automated dashboards make communication and collaboration easier

Major building and infrastructure projects hold regular stakeholder meetings to enable the various design disciplines, owners and project management teams to communicate and collaborate. These meetings are essential to delivering a large design project. In the past, supporting information raised during these meetings was typically manually collated, added to a pdf 'dashboard' and distributed to teams. Whilst this was once cutting edge, it can now be automated and distributed online, reducing the time spent collating data and allowing more time to be spent on collaboration and stakeholder engagement.

Reason 6: More-efficient handovers

Accurate, 'as-built,' digital models with resolved issues and clashes, full data compliance and optimized quantities improves construction efficiency and reduces errors and time delays. After construction, digital models can be transitioned to operations and maintenance teams to set up enterprise asset-management systems and building-management systems to support ongoing operations and maintenance.

Using digital to drive this process makes it easier to manage the required data handover process and ensuring the data is compliant with standards at key points of handover, thereby streamlining the asset-handover process.

Typical Engineering Details

In addition to the Engineering Features, Typical Engineering Details are included. These are indicative schematics to assist in the compilation of costing exercises. The user should note that these are only examples and cannot be construed to reflect the design for each and every situation. They are merely provided to assist the user with gaining an understanding of the Engineering concepts and elements making up such.

ELECTRICAL

- Urban Network Mainly Underground
- Urban Network Mainly Underground with Reinforcement
- Urban Network Mainly Underground with Substation Reinforcement
- Typical Simple 11 kV Network Connection For LV Intakes Up To 1000 kVA
- Typical 11 kV Network Connections For MV Intakes 1000 kVA To 6000 kVA
- Static UPS System – Simplified Single Line Schematic For a Single Module
- Typical Data Transmission (Structured Cabling)
- Typical Networked Lighting Control System
- Typical Standby Power System, Single Line Schematic
- Typical Fire Detection and Alarm Schematic
- Typical Block Diagram – Access Control System (ACS)
- Typical Block Diagram – Intruder Detection System (IDS)
- Typical Block Diagram – Digital CCTV

MECHANICAL

- Fan Coil Unit System
- Displacement Ventilation System
- Chilled Ceiling System (Passive System)
- Chilled Beam System (Passive or Active System)
- Variable Air Volume (VAV)
- Variable Refrigerant Volume System (VRV)
- Reverse cycle heat pump

Urban Network Mainly Underground

Details: Connection to small housing development 10 houses, 60 m of LV cable from local 11 kV/400 V substation route in footpath and verge, 10 m of service cable to each plot in verge

Supply Capacity: 200 kVA

Connection Voltage: LV

3 phase supply

	Labour	Plant	Materials	Overheads	Total
Mains Cable	£330	£110	£ 1,320	£350	£2,110
Service Cable	£330	£110	£550	£200	£1,190
Jointing	£1,320	£440	£1,650	£680	£4.090
Termination	Incl.	Incl.	Incl.	Incl.	Incl.
Trench/Reinstate	£1,980	£660	£1,650	£860	£5,150
Substation	–	–	–	–	–
HV Trench & Joint	–	–	–	–	–
HV Cable Install	–	–	–	–	–
Total Calculated Price	£3,960	£1,320	£5,170	£2,090	£12,540
Total Non-contestable Elements and Associated Charges					£3,300
Grand Total Calculated Price excl. VAT					£15,840

Urban Network Mainly Underground with Reinforcement

Details: Connection to small housing development 10 houses, 60 m of LV cable from local 11 kV/400 V substation route in footpath and verge, 10 m of service cable to each plot in unmade ground. Scheme includes reinforcement of LV distribution board at substation

Supply Capacity: 200 kVA

Connection Voltage: LV

Single phase supply

Breakdown of Detailed Cost Information

	Labour	Plant	Materials	Overheads	Total
Mains Cable	£330	£110	£ 1,320	£350	£2,110
Service Cable	£330	£110	£550	£200	£1,190
Jointing	£1,320	£440	£1,650	£680	£4.090
Termination	Incl.	Incl.	Incl.	Incl.	Incl.
Trench/Reinstate	£1,980	£660	£1,650	£860	£5,150
Substation	–	–	–	–	–
HV Trench & Joint	–	–	–	–	–
HV Cable Install	–	–	–	–	–
Total Calculated Price	£3,960	£1,320	£5,170	£2,090	£12,540
Total Non-contestable Elements and Associated Charges					£14,300
Grand Total Calculated Price excl. VAT					£26,840

Urban Network Mainly Underground with Substation Reinforcement

Details: Connection to small housing development 10 houses, 60 m of LV cable from local 11 kV/400 V substation route in footpath and verge, 10 m of service cable to each plot in verge. Scheme includes reinforcement of LV distribution board and new substation and 20 m of HV cable

Supply Capacity: 200 kVA

Connection Voltage: LV

Single phase

Breakdown of Detailed Cost Information

	Labour	Plant	Materials	Overheads	Total
Mains Cable	£330	£110	£ 1,650	£420	£2,510
Service Cable	£330	£110	£550	£200	£1,190
Jointing	£1,320	£440	£1,650	£680	£4.090
Termination	Incl.	Incl.	Incl.	Incl.	Incl.
Trench/Reinstate	£1,980	£660	£1,650	£860	£5,150
Substation	£11,000	£3,300	£88,000	£20,460	£122,760
HV trench & joint	£1,980	£660	£1,100	£750	£4,490
HV cable install	£330	£110	£1,240	£340	£2,020
Total Calculated Price	£17,270	£5,390	£95,840	£23,710	£142,210

Total Non-contestable Elements and Associated Charges	£16,500
Grand Total Calculated Price excl. VAT	£158,710

Client LV Intake Switchboard (400 V)

DNO metering air circuit breaker (ACB)

DNO Transformer (typically iol filled, 500 to 1000 kVA, 11 kV/400 V)

DNO Ring main (RMU)

DNO 11 kV network

Client Demise

DNO* Demise

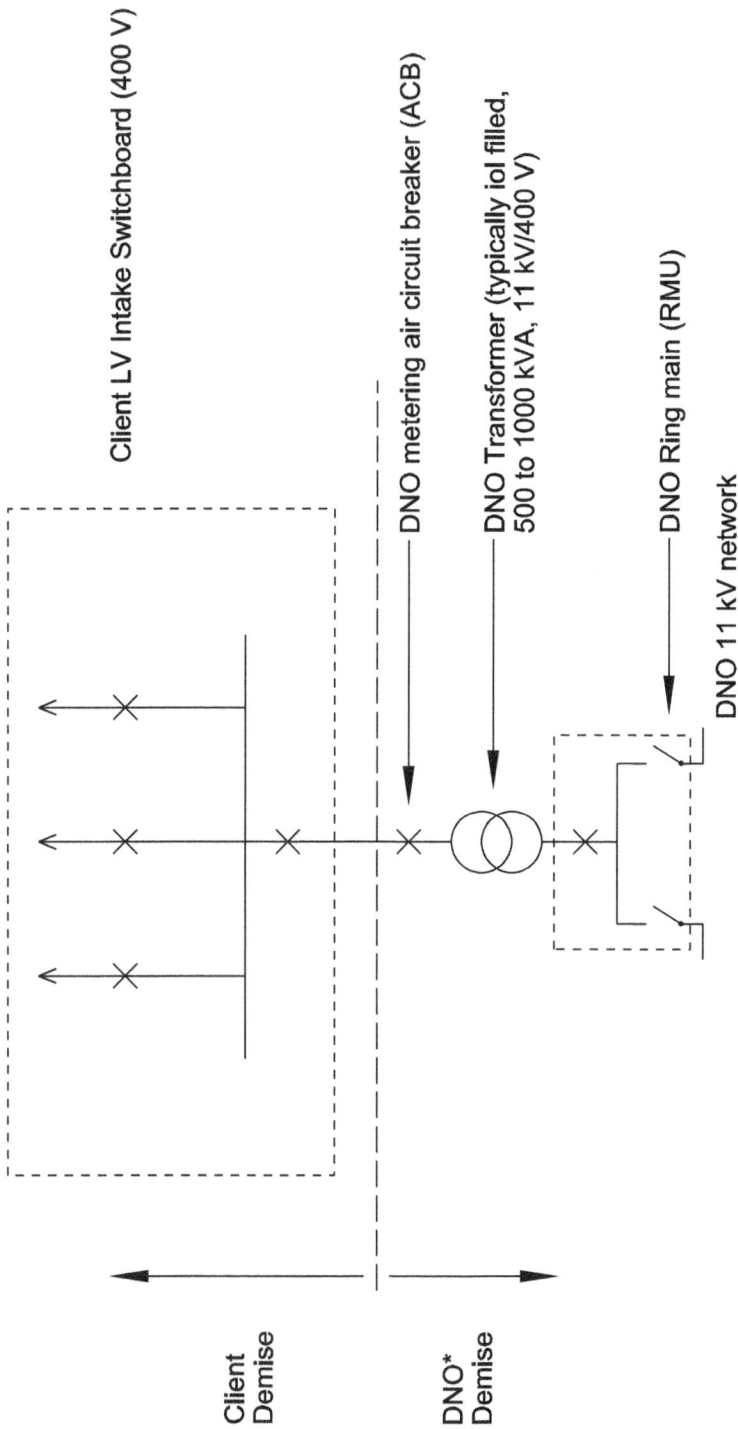

Note: *DNO – Distribution Network Operator

Typical Simple 11 kV Network Connection for LV Intakes up to 1000 kVA

109

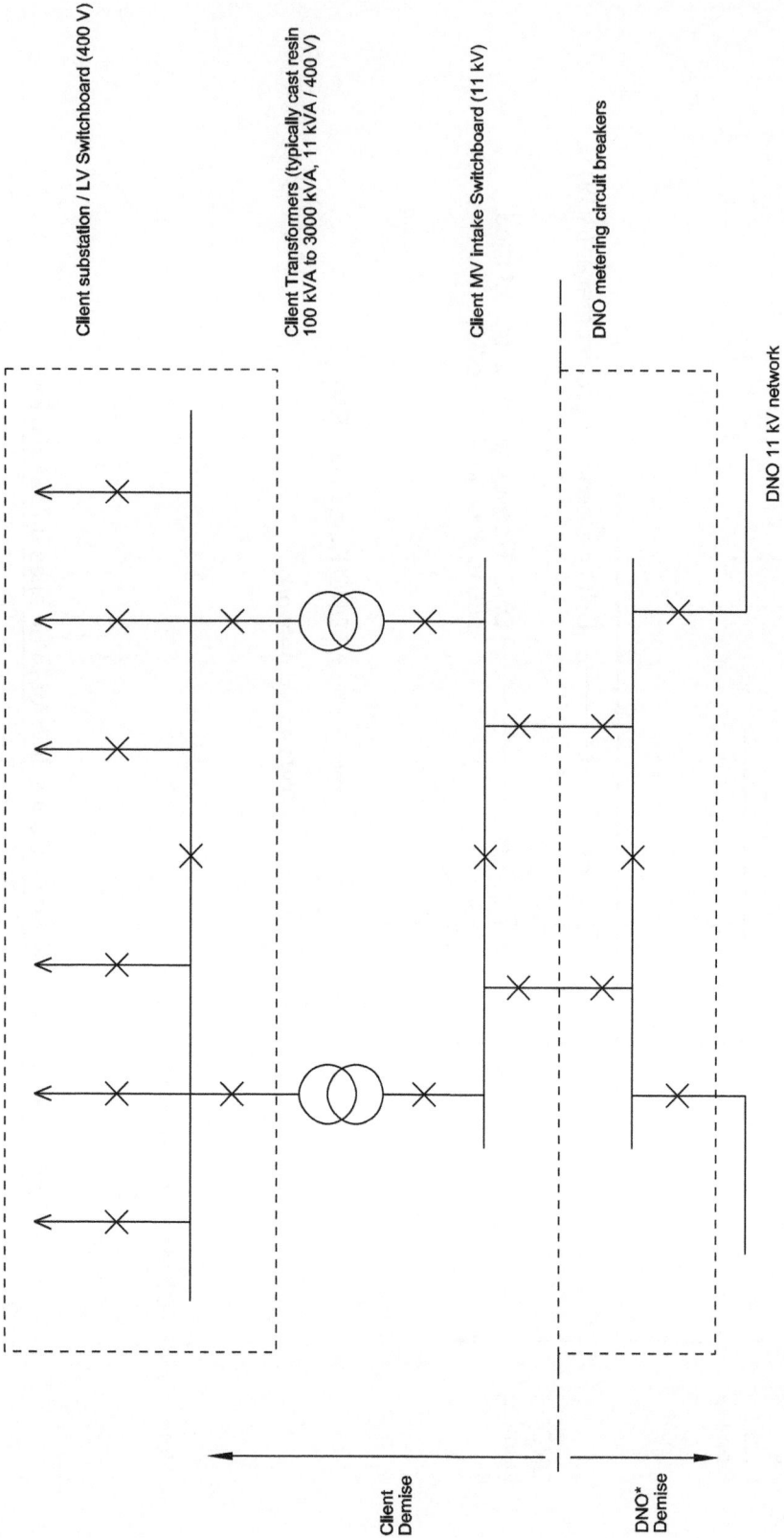

Client substation / LV Switchboard (400 V)

Client Transformers (typically cast resin 100 kVA to 3000 kVA, 11 kVA / 400 V)

Client MV Intake Switchboard (11 kV)

DNO metering circuit breakers

DNO 11 kV network

Client Demise

DNO* Demise

Note: *DNO - Distribution Network Operator

Typical 11 kV Network Connection for MV Intakes 1000 kVA up to 6000 kVA

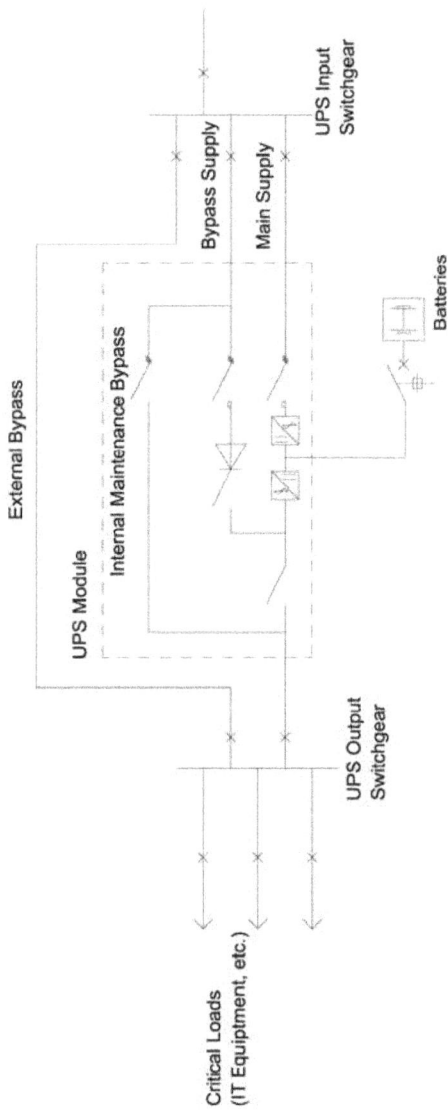

External Bypass

UPS Module

Internal Maintenance Bypass

Bypass Supply

Main Supply

UPS Input
Switchgear

Batteries

UPS Output
Switchgear

Critical Loads
(IT Equiptment, etc.)

Static UPS System – Simplified Schematic For Single Module

111

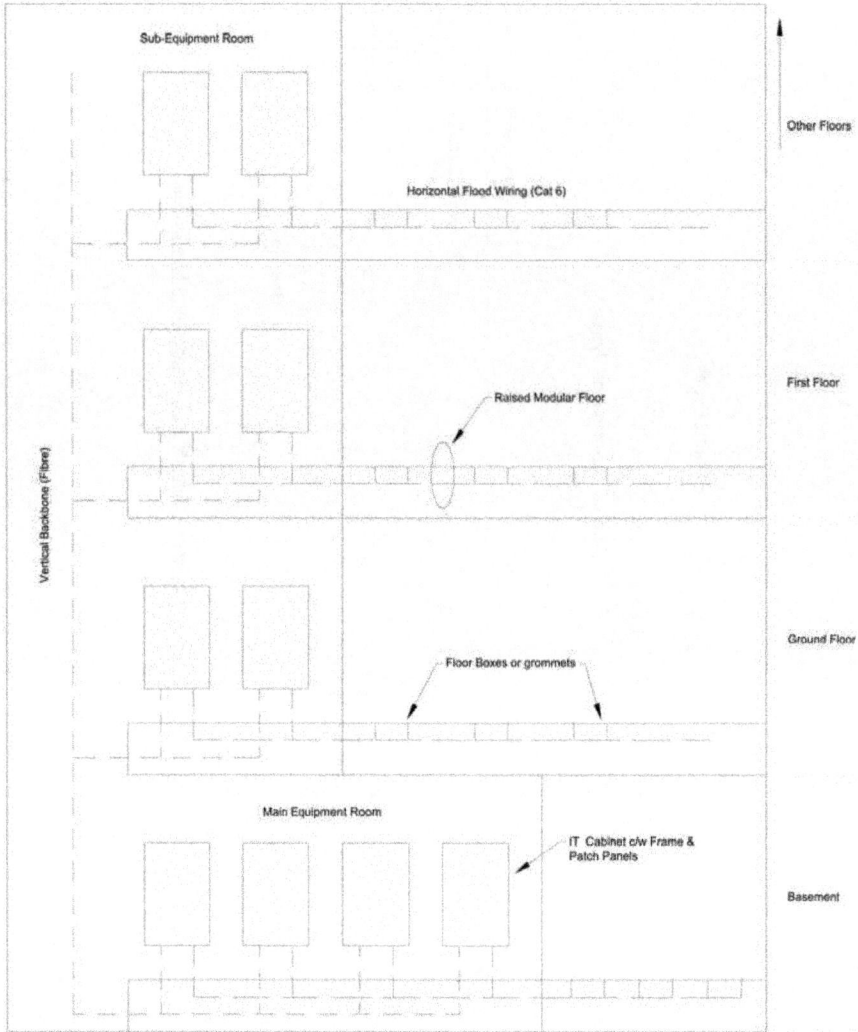

Sub-Equipment Room

Horizontal Flood Wiring (Cat 6)

Other Floors

Raised Modular Floor

First Floor

Floor Boxes or grommets

Ground Floor

Main Equipment Room

IT Cabinet c/w Frame & Patch Panels

Basement

Vertical Backbone (Fibre)

Typical Data Transmission (Structured Cabling)

112

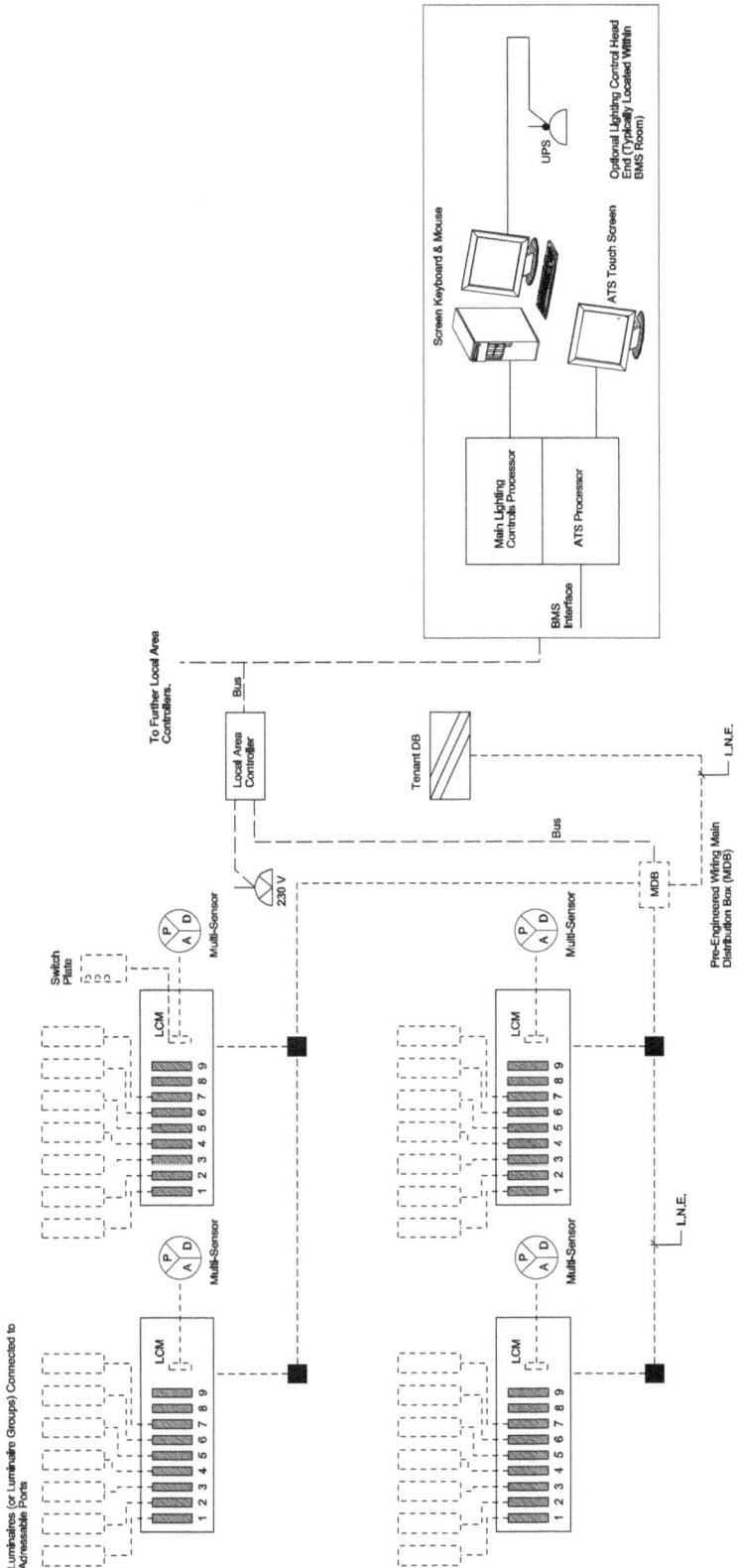

Typical Networked Lighting Control System

113

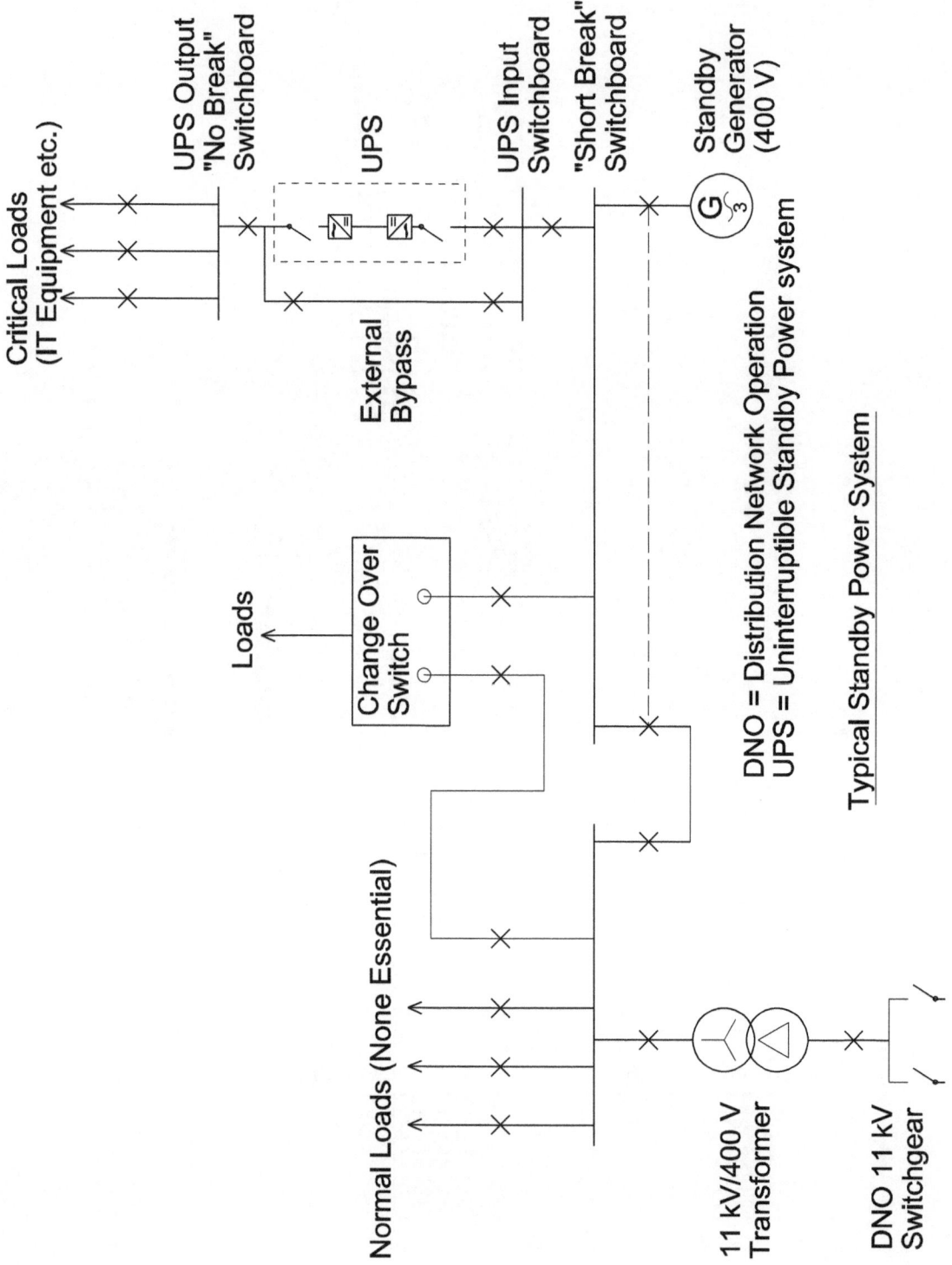

Critical Loads
(IT Equipment etc.)

UPS Output "No Break" Switchboard

UPS

External Bypass

UPS Input Switchboard

"Short Break" Switchboard

Standby Generator (400 V)

Loads

Change Over Switch

Normal Loads (None Essential)

11 kV/400 V Transformer

DNO 11 kV Switchgear

DNO = Distribution Network Operation
UPS = Uninterruptible Standby Power system

Typical Standby Power System

114

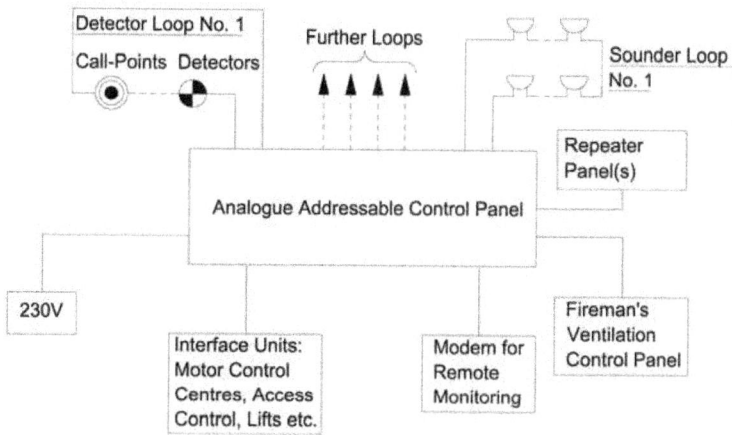

Detector Loop No. 1

Call-Points Detectors

Further Loops

Sounder Loop
No. 1

Repeater
Panel(s)

Analogue Addressable Control Panel

230V

Interface Units:
Motor Control
Centres, Access
Control, Lifts etc.

Modem for
Remote
Monitoring

Fireman's
Ventilation
Control Panel

Typical Fire Detection and Alarm Schematic

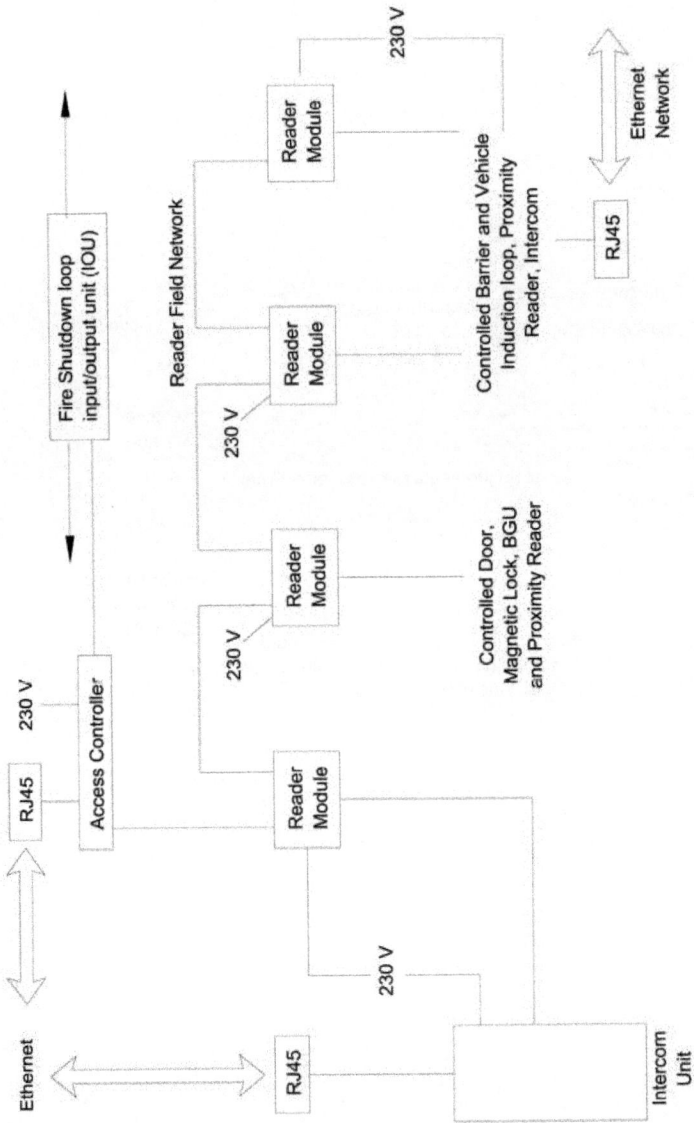

Typical Block Diagram – Access Control System

Passive Infared (PIR) and volumetric detectors

Personal attack buttons

External Sounder

Internal Loudspeaker

Multi-zone Control Panel including keypad and battery back-up

230 V

Remote Keypad

Contact loops for doors and windows

Vibration Detectors

Final exit set button

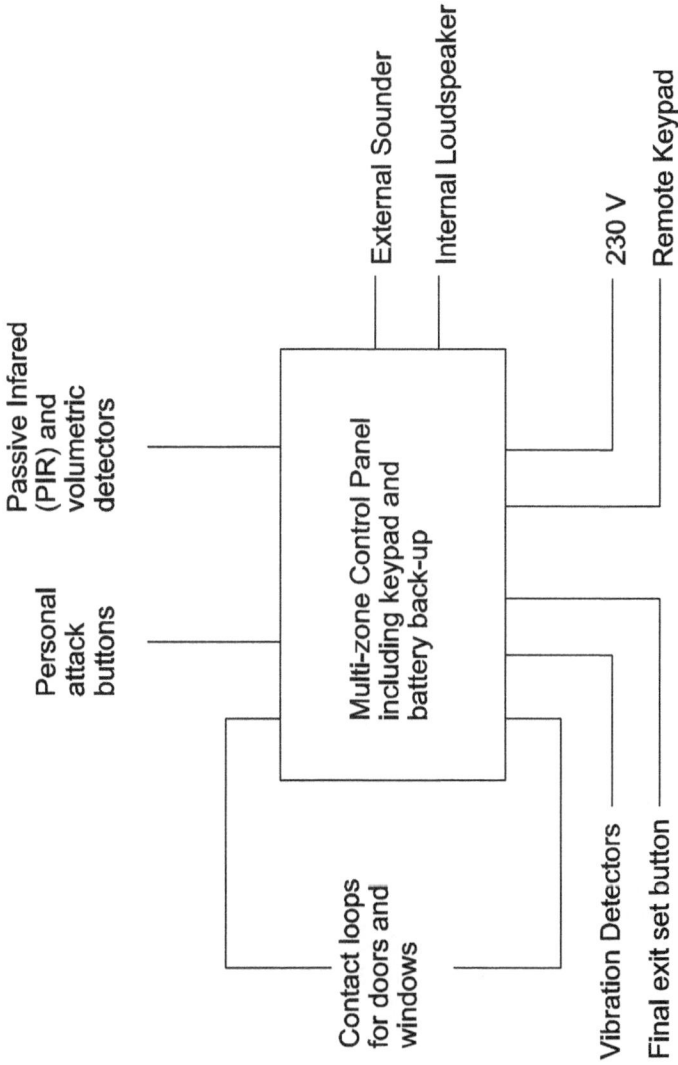

Typical Block Diagram – Intruder Detection System (IDS)

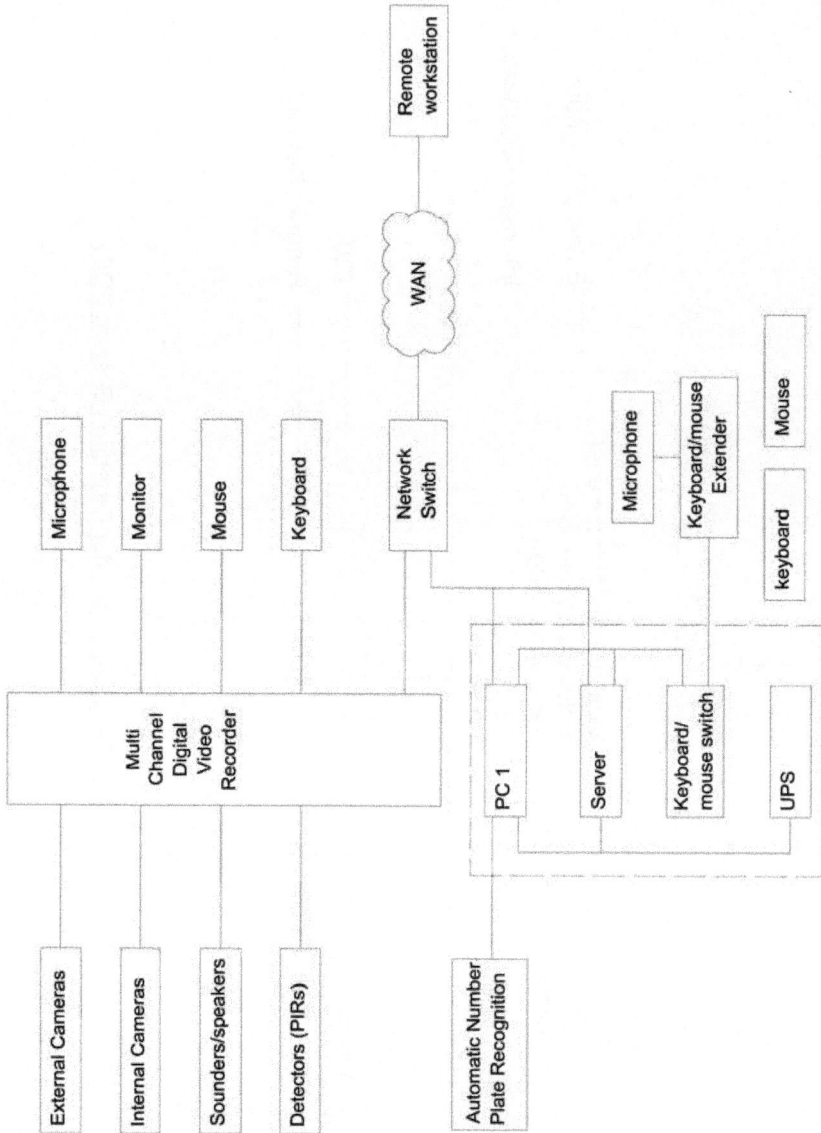

Typical Block Diagram – Digital CCTV

External Cameras
Internal Cameras
Sounders/speakers
Detectors (PIRs)

Multi Channel Digital Video Recorder

Microphone
Monitor
Mouse
Keyboard

Network Switch

WAN

Remote workstation

Automatic Number Plate Recognition

PC 1
Server
Keyboard/ mouse switch
UPS

Microphone
Keyboard/mouse Extender
Mouse
keyboard

FAN COIL UNIT SYSTEM

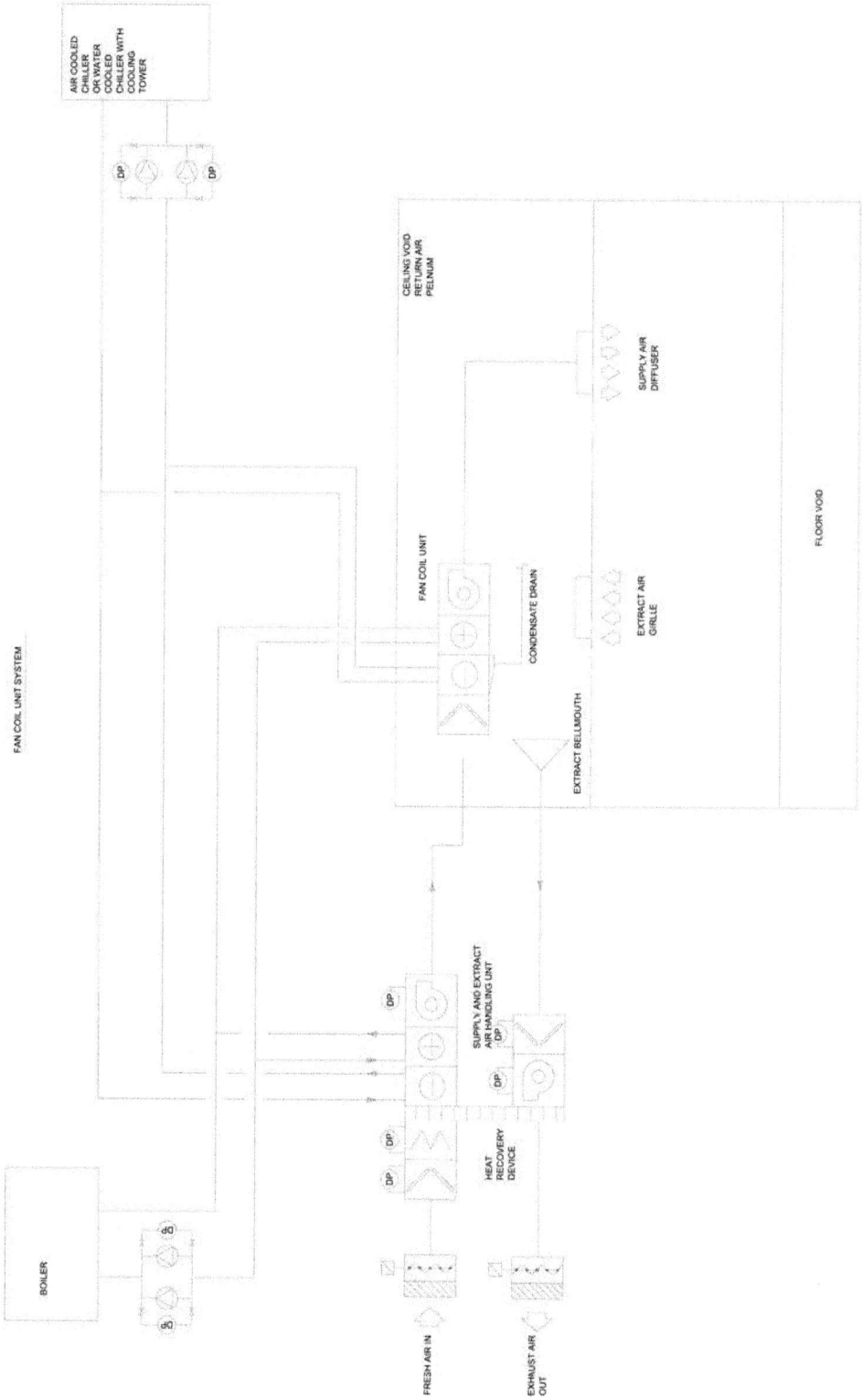

AIR COOLED
CHILLER
OR WATER
COOLED
CHILLER WITH
COOLING
TOWER

DP

DP

BOILER

DP

DP

CEILING VOID
RETURN AIR
PELNUM

SUPPLY AIR
DIFFUSER

FAN COIL UNIT

CONDENSATE DRAIN

EXTRACT AIR
GIRLLE

EXTRACT BELLMOUTH

FLOOR VOID

SUPPLY AND EXTRACT
AIR HANDLING UNIT

DP

DP

DP

DP

HEAT
RECOVERY
DEVICE

FRESH AIR IN

EXHAUST AIR
OUT

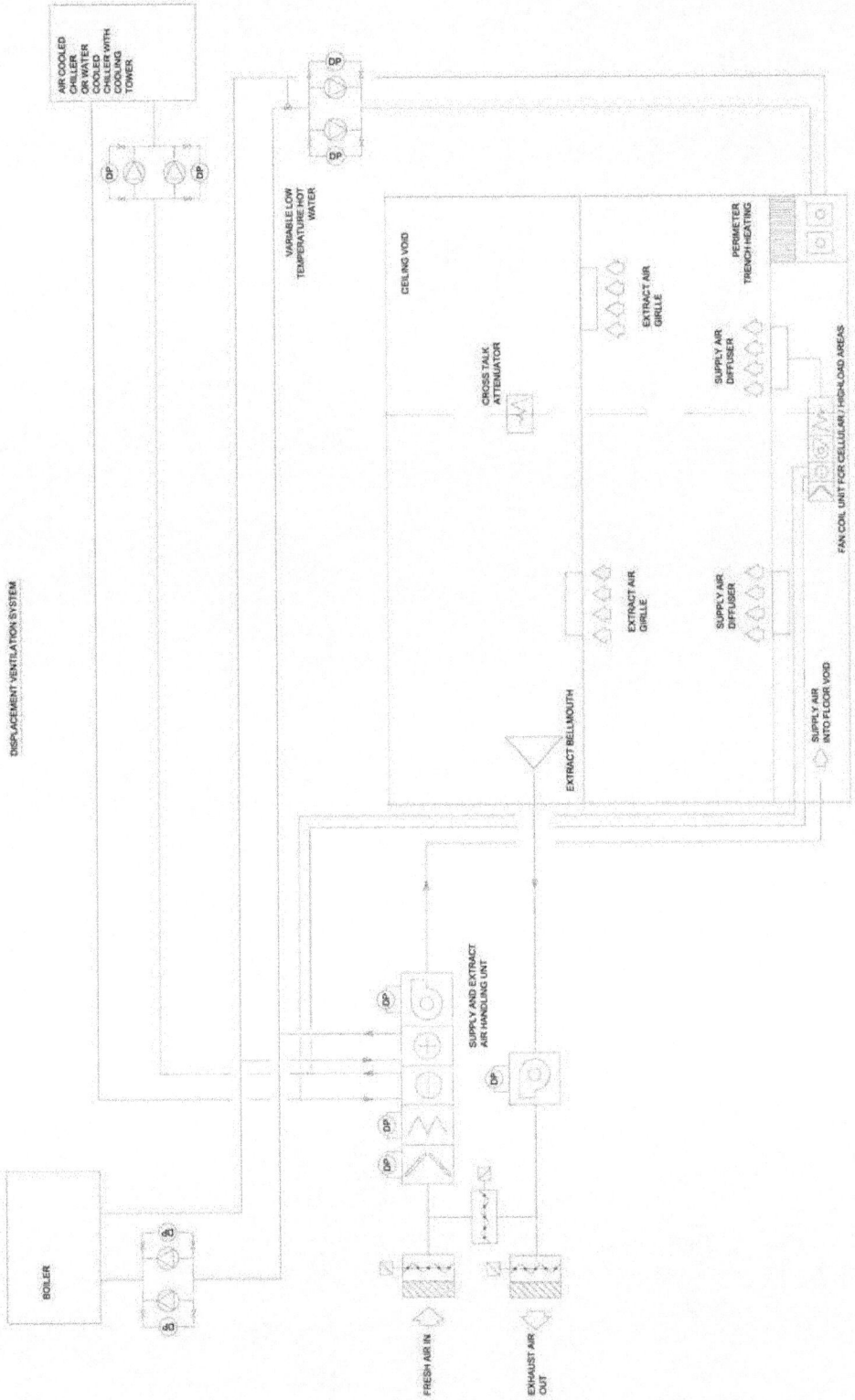

DISPLACEMENT VENTILATION SYSTEM

AIR COOLED CHILLER OR WATER COOLED CHILLER WITH COOLING TOWER

VARIABLE LOW TEMPERATURE HOT WATER

CEILING VOID

EXTRACT AIR GRILLE

CROSS TALK ATTENUATOR

SUPPLY AIR DIFFUSER

PERIMETER TRENCH HEATING

EXTRACT AIR GRILLE

SUPPLY AIR DIFFUSER

FAN COIL UNIT FOR CELLULAR / HIGH LOAD AREAS

EXTRACT BELLMOUTH

SUPPLY AIR INTO FLOOR VOID

SUPPLY AND EXTRACT AIR HANDLING UNIT

BOILER

FRESH AIR IN

EXHAUST AIR OUT

DP

CHILLED CEILING SYSTEM

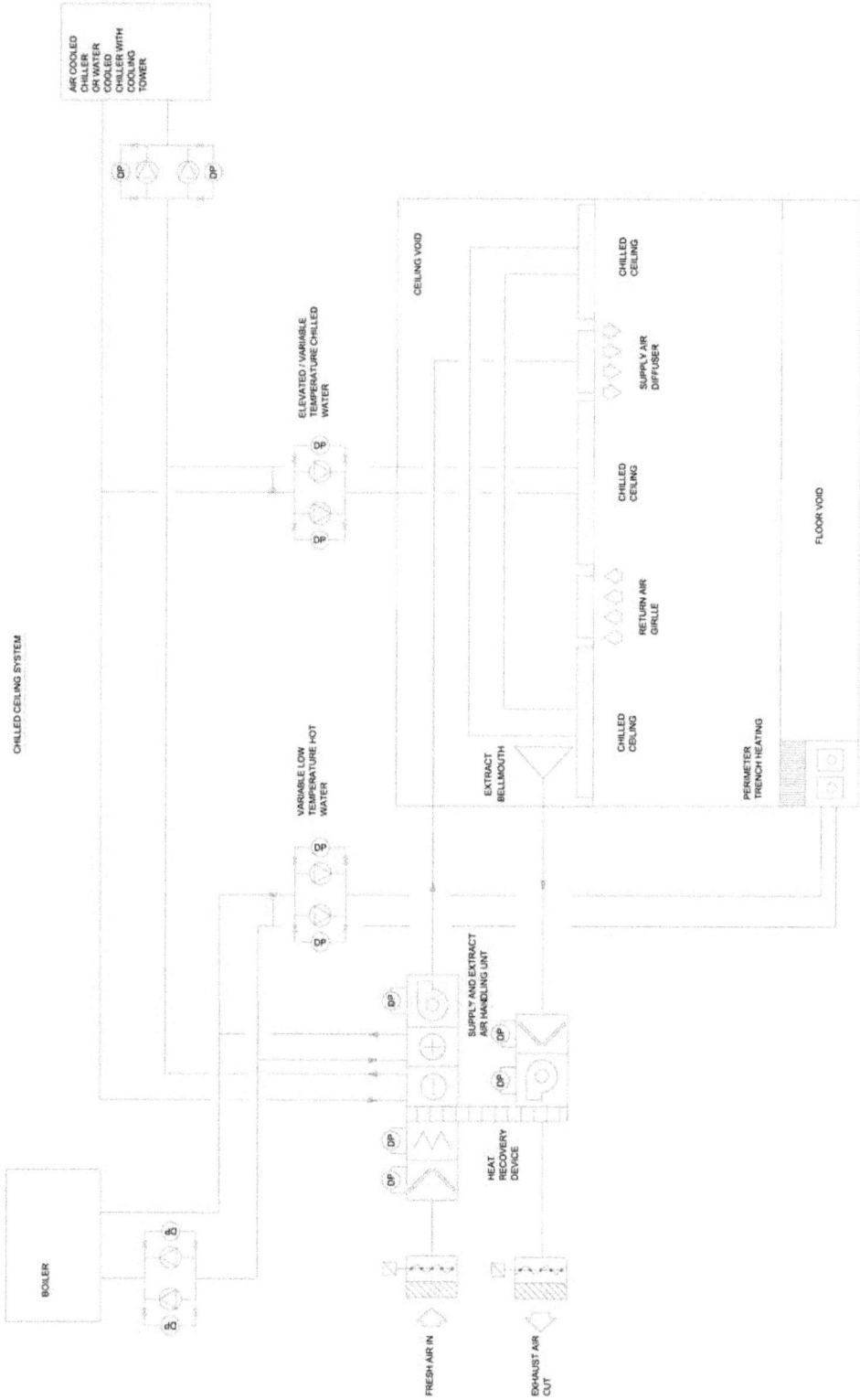

AIR COOLED
CHILLER
OR WATER
COOLED
CHILLER WITH
COOLING
TOWER

BOILER

DP

DP

DP

DP

DP

DP

ELEVATED / VARIABLE
TEMPERATURE CHILLED
WATER

VARIABLE LOW
TEMPERATURE HOT
WATER

DP

DP

DP

DP

CEILING VOID

CHILLED
CEILING

CHILLED
CEILING

SUPPLY AIR
DIFFUSER

RETURN AIR
GRILLE

CHILLED
CEILING

EXTRACT
BELLMOUTH

FLOOR VOID

PERIMETER
TRENCH HEATING

SUPPLY AND EXTRACT
AIR HANDLING UNIT

DP

DP

DP

DP

DP

HEAT
RECOVERY
DEVICE

FRESH AIR IN

EXHAUST AIR
OUT

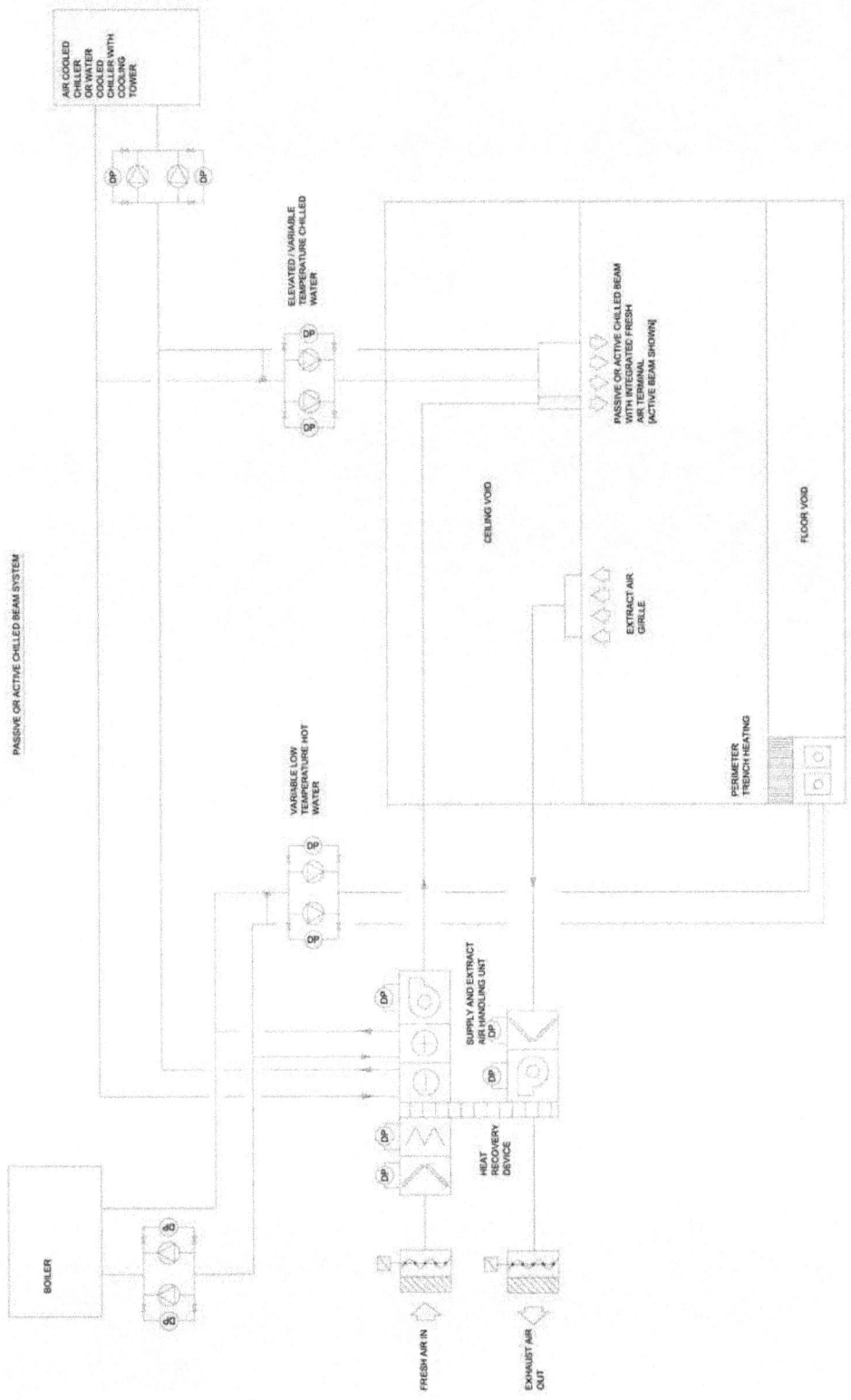

PASSIVE OR ACTIVE CHILLED BEAM SYSTEM

AIR COOLED CHILLER OR WATER COOLED CHILLER WITH COOLING TOWER

ELEVATED / VARIABLE TEMPERATURE CHILLED WATER

PASSIVE OR ACTIVE CHILLED BEAM WITH INTEGRATED FRESH AIR TERMINAL (ACTIVE BEAM SHOWN)

CEILING VOID

FLOOR VOID

EXTRACT AIR GRILLE

PERIMETER TRENCH HEATING

VARIABLE LOW TEMPERATURE HOT WATER

SUPPLY AND EXTRACT AIR HANDLING UNIT

HEAT RECOVERY DEVICE

BOILER

FRESH AIR IN

EXHAUST AIR OUT

VARIABLE AIR VOLUME SYSTEM

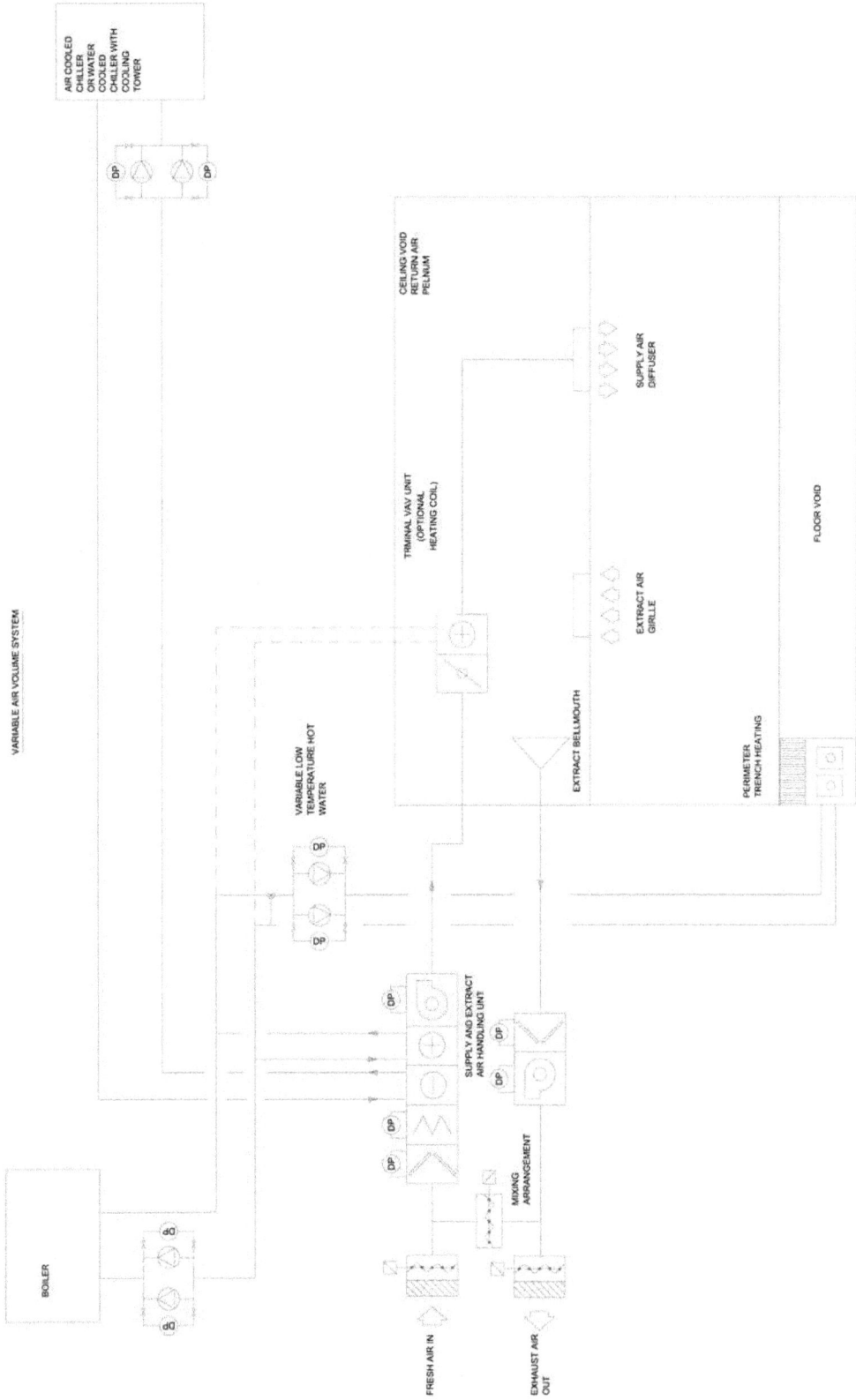

AIR COOLED CHILLER OR WATER COOLED CHILLER WITH COOLING TOWER

CEILING VOID RETURN AIR PELNUM

SUPPLY AIR DIFFUSER

TRIMINAL VAV UNIT (OPTIONAL HEATING COIL)

EXTRACT AIR GIRLLE

FLOOR VOID

VARIABLE LOW TEMPERATURE HOT WATER

EXTRACT BELLMOUTH

PERIMETER TRENCH HEATING

SUPPLY AND EXTRACT AIR HANDLING UNIT

MIXING ARRANGEMENT

BOILER

FRESH AIR IN

EXHAUST AIR OUT

DP

VARIABLE REFRIGERANT VOLUME SYSTEM

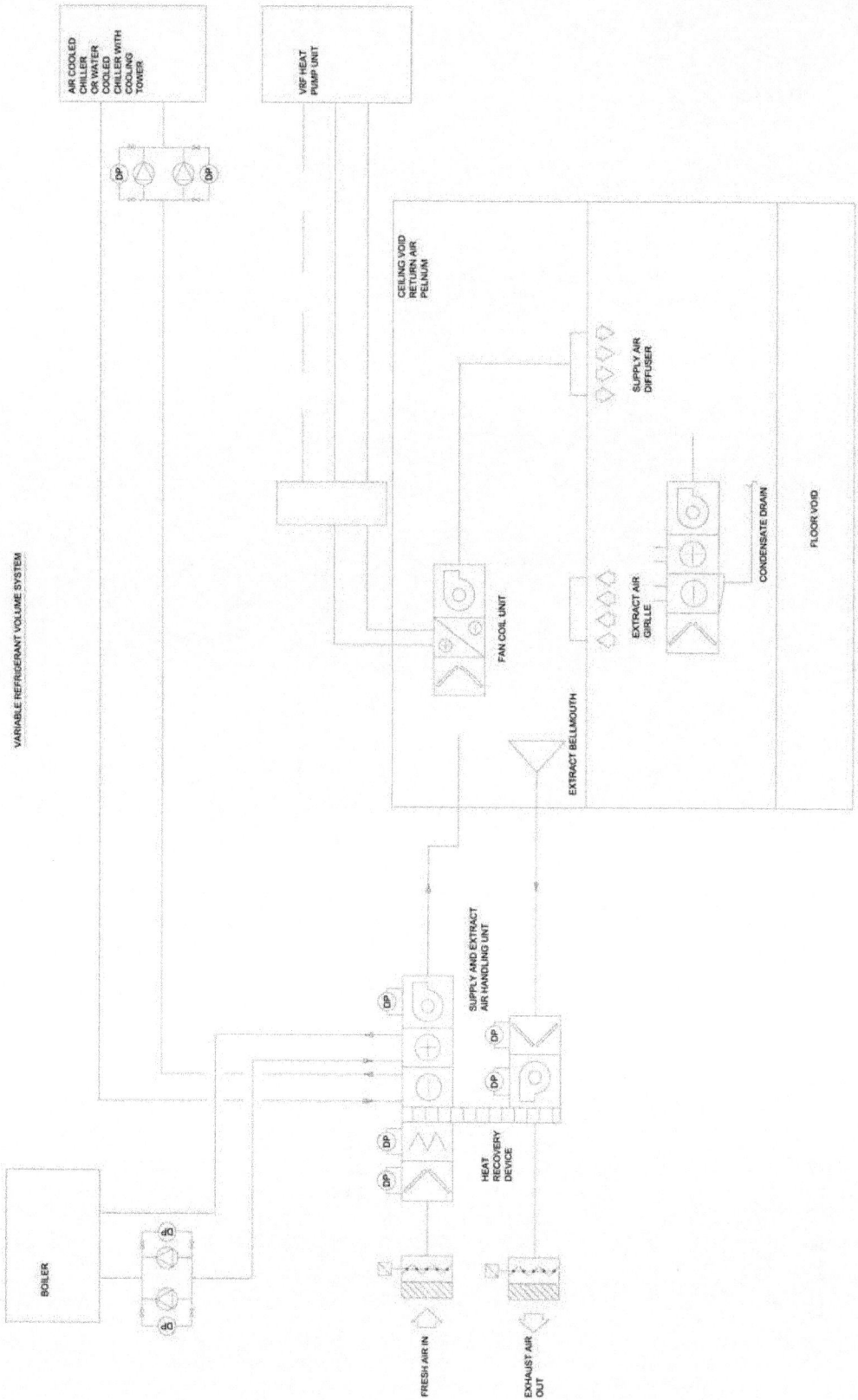

AIR COOLED
CHILLER
OR WATER
COOLED
CHILLER WITH
COOLING
TOWER

VRF HEAT
PUMP UNIT

CEILING VOID
RETURN AIR
PLENUM

SUPPLY AIR
DIFFUSER

FAN COIL UNIT

EXTRACT AIR
GRILLE

CONDENSATE DRAIN

FLOOR VOID

EXTRACT BELLMOUTH

BOILER

SUPPLY AND EXTRACT
AIR HANDLING UNIT

HEAT
RECOVERY
DEVICE

DP

FRESH AIR IN

EXHAUST AIR
OUT

124

REVERSE CYCLE HEAT PUMP

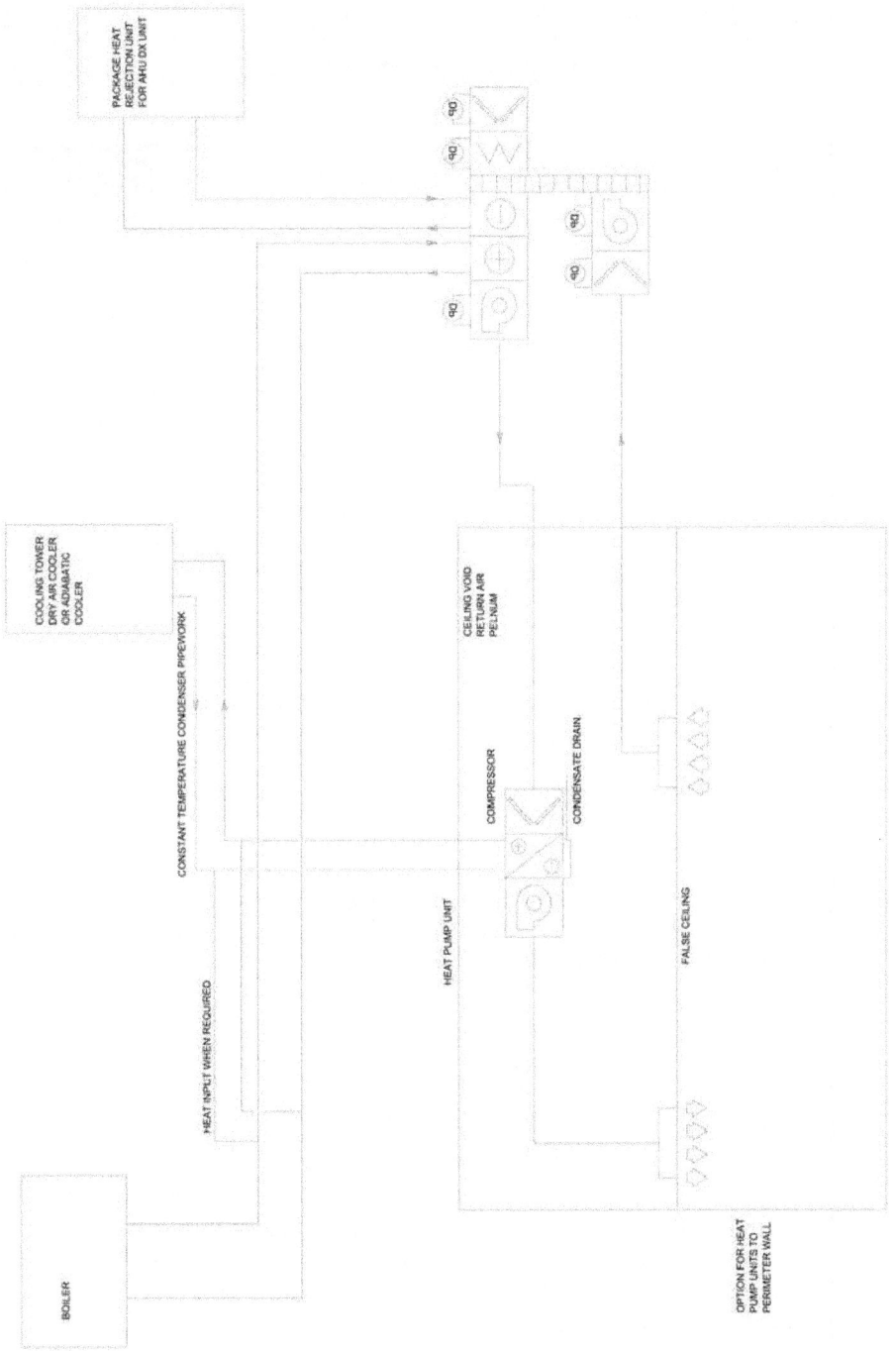

PACKAGE HEAT
REJECTION UNIT
FOR AHU DX UNIT

COOLING TOWER
DRY AIR COOLER
OR ADIABATIC
COOLER

CONSTANT TEMPERATURE CONDENSER PIPEWORK

HEAT INPUT WHEN REQUIRED

BOILER

CEILING VOID
RETURN AIR
PELNUM

COMPRESSOR

CONDENSATE DRAIN

HEAT PUMP UNIT

FALSE CEILING

OPTION FOR HEAT
PUMP UNITS TO
PERIMETER WALL

BIM and Quantity Surveying

S Pittard *et al.*

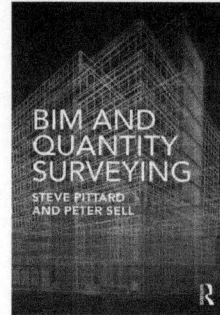

The sudden arrival of Building Information Modelling (BIM) as a key part of the building industry is redefining the roles and working practices of its stakeholders. Many clients, designers, contractors, quantity surveyors, and building managers are still finding their feet in an industry where BIM compliance can bring great rewards.

This guide is designed to help quantity surveying practitioners and students understand what BIM means for them, and how they should prepare to work successfully on BIM compliant projects. The case studies show how firms at the forefront of this technology have integrated core quantity surveying responsibilities like cost estimating, tendering, and development appraisal into high profile BIM projects. In addition to this, the implications for project management, facilities management, contract administration and dispute resolution are also explored through case studies, making this a highly valuable guide for those in a range of construction project management roles.

Featuring a chapter describing how the role of the quantity surveyor is likely to permanently shift as a result of this development, as well as descriptions of tools used, this covers both the organisational and practical aspects of a crucial topic.

December 2015: 234 x 156 mm: 258 pp
Pbk: 978-0-415-87043-6; £32.99

To Order: Tel: +44 (0) 1235 400524 Fax: +44 (0) 1235 400525
or Post: Taylor and Francis Customer Services,
Bookpoint Ltd, Unit T1, 200 Milton Park, Abingdon, Oxon, OX14 4TA UK
Email: book.orders@tandf.co.uk

For a complete listing of all our titles visit:
www.tandf.co.uk

Taylor & Francis
Taylor & Francis Group

PART 2

Approximate Estimating

DIRECTIONS

The prices shown in this section of the book are average prices on a fixed price basis for typical buildings tendered during the second quarter of 2020. Unless otherwise noted, they exclude external services and professional fees.

The information in this section has been arranged to follow more closely the order in which estimates may be developed, in accordance with RIBA work stages.

a)   Cost Indices and Regional Variations – These provide information regarding the adjustments to be made to estimates taking into account current pricing levels for different locations in the UK.

b)   Feasibility Costs – These provide a range of data (based on a rate per square metre) for all-in engineering costs, excluding lifts, associated with a wide variety of building types. These would typically be used at work stage 0/1 (feasibility) of a project.

c)   Elemental Rates – The outline costs for offices have been developed further to provide rates for the alternative solutions for each of the services elements. These would typically be used at work stage 2, outline proposal.

Where applicable, costs have been identified as Shell and Core and Fit-Out to reflect projects where the choice of procurement has dictated that the project is divided into two distinctive contractual parts.

Such detail would typically be required at work stage 3, detailed proposals.

d)   All-in-Rates – These are provided for a number of items and complete parts of a system, i.e. boiler plant, ductwork, pipework, electrical switchgear and small power distribution, together with lifts and escalators. Refer to the relevant section for further guidance notes.

e)   Elemental Costs – These are provided for a diverse range of building types: offices, laboratory, shopping mall, airport terminal building, supermarket, performing arts centre, sports hall, luxury hotel, hospital and secondary school. Also included Is a separate analysis of a building management system for an office block. In each case, a full analysis of engineering services costs is given to show the division between all elements and their relative costs to the total building area. A regional variation factor has been applied to bring these analyses to a common London base.

Prices should be applied to the total floor area of all storeys of the building under consideration. The area should be measured between the external walls without deduction for internal walls and staircases/lift shafts, i.e. GIA (Gross Internal Area).

Although prices are reviewed in the light of recent tenders it has only been possible to provide a range of prices for each building type. This should serve to emphasize that these can only be average prices for typical requirements and that such prices can vary widely depending on variations in size, location, phasing, specification, site conditions, procurement route, programme, market conditions and net to gross area efficiencies. Rates per square metre should not therefore be used indiscriminately and each case needs to be assessed on its own merits.

The prices do not include for incidental builder's work nor for profit and attendance by a Main Contractor where the work is executed as a subcontract: they do however include for preliminaries, profit and overheads for the services contractor. Capital contributions to statutory authorities and public undertakings and the cost of work carried out by them have been excluded.

Where services works are procured indirectly, i.e. ductwork via a mechanical subcontractor, the reader should make due allowance for the addition of a further level of profit etc.

COST INDICES

The following tables reflect the major changes in cost to contractors but do not necessarily reflect changes in tender levels. In addition to changes in labour and materials costs, tenders are affected by other factors such as the degree of competition in the particular industry, the area where the work is to be carried out, the availability of labour and the prevailing economic conditions. This has meant in years when there has been an abundance of work, tender levels have tended to increase at a greater rate than can be accounted for solely by increases in basic labour and material costs and, conversely, when there is a shortage of work this has tended to result in keener tenders. Allowances for these factors are impossible to assess on a general basis and can only be based on experience and knowledge of the particular circumstances.

In compiling the tables, the cost of labour has been calculated on the basis of a notional gang as set out elsewhere in the book. The proportion of labour to materials has been assumed as follows:

Mechanical Services – 30:70, Electrical Services – 50:50 (2015 = 100)

Mechanical Services

Year	First Quarter	Second Quarter	Third Quarter	Fourth Quarter	Annual Average	Annual Change
2015	99.9	99.8	99.3	101.0	100.0	1.1%
2016	101.1	101.9	102.5	104.4	102.5	2.5%
2017	105.6	105.8	106.5	108.5	106.6	4.0%
2018	109.6	110.7	110.7	112.3	110.8	4.0%
2019	112.6	114.0	114.2	114.6	113.9	2.7%
2020	114.7 (P)	114.9 (F)	114.9	115.4	115.0 (F)	1.0%
2021	116.0	116.3	116.6	117.4	116.6	1.4%
2022	118.4	119.1	119.7	120.8	119.5	2.5%
2023	119.1	119.7	120.8	122.0	120.4	0.8%

Electrical Services

Year	First Quarter	Second Quarter	Third Quarter	Fourth Quarter	Annual Average	Annual Change
2015	100.2	100.2	99.9	99.7	100.0	1.8%
2016	102.3	102.5	102.5	102.9	102.6	2.6%
2017	104.9	105.1	105.0	105.3	105.1	2.4%
2018	107.8	108.6	109.0	107.9	108.3	3.1%
2019	108.7	110.3	111.3	110.5	110.2	1.7%
2020	110.7 (P)	110.9 (F)	111.0	111.0	110.9 (F)	0.6%
2021	112.1	112.4	112,7	112.9	112.5	1.5%
2022	114.5	115.1	115.7	116.3	115.4	2.6%
2023	115.1	115.1	116.3	118.0	116.3	0.8%

(P = Provisional)

(F = Forecast)

COST INDICES

Regional Variations

Prices throughout this Book apply to work in the London area (see Directions at the beginning of the Mechanical Installations and Electrical Installations sections). However, prices for mechanical and electrical services installations will of course vary from region to region, largely as a result of differing labour costs but also depending on the degree of accessibility, urbanization and local market conditions.

The following table of regional factors is intended to provide readers with indicative adjustments that may be made to the prices in the Book for locations outside of London. The figures are of necessity averages for regions and further adjustments should be considered for city centre or very isolated locations, or other known local factors.

Inner London	1.06	North East	0.83
Outer London	1.00	North West	0.88
South East	0.96 (Excl GL)	Yorkshire and Humberside	0.86
South West	0.90	Scotland	0.88
East of England	0.93	Wales	0.86
East Midlands	0.88	Northern Ireland	0.74
West Midlands	0.86		

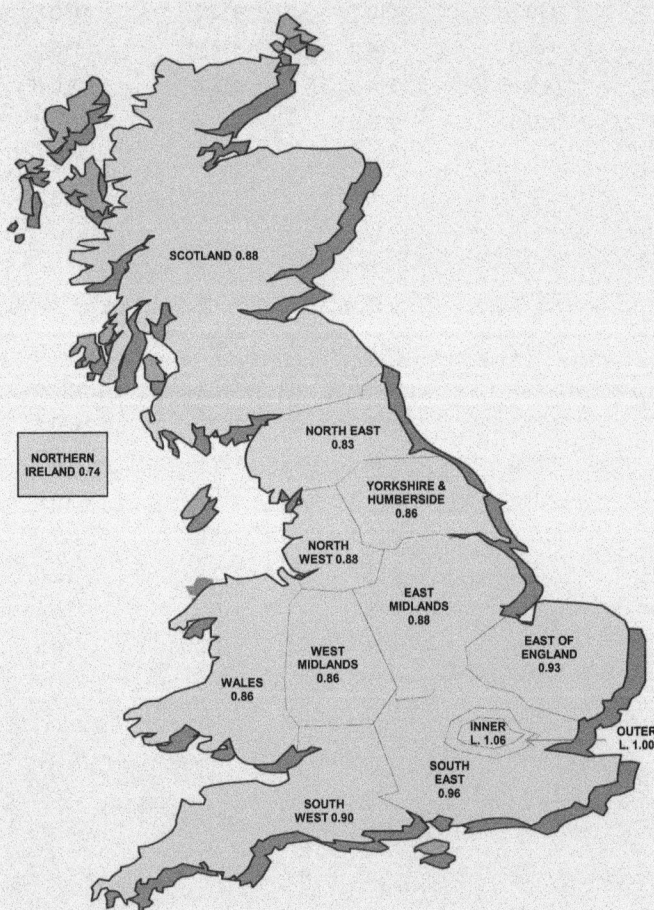

RIBA STAGE 0/1 PREPARATION

Item	Unit	Range £	
Typical Square Metre Rates for Engineering Services			
The following examples indicate the range of rates within each building type for engineering services, excluding lifts etc., utilities services and professional fees. Based on Gross Internal Area (GIA).			
Industrial Buildings			
Factories			
Owner occupation: Includes for rainwater, soil/waste, LTHW heating via HL radiant heaters, BMS, LV installations, lighting, fire alarms, security, earthing	m²	129.00 to	160.00
Owner occupation: Includes for rainwater, soil/waste, sprinklers, LTHW heating via HL gas fired heaters, local air conditioning, BMS, HV/LV installations, lighting, fire alarms, security, earthing	m²	185.00 to	232.00
Warehouses			
High bay for owner occupation: Includes for rainwater, soil/waste, LTHW heating via HL gas fired heaters, BMS, HV/LV installations, lighting, fire alarms, security, earthing	m²	108.00 to	129.00
High bay for owner occupation: Includes for rainwater, soil/waste, sprinklers, LTHW heating via HL radiant heaters, local air conditioning, BMS, HV/LV installations, lighting, fire alarms, security, earthing	m²	210.00 to	252.00
Distribution Centres			
High bay for letting: Includes for rainwater, soil/waste, LTHW heating via HL gas fired heaters, BMS, HV/LV installations, lighting, fire alarms, security, earthing	m²	134.00 to	170.00
High bay for owner occupation: Includes for rainwater, soil/waste, sprinklers, LTHW heating via HL radiant heaters, local air conditioning, BMS, HV/LV installations, lighting, fire alarms, security	m²	247.00 to	275.00
Office Buildings 5,000 m² to 15,000 m²			
Offices for Letting			
Shell & Core and Cat A non-air-conditioned: Includes for rainwater, soil/waste, cold water, hot water via local electrical heaters, LTHW heating via radiator heaters, toilet extract, LV installations, lighting, small power (landlords), fire alarms, earthing, security, IT wireways & controls	m²	325.00 to	385.00
Shell & Core and Cat A non-air-conditioned: Includes for rainwater, soil/waste, cold water, hot water, LTHW heating via perimeter heaters, toilet extract, LV installations, lighting, small power (landlords), fire alarms, earthing, security wireways, IT wireways & controls	m²	340.00 to	400.00
Shell & Core and Cat A air-conditioned: Includes for rainwater, soil/waste, cold water, hot water, VRV 3 pipe heat pumps, toilet extract, LV installations, lighting, small power (landlords), fire alarms, earthing, security, IT wireways & controls	m²	505.00 to	645.00
Shell & Core and Cat A air-conditioned: Includes for rainwater, soil/waste, cold water, hot water via local electrical heaters, LTHW heating via perimeter heaters, 2 pipe, toilet extract, BMS, LV installations, lighting, small power (landlords), fire alarms, earthing, security & IT wireways	m²	850.00 to	975.00

RIBA STAGE 0/1 PREPARATION

Item	Unit	Range £
Typical Square Metre Rates for Engineering Services – cont		
Office Buildings 5,000 m² to 15,000 m² – cont		
Offices for Owner Occupation		
Non-air-conditioned: Includes for rainwater, soil/waste, cold water, hot water via local electrical heaters, LTHW heating via radiator heaters, toilet extract, LV installations, lighting, small power (landlords), dry risers, fire alarms, earthing, security, IT wireways & controls	m²	340.00 to 405.00
Non-air-conditioned: Includes for rainwater, soil/waste, cold water, hot water, dry risers, LTHW heating via perimeter heaters, toilet extract, LV installations, lighting, small power (landlords), dry risers, fire alarms, earthing, security, IT wireways & controls	m²	355.00 to 425.00
Air-conditioned: Includes for rainwater, soil/waste, cold water, hot water via centralized system, dry risers, 4 pipe air conditioning, toilet extract, BMS, LV installations, life safety standby generators, lighting, small power, sprinkler protection, fire alarms, earthing, lightning protection, security, IT wireways, fire alarms L1/P1 & controls	m²	900.00 to 1200.00
Health and Welfare Facilities		
District General Hospitals		
Natural ventilation: Includes for rainwater, soil/waste, cold water, hot water, dry risers, sprinklers, medical gases, LTHW heating, WC/kitchen extract, BMS, LV installations, standby generation, lighting, small power, fire alarms, earthing/lightning protection, nurse call systems, security, IT wireways	m²	1150.00 to 1245.00
Natural ventilation: Includes for rainwater, soil/waste, cold water, hot water, dry risers, sprinklers, medical gases, LTHW heating, localized VAV air conditioning and mechanical ventilation, WC/kitchen extract, BMS, LV installations, standby generation, lighting, small power, fire alarms, earthing/lightning protection, nurse call systems, security, IT wireways	m²	1250.00 to 1350.00
Private Hospitals		
Natural ventilation: Includes for rainwater, soil/waste, cold water, hot water, dry risers, sprinklers, medical gases, LTHW heating, localized VAV air conditioning, WC/kitchen extract, BMS, LV installations, standby generation, lighting, small power, fire alarms, earthing/lightning protection, nurse call systems, security, IT wireways	m²	1270.00 to 1385.00
Air-conditioned: Includes for rainwater, soil/waste, cold water, hot water, dry risers, sprinklers, medical gases, LTHW heating, supply and extract ventilation, 4 pipe air conditioning, WC/kitchen extract, BMS, LV installations, standby generation, lighting, small power, fire alarms, earthing/lightning protection, nurse call systems, security, IT wireways	m²	1450.00 to 1560.00
GP Surgery		
Natural ventilation: Includes for rainwater, soil/waste, cold water, hot water, LTHW heating, WC/kitchen extract, BMS, LV installations, lighting, small power, fire alarms, earthing/lightning protection, nurse call systems, security, IT wireways	m²	800.00 to 915.00
Air-conditioned: Includes for rainwater, soil/waste, cold water, hot water, LTHW heating, supply and extract, BMS, LV installations, lighting, small power, fire alarms, earthing/lightning protection, nurse call systems, security, IT wireways	m²	885.00 to 1125.00

RIBA STAGE 0/1 PREPARATION

Item	Unit	Range £	
Entertainment and Recreation Buildings			
Non-Performing			
Natural ventilation: Includes for rainwater, soil/waste, cold water, central hot water, dry risers, LTHW heating, toilet extract, kitchen extract, controls, LV installations, lighting, small power, fire alarms, earthing, security, IT wireways	m²	360.00 to	445.00
Comfort cooled: Includes for rainwater, soil/waste, cold water, hot water via electrical heaters, sprinklers/dry risers, LTHW heating, DX air conditioning, kitchen/toilet extract, BMS, LV installations, lighting, small power, fire alarms, earthing, security, IT wireways	m²	660.00 to	790.00
Performing Arts (with Theatre)			
Natural ventilation: Includes for rainwater, soil/waste, cold water, central hot water, sprinklers/dry risers, LTHW heating, toilet extract, kitchen extract, controls, LV installations, lighting, small power, fire alarms, earthing, security, IT wireways	m²	600.00 to	720.00
Comfort cooled: Includes for rainwater, soil/waste, cold water, central hot water, sprinklers/dry risers, LTHW heating, air conditioning, kitchen/toilet extract, BMS, LV installations, lighting including enhanced dimming/scene setting, small power, fire alarms, earthing, security, IT wireways including production, audio and video recording	m²	820.00 to	990.00
Sports Halls			
Natural ventilation: Includes for rainwater, soil/waste, cold water, hot water gas fired heaters, LTHW heating, toilet extract, BMS, LV installations, lighting, small power, fire alarms, earthing, security	m²	275.00 to	325.00
Comfort cooled: Includes for rainwater, soil/waste, cold water, hot water via LTHW heat exchangers, LTHW heating, air conditioning via AHUs to limited areas, toilet extract, BMS, LV installations, lighting, small power, fire alarms, earthing, security	m²	380.00 to	465.00
Multi-Purpose Leisure Centre			
Natural ventilation: Includes for rainwater, soil/waste, cold water, hot water gas fired heaters, LTHW heating, toilet extract, BMS, LV installations, lighting, small power, fire alarms, earthing, security, IT wireways	m²	380.00 to	465.00
Comfort cooled: Includes for rainwater, soil/waste, cold water, hot water LTHW heat exchangers, LTHW heating, air conditioning via AHU to limited areas, pool hall supply/extract, kitchen/toilet extract, BMS, LV installations, lighting, small power, fire alarms, earthing, security, IT wireways	m²	610.00 to	770.00
Retail Buildings			
Open Arcade			
Natural ventilation: Includes for rainwater, soil/waste, cold water, hot water, sprinklers/dry risers, LTHW heating, toilet extract, smoke extract, BMS, LV installations, life safety standby generators, lighting, small power, fire alarms, public address, earthing/lightning protection, security, IT wireways	m²	355.00 to	445.00
Enclosed Shopping Mall			
Air-conditioned: Includes for rainwater, soil/waste, cold water, hot water, sprinklers/dry risers, LTHW heating, air conditioning via AHUs, toilet extract, smoke extract, BMS, LV installations, life safety standby generators, lighting, small power, fire alarms, public address, earthing/lightning protection, CCTV/ security, IT wireways people counting systems	m²	545.00 to	660.00

Approximate Estimating

RIBA STAGE 0/1 PREPARATION

Item	Unit	Range £	
Typical Square Metre Rates for Engineering Services – cont			
Retail Buildings – cont			
Department Stores			
Air-conditioned: Includes for sanitaryware, soil/waste, cold water, hot water, sprinklers/dry risers, LTHW heating, air conditioning via AHUs, toilet extract, smoke extract, BMS, LV installations, life safety standby generators, lighting, small power, fire alarms, public address, earthing, lightning protection, CCTV/security, IT installation wireways	m²	400.00 to	495.00
Supermarkets			
Air-conditioned: Includes for rainwater, soil/waste, cold water, hot water, sprinklers, LTHW heating, air conditioning via AHUs, toilet extract, BMS, LV installations, lighting, small power, fire alarms, earthing, lightning protection, security, IT wireways, refrigeration	m²	590.00 to	725.00
Educational Buildings			
Secondary Schools (Academy)			
Natural Ventilation: Includes for rainwater, soil/waste, cold water, central hot water, LTHW heating, toilet extract, BMS, LV installations, lighting, small power, fire alarms, earthing, security, IT wireways	m²	425.00 to	490.00
Natural vent with comfort cooling to selected areas (BB93 compliant): Includes for rainwater, soil/waste, cold water, central hot water, LTHW heating, DX air conditioning, general supply/extract, toilet extract, BMS, LV installations, lighting, small power, fire alarms, earthing, security, IT wireways	m²	490.00 to	560.00
Scientific Buildings			
Educational Research			
Comfort cooled: Includes for rainwater, soil/waste, cold water, central hot water, dry risers, compressed air, medical gasses, LTHW heating, 4 pipe air conditioning, toilet extract, fume, BMS, LV installations, lighting, small power, fire alarms, earthing, lightning protection, security, IT wireways	m²	770.00 to	940.00
Air-conditioned: Includes for rainwater, soil/waste, laboratory waste, cold water, central hot water, specialist water, dry risers, compressed air, medical gasses, steam, LTHW heating, VAV air conditioning, Comms room cooling, toilet extract, fume extract, BMS, LV installations, UPS, standby generators, lighting, small power, fire alarms, earthing, lightning protection, security, IT wireways	m²	1370.00 to	1675.00
Commercial Research			
Air-conditioned; Includes for rainwater, soil/waste, laboratory waste, cold water, central hot water, specialist water, dry risers, compressed air, medical gasses, steam, LTHW heating, VAV air conditioning, Comm room cooling, toilet extract, fume extract, BMS, LV installations, UPS, standby generators, lighting, small power, fire alarms, earthing, lightning protection, security, IT wireways	m²	1500.00 to	1800.00

RIBA STAGE 0/1 PREPARATION

Item	Unit	Range £
Residential Facilities		
Hotels		
1 to 3 Star: Includes for rainwater, soil/waste, cold water, hot water, dry risers, LTHW heating via radiators, refrigerate cooling to bedrooms, toilet/ bathroom extract, kitchen extract, BMS, LV installations, lighting, small power, fire alarms, earthing, lightning protection, security, IT wireways	m²	490.00 to 650.00
4 to 5 Star: Includes for rainwater, soil/waste, cold water, hot water, sprinklers, dry risers, 4 pipe air conditioning, kitchen extract, toilet/bathroom extract, BMS, LV installations, life safety standby generators, lighting, small power, fire alarms, earthing, security, IT wireways	m²	725.00 to 950.00

RIBA STAGE 2/3 DESIGN

Item	Unit	Range £	
Elemental Rates for Alternative Engineering Services Solutions – Offices			
The following examples of building types indicate the range of rates for alternative design solutions for each of the engineering services elements based on Gross Internal Area for the Shell and Core and Net Internal Area for the Fit-Out. Fit-Out is assumed to be to Cat A Standard.			
Consideration should be made for the size of the building, which may affect the economies of scale for rates, i.e. the larger the building the lower the rates.			
5 Services			
Shell & Core			
5.1 Sanitary Installations			
Building up to 5,000 m²	m² GIA	12.00 to	17.00
Building over 5,000 m² up to 15,000 m² (low rise)	m² GIA	10.00 to	15.00
5.3 Disposal Installations			
Building up to 5,000 m²	m² GIA	21.00 to	25.00
Building over 5,000 m² up to 15,000 m²	m² GIA	20.00 to	23.00
5.4 Water Installations			
Building up to 5,000 m²	m² GIA	27.00 to	30.00
Building over 5,000 m² up to 15,000 m²	m² GIA	24.00 to	28.00
5.6 Space Heating and Air Conditioning			
LPHW heating installation; including gas installations			
Building up to 5,000 m²	m² GIA	45.00 to	50.00
Building over 5,000 m² up to 15,000 m²	m² GIA	38.00 to	42.00
Comfort cooling			
2 pipe fan coil for building up to 5,000 m²	m² GIA	90.00 to	99.00
2 pipe fan coil for building over 5,000 m² up to 15,000 m²	m² GIA	81.00 to	89.00
2 pipe variable refrigerant volume (VRV) for building up to 5,000 m²	m² GIA	65.00 to	77.00
Full air conditioning			
4 pipe fan coil for building up to 5,000 m²	m² GIA	120.00 to	160.00
4 pipe fan coil for building over 5,000 m² up to 15,000 m²	m² GIA	110.00 to	150.00
4 pipe variable refrigerant volume for building up to 5,000 m²	m² GIA	105.00 to	119.00
Ventilated (active) chilled beams for building over 5,000 m² up to 15,000 m²	m² GIA	133.00 to	155.00
Chilled beam exposed services for building over 5,000 m² to 15,000 m²	m² GIA	141.00 to	165.00
Concealed passive chilled beams for building over 5,000 m² up to 15,000 m²	m² GIA	129.00 to	150.00
Chilled ceiling for building over 5,000 m² to 15,000 m²	m² GIA	131.00 to	155.00
Chilled ceiling/perimeter beams for building over 5,000 m² up to 15,000 m²	m² GIA	140.00 to	160.00
Displacement for building over 5,000 m² up to 15,000 m²	m² GIA	150.00 to	171.00
5.7 Ventilation Systems (excluding smoke extract)			
Building up to 5,000 m²	m² GIA	33.00 to	51.00
Building over 5,000 m² up to 15,000 m²	m² GIA	30.00 to	43.00
5.8 Electrical Installations			
LV installations			
Standby generators (life safety only)			
Buildings over 5,000 m² up to 15,000 m²	m² GIA	15.00 to	16.00
LV distribution			
Buildings up to 5,000 m²	m² GIA	30.00 to	37.00
Buildings over 5,000 m² up to 15,000 m²	m² GIA	38.00 to	48.00

RIBA STAGE 2/3 DESIGN

Item	Unit	Range £	
Lighting installations (including lighting controls and luminaires)			
Buildings up to 5,000 m²	m² GIA	21.00 to	27.00
Buildings over 5,000 m² up to 15,000 m²	m² GIA	20.00 to	26.00
Small power			
Buildings up to 5,000 m²	m² GIA	9.00 to	10.00
Buildings over 5,000 m² to up to 15,000 m²	m² GIA	7.00 to	9.00
Electrical installations for mechanical plant			
Buildings up to 5,000 m²	m² GIA	8.00 to	10.00
Buildings over 5,000 m² to up to 15,000 m²	m² GIA	9.00 to	11.00
5.11 Fire and Lightning Protection			
Fire protection over 5,000 m² to 15,000 m²			
Dry risers	m² GIA	3.00 to	6.00
Sprinkler installation	m² GIA	22.00 to	28.00
Protective installations			
Earthing			
Buildings up to 5,000 m²	m² GIA	2.00 to	3.00
Buildings over 5,000 m² to 15,000 m²	m² GIA	2.00 to	3.00
Lightning protection			
Buildings up to 5,000 m²	m² GIA	2.00 to	3.00
Buildings over 5,000 m² to 15,000 m²	m² GIA	2.00 to	3.00
5.12 Communication, Security and Control Systems			
Fire alarms (single stage)			
Buildings up to 5,000 m²	m² GIA	8.00 to	15.00
Buildings over 5,000 m² to 15,000 m²	m² GIA	12.00 to	18.00
Fire alarms (phased evacuation)			
Buildings over 5,000 m² to 15,000 m²	m² GIA	14.00 to	22.00
IT (wireways only)			
Buildings up to 5,000 m²	m² GIA	3.00 to	6.00
Buildings over 5,000 m² to 15,000 m²	m² GIA	3.00 to	6.00
Security			
Buildings up to 5,000 m²	m² GIA	6.00 to	9.00
Buildings over 5,000 m² to 15,000 m²	m² GIA	8.00 to	11.00
BMS controls; including MCC panels and control cabling			
Full air conditioning			
Buildings up to 5,000 m²	m² GIA	20.00 to	28.00
Buildings up to 15,000 m²	m² GIA	22.00 to	31.00
Fit-Out			
5.6 Space Heating and Air Conditioning			
LPHW Heating Installation			
Building up to 5,000 m²	m² NIA	53.00 to	65.00
Building over 5,000 m² to 15,000 m²	m² NIA	49.00 to	61.00
Comfort cooling			
2 pipe fan coil for building up to 5,000 m²	m² NIA	110.00 to	130.00
2 pipe fan coil for building over 5,000 m² to 15,000 m²	m² NIA	95.00 to	110.00
2 pipe variable refrigerant volume (VRV) for building up to 5,000 m²	m² NIA	85.00 to	100.00

Approximate Estimating

RIBA STAGE 2/3 DESIGN

Item	Unit	Range £	
Elemental Rates for Alternative Engineering Services Solutions – Offices – cont			
Fit-Out – cont			
Full air conditioning			
4 pipe fan coil for building up to 5,000 m²	m² NIA	180.00 to	210.00
4 pipe fan coil for building over 5,000 m² to 15,000 m²	m² NIA	180.00 to	210.00
4 pipe variable refrigerant volume for building up to 5,000 m²	m² NIA	160.00 to	180.00
Ventilated (active) chilled beams for building over 5,000 m² to 15,000 m²	m² NIA	190.00 to	215.00
Chilled beam exposed services for building over 5,000 m² to 15,000 m²	m² NIA	210.00 to	240.00
Concealed passive chilled beams for building over 5,000 m² to 15,000 m²	m² NIA	160.00 to	180.00
Chilled ceiling for building over 5,000 m² to 15,000 m²	m² NIA	230.00 to	275.00
Chilled ceiling/perimeter beams for building over 5,000 m² to 15,000 m²	m² NIA	250.00 to	300.00
Displacement for building over 5,000 m² to 15,000 m²	m² NIA	100.00 to	130.00
5.8 Electrical Installations			
Lighting installations (including lighting controls and luminaires)			
Buildings up to 5,000 m²	m² NIA	75.00 to	85.00
Buildings over 5,000 m² to 15,000 m²	m² NIA	85.00 to	100.00
Electrical installations for mechanical plant			
Buildings up to 5,000 m²	m² NIA	7.00 to	10.00
Buildings over 5,000 m² to 15,000 m²	m² NIA	7.00 to	10.00
5.11 Fire and Lightning Protection			
Fire protection over 5,000 m² to 15,000 m²			
Sprinkler installation	m² NIA	22.00 to	30.00
Earthing			
Buildings up to 5,000 m²	m² NIA	2.00 to	3.00
Buildings over 5,000 m² to 15,000 m²	m² NIA	2.00 to	3.00
5.12 Communication, Security and Control Systems			
Fire alarms (single stage)			
Buildings up to 5,000 m²	m² NIA	12.00 to	18.00
Buildings over 5,000 m² to 15,000 m²	m² NIA	12.00 to	18.00
Fire alarms (phased evacuation)			
Buildings over 5,000 m² to 15,000 m²	m² NIA	15.00 to	25.00
BMS controls; including MCC panels and control cabling			
Full air conditioning			
Buildings up to 5,000 m²	m² NIA	22.00 to	30.00
Buildings up to 15,000 m²	m² NIA	21.00 to	35.00

RIBA STAGE 2/3 DESIGN

Item	Unit	Range £	
Elemental Rates for Alternative Engineering Services Solutions – Hotels The following examples of building types indicate the range of rates for alternative design solutions for each of the engineering services elements based on Gross Internal Area for the Shell and Core and Net Internal Area for the Fit- Out. Fit-Out is assumed to be to Cat A Standard. Consideration should be made for the size of the building, which may affect the economies of scale for rates, i.e. the larger the building the lower the rates.			
5 Services			
Shell & Core			
5.1 Sanitary Installations			
2 to 3 star	m² GIA	30.00 to	35.00
4 to 5 star	m² GIA	45.00 to	60.00
5.3 Disposal Installations			
2 to 3 star	m² GIA	30.00 to	36.00
4 to 5 star	m² GIA	33.00 to	45.00
5.4 Water Installations			
2 to 3 star	m² GIA	43.00 to	52.00
4 to 5 star	m² GIA	60.00 to	80.00
5.6 Space Heating and Air Conditioning			
LPHW heating installation; including gas installations			
2 to 3 star	m² GIA	45.00 to	52.00
4 to 5 star	m² GIA	45.00 to	55.00
Air conditioning; including ventilation			
2 to 3 star – 4 pipe fan coil	m² GIA	245.00 to	320.00
4 to 5 star – 4 pipe fan coil	m² GIA	255.00 to	325.00
2 to 3 star – 3 pipe variable refrigerant volume	m² GIA	155.00 to	200.00
4 to 5 star – 3 pipe variable refrigerant volume	m² GIA	155.00 to	200.00
5.8 Electrical Installations			
LV installations			
Standby generators (life safety only)			
2 to 3 star	m² GIA	17.00 to	20.00
4 to 5 star	m² GIA	17.00 to	23.00
LV distribution			
2 to 3 star	m² GIA	35.00 to	45.00
4 to 5 star	m² GIA	50.00 to	61.00
Lighting installations			
2 to 3 star	m² GIA	25.00 to	32.00
4 to 5 star	m² GIA	40.00 to	51.00
Small power			
2 to 3 star	m² GIA	9.00 to	12.00
4 to 5 star	m² GIA	14.00 to	20.00
Electrical installations for mechanical plant			
2 to 3 star	m² GIA	7.00 to	9.00
4 to 5 star	m² GIA	7.00 to	9.00

RIBA STAGE 2/3 DESIGN

Item	Unit	Range £	
Elemental Rates for Alternative Engineering Services Solutions – Hotels – cont			
Shell & Core – cont			
5.11 Fire and Lightning Protection			
Fire protection			
2 to 3 star – Dry risers	m² GIA	11.00 to	14.00
4 to 5 star – Dry risers	m² GIA	11.00 to	14.00
2 to 3 star – Sprinkler installation	m² GIA	25.00 to	32.00
4 to 5 star – Sprinkler installation	m² GIA	28.00 to	37.00
Earthing			
2 to 3 star	m² GIA	2.00 to	2.50
4 to 5 star	m² GIA	2.00 to	2.50
Lightning protection			
2 to 3 star	m² GIA	2.00 to	3.00
4 to 5 star	m² GIA	2.00 to	3.00
5.12 Communication, Security and Control Systems			
Fire alarms			
2 to 3 star	m² GIA	16.50 to	23.00
4 to 5 star	m² GIA	16.50 to	26.00
IT			
2 to 3 star	m² GIA	14.00 to	17.00
4 to 5 star	m² GIA	14.00 to	20.00
Security			
2 to 3 star	m² GIA	20.00 to	27.00
4 to 5 star	m² GIA	21.00 to	28.00
BMS controls; including MCC panels and control cabling			
2 to 3 star	m² GIA	13.00 to	17.00
4 to 5 star	m² GIA	32.00 to	40.00

RIBA STAGE 2/3 DESIGN

Item	Unit	Range £	
Elemental Rates for Alternative Engineering Services Solutions – Apartments			
The following examples of building types indicate the range of rates for alternative design solutions for each of the engineering services elements based on Gross Internal Area for the Shell and Core and Net Internal Area for the Fit-Out. Fit-Out is assumed to be to Cat A Standard.			
Consideration should be made for the size of the building height, sales point (£/ft²) and location, which may affect the economies of scale for rates, i.e. the larger the building the lower the rates.			
5 Services			
Shell & Core			
5.1 Sanitary Installations			
Affordable	m² GIA	0.75 to	1.00
Private	m² GIA	3.00 to	3.50
5.3 Disposal Installations			
Affordable	m² GIA	21.00 to	26.00
Private	m² GIA	25.00 to	32.00
5.4 Water Installations			
Affordable	m² GIA	20.00 to	25.00
Private	m² GIA	34.00 to	44.00
5.5 Heat Source			
Affordable	m² GIA	10.00 to	12.00
Private	m² GIA	10.00 to	15.00
5.6 Space Heating and Air Conditioning			
Affordable	m² GIA	28.00 to	33.00
Private	m² GIA	80.00 to	110.00
5.7 Ventilation Systems			
Affordable	m² GIA	20.00 to	21.00
Private	m² GIA	23.00 to	28.00
5.8 Electrical Installations			
Affordable	m² GIA	45.00 to	55.00
Private	m² GIA	67.00 to	85.00
5.9 Fuel Installations (Gas)			
Affordable	m² GIA	3.00 to	3.50
Private	m² GIA	4.50 to	5.50
5.11 Fire and Lightning Protection			
Affordable	m² GIA	23.00 to	28.00
Private	m² GIA	33.00 to	41.00
5.12 Communication, Security and Control Systems			
Affordable	m² GIA	35.00 to	43.00
Private	m² GIA	48.00 to	60.00
5.13 Specialist Installations			
Affordable	m² GIA	10.00 to	12.00
Private	m² GIA	40.00 to	55.00

RIBA STAGE 2/3 DESIGN

Item	Unit	Range £
Elemental Rates for Alternative Engineering Services Solutions – Apartments – cont		
Fit-Out		
5.1 Sanitary Installations		
Affordable	m² NIA	34.00 to 41.00
Private	m² NIA	100.00 to 120.00
5.3 Disposal Installations		
Affordable	m² NIA	12.00 to 15.00
Private	m² NIA	25.00 to 35.00
5.4 Water Installations		
Affordable	m² NIA	30.00 to 35.00
Private	m² NIA	52.00 to 70.00
5.6 Space Heating and Air Conditioning		
Affordable	m² NIA	78.00 to 94.00
Private	m² NIA	210.00 to 260.00
5.7 Ventilation Systems		
Affordable (whole house vent)	m² NIA	38.00 to 46.00
Private (whole house vent)	m² NIA	50.00 to 60.00
5.8 Electrical Installations		
Affordable	m² NIA	60.00 to 75.00
Private	m² NIA	130.00 to 160.00
5.11 Fire and Lightning Protection		
Affordable	m² NIA	24.00 to 29.00
Private	m² NIA	30.00 to 38.00
5.12 Communication, Security and Control Systems		
Affordable	m² NIA	20.00 to 25.00
Private	m² NIA	105.00 to 130.00
5.13 Specialist Installations		
Affordable	m² NIA	12.00 to 14.00
Private	m² NIA	33.00 to 40.00

Note: The range in cost differs due to the vast diversity in services strategies available. The lower end of the scale reflects no comfort cooling, radiator heating in lieu of underfloor, etc. The high end of the scale is based on central plant installations with good quality apartments, which includes comfort cooling, sprinklers, home network installations, video entry, higher quality of sanitaryware and good quality LED lighting and lighting control.

APPROXIMATE ESTIMATING RATES – 5 SERVICES

Item	Unit	Range £	
5.3 Disposal Installations			
Above Ground Drainage			
Soil and waste	point	400.00 to	500.00
5.4 Water Installations			
Cold Water	point	400.00 to	500.00
Hot Water	point	500.00 to	600.00
Pipework			
Hot and Cold water			
Excludes insulation, valves and ancillaries, etc.			
Light gauge copper tube to EN1057 R250 (TX) formerly BS 2871 Part 1 Table X			
with joints as described including allowance for waste, fittings and supports			
assuming average runs with capillary joints up to 54 mm and bronze welded			
thereafter			
Horizontal High Level Distribution			
15 mm	m	38.00 to	45.00
22 mm	m	39.00 to	46.00
28 mm	m	52.00 to	63.00
35 mm	m	60.00 to	72.00
42 mm	m	71.00 to	89.00
54 mm	m	91.00 to	111.00
67 mm	m	109.00 to	138.00
76 mm	m	127.00 to	155.00
108 mm	m	190.00 to	242.00
Risers			
15 mm	m	22.00 to	28.00
22 mm	m	26.00 to	31.00
28 mm	m	38.00 to	45.00
35 mm	m	43.00 to	52.00
42 mm	m	54.00 to	65.00
54 mm	m	59.00 to	76.00
67 mm	m	108.00 to	124.00
76 mm	m	118.00 to	145.00
108 mm	m	151.00 to	178.00
Toilet Areas, etc., at Low Level			
15 mm	m	62.00 to	74.00
22 mm	m	74.00 to	89.00
28 mm	m	104.00 to	125.00

Approximate Estimating

APPROXIMATE ESTIMATING RATES – 5 SERVICES

Item	Unit	Range £	
5.5 Heat Source			
Gas fired boilers including gas train and controls	kW	39.00 to	46.00
Gas fired boilers including gas train, controls, flue, plantroom pipework, valves and insulation, pumps and pressurization unit	kW	103.00 to	129.00
5.6 Space Heating and Air Conditioning			
Pipework LTHW and Chilled Water			
Excludes insulation, valves and ancillaries, etc.			
Black heavy weight mild steel tube to BS 1387 with joints in the running length, allowance for waste, fittings and supports assuming average runs			
Horizontal Distribution – Basements, etc.			
15 mm	m	59.00 to	71.00
20 mm	m	62.00 to	75.00
25 mm	m	70.00 to	86.00
32 mm	m	82.00 to	100.00
40 mm	m	93.00 to	114.00
50 mm	m	116.00 to	139.00
65 mm	m	114.00 to	139.00
80 mm	m	152.00 to	183.00
100 mm	m	189.00 to	226.00
125 mm	m	250.00 to	307.00
150 mm	m	295.00 to	349.00
200 mm	m	435.00 to	530.00
250 mm	m	540.00 to	661.00
300 mm	m	790.00 to	953.00
Risers			
15 mm	m	41.00 to	50.00
20 mm	m	43.00 to	51.00
25 mm	m	48.00 to	58.00
32 mm	m	56.00 to	67.00
40 mm	m	62.00 to	73.00
50 mm	m	76.00 to	91.00
65 mm	m	84.00 to	102.00
80 mm	m	96.00 to	117.00
100 mm	m	124.00 to	151.00
125 mm	m	154.00 to	181.00
150 mm	m	196.00 to	238.00
200 mm	m	296.00 to	369.00
250 mm	m	360.00 to	438.00
300 mm	m	560.00 to	695.00
On Floor Distribution			
15 mm	m	66.00 to	79.00
20 mm	m	70.00 to	83.00
25 mm	m	78.00 to	96.00
32 mm	m	88.00 to	108.00
40 mm	m	98.00 to	120.00
50 mm	m	122.00 to	149.00

APPROXIMATE ESTIMATING RATES – 5 SERVICES

Item	Unit	Range £	
Plantroom Areas, etc.			
15 mm	m	62.00 to	72.00
20 mm	m	64.00 to	77.00
25 mm	m	70.00 to	82.00
32 mm	m	80.00 to	98.00
40 mm	m	90.00 to	108.00
50 mm	m	110.00 to	129.00
65 mm	m	120.00 to	139.00
80 mm	m	158.00 to	192.00
100 mm	m	198.00 to	241.00
125 mm	m	260.00 to	315.00
150 mm	m	306.00 to	363.00
200 mm	m	450.00 to	549.00
250 mm	m	555.00 to	662.00
300 mm	m	812.00 to	972.00
Chilled Water			
Air cooled R134a refrigerant chiller including control panel, anti-vibration mountings	kW	165.00 to	216.00
Air cooled R134a refrigerant chiller including control panel, anti-vibration mountings, plantroom pipework, valves, insulation, pumps and pressurization units	kW	252.00 to	309.00
Water cooled R134a refrigerant chiller including control panel, anti-vibration mountings	kW	93.00 to	139.00
Water cooled R134a refrigerant chiller including control panel, anti-vibration mountings, plantroom pipework, valves, insulation, pumps and pressurization units	kW	191.00 to	221.00
Absorption steam medium chiller including control panel, anti-vibration mountings, plantroom pipework, valves, insulation, pumps and pressurization units	kW	330.00 to	397.00
Heat Rejection			
Open circuit, forced draft cooling tower	kW	105.00 to	129.00
Closed circuit, forced draft cooling tower	kW	98.00 to	119.00
Dry Air Cooler	kW	60.00 to	65.00
Pumps			
Pumps including flexible connections, anti-vibration mountings	kPa	42.00 to	53.00
Pumps including flexible connections, anti-vibration mountings, plantroom pipework, valves, insulation and accessories	kPa	103.00 to	124.00
Ductwork			
The rates below allow for ductwork and for all other labour and material in fabrication, fittings, supports and jointing to equipment, stop and capped ends, elbows, bends, diminishing and transition pieces, regular and reducing couplings, volume control dampers, branch diffuser and 'snap on' grille connections, ties, 'Ys', crossover spigots, etc., turning vanes, regulating dampers, access doors and openings, hand-holes, test holes and covers, blanking plates, flanges, stiffeners, tie rods and all supports and brackets fixed to structure.			
Rectangular galvanized mild steel ductwork as HVCA DW 144 up to 1000 mm longest side	m²	67.00 to	80.00
Rectangular galvanized mild steel ductwork as HVCA DW 144 up to 2500 mm longest side	m²	72.00 to	88.00

APPROXIMATE ESTIMATING RATES – 5 SERVICES

Item	Unit	Range £	
5.6 Space Heating and Air Conditioning – cont			
Ductwork – cont			
Rectangular galvanized mild steel ductwork as HVCA DW 144 up to 3000 mm longest side and above	m²	98.00 to	118.00
Circular galvanized mild steel ductwork as HVCA DW 144	m²	72.00 to	88.00
Flat oval galvanized mild steel ductwork as HVCA DW 144 up to 545 mm wide	m²	72.00 to	88.00
Flat oval galvanized mild steel ductwork as HVCA DW 144 up to 880 mm wide	m²	79.00 to	98.00
Flat oval galvanized mild steel ductwork as HVCA DW 144 up to 1785 mm wide	m²	93.00 to	113.00
Packaged Air Handling Units			
Air handling unit including LPHW preheater coil, pre-filter panel, LPHW heater coils, chilled water coil, filter panels, inverter drive, motorized volume control dampers, sound attenuation, flexible connections to ductwork and all anti-vibration mountings	m³/s	8600.00 to	10700.00
5.7 Ventilating Systems			
Extract Fans			
Extract fan including inverter drive, sound attenuation, flexible connections to ductwork and all anti-vibration mountings	m³/s	3000.00 to	3500.00
5.8 Electrical Installations			
HV/LV Installations			
The cost of HV/LV equipment will vary according to the electricity supplier's requirements, the duty required and the actual location of the site. For estimating purposes, the items indicated below are typical of the equipment required in a HV substation incorporated into a building.			
Ring Main Unit			
Ring Main Unit, 11 kV including electrical terminations	point	15350.00 to	18600.00
Transfomers			
Oil filled transformers, 11 kV to 415 kV including electrical terminations	kVA	19.00 to	23.00
Cast resin transformers, 11 kV to 415 kV including electrical terminations	kVA	21.00 to	24.00
Midal filled transformers, 11 kV to 415 kV including electrical terminations	kVA	22.00 to	26.00
HV Switchgear			
Cubicle section HV switchpanel, Form 4 type 6 including air circuit breakers, meters and electrical terminations	section	17300.00 to	25500.00
LV Switchgear			
LV switchpanel, Form 3 including all isolators, fuses, meters and electrical terminations	isolator	2600.00 to	3300.00
LV switchpanel, Form 4 type 5 including all isolators, fuses, meters and electrical terminations	isolator	4100.00 to	5100.00
External Packaged Substation			
Extra over cost for prefabricated packaged substation housing, excludes base and protective security fencing	each	27500.00 to	33700.00

APPROXIMATE ESTIMATING RATES – 5 SERVICES

Item	Unit	Range £	
Standby Generating Sets			
Diesel powered including control panel, flue, oil day tank and attenuation			
Approximate installed cost, LV	kVA	320.00 to	380.00
Approximate installed cost, HV	kVA	370.00 to	440.00
Uninterruptible Power Supply			
Rotary UPS including control panel and choke transformer (excludes distribution)			
Approximate installed cost (range 1000 kVA to 2500 kVA)	kVA	495.00 to	600.00
Static UPS including control panel, automatic bypass, DC isolator and batteries for 10 minutes standby (excludes distribution)			
Approximate installed cost (range 500 kVA to 1000 kVA)	kVA	210.00 to	290.00
Small Power			
Approximate prices for wiring of power points of length not exceeding 20 m, including accessories, wireways but excluding distribution boards			
13 amp Accessories			
Wired in PVC insulated twin and earth cable in ring main circuit			
Domestic properties	point	62.00 to	75.00
Commercial properties	point	83.00 to	100.00
Industrial properties	point	83.00 to	100.00
Wired in PVC insulated twin and earth cable in radial circuit			
Domestic properties	point	83.00 to	100.00
Commercial properties	point	100.00 to	125.00
Industrial properties	point	100.00 to	125.00
Wired in LSF insulated single cable in ring main circuit			
Commercial properties	point	100.00 to	125.00
Industrial properties	point	100.00 to	125.00
Wired in LSF insulated single cable in radial circuit			
Commercial properties	point	120.00 to	145.00
Industrial properties	point	120.00 to	145.00
45 amp Wired in PVC Insulated Twin and Earth Cable			
Domestic properties	point	110.00 to	140.00
Low Voltage Power Circuits			
Three phase four wire radial circuit feeding an individual load, wired in LSF insulated single cable including all wireways, isolator, not exceeding 10 metres; in commercial properties			
Cable size:			
1.5 mm²	point	200.00 to	250.00
2.5 mm²	point	220.00 to	265.00
4 mm²	point	230.00 to	275.00
6 mm²	point	250.00 to	320.00
10 mm²	point	300.00 to	370.00
16 mm²	point	330.00 to	400.00

APPROXIMATE ESTIMATING RATES – 5 SERVICES

Item	Unit	Range £	
5.8 Electrical Installations – cont			
Low Voltage Power Circuits – cont			
Three phase four core radial circuit feeding an			
individual load item, wired in LSF/SWA/XLPE insulated			
cable including terminations, isolator, clipped to surface,			
not exceeding 10 metres; in commercial properties			
Cable size			
1.5 mm²	point	150.00 to	185.00
2.5 mm²	point	170.00 to	200.00
4 mm²	point	175.00 to	220.00
6 mm²	point	200.00 to	245.00
10 mm²	point	295.00 to	366.00
16 mm²	point	375.00 to	460.00
Lighting			
Approximate prices for wiring of lighting points including rose, wireways but			
excluding distribution boards, luminaires and switches			
Final Circuits			
Wired in PVC insulated twin and earth cable			
Domestic properties	point	46.00 to	56.00
Commercial properties	point	58.00 to	71.00
Industrial properties	point	58.00 to	71.00
Wired in LSF insulated single cable			
Commercial properties	point	77.00 to	92.00
Industrial properties	point	77.00 to	92.00
Electrical Works in Connection with Mechanical Services			
The cost of electrical connections to mechanical services equipment will vary			
depending on the type of building and complexity of the equipment.			
Typical rate for power wiring, isolators and associated wireways	m²	8.00 to	11.00
Fire Alarms			
Cost per point for two core FP200 wired system including all terminations,			
supports and wireways			
Call point	point	280.00 to	340.00
Smoke detector	point	240.00 to	290.00
Smoke/heat detector	point	280.00 to	340.00
Heat detector	point	260.00 to	320.00
Heat detector and sounder	point	240.00 to	280.00
Input/output/relay units	point	340.00 to	400.00
Alarm sounder	point	260.00 to	320.00
Alarm sounder/beacon	point	320.00 to	390.00
Speakers/voice sounders	point	320.00 to	390.00
Speakers/voice sounders (weatherproof)	point	340.00 to	400.00
Beacon/strobe	point	240.00 to	290.00
Beacon/strobe (weatherproof)	point	350.00 to	430.00
Door release units	point	350.00 to	430.00
Beam detector	point	990.00 to	1200.00

APPROXIMATE ESTIMATING RATES – 5 SERVICES

Item	Unit	Range £	
For costs for zone control panel, battery chargers and batteries, see 'Prices for Measured Work' section.			
External Lighting			
Estate Road Lighting			
Post type road lighting lantern 70 watt CDM-T 3000 complete with 5 m high column with hinged lockable door, control gear and cut-out including 2.5 mm two core butyl cable internal wiring, interconnections and earthing fed by 16 mm² four core XLPE/SWA/LSF cable and terminations. Approximate installed price per metre road length (based on 300 metres run) including time switch but excluding builder's work in connection			
Columns erected at 30 m intervals along road (cost per m of road)	m	78.00 to	97.00
Bollard Lighting			
Bollard lighting fitting 26 watt TC-D 3500k including control gear, all internal wiring, interconnections, earthing and 25 metres of 2.5 mm² three core XLPE/ SWA/LSF cable			
Approximate installed price excluding builder's work in connection	each	1120.00 to	1380.00
Outdoor Flood Lighting			
Wall mounted outdoor flood light fitting complete with tungsten halogen lamp, mounting bracket, wire guard and all internal wiring and containment, fixed to brickwork or concrete and connected			
Installed price 500 watt	point	230.00 to	290.00
Installed price 1000 watt	point	280.00 to	340.00
Pedestal mounted outdoor flood light fitting complete with 1000 watt MBF/U lamp, control gear, contained in weatherproof steel box, all internal wiring and containment, interconnections and earthing, fixed to brickwork or concrete and connected			
Approximate installed price excluding builder's work in connection	each	1120.00 to	1380.00
Building Management Installations			
Category A Fit-Out			
Option 1 –185 Nr four pipe fan coil – 740 points			
1.0 Field Equipment			
Network devices; Valves/actuators; Sensing devices	point	77.00 to	79.00
2.0 Cabling			
Power – from local isolator to DDC controller; Control – from DDC controller to field equipment	point	46.00 to	48.00
3.0 Programming			
Software – central facility; Software – network devices; Graphics	point	23.00 to	26.00
4.0 On site testing and commissioning			
Equipment; Programming/graphics; Power and control cabling	point	24.00 to	27.00
Total Option 1 – Four pipefan coil	point	160.00 to	170.00
Cost/FCU	each	690.00 to	710.00

APPROXIMATE ESTIMATING RATES – 5 SERVICES

Item	Unit	Range £	
5.8 Electrical Installations – cont			
Option 2 –185 Nr two pipe fan coil system with electric heating – 740 points			
1.0 Field Equipment			
Network devices; Valves/actuators/thyristors; Sensing devices	point	100.00 to	105.00
2.0 Cabling			
Power – from local isolator to DDC controller; Control – from DDC controller			
to field equipment	point	46.00 to	48.00
3.0 Programming			
Software – central facility; Software – network devices; Graphics	point	24.00 to	27.00
4.0 On site testing and commissioning			
Equipment; Programming/graphics; Power and control cabling	point	26.00 to	28.00
Total Option 2 – Two pipe fan coil with electric heating	point	190.00 to	200.00
Cost/FCU	each	790.00 to	810.00
Option 3 –180 Nr chilled beams with perimeter heating – 567 points			
1.0 Field Equipment			
Network devices; Valves/actuators; Sensing devices	point	95.00 to	97.00
2.0 Cabling			
Power – from local isolator to DDC controller; Control – from DDC controller			
to field equipment	point	64.00 to	66.00
3.0 Programming			
Cost/Point	point	34.00 to	36.00
4.0 On site testing and commissioning			
Equipment; Programming/Graphics; Power and control cabling	point	36.00 to	38.00
Total Option 3 – Chilled beams with perimeter heating	point	230.00 to	230.00
Cost/Chilled Beam	each	720.00 to	740.00
Shell & Core Only			
Main Plant – Cost/BMS Points	each	740.00 to	750.00
Landlord FCUs – Cost/Terminal Units	each	740.00 to	750.00
Trade Contract Preliminaries	%	20.00 to	30.00

Notes: The following are included in points rates
- DDC Controllers/Control Enclosures/Control Panels
- Motor Control Centre (MCC)
- Field Devices
- Control and Power Cabling from DDC Controllers/MCC
- Programming
- On Site Testing and Commissioning

APPROXIMATE ESTIMATING RATES – 5 SERVICES

Item	Unit	Range £
5.10 Lift and Conveyor Systems		
Lift Installations		
The cost of lift installations will vary depending upon a variety of circumstances. The following prices assume a car height of 2.2 metres, manufacturer's standard car finish, brushed stainless steel 2 panel centre opening doors to BSEN81 –20 and Lift Regulations 2016.		
Passenger Lifts, machine room above		
Electrically operated AC drive serving 2 levels with directional collective controls and a speed of 1.0 m/s		
8 Person	item	68340.00 to 83640.00
10 Person	item	73440.00 to 88740.00
13 Person	item	77520.00 to 94860.00
17 Person	item	87720.00 to 106080.00
21 Person	item	97920.00 to 119340.00
26 Person	item	109140.00 to 136680.00
Electrically operated AC drive serving 4 levels and a speed of 1.0 m/s		
8 Person	item	79560.00 to 97920.00
10 Person	item	84660.00 to 102000.00
13 Person	item	88740.00 to 109140.00
17 Person	item	98940.00 to 119340.00
21 Person	item	110160.00 to 136680.00
26 Person	item	130560.00 to 159120.00
Electrically operated AC drive serving 6 levels and a speed of 1.0 m/s		
8 Person	item	89760.00 to 110160.00
10 Person	item	94860.00 to 113220.00
13 Person	item	98940.00 to 119340.00
17 Person	item	110160.00 to 136680.00
21 Person	item	130560.00 to 159120.00
26 Person	item	141780.00 to 170340.00
Electrically operated AC drive serving 8 levels and a speed of 1.0 m/s		
8 Person	item	99960.00 to 125460.00
10 Person	item	106080.00 to 125460.00
13 Person	item	110160.00 to 136680.00
17 Person	item	130560.00 to 159120.00
21 Person	item	136680.00 to 164220.00
26 Person	item	159120.00 to 192780.00
Electrically operated AC drive serving 10 levels and a speed of 1.0 m/s		
8 Person	item	110160.00 to 136680.00
10 Person	item	113220.00 to 141780.00
13 Person	item	113220.00 to 141780.00
17 Person	item	136680.00 to 164220.00
21 Person	item	153000.00 to 181560.00
26 Person	item	176460.00 to 215220.00
Electrically operated AC drive serving 12 levels and a speed of 1.0 m/s		
8 Person	item	113220.00 to 141780.00
10 Person	item	130560.00 to 159120.00
13 Person	item	130560.00 to 159120.00
17 Person	item	147900.00 to 181560.00
21 Person	item	164220.00 to 204000.00
26 Person	item	187680.00 to 226440.00

APPROXIMATE ESTIMATING RATES – 5 SERVICES

Item	Unit	Range £
5.10 Lift and Conveyor Systems – cont		
Passenger Lifts, machine room above – cont		
Electrically operated AC drive serving 14 levels and a speed of 1.0 m/s		
8 Person	item	136680.00 to 164220.00
10 Person	item	136680.00 to 164220.00
13 Person	item	147900.00 to 181560.00
17 Person	item	164220.00 to 204000.00
21 Person	item	176460.00 to 215220.00
26 Person	item	204000.00 to 249900.00
Add to above for		
Increase speed from 1.0 m/s to 1.6 m/s		
8 Person	item	4590.00 to 5610.00
10 Person	item	4790.00 to 5920.00
13 Person	item	4900.00 to 6020.00
17 Person	item	4900.00 to 6020.00
21 Person	item	4900.00 to 6020.00
26 Person	item	4900.00 to 6020.00
Increase speed from 1.6 m/s to 2.0 m/s		
8 Person	item	6530.00 to 7850.00
10 Person	item	6940.00 to 8570.00
13 Person	item	7650.00 to 9180.00
17 Person	item	3470.00 to 4080.00
21 Person	item	3470.00 to 4280.00
26 Person	item	3470.00 to 4280.00
Increase speed from 2.0 m/s to 2.5 m/s		
8 Person	item	2450.00 to 2860.00
10 Person	item	2450.00 to 2860.00
13 Person	item	2860.00 to 3470.00
17 Person	item	2860.00 to 3470.00
21 Person	item	3470.00 to 4080.00
26 Person	item	3470.00 to 4080.00
Enhanced finish to car – Centre mirror, flat ceiling, carpet		
8 Person	item	3370.00 to 3980.00
10 Person	item	3570.00 to 4390.00
13 Person	item	3470.00 to 4390.00
17 Person	item	3980.00 to 4900.00
21 Person	item	4690.00 to 5710.00
26 Person	item	5410.00 to 6630.00
Bottom motor room		
8 Person	item	8060.00 to 9790.00
10 Person	item	8060.00 to 9790.00
13 Person	item	8060.00 to 9790.00
17 Person	item	9890.00 to 11930.00
21 Person	item	9890.00 to 11930.00
26 Person	item	10000.00 to 12550.00

APPROXIMATE ESTIMATING RATES – 5 SERVICES

Item	Unit	Range £	
Firefighting control			
8 Person	item	6530.00 to	7850.00
10 Person	item	6530.00 to	7850.00
13 Person	item	6530.00 to	7850.00
17 Person	item	6530.00 to	7850.00
21 Person	item	6530.00 to	7850.00
26 Person	item	6530.00 to	7850.00
Glass back			
8 Person	item	2960.00 to	3570.00
10 Person	item	3370.00 to	3980.00
13 Person	item	3880.00 to	4790.00
17 Person	item	4690.00 to	5710.00
21 Person	item	4690.00 to	5710.00
26 Person	item	4690.00 to	5710.00
Glass doors			
8 Person	item	22640.00 to	27230.00
10 Person	item	22640.00 to	27230.00
13 Person	item	24380.00 to	29480.00
17 Person	item	26110.00 to	32330.00
21 Person	item	24380.00 to	29480.00
26 Person	item	24380.00 to	29480.00
Painting to entire pit			
8 Person	item	2350.00 to	2860.00
10 Person	item	2350.00 to	2860.00
13 Person	item	2350.00 to	2860.00
17 Person	item	2350.00 to	2860.00
21 Person	item	2350.00 to	2860.00
26 Person	item	2350.00 to	2860.00
Dual seal shaft			
8 Person	item	4690.00 to	5710.00
10 Person	item	4690.00 to	5710.00
13 Person	item	4690.00 to	5710.00
17 Person	item	5610.00 to	6940.00
21 Person	item	5610.00 to	6940.00
26 Person	item	5610.00 to	6940.00
Dust sealing machine room			
8 Person	item	920.00 to	1120.00
10 Person	item	920.00 to	1120.00
13 Person	item	1530.00 to	1840.00
17 Person	item	1530.00 to	1840.00
21 Person	item	1530.00 to	1840.00
26 Person	item	1530.00 to	1840.00
Intercom to reception desk and security room			
8 Person	item	460.00 to	560.00
10 Person	item	460.00 to	560.00
13 Person	item	460.00 to	560.00
17 Person	item	460.00 to	560.00
21 Person	item	460.00 to	560.00
26 Person	item	460.00 to	560.00

Approximate Estimating

APPROXIMATE ESTIMATING RATES – 5 SERVICES

Item	Unit	Range £
5.10 Lift and Conveyor Systems – cont		
Add to above for – cont		
Heating, cooling and ventilation to machine room		
8 Person	item	1160.00 to 1450.00
10 Person	item	1160.00 to 1450.00
13 Person	item	1160.00 to 1450.00
17 Person	item	1160.00 to 1450.00
21 Person	item	1160.00 to 1450.00
26 Person	item	1160.00 to 1450.00
Shaft lighting/small power		
8 Person	item	4030.00 to 4930.00
10 Person	item	4030.00 to 4930.00
13 Person	item	4030.00 to 4930.00
17 Person	item	4030.00 to 4930.00
21 Person	item	4030.00 to 4930.00
26 Person	item	4030.00 to 4930.00
Motor room lighting/small power		
8 Person	item	1480.00 to 1790.00
10 Person	item	1480.00 to 1790.00
13 Person	item	1850.00 to 2240.00
17 Person	item	1850.00 to 2240.00
21 Person	item	1850.00 to 2240.00
26 Person	item	1850.00 to 2240.00
Lifting beams		
8 Person	item	1760.00 to 2130.00
10 Person	item	1760.00 to 2130.00
13 Person	item	1760.00 to 2130.00
17 Person	item	1960.00 to 2410.00
21 Person	item	1960.00 to 2410.00
26 Person	item	1960.00 to 2410.00
10 mm equipotential bonding of all entrance metalwork		
8 Person	item	920.00 to 1120.00
10 Person	item	920.00 to 1120.00
13 Person	item	920.00 to 1120.00
17 Person	item	920.00 to 1120.00
21 Person	item	920.00 to 1120.00
26 Person	item	920.00 to 1120.00
Shaft secondary steelwork		
8 Person	item	6730.00 to 8360.00
10 Person	item	6940.00 to 8470.00
13 Person	item	7140.00 to 8770.00
17 Person	item	7340.00 to 8980.00
21 Person	item	7340.00 to 8980.00
26 Person	item	7340.00 to 8980.00
Independent insurance inspection		
8 Person	item	2210.00 to 2690.00
10 Person	item	2210.00 to 2690.00
13 Person	item	2210.00 to 2690.00
17 Person	item	2210.00 to 2690.00
21 Person	item	2210.00 to 2690.00
26 Person	item	2210.00 to 2690.00

APPROXIMATE ESTIMATING RATES – 5 SERVICES

Item	Unit	Range £
12 month warranty service		
8 Person	item	1190.00 to 1450.00
10 Person	item	1190.00 to 1450.00
13 Person	item	1190.00 to 1450.00
17 Person	item	1190.00 to 1450.00
21 Person	item	1190.00 to 1450.00
26 Person	item	1190.00 to 1450.00
Passenger Lifts, machine room less		
Electrically operated AC drive serving 2 levels with directional collective controls and a speed of 1.0 m/s		
8 Person	item	61200.00 to 74460.00
10 Person	item	68340.00 to 84660.00
13 Person	item	71400.00 to 87720.00
17 Person	item	86700.00 to 105060.00
21 Person	item	94860.00 to 113220.00
26 Person	item	105060.00 to 125460.00
Electrically operated AC drive serving 4 levels and a speed of 1.0 m/s		
8 Person	item	70380.00 to 86700.00
10 Person	item	77520.00 to 94860.00
13 Person	item	81600.00 to 99960.00
17 Person	item	97920.00 to 119340.00
21 Person	item	106080.00 to 125460.00
26 Person	item	112200.00 to 141780.00
Electrically operated AC drive serving 6 levels and a speed of 1.0 m/s		
8 Person	item	81600.00 to 99960.00
10 Person	item	87720.00 to 106080.00
13 Person	item	91800.00 to 112200.00
17 Person	item	106080.00 to 125460.00
21 Person	item	112200.00 to 141780.00
26 Person	item	125460.00 to 147900.00
Electrically operated AC drive serving 8 levels and a speed of 1.0 m/s		
8 Person	item	90780.00 to 111180.00
10 Person	item	96900.00 to 113220.00
13 Person	item	102000.00 to 125460.00
17 Person	item	112200.00 to 141780.00
21 Person	item	125460.00 to 147900.00
26 Person	item	136680.00 to 170340.00
Electrically operated AC drive serving 10 levels and a speed of 1.0 m/s		
8 Person	item	100980.00 to 125460.00
10 Person	item	105060.00 to 125460.00
13 Person	item	112200.00 to 136680.00
17 Person	item	125460.00 to 147900.00
21 Person	item	136680.00 to 170340.00
26 Person	item	159120.00 to 192780.00
Electrically operated AC drive serving 12 levels and a speed of 1.0 m/s		
8 Person	item	113220.00 to 136680.00
10 Person	item	112200.00 to 136680.00
13 Person	item	125460.00 to 147900.00
17 Person	item	136680.00 to 170340.00
21 Person	item	147900.00 to 181560.00
26 Person	item	164220.00 to 204000.00

APPROXIMATE ESTIMATING RATES – 5 SERVICES

Item	Unit	Range £
5.10 Lift and Conveyor Systems – cont		
Passenger Lifts, machine room less – cont		
Electrically operated AC drive serving 14 levels and a speed of 1.0 m/s		
8 Person	item	125460.00 to 147900.00
10 Person	item	125460.00 to 147900.00
13 Person	item	130560.00 to 159120.00
17 Person	item	159120.00 to 192780.00
21 Person	item	164220.00 to 204000.00
26 Person	item	187680.00 to 226440.00
Add to above for		
Increase speed from 1.0 m/s to 1.6 m/s		
8 Person	item	3010.00 to 3680.00
10 Person	item	3010.00 to 3680.00
13 Person	item	3170.00 to 3860.00
17 Person	item	4880.00 to 5900.00
21 Person	item	5610.00 to 6920.00
26 Person	item	7140.00 to 8720.00
Enhanced finish to car – Centre mirror, flat ceiling, carpet		
8 Person	item	3280.00 to 3970.00
10 Person	item	3460.00 to 4250.00
13 Person	item	3460.00 to 4250.00
17 Person	item	3970.00 to 4880.00
21 Person	item	4590.00 to 5610.00
26 Person	item	5440.00 to 6570.00
Firefighting control		
8 Person	item	6340.00 to 7700.00
10 Person	item	6340.00 to 7700.00
13 Person	item	6340.00 to 7700.00
17 Person	item	6340.00 to 7700.00
21 Person	item	6340.00 to 7700.00
26 Person	item	6340.00 to 7700.00
Painting to entire pit		
8 Person	item	2300.00 to 2810.00
10 Person	item	2300.00 to 2810.00
13 Person	item	2300.00 to 2810.00
17 Person	item	2300.00 to 2810.00
21 Person	item	2300.00 to 2810.00
26 Person	item	2300.00 to 2810.00
Dual seal shaft		
8 Person	item	4650.00 to 5660.00
10 Person	item	4650.00 to 5660.00
13 Person	item	4650.00 to 5660.00
17 Person	item	5610.00 to 6920.00
21 Person	item	5610.00 to 6920.00
26 Person	item	5610.00 to 6920.00

APPROXIMATE ESTIMATING RATES – 5 SERVICES

Item	Unit	Range £	
Shaft lighting/small power			
8 Person	item	4030.00 to	4930.00
10 Person	item	4030.00 to	4930.00
13 Person	item	4030.00 to	4930.00
17 Person	item	4030.00 to	4930.00
21 Person	item	4030.00 to	4930.00
26 Person	item	4030.00 to	4930.00
Intercom to reception desk and security room			
8 Person	item	460.00 to	560.00
10 Person	item	460.00 to	560.00
13 Person	item	460.00 to	560.00
17 Person	item	460.00 to	560.00
21 Person	item	460.00 to	560.00
26 Person	item	460.00 to	560.00
Lifting beams			
8 Person	item	1730.00 to	2130.00
10 Person	item	1730.00 to	2130.00
13 Person	item	1730.00 to	2130.00
17 Person	item	1960.00 to	2410.00
21 Person	item	1960.00 to	2410.00
26 Person	item	1960.00 to	2410.00
10 mm equipotential bonding of all entrance metalwork			
8 Person	item	920.00 to	1120.00
10 Person	item	920.00 to	1120.00
13 Person	item	920.00 to	1120.00
17 Person	item	920.00 to	1120.00
21 Person	item	920.00 to	1120.00
26 Person	item	920.00 to	1120.00
Shaft secondary steelwork			
8 Person	item	6570.00 to	7930.00
10 Person	item	6680.00 to	8270.00
13 Person	item	6920.00 to	8500.00
17 Person	item	7140.00 to	8720.00
21 Person	item	7140.00 to	8720.00
26 Person	item	7140.00 to	8720.00
Independent insurance inspection			
8 Person	item	2130.00 to	2580.00
10 Person	item	2130.00 to	2580.00
13 Person	item	2130.00 to	2580.00
17 Person	item	2130.00 to	2580.00
21 Person	item	2130.00 to	2580.00
26 Person	item	2130.00 to	2580.00
12 month warranty service			
8 Person	item	1120.00 to	1370.00
10 Person	item	1120.00 to	1370.00
13 Person	item	1120.00 to	1370.00
17 Person	item	1120.00 to	1370.00
21 Person	item	1120.00 to	1370.00
26 Person	item	1120.00 to	1370.00

APPROXIMATE ESTIMATING RATES – 5 SERVICES

Item	Unit	Range £
5.10 Lift and Conveyor Systems – cont		
Goods Lifts, machine room above		
Electrically operated two speed serving 2 levels to take 1000 kg load, prime coated internal finish and a speed of 1.0 m/s		
2000 kg	item	113220.00 to 141780.00
2250 kg	item	125460.00 to 147900.00
2500 kg	item	125460.00 to 147900.00
3000 kg	item	136680.00 to 170340.00
Electrically operated two speed serving 4 levels and a speed of 1.0 m/s		
2000 kg	item	125460.00 to 147900.00
2250 kg	item	136680.00 to 170340.00
2500 kg	item	141780.00 to 170340.00
3000 kg	item	159120.00 to 192780.00
Electrically operated two speed serving 6 levels and a speed of 1.0 m/s		
2000 kg	item	141780.00 to 170340.00
2250 kg	item	159120.00 to 192780.00
2500 kg	item	159120.00 to 192780.00
3000 kg	item	176460.00 to 215220.00
Electrically operated two speed serving 8 levels and a speed of 1.0 m/s		
2000 kg	item	159120.00 to 192780.00
2250 kg	item	164220.00 to 204000.00
2500 kg	item	176460.00 to 215220.00
3000 kg	item	198900.00 to 243780.00
Electrically operated two speed serving 10 levels and a speed of 1.0 m/s		
2000 kg	item	176460.00 to 215220.00
2250 kg	item	187680.00 to 226440.00
2500 kg	Item	187680.00 to 226440.00
3000 kg	item	215220.00 to 261120.00
Electrically operated two speed serving 12 levels and a speed of 1.0 m/s		
2000 kg	item	187680.00 to 226440.00
2250 kg	item	198900.00 to 243780.00
2500 kg	item	198900.00 to 243780.00
3000 kg	item	238680.00 to 283560.00
Electrically operated two speed serving 14 levels and a speed of 1.0 m/s		
2000 kg	item	204000.00 to 249900.00
2250 kg	item	221340.00 to 272340.00
2500 kg	item	221340.00 to 272340.00
3000 kg	item	249900.00 to 306000.00
Add to above for		
Increase speed of travel from 1.0 m/s to 1.6 m/s		
2000 kg	item	1480.00 to 1790.00
Enhanced finish to car – Centre mirror, flat ceiling, carpet		
2000 kg	item	4030.00 to 4930.00
2250 kg	item	4030.00 to 4930.00
2500 kg	item	4030.00 to 4930.00

APPROXIMATE ESTIMATING RATES – 5 SERVICES

Item	Unit	Range £	
Bottom motor room			
2000 kg	item	9850.00 to	11890.00
Painting to entire pit			
2000 kg	item	2300.00 to	2810.00
2250 kg	item	2300.00 to	2810.00
2500 kg	item	2300.00 to	2810.00
3000 kg	item	2300.00 to	2810.00
Dual seal shaft			
2000 kg	item	4650.00 to	5660.00
2250 kg	item	4650.00 to	5660.00
2500 kg	item	4650.00 to	5660.00
3000 kg	item	4650.00 to	5660.00
Intercom to reception desk and security room			
2000 kg	item	470.00 to	570.00
2250 kg	item	470.00 to	570.00
2500 kg	item	470.00 to	570.00
3000 kg	item	470.00 to	570.00
Heating, cooling and ventilation to machine room			
2000 kg	item	1280.00 to	1560.00
2250 kg	item	1280.00 to	1560.00
2500 kg	item	1280.00 to	1560.00
3000 kg	item	1280.00 to	1560.00
Lifting beams			
2000 kg	item	1700.00 to	2100.00
2250 kg	item	1700.00 to	2100.00
2500 kg	item	1700.00 to	2100.00
3000 kg	item	1700.00 to	2100.00
10 mm equipotential bonding of all entrance metalwork			
2000 kg	item	920.00 to	1120.00
2250 kg	item	920.00 to	1120.00
2500 kg	item	920.00 to	1120.00
3000 kg	item	920.00 to	1120.00
Independent insurance inspection			
2000 kg	item	2180.00 to	2690.00
2250 kg	item	2180.00 to	2690.00
2500 kg	item	2180.00 to	2690.00
3000 kg	item	2180.00 to	2690.00
12 month warranty service			
2000 kg	item	1190.00 to	1450.00
2250 kg	item	1390.00 to	1670.00
2500 kg	item	1390.00 to	1670.00
3000 kg	item	1390.00 to	1670.00

APPROXIMATE ESTIMATING RATES – 5 SERVICES

Item	Unit	Range £
5.10 Lift and Conveyor Systems – cont		
Goods Lift, machine room less		
Electrically operated two speed serving 2 levels to take 1000 kg load, prime coated internal finish and a speed of 1.0 m/s		
2000 kg	item	100980.00 to 125460.00
2250 kg	item	106080.00 to 125460.00
2500 kg	item	113220.00 to 141780.00
Electrically operated two speed serving 4 levels and a speed of 1.0 m/s		
2000 kg	item	113220.00 to 141780.00
2250 kg	item	119340.00 to 147900.00
2500 kg	item	130560.00 to 159120.00
Electrically operated two speed serving 6 levels and a speed of 1.0 m/s		
2000 kg	item	130560.00 to 159120.00
2250 kg	item	130560.00 to 159120.00
2500 kg	item	147900.00 to 181560.00
Electrically operated two speed serving 8 levels and a speed of 1.0 m/s		
2000 kg	item	136680.00 to 170340.00
2250 kg	item	147900.00 to 181560.00
2500 kg	item	159120.00 to 192780.00
Electrically operated two speed serving 10 levels and a speed of 1.0 m/s		
2000 kg	item	153000.00 to 187680.00
2250 kg	item	159120.00 to 192780.00
2500 kg	item	164220.00 to 204000.00
Electrically operated two speed serving 12 levels and a speed of 1.0 m/s		
2000 kg	item	136680.00 to 170340.00
2250 kg	item	176460.00 to 215220.00
2500 kg	item	187680.00 to 226440.00
Electrically operated two speed serving 14 levels and a speed of 1.0 m/s		
2000 kg	item	176460.00 to 215220.00
2250 kg	item	187680.00 to 226440.00
2500 kg	item	198900.00 to 243780.00
Add to above for		
Increase speed of travel from 1.0 m/s to 1.6 m/s		
2000 kg	item	5610.00 to 6940.00
2250 kg	item	8950.00 to 11020.00
Enhanced finish to car – Centre mirror, flat ceiling, carpet		
2000 kg	item	4760.00 to 5810.00
2250 kg	item	6230.00 to 7650.00
2500 kg	item	9180.00 to 11220.00
Painting to entire pit		
2000 kg	item	2330.00 to 2830.00
2250 kg	item	2330.00 to 2830.00
2500 kg	item	2330.00 to 2830.00
Dual seal shaft		
2000 kg	item	4650.00 to 5660.00
2250 kg	item	4650.00 to 5660.00
2500 kg	item	4650.00 to 5660.00

APPROXIMATE ESTIMATING RATES – 5 SERVICES

Item	Unit	Range £
Intercom to reception desk and security room		
2000 kg	item	470.00 to 570.00
2250 kg	item	470.00 to 570.00
2500 kg	item	470.00 to 570.00
Lifting beams		
2000 kg	item	1760.00 to 2130.00
2250 kg	item	1760.00 to 2130.00
2500 kg	item	1760.00 to 2130.00
10 mm equipotential bonding of all entrance metalwork		
2000 kg	item	920.00 to 1120.00
2250 kg	item	920.00 to 1120.00
2500 kg	item	920.00 to 1120.00
Independent insurance inspection		
2000 kg	item	2210.00 to 2690.00
2250 kg	item	2210.00 to 2690.00
2500 kg	item	2210.00 to 2690.00
12 month warranty service		
2000 kg	item	1190.00 to 1450.00
2250 kg	item	1190.00 to 1450.00
2500 kg	item	1190.00 to 1450.00
Escalator Installations		
30Ø Pitch escalator with a rise of 3 to 6 metres with standard balustrades		
1000 mm step width	item	91800.00 to 112200.00
Add to above for		
Balustrade lighting	item	2860.00 to 3570.00
Skirting lighting	item	11320.00 to 14180.00
Emergency stop button pedestals	item	5100.00 to 6320.00
Truss cladding – stainless steel	item	28970.00 to 35190.00
Truss cladding – spray painted steel	item	25500.00 to 31210

APPROXIMATE ESTIMATING RATES – 5 SERVICES

Item	Unit	Range £
5.11 Fire and Lightning Protection		
Sprinkler Installations		
Recommended maximum area coverage per sprinkler head		
Extra light hazard, 21 m² of floor area		
Ordinary hazard, 12 m² of floor area		
Extra high hazard, 9 m² of floor area		
Equipment		
Sprinkler equipment installation, pipework, valve sets, booster pumps and water storage	item	81600.00 to 99600.00
Price per sprinkler head; including pipework, valves and supports	point	210.00 to 250.00
Hose Reels and Dry Risers		
Wall mounted concealed hose reel with 36 m hose including approximately 15 m of pipework and isolating valve		
Price per hose reel	point	2040.00 to 2550.00
100 mm dry riser main including 2 way breeching valve and box, 65 mm landing valve, complete with padlock and leather strap and automatic air vent and drain valve		
Price per landing	point	2450.00 to 2960.00
5.12 Communications and Security Installations		
Access Control Systems		
Door mounted access control unit inclusive of door furniture, lock plus software; including up to 50 m of cable and termination; including documentation testing and commissioning		
Internal single leaf door	point	1260.00 to 1520.00
Internal double door	point	1370.00 to 1680.00
External single leaf door	point	1470.00 to 1790.00
External double leaf door	point	1680.00 to 2100.00
Management control PC with printer software and commissioning up to 1000 users	point	17850.00 to 21000.00
CCTV Installations		
CCTV equipment inclusive of 50 m of cable including testing and commissioning		
Internal camera with bracket	point	1100.00 to 1310.00
Internal camera with housing	point	1340.00 to 1630.00
Internal PTZ camera with bracket	point	2050.00 to 2520.00
External fixed camera with housing	point	1470.00 to 1790.00
External PTZ camera dome	point	2890.00 to 3470.00
External PTZ camera dome with power	point	3520.00 to 4250.00

APPROXIMATE ESTIMATING RATES – 5 SERVICES

Item	Unit	Range £
Turnstiles		
Physical Access Control Barrier system – standard security level comprising unit caseworks in stainless steel finish, standard lane width – 650 mm, restricting panels standard 900 mm high, standard level detection sensor system, including provision to integrate Access Control Card Readers (issued by others),including LED Pictogram – Green Arrow/Red Cross, closed base for ease of installation and cable management		
Including delivery, installation and commissioning to a site in London; budget cost land configurations		
Single lane	item	10710.00 to 11730.00
Double lane	item	17340.00 to 18360.00
Triple lane	item	23970.00 to 24990.00
Physical Access Control Barrier system – medium security level comprising unit caseworks in stainless steel finish, standard lane width – 650 mm, restricting panels standard 900 mm high, medium level detection sensor system, including provision to integrate Access Control Card Readers (issued by others),including LED Pictogram – Green Arrow/Red Cross, closed base for ease of installation and cable management		
Including delivery, installation and commissioning to a site in London; budget cost land configurations		
Single lane	item	12750.00 to 13770.00
Double lane	item	18870.00 to 19890.00
Triple lane	item	26520.00 to 27540.00
Physical Access Control Barrier system – high security level comprising unit caseworks in stainless steel finish, standard lane width – 650 mm, restricting panels standard 1600 mm or 1800 mm high (same cost), higher level detection sensor system, including provision to integrate Access Control Card Readers (issued by others), including LED Pictogram – Green Arrow/Red Cross, closed base for ease of installation and cable management		
Including delivery, installation and commissioning to a site in London; budget cost land configurations		
Single lane	item	12750.00 to 13770.00
Double lane	item	21420.00 to 23460.00
Triple lane	item	30600.00 to 32130.00
High security full height security revolving door – 2300 mm high, four wing T25 sections, no centre column, positioning drive with horizontal and vertical safety strips at door leaves, controlled via electronic Access Control Card Readers (supplied by others), sensor system in ceiling monitors door segments, secure simultaneous bi-directional use possible, finish – standard finish – PPC to RAL colour, card reader mounting boxes, 4 LED lights in ceiling, rubber matting		
Including delivery, installation and commissioning to a site in London		
Budget cost – 1800 mm dia., per door	item	28050.00 to 29580.00
Budget cost – 2000 mm dia., per door	item	28050.00 to 29580.00
Additional cost – stainless steel finish	item	2000.00

APPROXIMATE ESTIMATING RATES – 5 SERVICES

Item	Unit	Range £	
5.12 Communications and Security Installations – cont			
IT Installations			
Data Cabling			
Complete channel link including patch leads, cable, panels, testing and documentation (excludes cabinets and/or frames, patch cords, backbone/ harness connectivity as well as containment)			
Low Level			
Cat 5e (up to 5,000 outlets)	point	60.00 to	70.00
Cat 5e (5,000 to 15,000 outlets)	point	40.00 to	50.00
Cat 6 (up to 5,000 outlets)	point	70.00 to	80.00
Cat 6 (5,000 to 15,000 outlets)	point	60.00 to	70.00
Cat 6a (up to 5,000 outlets)	point	80.00 to	100.00
Cat 6a (5,000 to 15,000 outlets)	point	80.00 to	90.00
Cat 7 (up to 5,000 outlets)	point	100.00 to	120.00
Cat 7 (5,000 to 15,000 outlets)	point	90.00 to	110.00
Note: LSZH cable based on average of 50 metres false floor low level installation assuming 1 workstation in 2.5 m × 2.5 m (to 3.2 m × 3.2 m) density, with 4 data points per workstation. Not applicable for installations with less than 250 No. outlets.			
High Level			
Cat 5e (up to 500 outlets)	point	70.00 to	80.00
Cat 5e (over 500 outlets)	point	60.00 to	70.00
Cat 6 (up to 500 outlets)	point	80.00 to	90.00
Cat 6 (over 500 outlets)	point	70.00 to	90.00
Cat 6a (up to 500 outlets)	point	90.00 to	110.00
Cat 6a (over 500 outlets)	point	90.00 to	110.00
Cat 7 (up to 500 outlets)	point	110.00 to	130.00
Cat 7 (over 500 outlets)	point	110.00 to	130.00
Note: High level at 10 × 10 m grid.			

APPROXIMATE ESTIMATING RATES – 5 SERVICES

Item	Unit	Range £	
5.13 Specialist Installations			
Photovoltaic Panels			
Assuming roof mounted array, rate includes panels, associated framework, cabling, fixings, controls, delivery and commissioning			
High efficiency	m^2	250.00 to	300.00
Standard efficiency	m^2	200.00 to	250.00
Electric Vehicle Chargers			
Electric vehicle charging units, PAYG including array charging, data, power, wall signage and associated cabling, controls, delivery and commissioning			
7.0 kW charging capacity	nr	1700.00 to	2000.00
22.0 kW charging capacity	nr	2500.00 to	3000.00

BUILDING MODELS – ELEMENTAL COST SUMMARIES

Item	Unit	Range £	
AIRPORT TERMINAL BUILDING			
New build airport terminal building, premium quality, located in the South East of England, handling both domestic and international flights with a gross internal floor area (GIFA) of 25,000 m². These costs exclude baggage handling, check-in systems, pre-check in and boarding security systems, vertical transportation and travellators, pre-conditioned air systems to aircraft, services to stands and visual docking systems. Costs assume that the works are undertaken under landside access/logistics environment.			
5 Services			
5.1 Sanitary Installations	m²	6.50 to	10.0
5.3 Disposal Installations			
rainwater	m²	7.00 to	9.00
soil and waste	m²	9.00 to	13.50
condensate	m²	1.50 to	3.00
5.4 Water Installations			
domestic hot and cold water services	m²	18.50 to	22.00
5.5 Heat Source	m²	6.50 to	8.50
5.6 Space Heating and Air Conditioning			
LTHW heating system	m²	83.00 to	95.00
chilled water system	m²	91.50 to	105.00
supply and extract air conditioning system	m²	114.00 to	139.00
local cooling; DX systems to IT rooms	m²	6.00 to	9.00
5.7 Ventilation Systems			
mechanical ventilation to baggage handling and plantrooms	m²	28.00 to	33.00
toilet extract ventilation	m²	9.50 to	12.50
smoke extract installation	m²	13.00 to	18.00
kitchen extract system	m²	3.00 to	5.00
5.8 Electrical Installations			
main HV/MV Installations including switchgear, transformers (Cast Resin)	m²	67.00 to	85.00
low Voltage (LV) – Incoming LV switchgear, distribution boards and distribution systems	m²	47.00 to	58.00
generators, life safety – Containerized standby diesel generators and 24 hour capacity belly tanks	m²	36.00 to	47.00
small power installations	m²	47.00 to	58.00
lighting installations; warehouse, office, ancillary and plant spaces	m²	125.00 to	155.00
emergency lighting installations	m²	12.50 to	20.00
power to mechanical services	m²	7.00 to	10.00
5.9 Fuel Installations			
gas mains to services	m²	2.50 to	4.00
5.11 Fire and Lightning Protection			
lightning protection	m²	0.75 to	1.00
earthing and bonding	m²	0.50 to	0.75
sprinkler installations	m²	36.00 to	46.00
dry riser and hose reel installations	m²	7.00 to	8.50
fire suppression to IT rooms	m²	5.00 to	6.00

BUILDING MODELS – ELEMENTAL COST SUMMARIES

Item	Unit	Range £	
5.12 Communication, Security and Control Systems			
fire alarm systems; DDI	m²	36.50 to	46.00
voice/public address systems	m²	23.00 to	26.00
other alarm systems, e.g. disabled refuge	m²	2.00 to	3.50
security installations; CCTV, access control and intruder detection	m²	39.00 to	48.00
wireways for IT, comms and FA systems	m²	15.00 to	19.00
structured IT cabling; fibre optic backbone and copper Cat 6 to office and warehouse	m²	13.00 to	16.00
BMS systems	m²	60.00 to	68.00
5.13 Specialist Installations			
flight information display systems		29.00 to	37.00
Total Cost/m² (based on GIFA of 25,000 m²)	m²	967.00 to	1170.00

Notes:
The above includes MEP preliminaries and OH&P allowances
The above excludes Main Contractor's preliminaries and OH&P, Utility
connections; electric, water, drainage and fibre connections, Contingency/risk
allowances, Foreign exchange fluctuations, VAT at prevailing rates

BUILDING MODELS – ELEMENTAL COST SUMMARIES

Item	Unit	Range £	
SHOPPING MALL (TENANT'S FIT-OUT EXCLUDED)			
Natural ventilation shopping mall with approximately 33,000 m² two storey retail area and a 13,000 m² above ground, mechanically ventilated, covered car park, situated in a town centre in South East England.			
5 Services			
5.1 Sanitary Installations	m²	1.10 to	1.20
5.3 Disposal Installations			
rainwater	m²	6.80 to	7.20
soil, waste and vent	m²	6.80 to	7.20
5.4 Water Installations			
cold water installation	m²	6.80 to	7.20
hot water installation	m²	5.70 to	6.00
5.6 Space Heating and Air Conditioning			
condenser water system	m²	34.00 to	35.60
LTHW installation	m²	4.80 to	5.00
air conditioning system	m²	34.00 to	35.60
over door heaters at entrances	m²	1.10 to	1.20
5.7 Ventilation Systems			
public toilet ventilation	m²	1.10 to	1.20
plantroom ventilation	m²	4.80 to	5.00
supply and extract systems to shop units	m²	14.90 to	15.60
toilet extract systems to shop units	m²	2.30 to	2.40
smoke ventilation to Mall area	m²	11.70 to	12.30
service corridor ventilation	m²	2.30 to	2.40
other miscellaneous ventilation	m²	19.60 to	20.70
5.8 Electrical Installations			
LV distribution	m²	25.50 to	27.00
standby power	m²	4.80 to	5.10
general lighting	m²	70.10 to	73.20
external lighting	m²	5.90 to	6.10
emergency lighting	m²	14.10 to	14.90
small power	m²	11.50 to	12.10
mechanical services power supplies	m²	3.80 to	3.90
general earthing	m²	1.10 to	1.20
UPS for security and CCTV equipment	m²	1.10 to	1.20
5.9 Fuel Installations (gas)			
gas supply and boilers	m²	2.25 to	2.35
gas supplies to anchor (major) stores	m²	1.10 to	1.20
5.11 Fire and Lightning Protection			
lightning protection	m²	1.10 to	1.20
sprinkler installations	m²	15.90 to	16.80

BUILDING MODELS – ELEMENTAL COST SUMMARIES

Item	Unit	Range £	
5.12 Communication, Security and Control Systems			
fire alarm installation	m²	9.20 to	9.80
public address/voice alarm	m²	5.90 to	6.10
security installation	m²	12.50 to	13.10
general containment	m²	13.80 to	14.50
5.13 Specialist Installations			
BMS/Controls	m²	19.60 to	20.50
Total Cost/m² (based on GIFA of 33,000 m²)	m²	377.00 to	396.00
CAR PARK – 13,000 m²			
5 Services			
5.3 Disposal Installations			
car park drainage	m²	5.60 to	5.80
5.7 Ventilation Systems			
car park ventilation (impulse fans)	m²	33.70 to	35.70
5.8 Electrical Installations			
LV distribution	m²	11.50 to	12.00
general lighting	m²	23.40 to	24.50
emergency lighting	m²	5.90 to	6.10
small power	m²	3.50 to	3.60
mechanical services power supplies	m²	5.60 to	6.00
general earthing	m²	1.10 to	1.20
ramp frost protection	m²	2.20 to	2.30
5.11 Fire and Lightning Protection			
sprinkler installation	m²	22.40 to	24.00
fire alarm installations	m²	22.40 to	24.00
5.12 Communication, Security and Control Systems			
security installation	m²	10.20 to	10.70
BMS/Controls	m²	6.80 to	7.10
5.13 Specialist Installations			
entry/exit barriers, pay stations	m²	6.60 to	7.00
Total Cost/m² (based on GIFA of 13,000 m²)	m²	160.00 to	170.00

Approximate Estimating

BUILDING MODELS – ELEMENTAL COST SUMMARIES

Item	Unit	Range £	
SUPERMARKET			
Supermarket located in the South East with a total gross floor area of 4,000 m², including a sales area of 2,350 m². The building is on one level and incorporates a main sales, coffee shop, bakery, offices and amenities areas and warehouse.			
5 Services			
5.1 Sanitary Installations	m²	1.60 to	1.70
5.3 Disposal Installations	m²	4.00 to	4.20
5.4 Water Installations			
hot and cold water services	m²	28.70 to	29.70
5.6 Space Heating and Air Conditioning			
heating and ventilation with cooling via DX units	m²	11.40 to	12.00
5.7 Ventilation Systems			
supply and extract systems	m²	2.90 to	3.10
5.8 Electrical Installations			
panels/boards	m²	32.30 to	34.00
containment	m²	1.70 to	1.80
general lighting	m²	21.20 to	22.30
small power	m²	11.70 to	12.20
mechanical services wiring	m²	2.40 to	2.60
5.11 Fire and Lightning Protection			
lightning protection	m²	1.30 to	1.40
5.12 Communication, Security and Control Systems			
fire alarms, detection and public address	m²	2.90 to	3.10
CCTV	m²	2.90 to	3.10
intruder alarm, detection and store security	m²	2.90 to	3.10
telecom and structured cabling	m²	0.80 to	0.90
BMS	m²	8.00 to	8.40
data cabinet	m²	2.90 to	3.10
controls wiring	m²	4.30 to	4.50
5.13 Specialist Installations			
Refrigeration			
installation	m²	31.80 to	33.50
plant	m²	31.80 to	33.50
cold store	m²	11.70 to	12.30
cabinets	m²	85.90 to	90.20
Total Cost/m² (based on GIFA of 4,000 m²)	m²	305.00 to	320.00

BUILDING MODELS – ELEMENTAL COST SUMMARIES

Item	Unit	Range £	
OFFICE BUILDING Speculative 15 storey office in Central London for multiple tenant occupancy with gross mounted water cooled chillers, located in basement. 1 level of basement of approx. 1,600 m².			
SHELL & CORE – 20,000 m²			
5 Services			
5.1 Sanitary Installations	m²	10.00 to	14.00
5.3 Disposal Installations			
rainwater/soil and waste	m²	17.00 to	22.00
condensate	m²	2.00 to	3.00
5.4 Water Installations			
hot and cold water services	m²	19.00 to	28.00
5.5 Heat Source	m²	8.00 to	12.00
5.6 Space Heating and Air Conditioning			
LTHW, plant and distribution	m²	16.00 to	20.00
chilled water, plant and distribution	m²	52.00 to	68.00
air, plant and distribution	m²	38.00 to	48.00
5.7 Ventilation Systems			
toilet extract ventilation	m²	7.00 to	10.00
basement extract; smoke extract system, fire-rated	m²	18.00 to	27.00
miscellaneous ventilation systems	m²	17.00 to	18.00
5.8 Electrical Installations			
life safety generator, fuel and flue	m²	12.00 to	18.00
HV/LV supply/distribution	m²	45.00 to	56.00
general lighting (excluding external lighting) and lighting control	m²	30.00 to	40.00
general power	m²	6.00 to	9.00
electrical services for mechanical equipment	m²	3.00 to	5.00
voice and data (wireways)	m²	2.00 to	4.00
security (wireways)	m²	1.00 to	2.00
5.9 Fuel Installations (gas)	m²	2.00 to	3.00
5.11 Fire and Lightning Protection			
wet risers	m²	9.00 to	13.00
sprinklers	m²	18.00 to	22.00
earthing and bonding	m²	2.00 to	3.00
lightning protection	m²	2.00 to	3.00
5.12 Communication, Security and Control Systems			
fire and voice alarms	m²	12.00 to	16.00
disabled/refuge alarms	m²	2.00 to	4.00
CCTV/Access control/intruder detection	m²	3.00 to	5.00
BMS	m²	25.00 to	28.00
Total Cost/m² (based on GIFA of 20,000 m²)	m²	378.00 to	501.00

BUILDING MODELS – ELEMENTAL COST SUMMARIES

Item	Unit	Range £	
OFFICE BUILDING – cont			
CATEGORY 'A' FIT-OUT – 13,000 m² NIA			
Fan Coil Solution: 4 pipe FCUs to perimeter zone at 2 pipe FCU to internal			
5 Services			
5.6 Space Heating and Air Conditioning			
4 pipe/2 pipe fan coil units	m²	20.00 to	24.00
LTHW heating	m²	30.00 to	35.00
chilled water	m²	38.00 to	44.00
condensate	m²	8.00 to	12.00
ductwork distribution including griller/diffusers	m²	65.00 to	80.00
5.8 Electrical Installations			
tenant distribution boards	m²	6.00 to	8.00
lighting installation, lighting control and emergency	m²	75.00 to	95.00
supply only floor boxes	m²	5.00 to	7.00
earthing and bonding	m²	2.00 to	3.00
electrical services in connection with mechanical	m²	3.00 to	5.00
5.11 Fire and Lightning Protection			
sprinkler installation	m²	20.00 to	25.00
5.12 Communication, Security and Control Systems			
fire and voice alarms	m²	13.00 to	16.00
BMS	m²	24.00 to	28.00
Total Cost/m² (based on NIFA of 13,000 m²)	m²	309.00 to	382.00

BUILDING MODELS – ELEMENTAL COST SUMMARIES

Item	Unit	Range £	
BUSINESS PARK New build office in the South East within the M25, part of a speculative business park with a gross floor area of 10,000 m². A full air displacement system with roof mounted air cooled chillers, gas fired boilers and air handling plant. Total M&E services value approximately £2,500,000 (shell & core) and £1,000,000 (Cat A Fit-out). Excluding lifts.			
SHELL & CORE – 10,000 m² GIA			
5 Services			
5.1 Sanitary Installations	m²	8.00 to	11.00
5.3 Disposal Installations			
condensate	m²	2.00 to	3.00
rainwater, soil and waste	m²	12.00 to	15.00
5.4 Water Installations			
hot and cold water services	m²	12.00 to	16.00
5.5 Heat Source	m²	8.00 to	11.00
5.6 Space Heating and Air Conditioning			
LTHW heating; plantroom and risers	m²	12.00 to	16.00
chilled water; plantroom and risers	m²	32.00 to	37.00
ductwork; plantroom and risers	m²	60.00 to	70.00
5.7 Ventilation Systems			
toilet and miscellaneous ventilation	m²	12.00 to	16.00
5.8 Electrical Installations			
LV supply/distribution	m²	26.00 to	32.00
general lighting	m²	21.00 to	26.00
general power	m²	4.00 to	6.00
electrical services in connection with mechanical services	m²	2.00 to	4.00
security (wireways)	m²	1.00 to	3.00
voice and data (wireways)	m²	1.00 to	3.00
5.9 Fuel Installations (gas)	m²	1.00 to	3.00
5.11 Fire and Lightning Protection			
earthing and bonding	m²	2.00 to	4.00
lightning protection	m²	2.00 to	4.00
dry risers	m²	1.00 to	3.00
5.12 Communication, Security and Control Systems			
fire alarms	m²	10.00 to	13.00
BMS	m²	22.00 to	26.00
Total Cost/m² (based on GIFA of 10,000 m²)	m²	251.00 to	322.00

BUILDING MODELS – ELEMENTAL COST SUMMARIES

Item	Unit	Range £	
BUSINESS PARK – cont			
CATEGORY 'A' FIT-OUT – 8,000 m² NIA			
Displacement Ventilation, LTHW heating only			
5 Services			
5.6 Space Heating and Air Conditioning			
LTHW heating and perimeter heaters	m²	38.00 to	47.00
floor swirl diffusers and supply ductwork	m²	23.00 to	27.50
5.8 Electrical Installations			
distribution boards	m²	1.00 to	3.00
general lighting, recessed including lighting controls	m²	50.00 to	60.00
5.11 Fire and Lightning Protection			
earthing and bonding	m²	1.00 to	2.00
5.12 Communication, Security and Control Systems			
fire alarms	m²	7.00 to	9.10
BMS	m²	5.00 to	7.00
Total Cost/m² (based on NIFA of 8,000 m²)	each	125.00 to	155.00

BUILDING MODELS – ELEMENTAL COST SUMMARIES

Item	Unit	Range £	
PERFORMING ARTS CENTRE (MEDIUM SPECIFICATION)			
Performing Arts centre with a Gross Internal Area (GIA) of approximately 8,000 m², based on a medium specification with cooling to the Auditorium.			
The development comprises dance studios and a theatre auditorium in the outer London area. The theatre would require all the necessary stage lighting, machinery and equipment installed in a modern professional theatre (these are excluded from the model, as assumed to be FF&E, but the containment and power wiring is included).			
5 Services			
5.1 Sanitary Installations	m²	10.00 to	13.00
5.3 Disposal Installations			
soil, waste and rainwater	m²	16.00 to	20.00
5.4 Water Installations			
cold water installation	m²	15.00 to	20.00
hot water installation	m²	13.00 to	16.00
5.5 Heat Source	m²	12.00 to	15.00
5.6 Space Heating and Air Conditioning			
heating	m²	74.00 to	79.00
chilled water system	m²	63.00 to	84.00
supply and extract air systems	m²	115.00 to	137.00
5.7 Ventilation Systems			
ventilation and extract systems to toilets, kitchen and workshop	m²	22.00 to	27.00
5.8 Electrical Installations			
LV supply/distribution	m²	42.00 to	47.00
general lighting	m²	105.00 to	127.00
small power	m²	32.00 to	37.00
power to mechanical plant	m²	9.00 to	10.00
5.9 Fuel Installations (gas)	m²	4.00 to	6.00
5.11 Fire and Lightning Protection			
lightning protection	m²	4.00 to	5.00
5.12 Communication, Security and Control Systems			
fire alarms and detection	m²	27.00 to	32.00
voice and data complete installation (excluding active equipment)	m²	32.00 to	42.00
security; access; control; disabled alarms; staff paging	m²	30.00 to	42.00
BMS	m²	47.00 to	58.00
5.13 Specialist Installations			
theatre systems includes for containment and power wiring	m²	27.00 to	32.00
Total Cost/m² (based on GIFA of 8,000 m²)	each	125.00 to	155.00

BUILDING MODELS – ELEMENTAL COST SUMMARIES

Item	Unit	Range £	
SPORTS HALL Single storey sports hall, located in the South East, with a gross internal area of 1,200 m² (40 m × 30 m).			
5 Services			
5.1 Sanitary Installations	m²	12.00 to	13.00
5.3 Disposal Installations			
rainwater	m²	4.00 to	5.00
soil and waste	m²	8.00 to	9.00
5.4 Water Installations			
hot and cold water services	m²	16.00 to	17.00
5.5 Heat Source			
boilers, flues, pumps and controls	m²	16.00 to	17.00
5.6 Space Heating and Air Conditioning			
warm air heating to sports hall area	m²	20.00 to	21.00
radiator heating to ancillary areas	m²	23.00 to	24.00
5.7 Ventilation Systems			
ventilation to changing, fitness and sports hall areas	m²	22.00 to	23.00
5.8 Electrical Installations			
main switchgear and sub-mains	m²	13.00 to	14.00
small power	m²	12.00 to	13.00
lighting and luminaires to sports areas	m²	27.00 to	28.00
lighting and luminaires to ancillary areas	m²	32.00 to	33.00
5.11 Fire and Lightning Protection			
lightning protection	m²	5.00 to	6.00
5.12 Communication, Security and Control Systems			
fire, smoke detection and alarm system, intruder detection	m²	13.00 to	14.00
CCTV installation	m²	15.00 to	16.00
public address and music systems	m²	8.00 to	9.00
wireways for voice and data	m²	4.00 to	5.00
Total Cost/m² (based on GIFA of 1,200 m²)	m²	250.00 to	267.00

BUILDING MODELS – ELEMENTAL COST SUMMARIES

Item	Unit	Range £	
STADIUM – NEW			
A three storey stadium, located in Greater London with a gross internal area of 85,000 m² and incorporating 60,000 spectator seats.			
5 Services			
5.1 Sanitary Installations	m²	11.40 to	12.10
5.3 Disposal Installations			
rainwater	m²	4.80 to	5.10
above ground drainage	m²	12.50 to	13.30
5.4 Water Installations			
hot and cold water services	m²	22.20 to	23.40
5.5 Heat Source	m²	10.30 to	10.90
5.6 Space Heating and Air Conditioning			
heating	m²	6.90 to	7.10
cooling	m²	15.90 to	16.70
5.7 Ventilation Systems	m²	63.60 to	66.80
5.8 Electrical Installations			
HV/LV supply	m²	11.40 to	12.10
LV distribution	m²	27.00 to	28.60
general lighting	m²	66.80 to	70.00
small power	m²	21.20 to	22.20
earthing and bonding	m²	1.10 to	1.20
power supply to mechanical equipment	m²	1.10 to	1.20
pitch lighting	m²	11.60 to	12.20
5.9 Fuel Installations (gas)	m²	5.30 to	5.60
5.11 Fire and Lightning Protection			
lightning protection	m²	1.10 to	1.20
hydrants	m²	2.30 to	2.40
5.12 Communication, Security and Control Systems			
wireways for data, TV, telecom and PA	m²	9.20 to	9.70
public address	m²	17.30 to	18.20
security	m²	16.10 to	16.90
data voice installations	m²	34.50 to	36.10
fire alarms	m²	9.30 to	9.80
disabled/refuse alarm/call systems	m²	3.50 to	3.60
BMS	m²	16.10 to	16.90
Total Cost/m² (based on GIFA of 85,000 m²)	m²	402.00 to	423.00
Total Cost/seat (based on 60,000 seats)	each	570.00 to	600.00

BUILDING MODELS – ELEMENTAL COST SUMMARIES

Item	Unit	Range £	
HOTELS			
200 bedroom, four star hotel, situated in Central London, with a gross internal floor area of 16,500 m².The development comprises a ten storey building with large suites on each guest floor, together with banqueting, meeting rooms and leisure facilities.			
5 Services			
5.1 Sanitary Installations	m²	45.00 to	53.50
5.3 Disposal Installations			
rainwater	m²	4.50 to	5.50
soil and waste	m²	28.00 to	33.00
5.4 Water Installations			
hot and cold water services	m²	56.50 to	73.00
5.5 Heat Source			
condensing boiler, CHP and flues	m²	16.00 to	21.50
5.6 Space Heating and Air Conditioning			
air conditioning and space heating system; chillers, pumps, CHW and LTHW pipework, insulation, 4 pipe FCU, ductwork, grilles and diffusers, to guest rooms, public areas, meeting and banquet rooms	m²	235.50 to	278.00
5.7 Ventilation Systems			
general bathroom extract from guest suites, ventilation to kitchens and bathrooms, etc.	m²	48.00 to	64.00
smoke extract	m²	13.00 to	17.00
5.8 Electrical Installations			
HV/LV installation, standby power, lighting, emergency lighting and small power to guest rooms and public areas, including earthing and bonding	m²	134.00 to	165.00
5.11 Fire and Lightning Protection			
dry risers and sprinkler installation	m²	41.50 to	51.50
lightning protection	m²	2.50 to	3.50
5.12 Communication, Security and Control Systems			
fire/smoke detection and fire alarm system	m²	18.00 to	25.00
security/access control	m²	21.50 to	29.00
integrated sound and AV system	m²	22.50 to	27.00
telephone and data and TV installation	m²	17.00 to	20.50
containment	m²	16.00 to	19.50
wire ways for voice and data (no hotel management and head end equipment)	m²	11.00 to	13.00
BMS	m²	34.50 to	41.00
Total Cost/m² (based on GIFA of 16,500 m²)	m²	765.00 to	940.50

BUILDING MODELS – ELEMENTAL COST SUMMARIES

Item	Unit	Range £	
PRIVATE HOSPITAL			
New build project building. The works consist of a new 80 bed hospital of approximately 15,000 m², eight storey with a plant room.			
All heat is provided from existing steam boiler plant, medical gases are also served from existing plant. The project includes the provision of additional standby electrical generation to serve the wider site requirements.			
This hospital has six operating theatres, ITU/HDU department, pathology facilities, diagnostic imaging, outpatient facilities and physiotherapy.			
5 Services			
5.1 Sanitary Installations	m²	35.00 to	37.00
5.3 Disposal Installations			
rainwater	m²	4.00 to	6.00
soil and waste	m²	38.00 to	40.00
5.4 Water Installations			
hot and cold water services	m²	87.00 to	90.00
5.5 Heat Source (included in 5.6)			
5.6 Space Heating and Air Conditioning			
LPHW heating	m²	72.00 to	74.00
chilled water	m²	55.00 to	58.00
steam and condensate	m²	35.00 to	38.00
ventilation, comfort cooling and air conditioning	m²	184.00 to	211.00
5.7 Ventilation Systems (included in 5.6)			
5.8 Electrical Installations			
HV distribution	m²	47.00 to	50.00
LV supply/distribution	m²	74.00 to	79.00
standby power	m²	44.00 to	48.00
UPS	m²	31.00 to	33.00
general lighting	m²	84.00 to	95.00
general power	m²	65.00 to	68.00
emergency lighting	m²	25.00 to	27.00
theatre lighting	m²	19.00 to	21.00
specialist lighting	m²	19.00 to	21.00
external lighting	m²	4.00 to	6.00
electrical supplies for mechanical services	m²	14.00 to	16.00
earthing and bonding	m²	5.00 to	7.00
5.9 Fuel Installations			
gas installations	m²	6.00 to	8.00
oil installations	m²	9.00 to	11.00
5.11 Fire and Lightning Protection			
dry risers	m²	4.00 to	6.00
sprinkler systems/gaseous fire suppression	m²	58.00 to	68.00
lightning protection	m²	2.00 to	4.00

BUILDING MODELS – ELEMENTAL COST SUMMARIES

Item	Unit	Range £	
PRIVATE HOSPITAL – cont			
5 Services – cont			
5.12 Communication, Security and Control Systems			
fire alarms and detection	m²	35.00 to	37.00
voice and data	m²	53.00 to	55.00
data containment	m²	10.00 to	12.00
security and CCTV	m²	30.00 to	32.00
nurse call and cardiac alarm system	m²	46.00 to	48.00
TV systems	m²	19.00 to	21.00
disabled WC alarm	m²	5.00 to	7.00
BMS	m²	80.00 to	82.00
5.13 Specialist Installations			
pneumatic tube conveying system	m²	7.00 to	10.00
medical gases	m²	82.00 to	84.00
Total Cost/m² (based on GIFA of 15,000 m²)	m²	1375.00 to	1500.00

BUILDING MODELS – ELEMENTAL COST SUMMARIES

Item	Unit	Range £	
SCHOOL New build secondary school (Academy) located in Southern England, with a gross internal floor area of 10,000 m². Total M&E services value approximately £5,000,000. The building comprises a three storey teaching block, including provision for music, drama, catering, sports hall, science laboratories, food technology, workshops and reception/admin (BB93 compliant). Excludes IT cabling and sprinkler protection.			
5 Services			
5.1 Sanitary Installations			
toilet cores and changing facilities and lab sinks	m²	14.00 to	16.00
5.3 Disposal Installations			
rainwater installations	m²	5.00 to	6.00
soil and waste	m²	13.00 to	16.00
5.4 Water Installations			
potable hot and cold water services	m²	30.00 to	32.00
non-potable hot and cold water services to labs and art rooms	m²	10.00 to	11.00
5.5 Heat Source			
gas fired boiler installation	m²	12.00 to	15.00
5.6 Space Heating and Air Conditioning			
LTHW heating system (primary)	m²	37.00 to	40.00
LTHW heating system (secondary)	m²	10.00 to	12.00
DX cooling system to ICT server rooms	m²	5.00 to	7.00
mechanical supply and extract ventilation including DX type cooling to Music, Drama, Kitchen/Dining and Sports Hall	m²	63.00 to	68.00
5.7 Ventilation Systems			
toilet extract systems	m²	8.00 to	10.00
changing area extract systems	m²	5.00 to	7.00
extract ventilation from design/food technology and science labs	m²	10.00 to	12.00
5.8 Electrical Installations			
mains and sub-mains distribution	m²	37.00 to	41.00
lighting and luminaires including emergency fittings	m²	79.00 to	84.00
small power installation	m²	39.00 to	42.00
earthing and bonding	m²	2.00 to	3.00
5.9 Fuel Installations (gas)	m²	8.00 to	10.00
5.11 Fire and Lightning Protection			
lightning protection	m²	4.00 to	4.50
5.12 Communication, Security and Control Systems			
containment for telephone, IT data, AV and security systems	m²	5.00 to	7.00
fire, smoke detection and alarm system	m²	19.00 to	21.00
security installations including CCTV, access control and intruder alarm	m²	21.00 to	22.00
disabled toilet, refuge and induction loop systems	m²	3.00 to	5.00
BMS – to plant	m²	24.00 to	28.00
BMS – to opening vents/windows	m²	10.00 to	12.00
Total Cost/m² (based on GIFA of 10,000 m²)	m²	473.00 to	532.00

BUILDING MODELS – ELEMENTAL COST SUMMARIES

Item	Unit	Range £	
AFFORDABLE RESIDENTIAL DEVELOPMENT			
An eight storey, 117 affordable residential development with a gross internal area of 11,400 m² and a net internal area of 8,400 m², situated within the Central London area. The development has no car park or communal facilities and achieves a net to gross efficiency of 74%.			
Based upon radiator LTHW heating within each apartment, with plate heat exchanger, whole house ventilation, plastic cold and hot water services pipework, sprinkler protection to apartments. Excludes remote metering. Based upon 2 bed units with 1 bathroom.			
SHELL & CORE			
5 Services			
5.1 Sanitary Installations	m²	0.75 to	1.00
5.3 Disposal Installations	m²	20.00 to	25.00
5.4 Water Installations	m²	20.00 to	24.00
5.5 Heat Source	m²	9.50 to	11.50
5.6 Space Heating and Air Conditioning	m²	24.00 to	29.50
5.7 Ventilation Systems	m²	19.00 to	21.00
5.8 Electrical Installations	m²	43.00 to	53.50
5.9 Fuel Installations – Gas	m²	3.00 to	3.50
5.11 Fire and Lightning Protection	m²	22.00 to	2600
5.12 Communication, Security and Control Systems	m²	34.00 to	41.00
5.13 Specialist Installations	m²	9.50 to	11.50
Total Cost/m² (based on GIFA of 11,400 m²)	m²	205.00 to	248.00
FITTING OUT			
5 Services			
5.1 Sanitary Installations	m²	34.00 to	41.00
5.3 Disposal Installations	m²	11.00 to	13.00
5.4 Water Installations	m²	28.50 to	33.50
5.5 Heat Source	m²	35.00 to	42.00
5.6 Space Heating and Air Conditioning	m²	42.00 to	52.50
5.7 Ventilation Systems	m²	37.00 to	44.00
5.8 Electrical Installations	m²	60.00 to	73.50
5.11 Fire and Lightning Protection	m²	22.00 to	27.50
5.12 Communication, Security and Control Systems	m²	19.00 to	24.00
5.13 Specialist Installations	m²	11.50 to	13.50
Total Cost/m² (based on NIA of 8,400 m²)	m²	300.00 to	365.00
Total cost per apartment – Shell & Core and Fit-Out	each	40000.00 to	50000.00

BUILDING MODELS – ELEMENTAL COST SUMMARIES

Item	Unit	Range £	
PRIVATE RESIDENTIAL DEVELOPMENT			
A 24 storey, 203 apartment private residential development with a gross internal area of 50,000 m² and a net internal area of 33,000 m², situated within London's prime residential area. The development has limited car park facilities, residents' communal areas such as gym, cinema room, self-stimulated, etc.			
Based upon LTHW underfloor heating, 4 pipe fan coils to principal rooms, with plate heat and cooling exchangers, whole house ventilation with summer house facility, copper hot and cold water services pipework, sprinkler protection to apartments, lighting control, TV/data outlets. Flood wiring for apartment, home automation and sound system only. Based upon a combination of 2/3 bedroom apartments with family bathroom and ensuite.			
SHELL & CORE			
5 Services			
5.1 Sanitary Installations	m²	3.00 to	3.50
5.3 Disposal Installations	m²	24.50 to	31.00
5.4 Water Installations	m²	34.00 to	43.50
5.5 Heat Source	m²	10.00 to	15.00
5.6 Space Heating and Air Conditioning	m²	78.50 to	100.00
5.7 Ventilation Systems	m²	23.00 to	27.00
5.8 Electrical Installations	m²	67.00 to	85.00
5.9 Fuel Installations – Gas	m²	4.50 to	5.50
5.11 Fire and Lightning Protection	m²	32.00 to	40.00
5.12 Communication, Security and Control Systems	m²	48.00 to	58.00
5.13 Specialist Installations	m²	40.00 to	53.00
Total Cost/m² (based on GIFA of 50,000 m²)	m²	365.00 to	460.00
FITTING OUT			
5 Services			
5.1 Sanitary Installations	m²	100.00 to	120.00
5.3 Disposal Installations	m²	25.00 to	32.00
5.4 Water Installations	m²	52.00 to	65.00
5.5 Heat Source	m²	52.00 to	63.00
5.6 Space Heating and Air Conditioning	m²	158.00 to	195.00
5.7 Ventilation Systems	m²	37.00 to	44.00
5.8 Electrical Installations	m²	126.00 to	158.00
5.11 Fire and Lightning Protection	m²	28.00 to	36.00
5.12 Communication, Security and Control Systems	m²	103.00 to	126.00
5.13 Specialist Installations	m²	32.00 to	39.00
Total Cost/m² (based on NIA of 33,000 m²)	m²	750.00 to	880.00
Total cost per apartment – Shell & Core and Fit-Out	each	96000.00 to	122,000.00

BUILDING MODELS – ELEMENTAL COST SUMMARIES

Item	Unit	Range £	
DISTRIBUTION CENTRE			
New build distribution centre located in the South East of England with a total gross floor area of 75,000 m², including a refrigerated cold box of 17,500 m². The building is on one level (no mezzanine decks) and incorporates a small office area at ground level, vehicle recovery unit, electric fork lift truck docking bay, gate house and associated plantrooms.			
5 Services			
5.3 Disposal Installations			
soil and waste	m²	7.00 to	7.50
rainwater	m²	2.50 to	3.00
5.4 Water Installations			
domestic hot and cold water services	m²	2.50 to	3.50
emergency drench showers & Cat 5 vehicle wash downs	m²	0.50 to	0.75
5.6 Space Heating and Air Conditioning			
heating with ventilation to offices, displacement system to main warehouse	m²	28.00 to	37.00
local cooling; DX systems to refrigerated cold box	m²	38.00 to	45.00
5.7 Ventilation Systems			
smoke extract system	m²	9.00 to	10.00
specialist extract to battery charging and maintenance areas	m²	2.00 to	3.00
5.8 Electrical Installations			
generator, life safety – containerized standby diesel generator and 24 hour capacity belly tanks	m²	5.00 to	6.00
main HV/MV installations including switchgear, transformers (Cast Resin)	m²	23.00 to	24.00
low voltage (LV) – incoming LV switchgear, distribution boards and distribution systems	m²	18.00 to	19.00
lighting installations; warehouse, office, ancillary and plant spaces	m²	15.00 to	19.00
small power installations inc. mech/HVAC plant power	m²	12.00 to	15.00
5.9 Fuel Installations			
gas mains to services	m²	0.75 to	1.00
5.11 Fire and Lightning Protection			
sprinklers (ESFR) including rack protection	m²	50.00 to	54.00
lightning protection	m²	0.75 to	1.00
earthing and bonding	m²	0.50 to	0.75
5.12 Communication, Security and Control Systems			
fire alarm systems; DDI	m²	13.00 to	15.00
security installations; CCTV, access control and intruder detection	m²	8.25 to	9.00
BMS systems	m²	6.00 to	7.00
other alarm systems, e.g. disabled refuge	m²	0.50 to	0.50
structured IT cabling; fibre optic backbone and copper Cat 6 to office and warehouse	m²	13.00 to	16.00
5.13 Specialist Installations			
testing and commissioning	m²	2.50 to	3.00
Total Cost/m² (based on GIFA of 75,000 m²)	m²	258.00 to	300.00

BUILDING MODELS – ELEMENTAL COST SUMMARIES

Item	Unit	Range £	
DATA CENTRE New build data centre in a single storey 'warehouse' type construction in the Greater London area/home counties proximity to the M25. Total Net Technical Area (NTA) of 2,000 m² (2 Nr data halls of 1,000 m² NTA each) with typically other areas of 300 m² for NOC/Security Room, office space, 350 m² ancillary space and 1,000 m² of internal plant areas. Total GIA of 3,650 m². Power and cooling to Technical Spaces designed to 1,500 W/m² of NTA = 3,000 kW IT load. Critical infrastructure built to UTI Tier III resilience standard (HVAC generally N +1 and Electrical on a dual 'A' and 'B' string supplies from MV/LV switchgear to IT cab (Critical UPSs N+1 on each string) with all heat rejection and standby generators (N+1) located in a secure, external plant farm. Total MEP services value of approximately £18,520,000 to £23,090,000. All costs expressed as £/m² against Net Technical Area – NTA.			
5 Services			
5.3 Disposal Installations			
soil and waste	m²	15.00 to	17.00
rainwater	m²	11.00 to	13.00
condensate	m²	11.00 to	15.00
5.4 Water Installations			
domestic hot and cold water services	m²	23.00 to	30.50
5.5 Heat Source			
for office and ancillary spaces only	m²	17.00 to	25.00
5.6 Space Heating and Air Conditioning			
chilled water plant with redundancy of N+1 to technical space; packaged air-cooled Turbocor chillers with free cooling capability	m²	515.00 to	675.00
chilled water distribution systems; plant, header and primary distribution to technical space CRAC units, office and ancillary space cooling units; single distribution ring with redundancy N+1 on pumps, pressurization, dosing and buffer vessels	m²	1040.00 to	1190.00
chilled water final connections to technical space CRAC office/ancillary space cooling units	m²	136.00 to	168.00
computer room air conditioning (CRAC) units; 100 kW single coil units with 30% of units with humidification; N+2 redundancy	m²	250.00 to	315.00
CRAC units to critical ancillary space, e.g. UPS module and battery rooms	m²	44.00 to	63.00
cooling units to office, support and other spaces; single coil FCUs	m²	35.00 to	55.00
5.7 Ventilation Systems			
supply and extract ventilation systems to data halls, switchrooms and ancillary spaces including dedicated gas extract to gaseous fire suppression protected spaces	m²	95.00 to	140.00
computer room floor grilles; heavy duty 600 × 600 mm metal grilles with dampers	m²	135.00 to	160.00
General supply and extract to office, support and ancillary spaces	m²	53.00 to	85.00

BUILDING MODELS – ELEMENTAL COST SUMMARIES

Item	Unit	Range £	
DATA CENTRE – cont			
5 Services – cont			
5.8 Electrical Installations			
main HV/MV installations including switchgear, transformers (Cast Resin); dual MV supply from local authority (REC supply cost excluded)	m²	195.00 to	245.00
Low Voltage (LV) switchgear – incoming LV switchgear, UPS Input and Output LV boards, mechanical services and support/ancillary area panels and distribution boards	m²	1010.00 to	1110.00
LV Distribution including busbars, cabling and containment systems to provide full dual 'A' and 'B' supplies to data halls and supplies to mechanical services, office and ancillary areas	m²	765.00 to	890.00
generators – containerized standby diesel generators, DC rated with acoustic treatment and exhaust systems; on an N+1 redundancy for critical IT electrical and cooling loads including synchronization panel and 48 hour capacity belly tanks	m²	950.00 to	1020.00
Uninterruptible Power Supplies (UPS) – Static UPS systems to provide 2 × (N +1) redundancy; critical IT loads – 600 kVA modules with 10 minute battery autonomy, 0.9 PFC, harmonic filtration and static by-pass; 500 kVA Static UPS with 10 minute battery autonomy for mechanical loads on N load basis	m²	795.00 to	925.00
Critical Power Distribution Units (PDUs) to data halls; dual 'A' and 'B' units to provide diverse supplies to each IT cabinet	m²	495.00 to	780.00
final critical IT power supplies. 2 Nr supplies to each cabinet from 'A' and 'B' PDUs	m²	420.00 to	570.00
lighting installations; data halls, office, support, ancillary and plant supplies	m²	136.00 to	170.00
5.9 Fuel Installations			
manual fill point and distribution pipework to generator belly tanks	m²	42.00 to	63.00
5.10 Lift and Conveyor Systems			
dock levellers and DDA access platform lifts only	m²	40.00 to	55.00
5.11 Fire and Lightning Protection			
gaseous fire suppression to technical, UPS and switchgear rooms	m²	310.00 to	350.00
lightning protection	m²	11.00 to	22.00
earthing and clean earth	m²	22.00 to	32.00
leak detection; to chilled water circuits within data hall service corridor	m²	55.00 to	75.00
5.12 Communication, Security and Control Systems			
fire alarm systems; DDI	m²	55.00 to	75.00
Very Early Smoke Detection Alarm (VESDA) to technical areas	m²	125.00 to	160.00
other alarm systems, e.g. disabled refuge	m²	10.00 to	20.00
Structured IT cabling; fibre optic backbone and copper Cat 6 to office and support spaces only. IT cabling to data halls EXCLUDED (Client fit-out item).	m²	120.00 to	175.00
Security installations; CCTV, Access Control and Intruder Detection	m²	525.00 to	730.00
BMS/EMS systems	m²	620.00 to	900.00
5.13 Specialist Installations			
Factory Acceptance Testing, Site Acceptance Testing, Integrated Systems Testing and 'Black Building' Testing @ 2% of MEP value	m²	180.00 to	225.00
Total Cost/m² (based on NTA of 2,000 m²)	m²	9261.00 to	11544.00
Total Cost/kW IT Load (based on 3,000 kW IT Load)	each	6174.00 to	7697.00
Notes: The above includes: MEP preliminaries and OH&P allowances			
The above excludes: Main contractor's preliminaries and OH&P; Utility connections; Electric, water, drainage and Fibre Connections; Contingency/Risk Allowances; Foreign Exchange fluctuations; VAT at prevailing rates			

BUILDING MODELS – ELEMENTAL COST SUMMARIES

Item	Unit	Range £	
GYM Gym located in Central London on the ground floor of a mixed use development. This model is based on a total of 2,000 m², with costs based on connection to central plant within plant rooms. The model includes for an AHU dedicated to providing ventilation to the gym. Services provided also include for 4 Pipe Fan Coil Units providing Heating and Cooling, additional underfloor heating to changing rooms. Excluded from the gym fit-out costs are FFE, active IT equipment and gym equipment.			
5 Services			
5.1 Sanitary Installations			
sanitary appliances including WCs, wash hand basins, cleaner's sinks, water fountains, urinals, showers and the provision of disabled toilets and accessible showers	m²	37.00 to	42.00
5.3 Disposal Installations			
soil, waste and vent installation to all sanitaryware points, provision of stub stacks, condensate drainage for fan coil units including insulation, rainwater disposal excluded as part of base build installations	m²	16.00 to	21.00
5.4 Water Installations			
installation of mains cold water services including meter, storage tanks, pumps, electromagnetic water conditioner and connections to sanitaryware, hot water including connection to base built plate heat exchanger for hot water generation, bulk storage, distribution, pump sets and connections to sanitaryware, miscellaneous water points to drinking fountains and plant supplies	m²	26.00 to	31.00
5.5 Heat Source			
connection to base build plate heat exchanger, energy meter, buffer vessels, associated pump sets and primary distribution pipework and insulation	m²	21.00 to	26.00
5.6 Space Heating and Air Conditioning			
connection to base build plate heat exchanger, energy meter, buffer vessels, associated pump sets and CHW distribution pipework to AHU and FCUs, secondary LTHW distribution to underfloor heating and FCUs, Intake and exhaust air ductwork to AHU. Air handling unit and associated supply and extract ductwork distribution	m²	240.00 to	285.00
5.7 Ventilation Systems			
Intake and exhaust air ductwork to fans, toilet and shower supply and extract system. Extract to stores and cleaner's cupboards. MVHR unit to office/staff rest room	m²	32.00 to	37.00
5.8 Electrical Installations			
LV distribution system including switchgear, containment and cabling, small power and lighting including emergency lighting and controls, floor boxes to gym equipment, enhanced lighting to lobbies, reception areas and toilet/ changing areas, earthing and bonding	m²	190.00 to	220.00
5.9 Fuel Installations			
5.11 Fire and Lightning Protection			
connection to base build installations, zone valve, concealed sprinkler heads throughout	m²	26.00 to	31.00

Approximate Estimating

BUILDING MODELS – ELEMENTAL COST SUMMARIES

Item	Unit	Range £
GYM – cont		
5 Services – cont		
5.12 Communication, Security and Control Systems		
fire alarm system, PA/VA installation, access control including card reader turnstiles at reception controlling entrance to the gym, security system comprising of security cameras, access control and intruder alarms to back of houses, TV and data installations to gym, Wi-Fi installation, complete sound system including speakers, cabling and central music generation and smart device docks, installation of central building management system including central control panels and BMS to plant and equipment	m²	210.00 to 240.00
5.13 Specialist Installations		
Total Cost/m² (based on GIFA of 2,000 m²)	m²	800.00 to 930.00

BUILDING MODELS – ELEMENTAL COST SUMMARIES

Item	Unit	Range £	

BAR

Bar located in Central London, 15th floor of a mixed use development. Model is based on a total GIA of 500 m², with costs based on complete stand-alone plant. Model includes a displacement ventilation installation, heating and cooling provided by a VRF installation via terminal units and an AHU. The cost model excludes the bar and back bar unit, lifts with connections only provided for electrical, comms, water and drainage, FFE and active IT equipment.

5 Services

Item	Unit	Range £	
5.1 Sanitary Installations			
sanitary appliances including WCs, wash hand basins, urinals and the provision of disabled toilets	m²	60.00 to	70.00
5.3 Disposal Installations			
soil, waste and vent installation to all sanitaryware points, provision of stub stacks, capped connections to bar areas, condensate drainage for fan coil units including insulation, rainwater gullies to terrace areas	m²	26.00 to	31.00
5.4 Water Installations			
installation of mains cold water services including meter, storage tanks, pumps, electromagnetic water conditioner and connections to sanitaryware, hot water generation, plant bulk storage, distribution, pump sets and connections to sanitaryware, capped connections to bar areas, miscellaneous water points and plant supplies	m²	140.00 to	160.00
5.6 Space Heating and Air Conditioning			
VRF installation serving the AHU and terminal units provide heating and cooling to bar areas, stand-alone cooling only VRF units to bar cellar and stores, specialist temperature control and humidity control units to wine cellar, Intake and exhaust air ductwork to AHU. Air handling unit and associated supply and extract ductwork distribution	m²	265.00 to	305.00
5.7 Ventilation Systems			
Supply and extract installations to WC areas and back of house areas. Extract to cleaner's cupboard, stores and the like	m²	60.00 to	70.00
5.8 Electrical Installations			
LV distribution system including switchgear, containment and cabling, small power and lighting including emergency lighting and controls, provision of enhanced lighting to lobbies, bar areas and toilet areas, Allowance for statement lighting, full lighting control and scene setting, capped electrical supply to bar area, earthing and bonding	m²	470.00 to	620.00
5.11 Fire and Lightning Protection			
connection to base build installations, zone valve, concealed sprinkler heads throughout	m²	32.00 to	37.00
5.12 Communication, Security and Control Systems			
fire alarm system, PA/VA installation, security system comprising of security cameras, access control and intruder alarms to back of houses, TV and data installations complete sound system including speakers, cabling and central music generation and smart device docks, Wi-Fi installation, installation of central building management system including central control panels and BMS to plant and equipment	m²	350.00 to	395.00
Total Cost/m² (based on GIFA of 500 m²)	m²	1410.00 to	1690.00

BUILDING MODELS – ELEMENTAL COST SUMMARIES

Item	Unit	Range £	

SPA

High end luxury spa located in Central London within the basement level of a mixed use development. This model is based on a total GIA of 1,050 m², with costs based on connection to central plant within plant rooms. The model includes for an AHU which is only providing ventilation to the spa area and swimming pool hall. Services provided also include for 4 pipe fan coil units providing heating and cooling, displacement ventilation to the pool hall, additional underfloor heating to changing rooms. Included within the cost plan are costs for swimming pool plant, steam rooms and a sauna. Excluded from the spa fit-out costs are FFE and active IT equipment. Locating elements such as spas within the basements of developments will attract a cost premium due to the difficulty associated with bringing services into these areas. A holistic approach to costing must be considered for spa areas, as basement digs to accommodate swimming pools and the associated structural and water proofing required.

5 Services

Item	Unit	Range £	
5.1 Sanitary Installations			
sanitary appliances including WCs, wash hand basins, cleaner's sinks, water fountains, urinals, showers, ice fountain and the provision of disabled toilets and accessible showers	m²	95.00 to	115.00
5.3 Disposal Installations			
soil, waste and vent installation to all sanitaryware points, provision of stub stacks, condensate drainage for fan coil units including insulation, back wash drainage to pool plant, rainwater gullies to terrace areas	m²	26.00 to	31.00
5.4 Water Installations			
installation of mains cold water services including meter, storage tanks, pumps, electromagnetic water conditioner and connections to sanitaryware, hot water including connection to base built plate heat exchanger for hot water generation, bulk storage, distribution, pump sets and connections to sanitaryware, softened water to experience shower, saunas and steam rooms, miscellaneous water points to drinking fountains and plant supplies	m²	60.00 to	70.00
5.5 Heat Source			
connection to base build plate heat exchanger, energy meter, buffer vessels, associated pump sets and primary distribution pipework and insulation	m²	32.00 to	37.00
5.6 Space Heating and Air Conditioning			
connection to base build plate heat exchanger for chilled water, energy meter, buffer vessels, associated pump sets and CHW distribution pipework to AHU and FCUs, VRF installation to comms/media rooms, secondary LTHW distribution to underfloor heating and FCUs, supply and extract installation to pool hall, Intake and exhaust air ductwork to AHU. Air handling unit and associated supply and extract ductwork distribution	m²	625.00 to	695.00
5.7 Ventilation Systems			
intake and exhaust air ductwork to fans, toilet and shower supply and extract system. Extract to stores and cleaner's cupboards. MVHR unit to office/staff rest room. Extra over for smoke extract	m²	125.00 to	150.00

BUILDING MODELS – ELEMENTAL COST SUMMARIES

Item	Unit	Range £
5.8 Electrical Installations		
LV distribution system including switchgear, containment and cabling, small power and lighting including emergency lighting and controls, floor boxes, enhanced lighting to lobbies, reception areas and toilet/changing areas, allowance for statement lighting, full lighting control, fibre optic lighting with pool and scene setting, earthing and bonding	m²	345.00 to 400.00
5.9 Fuel Installations		
5.11 Fire and Lightning Protection		
connection to base build installations, zone valve, concealed sprinkler heads throughout	m²	26.00 to 31.00
5.12 Communication, Security and Control Systems		
fire alarm system, PA/VA installation, access control including card reader turnstiles at reception controlling entrance to the spa, security system comprising of security cameras, access control and intruder alarms to back of houses, TV and data installations, Wi-Fi installation, complete sound system including speakers, cabling and central music generation and smart device docks, installation of central building management system including central control panels and BMS to plant and equipment	m²	345.00 to 390.00
5.13 Specialist Installations		
saunas and steam rooms, pool plant, including all pumps, water treatment, filters and controls	m²	530.00 to 575.00
Total Cost/m² (based on GIFA of 1,050 m²)	m²	2210.00 to 2495.00

Building Regulations Pocket Book

Ray Tricker and Samantha Alford

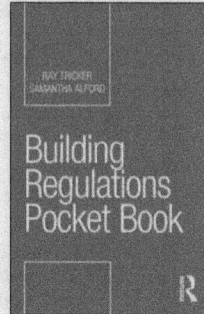

This handy guide provides you with all the information you need to comply with the UK Building Regulations and Approved Documents. On site, in the van, in the office, wherever you are, this is the book you'll refer to time and time again to double check the regulations on your current job.

The *Building Regulations Pocket Book* is the must have reliable and portable guide to compliance with the Building Regulations.

- Part 1 provides an overview of the Building Act
- Part 2 offers a handy guide to the dos and don'ts of gaining the Local Council's approval for Planning Permission and Building Regulations Approval
- Part 3 presents an overview of the requirements of the Approved Documents associated with the Building Regulations
- Part 4 is an easy to read explanation of the essential requirements of the Building Regulations that any architect, builder or DIYer needs to know to keep their work safe and compliant on both domestic or non-domestic jobs

This book is essential reading for all building contractors and sub-contractors, site engineers, building engineers, building control officers, building surveyors, architects, construction site managers and DIYers. Homeowners will also find it useful to understand what they are responsible for when they have work done on their home (ignorance of the regulations is no defence when it comes to compliance!).

February 2018: 186 x 123 mm: 456 pp
Pb: 978-0-8153-6838-0 : £21.99

To Order: Tel: +44 (0) 1235 400524 Fax: +44 (0) 1235 400525
or Post: Taylor and Francis Customer Services,
Bookpoint Ltd, Unit T1, 200 Milton Park, Abingdon, Oxon, OX14 4TA UK
Email: book.orders@tandf.co.uk

For a complete listing of all our titles visit:
www.tandf.co.uk

Taylor & Francis
Taylor & Francis Group

PART 3

Material Costs/Measured Work Prices – Mechanical Installations

New Aspects of Quantity Surveying Practice, 4th edition

Duncan Cartlidge

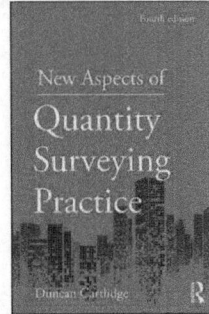

In this fourth edition of New Aspects of Quantity Surveying Practice, renowned quantity surveying author Duncan Cartlidge reviews the history of the quantity surveyor, examines and reflects on the state of current practice with a concentration on new and innovative practice, and attempts to predict the future direction of quantity surveying practice in the UK and worldwide.

The book champions the adaptability and flexibility of the quantity surveyor, whilst covering the hot topics which have emerged since the previous edition's publication, including:

- the RICS 'Futures' publication;
- Building Information Modelling (BIM);
- mergers and acquisitions;
- a more informed and critical evaluation of the NRM;
- greater discussion of ethics to reflect on the renewed industry interest;
- and a new chapter on Dispute Resolution.

As these issues create waves throughout the industry whilst it continues its global growth in emerging markets, such reflections on QS practice are now more important than ever. The book is essential reading for all Quantity Surveying students, teachers and professionals. It is particularly suited to undergraduate professional skills courses and non-cognate postgraduate students looking for an up to date understanding of the industry and the role.

December 2017: 234 x 156 mm: 282 pp
Pb: 978-1-138-67376-2: £50.99

To Order: Tel: +44 (0) 1235 400524 Fax: +44 (0) 1235 400525
or Post: Taylor and Francis Customer Services,
Bookpoint Ltd, Unit T1, 200 Milton Park, Abingdon, Oxon, OX14 4TA UK
Email: book.orders@tandf.co.uk

For a complete listing of all our titles visit:
www.tandf.co.uk

Taylor & Francis
Taylor & Francis Group

Material Costs/Measured Work Prices

DIRECTIONS

The following explanations are given for each of the column headings and letter codes.

Unit	Prices for each unit are given as singular (i.e. 1 metre, 1 nr) unless stated otherwise.
Net price	Industry tender prices, plus nominal allowance for fixings (unless measured separately), waste and applicable trade discounts.
Material cost	Net price plus percentage allowance for overheads (7%), profit (5%) and preliminaries (13%).
Labour norms	In man-hours for each operation.
Labour cost	Labour constant multiplied by the appropriate all-in man-hour cost based on gang rate (See also relevant Rates of Wages Section) plus percentage allowance for overheads, profit and preliminaries.
Measured work	Material cost plus Labour cost.
Price (total rate)	

MATERIAL COSTS

The Material Costs given are based at Second Quarter 2020 but exclude any charges in respect of VAT. The average rate of copper during this quarter is US $5179/UK £4,096 per tonne. Users of the book are advised to register on the Spon's website www.pricebooks.co.uk/updates to receive the free quarterly updates – alerts will then be provided by e-mail as changes arise.

MEASURED WORK PRICES

These prices are intended to apply to new work in the London area. The prices are for reasonable quantities of work and the user should make suitable adjustments if the quantities are especially small or especially large. Adjustments may also be required for locality (e.g. outside London – refer to cost indices in Approximate Estimating section for details of adjustment factors) and for the market conditions (e.g. volume of work secured or being tendered) at the time of use.

MECHANICAL INSTALLATIONS

The labour rate has been based on average gang rates per man hour effective from 7th October 2019. To this rate has been added 13% and 7% to cover preliminary items, site and head office overheads together with 5% for profit, resulting in an inclusive rate of £33.69 per man hour. The rate has been calculated on a working year of 2,016 hours; a detailed build-up of the rate is given at the end of these directions.

The rates and allowances will be subject to increases as covered by the National Agreements. Future changes will be published in the free Spon quarterly update by registering on their website.

DIRECTIONS

DUCTWORK INSTALLATIONS

The labour rate basis is as per Mechanical above and to this rate has been added 13% plus 7% to cover site and head office overheads only (factory overhead is included in the material rate) and preliminary items together with 5% for profit, resulting in an inclusive rate of £32.27 per man hour. The rate has been calculated on a working year of 2,016 hours; a detailed build-up of the rate is given at the end of these directions.

In calculating the 'Measured Work Prices' the following assumptions have been made:

(a)   That the work is carried out as a subcontract under the Standard Form of Building Contract.
(b)   That, unless otherwise stated, the work is being carried out in open areas at a height which would not require more than simple scaffolding.
(c)   That the building in which the work is being carried out is no more than six storeys high.

Where these assumptions are not valid, as for example where work is carried out in ducts and similar confined spaces or in multi-storey structures when additional time is needed to get to and from upper floors, then an appropriate adjustment must be made to the prices. Such adjustment will normally be to the labour element only.

Note The rates do not include for any uplift applied if the ductwork package is procured via the Mechanical Subcontractor

DIRECTIONS

LABOUR RATE – MECHANICAL & PUBLIC HEALTH

The annual cost of a notional twelve man gang

	FOREMAN	SENIOR CRAFTSMAN (+2 Welding skill)	SENIOR CRAFTSMAN	CRAFTSMAN	INSTALLER	MATE (Over 18)	SUB TOTALS
	1 NR	1 NR	2 NR	4 NR	2 NR	2 NR	
Hourly Rate from 7 October 2019	17.70	15.79	14.63	13.47	12.16	10.25	
Working hours per annum per man	1,680.00	1,680.00	1,680.00	1,680.00	1,680.00	1,680.00	
x Hourly rate × nr of men = £ per annum	29,736.00	25,746.40	49,156.80	90,518.40	40,857.60	34,440.00	271,236.00
Overtime Rate	24.94	22.29	20.63	18.98	17.15	14.44	
Overtime hours per annum per man	336.00	336.00	336.00	336.00	336.00	336.00	
x Hourly rate × nr of men = £ per annum	8,379.84	7,489.44	13,863.36	25,509.12	11,524.80	9,703.68	76,470.24
Total	38,115.84	34,016.64	63,020.16	116,027.52	52,382.40	44,143.68	347,706.24
Incentive schemes 0.00%	0.00	0.00	0.00	0.00	0.00	0.00	0.00
Daily Travel Time Allowance (15–20 miles each way)	10.51	10.51	10.51	10.51	10.51	10.51	
Days per annum per man	224.00	224.00	224.00	224.00	224.00	224.00	
x nr of men = £ per annum	2,354.24	2,354.24	4,708.48	9,416.96	4,708.48	4,708.48	28,250.88
Daily Travel Fare (15–20 miles each way)	15.60	15.60	15.60	15.60	15.60	15.60	
Days per annum per man	224.00	224.00	224.00	224.00	224.00	224.00	
x nr of men = £ per annum	3,494.40	3,494.40	6,988.80	13,977.60	6,988.80	6,988.80	41,932.80
Employer's Pension contributions at 7 October 2019:							
£ Contributions/annum	1,725.88	1,539.72	2,853.76	5,707.52	2,371.20	1,998.88	16,196.96
National Insurance Contributions:							
Gross pay – subject to NI	49,741.24	45,215.14	84,442.58	160,288.42	72,332.63	62,953.73	
% of NI Contributions	13.80	13.80	13.80	13.80	13.80	13.80	
£ Contributions/annum	5,379.10	4,754.50	8,682.70	16,179.05	7,011.52	5,717.24	47,724.11
Holiday Credit and Welfare contributions:							
Number of weeks	52	52	52	52	52	52	
Total weekly £ contribution each	92.08	85.90	77.77	85.90	66.28	57.39	
X nr of men = £ Contributions/annum	4,788.16	4,466.80	8,088.08	17,867.20	6,893.12	5,968.56	48,071.92
Holiday Top-up Funding including overtime	19.01	16.98	15.74	14.42	13.08	11.00	
Cost	988.60	883.06	1,637.06	2,999.14	1,359.83	1,144.21	9,011.90

SUB-TOTAL		538,894.81
TRAINING (INCLUDING ANY TRADE REGISTRATIONS) – SAY	1.00%	5,388.95
SEVERANCE PAY AND SUNDRY COSTS – SAY	1.50%	8,164.26
EMPLOYER'S LIABILITY AND THIRD PARTY INSURANCE – SAY	2.00%	11,048.96
ANNUAL COST OF NOTIONAL GANG		563,496.97
MEN ACTUALLY WORKING = 10.5 THEREFORE ANNUAL COST PER PRODUCTIVE MAN		53,666.38
AVERAGE NR OF HOURS WORKED PER MAN = 2016		
THEREFORE ALL-IN MAN HOURS		26.62
PRELIMINARY ITEMS – SAY	13%	3.46
SITE AND HEAD OFFICE OVERHEADS AND PROFIT (7% & 5% RESPECTIVELY) – SAY	12%	3.61
THEREFORE INCLUSIVE MAN-HOUR RATE		33.69

DIRECTIONS

Notes:

1)   The following assumptions have been made in the above calculations:
   a)   Increase in Hourly Rate from 7 October 2019, in accordance with the latest BESA Wage Agreement.
   b)   The working week of 37.5 hours i.e. the normal working week as defined by the National Agreement.
   c)   The actual hours worked are five days of 9 hours each.
   d)   A working year of 2,016 hours.
   e)   Five days in the year are lost through sickness or similar reason.
2)   National insurance contributions are those effective from 1st April 2020.
3)   Weekly Holiday Credit/Welfare Stamp values are those effective from 7 October 2019.
4)   Rates are based from 7 October 2019.
5)   Overtime rates are based on Premium Rate 1.
6)   Fares with Oyster Card (New Malden to Waterloo + Zone 1 – Anytime fare) current at June 2020 (TfL – 0343 222 1234).

DIRECTIONS

LABOUR RATE – DUCTWORK

The annual cost of notional eight man gang

	FOREMAN	SENIOR CRAFTSMAN	CRAFTSMAN	INSTALLER	SUB TOTALS
	1 NR	1 NR	4 NR	2 NR	
Hourly Rate from 7 October 2019	17.70	14.63	13.47	12.16	
Working hours per annum per man	1,680.00	1,680.00	1,680.00	1,680.00	
x Hourly rate × nr of men = £ per annum	29,736.00	24,578.40	90,518.40	40,857.60	185,690.40
Overtime Rate	24.94	20.63	18.98	17.15	
Overtime hours per annum per man	336.00	336.00	336.00	336.00	
x Hourly rate × nr of men = £ per annum	8,379.84	6,931.68	25,509.12	11,524.80	52,345.44
Total	38,115.84	31,510.08	116,027.52	52,382.40	238,035.84
Incentive schemes 0.00%	0.00	0.00	0.00	0.00	0.00
Daily Travel Time Allowance (15–20 miles each way)	10.51	10.51	10.51	10.51	
Days per annum per man	224.00	224.00	224.00	224.00	
x nr of men = £ per annum	2,354.24	2,354.24	9,416.96	4,708.48	18,833.92
Daily Travel Fare (15–20 miles each way)	15.60	15.60	15.60	15.60	
Days per annum per man	224.00	224.00	224.00	224.00	
x nr of men = £ per annum	3,494.40	3,494.40	13,977.60	6,988.80	27,955.20
Employer's Pension Contributions from 7th October 2019					
£ Contributions/annum	1,725.88	1,426.88	5,707.52	2,371.20	11,231.48
National Insurance Contributions:					
Gross pay – subject to NI	49,741.24	42,221.29	160,288.42	72,332.63	
% of NI Contributions	13.8	13.8	13.8	13.8	
£ Contributions/annum	5,379.10	4,341.35	16,179.05	7,011.52	32,911.02
Holiday Credit and Welfare contributions:					
Number of weeks	52	52	52	52	
Total weekly £ contribution each	92.08	77.77	85.90	66.28	
x nr of men = £ Contributions/annum	4,788.16	4,044.04	17,867.20	6,893.12	33,592.52
Holiday Top-up Funding including overtime	19.01	15.74	14.42	13.08	
Cost	988.60	818.53	2,999.14	1,359.83	6,166.10

	SUB-TOTAL		368,726.08
TRAINING (INCLUDING ANY TRADE REGISTRATIONS) – SAY		1.00%	3,687.26
SEVERANCE PAY AND SUNDRY COSTS – SAY		1.50%	5,586.20
EMPLOYER'S LIABILITY AND THIRD PARTY INSURANCE – SAY		2.00%	7,559.99
ANNUAL COST OF NOTIONAL GANG			385,559.53
MEN ACTUALLY WORKING = 7.5 THEREFORE ANNUAL COST PER PRODUCTIVE MAN			51,407.94
AVERAGE NR OF HOURS WORKED PER MAN = 2016			
THEREFORE ALL-IN MAN HOURS			25.50
PRELIMINARY ITEMS – SAY		13%	3.31
SITE AND HEAD OFFICE OVERHEADS AND PROFIT (7% & 5% RESPECTIVELY) – SAY		12%	3.46
THEREFORE INCLUSIVE MAN-HOUR RATE			32.27

DIRECTIONS

Notes:

1)   The following assumptions have been made in the above calculations:
    a)   Increase in Hourly Rate from 7 October 2019, in accordance with the latest BESA Wage Agreement.
    b)   The working week of 37.5 hours, i.e. the normal working week as defined by the National Agreement.
    c)   The actual hours worked are five days of 9 hours each.
    d)   A working year of 2,016 hours.
    e)   Five days in the year are lost through sickness or similar reason.
2)   National insurance contributions are those effective from 1st April 2020.
3)   Weekly Holiday Credit/Welfare Stamp values are those effective from 7 October 2019.
4)   Rates are based from 7 October 2019.
5)   Overtime rates are based on Premium Rate 1.
6)   Fares with Oyster Card (New Malden to Waterloo + Zone 1 – Anytime fare) current at June 2020 (TfL – 0343 222 1234).

33 DRAINAGE ABOVE GROUND

Item	Net Price £	Material £	Labour hours	Labour £	Unit	Total rate £
RAINWATER PIPEWORK/GUTTERS						
PVC-u gutters: push fit joints; fixed with brackets to backgrounds; BS 4576 BS EN 607						
Half round gutter, with brackets measured separately						
75 mm	4.20	5.32	0.69	23.25	m	**28.57**
100 mm	8.45	10.70	0.64	21.57	m	**32.27**
150 mm	9.74	12.33	0.82	27.63	m	**39.96**
Brackets: including fixing to backgrounds. For minimum fixing distances, refer to the Tables and Memoranda at the rear of the book						
75 mm; Fascia	1.65	2.08	0.15	5.05	nr	**7.13**
100 mm; Jointing	2.63	3.33	0.16	5.39	nr	**8.72**
100 mm; Support	1.06	1.34	0.16	5.39	nr	**6.73**
150 mm; Fascia	3.19	4.03	0.16	5.39	nr	**9.42**
Bracket supports: including fixing to backgrounds. For minimum fixing distances, refer to the Tables and Memoranda at the rear of the book						
Side rafter	4.18	5.29	0.16	5.39	nr	**10.68**
Top rafter	4.18	5.29	0.16	5.39	nr	**10.68**
Rise and fall	4.52	5.72	0.16	5.39	nr	**11.11**
Extra over fittings half round PVC-u gutter						
Union						
75 mm	2.44	3.09	0.19	6.41	nr	**9.50**
100 mm	2.63	3.33	0.24	8.09	nr	**11.42**
150 mm	9.82	12.43	0.28	9.43	nr	**21.86**
Rainwater pipe outlets						
Running: 75 × 53 mm dia.	5.64	7.13	0.12	4.03	nr	**11.16**
Running: 100 × 68 mm dia.	3.85	4.87	0.12	4.03	nr	**8.90**
Running: 150 × 110 mm dia.	18.01	22.79	0.12	4.03	nr	**26.82**
Stop end: 100 × 68 mm dia.	4.32	5.47	0.12	4.03	nr	**9.50**
Internal stop ends: short						
75 mm	2.44	4.48	0.09	3.04	nr	**7.52**
100 mm	1.25	2.98	0.09	3.04	nr	**6.02**
150 mm	6.50	8.22	0.09	3.04	nr	**11.26**
External stop ends: short						
75 mm	4.96	6.27	0.09	3.04	nr	**9.31**
100 mm	6.71	8.49	0.09	3.04	nr	**11.53**
150 mm	6.50	8.22	0.09	3.04	nr	**11.26**
Angles						
75 mm; 45°	6.66	8.43	0.20	6.73	nr	**15.16**
75 mm; 90°	6.66	8.43	0.20	6.73	nr	**15.16**
100 mm; 90°	4.60	5.82	0.20	6.73	nr	**12.55**
100 mm; 120°	5.08	6.43	0.20	6.73	nr	**13.16**
100 mm; 135°	5.08	6.43	0.20	6.73	nr	**13.16**
100 mm; Prefabricated to special angle	28.69	36.31	0.23	7.75	nr	**44.06**
100 mm; Prefabricated to raked angle	31.04	39.29	0.23	7.75	nr	**47.04**
150 mm; 90°	16.48	20.85	0.20	6.73	nr	**27.58**

33 DRAINAGE ABOVE GROUND

Item	Net Price £	Material £	Labour hours	Labour £	Unit	Total rate £
RAINWATER PIPEWORK/GUTTERS – cont						
Extra over fittings half round PVC-u gutter – cont						
Gutter adaptors						
100 mm; Stainless steel clip	2.36	2.99	0.16	5.39	nr	**8.38**
100 mm; Cast iron spigot	6.48	8.20	0.23	7.75	nr	**15.95**
100 mm; Cast iron socket	6.48	8.20	0.23	7.75	nr	**15.95**
100 mm; Cast iron 'ogee' spigot	6.58	8.33	0.23	7.75	nr	**16.08**
100 mm; Cast iron 'ogee' socket	6.58	8.33	0.23	7.75	nr	**16.08**
100 mm; Half round to Square PVC-u	11.76	14.88	0.23	7.75	nr	**22.63**
100 mm; Gutter overshoot guard	14.54	18.40	0.58	19.54	nr	**37.94**
Square gutter, with brackets measured separately						
120 mm square	3.98	5.04	0.82	27.63	m	**32.67**
Brackets: including fixing to backgrounds. For minimum fixing distances, refer to the Tables and Memoranda at the rear of the book						
Jointing	2.74	3.47	0.16	5.39	nr	**8.86**
Support	1.19	1.50	0.16	5.39	nr	**6.89**
Bracket support: including fixing to backgrounds. For minimum fixing distances, refer to the Tables and Memoranda at the rear of the book						
Side rafter	4.18	5.29	0.16	5.39	nr	**10.68**
Top rafter	4.18	5.29	0.16	5.39	nr	**10.68**
Rise and fall	4.52	5.72	0.16	5.39	nr	**11.11**
Extra over fittings square PVC-u gutter						
Rainwater pipe outlets						
Running: 62 mm square	3.86	4.88	0.12	4.03	nr	**8.91**
Stop end: 62 mm square	4.38	5.54	0.12	4.03	nr	**9.57**
Stop ends: short						
External	2.00	2.53	0.09	3.04	nr	**5.57**
Angles						
90°	4.97	6.29	0.20	6.73	nr	**13.02**
120°	12.44	15.75	0.20	6.73	nr	**22.48**
135°	5.82	7.37	0.20	6.73	nr	**14.10**
Prefabricated to special angle	29.43	37.25	0.23	7.75	nr	**45.00**
Prefabricated to raked angle	36.27	45.91	0.23	7.75	nr	**53.66**
Gutter adaptors						
Cast iron	18.87	23.88	0.23	7.75	nr	**31.63**
High capacity square gutter, with brackets measured separately						
137 mm	9.07	11.48	0.82	27.63	m	**39.11**

33 DRAINAGE ABOVE GROUND

Item	Net Price £	Material £	Labour hours	Labour £	Unit	Total rate £
Brackets: including fixing to backgrounds. For minimum fixing distances, refer to the Tables and Memoranda at the rear of the book						
Jointing	8.62	10.91	0.16	5.39	nr	**16.30**
Support	3.60	4.56	0.16	5.39	nr	**9.95**
Overslung	3.34	4.22	0.16	5.39	nr	**9.61**
Bracket supports: including fixing to backgrounds. For minimum fixing distances, refer to the Tables and Memoranda at the rear of the book						
Side rafter	5.13	6.50	0.16	5.39	nr	**11.89**
Top rafter	5.13	6.50	0.16	5.39	nr	**11.89**
Rise and fall	8.24	10.43	0.16	5.39	nr	**15.82**
Extra over fittings high capacity square PVC-u						
Rainwater pipe outlets						
Running: 75 mm square	14.13	17.89	0.12	4.03	nr	**21.92**
Running: 82 mm dia.	14.13	17.89	0.12	4.03	nr	**21.92**
Running: 110 mm dia.	12.36	15.65	0.12	4.03	nr	**19.68**
Screwed outlet adaptor						
75 mm square pipe	8.99	11.38	0.23	7.75	nr	**19.13**
Stop ends: short						
External	4.96	6.27	0.09	3.04	nr	**9.31**
Angles						
90°	12.36	15.65	0.20	6.73	nr	**22.38**
135°	14.55	18.41	0.20	6.73	nr	**25.14**
Prefabricated to special angle	42.01	53.17	0.23	7.75	nr	**60.92**
Prefabricated to raked internal angle	73.15	92.58	0.23	7.75	nr	**100.33**
Prefabricated to raked external angle	73.15	92.58	0.23	7.75	nr	**100.33**
Deep eliptical gutter, with brackets measured separately						
137 mm	4.91	6.22	0.82	27.63	m	**33.85**
Brackets: including fixing to backgrounds. For minimum fixing distances, refer to the Tables and Memoranda at the rear of the book						
Jointing	3.89	4.93	0.16	5.39	nr	**10.32**
Support	1.67	2.12	0.16	5.39	nr	**7.51**
Bracket support: including fixing to backgrounds. For minimum fixing distances, refer to the Tables and Memoranda at the rear of the book						
Side rafter	5.39	6.82	0.16	5.39	nr	**12.21**
Top rafter	5.39	6.82	0.16	5.39	nr	**12.21**
Rise and fall	8.65	10.94	0.16	5.39	nr	**16.33**
Extra over fittings deep eliptical PVC-u gutter						
Rainwater pipe outlets						
Running: 68 mm dia.	6.19	7.83	0.12	4.03	nr	**11.86**
Running: 82 mm dia.	5.87	7.43	0.12	4.03	nr	**11.46**
Stop end: 68 mm dia.	6.28	7.95	0.12	4.03	nr	**11.98**

33 DRAINAGE ABOVE GROUND

Item	Net Price £	Material £	Labour hours	Labour £	Unit	Total rate £
RAINWATER PIPEWORK/GUTTERS – cont						
Extra over fittings deep eliptical PVC-u gutter – cont						
Stop ends: short						
External	3.03	3.83	0.09	3.04	nr	**6.87**
Angles						
90°	6.24	7.90	0.20	6.73	nr	**14.63**
135°	7.41	9.37	0.20	6.73	nr	**16.10**
Prefabricated to special angle	21.69	27.45	0.23	7.75	nr	**35.20**
Gutter adaptors						
Stainless steel clip	5.33	6.74	0.16	5.39	nr	**12.13**
Marley deepflow	4.31	5.45	0.23	7.75	nr	**13.20**
Ogee profile PVC-u gutter, with brackets measured separately						
122 mm	5.17	6.54	0.82	27.63	m	**34.17**
Brackets: including fixing to backgrounds. For minimum fixing distances, refer to the Tables and Memoranda at the rear of the book						
Jointing	4.11	5.20	0.16	5.39	nr	**10.59**
Support	1.59	2.02	0.16	5.39	nr	**7.41**
Overslung	1.59	2.02	0.16	5.39	nr	**7.41**
Extra over fittings Ogee profile PVC-u gutter						
Rainwater pipe outlets						
Running: 68 mm dia.	5.87	7.43	0.12	4.03	nr	**11.46**
Stop ends: short						
Internal/External: left or right hand	2.98	3.77	0.09	3.04	nr	**6.81**
Angles						
90°: internal or external	5.94	7.52	0.20	6.73	nr	**14.25**
135°: internal or external	5.94	7.52	0.20	6.73	nr	**14.25**
PVC-u rainwater pipe: dry push fit joints; fixed with brackets to backgrounds; BS 4576/BS EN 607						
Pipe: circular, with brackets measured separately						
53 mm	6.68	8.46	0.61	20.55	m	**29.01**
68 mm	6.96	8.80	0.61	20.55	m	**29.35**
Pipe clip: including fixing to backgrounds. For minimum fixing distances, refer to the Tables and Memoranda at the rear of the book						
68 mm	1.56	1.97	0.16	5.39	nr	**7.36**
Pipe clip adjustable: including fixing to backgrounds. For minimum fixing distances, refer to the Tables and Memoranda at the rear of the book						
53 mm	2.38	3.01	0.16	5.39	nr	**8.40**
68 mm	3.33	4.21	0.16	5.39	nr	**9.60**

33 DRAINAGE ABOVE GROUND

Item	Net Price £	Material £	Labour hours	Labour £	Unit	Total rate £
Pipe clip drive in: including fixing to backgrounds. For minimum fixing distances, refer to the Tables and Memoranda at the rear of the book						
68 mm	3.74	4.74	0.16	5.39	nr	**10.13**
Extra over fittings circular pipework PVC-u						
Pipe coupler: PVC-u to PVC-u						
68 mm	1.94	2.45	0.12	4.03	nr	**6.48**
Pipe coupler: PVC-u to Cast Iron						
68 mm: to 3" cast iron	7.61	9.63	0.17	5.73	nr	**15.36**
68 mm: to 3¾" cast iron	26.17	33.12	0.17	5.73	nr	**38.85**
Access pipe: single socket						
68 mm	12.15	15.38	0.15	5.05	nr	**20.43**
Bend: short radius						
53 mm: 67.5°	2.83	3.58	0.20	6.73	nr	**10.31**
68 mm: 92.5°	4.16	5.26	0.20	6.73	nr	**11.99**
68 mm: 112.5°	4.16	5.26	0.20	6.73	nr	**11.99**
Bend: long radius						
68 mm: 112°	3.94	4.98	0.20	6.73	nr	**11.71**
Branch						
68 mm: 92°	25.56	32.35	0.23	7.75	nr	**40.10**
68 mm: 112°	25.64	32.45	0.23	7.75	nr	**40.20**
Double branch						
68 mm: 112°	50.59	64.03	0.24	8.09	nr	**72.12**
Shoe						
53 mm	4.19	5.30	0.12	4.03	nr	**9.33**
68 mm	6.19	7.83	0.12	4.03	nr	**11.86**
Rainwater head: including fixing to backgrounds						
68 mm	12.15	15.38	0.29	9.77	nr	**25.15**
Pipe: square, with brackets measured separately						
62 mm	4.26	5.39	0.45	15.16	m	**20.55**
75 mm	8.10	10.25	0.45	15.16	m	**25.41**
Pipe clip: including fixing to backgrounds. For minimum fixing distances, refer to the Tables and Memoranda at the rear of the book						
62 mm	1.42	1.79	0.16	5.39	nr	**7.18**
75 mm	3.16	4.00	0.16	5.39	nr	**9.39**
Pipe clip adjustable: including fixing to backgrounds. For minimum fixing distances, refer to the Tables and Memoranda at the rear of the book						
62 mm	4.41	5.58	0.16	5.39	nr	**10.97**

33 DRAINAGE ABOVE GROUND

Item	Net Price £	Material £	Labour hours	Labour £	Unit	Total rate £
RAINWATER PIPEWORK/GUTTERS – cont						
Extra over fittings square pipework PVC-u						
Pipe coupler: PVC-u to PVC-u						
62 mm	2.09	2.64	0.20	6.73	nr	9.37
75 mm	3.24	4.10	0.20	6.73	nr	10.83
Square to circular adaptor: single socket						
62 mm to 68 mm	4.07	5.15	0.20	6.73	nr	11.88
Square to circular adaptor: single socket						
75 mm to 62 mm	5.26	6.65	0.20	6.73	nr	13.38
Access pipe						
62 mm	24.33	30.79	0.16	5.39	nr	36.18
75 mm	25.68	32.50	0.16	5.39	nr	37.89
Bends						
62 mm: 92.5°	3.76	4.76	0.20	6.73	nr	11.49
62 mm: 112.5°	2.74	3.47	0.20	6.73	nr	10.20
75 mm: 112.5°	6.86	8.68	0.20	6.73	nr	15.41
Bends: prefabricated special angle						
62 mm	24.45	30.95	0.23	7.75	nr	38.70
75 mm	33.97	43.00	0.23	7.75	nr	50.75
Offset						
62 mm	5.89	7.46	0.20	6.73	nr	14.19
75 mm	16.22	20.53	0.20	6.73	nr	27.26
Offset: prefabricated special angle						
62 mm	24.45	30.95	0.23	7.75	nr	38.70
Shoe						
62 mm	3.46	4.38	0.12	4.03	nr	8.41
75 mm	5.92	7.49	0.12	4.03	nr	11.52
Branch						
62 mm	8.55	10.82	0.23	7.75	nr	18.57
75 mm	41.68	52.75	0.23	7.75	nr	60.50
Double branch						
62 mm	45.86	58.04	0.24	8.09	nr	66.13
Rainwater head						
62 mm	38.46	48.68	0.29	9.77	nr	58.45
75 mm	40.46	51.21	3.45	116.17	nr	167.38
PVC-u rainwater pipe: solvent welded joints; fixed with brackets to backgrounds; BS 4576/BS EN 607						
Pipe: circular, with brackets measured separately						
82 mm	13.12	16.61	0.35	11.79	m	28.40
Pipe clip: galvanized; including fixing to backgrounds. For minimum fixing distances, refer to the Tables and Memoranda at the rear of the book						
82 mm	5.34	6.75	0.58	19.54	nr	26.29

33 DRAINAGE ABOVE GROUND

Item	Net Price £	Material £	Labour hours	Labour £	Unit	Total rate £
Pipe clip: galvanized plastic coated; including fixing to backgrounds. For minimum fixing distances, refer to the Tables and Memoranda at the rear of the book						
82 mm	7.39	9.35	0.58	19.54	nr	28.89
Pipe clip: PVC-u including fixing to backgrounds. For minimum fixing distances, refer to the Tables and Memoranda at the rear of the book						
82 mm	3.97	5.03	0.58	19.54	nr	24.57
Pipe clip: PVC-u adjustable: including fixing to backgrounds. For minimum fixing distances, refer to the Tables and Memoranda at the rear of the book						
82 mm	7.26	9.18	0.58	19.54	nr	28.72
Extra over fittings circular pipework PVC-u						
Pipe coupler: PVC-u to PVC-u						
82 mm	6.65	8.41	0.21	7.08	nr	15.49
Access pipe						
82 mm	40.96	51.83	0.23	7.75	nr	59.58
Bend						
82 mm: 92, 112.5 and 135°	16.72	21.16	0.29	9.77	nr	30.93
Shoe						
82 mm	10.32	13.06	0.29	9.77	nr	22.83
110 mm	12.95	16.39	0.32	10.79	nr	27.18
Branch						
82 mm: 92, 112.5 and 135°	24.92	31.54	0.35	11.79	nr	43.33
Rainwater head						
82 mm	21.87	27.68	0.58	19.54	nr	47.22
110 mm	19.91	25.20	0.58	19.54	nr	44.74
Roof outlets: 178 dia.; Flat						
50 mm	21.86	27.66	1.15	38.74	nr	66.40
82 mm	21.86	27.66	1.15	38.74	nr	66.40
Roof outlets: 178 mm dia.; Domed						
50 mm	21.86	27.66	1.15	38.74	nr	66.40
82 mm	21.86	27.66	1.15	38.74	nr	66.40
Roof outlets: 406 mm dia.; Flat						
82 mm	42.69	54.03	1.15	38.74	nr	92.77
110 mm	42.69	54.03	1.15	38.74	nr	92.77
Roof outlets: 406 mm dia.; Domed						
82 mm	42.69	54.03	1.15	38.74	nr	92.77
110 mm	42.69	54.03	1.15	38.74	nr	92.77
Roof outlets: 406 mm dia.; Inverted						
82 mm	94.01	118.98	1.15	38.74	nr	157.72
110 mm	94.01	118.98	1.15	38.74	nr	157.72
Roof outlets: 406 mm dia.; Vent Pipe						
82 mm	61.75	78.15	1.15	38.74	nr	116.89
110 mm	61.75	78.15	1.15	38.74	nr	116.89

33 DRAINAGE ABOVE GROUND

Item	Net Price £	Material £	Labour hours	Labour £	Unit	Total rate £
RAINWATER PIPEWORK/GUTTERS – cont						
Extra over fittings circular pipework PVC-u – cont						
Balcony outlets: screed						
82 mm	36.54	46.24	1.15	38.74	nr	**84.98**
Balcony outlets: asphalt						
82 mm	36.35	46.01	1.15	38.74	nr	**84.75**
Adaptors						
82 mm × 62 mm square pipe	4.89	6.19	0.21	7.08	nr	**13.27**
82 mm × 68 mm circular pipe	4.89	6.19	0.21	7.08	nr	**13.27**
For 110 mm dia. pipework and fittings refer to 33 Drainage Above Ground						
Cast iron gutters: mastic and bolted joints; BS 460; fixed with brackets to backgrounds						
Half round gutter, with brackets measured separately						
100 mm	16.87	21.35	0.85	28.64	m	**49.99**
115 mm	17.59	22.27	0.97	32.68	m	**54.95**
125 mm	20.58	26.05	0.97	32.68	m	**58.73**
150 mm	35.18	44.52	1.12	37.73	m	**82.25**
Brackets; fixed to backgrounds. For minimum fixing distances, refer to the Tables and Memoranda at the rear of the book						
Fascia						
100 mm	3.96	5.01	0.16	5.39	nr	**10.40**
115 mm	3.96	5.01	0.16	5.39	nr	**10.40**
125 mm	3.96	5.01	0.16	5.39	nr	**10.40**
150 mm	5.01	6.34	0.16	5.39	nr	**11.73**
Rise and fall						
100 mm	7.86	9.95	0.39	13.14	nr	**23.09**
115 mm	7.86	9.95	0.39	13.14	nr	**23.09**
125 mm	8.08	10.23	0.39	13.14	nr	**23.37**
150 mm	8.22	10.40	0.39	13.14	nr	**23.54**
Top rafter						
100 mm	4.84	6.13	0.16	5.39	nr	**11.52**
115 mm	4.84	6.13	0.16	5.39	nr	**11.52**
125 mm	6.55	8.29	0.16	5.39	nr	**13.68**
150 mm	8.90	11.27	0.16	5.39	nr	**16.66**
Side rafter						
100 mm	4.84	6.13	0.16	5.39	nr	**11.52**
115 mm	4.84	6.13	0.16	5.39	nr	**11.52**
125 mm	6.55	8.29	0.16	5.39	nr	**13.68**
150 mm	8.90	11.27	0.16	5.39	nr	**16.66**

33 DRAINAGE ABOVE GROUND

Item	Net Price £	Material £	Labour hours	Labour £	Unit	Total rate £
Extra over fittings half round gutter cast iron BS 460						
Union						
100 mm	9.28	11.75	0.39	13.14	nr	**24.89**
115 mm	11.58	14.66	0.48	16.17	nr	**30.83**
125 mm	13.08	16.55	0.48	16.17	nr	**32.72**
150 mm	14.68	18.58	0.55	18.52	nr	**37.10**
Stop end; internal						
100 mm	4.74	6.00	0.12	4.03	nr	**10.03**
115 mm	6.13	7.76	0.15	5.05	nr	**12.81**
125 mm	6.13	7.76	0.15	5.05	nr	**12.81**
150 mm	8.51	10.77	0.20	6.73	nr	**17.50**
Stop end; external						
100 mm	4.74	6.00	0.12	4.03	nr	**10.03**
115 mm	5.99	7.58	0.15	5.05	nr	**12.63**
125 mm	6.13	7.76	0.15	5.05	nr	**12.81**
150 mm	8.51	10.77	0.20	6.73	nr	**17.50**
90° angle; single socket						
100 mm	14.07	17.81	0.39	13.14	nr	**30.95**
115 mm	14.47	18.31	0.43	14.49	nr	**32.80**
125 mm	17.07	21.60	0.43	14.49	nr	**36.09**
150 mm	31.19	39.47	0.50	16.84	nr	**56.31**
90° angle; double socket						
100 mm	17.06	21.59	0.39	13.14	nr	**34.73**
115 mm	18.10	22.90	0.43	14.49	nr	**37.39**
125 mm	23.41	29.62	0.43	14.49	nr	**44.11**
135° angle; single socket						
100 mm	14.37	18.19	0.39	13.14	nr	**31.33**
115 mm	14.50	18.36	0.43	14.49	nr	**32.85**
125 mm	21.43	27.13	0.43	14.49	nr	**41.62**
150 mm	31.80	40.24	0.50	16.84	nr	**57.08**
Running outlet						
65 mm outlet						
100 mm	13.73	17.37	0.39	13.14	nr	**30.51**
115 mm	14.94	18.91	0.43	14.49	nr	**33.40**
125 mm	17.07	21.60	0.43	14.49	nr	**36.09**
75 mm outlet						
100 mm	13.73	17.37	0.39	13.14	nr	**30.51**
115 mm	14.94	18.91	0.43	14.49	nr	**33.40**
125 mm	17.07	21.60	0.43	14.49	nr	**36.09**
150 mm	29.56	37.41	0.50	16.84	nr	**54.25**
100 mm outlet						
150 mm	29.56	37.41	0.50	16.84	nr	**54.25**
Stop end outlet; socket						
65 mm outlet						
100 mm	16.09	20.36	0.39	13.14	nr	**33.50**
115 mm	18.04	22.84	0.43	14.49	nr	**37.33**
75 mm outlet						
125 mm	16.09	20.36	0.43	14.49	nr	**34.85**
150 mm	33.83	80.23	0.50	16.84	nr	**97.07**

33 DRAINAGE ABOVE GROUND

Item	Net Price £	Material £	Labour hours	Labour £	Unit	Total rate £
RAINWATER PIPEWORK/GUTTERS – cont						
Extra over fittings half round gutter cast iron BS 460 – cont						
Stop end outlet – cont						
100 mm outlet						
150 mm	29.56	37.41	0.50	16.84	nr	**54.25**
Stop end outlet; spigot						
65 mm outlet						
100 mm	16.09	20.36	0.39	13.14	nr	**33.50**
115 mm	18.04	22.84	0.43	14.49	nr	**37.33**
75 mm outlet						
125 mm	16.09	20.36	0.43	14.49	nr	**34.85**
150 mm	33.83	42.82	0.50	16.84	nr	**59.66**
100 mm outlet						
150 mm	33.83	42.82	0.50	16.84	nr	**59.66**
Half round; 3 mm thick double beaded gutter, with brackets measured separately						
100 mm	17.18	21.74	0.85	28.64	m	**50.38**
115 mm	17.93	22.69	0.85	28.64	m	**51.33**
125 mm	20.57	26.03	0.97	32.68	m	**58.71**
Brackets; fixed to backgrounds. For minimum fixing distances, refer to the Tables and Memoranda at the rear of the book						
Fascia						
100 mm	3.96	5.01	0.16	5.39	nr	**10.40**
115 mm	3.96	5.01	0.16	5.39	nr	**10.40**
125 mm	3.96	5.01	0.16	5.39	nr	**10.40**
Extra over fittings half round 3 mm thick gutter BS 460						
Union						
100 mm	9.28	11.75	0.38	12.81	nr	**24.56**
115 mm	11.31	14.31	0.38	12.81	nr	**27.12**
125 mm	13.08	16.55	0.43	14.49	nr	**31.04**
Stop end; internal						
100 mm	4.74	6.00	0.12	4.03	nr	**10.03**
115 mm	6.13	7.76	0.12	4.03	nr	**11.79**
125 mm	6.15	7.78	0.15	5.05	nr	**12.83**
Stop end; external						
100 mm	4.74	6.00	0.12	4.03	nr	**10.03**
115 mm	6.13	7.76	0.12	4.03	nr	**11.79**
125 mm	6.15	7.78	0.15	5.05	nr	**12.83**
90° angle; single socket						
100 mm	14.63	18.51	0.38	12.81	nr	**31.32**
115 mm	14.81	18.75	0.38	12.81	nr	**31.56**
125 mm	17.07	21.60	0.43	14.49	nr	**36.09**

33 DRAINAGE ABOVE GROUND

Item	Net Price £	Material £	Labour hours	Labour £	Unit	Total rate £
135° angle; single socket						
100 mm	14.37	18.19	0.38	12.81	nr	**31.00**
115 mm	14.47	18.31	0.38	12.81	nr	**31.12**
125 mm	18.05	22.85	0.43	14.49	nr	**37.34**
Running outlet						
65 mm outlet						
100 mm	14.74	18.66	0.38	12.81	nr	**31.47**
115 mm	15.23	19.28	0.38	12.81	nr	**32.09**
125 mm	20.17	25.52	0.43	14.49	nr	**40.01**
75 mm outlet						
115 mm	15.23	19.28	0.38	12.81	nr	**32.09**
125 mm	17.67	22.37	0.43	14.49	nr	**36.86**
Stop end outlet; socket						
65 mm outlet						
100 mm	16.09	20.36	0.38	12.81	nr	**33.17**
115 mm	18.04	22.84	0.38	12.81	nr	**35.65**
125 mm	20.17	25.52	0.43	14.49	nr	**40.01**
75 mm outlet						
125 mm	20.54	26.00	0.43	14.49	nr	**40.49**
Stop end outlet; spigot						
65 mm outlet						
100 mm	16.09	20.36	0.38	12.81	nr	**33.17**
115 mm	18.04	22.84	0.38	12.81	nr	**35.65**
125 mm	20.17	25.52	0.43	14.49	nr	**40.01**
Deep half round gutter, with brackets measured separately						
100 × 75 mm	28.26	35.76	0.85	28.64	m	**64.40**
125 × 75 mm	36.51	46.21	0.97	32.68	m	**78.89**
Brackets; fixed to backgrounds. For minimum fixing distances, refer to the Tables and Memoranda at the rear of the book						
Fascia						
100 × 75 mm	13.35	16.90	0.16	5.39	nr	**22.29**
125 × 75 mm	16.44	20.81	0.16	5.39	nr	**26.20**
Extra over fittings deep half round gutter BS 460						
Union						
100 × 75 mm	15.52	19.64	0.38	12.81	nr	**32.45**
125 × 75 mm	16.44	20.81	0.43	14.49	nr	**35.30**
Stop end; internal						
100 × 75 mm	13.59	17.20	0.12	4.03	nr	**21.23**
125 × 75 mm	16.76	21.21	0.15	5.05	nr	**26.26**
Stop end; external						
100 × 75 mm	13.59	17.20	0.12	4.03	nr	**21.23**
125 × 75 mm	16.76	21.21	0.15	5.05	nr	**26.26**
90° angle; single socket						
100 × 75 mm	38.73	49.01	0.38	12.81	nr	**61.82**
125 × 75 mm	49.18	62.24	0.43	14.49	nr	**76.73**

33 DRAINAGE ABOVE GROUND

Item	Net Price £	Material £	Labour hours	Labour £	Unit	Total rate £
RAINWATER PIPEWORK/GUTTERS – cont						
Extra over fittings deep half round gutter BS 460 – cont						
135° angle; single socket						
100 × 75 mm	38.73	49.01	0.38	12.81	nr	**61.82**
125 × 75 mm	49.18	62.24	0.43	14.49	nr	**76.73**
Running outlet						
65 mm outlet						
100 × 75 mm	38.73	49.01	0.38	12.81	nr	**61.82**
125 × 75 mm	49.18	62.24	0.43	14.49	nr	**76.73**
75 mm outlet						
100 × 75 mm	38.73	49.01	0.38	12.81	nr	**61.82**
125 × 75 mm	35.69	45.17	0.43	14.49	nr	**59.66**
Stop end outlet; socket						
65 mm outlet						
100 × 75 mm	52.32	66.21	0.38	12.81	nr	**79.02**
75 mm outlet						
100 × 75 mm	52.32	66.21	0.38	12.81	nr	**79.02**
125 × 75 mm	52.32	66.21	0.43	14.49	nr	**80.70**
Stop end outlet; spigot						
65 mm outlet						
100 × 75 mm	52.32	66.21	0.38	12.81	nr	**79.02**
75 mm outlet						
100 × 75 mm	52.32	66.21	0.38	12.81	nr	**79.02**
125 × 75 mm	52.32	66.21	0.43	14.49	nr	**80.70**
Ogee gutter, with brackets measured separately						
100 mm	18.82	23.82	0.85	28.64	m	**52.46**
115 mm	20.69	26.19	0.97	32.68	m	**58.87**
125 mm	21.71	27.47	0.97	32.68	m	**60.15**
Brackets; fixed to backgrounds. For minimum fixing distances, refer to the Tables and Memoranda at the rear of the book						
Fascia						
100 mm	4.31	5.45	0.16	5.39	nr	**10.84**
115 mm	4.31	5.45	0.16	5.39	nr	**10.84**
125 mm	4.85	6.14	0.16	5.39	nr	**11.53**
Extra over fittings Ogee cast iron gutter BS 460						
Union						
100 mm	9.29	11.76	0.38	12.81	nr	**24.57**
115 mm	11.31	14.31	0.43	14.49	nr	**28.80**
125 mm	13.08	16.55	0.43	14.49	nr	**31.04**
Stop end; internal						
100 mm	4.84	6.13	0.12	4.03	nr	**10.16**
115 mm	6.25	7.91	0.15	5.05	nr	**12.96**
125 mm	6.25	7.91	0.15	5.05	nr	**12.96**

33 DRAINAGE ABOVE GROUND

Item	Net Price £	Material £	Labour hours	Labour £	Unit	Total rate £
Stop end; external						
100 mm	4.84	6.13	0.12	4.03	nr	**10.16**
115 mm	6.25	7.91	0.15	5.05	nr	**12.96**
125 mm	6.25	7.91	0.15	5.05	nr	**12.96**
90° angle; internal						
100 mm	14.68	18.58	0.38	12.81	nr	**31.39**
115 mm	15.91	20.14	0.43	14.49	nr	**34.63**
125 mm	17.37	21.99	0.43	14.49	nr	**36.48**
90° angle; external						
100 mm	14.68	18.58	0.38	12.81	nr	**31.39**
115 mm	15.91	20.14	0.43	14.49	nr	**34.63**
125 mm	17.37	21.99	0.43	14.49	nr	**36.48**
135° angle; internal						
100 mm	15.24	19.29	0.38	12.81	nr	**32.10**
115 mm	16.24	20.55	0.43	14.49	nr	**35.04**
125 mm	21.39	27.07	0.43	14.49	nr	**41.56**
135° angle; external						
100 mm	15.24	19.29	0.38	12.81	nr	**32.10**
115 mm	16.24	20.55	0.43	14.49	nr	**35.04**
125 mm	21.39	27.07	0.43	14.49	nr	**41.56**
Running outlet						
65 mm outlet						
100 mm	14.95	18.92	0.38	12.81	nr	**31.73**
115 mm	15.92	20.15	0.43	14.49	nr	**34.64**
125 mm	17.37	21.99	0.43	14.49	nr	**36.48**
75 mm outlet						
125 mm	17.37	21.99	0.43	14.49	nr	**36.48**
Stop end outlet; socket						
65 mm outlet						
100 mm	23.61	29.88	0.38	12.81	nr	**42.69**
115 mm	23.61	29.88	0.43	14.49	nr	**44.37**
125 mm	23.61	29.88	0.43	14.49	nr	**44.37**
75 mm outlet						
125 mm	23.61	29.88	0.43	14.49	nr	**44.37**
Stop end outlet; spigot						
65 mm outlet						
100 mm	23.61	29.88	0.38	12.81	nr	**42.69**
115 mm	23.61	29.88	0.43	14.49	nr	**44.37**
125 mm	23.61	29.88	0.43	14.49	nr	**44.37**
75 mm outlet						
125 mm	23.61	29.88	0.43	14.49	nr	**44.37**
Notts Ogee Gutter, with brackets measured separately						
115 mm	33.43	42.31	0.85	28.64	m	**70.95**
Brackets; fixed to backgrounds. For minimum fixing distances, refer to the Tables and Memoranda at the rear of the book						
Fascia						
115 mm	13.04	16.51	0.16	5.39	nr	**21.90**

33 DRAINAGE ABOVE GROUND

Item	Net Price £	Material £	Labour hours	Labour £	Unit	Total rate £
RAINWATER PIPEWORK/GUTTERS – cont						
Extra over fittings Notts Ogee Cast Iron Gutter BS 460						
Union						
115 mm	15.83	20.04	0.38	12.81	nr	**32.85**
Stop end; internal						
115 mm	13.04	16.51	0.16	5.39	nr	**21.90**
Stop end; external						
115 mm	13.04	16.51	0.16	5.39	nr	**21.90**
90° angle; internal						
115 mm	37.82	47.87	0.43	14.49	nr	**62.36**
90° angle; external						
115 mm	37.82	47.87	0.43	14.49	nr	**62.36**
135° angle; internal						
115 mm	38.54	48.78	0.43	14.49	nr	**63.27**
135° angle; external						
115 mm	38.54	48.78	0.43	14.49	nr	**63.27**
Running outlet						
65 mm outlet						
115 mm	45.34	57.38	0.43	14.49	nr	**71.87**
75 mm outlet						
115 mm	45.34	57.38	0.43	14.49	nr	**71.87**
Stop end outlet; socket						
65 mm outlet						
115 mm	58.37	73.88	0.43	14.49	nr	**88.37**
Stop end outlet; spigot						
65 mm outlet						
115 mm	58.37	73.88	0.43	14.49	nr	**88.37**
No 46 moulded Gutter, with brackets measured separately						
100 × 75 mm	32.33	40.91	0.85	28.64	m	**69.55**
125 × 100 mm	47.40	59.99	0.97	32.68	m	**92.67**
Brackets; fixed to backgrounds. For minimum fixing distances, refer to the Tables and Memoranda at the rear of the book						
Fascia						
100 × 75 mm	7.21	9.13	0.16	5.39	nr	**14.52**
125 × 100 mm	7.21	9.13	0.16	5.39	nr	**14.52**
Extra over fittings						
Union						
100 × 75 mm	15.23	19.28	0.38	12.81	nr	**32.09**
125 × 100 mm	17.65	22.33	0.43	14.49	nr	**36.82**
Stop end; internal						
100 × 75 mm	13.64	17.26	0.12	4.03	nr	**21.29**
125 × 100 mm	17.65	22.33	0.15	5.05	nr	**27.38**

33 DRAINAGE ABOVE GROUND

Item	Net Price £	Material £	Labour hours	Labour £	Unit	Total rate £
Stop end; external						
100 × 75 mm	13.64	17.26	0.12	4.03	nr	**21.29**
125 × 100 mm	17.65	22.33	0.15	5.05	nr	**27.38**
90° angle; internal						
100 × 75 mm	35.71	45.19	0.38	12.81	nr	**58.00**
125 × 100 mm	51.34	64.97	0.43	14.49	nr	**79.46**
90° angle; external						
100 × 75 mm	35.71	45.19	0.38	12.81	nr	**58.00**
125 × 100 mm	51.34	64.97	0.43	14.49	nr	**79.46**
135° angle; internal						
100 × 75 mm	36.40	46.07	0.38	12.81	nr	**58.88**
125 × 100 mm	51.34	64.97	0.43	14.49	nr	**79.46**
135° angle; external						
100 × 75 mm	36.40	46.07	0.38	12.81	nr	**58.88**
125 × 100 mm	51.34	64.97	0.43	14.49	nr	**79.46**
Running outlet						
65 mm outlet						
100 × 75 mm	36.40	46.07	0.38	12.81	nr	**58.88**
125 × 100 mm	51.34	64.97	0.43	14.49	nr	**79.46**
75 mm outlet						
100 × 75 mm	36.40	46.07	0.38	12.81	nr	**58.88**
125 × 100 mm	51.34	64.97	0.43	14.49	nr	**79.46**
100 mm outlet						
100 × 75 mm	36.40	46.07	0.38	12.81	nr	**58.88**
125 × 100 mm	51.34	64.97	0.43	14.49	nr	**79.46**
100 × 75 mm outlet						
125 × 100 mm	36.40	46.07	0.43	14.49	nr	**60.56**
Stop end outlet; socket						
65 mm outlet						
100 × 75 mm	68.99	87.32	0.38	12.81	nr	**100.13**
75 mm outlet						
125 × 100 mm	68.99	87.32	0.43	14.49	nr	**101.81**
Stop end outlet; spigot						
65 mm outlet						
100 × 75 mm	68.99	87.32	0.38	12.81	nr	**100.13**
75 mm outlet						
125 × 100 mm	68.99	87.32	0.43	14.49	nr	**101.81**
Box gutter, with brackets measured separately						
100 × 75 mm	52.43	66.36	0.85	28.64	m	**95.00**
Brackets; fixed to backgrounds. For minimum fixing distances, refer to the Tables and Memoranda at the rear of the book						
Fascia						
100 × 75 mm	8.31	10.52	0.16	5.39	nr	**15.91**

33 DRAINAGE ABOVE GROUND

Item	Net Price £	Material £	Labour hours	Labour £	Unit	Total rate £
RAINWATER PIPEWORK/GUTTERS – cont						
Extra over fittings box cast iron gutter BS 460						
Union						
100 × 75 mm	11.02	13.94	0.38	12.81	nr	**26.75**
Stop end; external						
100 × 75 mm	8.31	10.52	0.12	4.03	nr	**14.55**
90° angle						
100 × 75 mm	41.90	53.03	0.38	12.81	nr	**65.84**
135° angle						
100 × 75 mm	41.90	53.03	0.38	12.81	nr	**65.84**
Running outlet						
65 mm outlet						
100 × 75 mm	41.90	53.03	0.38	12.81	nr	**65.84**
75 mm outlet						
100 × 75 mm	41.90	53.03	0.38	12.81	nr	**65.84**
100 × 75 mm outlet						
100 × 75 mm	41.90	53.03	0.38	12.81	nr	**65.84**
Cast iron rainwater pipe; dry joints; BS 460; fixed to backgrounds: Circular						
Plain socket pipe, with brackets measured separately						
65 mm	30.34	38.39	0.69	23.25	m	**61.64**
75 mm	30.34	38.39	0.69	23.25	m	**61.64**
100 mm	41.41	52.40	0.69	23.25	m	**75.65**
Bracket; fixed to backgrounds. For minimum fixing distances, refer to the Tables and Memoranda at the rear of the book						
65 mm	11.08	14.02	0.29	9.77	nr	**23.79**
75 mm	11.13	14.09	0.29	9.77	nr	**23.86**
100 mm	11.27	14.27	0.29	9.77	nr	**24.04**
Eared socket pipe, with wall spacers measured separately						
65 mm	32.42	41.03	0.62	20.89	m	**61.92**
75 mm	32.42	41.03	0.62	20.89	m	**61.92**
100 mm	43.51	55.07	0.62	20.89	m	**75.96**
Wall spacer plate; eared pipework						
65 mm	7.16	9.06	0.16	5.39	nr	**14.45**
75 mm	7.28	9.22	0.16	5.39	nr	**14.61**
100 mm	11.27	14.27	0.16	5.39	nr	**19.66**

33 DRAINAGE ABOVE GROUND

Item	Net Price £	Material £	Labour hours	Labour £	Unit	Total rate £
Extra over fittings circular cast iron pipework BS 460						
Loose sockets						
Plain socket						
65 mm	8.77	11.10	0.23	7.75	nr	**18.85**
75 mm	8.77	11.10	0.23	7.75	nr	**18.85**
100 mm	13.55	17.15	0.23	7.75	nr	**24.90**
Eared socket						
65 mm	12.67	16.04	0.29	9.77	nr	**25.81**
75 mm	12.67	16.04	0.29	9.77	nr	**25.81**
100 mm	17.12	21.67	0.29	9.77	nr	**31.44**
Shoe; front projection						
Plain socket						
65 mm	27.27	34.52	0.23	7.75	nr	**42.27**
75 mm	27.27	34.52	0.23	7.75	nr	**42.27**
100 mm	36.77	46.54	0.23	7.75	nr	**54.29**
Eared socket						
65 mm	31.61	40.01	0.29	9.77	nr	**49.78**
75 mm	31.61	40.01	0.29	9.77	nr	**49.78**
100 mm	41.96	53.10	0.29	9.77	nr	**62.87**
Access Pipe						
65 mm	49.24	62.32	0.23	7.75	nr	**70.07**
75 mm	51.69	65.42	0.23	7.75	nr	**73.17**
100 mm	91.69	116.04	0.23	7.75	nr	**123.79**
100 mm; eared	103.53	131.03	0.29	9.77	nr	**140.80**
Bends; any degree						
65 mm	19.70	24.93	0.23	7.75	nr	**32.68**
75 mm	23.47	29.70	0.23	7.75	nr	**37.45**
100 mm	33.17	41.98	0.23	7.75	nr	**49.73**
Branch						
92.5°						
65 mm	38.01	48.10	0.29	9.77	nr	**57.87**
75 mm	41.92	53.05	0.29	9.77	nr	**62.82**
100 mm	48.85	61.82	0.29	9.77	nr	**71.59**
112.5°						
65 mm	38.01	48.10	0.29	9.77	nr	**57.87**
75 mm	41.92	53.05	0.29	9.77	nr	**62.82**
135°						
65 mm	38.01	48.10	0.29	9.77	nr	**57.87**
75 mm	41.92	53.05	0.29	9.77	nr	**62.82**
Offsets						
75 to 150 mm projection						
65 mm	29.60	37.46	0.25	8.43	nr	**45.89**
75 mm	29.60	37.46	0.25	8.43	nr	**45.89**
100 mm	55.85	70.68	0.25	8.43	nr	**79.11**
225 mm projection						
65 mm	30.17	38.18	0.25	8.43	nr	**46.61**
75 mm	30.17	38.18	0.25	8.43	nr	**46.61**
100 mm	55.85	70.68	0.25	8.43	nr	**79.11**

33 DRAINAGE ABOVE GROUND

Item	Net Price £	Material £	Labour hours	Labour £	Unit	Total rate £
RAINWATER PIPEWORK/GUTTERS – cont						
Extra over fittings circular cast iron pipework BS 460 – cont						
305 mm projection						
65 mm	40.37	51.09	0.25	8.43	nr	59.52
75 mm	42.38	53.64	0.25	8.43	nr	62.07
100 mm	68.97	87.29	0.25	8.43	nr	95.72
380 mm projection						
65 mm	80.57	101.96	0.25	8.43	nr	110.39
75 mm	80.57	101.96	0.25	8.43	nr	110.39
100 mm	109.96	139.16	0.25	8.43	nr	147.59
455 mm projection						
65 mm	94.31	119.36	0.25	8.43	nr	127.79
75 mm	94.31	119.36	0.25	8.43	nr	127.79
100 mm	136.75	173.07	0.25	8.43	nr	181.50
Cast iron rainwater pipe; dry joints; BS 460; fixed to backgrounds: Rectangular						
Plain socket						
100 × 75 mm	85.72	108.48	1.04	35.03	m	143.51
Bracket; fixed to backgrounds. For minimum fixing distances, refer to the Tables and Memoranda at the rear of the book						
100 × 75 mm; build in holdabat	48.78	61.73	0.35	11.79	nr	73.52
100 × 75 mm; trefoil earband	38.87	49.19	0.29	9.77	nr	58.96
100 × 75 mm; plain earband	37.59	47.58	0.29	9.77	nr	57.35
Eared Socket, with wall spacers measured separately						
100 × 75 mm	87.14	110.29	1.16	39.08	m	149.37
Wall spacer plate; eared pipework						
100 × 75	11.27	14.27	0.16	5.39	nr	19.66
Extra over fittings rectangular cast iron pipework BS 460						
Loose socket						
100 × 75 mm; plain	36.36	46.02	0.23	7.75	nr	53.77
100 × 75 mm; eared	60.23	76.23	0.29	9.77	nr	86.00
Shoe; front						
100 × 75 mm; plain	96.44	122.06	0.23	7.75	nr	129.81
100 × 75 mm; eared	117.84	149.14	0.29	9.77	nr	158.91
Shoe; side						
100 × 75 mm; plain	117.01	148.09	0.23	7.75	nr	155.84
100 × 75 mm; eared	146.28	185.14	0.29	9.77	nr	194.91
Bends; side; any degree						
100 × 75 mm; plain	88.73	112.29	0.25	8.43	nr	120.72
100 × 75 mm; 135°; plain	90.63	114.70	0.25	8.43	nr	123.13
Bends; side; any degree						
100 × 75 mm; eared	112.20	142.00	0.25	8.43	nr	150.43

33 DRAINAGE ABOVE GROUND

Item	Net Price £	Material £	Labour hours	Labour £	Unit	Total rate £
Bends; front; any degree						
100 × 75 mm; plain	84.05	106.38	0.25	8.43	nr	**114.81**
100 × 75 mm; eared	95.23	120.52	0.25	8.43	nr	**128.95**
Offset; side;						
Plain socket						
75 mm projection	116.92	147.97	0.25	8.43	nr	**156.40**
115 mm projection	121.60	153.90	0.25	8.43	nr	**162.33**
225 mm projection	151.49	191.72	0.25	8.43	nr	**200.15**
305 mm projection	174.61	220.99	0.25	8.43	nr	**229.42**
Offset; front						
Plain socket						
75 mm projection	88.95	112.57	0.25	8.43	nr	**121.00**
150 mm projection	98.30	124.41	0.25	8.43	nr	**132.84**
225 mm projection	121.69	154.01	0.25	8.43	nr	**162.44**
305 mm projection	144.79	183.24	0.25	8.43	nr	**191.67**
Eared socket						
75 mm projection	113.78	144.00	0.25	8.43	nr	**152.43**
150 mm projection	122.70	155.29	0.25	8.43	nr	**163.72**
225 mm projection	145.15	183.70	0.25	8.43	nr	**192.13**
305 mm projection	168.88	213.73	0.25	8.43	nr	**222.16**
Offset; plinth						
115 mm projection; plain	91.81	116.20	0.25	8.43	nr	**124.63**
115 mm projection; eared	118.24	149.64	0.25	8.43	nr	**158.07**
Rainwater heads						
Flat hopper						
210 × 160 × 185 mm; 65 mm outlet	69.85	88.40	0.40	13.47	nr	**101.87**
210 × 160 × 185 mm; 75 mm outlet	69.85	88.40	0.40	13.47	nr	**101.87**
250 × 215 × 215 mm; 100 mm outlet	82.79	104.78	0.40	13.47	nr	**118.25**
Flat rectangular						
225 × 125 × 125 mm; 65 mm outlet	97.12	122.92	0.40	13.47	nr	**136.39**
225 × 125 × 125 mm; 75 mm outlet	97.12	122.92	0.40	13.47	nr	**136.39**
280 × 150 × 130 mm; 100 mm outlet	134.09	169.70	0.40	13.47	nr	**183.17**
Rectangular						
250 × 180 × 175 mm; 75 mm outlet	90.55	114.60	0.40	13.47	nr	**128.07**
250 × 180 × 175 mm; 100 mm outlet	90.55	114.60	0.40	13.47	nr	**128.07**
300 × 250 × 200 mm; 65 mm outlet	126.12	159.62	0.40	13.47	nr	**173.09**
300 × 250 × 200 mm; 75 mm outlet	126.12	159.62	0.40	13.47	nr	**173.09**
300 × 250 × 200 mm; 100 mm outlet	126.12	159.62	0.40	13.47	nr	**173.09**
300 × 250 × 200 mm; 100 × 75 mm outlet	126.12	159.62	0.40	13.47	nr	**173.09**
Castellated rectangular						
250 × 180 × 175 mm; 65 mm outlet	90.55	114.60	0.40	13.47	nr	**128.07**

33 DRAINAGE ABOVE GROUND

Item	Net Price £	Material £	Labour hours	Labour £	Unit	Total rate £
DISPOSAL SYSTEMS						
Pricing note: degree angles are only indicated where material prices differ						
PVC-u overflow pipe; solvent welded joints; fixed with clips to backgrounds						
Pipe, with brackets measured separately						
19 mm	1.46	1.85	0.21	7.08	m	**8.93**
Fixings						
Pipe clip: including fixing to backgrounds. For minimum fixing distances, refer to the Tables and Memoranda at the rear of the book						
19 mm	0.57	0.72	0.18	6.06	nr	**6.78**
Extra over fittings overflow pipework						
PVC-u						
Straight coupler						
19 mm	1.56	1.97	0.17	5.73	nr	**7.70**
Bend						
19 mm: 91.25°	1.83	2.32	0.17	5.73	nr	**8.05**
19 mm: 135°	1.84	2.33	0.17	5.73	nr	**8.06**
Tee						
19 mm	1.99	2.52	0.18	6.06	nr	**8.58**
Reverse nut connector						
19 mm	0.73	0.92	0.15	5.05	nr	**5.97**
BSP adaptor: solvent welded socket to threaded socket						
19 mm × ¾"	2.68	3.39	0.14	4.72	nr	**8.11**
Straight tank connector						
19 mm	2.43	3.08	0.21	7.08	nr	**10.16**
32 mm	1.18	1.49	0.28	9.43	nr	**10.92**
40 mm	1.18	1.49	0.30	10.11	nr	**11.60**
Bent tank connector						
19 mm	2.86	3.62	0.21	7.08	nr	**10.70**
Tundish						
19 mm	38.01	48.10	0.38	12.81	nr	**60.91**
MuPVC waste pipe; solvent welded joints; fixed with clips to backgrounds; BS 5255						
Pipe, with brackets measured separately						
32 mm	2.32	2.93	0.23	7.75	m	**10.68**
40 mm	2.86	3.62	0.23	7.75	m	**11.37**
50 mm	4.34	5.49	0.26	8.76	m	**14.25**
Fixings						
Pipe clip: including fixing to backgrounds. For minimum fixing distances, refer to the Tables and Memoranda at the rear of the book						
32 mm	0.48	0.60	0.13	4.38	nr	**4.98**
40 mm	0.54	0.68	0.13	4.38	nr	**5.06**
50 mm	0.73	0.92	0.13	4.38	nr	**5.30**

33 DRAINAGE ABOVE GROUND

Item	Net Price £	Material £	Labour hours	Labour £	Unit	Total rate £
Pipe clip: expansion: including fixing to backgrounds. For minimum fixing distances, refer to the Tables and Memoranda at the rear of the book						
32 mm	0.57	0.72	0.13	4.38	nr	**5.10**
40 mm	0.59	0.75	0.13	4.38	nr	**5.13**
50 mm	1.39	1.76	0.13	4.38	nr	**6.14**
Pipe clip: metal; including fixing to backgrounds. For minimum fixing distances, refer to the Tables and Memoranda at the rear of the book						
32 mm	2.08	2.63	0.13	4.38	nr	**7.01**
40 mm	2.47	3.12	0.13	4.38	nr	**7.50**
50 mm	3.15	3.99	0.13	4.38	nr	**8.37**
Extra over fittings waste pipework MuPVC						
Screwed access plug						
32 mm	1.35	1.71	0.18	6.06	nr	**7.77**
40 mm	1.35	1.71	0.18	6.06	nr	**7.77**
50 mm	1.94	2.45	0.25	8.43	nr	**10.88**
Straight coupling						
32 mm	1.44	1.83	0.27	9.09	nr	**10.92**
40 mm	1.44	1.83	0.27	9.09	nr	**10.92**
50 mm	2.65	3.35	0.27	9.09	nr	**12.44**
Expansion coupling						
32 mm	2.56	3.24	0.27	9.09	nr	**12.33**
40 mm	3.07	3.89	0.27	9.09	nr	**12.98**
50 mm	4.17	5.28	0.27	9.09	nr	**14.37**
MuPVC to copper coupling						
32 mm	2.56	3.24	0.27	9.09	nr	**12.33**
40 mm	3.07	3.89	0.27	9.09	nr	**12.98**
50 mm	4.17	5.28	0.27	9.09	nr	**14.37**
Spigot and socket coupling						
32 mm	2.56	3.24	0.27	9.09	nr	**12.33**
40 mm	3.07	3.89	0.27	9.09	nr	**12.98**
50 mm	4.17	5.28	0.27	9.09	nr	**14.37**
Union						
32 mm	6.07	7.68	0.28	9.43	nr	**17.11**
40 mm	7.95	10.06	0.28	9.43	nr	**19.49**
50 mm	9.05	11.46	0.28	9.43	nr	**20.89**
Reducer: socket						
32 × 19 mm	2.19	2.77	0.27	9.09	nr	**11.86**
40 × 32 mm	1.44	1.83	0.27	9.09	nr	**10.92**
50 × 32 mm	2.09	2.64	0.27	9.09	nr	**11.73**
50 × 40 mm	2.56	3.24	0.27	9.09	nr	**12.33**
Reducer: level invert						
40 × 32 mm	1.80	2.27	0.27	9.09	nr	**11.36**
50 × 32 mm	2.22	2.81	0.27	9.09	nr	**11.90**
50 × 40 mm	2.22	2.81	0.27	9.09	nr	**11.90**

33 DRAINAGE ABOVE GROUND

Item	Net Price £	Material £	Labour hours	Labour £	Unit	Total rate £
DISPOSAL SYSTEMS – cont						
Extra over fittings waste pipework						
MuPVC – cont						
Swept bend						
32 mm	1.48	1.87	0.27	9.09	nr	**10.96**
32 mm: 165°	1.53	1.94	0.27	9.09	nr	**11.03**
40 mm	1.65	2.08	0.27	9.09	nr	**11.17**
40 mm: 165°	2.89	3.66	0.27	9.09	nr	**12.75**
50 mm	2.86	3.62	0.30	10.11	nr	**13.73**
50 mm: 165°	3.82	4.84	0.30	10.11	nr	**14.95**
Knuckle bend						
32 mm	1.35	1.71	0.27	9.09	nr	**10.80**
40 mm	1.49	1.88	0.27	9.09	nr	**10.97**
Spigot and socket bend						
32 mm	2.41	3.05	0.27	9.09	nr	**12.14**
32 mm: 150°	2.50	3.17	0.27	9.09	nr	**12.26**
40 mm	2.76	3.49	0.27	9.09	nr	**12.58**
50 mm	3.93	4.97	0.30	10.11	nr	**15.08**
Swept tee						
32 mm: 91.25°	1.97	2.50	0.31	10.44	nr	**12.94**
32 mm: 135°	2.37	3.00	0.31	10.44	nr	**13.44**
40 mm: 91.25°	2.51	3.18	0.31	10.44	nr	**13.62**
40 mm: 135°	3.13	3.96	0.31	10.44	nr	**14.40**
50 mm	4.89	6.19	0.31	10.44	nr	**16.63**
Swept cross						
40 mm: 91.25°	6.10	7.72	0.31	10.44	nr	**18.16**
50 mm: 91.25°	6.38	8.08	0.43	14.49	nr	**22.57**
50 mm: 135°	8.05	10.19	0.31	10.44	nr	**20.63**
Male iron adaptor						
32 mm	2.20	2.79	0.28	9.43	nr	**12.22**
40 mm	2.59	3.28	0.28	9.43	nr	**12.71**
Female iron adaptor						
32 mm	2.59	3.28	0.28	9.43	nr	**12.71**
40 mm	2.59	3.28	0.28	9.43	nr	**12.71**
50 mm	3.72	4.70	0.31	10.44	nr	**15.14**
Reverse nut adaptor						
32 mm	3.29	4.17	0.20	6.73	nr	**10.90**
40 mm	3.29	4.17	0.20	6.73	nr	**10.90**
Automatic air admittance valve						
32 mm	18.49	23.40	0.27	9.09	nr	**32.49**
40 mm	18.49	23.40	0.28	9.43	nr	**32.83**
50 mm	18.49	23.40	0.31	10.44	nr	**33.84**
MuPVC to metal adpator: including heat shrunk joint to metal						
50 mm	7.97	10.09	0.38	12.81	nr	**22.90**
Caulking bush: including joint to metal						
32 mm	3.47	4.39	0.31	10.44	nr	**14.83**
40 mm	3.47	4.39	0.31	10.44	nr	**14.83**
50 mm	3.47	4.39	0.32	10.79	nr	**15.18**

33 DRAINAGE ABOVE GROUND

Item	Net Price £	Material £	Labour hours	Labour £	Unit	Total rate £
Weathering apron						
50 mm	3.05	3.86	0.65	21.90	nr	**25.76**
Vent cowl						
50 mm	3.48	4.40	0.19	6.41	nr	**10.81**
ABS waste pipe; solvent welded joints; fixed with clips to backgrounds; BS 5255						
Pipe, with brackets measured separately						
32 mm	2.12	2.69	0.23	7.75	m	**10.44**
40 mm	2.65	3.35	0.23	7.75	m	**11.10**
50 mm	3.33	4.21	0.26	8.76	m	**12.97**
Fixings						
Pipe clip: including fixing to backgrounds. For minimum fixing distances, refer to the Tables and Memoranda at the rear of the book						
32 mm	0.36	0.46	0.17	5.73	nr	**6.19**
40 mm	0.45	0.57	0.17	5.73	nr	**6.30**
50 mm	1.35	1.71	0.17	5.73	nr	**7.44**
Pipe clip: expansion: including fixing to backgrounds. For minimum fixing distances, refer to the Tables and Memoranda at the rear of the book						
32 mm	0.36	0.46	0.17	5.73	nr	**6.19**
40 mm	0.45	0.57	0.17	5.73	nr	**6.30**
50 mm	1.35	1.71	0.17	5.73	nr	**7.44**
Pipe clip: metal; including fixing to backgrounds. For minimum fixing distances, refer to the Tables and Memoranda at the rear of the book						
32 mm	2.05	2.60	0.17	5.73	nr	**8.33**
40 mm	2.43	3.08	0.17	5.73	nr	**8.81**
50 mm	3.10	3.92	0.17	5.73	nr	**9.65**
Extra over fittings waste pipework ABS						
Screwed access plug						
32 mm	1.40	1.77	0.18	6.06	nr	**7.83**
40 mm	1.40	1.77	0.18	6.06	nr	**7.83**
50 mm	2.89	3.66	0.25	8.43	nr	**12.09**
Straight coupling						
32 mm	1.40	1.77	0.27	9.09	nr	**10.86**
40 mm	1.40	1.77	0.27	9.09	nr	**10.86**
50 mm	2.89	3.66	0.27	9.09	nr	**12.75**
Expansion coupling						
32 mm	2.89	3.66	0.27	9.09	nr	**12.75**
40 mm	2.89	3.66	0.27	9.09	nr	**12.75**
50 mm	4.43	5.61	0.27	9.09	nr	**14.70**
ABS to copper coupling						
32 mm	2.74	3.47	0.27	9.09	nr	**12.56**
40 mm	2.74	3.47	0.27	9.09	nr	**12.56**
50 mm	4.19	5.30	0.27	9.09	nr	**14.39**

33 DRAINAGE ABOVE GROUND

Item	Net Price £	Material £	Labour hours	Labour £	Unit	Total rate £
DISPOSAL SYSTEMS – cont						
Extra over fittings waste pipework ABS – cont						
Reducer: socket						
40 × 32 mm	1.40	1.77	0.27	9.09	nr	**10.86**
50 × 32 mm	3.18	4.02	0.27	9.09	nr	**13.11**
50 × 40 mm	3.18	4.02	0.27	9.09	nr	**13.11**
Swept bend						
32 mm	1.40	1.77	0.27	9.09	nr	**10.86**
40 mm	1.40	1.77	0.27	9.09	nr	**10.86**
50 mm	2.89	3.66	0.30	10.11	nr	**13.77**
Knuckle bend						
32 mm	1.40	1.77	0.27	9.09	nr	**10.86**
40 mm	1.40	1.77	0.27	9.09	nr	**10.86**
Swept tee						
32 mm	2.00	2.53	0.31	10.44	nr	**12.97**
40 mm	2.00	2.53	0.31	10.44	nr	**12.97**
50 mm	5.29	6.70	0.31	10.44	nr	**17.14**
Swept cross						
40 mm	6.74	8.53	0.23	7.75	nr	**16.28**
50 mm	7.70	9.74	0.43	14.49	nr	**24.23**
Male iron adaptor						
32 mm	2.89	3.66	0.28	9.43	nr	**13.09**
40 mm	2.89	3.66	0.28	9.43	nr	**13.09**
Female iron adapator						
32 mm	2.89	3.66	0.28	9.43	nr	**13.09**
40 mm	2.89	3.66	0.28	9.43	nr	**13.09**
50 mm	4.33	5.48	0.31	10.44	nr	**15.92**
Tank connectors						
32 mm	1.99	2.52	0.29	9.77	nr	**12.29**
40 mm	2.20	2.79	0.29	9.77	nr	**12.56**
Caulking bush: including joint to pipework						
50 mm	3.42	4.32	0.50	16.84	nr	**21.16**
Polypropylene waste pipe; push fit joints; fixed with clips to backgrounds; BS 5254. Pipe, with brackets measured separately						
32 mm	1.24	1.57	0.21	7.08	m	**8.65**
40 mm	1.51	1.92	0.21	7.08	m	**9.00**
50 mm	2.52	3.19	0.38	12.81	m	**16.00**
Fixings						
Pipe clip: saddle; including fixing to backgrounds. For minimum fixing distances, refer to the Tables and Memoranda at the rear of the book						
32 mm	0.42	0.53	0.17	5.73	nr	**6.26**
40 mm	0.42	0.53	0.17	5.73	nr	**6.26**

33 DRAINAGE ABOVE GROUND

Item	Net Price £	Material £	Labour hours	Labour £	Unit	Total rate £
Pipe clip: including fixing to backgrounds. For minimum fixing distances, refer to the Tables and Memoranda at the rear of the book						
50 mm	0.99	1.25	0.17	5.73	nr	6.98
Extra over fittings waste pipework polypropylene						
Screwed access plug						
32 mm	1.17	1.48	0.16	5.39	nr	6.87
40 mm	1.17	1.48	0.16	5.39	nr	6.87
50 mm	2.00	2.53	0.20	6.73	nr	9.26
Straight coupling						
32 mm	1.17	1.48	0.19	6.41	nr	7.89
40 mm	1.17	1.48	0.19	6.41	nr	7.89
50 mm	2.00	2.53	0.20	6.73	nr	9.26
Universal waste pipe coupler						
32 mm dia.	1.98	2.51	0.20	6.73	nr	9.24
40 mm dia.	2.25	2.84	0.20	6.73	nr	9.57
Reducer						
40 × 32 mm	2.00	2.53	0.19	6.41	nr	8.94
50 × 32 mm	2.11	2.67	0.19	6.41	nr	9.08
50 × 40 mm	2.17	2.74	0.20	6.73	nr	9.47
Swept bend						
32 mm	1.17	1.48	0.19	6.41	nr	7.89
40 mm	1.17	1.48	0.19	6.41	nr	7.89
50 mm	2.00	2.53	0.20	6.73	nr	9.26
Knuckle bend						
32 mm	1.17	1.48	0.19	6.41	nr	7.89
40 mm	1.17	1.48	0.19	6.41	nr	7.89
50 mm	2.00	2.53	0.20	6.73	nr	9.26
Spigot and socket bend						
32 mm	1.17	1.48	0.19	6.41	nr	7.89
40 mm	1.17	1.48	0.19	6.41	nr	7.89
Swept tee						
32 mm	1.29	1.64	0.22	7.41	nr	9.05
40 mm	1.29	1.64	0.22	7.41	nr	9.05
50 mm	2.17	2.74	0.23	7.75	nr	10.49
Male iron adaptor						
32 mm	1.17	1.48	0.13	4.38	nr	5.86
40 mm	1.17	1.48	0.19	6.41	nr	7.89
50 mm	2.00	2.53	0.15	5.05	nr	7.58
Tank connector						
32 mm	1.17	1.48	0.24	8.09	nr	9.57
40 mm	1.17	1.48	0.24	8.09	nr	9.57
50 mm	2.00	2.53	0.35	11.79	nr	14.32

33 DRAINAGE ABOVE GROUND

Item	Net Price £	Material £	Labour hours	Labour £	Unit	Total rate £
DISPOSAL SYSTEMS – cont						
Polypropylene traps; including fixing to appliance and connection to pipework; BS 3943						
Tubular P trap; 75 mm seal						
32 mm dia.	4.77	6.04	0.20	6.73	nr	**12.77**
40 mm dia.	5.50	6.96	0.20	6.73	nr	**13.69**
Tubular S trap; 75 mm seal						
32 mm dia.	5.99	7.58	0.20	6.73	nr	**14.31**
40 mm dia.	7.08	8.96	0.20	6.73	nr	**15.69**
Running tubular P trap; 75 mm seal						
32 mm dia.	7.30	9.24	0.20	6.73	nr	**15.97**
40 mm dia.	7.98	10.10	0.20	6.73	nr	**16.83**
Running tubular S trap; 75 mm seal						
32 mm dia.	8.80	11.13	0.20	6.73	nr	**17.86**
40 mm dia.	9.47	11.98	0.20	6.73	nr	**18.71**
Spigot and socket bend; converter from P to S Trap						
32 mm	1.87	2.36	0.20	6.73	nr	**9.09**
40 mm	1.99	2.52	0.21	7.08	nr	**9.60**
Bottle P trap; 75 mm seal						
32 mm dia.	5.31	6.72	0.20	6.73	nr	**13.45**
40 mm dia.	6.36	8.05	0.20	6.73	nr	**14.78**
Bottle S trap; 75 mm seal						
32 mm dia.	6.41	8.11	0.20	6.73	nr	**14.84**
40 mm dia.	7.78	9.84	0.25	8.43	nr	**18.27**
Bottle P trap; resealing; 75 mm seal						
32 mm dia.	6.61	8.37	0.20	6.73	nr	**15.10**
40 mm dia.	7.72	9.77	0.25	8.43	nr	**18.20**
Bottle S trap; resealing; 75 mm seal						
32 mm dia.	7.56	9.56	0.20	6.73	nr	**16.29**
40 mm dia.	8.78	11.11	0.25	8.43	nr	**19.54**
Bath trap, low level; 38 mm seal						
40 mm dia.	6.63	8.39	0.25	8.43	nr	**16.82**
Bath trap, low level; 38 mm seal complete with overflow hose						
40 mm dia.	10.23	12.95	0.25	8.43	nr	**21.38**
Bath trap; 75 mm seal complete with overlow hose						
40 mm dia.	10.20	12.91	0.25	8.43	nr	**21.34**
Bath trap; 75 mm seal complete with overflow hose and overflow outlet						
40 mm dia.	17.41	22.03	0.20	6.73	nr	**28.76**
Bath trap; 75 mm seal complete with overflow hose, overflow outlet and ABS chrome waste						
40 mm dia.	24.05	30.44	0.20	6.73	nr	**37.17**
Washing machine trap; 75 mm seal including stand pipe						
40 mm dia.	13.84	17.52	0.25	8.43	nr	**25.95**

33 DRAINAGE ABOVE GROUND

Item	Net Price £	Material £	Labour hours	Labour £	Unit	Total rate £
Washing machine standpipe						
40 mm dia.	7.22	9.14	0.25	8.43	nr	**17.57**
Plastic unslotted chrome plated basin/sink waste including plug						
32 mm	8.69	11.00	0.34	11.46	nr	**22.46**
40 mm	11.63	14.72	0.34	11.46	nr	**26.18**
Plastic slotted chrome plated basin/sink waste including plug						
32 mm	6.90	8.74	0.34	11.46	nr	**20.20**
40 mm	11.56	14.63	0.34	11.46	nr	**26.09**
Bath overflow outlet; plastic; white						
42 mm	6.07	7.68	0.37	12.47	nr	**20.15**
Bath overlow outlet; plastic; chrome plated						
42 mm	8.00	10.12	0.37	12.47	nr	**22.59**
Combined cistern and bath overflow outlet; plastic; white						
42 mm	12.43	15.74	0.39	13.14	nr	**28.88**
Combined cistern and bath overlow outlet; plastic; chrome plated						
42 mm	12.43	15.74	0.39	13.14	nr	**28.88**
Cistern overflow outlet; plastic; white						
42 mm	10.02	12.68	0.15	5.05	nr	**17.73**
Cistern overlow outlet; plastic; chrome plated						
42 mm	9.03	11.42	0.15	5.05	nr	**16.47**
PVC-u soil and waste pipe; solvent welded joints; fixed with clips to backgrounds; BS 4514/BS EN 607						
Pipe, with brackets measured separately						
82 mm	10.52	13.32	0.35	11.79	m	**25.11**
110 mm	10.73	13.57	0.41	13.81	m	**27.38**
160 mm	27.80	35.18	0.51	17.19	m	**52.37**
Fixings						
Galvanized steel pipe clip: including fixing to backgrounds. For minimum fixing distances, refer to the Tables and Memoranda at the rear of the book						
82 mm	4.37	5.53	0.18	6.06	nr	**11.59**
110 mm	4.53	5.73	0.18	6.06	nr	**11.79**
160 mm	10.92	13.82	0.18	6.06	nr	**19.88**
Plastic coated steel pipe clip: including fixing to backgrounds. For minimum fixing distances, refer to the Tables and Memoranda at the rear of the book						
82 mm	6.12	7.75	0.18	6.06	nr	**13.81**
110 mm	8.33	10.54	0.18	6.06	nr	**16.60**
160 mm	10.51	13.31	0.18	6.06	nr	**19.37**

33 DRAINAGE ABOVE GROUND

Item	Net Price £	Material £	Labour hours	Labour £	Unit	Total rate £
DISPOSAL SYSTEMS – cont						
Fixings – cont						
Plastic pipe clip: including fixing to backgrounds. For minimum fixing distances, refer to the Tables and Memoranda at the rear of the book						
82 mm	3.23	4.09	0.18	6.06	nr	**10.15**
110 mm	6.22	7.87	0.18	6.06	nr	**13.93**
Plastic coated steel pipe clip: adjustable; including fixing to backgrounds. For minimum fixing distances, refer to the Tables and Memoranda at the rear of the book						
82 mm	4.93	6.24	0.20	6.73	nr	**12.97**
110 mm	6.17	7.81	0.20	6.73	nr	**14.54**
Galvanized steel pipe clip: drive in; including fixing to backgrounds. For minimum fixing distances, refer to the Tables and Memoranda at the rear of the book						
110 mm	9.42	11.92	0.22	7.41	nr	**19.33**
Extra over fittings solvent welded pipework PVC-u						
Straight coupling						
82 mm	5.58	7.07	0.21	7.08	nr	**14.15**
110 mm	6.98	8.84	0.22	7.41	nr	**16.25**
160 mm	20.12	25.47	0.24	8.09	nr	**33.56**
Expansion coupling						
82 mm	8.35	10.57	0.21	7.08	nr	**17.65**
110 mm	8.55	10.82	0.22	7.41	nr	**18.23**
160 mm	25.70	32.52	0.24	8.09	nr	**40.61**
Slip coupling; double ring socket						
82 mm	17.36	21.97	0.21	7.08	nr	**29.05**
110 mm	21.73	27.50	0.22	7.41	nr	**34.91**
160 mm	34.51	43.68	0.24	8.09	nr	**51.77**
Puddle flanges						
110 mm	169.56	214.59	0.45	15.16	nr	**229.75**
160 mm	274.11	346.91	0.55	18.52	nr	**365.43**
Socket reducer						
82 to 50 mm	8.50	10.75	0.18	6.06	nr	**16.81**
110 to 50 mm	10.59	13.41	0.18	6.06	nr	**19.47**
110 to 82 mm	10.92	13.82	0.22	7.41	nr	**21.23**
160 to 110 mm	22.10	27.97	0.26	8.76	nr	**36.73**
Socket plugs						
82 mm	6.58	8.33	0.15	5.05	nr	**13.38**
110 mm	9.62	12.17	0.20	6.73	nr	**18.90**
160 mm	17.71	22.41	0.27	9.09	nr	**31.50**
Access door; including cutting into pipe						
82 mm	18.66	23.62	0.28	9.43	nr	**33.05**
110 mm	18.66	23.62	0.34	11.46	nr	**35.08**
160 mm	33.33	42.18	0.46	15.50	nr	**57.68**

33 DRAINAGE ABOVE GROUND

Item	Net Price £	Material £	Labour hours	Labour £	Unit	Total rate £
Screwed access cap						
82 mm	13.20	16.71	0.15	5.05	nr	21.76
110 mm	15.56	19.69	0.20	6.73	nr	26.42
160 mm	29.30	37.08	0.27	9.09	nr	46.17
Access pipe: spigot and socket						
110 mm	23.01	29.12	0.22	7.41	nr	36.53
Access pipe: double socket						
110 mm	23.01	29.12	0.22	7.41	nr	36.53
Swept bend						
82 mm	14.00	17.72	0.29	9.77	nr	27.49
110 mm	16.40	20.75	0.32	10.79	nr	31.54
160 mm	40.86	51.71	0.49	16.51	nr	68.22
Bend; special angle						
82 mm	26.99	34.16	0.29	9.77	nr	43.93
110 mm	32.18	40.72	0.32	10.79	nr	51.51
160 mm	54.34	68.77	0.49	16.51	nr	85.28
Spigot and socket bend						
82 mm	13.55	17.15	0.26	8.76	nr	25.91
110 mm	15.88	20.09	0.32	10.79	nr	30.88
110 mm: 135°	17.85	22.59	0.32	10.79	nr	33.38
160 mm: 135°	39.52	50.02	0.44	14.82	nr	64.84
Variable bend: single socket						
110 mm	30.76	38.93	0.33	11.11	nr	50.04
Variable bend: double socket						
110 mm	30.69	38.84	0.33	11.11	nr	49.95
Access bend						
110 mm	45.53	57.62	0.33	11.11	nr	68.73
Single branch: two bosses						
82 mm	19.58	24.79	0.35	11.79	nr	36.58
82 mm: 104°	20.88	26.42	0.35	11.79	nr	38.21
110 mm	21.70	27.46	0.42	14.15	nr	41.61
110 mm: 135°	22.64	28.65	0.42	14.15	nr	42.80
160 mm	46.08	58.32	0.50	16.84	nr	75.16
160 mm: 135°	47.12	59.64	0.50	16.84	nr	76.48
Single branch; four bosses						
110 mm	26.42	33.43	0.42	14.15	nr	47.58
Single access branch						
82 mm	63.49	80.35	0.35	11.79	nr	92.14
110 mm	37.15	47.02	0.42	14.15	nr	61.17
Unequal single branch						
160 × 160 × 110 mm	52.02	65.83	0.50	16.84	nr	82.67
160 × 160 × 110 mm: 135°	55.91	70.76	0.50	16.84	nr	87.60
Double branch						
110 mm	56.06	70.95	0.42	14.15	nr	85.10
110 mm: 135°	53.62	67.86	0.42	14.15	nr	82.01
Corner branch						
110 mm	94.73	119.88	0.42	14.15	nr	134.03
Unequal double branch						
160 × 160 × 110 mm	96.89	122.63	0.50	16.84	nr	139.47

33 DRAINAGE ABOVE GROUND

Item	Net Price £	Material £	Labour hours	Labour £	Unit	Total rate £
DISPOSAL SYSTEMS – cont						
Extra over fittings solvent welded pipework PVC-u – cont						
Single boss pipe; single socket						
110 × 110 × 32 mm	5.86	7.41	0.24	8.09	nr	**15.50**
110 × 110 × 40 mm	5.86	7.41	0.24	8.09	nr	**15.50**
110 × 110 × 50 mm	6.19	7.83	0.24	8.09	nr	**15.92**
Single boss pipe; triple socket						
110 × 110 × 40 mm	9.52	12.05	0.24	8.09	nr	**20.14**
Waste boss; including cutting into pipe						
82 to 32 mm	7.67	9.71	0.29	9.77	nr	**19.48**
82 to 40 mm	7.67	9.71	0.29	9.77	nr	**19.48**
110 to 32 mm	7.67	9.71	0.29	9.77	nr	**19.48**
110 to 40 mm	7.67	9.71	0.29	9.77	nr	**19.48**
110 to 50 mm	7.95	10.06	0.29	9.77	nr	**19.83**
160 to 32 mm	10.84	13.72	0.30	10.11	nr	**23.83**
160 to 40 mm	10.84	13.72	0.35	11.79	nr	**25.51**
160 to 50 mm	10.84	13.72	0.40	13.47	nr	**27.19**
Self locking waste boss; including cutting into pipe						
110 to 32 mm	10.24	12.96	0.30	10.11	nr	**23.07**
110 to 40 mm	10.70	13.54	0.30	10.11	nr	**23.65**
110 to 50 mm	13.83	17.51	0.30	10.11	nr	**27.62**
Adaptor saddle; including cutting to pipe						
82 to 32 mm	5.10	6.45	0.29	9.77	nr	**16.22**
110 to 40 mm	6.30	7.97	0.29	9.77	nr	**17.74**
160 to 50 mm	11.41	14.44	0.29	9.77	nr	**24.21**
Branch boss adaptor						
32 mm	3.05	3.86	0.26	8.76	nr	**12.62**
40 mm	3.05	3.86	0.26	8.76	nr	**12.62**
50 mm	4.36	5.52	0.26	8.76	nr	**14.28**
Branch boss adaptor bend						
32 mm	4.17	5.28	0.26	8.76	nr	**14.04**
40 mm	4.54	5.75	0.26	8.76	nr	**14.51**
50 mm	5.44	6.89	0.26	8.76	nr	**15.65**
Automatic air admittance valve						
82 to 110 mm	47.83	60.54	0.19	6.41	nr	**66.95**
PVC-u to metal adpator: including heat shrunk joint to metal						
110 mm	13.32	16.86	0.57	19.20	nr	**36.06**
Caulking bush: including joint to pipework						
82 mm	13.11	16.59	0.46	15.50	nr	**32.09**
110 mm	13.11	16.59	0.46	15.50	nr	**32.09**
Vent cowl						
82 mm	3.96	5.01	0.13	4.38	nr	**9.39**
110 mm	3.99	5.05	0.13	4.38	nr	**9.43**
160 mm	10.44	13.22	0.13	4.38	nr	**17.60**

33 DRAINAGE ABOVE GROUND

Item	Net Price £	Material £	Labour hours	Labour £	Unit	Total rate £
Weathering apron; to lead slates						
82 mm	3.96	5.01	1.15	38.74	nr	**43.75**
110 mm	4.53	5.73	1.15	38.74	nr	**44.47**
160 mm	13.63	17.25	1.15	38.74	nr	**55.99**
Weathering apron; to asphalt						
82 mm	16.64	21.06	1.10	37.06	nr	**58.12**
110 mm	16.64	21.06	1.10	37.06	nr	**58.12**
Weathering slate; flat; 406 × 406 mm						
82 mm	46.95	59.42	1.04	35.03	nr	**94.45**
110 mm	46.95	59.42	1.04	35.03	nr	**94.45**
Weathering slate; flat; 457 × 457 mm						
82 mm	48.17	60.96	1.04	35.03	nr	**95.99**
110 mm	48.17	60.96	1.04	35.03	nr	**95.99**
Weathering slate; angled; 610 × 610 mm						
82 mm	65.09	82.38	1.04	35.03	nr	**117.41**
110 mm	65.09	82.38	1.04	35.03	nr	**117.41**
PVC-u soil and waste pipe; ring seal joints; fixed with clips to backgrounds; BS 4514/ BS EN 607						
Pipe, with brackets measured separately						
82 mm dia.	9.28	11.75	0.35	11.79	m	**23.54**
110 mm dia.	9.35	11.84	0.41	13.81	m	**25.65**
160 mm dia.	33.53	42.44	0.51	17.19	m	**59.63**
Fixings						
Galvanized steel pipe clip: including fixing to backgrounds. For minimum fixing distances, refer to the Tables and Memoranda at the rear of the book						
82 mm	4.37	5.53	0.18	6.06	nr	**11.59**
110 mm	4.53	5.73	0.18	6.06	nr	**11.79**
160 mm	10.92	13.82	0.18	6.06	nr	**19.88**
Plastic coated steel pipe clip: including fixing to backgrounds. For minimum fixing distances, refer to the Tables and Memoranda at the rear of the book						
82 mm	6.12	7.75	0.18	6.06	nr	**13.81**
110 mm	8.33	10.54	0.18	6.06	nr	**16.60**
160 mm	10.51	13.31	0.18	6.06	nr	**19.37**
Plastic pipe clip: including fixing to backgrounds. For minimum fixing distances, refer to the Tables and Memoranda at the rear of the book						
82 mm	3.23	4.09	0.18	6.06	nr	**10.15**
110 mm	6.22	7.87	0.18	6.06	nr	**13.93**
Plastic coated steel pipe clip: adjustable; including fixing to backgrounds. For minimum fixing distances, refer to the Tables and Memoranda at the rear of the book						
82 mm	4.93	6.24	0.20	6.73	nr	**12.97**
110 mm	6.17	7.81	0.20	6.73	nr	**14.54**

33 DRAINAGE ABOVE GROUND

Item	Net Price £	Material £	Labour hours	Labour £	Unit	Total rate £
DISPOSAL SYSTEMS – cont						
Fixings – cont						
Galvanized steel pipe clip: drive in; including fixing to backgrounds. For minimum fixing distances, refer to the Tables and Memoranda at the rear of the book						
110 mm	9.42	11.92	0.22	7.41	nr	**19.33**
Extra over fittings ring seal pipework						
PVC-u						
Straight coupling						
82 mm	6.08	7.69	0.21	7.08	nr	**14.77**
110 mm	6.81	8.62	0.22	7.41	nr	**16.03**
160 mm	23.01	29.12	0.24	8.09	nr	**37.21**
Straight coupling; double socket						
82 mm	10.73	13.57	0.21	7.08	nr	**20.65**
110 mm	9.73	12.31	0.22	7.41	nr	**19.72**
160 mm	23.01	29.12	0.24	8.09	nr	**37.21**
Reducer; socket						
82 to 50 mm	10.00	12.66	0.15	5.05	nr	**17.71**
110 to 50 mm	13.95	17.65	0.15	5.05	nr	**22.70**
110 to 82 mm	13.95	17.65	0.19	6.41	nr	**24.06**
160 to 110 mm	20.10	25.44	0.31	10.44	nr	**35.88**
Access cap						
82 mm	10.89	13.79	0.15	5.05	nr	**18.84**
110 mm	12.83	16.24	0.17	5.73	nr	**21.97**
Access cap; pressure plug						
160 mm	30.17	38.18	0.33	11.11	nr	**49.29**
Access pipe						
82 mm	26.73	33.82	0.22	7.41	nr	**41.23**
110 mm	26.68	33.77	0.22	7.41	nr	**41.18**
160 mm	56.01	70.88	0.24	8.09	nr	**78.97**
Bend						
82 mm	13.59	17.20	0.29	9.77	nr	**26.97**
82 mm; adjustable radius	30.65	38.79	0.29	9.77	nr	**48.56**
110 mm	14.63	18.51	0.32	10.79	nr	**29.30**
110 mm; adjustable radius	27.28	34.53	0.32	10.79	nr	**45.32**
160 mm	43.91	55.57	0.49	16.51	nr	**72.08**
160 mm; adjustable radius	57.60	72.90	0.49	16.51	nr	**89.41**
Bend; spigot and socket						
110 mm	14.76	18.68	0.32	10.79	nr	**29.47**
Bend; offset						
82 mm	11.97	15.15	0.21	7.08	nr	**22.23**
110 mm	19.20	24.30	0.32	10.79	nr	**35.09**
160 mm	42.35	53.60	0.32	10.79	nr	**64.39**
Bend; access						
110 mm	40.28	50.98	0.33	11.11	nr	**62.09**

33 DRAINAGE ABOVE GROUND

Item	Net Price £	Material £	Labour hours	Labour £	Unit	Total rate £
Single branch						
82 mm	21.34	27.00	0.35	11.79	nr	38.79
110 mm	19.73	24.96	0.42	14.15	nr	39.11
110 mm; 45°	20.30	25.69	0.31	10.44	nr	36.13
160 mm	50.88	64.39	0.50	16.84	nr	81.23
Single branch; access						
82 mm	31.98	40.48	0.35	11.79	nr	52.27
110 mm	46.30	58.60	0.42	14.15	nr	72.75
Unequal single branch						
160 × 160 × 110 mm	51.99	65.80	0.50	16.84	nr	82.64
160 × 160 × 110 mm; 45°	55.91	70.76	0.50	16.84	nr	87.60
Double branch; 4 bosses						
110 mm	51.65	65.36	0.49	16.51	nr	81.87
Corner branch; 2 bosses						
110 mm	94.71	119.86	0.49	16.51	nr	136.37
Multibranch; 4 bosses						
110 mm	28.28	35.80	0.52	17.52	nr	53.32
Boss branch						
110 × 32 mm	5.85	7.40	0.34	11.46	nr	18.86
110 × 40 mm	5.85	7.40	0.34	11.46	nr	18.86
Strap on boss						
110 × 32 mm	6.52	8.25	0.30	10.11	nr	18.36
110 × 40 mm	6.52	8.25	0.30	10.11	nr	18.36
110 × 50 mm	6.57	8.31	0.30	10.11	nr	18.42
Patch boss						
82 × 32 mm	7.67	9.71	0.31	10.44	nr	20.15
82 × 40 mm	7.67	9.71	0.31	10.44	nr	20.15
82 × 50 mm	7.67	9.71	0.31	10.44	nr	20.15
Boss pipe; collar 4 boss						
110 mm	9.52	12.05	0.35	11.79	nr	23.84
Boss adaptor; rubber; push fit						
32 mm	3.56	4.50	0.26	8.76	nr	13.26
40 mm	3.56	4.50	0.26	8.76	nr	13.26
50 mm	4.03	5.10	0.26	8.76	nr	13.86
WC connector; cap and seal; solvent socket						
110 mm	9.92	12.56	0.23	7.75	nr	20.31
110 mm; 90°	15.94	20.17	0.27	9.09	nr	29.26
Vent terminal						
82 mm	3.96	5.01	0.13	4.38	nr	9.39
110 mm	3.99	5.05	0.13	4.38	nr	9.43
160 mm	10.46	13.24	0.13	4.38	nr	17.62
Weathering slate; inclined; 610 × 610 mm						
82 mm	65.09	82.38	1.04	35.03	nr	117.41
110 mm	65.09	82.38	1.04	35.03	nr	117.41
Weathering slate; inclined; 450 × 450 mm						
82 mm	48.17	60.96	1.04	35.03	nr	95.99
110 mm	48.17	60.96	1.04	35.03	nr	95.99
Weathering slate; flat; 400 × 400 mm						
82 mm	46.97	59.45	1.04	35.03	nr	94.48
110 mm	46.97	59.45	1.04	35.03	nr	94.48

33 DRAINAGE ABOVE GROUND

Item	Net Price £	Material £	Labour hours	Labour £	Unit	Total rate £
DISPOSAL SYSTEMS – cont						
Extra over fittings ring seal pipework PVC-u – cont						
Air admittance valve						
82 mm	47.83	60.54	0.19	6.41	nr	**66.95**
110 mm	47.83	60.54	0.19	6.41	nr	**66.95**
Cast iron pipe; nitrile rubber gasket joint with continuity clip BSEN 877; fixed vertically to backgrounds						
Pipe, with brackets and jointing couplings measured separately						
50 mm	23.68	29.97	0.31	10.44	m	**40.41**
75 mm	27.59	34.92	0.34	11.46	m	**46.38**
100 mm	30.65	38.79	0.37	12.47	m	**51.26**
150 mm	64.02	81.02	0.60	20.22	m	**101.24**
Fixings						
Brackets; fixed to backgrounds. For minimum fixing distances, refer to the Tables and Memoranda at the rear of the book						
50 mm	8.83	11.18	0.15	5.05	nr	**16.23**
75 mm	9.37	11.86	0.18	6.06	nr	**17.92**
100 mm	9.60	12.15	0.18	6.06	nr	**18.21**
150 mm	17.79	22.51	0.20	6.73	nr	**29.24**
Extra over fittings nitrile gasket cast iron pipework BS 416/6087, with jointing couplings measured separately						
Standard coupling						
50 mm	11.45	14.49	0.17	5.73	nr	**20.22**
75 mm	12.54	15.87	0.17	5.73	nr	**21.60**
100 mm	16.37	20.72	0.17	5.73	nr	**26.45**
150 mm	32.67	41.35	0.17	5.73	nr	**47.08**
Conversion coupling						
65 × 75 mm	13.29	16.82	0.60	20.23	nr	**37.05**
70 × 75 mm	13.29	16.82	0.60	20.23	nr	**37.05**
90 × 100 mm	17.11	21.65	0.67	22.58	nr	**44.23**
Access pipe; round door						
50 mm	33.48	42.37	0.41	13.81	nr	**56.18**
75 mm	48.13	60.92	0.46	15.50	nr	**76.42**
100 mm	50.59	64.03	0.67	22.60	nr	**86.63**
150 mm	84.15	106.50	0.83	27.98	nr	**134.48**
Access pipe; square door						
100 mm	99.69	126.17	0.67	22.60	nr	**148.77**
150 mm	152.60	193.13	0.83	27.98	nr	**221.11**
Taper reducer						
75 mm	27.51	34.82	0.60	20.23	nr	**55.05**
100 mm	34.92	44.20	0.67	22.58	nr	**66.78**
150 mm	68.00	86.06	0.83	27.98	nr	**114.04**

33 DRAINAGE ABOVE GROUND

Item	Net Price £	Material £	Labour hours	Labour £	Unit	Total rate £
Blank cap						
50 mm	6.95	8.79	0.24	8.09	nr	16.88
75 mm	8.66	10.96	0.26	8.76	nr	19.72
100 mm	8.42	10.65	0.32	10.79	nr	21.44
150 mm	12.16	15.39	0.40	13.47	nr	28.86
Blank cap; 50 mm screwed tapping						
75 mm	17.10	21.64	0.26	8.76	nr	30.40
100 mm	19.84	25.11	0.32	10.79	nr	35.90
150 mm	22.03	27.88	0.40	13.47	nr	41.35
Universal connector						
50 × 56/48/40 mm	10.33	13.07	0.33	11.11	nr	24.18
Change piece; BS416						
100 mm	22.03	27.88	0.47	15.84	nr	43.72
WC connector						
100 mm	35.44	44.86	0.49	16.51	nr	61.37
Boss pipe; 2" BSPT socket						
50 mm	28.30	35.82	0.58	19.54	nr	55.36
75 mm	41.42	52.42	0.65	21.90	nr	74.32
100 mm	49.48	62.62	0.79	26.61	nr	89.23
150 mm	77.09	97.56	0.86	28.97	nr	126.53
Boss pipe; 2" BSPT socket; 135°						
100 mm	59.62	75.45	0.79	26.61	nr	102.06
Boss pipe; 2 × 2" BSPT socket; opposed						
75 mm	54.88	69.45	0.65	21.90	nr	91.35
100 mm	60.77	76.91	0.79	26.61	nr	103.52
Boss pipe; 2 × 2" BSPT socket; in line						
100 mm	64.61	81.77	0.79	26.61	nr	108.38
Boss pipe; 2 × 2" BSPT socket; 90°						
100 mm	63.98	80.98	0.79	26.61	nr	107.59
Bend; short radius						
50 mm	20.02	25.33	0.50	16.84	nr	42.17
75 mm	22.65	28.66	0.60	20.23	nr	48.89
100 mm	27.69	35.04	0.67	22.58	nr	57.62
100 mm; 11°	23.87	30.21	0.67	22.58	nr	52.79
100 mm; 67°	27.69	35.04	0.67	22.58	nr	57.62
150 mm	49.48	62.62	0.83	27.98	nr	90.60
Access bend; short radius						
50 mm	49.31	62.41	0.50	16.84	nr	79.25
75 mm	53.49	67.69	0.60	20.23	nr	87.92
100 mm	58.60	74.17	0.67	22.58	nr	96.75
100 mm; 45°	58.60	74.17	0.67	22.58	nr	96.75
150 mm	83.19	105.28	0.83	27.98	nr	133.26
150 mm; 45°	83.19	105.28	0.83	27.98	nr	133.26
Long radius bend						
75 mm	37.78	47.81	0.60	20.23	nr	68.04
100 mm	44.87	56.78	0.67	22.58	nr	79.36
100 mm; 5°	27.69	35.04	0.67	22.58	nr	57.62
150 mm	97.78	123.75	0.83	27.98	nr	151.73
150 mm; 22.5°	102.51	129.74	0.83	27.98	nr	157.72

33 DRAINAGE ABOVE GROUND

Item	Net Price £	Material £	Labour hours	Labour £	Unit	Total rate £
DISPOSAL SYSTEMS – cont						
Extra over fittings nitrile gasket cast iron pipework BS 416/6087, with jointing couplings measured separately – cont						
Access bend; long radius						
75 mm	67.04	84.85	0.60	20.23	nr	**105.08**
100 mm	75.78	95.91	0.67	22.58	nr	**118.49**
150 mm	133.45	168.90	0.83	27.98	nr	**196.88**
Long tail bend						
100 × 250 mm long	35.78	45.28	0.70	23.59	nr	**68.87**
100 × 815 mm long	113.74	143.95	0.70	23.59	nr	**167.54**
Offset						
75 mm projection						
75 mm	29.53	37.37	0.53	17.85	nr	**55.22**
100 mm	29.14	36.88	0.66	22.23	nr	**59.11**
115 mm projection						
75 mm	32.45	41.07	0.53	17.85	nr	**58.92**
100 mm	36.39	46.05	0.66	22.23	nr	**68.28**
150 mm projection						
75 mm	38.13	48.26	0.53	17.85	nr	**66.11**
100 mm	38.13	48.26	0.66	22.23	nr	**70.49**
225 mm projection						
100 mm	41.69	52.76	0.66	22.23	nr	**74.99**
300 mm projection						
100 mm	44.87	56.78	0.66	22.23	nr	**79.01**
Branch; equal and unequal						
50 mm	30.10	38.09	0.78	26.28	nr	**64.37**
75 mm	34.07	43.12	0.85	28.65	nr	**71.77**
100 mm	42.81	54.19	1.00	33.69	nr	**87.88**
150 mm	106.12	134.31	1.20	40.44	nr	**174.75**
150 × 100 mm; 87.5°	81.21	102.78	1.21	40.60	nr	**143.38**
150 × 100 mm; 45°	119.12	150.76	1.21	40.60	nr	**191.36**
Branch; 2" BSPT screwed socket						
100 mm	57.35	72.59	1.00	33.69	nr	**106.28**
Branch; long tail						
100 × 915 mm long	118.58	150.08	1.00	33.69	nr	**183.77**
Access branch; equal and unequal						
50 mm	65.57	82.98	0.78	26.28	nr	**109.26**
75 mm	65.57	82.98	0.85	28.65	nr	**111.63**
100 mm	73.72	93.30	1.02	34.37	nr	**127.67**
150 mm	156.91	198.59	1.20	40.44	nr	**239.03**
150 × 100 mm; 87.5°	131.87	166.89	1.20	40.44	nr	**207.33**
150 × 100 mm; 45°	142.95	180.91	1.20	40.44	nr	**221.35**
Parallel branch						
100 mm	44.87	56.78	1.00	33.69	nr	**90.47**
Double branch						
75 mm	50.59	64.03	0.95	32.01	nr	**96.04**
100 mm	52.95	67.01	1.30	43.81	nr	**110.82**
150 × 100 mm	149.11	188.71	1.56	52.56	nr	**241.27**

33 DRAINAGE ABOVE GROUND

Item	Net Price £	Material £	Labour hours	Labour £	Unit	Total rate £
Double access branch						
100 mm	83.85	106.12	1.43	48.19	nr	154.31
Corner branch						
100 mm	75.28	95.28	1.30	43.81	nr	139.09
Puddle flange; grey epoxy coated						
100 mm	50.34	63.71	1.00	33.69	nr	97.40
Roof vent connector; asphalt						
75 mm	61.26	77.53	0.90	30.32	nr	107.85
100 mm	47.84	60.55	0.97	32.68	nr	93.23
P trap						
100 mm	44.39	56.18	1.00	33.69	nr	89.87
P trap with access						
50 mm	67.94	85.98	0.77	25.95	nr	111.93
75 mm	67.94	85.98	0.90	30.32	nr	116.30
100 mm	75.30	95.30	1.16	39.08	nr	134.38
150 mm	131.48	166.40	1.77	59.64	nr	226.04
Bellmouth gully inlet						
100 mm	64.25	81.31	1.08	36.39	nr	117.70
Balcony gully inlet						
100 mm	193.99	245.52	1.08	36.39	nr	281.91
Roof outlet						
Flat grate						
75 mm	119.75	151.56	0.83	27.96	nr	179.52
100 mm	168.47	213.21	1.08	36.39	nr	249.60
Dome grate						
75 mm	119.75	151.56	0.83	27.96	nr	179.52
100 mm	188.97	239.16	1.08	36.39	nr	275.55
Top hat						
100 mm	255.83	323.78	1.08	36.39	nr	360.17
Cast iron pipe; EPDM rubber gasket joint with continuity clip; BS EN877; fixed to backgrounds						
Pipe, with brackets and jointing couplings measured separately						
50 mm	24.29	30.74	0.31	10.44	m	41.18
70 mm	28.39	35.93	0.34	11.46	m	47.39
100 mm	33.79	42.76	0.37	12.47	m	55.23
125 mm	54.22	68.62	0.65	21.90	m	90.52
150 mm	66.96	84.74	0.70	23.58	m	108.32
200 mm	111.90	141.62	1.14	38.42	m	180.04
250 mm	156.35	197.88	1.25	42.11	m	239.99
300 mm	194.70	246.41	1.53	51.54	m	297.95

33 DRAINAGE ABOVE GROUND

Item	Net Price £	Material £	Labour hours	Labour £	Unit	Total rate £
DISPOSAL SYSTEMS – cont						
Fixings						
Brackets; fixed to backgrounds. For minimum fixing distances, refer to the Tables and Memoranda at the rear of the book						
Ductile iron						
50 mm	10.32	13.06	0.10	3.37	nr	**16.43**
70 mm	10.32	13.06	0.10	3.37	nr	**16.43**
100 mm	11.93	15.10	0.15	5.05	nr	**20.15**
150 mm	22.08	27.94	0.20	6.73	nr	**34.67**
200 mm	81.92	103.68	0.25	8.43	nr	**112.11**
Mild steel; vertical						
125 mm	20.11	25.45	0.15	5.05	nr	**30.50**
Mild steel; stand off						
250 mm	43.77	55.40	0.25	8.43	nr	**63.83**
300 mm	48.22	61.03	0.25	8.43	nr	**69.46**
Stack support; rubber seal						
70 mm	40.94	51.81	0.55	18.52	nr	**70.33**
100 mm	45.55	57.65	0.65	21.90	nr	**79.55**
125 mm	50.52	63.94	0.74	24.93	nr	**88.87**
150 mm	72.09	91.24	0.86	28.97	nr	**120.21**
Wall spacer plate; cast iron (eared sockets)						
50 mm	8.79	11.12	0.10	3.37	nr	**14.49**
70 mm	8.79	11.12	0.10	3.37	nr	**14.49**
100 mm	8.79	11.12	0.10	3.37	nr	**14.49**
Extra over fittings EPDM rubber jointed cast iron pipework BS EN 877, with jointing couplings measured separately						
Coupling						
50 mm	11.03	13.96	0.10	3.37	nr	**17.33**
70 mm	12.14	15.37	0.10	3.37	nr	**18.74**
100 mm	15.81	20.01	0.10	3.37	nr	**23.38**
125 mm	19.64	24.85	0.15	5.05	nr	**29.90**
150 mm	31.68	40.10	0.30	10.11	nr	**50.21**
200 mm	70.87	89.69	0.35	11.79	nr	**101.48**
250 mm	101.52	128.49	0.40	13.47	nr	**141.96**
300 mm	117.47	148.67	0.50	16.84	nr	**165.51**
Plain socket						
50 mm	27.73	35.09	0.10	3.37	nr	**38.46**
70 mm	27.73	35.09	0.10	3.37	nr	**38.46**
100 mm	31.83	40.29	0.10	3.37	nr	**43.66**
150 mm	53.68	67.94	0.10	3.37	nr	**71.31**
Eared socket						
50 mm	28.55	36.13	0.25	8.43	nr	**44.56**
70 mm	28.55	36.13	0.25	8.43	nr	**44.56**
100 mm	34.59	43.78	0.25	8.43	nr	**52.21**
150 mm	56.72	71.78	0.25	8.43	nr	**80.21**

33 DRAINAGE ABOVE GROUND

Item	Net Price £	Material £	Labour hours	Labour £	Unit	Total rate £
Slip socket						
50 mm	35.91	45.45	0.25	8.43	nr	53.88
70 mm	35.91	45.45	0.25	8.43	nr	53.88
100 mm	41.94	53.08	0.25	8.43	nr	61.51
150 mm	63.12	79.89	0.25	8.43	nr	88.32
Stack support pipe						
70 mm	26.49	33.52	0.74	24.93	nr	58.45
100 mm	36.39	46.05	0.88	29.66	nr	75.71
125 mm	38.92	49.26	1.00	33.69	nr	82.95
150 mm	52.76	66.77	1.19	40.10	nr	106.87
Access pipe; round door						
50 mm	50.88	64.39	0.27	9.09	nr	73.48
70 mm	53.83	68.13	0.30	10.11	nr	78.24
100 mm	59.17	74.88	0.32	10.79	nr	85.67
150 mm	107.10	135.54	0.71	23.92	nr	159.46
Access pipe; square door						
100 mm	114.41	144.79	0.32	10.79	nr	155.58
125 mm	119.07	150.70	0.67	22.58	nr	173.28
150 mm	179.14	226.72	0.71	23.92	nr	250.64
200 mm	355.85	450.36	1.21	40.77	nr	491.13
250 mm	559.91	708.62	1.31	44.13	nr	752.75
300 mm	698.44	883.95	1.41	47.50	nr	931.45
Taper reducer						
70 mm	29.38	37.18	0.30	10.11	nr	47.29
100 mm	34.54	43.71	0.32	10.79	nr	54.50
125 mm	34.73	43.95	0.64	21.57	nr	65.52
150 mm	66.31	83.92	0.67	22.58	nr	106.50
200 mm	107.68	136.28	1.15	38.74	nr	175.02
250 mm	222.42	281.49	1.25	42.11	nr	323.60
300 mm	305.76	386.97	1.37	46.16	nr	433.13
Blank cap						
50 mm	7.69	9.73	0.24	8.09	nr	17.82
70 mm	8.12	10.28	0.26	8.76	nr	19.04
100 mm	9.43	11.94	0.32	10.79	nr	22.73
125 mm	13.35	16.90	0.35	11.79	nr	28.69
150 mm	13.65	17.27	0.40	13.47	nr	30.74
200 mm	60.26	76.26	0.60	20.22	nr	96.48
250 mm	123.14	155.85	0.65	21.90	nr	177.75
300 mm	175.49	222.10	0.72	24.26	nr	246.36
Blank cap; 50 mm screwed tapping						
70 mm	18.10	22.90	0.26	8.76	nr	31.66
100 mm	19.55	24.74	0.32	10.79	nr	35.53
150 mm	23.48	29.71	0.40	13.47	nr	43.18
Universal connector; EPDM rubber						
50 × 56/48/40 mm	15.70	19.87	0.30	10.11	nr	29.98
Blank end; push fit						
100 × 38/32 mm	17.52	22.18	0.32	10.79	nr	32.97
Boss pipe; 2" BSPT socket						
50 mm	43.02	54.44	0.27	9.09	nr	63.53
75 mm	43.02	54.44	0.30	10.11	nr	64.55
100 mm	52.57	66.53	0.32	10.79	nr	77.32
150 mm	85.72	108.48	0.71	23.92	nr	132.40

33 DRAINAGE ABOVE GROUND

Item	Net Price £	Material £	Labour hours	Labour £	Unit	Total rate £
DISPOSAL SYSTEMS – cont						
Extra over fittings EPDM rubber jointed cast iron pipework BS EN 877, with jointing couplings measured separately – cont						
Boss pipe; 2 × 2" BSPT socket; opposed						
100 mm	67.93	85.97	0.32	10.79	nr	**96.76**
Boss pipe; 2 × 2" BSPT socket; 90°						
100 mm	67.93	85.97	0.32	10.79	nr	**96.76**
Manifold connector						
100 mm	104.87	132.72	0.64	21.57	nr	**154.29**
150 mm	146.13	184.95	1.00	33.69	nr	**218.64**
Bend; short radius						
50 mm	19.10	24.17	0.27	9.09	nr	**33.26**
70 mm	21.50	27.20	0.30	10.11	nr	**37.31**
100 mm	25.44	32.20	0.32	10.79	nr	**42.99**
125 mm	45.14	57.13	0.62	20.89	nr	**78.02**
150 mm	45.70	57.84	0.67	22.58	nr	**80.42**
200 mm; 45°	136.05	172.19	1.21	40.77	nr	**212.96**
250 mm; 45°	265.47	335.98	1.31	44.13	nr	**380.11**
300 mm; 45°	373.44	472.63	1.43	48.18	nr	**520.81**
Access bend; short radius						
70 mm	41.66	52.73	0.30	10.11	nr	**62.84**
100 mm	60.85	77.01	0.32	10.79	nr	**87.80**
150 mm	107.09	135.53	0.67	22.58	nr	**158.11**
Bend; long radius bend						
100 mm; 88°	64.66	81.84	0.32	10.79	nr	**92.63**
100 mm; 22°	47.69	60.36	0.32	10.79	nr	**71.15**
150 mm; 88°	185.38	234.62	0.67	22.58	nr	**257.20**
Access bend; long radius						
100 mm	78.72	99.62	0.32	10.79	nr	**110.41**
150 mm	191.24	242.03	0.32	10.79	nr	**252.82**
Bend; long tail						
100 mm	44.60	56.45	0.32	10.79	nr	**67.24**
Bend; long tail double						
70 mm	69.10	87.45	0.30	10.11	nr	**97.56**
100 mm	76.23	96.48	0.32	10.79	nr	**107.27**
Offset; 75 mm projection						
100 mm	39.11	49.49	0.32	10.79	nr	**60.28**
Offset; 130 mm projection						
50 mm	32.35	40.95	0.27	9.09	nr	**50.04**
70 mm	49.09	62.13	0.30	10.11	nr	**72.24**
100 mm	64.48	81.60	0.32	10.79	nr	**92.39**
125 mm	81.71	103.41	0.67	22.58	nr	**125.99**
Branch; equal and unequal						
50 mm	30.62	38.75	0.37	12.47	nr	**51.22**
70 mm	32.34	40.92	0.40	13.47	nr	**54.39**
100 mm	44.37	56.16	0.42	14.15	nr	**70.31**
125 mm	88.93	112.55	0.76	25.60	nr	**138.15**
150 mm	96.76	122.46	0.97	32.68	nr	**155.14**
200 mm	248.21	314.14	1.51	50.88	nr	**365.02**

33 DRAINAGE ABOVE GROUND

Item	Net Price £	Material £	Labour hours	Labour £	Unit	Total rate £
250 mm	317.20	401.45	1.63	54.91	nr	**456.36**
300 mm	523.39	662.40	1.77	59.64	nr	**722.04**
Branch; radius; equal and unequal						
70 mm	39.40	49.86	0.40	13.47	nr	**63.33**
100 mm	51.87	65.64	0.42	14.15	nr	**79.79**
150 mm	112.18	141.97	1.37	46.16	nr	**188.13**
200 mm	315.67	399.52	1.51	50.88	nr	**450.40**
Branch; long tail						
100 mm	144.69	183.12	0.52	17.52	nr	**200.64**
Access branch; radius; equal and unequal						
70 mm	57.78	73.12	0.40	13.47	nr	**86.59**
100 mm	78.32	99.12	0.42	14.15	nr	**113.27**
150 mm	186.99	236.66	0.97	32.68	nr	**269.34**
Double branch; equal and unequal						
100 mm	43.68	55.28	0.52	17.52	nr	**72.80**
100 mm; 70°	65.16	82.47	0.52	17.52	nr	**99.99**
150 mm	186.99	236.66	1.37	46.16	nr	**282.82**
200 mm	319.80	404.73	1.51	50.88	nr	**455.61**
Double branch; radius; equal and unequal						
100 mm	56.34	71.30	0.52	17.52	nr	**88.82**
150 mm	230.66	291.93	1.37	46.16	nr	**338.09**
Corner branch						
100 mm	113.97	144.24	0.52	17.52	nr	**161.76**
150 mm	126.01	159.48	0.52	17.52	nr	**177.00**
Corner branch; long tail						
100 mm	170.03	215.19	0.52	17.52	nr	**232.71**
Roof vent connector; asphalt						
100 mm	71.70	90.74	0.32	10.79	nr	**101.53**
Movement connector						
100 mm	91.07	115.26	0.32	10.79	nr	**126.05**
150 mm	168.59	213.37	0.67	22.58	nr	**235.95**
Expansion plugs						
70 mm	20.26	25.64	0.32	10.79	nr	**36.43**
100 mm	25.25	31.95	0.39	13.14	nr	**45.09**
150 mm	45.32	57.36	0.55	18.52	nr	**75.88**
P trap						
100 mm dia.	47.23	59.77	0.32	10.79	nr	**70.56**
P trap with access						
50 mm	72.13	91.29	0.27	9.09	nr	**100.38**
70 mm	72.13	91.29	0.30	10.11	nr	**101.40**
100 mm	78.13	98.88	0.32	10.79	nr	**109.67**
150 mm	139.69	176.79	0.67	22.58	nr	**199.37**
Branch trap						
100 mm	171.19	216.65	0.42	14.15	nr	**230.80**
Stench trap						
100 mm	332.91	421.33	0.42	14.15	nr	**435.48**
Balcony gully inlet						
100 mm	188.96	239.14	1.00	33.69	nr	**272.83**

33 DRAINAGE ABOVE GROUND

Item	Net Price £	Material £	Labour hours	Labour £	Unit	Total rate £
DISPOSAL SYSTEMS – cont						
Extra over fittings EPDM rubber jointed cast iron pipework BS EN 877, with jointing couplings measured separately – cont						
Roof outlet						
Flat grate						
70 mm	116.65	147.63	1.00	33.69	nr	**181.32**
100 mm	164.09	207.67	1.00	33.69	nr	**241.36**
Dome grate						
70 mm	116.65	147.63	1.00	33.69	nr	**181.32**
100 mm	184.04	232.93	1.00	33.69	nr	**266.62**
Top hat						
100 mm	249.18	315.36	1.00	33.69	nr	**349.05**
Floor drains; for cast iron pipework BS 416 and BS EN877						
Adjustable clamp plate body						
100 mm; 165 mm nickel bronze grate and frame	111.08	140.58	0.50	16.84	nr	**157.42**
100 mm; 165 mm nickel bronze rodding eye	129.26	163.59	0.50	16.84	nr	**180.43**
100 mm; 150 × 150 mm nickel bronze grate and frame	124.63	157.73	0.50	16.84	nr	**174.57**
100 mm; 150 × 150 mm nickel bronze rodding eye	129.26	163.59	0.50	16.84	nr	**180.43**
Deck plate body						
100 mm; 165 mm nickel bronze grate and frame	111.08	140.58	0.50	16.84	nr	**157.42**
100 mm; 165 mm nickel bronze rodding eye	129.26	163.59	0.50	16.84	nr	**180.43**
100 mm; 150 × 150 mm nickel bronze grate and frame	124.63	157.73	0.50	16.84	nr	**174.57**
100 mm; 150 × 150 mm nickel bronze rodding eye	129.26	163.59	0.50	16.84	nr	**180.43**
Extra for						
100 mm; Screwed extension piece	43.32	54.82	0.30	10.11	nr	**64.93**
100 mm; Grating extension piece; screwed or spigot	32.86	41.59	0.30	10.11	nr	**51.70**
100 mm; Brewary trap	1195.46	1512.97	2.00	67.38	nr	**1580.35**
High density polyethylene (HDPE) pipes and fittings						
Pipe, with brackets and jointing couplings measured separately						
40 mm dia.	3.32	4.20	0.47	15.84	m	**20.04**
50 mm dia.	3.94	4.98	0.53	17.85	m	**22.83**
56 mm dia.	4.73	5.98	0.60	20.23	m	**26.21**

33 DRAINAGE ABOVE GROUND

Item	Net Price £	Material £	Labour hours	Labour £	Unit	Total rate £
63 mm dia.	5.43	6.88	0.60	20.23	m	**27.11**
75 mm dia.	6.25	7.91	0.90	30.32	m	**38.23**
90 mm dia.	9.29	11.76	0.90	30.32	m	**42.08**
110 mm dia.	11.76	14.88	1.10	37.06	m	**51.94**
125 mm dia.	16.69	21.12	1.20	40.44	m	**61.56**
160 mm dia.	26.94	34.09	1.48	49.86	m	**83.95**
200 mm dia.	34.91	44.18	1.77	59.64	m	**103.82**
250 mm dia.	55.96	70.82	1.75	59.00	m	**129.82**
315 mm dia.	86.47	109.44	1.90	64.02	m	**173.46**
Extra over fittings						
Coupler						
40 mm dia.	5.03	6.36	0.48	16.17	nr	**22.53**
50 mm dia.	5.18	6.55	0.52	17.52	nr	**24.07**
56 mm dia.	5.21	6.60	0.58	19.54	nr	**26.14**
63 mm dia.	6.95	8.79	0.58	19.54	nr	**28.33**
90 mm dia.	7.27	9.21	0.67	22.58	nr	**31.79**
110 mm dia.	7.77	9.83	0.74	24.93	nr	**34.76**
125 mm dia.	13.38	16.93	0.83	27.96	nr	**44.89**
160 mm dia.	18.61	23.55	1.00	33.69	nr	**57.24**
200 mm dia.	129.58	164.00	1.35	45.48	nr	**209.48**
250 mm dia.	214.22	271.12	1.50	50.53	nr	**321.65**
315 mm dia.	242.98	307.52	1.80	60.65	nr	**368.17**
Cap						
40 mm dia.	2.88	3.64	0.24	8.09	nr	**11.73**
50 mm dia.	2.46	3.11	0.26	8.76	nr	**11.87**
56 mm dia.	2.46	3.11	0.30	10.11	nr	**13.22**
63 mm dia.	2.46	3.11	0.32	10.79	nr	**13.90**
75 mm dia.	2.46	3.11	0.35	11.79	nr	**14.90**
90 mm dia.	3.11	3.93	0.37	12.47	nr	**16.40**
110 mm dia.	5.82	7.37	0.40	13.47	nr	**20.84**
125 mm dia.	7.31	9.25	0.46	15.50	nr	**24.75**
160 mm dia.	23.37	29.58	0.50	16.84	nr	**46.42**
200 mm dia.	33.11	41.90	0.68	22.90	nr	**64.80**
250 mm dia.	42.87	54.25	0.75	25.26	nr	**79.51**
315 mm dia.	48.72	61.66	0.90	30.32	nr	**91.98**
Reducer						
50 mm dia.	2.57	3.25	0.54	18.19	nr	**21.44**
56 mm dia.	2.81	3.56	0.60	20.22	nr	**23.78**
63 mm dia.	2.81	3.56	0.67	22.58	nr	**26.14**
75 mm dia.	3.29	4.17	0.67	22.58	nr	**26.75**
90 mm dia.	4.21	5.33	0.67	22.58	nr	**27.91**
110 mm dia.	5.16	6.53	0.74	24.93	nr	**31.46**
125 mm dia.	8.19	10.36	0.83	27.96	nr	**38.32**
160 mm dia.	9.17	11.60	1.10	37.06	nr	**48.66**
200 mm dia.	58.35	73.85	1.40	47.17	nr	**121.02**
Bend; 45°						
40 mm dia.	2.30	2.91	0.45	15.16	nr	**18.07**
50 mm dia.	2.31	2.92	0.50	16.84	nr	**19.76**
56 mm dia.	2.71	3.43	0.55	18.52	nr	**21.95**

33 DRAINAGE ABOVE GROUND

Item	Net Price £	Material £	Labour hours	Labour £	Unit	Total rate £
DISPOSAL SYSTEMS – cont						
Extra over fittings – cont						
63 mm dia.	3.48	4.40	0.58	19.54	nr	23.94
75 mm dia.	4.46	5.64	0.67	22.58	nr	28.22
90 mm dia.	6.39	8.09	0.67	22.58	nr	30.67
110 mm dia.	6.39	8.09	0.74	24.93	nr	33.02
125 mm dia.	10.30	13.04	0.83	27.96	nr	41.00
160 mm dia.	27.05	34.24	1.00	33.69	nr	67.93
200 mm dia.	74.60	94.42	1.40	47.15	nr	141.57
250 mm dia.	121.36	153.60	1.80	60.65	nr	214.25
315 mm dia.	168.31	213.01	2.40	80.86	nr	293.87
Bend; 90 °						
40 mm dia.	2.30	46.42	0.40	13.47	nr	59.89
50 mm dia.	2.71	43.51	0.50	16.84	nr	60.35
56 mm dia.	3.29	57.55	0.45	15.16	nr	72.71
63 mm dia.	3.87	53.38	0.58	19.54	nr	72.92
75 mm dia.	5.82	82.45	0.67	22.58	nr	105.03
90 mm dia.	7.79	119.95	0.64	21.57	nr	141.52
110 mm dia.	7.79	119.95	0.74	24.93	nr	144.88
125 mm dia.	12.05	138.30	0.83	27.96	nr	166.26
160 mm dia.	26.27	261.27	1.00	33.69	nr	294.96
200 mm dia.	128.37	722.46	1.40	47.17	nr	769.63
250 mm dia.	214.32	883.78	1.80	60.65	nr	944.43
315 mm dia.	311.74	1161.35	2.40	80.86	nr	1242.21
Equal tee						
40 mm dia.	4.59	5.81	0.60	20.22	nr	26.03
50 mm dia.	5.82	7.37	0.70	23.58	nr	30.95
56 mm dia.	5.82	7.37	0.70	23.58	nr	30.95
63 mm dia.	6.05	7.66	0.75	25.26	nr	32.92
75 mm dia.	7.16	9.06	0.83	27.96	nr	37.02
90 mm dia.	9.12	11.55	0.87	29.31	nr	40.86
110 mm dia.	10.49	13.27	1.00	33.69	nr	46.96
125 mm dia.	15.54	19.67	1.08	36.39	nr	56.06
160 mm dia.	58.43	73.95	1.35	45.48	nr	119.43
200 mm dia.	115.89	146.68	1.90	64.02	nr	210.70
250 mm dia.	165.61	209.60	2.70	90.96	nr	300.56
315 mm dia.	233.80	295.89	3.60	121.28	nr	417.17
Reducing tee						
50 mm dia.	5.82	7.37	0.70	23.58	nr	30.95
56 mm dia.	5.82	7.37	0.70	23.58	nr	30.95
63 mm dia.	6.05	7.66	0.75	25.26	nr	32.92
75 mm dia.	7.16	9.06	0.83	27.96	nr	37.02
90 mm dia.	9.12	11.55	0.87	29.31	nr	40.86
110 mm dia.	10.49	13.27	1.00	33.69	nr	46.96
125 mm dia.	15.54	19.67	1.08	36.39	nr	56.06
160 mm dia.	58.43	73.95	1.35	45.48	nr	119.43
200 mm dia.	115.89	146.68	1.90	64.02	nr	210.70
250 mm dia.	165.61	209.60	2.70	90.96	nr	300.56
315 mm dia.	233.80	295.89	3.60	121.28	nr	417.17

33 DRAINAGE ABOVE GROUND

Item	Net Price £	Material £	Labour hours	Labour £	Unit	Total rate £
Y-branch 45°						
40 mm dia.	5.61	7.10	0.65	21.90	nr	**29.00**
50 mm dia.	5.82	7.37	0.70	23.58	nr	**30.95**
56 mm dia.	5.82	7.37	0.70	23.58	nr	**30.95**
63 mm dia.	6.05	7.66	0.75	25.26	nr	**32.92**
75 mm dia.	6.79	8.59	0.83	27.96	nr	**36.55**
90 mm dia.	9.70	12.28	0.87	29.31	nr	**41.59**
110 mm dia.	11.65	14.74	1.00	33.69	nr	**48.43**
125 mm dia.	15.54	19.67	1.08	36.39	nr	**56.06**
160 mm dia.	53.76	68.04	1.35	45.48	nr	**113.52**
200 mm dia.	173.39	219.44	1.90	64.02	nr	**283.46**
250 mm dia.	292.26	369.88	2.70	90.96	nr	**460.84**
315 mm dia.	428.65	542.49	3.60	121.28	nr	**663.77**
Reducing Y-branch 45°						
50 mm dia.	5.82	7.37	0.70	23.58	nr	**30.95**
56 mm dia.	5.82	7.37	0.70	23.58	nr	**30.95**
63 mm dia.	6.05	7.66	0.75	25.26	nr	**32.92**
75 mm dia.	6.79	8.59	0.83	27.96	nr	**36.55**
90 mm dia.	10.11	12.79	0.87	29.31	nr	**42.10**
110 mm dia.	11.65	14.74	1.00	33.69	nr	**48.43**
125 mm dia.	15.54	19.67	1.08	36.39	nr	**56.06**
160 mm dia.	72.46	91.71	1.35	45.48	nr	**137.19**
200 mm dia.	115.89	146.68	1.90	64.02	nr	**210.70**
250 mm dia.	214.32	271.24	2.70	90.96	nr	**362.20**
315 mm dia.	360.44	456.18	3.60	121.28	nr	**577.46**

38 PIPED SUPPLY SYSTEMS

Item	Net Price £	Material £	Labour hours	Labour £	Unit	Total rate £
COLD WATER PIPELINES: COPPER PIPEWORK						
Copper pipe; capillary or compression joints in the running length; EN1057 R250 (TX) formerly BS 2871 Table X						
Fixed vertically or at low level, with brackets measured separately						
12 mm dia.	1.77	2.24	0.39	13.14	m	**15.38**
15 mm dia.	1.99	2.52	0.40	13.47	m	**15.99**
22 mm dia.	4.00	5.06	0.47	15.84	m	**20.90**
28 mm dia.	5.07	6.42	0.51	17.19	m	**23.61**
35 mm dia.	12.02	15.21	0.58	19.54	m	**34.75**
42 mm dia.	14.64	18.52	0.66	22.23	m	**40.75**
54 mm dia.	18.82	23.82	0.72	24.26	m	**48.08**
67 mm dia.	24.62	31.16	0.75	25.26	m	**56.42**
76 mm dia.	34.80	44.04	0.76	25.60	m	**69.64**
108 mm dia.	50.08	63.38	0.78	26.28	m	**89.66**
133 mm dia.	64.97	82.23	1.05	35.37	m	**117.60**
159 mm dia.	102.98	130.33	1.15	38.74	m	**169.07**
Fixed horizontally at high level or suspended, with brackets measured separately						
12 mm dia.	1.77	2.24	0.45	15.16	m	**17.40**
15 mm dia.	1.99	2.52	0.46	15.50	m	**18.02**
22 mm dia.	4.00	5.06	0.54	18.19	m	**23.25**
28 mm dia.	5.07	12.84	0.59	19.88	m	**32.72**
35 mm dia.	12.02	15.21	0.67	22.58	m	**37.79**
42 mm dia.	14.64	18.52	0.76	25.60	m	**44.12**
54 mm dia.	18.82	23.82	0.83	27.96	m	**51.78**
67 mm dia.	24.62	31.16	0.86	28.97	m	**60.13**
76 mm dia.	34.80	44.04	0.87	29.31	m	**73.35**
108 mm dia.	50.08	63.38	0.90	30.32	m	**93.70**
133 mm dia.	64.97	82.23	1.21	40.77	m	**123.00**
159 mm dia.	102.98	130.33	1.32	44.48	m	**174.81**
Copper pipe; capillary or compression joints in the running length; EN1057 R250 (TY) formerly BS 2871 Table Y						
Fixed vertically or at low level with brackets measured separately (Refer to Copper Pipe Table X Section)						
12 mm dia.	2.29	2.90	0.41	13.81	m	**16.71**
15 mm dia.	3.27	4.14	0.43	14.49	m	**18.63**
22 mm dia.	5.76	7.29	0.50	16.84	m	**24.13**
28 mm dia.	7.44	9.42	0.54	18.19	m	**27.61**
35 mm dia.	10.86	13.74	0.62	20.89	m	**34.63**
42 mm dia.	13.16	16.65	0.71	23.92	m	**40.57**
54 mm dia.	22.44	28.40	0.78	26.28	m	**54.68**
67 mm dia.	29.81	37.73	0.82	27.63	m	**65.36**
76 mm dia.	43.56	55.13	0.60	20.22	m	**75.35**
108 mm dia.	60.48	76.54	0.88	29.66	m	**106.20**

38 PIPED SUPPLY SYSTEMS

Item	Net Price £	Material £	Labour hours	Labour £	Unit	Total rate £
Copper pipe; capillary or compression joints in the running length; EN1057 R250 (TX) formerly BS 2871 Table X						
Plastic coated gas and cold water service pipe for corrosive environments, fixed vertically or at low level with brackets measured separately						
15 mm dia. (white)	4.57	5.78	0.59	19.88	m	**25.66**
22 mm dia. (white)	8.11	10.26	0.68	22.92	m	**33.18**
28 mm dia. (white)	8.37	10.60	0.74	24.93	m	**35.53**
Fixings						
Saddle band						
6 mm dia.	0.06	0.08	0.05	1.72	nr	**1.80**
8 mm dia.	0.06	0.08	0.07	2.31	nr	**2.39**
10 mm dia.	0.09	0.11	0.09	2.89	nr	**3.00**
12 mm dia.	0.09	0.11	0.11	3.61	nr	**3.72**
15 mm dia.	0.09	0.11	0.13	4.38	nr	**4.49**
22 mm dia.	0.19	0.24	0.13	4.38	nr	**4.62**
28 mm dia.	0.09	0.11	0.16	5.39	nr	**5.50**
35 mm dia.	0.34	0.43	0.18	6.06	nr	**6.49**
42 mm dia.	0.71	0.90	0.21	7.08	nr	**7.98**
54 mm dia.	0.95	1.20	0.21	7.08	nr	**8.28**
Single spacing clip						
15 mm dia.	0.20	0.26	0.14	4.72	nr	**4.98**
22 mm dia.	0.20	0.26	0.15	5.05	nr	**5.31**
28 mm dia.	0.47	0.59	0.17	5.73	nr	**6.32**
Two piece spacing clip						
8 mm dia. Bottom	0.16	0.20	0.11	3.71	nr	**3.91**
8 mm dia. Top	0.16	0.20	0.11	3.71	nr	**3.91**
12 mm dia. Bottom	0.16	0.20	0.13	4.38	nr	**4.58**
12 mm dia. Top	0.16	0.20	0.13	4.38	nr	**4.58**
15 mm dia. Bottom	0.19	0.24	0.13	4.38	nr	**4.62**
15 mm dia. Top	0.19	0.24	0.13	4.38	nr	**4.62**
22 mm dia. Bottom	0.19	0.24	0.14	4.72	nr	**4.96**
22 mm dia. Top	0.20	0.26	0.14	4.72	nr	**4.98**
28 mm dia. Bottom	0.19	0.24	0.16	5.39	nr	**5.63**
28 mm dia. Top	0.33	0.41	0.16	5.39	nr	**5.80**
35 mm dia. Bottom	0.20	0.26	0.21	7.08	nr	**7.34**
35 mm dia. Top	0.46	0.58	0.21	7.08	nr	**7.66**
Single pipe bracket						
15 mm dia.	1.66	2.11	0.14	4.72	nr	**6.83**
22 mm dia.	1.68	2.13	0.14	4.72	nr	**6.85**
28 mm dia.	2.31	2.92	0.17	5.73	nr	**8.65**
Single pipe ring						
15 mm dia.	0.35	0.45	0.26	8.76	nr	**9.21**
22 mm dia.	0.42	0.53	0.26	8.76	nr	**9.29**
28 mm dia.	0.46	0.58	0.31	10.44	nr	**11.02**
35 mm dia.	0.62	0.78	0.32	10.79	nr	**11.57**
42 mm dia.	0.93	1.18	0.32	10.79	nr	**11.97**
54 mm dia.	1.43	1.81	0.34	11.46	nr	**13.27**
67 mm dia.	1.64	2.07	0.35	11.79	nr	**13.86**
76 mm dia.	1.92	2.43	0.42	14.15	nr	**16.58**
108 mm dia.	3.88	4.91	0.42	14.15	nr	**19.06**

38 PIPED SUPPLY SYSTEMS

Item	Net Price £	Material £	Labour hours	Labour £	Unit	Total rate £
COLD WATER PIPELINES: COPPER PIPEWORK – cont						
Fixings – cont						
Double pipe ring						
15 mm dia.	0.37	0.47	0.26	8.76	nr	**9.23**
22 mm dia.	0.67	0.85	0.26	8.76	nr	**9.61**
28 mm dia.	0.60	0.76	0.31	10.44	nr	**11.20**
35 mm dia.	0.64	0.81	0.32	10.79	nr	**11.60**
42 mm dia.	1.40	1.77	0.32	10.79	nr	**12.56**
54 mm dia.	1.50	1.90	0.34	11.46	nr	**13.36**
67 mm dia.	2.10	2.65	0.35	11.79	nr	**14.44**
76 mm dia.	2.76	3.49	0.42	14.15	nr	**17.64**
108 mm dia.	4.85	6.14	0.42	14.15	nr	**20.29**
Wall bracket						
15 mm dia.	3.28	4.16	0.05	1.68	nr	**5.84**
22 mm dia.	4.31	5.45	0.05	1.68	nr	**7.13**
28 mm dia.	5.14	6.51	0.05	1.68	nr	**8.19**
35 mm dia.	7.69	9.73	0.05	1.68	nr	**11.41**
42 mm dia.	10.19	12.89	0.05	1.68	nr	**14.57**
54 mm dia.	15.87	20.08	0.05	1.68	nr	**21.76**
Hospital bracket						
15 mm dia.	5.42	6.85	0.26	8.76	nr	**15.61**
22 mm dia.	6.42	8.12	0.26	8.76	nr	**16.88**
28 mm dia.	7.53	9.53	0.31	10.44	nr	**19.97**
35 mm dia.	7.97	10.09	0.32	10.79	nr	**20.88**
42 mm dia.	11.39	14.41	0.32	10.79	nr	**25.20**
54 mm dia.	15.24	19.29	0.34	11.46	nr	**30.75**
Screw on backplate, female						
All sizes 15 mm to 54 mm × 10 mm	1.73	2.18	0.10	3.37	nr	**5.55**
Screw on backplate, male						
All sizes 15 mm to 54 mm × 10 mm	2.32	2.93	0.10	3.37	nr	**6.30**
Pipe joist clips, single						
15 mm dia.	1.10	1.39	0.08	2.70	nr	**4.09**
22 mm dia.	1.10	1.39	0.08	2.70	nr	**4.09**
Pipe joist clips, double						
15 mm dia.	1.53	1.94	0.08	2.70	nr	**4.64**
22 mm dia.	0.83	1.05	0.08	2.70	nr	**3.75**
Extra over channel sections for fabricated hangers and brackets						
Galvanized steel; including inserts, bolts, nuts, washers; fixed to backgrounds						
41 × 21 mm	6.79	8.59	0.29	9.77	m	**18.36**
41 × 41 mm	8.14	10.30	0.29	9.77	m	**20.07**
Threaded rods; metric thread; including nuts, washers etc.						
10 mm dia. × 600 mm long for ring clips up to 54 mm	2.19	2.77	0.18	6.06	nr	**8.83**
12 mm dia. × 600 mm long for ring clips from 54 mm	3.43	4.35	0.18	6.06	nr	**10.41**

38 PIPED SUPPLY SYSTEMS

Item	Net Price £	Material £	Labour hours	Labour £	Unit	Total rate £
Extra over copper pipes; capillary fittings; BS 864						
Stop end						
15 mm dia.	1.61	2.04	0.13	4.38	nr	6.42
22 mm dia.	2.97	3.76	0.14	4.72	nr	8.48
28 mm dia.	5.33	6.74	0.17	5.73	nr	12.47
35 mm dia.	11.75	14.87	0.19	6.41	nr	21.28
42 mm dia.	20.25	25.63	0.22	7.41	nr	33.04
54 mm dia.	28.27	35.78	0.23	7.75	nr	43.53
Straight coupling; copper to copper						
6 mm dia.	1.68	2.13	0.23	7.75	nr	9.88
8 mm dia.	1.71	2.16	0.23	7.75	nr	9.91
10 mm dia.	0.88	1.11	0.23	7.75	nr	8.86
15 mm dia.	0.19	0.24	0.23	7.75	nr	7.99
22 mm dia.	0.50	0.64	0.26	8.76	nr	9.40
28 mm dia.	1.48	1.87	0.30	10.11	nr	11.98
35 mm dia.	4.81	6.09	0.34	11.46	nr	17.55
42 mm dia.	8.03	10.16	0.38	12.81	nr	22.97
54 mm dia.	14.82	18.76	0.42	14.15	nr	32.91
67 mm dia.	43.97	55.65	0.53	17.85	nr	73.50
Adaptor coupling; imperial to metric						
½" × 15 mm dia.	3.49	4.41	0.27	9.09	nr	13.50
¾" × 22 mm dia.	3.06	3.88	0.31	10.44	nr	14.32
1" × 28 mm dia.	5.99	7.58	0.36	12.13	nr	19.71
1¼" × 35 mm dia.	10.01	12.67	0.41	13.81	nr	26.48
1½" × 42 mm dia.	12.75	16.14	0.46	15.50	nr	31.64
Reducing coupling						
15 × 10 mm dia.	3.66	4.64	0.23	7.75	nr	12.39
22 × 10 mm dia.	5.37	6.80	0.26	8.76	nr	15.56
22 × 15 mm dia.	5.74	7.27	0.27	9.09	nr	16.36
28 × 15 mm dia.	6.60	8.36	0.28	9.43	nr	17.79
28 × 22 mm dia.	6.65	8.41	0.30	10.11	nr	18.52
35 × 28 mm dia.	9.54	12.07	0.34	11.46	nr	23.53
42 × 35 mm dia.	14.00	17.72	0.38	12.81	nr	30.53
54 × 35 mm dia.	24.53	31.05	0.42	14.15	nr	45.20
54 × 42 mm dia.	26.78	33.89	0.42	14.15	nr	48.04
Straight female connector						
15 mm × ½" dia.	3.64	4.60	0.27	9.09	nr	13.69
22 mm × ¾" dia.	5.26	6.65	0.31	10.44	nr	17.09
28 mm × 1" dia.	9.95	12.59	0.36	12.13	nr	24.72
35 mm × 1¼" dia.	17.21	21.78	0.41	13.81	nr	35.59
42 mm × 1½" dia.	22.32	28.25	0.46	15.50	nr	43.75
54 mm × 2" dia.	35.42	44.82	0.52	17.52	nr	62.34
Straight male connector						
15 mm × ½" dia.	3.10	3.92	0.27	9.09	nr	13.01
22 mm × ¾" dia.	5.54	7.01	0.31	10.44	nr	17.45
28 mm × 1" dia.	8.92	11.29	0.36	12.13	nr	23.42
35 mm × 1¼" dia.	15.68	19.85	0.41	13.81	nr	33.66
42 mm × 1½" dia.	20.20	25.57	0.46	15.50	nr	41.07
54 mm × 2" dia.	30.66	38.81	0.52	17.52	nr	56.33
67 mm × 2½" dia.	48.96	61.96	0.63	21.22	nr	83.18

38 PIPED SUPPLY SYSTEMS

Item	Net Price £	Material £	Labour hours	Labour £	Unit	Total rate £
COLD WATER PIPELINES: COPPER PIPEWORK – cont						
Extra over copper pipes – cont						
Female reducing connector						
15 mm × ¾" dia.	9.03	11.42	0.27	9.09	nr	20.51
Male reducing connector						
15 mm × ¾" dia.	8.08	10.23	0.27	9.09	nr	19.32
22 mm × 1" dia.	12.29	15.56	0.31	10.44	nr	26.00
Flanged connector						
28 mm dia.	57.73	73.06	0.36	12.13	nr	85.19
35 mm dia.	73.09	92.50	0.41	13.81	nr	106.31
42 mm dia.	87.35	110.56	0.46	15.50	nr	126.06
54 mm dia.	132.06	167.14	0.52	17.52	nr	184.66
67 mm dia.	155.29	196.54	0.61	20.55	nr	217.09
Tank connector						
15 mm × ½" dia.	7.74	9.80	0.25	8.43	nr	18.23
22 mm × ¾" dia.	11.78	14.91	0.28	9.43	nr	24.34
28 mm × 1" dia.	15.49	19.60	0.32	10.79	nr	30.39
35 mm × 1¼" dia.	19.87	25.14	0.37	12.47	nr	37.61
42 mm × 1½" dia.	26.05	32.97	0.43	14.49	nr	47.46
54 mm × 2" dia.	39.82	50.40	0.46	15.50	nr	65.90
Tank connector with long thread						
15 mm × ½" dia.	10.01	12.67	0.30	10.11	nr	22.78
22 mm × ¾" dia.	14.27	18.07	0.33	11.11	nr	29.18
28 mm × 1" dia.	17.65	22.33	0.39	13.14	nr	35.47
Reducer						
15 × 10 mm dia.	1.23	1.56	0.23	7.75	nr	9.31
22 × 15 mm dia.	0.87	1.10	0.26	8.76	nr	9.86
28 × 15 mm dia.	3.25	4.11	0.28	9.43	nr	13.54
28 × 22 mm dia.	2.49	3.15	0.30	10.11	nr	13.26
35 × 22 mm dia.	9.25	11.70	0.34	11.46	nr	23.16
42 × 22 mm dia.	16.71	21.15	0.36	12.13	nr	33.28
42 × 35 mm dia.	12.93	16.36	0.38	12.81	nr	29.17
54 × 35 mm dia.	27.14	34.35	0.40	13.47	nr	47.82
54 × 42 mm dia.	23.40	29.61	0.42	14.15	nr	43.76
67 × 54 mm dia.	31.82	40.28	0.53	17.85	nr	58.13
Adaptor; copper to female iron						
15 mm × ½" dia.	6.27	7.94	0.27	9.09	nr	17.03
22 mm × ¾" dia.	9.55	12.08	0.31	10.44	nr	22.52
28 mm × 1" dia.	13.45	17.02	0.36	12.13	nr	29.15
35 mm × 1¼" dia.	24.35	30.82	0.41	13.81	nr	44.63
42 mm × 1½" dia.	30.66	38.81	0.46	15.50	nr	54.31
54 mm × 2" dia.	36.88	46.67	0.52	17.52	nr	64.19
Adaptor; copper to male iron						
15 mm × ½" dia.	6.39	8.09	0.27	9.09	nr	17.18
22 mm × ¾" dia.	8.18	10.35	0.31	10.44	nr	20.79
28 mm × 1" dia.	13.67	17.30	0.36	12.13	nr	29.43
35 mm × 1¼" dia.	19.88	25.16	0.41	13.81	nr	38.97
42 mm × 1½" dia.	27.47	34.76	0.46	15.50	nr	50.26
54 mm × 2" dia.	36.88	46.67	0.52	17.52	nr	64.19

38 PIPED SUPPLY SYSTEMS

Item	Net Price £	Material £	Labour hours	Labour £	Unit	Total rate £
Union coupling						
15 mm dia.	8.64	10.93	0.41	13.81	nr	24.74
22 mm dia.	13.84	17.52	0.45	15.16	nr	32.68
28 mm dia.	20.20	25.57	0.51	17.19	nr	42.76
35 mm dia.	26.51	33.56	0.64	21.57	nr	55.13
42 mm dia.	38.72	49.00	0.68	22.90	nr	71.90
54 mm dia.	73.69	93.26	0.78	26.28	nr	119.54
67 mm dia.	124.78	157.92	0.96	32.35	nr	190.27
Elbow						
15 mm dia.	0.34	0.43	0.23	7.75	nr	8.18
22 mm dia.	0.90	1.14	0.26	8.76	nr	9.90
28 mm dia.	2.37	3.00	0.31	10.44	nr	13.44
35 mm dia.	10.30	13.04	0.35	11.79	nr	24.83
42 mm dia.	17.01	21.53	0.41	13.81	nr	35.34
54 mm dia.	35.14	44.48	0.44	14.82	nr	59.30
67 mm dia.	91.21	115.44	0.54	18.19	nr	133.63
Backplate elbow						
15 mm dia.	6.49	8.21	0.51	17.19	nr	25.40
22 mm dia.	13.96	17.66	0.54	18.19	nr	35.85
Overflow bend						
22 mm dia.	19.66	24.89	0.26	8.76	nr	33.65
Return bend						
15 mm dia.	9.73	12.31	0.23	7.75	nr	20.06
22 mm dia.	19.12	24.20	0.26	8.76	nr	32.96
28 mm dia.	24.43	30.92	0.31	10.44	nr	41.36
Obtuse elbow						
15 mm dia.	1.26	1.59	0.23	7.75	nr	9.34
22 mm dia.	2.60	3.29	0.26	8.76	nr	12.05
28 mm dia.	4.99	6.32	0.31	10.44	nr	16.76
35 mm dia.	15.55	19.68	0.36	12.13	nr	31.81
42 mm dia.	27.68	35.03	0.41	13.81	nr	48.84
54 mm dia.	50.03	63.31	0.44	14.82	nr	78.13
67 mm dia.	90.78	114.89	0.54	18.19	nr	133.08
Straight tap connector						
15 mm × ½" dia.	1.67	2.12	0.13	4.38	nr	6.50
22 mm × ¾" dia.	2.48	3.14	0.14	4.72	nr	7.86
Bent tap connector						
15 mm × ½" dia.	2.47	3.12	0.13	4.38	nr	7.50
22 mm × ¾" dia.	7.58	9.60	0.14	4.72	nr	14.32
Bent male union connector						
15 mm × ½" dia.	12.67	16.04	0.41	13.81	nr	29.85
22 mm × ¾" dia.	16.44	20.81	0.45	15.16	nr	35.97
28 mm × 1" dia.	23.55	29.80	0.51	17.19	nr	46.99
35 mm × 1¼" dia.	38.41	48.61	0.64	21.57	nr	70.18
42 mm × 1½" dia.	62.43	79.02	0.68	22.90	nr	101.92
54 mm × 2" dia.	98.61	124.80	0.78	26.28	nr	151.08
Bent female union connector						
15 mm dia.	12.67	16.04	0.41	13.81	nr	29.85
22 mm × ¾" dia.	16.44	20.81	0.45	15.16	nr	35.97
28 mm × 1" dia.	23.55	29.80	0.51	17.19	nr	46.99

38 PIPED SUPPLY SYSTEMS

Item	Net Price £	Material £	Labour hours	Labour £	Unit	Total rate £
COLD WATER PIPELINES: COPPER PIPEWORK – cont						
Extra over copper pipes – cont						
35 mm × 1¼" dia.	38.41	48.61	0.64	21.57	nr	70.18
42 mm × 1½" dia.	62.43	79.02	0.68	22.90	nr	101.92
54 mm × 2" dia.	98.61	124.80	0.78	26.28	nr	151.08
Straight union adaptor						
15 mm × ¾" dia.	5.41	6.84	0.41	13.81	nr	20.65
22 mm × 1" dia.	7.68	9.72	0.45	15.16	nr	24.88
28 mm × 1¼" dia.	12.42	15.71	0.51	17.19	nr	32.90
35 mm × 1½" dia.	19.12	24.20	0.64	21.57	nr	45.77
42 mm × 2" dia.	24.14	30.55	0.68	22.90	nr	53.45
54 mm × 2½" dia.	37.27	47.17	0.78	26.28	nr	73.45
Straight male union connector						
15 mm × ½" dia.	10.79	13.65	0.41	13.81	nr	27.46
22 mm × ¾" dia.	14.00	17.72	0.45	15.16	nr	32.88
28 mm × 1" dia.	20.84	26.38	0.51	17.19	nr	43.57
35 mm × 1¼" dia.	30.02	37.99	0.64	21.57	nr	59.56
42 mm × 1½" dia.	47.17	59.70	0.68	22.90	nr	82.60
54 mm × 2" dia.	67.78	85.78	0.78	26.28	nr	112.06
Straight female union connector						
15 mm × ½" dia.	10.79	13.65	0.41	13.81	nr	27.46
22 mm × ¾" dia.	14.00	17.72	0.45	15.16	nr	32.88
28 mm × 1" dia.	20.84	26.38	0.51	17.19	nr	43.57
35 mm × 1¼" dia.	30.02	37.99	0.64	21.57	nr	59.56
42 mm × 1½" dia.	47.17	59.70	0.68	22.90	nr	82.60
54 mm × 2" dia.	67.78	85.78	0.78	26.28	nr	112.06
Male nipple						
¾ × ½" dia.	2.54	3.21	0.24	7.99	nr	11.20
1 × ¾" dia.	3.06	3.88	0.32	10.79	nr	14.67
1¼ × 1" dia.	3.35	4.24	0.37	12.47	nr	16.71
1½ × 1¼" dia.	12.46	15.77	0.42	14.15	nr	29.92
2 × 1½" dia.	25.50	32.27	0.46	15.50	nr	47.77
2½ × 2" dia.	33.22	42.04	0.56	18.87	nr	60.91
Female nipple						
¾ × ½" dia.	3.94	4.98	0.19	6.41	nr	11.39
1 × ¾" dia.	6.16	7.80	0.32	10.79	nr	18.59
1¼ × 1" dia.	8.42	10.65	0.37	12.47	nr	23.12
1½ × 1¼" dia.	12.46	15.77	0.42	14.15	nr	29.92
2 × 1½" dia.	25.50	32.27	0.46	15.50	nr	47.77
2½ × 2" dia.	33.22	42.04	0.56	18.87	nr	60.91
Equal tee						
10 mm dia.	3.38	4.28	0.25	8.43	nr	12.71
15 mm dia.	0.34	0.43	0.36	12.13	nr	12.56
22 mm dia.	0.90	1.14	0.39	13.14	nr	14.28
28 mm dia.	6.58	8.33	0.43	14.49	nr	22.82
35 mm dia.	16.77	21.22	0.57	19.20	nr	40.42
42 mm dia.	26.90	34.05	0.60	20.22	nr	54.27
54 mm dia.	54.24	68.64	0.65	21.90	nr	90.54
67 mm dia.	73.80	93.40	0.78	26.28	nr	119.68

38 PIPED SUPPLY SYSTEMS

Item	Net Price £	Material £	Labour hours	Labour £	Unit	Total rate £
Female tee, reducing branch FI						
15 × 15 mm × ¼" dia.	8.12	10.28	0.36	12.13	nr	**22.41**
22 × 22 mm × ½" dia.	9.93	12.57	0.39	13.16	nr	**25.73**
28 × 28 mm × ¾" dia.	19.45	24.62	0.43	14.49	nr	**39.11**
35 × 35 mm × ¾" dia.	28.08	35.54	0.47	15.84	nr	**51.38**
42 × 42 mm × ½" dia.	33.73	42.68	0.60	20.22	nr	**62.90**
Backplate tee						
15 × 15 mm × ½" dia.	15.35	19.43	0.62	20.89	nr	**40.32**
Heater tee						
½ × ½" × 15 mm dia.	13.80	17.46	0.36	12.13	nr	**29.59**
Union heater tee						
½ × ½" × 15 mm dia.	13.80	17.46	0.36	12.13	nr	**29.59**
Sweep tee – equal						
15 mm dia.	10.97	13.89	0.36	12.13	nr	**26.02**
22 mm dia.	14.12	17.88	0.39	13.14	nr	**31.02**
28 mm dia.	23.77	30.08	0.43	14.49	nr	**44.57**
35 mm dia.	33.71	42.66	0.57	19.20	nr	**61.86**
42 mm dia.	49.99	63.27	0.60	20.22	nr	**83.49**
54 mm dia.	55.36	70.07	0.65	21.90	nr	**91.97**
67 mm dia.	75.47	95.51	0.78	26.28	nr	**121.79**
Sweep tee – reducing						
22 × 22 × 15 mm dia.	11.83	14.97	0.39	13.14	nr	**28.11**
28 × 28 × 22 mm dia.	20.12	25.47	0.43	14.49	nr	**39.96**
35 × 35 × 22 mm dia.	33.71	42.66	0.57	19.20	nr	**61.86**
Sweep tee – double						
15 mm dia.	12.42	15.71	0.36	12.13	nr	**27.84**
22 mm dia.	16.90	21.39	0.39	13.14	nr	**34.53**
28 mm dia.	25.69	32.51	0.43	14.49	nr	**47.00**
Cross						
15 mm dia.	16.43	20.80	0.48	16.17	nr	**36.97**
22 mm dia.	18.34	23.21	0.53	17.85	nr	**41.06**
28 mm dia.	26.34	33.33	0.61	20.55	nr	**53.88**
Extra over copper pipes; high duty capillary fittings; BS 864						
Stop end						
15 mm dia.	7.69	9.73	0.16	5.39	nr	**15.12**
Straight coupling; copper to copper						
15 mm dia.	3.51	4.45	0.27	9.09	nr	**13.54**
22 mm dia.	5.64	7.13	0.32	10.79	nr	**17.92**
28 mm dia.	7.54	9.54	0.37	12.47	nr	**22.01**
35 mm dia.	14.08	17.82	0.43	14.49	nr	**32.31**
42 mm dia.	15.40	19.49	0.50	16.84	nr	**36.33**
54 mm dia.	22.65	28.66	0.54	18.19	nr	**46.85**
Reducing coupling						
15 × 12 mm dia.	6.63	8.39	0.27	9.09	nr	**17.48**
22 × 15 mm dia.	7.69	9.73	0.32	10.79	nr	**20.52**
28 × 22 mm dia.	10.59	13.41	0.37	12.47	nr	**25.88**

38 PIPED SUPPLY SYSTEMS

Item	Net Price £	Material £	Labour hours	Labour £	Unit	Total rate £
COLD WATER PIPELINES: COPPER PIPEWORK – cont						
Extra over copper pipes – cont						
Straight female connector						
15 mm × ½" dia.	8.64	10.93	0.32	10.79	nr	**21.72**
22 mm × ¾" dia.	9.76	12.35	0.36	12.13	nr	**24.48**
28 mm × 1" dia.	14.40	18.22	0.42	14.15	nr	**32.37**
Straight male connector						
15 mm × ½" dia.	8.42	10.65	0.32	10.79	nr	**21.44**
22 mm × ¾" dia.	9.76	12.35	0.36	12.13	nr	**24.48**
28 mm × 1" dia.	14.40	18.22	0.42	14.15	nr	**32.37**
42 mm × 1½" dia.	28.09	35.55	0.53	17.85	nr	**53.40**
54 mm × 2" dia.	45.66	57.79	0.62	20.89	nr	**78.68**
Reducer						
15 × 12 mm dia.	4.36	5.52	0.27	9.09	nr	**14.61**
22 × 15 mm dia.	4.25	5.38	0.32	10.79	nr	**16.17**
28 × 22 mm dia.	7.69	9.73	0.37	12.47	nr	**22.20**
35 × 28 mm dia.	9.76	12.35	0.43	14.49	nr	**26.84**
42 × 35 mm dia.	12.58	15.93	0.50	16.84	nr	**32.77**
54 × 42 mm dia.	20.27	25.66	0.39	13.14	nr	**38.80**
Straight union adaptor						
15 mm × ¾" dia.	7.03	8.89	0.27	9.09	nr	**17.98**
22 mm × 1" dia.	9.53	12.06	0.32	10.79	nr	**22.85**
28 mm × 1¼" dia.	12.58	15.93	0.37	12.47	nr	**28.40**
35 mm × 1½" dia.	22.80	28.85	0.43	14.49	nr	**43.34**
42 mm × 2" dia.	28.89	36.57	0.50	16.84	nr	**53.41**
Bent union adaptor						
15 mm × ¾" dia.	18.28	23.14	0.27	9.09	nr	**32.23**
22 mm × 1" dia.	24.68	31.24	0.32	10.79	nr	**42.03**
28 mm × 1¼" dia.	33.22	42.04	0.37	12.47	nr	**54.51**
Adaptor; male copper to FI						
15 mm × ½" dia.	8.23	10.42	0.27	9.09	nr	**19.51**
22 mm × ¾" dia.	14.12	17.88	0.32	10.79	nr	**28.67**
Union coupling						
15 mm dia.	15.85	20.06	0.54	18.19	nr	**38.25**
22 mm dia.	20.27	25.66	0.60	20.22	nr	**45.88**
28 mm dia.	28.17	35.65	0.68	22.90	nr	**58.55**
35 mm dia.	49.16	62.22	0.83	27.96	nr	**90.18**
42 mm dia.	57.90	73.28	0.89	29.98	nr	**103.26**
Elbow						
15 mm dia.	10.20	12.91	0.27	9.09	nr	**22.00**
22 mm dia.	10.91	13.81	0.32	10.79	nr	**24.60**
28 mm dia.	16.20	20.51	0.37	12.47	nr	**32.98**
35 mm dia.	25.33	32.05	0.43	14.49	nr	**46.54**
42 mm dia.	31.55	39.93	0.50	16.84	nr	**56.77**
54 mm dia.	54.87	69.44	0.52	17.52	nr	**86.96**
Return bend						
28 mm dia.	24.43	30.92	0.37	12.47	nr	**43.39**
35 mm dia.	30.11	38.10	0.43	14.49	nr	**52.59**

38 PIPED SUPPLY SYSTEMS

Item	Net Price £	Material £	Labour hours	Labour £	Unit	Total rate £
Bent male union connector						
15 mm × ½" dia.	23.66	29.95	0.54	18.19	nr	**48.14**
22 mm × ¾" dia.	31.87	40.33	0.60	20.22	nr	**60.55**
28 mm × 1" dia.	57.90	73.28	0.68	22.90	nr	**96.18**
Composite flange						
35 mm dia.	4.77	6.04	0.38	12.81	nr	**18.85**
42 mm dia.	6.20	7.85	0.41	13.81	nr	**21.66**
54 mm dia.	7.91	10.01	0.43	14.49	nr	**24.50**
Equal tee						
15 mm dia.	11.72	14.83	0.44	14.82	nr	**29.65**
22 mm dia.	14.75	18.67	0.47	15.84	nr	**34.51**
28 mm dia.	19.44	24.61	0.53	17.85	nr	**42.46**
35 mm dia.	33.22	42.04	0.70	23.58	nr	**65.62**
42 mm dia.	42.30	53.54	0.84	28.30	nr	**81.84**
54 mm dia.	66.60	84.29	0.79	26.61	nr	**110.90**
Reducing tee						
15 × 12 mm dia.	16.07	20.34	0.44	14.82	nr	**35.16**
22 × 15 mm dia.	18.98	24.02	0.47	15.84	nr	**39.86**
28 × 22 mm dia.	27.12	34.33	0.53	17.85	nr	**52.18**
35 × 28 mm dia.	42.99	54.41	0.73	24.60	nr	**79.01**
42 × 28 mm dia.	55.03	69.64	0.84	28.30	nr	**97.94**
54 × 28 mm dia.	86.91	110.00	1.01	34.04	nr	**144.04**
Extra over copper pipes; compression fittings; BS 864						
Stop end						
15 mm dia.	1.80	2.27	0.10	3.37	nr	**5.64**
22 mm dia.	2.14	2.71	0.12	4.03	nr	**6.74**
28 mm dia.	2.45	3.10	0.15	5.05	nr	**8.15**
Straight connector; copper to copper						
15 mm dia.	3.73	4.72	0.18	6.06	nr	**10.78**
22 mm dia.	4.94	6.25	0.21	7.08	nr	**13.33**
28 mm dia.	6.57	8.31	0.24	8.09	nr	**16.40**
Straight connector; copper to imperial copper						
22 mm dia.	6.15	7.78	0.21	7.08	nr	**14.86**
Male coupling; copper to MI (BSP)						
15 mm dia.	1.06	1.34	0.19	6.41	nr	**7.75**
22 mm dia.	1.67	2.12	0.23	7.75	nr	**9.87**
28 mm dia.	3.88	4.91	0.26	8.76	nr	**13.67**
Male coupling with long thread and backnut						
15 mm dia.	6.83	8.65	0.19	6.41	nr	**15.06**
22 mm dia.	8.67	10.98	0.23	7.75	nr	**18.73**
Female coupling; copper to FI (BSP)						
15 mm dia.	1.31	1.66	0.19	6.41	nr	**8.07**
22 mm dia.	1.87	2.36	0.23	7.75	nr	**10.11**
28 mm dia.	5.42	6.85	0.27	9.09	nr	**15.94**
Elbow						
15 mm dia.	1.39	1.76	0.18	6.06	nr	**7.82**
22 mm dia.	2.38	3.01	0.21	7.08	nr	**10.09**
28 mm dia.	8.02	10.15	0.24	8.09	nr	**18.24**

38 PIPED SUPPLY SYSTEMS

Item	Net Price £	Material £	Labour hours	Labour £	Unit	Total rate £
COLD WATER PIPELINES: COPPER PIPEWORK – cont						
Extra over copper pipes – cont						
Male elbow; copper to FI (BSP)						
15 mm × ½" dia.	2.68	3.39	0.19	6.41	nr	9.80
22 mm × ¾" dia.	3.48	4.40	0.23	7.75	nr	12.15
28 mm × 1" dia.	8.37	10.60	0.27	9.09	nr	19.69
Female elbow; copper to FI (BSP)						
15 mm × ½" dia.	4.10	5.19	0.19	6.41	nr	11.60
22 mm × ¾" dia.	5.93	7.50	0.23	7.75	nr	15.25
28 mm × 1" dia.	10.45	13.23	0.27	9.09	nr	22.32
Backplate elbow						
15 mm × ½" dia.	5.93	7.50	0.50	16.84	nr	24.34
Tank coupling; long thread						
22 mm dia.	8.67	10.98	0.46	15.50	nr	26.48
Tee equal						
15 mm dia.	1.98	2.51	0.28	9.43	nr	11.94
22 mm dia.	3.31	4.19	0.30	10.11	nr	14.30
28 mm dia.	15.10	19.11	0.34	11.46	nr	30.57
Tee reducing						
22 mm dia.	8.54	10.81	0.30	10.11	nr	20.92
Backplate tee						
15 mm dia.	20.86	26.40	0.62	20.89	nr	47.29
Extra over fittings; silver brazed welded joints						
Reducer						
76 × 67 mm dia.	34.42	43.56	1.40	47.17	nr	90.73
108 × 76 mm dia.	72.15	91.31	1.80	60.65	nr	151.96
133 × 108 mm dia.	143.87	182.08	2.20	74.11	nr	256.19
159 × 133 mm dia.	184.00	232.87	2.60	87.60	nr	320.47
90° elbow						
76 mm dia.	83.16	105.25	1.60	53.91	nr	159.16
108 mm dia.	153.10	193.76	2.00	67.38	nr	261.14
133 mm dia.	323.64	409.60	2.40	80.86	nr	490.46
159 mm dia.	401.79	508.50	2.80	94.34	nr	602.84
45° elbow						
76 mm dia.	77.93	98.63	1.60	53.91	nr	152.54
108 mm dia.	122.21	154.67	2.00	67.38	nr	222.05
133 mm dia.	312.50	395.51	2.40	80.86	nr	476.37
159 mm dia.	449.81	569.28	2.80	94.34	nr	663.62
Equal tee						
76 mm dia.	101.28	128.18	2.40	80.86	nr	209.04
108 mm dia.	157.33	199.11	3.00	101.07	nr	300.18
133 mm dia.	370.51	468.92	3.60	121.28	nr	590.20
159 mm dia.	441.89	559.26	4.20	141.49	nr	700.75

38 PIPED SUPPLY SYSTEMS

Item	Net Price £	Material £	Labour hours	Labour £	Unit	Total rate £
Extra over copper pipes; dezincification resistant compression fittings; BS 864						
Stop end						
15 mm dia.	2.58	3.27	0.10	3.37	nr	6.64
22 mm dia.	3.73	4.72	0.13	4.38	nr	9.10
28 mm dia.	7.99	10.11	0.15	5.05	nr	15.16
35 mm dia.	12.51	15.84	0.18	6.06	nr	21.90
42 mm dia.	20.86	26.40	0.20	6.73	nr	33.13
Straight coupling; copper to copper						
15 mm dia.	2.06	2.61	0.18	6.06	nr	8.67
22 mm dia.	3.38	4.28	0.21	7.08	nr	11.36
28 mm dia.	7.64	9.67	0.24	8.09	nr	17.76
35 mm dia.	16.17	20.46	0.29	9.77	nr	30.23
42 mm dia.	21.26	26.90	0.33	11.11	nr	38.01
54 mm dia.	31.79	40.23	0.38	12.81	nr	53.04
Straight swivel connector; copper to imperial copper						
22 mm dia.	7.61	9.63	0.20	6.73	nr	16.36
Male coupling; copper to MI (BSP)						
15 mm × ½" dia.	1.84	2.33	0.19	6.41	nr	8.74
22 mm × ¾" dia.	2.79	3.53	0.23	7.75	nr	11.28
28 mm × 1" dia.	5.41	6.84	0.26	8.76	nr	15.60
35 mm × 1¼" dia.	12.28	15.55	0.32	10.79	nr	26.34
42 mm × 1½" dia.	18.44	23.34	0.37	12.47	nr	35.81
54 mm × 2" dia.	27.20	34.43	0.57	19.20	nr	53.63
Male coupling with long thread and backnuts						
22 mm dia.	10.45	13.23	0.23	7.75	nr	20.98
28 mm dia.	11.58	14.66	0.24	8.09	nr	22.75
Female coupling; copper to FI (BSP)						
15 mm × ½" dia.	2.20	2.79	0.19	6.41	nr	9.20
22 mm × ¾" dia.	3.22	4.08	0.23	7.75	nr	11.83
28 mm × 1" dia.	7.01	8.87	0.27	9.09	nr	17.96
35 mm × 1¼" dia.	14.77	18.69	0.32	10.79	nr	29.48
42 mm × 1½" dia.	19.83	25.10	0.37	12.47	nr	37.57
54 mm × 2" dia.	29.10	36.83	0.42	14.15	nr	50.98
Elbow						
15 mm dia.	2.48	3.14	0.18	6.06	nr	9.20
22 mm dia.	3.97	5.03	0.21	7.08	nr	12.11
28 mm dia.	9.86	12.48	0.24	8.09	nr	20.57
35 mm dia.	21.82	27.62	0.29	9.77	nr	37.39
42 mm dia.	29.55	37.40	0.33	11.11	nr	48.51
54 mm dia.	50.86	64.37	0.38	12.81	nr	77.18
Male elbow; copper to MI (BSP)						
15 mm × ½" dia.	4.31	5.45	0.19	6.41	nr	11.86
22 mm × ¾" dia.	4.82	6.10	0.23	7.75	nr	13.85
28 mm × 1" dia.	9.03	11.42	0.27	9.09	nr	20.51
Female elbow; copper to FI (BSP)						
15 mm × ½" dia.	4.60	5.82	0.19	6.41	nr	12.23
22 mm × ¾" dia.	6.64	8.40	0.23	7.75	nr	16.15
28 mm × 1" dia.	11.01	13.93	0.27	9.09	nr	23.02

38 PIPED SUPPLY SYSTEMS

Item	Net Price £	Material £	Labour hours	Labour £	Unit	Total rate £
COLD WATER PIPELINES: COPPER PIPEWORK – cont						
Extra over copper pipes – cont						
Backplate elbow						
15 mm × ½" dia.	6.72	8.50	0.50	16.84	nr	**25.34**
Straight tap connector						
15 mm dia.	3.81	4.83	0.13	4.38	nr	**9.21**
22 mm dia.	8.26	10.45	0.15	5.05	nr	**15.50**
Tank coupling						
15 mm dia.	5.25	6.64	0.19	6.41	nr	**13.05**
22 mm dia.	5.80	7.34	0.23	7.75	nr	**15.09**
28 mm dia.	12.26	15.51	0.27	9.09	nr	**24.60**
35 mm dia.	21.68	27.44	0.32	10.79	nr	**38.23**
42 mm dia.	35.23	44.59	0.37	12.47	nr	**57.06**
54 mm dia.	45.38	57.43	0.31	10.44	nr	**67.87**
Tee equal						
15 mm dia.	3.49	4.41	0.28	9.43	nr	**13.84**
22 mm dia.	5.76	7.29	0.30	10.11	nr	**17.40**
28 mm dia.	15.72	19.89	0.34	11.46	nr	**31.35**
35 mm dia.	28.38	35.92	0.43	14.49	nr	**50.41**
42 mm dia.	44.63	56.48	0.46	15.50	nr	**71.98**
54 mm dia.	71.69	90.73	0.54	18.19	nr	**108.92**
Tee reducing						
22 mm dia.	9.21	11.66	0.30	10.11	nr	**21.77**
28 mm dia.	15.19	19.22	0.34	11.46	nr	**30.68**
35 mm dia.	27.73	35.09	0.43	14.49	nr	**49.58**
42 mm dia.	42.87	54.25	0.46	15.50	nr	**69.75**
54 mm dia.	71.69	90.73	0.54	18.19	nr	**108.92**
Extra over copper pipes; bronze one piece brazing flanges; metric, including jointing ring and bolts						
Bronze flange; PN6						
15 mm dia.	32.58	41.24	0.27	9.09	nr	**50.33**
22 mm dia.	38.82	49.13	0.32	10.79	nr	**59.92**
28 mm dia.	42.65	53.97	0.36	12.13	nr	**66.10**
35 mm dia.	50.95	64.48	0.47	15.84	nr	**80.32**
42 mm dia.	55.59	70.36	0.54	18.19	nr	**88.55**
54 mm dia.	64.48	81.60	0.63	21.22	nr	**102.82**
67 mm dia.	76.01	96.20	0.77	25.95	nr	**122.15**
76 mm dia.	91.61	115.94	0.93	31.34	nr	**147.28**
108 mm dia.	121.04	153.19	1.14	38.42	nr	**191.61**
133 mm dia.	165.85	209.90	1.41	47.50	nr	**257.40**
159 mm dia.	239.26	302.80	1.74	58.62	nr	**361.42**
Bronze flange; PN10						
15 mm dia.	37.30	47.21	0.27	9.09	nr	**56.30**
22 mm dia.	39.70	50.24	0.32	10.79	nr	**61.03**
28 mm dia.	43.66	55.26	0.38	12.81	nr	**68.07**
35 mm dia.	52.79	66.81	0.47	15.84	nr	**82.65**
42 mm dia.	59.17	74.88	0.54	18.19	nr	**93.07**
54 mm dia.	64.75	81.95	0.63	21.22	nr	**103.17**

38 PIPED SUPPLY SYSTEMS

Item	Net Price £	Material £	Labour hours	Labour £	Unit	Total rate £
67 mm dia.	76.08	96.29	0.77	25.95	nr	**122.24**
76 mm dia.	91.61	115.94	0.93	31.34	nr	**147.28**
108 mm dia.	115.78	146.53	1.14	38.42	nr	**184.95**
133 mm dia.	154.59	195.65	1.41	47.50	nr	**243.15**
159 mm dia.	195.95	247.99	1.74	58.62	nr	**306.61**
Bronze flange; PN16						
15 mm dia.	41.04	51.95	0.27	9.09	nr	**61.04**
22 mm dia.	43.69	55.29	0.32	10.79	nr	**66.08**
28 mm dia.	48.03	60.78	0.38	12.81	nr	**73.59**
35 mm dia.	58.07	73.49	0.47	15.84	nr	**89.33**
42 mm dia.	65.10	82.39	0.54	18.19	nr	**100.58**
54 mm dia.	71.21	90.13	0.63	21.22	nr	**111.35**
67 mm dia.	83.68	105.91	0.77	25.95	nr	**131.86**
76 mm dia.	100.76	127.52	0.93	31.34	nr	**158.86**
108 mm dia.	127.36	161.19	1.14	38.42	nr	**199.61**
133 mm dia.	170.05	215.22	1.41	47.50	nr	**262.72**
159 mm dia.	215.53	272.78	1.74	58.62	nr	**331.40**
Extra over copper pipes; bronze blank flanges; metric, including jointing ring and bolts						
Gunmetal blank flange; PN6						
15 mm dia.	27.35	34.62	0.27	9.09	nr	**43.71**
22 mm dia.	34.89	44.16	0.27	9.09	nr	**53.25**
28 mm dia.	35.86	45.38	0.27	9.09	nr	**54.47**
35 mm dia.	58.69	74.28	0.32	10.79	nr	**85.07**
42 mm dia.	79.18	100.21	0.32	10.79	nr	**111.00**
54 mm dia.	87.30	110.49	0.34	11.46	nr	**121.95**
67 mm dia.	107.42	135.95	0.36	12.13	nr	**148.08**
76 mm dia.	138.38	175.13	0.37	12.47	nr	**187.60**
108 mm dia.	219.86	278.25	0.41	13.81	nr	**292.06**
133 mm dia.	259.36	328.25	0.58	19.54	nr	**347.79**
159 mm dia.	324.18	410.28	0.61	20.55	nr	**430.83**
Gunmetal blank flange; PN10						
15 mm dia.	33.07	41.85	0.27	9.09	nr	**50.94**
22 mm dia.	42.80	54.16	0.27	9.09	nr	**63.25**
28 mm dia.	47.38	59.96	0.27	9.09	nr	**69.05**
35 mm dia.	58.69	74.28	0.32	10.79	nr	**85.07**
42 mm dia.	110.00	139.22	0.32	10.79	nr	**150.01**
54 mm dia.	125.41	158.72	0.34	11.46	nr	**170.18**
67 mm dia.	134.53	170.26	0.46	15.50	nr	**185.76**
76 mm dia.	178.78	226.26	0.47	15.84	nr	**242.10**
108 mm dia.	209.81	265.54	0.51	17.19	nr	**282.73**
133 mm dia.	278.73	352.76	0.58	19.54	nr	**372.30**
159 mm dia.	379.60	480.42	0.71	23.92	nr	**504.34**
Gunmetal blank flange; PN16						
15 mm dia.	33.07	41.85	0.27	9.09	nr	**50.94**
22 mm dia.	43.45	54.99	0.27	9.09	nr	**64.08**
28 mm dia.	47.38	59.96	0.27	9.09	nr	**69.05**
35 mm dia.	58.69	74.28	0.32	10.79	nr	**85.07**
42 mm dia.	110.00	139.22	0.32	10.79	nr	**150.01**
54 mm dia.	125.41	158.72	0.34	11.46	nr	**170.18**

38 PIPED SUPPLY SYSTEMS

Item	Net Price £	Material £	Labour hours	Labour £	Unit	Total rate £
COLD WATER PIPELINES: COPPER PIPEWORK – cont						
Extra over copper pipes – cont						
67 mm dia.	156.26	197.76	0.46	15.50	nr	**213.26**
76 mm dia.	178.78	226.26	0.47	15.84	nr	**242.10**
108 mm dia.	209.81	265.54	0.51	17.19	nr	**282.73**
133 mm dia.	481.90	609.90	0.58	19.54	nr	**629.44**
159 mm dia.	575.61	728.49	0.71	23.92	nr	**752.41**
Extra over copper pipes; bronze screwed flanges; metric, including jointing ring and bolts						
Gunmetal screwed flange; 6 BSP						
15 mm dia.	27.35	34.62	0.35	11.79	nr	**46.41**
22 mm dia.	31.61	40.01	0.47	15.84	nr	**55.85**
28 mm dia.	32.95	41.70	0.52	17.52	nr	**59.22**
35 mm dia.	44.42	56.21	0.62	20.89	nr	**77.10**
42 mm dia.	53.32	67.48	0.70	23.58	nr	**91.06**
54 mm dia.	72.47	91.72	0.84	28.30	nr	**120.02**
67 mm dia.	90.94	115.09	1.03	34.70	nr	**149.79**
76 mm dia.	109.77	138.92	1.22	41.10	nr	**180.02**
108 mm dia.	173.84	220.01	1.41	47.50	nr	**267.51**
133 mm dia.	206.08	260.81	1.75	58.97	nr	**319.78**
159 mm dia.	262.33	332.00	2.21	74.46	nr	**406.46**
Gunmetal screwed flange; 10 BSP						
15 mm dia.	33.12	41.92	0.35	11.79	nr	**53.71**
22 mm dia.	38.57	48.81	0.47	15.84	nr	**64.65**
28 mm dia.	42.59	53.91	0.52	17.52	nr	**71.43**
35 mm dia.	60.70	76.82	0.62	20.89	nr	**97.71**
42 mm dia.	74.12	93.81	0.70	23.58	nr	**117.39**
54 mm dia.	105.39	133.38	0.84	28.30	nr	**161.68**
67 mm dia.	123.38	156.15	1.03	34.70	nr	**190.85**
76 mm dia.	139.28	176.28	1.22	41.10	nr	**217.38**
108 mm dia.	184.88	233.98	1.41	47.50	nr	**281.48**
133 mm dia.	224.07	283.58	1.75	58.97	nr	**342.55**
159 mm dia.	397.62	503.23	2.21	74.46	nr	**577.69**
Gunmetal screwed flange; 16 BSP						
15 mm dia.	26.02	32.93	0.35	11.79	nr	**44.72**
22 mm dia.	32.90	41.64	0.47	15.84	nr	**57.48**
28 mm dia.	38.08	48.19	0.52	17.52	nr	**65.71**
35 mm dia.	58.20	73.66	0.62	20.89	nr	**94.55**
42 mm dia.	70.05	88.66	0.70	23.58	nr	**112.24**
54 mm dia.	90.09	114.02	0.84	28.30	nr	**142.32**
67 mm dia.	130.22	164.81	1.03	34.70	nr	**199.51**
76 mm dia.	149.33	188.99	1.22	41.10	nr	**230.09**
108 mm dia.	174.73	221.13	1.41	47.50	nr	**268.63**
133 mm dia.	283.13	358.33	1.75	58.97	nr	**417.30**
159 mm dia.	350.21	443.23	2.21	74.46	nr	**517.69**

38 PIPED SUPPLY SYSTEMS

Item	Net Price £	Material £	Labour hours	Labour £	Unit	Total rate £
Extra over copper pipes; labour						
Made bend						
15 mm dia.	–	–	0.26	8.76	nr	**8.76**
22 mm dia.	–	–	0.28	9.43	nr	**9.43**
28 mm dia.	–	–	0.31	10.44	nr	**10.44**
35 mm dia.	–	–	0.42	14.15	nr	**14.15**
42 mm dia.	–	–	0.51	17.19	nr	**17.19**
54 mm dia.	–	–	0.58	19.54	nr	**19.54**
67 mm dia.	–	–	0.69	23.25	nr	**23.25**
76 mm dia.	–	–	0.80	26.96	nr	**26.96**
Bronze butt weld						
15 mm dia.	–	–	0.25	8.43	nr	**8.43**
22 mm dia.	–	–	0.31	10.44	nr	**10.44**
28 mm dia.	–	–	0.37	12.47	nr	**12.47**
35 mm dia.	–	–	0.49	16.51	nr	**16.51**
42 mm dia.	–	–	0.58	19.54	nr	**19.54**
54 mm dia.	–	–	0.72	24.26	nr	**24.26**
67 mm dia.	–	–	0.88	29.66	nr	**29.66**
76 mm dia.	–	–	1.08	36.39	nr	**36.39**
108 mm dia.	–	–	1.37	46.16	nr	**46.16**
133 mm dia.	–	–	1.73	58.28	nr	**58.28**
159 mm dia.	–	–	2.03	68.40	nr	**68.40**
PRESS FIT (copper fittings); Mechanical press fit joints; butyl rubber O ring						
Coupler						
15 mm dia.	1.16	1.47	0.36	12.13	nr	**13.60**
22 mm dia.	1.84	2.33	0.36	12.13	nr	**14.46**
28 mm dia.	3.77	4.77	0.44	14.82	nr	**19.59**
35 mm dia.	4.76	6.03	0.44	14.82	nr	**20.85**
42 mm dia.	8.56	10.83	0.52	17.52	nr	**28.35**
54 mm dia.	10.92	13.82	0.60	20.22	nr	**34.04**
Stop end						
22 mm dia.	2.95	3.73	0.18	6.06	nr	**9.79**
28 mm dia.	4.67	5.91	0.22	7.41	nr	**13.32**
35 mm dia.	8.01	10.14	0.22	7.41	nr	**17.55**
42 mm dia.	12.03	15.22	0.26	8.76	nr	**23.98**
54 mm dia.	14.49	18.33	0.30	10.11	nr	**28.44**
Reducer						
22 × 15 mm dia.	1.36	1.72	0.36	12.13	nr	**13.85**
28 × 15 mm dia.	3.59	4.55	0.40	13.47	nr	**18.02**
28 × 22 mm dia.	3.73	4.72	0.40	13.47	nr	**18.19**
35 × 22 mm dia.	4.47	5.66	0.40	13.47	nr	**19.13**
35 × 28 mm dia.	4.93	6.24	0.44	14.82	nr	**21.06**
42 × 22 mm dia.	7.71	9.76	0.44	14.82	nr	**24.58**
42 × 28 mm dia.	7.35	9.31	0.48	16.17	nr	**25.48**
42 × 35 mm dia.	7.35	9.31	0.48	16.17	nr	**25.48**
54 × 35 mm dia.	9.95	12.59	0.52	17.52	nr	**30.11**
54 × 42 mm dia.	9.95	12.59	0.56	18.87	nr	**31.46**

38 PIPED SUPPLY SYSTEMS

Item	Net Price £	Material £	Labour hours	Labour £	Unit	Total rate £
COLD WATER PIPELINES: COPPER PIPEWORK – cont						
PRESS FIT (copper fittings) – cont						
90° elbow						
15 mm dia.	1.27	1.61	0.36	12.13	nr	**13.74**
22 mm dia.	2.12	2.69	0.36	12.13	nr	**14.82**
28 mm dia.	4.55	5.76	0.44	14.82	nr	**20.58**
35 mm dia.	9.40	11.89	0.44	14.82	nr	**26.71**
42 mm dia.	17.66	22.36	0.52	17.52	nr	**39.88**
54 mm dia.	24.52	31.04	0.60	20.22	nr	**51.26**
45° elbow						
15 mm dia.	1.56	1.97	0.36	12.13	nr	**14.10**
22 mm dia.	2.15	2.72	0.36	12.13	nr	**14.85**
28 mm dia.	6.46	8.18	0.44	14.82	nr	**23.00**
35 mm dia.	9.21	11.66	0.44	14.82	nr	**26.48**
42 mm dia.	15.33	19.40	0.52	17.52	nr	**36.92**
54 mm dia.	21.81	27.61	0.60	20.22	nr	**47.83**
Equal tee						
15 mm dia.	2.01	2.54	0.54	18.19	nr	**20.73**
22 mm dia.	3.66	4.64	0.54	18.19	nr	**22.83**
28 mm dia.	6.58	8.33	0.66	22.23	nr	**30.56**
35 mm dia.	11.38	14.40	0.66	22.23	nr	**36.63**
42 mm dia.	22.47	28.44	0.78	26.28	nr	**54.72**
54 mm dia.	28.03	35.47	0.90	30.32	nr	**65.79**
Reducing tee						
22 × 15 mm dia.	2.97	3.76	0.54	18.19	nr	**21.95**
28 × 15 mm dia.	5.79	7.32	0.62	20.89	nr	**28.21**
28 × 22 mm dia.	7.77	9.83	0.62	20.89	nr	**30.72**
35 × 22 mm dia.	10.14	12.84	0.62	20.89	nr	**33.73**
35 × 28 mm dia.	11.28	14.28	0.62	20.89	nr	**35.17**
42 × 28 mm dia.	20.40	25.82	0.70	23.58	nr	**49.40**
42 × 35 mm dia.	20.40	25.82	0.70	23.58	nr	**49.40**
54 × 35 mm dia.	34.46	43.61	0.82	27.63	nr	**71.24**
54 × 42 mm dia.	34.46	43.61	0.82	27.63	nr	**71.24**
Male iron connector; BSP thread						
15 mm dia.	4.25	5.38	0.18	6.06	nr	**11.44**
22 mm dia.	6.32	8.00	0.18	6.06	nr	**14.06**
28 mm dia.	8.44	10.68	0.22	7.41	nr	**18.09**
35 mm dia.	15.27	19.33	0.22	7.41	nr	**26.74**
42 mm dia.	20.48	25.92	0.26	8.76	nr	**34.68**
54 mm dia.	39.50	49.99	0.30	10.11	nr	**60.10**
90° elbow; male iron BSP thread						
15 mm dia.	6.88	8.70	0.36	12.13	nr	**20.83**
22 mm dia.	10.79	13.65	0.36	12.13	nr	**25.78**
28 mm dia.	16.52	20.91	0.44	14.82	nr	**35.73**
35 mm dia.	21.49	27.19	0.44	14.82	nr	**42.01**
42 mm dia.	28.01	35.45	0.52	17.52	nr	**52.97**
54 mm dia.	40.92	51.79	0.60	20.22	nr	**72.01**

38 PIPED SUPPLY SYSTEMS

Item	Net Price £	Material £	Labour hours	Labour £	Unit	Total rate £
Female iron connector; BSP thread						
15 mm dia.	4.86	6.15	0.18	6.06	nr	**12.21**
22 mm dia.	6.42	8.12	0.18	6.06	nr	**14.18**
28 mm dia.	8.63	10.92	0.22	7.41	nr	**18.33**
35 mm dia.	16.90	21.39	0.22	7.41	nr	**28.80**
42 mm dia.	24.15	30.56	0.26	8.76	nr	**39.32**
54 mm dia.	41.43	52.44	0.30	10.11	nr	**62.55**
90° elbow; female iron BSP thread						
15 mm dia.	5.80	7.34	0.36	12.13	nr	**19.47**
22 mm dia.	8.56	10.83	0.36	12.13	nr	**22.96**
28 mm dia.	14.12	17.88	0.44	14.82	nr	**32.70**
35 mm dia.	18.26	23.11	0.44	14.82	nr	**37.93**
42 mm dia.	24.85	31.45	0.52	17.52	nr	**48.97**
54 mm dia.	36.54	46.24	0.60	20.22	nr	**66.46**

38 PIPED SUPPLY SYSTEMS

Item	Net Price £	Material £	Labour hours	Labour £	Unit	Total rate £
COLD WATER PIPELINES: STAINLESS STEEL PIPEWORK						
Stainless steel pipes; capillary or compression joints; BS 4127, vertical or at low level, with brackets measured separately						
Grade 304; satin finish						
15 mm dia.	4.24	5.36	0.41	13.81	m	**19.17**
22 mm dia.	5.95	7.53	0.51	17.19	m	**24.72**
28 mm dia.	8.10	10.25	0.58	19.54	m	**29.79**
35 mm dia.	12.24	15.49	0.65	21.91	m	**37.40**
42 mm dia.	15.52	19.64	0.71	23.93	m	**43.57**
54 mm dia.	21.64	27.38	0.80	26.96	m	**54.34**
Grade 316; satin finish						
15 mm dia.	5.44	6.89	0.61	20.55	m	**27.44**
22 mm dia.	10.19	12.89	0.76	25.61	m	**38.50**
28 mm dia.	12.08	15.29	0.87	29.32	m	**44.61**
35 mm dia.	21.92	27.74	0.98	33.03	m	**60.77**
42 mm dia.	28.38	35.92	1.06	35.73	m	**71.65**
54 mm dia.	33.07	41.85	1.16	39.08	m	**80.93**
Fixings						
Single pipe ring						
15 mm dia.	13.21	16.72	0.26	8.76	nr	**25.48**
22 mm dia.	15.39	19.48	0.26	8.76	nr	**28.24**
28 mm dia.	16.12	20.41	0.31	10.44	nr	**30.85**
35 mm dia.	18.31	23.17	0.32	10.79	nr	**33.96**
42 mm dia.	20.81	26.34	0.32	10.79	nr	**37.13**
54 mm dia.	23.75	30.06	0.34	11.46	nr	**41.52**
Screw on backplate, female						
All sizes 15 mm to 54 mm dia.	11.92	15.09	0.10	3.37	nr	**18.46**
Screw on backplate, male						
All sizes 15 mm to 54 mm dia.	13.57	17.17	0.10	3.37	nr	**20.54**
Stainless steel threaded rods; metric thread; including nuts, washers etc.						
10 mm dia. × 600 mm long	13.93	17.63	0.18	6.06	nr	**23.69**
Extra over stainless steel pipes; capillary fittings						
Straight coupling						
15 mm dia.	1.56	1.97	0.25	8.43	nr	**10.40**
22 mm dia.	2.51	3.18	0.28	9.43	nr	**12.61**
28 mm dia.	3.32	4.20	0.33	11.11	nr	**15.31**
35 mm dia.	7.67	9.71	0.37	12.47	nr	**22.18**
42 mm dia.	8.85	11.20	0.42	14.15	nr	**25.35**
54 mm dia.	13.31	16.84	0.45	15.16	nr	**32.00**
45° bend						
15 mm dia.	8.37	10.60	0.25	8.43	nr	**19.03**
22 mm dia.	10.99	13.91	0.30	10.00	nr	**23.91**
28 mm dia.	13.49	17.07	0.33	11.11	nr	**28.18**

38 PIPED SUPPLY SYSTEMS

Item	Net Price £	Material £	Labour hours	Labour £	Unit	Total rate £
35 mm dia.	16.00	20.25	0.37	12.47	nr	**32.72**
42 mm dia.	20.59	26.06	0.42	14.15	nr	**40.21**
54 mm dia.	25.46	32.22	0.45	15.16	nr	**47.38**
90° bend						
15 mm dia.	4.34	5.49	0.28	9.43	nr	**14.92**
22 mm dia.	5.84	7.39	0.28	9.43	nr	**16.82**
28 mm dia.	8.23	10.42	0.33	11.11	nr	**21.53**
35 mm dia.	20.07	25.40	0.37	12.47	nr	**37.87**
42 mm dia.	27.62	34.96	0.42	14.15	nr	**49.11**
54 mm dia.	37.47	47.42	0.45	15.16	nr	**62.58**
Reducer						
22 × 15 mm dia.	9.97	12.62	0.28	9.43	nr	**22.05**
28 × 22 mm dia.	11.12	14.08	0.33	11.11	nr	**25.19**
35 × 28 mm dia.	13.63	17.25	0.37	12.47	nr	**29.72**
42 × 35 mm dia.	14.66	18.56	0.42	14.15	nr	**32.71**
54 × 42 mm dia.	43.51	55.07	0.48	16.20	nr	**71.27**
Tap connector						
15 mm dia.	21.05	26.64	0.13	4.38	nr	**31.02**
22 mm dia.	27.83	35.22	0.14	4.72	nr	**39.94**
28 mm dia.	38.63	48.89	0.17	5.73	nr	**54.62**
Tank connector						
15 mm dia.	27.18	34.40	0.13	4.38	nr	**38.78**
22 mm dia.	40.45	51.20	0.13	4.38	nr	**55.58**
28 mm dia.	49.81	63.04	0.15	5.05	nr	**68.09**
35 mm dia.	72.03	91.16	0.18	6.06	nr	**97.22**
42 mm dia.	95.12	120.39	0.21	7.08	nr	**127.47**
54 mm dia.	144.00	182.25	0.24	8.09	nr	**190.34**
Tee equal						
15 mm dia.	7.75	9.81	0.37	12.47	nr	**22.28**
22 mm dia.	9.64	12.20	0.40	13.47	nr	**25.67**
28 mm dia.	11.66	14.76	0.45	15.16	nr	**29.92**
35 mm dia.	27.99	35.43	0.59	19.88	nr	**55.31**
42 mm dia.	34.56	43.74	0.62	20.90	nr	**64.64**
54 mm dia.	69.77	88.30	0.67	22.58	nr	**110.88**
Unequal tee						
22 × 15 mm dia.	15.71	19.88	0.37	12.47	nr	**32.35**
28 × 15 mm dia.	17.70	22.40	0.45	15.16	nr	**37.56**
28 × 22 mm dia.	17.70	22.40	0.45	15.16	nr	**37.56**
35 × 22 mm dia.	30.93	39.14	0.59	19.88	nr	**59.02**
35 × 28 mm dia.	30.93	39.14	0.59	19.88	nr	**59.02**
42 × 28 mm dia.	38.03	48.13	0.62	20.90	nr	**69.03**
42 × 35 mm dia.	38.03	48.13	0.62	20.90	nr	**69.03**
54 × 35 mm dia.	78.72	99.62	0.67	22.58	nr	**122.20**
54 × 42 mm dia.	78.72	99.62	0.67	22.58	nr	**122.20**
Union, conical seat						
15 mm dia.	34.86	44.12	0.25	8.43	nr	**52.55**
22 mm dia.	54.89	69.47	0.28	9.43	nr	**78.90**
28 mm dia.	70.97	89.82	0.33	11.11	nr	**100.93**
35 mm dia.	93.15	117.89	0.37	12.47	nr	**130.36**
42 mm dia.	117.49	148.69	0.42	14.15	nr	**162.84**
54 mm dia.	155.45	196.74	0.45	15.16	nr	**211.90**

38 PIPED SUPPLY SYSTEMS

Item	Net Price £	Material £	Labour hours	Labour £	Unit	Total rate £
COLD WATER PIPELINES: STAINLESS STEEL PIPEWORK – cont						
Extra over stainless steel pipes – cont						
Union, flat seat						
15 mm dia.	36.40	46.07	0.25	8.43	nr	**54.50**
22 mm dia.	56.67	71.72	0.28	9.43	nr	**81.15**
28 mm dia.	73.26	92.71	0.33	11.11	nr	**103.82**
35 mm dia.	95.73	121.15	0.37	12.47	nr	**133.62**
42 mm dia.	120.61	152.64	0.42	14.15	nr	**166.79**
54 mm dia.	161.71	204.66	0.45	15.16	nr	**219.82**
Extra over stainless steel pipes; compression fittings						
Straight coupling						
15 mm dia.	30.30	38.35	0.18	6.06	nr	**44.41**
22 mm dia.	57.72	73.05	0.22	7.41	nr	**80.46**
28 mm dia.	77.71	98.35	0.25	8.43	nr	**106.78**
35 mm dia.	119.96	151.82	0.30	10.11	nr	**161.93**
42 mm dia.	140.02	177.21	0.40	13.47	nr	**190.68**
90° bend						
15 mm dia.	38.19	48.33	0.18	6.06	nr	**54.39**
22 mm dia.	75.88	96.03	0.22	7.41	nr	**103.44**
28 mm dia.	103.47	130.95	0.25	8.43	nr	**139.38**
35 mm dia.	209.53	265.18	0.33	11.11	nr	**276.29**
42 mm dia.	306.21	387.54	0.35	11.79	nr	**399.33**
Reducer						
22 × 15 mm dia.	54.97	69.57	0.28	9.43	nr	**79.00**
28 × 22 mm dia.	75.30	95.30	0.28	9.43	nr	**104.73**
35 × 28 mm dia.	110.02	139.24	0.30	10.11	nr	**149.35**
42 × 35 mm dia.	146.36	185.24	0.37	12.47	nr	**197.71**
Stud coupling						
15 mm dia.	31.49	39.85	0.42	14.15	nr	**54.00**
22 mm dia.	53.26	67.40	0.25	8.43	nr	**75.83**
28 mm dia.	73.95	93.59	0.25	8.43	nr	**102.02**
35 mm dia.	118.46	149.92	0.37	12.47	nr	**162.39**
42 mm dia.	140.02	177.21	0.42	14.15	nr	**191.36**
Equal tee						
15 mm dia.	53.78	68.06	0.37	12.47	nr	**80.53**
22 mm dia.	111.10	140.60	0.40	13.47	nr	**154.07**
28 mm dia.	151.97	192.34	0.45	15.16	nr	**207.50**
35 mm dia.	302.17	382.42	0.59	19.88	nr	**402.30**
42 mm dia.	419.00	530.29	0.62	20.90	nr	**551.19**
Running tee						
15 mm dia.	66.21	83.80	0.37	12.47	nr	**96.27**
22 mm dia.	119.27	150.95	0.40	13.47	nr	**164.42**
28 mm dia.	201.94	255.57	0.59	19.88	nr	**275.45**

38 PIPED SUPPLY SYSTEMS

Item	Net Price £	Material £	Labour hours	Labour £	Unit	Total rate £
PRESS FIT (stainless steel); Press fit jointing system; butyl rubber O ring mechanical joint						
Pipework						
15 mm dia.	5.21	6.60	0.46	15.50	m	**22.10**
22 mm dia.	8.33	10.54	0.48	16.17	m	**26.71**
28 mm dia.	10.26	12.98	0.52	17.52	m	**30.50**
35 mm dia.	15.12	19.14	0.56	18.87	m	**38.01**
42 mm dia.	18.60	23.54	0.58	19.54	m	**43.08**
54 mm dia.	23.69	29.98	0.66	22.23	m	**52.21**
Fixings						
For stainless steel pipes						
Refer to fixings for stainless steel pipes; capillary or compression joints; BS 4127						
Extra over stainless steel pipes; Press fit jointing system;						
Coupling						
15 mm dia.	6.12	7.75	0.36	12.13	nr	**19.88**
22 mm dia.	7.71	9.76	0.36	12.13	nr	**21.89**
28 mm dia.	8.67	10.98	0.44	14.82	nr	**25.80**
35 mm dia.	10.77	13.63	0.44	14.82	nr	**28.45**
42 mm dia.	14.71	18.61	0.52	17.52	nr	**36.13**
54 mm dia.	17.70	22.40	0.60	20.22	nr	**42.62**
Stop end						
22 mm dia.	5.81	7.36	0.18	6.06	nr	**13.42**
28 mm dia.	6.77	8.57	0.22	7.41	nr	**15.98**
35 mm dia.	11.07	14.01	0.22	7.41	nr	**21.42**
42 mm dia.	15.56	19.69	0.26	8.76	nr	**28.45**
54 mm dia.	18.00	22.78	0.30	10.11	nr	**32.89**
Reducer						
22 × 15 mm dia.	7.29	9.23	0.36	12.13	nr	**21.36**
28 × 15 mm dia.	8.26	10.45	0.40	13.47	nr	**23.92**
28 × 22 mm dia.	8.55	10.82	0.40	13.47	nr	**24.29**
35 × 22 mm dia.	10.44	13.22	0.40	13.47	nr	**26.69**
35 × 28 mm dia.	12.93	16.36	0.44	14.82	nr	**31.18**
42 × 35 mm dia.	13.64	17.26	0.48	16.17	nr	**33.43**
54 × 42 mm dia.	15.59	19.73	0.56	18.87	nr	**38.60**
90° bend						
15 mm dia.	8.76	11.09	0.36	12.13	nr	**23.22**
22 mm dia.	12.24	15.49	0.36	12.13	nr	**27.62**
28 mm dia.	15.44	19.54	0.44	14.82	nr	**34.36**
35 mm dia.	24.30	30.76	0.44	14.82	nr	**45.58**
42 mm dia.	40.57	51.34	0.52	17.52	nr	**68.86**
54 mm dia.	56.03	70.91	0.60	20.22	nr	**91.13**
45° bend						
15 mm dia.	11.88	15.03	0.36	12.13	nr	**27.16**
22 mm dia.	14.77	18.69	0.36	12.13	nr	**30.82**
28 mm dia.	17.19	21.75	0.44	14.82	nr	**36.57**

38 PIPED SUPPLY SYSTEMS

Item	Net Price £	Material £	Labour hours	Labour £	Unit	Total rate £
COLD WATER PIPELINES: STAINLESS STEEL PIPEWORK – cont						
Extra over stainless steel pipes – cont						
45° bend – cont						
35 mm dia.	20.17	25.52	0.44	14.82	nr	**40.34**
42 mm dia.	32.45	41.07	0.52	17.52	nr	**58.59**
54 mm dia.	42.16	53.36	0.60	20.22	nr	**73.58**
Equal tee						
15 mm dia.	14.36	18.18	0.54	18.19	nr	**36.37**
22 mm dia.	17.62	22.30	0.54	18.19	nr	**40.49**
28 mm dia.	20.57	26.03	0.66	22.23	nr	**48.26**
35 mm dia.	26.07	33.00	0.66	22.23	nr	**55.23**
42 mm dia.	37.00	46.83	0.78	26.28	nr	**73.11**
54 mm dia.	44.27	56.03	0.90	30.32	nr	**86.35**
Reducing tee						
22 × 15 mm dia.	15.06	19.06	0.54	18.19	nr	**37.25**
28 × 15 mm dia.	18.25	23.09	0.62	20.89	nr	**43.98**
28 × 22 mm dia.	19.76	25.01	0.62	20.89	nr	**45.90**
35 × 22 mm dia.	23.42	29.64	0.62	20.89	nr	**50.53**
35 × 28 mm dia.	24.44	30.93	0.62	20.89	nr	**51.82**
42 × 28 mm dia.	34.75	43.98	0.70	23.58	nr	**67.56**
42 × 35 mm dia.	35.80	45.30	0.70	23.58	nr	**68.88**
54 × 35 mm dia.	40.44	51.18	0.82	27.63	nr	**78.81**
54 × 42 mm dia.	41.58	52.63	0.82	27.63	nr	**80.26**
Fixings						
For stainless steel pipes						
Refer to fixings for stainless steel pipes;						
capillary or compression joints; BS 4127						

38 PIPED SUPPLY SYSTEMS

Item	Net Price £	Material £	Labour hours	Labour £	Unit	Total rate £
COLD WATER PIPELINES: MEDIUM DENSITY POLYETHYLENE – BLUE PIPEWORK						
Note: MDPE is sized on Outside Dia., i.e. OD not ID						
Pipes for water distribution; laid underground; electrofusion joints in the running length; BS 6572						
Coiled service pipe						
20 mm dia.	0.95	1.20	0.37	12.47	m	**13.67**
25 mm dia.	1.24	1.57	0.41	13.81	m	**15.38**
32 mm dia.	2.09	2.64	0.47	15.84	m	**18.48**
50 mm dia.	5.03	6.36	0.53	17.85	m	**24.21**
63 mm dia.	7.88	9.97	0.60	20.23	m	**30.20**
Mains service pipe						
90 mm dia.	11.91	15.08	0.90	30.32	m	**45.40**
110 mm dia.	17.82	22.56	1.10	37.06	m	**59.62**
125 mm dia.	22.54	28.53	1.20	40.44	m	**68.97**
160 mm dia.	35.93	45.47	1.48	49.86	m	**95.33**
180 mm dia.	46.80	59.23	1.50	50.59	m	**109.82**
225 mm dia.	71.11	89.99	1.77	59.64	m	**149.63**
250 mm dia.	89.75	113.59	1.75	59.00	m	**172.59**
315 mm dia.	138.58	175.39	1.90	64.02	m	**239.41**
Extra over fittings; MDPE blue; electrofusion joints						
Coupler						
20 mm dia.	8.10	10.25	0.36	12.13	nr	**22.38**
25 mm dia.	8.10	10.25	0.40	13.47	nr	**23.72**
32 mm dia.	8.10	10.25	0.44	14.82	nr	**25.07**
40 mm dia.	11.95	15.12	0.48	16.17	nr	**31.29**
50 mm dia.	11.47	14.52	0.52	17.52	nr	**32.04**
63 mm dia.	15.00	18.98	0.58	19.54	nr	**38.52**
90 mm dia.	22.05	27.91	0.67	22.58	nr	**50.49**
110 mm dia.	35.47	44.89	0.74	24.93	nr	**69.82**
125 mm dia.	40.02	50.65	0.83	27.96	nr	**78.61**
160 mm dia.	63.65	80.55	1.00	33.69	nr	**114.24**
180 mm dia.	74.84	94.72	1.25	42.11	nr	**136.83**
225 mm dia.	119.60	151.37	1.35	45.48	nr	**196.85**
250 mm dia.	175.13	221.65	1.50	50.53	nr	**272.18**
315 mm dia.	288.81	365.52	1.80	60.65	nr	**426.17**
Extra over fittings; MDPE blue; butt fused joints						
Cap						
25 mm dia.	15.36	19.44	0.20	6.73	nr	**26.17**
32 mm dia.	15.36	19.44	0.22	7.41	nr	**26.85**
40 mm dia.	16.27	20.60	0.24	8.09	nr	**28.69**

38 PIPED SUPPLY SYSTEMS

Item	Net Price £	Material £	Labour hours	Labour £	Unit	Total rate £
COLD WATER PIPELINES: MEDIUM DENSITY POLYETHYLENE – BLUE PIPEWORK – cont						
Extra over fittings – cont						
Cap – cont						
50 mm dia.	23.73	30.03	0.26	8.76	nr	**38.79**
63 mm dia.	27.22	34.45	0.32	10.79	nr	**45.24**
90 mm dia.	44.91	56.84	0.37	12.47	nr	**69.31**
110 mm dia.	90.44	114.46	0.40	13.47	nr	**127.93**
125 mm dia.	72.20	91.38	0.46	15.50	nr	**106.88**
160 mm dia.	82.54	104.46	0.50	16.84	nr	**121.30**
180 mm dia.	138.06	174.73	0.60	20.22	nr	**194.95**
225 mm dia.	163.08	206.39	0.68	22.90	nr	**229.29**
250 mm dia.	239.37	302.95	0.75	25.26	nr	**328.21**
315 mm dia.	308.66	390.64	0.90	30.32	nr	**420.96**
Reducer						
63 × 32 mm dia.	20.75	26.26	0.54	18.19	nr	**44.45**
63 × 50 mm dia.	24.27	30.72	0.60	20.22	nr	**50.94**
90 × 63 mm dia.	30.71	38.86	0.67	22.58	nr	**61.44**
110 × 90 mm dia.	42.32	53.56	0.74	24.93	nr	**78.49**
125 × 90 mm dia.	61.55	77.90	0.83	27.96	nr	**105.86**
125 × 110 mm dia.	67.82	85.84	1.00	33.69	nr	**119.53**
160 × 110 mm dia.	104.18	131.85	1.10	37.06	nr	**168.91**
180 × 125 mm dia.	113.16	143.21	1.25	42.11	nr	**185.32**
225 × 160 mm dia.	184.52	233.53	1.40	47.17	nr	**280.70**
250 × 180 mm dia.	141.77	179.42	1.80	60.65	nr	**240.07**
315 × 250 mm dia.	163.09	206.40	2.40	80.86	nr	**287.26**
Bend; 45°						
50 mm dia.	31.67	40.08	0.50	16.84	nr	**56.92**
63 mm dia.	38.31	48.48	0.58	19.54	nr	**68.02**
90 mm dia.	59.33	75.08	0.67	22.58	nr	**97.66**
110 mm dia.	86.99	110.10	0.74	24.93	nr	**135.03**
125 mm dia.	97.22	123.04	0.83	27.96	nr	**151.00**
160 mm dia.	180.17	228.02	1.00	33.69	nr	**261.71**
180 mm dia.	205.10	259.57	1.25	42.11	nr	**301.68**
225 mm dia.	261.43	330.87	1.40	47.15	nr	**378.02**
250 mm dia.	282.90	358.04	1.80	60.65	nr	**418.69**
315 mm dia.	352.46	446.07	2.40	80.86	nr	**526.93**
Bend; 90°						
50 mm dia.	31.67	40.08	0.50	16.84	nr	**56.92**
63 mm dia.	38.31	48.48	0.58	19.54	nr	**68.02**
90 mm dia.	59.33	75.08	0.67	22.58	nr	**97.66**
110 mm dia.	86.99	110.10	0.74	24.93	nr	**135.03**
125 mm dia.	97.22	123.04	0.83	27.96	nr	**151.00**
160 mm dia.	180.17	228.02	1.00	33.69	nr	**261.71**
180 mm dia.	349.95	442.89	1.25	42.11	nr	**485.00**
225 mm dia.	442.47	559.99	1.40	47.17	nr	**607.16**
250 mm dia.	483.99	612.54	1.80	60.65	nr	**673.19**
315 mm dia.	605.89	766.82	2.40	80.86	nr	**847.68**

38 PIPED SUPPLY SYSTEMS

Item	Net Price £	Material £	Labour hours	Labour £	Unit	Total rate £
Equal tee						
50 mm dia.	34.63	43.83	0.70	23.58	nr	**67.41**
63 mm dia.	37.79	47.82	0.75	25.26	nr	**73.08**
90 mm dia.	69.02	87.35	0.87	29.31	nr	**116.66**
110 mm dia.	102.93	130.27	1.00	33.69	nr	**163.96**
125 mm dia.	132.88	168.17	1.08	36.39	nr	**204.56**
160 mm dia.	218.49	276.52	1.35	45.48	nr	**322.00**
180 mm dia.	223.58	282.97	1.63	54.91	nr	**337.88**
225 mm dia.	270.96	342.92	1.90	64.02	nr	**406.94**
250 mm dia.	373.88	473.18	2.70	90.96	nr	**564.14**
315 mm dia.	932.07	1179.63	3.60	121.28	nr	**1300.91**
Extra over plastic fittings, compression joints						
Straight connector						
20 mm dia.	3.58	4.54	0.38	12.81	nr	**17.35**
25 mm dia.	3.75	4.75	0.45	15.16	nr	**19.91**
32 mm dia.	8.91	11.28	0.50	16.84	nr	**28.12**
50 mm dia.	20.52	25.97	0.68	22.92	nr	**48.89**
63 mm dia.	30.90	39.11	0.85	28.65	nr	**67.76**
Reducing connector						
25 mm dia.	7.36	9.32	0.38	12.81	nr	**22.13**
32 mm dia.	11.88	15.03	0.45	15.16	nr	**30.19**
50 mm dia.	32.91	41.65	0.50	16.84	nr	**58.49**
63 mm dia.	45.93	58.13	0.62	20.90	nr	**79.03**
Straight connector; polyethylene to MI						
20 mm dia.	3.24	4.10	0.31	10.44	nr	**14.54**
25 mm dia.	5.52	6.99	0.35	11.79	nr	**18.78**
32 mm dia.	5.98	7.57	0.40	13.47	nr	**21.04**
50 mm dia.	15.26	19.31	0.55	18.52	nr	**37.83**
63 mm dia.	21.52	27.24	0.65	21.91	nr	**49.15**
Straight connector; polyethylene to FI						
20 mm dia.	4.36	5.52	0.31	10.44	nr	**15.96**
25 mm dia.	4.72	5.97	0.35	11.79	nr	**17.76**
32 mm dia.	5.64	7.13	0.40	13.47	nr	**20.60**
50 mm dia.	17.93	22.69	0.55	18.52	nr	**41.21**
63 mm dia.	25.11	31.77	0.75	25.28	nr	**57.05**
Elbow						
20 mm dia.	4.77	6.04	0.38	12.81	nr	**18.85**
25 mm dia.	7.05	8.93	0.45	15.16	nr	**24.09**
32 mm dia.	10.28	13.01	0.50	16.84	nr	**29.85**
50 mm dia.	23.84	30.17	0.68	22.92	nr	**53.09**
63 mm dia.	32.41	41.01	0.80	26.96	nr	**67.97**
Elbow; polyethylene to MI						
25 mm dia.	6.07	7.68	0.35	11.79	nr	**19.47**
Elbow; polyethylene to FI						
20 mm dia.	4.34	5.49	0.31	10.44	nr	**15.93**
25 mm dia.	5.90	7.47	0.35	11.79	nr	**19.26**
32 mm dia.	8.84	11.19	0.42	14.15	nr	**25.34**
50 mm dia.	20.99	26.57	0.50	16.84	nr	**43.41**
63 mm dia.	27.51	34.82	0.55	18.52	nr	**53.34**

38 PIPED SUPPLY SYSTEMS

Item	Net Price £	Material £	Labour hours	Labour £	Unit	Total rate £
COLD WATER PIPELINES: MEDIUM DENSITY POLYETHYLENE – BLUE PIPEWORK – cont						
Extra over plastic fittings, compression joints – cont						
Tank coupling						
25 mm dia.	9.11	11.52	0.42	14.15	nr	**25.67**
Equal tee						
20 mm dia.	6.42	8.12	0.53	17.85	nr	**25.97**
25 mm dia.	10.04	12.71	0.55	18.52	nr	**31.23**
32 mm dia.	12.58	15.93	0.64	21.57	nr	**37.50**
50 mm dia.	29.33	37.12	0.75	25.28	nr	**62.40**
63 mm dia.	45.43	57.50	0.87	29.32	nr	**86.82**
Equal tee; FI branch						
20 mm dia.	6.22	7.87	0.45	15.16	nr	**23.03**
25 mm dia.	9.91	12.54	0.50	16.84	nr	**29.38**
32 mm dia.	12.06	15.27	0.60	20.23	nr	**35.50**
50 mm dia.	27.79	35.17	0.68	22.92	nr	**58.09**
63 mm dia.	39.02	49.38	0.81	27.29	nr	**76.67**
Equal tee; MI branch						
25 mm dia.	9.74	12.33	0.50	16.84	nr	**29.17**

38 PIPED SUPPLY SYSTEMS

Item	Net Price £	Material £	Labour hours	Labour £	Unit	Total rate £
COLD WATER PIPELINES: ABS PIPEWORK						
Pipes; solvent welded joints in the running length, brackets measured separately						
Class C (9 bar pressure)						
1" dia.	6.44	8.15	0.30	10.11	m	18.26
1¼" dia.	10.83	13.71	0.33	11.11	m	24.82
1½" dia.	13.74	17.39	0.36	12.13	m	29.52
2" dia.	18.51	23.43	0.39	13.14	m	36.57
3" dia.	38.14	48.27	0.46	15.50	m	63.77
4" dia.	62.86	79.55	0.53	17.85	m	97.40
6" dia.	124.27	157.28	0.76	25.60	m	182.88
8" dia.	213.43	270.12	0.97	32.68	m	302.80
Class E (15 bar pressure)						
½" dia.	4.91	6.22	0.24	8.09	m	14.31
¾" dia.	7.56	9.56	0.27	9.09	m	18.65
1" dia.	9.96	12.60	0.30	10.11	m	22.71
1¼" dia.	14.87	18.82	0.33	11.11	m	29.93
1½" dia.	19.60	24.81	0.36	12.13	m	36.94
2" dia.	24.54	31.06	0.39	13.14	m	44.20
3" dia.	49.36	62.47	0.49	16.51	m	78.98
4" dia.	79.40	100.49	0.57	19.20	m	119.69
Fixings						
Refer to steel pipes; galvanized iron. For minimum fixing dimensions, refer to the Tables and Memoranda at the rear of the book						
Extra over fittings; solvent welded joints						
Cap						
½" dia.	2.10	2.65	0.16	5.39	nr	8.04
¾" dia.	2.43	3.08	0.19	6.41	nr	9.49
1" dia.	2.79	3.53	0.22	7.41	nr	10.94
1¼" dia.	4.66	5.90	0.25	8.43	nr	14.33
1½" dia.	7.18	9.08	0.28	9.43	nr	18.51
2" dia.	9.09	11.50	0.31	10.44	nr	21.94
3" dia.	27.32	34.57	0.36	12.13	nr	46.70
4" dia.	41.78	52.88	0.44	14.82	nr	67.70
Elbow 90°						
½" dia.	2.93	3.71	0.29	9.77	nr	13.48
¾" dia.	3.51	4.45	0.34	11.46	nr	15.91
1" dia.	4.91	6.22	0.40	13.47	nr	19.69
1¼" dia.	8.29	10.49	0.45	15.16	nr	25.65
1½" dia.	10.77	13.63	0.51	17.19	nr	30.82
2" dia.	16.40	20.75	0.56	18.87	nr	39.62
3" dia.	47.11	59.62	0.65	21.90	nr	81.52
4" dia.	70.36	89.05	0.80	26.96	nr	116.01
6" dia.	283.24	358.47	1.21	40.77	nr	399.24
8" dia.	432.26	547.06	1.45	48.85	nr	595.91

38 PIPED SUPPLY SYSTEMS

Item	Net Price £	Material £	Labour hours	Labour £	Unit	Total rate £
COLD WATER PIPELINES: ABS PIPEWORK – cont						
Extra over fittings – cont						
Elbow 45°						
½" dia.	5.68	7.19	0.29	9.77	nr	**16.96**
¾" dia.	5.75	7.28	0.34	11.46	nr	**18.74**
1" dia.	7.18	9.08	0.40	13.47	nr	**22.55**
1¼" dia.	10.52	13.32	0.45	15.16	nr	**28.48**
1½" dia.	13.02	16.48	0.51	17.19	nr	**33.67**
2" dia.	18.09	22.89	0.56	18.87	nr	**41.76**
3" dia.	42.60	53.92	0.65	21.90	nr	**75.82**
4" dia.	88.33	111.79	0.80	26.96	nr	**138.75**
6" dia.	183.09	231.72	1.21	40.77	nr	**272.49**
8" dia.	393.98	498.62	1.45	48.85	nr	**547.47**
Reducing bush						
¾" × ½" dia.	2.16	2.73	0.42	14.15	nr	**16.88**
1" × ½" dia.	2.79	3.53	0.45	15.16	nr	**18.69**
1" × ¾" dia.	2.79	3.53	0.45	15.16	nr	**18.69**
1¼" × 1" dia.	3.75	4.75	0.48	16.17	nr	**20.92**
1½" × ¾" dia.	4.91	6.22	0.51	17.19	nr	**23.41**
1½" × 1" dia.	4.91	6.22	0.51	17.19	nr	**23.41**
1½" × 1¼" dia.	4.91	6.22	0.51	17.19	nr	**23.41**
2" × 1" dia.	6.44	8.15	0.56	18.87	nr	**27.02**
2" × 1¼" dia.	6.44	8.15	0.56	18.87	nr	**27.02**
2" × 1½" dia.	6.44	8.15	0.56	18.87	nr	**27.02**
3" × 1½" dia.	18.09	22.89	0.65	21.90	nr	**44.79**
3" × 2" dia.	18.09	22.89	0.65	21.90	nr	**44.79**
4" × 3" dia.	24.96	31.58	0.80	26.96	nr	**58.54**
6" × 4" dia.	76.84	97.25	1.21	40.77	nr	**138.02**
Union						
½" dia.	11.66	14.76	0.34	11.46	nr	**26.22**
¾" dia.	12.58	15.93	0.39	13.14	nr	**29.07**
1" dia.	16.94	21.44	0.43	14.49	nr	**35.93**
1¼" dia.	20.76	26.28	0.50	16.84	nr	**43.12**
1½" dia.	28.60	36.20	0.57	19.20	nr	**55.40**
2" dia.	37.30	47.21	0.62	20.89	nr	**68.10**
Sockets						
½" dia.	2.16	2.73	0.34	11.46	nr	**14.19**
¾" dia.	2.43	3.08	0.39	13.14	nr	**16.22**
1" dia.	2.79	3.53	0.43	14.49	nr	**18.02**
1¼" dia.	4.91	6.22	0.50	16.84	nr	**23.06**
1½" dia.	5.92	7.49	0.57	19.20	nr	**26.69**
2" dia.	8.29	10.49	0.62	20.89	nr	**31.38**
3" dia.	33.37	42.24	0.70	23.58	nr	**65.82**
4" dia.	47.35	59.93	0.70	23.58	nr	**83.51**
6" dia.	118.33	149.76	1.26	42.45	nr	**192.21**
8" dia.	236.33	299.10	1.55	52.21	nr	**351.31**

38 PIPED SUPPLY SYSTEMS

Item	Net Price £	Material £	Labour hours	Labour £	Unit	Total rate £
Barrel nipple						
½" dia.	4.08	5.16	0.34	11.46	nr	16.62
¾" dia.	5.30	6.71	0.39	13.14	nr	19.85
1" dia.	6.87	8.69	0.43	14.49	nr	23.18
1¼" dia.	9.53	12.06	0.50	16.84	nr	28.90
1½" dia.	11.24	14.22	0.57	19.20	nr	33.42
2" dia.	13.60	17.21	0.62	20.89	nr	38.10
3" dia.	36.16	45.76	0.70	23.58	nr	69.34
Tee, 90°						
½" dia.	3.34	4.22	0.41	13.81	nr	18.03
¾" dia.	4.66	5.90	0.47	15.84	nr	21.74
1" dia.	6.44	8.15	0.55	18.52	nr	26.67
1¼" dia.	9.26	11.72	0.64	21.57	nr	33.29
1½" dia.	13.60	17.21	0.71	23.92	nr	41.13
2" dia.	20.76	26.28	0.78	26.28	nr	52.56
3" dia.	60.56	76.64	0.91	30.65	nr	107.29
4" dia.	88.90	112.52	1.12	37.73	nr	150.25
6" dia.	310.71	393.23	1.69	56.94	nr	450.17
8" dia.	484.47	613.14	2.03	68.40	nr	681.54
Full face flange						
½" dia.	36.19	45.80	0.10	3.37	nr	49.17
¾" dia.	37.04	46.88	0.13	4.38	nr	51.26
1" dia.	40.12	50.78	0.15	5.05	nr	55.83
1¼" dia.	44.60	56.45	0.18	6.06	nr	62.51
1½" dia.	53.63	67.87	0.21	7.08	nr	74.95
2" dia.	72.59	91.87	0.29	9.77	nr	101.64
3" dia.	124.47	157.53	0.37	12.47	nr	170.00
4" dia.	163.13	206.46	0.41	13.81	nr	220.27

38 PIPED SUPPLY SYSTEMS

Item	Net Price £	Material £	Labour hours	Labour £	Unit	Total rate £
COLD WATER PIPELINES: PVC-u PIPEWORK						
Pipes; solvent welded joints in the running length, brackets measured separately						
Class C (9 bar pressure)						
2" dia.	15.26	19.31	0.41	13.81	m	33.12
3" dia.	29.25	37.02	0.47	15.84	m	52.86
4" dia.	51.89	65.68	0.50	16.84	m	82.52
6" dia.	112.29	142.12	1.76	59.29	m	201.41
Class D (12 bar pressure)						
1¼" dia.	8.89	11.26	0.41	13.81	m	25.07
1½" dia.	12.22	15.47	0.42	14.15	m	29.62
2" dia.	18.95	23.98	0.45	15.16	m	39.14
3" dia.	40.61	51.40	0.48	16.17	m	67.57
4" dia.	67.96	86.00	0.53	17.85	m	103.85
6" dia.	126.13	159.63	0.58	19.54	m	179.17
Class E (15 bar pressure)						
½" dia.	4.35	5.51	0.38	12.81	m	18.32
¾" dia.	6.22	7.87	0.40	13.47	m	21.34
1" dia.	7.23	9.15	0.41	13.81	m	22.96
1¼" dia.	10.61	13.43	0.41	13.81	m	27.24
1½" dia.	13.81	17.48	0.42	14.15	m	31.63
2" dia.	21.58	27.32	0.45	15.16	m	42.48
3" dia.	46.76	59.18	0.47	15.84	m	75.02
4" dia.	76.77	97.16	0.50	16.84	m	114.00
6" dia.	166.29	210.46	0.53	17.85	m	228.31
Class 7						
½" dia.	7.69	9.73	0.32	10.79	m	20.52
¾" dia.	10.80	13.66	0.33	11.11	m	24.77
1" dia.	16.46	20.83	0.40	13.47	m	34.30
1¼" dia.	22.61	28.62	0.40	13.47	m	42.09
1½" dia.	27.99	35.43	0.41	13.81	m	49.24
2" dia.	46.53	58.89	0.43	14.49	m	73.38
Fixings						
Refer to steel pipes; galvanized iron. For minimum fixing dimensions, refer to the Tables and Memoranda at the rear of the book						
Extra over fittings; solvent welded joints						
End cap						
½" dia.	1.48	1.87	0.17	5.73	nr	7.60
¾" dia.	1.73	2.18	0.19	6.41	nr	8.59
1" dia.	1.96	2.48	0.22	7.41	nr	9.89
1¼" dia.	3.06	3.88	0.25	8.43	nr	12.31
1½" dia.	5.11	6.46	0.28	9.43	nr	15.89
2" dia.	6.27	7.94	0.31	10.44	nr	18.38
3" dia.	19.19	24.28	0.36	12.13	nr	36.41
4" dia.	29.63	37.50	0.44	14.82	nr	52.32
6" dia.	71.62	90.64	0.67	22.58	nr	113.22

38 PIPED SUPPLY SYSTEMS

Item	Net Price £	Material £	Labour hours	Labour £	Unit	Total rate £
Socket						
½" dia.	1.57	1.98	0.31	10.44	nr	**12.42**
¾" dia.	1.73	2.18	0.35	11.79	nr	**13.97**
1" dia.	2.02	2.55	0.42	14.15	nr	**16.70**
1¼" dia.	3.64	4.60	0.45	15.16	nr	**19.76**
1½" dia.	4.27	5.41	0.51	17.19	nr	**22.60**
2" dia.	6.07	7.68	0.56	18.87	nr	**26.55**
3" dia.	23.20	29.37	0.65	21.90	nr	**51.27**
4" dia.	33.63	42.56	0.80	26.96	nr	**69.52**
6" dia.	84.36	106.77	1.21	40.77	nr	**147.54**
Reducing socket						
¾ × ½" dia.	1.83	2.32	0.31	10.44	nr	**12.76**
1 × ¾" dia.	2.31	2.92	0.35	11.79	nr	**14.71**
1¼ × 1" dia.	4.40	5.57	0.42	14.15	nr	**19.72**
1½ × 1¼" dia.	4.91	6.22	0.45	15.16	nr	**21.38**
2 × 1½" dia.	7.41	9.37	0.51	17.19	nr	**26.56**
3 × 2" dia.	22.57	28.56	0.56	18.87	nr	**47.43**
4 × 3" dia.	33.39	42.26	0.65	21.90	nr	**64.16**
6 × 4" dia.	121.74	154.08	0.80	26.96	nr	**181.04**
8 × 6" dia.	188.54	238.62	1.21	40.77	nr	**279.39**
Elbow, 90°						
½" dia.	2.07	2.62	0.31	10.44	nr	**13.06**
¾" dia.	2.49	3.15	0.35	11.79	nr	**14.94**
1" dia.	3.48	4.40	0.42	14.15	nr	**18.55**
1¼" dia.	6.07	7.68	0.45	15.16	nr	**22.84**
1½" dia.	7.81	9.89	0.45	15.16	nr	**25.05**
2" dia.	11.59	14.67	0.56	18.87	nr	**33.54**
3" dia.	33.39	42.26	0.65	21.90	nr	**64.16**
4" dia.	50.31	63.67	0.80	26.96	nr	**90.63**
6" dia.	199.19	252.09	1.21	40.77	nr	**292.86**
Elbow 45°						
½" dia.	3.96	5.01	0.31	10.44	nr	**15.45**
¾" dia.	4.21	5.33	0.35	11.79	nr	**17.12**
1" dia.	5.11	6.46	0.45	15.16	nr	**21.62**
1¼" dia.	7.32	9.26	0.45	15.16	nr	**24.42**
1½" dia.	9.19	11.63	0.51	17.19	nr	**28.82**
2" dia.	12.95	16.39	0.56	18.87	nr	**35.26**
3" dia.	30.48	38.57	0.65	21.90	nr	**60.47**
4" dia.	62.63	79.26	0.80	26.96	nr	**106.22**
6" dia.	129.22	163.54	1.21	40.77	nr	**204.31**
Bend 90° (long radius)						
3" dia.	93.12	117.86	0.65	21.90	nr	**139.76**
4" dia.	188.11	238.07	0.80	26.96	nr	**265.03**
6" dia.	413.38	523.17	1.21	40.77	nr	**563.94**
Bend 45° (long radius)						
1½" dia.	22.11	27.98	0.51	17.19	nr	**45.17**
2" dia.	36.13	45.73	0.56	18.87	nr	**64.60**
3" dia.	77.23	97.74	0.65	21.90	nr	**119.64**
4" dia.	150.31	190.23	0.80	26.96	nr	**217.19**

38 PIPED SUPPLY SYSTEMS

Item	Net Price £	Material £	Labour hours	Labour £	Unit	Total rate £
COLD WATER PIPELINES: PVC-u PIPEWORK – cont						
Extra over fittings – cont						
Socket union						
½" dia.	8.03	10.16	0.34	11.46	nr	21.62
¾" dia.	9.19	11.63	0.39	13.14	nr	24.77
1" dia.	11.93	15.10	0.45	15.16	nr	30.26
1¼" dia.	14.84	18.78	0.50	16.84	nr	35.62
1½" dia.	20.32	25.72	0.57	19.20	nr	44.92
2" dia.	26.30	33.29	0.62	20.89	nr	54.18
3" dia.	97.92	123.93	0.70	23.58	nr	147.51
4" dia.	132.58	167.80	0.89	29.98	nr	197.78
Saddle plain						
2" × 1¼" dia.	20.66	26.15	0.42	14.15	nr	40.30
3" × 1½" dia.	29.04	36.76	0.48	16.17	nr	52.93
4" × 2" dia.	32.70	41.38	0.68	22.90	nr	64.28
6" × 2" dia.	38.41	48.61	0.91	30.65	nr	79.26
Straight tank connector						
½" dia.	5.34	6.75	0.13	4.38	nr	11.13
¾" dia.	6.04	7.65	0.14	4.72	nr	12.37
1" dia.	12.82	16.23	0.14	4.72	nr	20.95
1¼" dia.	32.58	41.24	0.16	5.39	nr	46.63
1½" dia.	35.72	45.20	0.18	6.06	nr	51.26
2" dia.	42.78	54.14	0.24	8.09	nr	62.23
3" dia.	43.87	55.52	0.29	9.77	nr	65.29
Equal tee						
½" dia.	2.41	3.05	0.44	14.82	nr	17.87
¾" dia.	3.06	3.88	0.48	16.17	nr	20.05
1" dia.	4.58	5.80	0.54	18.19	nr	23.99
1¼" dia.	6.49	8.21	0.70	23.58	nr	31.79
1½" dia.	9.35	11.84	0.74	24.93	nr	36.77
2" dia.	14.84	18.78	0.80	26.96	nr	45.74
3" dia.	43.01	54.43	1.04	35.03	nr	89.46
4" dia.	63.05	79.80	1.28	43.12	nr	122.92
6" dia.	219.64	277.97	1.93	65.03	nr	343.00

38 PIPED SUPPLY SYSTEMS

Item	Net Price £	Material £	Labour hours	Labour £	Unit	Total rate £
COLD WATER PIPELINES: PVC-C PIPEWORK						
Pipes; solvent welded in the running length, brackets measured separately						
Pipe; 3 m long; PN25						
16 × 2.0 mm	4.84	6.13	0.20	6.73	m	**12.86**
20 × 2.3 mm	7.31	9.25	0.20	6.73	m	**15.98**
25 × 2.8 mm	9.48	12.00	0.20	6.73	m	**18.73**
32 × 3.6 mm	14.08	17.82	0.20	6.73	m	**24.55**
Pipe; 5 m long; PN25						
40 × 4.5 mm	17.17	21.73	0.20	6.73	m	**28.46**
50 × 5.6 mm	25.84	32.70	0.20	6.73	m	**39.43**
63 × 7.0 mm	39.90	50.50	0.20	6.73	m	**57.23**
Fixings						
Refer to steel pipes; galvanized iron. For minimum fixing dimensions, refer to the Tables and Memoranda at the rear of the book						
Extra over fittings; solvent welded joints						
Straight coupling; PN25						
16 mm	0.75	0.95	0.20	6.73	nr	**7.68**
20 mm	1.06	1.34	0.20	6.73	nr	**8.07**
25 mm	1.32	1.67	0.20	6.73	nr	**8.40**
32 mm	4.02	5.08	0.20	6.73	nr	**11.81**
40 mm	5.16	6.53	0.20	6.73	nr	**13.26**
50 mm	6.91	8.75	0.20	6.73	nr	**15.48**
63 mm	12.21	15.46	0.20	6.73	nr	**22.19**
Elbow; 90°; PN25						
16 mm	1.21	1.53	0.20	6.73	nr	**8.26**
20 mm	1.85	2.34	0.20	6.73	nr	**9.07**
25 mm	2.31	2.92	0.20	6.73	nr	**9.65**
32 mm	4.81	6.09	0.20	6.73	nr	**12.82**
40 mm	7.41	9.37	0.20	6.73	nr	**16.10**
50 mm	10.25	12.97	0.20	6.73	nr	**19.70**
63 mm	17.53	22.19	0.20	6.73	nr	**28.92**
Elbow; 45°; PN25						
20 mm	1.85	2.34	0.20	6.73	nr	**9.07**
25 mm	2.31	2.92	0.20	6.73	nr	**9.65**
32 mm	4.81	6.09	0.20	6.73	nr	**12.82**
40 mm	7.41	9.37	0.20	6.73	nr	**16.10**
50 mm	10.25	12.97	0.20	6.73	nr	**19.70**
63 mm	17.53	22.19	0.20	6.73	nr	**28.92**
Reducer fitting; single stage reduction						
20/16 mm	1.32	1.67	0.20	6.73	nr	**8.40**
25/20 mm	1.58	2.00	0.20	6.73	nr	**8.73**
32/25 mm	3.17	4.01	0.20	6.73	nr	**10.74**
40/32 mm	4.15	5.25	0.20	6.73	nr	**11.98**
50/40 mm	4.81	6.09	0.20	6.73	nr	**12.82**
63/50 mm	7.28	9.22	0.20	6.73	nr	**15.95**

38 PIPED SUPPLY SYSTEMS

Item	Net Price £	Material £	Labour hours	Labour £	Unit	Total rate £
COLD WATER PIPELINES: PVC-C PIPEWORK – cont						
Extra over fittings – cont						
Equal tee; 90°; PN25						
16 mm	2.01	2.54	0.20	6.73	nr	9.27
20 mm	2.77	3.51	0.20	6.73	nr	10.24
25 mm	3.51	4.45	0.20	6.73	nr	11.18
32 mm	5.71	7.22	0.20	6.73	nr	13.95
40 mm	9.87	12.49	0.20	6.73	nr	19.22
50 mm	14.78	18.70	0.20	6.73	nr	25.43
63 mm	24.91	31.53	0.20	6.73	nr	38.26
Cap; PN25						
20 mm	1.38	1.75	0.20	6.73	nr	8.48
25 mm	1.85	2.34	0.20	6.73	nr	9.07
32 mm	2.69	3.40	0.20	6.73	nr	10.13
40 mm	3.69	4.67	0.20	6.73	nr	11.40
50 mm	5.16	6.53	0.20	6.73	nr	13.26
63 mm	8.22	10.40	0.20	6.73	nr	17.13

38 PIPED SUPPLY SYSTEMS

Item	Net Price £	Material £	Labour hours	Labour £	Unit	Total rate £
COLD WATER PIPELINES: SCREWED STEEL PIPEWORK						
Galvanized steel pipes; screwed and socketed joints; BS 1387: 1985						
Galvanized; medium, fixed vertically, with brackets measured separately, screwed joints are within the running length, but any flanges are additional						
10 mm dia.	5.16	6.53	0.51	17.19	m	**23.72**
15 mm dia.	4.65	5.88	0.52	17.52	m	**23.40**
20 mm dia.	5.25	6.64	0.55	18.52	m	**25.16**
25 mm dia.	7.34	9.28	0.60	20.22	m	**29.50**
32 mm dia.	9.09	11.50	0.67	22.58	m	**34.08**
40 mm dia.	10.56	13.36	0.75	25.26	m	**38.62**
50 mm dia.	14.81	18.75	0.85	28.64	m	**47.39**
65 mm dia.	20.07	25.40	0.93	31.34	m	**56.74**
80 mm dia.	25.99	32.89	1.07	36.04	m	**68.93**
100 mm dia.	36.78	46.55	1.46	49.19	m	**95.74**
125 mm dia.	58.48	74.01	1.72	57.95	m	**131.96**
150 mm dia.	67.91	85.95	1.96	66.04	m	**151.99**
Galvanized; heavy, fixed vertically, with brackets measured separately, screwed joints are within the running length, but any flanges are additional						
15 mm dia.	5.52	6.99	0.52	17.52	m	**24.51**
20 mm dia.	6.23	7.88	0.55	18.52	m	**26.40**
25 mm dia.	8.90	11.27	0.60	20.22	m	**31.49**
32 mm dia.	11.04	13.98	0.67	22.58	m	**36.56**
40 mm dia.	12.89	16.32	0.75	25.26	m	**41.58**
50 mm dia.	17.87	22.61	0.85	28.64	m	**51.25**
65 mm dia.	24.27	30.72	0.93	31.34	m	**62.06**
80 mm dia.	30.82	39.01	1.07	36.04	m	**75.05**
100 mm dia.	42.97	54.39	1.46	49.19	m	**103.58**
125 mm dia.	62.26	78.79	1.72	57.95	m	**136.74**
150 mm dia.	72.82	92.16	1.96	66.04	m	**158.20**
Galvanized; medium, fixed horizontally or suspended at high level, with brackets measured separately, screwed joints are within the running length, but any flanges are additional						
10 mm dia.	5.16	6.53	0.51	17.19	m	**23.72**
15 mm dia.	4.65	5.88	0.52	17.52	m	**23.40**
20 mm dia.	5.25	6.64	0.55	18.52	m	**25.16**
25 mm dia.	7.34	9.28	0.60	20.22	m	**29.50**
32 mm dia.	9.09	11.50	0.67	22.58	m	**34.08**
40 mm dia.	10.56	13.36	0.75	25.26	m	**38.62**
50 mm dia.	14.81	18.75	0.85	28.64	m	**47.39**
65 mm dia.	20.07	25.40	0.93	31.34	m	**56.74**
80 mm dia.	25.99	32.89	1.07	36.04	m	**68.93**
100 mm dia.	36.78	46.55	1.46	49.19	m	**95.74**
125 mm dia.	58.48	74.01	1.72	57.95	m	**131.96**
150 mm dia.	67.91	85.95	1.96	66.04	m	**151.99**

38 PIPED SUPPLY SYSTEMS

Item	Net Price £	Material £	Labour hours	Labour £	Unit	Total rate £
COLD WATER PIPELINES: SCREWED STEEL PIPEWORK – cont						
Galvanized steel pipes – cont						
Galvanized; heavy, fixed horizontally or suspended at high level, with brackets measured separately, screwed joints are within the running length, but any flanges are additional						
15 mm dia.	5.52	6.99	0.52	17.52	m	**24.51**
20 mm dia.	6.23	7.88	0.55	18.52	m	**26.40**
25 mm dia.	8.90	11.27	0.60	20.22	m	**31.49**
32 mm dia.	11.04	13.98	0.67	22.58	m	**36.56**
40 mm dia.	12.89	16.32	0.75	25.26	m	**41.58**
50 mm dia.	17.87	22.61	0.85	28.64	m	**51.25**
65 mm dia.	24.27	30.72	0.93	31.34	m	**62.06**
80 mm dia.	30.82	39.01	1.07	36.04	m	**75.05**
100 mm dia.	42.97	54.39	1.46	49.19	m	**103.58**
125 mm dia.	62.26	78.79	1.72	57.95	m	**136.74**
150 mm dia.	72.82	92.16	1.96	66.04	m	**158.20**
Fixings						
For steel pipes; galvanized iron. For minimum fixing dimensions, refer to the Tables and Memoranda at the rear of the book						
Single pipe bracket, screw on, galvanized iron; screwed to wood						
15 mm dia.	1.35	1.71	0.14	4.72	nr	**6.43**
20 mm dia.	1.50	1.90	0.14	4.72	nr	**6.62**
25 mm dia.	1.75	2.22	0.17	5.73	nr	**7.95**
32 mm dia.	2.39	3.02	0.19	6.41	nr	**9.43**
40 mm dia.	3.55	4.49	0.22	7.41	nr	**11.90**
50 mm dia.	4.71	5.96	0.22	7.41	nr	**13.37**
65 mm dia.	5.56	7.03	0.28	9.43	nr	**16.46**
80 mm dia.	8.71	11.02	0.32	10.79	nr	**21.81**
100 mm dia.	12.59	15.94	0.35	11.79	nr	**27.73**
Single pipe bracket, screw on, galvanized iron; plugged and screwed						
15 mm dia.	1.35	1.71	0.25	8.43	nr	**10.14**
20 mm dia.	1.50	1.90	0.25	8.43	nr	**10.33**
25 mm dia.	1.75	2.22	0.30	10.11	nr	**12.33**
32 mm dia.	2.39	3.02	0.32	10.79	nr	**13.81**
40 mm dia.	3.55	4.49	0.32	10.79	nr	**15.28**
50 mm dia.	4.71	5.96	0.32	10.79	nr	**16.75**
65 mm dia.	5.56	7.03	0.35	11.79	nr	**18.82**
80 mm dia.	8.71	11.02	0.42	14.15	nr	**25.17**
100 mm dia.	12.59	15.94	0.42	14.15	nr	**30.09**

38 PIPED SUPPLY SYSTEMS

Item	Net Price £	Material £	Labour hours	Labour £	Unit	Total rate £
Single pipe bracket for building in, galvanized iron						
15 mm dia.	1.43	1.81	0.10	3.37	nr	**5.18**
20 mm dia.	1.61	2.04	0.11	3.71	nr	**5.75**
25 mm dia.	1.75	2.22	0.12	4.03	nr	**6.25**
32 mm dia.	1.85	2.34	0.14	4.72	nr	**7.06**
40 mm dia.	2.39	3.02	0.15	5.05	nr	**8.07**
50 mm dia.	2.88	3.64	0.16	5.39	nr	**9.03**
Pipe ring, single socket, galvanized iron						
15 mm dia.	1.43	1.81	0.10	3.37	nr	**5.18**
20 mm dia.	1.61	2.04	0.11	3.71	nr	**5.75**
25 mm dia.	1.75	2.22	0.12	4.03	nr	**6.25**
32 mm dia.	1.76	2.23	0.15	5.05	nr	**7.28**
40 mm dia.	2.31	2.92	0.15	5.05	nr	**7.97**
50 mm dia.	3.02	3.82	0.16	5.39	nr	**9.21**
65 mm dia.	4.20	5.32	0.30	10.11	nr	**15.43**
80 mm dia.	4.94	6.25	0.35	11.79	nr	**18.04**
100 mm dia.	7.94	10.05	0.40	13.47	nr	**23.52**
125 mm dia.	16.06	20.33	0.60	20.23	nr	**40.56**
150 mm dia.	19.39	24.54	0.77	25.95	nr	**50.49**
Pipe ring, double socket, galvanized iron						
15 mm dia.	12.64	15.99	0.10	3.37	nr	**19.36**
20 mm dia.	14.54	18.40	0.11	3.71	nr	**22.11**
25 mm dia.	16.03	20.28	0.12	4.03	nr	**24.31**
32 mm dia.	17.67	22.37	0.14	4.72	nr	**27.09**
40 mm dia.	22.32	28.25	0.15	5.05	nr	**33.30**
50 mm dia.	25.40	32.14	0.16	5.39	nr	**37.53**
Screw on backplate (Male), galvanized iron; plugged and screwed						
All sizes 15 mm to 50 mm × M12	1.12	1.42	0.10	3.37	nr	**4.79**
Screw on backplate (Female), galvanized iron; plugged and screwed						
All sizes 15 mm to 50 mm × M12	1.12	1.42	0.10	3.37	nr	**4.79**
Extra over channel sections for fabricated hangers and brackets						
Galvanized steel; including inserts, bolts, nuts, washers; fixed to backgrounds						
41 × 21 mm	6.79	8.59	0.29	9.77	m	**18.36**
41 × 41 mm	8.14	10.30	0.29	9.77	m	**20.07**
Threaded rods; metric thread; including nuts, washers etc.						
10 mm dia. × 600 mm long	2.19	2.77	0.18	6.06	nr	**8.83**
12 mm dia. × 600 mm long	3.43	4.35	0.18	6.06	nr	**10.41**

38 PIPED SUPPLY SYSTEMS

Item	Net Price £	Material £	Labour hours	Labour £	Unit	Total rate £
COLD WATER PIPELINES: SCREWED STEEL PIPEWORK – cont						
Extra over steel flanges, screwed and drilled; metric; BS 4504						
Screwed flanges; PN6						
15 mm dia.	17.17	21.73	0.35	11.79	nr	33.52
20 mm dia.	17.17	21.73	0.47	15.84	nr	37.57
25 mm dia.	17.17	21.73	0.53	17.85	nr	39.58
32 mm dia.	17.17	21.73	0.62	20.89	nr	42.62
40 mm dia.	17.17	21.73	0.70	23.58	nr	45.31
50 mm dia.	18.31	23.17	0.84	28.30	nr	51.47
65 mm dia.	25.46	32.22	1.03	34.70	nr	66.92
80 mm dia.	35.96	45.51	1.23	41.44	nr	86.95
100 mm dia.	42.52	53.82	1.41	47.50	nr	101.32
125 mm dia.	77.99	98.71	1.77	59.64	nr	158.35
150 mm dia.	77.99	98.71	2.21	74.46	nr	173.17
Screwed flanges; PN16						
15 mm dia.	21.84	27.64	0.35	11.79	nr	39.43
20 mm dia.	21.84	27.64	0.47	15.84	nr	43.48
25 mm dia.	21.84	27.64	0.53	17.85	nr	45.49
32 mm dia.	23.32	29.51	0.62	20.89	nr	50.40
40 mm dia.	23.32	29.51	0.70	23.58	nr	53.09
50 mm dia.	28.45	36.01	0.84	28.30	nr	64.31
65 mm dia.	35.52	44.96	1.03	34.70	nr	79.66
80 mm dia.	43.26	54.75	1.23	41.44	nr	96.19
100 mm dia.	48.52	61.41	1.41	47.50	nr	108.91
125 mm dia.	84.03	107.49	1.77	59.64	nr	167.13
150 mm dia.	83.56	105.75	2.21	74.46	nr	180.21
Extra over steel flanges, screwed and drilled; imperial; BS 10						
Screwed flanges; Table E						
½" dia.	29.33	37.12	0.35	11.79	nr	48.91
¾" dia.	29.33	37.12	0.47	15.84	nr	52.96
1" dia.	29.33	37.12	0.53	17.85	nr	54.97
1¼" dia.	29.33	37.12	0.62	20.89	nr	58.01
1½" dia.	29.33	37.12	0.70	23.58	nr	60.70
2" dia.	29.33	37.12	0.84	28.30	nr	65.42
2½" dia.	34.92	44.20	1.03	34.70	nr	78.90
3" dia.	42.07	53.24	1.23	41.44	nr	94.68
4" dia.	53.52	67.74	1.41	47.50	nr	115.24
5" dia.	113.44	143.57	1.77	59.64	nr	203.21
Extra over steel flange connections						
Bolted connection between pair of flanges; including gasket, bolts, nuts and washers						
50 mm dia.	52.13	65.98	0.53	17.85	nr	83.83
65 mm dia.	65.68	83.13	0.53	17.85	nr	100.98
80 mm dia.	75.35	95.37	0.53	17.85	nr	113.22
100 mm dia.	89.92	113.80	0.53	17.85	nr	131.65
125 mm dia.	169.84	214.95	0.61	20.55	nr	235.50
150 mm dia.	174.69	221.09	0.90	30.32	nr	251.41

38 PIPED SUPPLY SYSTEMS

Item	Net Price £	Material £	Labour hours	Labour £	Unit	Total rate £
Extra over heavy steel tubular fittings; BS 1387						
Long screw connection with socket and backnut						
15 mm dia.	6.77	8.57	0.63	21.22	nr	**29.79**
20 mm dia.	8.50	10.75	0.84	28.30	nr	**39.05**
25 mm dia.	11.11	14.06	0.95	32.01	nr	**46.07**
32 mm dia.	14.63	18.51	1.11	37.40	nr	**55.91**
40 mm dia.	17.81	22.55	1.28	43.12	nr	**65.67**
50 mm dia.	26.28	33.26	1.53	51.54	nr	**84.80**
65 mm dia.	60.37	76.41	1.87	63.00	nr	**139.41**
80 mm dia.	78.07	98.81	2.21	74.46	nr	**173.27**
100 mm dia.	88.57	112.09	3.05	102.75	nr	**214.84**
Running nipple						
15 mm dia.	1.70	2.15	0.50	16.84	nr	**18.99**
20 mm dia.	2.11	2.67	0.68	22.90	nr	**25.57**
25 mm dia.	2.27	2.88	0.77	25.95	nr	**28.83**
32 mm dia.	3.66	4.64	0.90	30.32	nr	**34.96**
40 mm dia.	4.94	6.25	1.03	34.70	nr	**40.95**
50 mm dia.	7.51	9.51	1.23	41.44	nr	**50.95**
65 mm dia.	16.14	20.43	1.50	50.53	nr	**70.96**
80 mm dia.	25.17	31.85	1.78	59.96	nr	**91.81**
100 mm dia.	39.45	49.93	2.38	80.19	nr	**130.12**
Barrel nipple						
15 mm dia.	1.42	1.79	0.50	16.84	nr	**18.63**
20 mm dia.	2.14	2.71	0.68	22.90	nr	**25.61**
25 mm dia.	2.40	3.04	0.77	25.95	nr	**28.99**
32 mm dia.	3.98	5.04	0.90	30.32	nr	**35.36**
40 mm dia.	4.43	5.61	1.03	34.70	nr	**40.31**
50 mm dia.	6.33	8.01	1.23	41.44	nr	**49.45**
65 mm dia.	13.53	17.12	1.50	50.53	nr	**67.65**
80 mm dia.	18.89	23.91	1.78	59.96	nr	**83.87**
100 mm dia.	34.17	43.24	2.38	80.19	nr	**123.43**
125 mm dia.	63.47	80.33	2.87	96.69	nr	**177.02**
150 mm dia.	99.99	126.55	3.39	114.21	nr	**240.76**
Close taper nipple						
15 mm dia.	2.01	2.54	0.50	16.84	nr	**19.38**
20 mm dia.	2.61	3.30	0.68	22.90	nr	**26.20**
25 mm dia.	3.42	4.32	0.77	25.95	nr	**30.27**
32 mm dia.	5.10	6.45	0.90	30.32	nr	**36.77**
40 mm dia.	6.31	7.99	1.03	34.70	nr	**42.69**
50 mm dia.	9.70	12.28	1.23	41.44	nr	**53.72**
65 mm dia.	15.32	19.39	1.50	50.53	nr	**69.92**
80 mm dia.	25.11	31.77	1.78	59.96	nr	**91.73**
100 mm dia.	47.74	60.42	2.38	80.19	nr	**140.61**
90° bend with socket						
15 mm dia.	5.54	7.01	0.64	21.57	nr	**28.58**
20 mm dia.	7.46	9.44	0.85	28.64	nr	**38.08**
25 mm dia.	11.41	14.44	0.97	32.68	nr	**47.12**

38 PIPED SUPPLY SYSTEMS

Item	Net Price £	Material £	Labour hours	Labour £	Unit	Total rate £
COLD WATER PIPELINES: SCREWED STEEL PIPEWORK – cont						
Extra over heavy steel tubular fittings – cont						
90° bend with socket – cont						
32 mm dia.	16.37	20.72	1.12	37.73	nr	58.45
40 mm dia.	19.99	25.30	1.29	43.46	nr	68.76
50 mm dia.	31.09	39.35	1.55	52.21	nr	91.56
65 mm dia.	62.68	79.33	1.89	63.67	nr	143.00
80 mm dia.	93.03	117.73	2.24	75.47	nr	193.20
100 mm dia.	164.96	208.77	3.09	104.10	nr	312.87
125 mm dia.	404.06	511.38	3.92	132.07	nr	643.45
150 mm dia.	606.63	767.75	4.74	159.69	nr	927.44
Extra over heavy steel fittings; BS 1740						
Plug						
15 mm dia.	1.53	1.94	0.28	9.43	nr	11.37
20 mm dia.	2.39	3.02	0.38	12.81	nr	15.83
25 mm dia.	4.20	5.32	0.44	14.82	nr	20.14
32 mm dia.	6.52	8.25	0.51	17.19	nr	25.44
40 mm dia.	7.19	9.09	0.59	19.88	nr	28.97
50 mm dia.	10.28	13.01	0.70	23.58	nr	36.59
65 mm dia.	24.58	31.11	0.85	28.64	nr	59.75
80 mm dia.	45.99	58.21	1.00	33.69	nr	91.90
100 mm dia.	88.29	111.74	1.44	48.51	nr	160.25
Socket						
15 mm dia.	1.70	2.15	0.64	21.57	nr	23.72
20 mm dia.	1.93	2.44	0.85	28.64	nr	31.08
25 mm dia.	2.73	3.45	0.97	32.68	nr	36.13
32 mm dia.	3.93	4.97	1.12	37.73	nr	42.70
40 mm dia.	4.77	6.04	1.29	43.46	nr	49.50
50 mm dia.	7.36	9.32	1.55	52.21	nr	61.53
65 mm dia.	14.62	18.50	1.89	63.67	nr	82.17
80 mm dia.	18.91	23.93	2.24	75.47	nr	99.40
100 mm dia.	35.60	45.06	3.09	104.10	nr	149.16
150 mm dia.	84.97	107.54	4.74	159.69	nr	267.23
Elbow, female/female						
15 mm dia.	9.77	12.36	0.64	21.57	nr	33.93
20 mm dia.	12.74	16.13	0.85	28.64	nr	44.77
25 mm dia.	17.29	21.88	0.97	32.68	nr	54.56
32 mm dia.	32.19	40.73	1.12	37.73	nr	78.46
40 mm dia.	38.39	48.59	1.29	43.46	nr	92.05
50 mm dia.	62.93	79.64	1.55	52.21	nr	131.85
65 mm dia.	153.76	194.60	1.89	63.67	nr	258.27
80 mm dia.	183.34	232.03	2.24	75.47	nr	307.50
100 mm dia.	317.51	401.84	3.09	104.10	nr	505.94
Equal tee						
15 mm dia.	12.15	15.38	0.91	30.65	nr	46.03
20 mm dia.	14.11	17.85	1.22	41.10	nr	58.95
25 mm dia.	20.80	26.32	1.40	47.17	nr	73.49

38 PIPED SUPPLY SYSTEMS

Item	Net Price £	Material £	Labour hours	Labour £	Unit	Total rate £
32 mm dia.	42.96	54.36	1.62	54.58	nr	**108.94**
40 mm dia.	46.75	59.17	1.86	62.66	nr	**121.83**
50 mm dia.	76.02	96.21	2.21	74.46	nr	**170.67**
65 mm dia.	183.85	232.68	2.72	91.64	nr	**324.32**
80 mm dia.	197.32	249.73	3.21	108.15	nr	**357.88**
100 mm dia.	317.56	401.90	4.44	149.58	nr	**551.48**
Extra over malleable iron fittings; BS 143						
Cap						
15 mm dia.	3.92	4.96	0.32	10.79	nr	**15.75**
20 mm dia.	4.10	5.19	0.43	14.49	nr	**19.68**
25 mm dia.	7.42	9.39	0.49	16.51	nr	**25.90**
32 mm dia.	14.08	17.82	0.58	19.54	nr	**37.36**
40 mm dia.	15.11	19.12	0.66	22.23	nr	**41.35**
50 mm dia.	24.02	30.40	0.78	26.28	nr	**56.68**
65 mm dia.	41.28	52.25	0.96	32.35	nr	**84.60**
80 mm dia.	68.10	86.18	1.13	38.07	nr	**124.25**
100 mm dia.	116.09	146.92	1.70	57.27	nr	**204.19**
Plain plug, hollow						
15 mm dia.	1.91	2.42	0.28	9.43	nr	**11.85**
20 mm dia.	2.96	3.74	0.38	12.81	nr	**16.55**
25 mm dia.	5.19	6.56	0.44	14.82	nr	**21.38**
32 mm dia.	8.04	10.18	0.51	17.19	nr	**27.37**
40 mm dia.	8.88	11.23	0.59	19.88	nr	**31.11**
50 mm dia.	12.72	16.09	0.70	23.58	nr	**39.67**
65 mm dia.	30.37	38.44	0.85	28.64	nr	**67.08**
80 mm dia.	56.85	71.95	1.00	33.69	nr	**105.64**
100 mm dia.	109.15	138.14	1.44	48.51	nr	**186.65**
Plain plug, solid						
15 mm dia.	2.37	3.00	0.29	9.77	nr	**12.77**
20 mm dia.	2.51	3.18	0.38	12.81	nr	**15.99**
25 mm dia.	3.34	4.22	0.44	14.82	nr	**19.04**
32 mm dia.	5.15	6.52	0.51	17.19	nr	**23.71**
40 mm dia.	6.93	8.77	0.59	19.88	nr	**28.65**
50 mm dia.	9.11	11.52	0.70	23.58	nr	**35.10**
Elbow, male/female						
15 mm dia.	1.23	1.56	0.64	21.57	nr	**23.13**
20 mm dia.	1.66	2.11	0.85	28.64	nr	**30.75**
25 mm dia.	2.75	3.48	0.97	32.68	nr	**36.16**
32 mm dia.	6.23	7.88	1.12	37.73	nr	**45.61**
40 mm dia.	8.61	10.90	1.29	43.46	nr	**54.36**
50 mm dia.	11.10	14.04	1.55	52.21	nr	**66.25**
65 mm dia.	24.74	31.32	1.89	63.67	nr	**94.99**
80 mm dia.	33.84	42.83	2.24	75.47	nr	**118.30**
100 mm dia.	59.15	74.86	3.09	104.10	nr	**178.96**
Elbow						
15 mm dia.	1.10	1.39	0.64	21.57	nr	**22.96**
20 mm dia.	1.51	1.92	0.85	28.64	nr	**30.56**
25 mm dia.	2.35	2.98	0.97	32.68	nr	**35.66**
32 mm dia.	4.89	6.19	1.12	37.73	nr	**43.92**

38 PIPED SUPPLY SYSTEMS

Item	Net Price £	Material £	Labour hours	Labour £	Unit	Total rate £
COLD WATER PIPELINES: SCREWED STEEL PIPEWORK – cont						
Extra over malleable iron fittings – cont						
Elbow – cont						
40 mm dia.	7.31	9.25	1.29	43.46	nr	**52.71**
50 mm dia.	8.57	10.84	1.55	52.21	nr	**63.05**
65 mm dia.	19.11	24.18	1.89	63.67	nr	**87.85**
80 mm dia.	28.06	35.52	2.24	75.47	nr	**110.99**
100 mm dia.	48.17	60.96	3.09	104.10	nr	**165.06**
125 mm dia.	115.61	146.32	4.44	149.58	nr	**295.90**
150 mm dia.	215.24	272.41	5.79	195.07	nr	**467.48**
45° elbow						
15 mm dia.	2.84	3.60	0.64	21.57	nr	**25.17**
20 mm dia.	3.52	4.46	0.85	28.64	nr	**33.10**
25 mm dia.	4.84	6.13	0.97	32.68	nr	**38.81**
32 mm dia.	11.26	14.25	1.12	37.73	nr	**51.98**
40 mm dia.	13.25	16.77	1.29	43.46	nr	**60.23**
50 mm dia.	18.15	22.97	1.55	52.21	nr	**75.18**
65 mm dia.	25.51	32.29	1.89	63.67	nr	**95.96**
80 mm dia.	38.39	48.59	2.24	75.47	nr	**124.06**
100 mm dia.	74.00	93.65	3.09	104.10	nr	**197.75**
150 mm dia.	225.20	285.02	5.79	195.07	nr	**480.09**
Bend, male/female						
15 mm dia.	2.17	2.74	0.64	21.57	nr	**24.31**
20 mm dia.	3.57	4.51	0.85	28.64	nr	**33.15**
25 mm dia.	5.03	6.36	0.97	32.68	nr	**39.04**
32 mm dia.	8.50	10.75	1.12	37.73	nr	**48.48**
40 mm dia.	12.46	15.77	1.29	43.46	nr	**59.23**
50 mm dia.	23.42	29.64	1.55	52.21	nr	**81.85**
65 mm dia.	35.84	45.36	1.89	63.67	nr	**109.03**
80 mm dia.	55.20	69.87	2.24	75.47	nr	**145.34**
100 mm dia.	136.69	173.00	3.09	104.10	nr	**277.10**
Bend, male						
15 mm dia.	4.99	6.32	0.64	21.57	nr	**27.89**
20 mm dia.	5.60	7.09	0.85	28.64	nr	**35.73**
25 mm dia.	8.22	10.40	0.97	32.68	nr	**43.08**
32 mm dia.	18.18	23.00	1.12	37.73	nr	**60.73**
40 mm dia.	25.50	32.27	1.29	43.46	nr	**75.73**
50 mm dia.	38.69	48.97	1.55	52.21	nr	**101.18**
Bend, female						
15 mm dia.	2.24	2.83	0.64	21.57	nr	**24.40**
20 mm dia.	3.19	4.03	0.85	28.64	nr	**32.67**
25 mm dia.	4.48	5.67	0.97	32.68	nr	**38.35**
32 mm dia.	8.73	11.04	1.12	37.73	nr	**48.77**
40 mm dia.	10.39	13.15	1.29	43.46	nr	**56.61**
50 mm dia.	16.37	20.72	1.55	52.21	nr	**72.93**
65 mm dia.	35.84	45.36	1.89	63.67	nr	**109.03**
80 mm dia.	53.15	67.27	2.24	75.47	nr	**142.74**
100 mm dia.	111.52	141.14	3.09	104.10	nr	**245.24**
125 mm dia.	224.99	284.75	4.44	149.58	nr	**434.33**
150 mm dia.	494.29	625.58	5.79	195.07	nr	**820.65**

38 PIPED SUPPLY SYSTEMS

Item	Net Price £	Material £	Labour hours	Labour £	Unit	Total rate £
Return bend						
15 mm dia.	11.41	14.44	0.64	21.57	nr	36.01
20 mm dia.	18.45	23.35	0.85	28.64	nr	51.99
25 mm dia.	23.02	29.13	0.97	32.68	nr	61.81
32 mm dia.	31.96	40.44	1.12	37.73	nr	78.17
40 mm dia.	43.29	54.79	1.29	43.46	nr	98.25
50 mm dia.	66.10	83.65	1.55	52.21	nr	135.86
Equal socket, parallel thread						
15 mm dia.	1.07	1.36	0.64	21.57	nr	22.93
20 mm dia.	1.40	1.77	0.85	28.64	nr	30.41
25 mm dia.	1.77	2.24	0.97	32.68	nr	34.92
32 mm dia.	3.94	4.98	1.12	37.73	nr	42.71
40 mm dia.	5.35	6.78	1.29	43.46	nr	50.24
50 mm dia.	7.74	9.80	1.55	52.21	nr	62.01
65 mm dia.	12.28	15.55	1.89	63.67	nr	79.22
80 mm dia.	17.39	22.01	2.24	75.47	nr	97.48
100 mm dia.	28.71	36.33	3.09	104.10	nr	140.43
Concentric reducing socket						
20 × 15 mm dia.	1.71	2.16	0.76	25.60	nr	27.76
25 × 15 mm dia.	2.24	2.83	0.86	28.97	nr	31.80
25 × 20 mm dia.	2.11	2.67	0.86	28.97	nr	31.64
32 × 25 mm dia.	4.05	5.13	1.01	34.04	nr	39.17
40 × 25 mm dia.	5.35	6.78	1.16	39.08	nr	45.86
40 × 32 mm dia.	5.93	7.50	1.16	39.08	nr	46.58
50 × 25 mm dia.	10.26	12.98	1.38	46.50	nr	59.48
50 × 40 mm dia.	8.30	10.51	1.38	46.50	nr	57.01
65 × 50 mm dia.	14.49	18.33	1.69	56.94	nr	75.27
80 × 50 mm dia.	18.08	22.88	2.00	67.38	nr	90.26
100 × 50 mm dia.	36.03	45.60	2.75	92.65	nr	138.25
100 × 80 mm dia.	107.79	136.42	2.75	92.65	nr	229.07
150 × 100 mm dia.	100.16	126.76	4.10	138.13	nr	264.89
Eccentric reducing socket						
20 × 15 mm dia.	3.88	4.91	0.76	25.60	nr	30.51
25 × 15 mm dia.	11.00	13.92	0.86	28.97	nr	42.89
25 × 20 mm dia.	12.49	15.80	0.86	28.97	nr	44.77
32 × 25 mm dia.	16.15	20.44	1.01	34.04	nr	54.48
40 × 25 mm dia.	18.50	23.41	1.16	39.08	nr	62.49
40 × 32 mm dia.	20.15	25.50	1.16	39.08	nr	64.58
50 × 25 mm dia.	20.20	25.57	1.18	39.75	nr	65.32
50 × 40 mm dia.	20.25	25.63	1.28	43.12	nr	68.75
65 × 50 mm dia.	20.52	25.97	1.69	56.94	nr	82.91
80 × 50 mm dia.	21.86	27.66	2.00	67.38	nr	95.04
Hexagon bush						
20 × 15 mm dia.	0.96	1.21	0.37	12.47	nr	13.68
25 × 15 mm dia.	1.32	1.67	0.43	14.49	nr	16.16
25 × 20 mm dia.	1.23	1.56	0.43	14.49	nr	16.05
32 × 25 mm dia.	1.67	2.12	0.51	17.19	nr	19.31
40 × 25 mm dia.	2.49	3.15	0.58	19.54	nr	22.69
40 × 32 mm dia.	2.49	3.15	0.58	19.54	nr	22.69
50 × 25 mm dia.	5.27	6.68	0.71	23.92	nr	30.60
50 × 40 mm dia.	4.91	6.22	0.71	23.92	nr	30.14

38 PIPED SUPPLY SYSTEMS

Item	Net Price £	Material £	Labour hours	Labour £	Unit	Total rate £
COLD WATER PIPELINES: SCREWED STEEL PIPEWORK – cont						
Extra over malleable iron fittings – cont						
Hexagon bush – cont						
65 × 50 mm dia.	9.07	11.48	0.84	28.30	nr	39.78
80 × 50 mm dia.	13.69	17.33	1.00	33.69	nr	51.02
100 × 50 mm dia.	31.71	40.13	1.52	51.21	nr	91.34
100 × 80 mm dia.	26.38	33.39	1.52	51.21	nr	84.60
150 × 100 mm dia.	94.78	119.95	2.48	83.55	nr	203.50
Hexagon nipple						
15 mm dia.	1.05	1.33	0.28	9.43	nr	10.76
20 mm dia.	1.17	1.48	0.38	12.81	nr	14.29
25 mm dia.	1.66	2.11	0.44	14.82	nr	16.93
32 mm dia.	3.53	4.47	0.51	17.19	nr	21.66
40 mm dia.	4.05	5.13	0.59	19.88	nr	25.01
50 mm dia.	7.39	9.35	0.70	23.58	nr	32.93
65 mm dia.	12.37	15.66	0.85	28.64	nr	44.30
80 mm dia.	17.10	21.64	1.00	33.69	nr	55.33
100 mm dia.	30.35	38.42	1.44	48.51	nr	86.93
150 mm dia.	84.02	106.33	2.32	78.16	nr	184.49
Union, male/female						
15 mm dia.	5.19	6.56	0.64	21.57	nr	28.13
20 mm dia.	6.37	8.06	0.85	28.64	nr	36.70
25 mm dia.	7.39	9.35	0.97	32.68	nr	42.03
32 mm dia.	13.06	16.53	1.12	37.73	nr	54.26
40 mm dia.	16.71	21.15	1.29	43.46	nr	64.61
50 mm dia.	26.27	33.25	1.55	52.21	nr	85.46
65 mm dia.	58.79	74.40	1.89	63.67	nr	138.07
Union, female						
15 mm dia.	4.91	6.22	0.64	21.57	nr	27.79
20 mm dia.	5.40	6.83	0.85	28.64	nr	35.47
25 mm dia.	6.33	8.01	0.97	32.68	nr	40.69
32 mm dia.	10.92	13.82	1.12	37.73	nr	51.55
40 mm dia.	12.33	15.60	1.29	43.46	nr	59.06
50 mm dia.	18.38	23.26	1.55	52.21	nr	75.47
65 mm dia.	47.07	59.57	1.89	63.67	nr	123.24
80 mm dia.	62.22	78.75	2.24	75.47	nr	154.22
100 mm dia.	118.48	149.95	3.09	104.10	nr	254.05
Union elbow, male/female						
15 mm dia.	7.83	9.91	0.64	21.57	nr	31.48
20 mm dia.	9.82	12.43	0.85	28.64	nr	41.07
25 mm dia.	13.79	17.45	0.97	32.68	nr	50.13
Twin elbow						
15 mm dia.	7.53	9.53	0.91	30.65	nr	40.18
20 mm dia.	8.33	10.54	1.22	41.10	nr	51.64
25 mm dia.	13.49	17.07	1.39	46.83	nr	63.90
32 mm dia.	27.03	34.20	1.62	54.58	nr	88.78
40 mm dia.	34.22	43.31	1.86	62.66	nr	105.97
50 mm dia.	43.98	55.66	2.21	74.46	nr	130.12
65 mm dia.	71.08	89.96	2.72	91.64	nr	181.60
80 mm dia.	80.67	102.10	3.21	108.15	nr	210.25

38 PIPED SUPPLY SYSTEMS

Item	Net Price £	Material £	Labour hours	Labour £	Unit	Total rate £
Equal tee						
15 mm dia.	1.51	1.92	0.91	30.65	nr	**32.57**
20 mm dia.	2.20	2.79	1.22	41.10	nr	**43.89**
25 mm dia.	3.17	4.01	1.39	46.83	nr	**50.84**
32 mm dia.	6.71	8.49	1.62	54.58	nr	**63.07**
40 mm dia.	9.16	11.59	1.86	62.66	nr	**74.25**
50 mm dia.	13.18	16.68	2.21	74.46	nr	**91.14**
65 mm dia.	30.92	39.13	2.72	91.64	nr	**130.77**
80 mm dia.	36.03	45.60	3.21	108.15	nr	**153.75**
100 mm dia.	65.33	82.68	4.44	149.58	nr	**232.26**
125 mm dia.	121.19	153.37	5.38	181.26	nr	**334.63**
150 mm dia.	289.81	366.79	6.31	212.59	nr	**579.38**
Tee reducing on branch						
20 × 15 mm dia.	2.26	2.86	1.22	41.10	nr	**43.96**
25 × 15 mm dia.	3.09	3.91	1.39	46.83	nr	**50.74**
25 × 20 mm dia.	3.52	4.46	1.39	46.83	nr	**51.29**
32 × 25 mm dia.	6.79	8.59	1.62	54.58	nr	**63.17**
40 × 25 mm dia.	8.61	10.90	1.86	62.66	nr	**73.56**
40 × 32 mm dia.	12.64	15.99	1.86	62.66	nr	**78.65**
50 × 25 mm dia.	11.44	14.48	2.21	74.46	nr	**88.94**
50 × 40 mm dia.	17.79	22.51	2.21	74.46	nr	**96.97**
65 × 50 mm dia.	27.45	34.74	2.72	91.64	nr	**126.38**
80 × 50 mm dia.	37.11	46.96	3.21	108.15	nr	**155.11**
100 × 50 mm dia.	61.43	77.75	4.44	149.58	nr	**227.33**
100 × 80 mm dia.	94.78	119.95	4.44	149.58	nr	**269.53**
150 × 100 mm dia.	153.74	194.58	6.28	211.57	nr	**406.15**
Equal pitcher tee						
15 mm dia.	5.25	6.64	0.91	30.65	nr	**37.29**
20 mm dia.	6.47	8.19	1.22	41.10	nr	**49.29**
25 mm dia.	9.68	12.25	1.39	46.83	nr	**59.08**
32 mm dia.	15.01	19.00	1.62	54.58	nr	**73.58**
40 mm dia.	23.24	29.41	1.86	62.66	nr	**92.07**
50 mm dia.	32.60	41.26	2.21	74.46	nr	**115.72**
65 mm dia.	46.40	58.72	2.72	91.64	nr	**150.36**
80 mm dia.	72.40	91.63	3.21	108.15	nr	**199.78**
100 mm dia.	162.90	206.17	4.44	149.58	nr	**355.75**
Cross						
15 mm dia.	4.35	5.51	1.00	33.69	nr	**39.20**
20 mm dia.	6.82	8.64	1.33	44.80	nr	**53.44**
25 mm dia.	8.63	10.92	1.51	50.88	nr	**61.80**
32 mm dia.	12.84	16.25	1.77	59.64	nr	**75.89**
40 mm dia.	17.29	21.88	2.02	68.05	nr	**89.93**
50 mm dia.	26.88	34.01	2.42	81.52	nr	**115.53**
65 mm dia.	38.39	48.59	2.97	100.06	nr	**148.65**
80 mm dia.	51.03	64.58	3.50	117.91	nr	**182.49**
100 mm dia.	105.32	133.29	4.84	163.06	nr	**296.35**

38 PIPED SUPPLY SYSTEMS

Item	Net Price £	Material £	Labour hours	Labour £	Unit	Total rate £
COLD WATER PIPELINES: PIPEWORK ANCILLARIES – VALVES						
Regulators						
Gunmetal; self-acting two port thermostat; single seat; screwed; normally closed; with adjustable or fixed bleed device						
25 mm dia.	778.13	984.80	1.46	49.19	nr	**1033.99**
32 mm dia.	800.58	1013.22	1.45	48.85	nr	**1062.07**
40 mm dia.	855.72	1083.00	1.55	52.24	nr	**1135.24**
50 mm dia.	1030.12	1303.72	1.68	56.59	nr	**1360.31**
Self acting temperature regulator for storage calorifier; integral sensing element and pocket; screwed ends						
15 mm dia.	752.20	951.99	1.32	44.48	nr	**996.47**
25 mm dia.	825.66	1044.96	1.52	51.21	nr	**1096.17**
32 mm dia.	1066.21	1349.40	1.79	60.30	nr	**1409.70**
40 mm dia.	1304.31	1650.73	1.99	67.04	nr	**1717.77**
50 mm dia.	1524.65	1929.59	2.26	76.14	nr	**2005.73**
Self acting temperature regulator for storage calorifier; integral sensing element and pocket; flanged ends; bolted connection						
15 mm dia.	1104.22	1397.50	0.61	20.55	nr	**1418.05**
25 mm dia.	1263.78	1599.44	0.72	24.26	nr	**1623.70**
32 mm dia.	1593.04	2016.16	0.94	31.66	nr	**2047.82**
40 mm dia.	1886.83	2387.97	1.03	34.70	nr	**2422.67**
50 mm dia.	2190.75	2772.62	1.18	39.75	nr	**2812.37**
Chrome plated thermostatic mixing valves including non-return valves and inlet swivel connections with strainers; copper compression fittings						
15 mm dia.	121.88	154.25	0.69	23.25	nr	**177.50**
Chrome plated thermostatic mixing valves including non-return valves and inlet swivel connections with angle pattern combined isolating valves and strainers; copper compression fittings						
15 mm dia.	200.01	253.13	0.69	23.25	nr	**276.38**
Gunmetal thermostatic mixing valves including non-return valves and inlet swivel connections with strainers; copper compression fittings						
15 mm dia.	252.95	320.13	0.69	23.25	nr	**343.38**
Gunmetal thermostatic mixing valves including non-return valves and inlet swivel connections with angle pattern combined isolating valves and strainers; copper compression fittings						
15 mm dia.	265.94	336.57	0.69	23.25	nr	**359.82**

38 PIPED SUPPLY SYSTEMS

Item	Net Price £	Material £	Labour hours	Labour £	Unit	Total rate £
Ball float valves						
Bronze, equilibrium; copper float; working pressure cold services up to 16 bar; flanged ends; BS 4504 Table 16/21; bolted connections						
25 mm dia.	165.06	208.90	1.04	35.03	nr	243.93
32 mm dia.	228.98	289.80	1.22	41.10	nr	330.90
40 mm dia.	310.58	393.08	1.38	46.50	nr	439.58
50 mm dia.	488.03	617.65	1.66	55.92	nr	673.57
65 mm dia.	523.35	662.36	1.93	65.03	nr	727.39
80 mm dia.	583.56	738.55	2.16	72.77	nr	811.32
Heavy, equilibrium; with long tail and backnut; copper float; screwed for iron						
25 mm dia.	147.61	186.82	1.58	53.23	nr	240.05
32 mm dia.	202.65	256.47	1.78	59.96	nr	316.43
40 mm dia.	285.70	361.58	1.90	64.02	nr	425.60
50 mm dia.	375.18	474.82	2.65	89.28	nr	564.10
Brass, ball valve; BS 1212; copper float; screwed						
15 mm dia.	8.22	10.40	0.25	8.43	nr	18.83
22 mm dia.	14.94	18.91	0.29	9.77	nr	28.68
28 mm dia.	65.08	82.36	0.35	11.79	nr	94.15
Gate valves						
DZR copper alloy wedge non-rising stem; capillary joint to copper						
15 mm dia.	17.26	21.84	0.84	28.30	nr	50.14
22 mm dia.	21.18	26.80	1.01	34.04	nr	60.84
28 mm dia.	28.64	36.24	1.19	40.10	nr	76.34
35 mm dia.	51.41	65.06	1.38	46.50	nr	111.56
42 mm dia.	87.34	110.53	1.62	54.58	nr	165.11
54 mm dia.	122.38	154.88	1.94	65.35	nr	220.23
Cocks; capillary joints to copper						
Stopcock; brass head with gun metal body						
15 mm dia.	7.56	9.56	0.45	15.16	nr	24.72
22 mm dia.	14.10	17.84	0.46	15.50	nr	33.34
28 mm dia.	40.09	50.74	0.54	18.19	nr	68.93
Lockshield stop cocks; brass head with gun metal body						
15 mm dia.	8.72	11.03	0.45	15.16	nr	26.19
22 mm dia.	8.72	11.03	0.46	15.50	nr	26.53
28 mm dia.	8.72	11.03	0.54	18.19	nr	29.22
DZR stopcock; brass head with gun metal body						
15 mm dia.	20.69	26.19	0.45	15.16	nr	41.35
22 mm dia.	35.83	45.35	0.46	15.50	nr	60.85
28 mm dia.	59.72	75.58	0.54	18.19	nr	93.77

38 PIPED SUPPLY SYSTEMS

Item	Net Price £	Material £	Labour hours	Labour £	Unit	Total rate £
COLD WATER PIPELINES: PIPEWORK ANCILLARIES – VALVES – cont						
Cocks – cont						
Gunmetal stopcock						
35 mm dia.	78.05	98.78	0.69	23.25	nr	122.03
42 mm dia.	103.66	131.20	0.71	23.92	nr	155.12
54 mm dia.	154.83	195.96	0.81	27.28	nr	223.24
Double union stopcock						
15 mm dia.	22.24	28.15	0.60	20.22	nr	48.37
22 mm dia.	31.27	39.58	0.60	20.22	nr	59.80
28 mm dia.	55.63	70.40	0.69	23.25	nr	93.65
Double union DZR stopcock						
15 mm dia.	36.30	45.94	0.60	20.22	nr	66.16
22 mm dia.	44.63	56.48	0.61	20.55	nr	77.03
28 mm dia.	82.48	104.38	0.69	23.25	nr	127.63
Double union gun metal stopcock						
35 mm dia.	137.42	173.91	0.63	21.22	nr	195.13
42 mm dia.	188.60	238.69	0.67	22.58	nr	261.27
54 mm dia.	296.70	375.50	0.85	28.64	nr	404.14
Double union stopcock with easy clean cover						
15 mm dia.	38.53	48.76	0.60	20.22	nr	68.98
22 mm dia.	48.08	60.85	0.61	20.55	nr	81.40
28 mm dia.	89.36	113.10	0.69	23.25	nr	136.35
Combined stopcock and drain						
15 mm dia.	38.22	48.37	0.67	22.58	nr	70.95
22 mm dia.	46.63	59.01	0.68	22.90	nr	81.91
Combined DZR stopcock and drain						
15 mm dia.	50.51	63.93	0.67	22.58	nr	86.51
Gate valve						
DZR copper alloy wedge non-rising stem; compression joint to copper						
15 mm dia.	17.26	21.84	0.84	28.30	nr	50.14
22 mm dia.	21.18	26.80	1.01	34.04	nr	60.84
28 mm dia.	28.64	36.24	1.19	40.10	nr	76.34
35 mm dia.	51.41	65.06	1.38	46.50	nr	111.56
42 mm dia.	87.34	110.53	1.62	54.58	nr	165.11
54 mm dia.	122.38	154.88	1.94	65.35	nr	220.23
Cocks; compression joints to copper						
Stopcock; brass head gun metal body						
15 mm dia.	6.87	8.69	0.42	14.15	nr	22.84
22 mm dia.	12.81	16.22	0.42	14.15	nr	30.37
28 mm dia.	36.44	46.12	0.45	15.16	nr	61.28
Lockshield stopcock; brass head gun metal body						
15 mm dia.	7.27	9.21	0.42	14.15	nr	23.36
22 mm dia.	10.41	13.17	0.42	14.15	nr	27.32
28 mm dia.	18.52	23.44	0.45	15.16	nr	38.60

38 PIPED SUPPLY SYSTEMS

Item	Net Price £	Material £	Labour hours	Labour £	Unit	Total rate £
DZR Stopcock						
15 mm dia.	17.24	21.82	0.38	12.81	nr	**34.63**
22 mm dia.	29.87	37.80	0.39	13.14	nr	**50.94**
28 mm dia.	49.75	62.97	0.40	13.47	nr	**76.44**
35 mm dia.	78.05	98.78	0.52	17.52	nr	**116.30**
42 mm dia.	103.66	131.20	0.54	18.19	nr	**149.39**
54 mm dia.	154.83	195.96	0.63	21.22	nr	**217.18**
DZR lockshield stopcock						
15 mm dia.	7.27	9.21	0.38	12.81	nr	**22.02**
22 mm dia.	11.47	14.52	0.39	13.14	nr	**27.66**
Combined stop/draincock						
15 mm dia.	38.22	48.37	0.22	7.41	nr	**55.78**
22 mm dia.	49.18	62.24	0.45	15.16	nr	**77.40**
DZR combined stop/draincock						
15 mm dia.	50.51	63.93	0.41	13.81	nr	**77.74**
22 mm dia.	65.45	82.84	0.42	14.15	nr	**96.99**
Stopcock to polyethylene						
15 mm dia.	26.41	33.42	0.38	12.81	nr	**46.23**
20 mm dia.	33.26	42.09	0.39	13.14	nr	**55.23**
25 mm dia.	44.24	55.99	0.40	13.47	nr	**69.46**
Draw off coupling						
15 mm dia.	16.04	20.31	0.38	12.81	nr	**33.12**
DZR draw off coupling						
15 mm dia.	16.04	20.31	0.38	12.81	nr	**33.12**
22 mm dia.	18.56	23.49	0.39	13.14	nr	**36.63**
Draw off elbow						
15 mm dia.	17.60	22.28	0.38	12.81	nr	**35.09**
22 mm dia.	21.68	27.44	0.39	13.14	nr	**40.58**
Lockshield drain cock						
15 mm dia.	18.41	23.30	0.41	13.81	nr	**37.11**
Check valves						
DZR copper alloy and bronze, WRC approved cartridge double check valve; BS 6282; working pressure cold services up to 10 bar at 65°C; screwed ends						
32 mm dia.	110.81	140.25	1.38	46.50	nr	**186.75**
40 mm dia.	126.29	159.84	1.62	54.58	nr	**214.42**
50 mm dia.	207.06	262.06	1.94	65.35	nr	**327.41**

38 PIPED SUPPLY SYSTEMS

Item	Net Price £	Material £	Labour hours	Labour £	Unit	Total rate £
COLD WATER PIPELINES: PIPEWORK ANCILLARIES – PUMPS						
Packaged cold water pressure booster set; fully automatic; 3 phase supply; includes fixing in position; electrical work elsewhere						
Pressure booster set						
0.75 l/s @ 30 m head	4852.99	6141.95	9.38	316.02	nr	**6457.97**
1.5 l/s @ 30 m head	5762.94	7293.57	9.38	316.02	nr	**7609.59**
3 l/s @ 30 m head	7041.17	8911.30	10.38	349.71	nr	**9261.01**
6 l/s @ 30 m head	15902.24	20125.87	10.38	349.71	nr	**20475.58**
12 l/s @ 30 m head	19866.94	25143.60	12.38	417.09	nr	**25560.69**
0.75 l/s @ 50 m head	5524.63	6991.97	9.38	316.02	nr	**7307.99**
1.5 l/s @ 50 m head	7041.17	8911.30	9.38	316.02	nr	**9227.32**
3 l/s @ 50 m head	7777.80	9843.58	10.38	349.71	nr	**10193.29**
6 l/s @ 50 m head	17761.10	22478.44	10.38	349.71	nr	**22828.15**
12 l/s @ 50 m head	21816.79	27611.33	12.38	417.09	nr	**28028.42**
0.75 l/s @ 70 m head	6044.58	7650.03	9.38	316.02	nr	**7966.05**
1.5 l/s @ 70 m head	7773.48	9838.11	9.38	316.02	nr	**10154.13**
3 l/s @ 70 m head	8276.09	10474.22	10.38	349.71	nr	**10823.93**
6 l/s @ 70 m head	19390.32	24540.39	10.38	349.71	nr	**24890.10**
12 l/s @ 70 m head	23571.66	29832.30	12.38	417.09	nr	**30249.39**
Automatic sump pump for clear and drainage water; single stage centrifugal pump, presure tight electric motor; single phase supply; includes fixing in position; electrical work elsewhere						
Single pump						
1 l/s @ 2.68 m total head	264.78	335.10	3.50	117.91	nr	**453.01**
1 l/s @ 4.68 m total head	291.46	368.87	3.50	117.91	nr	**486.78**
1 l/s @ 6.68 m total head	384.78	486.98	3.50	117.91	nr	**604.89**
2 l/s @ 4.38 m total head	384.78	486.98	4.00	134.76	nr	**621.74**
2 l/s @ 6.38 m total head	384.78	486.98	4.00	134.76	nr	**621.74**
2 l/s @ 8.38 m total head	495.26	626.80	4.00	134.76	nr	**761.56**
3 l/s @ 3.7 m total head	384.78	486.98	4.50	151.60	nr	**638.58**
3 l/s @ 5.7 m total head	495.26	626.80	4.50	151.60	nr	**778.40**
4 l/s @ 2.9 m total head	384.78	486.98	5.00	168.45	nr	**655.43**
4 l/s @ 4.9 m total head	495.26	626.80	5.00	168.45	nr	**795.25**
4 l/s @ 6.9 m total head	1373.41	1738.18	5.00	168.45	nr	**1906.63**
Extra for high level alarm box with single float switch, local alarm and volt free contacts for remote alarm.	403.01	510.05	–	–	nr	**510.05**
Duty/standby pump unit						
1 l/s @ 2.68 m total head	502.88	636.44	5.00	168.45	nr	**804.89**
1 l/s @ 4.68 m total head	552.43	699.16	5.00	168.45	nr	**867.61**
1 l/s @ 6.68 m total head	746.71	945.03	5.00	168.45	nr	**1113.48**

38 PIPED SUPPLY SYSTEMS

Item	Net Price £	Material £	Labour hours	Labour £	Unit	Total rate £
2 l/s @ 4.38 m total head	746.71	945.03	5.50	185.29	nr	**1130.32**
2 l/s @ 6.38 m total head	746.71	945.03	5.50	185.29	nr	**1130.32**
2 l/s @ 8.38 m total head	956.26	1210.24	5.50	185.29	nr	**1395.53**
3 l/s @ 3.7 m total head	746.71	945.03	6.00	202.14	nr	**1147.17**
3 l/s @ 5.7 m total head	956.26	1210.24	6.00	202.14	nr	**1412.38**
4 l/s @ 2.9 m total head	746.71	945.03	6.50	218.98	nr	**1164.01**
4 l/s @ 4.9 m total head	956.26	1210.24	6.50	218.98	nr	**1429.22**
4/s @ 6.9 m total head	2541.09	3216.00	7.00	235.83	nr	**3451.83**
Extra for 4 nr float switches to give pump on, off and high level alarm	421.07	532.91	–	–	nr	**532.91**
Extra for dual pump control panel, internal wall mounted IP54, including volt free contacts	2005.13	2537.70	4.00	134.76	nr	**2672.46**

38 PIPED SUPPLY SYSTEMS

Item	Net Price £	Material £	Labour hours	Labour £	Unit	Total rate £
COLD WATER PIPELINES: PIPEWORK ANCILLARIES – TANKS						
Cisterns; fibreglass; complete with ball valve, screened fittings, outlets, drain provisions, fixing plate and fitted covers						
Rectangular						
60 litre capacity	416.00	526.49	1.33	44.80	nr	**571.29**
100 litre capacity	474.00	599.89	1.40	47.17	nr	**647.06**
150 litre capacity	542.00	685.96	1.61	54.24	nr	**740.20**
250 litre capacity	584.00	739.11	1.61	54.24	nr	**793.35**
420 litre capacity	723.00	915.03	1.99	67.04	nr	**982.07**
730 litre capacity	1021.00	1292.18	3.31	111.51	nr	**1403.69**
800 litre capacity	1054.00	1333.94	3.60	121.28	nr	**1455.22**
1700 litre capacity	1492.00	1888.28	13.32	448.76	nr	**2337.04**
2250 litre capacity	1756.00	2222.39	20.18	679.86	nr	**2902.25**
3400 litre capacity	2376.00	3007.07	24.50	825.41	nr	**3832.48**
4500 litre capacity	2750.00	3480.40	29.91	1007.68	nr	**4488.08**
Cisterns; polypropylene; complete with ball valve, fixing plate and cover; includes placing in position						
Rectangular						
18 litre capacity	78.60	99.48	1.00	33.69	nr	**133.17**
68 litre capacity	105.92	134.05	1.00	33.69	nr	**167.74**
91 litre capacity	106.60	134.92	1.00	33.69	nr	**168.61**
114 litre capacity	118.61	150.11	1.00	33.69	nr	**183.80**
182 litre capacity	156.63	198.23	1.00	33.69	nr	**231.92**
227 litre capacity	199.50	252.48	1.00	33.69	nr	**286.17**
Circular						
114 litre capacity	110.94	140.40	1.00	33.69	nr	**174.09**
227 litre capacity	113.14	143.19	1.00	33.69	nr	**176.88**
318 litre capacity	188.18	238.16	1.00	33.69	nr	**271.85**
455 litre capacity	216.28	273.73	1.00	33.69	nr	**307.42**
Steel sectional water storage tank; hot pressed steel tank to BS 1564 TYPE 1; 5 mm plate; pre-insulated and complete with all connections and fittings to comply with BSEN 13280; 2001 and WRAS water supply (water fittings) regulations 1999; externally flanged base and sides; cost of erection (on prepared base) is included within the net price, labour cost allows for offloading and positioning materials						
Note: Prices are based on the most economical tank size for each volume, and the cost will vary with differing tank dimensions, for the same volume						

38 PIPED SUPPLY SYSTEMS

Item	Net Price £	Material £	Labour hours	Labour £	Unit	Total rate £
Volume, size						
4,900 litres, 3.66 m × 1.22 m × 1,22 m (h)	6415.55	8119.52	6.00	202.14	nr	8321.66
20,300 litres, 3.66 m × 2.4 m × 2.4 m (h)	13668.45	17298.79	12.00	404.29	nr	17703.08
52,000 litres, 6.1 m × 3.6 m × 2.4 m (h)	24368.74	30841.08	19.00	640.11	nr	31481.19
94,000 litres, 7.3 m × 3.6 m × 3.6 m (h)	42905.16	54300.77	28.00	943.33	nr	55244.10
140,000 litres, 9.7 m × 6.1 m × 2.44 m (h)	47871.31	60585.93	28.00	943.33	nr	61529.26
GRP sectional water storage tank; pre-insulated and complete with twin compartment and basic connections and fittings to comply with BSEN 13280; 2001 and WRAS water supply (water fittings) regulations 1999; externally flanged base and sides; cost of erection (on prepared base) is included within the net price, labour cost allows for offloading and positioning materials						
Note: Prices are based on the most economical tank size for each volume, and the cost will vary with differing tank dimensions, for the same volume						
Volume, size						
4,500 litres, 3 m × 1 m × 1.5 m (h)	7006.00	8866.79	5.00	168.45	nr	9035.24
10,000 litres, 2.5 m × 2 m × 2 m (h)	10020.00	12681.31	7.00	235.83	nr	12917.14
20,000 litres, 4 m × 2.5 m × 2 m (h)	12970.00	16414.83	10.00	336.91	nr	16751.74
30,000 litres 5 m × 3 m × 2 m (h)	16156.00	20447.03	12.00	404.29	nr	20851.32
40,000 litres, 5 m × 4 m × 2 m (h)	18858.00	23866.68	12.00	404.29	nr	24270.97
50,000 litres, 5 m × 4 m × 2.5 m (h)	23065.00	29191.06	14.00	471.67	nr	29662.73
60,000 litres, 6 m × 4 m × 2.5 m (h)	24534.00	31050.23	16.00	539.04	nr	31589.27
70,000 litres, 7 m × 4 m × 2.5 m (h)	27728.00	35092.56	16.00	539.04	nr	35631.60
80,000 litres, 8 m × 4 m × 2.5 m (h)	29336.00	37127.64	16.00	539.04	nr	37666.68
90,000 litres, 6 m × 5 m × 3 m (h)	31609.00	40004.35	16.00	539.04	nr	40543.39
105,000 litres, 7 m × 5 m × 3 m (h)	35415.00	44821.22	24.00	808.56	nr	45629.78
120,000 litres, 8 m × 5 m × 3 m (h)	37014.00	46844.92	24.00	808.56	nr	47653.48
135,000 litres, 9 m × 6 m × 2.5 m (h)	41070.00	51978.19	24.00	808.56	nr	52786.75
144,000 litres, 8 m × 6 m × 3 m (h)	41126.00	52049.07	24.00	808.56	nr	52857.63
CLEANING AND CHEMICAL TREATMENT						
Electromagnetic water conditioner; WRAS approved; complete with control box; maximum inlet pressure 16 bar; electrical work elsewhere						
Connection size, nominal flow rate at 50 mbar						
20 mm dia., 0.3 l/s	1645.00	2081.91	1.25	42.11	nr	2124.02
25 mm dia., 0.6 l/s	2160.00	2733.70	1.45	48.85	nr	2782.55
32 mm dia., 1.2 l/s	3080.00	3898.05	1.55	52.21	nr	3950.26
40 mm dia., 1.7 l/s	3680.00	4657.41	1.65	55.59	nr	4713.00
50 mm dia., 3.4 l/s	4820.00	6100.19	1.75	58.97	nr	6159.16
65 mm dia., 5.2 l/s	5275.00	6676.04	1.90	64.02	nr	6740.06
100 mm dia., 30.5 l/s, 595 mbar	9850.00	12466.16	3.00	101.07	nr	12567.23

38 PIPED SUPPLY SYSTEMS

Item	Net Price £	Material £	Labour hours	Labour £	Unit	Total rate £
COLD WATER PIPELINES: PIPEWORK ANCILLARIES – TANKS – cont						
CLEANING AND CHEMICAL TREATMENT – cont						
Ultraviolet water sterillizing unit; WRAS approved; complete with control unit; UV lamp housed in quartz tube; unit complete with UV intensity sensor, flushing and discharge valve and facilities for remote alarm; electrical work elsewhere						
Maximum flow rate (@ 400 J/m^2 exposure in accordance with BSEN 14897), connection size						
0.8 l/s, 25 mm dia.	3065.00	3879.06	1.98	66.71	nr	3945.77
1.25 l/s, 40 mm dia.	3875.00	4904.20	1.98	66.71	nr	4970.91
2.57 l/s, 50 mm dia.	4410.00	5581.30	1.98	66.71	nr	5648.01
4.46 l/s, 80 mm dia.	10070.00	12744.59	1.98	66.71	nr	12811.30
10.5 l/s, 100 mm dia.	12510.00	15832.66	1.98	66.71	nr	15899.37
20.1 l/s, 100 mm dia.	14070.00	17806.99	3.60	121.28	nr	17928.27
26.9 l/s 125 mm dia.	15800.00	19996.48	3.60	121.28	nr	20117.76
33.7 l/s 125 mm dia.	17830.00	22565.65	3.60	121.28	nr	22686.93
Base exchange water softener; WRAS approved; complete with resin tank, brine tank and consumption data monitoring facilities						
Capacities of softeners are based on 300 ppm hardness and quoted in m³ of softened water produced. Design flow rates are recommended for continuous use						
Simplex configuration						
Design flow rate, min-max softend water produced						
0.97 l/s, 7.7 m³	2975.00	3765.16	10.00	336.91	nr	4102.07
1.25 l/s, 11.7 m³	3050.00	3860.08	12.00	404.29	nr	4264.37
1.8 l/s, 19.8 m³	4545.00	5752.15	12.00	404.29	nr	6156.44
2.23 l/s, 23.4 m³	5100.00	6454.56	12.00	404.29	nr	6858.85
2.92 l/s, 31.6 m³	5275.00	6676.04	15.00	505.36	nr	7181.40
3.75 l/s, 39.8 m³	6300.00	7973.28	15.00	505.36	nr	8478.64
4.72 l/s, 60.3 m³	7155.00	9055.37	18.00	606.42	nr	9661.79
6.66 l/s, 96.6 m³	8450.00	10694.32	20.00	673.80	nr	11368.12
Duplex configuration						
Design flow rate, min-max softend water produced						
0.97 l/s, 7.7 m³	4390.00	5555.98	12.00	404.29	nr	5960.27
1.25 l/s, 11.7 m³	4570.00	5783.79	12.00	404.29	nr	6188.08
1.81 l/s, 19.8 m³	7665.00	9700.82	15.00	505.36	nr	10206.18

38 PIPED SUPPLY SYSTEMS

Item	Net Price £	Material £	Labour hours	Labour £	Unit	Total rate £
2.23 l/s, 23.4 m³	8785.00	11118.30	15.00	505.36	nr	**11623.66**
2.29 l/s, 31.6 m³	9055.00	11460.01	18.00	606.42	nr	**12066.43**
3.75 l/s, 39.8 m³	11260.00	14250.66	18.00	606.42	nr	**14857.08**
4.72 l/s, 60.3 m³	12850.00	16262.96	18.00	606.42	nr	**16869.38**
6.66/s, 96.6 m³	12850.00	16262.96	23.00	774.87	nr	**17037.83**
THERMAL INSULATION						
Flexible closed cell walled insulation; Class 1/Class O; adhesive joints; including around fittings						
6 mm wall thickness						
15 mm dia.	1.53	1.94	0.15	5.05	m	**6.99**
22 mm dia.	1.81	2.30	0.15	5.05	m	**7.35**
28 mm dia.	2.29	2.90	0.15	5.05	m	**7.95**
9 mm wall thickness						
15 mm dia.	1.76	2.23	0.15	5.05	m	**7.28**
22 mm dia.	2.20	2.79	0.15	5.05	m	**7.84**
28 mm dia.	2.36	2.99	0.15	5.05	m	**8.04**
35 mm dia.	2.67	3.38	0.15	5.05	m	**8.43**
42 mm dia.	3.05	3.86	0.15	5.05	m	**8.91**
54 mm dia.	3.16	4.00	0.15	5.05	m	**9.05**
13 mm wall thickness						
15 mm dia.	2.25	2.84	0.15	5.05	m	**7.89**
22 mm dia.	2.76	3.49	0.15	5.05	m	**8.54**
28 mm dia.	3.28	4.16	0.15	5.05	m	**9.21**
35 mm dia.	3.54	4.48	0.15	5.05	m	**9.53**
42 mm dia.	4.16	5.26	0.15	5.05	m	**10.31**
54 mm dia.	5.80	7.34	0.15	5.05	m	**12.39**
67 mm dia.	8.57	10.84	0.15	5.05	m	**15.89**
76 mm dia.	10.08	12.76	0.15	5.05	m	**17.81**
108 mm dia.	10.57	13.37	0.15	5.05	m	**18.42**
19 mm wall thickness						
15 mm dia.	3.65	4.61	0.15	5.05	m	**9.66**
22 mm dia.	4.45	5.63	0.15	5.05	m	**10.68**
28 mm dia.	6.04	7.65	0.15	5.05	m	**12.70**
35 mm dia.	6.99	8.85	0.15	5.05	m	**13.90**
42 mm dia.	8.25	10.44	0.15	5.05	m	**15.49**
54 mm dia.	10.44	13.22	0.15	5.05	m	**18.27**
67 mm dia.	12.51	15.84	0.15	5.05	m	**20.89**
76 mm dia.	14.46	18.30	0.22	7.41	m	**25.71**
108 mm dia.	19.72	24.95	0.22	7.41	m	**32.36**
25 mm wall thickness						
15 mm dia.	7.17	9.07	0.15	5.05	m	**14.12**
22 mm dia.	7.84	9.92	0.15	5.05	m	**14.97**
28 mm dia.	8.88	11.23	0.15	5.05	m	**16.28**
35 mm dia.	9.84	12.45	0.15	5.05	m	**17.50**
42 mm dia.	10.50	13.28	0.15	5.05	m	**18.33**
54 mm dia.	12.33	15.60	0.15	5.05	m	**20.65**
67 mm dia.	16.62	21.03	0.15	5.05	m	**26.08**
76 mm dia.	18.74	23.72	0.22	7.41	m	**31.13**

38 PIPED SUPPLY SYSTEMS

Item	Net Price £	Material £	Labour hours	Labour £	Unit	Total rate £
COLD WATER PIPELINES: PIPEWORK ANCILLARIES – TANKS – cont						
Flexible closed cell walled insulation – cont						
32 mm wall thickness						
15 mm dia.	8.97	11.36	0.15	5.05	m	**16.41**
22 mm dia.	9.84	12.45	0.15	5.05	m	**17.50**
28 mm dia.	11.41	14.44	0.15	5.05	m	**19.49**
35 mm dia.	11.78	14.91	0.15	5.05	m	**19.96**
42 mm dia.	13.72	17.36	0.15	5.05	m	**22.41**
54 mm dia.	17.26	21.84	0.15	5.05	m	**26.89**
76 mm dia.	25.81	32.67	0.22	7.41	m	**40.08**

Note: For mineral fibre sectional insulation; bright class O foil faced; bright class O foil taped joints; 19 mm aluminium bands rates, refer to section – Thermal Insulation

Note: For mineral fibre sectional insulation; bright class O foil faced; bright class O foil taped joints; 22 swg plain/embossed aluminium cladding; pop riveted rates, refer to section – Thermal Insulation

Note: For mineral fibre sectional insulation; bright class O foil faced; bright class O foil taped joints; 0.8 mm polyisobutylene sheeting; welded joints rates, refer to section – Thermal Insulation

38 PIPED SUPPLY SYSTEMS

Item	Net Price £	Material £	Labour hours	Labour £	Unit	Total rate £
HOT WATER						
PIPELINES						
Note: For pipework prices refer to section – Cold Water						
PIPELINE ANCILLARIES						
Note: For prices for ancillaries refer to section – Cold Water						
STORAGE CYLINDERS/CALORIFIERS						
Insulated copper storage cylinders; BS 699; includes placing in position						
Grade 3 (maximum 10 m working head)						
BS size 6; 115 litre capacity; 400 mm dia.; 1050 mm high	197.65	250.14	1.50	50.59	nr	**300.73**
BS size 7; 120 litre capacity; 450 mm dia.; 900 mm high	233.69	295.76	2.00	67.38	nr	**363.14**
BS size 8; 144 litre capacity; 450 mm dia.; 1050 mm high	248.42	314.40	2.80	94.37	nr	**408.77**
Grade 4 (maximum 6 m working head)						
BS size 2; 96 litre capacity; 400 mm dia.; 900 mm high	153.04	193.69	1.50	50.59	nr	**244.28**
BS size 7; 120 litre capacity; 450 mm dia.; 900 mm high	160.83	203.55	1.50	50.59	nr	**254.14**
BS size 8; 144 litre capacity; 450 mm dia.; 1050 mm high	191.98	242.97	1.50	50.59	nr	**293.56**
BS size 9; 166 litre capacity; 450 mm dia.; 1200 mm high	280.71	355.26	1.50	50.59	nr	**405.85**
Storage cylinders; brazed copper construction; to BS 699; screwed bosses; includes placing in position						
Tested to 2.2 bar, 15 m maximum head						
144 litres	870.02	1101.09	3.00	101.17	nr	**1202.26**
160 litres	983.53	1244.76	3.00	101.17	nr	**1345.93**
200 litres	1013.76	1283.02	3.76	126.66	nr	**1409.68**
255 litres	1153.71	1460.13	3.76	126.66	nr	**1586.79**
290 litres	1558.46	1972.39	3.76	126.66	nr	**2099.05**
370 litres	1785.46	2259.68	4.50	151.76	nr	**2411.44**
450 litres	2432.15	3078.13	5.00	168.45	nr	**3246.58**
Tested to 2.55 bar, 17 m maximum head						
550 litres	2633.52	3332.99	5.00	168.45	nr	**3501.44**
700 litres	3078.85	3896.59	6.02	202.96	nr	**4099.55**
800 litres	3561.07	4506.89	6.54	220.20	nr	**4727.09**
900 litres	3853.44	4876.92	8.00	269.52	nr	**5146.44**
1000 litres	4066.46	5146.51	8.00	269.52	nr	**5416.03**
1250 litres	4453.71	5636.61	13.16	443.28	nr	**6079.89**
1500 litres	6821.67	8633.51	15.15	510.45	nr	**9143.96**
2000 litres	8186.61	10360.97	17.24	580.88	nr	**10941.85**
3000 litres	11499.37	14553.60	24.39	821.72	nr	**15375.32**

38 PIPED SUPPLY SYSTEMS

Item	Net Price £	Material £	Labour hours	Labour £	Unit	Total rate £
HOT WATER – cont						
STORAGE CYLINDERS/CALORIFIERS – cont						
Indirect cylinders; copper; bolted top; up to 5 tappings for connections; BS 1586; includes placing in position						
Grade 3, tested to 1.45 bar, 10 m maximum head						
74 litre capacity	398.90	504.85	1.50	50.59	nr	**555.44**
96 litre capacity	406.11	513.97	1.50	50.59	nr	**564.56**
114 litre capacity	417.00	527.76	1.50	50.59	nr	**578.35**
117 litre capacity	432.81	547.77	2.00	67.38	nr	**615.15**
140 litre capacity	445.99	564.45	2.50	84.22	nr	**648.67**
162 litre capacity	623.69	789.34	3.00	101.17	nr	**890.51**
190 litre capacity	681.74	862.81	3.51	118.20	nr	**981.01**
245 litre capacity	797.77	1009.66	3.80	128.11	nr	**1137.77**
280 litre capacity	1414.18	1789.78	4.00	134.76	nr	**1924.54**
360 litre capacity	1530.24	1936.67	4.50	151.76	nr	**2088.43**
440 litre capacity	1776.81	2248.74	4.50	151.76	nr	**2400.50**
Grade 2, tested to 2.2 bar, 15 m maximum head						
117 litre capacity	576.53	729.66	2.00	67.38	nr	**797.04**
140 litre capacity	627.34	793.96	2.50	84.22	nr	**878.18**
162 litre capacity	717.97	908.67	2.80	94.37	nr	**1003.04**
190 litre capacity	834.06	1055.59	3.00	101.17	nr	**1156.76**
245 litre capacity	1008.09	1275.84	4.00	134.76	nr	**1410.60**
280 litre capacity	1610.00	2037.62	4.00	134.76	nr	**2172.38**
360 litre capacity	1776.81	2248.74	4.50	151.76	nr	**2400.50**
440 litre capacity	2103.14	2661.74	4.50	151.76	nr	**2813.50**
Grade 1, tested 3.65 bar, 25 m maximum head						
190 litre capacity	1240.17	1569.56	3.00	101.17	nr	**1670.73**
245 litre capacity	1410.52	1785.16	3.00	101.17	nr	**1886.33**
280 litre capacity	1994.39	2524.10	4.00	134.76	nr	**2658.86**
360 litre capacity	2523.84	3194.17	4.50	151.76	nr	**3345.93**
440 litre capacity	3064.07	3877.89	4.50	151.76	nr	**4029.65**
Indirect cylinders, including manhole; BS 853						
Grade 3, tested to 1.5 bar, 10 m maximum head						
550 litre capacity	2579.21	3264.25	5.21	175.46	nr	**3439.71**
700 litre capacity	2855.58	3614.03	6.02	202.96	nr	**3816.99**
800 litre capacity	3316.14	4196.91	6.54	220.20	nr	**4417.11**
1000 litre capacity	4145.21	5246.18	7.04	237.26	nr	**5483.44**
1500 litre capacity	4790.00	6062.22	10.00	336.91	nr	**6399.13**
2000 litre capacity	6632.34	8393.88	16.13	543.39	nr	**8937.27**

38 PIPED SUPPLY SYSTEMS

Item	Net Price £	Material £	Labour hours	Labour £	Unit	Total rate £
Grade 2, tested to 2.55 bar, 15 m maximum head						
550 litre capacity	2859.54	3619.03	5.21	175.46	nr	**3794.49**
700 litre capacity	3578.91	4529.47	6.02	202.96	nr	**4732.43**
800 litre capacity	3776.73	4779.82	6.54	220.20	nr	**5000.02**
1000 litre capacity	4675.97	5917.91	7.04	237.26	nr	**6155.17**
1500 litre capacity	5755.03	7283.56	10.00	336.91	nr	**7620.47**
2000 litre capacity	7193.81	9104.49	16.13	543.39	nr	**9647.88**
Grade 1, tested to 4 bar, 25 m maximum head						
550 litre capacity	3327.12	4210.81	5.21	175.46	nr	**4386.27**
700 litre capacity	3776.73	4779.82	6.02	202.96	nr	**4982.78**
800 litre capacity	4046.52	5121.28	6.54	220.20	nr	**5341.48**
1000 litre capacity	5395.37	6828.38	7.04	237.26	nr	**7065.64**
1500 litre capacity	6474.38	8193.98	10.00	336.91	nr	**8530.89**
2000 litre capacity	7913.16	10014.89	16.13	543.39	nr	**10558.28**
Storage calorifiers; copper; heater battery capable of raising temperature of contents from 10°C to 65°C in one hour; static head not exceeding 1.35 bar; BS 853; includes fixing in position on cradles or legs.						
Horizontal; primary LPHW at 82°C/71°C						
400 litre capacity	3764.33	4764.13	7.04	237.26	nr	**5001.39**
1000 litre capacity	6022.89	7622.57	8.00	269.52	nr	**7892.09**
2000 litre capacity	12045.86	15245.24	14.08	474.51	nr	**15719.75**
3000 litre capacity	14869.08	18818.31	25.00	842.26	nr	**19660.57**
4000 litre capacity	18068.73	22867.78	40.00	1347.61	nr	**24215.39**
4500 litre capacity	20362.68	25771.01	50.00	1684.51	nr	**27455.52**
Vertical; primary LPHW at 82°C/71°C						
400 litre capacity	4242.41	5369.19	7.04	237.26	nr	**5606.45**
1000 litre capacity	6818.18	8629.08	8.00	269.52	nr	**8898.60**
2000 litre capacity	12986.94	16436.27	14.08	474.51	nr	**16910.78**
3000 litre capacity	16233.68	20545.35	25.00	842.26	nr	**21387.61**
4000 litre capacity	19913.24	25202.20	40.00	1347.61	nr	**26549.81**
4500 litre capacity	22510.66	28489.50	50.00	1684.51	nr	**30174.01**
Storage calorifiers; galvanized mild steel; heater battery capable of raising temperature of contents from 10°C to 65°C in one hour; static head not exceeding 1.35 bar; BS 853; includes fixing in position on cradles or legs						
Horizontal; primary LPHW at 82°C/71°C						
400 litre capacity	3764.33	4764.13	7.04	237.26	nr	**5001.39**
1000 litre capacity	6022.89	7622.57	8.00	269.52	nr	**7892.09**
2000 litre capacity	12045.86	15245.24	14.08	474.51	nr	**15719.75**
3000 litre capacity	14869.08	18818.31	25.00	842.26	nr	**19660.57**
4000 litre capacity	18068.73	22867.78	40.00	1347.61	nr	**24215.39**
4500 litre capacity	20362.68	25771.01	50.00	1684.51	nr	**27455.52**

38 PIPED SUPPLY SYSTEMS

Item	Net Price £	Material £	Labour hours	Labour £	Unit	Total rate £
HOT WATER – cont						
STORAGE CYLINDERS/CALORIFIERS – cont						
Storage calorifiers – cont						
Vertical; primary LPHW at 82°C/71°C						
400 litre capacity	4242.41	5369.19	7.04	237.26	nr	**5606.45**
1000 litre capacity	6818.18	8629.08	8.00	269.52	nr	**8898.60**
2000 litre capacity	12986.94	16436.27	14.08	474.51	nr	**16910.78**
3000 litre capacity	16233.68	20545.35	25.00	842.26	nr	**21387.61**
4000 litre capacity	19913.24	25202.20	40.00	1347.61	nr	**26549.81**
4500 litre capacity	22510.66	28489.50	50.00	1684.51	nr	**30174.01**
Indirect cylinders; mild steel, welded throughout, galvanized, with connections. Tested to 4 bar, 95C. Includes sensors, with full insulation and cases, includes delivery.						
222 litre nominal content	1076.00	1361.79	2.50	84.22	nr	**1446.01**
278 litre nominal content	1195.00	1512.39	2.80	94.37	nr	**1606.76**
474 litre nominal content	1494.00	1890.81	3.00	101.17	nr	**1991.98**
765 litre nominal content	1673.00	2117.35	3.00	101.14	nr	**2218.49**
956 litre nominal content	1972.00	2495.76	3.00	101.17	nr	**2596.93**
1365 litre nominal content	2557.00	3236.14	4.00	134.76	nr	**3370.90**
2039 litre nominal content	3262.00	4128.39	5.00	168.45	nr	**4296.84**
2361 litre nominal content	4218.00	5338.30	6.02	202.96	nr	**5541.26**
Indirect cylinders; mild steel welded throughout, galvanized, with connections. Tested to 4 bar, 95C. Includes sensors. Includes delivery.						
4021 litre nominal content	5443.00	6888.66	1.50	50.53	nr	**6939.19**
5897 litre nominal content	6453.00	8166.92	1.50	50.53	nr	**8217.45**
8000 litre nominal content	10156.00	12853.43	1.50	50.53	nr	**12903.96**
10170 litre nominal content	11932.00	15101.14	2.00	67.38	nr	**15168.52**
Local electric hot water heaters						
Unvented multipoint water heater; providing hot water for one or more outlets; used with conventional taps or mixers; factory fitted temperature and pressure relief valve; externally adjustable thermostat; elemental 'on' indicator; fitted with 1 m of 3 core cable; electrical supply and connection excluded						
5 litre capacity; 2.2 kW rating	219.86	278.25	1.50	50.53	nr	**328.78**
10 litre capacity; 2.2 kW rating	272.46	344.83	1.50	50.53	nr	**395.36**
15 litre capacity; 2.2 kW rating	485.59	614.57	1.50	50.53	nr	**665.10**
30 litre capacity; 2.2 kW rating	485.59	614.57	2.00	67.38	nr	**681.95**
50 litre capacity; 2.2 kW rating	511.24	647.02	2.00	67.38	nr	**714.40**
80 litre capacity; 2.2 kW rating	1005.98	1273.17	2.00	67.38	nr	**1340.55**
100 litre capacity; 2.2 kW rating	1079.98	1366.83	2.00	67.38	nr	**1434.21**

38 PIPED SUPPLY SYSTEMS

Item	Net Price £	Material £	Labour hours	Labour £	Unit	Total rate £
Accessories						
Pressure reducing valve and expansion kit	175.58	222.22	2.00	67.38	nr	**289.60**
Thermostatic blending valve	93.44	118.26	1.00	33.69	nr	**151.95**
SOLAR THERMAL PACKAGES						
Packages include solar collectors, mild steel storage vessel, fresh water module (including plate heat exchanger), roof fixing kit, glycol, stratified pump station (including plate heat exchanger), solar expansion vessel, controls, anti-legionella protection; delivery and commissioning						
No. of panels; storage size						
6 Solar Collectors, 1000 l storage vessel	8258.00	10451.32	11.00	370.60	nr	**10821.92**
8 Solar Collectors, 1500 l storage vessel	9548.00	12083.95	13.00	437.98	nr	**12521.93**
10 Solar Collectors, 2000 l storage vessel	10023.00	12685.11	15.00	505.36	nr	**13190.47**
12 Solar Collectors, 2000 l storage vessel	10990.00	13908.94	17.00	572.73	nr	**14481.67**
14 Solar Collectors, 2000 l storage vessel	11964.00	15141.64	18.00	606.42	nr	**15748.06**
16 Solar Collectors, 2500 l storage vessel	12963.00	16405.97	20.00	673.80	nr	**17079.77**

38 PIPED SUPPLY SYSTEMS

Item	Net Price £	Material £	Labour hours	Labour £	Unit	Total rate £
NATURAL GAS: PIPELINES: MEDIUM DENSITY POLYETHYLENE – YELLOW						
Pipe; laid underground; electrofusion joints in the running length; BS 6572; BGT PL2 standards						
Coiled service pipe						
20 mm dia.	1.76	2.23	0.37	12.47	m	14.70
25 mm dia.	2.29	2.90	0.41	13.81	m	16.71
32 mm dia.	3.78	4.78	0.47	15.84	m	20.62
63 mm dia.	14.42	18.24	0.60	20.22	m	38.46
90 mm dia.	18.91	23.93	0.90	30.32	m	54.25
Mains service pipe						
63 mm dia.	14.00	17.72	0.60	20.22	m	37.94
90 mm dia.	18.35	23.23	0.90	30.32	m	53.55
125 mm dia.	35.44	44.86	1.20	40.42	m	85.28
180 mm dia.	73.19	92.62	1.50	50.53	m	143.15
250 mm dia.	134.71	170.49	1.75	58.97	m	229.46
Extra over fittings, electrofusion joints						
Straight connector						
32 mm dia.	10.97	13.89	0.47	15.84	nr	29.73
63 mm dia.	20.59	26.06	0.58	19.54	nr	45.60
90 mm dia.	30.37	38.44	0.67	22.58	nr	61.02
125 mm dia.	56.88	71.98	0.83	27.96	nr	99.94
180 mm dia.	102.44	129.65	1.25	42.11	nr	171.76
Reducing connector						
90 × 63 mm dia.	42.42	53.68	0.67	22.58	nr	76.26
125 × 90 mm dia.	85.07	107.67	0.83	27.96	nr	135.63
180 × 125 mm dia.	156.07	197.52	1.25	42.11	nr	239.63
Bend; 45°						
90 mm dia.	81.98	103.76	0.67	22.58	nr	126.34
125 mm dia.	134.11	169.72	0.83	27.96	nr	197.68
180 mm dia.	304.21	385.01	1.25	42.11	nr	427.12
Bend; 90°						
63 mm dia.	53.11	67.21	0.58	19.54	nr	86.75
90 mm dia.	81.98	103.76	0.67	22.58	nr	126.34
125 mm dia.	134.11	169.72	0.83	27.96	nr	197.68
180 mm dia.	304.21	385.01	1.25	42.11	nr	427.12
Extra over malleable iron fittings, compression joints						
Straight connector						
20 mm dia.	15.64	19.79	0.38	12.81	nr	32.60
25 mm dia.	17.04	21.57	0.45	15.16	nr	36.73
32 mm dia.	19.13	24.21	0.50	16.84	nr	41.05
63 mm dia.	38.41	48.61	0.85	28.65	nr	77.26
Straight connector; polyethylene to MI						
20 mm dia.	13.29	16.82	0.31	10.44	nr	27.26
25 mm dia.	14.50	18.36	0.35	11.79	nr	30.15
32 mm dia.	16.18	20.47	0.40	13.47	nr	33.94
63 mm dia.	27.12	34.33	0.65	21.91	nr	56.24

38 PIPED SUPPLY SYSTEMS

Item	Net Price £	Material £	Labour hours	Labour £	Unit	Total rate £
Straight connector; polyethylene to FI						
20 mm dia.	13.29	16.82	0.31	10.44	nr	**27.26**
25 mm dia.	16.18	20.47	0.35	11.79	nr	**32.26**
32 mm dia.	16.18	20.47	0.40	13.47	nr	**33.94**
63 mm dia.	27.12	34.33	0.75	25.28	nr	**59.61**
Elbow						
20 mm dia.	20.34	25.74	0.38	12.81	nr	**38.55**
25 mm dia.	22.19	28.08	0.45	15.16	nr	**43.24**
32 mm dia.	24.85	31.45	0.50	16.84	nr	**48.29**
63 mm dia.	49.94	63.20	0.80	26.96	nr	**90.16**
Equal tee						
20 mm dia.	27.12	34.33	0.53	17.85	nr	**52.18**
25 mm dia.	31.69	40.11	0.55	18.52	nr	**58.63**
32 mm dia.	39.93	50.53	0.64	21.57	nr	**72.10**
GAS BOOSTERS						
Complete skid mounted gas booster set, including AV mounts, flexible connections, low pressure switch, control panel and NRV (for run/standby unit); 3 phase supply; in accordance with IGE/UP/2; includes delivery, offloading and positioning						
Single unit						
Flow, pressure range						
0–200 m³/hour, 0.1–2.6 kPa	4107.02	5197.84	10.00	336.91	nr	**5534.75**
0–200 m³/hour, 0.1–4.0 kPa	4686.22	5930.88	10.00	336.91	nr	**6267.79**
0–200 m³/hour, 0.1–7 kPa	5453.43	6901.87	10.00	336.91	nr	**7238.78**
0–200 m³/hour, 0.1–9.5 kPa	5679.12	7187.50	10.00	336.91	nr	**7524.41**
0–200 m³/hour 0.1–11.0 kPa	6344.83	8030.02	10.00	336.91	nr	**8366.93**
0–400 m³/hour, 0.1–4.0 kPa	6901.01	8733.92	10.00	336.91	nr	**9070.83**
0–1000 m³/hour, 0.1–7.4 kPa	7168.44	9072.38	10.00	336.91	nr	**9409.29**
50–1000 m³/hour, 0.1–16.0 kPa	13530.37	17124.04	20.00	673.80	nr	**17797.84**
50–1000 m³/hour, 0.1–24.5 kPa	15141.73	19163.37	20.00	673.80	nr	**19837.17**
50–1000 m³/hour, 0.1–31.0 kPa	17221.60	21795.66	20.00	673.80	nr	**22469.46**
50–1000 m³/hour, 0.1–41.0 kPa	19008.10	24056.65	20.00	673.80	nr	**24730.45**
50–1000 m³/hour, 0.1–51.0 kPa	19722.62	24960.95	20.00	673.80	nr	**25634.75**
100–1800 m³/hour, 3.5–23.5 kPa	20237.94	25613.13	20.00	673.80	nr	**26286.93**
100–1800 m³/hour, 4.5–27.0 kPa	21828.83	27626.57	20.00	673.80	nr	**28300.37**
100–1800 m³/hour, 6.0–32.5 kPa	24175.68	30596.74	20.00	673.80	nr	**31270.54**
100–1800 m³/hour, 7.2–39.0 kPa	27812.56	35199.57	20.00	673.80	nr	**35873.37**
100–1800 m³/hour, 9.0–42.0 kPa	29234.23	36998.84	20.00	673.80	nr	**37672.64**
Run/Standby unit						
Flow, pressure range						
0–200 m³/hour, 0.1–2.6 kPa	20214.37	25583.31	16.00	539.04	nr	**26122.35**
0–200 m³/hour, 0.1–4.0 kPa	20830.02	26362.47	16.00	539.04	nr	**26901.51**
0–200 m³/hour, 0.1–7 kPa	21312.47	26973.06	16.00	539.04	nr	**27512.10**
0–200 m³/hour, 0.1–9.5 kPa	21840.57	27641.42	16.00	539.04	nr	**28180.46**
0–200 m³/hour 0.1–11.0 kPa	22202.41	28099.37	16.00	539.04	nr	**28638.41**
0–400 m³/hour, 0.1–4.0 kPa	24484.26	30987.28	16.00	539.04	nr	**31526.32**
0–1000 m³/hour, 0.1–7.4 kPa	31209.17	39498.32	25.00	842.26	nr	**40340.58**

38 PIPED SUPPLY SYSTEMS

Item	Net Price £	Material £	Labour hours	Labour £	Unit	Total rate £
NATURAL GAS: PIPELINES: MEDIUM DENSITY POLYETHYLENE – YELLOW – cont						
Complete skid mounted gas booster set, including AV mounts, flexible connections, low pressure switch, control panel and NRV (for run/standby unit) – cont						
Flow, pressure range – cont						
50–1000 m³/hour, 0.1–16.0 kPa	42774.89	54135.91	25.00	842.26	nr	**54978.17**
50–1000 m³/hour, 0.1–24.5 kPa	48358.92	61203.05	25.00	842.26	nr	**62045.31**
50–1000 m³/hour, 0.1–31.0 kPa	53929.86	68253.63	25.00	842.26	nr	**69095.89**
50–1000 m³/hour, 0.1–41.0 kPa	59517.15	75324.91	25.00	842.26	nr	**76167.17**
50–1000 m³/hour, 0.1–51.0 kPa	59784.43	75663.18	25.00	842.26	nr	**76505.44**
100–1800 m³/hour, 3.5–23.5 kPa	62201.28	78721.94	25.00	842.26	nr	**79564.20**
100–1800 m³/hour, 4.5–27.0 kPa	63906.62	80880.22	25.00	842.26	nr	**81722.48**
100–1800 m³/hour, 6.0–32.5 kPa	67894.74	85927.59	25.00	842.26	nr	**86769.85**
100–1800 m³/hour, 7.2–39.0 kPa	72165.08	91332.12	25.00	842.26	nr	**92174.38**
100–1800 m³/hour, 9.0–42.0 kPa	75855.18	96002.31	25.00	842.26	nr	**96844.57**

38 PIPED SUPPLY SYSTEMS

Item	Net Price £	Material £	Labour hours	Labour £	Unit	Total rate £
NATURAL GAS: PIPELINES: SCREWED STEEL						
For prices for steel pipework refer to section – Low Temperature Hot Water Heating						
PIPE IN PIPE						
Note: For pipe in pipe, a sleeve size two pipe sizes bigger than actual pipe size has been allowed. All rates refer to actual pipe size. Black steel pipes – Screwed and socketed joints; BS 1387: 1985 upto 50 mm pipe size. Butt welded joints; BS 1387: 1985 65 mm pipe size and above.						
Pipe dia.						
25 mm	16.50	20.89	1.73	58.28	m	**79.17**
32 mm	21.91	27.73	1.95	65.70	m	**93.43**
40 mm	28.18	35.66	2.16	72.77	m	**108.43**
50 mm	36.97	46.79	2.44	82.20	m	**128.99**
65 mm	50.71	64.18	2.95	99.39	m	**163.57**
80 mm	61.30	77.58	3.42	115.23	m	**192.81**
100 mm	76.55	96.88	4.00	134.76	m	**231.64**
Extra over black steel pipes – Screwed pipework; black malleable iron fittings; BS 143. Welded pipework; butt welded steel fittings; BS 1965						
Bend, 90°						
25 mm	9.52	12.05	2.91	98.03	m	**110.08**
32 mm	14.49	18.33	3.45	116.23	m	**134.56**
40 mm	20.30	25.69	5.34	179.91	m	**205.60**
50 mm	22.88	28.95	6.53	220.00	m	**248.95**
65 mm	30.90	39.11	8.84	297.82	m	**336.93**
80 mm	52.68	66.67	10.73	361.49	m	**428.16**
100 mm	69.96	88.54	12.76	429.89	m	**518.43**
Bend, 45°						
25 mm	12.03	15.22	2.91	98.03	m	**113.25**
32 mm	15.58	19.72	3.45	116.23	m	**135.95**
40 mm	17.42	22.04	5.34	179.91	m	**201.95**
50 mm	20.78	26.30	6.53	220.00	m	**246.30**
65 mm	27.14	34.35	8.84	297.82	m	**332.17**
80 mm	46.42	58.74	10.73	361.49	m	**420.23**
100 mm	59.46	75.25	12.76	429.89	m	**505.14**
Equal tee						
25 mm	80.09	101.36	4.18	140.83	m	**242.19**
32 mm	82.24	104.08	4.94	166.42	m	**270.50**
40 mm	101.17	128.04	7.28	245.26	m	**373.30**
50 mm	103.80	131.36	8.48	285.70	m	**417.06**
65 mm	138.20	174.91	11.47	386.42	m	**561.33**
80 mm	237.11	300.08	14.23	479.40	m	**779.48**
100 mm	268.35	339.63	17.92	603.72	m	**943.35**

38 PIPED SUPPLY SYSTEMS

Item	Net Price £	Material £	Labour hours	Labour £	Unit	Total rate £
NATURAL GAS: PIPELINES: SCREWED STEEL – cont						
Copper pipe; capillary or compression joints in the running length; EN1057 R250 (TX) formerly BS 2871 Table X						
Plastic coated gas service pipe for corrosive environments, fixed vertically or at low level with brackets measured separately						
15 mm dia. (yellow)	9.45	11.96	0.85	28.65	m	**40.61**
22 mm dia. (yellow)	18.62	23.56	0.96	32.36	m	**55.92**
28 mm dia. (yellow)	23.65	29.93	1.06	35.72	m	**65.65**
Copper pipe; capillary or compression joints in the running length; EN1057 R250 (TY) formerly BS 2871 Table Y						
Plastic coated gas and cold water service pipe for corrosive environments, fixed vertically or at low level with brackets measured separately (Refer to Copper Pipe Table X Section)						
15 mm dia. (yellow)	10.92	13.82	0.61	20.55	m	**34.37**
22 mm dia. (yellow)	19.84	25.11	0.69	23.25	m	**48.36**
Fixings						
Refer to section – Cold Water						
Extra over copper pipes; capillary fittings; BS 864						
Refer to section – Cold Water						

38 PIPED SUPPLY SYSTEMS

Item	Net Price £	Material £	Labour hours	Labour £	Unit	Total rate £
FUEL OIL STORAGE/DISTRIBUTION						
PIPELINES						
For pipework prices refer to section – Low Temperature Hot Water Heating						
TANKS						
Fuel storage tanks; mild steel; with all necessary screwed bosses; oil resistant joint rings; includes placing in position. Rectangular shape						
1360 litre (300 gallon) capacity; 2 mm plate	453.84	574.38	12.03	405.29	nr	979.67
2730 litre (600 gallon) capacity; 2.5 mm plate	606.45	767.52	18.60	626.64	nr	1394.16
4550 litre (1000 gallon) capacity; 3 mm plate	1271.48	1609.18	25.00	842.26	nr	2451.44
Fuel storage tanks; 5 mm plate mild steel to BS 799 type J; complete with raised neck manhole with bolted cover, screwed connections, vent and fill connections, drain valve, gauge and overfill alarm; includes placing in position; excludes pumps and control panel. Nominal capacity size						
5,600 litres, 2.5 m × 1.5 m × 1.5 m high	2967.71	3755.93	20.00	673.80	nr	4429.73
Extra for bund unit (internal use)	1706.65	2159.93	30.00	1010.71	nr	3170.64
Extra for external use with bund (watertight)	1080.87	1367.95	2.00	67.38	nr	1435.33
10,200 litres, 3.05 m × 1.83 m × 1.83 m high	3688.29	4667.90	30.00	1010.71	nr	5678.61
Extra for bund unit (internal use)	2484.14	3143.93	40.00	1347.61	nr	4491.54
Extra for external use with bund (watertight)	1327.40	1679.96	2.00	67.38	nr	1747.34
15,000 litres, 3.75 m × 2 m × 2 m high	4664.88	5903.87	40.00	1347.61	nr	7251.48
Extra for bund unit (internal use)	3337.49	4223.92	55.00	1852.96	nr	6076.88
Extra for external use with bund (watertight)	1554.97	1967.97	2.00	67.38	nr	2035.35
20,000 litres, 4 m × 2.5 m × 2 m high	6153.48	7787.84	50.00	1684.51	nr	9472.35
Extra for bund unit (internal use)	4048.62	5123.93	65.00	2189.87	nr	7313.80
Extra for external use with bund (watertight)	1839.40	2327.94	2.00	67.38	nr	2395.32
Extra for BMS output (all tank sizes)	644.74	815.99	–	–	nr	815.99
Fuel storage tanks; plastic; with all necessary screwed bosses; oil resistant joint rings; includes placing in position Cylindrical; horizontal						
1350 litre (300 gallon) capacity	419.00	530.29	4.30	144.87	nr	675.16
2500 litre (550 gallon) capacity	729.00	922.62	4.88	164.42	nr	1087.04

38 PIPED SUPPLY SYSTEMS

Item	Net Price £	Material £	Labour hours	Labour £	Unit	Total rate £
FUEL OIL STORAGE/DISTRIBUTION – cont						
Fuel storage tanks – cont						
Cylindrical; vertical						
1365 litre (300 gallon) capacity	329.00	416.38	3.73	125.66	nr	**542.04**
2600 litre (570 gallon) capacity	489.00	618.88	4.88	164.42	nr	**783.30**
Bunded tanks						
1135 litre (250 gallon) capacity	985.00	1246.62	4.30	144.87	nr	**1391.49**
1590 litre (350 gallon) capacity	1110.00	1404.82	4.88	164.42	nr	**1569.24**
2500 litre (550 gallon) capacity	1269.00	1606.05	5.95	200.46	nr	**1806.51**

38 PIPED SUPPLY SYSTEMS

Item	Net Price £	Material £	Labour hours	Labour £	Unit	Total rate £
FIRE HOSE REELS: PIPEWORK ANCILLARIES						
PIPELINES						
For pipework prices refer to section – Cold Water						
PIPELINE ANCILLARIES						
For prices for ancillaries refer to section – Cold Water						
Hose reels; automatic; connection to 25 mm screwed joint; reel with 30.5 metres, 19 mm rubber hose; suitable for working pressure up to 7 bar						
Reels						
Non-swing pattern	194.38	246.01	3.75	126.35	nr	**372.36**
Recessed non-swing pattern	349.39	442.19	3.75	126.35	nr	**568.54**
Swinging pattern	226.23	286.32	3.75	126.35	nr	**412.67**
Recessed swinging pattern	226.23	286.32	3.75	126.35	nr	**412.67**
Hose reels; manual; connection to 25 mm screwed joint; reel with 30.5 metres, 19 mm rubber hose; suitable for working pressure up to 7 bar						
Reels						
Non-swing pattern	157.08	198.80	3.25	109.50	nr	**308.30**
Recessed non-swing pattern	349.39	442.19	3.25	109.50	nr	**551.69**
Swinging pattern	186.14	235.58	3.25	109.50	nr	**345.08**
Recessed swinging pattern	186.14	235.58	3.25	109.50	nr	**345.08**

38 PIPED SUPPLY SYSTEMS

Item	Net Price £	Material £	Labour hours	Labour £	Unit	Total rate £
DRY RISERS: PIPEWORK ANCILLARIES						
PIPELINES						
For pipework prices refer to section – Cold Water						
VALVES (BS 5041, parts 2 and 3)						
Bronze/gunmetal inlet breeching for pumping in with 65 mm dia. instantaneous male coupling; with cap, chain and 25 mm drain valve						
Double inlet with back pressure valve, flanged to steel	298.23	377.44	1.75	58.97	nr	**436.41**
Quadruple inlet with back pressure valve, flanged to steel	667.75	845.11	1.75	58.97	nr	**904.08**
Bronze/gunmetal gate type outlet valve with 65 mm dia. instantaneous female coupling; cap and chain; wheel head secured by padlock and leather strap						
Flanged to BS 4504 PN6 (bolted connection to counter flanges measured separately)	239.17	302.69	1.75	58.97	nr	**361.66**
Bronze/gunmetal landing type outlet valve, with 65 mm dia. instantaneous female coupling; cap and chain; wheelhead secured by padlock and leather strap; bolted connections to counter flanges measured separately						
Horizontal, flanged to BS 4504 PN6	260.48	329.66	1.50	50.53	nr	**380.19**
Oblique, flanged to BS 4504 PN6	260.48	329.66	1.50	50.53	nr	**380.19**
Air valve, screwed joints to steel						
25 mm dia.	51.77	65.52	0.55	18.52	nr	**84.04**
INLET BOXES (BS 5041, part 5)						
Steel dry riser inlet box with hinged wire glazed door suitably lettered (fixing by others)						
610 × 460 × 325 mm; double inlet	317.42	401.72	3.00	101.07	nr	**502.79**
610 × 610 × 356 mm; quadruple inlet	592.38	749.72	3.00	101.07	nr	**850.79**
OUTLET BOXES (BS 5041, part 5)						
Steel dry riser outlet box with hinged wire glazed door suitably lettered (fixing by others)						
610 × 460 × 325 mm; single outlet	309.95	392.27	3.00	101.07	nr	**493.34**

38 PIPED SUPPLY SYSTEMS

Item	Net Price £	Material £	Labour hours	Labour £	Unit	Total rate £
SPRINKLERS						
PIPELINES						
Prefabricated black steel pipework; screwed joints, including all coupliings, unions and the like to BS 1387:1985; includes fixing to backgrounds, with brackets measured separately						
Heavy weight						
25 mm dia. – pipe plus allowance for coupling every 6 m	6.89	8.72	0.47	15.84	m	**24.56**
32 mm dia. – pipe plus allowance for coupling every 6 m	8.56	10.83	0.53	17.85	m	**28.68**
40 mm dia. – pipe plus allowance for coupling every 6 m	9.96	12.60	0.58	19.54	m	**32.14**
50 mm dia. – pipe plus allowance for coupling every 6 m	13.83	17.51	0.63	21.22	m	**38.73**
Fixings						
For steel pipes; black malleable iron. For minimum fixing distances, refer to the Tables and Memoranda at the rear of the book						
Pipe ring, single socket, black malleable iron						
25 mm dia.	1.23	1.56	0.12	4.03	nr	**5.59**
32 mm dia.	1.31	1.66	0.14	4.72	nr	**6.38**
40 mm dia.	1.68	2.13	0.15	5.05	nr	**7.18**
50 mm dia.	2.12	2.69	0.16	5.39	nr	**8.08**
Extra over channel sections for fabricated hangers and brackets						
Galvanized steel; including inserts, bolts, nuts, washers; fixed to backgrounds						
41 × 21 mm	6.79	8.59	0.29	9.77	m	**18.36**
41 × 41 mm	8.14	10.30	0.29	9.77	m	**20.07**
Threaded rods; metric thread; including nuts, washers etc.						
12 mm dia. × 600 mm long	3.43	4.35	0.18	6.06	nr	**10.41**
Extra over for black malleable iron fittings; BS 143						
Plain plug, solid						
25 mm dia.	2.45	3.10	0.40	13.47	nr	**16.57**
32 mm dia.	3.53	4.47	0.44	14.82	nr	**19.29**
40 mm dia.	4.97	6.29	0.48	16.17	nr	**22.46**
50 mm dia.	6.53	8.27	0.56	18.87	nr	**27.14**

38 PIPED SUPPLY SYSTEMS

Item	Net Price £	Material £	Labour hours	Labour £	Unit	Total rate £
SPRINKLERS – cont						
Extra over for black malleable iron fittings – cont						
Concentric reducing socket						
32 mm dia.	2.91	3.68	0.48	16.17	nr	**19.85**
40 mm dia.	3.79	4.79	0.55	18.52	nr	**23.31**
50 mm dia.	5.33	6.74	0.60	20.22	nr	**26.96**
Elbow; 90 Degree female/female						
25 mm dia.	1.70	2.15	0.44	14.82	nr	**16.97**
32 mm dia.	3.14	3.98	0.53	17.85	nr	**21.83**
40 mm dia.	5.25	6.64	0.60	20.22	nr	**26.86**
50 mm dia.	6.16	7.80	0.65	21.90	nr	**29.70**
Tee						
25 mm dia. equal	2.34	2.96	0.51	17.19	nr	**20.15**
32 mm dia. reducing to 25 mm dia.	4.25	5.38	0.54	18.19	nr	**23.57**
40 mm dia.	6.53	8.27	0.65	21.90	nr	**30.17**
50 mm dia.	9.39	11.88	0.78	26.28	nr	**38.16**
Cross tee						
25 mm dia. equal	6.36	8.05	1.16	39.08	nr	**47.13**
32 mm dia.	9.25	11.70	1.40	47.17	nr	**58.87**
40 mm dia.	11.88	15.03	1.60	53.91	nr	**68.94**
50 mm dia.	19.30	24.43	1.68	56.59	nr	**81.02**
Prefabricated black steel pipework; welded joints, including all coupliings, unions and the like to BS 1387:1985; fixing to backgrounds						
Heavy weight						
65 mm dia. – pipe length only, welded joints on runs assumed	18.09	22.89	0.65	21.90	m	**44.79**
80 mm dia. – pipe length only, welded joints on runs assumed	23.03	29.14	0.70	23.58	m	**52.72**
100 mm dia. – pipe length only, welded joints on runs assumed	32.13	40.67	0.85	28.64	m	**69.31**
150 mm dia. – pipe length only, welded joints on runs assumed	43.47	55.01	1.15	38.74	m	**93.75**
Fixings						
For steel pipes; black malleable iron. For minimum fixing distances, refer to the Tables and Memoranda at the rear of the book						
Pipe ring, single socket, black malleable iron						
65 mm dia.	3.07	3.89	0.30	10.11	nr	**14.00**
80 mm dia.	3.68	4.66	0.35	11.79	nr	**16.45**
100 mm dia.	5.57	7.04	0.40	13.47	nr	**20.51**
150 mm dia.	12.61	15.96	0.77	25.95	nr	**41.91**

38 PIPED SUPPLY SYSTEMS

Item	Net Price £	Material £	Labour hours	Labour £	Unit	Total rate £
Extra over fittings						
Reducer (one size down)						
65 mm dia.	9.29	11.76	2.70	90.96	nr	**102.72**
80 mm dia.	12.08	15.29	2.86	96.35	nr	**111.64**
100 mm dia.	24.09	30.49	3.22	108.48	nr	**138.97**
150 mm dia.	66.93	84.71	4.20	141.49	nr	**226.20**
Elbow; 90°						
65 mm dia.	12.76	16.15	3.06	103.10	nr	**119.25**
80 mm dia.	18.76	23.74	3.40	114.55	nr	**138.29**
100 mm dia.	23.46	29.69	3.70	124.64	nr	**154.33**
150 mm dia.	35.16	44.50	5.20	175.18	nr	**219.68**
Branch bend						
65 mm dia.	13.00	16.45	3.60	121.28	nr	**137.73**
80 mm dia.	15.22	19.26	3.80	128.03	nr	**147.29**
100 mm dia.	23.41	29.62	5.10	171.82	nr	**201.44**
150 mm dia.	60.58	76.68	7.50	252.67	nr	**329.35**
Prefabricated black steel pipe; Victaulic Firelock joints; including all couplings and the like to BS 1387: 1985; fixing to backgrounds						
Heavy weight						
65 mm dia. – pipe plus allowance for Victaulic coupling every 6 m	15.59	19.73	0.70	23.58	m	**43.31**
80 mm dia. – pipe plus allowance for Victaulic coupling every 6 m	20.27	25.66	0.78	26.28	m	**51.94**
100 mm dia. – pipe plus allowance for Victaulic coupling every 6 m	29.55	37.40	0.93	31.34	m	**68.74**
150 mm dia. – pipe plus allowance for Victaulic coupling every 6 m	58.83	74.46	1.25	42.11	m	**116.57**
Fixings						
For fixings refer to For steel pipes; black malleable iron.						
Extra over fittings						
Coupling						
65 mm dia. – Victaulic coupling	11.42	14.45	0.26	8.76	nr	**23.21**
80 mm dia. – Victaulic coupling	12.24	15.49	0.26	8.76	nr	**24.25**
100 mm dia. – Victaulic coupling	16.04	20.31	0.32	10.79	nr	**31.10**
150 mm dia. – Victaulic coupling	28.23	35.73	0.35	11.79	nr	**47.52**
Reducer						
65 mm dia. – Victaulic excl couplings	9.29	11.76	0.48	16.17	nr	**27.93**
80 mm dia. – Victaulic excl couplings	12.08	15.29	0.43	14.49	nr	**29.78**
100 mm dia. – Victaulic excl couplings	24.09	30.49	0.46	15.50	nr	**45.99**
150 mm dia. – Victaulic excl couplings	36.14	45.74	0.45	15.16	nr	**60.90**
Elbow; any degree						
65 mm dia. – Victaulic excl couplings	12.76	16.15	0.56	18.87	nr	**35.02**
80 mm dia. – Victaulic excl couplings	18.76	23.74	0.63	21.22	nr	**44.96**
100 mm dia. – Victaulic excl couplings	36.18	45.79	0.71	23.92	nr	**69.71**
150 mm dia. – Victaulic excl couplings	54.27	68.69	0.80	26.96	nr	**95.65**

38 PIPED SUPPLY SYSTEMS

Item	Net Price £	Material £	Labour hours	Labour £	Unit	Total rate £
SPRINKLERS – cont						
Extra over fittings – cont						
Equal tee						
65 mm dia. – Victaulic excl couplings	19.79	25.04	0.74	24.93	nr	**49.97**
80 mm dia. – Victaulic excl couplings	23.82	30.15	0.83	27.96	nr	**58.11**
100 mm dia. – Victaulic excl couplings	43.66	55.26	0.94	31.66	nr	**86.92**
150 mm dia. – Victaulic excl couplings	65.51	82.91	1.05	35.37	nr	**118.28**
Y11 – PIPELINE ANCILLARIES						
SPRINKLER HEADS						
Sprinkler heads; brass body; frangible glass bulb; manufactured to standard operating temperature of 57–141°C; quick response; RTI<50						
Conventional pattern; 15 mm dia.	4.97	6.29	0.15	5.05	nr	**11.34**
Sidewall pattern; 15 mm dia.	7.29	9.23	0.15	5.05	nr	**14.28**
Conventional pattern; 15 mm dia.; satin chrome plated	5.83	7.38	0.15	5.05	nr	**12.43**
Sidewall pattern; 15 mm dia.; satin chrome plated	7.57	9.58	0.15	5.05	nr	**14.63**
Fully concealed; fusible link; 15 mm dia.	18.20	23.04	0.15	5.05	nr	**28.09**
VALVES						
Wet system alarm valves; including internal non-return valve; working pressure up to 12.5 bar; BS4504 PN16 flanged ends; bolted connections						
100 mm dia.	1601.72	2027.13	25.00	842.26	nr	**2869.39**
150 mm dia.	1956.55	2476.21	25.00	842.26	nr	**3318.47**
Wet system by-pass alarm valves; including internal non-return valve; working pressure up to 12.5 bar; BS4504 PN16 flanged ends; bolted connections						
100 mm dia.	2823.39	3573.28	25.00	842.26	nr	**4415.54**
150 mm dia.	3569.73	4517.84	25.00	842.26	nr	**5360.10**
Alternate system wet/dry alarm station; including butterfly valve, wet alarm valve, dry pipe differential pressure valve and pressure gauges; working pressure up to 10.5 bar; BS4505 PN16 flanged ends; bolted connections						
100 mm dia.	3499.18	4428.56	40.00	1347.61	nr	**5776.17**
150 mm dia.	4077.88	5160.96	40.00	1347.61	nr	**6508.57**
Alternate system wet/dry alarm station; including electrically operated butterfly valve, water supply accelerator set, wet alarm valve, dry pipe differential pressure valve and pressure gauges; working pressure up to 10.5 bar; BS4505 PN16 flanged ends; bolted connections						
100 mm dia.	3988.63	5048.01	45.00	1516.07	nr	**6564.08**
150 mm dia.	4572.85	5787.40	45.00	1516.07	nr	**7303.47**

38 PIPED SUPPLY SYSTEMS

Item	Net Price £	Material £	Labour hours	Labour £	Unit	Total rate £
ALARM/GONGS						
Water operated motor alarm and gong; stainless steel and aluminum body and gong; screwed connections						
Connection to sprinkler system and drain pipework	488.23	617.90	6.00	202.14	nr	**820.04**
Y21 – WATER TANKS						
Note: Prices are based on the most economical tank size for each volume, and the cost will vary with differing tank dimensions, for the same volume						
Steel sectional sprinkler tank; ordinary hazard, life safety classification; two compartment tank, complete with all fittings and accessories to comply with LPCB type A requirements; cost of erection (on prepared supports) is included within net price, labour cost allows for offloading and positioning of materials						
Volume, size						
70 m³, 6.1 m × 4.88 m × 2.44 m (h)	30497.55	38597.70	24.00	808.56	nr	**39406.26**
105 m³, 7.3 m × 6.1 m × 2.44 m (h)	35787.13	45292.20	28.00	943.33	nr	**46235.53**
168 m³, 9.76 m × 4.88 mx 3.66 m (h)	44215.94	55959.69	28.00	943.33	nr	**56903.02**
211 m³, 9.76 m × 6.1 m × 3.66 m (h)	48814.90	61780.14	32.00	1078.09	nr	**62858.23**
Steel sectional sprinkler tank; ordinary hazard, property protection classification; single compartment tank, complete with all fittings and accessories to comply with LPCB type A requirements; cost of erection (on prepared supports) is included within net price, labour cost allows for offloading and positioning of materials						
Volume, size						
70 m³, 6.1 m × 4.88 m × 2.44 m (h)	21503.66	27215.04	24.00	808.56	nr	**28023.60**
105 m³, 7.3 m × 6.1 m × 2.44 m (h)	26565.68	33621.53	28.00	943.33	nr	**34564.86**
168 m³, 9.76 m × 4.88 m × 3.66 m (h)	34327.39	43444.74	28.00	943.33	nr	**44388.07**
211 m³, 9.76 m × 6.1 m × 3.66 m (h)	38518.25	48748.69	32.00	1078.09	nr	**49826.78**
GRP sectional sprinkler tank; ordinary hazard, life safety classification; two compartment tank, complete with all fittings and accessories to comply with LPCB type A requirements; cost of erection (on prepared supports) is included within net price, labour cost allows for offloading and positioning of materials						
Volume, size						
55 m³, 6 m × 4 m × 3 m (h)	32126.79	40659.66	14.00	471.67	nr	**41131.33**
70 m³, 6 m × 5 m × 3 m (h)	35920.09	45460.46	16.00	539.04	nr	**45999.50**
80 m³, 8 m × 4 m × 3 m (h)	37226.07	47113.32	16.00	539.04	nr	**47652.36**

38 PIPED SUPPLY SYSTEMS

Item	Net Price £	Material £	Labour hours	Labour £	Unit	Total rate £
SPRINKLERS – cont						
Y21 – WATER TANKS – cont						
105 m³, 10 m × 4 m × 3 m (h)	42124.19	53312.37	24.00	808.56	nr	**54120.93**
125 m³, 10 m × 5 m × 3 m (h)	47506.75	60124.55	24.00	808.56	nr	**60933.11**
140 m³, 8 m × 7 m × 3 m (h)	49474.24	62614.60	24.00	808.56	nr	**63423.16**
135 m³, 9 m × 6 m × 3 m (h)	50970.00	64507.63	24.00	808.56	nr	**65316.19**
160 m³, 13 m × 5 m × 3 m (h)	55960.20	70823.23	24.00	808.56	nr	**71631.79**
185 m³, 12 m × 6 m × 3 m (h)	59195.39	74917.68	24.00	808.56	nr	**75726.24**
GRP sectional sprinkler tank; ordinary hazard, property protection classification; single compartment tank, complete with all fittings and accessories to comply with LPCB type A requirements; cost of erection (on prepared supports) is within net price, labour cost allows for offloading and positioning of materials						
Volume, size						
55 m³, 6 m × 4 m × 3 m (h)	24715.87	31280.40	14.00	471.67	nr	**31752.07**
70 m³, 6 m × 5 m × 3 m (h)	27922.74	35339.02	16.00	539.04	nr	**35878.06**
80 m³, 8 m × 4 m × 3 m (h)	29803.79	37719.67	16.00	539.04	nr	**38258.71**
105 m³, 10 m × 4 m × 3 m (h)	34915.82	44189.47	24.00	808.56	nr	**44998.03**
125 m³, 10 m × 5 m × 3 m (h)	39482.49	49969.04	24.00	808.56	nr	**50777.60**
140 m³, 8 m × 7 m × 3 m (h)	40873.45	51729.44	24.00	808.56	nr	**52538.00**
135 m³, 9 m × 6 m × 3 m (h)	42944.53	54350.60	24.00	808.56	nr	**55159.16**
160 m³, 13 m × 5 m × 3 m (h)	48144.16	60931.25	24.00	808.56	nr	**61739.81**
185 m³, 12 m × 6 m × 3 m (h)	50576.23	64009.28	24.00	808.56	nr	**64817.84**

38 PIPED SUPPLY SYSTEMS

Item	Net Price £	Material £	Labour hours	Labour £	Unit	Total rate £
WATERMIST SYSTEM						
PIPELINES						
For pipework prices refer to section – Cold Water: Stainless steel						
PIPELINE ANCILLARIES						
Nozzles						
Semco CJ nozzles	105.13	133.06	0.75	25.26	nr	**158.32**
extra over for rosettes (1 per nozzle)	13.79	17.45	0.20	6.73	nr	**24.18**
extra over for protection caps (1 per nozzle)	1.28	1.62	0.10	3.37	nr	**4.99**
PUMP SETS						
Automatic pump units (APU); includes fixing in position; electrical work elsewhere						
15 l/s pump unit	1923.07	2433.84	16.00	539.04	nr	**2972.88**
18 l/s pump unit	2243.59	2839.49	16.00	539.04	nr	**3378.53**
extra over for control module to duty pump	320.00	404.99	2.00	67.38	nr	**472.37**

38 PIPED SUPPLY SYSTEMS

Item	Net Price £	Material £	Labour hours	Labour £	Unit	Total rate £
FIRE EXTINGUISHERS AND HYDRANTS						
EXTINGUISHERS						
Fire extinguishers; hand held; BS 5423; placed in position						
Water type; cartridge operated; for Class A fires						
Water type, 9 litre capacity; 55 gm CO_2 cartridge; Class A fires (fire-rating 13 A)	36.76	46.52	1.00	33.69	nr	**80.21**
Foam type, 9 litre capacity; 75 gm CO_2 cartridge; Class A & B fires (fire-rating 13 A:183B)	47.88	60.59	1.00	33.69	nr	**94.28**
Dry powder type; cartridge operated; for Class A, B & C fires and electrical equipment fires						
Dry powder type, 1 kg capacity; 12 gm CO_2 cartridge; Class A, B & C fires (fire-rating 5 A:34B)	37.39	47.32	1.00	33.69	nr	**81.01**
Dry powder type, 2 kg capacity; 28 gm CO_2 cartridge; Class A, B & C fires (fire-rating 13 A:55B)	50.35	63.73	1.00	33.69	nr	**97.42**
Dry powder type, 4 kg capacity; 90 gm CO_2 cartridge; Class A, B & C fires (fire-rating 21 A:183B)	89.69	113.51	1.00	33.69	nr	**147.20**
Dry powder type, 9 kg capacity; 190 gm CO_2 cartridge; Class A, B & C fires (fire-rating 43 A:233B)	120.18	152.10	1.00	33.69	nr	**185.79**
Dry powder type; stored pressure type; for Class A, B & C fires and electrical equipment fires						
Dry powder type, 1 kg capacity; Class A, B & C fires (fire-rating 5 A:34B)	16.26	20.57	1.00	33.69	nr	**54.26**
Dry powder type, 2 kg capacity; Class A, B & C fires (fire-rating 13 A:55B)	14.92	18.88	1.00	33.69	nr	**52.57**
Dry powder type, 4 kg capacity; Class A, B & C fires (fire-rating 21 A:183B)	30.89	39.10	1.00	33.69	nr	**72.79**
Dry powder type, 9 kg capacity; Class A, B & C fires (fire-rating 43 A:233B)	32.74	41.44	1.00	33.69	nr	**75.13**
Carbon dioxide type; for Class B fires and electrical equipment fires						
CO_2 type with hose and horn, 2 kg capacity, Class B fires (fire-rating 34B)	32.13	40.67	1.00	33.69	nr	**74.36**
CO_2 type with hose and horn, 5 kg capacity, Class B fires (fire-rating 55B)	64.36	81.46	1.00	33.69	nr	**115.15**
Glass fibre blanket, in GRP container						
1100 × 1100 mm	12.35	15.64	0.50	16.84	nr	**32.48**
1200 × 1200 mm	14.41	18.23	0.50	16.84	nr	**35.07**
1800 × 1200 mm	18.12	22.94	0.50	16.84	nr	**39.78**

38 PIPED SUPPLY SYSTEMS

Item	Net Price £	Material £	Labour hours	Labour £	Unit	Total rate £
HYDRANTS						
Fire hydrants; bolted connections						
Underground hydrants, complete with frost plug to BS 750						
Sluice valve pattern type 1	246.88	312.45	4.50	151.60	nr	**464.05**
Screw down pattern type 2	179.32	226.95	4.50	151.60	nr	**378.55**
Stand pipe for underground hydrant; screwed base; light alloy						
Single outlet	141.99	179.70	1.00	33.69	nr	**213.39**
Double outlet	207.76	262.94	1.00	33.69	nr	**296.63**
64 mm dia. bronze/gunmetal outlet valves						
Oblique flanged landing valve	155.42	196.69	1.00	33.69	nr	**230.38**
Oblique screwed landing valve	155.42	196.69	1.00	33.69	nr	**230.38**
Cast iron surface box; fixing by others						
400 × 200 × 100 mm	136.08	172.22	1.00	33.69	nr	**205.91**
500 × 200 × 150 mm	183.35	232.05	1.00	33.69	nr	**265.74**
Frost plug	33.02	41.79	0.25	8.43	nr	**50.22**

38 MECHANICAL/COOLING/HEATING SYSTEMS

Item	Net Price £	Material £	Labour hours	Labour £	Unit	Total rate £
GAS/OIL FIRED BOILERS – DOMESTIC						
Domestic water boilers; stove enamelled casing; electric controls; placing in position; assembling and connecting; electrical work elsewhere						
Gas fired; floor standing; connected to conventional flue						
9 to 12 kW	647.88	819.95	8.59	289.41	nr	**1109.36**
12 to 15 kW	681.56	862.58	8.59	289.41	nr	**1151.99**
15 to 18 kW	725.93	918.74	8.88	299.17	nr	**1217.91**
18 to 21 kW	843.38	1067.38	9.92	334.21	nr	**1401.59**
21 to 23 kW	934.97	1183.30	10.66	359.14	nr	**1542.44**
23 to 29 kW	1212.64	1534.71	11.81	397.88	nr	**1932.59**
29 to 37 kW	1434.26	1815.20	11.81	397.88	nr	**2213.08**
37 to 41 kW	1491.57	1887.73	12.68	427.19	nr	**2314.92**
Gas fired; wall hung; connected to conventional flue						
9 to 12 kW	569.43	720.68	8.59	289.41	nr	**1010.09**
12 to 15 kW	735.15	930.41	8.59	289.41	nr	**1219.82**
13 to 18 kW	932.97	1180.77	8.59	289.41	nr	**1470.18**
Gas fired; floor standing; connected to balanced flue						
9 to 12 kW	809.30	1024.25	9.16	308.60	nr	**1332.85**
12 to 15 kW	842.05	1065.70	10.95	368.91	nr	**1434.61**
15 to 18 kW	906.39	1147.13	11.98	403.61	nr	**1550.74**
18 to 21 kW	1068.10	1351.78	12.78	430.56	nr	**1782.34**
21 to 23 kW	1231.90	1559.10	12.78	430.56	nr	**1989.66**
23 to 29 kW	1570.21	1987.26	15.45	520.52	nr	**2507.78**
29 to 37 kW	3018.56	3820.29	17.65	594.63	nr	**4414.92**
Gas fired; wall hung; connected to balanced flue						
6 to 9 kW	611.06	773.36	9.16	308.60	nr	**1081.96**
9 to 12 kW	699.25	884.97	9.16	308.60	nr	**1193.57**
12 to 15 kW	791.80	1002.10	9.45	318.37	nr	**1320.47**
15 to 18 kW	949.41	1201.57	9.74	328.15	nr	**1529.72**
18 to 22 kW	1007.77	1275.43	9.74	328.15	nr	**1603.58**
Gas fired; wall hung; connected to fan flue (including flue kit)						
6 to 9 kW	698.87	884.49	9.16	308.60	nr	**1193.09**
9 to 12 kW	778.82	985.68	9.16	308.60	nr	**1294.28**
12 to 15 kW	845.43	1069.98	10.95	368.91	nr	**1438.89**
15 to 18 kW	910.67	1152.55	11.98	403.61	nr	**1556.16**
18 to 23 kW	1186.58	1501.74	12.78	430.56	nr	**1932.30**
23 to 29 kW	1536.60	1944.72	15.45	520.52	nr	**2465.24**
29 to 35 kW	1886.50	2387.56	17.65	594.63	nr	**2982.19**

38 MECHANICAL/COOLING/HEATING SYSTEMS

Item	Net Price £	Material £	Labour hours	Labour £	Unit	**Total rate £**
Oil fired; floor standing; connected to conventional flue						
12 to 15 kW	1213.44	1535.73	10.38	349.71	nr	**1885.44**
15 to 19 kW	1265.80	1601.99	12.20	411.02	nr	**2013.01**
21 to 25 kW	1444.64	1828.33	14.30	481.78	nr	**2310.11**
26 to 32 kW	1583.86	2004.53	15.80	532.31	nr	**2536.84**
35 to 50 kW	1788.23	2263.18	20.46	689.30	nr	**2952.48**
Fire place mounted natural gas fire and back boiler; cast iron water boiler; electric control box; fire output 3 kW with wood surround						
10.50 kW	354.01	448.03	8.88	299.17	nr	**747.20**

38 MECHANICAL/COOLING/HEATING SYSTEMS

Item	Net Price £	Material £	Labour hours	Labour £	Unit	Total rate £
GAS/OIL FIRED BOILERS – COMMERCIAL; FORCED DRAFT						
Commercial steel shell floor standing boilers; pressure jet burner; 8 bar max working pressure; including controls, enamelled jacket, insulation, placing in position and commissioning; electrical work elsewhere						
Natural gas; burner and boiler (high/low type), connected to conventional flue						
411–500 kW	11692.00	14797.40	12.00	404.29	nr	**15201.69**
601–750 kW	13802.00	17467.81	12.00	404.29	nr	**17872.10**
1151–1250 kW	18904.00	23924.90	14.00	471.67	nr	**24396.57**
1426–1500 kW	23146.00	29293.58	14.00	471.67	nr	**29765.25**
1625–1760 kW	27274.00	34517.97	14.00	471.67	nr	**34989.64**
Natural gas; burner and boiler (modulating type), connected to conventional flue						
411–500 kW	12531.00	15859.23	12.00	404.29	nr	**16263.52**
601–750 kW	14479.00	18324.62	12.00	404.29	nr	**18728.91**
1151–1250 kW	21198.00	26828.19	12.00	404.29	nr	**27232.48**
1426–1500 kW	25304.00	32024.74	14.00	471.67	nr	**32496.41**
1625–1760 kW	30433.00	38516.00	14.00	471.67	nr	**38987.67**
1900–2050 kW	31931.00	40411.87	14.00	471.67	nr	**40883.54**
2450–2650 kW	37478.00	47432.16	14.00	471.67	nr	**47903.83**
Natural gas; low Nox burner and boiler (modulating type), connected to conventional flue						
420 kW	16334.00	20672.31	12.00	404.29	nr	**21076.60**
530 kW	17607.00	22283.42	12.00	404.29	nr	**22687.71**
620 kW	19923.00	25214.55	12.00	404.29	nr	**25618.84**
750 kW	20131.00	25477.79	14.00	471.67	nr	**25949.46**
1000 kW	25374.00	32113.33	14.00	471.67	nr	**32585.00**
1250 kW	27963.00	35389.97	14.00	471.67	nr	**35861.64**
1500 kW	37187.00	47063.87	14.00	471.67	nr	**47535.54**
1800 kW	42406.00	53669.03	14.00	471.67	nr	**54140.70**
2200 kW	46068.00	58303.66	14.00	471.67	nr	**58775.33**
2700 kW	55220.00	69886.43	14.00	471.67	nr	**70358.10**
Oil; burner and boiler (high/low type), connected to conventional flue						
411–500 kW	11292.00	14291.16	12.00	404.29	nr	**14695.45**
601–750 kW	12745.00	16130.07	12.00	404.29	nr	**16534.36**
1151–1250 kW	17368.00	21980.94	12.00	404.29	nr	**22385.23**
1426–1500 kW	21149.00	26766.17	14.00	471.67	nr	**27237.84**
1625–1760 kW	25194.00	31885.53	14.00	471.67	nr	**32357.20**
Oil; burner and boiler (modulating type), connected to conventional flue						
420 kW	15237.00	19283.95	12.00	404.29	nr	**19688.24**
530 kW	15350.00	19426.96	12.00	404.29	nr	**19831.25**
620 kW	16873.00	21354.47	12.00	404.29	nr	**21758.76**

38 MECHANICAL/COOLING/HEATING SYSTEMS

Item	Net Price £	Material £	Labour hours	Labour £	Unit	Total rate £
750 kW	18335.00	23204.78	14.00	471.67	nr	23676.45
1000 kW	22779.00	28829.10	14.00	471.67	nr	29300.77
1250 kW	32768.00	41471.18	14.00	471.67	nr	41942.85
1500 kW	31073.00	39325.99	14.00	471.67	nr	39797.66
1800 kW	36316.00	45961.53	14.00	471.67	nr	46433.20
2200 kW	56230.00	71164.69	14.00	471.67	nr	71636.36
2700 kW	63048.00	79793.55	14.00	471.67	nr	80265.22
Dual fuel; burner and boiler (high/low oil and modulating gast type), connected to conventional flue						
411–500 kW	14479.00	18324.62	12.00	404.29	nr	18728.91
601–750 kW	16263.00	20582.45	12.00	404.29	nr	20986.74
1151–1250 kW	24655.00	31203.37	12.00	404.29	nr	31607.66
1426–1500 kW	29653.00	37528.84	14.00	471.67	nr	38000.51
1625–1760 kW	34325.00	43441.72	14.00	471.67	nr	43913.39
2450–2650-kW	44671.00	56535.62	14.00	471.67	nr	57007.29
Dual fuel; low NOX burner and boiler (high/low oil and modulating gast type), connected to conventional flue						
420 kW	17716.00	22421.37	12.00	404.29	nr	22825.66
530 kW	19501.00	24680.47	12.00	404.29	nr	25084.76
620 kW	20172.00	25529.68	12.00	404.29	nr	25933.97
750 kW	21915.00	27735.62	14.00	471.67	nr	28207.29
1000 kW	29373.00	37174.47	14.00	471.67	nr	37646.14
1250 kW	41178.00	52114.88	14.00	471.67	nr	52586.55
1500 kW	42110.00	53294.42	14.00	471.67	nr	53766.09
1800 kW	49960.00	63229.38	14.00	471.67	nr	63701.05
2200 kW	52324.00	66221.25	14.00	471.67	nr	66692.92
2700 kW	56169.00	71087.49	14.00	471.67	nr	71559.16

38 MECHANICAL/COOLING/HEATING SYSTEMS

Item	Net Price £	Material £	Labour hours	Labour £	Unit	Total rate £
GAS/OIL FIRED BOILERS – COMMERCIAL; CONDENSING						
Gas boiler; Low Nox wall mounted condensing boiler, with high efficiency modulating premix burner; Aluminium heat exchanger; includes integral control panel; delivery included						
Maximum output						
35 kW	2320.00	2936.19	11.00	370.60	nr	**3306.79**
45 kW	2675.00	3385.48	11.00	370.60	nr	**3756.08**
60 kW	3322.00	4204.32	11.00	370.60	nr	**4574.92**
80 kW	3916.00	4956.09	11.00	370.60	nr	**5326.69**
100 kW	4380.00	5543.33	11.00	370.60	nr	**5913.93**
120 kW	4519.00	5719.25	11.00	370.60	nr	**6089.85**
Commissioning Rate	250.00	316.40	–	–	nr	**316.40**
Gas boiler; Low Nox floor standing condensing boiler with high efficiency modulating premix burner; Stainless steel heat exchanger; includes controls; delivery and commissioning						
Maximum output						
50 kW	4595.86	5816.52	8.00	269.52	nr	**6086.04**
70 kW	5076.87	6425.28	8.00	269.52	nr	**6694.80**
100 kW	5494.02	6953.23	8.00	269.52	nr	**7222.75**
125 kW	5673.24	7180.05	10.00	336.91	nr	**7516.96**
150 kW	6328.32	8009.12	10.00	336.91	nr	**8346.03**
200 kW	7284.16	9218.83	10.00	336.91	nr	**9555.74**
250 kW	8351.24	10569.33	10.00	336.91	nr	**10906.24**
300 kW	9312.23	11785.56	10.00	336.91	nr	**12122.47**
350 kW	10507.03	13297.69	10.00	336.91	nr	**13634.60**
400 kW	11822.34	14962.35	10.00	336.91	nr	**15299.26**
450 kW	13126.32	16612.67	10.00	336.91	nr	**16949.58**
500 kW	14325.24	18130.02	10.00	336.91	nr	**18466.93**
575 kW	16116.41	20396.92	10.00	336.91	nr	**20733.83**
650 kW	17193.79	21760.46	10.00	336.91	nr	**22097.37**
720 kW	20057.19	25384.37	10.00	336.91	nr	**25721.28**
850 kW	24773.56	31353.41	12.00	404.29	nr	**31757.70**
1000 kW	27608.12	34940.84	12.00	404.29	nr	**35345.13**
1150 kW	28726.70	36356.51	12.00	404.29	nr	**36760.80**
1300 kW	36381.66	46044.63	14.00	471.67	nr	**46516.30**
1440 kW	41394.67	52389.10	14.00	471.67	nr	**52860.77**
1550 kW	48925.00	61919.48	14.00	471.67	nr	**62391.15**
1700 kW	51425.84	65084.54	14.00	471.67	nr	**65556.21**
2000 kW	57690.30	73012.84	14.00	471.67	nr	**73484.51**
2300 kW	59967.63	75895.03	14.00	471.67	nr	**76366.70**
3100 kW	76849.00	97260.09	14.00	471.67	nr	**97731.76**

38 MECHANICAL/COOLING/HEATING SYSTEMS

Item	Net Price £	Material £	Labour hours	Labour £	Unit	Total rate £
Gas boilers, wall hung cascade; Low Nox frame mounted, hung condensing boiler with high efficiency modulating premix burner; Aluminium heat exchanger; includes integral control panel; pumps and headers included, delivery included						
Maximum cascade output						
160 kW (2 units)	9349.00	11832.09	10.00	336.91	nr	**12169.00**
200 kW (2 units)	11654.00	14749.30	10.00	336.91	nr	**15086.21**
240 kW (3 units)	14882.00	18834.66	10.00	336.91	nr	**19171.57**
300 kW (3 units)	17921.00	22680.82	12.00	404.29	nr	**23085.11**
360 kW (3 units)	18339.00	23209.84	12.00	404.29	nr	**23614.13**
400 kW (4 units)	23277.00	29459.37	12.00	404.29	nr	**29863.66**
480 kW (4 units)	23834.00	30164.31	12.00	404.29	nr	**30568.60**
600 kW (5 units)	30270.00	38309.71	14.00	471.67	nr	**38781.38**
720 kW (6 units)	35736.00	45227.48	14.00	471.67	nr	**45699.15**
Commissioning Rate	190.00	240.46	–	–	nr	**240.46**
Oil boiler; Low Nox floor standing condensing boiler with high efficiency modulating premix burner; Stainless steel heat exchanger; includes controls; delivery and commissioning						
50 kW	5404.41	6839.82	8.00	269.52	nr	**7109.34**
80 kW	7154.38	9054.58	8.00	269.52	nr	**9324.10**
110 kW	9696.42	12271.78	8.00	269.52	nr	**12541.30**
130 kW	9988.94	12642.00	10.00	336.91	nr	**12978.91**
160 kW	10522.48	13317.25	10.00	336.91	nr	**13654.16**
200 kW	11228.03	14210.19	10.00	336.91	nr	**14547.10**
250 kW	14125.42	17877.13	10.00	336.91	nr	**18214.04**
300 kW	15898.05	20120.58	10.00	336.91	nr	**20457.49**
320 kW	22340.70	28274.39	10.00	336.91	nr	**28611.30**
400 kW	24819.91	31412.08	12.00	404.29	nr	**31816.37**
500 kW	30733.14	38895.86	12.00	404.29	nr	**39300.15**
600 kW	34274.28	43377.53	12.00	404.29	nr	**43781.82**

38 MECHANICAL/COOLING/HEATING SYSTEMS

Item	Net Price £	Material £	Labour hours	Labour £	Unit	Total rate £
GAS/OIL BOILERS – FLUE SYSTEMS						
Flues; suitable for domestic, medium sized industrial and commercial oil and gas appliances; stainless steel, twin wall, insulated; for use internally or externally						
Straight length; 120 mm long; including one locking band						
127 mm dia.	46.69	59.09	0.49	16.51	nr	**75.60**
152 mm dia.	52.31	66.20	0.51	17.19	nr	**83.39**
178 mm dia.	60.72	76.84	0.54	18.19	nr	**95.03**
203 mm dia.	69.08	87.43	0.58	19.54	nr	**106.97**
254 mm dia.	81.70	103.40	0.70	23.58	nr	**126.98**
304 mm dia.	102.36	129.55	0.74	24.93	nr	**154.48**
355 mm dia.	146.58	185.52	0.80	26.96	nr	**212.48**
Straight length; 300 mm long; including one locking band						
127 mm dia.	72.02	91.15	0.52	17.52	nr	**108.67**
152 mm dia.	81.41	103.03	0.52	17.52	nr	**120.55**
178 mm dia.	93.55	118.40	0.55	18.52	nr	**136.92**
203 mm dia.	105.86	133.97	0.64	21.57	nr	**155.54**
254 mm dia.	117.40	148.58	0.79	26.61	nr	**175.19**
304 mm dia.	140.84	178.25	0.86	28.97	nr	**207.22**
355 mm dia.	154.55	195.60	0.94	31.66	nr	**227.26**
400 mm dia.	165.38	209.31	1.03	34.70	nr	**244.01**
450 mm dia.	189.18	239.42	1.03	34.70	nr	**274.12**
500 mm dia.	202.90	256.79	1.10	37.06	nr	**293.85**
550 mm dia.	223.91	283.38	1.10	37.06	nr	**320.44**
600 mm dia.	247.07	312.69	1.10	37.06	nr	**349.75**
Straight length; 500 mm long; including one locking band						
127 mm dia.	84.64	107.12	0.55	18.52	nr	**125.64**
152 mm dia.	94.36	119.43	0.55	18.52	nr	**137.95**
178 mm dia.	106.26	134.48	0.63	21.22	nr	**155.70**
203 mm dia.	124.21	157.20	0.63	21.22	nr	**178.42**
254 mm dia.	144.28	182.60	0.86	28.97	nr	**211.57**
304 mm dia.	172.84	218.75	0.95	32.01	nr	**250.76**
355 mm dia.	194.11	245.66	1.03	34.70	nr	**280.36**
400 mm dia.	211.94	268.23	1.12	37.73	nr	**305.96**
450 mm dia.	244.77	309.78	1.12	37.73	nr	**347.51**
500 mm dia.	263.74	333.79	1.19	40.10	nr	**373.89**
550 mm dia.	290.74	367.96	1.19	40.10	nr	**408.06**
600 mm dia.	301.17	381.16	1.19	40.10	nr	**421.26**
Straight length; 1000 mm long; including one locking band						
127 mm dia.	151.29	191.48	0.62	20.89	nr	**212.37**
152 mm dia.	168.63	213.42	0.68	22.90	nr	**236.32**
178 mm dia.	189.73	240.12	0.74	24.93	nr	**265.05**
203 mm dia.	223.48	282.83	0.80	26.96	nr	**309.79**
254 mm dia.	253.26	320.52	0.87	29.31	nr	**349.83**
304 mm dia.	292.34	369.98	1.06	35.72	nr	**405.70**
355 mm dia.	335.46	424.56	1.16	39.08	nr	**463.64**

38 MECHANICAL/COOLING/HEATING SYSTEMS

Item	Net Price £	Material £	Labour hours	Labour £	Unit	Total rate £
400 mm dia.	359.25	454.66	1.26	42.45	nr	**497.11**
450 mm dia.	379.07	479.75	1.26	42.45	nr	**522.20**
500 mm dia.	411.37	520.63	1.33	44.80	nr	**565.43**
550 mm dia.	452.42	572.58	1.33	44.80	nr	**617.38**
600 mm dia.	475.30	601.54	1.33	44.80	nr	**646.34**
Adjustable length; boiler removal; internal use only; including one locking band						
127 mm dia.	69.36	87.79	0.52	17.52	nr	**105.31**
152 mm dia.	78.36	99.18	0.55	18.52	nr	**117.70**
178 mm dia.	88.82	112.41	0.59	19.88	nr	**132.29**
203 mm dia.	101.86	128.91	0.64	21.57	nr	**150.48**
254 mm dia.	151.02	191.13	0.79	26.61	nr	**217.74**
304 mm dia.	180.89	228.94	0.86	28.97	nr	**257.91**
355 mm dia.	202.84	256.72	0.99	33.35	nr	**290.07**
400 mm dia.	354.95	449.22	0.91	30.65	nr	**479.87**
450 mm dia.	379.94	480.85	0.91	30.65	nr	**511.50**
500 mm dia.	414.39	524.45	0.99	33.35	nr	**557.80**
550 mm dia.	452.07	572.14	0.99	33.35	nr	**605.49**
600 mm dia.	473.25	598.94	0.99	33.35	nr	**632.29**
Inspection length; 500 mm long; including one locking band						
127 mm dia.	178.26	225.60	0.55	18.52	nr	**244.12**
152 mm dia.	184.59	233.62	0.55	18.52	nr	**252.14**
178 mm dia.	194.42	246.05	0.63	21.22	nr	**267.27**
203 mm dia.	205.88	260.56	0.63	21.22	nr	**281.78**
254 mm dia.	266.19	336.88	0.86	28.97	nr	**365.85**
304 mm dia.	288.49	365.11	0.95	32.01	nr	**397.12**
355 mm dia.	325.74	412.26	1.03	34.70	nr	**446.96**
400 mm dia.	522.59	661.39	1.12	37.73	nr	**699.12**
450 mm dia.	536.81	679.39	1.12	37.73	nr	**717.12**
500 mm dia.	589.05	745.51	1.19	40.10	nr	**785.61**
550 mm dia.	615.96	779.55	1.19	40.10	nr	**819.65**
600 mm dia.	626.42	792.79	1.19	40.10	nr	**832.89**
Adapters						
127 mm dia.	14.91	18.87	0.49	16.51	nr	**35.38**
152 mm dia.	16.31	20.64	0.51	17.19	nr	**37.83**
178 mm dia.	17.45	22.09	0.54	18.19	nr	**40.28**
203 mm dia.	19.86	25.13	0.58	19.54	nr	**44.67**
254 mm dia.	21.85	27.65	0.70	23.58	nr	**51.23**
304 mm dia.	27.26	34.50	0.74	24.93	nr	**59.43**
355 mm dia.	33.10	41.89	0.80	26.96	nr	**68.85**
400 mm dia.	37.13	47.00	0.89	29.98	nr	**76.98**
450 mm dia.	39.66	50.20	0.89	29.98	nr	**80.18**
500 mm dia.	42.10	53.28	0.96	32.35	nr	**85.63**
550 mm dia.	48.94	61.94	0.96	32.35	nr	**94.29**
600 mm dia.	58.82	74.45	0.96	32.35	nr	**106.80**

38 MECHANICAL/COOLING/HEATING SYSTEMS

Item	Net Price £	Material £	Labour hours	Labour £	Unit	Total rate £
GAS/OIL BOILERS – FLUE SYSTEMS – cont						
Fittings for flue system						
90° insulated tee; including two locking bands						
127 mm dia.	173.31	219.34	1.89	63.67	nr	**283.01**
152 mm dia.	200.03	253.15	2.04	68.72	nr	**321.87**
178 mm dia.	218.63	276.70	2.39	80.52	nr	**357.22**
203 mm dia.	255.99	323.98	2.56	86.25	nr	**410.23**
254 mm dia.	259.32	328.19	2.95	99.39	nr	**427.58**
304 mm dia.	322.48	408.13	3.41	114.88	nr	**523.01**
355 mm dia.	411.12	520.32	3.77	127.02	nr	**647.34**
400 mm dia.	541.74	685.63	4.25	143.19	nr	**828.82**
450 mm dia.	564.65	714.62	4.76	160.36	nr	**874.98**
500 mm dia.	638.94	808.64	5.12	172.49	nr	**981.13**
550 mm dia.	686.82	869.24	5.61	189.00	nr	**1058.24**
600 mm dia.	714.64	904.44	5.98	201.47	nr	**1105.91**
135° insulated tee; including two locking bands						
127 mm dia.	224.78	284.48	1.89	63.67	nr	**348.15**
152 mm dia.	243.04	307.60	2.04	68.72	nr	**376.32**
178 mm dia.	265.72	336.29	2.39	80.52	nr	**416.81**
203 mm dia.	337.23	426.80	2.56	86.25	nr	**513.05**
254 mm dia.	383.19	484.96	2.95	99.39	nr	**584.35**
304 mm dia.	288.72	365.40	3.41	114.88	nr	**480.28**
355 mm dia.	568.10	718.98	3.77	127.02	nr	**846.00**
400 mm dia.	747.82	946.44	4.25	143.19	nr	**1089.63**
450 mm dia.	809.57	1024.59	4.76	160.36	nr	**1184.95**
500 mm dia.	943.13	1193.63	5.12	172.49	nr	**1366.12**
550 mm dia.	967.66	1224.68	5.61	189.00	nr	**1413.68**
600 mm dia.	1021.26	1292.50	5.98	201.47	nr	**1493.97**
Wall sleeve; for 135° tee through wall						
127 mm dia.	23.00	29.11	1.89	63.67	nr	**92.78**
152 mm dia.	30.87	39.07	2.04	68.72	nr	**107.79**
178 mm dia.	32.32	40.90	2.39	80.52	nr	**121.42**
203 mm dia.	36.37	46.03	2.56	86.25	nr	**132.28**
254 mm dia.	40.56	51.33	2.95	99.39	nr	**150.72**
304 mm dia.	47.27	59.83	3.41	114.88	nr	**174.71**
355 mm dia.	52.51	66.46	3.77	127.02	nr	**193.48**
15° insulated elbow; including two locking bands						
127 mm dia.	120.27	152.22	1.57	52.89	nr	**205.11**
152 mm dia.	133.67	169.18	1.79	60.30	nr	**229.48**
178 mm dia.	142.96	180.92	2.05	69.06	nr	**249.98**
203 mm dia.	151.56	191.81	2.33	78.49	nr	**270.30**
254 mm dia.	156.10	197.56	2.45	82.54	nr	**280.10**
304 mm dia.	197.34	249.75	3.43	115.56	nr	**365.31**
355 mm dia.	264.00	334.12	4.71	158.68	nr	**492.80**

38 MECHANICAL/COOLING/HEATING SYSTEMS

Item	Net Price £	Material £	Labour hours	Labour £	Unit	Total rate £
30° insulated elbow; including two locking bands						
127 mm dia.	120.27	152.22	1.44	48.51	nr	**200.73**
152 mm dia.	133.67	169.18	1.62	54.58	nr	**223.76**
178 mm dia.	142.96	180.92	1.89	63.67	nr	**244.59**
203 mm dia.	151.56	191.81	2.17	73.11	nr	**264.92**
254 mm dia.	156.10	197.56	2.16	72.77	nr	**270.33**
304 mm dia.	197.34	249.75	2.74	92.31	nr	**342.06**
355 mm dia.	263.50	333.49	3.17	106.80	nr	**440.29**
400 mm dia.	264.00	334.12	3.53	118.93	nr	**453.05**
450 mm dia.	305.97	387.24	3.88	130.73	nr	**517.97**
500 mm dia.	320.78	405.98	4.24	142.84	nr	**548.82**
550 mm dia.	344.26	435.69	4.61	155.31	nr	**591.00**
600 mm dia.	375.38	475.08	4.96	167.12	nr	**642.20**
45° insulated elbow; including two locking bands						
127 mm dia.	120.27	152.22	1.44	48.51	nr	**200.73**
152 mm dia.	133.67	169.18	1.51	50.88	nr	**220.06**
178 mm dia.	142.96	180.92	1.58	53.23	nr	**234.15**
203 mm dia.	151.56	191.81	1.66	55.92	nr	**247.73**
254 mm dia.	156.10	197.56	1.72	57.95	nr	**255.51**
304 mm dia.	197.34	249.75	1.80	60.65	nr	**310.40**
355 mm dia.	264.00	334.12	1.94	65.35	nr	**399.47**
400 mm dia.	336.67	426.09	2.01	67.73	nr	**493.82**
450 mm dia.	352.64	446.30	2.09	70.41	nr	**516.71**
500 mm dia.	378.27	478.74	2.16	72.77	nr	**551.51**
550 mm dia.	412.56	522.13	2.23	75.13	nr	**597.26**
600 mm dia.	423.01	535.36	2.30	77.49	nr	**612.85**
Flue supports						
Wall support, galvanized; including plate and brackets						
127 mm dia.	73.02	92.41	2.24	75.47	nr	**167.88**
152 mm dia.	80.28	101.61	2.44	82.20	nr	**183.81**
178 mm dia.	88.63	112.17	2.52	84.90	nr	**197.07**
203 mm dia.	92.29	116.80	2.77	93.33	nr	**210.13**
254 mm dia.	110.06	139.29	2.98	100.40	nr	**239.69**
304 mm dia.	124.44	157.49	3.46	116.57	nr	**274.06**
355 mm dia.	166.57	210.81	4.08	137.46	nr	**348.27**
400 mm dia.; with 300 mm support length and collar	427.86	541.50	4.80	161.72	nr	**703.22**
450 mm dia.; with 300 mm support length and collar	456.16	577.32	5.62	189.34	nr	**766.66**
500 mm dia.; with 300 mm support length and collar	498.35	630.72	6.24	210.22	nr	**840.94**
550 mm dia.; with 300 mm support length and collar	545.23	690.04	6.97	234.82	nr	**924.86**
600 mm dia.; with 300 mm support length and collar	575.56	728.43	7.49	252.34	nr	**980.77**

38 MECHANICAL/COOLING/HEATING SYSTEMS

Item	Net Price £	Material £	Labour hours	Labour £	Unit	Total rate £
GAS/OIL BOILERS – FLUE SYSTEMS – cont						
Flue supports – cont						
Ceiling/floor support						
127 mm dia.	22.61	28.62	1.86	62.66	nr	91.28
152 mm dia.	25.13	31.81	2.14	72.11	nr	103.92
178 mm dia.	30.09	38.08	1.93	65.03	nr	103.11
203 mm dia.	45.30	57.33	2.74	92.31	nr	149.64
254 mm dia.	51.68	65.41	3.21	108.15	nr	173.56
304 mm dia.	59.06	74.75	3.68	123.97	nr	198.72
355 mm dia.	70.39	89.08	4.28	144.19	nr	233.27
400 mm dia.	89.91	113.79	4.86	163.73	nr	277.52
450 mm dia.	97.64	123.57	5.46	183.96	nr	307.53
500 mm dia.	118.54	150.02	6.04	203.48	nr	353.50
550 mm dia.	145.02	183.53	6.65	224.03	nr	407.56
600 mm dia.	208.10	263.37	7.24	243.91	nr	507.28
Ceiling/floor firestop spacer						
127 mm dia.	4.40	5.57	0.66	22.23	nr	27.80
152 mm dia.	4.90	6.20	0.69	23.25	nr	29.45
178 mm dia.	5.53	7.00	0.70	23.58	nr	30.58
203 mm dia.	6.56	8.30	0.87	29.31	nr	37.61
254 mm dia.	6.78	8.58	0.91	30.65	nr	39.23
304 mm dia.	8.14	10.30	0.95	32.01	nr	42.31
355 mm dia.	14.82	18.76	0.99	33.35	nr	52.11
Wall band; internal or external use						
127 mm dia.	26.61	33.68	1.03	34.70	nr	68.38
152 mm dia.	27.79	35.17	1.07	36.04	nr	71.21
178 mm dia.	28.89	36.57	1.11	37.40	nr	73.97
203 mm dia.	30.50	38.60	1.18	39.75	nr	78.35
254 mm dia.	31.74	40.17	1.30	43.80	nr	83.97
304 mm dia.	34.30	43.41	1.45	48.85	nr	92.26
355 mm dia.	36.49	46.18	1.65	55.59	nr	101.77
400 mm dia.	45.26	57.28	1.85	62.33	nr	119.61
450 mm dia.	48.22	61.03	2.39	80.52	nr	141.55
500 mm dia.	57.97	73.37	2.25	75.81	nr	149.18
550 mm dia.	60.72	76.84	2.45	82.54	nr	159.38
600 mm dia.	64.19	81.23	2.66	89.62	nr	170.85
Flashings and terminals						
Insulated top stub; including one locking band						
127 mm dia.	65.57	82.98	1.49	50.20	nr	133.18
152 mm dia.	74.15	93.84	1.90	64.02	nr	157.86
178 mm dia.	79.83	101.04	1.92	64.68	nr	165.72
203 mm dia.	84.95	107.51	2.20	74.11	nr	181.62
254 mm dia.	90.20	114.16	2.49	83.89	nr	198.05
304 mm dia.	123.29	156.04	2.79	94.00	nr	250.04
355 mm dia.	162.77	206.00	3.19	107.48	nr	313.48
400 mm dia.	156.95	198.63	3.59	120.95	nr	319.58
450 mm dia.	170.51	215.80	3.97	133.75	nr	349.55
500 mm dia.	197.25	249.64	4.38	147.57	nr	397.21
550 mm dia.	206.40	261.22	4.78	161.03	nr	422.25
600 mm dia.	213.60	270.33	5.17	174.18	nr	444.51

38 MECHANICAL/COOLING/HEATING SYSTEMS

Item	Net Price £	Material £	Labour hours	Labour £	Unit	Total rate £
Rain cap; including one locking band						
127 mm dia.	35.06	44.37	1.49	50.20	nr	**94.57**
152 mm dia.	36.64	46.37	1.54	51.88	nr	**98.25**
178 mm dia.	40.34	51.05	1.72	57.95	nr	**109.00**
203 mm dia.	48.24	61.05	2.00	67.38	nr	**128.43**
254 mm dia.	63.43	80.28	2.49	83.89	nr	**164.17**
304 mm dia.	43.43	54.97	2.80	94.34	nr	**149.31**
355 mm dia.	114.58	145.02	3.19	107.48	nr	**252.50**
400 mm dia.	114.31	144.67	3.45	116.23	nr	**260.90**
450 mm dia.	124.32	157.34	3.97	133.75	nr	**291.09**
500 mm dia.	134.41	170.11	4.38	147.57	nr	**317.68**
550 mm dia.	144.22	182.53	4.78	161.03	nr	**343.56**
600 mm dia.	154.20	195.16	5.17	174.18	nr	**369.34**
Round top; including one locking band						
127 mm dia.	67.03	84.83	1.49	50.20	nr	**135.03**
152 mm dia.	73.08	92.49	1.65	55.59	nr	**148.08**
178 mm dia.	83.30	105.43	1.92	64.68	nr	**170.11**
203 mm dia.	98.04	124.08	2.20	74.11	nr	**198.19**
254 mm dia.	115.92	146.71	2.49	83.89	nr	**230.60**
304 mm dia.	152.25	192.68	2.80	94.34	nr	**287.02**
355 mm dia.	203.11	257.05	3.19	107.48	nr	**364.53**
Coping cap; including one locking band						
127 mm dia.	37.54	47.51	1.49	50.20	nr	**97.71**
152 mm dia.	39.32	49.76	1.65	55.59	nr	**105.35**
178 mm dia.	43.26	54.75	1.92	64.68	nr	**119.43**
203 mm dia.	51.92	65.71	2.20	74.11	nr	**139.82**
254 mm dia.	63.43	80.28	2.49	83.89	nr	**164.17**
304 mm dia.	85.52	108.24	2.79	94.00	nr	**202.24**
355 mm dia.	114.58	145.02	3.19	107.48	nr	**252.50**
Storm collar						
127 mm dia.	7.11	8.99	0.52	17.52	nr	**26.51**
152 mm dia.	7.61	9.63	0.55	18.52	nr	**28.15**
178 mm dia.	8.44	10.68	0.57	19.20	nr	**29.88**
203 mm dia.	8.84	11.19	0.66	22.23	nr	**33.42**
254 mm dia.	11.08	14.02	0.66	22.23	nr	**36.25**
304 mm dia.	11.53	14.59	0.72	24.26	nr	**38.85**
355 mm dia.	12.33	15.60	0.77	25.95	nr	**41.55**
400 mm dia.	32.70	41.38	0.82	27.63	nr	**69.01**
450 mm dia.	35.99	45.55	0.87	29.31	nr	**74.86**
500 mm dia.	39.24	49.66	0.92	30.99	nr	**80.65**
550 mm dia.	42.53	53.83	0.98	33.02	nr	**86.85**
600 mm dia.	45.79	57.95	1.03	34.70	nr	**92.65**
Flat flashing; including storm collar and sealant						
127 mm dia.	39.97	50.59	1.49	50.20	nr	**100.79**
152 mm dia.	41.30	52.27	1.65	55.59	nr	**107.86**
178 mm dia.	43.33	54.84	1.92	64.68	nr	**119.52**
203 mm dia.	47.46	60.07	2.20	74.11	nr	**134.18**
254 mm dia.	64.84	82.06	2.49	83.89	nr	**165.95**
304 mm dia.	77.47	98.04	2.80	94.34	nr	**192.38**
355 mm dia.	122.09	154.52	3.20	107.80	nr	**262.32**

38 MECHANICAL/COOLING/HEATING SYSTEMS

Item	Net Price £	Material £	Labour hours	Labour £	Unit	Total rate £
GAS/OIL BOILERS – FLUE SYSTEMS – cont						
Flashings and terminals – cont						
400 mm dia.	170.31	215.54	3.59	120.95	nr	**336.49**
450 mm dia.	195.66	247.63	3.97	133.75	nr	**381.38**
500 mm dia.	211.95	268.24	4.38	147.57	nr	**415.81**
550 mm dia.	225.06	284.84	4.78	161.03	nr	**445.87**
600 mm dia.	233.19	295.12	5.17	174.18	nr	**469.30**
5°–30° rigid adjustable flashing; including storm collar and sealant						
127 mm dia.	68.07	86.15	1.49	50.20	nr	**136.35**
152 mm dia.	71.72	90.76	1.65	55.59	nr	**146.35**
178 mm dia.	76.26	96.51	1.92	64.68	nr	**161.19**
203 mm dia.	80.19	101.48	2.20	74.11	nr	**175.59**
254 mm dia.	84.43	106.86	2.49	83.89	nr	**190.75**
304 mm dia.	104.47	132.22	2.80	94.34	nr	**226.56**
355 mm dia.	118.82	150.38	3.19	107.48	nr	**257.86**
400 mm dia.	383.05	484.79	3.59	120.95	nr	**605.74**
450 mm dia.	447.04	565.78	3.97	133.75	nr	**699.53**
500 mm dia.	477.25	604.00	4.38	147.57	nr	**751.57**
550 mm dia.	499.42	632.06	4.77	160.71	nr	**792.77**
600 mm dia.	543.36	687.68	5.17	174.18	nr	**861.86**
Domestic and small commercial; twin walled gas vent system suitable for gas fired appliances; domestic gas boilers; small commercial boilers with internal or external flues						
152 mm long						
100 mm dia.	7.28	9.22	0.52	17.52	nr	**26.74**
125 mm dia.	8.94	11.31	0.52	17.52	nr	**28.83**
150 mm dia.	9.69	12.26	0.52	17.52	nr	**29.78**
305 mm long						
100 mm dia.	11.06	14.00	0.52	17.52	nr	**31.52**
125 mm dia.	12.97	16.42	0.52	17.52	nr	**33.94**
150 mm dia.	15.38	19.47	0.52	17.52	nr	**36.99**
457 mm long						
100 mm dia.	12.22	15.47	0.55	18.52	nr	**33.99**
125 mm dia.	13.74	17.39	0.55	18.52	nr	**35.91**
150 mm dia.	17.00	21.52	0.55	18.52	nr	**40.04**
914 mm long						
100 mm dia.	21.84	27.64	0.62	20.89	nr	**48.53**
125 mm dia.	25.46	32.22	0.62	20.89	nr	**53.11**
150 mm dia.	29.19	36.94	0.62	20.89	nr	**57.83**
1524 mm long						
100 mm dia.	31.54	39.92	0.82	27.63	nr	**67.55**
125 mm dia.	38.82	49.13	0.84	28.30	nr	**77.43**
150 mm dia.	41.68	52.75	0.84	28.30	nr	**81.05**

38 MECHANICAL/COOLING/HEATING SYSTEMS

Item	Net Price £	Material £	Labour hours	Labour £	Unit	Total rate £
Adjustable length 305 mm long						
100 mm dia.	13.99	17.71	0.56	18.87	nr	**36.58**
125 mm dia.	15.70	19.87	0.56	18.87	nr	**38.74**
150 mm dia.	19.79	25.04	0.56	18.87	nr	**43.91**
Adjustable length 457 mm long						
100 mm dia.	18.86	23.87	0.56	18.87	nr	**42.74**
125 mm dia.	22.87	28.94	0.56	18.87	nr	**47.81**
150 mm dia.	25.44	32.20	0.56	18.87	nr	**51.07**
Adjustable elbow 0°–90°						
100 mm dia.	15.95	20.18	0.48	16.17	nr	**36.35**
125 mm dia.	18.86	23.87	0.48	16.17	nr	**40.04**
150 mm dia.	23.61	29.88	0.48	16.17	nr	**46.05**
Draughthood connector						
100 mm dia.	4.92	6.23	0.48	16.17	nr	**22.40**
125 mm dia.	5.54	7.01	0.48	16.17	nr	**23.18**
150 mm dia.	6.02	7.62	0.48	16.17	nr	**23.79**
Adaptor						
100 mm dia.	11.94	15.11	0.48	16.17	nr	**31.28**
125 mm dia.	12.19	15.42	0.48	16.17	nr	**31.59**
150 mm dia.	12.45	15.76	0.48	16.17	nr	**31.93**
Support plate						
100 mm dia.	8.61	10.90	0.48	16.17	nr	**27.07**
125 mm dia.	9.16	11.59	0.48	16.17	nr	**27.76**
150 mm dia.	9.81	12.42	0.48	16.17	nr	**28.59**
Wall band						
100 mm dia.	7.79	9.86	0.48	16.17	nr	**26.03**
125 mm dia.	8.30	10.51	0.48	16.17	nr	**26.68**
150 mm dia.	10.52	13.32	0.48	16.17	nr	**29.49**
Firestop						
100 mm dia.	3.37	4.27	0.48	16.17	nr	**20.44**
125 mm dia.	3.37	4.27	0.48	16.17	nr	**20.44**
150 mm dia.	3.88	4.91	0.48	16.17	nr	**21.08**
Flat flashing						
125 mm dia.	22.95	29.04	0.55	18.52	nr	**47.56**
150 mm dia.	31.95	40.43	0.55	18.52	nr	**58.95**
Adjustable flashing 5°–30°						
100 mm dia.	59.83	75.72	0.55	18.52	nr	**94.24**
125 mm dia.	93.40	118.20	0.55	18.52	nr	**136.72**
Storm collar						
100 mm dia.	4.77	6.04	0.55	18.52	nr	**24.56**
125 mm dia.	4.88	6.17	0.55	18.52	nr	**24.69**
150 mm dia.	5.00	6.33	0.55	18.52	nr	**24.85**
Gas vent terminal						
100 mm dia.	18.21	23.05	0.55	18.52	nr	**41.57**
125 mm dia.	20.01	25.32	0.55	18.52	nr	**43.84**
150 mm dia.	25.67	32.49	0.55	18.52	nr	**51.01**
Twin wall galvanized steel flue box, 125 mm dia.; fitted for gas fire, where no chimney exists						
Free-standing	115.70	146.43	2.15	72.43	nr	**218.86**
Recess	115.70	146.43	2.15	72.43	nr	**218.86**
Back boiler	85.24	107.88	2.40	80.86	nr	**188.74**

38 MECHANICAL/COOLING/HEATING SYSTEMS

Item	Net Price £	Material £	Labour hours	Labour £	Unit	Total rate £
PACKAGED STEAM GENERATORS						
Packaged steam boilers; boiler mountings centrifugal water feed pump; insulation; and sheet steel wrap around casing; plastic coated						
Gas fired						
276 kW	14921.00	18884.02	86.00	2897.36	nr	**21781.38**
1384 kW	56984.00	72118.95	148.00	4986.16	nr	**77105.11**
2940 KW	64414.00	81522.36	207.00	6973.88	nr	**88496.24**
4843 kW	96373.00	121969.67	295.00	9938.63	nr	**131908.30**
5213 kW	111631.00	141280.19	310.00	10443.99	nr	**151724.18**
Gas oil fired						
276 kW	11883.00	15039.12	86.00	2897.36	nr	**17936.48**
1384 kW	59456.00	75247.51	148.00	4986.16	nr	**80233.67**
2940 kW	71589.00	90603.04	207.00	6973.88	nr	**97576.92**
4843 kW	103580.00	131090.85	295.00	9938.63	nr	**141029.48**
5213 kW	124987.00	158183.55	310.00	10443.99	nr	**168627.54**

38 MECHANICAL/COOLING/HEATING SYSTEMS

Item	Net Price £	Material £	Labour hours	Labour £	Unit	Total rate £
LOW TEMPERATURE HOT WATER HEATING; PIPELINE: SCREWED STEEL						
Black steel pipes; screwed and socketed joints; BS 1387: 1985. Fixed vertically, brackets measured separately. Screwed joints are within the running length, but any flanges are additional						
Medium weight						
10 mm dia.	5.19	6.56	0.37	12.47	m	**19.03**
15 mm dia.	3.27	4.14	0.37	12.47	m	**16.61**
20 mm dia.	3.85	4.87	0.37	12.47	m	**17.34**
25 mm dia.	5.54	7.01	0.41	13.81	m	**20.82**
32 mm dia.	6.85	8.67	0.48	16.17	m	**24.84**
40 mm dia.	7.96	10.07	0.52	17.52	m	**27.59**
50 mm dia.	11.19	14.16	0.62	20.89	m	**35.05**
65 mm dia.	15.19	19.22	0.65	21.90	m	**41.12**
80 mm dia.	19.75	25.00	1.10	37.06	m	**62.06**
100 mm dia.	27.96	35.38	1.31	44.13	m	**79.51**
125 mm dia.	35.59	45.05	1.66	55.92	m	**100.97**
150 mm dia.	41.32	52.29	1.88	63.35	m	**115.64**
Heavy weight						
15 mm dia.	3.89	4.93	0.37	12.47	m	**17.40**
20 mm dia.	4.61	5.84	0.37	12.47	m	**18.31**
25 mm dia.	6.76	8.56	0.41	13.81	m	**22.37**
32 mm dia.	8.39	10.62	0.48	16.17	m	**26.79**
40 mm dia.	9.79	12.39	0.52	17.52	m	**29.91**
50 mm dia.	13.57	17.17	0.62	20.89	m	**38.06**
65 mm dia.	18.46	23.36	0.64	21.57	m	**44.93**
80 mm dia.	23.50	29.75	1.10	37.06	m	**66.81**
100 mm dia.	32.80	41.51	1.31	44.13	m	**85.64**
125 mm dia.	37.96	48.04	1.66	55.92	m	**103.96**
150 mm dia.	44.36	56.15	1.88	63.35	m	**119.50**
200 mm dia.	69.93	88.50	2.99	100.73	m	**189.23**
250 mm dia.	87.67	110.96	3.49	117.58	m	**228.54**
300 mm dia.	104.39	132.12	3.91	131.72	m	**263.84**
Black steel pipes; screwed and socketed joints; BS 1387: 1985. Fixed at high level or suspended, brackets measured separately. Screwed joints are within the running length, but any flanges are additional						
Medium weight						
10 mm dia.	5.19	6.56	0.58	19.54	m	**26.10**
15 mm dia.	3.27	4.14	0.58	19.54	m	**23.68**
20 mm dia.	3.85	4.87	0.58	19.54	m	**24.41**
25 mm dia.	5.54	7.01	0.60	20.22	m	**27.23**
32 mm dia.	6.85	8.67	0.68	22.90	m	**31.57**
40 mm dia.	7.96	10.07	0.73	24.60	m	**34.67**
50 mm dia.	11.19	14.16	0.85	28.64	m	**42.80**
65 mm dia.	15.19	19.22	0.88	29.66	m	**48.88**

38 MECHANICAL/COOLING/HEATING SYSTEMS

Item	Net Price £	Material £	Labour hours	Labour £	Unit	Total rate £
LOW TEMPERATURE HOT WATER HEATING; PIPELINE: SCREWED STEEL – cont						
Black steel pipes – cont						
80 mm dia.	19.75	25.00	1.45	48.85	m	**73.85**
100 mm dia.	27.96	35.38	1.74	58.62	m	**94.00**
125 mm dia.	35.59	45.05	2.21	74.46	m	**119.51**
150 mm dia.	41.32	52.29	2.50	84.22	m	**136.51**
Heavy weight						
15 mm dia.	3.89	4.93	0.58	19.54	m	**24.47**
20 mm dia.	4.61	5.84	0.58	19.54	m	**25.38**
25 mm dia.	6.76	8.56	0.60	20.22	m	**28.78**
32 mm dia.	8.39	10.62	0.68	22.90	m	**33.52**
40 mm dia.	9.79	12.39	0.73	24.60	m	**36.99**
50 mm dia.	13.57	17.17	0.85	28.64	m	**45.81**
65 mm dia.	18.46	23.36	0.88	29.66	m	**53.02**
80 mm dia.	23.50	29.75	1.45	48.85	m	**78.60**
100 mm dia.	32.80	41.51	1.74	58.62	m	**100.13**
125 mm dia.	37.96	48.04	2.21	74.46	m	**122.50**
150 mm dia.	44.36	56.15	2.50	84.22	m	**140.37**
200 mm dia.	69.93	88.50	2.99	100.73	m	**189.23**
250 mm dia.	87.67	110.96	3.49	117.58	m	**228.54**
300 mm dia.	104.39	132.12	3.91	131.72	m	**263.84**
Fixings						
For steel pipes; black malleable iron. For minimum fixing distances, refer to the Tables and Memoranda at the rear of the book						
Single pipe bracket, screw on, black malleable iron; screwed to wood						
15 mm dia.	0.85	1.08	0.14	4.72	nr	**5.80**
20 mm dia.	0.95	1.20	0.14	4.72	nr	**5.92**
25 mm dia.	1.11	1.40	0.17	5.73	nr	**7.13**
32 mm dia.	2.02	2.55	0.19	6.41	nr	**8.96**
40 mm dia.	1.93	2.44	0.22	7.41	nr	**9.85**
50 mm dia.	2.70	3.42	0.22	7.41	nr	**10.83**
65 mm dia.	3.54	4.48	0.28	9.43	nr	**13.91**
80 mm dia.	4.84	6.13	0.32	10.79	nr	**16.92**
100 mm dia.	7.10	8.98	0.35	11.79	nr	**20.77**
Single pipe bracket, screw on, black malleable iron; plugged and screwed						
15 mm dia.	0.85	1.08	0.25	8.43	nr	**9.51**
20 mm dia.	0.95	1.20	0.25	8.43	nr	**9.63**
25 mm dia.	1.11	1.40	0.30	10.11	nr	**11.51**
32 mm dia.	2.02	2.55	0.32	10.79	nr	**13.34**
40 mm dia.	1.93	2.44	0.32	10.79	nr	**13.23**
50 mm dia.	2.70	3.42	0.32	10.79	nr	**14.21**
65 mm dia.	3.54	4.48	0.35	11.79	nr	**16.27**

38 MECHANICAL/COOLING/HEATING SYSTEMS

Item	Net Price £	Material £	Labour hours	Labour £	Unit	Total rate £
80 mm dia.	4.84	6.13	0.42	14.15	nr	**20.28**
100 mm dia.	7.10	8.98	0.42	14.15	nr	**23.13**
Single pipe bracket for building in, black malleable iron						
15 mm dia.	2.49	3.15	0.10	3.37	nr	**6.52**
20 mm dia.	2.49	3.15	0.11	3.71	nr	**6.86**
25 mm dia.	2.49	3.15	0.12	4.03	nr	**7.18**
32 mm dia.	2.83	3.58	0.14	4.72	nr	**8.30**
40 mm dia.	2.85	3.61	0.15	5.05	nr	**8.66**
50 mm dia.	2.85	3.61	0.16	5.39	nr	**9.00**
Pipe ring, single socket, black malleable iron						
15 mm dia.	0.99	1.25	0.10	3.37	nr	**4.62**
20 mm dia.	1.10	1.39	0.11	3.71	nr	**5.10**
25 mm dia.	1.23	1.56	0.12	4.03	nr	**5.59**
32 mm dia.	1.31	1.66	0.14	4.72	nr	**6.38**
40 mm dia.	1.68	2.13	0.15	5.05	nr	**7.18**
50 mm dia.	2.12	2.69	0.16	5.39	nr	**8.08**
65 mm dia.	3.07	3.89	0.30	10.11	nr	**14.00**
80 mm dia.	3.68	4.66	0.35	11.79	nr	**16.45**
100 mm dia.	5.57	7.04	0.40	13.47	nr	**20.51**
125 mm dia.	11.04	13.98	0.60	20.23	nr	**34.21**
150 mm dia.	12.61	15.96	0.77	25.95	nr	**41.91**
200 mm dia.	34.87	44.13	0.90	30.32	nr	**74.45**
250 mm dia.	43.59	55.17	1.10	37.06	nr	**92.23**
300 mm dia.	52.43	66.36	1.25	42.11	nr	**108.47**
350 mm dia.	61.04	77.26	1.50	50.53	nr	**127.79**
400 mm dia.	72.23	91.41	1.75	58.97	nr	**150.38**
Pipe ring, double socket, black malleable iron						
15 mm dia.	1.14	1.44	0.10	3.37	nr	**4.81**
20 mm dia.	1.32	1.67	0.11	3.71	nr	**5.38**
25 mm dia.	1.47	1.86	0.12	4.03	nr	**5.89**
32 mm dia.	1.74	2.21	0.14	4.72	nr	**6.93**
40 mm dia.	1.90	2.41	0.15	5.05	nr	**7.46**
50 mm dia.	2.96	3.74	0.16	5.39	nr	**9.13**
Screw on backplate (Male), black malleable iron; plugged and screwed						
M12	0.60	0.76	0.10	3.37	nr	**4.13**
Screw on backplate (Female), black malleable iron; plugged and screwed						
M12	0.60	0.76	0.10	3.37	nr	**4.13**
Extra over channel sections for fabricated hangers and brackets						
Galvanized steel; including inserts, bolts, nuts, washers; fixed to backgrounds						
41 × 21 mm	6.79	8.59	0.29	9.77	m	**18.36**
41 × 41 mm	8.14	10.30	0.29	9.77	m	**20.07**
Threaded rods; metric thread; including nuts, washers etc.						
12 mm dia. × 600 mm long	3.43	4.35	0.18	6.06	nr	**10.41**

38 MECHANICAL/COOLING/HEATING SYSTEMS

Item	Net Price £	Material £	Labour hours	Labour £	Unit	Total rate £
LOW TEMPERATURE HOT WATER HEATING; PIPELINE: SCREWED STEEL – cont						
Pipe roller and chair						
Roller and chair; black malleable						
Up to 50 mm dia.	18.01	22.79	0.20	6.73	nr	**29.52**
65 mm dia.	18.55	23.48	0.20	6.73	nr	**30.21**
80 mm dia.	20.15	25.50	0.20	6.73	nr	**32.23**
100 mm dia.	20.51	25.96	0.20	6.73	nr	**32.69**
125 mm dia.	28.73	36.36	0.20	6.73	nr	**43.09**
150 mm dia.	28.73	36.36	0.30	10.11	nr	**46.47**
175 mm dia.	48.47	61.34	0.30	10.11	nr	**71.45**
200 mm dia.	74.24	93.96	0.30	10.11	nr	**104.07**
250 mm dia.	96.84	122.56	0.30	10.11	nr	**132.67**
300 mm dia.	121.95	154.34	0.30	10.11	nr	**164.45**
Roller bracket; black malleable						
25 mm dia.	3.69	4.67	0.20	6.73	nr	**11.40**
32 mm dia.	3.86	4.88	0.20	6.73	nr	**11.61**
40 mm dia.	4.14	5.24	0.20	6.73	nr	**11.97**
50 mm dia.	4.36	5.52	0.20	6.73	nr	**12.25**
65 mm dia.	5.75	7.28	0.20	6.73	nr	**14.01**
80 mm dia.	8.27	10.47	0.20	6.73	nr	**17.20**
100 mm dia.	9.21	11.66	0.20	6.73	nr	**18.39**
125 mm dia.	15.19	19.22	0.20	6.73	nr	**25.95**
150 mm dia.	15.19	19.22	0.30	10.11	nr	**29.33**
175 mm dia.	33.85	42.84	0.30	10.11	nr	**52.95**
200 mm dia.	33.85	42.84	0.30	10.11	nr	**52.95**
250 mm dia.	45.01	56.96	0.30	10.11	nr	**67.07**
300 mm dia.	55.27	69.96	0.30	10.11	nr	**80.07**
350 mm dia.	88.88	112.48	0.30	10.11	nr	**122.59**
400 mm dia.	101.59	128.58	0.30	10.11	nr	**138.69**
Extra over black steel screwed pipes; black steel flanges, screwed and drilled; metric; BS 4504						
Screwed flanges; PN6						
15 mm dia.	9.40	11.89	0.35	11.79	nr	**23.68**
20 mm dia.	9.40	11.89	0.47	15.84	nr	**27.73**
25 mm dia.	9.40	11.89	0.53	17.85	nr	**29.74**
32 mm dia.	9.40	11.89	0.62	20.89	nr	**32.78**
40 mm dia.	9.40	11.89	0.70	23.58	nr	**35.47**
50 mm dia.	10.01	12.67	0.84	28.30	nr	**40.97**
65 mm dia.	13.93	17.63	1.03	34.70	nr	**52.33**
80 mm dia.	19.65	24.86	1.23	41.44	nr	**66.30**
100 mm dia.	23.23	29.40	1.41	47.50	nr	**76.90**
125 mm dia.	42.63	53.95	1.77	59.64	nr	**113.59**
150 mm dia.	42.63	53.95	2.21	74.46	nr	**128.41**

38 MECHANICAL/COOLING/HEATING SYSTEMS

Item	Net Price £	Material £	Labour hours	Labour £	Unit	Total rate £
Screwed flanges; PN16						
15 mm dia.	11.93	15.10	0.35	11.79	nr	26.89
20 mm dia.	11.93	15.10	0.47	15.84	nr	30.94
25 mm dia.	11.93	15.10	0.53	17.85	nr	32.95
32 mm dia.	12.75	16.14	0.62	20.89	nr	37.03
40 mm dia.	12.75	16.14	0.70	23.58	nr	39.72
50 mm dia.	15.55	19.68	0.84	28.30	nr	47.98
65 mm dia.	19.43	24.60	1.03	34.70	nr	59.30
80 mm dia.	23.66	29.95	1.23	41.44	nr	71.39
100 mm dia.	26.52	33.57	1.41	47.50	nr	81.07
125 mm dia.	46.41	58.73	1.77	59.64	nr	118.37
150 mm dia.	45.67	57.80	2.21	74.46	nr	132.26
Extra over black steel screwed pipes; black steel flanges, screwed and drilled; imperial; BS 10						
Screwed flanges; Table E						
½" dia.	17.83	22.57	0.35	11.79	nr	34.36
¾" dia.	17.83	22.57	0.47	15.84	nr	38.41
1" dia.	17.81	22.55	0.53	17.85	nr	40.40
1¼" dia.	17.83	22.57	0.62	20.89	nr	43.46
1½" dia.	17.83	22.57	0.70	23.58	nr	46.15
2" dia.	17.83	22.57	0.84	28.30	nr	50.87
2½" dia.	21.21	26.85	1.03	34.70	nr	61.55
3" dia.	24.03	30.41	1.23	41.44	nr	71.85
4" dia.	32.52	41.16	1.41	47.50	nr	88.66
5" dia.	68.89	87.19	1.77	59.64	nr	146.83
6" dia.	68.89	87.19	2.21	74.46	nr	161.65
Extra over black steel screwed pipes; black steel flange connections						
Bolted connection between pair of flanges; including gasket, bolts, nuts and washers						
50 mm dia.	8.96	11.33	0.53	17.85	nr	29.18
65 mm dia.	10.62	13.44	0.53	17.85	nr	31.29
80 mm dia.	15.93	20.16	0.53	17.85	nr	38.01
100 mm dia.	17.30	21.90	0.61	20.55	nr	42.45
125 mm dia.	20.36	25.77	0.61	20.55	nr	46.32
150 mm dia.	29.09	36.81	0.90	30.32	nr	67.13
Extra over black steel screwed pipes; black heavy steel tubular fittings; BS 1387						
Long screw connection with socket and backnut						
15 mm dia.	6.17	7.81	0.63	21.22	nr	29.03
20 mm dia.	7.77	9.83	0.84	28.30	nr	38.13
25 mm dia.	10.20	12.91	0.95	32.01	nr	44.92
32 mm dia.	13.67	17.30	1.11	37.40	nr	54.70
40 mm dia.	16.44	20.81	1.28	43.12	nr	63.93
50 mm dia.	24.22	30.65	1.53	51.54	nr	82.19
65 mm dia.	33.87	42.86	1.87	63.00	nr	105.86

38 MECHANICAL/COOLING/HEATING SYSTEMS

Item	Net Price £	Material £	Labour hours	Labour £	Unit	Total rate £
LOW TEMPERATURE HOT WATER HEATING; PIPELINE: SCREWED STEEL – cont						
Extra over black steel screwed pipes – cont						
Long screw connection with socket and backnut – cont						
80 mm dia.	77.45	98.02	2.21	74.46	nr	172.48
100 mm dia.	125.47	158.79	3.05	102.75	nr	261.54
Running nipple						
15 mm dia.	1.60	2.03	0.50	16.84	nr	18.87
20 mm dia.	2.02	2.55	0.68	22.90	nr	25.45
25 mm dia.	2.47	3.12	0.77	25.95	nr	29.07
32 mm dia.	3.49	4.41	0.90	30.32	nr	34.73
40 mm dia.	4.70	5.95	1.03	34.70	nr	40.65
50 mm dia.	7.15	9.05	1.23	41.44	nr	50.49
65 mm dia.	15.36	19.44	1.50	50.53	nr	69.97
80 mm dia.	23.97	30.34	1.78	59.96	nr	90.30
100 mm dia.	37.53	47.50	2.38	80.19	nr	127.69
Barrel nipple						
15 mm dia.	0.91	1.15	0.50	16.84	nr	17.99
20 mm dia.	1.45	1.84	0.68	22.90	nr	24.74
25 mm dia.	1.92	2.43	0.77	25.95	nr	28.38
32 mm dia.	2.85	3.61	0.90	30.32	nr	33.93
40 mm dia.	3.52	4.46	1.03	34.70	nr	39.16
50 mm dia.	5.03	6.36	1.23	41.44	nr	47.80
65 mm dia.	10.74	13.60	1.50	50.53	nr	64.13
80 mm dia.	14.98	18.96	1.78	59.96	nr	78.92
100 mm dia.	27.13	34.34	2.38	80.19	nr	114.53
125 mm dia.	50.34	63.71	2.87	96.69	nr	160.40
150 mm dia.	79.34	100.41	3.39	114.21	nr	214.62
Close taper nipple						
15 mm dia.	1.92	2.43	0.50	16.84	nr	19.27
20 mm dia.	2.47	3.12	0.68	22.90	nr	26.02
25 mm dia.	3.23	4.09	0.77	25.95	nr	30.04
32 mm dia.	4.84	6.13	0.90	30.32	nr	36.45
40 mm dia.	6.00	7.59	1.03	34.70	nr	42.29
50 mm dia.	9.26	11.72	1.23	41.44	nr	53.16
65 mm dia.	17.95	22.71	1.50	50.53	nr	73.24
80 mm dia.	22.12	28.00	1.78	59.96	nr	87.96
100 mm dia.	45.46	57.53	2.38	80.19	nr	137.72
Extra over black steel screwed pipes; black malleable iron fittings; BS 143						
Cap						
15 mm dia.	0.76	0.96	0.32	10.79	nr	11.75
20 mm dia.	0.88	1.11	0.43	14.49	nr	15.60
25 mm dia.	1.11	1.40	0.49	16.51	nr	17.91

38 MECHANICAL/COOLING/HEATING SYSTEMS

Item	Net Price £	Material £	Labour hours	Labour £	Unit	Total rate £
32 mm dia.	1.94	2.45	0.58	19.54	nr	**21.99**
40 mm dia.	2.30	2.91	0.66	22.23	nr	**25.14**
50 mm dia.	4.78	6.05	0.78	26.28	nr	**32.33**
65 mm dia.	7.98	10.10	0.96	32.35	nr	**42.45**
80 mm dia.	9.03	11.42	1.13	38.07	nr	**49.49**
100 mm dia.	19.80	25.05	1.70	57.27	nr	**82.32**
Plain plug, hollow						
15 mm dia.	0.54	0.68	0.28	9.43	nr	**10.11**
20 mm dia.	0.69	0.87	0.38	12.81	nr	**13.68**
25 mm dia.	0.94	1.19	0.44	14.82	nr	**16.01**
32 mm dia.	1.30	1.65	0.51	17.19	nr	**18.84**
40 mm dia.	2.46	3.11	0.59	19.88	nr	**22.99**
50 mm dia.	3.45	4.37	0.70	23.58	nr	**27.95**
65 mm dia.	5.62	7.11	0.85	28.64	nr	**35.75**
80 mm dia.	8.78	11.11	1.00	33.69	nr	**44.80**
100 mm dia.	16.15	20.44	1.44	48.51	nr	**68.95**
Plain plug, solid						
15 mm dia.	1.67	2.12	0.28	9.43	nr	**11.55**
20 mm dia.	1.62	2.05	0.38	12.81	nr	**14.86**
25 mm dia.	2.41	3.05	0.44	14.82	nr	**17.87**
32 mm dia.	3.47	4.39	0.51	17.19	nr	**21.58**
40 mm dia.	4.87	6.16	0.59	19.88	nr	**26.04**
50 mm dia.	6.41	8.11	0.70	23.58	nr	**31.69**
90° elbow, male/female						
15 mm dia.	0.88	1.11	0.64	21.57	nr	**22.68**
20 mm dia.	1.19	1.50	0.85	28.64	nr	**30.14**
25 mm dia.	1.98	2.51	0.97	32.68	nr	**35.19**
32 mm dia.	3.63	4.59	1.12	37.73	nr	**42.32**
40 mm dia.	6.08	7.69	1.29	43.46	nr	**51.15**
50 mm dia.	7.82	9.90	1.55	52.21	nr	**62.11**
65 mm dia.	16.88	21.36	1.89	63.67	nr	**85.03**
80 mm dia.	23.07	29.20	2.24	75.47	nr	**104.67**
100 mm dia.	40.35	51.07	3.09	104.10	nr	**155.17**
90° elbow						
15 mm dia.	0.80	1.01	0.64	21.57	nr	**22.58**
20 mm dia.	1.09	1.38	0.85	28.64	nr	**30.02**
25 mm dia.	1.67	2.12	0.97	32.68	nr	**34.80**
32 mm dia.	3.08	3.90	1.12	37.73	nr	**41.63**
40 mm dia.	5.15	6.52	1.29	43.46	nr	**49.98**
50 mm dia.	6.05	7.66	1.55	52.21	nr	**59.87**
65 mm dia.	13.04	16.51	1.89	63.67	nr	**80.18**
80 mm dia.	19.14	24.23	2.24	75.47	nr	**99.70**
100 mm dia.	36.93	46.74	3.09	104.10	nr	**150.84**
125 mm dia.	89.83	113.69	4.44	149.58	nr	**263.27**
150 mm dia.	167.20	211.61	5.79	195.07	nr	**406.68**
45° elbow						
15 mm dia.	1.91	2.42	0.64	21.57	nr	**23.99**
20 mm dia.	2.33	2.95	0.85	28.64	nr	**31.59**
25 mm dia.	3.47	4.39	0.97	32.68	nr	**37.07**

38 MECHANICAL/COOLING/HEATING SYSTEMS

Item	Net Price £	Material £	Labour hours	Labour £	Unit	Total rate £
LOW TEMPERATURE HOT WATER HEATING; PIPELINE: SCREWED STEEL – cont						
Extra over black steel screwed pipes – cont						
45° elbow – cont						
32 mm dia.	7.09	8.97	1.12	37.73	nr	**46.70**
40 mm dia.	8.69	11.00	1.29	43.46	nr	**54.46**
50 mm dia.	11.91	15.08	1.55	52.21	nr	**67.29**
65 mm dia.	17.43	22.06	1.89	63.67	nr	**85.73**
80 mm dia.	26.18	33.13	2.24	75.47	nr	**108.60**
100 mm dia.	57.41	72.65	3.09	104.10	nr	**176.75**
150 mm dia.	160.33	202.91	5.79	195.07	nr	**397.98**
90° bend, male/female						
15 mm dia.	1.56	1.97	0.64	21.57	nr	**23.54**
20 mm dia.	2.28	2.89	0.85	28.64	nr	**31.53**
25 mm dia.	3.35	4.24	0.97	32.68	nr	**36.92**
32 mm dia.	5.33	6.74	1.12	37.73	nr	**44.47**
40 mm dia.	8.42	10.65	1.29	43.46	nr	**54.11**
50 mm dia.	14.71	18.61	1.55	52.21	nr	**70.82**
65 mm dia.	24.46	30.96	1.89	63.67	nr	**94.63**
80 mm dia.	37.64	47.63	2.24	75.47	nr	**123.10**
100 mm dia.	93.28	118.06	3.09	104.10	nr	**222.16**
90° bend, male						
15 mm dia.	3.59	4.55	0.64	21.57	nr	**26.12**
20 mm dia.	4.02	5.08	0.85	28.64	nr	**33.72**
25 mm dia.	5.89	7.46	0.97	32.68	nr	**40.14**
32 mm dia.	12.79	16.18	1.12	37.73	nr	**53.91**
40 mm dia.	17.96	22.72	1.29	43.46	nr	**66.18**
50 mm dia.	27.27	34.52	1.55	52.21	nr	**86.73**
90° bend, female						
15 mm dia.	1.43	1.81	0.64	21.57	nr	**23.38**
20 mm dia.	2.04	2.59	0.85	28.64	nr	**31.23**
25 mm dia.	2.86	3.62	0.97	32.68	nr	**36.30**
32 mm dia.	5.44	6.89	1.12	37.73	nr	**44.62**
40 mm dia.	7.24	9.16	1.29	43.46	nr	**52.62**
50 mm dia.	10.21	12.92	1.55	52.21	nr	**65.13**
65 mm dia.	21.78	27.56	1.89	63.67	nr	**91.23**
80 mm dia.	34.71	43.93	2.24	75.47	nr	**119.40**
100 mm dia.	76.10	96.31	3.09	104.10	nr	**200.41**
125 mm dia.	179.51	227.19	4.44	149.58	nr	**376.77**
150 mm dia.	326.46	413.17	5.79	195.07	nr	**608.24**
Return bend						
15 mm dia.	8.19	10.36	0.64	21.57	nr	**31.93**
20 mm dia.	13.25	16.77	0.85	28.64	nr	**45.41**
25 mm dia.	16.52	20.91	0.97	32.68	nr	**53.59**
32 mm dia.	25.62	32.42	1.12	37.73	nr	**70.15**
40 mm dia.	30.50	38.60	1.29	43.46	nr	**82.06**
50 mm dia.	46.55	58.91	1.55	52.21	nr	**111.12**

38 MECHANICAL/COOLING/HEATING SYSTEMS

Item	Net Price £	Material £	Labour hours	Labour £	Unit	Total rate £
Equal socket, parallel thread						
15 mm dia.	0.76	0.96	0.64	21.57	nr	**22.53**
20 mm dia.	0.94	1.19	0.85	28.64	nr	**29.83**
25 mm dia.	1.29	1.64	0.97	32.68	nr	**34.32**
32 mm dia.	2.59	3.28	1.12	37.73	nr	**41.01**
40 mm dia.	3.76	4.76	1.29	43.46	nr	**48.22**
50 mm dia.	5.45	6.90	1.55	52.21	nr	**59.11**
65 mm dia.	8.37	10.60	1.89	63.67	nr	**74.27**
80 mm dia.	11.88	15.03	2.24	75.47	nr	**90.50**
100 mm dia.	19.57	24.76	3.09	104.10	nr	**128.86**
Concentric reducing socket						
20 × 15 mm dia.	1.22	1.55	0.76	25.60	nr	**27.15**
25 × 15 mm dia.	1.43	1.81	0.85	28.64	nr	**30.45**
25 × 20 mm dia.	1.52	1.93	0.86	28.97	nr	**30.90**
32 × 25 mm dia.	2.85	3.61	1.01	34.04	nr	**37.65**
40 × 25 mm dia.	3.59	4.55	1.16	39.08	nr	**43.63**
40 × 32 mm dia.	3.72	4.70	1.16	39.08	nr	**43.78**
50 × 25 mm dia.	5.35	6.78	1.38	46.50	nr	**53.28**
50 × 40 mm dia.	5.21	6.60	1.38	46.50	nr	**53.10**
65 × 50 mm dia.	9.48	12.00	1.69	56.94	nr	**68.94**
80 × 50 mm dia.	12.33	15.60	2.00	67.38	nr	**82.98**
100 × 50 mm dia.	24.61	31.15	2.75	92.65	nr	**123.80**
100 × 80 mm dia.	33.93	42.94	2.75	92.65	nr	**135.59**
150 × 100 mm dia.	68.32	86.46	4.10	138.13	nr	**224.59**
Eccentric reducing socket						
20 × 15 mm dia.	2.16	2.73	0.73	24.60	nr	**27.33**
25 × 15 mm dia.	6.23	7.88	0.85	28.64	nr	**36.52**
25 × 20 mm dia.	7.05	8.93	0.85	28.64	nr	**37.57**
32 × 25 mm dia.	9.61	12.16	1.01	34.04	nr	**46.20**
40 × 25 mm dia.	11.01	13.93	1.16	39.08	nr	**53.01**
40 × 32 mm dia.	5.53	7.00	1.16	39.08	nr	**46.08**
50 × 25 mm dia.	7.14	9.04	1.38	46.50	nr	**55.54**
50 × 40 mm dia.	9.85	12.47	1.38	46.50	nr	**58.97**
65 × 50 mm dia.	11.45	14.49	1.69	56.94	nr	**71.43**
80 × 50 mm dia.	20.04	25.37	2.00	67.38	nr	**92.75**
Hexagon bush						
20 × 15 mm dia.	0.69	0.87	0.37	12.47	nr	**13.34**
25 × 15 mm dia.	0.84	1.06	0.43	14.49	nr	**15.55**
25 × 20 mm dia.	0.88	1.11	0.43	14.49	nr	**15.60**
32 × 25 mm dia.	1.13	1.43	0.51	17.19	nr	**18.62**
40 × 25 mm dia.	1.80	2.27	0.58	19.54	nr	**21.81**
40 × 32 mm dia.	1.65	2.08	0.58	19.54	nr	**21.62**
50 × 25 mm dia.	3.72	4.70	0.71	23.92	nr	**28.62**
50 × 40 mm dia.	3.45	4.37	0.71	23.92	nr	**28.29**
65 × 50 mm dia.	5.76	7.29	0.85	28.64	nr	**35.93**
80 × 50 mm dia.	9.34	11.82	1.00	33.69	nr	**45.51**
100 × 50 mm dia.	21.61	27.35	1.52	51.21	nr	**78.56**
100 × 80 mm dia.	29.82	37.74	1.52	51.21	nr	**88.95**
150 × 100 mm dia.	64.66	81.84	2.57	86.58	nr	**168.42**

38 MECHANICAL/COOLING/HEATING SYSTEMS

Item	Net Price £	Material £	Labour hours	Labour £	Unit	Total rate £
LOW TEMPERATURE HOT WATER HEATING; PIPELINE: SCREWED STEEL – cont						
Extra over black steel screwed pipes – cont						
Hexagon nipple						
15 mm dia.	0.74	0.94	0.28	9.43	nr	**10.37**
20 mm dia.	0.84	1.06	0.38	12.81	nr	**13.87**
25 mm dia.	1.19	1.50	0.44	14.82	nr	**16.32**
32 mm dia.	2.46	3.11	0.51	17.19	nr	**20.30**
40 mm dia.	2.85	3.61	0.59	19.88	nr	**23.49**
50 mm dia.	5.18	6.55	0.70	23.58	nr	**30.13**
65 mm dia.	8.44	10.68	0.85	28.64	nr	**39.32**
80 mm dia.	12.20	15.44	1.00	33.69	nr	**49.13**
100 mm dia.	20.70	26.20	1.44	48.51	nr	**74.71**
150 mm dia.	46.21	58.49	2.32	78.16	nr	**136.65**
Union, male/female						
15 mm dia.	3.89	4.93	0.64	21.57	nr	**26.50**
20 mm dia.	4.76	6.03	0.85	28.64	nr	**34.67**
25 mm dia.	5.95	7.53	0.97	32.68	nr	**40.21**
32 mm dia.	9.17	11.60	1.12	37.73	nr	**49.33**
40 mm dia.	11.73	14.84	1.29	43.46	nr	**58.30**
50 mm dia.	18.49	23.40	1.55	52.21	nr	**75.61**
65 mm dia.	35.28	44.65	1.89	63.67	nr	**108.32**
80 mm dia.	48.51	61.40	2.24	75.47	nr	**136.87**
Union, female						
15 mm dia.	3.20	4.05	0.64	21.57	nr	**25.62**
20 mm dia.	3.47	4.39	0.85	28.64	nr	**33.03**
25 mm dia.	4.04	5.12	0.97	32.68	nr	**37.80**
32 mm dia.	7.61	9.63	1.12	37.73	nr	**47.36**
40 mm dia.	8.61	10.90	1.29	43.46	nr	**54.36**
50 mm dia.	14.26	18.04	1.55	52.21	nr	**70.25**
65 mm dia.	30.75	38.92	1.89	63.67	nr	**102.59**
80 mm dia.	42.46	53.74	2.24	75.47	nr	**129.21**
100 mm dia.	80.83	102.30	3.09	104.10	nr	**206.40**
Union elbow, male/female						
15 mm dia.	5.63	7.12	0.55	18.52	nr	**25.64**
20 mm dia.	7.06	8.94	0.85	28.64	nr	**37.58**
25 mm dia.	9.90	12.53	0.97	32.68	nr	**45.21**
Twin elbow						
15 mm dia.	4.56	5.77	0.91	30.65	nr	**36.42**
20 mm dia.	5.06	6.41	1.22	41.10	nr	**47.51**
25 mm dia.	8.17	10.34	1.39	46.83	nr	**57.17**
32 mm dia.	16.77	21.22	1.62	54.58	nr	**75.80**
40 mm dia.	21.23	26.87	1.86	62.66	nr	**89.53**
50 mm dia.	27.30	34.55	2.21	74.46	nr	**109.01**
65 mm dia.	48.49	61.36	2.72	91.64	nr	**153.00**
80 mm dia.	82.64	104.59	3.21	108.15	nr	**212.74**

38 MECHANICAL/COOLING/HEATING SYSTEMS

Item	Net Price £	Material £	Labour hours	Labour £	Unit	Total rate £
Equal tee						
15 mm dia.	1.09	1.38	0.91	30.65	nr	**32.03**
20 mm dia.	1.59	2.02	1.22	41.10	nr	**43.12**
25 mm dia.	2.29	2.90	1.39	46.83	nr	**49.73**
32 mm dia.	4.16	5.26	1.62	54.58	nr	**59.84**
40 mm dia.	6.41	8.11	1.86	62.66	nr	**70.77**
50 mm dia.	9.22	11.67	2.21	74.46	nr	**86.13**
65 mm dia.	20.19	25.55	2.72	91.64	nr	**117.19**
80 mm dia.	24.61	31.15	3.21	108.15	nr	**139.30**
100 mm dia.	44.59	56.44	4.44	149.58	nr	**206.02**
125 mm dia.	129.50	163.90	5.38	181.26	nr	**345.16**
150 mm dia.	206.33	261.13	6.31	212.59	nr	**473.72**
Tee reducing on branch						
20 × 15 mm dia.	1.43	1.81	1.22	41.10	nr	**42.91**
25 × 15 mm dia.	1.98	2.51	1.39	46.83	nr	**49.34**
25 × 20 mm dia.	2.35	2.98	1.39	46.83	nr	**49.81**
32 × 25 mm dia.	4.09	5.17	1.62	54.58	nr	**59.75**
40 × 25 mm dia.	5.41	6.84	1.86	62.66	nr	**69.50**
40 × 32 mm dia.	7.11	8.99	1.86	62.66	nr	**71.65**
50 × 25 mm dia.	8.04	10.18	2.21	74.46	nr	**84.64**
50 × 40 mm dia.	12.00	15.19	2.21	74.46	nr	**89.65**
65 × 50 mm dia.	18.74	23.72	2.72	91.64	nr	**115.36**
80 × 50 mm dia.	25.32	32.04	3.21	108.15	nr	**140.19**
100 × 50 mm dia.	41.93	53.07	4.44	149.58	nr	**202.65**
100 × 80 mm dia.	64.66	81.84	4.44	149.58	nr	**231.42**
150 × 100 mm dia.	151.99	192.36	6.31	212.59	nr	**404.95**
Equal pitcher tee						
15 mm dia.	3.33	4.21	0.91	30.65	nr	**34.86**
20 mm dia.	4.14	5.24	1.22	41.10	nr	**46.34**
25 mm dia.	6.23	7.88	1.39	46.83	nr	**54.71**
32 mm dia.	9.45	11.96	1.62	54.58	nr	**66.54**
40 mm dia.	14.57	18.44	1.86	62.66	nr	**81.10**
50 mm dia.	20.47	25.91	2.21	74.46	nr	**100.37**
65 mm dia.	31.65	40.05	2.72	91.64	nr	**131.69**
80 mm dia.	49.41	62.53	3.21	108.15	nr	**170.68**
100 mm dia.	111.14	140.66	4.44	149.58	nr	**290.24**
Cross						
15 mm dia.	3.25	4.11	1.00	33.69	nr	**37.80**
20 mm dia.	4.89	6.19	1.33	44.80	nr	**50.99**
25 mm dia.	6.23	7.88	1.51	50.88	nr	**58.76**
32 mm dia.	9.05	11.46	1.77	59.64	nr	**71.10**
40 mm dia.	11.65	14.74	2.02	68.05	nr	**82.79**
50 mm dia.	18.93	23.96	2.42	81.52	nr	**105.48**
65 mm dia.	26.18	33.13	2.97	100.06	nr	**133.19**
80 mm dia.	34.81	44.06	3.50	117.91	nr	**161.97**
100 mm dia.	71.85	90.93	4.84	163.06	nr	**253.99**

38 MECHANICAL/COOLING/HEATING SYSTEMS

Item	Net Price £	Material £	Labour hours	Labour £	Unit	Total rate £
LOW TEMPERATURE HOT WATER HEATING; PIPELINE: BLACK WELDED STEEL						
Black steel pipes; butt welded joints; BS 1387: 1985; including protective painting. Fixed Vertical with brackets measured separately (Refer to Screwed Steel Section). Welded butt joints are within the running length, but any flanges are additional						
Medium weight						
10 mm dia.	3.07	3.89	0.37	12.47	m	**16.36**
15 mm dia.	3.27	4.14	0.37	12.47	m	**16.61**
20 mm dia.	3.85	4.87	0.37	12.47	m	**17.34**
25 mm dia.	5.54	7.01	0.41	13.81	m	**20.82**
32 mm dia.	6.85	8.67	0.48	16.17	m	**24.84**
40 mm dia.	7.96	10.07	0.52	17.52	m	**27.59**
50 mm dia.	11.19	14.16	0.62	20.89	m	**35.05**
65 mm dia.	15.19	19.22	0.64	21.57	m	**40.79**
80 mm dia.	19.75	25.00	1.10	37.06	m	**62.06**
100 mm dia.	27.96	35.38	1.31	44.13	m	**79.51**
125 mm dia.	35.59	45.05	1.66	55.92	m	**100.97**
150 mm dia.	41.32	52.29	1.88	63.35	m	**115.64**
Heavy weight						
15 mm dia.	3.89	4.93	0.37	12.47	m	**17.40**
20 mm dia.	4.61	5.84	0.37	12.47	m	**18.31**
25 mm dia.	6.76	8.56	0.41	13.81	m	**22.37**
32 mm dia.	8.39	10.62	0.48	16.17	m	**26.79**
40 mm dia.	9.79	12.39	0.52	17.52	m	**29.91**
50 mm dia.	13.57	17.17	0.62	20.89	m	**38.06**
65 mm dia.	18.46	23.36	0.64	21.57	m	**44.93**
80 mm dia.	23.50	29.75	1.10	37.06	m	**66.81**
100 mm dia.	32.80	41.51	1.31	44.13	m	**85.64**
125 mm dia.	37.96	48.04	1.66	55.92	m	**103.96**
150 mm dia.	44.36	56.15	1.88	63.35	m	**119.50**
Black steel pipes; butt welded joints; BS 1387: 1985; including protective painting. Fixed at High Level or Suspended with brackets measured separately (Refer to Screwed Steel Section). Welded butt joints are within the running length, but any flanges are additional						
Medium weight						
10 mm dia.	3.07	3.89	0.58	19.54	m	**23.43**
15 mm dia.	3.27	4.14	0.58	19.54	m	**23.68**
20 mm dia.	3.85	4.87	0.58	19.54	m	**24.41**
25 mm dia.	5.54	7.01	0.60	20.22	m	**27.23**
32 mm dia.	6.85	8.67	0.68	22.90	m	**31.57**
40 mm dia.	7.96	10.07	0.73	24.60	m	**34.67**
50 mm dia.	11.19	14.16	0.85	28.64	m	**42.80**

38 MECHANICAL/COOLING/HEATING SYSTEMS

Item	Net Price £	Material £	Labour hours	Labour £	Unit	Total rate £
65 mm dia.	15.19	19.22	0.88	29.66	m	**48.88**
80 mm dia.	19.75	25.00	1.45	48.85	m	**73.85**
100 mm dia.	27.96	35.38	1.74	58.62	m	**94.00**
125 mm dia.	35.59	45.05	2.21	74.46	m	**119.51**
150 mm dia.	41.32	52.29	2.50	84.22	m	**136.51**
Heavy weight						
15 mm dia.	3.89	4.93	0.58	19.54	m	**24.47**
20 mm dia.	4.61	5.84	0.58	19.54	m	**25.38**
25 mm dia.	6.76	8.56	0.60	20.22	m	**28.78**
32 mm dia.	8.39	10.62	0.68	22.90	m	**33.52**
40 mm dia.	9.79	12.39	0.73	24.60	m	**36.99**
50 mm dia.	13.57	17.17	0.85	28.64	m	**45.81**
65 mm dia.	18.46	23.36	0.88	29.66	m	**53.02**
80 mm dia.	23.50	29.75	1.45	48.85	m	**78.60**
100 mm dia.	32.80	41.51	1.74	58.62	m	**100.13**
125 mm dia.	37.96	48.04	2.21	74.46	m	**122.50**
150 mm dia.	44.36	56.15	2.50	84.22	m	**140.37**

Fixings
Refer to steel pipes; black malleable iron. For minimum fixing distances, refer to the Tables and Memoranda to the rear of the book

Extra over black steel butt welded pipes; black steel flanges, welded and drilled; metric; BS 4504

Item	Net Price £	Material £	Labour hours	Labour £	Unit	Total rate £
Welded flanges; PN6						
15 mm dia.	4.06	5.14	0.59	19.88	nr	**25.02**
20 mm dia.	4.06	5.14	0.69	23.25	nr	**28.39**
25 mm dia.	4.06	5.14	0.84	28.30	nr	**33.44**
32 mm dia.	6.73	8.51	1.00	33.69	nr	**42.20**
40 mm dia.	7.19	9.09	1.11	37.40	nr	**46.49**
50 mm dia.	7.39	9.35	1.37	46.16	nr	**55.51**
65 mm dia.	10.86	13.74	1.54	51.88	nr	**65.62**
80 mm dia.	16.32	20.65	1.67	56.27	nr	**76.92**
100 mm dia.	17.82	22.56	2.22	74.79	nr	**97.35**
125 mm dia.	23.35	29.56	2.61	87.93	nr	**117.49**
150 mm dia.	23.60	29.87	2.99	100.73	nr	**130.60**
Welded flanges; PN16						
15 mm dia.	8.13	10.29	0.59	19.88	nr	**30.17**
20 mm dia.	8.13	10.29	0.69	23.25	nr	**33.54**
25 mm dia.	8.13	10.29	0.84	28.30	nr	**38.59**
32 mm dia.	10.96	13.87	1.00	33.69	nr	**47.56**
40 mm dia.	10.96	13.87	1.11	37.40	nr	**51.27**
50 mm dia.	14.40	18.22	1.37	46.16	nr	**64.38**
65 mm dia.	16.62	21.03	1.54	51.88	nr	**72.91**
80 mm dia.	21.13	26.75	1.67	56.27	nr	**83.02**
100 mm dia.	21.57	27.29	2.22	74.79	nr	**102.08**
125 mm dia.	34.23	43.32	2.61	87.93	nr	**131.25**
150 mm dia.	40.87	51.72	2.99	100.73	nr	**152.45**

38 MECHANICAL/COOLING/HEATING SYSTEMS

Item	Net Price £	Material £	Labour hours	Labour £	Unit	Total rate £
LOW TEMPERATURE HOT WATER HEATING; PIPELINE: BLACK WELDED STEEL – cont						
Extra over black steel butt welded pipes – cont						
Blank flanges, slip on for welding; PN6						
15 mm dia.	2.54	3.21	0.48	16.17	nr	19.38
20 mm dia.	2.54	3.21	0.55	18.52	nr	21.73
25 mm dia.	2.54	3.21	0.64	21.57	nr	24.78
32 mm dia.	5.22	6.61	0.76	25.60	nr	32.21
40 mm dia.	5.22	6.61	0.84	28.30	nr	34.91
50 mm dia.	4.75	6.01	1.01	34.04	nr	40.05
65 mm dia.	9.41	11.91	1.30	43.80	nr	55.71
80 mm dia.	9.82	12.43	1.41	47.50	nr	59.93
100 mm dia.	10.09	12.77	1.78	59.96	nr	72.73
125 mm dia.	20.15	25.50	2.06	69.41	nr	94.91
150 mm dia.	19.07	24.14	2.35	79.17	nr	103.31
Blank flanges, slip on for welding; PN16						
15 mm dia.	2.31	2.92	0.48	16.17	nr	19.09
20 mm dia.	3.54	4.48	0.55	18.52	nr	23.00
25 mm dia.	4.31	5.45	0.64	21.57	nr	27.02
32 mm dia.	4.41	5.58	0.76	25.60	nr	31.18
40 mm dia.	5.71	7.22	0.84	28.30	nr	35.52
50 mm dia.	8.85	11.20	1.01	34.04	nr	45.24
65 mm dia.	10.86	13.74	1.30	43.80	nr	57.54
80 mm dia.	14.09	17.83	1.41	47.50	nr	65.33
100 mm dia.	17.46	22.10	1.78	59.96	nr	82.06
125 mm dia.	25.63	32.44	2.06	69.41	nr	101.85
150 mm dia.	33.07	41.85	2.35	79.17	nr	121.02
Extra over black steel butt welded pipes; black steel flanges, welding and drilled; imperial; BS 10						
Welded flanges; Table E						
15 mm dia.	8.34	10.55	0.59	19.88	nr	30.43
20 mm dia.	8.34	10.55	0.69	23.25	nr	33.80
25 mm dia.	8.34	10.55	0.84	28.30	nr	38.85
32 mm dia.	8.34	10.55	1.00	33.69	nr	44.24
40 mm dia.	8.56	10.83	1.11	37.40	nr	48.23
50 mm dia.	10.95	13.85	1.37	46.16	nr	60.01
65 mm dia.	13.32	16.86	1.54	51.88	nr	68.74
80 mm dia.	17.51	22.16	1.67	56.27	nr	78.43
100 mm dia.	25.86	32.73	2.22	74.79	nr	107.52
125 mm dia.	50.32	63.68	2.61	87.93	nr	151.61
150 mm dia.	50.32	63.68	2.99	100.73	nr	164.41
Blank flanges, slip on for welding; Table E						
15 mm dia.	5.74	7.27	0.48	16.17	nr	23.44
20 mm dia.	5.74	7.27	0.55	18.52	nr	25.79
25 mm dia.	5.74	7.27	0.64	21.57	nr	28.84

38 MECHANICAL/COOLING/HEATING SYSTEMS

Item	Net Price £	Material £	Labour hours	Labour £	Unit	Total rate £
32 mm dia.	7.21	9.13	0.76	25.60	nr	34.73
40 mm dia.	7.85	9.93	0.84	28.30	nr	38.23
50 mm dia.	8.66	10.96	1.01	34.04	nr	45.00
65 mm dia.	9.99	12.64	1.30	43.80	nr	56.44
80 mm dia.	21.79	27.57	1.41	47.50	nr	75.07
100 mm dia.	24.75	31.33	1.78	59.96	nr	91.29
125 mm dia.	40.71	51.52	2.06	69.41	nr	120.93
150 mm dia.	52.34	66.24	2.35	79.17	nr	145.41
Extra over black steel butt welded pipes; black steel flange connections						
Bolted connection between pair of flanges; including gasket, bolts, nuts and washers						
50 mm dia.	8.96	11.33	0.50	16.84	nr	28.17
65 mm dia.	10.62	13.44	0.50	16.84	nr	30.28
80 mm dia.	15.93	20.16	0.50	16.84	nr	37.00
100 mm dia.	17.30	21.90	0.50	16.84	nr	38.74
125 mm dia.	20.36	25.77	0.50	16.84	nr	42.61
150 mm dia.	29.09	36.81	0.88	29.66	nr	66.47
Extra over fittings; BS 1965; butt welded						
Cap						
25 mm dia.	19.66	24.89	0.47	15.84	nr	40.73
32 mm dia.	19.66	24.89	0.59	19.88	nr	44.77
40 mm dia.	19.78	25.03	0.70	23.58	nr	48.61
50 mm dia.	23.12	29.27	0.99	33.35	nr	62.62
65 mm dia.	27.17	34.38	1.35	45.48	nr	79.86
80 mm dia.	27.66	35.01	1.66	55.92	nr	90.93
100 mm dia.	36.15	45.75	2.23	75.13	nr	120.88
125 mm dia.	50.84	64.34	3.03	102.09	nr	166.43
150 mm dia.	58.81	74.44	3.79	127.69	nr	202.13
Concentric reducer						
20 × 15 mm dia.	10.95	13.85	0.69	23.25	nr	37.10
25 × 15 mm dia.	13.67	17.30	0.87	29.31	nr	46.61
25 × 20 mm dia.	10.57	13.37	0.87	29.31	nr	42.68
32 × 25 mm dia.	14.56	18.42	1.08	36.39	nr	54.81
40 × 25 mm dia.	18.99	24.04	1.38	46.50	nr	70.54
40 × 32 mm dia.	13.01	16.46	1.38	46.50	nr	62.96
50 × 25 mm dia.	18.06	22.86	1.82	61.32	nr	84.18
50 × 40 mm dia.	12.79	16.18	1.82	61.32	nr	77.50
65 × 50 mm dia.	17.66	22.36	2.52	84.90	nr	107.26
80 × 50 mm dia.	17.96	22.72	3.24	109.16	nr	131.88
100 × 50 mm dia.	29.76	37.67	4.08	137.46	nr	175.13
100 × 80 mm dia.	20.47	25.91	4.08	137.46	nr	163.37
125 × 80 mm dia.	44.71	56.58	4.71	158.68	nr	215.26
150 × 100 mm dia.	48.36	61.21	5.33	179.56	nr	240.77
Eccentric reducer						
20 × 15 mm dia.	16.29	20.62	0.69	23.25	nr	43.87
25 × 15 mm dia.	21.32	26.98	0.87	29.31	nr	56.29
25 × 20 mm dia.	17.90	22.66	0.87	29.31	nr	51.97

38 MECHANICAL/COOLING/HEATING SYSTEMS

Item	Net Price £	Material £	Labour hours	Labour £	Unit	Total rate £
LOW TEMPERATURE HOT WATER HEATING; PIPELINE: BLACK WELDED STEEL – cont						
Extra over fittings – cont						
Eccentric reducer – cont						
32 × 25 mm dia.	19.78	25.03	1.08	36.39	nr	**61.42**
40 × 25 mm dia.	24.42	30.90	1.38	46.50	nr	**77.40**
40 × 32 mm dia.	23.32	29.51	1.38	46.50	nr	**76.01**
50 × 25 mm dia.	27.83	35.22	1.82	61.32	nr	**96.54**
50 × 40 mm dia.	20.05	25.38	1.82	61.32	nr	**86.70**
65 × 50 mm dia.	23.66	29.95	2.52	84.90	nr	**114.85**
80 × 50 mm dia.	28.96	36.65	3.24	109.16	nr	**145.81**
100 × 50 mm dia.	49.07	62.10	4.08	137.46	nr	**199.56**
100 × 80 mm dia.	36.20	45.82	4.08	137.46	nr	**183.28**
125 × 80 mm dia.	96.71	122.39	4.71	158.68	nr	**281.07**
150 × 100 mm dia.	73.23	92.68	5.33	179.56	nr	**272.24**
45° elbow, long radius						
15 mm dia.	4.11	5.20	0.56	18.87	nr	**24.07**
20 mm dia.	4.16	5.26	0.75	25.26	nr	**30.52**
25 mm dia.	5.41	6.84	0.93	31.34	nr	**38.18**
32 mm dia.	6.57	8.31	1.17	39.42	nr	**47.73**
40 mm dia.	6.65	8.41	1.46	49.19	nr	**57.60**
50 mm dia.	9.09	11.50	1.97	66.37	nr	**77.87**
65 mm dia.	11.71	14.82	2.70	90.96	nr	**105.78**
80 mm dia.	10.83	13.71	3.32	111.85	nr	**125.56**
100 mm dia.	16.37	20.72	4.09	137.79	nr	**158.51**
125 mm dia.	34.81	44.06	4.94	166.42	nr	**210.48**
150 mm dia.	43.25	54.73	5.78	194.72	nr	**249.45**
90° elbow, long radius						
15 mm dia.	4.25	5.38	0.56	18.87	nr	**24.25**
20 mm dia.	4.15	5.25	0.75	25.26	nr	**30.51**
25 mm dia.	5.52	6.99	0.93	31.34	nr	**38.33**
32 mm dia.	6.59	8.34	1.17	39.42	nr	**47.76**
40 mm dia.	6.66	8.43	1.46	49.19	nr	**57.62**
50 mm dia.	9.09	11.50	1.97	66.37	nr	**77.87**
65 mm dia.	11.71	14.82	2.70	90.96	nr	**105.78**
80 mm dia.	12.72	16.09	3.32	111.85	nr	**127.94**
100 mm dia.	19.26	24.37	4.09	137.79	nr	**162.16**
125 mm dia.	40.08	50.72	4.94	166.42	nr	**217.14**
150 mm dia.	50.86	64.37	5.78	194.72	nr	**259.09**
Branch bend (based on branch and pipe sizes being the same)						
15 mm dia.	16.93	21.43	0.85	28.64	nr	**50.07**
20 mm dia.	16.78	21.24	0.85	28.64	nr	**49.88**
25 mm dia.	16.95	21.45	1.02	34.36	nr	**55.81**
32 mm dia.	16.02	20.27	1.11	37.40	nr	**57.67**
40 mm dia.	15.81	20.01	1.36	45.82	nr	**65.83**
50 mm dia.	15.36	19.44	1.70	57.27	nr	**76.71**
65 mm dia.	22.71	28.74	1.78	59.96	nr	**88.70**

38 MECHANICAL/COOLING/HEATING SYSTEMS

Item	Net Price £	Material £	Labour hours	Labour £	Unit	Total rate £
80 mm dia.	35.48	44.90	1.82	61.32	nr	106.22
100 mm dia.	46.28	58.58	1.87	63.00	nr	121.58
125 mm dia.	81.55	103.21	2.21	74.46	nr	177.67
150 mm dia.	125.40	158.70	2.65	89.28	nr	247.98
Equal tee						
15 mm dia.	40.13	50.79	0.82	27.63	nr	78.42
20 mm dia.	40.13	50.79	1.10	37.06	nr	87.85
25 mm dia.	40.13	50.79	1.35	45.48	nr	96.27
32 mm dia.	40.13	50.79	1.63	54.91	nr	105.70
40 mm dia.	40.13	50.79	2.14	72.11	nr	122.90
50 mm dia.	42.27	53.50	3.02	101.74	nr	155.24
65 mm dia.	61.67	78.05	3.61	121.62	nr	199.67
80 mm dia.	61.79	78.20	4.18	140.83	nr	219.03
100 mm dia.	76.85	97.26	5.24	176.53	nr	273.79
125 mm dia.	175.90	222.62	6.70	225.72	nr	448.34
150 mm dia.	192.16	243.20	8.45	284.68	nr	527.88
Extra over black steel butt welded pipes; labour						
Made bend						
15 mm dia.	–	–	0.42	14.15	nr	14.15
20 mm dia.	–	–	0.42	14.15	nr	14.15
25 mm dia.	–	–	0.50	16.84	nr	16.84
32 mm dia.	–	–	0.62	20.89	nr	20.89
40 mm dia.	–	–	0.74	24.93	nr	24.93
50 mm dia.	–	–	0.89	29.98	nr	29.98
65 mm dia.	–	–	1.05	35.37	nr	35.37
80 mm dia.	–	–	1.13	38.07	nr	38.07
100 mm dia.	–	–	2.90	97.71	nr	97.71
125 mm dia.	–	–	3.56	119.94	nr	119.94
150 mm dia.	–	–	4.18	140.83	nr	140.83
Splay cut end						
15 mm dia.	–	–	0.14	4.72	nr	4.72
20 mm dia.	–	–	0.16	5.39	nr	5.39
25 mm dia.	–	–	0.18	6.06	nr	6.06
32 mm dia.	–	–	0.25	8.43	nr	8.43
40 mm dia.	–	–	0.27	9.09	nr	9.09
50 mm dia.	–	–	0.31	10.44	nr	10.44
65 mm dia.	–	–	0.35	11.79	nr	11.79
80 mm dia.	–	–	0.40	13.47	nr	13.47
100 mm dia.	–	–	0.48	16.17	nr	16.17
125 mm dia.	–	–	0.56	18.87	nr	18.87
150 mm dia.	–	–	0.64	21.57	nr	21.57
Screwed joint to fitting						
15 mm dia.	–	–	0.30	10.11	nr	10.11
20 mm dia.	–	–	0.40	13.47	nr	13.47
25 mm dia.	–	–	0.46	15.50	nr	15.50
32 mm dia.	–	–	0.53	17.85	nr	17.85
40 mm dia.	–	–	0.61	20.55	nr	20.55
50 mm dia.	–	–	0.73	24.60	nr	24.60
65 mm dia.	–	–	0.89	29.98	nr	29.98

38 MECHANICAL/COOLING/HEATING SYSTEMS

Item	Net Price £	Material £	Labour hours	Labour £	Unit	Total rate £
LOW TEMPERATURE HOT WATER HEATING; PIPELINE: BLACK WELDED STEEL – cont						
Extra over black steel butt welded pipes – cont						
Screwed joint to fitting – cont						
80 mm dia.	–	–	1.05	35.37	nr	**35.37**
100 mm dia.	–	–	1.46	49.19	nr	**49.19**
125 mm dia.	–	–	2.10	70.75	nr	**70.75**
150 mm dia.	–	–	2.73	91.97	nr	**91.97**
Straight butt weld						
15 mm dia.	–	–	0.31	10.44	nr	**10.44**
20 mm dia.	–	–	0.42	14.15	nr	**14.15**
25 mm dia.	–	–	0.52	17.52	nr	**17.52**
32 mm dia.	–	–	0.69	23.25	nr	**23.25**
40 mm dia.	–	–	0.83	27.96	nr	**27.96**
50 mm dia.	–	–	1.22	41.10	nr	**41.10**
65 mm dia.	–	–	1.57	52.89	nr	**52.89**
80 mm dia.	–	–	1.95	65.70	nr	**65.70**
100 mm dia.	–	–	2.38	80.19	nr	**80.19**
125 mm dia.	–	–	2.83	95.33	nr	**95.33**
150 mm dia.	–	–	3.27	110.17	nr	**110.17**

38 MECHANICAL/COOLING/HEATING SYSTEMS

Item	Net Price £	Material £	Labour hours	Labour £	Unit	Total rate £
LOW TEMPERATURE HOT WATER HEATING; PIPELINE: CARBON WELDED STEEL						
Hot finished seamless carbon steel pipe; BS 806 and BS 3601; wall thickness to BS 3600; butt welded joints; including protective painting, fixed vertically or at low level, brackets measured separately (Refer to Screwed Pipework Section). Welded butt joints are within the running length, but any flanges are additional						
Pipework						
200 mm dia.	101.18	128.05	2.04	68.72	m	196.77
250 mm dia.	91.09	115.28	2.59	87.26	m	202.54
300 mm dia.	214.15	271.03	2.99	100.73	m	371.76
350 mm dia.	226.01	286.04	3.52	118.59	m	404.63
400 mm dia.	394.05	498.71	4.08	137.46	m	636.17
Hot finished seamless carbon steel pipe; BS 806 and BS 3601; wall thickness to BS 3600; butt welded joints; including protective painting, fixed at high level or suspended, brackets measured separately (Refer to Screwed Pipework Section). Welded butt joints are within the running length, but any flanges are additional						
Pipework						
200 mm dia.	101.18	128.05	3.70	124.64	m	252.69
250 mm dia.	91.09	115.28	4.73	159.35	m	274.63
300 mm dia.	214.15	271.03	5.65	190.34	m	461.37
350 mm dia.	226.01	286.04	6.68	225.05	m	511.09
400 mm dia.	394.05	498.71	7.70	259.41	m	758.12
Fixings						
Refer to steel pipes; black malleable iron. For minimum fixing distances, refer to the Tables and Memoranda to the rear of the book						
Extra over fittings; BS 1965 part 1; butt welded						
Cap						
200 mm dia.	82.83	104.83	3.70	124.64	nr	229.47
250 mm dia.	158.96	201.17	4.73	159.35	nr	360.52
300 mm dia.	173.85	220.02	5.65	190.34	nr	410.36
350 mm dia.	267.85	338.99	6.68	225.05	nr	564.04
400 mm dia.	309.82	392.11	7.70	259.41	nr	651.52
Concentric reducer						
200 mm × 150 mm dia.	90.49	114.52	7.27	244.93	nr	359.45
250 mm × 150 mm dia.	131.22	166.07	9.05	304.90	nr	470.97
250 mm × 200 mm dia.	80.31	101.64	9.10	306.58	nr	408.22

38 MECHANICAL/COOLING/HEATING SYSTEMS

Item	Net Price £	Material £	Labour hours	Labour £	Unit	Total rate £
LOW TEMPERATURE HOT WATER HEATING; PIPELINE: CARBON WELDED STEEL – cont						
Extra over fittings – cont						
Concentric reducer – cont						
300 mm × 150 mm dia.	151.57	191.82	10.75	362.17	nr	**553.99**
300 mm × 200 mm dia.	156.97	198.67	10.75	362.17	nr	**560.84**
300 mm × 250 mm dia.	139.54	176.60	11.15	375.65	nr	**552.25**
350 mm × 200 mm dia.	238.04	301.27	12.50	421.13	nr	**722.40**
350 mm × 250 mm dia.	199.57	252.57	12.70	427.86	nr	**680.43**
350 mm × 300 mm dia.	199.57	252.57	13.00	437.98	nr	**690.55**
400 mm × 250 mm dia.	291.82	369.33	14.46	487.17	nr	**856.50**
400 mm × 300 mm dia.	361.39	457.37	14.51	488.85	nr	**946.22**
400 mm × 350 mm dia.	314.99	398.65	15.16	510.74	nr	**909.39**
Eccentric reducer						
200 mm × 150 mm dia.	163.43	206.84	7.27	244.93	nr	**451.77**
250 mm × 150 mm dia.	217.77	275.61	9.05	304.90	nr	**580.51**
250 mm × 200 mm dia.	142.49	180.33	9.10	306.58	nr	**486.91**
300 mm × 150 mm dia.	261.25	330.64	10.75	362.17	nr	**692.81**
300 mm × 200 mm dia.	301.32	381.35	10.75	362.17	nr	**743.52**
300 mm × 250 mm dia.	242.55	306.97	11.15	375.65	nr	**682.62**
350 mm × 200 mm dia.	276.75	350.26	12.50	421.13	nr	**771.39**
350 mm × 250 mm dia.	230.08	291.19	12.70	427.86	nr	**719.05**
350 mm × 300 mm dia.	230.08	291.19	13.00	437.98	nr	**729.17**
400 mm × 250 mm dia.	369.52	467.67	14.46	487.17	nr	**954.84**
400 mm × 300 mm dia.	346.04	437.95	14.51	488.85	nr	**926.80**
400 mm × 350 mm dia.	426.37	539.62	15.16	510.74	nr	**1050.36**
45° elbow						
200 mm dia.	86.55	109.54	7.75	261.11	nr	**370.65**
250 mm dia.	162.54	205.71	10.05	338.59	nr	**544.30**
300 mm dia.	237.49	300.56	12.20	411.02	nr	**711.58**
350 mm dia.	192.20	243.25	14.65	493.56	nr	**736.81**
400 mm dia.	319.43	404.28	17.12	576.77	nr	**981.05**
90° elbow						
200 mm dia.	101.83	128.88	7.75	261.11	nr	**389.99**
250 mm dia.	191.27	242.08	10.05	338.59	nr	**580.67**
300 mm dia.	279.42	353.63	12.20	411.02	nr	**764.65**
350 mm dia.	384.12	486.15	14.65	493.56	nr	**979.71**
400 mm dia.	490.64	620.95	17.12	576.77	nr	**1197.72**
Equal tee						
200 mm dia.	268.99	340.44	11.25	379.02	nr	**719.46**
250 mm dia.	461.25	583.76	14.53	489.52	nr	**1073.28**
300 mm dia.	572.43	724.47	17.55	591.26	nr	**1315.73**
350 mm dia.	768.00	971.98	20.98	706.82	nr	**1678.80**
400 mm dia.	877.71	1110.83	24.38	821.37	nr	**1932.20**

38 MECHANICAL/COOLING/HEATING SYSTEMS

Item	Net Price £	Material £	Labour hours	Labour £	Unit	Total rate £
Extra over black steel butt welded pipes; labour						
Straight butt weld						
200 mm dia.	–	–	4.08	137.46	nr	**137.46**
250 mm dia.	–	–	5.20	175.18	nr	**175.18**
300 mm dia.	–	–	6.22	209.56	nr	**209.56**
350 mm dia.	–	–	7.33	246.95	nr	**246.95**
400 mm dia.	–	–	8.41	283.33	nr	**283.33**
Branch weld						
100 mm dia.	–	–	3.46	116.57	nr	**116.57**
125 mm dia.	–	–	4.23	142.51	nr	**142.51**
150 mm dia.	–	–	5.00	168.45	nr	**168.45**
Extra over black steel butt welded pipes; black steel flanges, welding and drilled; metric; BS 4504						
Welded flanges; PN16						
200 mm dia.	55.38	70.09	4.10	138.13	nr	**208.22**
250 mm dia.	79.55	100.68	5.33	179.56	nr	**280.24**
300 mm dia.	111.44	141.04	6.40	215.62	nr	**356.66**
350 mm dia.	215.53	272.78	7.43	250.32	nr	**523.10**
400 mm dia.	284.57	360.15	8.45	284.68	nr	**644.83**
Welded flanges; PN25						
200 mm dia.	183.90	232.75	4.10	138.13	nr	**370.88**
250 mm dia.	220.44	278.99	5.33	179.56	nr	**458.55**
300 mm dia.	298.08	377.25	6.40	215.62	nr	**592.87**
Blank flanges, slip on for welding; PN16						
200 mm dia.	128.06	162.08	2.70	90.96	nr	**253.04**
250 mm dia.	194.63	246.32	3.48	117.24	nr	**363.56**
300 mm dia.	207.95	263.18	4.20	141.49	nr	**404.67**
350 mm dia.	325.44	411.88	4.78	161.03	nr	**572.91**
400 mm dia.	430.31	544.60	5.35	180.24	nr	**724.84**
Blank flanges, slip on for welding; PN25						
200 mm dia.	128.06	162.08	2.70	90.96	nr	**253.04**
250 mm dia.	194.63	246.32	3.48	117.24	nr	**363.56**
300 mm dia.	207.95	263.18	4.20	141.49	nr	**404.67**
Extra over black steel butt welded pipes; black steel flange connections						
Bolted connection between pair of flanges; including gasket, bolts, nuts and washers						
200 mm dia.	107.47	136.01	3.83	129.02	nr	**265.03**
250 mm dia.	166.32	210.49	4.93	166.10	nr	**376.59**
300 mm dia.	226.78	287.01	5.90	198.78	nr	**485.79**

38 MECHANICAL/COOLING/HEATING SYSTEMS

Item	Net Price £	Material £	Labour hours	Labour £	Unit	Total rate £
LOW TEMPERATURE HOT WATER HEATING; PIPELINE: PRESS FIT						
Press fit jointing system; operating temperature −20°C to +120°C; operating pressure 16 bar; butyl rubber 'O' ring mechanical joint. With brackets measured separately (Refer to Screwed Steel Section)						
Carbon steel						
Pipework						
15 mm dia.	1.74	2.21	0.46	15.50	m	**17.71**
20 mm dia.	2.80	3.54	0.48	16.17	m	**19.71**
25 mm dia.	3.96	5.01	0.52	17.52	m	**22.53**
32 mm dia.	5.11	6.46	0.56	18.87	m	**25.33**
40 mm dia.	6.96	8.80	0.58	19.54	m	**28.34**
50 mm dia.	9.04	11.45	0.66	22.23	m	**33.68**
Extra over for carbon steel press fit fittings						
Coupling						
15 mm dia.	1.18	1.49	0.36	12.13	nr	**13.62**
20 mm dia.	1.45	1.84	0.36	12.13	nr	**13.97**
25 mm dia.	1.83	2.32	0.44	14.82	nr	**17.14**
32 mm dia.	3.06	3.88	0.44	14.82	nr	**18.70**
40 mm dia.	4.09	5.17	0.52	17.52	nr	**22.69**
50 mm dia.	4.83	6.12	0.60	20.22	nr	**26.34**
Reducer						
20 × 15 mm dia.	1.10	1.39	0.36	12.13	nr	**13.52**
25 × 15 mm dia.	1.45	1.84	0.40	13.47	nr	**15.31**
25 × 20 mm dia.	1.51	1.92	0.40	13.47	nr	**15.39**
32 × 20 mm dia.	1.69	2.14	0.40	13.47	nr	**15.61**
32 × 25 mm dia.	1.80	2.27	0.44	14.82	nr	**17.09**
40 × 32 mm dia.	3.91	4.95	0.48	16.17	nr	**21.12**
50 × 20 mm dia.	11.31	14.31	0.48	16.17	nr	**30.48**
50 × 25 mm dia.	11.40	14.43	0.52	17.52	nr	**31.95**
50 × 40 mm dia.	11.99	15.18	0.56	18.87	nr	**34.05**
90° elbow						
15 mm dia.	1.71	2.16	0.36	12.13	nr	**14.29**
20 mm dia.	2.26	2.86	0.36	12.13	nr	**14.99**
25 mm dia.	3.10	3.92	0.44	14.82	nr	**18.74**
32 mm dia.	7.74	9.80	0.44	14.82	nr	**24.62**
40 mm dia.	12.37	15.66	0.52	17.52	nr	**33.18**
50 mm dia.	14.77	18.69	0.60	20.22	nr	**38.91**
45° elbow						
15 mm dia.	2.05	2.60	0.36	12.13	nr	**14.73**
20 mm dia.	2.29	2.90	0.36	12.13	nr	**15.03**
25 mm dia.	3.11	3.93	0.44	14.82	nr	**18.75**
32 mm dia.	6.14	7.77	0.44	14.82	nr	**22.59**
40 mm dia.	7.69	9.73	0.52	17.52	nr	**27.25**
50 mm dia.	8.70	11.01	0.60	20.22	nr	**31.23**

38 MECHANICAL/COOLING/HEATING SYSTEMS

Item	Net Price £	Material £	Labour hours	Labour £	Unit	Total rate £
Equal tee						
15 mm dia.	3.29	4.17	0.54	18.19	nr	**22.36**
20 mm dia.	3.81	4.83	0.54	18.19	nr	**23.02**
25 mm dia.	5.11	6.46	0.66	22.23	nr	**28.69**
32 mm dia.	7.95	10.06	0.66	22.23	nr	**32.29**
40 mm dia.	11.73	14.84	0.78	26.28	nr	**41.12**
50 mm dia.	14.07	17.81	0.90	30.32	nr	**48.13**
Reducing tee						
20 × 15 mm dia.	3.74	4.74	0.54	18.19	nr	**22.93**
25 × 15 mm dia.	5.06	6.41	0.62	20.89	nr	**27.30**
25 × 20 mm dia.	5.49	6.94	0.62	20.89	nr	**27.83**
32 × 15 mm dia.	7.41	9.37	0.62	20.89	nr	**30.26**
32 × 20 mm dia.	8.00	10.12	0.62	20.89	nr	**31.01**
32 × 25 mm dia.	8.13	10.29	0.62	20.89	nr	**31.18**
40 × 20 mm dia.	10.71	13.55	0.70	23.58	nr	**37.13**
40 × 25 mm dia.	11.11	14.06	0.70	23.58	nr	**37.64**
40 × 32 mm dia.	10.86	13.74	0.70	23.58	nr	**37.32**
50 × 20 mm dia.	12.77	16.16	0.82	27.63	nr	**43.79**
50 × 25 mm dia.	13.03	16.49	0.82	27.63	nr	**44.12**
50 × 32 mm dia.	13.42	16.98	0.82	27.63	nr	**44.61**
50 × 40 mm dia.	14.06	17.80	0.82	27.63	nr	**45.43**

38 MECHANICAL/COOLING/HEATING SYSTEMS

Item	Net Price £	Material £	Labour hours	Labour £	Unit	Total rate £
LOW TEMPERATURE HOT WATER HEATING; PIPELINE: MECHANICAL GROOVED						
MECHANICAL GROOVED						
Mechanical grooved jointing system; working temperature not exceeding 82°C BS 5750; pipework complete with grooved joints; painted finish. With brackets measured separately (Refer to Screwed Steel Section)						
Grooved joints						
65 mm	16.00	20.25	0.58	19.54	m	**39.79**
80 mm	16.77	21.22	0.68	22.90	m	**44.12**
100 mm	21.75	27.53	0.79	26.61	m	**54.14**
125 mm	34.15	43.22	1.02	34.36	m	**77.58**
150 mm	35.65	45.11	1.15	38.74	m	**83.85**
Extra over mechanical grooved system fittings						
Couplings						
65 mm	16.00	20.25	0.41	13.81	nr	**34.06**
80 mm	16.77	21.22	0.41	13.81	nr	**35.03**
100 mm	21.75	27.53	0.66	22.23	nr	**49.76**
125 mm	34.15	43.22	0.68	22.90	nr	**66.12**
150 mm	35.65	45.11	0.80	26.96	nr	**72.07**
Concentric reducers (one size down)						
80 mm	25.68	32.50	0.59	19.88	nr	**52.38**
100 mm	25.74	32.58	0.71	23.92	nr	**56.50**
125 mm	44.36	56.15	0.85	28.64	nr	**84.79**
150 mm	57.52	72.80	0.98	33.02	nr	**105.82**
Short radius elbow; 90°						
65 mm	29.41	37.22	0.53	17.85	nr	**55.07**
80 mm	30.00	37.97	0.61	20.55	nr	**58.52**
100 mm	40.17	50.84	0.80	26.96	nr	**77.80**
125 mm	66.25	83.84	0.90	30.32	nr	**114.16**
150 mm	85.98	108.82	0.94	31.66	nr	**140.48**
Short radius elbow; 45°						
65 mm	25.17	31.85	0.53	17.85	nr	**49.70**
80 mm	28.28	35.80	0.61	20.55	nr	**56.35**
100 mm	35.15	44.49	0.80	26.96	nr	**71.45**
125 mm	59.47	75.26	0.90	30.32	nr	**105.58**
150 mm	65.48	82.87	0.94	31.66	nr	**114.53**
Equal tee						
65 mm	52.93	66.99	0.83	27.96	nr	**94.95**
80 mm	56.07	70.96	0.93	31.34	nr	**102.30**
100 mm	62.70	79.35	1.18	39.75	nr	**119.10**
125 mm	166.30	210.47	1.37	46.16	nr	**256.63**
150 mm	154.19	195.14	1.43	48.18	nr	**243.32**

38 MECHANICAL/COOLING/HEATING SYSTEMS

Item	Net Price £	Material £	Labour hours	Labour £	Unit	Total rate £
LOW TEMPERATURE HOT WATER HEATING; PIPELINE: PLASTIC PIPEWORK						
Polypropylene PP-R 80 pipe, mechanically stabilized by fibre compound mixture in middle layer; suitable for continuous working temperatures of 0–90°C; thermally fused joints in the running length						
Pipe; 4 m long; PN 20						
20 mm dia.	3.18	4.02	0.35	11.79	m	**15.81**
25 mm dia.	4.79	6.06	0.39	13.14	m	**19.20**
32 mm dia.	5.47	6.92	0.43	14.49	m	**21.41**
40 mm dia.	7.30	9.24	0.47	15.84	m	**25.08**
50 mm dia.	10.62	13.44	0.51	17.19	m	**30.63**
63 mm dia.	17.53	22.19	0.52	17.52	m	**39.71**
75 mm dia.	22.68	28.71	0.60	20.22	m	**48.93**
90 mm dia.	34.99	44.28	0.69	23.25	m	**67.53**
110 mm dia.	52.59	66.56	0.69	23.25	m	**89.81**
125 mm dia.	56.34	71.30	0.85	28.64	m	**99.94**
Fixings						
Refer to steel pipes; black malleable iron. For minimum fixing distances, refer to the Tables and Memoranda to the rear of the book						
Extra over fittings; thermally fused joints						
Overbridge bow						
20 mm dia.	2.11	2.67	0.51	17.19	nr	**19.86**
25 mm dia.	3.88	4.91	0.56	18.87	nr	**23.78**
32 mm dia.	7.77	9.83	0.65	21.90	nr	**31.73**
Elbow 90°						
20 mm dia.	0.71	0.90	0.44	14.82	nr	**15.72**
25 mm dia.	0.94	1.19	0.52	17.52	nr	**18.71**
32 mm dia.	1.35	1.71	0.59	19.88	nr	**21.59**
40 mm dia.	2.13	2.70	0.66	22.23	nr	**24.93**
50 mm dia.	4.60	5.82	0.73	24.60	nr	**30.42**
63 mm dia.	7.03	8.89	0.85	28.64	nr	**37.53**
75 mm dia.	15.57	19.70	0.85	28.64	nr	**48.34**
90 mm dia.	28.77	36.41	1.04	35.03	nr	**71.44**
110 mm dia.	40.89	51.76	1.04	35.03	nr	**86.79**
125 mm dia.	63.02	79.76	1.30	43.80	nr	**123.56**
Long bend 90°						
20 mm dia.	3.78	4.78	0.48	16.17	nr	**20.95**
25 mm dia.	3.97	5.03	0.57	19.20	nr	**24.23**
32 mm dia.	4.54	5.75	0.65	21.90	nr	**27.65**
40 mm dia.	8.49	10.74	0.73	24.60	nr	**35.34**
Elbow 90°, female/male						
20 mm dia.	0.73	0.92	0.44	14.82	nr	**15.74**
25 mm dia.	0.96	1.21	0.52	17.52	nr	**18.73**
32 mm dia.	1.37	1.74	0.59	19.88	nr	**21.62**

38 MECHANICAL/COOLING/HEATING SYSTEMS

Item	Net Price £	Material £	Labour hours	Labour £	Unit	Total rate £
LOW TEMPERATURE HOT WATER HEATING; PIPELINE: PLASTIC PIPEWORK – cont						
Extra over fittings – cont						
Elbow 45°						
20 mm dia.	0.73	0.92	0.44	14.82	nr	**15.74**
25 mm dia.	0.96	1.21	0.52	17.52	nr	**18.73**
32 mm dia.	1.37	1.74	0.59	19.88	nr	**21.62**
40 mm dia.	2.14	2.71	0.66	22.23	nr	**24.94**
50 mm dia.	4.60	5.82	0.73	24.60	nr	**30.42**
63 mm dia.	7.05	8.93	0.85	28.64	nr	**37.57**
75 mm dia.	15.58	19.72	0.85	28.64	nr	**48.36**
90 mm dia.	28.77	36.41	1.04	35.03	nr	**71.44**
110 mm dia.	40.91	51.78	1.04	35.03	nr	**86.81**
125 mm dia.	63.03	79.77	1.30	43.80	nr	**123.57**
Elbow 45°, female/male						
20 mm dia.	0.73	0.92	0.44	14.82	nr	**15.74**
25 mm dia.	0.96	1.21	0.52	17.52	nr	**18.73**
32 mm dia.	1.37	1.74	0.59	19.88	nr	**21.62**
T-Piece 90°						
20 mm dia.	0.99	1.25	0.61	20.55	nr	**21.80**
25 mm dia.	1.34	1.69	0.72	24.26	nr	**25.95**
32 mm dia.	1.74	2.21	0.83	27.96	nr	**30.17**
40 mm dia.	2.70	3.42	0.92	30.99	nr	**34.41**
50 mm dia.	7.66	9.70	1.01	34.04	nr	**43.74**
63 mm dia.	10.97	13.89	1.11	37.40	nr	**51.29**
75 mm dia.	18.29	23.15	1.18	39.75	nr	**62.90**
90 mm dia.	33.67	42.62	1.46	49.19	nr	**91.81**
110 mm dia.	52.53	66.48	1.46	49.19	nr	**115.67**
125 mm dia.	69.81	88.36	1.82	61.32	nr	**149.68**
T-Piece 90° reducing						
25 × 20 × 20 mm	1.37	1.74	0.72	24.26	nr	**26.00**
25 × 20 × 25 mm	1.37	1.74	0.72	24.26	nr	**26.00**
32 × 20 × 32 mm	1.74	2.21	0.83	27.96	nr	**30.17**
32 × 25 × 32 mm	1.74	2.21	0.83	27.96	nr	**30.17**
40 × 20 × 40 mm	2.70	3.42	0.92	30.99	nr	**34.41**
40 × 25 × 40 mm	2.70	3.42	0.92	30.99	nr	**34.41**
40 × 32 × 40 mm	2.70	3.42	0.92	30.99	nr	**34.41**
50 × 25 × 50 mm	7.66	9.70	1.01	34.04	nr	**43.74**
50 × 32 × 50 mm	7.66	9.70	1.01	34.04	nr	**43.74**
50 × 40 × 50 mm	7.66	9.70	1.01	34.04	nr	**43.74**
63 × 20 × 63 mm	10.33	13.07	1.11	37.40	nr	**50.47**
63 × 25 × 63 mm	10.33	13.07	1.11	37.40	nr	**50.47**
63 × 32 × 63 mm	10.33	13.07	1.11	37.40	nr	**50.47**
63 × 40 × 63 mm	10.33	13.07	1.01	34.04	nr	**47.11**
63 × 50 × 63 mm	10.33	13.07	1.01	34.04	nr	**47.11**
75 × 20 × 75 mm	16.79	21.25	1.18	39.75	nr	**61.00**
75 × 25 × 75 mm	16.79	21.25	1.18	39.75	nr	**61.00**
75 × 32 × 75 mm	16.79	21.25	1.18	39.75	nr	**61.00**
75 × 40 × 75 mm	16.79	21.25	1.18	39.75	nr	**61.00**

38 MECHANICAL/COOLING/HEATING SYSTEMS

Item	Net Price £	Material £	Labour hours	Labour £	Unit	Total rate £
75 × 50 × 75 mm	16.79	21.25	1.18	39.75	nr	**61.00**
75 × 63 × 75 mm	16.79	21.25	1.18	39.75	nr	**61.00**
90 × 63 × 90 mm	33.67	42.62	1.46	49.19	nr	**91.81**
110 × 75 × 110 mm	52.53	66.48	1.46	49.19	nr	**115.67**
110 × 90 × 110 mm	52.53	66.48	1.46	49.19	nr	**115.67**
125 × 90 × 125 mm	62.49	79.08	1.82	61.32	nr	**140.40**
125 × 110 × 125 mm	63.77	80.71	1.82	61.32	nr	**142.03**
Reducer						
25 × 20 mm	0.79	1.00	0.59	19.88	nr	**20.88**
32 × 20 mm	1.02	1.29	0.62	20.89	nr	**22.18**
32 × 25 mm	1.02	1.29	0.62	20.89	nr	**22.18**
40 × 20 mm	1.61	2.04	0.66	22.23	nr	**24.27**
40 × 25 mm	1.61	2.04	0.66	22.23	nr	**24.27**
40 × 32 mm	1.61	2.04	0.66	22.23	nr	**24.27**
50 × 20 mm	2.58	3.27	0.73	24.60	nr	**27.87**
50 × 25 mm	2.58	3.27	0.73	24.60	nr	**27.87**
50 × 32 mm	2.58	3.27	0.73	24.60	nr	**27.87**
50 × 40 mm	2.58	3.27	0.73	24.60	nr	**27.87**
63 × 40 mm	5.21	6.60	0.78	26.28	nr	**32.88**
63 × 25 mm	5.21	6.60	0.78	26.28	nr	**32.88**
63 × 32 mm	5.21	6.60	0.78	26.28	nr	**32.88**
63 × 50 mm	5.21	6.60	0.78	26.28	nr	**32.88**
75 × 50 mm	5.82	7.37	0.85	28.64	nr	**36.01**
75 × 63 mm	5.82	7.37	0.85	28.64	nr	**36.01**
90 × 63 mm	13.00	16.45	1.04	35.03	nr	**51.48**
90 × 75 mm	13.00	16.45	1.04	35.03	nr	**51.48**
110 × 90 mm	21.00	26.58	1.17	39.42	nr	**66.00**
125 × 110 mm	32.81	41.53	1.43	48.18	nr	**89.71**
Socket						
20 mm dia.	0.70	0.88	0.51	17.19	nr	**18.07**
25 mm dia.	0.79	1.00	0.56	18.87	nr	**19.87**
32 mm dia.	1.01	1.28	0.65	21.90	nr	**23.18**
40 mm dia.	1.24	1.57	0.74	24.93	nr	**26.50**
50 mm dia.	2.56	3.24	0.81	27.28	nr	**30.52**
63 mm dia.	5.21	6.60	0.86	28.97	nr	**35.57**
75 mm dia.	5.82	7.37	0.91	30.65	nr	**38.02**
90 mm dia.	15.02	19.01	0.91	30.65	nr	**49.66**
110 mm dia.	25.52	32.30	0.91	30.65	nr	**62.95**
125 mm dia.	35.55	44.99	1.30	43.80	nr	**88.79**
End cap						
20 mm dia.	1.09	1.38	0.25	8.43	nr	**9.81**
25 mm dia.	1.34	1.69	0.29	9.77	nr	**11.46**
32 mm dia.	1.66	2.11	0.33	11.11	nr	**13.22**
40 mm dia.	2.66	3.37	0.36	12.13	nr	**15.50**
50 mm dia.	3.68	4.66	0.40	13.47	nr	**18.13**
63 mm dia.	6.19	7.83	0.44	14.82	nr	**22.65**
75 mm dia.	8.90	11.27	0.47	15.84	nr	**27.11**
90 mm dia.	20.17	25.52	0.57	19.20	nr	**44.72**
110 mm dia.	24.25	30.69	0.57	19.20	nr	**49.89**
125 mm dia.	36.94	46.75	0.85	28.64	nr	**75.39**

38 MECHANICAL/COOLING/HEATING SYSTEMS

Item	Net Price £	Material £	Labour hours	Labour £	Unit	Total rate £
LOW TEMPERATURE HOT WATER HEATING; PIPELINE: PLASTIC PIPEWORK – cont						
Extra over fittings – cont						
Stub flange with gasket						
32 mm dia.	24.50	31.01	0.23	7.75	nr	**38.76**
40 mm dia.	30.83	39.02	0.27	9.09	nr	**48.11**
50 mm dia.	37.29	47.20	0.38	12.81	nr	**60.01**
63 mm dia.	44.76	56.65	0.43	14.49	nr	**71.14**
75 mm dia.	52.53	66.48	0.48	16.17	nr	**82.65**
90 mm dia.	71.09	89.97	0.53	17.85	nr	**107.82**
110 mm dia.	99.56	126.00	0.53	17.85	nr	**143.85**
125 mm dia.	142.70	180.60	0.75	25.26	nr	**205.86**
Weld in saddle with female thread						
40–½"	1.76	2.23	0.36	12.13	nr	**14.36**
50–½"	1.76	2.23	0.36	12.13	nr	**14.36**
63–½"	1.76	2.23	0.40	13.47	nr	**15.70**
75–½"	1.76	2.23	0.40	13.47	nr	**15.70**
90–½"	1.76	2.23	0.46	15.50	nr	**17.73**
110–½"	1.76	2.23	0.46	15.50	nr	**17.73**
Weld in saddle with male thread						
50–½"	1.76	2.23	0.36	12.13	nr	**14.36**
63–½"	1.76	2.23	0.40	13.47	nr	**15.70**
75–½"	1.76	2.23	0.40	13.47	nr	**15.70**
90–½"	1.76	2.23	0.46	15.50	nr	**17.73**
110–½"	1.76	2.23	0.46	15.50	nr	**17.73**
Transition piece, round with female thread						
20 × ½"	4.12	5.22	0.29	9.77	nr	**14.99**
20 × ¾"	5.44	6.89	0.29	9.77	nr	**16.66**
25 × ½"	4.12	5.22	0.33	11.11	nr	**16.33**
25 × ¾"	5.44	6.89	0.33	11.11	nr	**18.00**
Transition piece, hexagon with female thread						
32 × 1"	15.33	19.40	0.36	12.13	nr	**31.53**
40 × 1¼"	24.27	30.72	0.36	12.13	nr	**42.85**
50 × 1½"	28.17	35.65	0.36	12.13	nr	**47.78**
63 × 2"	43.63	55.22	0.40	13.47	nr	**68.69**
75 × 2"	45.53	57.62	0.40	13.47	nr	**71.09**
125 × 5"	87.21	110.38	0.51	17.19	nr	**127.57**
Stop valve for surface assembly						
20 mm dia.	16.93	21.43	0.25	8.43	nr	**29.86**
25 mm dia.	16.93	21.43	0.29	9.77	nr	**31.20**
32 mm dia.	31.89	40.36	0.33	11.11	nr	**51.47**
Ball valve						
20 mm dia.	59.62	75.45	0.25	8.43	nr	**83.88**
25 mm dia.	63.89	80.86	0.29	9.77	nr	**90.63**
32 mm dia.	76.77	97.16	0.33	11.11	nr	**108.27**
40 mm dia.	97.98	124.01	0.36	12.13	nr	**136.14**
50 mm dia.	134.51	170.24	0.40	13.47	nr	**183.71**
63 mm dia.	151.67	191.96	0.44	14.82	nr	**206.78**

38 MECHANICAL/COOLING/HEATING SYSTEMS

Item	Net Price £	Material £	Labour hours	Labour £	Unit	Total rate £
Floor or ceiling cover plates						
Plastic						
15 mm dia.	0.55	0.69	0.16	5.39	nr	**6.08**
20 mm dia.	0.62	0.78	0.22	7.41	nr	**8.19**
25 mm dia.	0.66	0.84	0.22	7.41	nr	**8.25**
32 mm dia.	0.72	0.91	0.24	8.09	nr	**9.00**
40 mm dia.	1.61	2.04	0.26	8.76	nr	**10.80**
50 mm dia.	1.76	2.23	0.26	8.76	nr	**10.99**
Chromium plated						
15 mm dia.	3.73	4.72	0.16	5.39	nr	**10.11**
20 mm dia.	3.99	5.05	0.17	5.73	nr	**10.78**
25 mm dia.	4.12	5.22	0.21	7.08	nr	**12.30**
32 mm dia.	4.22	5.34	0.22	7.41	nr	**12.75**
40 mm dia.	4.77	6.04	0.26	8.76	nr	**14.80**
50 mm dia.	5.70	7.21	0.26	8.76	nr	**15.97**

38 MECHANICAL/COOLING/HEATING SYSTEMS

Item	Net Price £	Material £	Labour hours	Labour £	Unit	Total rate £
LOW TEMPERATURE HOT WATER HEATING; PIPELINE: PIPELINE ANCILLERIES						
EXPANSION JOINTS						
Axial movement bellows expansion joints; stainless steel						
Screwed ends for steel pipework; up to 6 bar G at 100°C						
15 mm dia.	91.49	115.79	0.68	22.90	nr	**138.69**
20 mm dia.	97.73	123.68	0.81	27.28	nr	**150.96**
25 mm dia.	101.52	128.49	0.93	31.34	nr	**159.83**
32 mm dia.	107.49	136.04	1.06	35.72	nr	**171.76**
40 mm dia.	113.85	144.09	1.16	39.08	nr	**183.17**
50 mm dia.	113.42	143.54	1.19	40.10	nr	**183.64**
Screwed ends for steel pipework; aluminium and steel outer sleeves; up to 16 bar G at 120°C						
20 mm dia.	99.67	126.15	1.32	44.48	nr	**170.63**
25 mm dia.	101.52	128.49	1.52	51.21	nr	**179.70**
32 mm dia.	107.49	136.04	1.80	60.65	nr	**196.69**
40 mm dia.	113.85	144.09	2.03	68.40	nr	**212.49**
50 mm dia.	115.68	146.41	2.26	76.14	nr	**222.55**
Flanged ends for steel pipework; aluminium and steel outer sleeves; up to 16 bar G at 120°C						
20 mm dia.	88.19	111.61	0.53	17.85	nr	**129.46**
25 mm dia.	94.48	119.57	0.64	21.57	nr	**141.14**
32 mm dia.	130.19	164.76	0.74	24.93	nr	**189.69**
40 mm dia.	140.68	178.05	0.82	27.63	nr	**205.68**
50 mm dia.	166.92	211.25	0.89	29.98	nr	**241.23**
Flanged ends for steel pipework; up to 16 bar G at 120°C						
65 mm dia.	154.32	195.31	1.10	37.06	nr	**232.37**
80 mm dia.	165.87	209.92	1.31	44.13	nr	**254.05**
100 mm dia.	190.00	240.46	1.78	59.96	nr	**300.42**
150 mm dia.	290.80	368.03	3.08	103.77	nr	**471.80**
Screwed ends for non-ferrous pipework; up to 6 bar G at 100°C						
20 mm dia.	99.67	126.15	0.72	24.26	nr	**150.41**
25 mm dia.	101.52	128.49	0.84	28.30	·nr	**156.79**
32 mm dia.	107.49	136.04	1.02	34.36	nr	**170.40**
40 mm dia.	113.85	144.09	1.11	37.40	nr	**181.49**
50 mm dia.	115.68	146.41	1.18	39.75	nr	**186.16**
Flanged ends for steel, copper or non-ferrous pipework; up to 16 bar G at 120°C						
65 mm dia.	198.42	251.12	0.87	29.31	nr	**280.43**
80 mm dia.	204.71	259.08	0.95	32.01	nr	**291.09**
100 mm dia.	236.22	298.96	1.15	38.74	nr	**337.70**
150 mm dia.	387.37	490.26	1.36	45.82	nr	**536.08**

38 MECHANICAL/COOLING/HEATING SYSTEMS

Item	Net Price £	Material £	Labour hours	Labour £	Unit	Total rate £
Angular movement bellows expansion joints; stainless steel						
Flanged ends for steel pipework; up to 16 bar G at 120°C						
50 mm dia.	228.72	289.46	0.71	23.92	nr	**313.38**
65 mm dia.	277.58	351.31	0.83	27.96	nr	**379.27**
80 mm dia.	324.38	410.54	0.91	30.65	nr	**441.19**
100 mm dia.	418.76	529.98	0.97	32.68	nr	**562.66**
125 mm dia.	553.14	700.06	1.16	39.08	nr	**739.14**
150 mm dia.	657.73	832.42	1.18	39.75	nr	**872.17**

38 MECHANICAL/COOLING/HEATING SYSTEMS

Item	Net Price £	Material £	Labour hours	Labour £	Unit	Total rate £
LOW TEMPERATURE HOT WATER HEATING; PIPELINE: VALVES						
Gate valves						
Bronze gate valve; non-rising stem; BS 5154, series B, PN 16; working pressure up to 16 bar from −10°C to 100°C; 7 bar for saturated steam; screwed ends to steel						
15 mm dia.	55.07	69.70	1.18	39.75	nr	**109.45**
20 mm dia.	75.22	95.20	1.24	41.78	nr	**136.98**
25 mm dia.	95.73	121.15	1.31	44.13	nr	**165.28**
32 mm dia.	130.40	165.03	1.43	48.18	nr	**213.21**
40 mm dia.	174.09	220.33	1.53	51.54	nr	**271.87**
50 mm dia.	240.20	304.00	1.63	54.91	nr	**358.91**
Bronze gate valve; non-rising stem; BS 5154, series B, PN 32; working pressure up to 14 bar for saturated steam, 32 bar from −10°C to 100°C; screwed ends to steel						
15 mm dia.	48.19	60.98	1.11	37.40	nr	**98.38**
20 mm dia.	62.51	79.12	1.28	43.12	nr	**122.24**
25 mm dia.	84.68	107.17	1.49	50.20	nr	**157.37**
32 mm dia.	116.96	148.02	1.88	63.35	nr	**211.37**
40 mm dia.	157.03	198.73	2.31	77.82	nr	**276.55**
50 mm dia.	220.65	279.25	2.80	94.34	nr	**373.59**
Bronze gate valve; non-rising stem; BS 5154, series B, PN 20; working pressure up to 9 bar for saturated steam, 20 bar from −10°C to 100°C; screwed ends to steel						
15 mm dia.	32.87	41.60	0.84	28.30	nr	**69.90**
20 mm dia.	46.70	59.10	1.01	34.04	nr	**93.14**
25 mm dia.	60.43	76.48	1.19	40.10	nr	**116.58**
32 mm dia.	86.23	109.13	1.38	46.50	nr	**155.63**
40 mm dia.	118.02	149.36	1.62	54.58	nr	**203.94**
50 mm dia.	168.65	213.44	1.94	65.35	nr	**278.79**
Bronze gate valve; non-rising stem; BS 5154, series B, PN 16; working pressure up to 7 bar for saturated steam, 16 bar from −10°C to 100°C; BS4504 flanged ends; bolted connections						
15 mm dia.	148.00	187.31	1.18	39.75	nr	**227.06**
20 mm dia.	163.20	206.55	1.24	41.78	nr	**248.33**
25 mm dia.	214.92	272.00	1.31	44.13	nr	**316.13**
32 mm dia.	269.64	341.25	1.43	48.18	nr	**389.43**
40 mm dia.	326.01	412.60	1.53	51.54	nr	**464.14**
50 mm dia.	450.70	570.40	1.63	54.91	nr	**625.31**
65 mm dia.	699.98	885.90	1.71	57.61	nr	**943.51**
80 mm dia.	990.68	1253.81	1.88	63.35	nr	**1317.16**
100 mm dia.	1800.37	2278.55	2.03	68.40	nr	**2346.95**

38 MECHANICAL/COOLING/HEATING SYSTEMS

Item	Net Price £	Material £	Labour hours	Labour £	Unit	Total rate £
Cast iron gate valve; bronze trim; non-rising stem; BS 5150, PN6; working pressure 6 bar from −10°C to 120°C; BS4504 flanged ends; bolted connections						
50 mm dia.	293.13	370.99	1.85	62.33	nr	**433.32**
65 mm dia.	306.38	387.76	2.00	67.38	nr	**455.14**
80 mm dia.	354.08	448.12	2.27	76.48	nr	**524.60**
100 mm dia.	447.91	566.88	2.76	92.98	nr	**659.86**
125 mm dia.	641.11	811.38	6.05	203.83	nr	**1015.21**
150 mm dia.	737.73	933.67	8.03	270.54	nr	**1204.21**
200 mm dia.	1405.19	1778.40	9.17	308.94	nr	**2087.34**
250 mm dia.	2162.17	2736.44	10.72	361.17	nr	**3097.61**
300 mm dia.	2564.49	3245.61	11.75	395.86	nr	**3641.47**
Cast iron gate valve; bronze trim; non-rising stem; BS 5150, PN10; working pressure up to 8.4 bar for saturated steam, 10 bar from −10°C to 120°C; BS4504 flanged ends; bolted connections						
50 mm dia.	307.47	389.13	1.85	62.33	nr	**451.46**
65 mm dia.	382.10	483.58	2.00	67.38	nr	**550.96**
80 mm dia.	425.42	538.41	2.27	76.48	nr	**614.89**
100 mm dia.	566.30	716.71	2.76	92.98	nr	**809.69**
125 mm dia.	791.75	1002.04	6.05	203.83	nr	**1205.87**
150 mm dia.	934.14	1182.25	8.03	270.54	nr	**1452.79**
200 mm dia.	1716.29	2172.14	9.17	308.94	nr	**2481.08**
250 mm dia.	2581.16	3266.72	10.72	361.17	nr	**3627.89**
300 mm dia.	2904.56	3676.01	11.75	395.86	nr	**4071.87**
350 mm dia.	3378.33	4275.61	12.67	426.87	nr	**4702.48**
Cast iron gate valve; bronze trim; non-rising stem; BS 5163 series A, PN16; working pressure for cold water services up to 16 bar; BS4504 flanged ends; bolted connections						
50 mm dia.	307.47	389.13	1.85	62.33	nr	**451.46**
65 mm dia.	382.10	483.58	2.00	67.38	nr	**550.96**
80 mm dia.	425.42	538.41	2.27	76.48	nr	**614.89**
100 mm dia.	566.30	716.71	2.76	92.98	nr	**809.69**
125 mm dia.	791.75	1002.04	6.05	203.83	nr	**1205.87**
150 mm dia.	934.14	1182.25	8.03	270.54	nr	**1452.79**
Ball valves						
Malleable iron body; lever operated stainless steel ball and stem; working pressure up to 12 bar; flanged ends to BS 4504 16/11; bolted connections						
40 mm dia.	408.66	517.20	1.54	51.88	nr	**569.08**
50 mm dia.	514.95	651.72	1.64	55.26	nr	**706.98**
80 mm dia.	832.98	1054.22	1.92	64.68	nr	**1118.90**
100 mm dia.	861.21	1089.95	2.80	94.34	nr	**1184.29**
150 mm dia.	1591.93	2014.75	12.05	405.97	nr	**2420.72**

38 MECHANICAL/COOLING/HEATING SYSTEMS

Item	Net Price £	Material £	Labour hours	Labour £	Unit	Total rate £
LOW TEMPERATURE HOT WATER HEATING; PIPELINE: VALVES – cont						
Ball valves – cont						
Malleable iron body; lever operated stainless steel ball and stem; working pressure up to 16 bar; screwed ends to steel						
15 mm dia.	36.68	46.42	1.34	45.16	nr	91.58
20 mm dia.	48.53	61.42	1.34	45.16	nr	106.58
25 mm dia.	49.92	63.18	1.40	47.19	nr	110.37
32 mm dia.	68.77	87.04	1.46	49.26	nr	136.30
40 mm dia.	68.77	87.04	1.54	51.91	nr	138.95
50 mm dia.	82.24	104.08	1.64	55.32	nr	159.40
Carbon steel body; lever operated stainless steel ball and stem; Class 150; working pressure up to 19 bar; screwed ends to steel						
15 mm dia.	89.81	113.67	0.84	28.31	nr	141.98
20 mm dia.	100.95	127.76	1.14	38.42	nr	166.18
25 mm dia.	109.30	138.33	1.30	43.81	nr	182.14
32 mm dia.	140.95	178.38	1.42	47.84	nr	226.22
40 mm dia.	176.18	222.97	1.56	52.56	nr	275.53
50 mm dia.	220.24	278.73	1.68	56.59	nr	335.32
Globe valves						
Bronze; rising stem; renewable disc; BS 5154 series B, PN32; working pressure up to 14 bar for saturated steam, 32 bar from −10°C to 100°C; screwed ends to steel						
15 mm dia.	48.91	61.90	0.77	25.95	nr	87.85
20 mm dia.	66.89	84.66	1.03	34.70	nr	119.36
25 mm dia.	95.97	121.46	1.19	40.10	nr	161.56
32 mm dia.	135.11	170.99	1.38	46.50	nr	217.49
40 mm dia.	188.15	238.12	1.62	54.58	nr	292.70
50 mm dia.	288.80	365.50	1.61	54.24	nr	419.74
Bronze; needle valve; rising stem; BS 5154, series B, PN32; working pressure up to 14 bar for saturated steam, 32 bar from −10°C to 100°C; screwed ends to steel						
15 mm dia.	37.19	47.06	1.07	36.04	nr	83.10
20 mm dia.	62.96	79.68	1.18	39.75	nr	119.43
25 mm dia.	88.77	112.35	1.27	42.80	nr	155.15
32 mm dia.	184.77	233.84	1.35	45.48	nr	279.32
40 mm dia.	291.09	368.40	1.47	49.53	nr	417.93
50 mm dia.	368.70	466.63	1.61	54.24	nr	520.87
Bronze; rising stem; renewable disc; BS 5154, series B, PN16; working pressure up to 7 bar for saturated steam, 16 bar from −10°C to 100°C; BS4504 flanged ends; bolted connections						
15 mm dia.	98.84	125.09	1.16	39.08	nr	164.17
20 mm dia.	114.33	144.69	1.26	42.45	nr	187.14
25 mm dia.	200.48	253.72	1.38	46.50	nr	300.22

38 MECHANICAL/COOLING/HEATING SYSTEMS

Item	Net Price £	Material £	Labour hours	Labour £	Unit	Total rate £
32 mm dia.	253.08	320.30	1.47	49.53	nr	369.83
40 mm dia.	287.67	364.08	1.56	52.56	nr	416.64
50 mm dia.	412.73	522.35	1.71	57.61	nr	579.96
Bronze; rising stem; renewable disc; BS 2060, class 250; working pressure up to 24 bar for saturated steam, 38 bar from −10°C to 100°C; flanged ends (BS 10 table H); bolted connections						
15 mm dia.	251.71	318.56	1.16	39.08	nr	357.64
20 mm dia.	292.54	370.24	1.26	42.45	nr	412.69
25 mm dia.	402.00	508.77	1.38	46.50	nr	555.27
32 mm dia.	525.16	664.64	1.47	49.53	nr	714.17
40 mm dia.	628.87	795.89	1.56	52.56	nr	848.45
50 mm dia.	977.87	1237.59	1.71	57.61	nr	1295.20
65 mm dia.	1169.44	1480.05	1.88	63.35	nr	1543.40
80 mm dia.	1486.06	1880.76	2.03	68.40	nr	1949.16
Check valves						
Bronze; swing pattern; BS 5154 series B, PN 25; working pressure up to 10.5 bar for saturated steam, 25 bar from −10°C to 100°C; screwed ends to steel						
15 mm dia.	22.16	28.04	0.77	25.95	nr	53.99
20 mm dia.	26.33	33.32	1.03	34.70	nr	68.02
25 mm dia.	36.49	46.18	1.19	40.10	nr	86.28
32 mm dia.	61.83	78.25	1.38	46.50	nr	124.75
40 mm dia.	76.91	97.34	1.62	54.58	nr	151.92
50 mm dia.	117.97	149.31	1.94	65.35	nr	214.66
65 mm dia.	226.96	287.24	2.45	82.54	nr	369.78
80 mm dia.	320.92	406.16	2.83	95.33	nr	501.49
Bronze; vertical lift pattern; BS 5154 series B, PN32; working pressure up to 14 bar for saturated steam, 32 bar from −10°C to 100°C; screwed ends to steel						
15 mm dia.	28.88	36.55	0.96	32.35	nr	68.90
20 mm dia.	41.07	51.98	1.07	36.04	nr	88.02
25 mm dia.	60.31	76.33	1.17	39.42	nr	115.75
32 mm dia.	91.87	116.27	1.33	44.80	nr	161.07
40 mm dia.	119.71	151.50	1.41	47.50	nr	199.00
50 mm dia.	179.35	226.99	1.55	52.21	nr	279.20
65 mm dia.	495.53	627.14	1.80	60.65	nr	687.79
80 mm dia.	740.82	937.59	1.99	67.04	nr	1004.63
Bronze; oblique swing pattern; BS 5154 series A, PN32; working pressure up to 14 bar for saturated steam, 32 bar from −10°C to 120°C; screwed connections to steel						
15 mm dia.	36.69	46.44	0.96	32.35	nr	78.79
20 mm dia.	43.58	55.16	1.07	36.04	nr	91.20
25 mm dia.	60.42	76.46	1.17	39.42	nr	115.88
32 mm dia.	90.35	114.35	1.33	44.80	nr	159.15
40 mm dia.	116.45	147.38	1.41	47.50	nr	194.88
50 mm dia.	192.92	244.16	1.55	52.21	nr	296.37

38 MECHANICAL/COOLING/HEATING SYSTEMS

Item	Net Price £	Material £	Labour hours	Labour £	Unit	Total rate £
LOW TEMPERATURE HOT WATER HEATING; PIPELINE: VALVES – cont						
Check valves – cont						
Cast iron; swing pattern; BS 5153 PN6; working pressure up to 6 bar from −10°C to 120°C; BS 4504 flanged ends; bolted connections						
50 mm dia.	633.53	801.80	1.86	62.66	nr	864.46
65 mm dia.	693.63	877.86	2.00	67.38	nr	945.24
80 mm dia.	780.75	988.12	2.56	86.25	nr	1074.37
100 mm dia.	1000.17	1265.81	2.76	92.98	nr	1358.79
125 mm dia.	1492.11	1888.41	6.05	203.83	nr	2092.24
150 mm dia.	1677.69	2123.28	8.11	273.24	nr	2396.52
200 mm dia.	3710.18	4695.60	9.26	311.98	nr	5007.58
250 mm dia.	4637.61	5869.36	10.72	361.17	nr	6230.53
300 mm dia.	5565.14	7043.24	11.75	395.86	nr	7439.10
Cast iron; horizontal lift pattern; BS 5153 PN16; working pressure up to 13 bar for saturated steam, 16 bar from −10°C to 120°C; BS 4504 flanged ends; bolted connections						
50 mm dia.	572.09	724.04	1.86	62.66	nr	786.70
65 mm dia.	572.09	724.04	2.00	67.38	nr	791.42
80 mm dia.	708.30	896.43	2.56	86.25	nr	982.68
100 mm dia.	824.84	1043.92	2.96	99.72	nr	1143.64
125 mm dia.	1230.65	1557.51	7.76	261.43	nr	1818.94
150 mm dia.	1383.61	1751.10	10.50	353.75	nr	2104.85
Cast iron; semi-lugged butterfly valve; BS5155 PN16; working pressure 16 bar from −10°C to 120°C; BS 4504 flanged ends; bolted connections						
50 mm dia.	190.93	241.64	2.20	74.11	nr	315.75
65 mm dia.	197.04	249.38	2.31	77.82	nr	327.20
80 mm dia.	232.10	293.74	2.88	97.04	nr	390.78
100 mm dia.	322.65	408.34	3.11	104.78	nr	513.12
125 mm dia.	470.92	596.00	5.02	169.12	nr	765.12
150 mm dia.	540.54	684.11	6.98	235.17	nr	919.28
200 mm dia.	862.88	1092.06	8.25	277.95	nr	1370.01
250 mm dia.	1273.58	1611.85	10.47	352.73	nr	1964.58
300 mm dia.	1910.42	2417.82	11.48	386.77	nr	2804.59
Cast iron; semi-lugged butterfly valve; BS5155 PN20; working pressure up to 20 bar from −10°C to 120°C; BS 4504 flanged ends; bolted connections						
50 mm dia.	1709.82	2163.95	2.20	74.11	nr	2238.06
65 mm dia.	1769.00	2238.85	2.31	77.82	nr	2316.67
80 mm dia.	1888.61	2390.23	2.88	97.04	nr	2487.27
100 mm dia.	2299.83	2910.67	3.11	104.78	nr	3015.45
125 mm dia.	2951.96	3736.00	5.02	169.12	nr	3905.12
150 mm dia.	3350.14	4239.94	5.02	169.12	nr	4409.06
200 mm dia.	3853.13	4876.52	8.25	277.95	nr	5154.47
250 mm dia.	6075.71	7689.42	10.47	352.73	nr	8042.15
300 mm dia.	7361.91	9317.24	11.48	386.77	nr	9704.01

38 MECHANICAL/COOLING/HEATING SYSTEMS

Item	Net Price £	Material £	Labour hours	Labour £	Unit	Total rate £
Cast iron; semi-lugged butterfly valve; BS5155 PN20; working pressure up to 30 bar from −10°C to 120°C; BS 4504 flanged ends; bolted connections						
50 mm dia.	1709.82	2163.95	2.20	74.11	nr	**2238.06**
65 mm dia.	1769.00	2238.85	2.31	77.82	nr	**2316.67**
80 mm dia.	1888.61	2390.23	2.88	97.04	nr	**2487.27**
100 mm dia.	2299.83	2910.67	3.11	104.78	nr	**3015.45**
125 mm dia.	2951.96	3736.00	5.02	169.12	nr	**3905.12**
150 mm dia.	3350.14	4239.94	5.02	169.12	nr	**4409.06**
200 mm dia.	3853.13	4876.52	8.25	277.95	nr	**5154.47**
250 mm dia.	6075.71	7689.42	10.47	352.73	nr	**8042.15**
300 mm dia.	7361.91	9317.24	11.48	386.77	nr	**9704.01**
Commissioning valves						
Bronze commissioning set; metering station; double regulating valve; BS5154 PN20 Series B; working pressure 20 bar from −10°C to 100°C; screwed ends to steel						
15 mm dia.	95.84	121.30	1.08	36.39	nr	**157.69**
20 mm dia.	154.00	194.90	1.46	49.19	nr	**244.09**
25 mm dia.	183.12	231.76	1.68	56.59	nr	**288.35**
32 mm dia.	245.52	310.73	1.95	65.70	nr	**376.43**
40 mm dia.	356.07	450.64	2.27	76.48	nr	**527.12**
50 mm dia.	537.38	680.11	2.73	91.97	nr	**772.08**
Cast iron commissioning set; metering station; double regulating valve; BS5152 PN16; working pressure 16 bar from −10°C to 90°C; flanged ends (BS 4504, Part 1, Table 16); bolted connections						
65 mm dia.	814.13	1030.37	1.80	60.65	nr	**1091.02**
80 mm dia.	974.81	1233.72	2.56	86.25	nr	**1319.97**
100 mm dia.	1357.12	1717.58	2.30	77.49	nr	**1795.07**
125 mm dia.	1987.83	2515.80	2.44	82.20	nr	**2598.00**
150 mm dia.	2618.59	3314.09	2.90	97.71	nr	**3411.80**
200 mm dia.	6702.53	8482.72	8.26	278.28	nr	**8761.00**
250 mm dia.	8384.79	10611.79	10.49	353.40	nr	**10965.19**
300 mm dia.	9106.05	11524.62	11.49	387.09	nr	**11911.71**
Cast iron variable orifice double regulating valve; orifice valve; BS5152 PN16; working pressure 16 bar from −10° to 90°C; flanged ends (BS 4504, Part 1, Table 16); bolted connections						
65 mm dia.	596.07	754.39	2.00	67.38	nr	**821.77**
80 mm dia.	729.58	923.36	2.56	86.25	nr	**1009.61**
100 mm dia.	1001.21	1267.13	2.96	99.72	nr	**1366.85**
125 mm dia.	1499.31	1897.53	7.76	261.43	nr	**2158.96**
150 mm dia.	1925.82	2437.32	10.50	353.75	nr	**2791.07**
200 mm dia.	5047.40	6387.99	8.26	278.28	nr	**6666.27**
250 mm dia.	7674.53	9712.89	10.49	353.40	nr	**10066.29**
300 mm dia.	13766.57	17422.97	11.49	387.09	nr	**17810.06**

38 MECHANICAL/COOLING/HEATING SYSTEMS

Item	Net Price £	Material £	Labour hours	Labour £	Unit	Total rate £
LOW TEMPERATURE HOT WATER HEATING; PIPELINE: VALVES – cont						
Commissioning valves – cont						
Cast iron globe valve with double regulating feature; BS5152 PN16; working pressure 16 bar from −10°C to 120°C; flanged ends (BS 4504, Part 1, Table 16); bolted connections						
65 mm dia.	710.06	898.65	2.00	67.38	nr	**966.03**
80 mm dia.	858.69	1086.76	2.56	86.25	nr	**1173.01**
100 mm dia.	1131.14	1431.57	2.96	99.72	nr	**1531.29**
125 mm dia.	1577.11	1995.99	7.76	261.43	nr	**2257.42**
150 mm dia.	1996.06	2526.22	10.50	353.75	nr	**2879.97**
200 mm dia.	5167.58	6540.09	8.26	278.28	nr	**6818.37**
250 mm dia.	7818.44	9895.02	10.49	353.40	nr	**10248.42**
300 mm dia.	12283.91	15546.52	11.49	387.09	nr	**15933.61**
Bronze autoflow commissioning valve; PN25; working pressure 25 bar up to 100°C; screwed ends to steel						
15 mm dia.	108.68	137.55	0.82	27.63	nr	**165.18**
20 mm dia.	115.07	145.63	1.08	36.39	nr	**182.02**
25 mm dia.	143.28	181.34	1.27	42.80	nr	**224.14**
32 mm dia.	223.78	283.21	1.50	50.53	nr	**333.74**
40 mm dia.	245.03	310.11	1.76	59.29	nr	**369.40**
50 mm dia.	362.30	458.53	2.13	71.76	nr	**530.29**
Ductile iron autoflow commissioning valves; PN16; working pressure 16 bar from −10°C to 120°C; for ANSI 150 flanged ends						
65 mm dia.	954.09	1207.49	2.31	77.82	nr	**1285.31**
80 mm dia.	1054.85	1335.02	2.88	97.04	nr	**1432.06**
100 mm dia.	1347.56	1705.47	3.11	104.78	nr	**1810.25**
150 mm dia.	2775.98	3513.28	6.98	235.17	nr	**3748.45**
200 mm dia.	4043.45	5117.39	8.26	278.28	nr	**5395.67**
250 mm dia.	5615.77	7107.32	10.49	353.40	nr	**7460.72**
300 mm dia.	7172.20	9077.14	11.49	387.09	nr	**9464.23**
Strainers						
Bronze strainer; Y type; PN32; working pressure 32 bar from −10°C to 100°C; screwed ends to steel						
15 mm dia.	43.56	55.13	0.82	27.63	nr	**82.76**
20 mm dia.	55.32	70.01	1.08	36.39	nr	**106.40**
25 mm dia.	73.40	92.89	1.27	42.80	nr	**135.69**
32 mm dia.	118.55	150.04	1.50	50.53	nr	**200.57**
40 mm dia.	153.33	194.05	1.76	59.29	nr	**253.34**
50 mm dia.	256.80	325.00	2.13	71.76	nr	**396.76**

38 MECHANICAL/COOLING/HEATING SYSTEMS

Item	Net Price £	Material £	Labour hours	Labour £	Unit	Total rate £
Cast iron strainer; Y type; PN16; working pressure 16 bar from −10°C to 120°C; BS 4504 flanged ends						
65 mm dia.	270.92	342.88	2.31	77.82	nr	**420.70**
80 mm dia.	320.66	405.83	2.88	97.04	nr	**502.87**
100 mm dia.	469.20	593.82	3.11	104.78	nr	**698.60**
125 mm dia.	864.94	1094.67	5.02	169.12	nr	**1263.79**
150 mm dia.	1109.73	1404.47	6.98	235.17	nr	**1639.64**
200 mm dia.	1862.63	2357.34	8.26	278.28	nr	**2635.62**
250 mm dia.	2693.46	3408.84	10.49	353.40	nr	**3762.24**
300 mm dia.	4386.24	5551.22	11.49	387.09	nr	**5938.31**
Regulators						
Gunmetal; self-acting two port thermostatic regulator; single seat; screwed ends; complete with sensing element, 2 m long capillary tube						
15 mm dia.	767.94	971.90	1.37	46.16	nr	**1018.06**
20 mm dia.	792.65	1003.17	1.24	41.78	nr	**1044.95**
25 mm dia.	813.45	1029.50	1.34	45.15	nr	**1074.65**
Gunmetal; self-acting two port thermostatic regulator; double seat; flanged ends (BS 4504 PN25); with sensing element, 2 m long capillary tube; steel body						
65 mm dia.	2847.97	3604.40	1.23	47.52	nr	**3651.92**
80 mm dia.	3361.39	4254.17	1.62	62.60	nr	**4316.77**
Control Valves						
Electrically operated (electrical work elsewhere)						
Pressure independent control valve (PICV)						
Brass; pressure independent control valve; rotary type; PN 25; maximum working pressure 25 bar at from −10°C to 120°C; screwed ends to steel; electrical work elsewhere						
15 mm dia.	88.16	111.57	1.08	36.39	nr	**147.96**
20 mm dia.	105.55	133.58	1.18	39.75	nr	**173.33**
25 mm dia.	155.43	196.72	1.35	45.48	nr	**242.20**
32 mm dia.	226.19	286.26	1.46	49.19	nr	**335.45**
40 mm dia.	250.54	317.08	1.53	51.54	nr	**368.62**
50 mm dia.	437.27	553.41	1.61	54.24	nr	**607.65**
Brass; pressure independent control valve; rotary type with electronic actuator; 24 V motor; PN 25; maximum working pressure 25 bar at from −10°C to 120°C; screwed ends to steel; electrical work elsewhere						
15 mm dia.	184.43	233.42	2.06	69.41	nr	**302.83**
20 mm dia.	201.82	255.43	2.15	72.43	nr	**327.86**
25 mm dia.	251.70	318.55	2.27	76.48	nr	**395.03**
32 mm dia.	354.93	449.20	2.35	79.17	nr	**528.37**
40 mm dia.	379.28	480.02	2.47	83.22	nr	**563.24**
50 mm dia.	578.78	732.50	2.55	85.90	nr	**818.40**

38 MECHANICAL/COOLING/HEATING SYSTEMS

Item	Net Price £	Material £	Labour hours	Labour £	Unit	Total rate £
LOW TEMPERATURE HOT WATER HEATING; PIPELINE: VALVES – cont						
Control Valves – cont						
Differential pressure control valves (DPCV)						
Brass; differential control valve; PN 25; maximum working pressure 25 bar at from −20°C to 120°C; screwed ends to steel						
15 mm dia.	242.43	306.82	1.99	67.04	nr	373.86
20 mm dia.	261.06	330.40	2.03	68.40	nr	398.80
25 mm dia.	290.09	367.14	2.11	71.09	nr	438.23
32 mm dia.	328.37	415.59	2.17	73.11	nr	488.70
40 m m dia.	566.00	716.33	2.26	76.14	nr	792.47
50 mm dia.	679.61	860.12	2.31	77.82	nr	937.94
Cast iron; butterfly type; two position electrically controlled 240 V motor and linkage mechanism; for low pressure hot water; maximum pressure 6 bar at 120°C; flanged ends; electrical work elsewhere						
25 mm dia.	919.58	1163.83	1.47	49.53	nr	1213.36
32 mm dia.	950.19	1202.56	1.52	51.21	nr	1253.77
40 mm dia.	1327.74	1680.39	1.61	54.24	nr	1734.63
50 mm dia.	1365.46	1728.13	1.71	57.61	nr	1785.74
65 mm dia.	1399.66	1771.41	2.51	84.57	nr	1855.98
80 mm dia.	1454.39	1840.68	2.69	90.63	nr	1931.31
100 mm dia.	1526.27	1931.65	2.81	94.66	nr	2026.31
125 mm dia.	1690.53	2139.54	2.94	99.04	nr	2238.58
150 mm dia.	1831.96	2318.52	3.33	112.18	nr	2430.70
200 mm dia.	2272.25	2875.76	3.67	123.65	nr	2999.41
Cast iron; three way 240 V motorized; for low pressure hot water; maximum pressure 6 bar 120°C; flanged ends, drilled (BS 10, Table F)						
25 mm dia.	1027.79	1300.77	1.99	67.04	nr	1367.81
40 mm dia.	1069.35	1353.37	2.13	71.76	nr	1425.13
50 mm dia.	1083.10	1370.77	3.21	108.15	nr	1478.92
65 mm dia.	1189.16	1505.00	3.23	108.82	nr	1613.82
80 mm dia.	1332.47	1686.37	3.50	117.91	nr	1804.28
2 port control valves						
Cast iron; 2 port; motorized control valve; normally open for heating applications; PN 16; screwed ends to steel						
15 mm dia.	41.75	52.84	1.24	41.78	nr	94.62
22 mm dia.	63.75	80.68	1.31	44.13	nr	124.81
28 mm dia.	76.24	96.49	1.37	46.16	nr	142.65
32 mm dia.	89.08	112.74	1.45	48.85	nr	161.59
40 mm dia.	111.35	140.93	1.52	51.21	nr	192.14
50 mm dia.	139.18	176.14	1.60	53.91	nr	230.05

38 MECHANICAL/COOLING/HEATING SYSTEMS

Item	Net Price £	Material £	Labour hours	Labour £	Unit	Total rate £
Two port normally closed motorized valve; electric actuator; spring return; domestic usage						
22 mm dia.	143.75	181.93	1.18	39.75	nr	**221.68**
28 mm dia.	202.85	256.73	1.35	45.48	nr	**302.21**
Two port on/off motorized valve; electric actuator; spring return; domestic usage						
22 mm dia.	143.75	181.93	1.18	39.75	nr	**221.68**
3 port control valves						
Cast iron; 3 port; motorized control valve with temperature control unit; PN 25; screwed ends to steel						
22 mm dia.	64.96	82.21	2.00	67.38	nr	**149.59**
25 mm dia.	73.64	93.20	2.13	71.76	nr	**164.96**
32 mm dia.	94.28	119.32	2.19	73.79	nr	**193.11**
40 mm dia.	117.84	149.14	2.26	76.14	nr	**225.28**
Three port motorized valve; electric actuator; spring return; domestic usage						
22 mm dia.	218.37	276.37	1.18	39.75	nr	**316.12**
4 port control valves						
Cast iron; 4 port; fixed orifice double regulating and control valve; PN 16; screwed ends to steel						
22 mm dia.	57.57	72.86	2.89	97.36	nr	**170.22**
Safety and relief valves						
Bronze relief valve; spring type; side outlet; working pressure up to 20.7 bar at 120°C; screwed ends to steel						
15 mm dia.	166.90	211.23	0.26	8.76	nr	**219.99**
20 mm dia.	183.04	231.66	0.36	12.13	nr	**243.79**
Bronze relief valve; spring type; side outlet; working pressure up to 17.2 bar at 120°C; screwed ends to steel						
25 mm dia.	250.34	316.83	0.38	12.81	nr	**329.64**
32 mm dia.	339.18	429.26	0.48	16.17	nr	**445.43**
Bronze relief valve; spring type; side outlet; working pressure up to 13.8 bar at 120°C; screwed ends to steel						
40 mm dia.	442.81	560.43	0.64	21.57	nr	**582.00**
50 mm dia.	577.39	730.74	0.76	25.60	nr	**756.34**
65 mm dia.	1069.06	1353.00	0.94	31.66	nr	**1384.66**
80 mm dia.	1402.59	1775.12	1.10	37.06	nr	**1812.18**

38 MECHANICAL/COOLING/HEATING SYSTEMS

Item	Net Price £	Material £	Labour hours	Labour £	Unit	Total rate £
LOW TEMPERATURE HOT WATER HEATING; PIPELINE: VALVES – cont						
Cocks; screwed joints to steel						
Bronze gland cock; complete with malleable iron lever; working pressure up to 10 bar at 100°C; screwed ends to steel						
15 mm dia.	58.66	74.24	0.77	25.95	nr	**100.19**
20 mm dia.	83.44	105.60	1.03	34.70	nr	**140.30**
25 mm dia.	112.43	142.30	1.19	40.10	nr	**182.40**
32 mm dia.	169.89	215.02	1.38	46.50	nr	**261.52**
40 mm dia.	237.23	300.24	1.62	54.58	nr	**354.82**
50 mm dia.	362.05	458.21	1.94	65.35	nr	**523.56**
Bronze three-way plug cock; complete with malleable iron lever; working pressure up to 10 bar at 100°C; screwed ends to steel						
15 mm dia.	118.90	150.48	0.77	25.95	nr	**176.43**
20 mm dia.	137.63	174.18	1.03	34.70	nr	**208.88**
25 mm dia.	192.38	243.48	1.19	40.10	nr	**283.58**
32 mm dia.	272.85	345.32	1.38	46.50	nr	**391.82**
40 mm dia.	328.98	416.36	1.62	54.58	nr	**470.94**
Air vents; including regulating, adjusting and testing						
Automatic air vent; maximum pressure up to 7 bar at 93°C; screwed ends to steel						
15 mm dia.	14.95	18.92	0.80	26.96	nr	**45.88**
Automatic air vent; maximum pressure up to 7 bar at 93°C; lockhead isolating valve; screwed ends to steel						
15 mm dia.	14.95	18.92	0.83	27.96	nr	**46.88**
Automatic air vent; maximum pressure up to 17 bar at 200°C; flanged ends (BS10, Table H); bolted connections to counter flange (measured separately)						
15 mm dia.	538.25	681.21	0.83	27.96	nr	**709.17**
Radiator valves						
Bronze; wheelhead or lockshield; chromium plated finish; screwed joints to steel						
Straight						
15 mm dia.	19.10	24.17	0.59	19.88	nr	**44.05**
20 mm dia.	29.72	37.61	0.73	24.60	nr	**62.21**
25 mm dia.	37.30	47.21	0.85	28.64	nr	**75.85**
Angled						
15 mm dia.	64.62	81.78	0.59	19.88	nr	**101.66**
20 mm dia.	85.19	107.81	0.73	24.60	nr	**132.41**
25 mm dia.	109.54	138.63	0.85	28.64	nr	**167.27**

38 MECHANICAL/COOLING/HEATING SYSTEMS

Item	Net Price £	Material £	Labour hours	Labour £	Unit	Total rate £
Bronze; wheelhead or lockshield; chromium plated finish; compression joints to copper						
Straight						
15 mm dia.	34.21	43.30	0.59	19.88	nr	**63.18**
20 mm dia.	43.11	54.56	0.73	24.60	nr	**79.16**
25 mm dia.	54.17	68.56	0.85	28.64	nr	**97.20**
Angled						
15 mm dia.	25.81	32.67	0.59	19.88	nr	**52.55**
20 mm dia.	27.09	34.28	0.73	24.60	nr	**58.88**
25 mm dia.	33.58	42.50	0.85	28.64	nr	**71.14**
Twin entry						
8 mm dia.	34.56	43.74	0.23	7.75	nr	**51.49**
10 mm dia.	39.22	49.64	0.23	7.75	nr	**57.39**
Bronze; thermostatic head; chromium plated finish; compression joints to copper						
Straight						
15 mm dia.	14.14	17.90	0.59	19.88	nr	**37.78**
20 mm dia.	20.74	26.25	0.73	24.60	nr	**50.85**
Angled						
15 mm dia.	13.15	16.64	0.59	19.88	nr	**36.52**
20 mm dia.	19.28	24.40	0.73	24.60	nr	**49.00**

38 MECHANICAL/COOLING/HEATING SYSTEMS

Item	Net Price £	Material £	Labour hours	Labour £	Unit	Total rate £
LOW TEMPERATURE HOT WATER HEATING; PIPELINE: EQUIPMENT						
GAUGES						
Thermometers and pressure gauges						
Dial thermometer; coated steel case and dial; glass window; brass pocket; BS 5235; pocket length 100 mm; screwed end Back/bottom entry						
100 mm dia. face	75.29	95.29	0.81	27.29	nr	**122.58**
150 mm dia. face	123.41	156.18	0.81	27.29	nr	**183.47**
Dial thermometer; coated steel case and dial; glass window; brass pocket; BS 5235; pocket length 100 mm; screwed end						
100 mm dia. face	111.62	141.27	0.81	27.29	nr	**168.56**
150 mm dia. face	125.00	158.20	0.81	27.29	nr	**185.49**
PRESSURIZATION UNITS						
LTHW pressurization unit complete with expansion vessel(s), interconnecting pipework and all necessary isolating and drain valves; includes placing in position; electrical work elsewhere. Selection based on a final working pressure of 4 bar, a 3 m static head and system operating temperatures of 82/71°C System volume						
2,400 litres	2254.21	2852.93	15.00	505.36	nr	**3358.29**
6,000–20,000 litres	4089.92	5176.20	22.00	741.18	nr	**5917.38**
25,000 litres	4659.78	5897.42	22.00	741.18	nr	**6638.60**
DIRT SEPARATORS						
Dirt separator; maximum operating temperature and pressure of 110°C and 10 bar; fitted with drain valve Bore size, flow rate (at 1.0 m/s velocity); threaded connections						
32 mm dia., 3.7 m³/h	107.54	136.10	2.29	77.15	nr	**213.25**
40 mm dia., 5.0 m³/h	128.02	162.02	2.45	82.54	nr	**244.56**
Bore size, flow rate (at 1.5 m/s velocity); flanged connections to PN16						
50 mm dia., 13.0 m³/h	723.75	915.98	3.00	101.07	nr	**1017.05**
65 mm dia., 21.0 m³/h	754.49	954.88	3.00	101.07	nr	**1055.95**
80 mm dia., 29.0 m³/h	1061.73	1343.72	3.84	129.37	nr	**1473.09**

38 MECHANICAL/COOLING/HEATING SYSTEMS

Item	Net Price £	Material £	Labour hours	Labour £	Unit	Total rate £
100 mm dia., 49.0 m³/h	1123.18	1421.49	4.44	149.58	nr	**1571.07**
125 mm dia., 74.0 m³/h	2152.48	2724.18	11.64	392.16	nr	**3116.34**
150 mm dia., 109.0 m³/h	2244.65	2840.82	15.75	530.61	nr	**3371.43**
200 mm dia., 181.0 m³/h	3383.18	4281.75	15.75	530.61	nr	**4812.36**
250 mm dia., 288.0 m³/h	5996.55	7589.23	15.75	530.61	nr	**8119.84**
300 mm dia., 407.0 m³/h	8765.21	11093.25	17.24	580.82	nr	**11674.07**
MICROBUBBLE DEAERATORS						
Microbubble deaerator; maximum operating temperature and pressure of 110°C and 10 bar; fitted with drain valve						
Bore size, flow rate (at 1.0 m/s velocity); threaded connections						
32 mm dia., 3.7 m³/h	99.00	125.29	2.29	77.15	nr	**202.44**
40 mm dia., 5.0 m³/h	117.79	149.07	2.45	82.54	nr	**231.61**
Bore size, flow rate (at 1.5 m/s velocity); flanged connections to PN16						
50 mm dia., 13.0 m³/h	723.75	915.98	3.00	101.07	nr	**1017.05**
65 mm dia., 21.0 m³/h	753.04	953.05	3.00	101.07	nr	**1054.12**
80 mm dia., 29.0 m³/h	1061.73	1343.72	3.84	129.37	nr	**1473.09**
100 mm dia., 49.0 m³/h	1123.18	1421.49	4.44	149.58	nr	**1571.07**
125 mm dia., 74.0 m³/h	2152.48	2724.18	11.64	392.16	nr	**3116.34**
150 mm dia., 109.0 m³/h	2244.65	2840.82	15.75	530.61	nr	**3371.43**
200 mm dia., 181.0 m³/h	3383.18	4281.75	15.75	530.61	nr	**4812.36**
250 mm dia., 288.0 m³/h	5996.55	7589.23	15.75	530.61	nr	**8119.84**
300 mm dia., 407.0 m³/h	8795.94	11132.14	17.24	580.82	nr	**11712.96**
COMBINED MICROBUBBLE DEAERATORS AND DIRT SEPARATORS						
Combined deaerator and dirt separators; maximum operating temperature and pressure of 110°C and 10 bar; fitted with drain valve						
Bore size, flow rate (at 1.5 m/s velocity); threaded connections						
25 mm dia., 2.0 m³/h	136.56	172.83	2.75	92.65	nr	**265.48**
Bore size, flow rate (at 1.5 m/s velocity); flanged connections to PN16						
50 mm dia., 13.0 m³/h	923.46	1168.73	3.60	121.28	nr	**1290.01**
65 mm dia., 21.0 m³/h	954.19	1207.62	3.60	121.28	nr	**1328.90**
80 mm dia., 29.0 m³/h	1292.17	1635.37	4.61	155.31	nr	**1790.68**
100 mm dia., 49.0 m³/h	1353.98	1713.60	5.33	179.56	nr	**1893.16**
125 mm dia., 74.0 m³/h	2582.63	3268.57	13.97	470.65	nr	**3739.22**
150 mm dia., 109.0 m³/h	2676.51	3387.40	18.90	636.75	nr	**4024.15**
200 mm dia., 181.0 m³/h	4151.32	5253.91	18.90	636.75	nr	**5890.66**
250 mm dia., 288.0 m³/h	7319.44	9263.49	18.90	636.75	nr	**9900.24**
300 mm dia., 407.0 m³/h	11224.94	14206.28	20.69	697.05	nr	**14903.33**

38 MECHANICAL/COOLING/HEATING SYSTEMS

Item	Net Price £	Material £	Labour hours	Labour £	Unit	Total rate £
LOW TEMPERATURE HOT WATER HEATING; PIPELINE: PUMPS						
Centrifugal heating and chilled water pump; belt drive; 3 phase, 1450 rpm motor; max. pressure 1000 kN/m²; max. temperature 125°C; bed plate; coupling guard; bolted connections; supply only mating flanges; includes fixing on prepared concrete base; electrical work elsewhere						
40 mm pump size; 4.0 l/s at 70 kPa max head; 0.25 kW max motor rating	1757.07	2223.75	7.59	255.72	nr	**2479.47**
40 mm pump size; 4.0 l/s at 130 kPa max head; 1.5 kW max motor rating	2198.95	2782.99	8.09	272.56	nr	**3055.55**
50 mm pump size; 8.5 l/s at 90 kPa max head; 2.2 kW max motor rating	2267.73	2870.03	8.67	292.10	nr	**3162.13**
50 mm pump size; 8.5 l/s at 190 kPa max head; 3 kW max motor rating	3941.42	4988.26	11.20	377.33	nr	**5365.59**
50 mm pump size; 8.5 l/s at 215 kPa max head; 4 kW max motor rating	4359.84	5517.81	11.70	394.17	nr	**5911.98**
65 mm pump size; 14.0 l/s at 90 kPa max head; 3 kW max motor rating	2259.41	2859.51	11.70	394.17	nr	**3253.68**
65 mm pump size; 14.0 l/s at 160 kPa max head; 4 kW max motor rating	2505.34	3170.75	11.70	394.17	nr	**3564.92**
80 mm pump size; 14.5 l/s at 210 kPa max head; 5.5 kW max motor rating	3843.47	4864.29	11.70	394.17	nr	**5258.46**
80 mm pump size; 22.0 l/s at 130 kPa max head; 5.5 kW max motor rating	3843.47	4864.29	13.64	459.54	nr	**5323.83**
80 mm pump size; 22.0 l/s at 200 kPa max head; 7.5 kW max motor rating	4024.80	5093.78	13.64	459.54	nr	**5553.32**
100 mm pump size; 22.0 l/s at 250 kPa max head; 11 kW max motor rating	5648.46	7148.69	13.64	459.54	nr	**7608.23**
100 mm pump size; 30.0 l/s at 100 kPa max head; 4.0 kW max motor rating	3161.90	4001.70	19.15	645.16	nr	**4646.86**
100 mm pump size; 36.0 l/s at 250 kPa max head; 15.0 kW max motor rating	5981.99	7570.81	19.15	645.16	nr	**8215.97**
100 mm pump size; 36.0 l/s at 550 kPa max head; 30.0 kW max motor rating	8118.38	10274.62	19.15	645.16	nr	**10919.78**
Centrifugal heating and chilled water pump; twin head belt drive; 3 phase, 1450 rpm motor; max. pressure 1000 kN/m²; max. temperature 125°C; bed plate; coupling guard; bolted connections; supply only mating flanges; includes fixing on prepared concrete base; electrical work elsewhere						
40 mm pump size; 4.0 l/s at 70 kPa max head; 0.75 kW max motor rating	3764.27	4764.07	7.59	255.72	nr	**5019.79**
40 mm pump size; 4.0 l/s at 130 kPa max head; 1.5 kW max motor rating	4531.30	5734.81	8.09	272.56	nr	**6007.37**

38 MECHANICAL/COOLING/HEATING SYSTEMS

Item	Net Price £	Material £	Labour hours	Labour £	Unit	Total rate £
50 mm pump size; 8.5 l/s at 90 kPa max head; 2.2 kW max motor rating	4629.27	5858.81	8.67	292.10	nr	**6150.91**
50 mm pump size; 8.5 l/s at 190 kPa max head; 4 kW max motor rating	5335.83	6753.03	11.20	377.33	nr	**7130.36**
65 mm pump size; 8.5 l/s at 215 kPa max head; 4 kW max motor rating	7332.55	9280.07	11.70	394.17	nr	**9674.24**
65 mm pump size; 14.0 l/s at 90 kPa max head; 3 kW max motor rating	7699.19	9744.09	11.70	394.17	nr	**10138.26**
65 mm pump size; 14.0 l/s at 160 kPa max head; 4 kW max motor rating	8084.16	10231.31	11.70	394.17	nr	**10625.48**
80 mm pump size; 14.5 l/s at 210 kPa max head; 7.5 kW max motor rating	8360.16	10580.62	13.64	459.54	nr	**11040.16**
Centrifugal heating and chilled water pump; close coupled; 3 phase, 1450 rpm motor; max. pressure 1000 kN/m²; max. temperature 110°C; bed plate; coupling guard; bolted connections; supply only mating flanges; includes fixing on prepared concrete base; electrical work elsewhere						
40 mm pump size; 4.0 l/s at 23 kPa max head; 0.55 kW max motor rating	1178.27	1491.22	7.31	246.28	nr	**1737.50**
50 mm pump size; 4.0 l/s at 75 kPa max head; 0.75 kW max motor rating	1306.08	1652.97	7.31	246.28	nr	**1899.25**
50 mm pump size; 7.0 l/s at 65 kPa max head; 0.75 kW max motor rating	1306.08	1652.97	8.01	269.86	nr	**1922.83**
65 mm pump size; 10.0 l/s at 33 kPa max head; 0.75 kW max motor rating	1480.42	1873.61	8.01	269.86	nr	**2143.47**
50 mm pump size; 4.0 l/s at 120 kPa max head; 1.5 kW max motor rating	1782.49	2255.92	8.01	269.86	nr	**2525.78**
80 mm pump size; 16.0 l/s at 80 kPa max head; 2.2 kW max motor rating	2307.73	2920.66	12.35	416.08	nr	**3336.74**
80 mm pump size; 16.0 l/s at 120 kPa max head; 4.0 kW max motor rating	2312.43	2926.61	12.35	416.08	nr	**3342.69**
100 mm pump size; 28.0 l/s at 40 kPa max head; 2.2 kW max motor rating	2279.89	2885.43	17.86	601.71	nr	**3487.14**
100 mm pump size; 28.0 l/s at 90 kPa max head; 4.0 kW max motor rating	2516.88	3185.36	17.86	601.71	nr	**3787.07**
125 mm pump size; 40.0 l/s at 50 kPa max head; 3.0 kW max motor rating	2400.70	3038.32	25.85	870.90	nr	**3909.22**
125 mm pump size; 40.0 l/s at 120 kPa max head; 7.5 kW max motor rating	2963.11	3750.11	25.85	870.90	nr	**4621.01**
150 mm pump size; 70.0 l/s at 75 kPa max head; 11 kW max motor rating	4355.18	5511.91	30.43	1025.20	nr	**6537.11**
150 mm pump size; 70.0 l/s at 120 kPa max head; 15.0 kW max motor rating	4654.99	5891.36	30.43	1025.20	nr	**6916.56**
150 mm pump size; 70.0 l/s at 150 kPa max head; 15.0 kW max motor rating	4654.99	5891.36	30.43	1025.20	nr	**6916.56**

38 MECHANICAL/COOLING/HEATING SYSTEMS

Item	Net Price £	Material £	Labour hours	Labour £	Unit	Total rate £
LOW TEMPERATURE HOT WATER HEATING; PIPELINE: PUMPS – cont						
Centrifugal heating & chilled water pump; close coupled; 3 phase, variable speed motor; max. system pressure 1000 kN/m²; max. temperature 110°C; bed plate; coupling guard; bolted connections; supply only mating flanges; includes fixing on prepared concrete base; electrical work elsewhere.						
40 mm pump size; 4.0 l/s at 23 kPa max head; 0.55 kW max motor rating	2126.47	2691.26	7.31	246.28	nr	**2937.54**
40 mm pump size; 4.0 l/s at 75 kPa max head; 0.75 kW max motor rating	2347.25	2970.68	7.31	246.28	nr	**3216.96**
50 mm pump size; 7.0 l/s at 65 kPa max head; 1.5 kW max motor rating	2879.44	3644.22	8.01	269.86	nr	**3914.08**
50 mm pump size; 10.0 l/s at 33 kPa max head; 1.5 kW max motor rating	2879.44	3644.22	8.01	269.86	nr	**3914.08**
50 mm pump size; 4.0 l/s at 120 kPa max head; 1.5 kW max motor rating	2879.44	3644.22	8.01	269.86	nr	**3914.08**
80 mm pump size; 16.0 l/s at 80 kPa max head; 2.2 kW max motor rating	3753.28	4750.16	12.35	416.08	nr	**5166.24**
80 mm pump size; 16.0 l/s at 120 kPa max head; 3.0 kW max motor rating	4048.42	5123.68	12.35	416.08	nr	**5539.76**
100 mm pump size; 28.0 l/s at 40 kPa max head; 2.2 kW max motor rating	3846.22	4867.78	17.86	601.71	nr	**5469.49**
100 mm pump size; 28.0 l/s at 90 kPa max head; 4.0 kW max motor rating	4587.59	5806.06	17.86	601.71	nr	**6407.77**
125 mm pump size; 40.0 l/s at 50 kPa max head; 3.0 kW max motor rating	4403.91	5573.59	25.85	870.90	nr	**6444.49**
125 mm pump size; 40.0 l/s at 120 kPa max head; 7.5 kW max motor rating	5435.61	6879.31	25.85	870.90	nr	**7750.21**
150 mm pump size; 70.0 l/s at 75 kPa max head; 7.5 kW max motor rating	7164.91	9067.91	30.43	1025.20	nr	**10093.11**
Glandless domestic heating pump; for low pressure domestic hot water heating systems; 240 volt; 50 Hz electric motor; max working pressure 1000 N/m² and max temperature of 130°C; includes fixing in position; electrical work elsewhere						
1" BSP unions – 2 speed	169.69	214.76	1.58	53.23	nr	**267.99**
1.25" BSP unions – 3 speed	250.05	316.47	1.58	53.23	nr	**369.70**
Glandless pumps; for hot water secondary supply; silent running; 3 phase; max pressure 1000 kN/m²; max temperature 130°C ; bolted connections; supply only mating flanges; including fixing in position; electrical elsewhere						
1" BSP unions – 3 speed	348.43	440.98	1.58	53.23	nr	**494.21**

38 MECHANICAL/COOLING/HEATING SYSTEMS

Item	Net Price £	Material £	Labour hours	Labour £	Unit	Total rate £
Pipeline mounted circulator; for heating and chilled water; silent running; 3 phase; 1450 rpm motor; max pressure 1000 kN/m²; max temperature 120°C; bolted connections; supply only mating flanges; includes fixing in position; electrical elsewhere						
32 mm pump size; 2.0 l/s at 17 kPa max head; 0.2 kW max motor rating	696.88	881.97	6.44	216.97	nr	**1098.94**
50 mm pump size; 3.0 l/s at 20 kPa max head; 0.2 kW max motor rating	696.88	881.97	6.86	231.11	nr	**1113.08**
65 mm pump size; 5.0 l/s at 30 kPa max head; 0.37 kW max motor rating	1102.39	1395.18	7.48	252.01	nr	**1647.19**
65 mm pump size; 8.0 l/s at 37 kPa max head; 0.75 kW max motor rating	1234.29	1562.12	7.48	252.01	nr	**1814.13**
80 mm pump size; 12.0 l/s at 42 kPa max head; 1.1 kW max motor rating	1435.10	1816.26	8.01	269.86	nr	**2086.12**
100 mm pump size; 25.0 l/s at 37 kPa max head; 2.2 kW max motor rating	2045.35	2588.60	9.11	306.92	nr	**2895.52**
Dual pipeline mounted circulator; for heating & chilled water; silent running; 3 phase; 1450 rpm motor; max pressure 1000 kN/m²; max temperature 120°C; bolted connections; supply only mating flanges; includes fixing in position; electrical work elsewhere						
40 mm pump size; 2.0 l/s at 17 kPa max head; 0.8 kW max motor rating	1285.49	1626.91	7.88	265.48	nr	**1892.39**
50 mm pump size; 3.0 l/s at 20 kPa max head; 0.2 kW max motor rating	1289.41	1631.87	8.01	269.86	nr	**1901.73**
65 mm pump size; 5.0 l/s at 30 kPa max head; 0.37 kW max motor rating	2092.59	2648.39	9.20	309.95	nr	**2958.34**
65 mm pump size; 8.0 l/s at 37 kPa max head; 0.75 kW max motor rating	2419.38	3061.97	9.20	309.95	nr	**3371.92**
100 mm pump size; 12.0 l/s at 42 kPa max head; 1.1 kW max motor rating	2736.30	3463.06	9.45	318.37	nr	**3781.43**
Glandless accelerator pumps; for low and medium pressure heating services; silent running; 3 phase; 1450 rpm motor; max pressure 1000 kN/m²; max temperature 130°C; bolted connections; supply only mating flanges; includes fixing in position; electrical work elsewhere						
40 mm pump size; 4.0 l/s at 15 kPa max head; 0.35 kW max motor rating	538.17	681.11	6.94	233.81	nr	**914.92**
50 mm pump size; 6.0 l/s at 20 kPa max head; 0.45 kW max motor rating	573.53	725.86	7.35	247.63	nr	**973.49**
80 mm pump size; 13.0 l/s at 28 kPa max head; 0.58 kW max motor rating	1130.15	1430.32	7.76	261.43	nr	**1691.75**

38 MECHANICAL/COOLING/HEATING SYSTEMS

Item	Net Price £	Material £	Labour hours	Labour £	Unit	Total rate £
LOW TEMPERATURE HOT WATER HEATING; PIPELINE: PUMPS – cont						
Air source heat pumps						
Two stage heat pump (air to water) in monoblock design, cascadable for outdoor installation. Providing heating and cooling. Unit and controller only.						
Heat output and cooling capacity						
17 kW heating, 18 kW cooling	10525.95	13321.64	12.00	404.29	nr	**13725.93**
24 kW heating, 24 kW cooling	11963.64	15141.18	12.00	404.29	nr	**15545.47**
32 kW heating, 31 kW cooling	13389.56	16945.82	12.00	404.29	nr	**17350.11**
64 kW heating, 62 kW cooling	19828.53	25094.99	12.00	404.29	nr	**25499.28**

38 MECHANICAL/COOLING/HEATING SYSTEMS

Item	Net Price £	Material £	Labour hours	Labour £	Unit	Total rate £
LOW TEMPERATURE HOT WATER HEATING; PIPELINE: HEAT EXCHANGERS						
Plate heat exchanger; for use in LTHW systems; painted carbon steel frame; stainless steel plates, nitrile rubber gaskets; design pressure of 10 bar and operating temperature of 110/135°C						
Primary side; 80°C in, 69°C out; secondary side; 82°C in, 71°C out						
107 kW, 2.38 l/s	2498.50	3162.11	10.00	336.91	nr	**3499.02**
245 kW, 5.46 l/s	3839.32	4859.04	10.00	336.91	nr	**5195.95**
287 kW, 6.38 l/s	4261.83	5393.77	10.00	336.91	nr	**5730.68**
328 kW, 7.31 l/s	4650.35	5885.49	10.00	336.91	nr	**6222.40**
364 kW, 8.11 l/s	5518.94	6984.77	10.00	336.91	nr	**7321.68**
403 kW, 8.96 l/s	5793.65	7332.44	10.00	336.91	nr	**7669.35**
453 kW, 10.09 l/s	6027.81	7628.80	10.00	336.91	nr	**7965.71**
490 kW, 10.89 l/s	6248.87	7908.57	10.00	336.91	nr	**8245.48**
1000 kW, 21.7 l/s	10901.20	13796.52	12.00	404.29	nr	**14200.81**
1500 kW, 32.6 l/s	14955.80	18928.06	12.00	404.29	nr	**19332.35**
2000 kW, 43.4 l/s	19306.13	24433.84	15.00	505.36	nr	**24939.20**
2500 kW, 54.3 l/s	23627.08	29902.43	15.00	505.36	nr	**30407.79**
Note: For temperature conditions different to those above, the cost of the units can vary significantly, and so manufacturer's advice should be sought.						
District heating HIUs						
District heating station; for conventional connection to a district or local heating network; offering primary F/R and secondary F/R connections, with stainless steel plate heat exchanger, integrated control panel, heat meter and combination valve to adjust output						
Heat output						
149 kW	5091.84	6444.23	12.00	404.29	nr	**6848.52**
194 kW	5677.36	7185.27	12.00	404.29	nr	**7589.56**
240 kW	5715.84	7233.97	12.00	404.29	nr	**7638.26**
400 kW	6834.88	8650.22	12.00	404.29	nr	**9054.51**
494 kW	7987.20	10108.60	12.00	404.29	nr	**10512.89**
572 kW	9205.04	11649.90	12.00	404.29	nr	**12054.19**
915 kW	11245.52	14232.33	14.00	471.67	nr	**14704.00**
1417 kW	14457.04	18296.84	14.00	471.67	nr	**18768.51**

38 MECHANICAL/COOLING/HEATING SYSTEMS

Item	Net Price £	Material £	Labour hours	Labour £	Unit	Total rate £
LOW TEMPERATURE HOT WATER HEATING; PIPELINE: CALORIFIERS						
Non-storage calorifiers; mild steel; heater battery duty 116°C/90°C to BS 853, maximum test on shell 11.55 bar, tubes 26.25 bar						
Horizontal or vertical; primary water at 116°C on, 90°C off						
40 kW capacity	1081.68	1368.98	3.00	101.07	nr	**1470.05**
88 kW capacity	1260.15	1594.85	5.00	168.45	nr	**1763.30**
176 kW capacity	1444.07	1827.62	7.04	237.26	nr	**2064.88**
293 kW capacity	2002.64	2534.54	9.01	303.52	nr	**2838.06**
586 kW capacity	2983.52	3775.95	22.22	748.66	nr	**4524.61**
879 kW capacity	4155.12	5258.72	28.57	962.57	nr	**6221.29**
1465 kW capacity	6675.44	8448.44	50.00	1684.51	nr	**10132.95**
2000 kW capacity	12150.17	15377.25	60.00	2021.42	nr	**17398.67**

38 MECHANICAL/COOLING/HEATING SYSTEMS

Item	Net Price £	Material £	Labour hours	Labour £	Unit	Total rate £
LOW TEMPERATURE HOT WATER HEATING; PIPELINE: HEAT EMITTERS						
Perimeter convector heating; metal casing with standard finish; aluminium extruded grille; including backplates						
Top/sloping/flat front outlet						
60 × 200 mm	36.76	46.52	2.00	67.38	m	**113.90**
60 × 300 mm	39.99	50.61	2.00	67.38	m	**117.99**
60 × 450 mm	49.57	62.73	2.00	67.38	m	**130.11**
60 × 525 mm	52.77	66.79	2.00	67.38	m	**134.17**
60 × 600 mm	57.56	72.84	2.00	67.38	m	**140.22**
90 × 260 mm	39.99	50.61	2.00	67.38	m	**117.99**
90 × 300 mm	41.57	52.61	2.00	67.38	m	**119.99**
90 × 450 mm	51.19	64.78	2.00	67.38	m	**132.16**
90 × 525 mm	55.97	70.84	2.00	67.38	m	**138.22**
90 × 600 mm	59.18	74.89	2.00	67.38	m	**142.27**
Extra over for dampers						
Damper	17.58	22.25	0.25	8.43	nr	**30.68**
Extra over for fittings						
60 mm end caps	14.07	17.81	0.25	8.43	nr	**26.24**
90 mm end caps	22.14	28.02	0.25	8.43	nr	**36.45**
60 mm corners	29.91	37.86	0.25	8.43	nr	**46.29**
90 mm corners	43.99	55.68	0.25	8.43	nr	**64.11**

38 MECHANICAL/COOLING/HEATING SYSTEMS

Item	Net Price £	Material £	Labour hours	Labour £	Unit	Total rate £
LOW TEMPERATURE HOT WATER HEATING; PIPELINE: RADIATORS						
Radiant strip heaters						
Suitable for connection to hot water system; aluminium sheet panels with steel pipe clamped to upper surface; including insulation, sliding brackets, cover plates, end closures; weld or screwed BSP ends						
One pipe						
1500 mm long	74.36	94.11	3.11	104.78	nr	**198.89**
3000 mm long	117.60	148.84	3.11	104.78	nr	**253.62**
4500 mm long	159.54	201.91	3.11	104.78	nr	**306.69**
6000 mm long	219.92	278.33	3.11	104.78	nr	**383.11**
Two pipe						
1500 mm long	138.82	175.69	4.15	139.81	nr	**315.50**
3000 mm long	219.27	277.51	4.15	139.81	nr	**417.32**
4500 mm long	299.43	378.96	4.15	139.81	nr	**518.77**
6000 mm long	403.40	510.54	4.15	139.81	nr	**650.35**
Pressed steel panel type radiators; fixed with and including brackets; taking down once for decoration; refixing						
300 mm high; single panel						
500 mm length	20.30	25.69	2.03	68.40	nr	**94.09**
1000 mm length	40.56	51.33	2.03	68.40	nr	**119.73**
1500 mm length	53.15	67.27	2.03	68.40	nr	**135.67**
2000 mm length	59.63	75.47	2.47	83.22	nr	**158.69**
2500 mm length	66.09	83.64	2.97	100.06	nr	**183.70**
3000 mm length	79.09	100.09	3.22	108.48	nr	**208.57**
300 mm high; double panel; convector						
500 mm length	39.02	49.38	2.13	71.76	nr	**121.14**
1000 mm length	78.09	98.83	2.13	71.76	nr	**170.59**
1500 mm length	117.12	148.23	2.13	71.76	nr	**219.99**
2000 mm length	156.19	197.67	2.57	86.58	nr	**284.25**
2500 mm length	195.20	247.05	3.07	103.42	nr	**350.47**
3000 mm length	234.26	296.48	3.31	111.51	nr	**407.99**
450 mm high; single panel						
500 mm length	18.93	23.96	2.08	70.08	nr	**94.04**
1000 mm length	37.90	47.97	2.08	70.08	nr	**118.05**
1600 mm length	60.63	76.73	2.53	85.24	nr	**161.97**
2000 mm length	75.81	95.95	2.97	100.06	nr	**196.01**
2400 mm length	90.94	115.09	3.47	116.91	nr	**232.00**
3000 mm length	113.67	143.86	3.82	128.70	nr	**272.56**
450 mm high; double panel; convector						
500 mm length	34.70	43.92	2.18	73.44	nr	**117.36**
1000 mm length	69.39	87.82	2.18	73.44	nr	**161.26**
1600 mm length	127.09	160.84	2.63	88.60	nr	**249.44**
2000 mm length	213.65	270.39	3.06	103.10	nr	**373.49**
2400 mm length	256.40	324.50	3.37	113.53	nr	**438.03**
3000 mm length	320.50	405.62	3.92	132.07	nr	**537.69**

38 MECHANICAL/COOLING/HEATING SYSTEMS

Item	Net Price £	Material £	Labour hours	Labour £	Unit	Total rate £
600 mm high; single panel						
500 mm length	25.38	32.12	2.18	73.44	nr	**105.56**
1000 mm length	50.75	64.23	2.43	81.87	nr	**146.10**
1600 mm length	81.19	102.75	3.13	105.45	nr	**208.20**
2000 mm length	101.49	128.44	3.77	127.02	nr	**255.46**
2400 mm length	121.80	154.15	4.07	137.11	nr	**291.26**
3000 mm length	152.26	192.70	5.11	172.16	nr	**364.86**
600 mm high; double panel; convector						
500 mm length	43.69	55.29	2.28	76.81	nr	**132.10**
1000 mm length	87.38	110.59	2.28	76.81	nr	**187.40**
1600 mm length	160.04	202.55	3.23	108.82	nr	**311.37**
2000 mm length	269.05	340.51	3.87	130.38	nr	**470.89**
2400 mm length	322.86	408.61	4.17	140.49	nr	**549.10**
3000 mm length	403.57	510.75	5.24	176.53	nr	**687.28**
700 mm high; single panel						
500 mm length	29.68	37.56	2.23	75.13	nr	**112.69**
1000 mm length	59.36	75.13	2.83	95.33	nr	**170.46**
1600 mm length	94.99	120.22	3.73	125.66	nr	**245.88**
2000 mm length	118.75	150.29	4.46	150.26	nr	**300.55**
2400 mm length	142.52	180.38	4.48	150.93	nr	**331.31**
3000 mm length	178.10	225.40	5.24	176.53	nr	**401.93**
700 mm high; double panel; convector						
500 mm length	56.78	71.86	2.33	78.49	nr	**150.35**
1000 mm length	152.78	193.36	3.08	103.77	nr	**297.13**
1600 mm length	244.42	309.33	3.83	129.02	nr	**438.35**
2000 mm length	305.53	386.68	4.17	140.49	nr	**527.17**
2400 mm length	366.65	464.03	4.37	147.22	nr	**611.25**
3000 mm length	458.31	580.04	4.82	162.39	nr	**742.43**
Flat panel type steel radiators; fixed with and including brackets; taking down once for decoration; refixing						
300 mm high; single panel (44 mm deep)						
500 mm length	208.97	264.48	2.03	68.40	nr	**332.88**
1000 mm length	328.21	415.39	2.03	68.40	nr	**483.79**
1500 mm length	447.46	566.31	2.03	68.40	nr	**634.71**
2000 mm length	566.71	717.23	2.47	83.22	nr	**800.45**
2400 mm length	662.11	837.96	2.97	100.06	nr	**938.02**
3000 mm length	952.61	1205.62	3.22	108.48	nr	**1314.10**
300 mm high; double panel (100 mm deep)						
500 mm length	414.52	524.62	2.03	68.40	nr	**593.02**
1000 mm length	649.60	822.14	2.03	68.40	nr	**890.54**
1500 mm length	884.70	1119.68	2.03	68.40	nr	**1188.08**
2000 mm length	1119.78	1417.19	2.47	83.22	nr	**1500.41**
2400 mm length	1307.85	1655.21	2.97	100.06	nr	**1755.27**
3000 mm length	1737.36	2198.81	3.22	108.48	nr	**2307.29**
500 mm high; single panel (44 mm deep)						
500 mm length	246.44	311.90	2.13	71.76	nr	**383.66**
1000 mm length	403.17	510.25	2.13	71.76	nr	**582.01**
1500 mm length	559.89	708.60	2.13	71.76	nr	**780.36**

38 MECHANICAL/COOLING/HEATING SYSTEMS

Item	Net Price £	Material £	Labour hours	Labour £	Unit	Total rate £
LOW TEMPERATURE HOT WATER HEATING; PIPELINE: RADIATORS – cont						
Flat panel type steel radiators – cont						
2000 mm length	716.61	906.94	2.57	86.58	nr	**993.52**
2400 mm length	841.99	1065.62	3.07	103.42	nr	**1169.04**
3000 mm length	1177.48	1490.22	3.31	111.51	nr	**1601.73**
500 mm high; double panel (100 mm deep)						
500 mm length	482.66	610.86	2.08	70.08	nr	**680.94**
1000 mm length	785.89	994.63	2.08	70.08	nr	**1064.71**
1500 mm length	1089.09	1378.35	2.53	85.24	nr	**1463.59**
2000 mm length	1392.34	1762.14	2.97	100.06	nr	**1862.20**
2400 mm length	1634.92	2069.16	3.47	116.91	nr	**2186.07**
3000 mm length	2146.21	2716.25	3.82	128.70	nr	**2844.95**
600 mm high; single panel (44 mm deep)						
500 mm length	278.24	352.14	2.18	73.44	nr	**425.58**
1000 mm length	455.41	576.36	2.18	73.44	nr	**649.80**
1500 mm length	632.57	1376.95	2.63	88.60	nr	**1465.55**
2000 mm length	809.74	1024.81	3.06	103.10	nr	**1127.91**
2400 mm length	951.48	1204.19	3.37	113.53	nr	**1317.72**
3000 mm length	1311.48	1659.81	3.92	132.07	nr	**1791.88**
600 mm high; double panel (100 mm deep)						
500 mm length	549.44	695.37	2.18	73.44	nr	**768.81**
1000 mm length	896.96	1135.19	2.43	81.87	nr	**1217.06**
1500 mm length	1244.47	1575.00	3.13	105.45	nr	**1680.45**
2000 mm length	1592.00	2014.84	3.77	127.02	nr	**2141.86**
2400 mm length	1870.01	2366.68	4.07	137.11	nr	**2503.79**
3000 mm length	2434.44	3081.03	5.11	172.16	nr	**3253.19**
700 mm high; single panel (44 mm deep)						
500 mm length	302.09	382.32	2.28	76.81	nr	**459.13**
1000 mm length	503.11	636.73	2.28	76.81	nr	**713.54**
1500 mm length	704.12	891.14	3.23	108.82	nr	**999.96**
2000 mm length	905.13	1145.54	3.87	130.38	nr	**1275.92**
2400 mm length	1066.27	1349.48	4.17	140.49	nr	**1489.97**
3000 mm length	1454.58	1840.92	5.24	176.53	nr	**2017.45**
700 mm high; double panel (100 mm deep)						
500 mm length	593.74	751.44	2.23	75.13	nr	**826.57**
1000 mm length	985.55	1247.31	2.83	95.33	nr	**1342.64**
1500 mm length	1377.36	1743.19	3.73	125.66	nr	**1868.85**
2000 mm length	1769.16	2239.05	4.46	150.26	nr	**2389.31**
2400 mm length	2082.61	2635.75	4.48	150.93	nr	**2786.68**

38 MECHANICAL/COOLING/HEATING SYSTEMS

Item	Net Price £	Material £	Labour hours	Labour £	Unit	Total rate £
Fan convector; sheet metal casing with lockable access panel; centrifugal fan; air filter; LPHW heating coil; extruded aluminium grilles; 3 speed; includes fixing in position; electrical work elsewhere						
Free-standing flat top, 695 mm high, medium speed rating						
Entering air temperature, 18°C						
695 mm long, 1 row 1.94 kW, 75 l/sec	706.79	894.51	2.73	91.97	nr	**986.48**
695 mm long, 2 row 2.64 kW, 75 l/sec	706.79	894.51	2.73	91.97	nr	**986.48**
895 mm long, 1 row 4.02 kW, 150 l/sec	796.44	1007.98	2.73	91.97	nr	**1099.95**
895 mm long, 2 row 5.62 kW, 150 l/sec	796.44	1007.98	2.73	91.97	nr	**1099.95**
1195 mm long, 1 row 6.58 kW, 250 l/sec	906.76	1147.60	3.00	101.07	nr	**1248.67**
1195 mm long, 2 row 9.27 kW, 250 l/sec	906.76	1147.60	3.00	101.07	nr	**1248.67**
1495 mm long, 1 row 9.04 kW, 340 l/sec	1011.93	1280.70	3.26	109.83	nr	**1390.53**
1495 mm long, 2 row 12.73 kW, 340 l/sec	1011.93	1280.70	3.26	109.83	nr	**1390.53**
Free-standing flat top, 695 mm high, medium speed rating, c/w floor plinth						
695 mm long, 1 row 1.94 kW, 75 l/sec	742.12	939.23	2.73	91.97	nr	**1031.20**
695 mm long, 2 row 2.64 kW, 75 l/sec	742.12	939.23	2.73	91.97	nr	**1031.20**
895 mm long, 1 row 4.02 kW, 150 l/sec	836.26	1058.37	2.73	91.97	nr	**1150.34**
895 mm long, 2 row 5.62 kW, 150 l/sec	836.26	1058.37	2.73	91.97	nr	**1150.34**
1195 mm long, 1 row 6.58 kW, 250 l/sec	952.12	1205.01	3.00	101.07	nr	**1306.08**
1195 mm long, 2 row 9.27 kW, 250 l/sec	952.12	1205.01	3.00	101.07	nr	**1306.08**
1495 mm long, 1 row 9.04 kW, 340 l/sec	1062.52	1344.73	3.26	109.83	nr	**1454.56**
1495 mm long, 2 row 12.73 kW, 340 l/sec	1062.52	1344.73	3.26	109.83	nr	**1454.56**
Free-standing sloping top, 695 mm high, medium speed rating, c/w floor plinth						
695 mm long, 1 row 1.94 kW, 75 l/sec	767.99	971.97	2.73	91.97	nr	**1063.94**
695 mm long, 2 row 2.64 kW, 75 l/sec	767.99	971.97	2.73	91.97	nr	**1063.94**
895 mm long, 1 row 4.02 kW, 150 l/sec	862.12	1091.10	2.73	91.97	nr	**1183.07**
895 mm long, 2 row 5.62 kW, 150 l/sec	862.12	1091.10	2.73	91.97	nr	**1183.07**
1195 mm long, 1 row 6.58 kW, 250 l/sec	977.97	1237.72	3.00	101.07	nr	**1338.79**
1195 mm long, 2 row 9.27 kW, 250 l/sec	977.97	1237.72	3.00	101.07	nr	**1338.79**
1495 mm long, 1 row 9.04 kW, 340 l/sec	1088.37	1377.44	3.26	109.83	nr	**1487.27**
1495 mm long, 2 row 12.73 kW, 340 l/sec	1088.37	1377.44	3.26	109.83	nr	**1487.27**
Wall mounted high level sloping discharge						
695 mm long, 1 row 1.94 kW, 75 l/sec	775.75	981.79	2.73	91.97	nr	**1073.76**
695 mm long, 2 row 2.64 kW, 75 l/sec	775.75	981.79	2.73	91.97	nr	**1073.76**
895 mm long, 1 row 4.02 kW, 150 l/sec	798.18	1010.17	2.73	91.97	nr	**1102.14**
895 mm long, 2 row 5.62 kW, 150 l/sec	798.18	1010.17	2.73	91.97	nr	**1102.14**
1195 mm long, 1 row 6.58 kW, 250 l/sec	971.39	1229.39	3.00	101.07	nr	**1330.46**
1195 mm long, 2 row 9.27 kW, 250 l/sec	971.39	1229.39	3.00	101.07	nr	**1330.46**
1495 mm long, 1 row 9.04 kW, 340 l/sec	1053.29	1333.05	3.26	109.83	nr	**1442.88**
1495 mm long, 2 row 12.73 kW, 340 l/sec	1053.29	1333.05	3.26	109.83	nr	**1442.88**
Ceiling mounted sloping inlet/outlet 665 mm wide						
895 mm long, 1 row 4.02 kW, 150 l/sec	877.45	1110.50	4.15	139.81	nr	**1250.31**
895 mm long, 2 row 5.62 kW, 150 l/sec	877.45	1110.50	4.15	139.81	nr	**1250.31**
1195 mm long, 1 row 6.58 kW, 250 l/sec	986.08	1247.98	4.15	139.81	nr	**1387.79**
1195 mm long, 2 row 9.27 kW, 250 l/sec	986.08	1247.98	4.15	139.81	nr	**1387.79**
1495 mm long, 1 row 9.04 kW, 340 l/sec	1082.60	1370.14	4.15	139.81	nr	**1509.95**
1495 mm long, 2 row 12.73 kW, 340 l/sec	1082.60	1370.14	4.15	139.81	nr	**1509.95**

38 MECHANICAL/COOLING/HEATING SYSTEMS

Item	Net Price £	Material £	Labour hours	Labour £	Unit	Total rate £
LOW TEMPERATURE HOT WATER HEATING; PIPELINE: RADIATORS – cont						
Fan convector – cont						
Free-standing unit, extended height 1700/ 1900/2100 mm						
895 mm long, 1 row 4.02 kW, 150 l/sec	1015.72	1285.49	3.11	104.78	nr	**1390.27**
895 mm long, 2 row 5.62 kW, 150 l/sec	1015.37	1285.05	3.11	104.78	nr	**1389.83**
1195 mm long, 1 row 6.58 kW, 250 l/sec	1186.04	1501.06	3.11	104.78	nr	**1605.84**
1195 mm long, 2 row 9.27 kW, 250 l/sec	1186.04	1501.06	3.11	104.78	nr	**1605.84**
1495 mm long, 1 row 9.04 kW, 340 l/sec	1305.00	1651.61	3.11	104.78	nr	**1756.39**
1495 mm long, 2 row 12.73 kW, 340 l/sec	1305.00	1651.61	3.11	104.78	nr	**1756.39**
LTHW trench heating; water temperatures 90°C/70°C; room air temperature 20°C; convector with copper tubes and aluminium fins within steel duct; Includes fixing within floor screed; electrical work elsewhere						
Natural convection type						
Normal capacity, 182 mm width, complete with linear, natural anodized aluminium grille (grille also costed separately below)						
90 mm deep						
1300 mm long, 287 W output	291.22	368.57	2.00	67.38	nr	**435.95**
2300 mm long, 576 W output	475.86	602.25	4.00	134.76	nr	**737.01**
3300 mm long, 864 W output	651.39	824.40	5.00	168.45	nr	**992.85**
4500 mm long, 1210 W output	891.63	1128.44	7.00	235.83	nr	**1364.27**
4900 mm long, 1325 W output	957.28	1211.54	8.00	269.52	nr	**1481.06**
110 mm deep						
1300 mm long, 331 W output	293.72	371.73	2.00	67.38	nr	**439.11**
2300 mm long, 662 W output	480.04	607.54	4.00	134.76	nr	**742.30**
3300 mm long, 993 W output	657.26	831.82	5.00	168.45	nr	**1000.27**
4500 mm long, 1389 W output	899.81	1138.80	7.00	235.83	nr	**1374.63**
4900 mm long, 1522 W output	965.76	1222.27	8.00	269.52	nr	**1491.79**
140 mm deep						
1300 mm long, 551 W output	345.82	437.67	2.00	67.38	nr	**505.05**
2300 mm long, 1100 W output	568.59	719.61	4.00	134.76	nr	**854.37**
3300 mm long, 1651 W output	781.00	988.43	5.00	168.45	nr	**1156.88**
4500 mm long, 2310 W output	1064.48	1347.20	7.00	235.83	nr	**1583.03**
4900 mm long, 2530 W output	1144.48	1448.45	8.00	269.52	nr	**1717.97**
190 mm deep						
1300 mm long, 625 W output	357.49	452.44	2.00	67.38	nr	**519.82**
2300 mm long, 1250 W output	589.62	746.22	4.00	134.76	nr	**880.98**
3300 mm long, 1874 W output	811.39	1026.89	5.00	168.45	nr	**1195.34**
4500 mm long, 2624 W output	1106.85	1400.83	7.00	235.83	nr	**1636.66**
4900 mm long, 2875 W output	1190.40	1506.57	8.00	269.52	nr	**1776.09**

38 MECHANICAL/COOLING/HEATING SYSTEMS

Item	Net Price £	Material £	Labour hours	Labour £	Unit	Total rate £
Fan assisted type (outputs assume fan at 50%)						
Normal capacity, 182 mm width, complete with natural anodized aluminium grille						
140 mm deep						
1300 mm long, 1439 W output	568.03	718.89	2.00	67.38	nr	**786.27**
2300 mm long, 3331 W output	939.37	1188.87	4.00	134.76	nr	**1323.63**
3300 mm long, 5225 W output	1296.67	1641.07	5.00	168.45	nr	**1809.52**
4300 mm long, 5753 W output	1521.94	1926.16	7.00	235.83	nr	**2161.99**
4900 mm long, 6070 W output	1641.43	2077.40	8.00	269.52	nr	**2346.92**
Linear grille anodized aluminium, 170 mm width (if supplied as a separate item)	154.62	195.69	–	–	m	**195.69**
Roll up grille, natural anodized aluminium	154.62	195.69	–	–	m	**195.69**
Thermostatic valve with remote regulator (c/w valve body)	63.84	80.80	4.00	134.76	nr	**215.56**
Fan speed controller	31.20	39.49	2.00	67.38	nr	**106.87**
Note: As an alternative to thermostatic control, the system can be controlled via two port valves. Refer to valve section for valve prices.						
LTHW underfloor heating; water flow and return temperatures of 60°C and 70°C; pipework at 300 mm centres; pipe fixings; flow and return manifolds and zone actuators; wiring block; insulation; includes fixing in position; excludes secondary pump, mixing valve, zone thermostats and floor finishes; electrical work elsewhere						
Note: All rates are expressed on a m² basis, for the following example areas						
Screeded floor with 15–25 mm stone/marble finish (producing 80–100 W/m²)						
250 m² area (single zone)	26.63	33.70	0.14	4.72	m²	**38.42**
1000 m² area (single zone)	26.94	34.09	0.12	4.03	m²	**38.12**
5000 m² area (multi-zone)	26.87	34.00	0.10	3.37	m²	**37.37**
Screeded floor with 10 mm carpet tile (producing 80–100 W/m²)						
250 m² area (single zone)	26.63	33.70	0.14	4.72	m²	**38.42**
1000 m² area (single zone)	26.94	34.09	0.12	4.03	m²	**38.12**
5000 m² area (multi-zone)	26.87	34.00	0.10	3.37	m²	**37.37**
Floating timber floor with 20 mm timber finish (producing 70–80 Wm²)						
250 m² are (single zone)	34.84	44.09	0.14	4.72	m²	**48.81**
1000 m² area(single zone)	37.35	47.28	0.12	4.03	m²	**51.31**
5000 m² area (multi-zone)	35.68	45.16	0.10	3.37	m²	**48.53**

38 MECHANICAL/COOLING/HEATING SYSTEMS

Item	Net Price £	Material £	Labour hours	Labour £	Unit	Total rate £
LOW TEMPERATURE HOT WATER HEATING; PIPELINE: RADIATORS – cont						
LTHW underfloor heating – cont						
Floating timber floor with 10 mm carpet tile (producing 70–80 W/m²)						
250 m² are (single zone)	45.35	57.40	0.16	5.39	m²	**62.79**
1000 m² area(single zone)	42.05	53.22	0.12	4.03	m²	**57.25**
5000 m² area (multi-zone)	40.23	50.92	0.10	3.37	m²	**54.29**

38 MECHANICAL/COOLING/HEATING SYSTEMS

Item	Net Price £	Material £	Labour hours	Labour £	Unit	Total rate £
LOW TEMPERATURE HOT WATER HEATING; PIPELINE: PIPE FREEZING						
Freeze isolation of carbon steel or copper pipelines containing static water, either side of work location, freeze duration not exceeding 4 hours assuming that flow and return circuits are treated concurrently and activities undertaken during normal working hours						
Up to 4 freezes						
50 mm dia.	482.86	611.11	1.00	33.69	nr	**644.80**
65 mm dia.	482.86	611.11	1.00	33.69	nr	**644.80**
80 mm dia.	552.01	698.62	1.00	33.69	nr	**732.31**
100 mm dia.	621.19	786.17	1.00	33.69	nr	**819.86**
150 mm dia.	1033.62	1308.15	1.00	33.69	nr	**1341.84**
200 mm dia.	1584.34	2005.14	1.00	33.69	nr	**2038.83**

38 MECHANICAL/COOLING/HEATING SYSTEMS

Item	Net Price £	Material £	Labour hours	Labour £	Unit	Total rate £
LOW TEMPERATURE HOT WATER HEATING; PIPELINE: ENERGY METERS						
Ultrasonic						
Energy meter for measuring energy use in LTHW systems; includes ultrasonic flow meter (with sensor and signal converter), energy calculator, pair of temperature sensors with brass pockets, and 3 m of interconnecting cable; includes fixing in position; electrical work elsewhere						
Pipe size (flanged connections to PN16); maximum flow rate						
50 mm, 36 m³/hr	1389.38	1758.40	1.80	60.65	nr	**1819.05**
65 mm, 60 m³/hr	1531.50	1938.26	2.32	78.16	nr	**2016.42**
80 mm, 100 m³/hr	1716.70	2172.65	2.56	86.25	nr	**2258.90**
125 mm, 250 m³/hr	1988.96	2517.22	3.60	121.28	nr	**2638.50**
150 mm, 360 m³/hr	2159.14	2732.61	4.80	161.72	nr	**2894.33**
200 mm, 600 m³/hr	2412.38	3053.11	6.24	210.22	nr	**3263.33**
250 mm, 1000 m³/hr	2783.74	3523.11	9.60	323.42	nr	**3846.53**
300 mm, 1500 m³/hr	3272.23	4141.33	10.80	363.87	nr	**4505.20**
350 mm, 2000 m³/hr	3942.89	4990.13	13.20	444.71	nr	**5434.84**
400 mm, 2500 m³/hr	4513.46	5712.24	15.60	525.57	nr	**6237.81**
500 mm, 3000 m³/hr	5122.05	6482.47	24.00	808.56	nr	**7291.03**
600 mm, 3500 m³/hr	5754.67	7283.11	28.00	943.33	nr	**8226.44**

38 MECHANICAL/COOLING/HEATING SYSTEMS

Item	Net Price £	Material £	Labour hours	Labour £	Unit	Total rate £
LOW TEMPERATURE HOT WATER HEATING: CONTROL COMPONENTS MECHANICAL						
Room thermostats; light and medium duty; installed and connected						
Range 3°C to 27°C; 240 volt						
1 amp; on/off type	30.95	39.17	0.30	10.11	nr	**49.28**
Range 0°C to +15°C; 240 volt						
6 amp; frost thermostat	20.87	26.41	0.30	10.11	nr	**36.52**
Range 3°C to 27°C; 250 volt						
2 amp; changeover type; dead zone	51.77	65.52	0.30	10.11	nr	**75.63**
2 amp; changeover type	24.03	30.41	0.30	10.11	nr	**40.52**
2 amp; changeover type; concealed setting	30.28	38.33	0.30	10.11	nr	**48.44**
6 amp; on/off type	18.69	23.65	0.30	10.11	nr	**33.76**
6 amp; temperature set-back	38.68	48.96	0.30	10.11	nr	**59.07**
16 amp; on/off type	28.80	36.44	0.30	10.11	nr	**46.55**
16 amp; on/off type; concealed setting	31.38	39.72	0.30	10.11	nr	**49.83**
20 amp; on/off type; concealed setting	26.74	33.85	0.30	10.11	nr	**43.96**
20 amp; indicated 'off' position	33.71	42.66	0.30	10.11	nr	**52.77**
20 amp; manual; double pole on/off and neon indicator	61.37	77.67	0.30	10.11	nr	**87.78**
20 amp; indicated 'off' position	40.16	50.83	0.30	10.11	nr	**60.94**
Range 10°C to 40°C; 240 volt						
20 amp; changeover contacts	36.60	46.32	0.30	10.11	nr	**56.43**
2 amp; 'heating-cooling' switch	79.15	100.17	0.30	10.11	nr	**110.28**
Surface thermostats						
Cylinder thermostat						
6 amp; changeover type; with cable	20.98	26.56	0.25	8.43	nr	**34.99**
Electrical thermostats; installed and connected						
Range 5°C to 30°C; 230 volt standard port single time						
10 amp with sensor	29.13	36.87	0.30	10.11	nr	**46.98**
Range 5°C to 30°C; 230 volt standard port double time						
10 amp with sensor	33.00	41.76	0.30	10.11	nr	**51.87**
10 amp with sensor and on/off switch	50.35	63.73	0.30	10.11	nr	**73.84**
Radiator thermostats						
Angled valve body; thermostatic head; built-in sensor						
15 mm; liquid filled	18.70	23.67	0.84	28.31	nr	**51.98**
15 mm; wax filled	18.70	23.67	0.84	28.31	nr	**51.98**

38 MECHANICAL/COOLING/HEATING SYSTEMS

Item	Net Price £	Material £	Labour hours	Labour £	Unit	Total rate £
LOW TEMPERATURE HOT WATER HEATING: CONTROL COMPONENTS MECHANICAL – cont						
Immersion thermostats; stem type; domestic water boilers; fitted; electrical work elsewhere						
Temperature range 0°C to 40°C						
Non standard; 280 mm stem	11.07	14.01	0.25	8.43	nr	**22.44**
Temperature range 18°C to 88°C						
13 amp; 178 mm stem	7.35	9.31	0.25	8.43	nr	**17.74**
20 amp; 178 mm stem	11.44	14.48	0.25	8.43	nr	**22.91**
Non standard; pocket clip; 280 mm stem	10.61	13.43	0.25	8.43	nr	**21.86**
Temperature range 40°C to 80°C						
13 amp; 178 mm stem	4.05	5.13	0.25	8.43	nr	**13.56**
20 amp; 178 mm stem	7.64	9.67	0.25	8.43	nr	**18.10**
Non standard; pocket clip; 280 mm stem	11.90	15.06	0.25	8.43	nr	**23.49**
13 amp; 457 mm stem	4.79	6.06	0.25	8.43	nr	**14.49**
20 amp; 457 mm stem	8.36	10.58	0.25	8.43	nr	**19.01**
Temperature range 50°C to 100°C						
Non standard; 1780 mm stem	10.00	12.66	0.25	8.43	nr	**21.09**
Non standard; 280 mm stem	10.43	13.20	0.25	8.43	nr	**21.63**
Pockets for thermostats						
For 178 mm stem	13.10	16.58	0.25	8.43	nr	**25.01**
For 280 mm stem	13.32	16.86	0.25	8.43	nr	**25.29**
Immersion thermostats; stem type; industrial installations; fitted; electrical work elsewhere						
Temperature range 5°C to 105°C						
For 305 mm stem	164.50	208.20	0.50	16.84	nr	**225.04**

38 MECHANICAL/COOLING/HEATING SYSTEMS

Item	Net Price £	Material £	Labour hours	Labour £	Unit	Total rate £
THERMAL INSULATION						
For flexible closed cell insulation see section – Cold Water						
Mineral fibre sectional insulation; bright class O foil faced; bright class O foil taped joints; 19 mm aluminium bands						
Concealed pipework						
20 mm thick						
15 mm dia.	4.24	5.36	0.15	5.05	m	**10.41**
20 mm dia.	4.52	5.72	0.15	5.05	m	**10.77**
25 mm dia.	4.84	6.13	0.15	5.05	m	**11.18**
32 mm dia.	5.42	6.85	0.15	5.05	m	**11.90**
40 mm dia.	5.80	7.34	0.15	5.05	m	**12.39**
50 mm dia.	6.63	8.39	0.15	5.05	m	**13.44**
Extra over for fittings concealed insulation						
Flange/union						
15 mm dia.	2.12	2.69	0.13	4.38	nr	**7.07**
20 mm dia.	2.26	2.86	0.13	4.38	nr	**7.24**
25 mm dia.	2.44	3.09	0.13	4.38	nr	**7.47**
32 mm dia.	2.71	3.43	0.13	4.38	nr	**7.81**
40 mm dia.	2.88	3.64	0.13	4.38	nr	**8.02**
50 mm dia.	3.31	4.19	0.13	4.38	nr	**8.57**
Valves						
15 mm dia.	4.24	5.36	0.15	5.05	nr	**10.41**
20 mm dia.	4.84	6.13	0.15	5.05	nr	**11.18**
25 mm dia.	4.84	6.13	0.15	5.05	nr	**11.18**
32 mm dia.	5.42	6.85	0.15	5.05	nr	**11.90**
40 mm dia.	5.80	7.34	0.15	5.05	nr	**12.39**
50 mm dia.	6.63	8.39	0.15	5.05	nr	**13.44**
Expansion bellows						
15 mm dia.	8.47	10.72	0.22	7.41	nr	**18.13**
20 mm dia.	9.05	11.46	0.22	7.41	nr	**18.87**
25 mm dia.	9.73	12.31	0.22	7.41	nr	**19.72**
32 mm dia.	10.82	13.70	0.22	7.41	nr	**21.11**
40 mm dia.	11.59	14.67	0.22	7.41	nr	**22.08**
50 mm dia.	13.24	16.76	0.22	7.41	nr	**24.17**
25 mm thick						
15 mm dia.	4.66	5.90	0.15	5.05	m	**10.95**
20 mm dia.	5.01	6.34	0.15	5.05	m	**11.39**
25 mm dia.	5.63	7.12	0.15	5.05	m	**12.17**
32 mm dia.	6.13	7.76	0.15	5.05	m	**12.81**
40 mm dia.	6.55	8.29	0.15	5.05	m	**13.34**
50 mm dia.	7.50	9.49	0.15	5.05	m	**14.54**
65 mm dia.	8.58	10.86	0.15	5.05	m	**15.91**
80 mm dia.	9.40	11.89	0.22	7.41	m	**19.30**
100 mm dia.	12.38	15.67	0.22	7.41	m	**23.08**
125 mm dia.	14.32	18.12	0.22	7.41	m	**25.53**
150 mm dia.	17.06	21.59	0.22	7.41	m	**29.00**

38 MECHANICAL/COOLING/HEATING SYSTEMS

Item	Net Price £	Material £	Labour hours	Labour £	Unit	Total rate £
THERMAL INSULATION – cont						
25 mm thick – cont						
200 mm dia.	24.11	30.51	0.25	8.43	m	38.94
250 mm dia.	28.86	36.52	0.25	8.43	m	44.95
300 mm dia.	30.87	39.07	0.25	8.43	m	47.50
Extra over for fittings concealed insulation						
Flange/union						
15 mm dia.	2.33	2.95	0.13	4.38	nr	7.33
20 mm dia.	2.50	3.17	0.13	4.38	nr	7.55
25 mm dia.	2.81	3.56	0.13	4.38	nr	7.94
32 mm dia.	3.06	3.88	0.13	4.38	nr	8.26
40 mm dia.	3.26	4.12	0.13	4.38	nr	8.50
50 mm dia.	3.74	4.74	0.13	4.38	nr	9.12
65 mm dia.	4.28	5.42	0.13	4.38	nr	9.80
80 mm dia.	4.71	5.96	0.18	6.06	nr	12.02
100 mm dia.	6.20	7.85	0.18	6.06	nr	13.91
125 mm dia.	7.16	9.06	0.18	6.06	nr	15.12
150 mm dia.	8.54	10.81	0.18	6.06	nr	16.87
200 mm dia.	12.05	15.25	0.22	7.41	nr	22.66
250 mm dia.	14.44	18.28	0.22	7.41	nr	25.69
300 mm dia.	15.42	19.51	0.22	7.41	nr	26.92
Valves						
15 mm dia.	4.66	5.90	0.15	5.05	nr	10.95
20 mm dia.	5.01	6.34	0.15	5.05	nr	11.39
25 mm dia.	5.63	7.12	0.15	5.05	nr	12.17
32 mm dia.	6.13	7.76	0.15	5.05	nr	12.81
40 mm dia.	6.55	8.29	0.15	5.05	nr	13.34
50 mm dia.	7.50	9.49	0.15	5.05	nr	14.54
65 mm dia.	8.58	10.86	0.15	5.05	nr	15.91
80 mm dia.	9.40	11.89	0.20	6.73	nr	18.62
100 mm dia.	12.38	15.67	0.20	6.73	nr	22.40
125 mm dia.	14.32	18.12	0.20	6.73	nr	24.85
150 mm dia.	17.06	21.59	0.20	6.73	nr	28.32
200 mm dia.	24.11	30.51	0.25	8.43	nr	38.94
250 mm dia.	28.86	36.52	0.25	8.43	nr	44.95
300 mm dia.	30.87	39.07	0.25	8.43	nr	47.50
Expansion bellows						
15 mm dia.	9.32	11.79	0.22	7.41	nr	19.20
20 mm dia.	10.07	12.75	0.22	7.41	nr	20.16
25 mm dia.	11.23	14.21	0.22	7.41	nr	21.62
32 mm dia.	12.23	15.48	0.22	7.41	nr	22.89
40 mm dia.	13.11	16.59	0.22	7.41	nr	24.00
50 mm dia.	15.01	19.00	0.22	7.41	nr	26.41
65 mm dia.	17.13	21.68	0.22	7.41	nr	29.09
80 mm dia.	18.78	23.77	0.29	9.77	nr	33.54
100 mm dia.	24.76	31.34	0.29	9.77	nr	41.11
125 mm dia.	28.66	36.28	0.29	9.77	nr	46.05
150 mm dia.	34.15	43.22	0.29	9.77	nr	52.99

38 MECHANICAL/COOLING/HEATING SYSTEMS

Item	Net Price £	Material £	Labour hours	Labour £	Unit	Total rate £
200 mm dia.	48.23	61.04	0.36	12.13	nr	**73.17**
250 mm dia.	57.76	73.10	0.36	12.13	nr	**85.23**
300 mm dia.	61.74	78.14	0.36	12.13	nr	**90.27**
30 mm thick						
15 mm dia.	6.06	7.67	0.15	5.05	m	**12.72**
20 mm dia.	6.49	8.21	0.15	5.05	m	**13.26**
25 mm dia.	6.86	8.68	0.15	5.05	m	**13.73**
32 mm dia.	7.47	9.45	0.15	5.05	m	**14.50**
40 mm dia.	7.93	10.04	0.15	5.05	m	**15.09**
50 mm dia.	9.06	11.47	0.15	5.05	m	**16.52**
65 mm dia.	10.26	12.98	0.15	5.05	m	**18.03**
80 mm dia.	11.21	14.19	0.22	7.41	m	**21.60**
100 mm dia.	14.47	18.31	0.22	7.41	m	**25.72**
125 mm dia.	16.70	21.13	0.22	7.41	m	**28.54**
150 mm dia.	19.59	24.80	0.22	7.41	m	**32.21**
200 mm dia.	27.37	34.64	0.25	8.43	m	**43.07**
250 mm dia.	32.54	41.18	0.25	8.43	m	**49.61**
300 mm dia.	34.51	43.68	0.25	8.43	m	**52.11**
350 mm dia.	37.92	47.99	0.25	8.43	m	**56.42**
Extra over for fittings concealed insulation						
Flange/union						
15 mm dia.	3.02	3.82	0.13	4.38	nr	**8.20**
20 mm dia.	3.25	4.11	0.13	4.38	nr	**8.49**
25 mm dia.	3.41	4.31	0.13	4.38	nr	**8.69**
32 mm dia.	3.74	4.74	0.13	4.38	nr	**9.12**
40 mm dia.	3.95	5.00	0.13	4.38	nr	**9.38**
50 mm dia.	4.54	5.75	0.13	4.38	nr	**10.13**
65 mm dia.	5.14	6.51	0.13	4.38	nr	**10.89**
80 mm dia.	5.58	7.07	0.18	6.06	nr	**13.13**
100 mm dia.	7.26	9.18	0.18	6.06	nr	**15.24**
125 mm dia.	8.32	10.53	0.18	6.06	nr	**16.59**
150 mm dia.	9.80	12.40	0.18	6.06	nr	**18.46**
200 mm dia.	13.71	17.35	0.22	7.41	nr	**24.76**
250 mm dia.	16.29	20.62	0.22	7.41	nr	**28.03**
300 mm dia.	17.27	21.86	0.22	7.41	nr	**29.27**
350 mm dia.	18.96	23.99	0.22	7.41	nr	**31.40**
Valves						
15 mm dia.	6.49	8.21	0.15	5.05	nr	**13.26**
20 mm dia.	6.49	8.21	0.15	5.05	nr	**13.26**
25 mm dia.	6.86	8.68	0.15	5.05	nr	**13.73**
32 mm dia.	7.47	9.45	0.15	5.05	nr	**14.50**
40 mm dia.	7.93	10.04	0.15	5.05	nr	**15.09**
50 mm dia.	9.06	11.47	0.15	5.05	nr	**16.52**
65 mm dia.	10.26	12.98	0.15	5.05	nr	**18.03**
80 mm dia.	11.21	14.19	0.20	6.73	nr	**20.92**
100 mm dia.	14.47	18.31	0.20	6.73	nr	**25.04**
125 mm dia.	16.70	21.13	0.20	6.73	nr	**27.86**
150 mm dia.	19.59	24.80	0.20	6.73	nr	**31.53**

38 MECHANICAL/COOLING/HEATING SYSTEMS

Item	Net Price £	Material £	Labour hours	Labour £	Unit	Total rate £
THERMAL INSULATION – cont						
Extra over for fittings concealed insulation – cont						
200 mm dia.	27.37	34.64	0.25	8.43	nr	**43.07**
250 mm dia.	32.54	41.18	0.25	8.43	nr	**49.61**
300 mm dia.	34.51	43.68	0.25	8.43	nr	**52.11**
350 mm dia.	37.92	47.99	0.25	8.43	nr	**56.42**
Expansion bellows						
15 mm dia.	12.13	15.36	0.22	7.41	nr	**22.77**
20 mm dia.	12.96	16.40	0.22	7.41	nr	**23.81**
25 mm dia.	13.73	17.37	0.22	7.41	nr	**24.78**
32 mm dia.	15.00	18.98	0.22	7.41	nr	**26.39**
40 mm dia.	15.87	20.08	0.22	7.41	nr	**27.49**
50 mm dia.	18.14	22.96	0.22	7.41	nr	**30.37**
65 mm dia.	20.52	25.97	0.22	7.41	nr	**33.38**
80 mm dia.	22.38	28.32	0.29	9.77	nr	**38.09**
100 mm dia.	28.99	36.69	0.29	9.77	nr	**46.46**
125 mm dia.	33.34	42.19	0.29	9.77	nr	**51.96**
150 mm dia.	39.17	49.57	0.29	9.77	nr	**59.34**
200 mm dia.	54.74	69.28	0.36	12.13	nr	**81.41**
250 mm dia.	65.14	82.44	0.36	12.13	nr	**94.57**
300 mm dia.	69.06	87.40	0.36	12.13	nr	**99.53**
350 mm dia.	75.82	95.96	0.36	12.13	nr	**108.09**
40 mm thick						
15 mm dia.	7.75	9.81	0.15	5.05	m	**14.86**
20 mm dia.	8.00	10.12	0.15	5.05	m	**15.17**
25 mm dia.	8.61	10.90	0.15	5.05	m	**15.95**
32 mm dia.	9.17	11.60	0.15	5.05	m	**16.65**
40 mm dia.	9.66	12.23	0.15	5.05	m	**17.28**
50 mm dia.	10.92	13.82	0.15	5.05	m	**18.87**
65 mm dia.	12.24	15.49	0.15	5.05	m	**20.54**
80 mm dia.	13.30	16.83	0.22	7.41	m	**24.24**
100 mm dia.	17.31	21.91	0.22	7.41	m	**29.32**
125 mm dia.	19.55	24.74	0.22	7.41	m	**32.15**
150 mm dia.	22.80	28.85	0.22	7.41	m	**36.26**
200 mm dia.	31.51	39.88	0.25	8.43	m	**48.31**
250 mm dia.	36.92	46.73	0.25	8.43	m	**55.16**
300 mm dia.	39.47	49.95	0.25	8.43	m	**58.38**
350 mm dia.	43.66	55.26	0.25	8.43	m	**63.69**
400 mm dia.	48.64	61.56	0.25	8.43	m	**69.99**
Extra over for fittings concealed insulation						
Flange/union						
15 mm dia.	3.90	4.94	0.13	4.38	nr	**9.32**
20 mm dia.	3.99	5.05	0.13	4.38	nr	**9.43**
25 mm dia.	4.29	5.43	0.13	4.38	nr	**9.81**
32 mm dia.	4.58	5.80	0.13	4.38	nr	**10.18**
40 mm dia.	4.81	6.09	0.13	4.38	nr	**10.47**
50 mm dia.	5.46	6.91	0.13	4.38	nr	**11.29**
65 mm dia.	6.13	7.76	0.13	4.38	nr	**12.14**

38 MECHANICAL/COOLING/HEATING SYSTEMS

Item	Net Price £	Material £	Labour hours	Labour £	Unit	Total rate £
80 mm dia.	6.67	8.44	0.18	6.06	nr	**14.50**
100 mm dia.	8.64	10.93	0.18	6.06	nr	**16.99**
125 mm dia.	9.76	12.35	0.18	6.06	nr	**18.41**
150 mm dia.	11.41	14.44	0.18	6.06	nr	**20.50**
200 mm dia.	15.76	19.95	0.22	7.41	nr	**27.36**
250 mm dia.	18.45	23.35	0.22	7.41	nr	**30.76**
300 mm dia.	19.71	24.94	0.22	7.41	nr	**32.35**
350 mm dia.	21.83	27.63	0.22	7.41	nr	**35.04**
400 mm dia.	24.35	30.82	0.22	7.41	nr	**38.23**
Valves						
15 mm dia.	7.75	9.81	0.15	5.05	nr	**14.86**
20 mm dia.	8.00	10.12	0.15	5.05	nr	**15.17**
25 mm dia.	8.61	10.90	0.15	5.05	nr	**15.95**
32 mm dia.	9.17	11.60	0.15	5.05	nr	**16.65**
40 mm dia.	9.66	12.23	0.15	5.05	nr	**17.28**
50 mm dia.	10.92	13.82	0.15	5.05	nr	**18.87**
65 mm dia.	12.24	15.49	0.15	5.05	nr	**20.54**
80 mm dia.	13.30	16.83	0.20	6.73	nr	**23.56**
100 mm dia.	17.31	21.91	0.20	6.73	nr	**28.64**
125 mm dia.	19.55	24.74	0.20	6.73	nr	**31.47**
150 mm dia.	22.80	28.85	0.20	6.73	nr	**35.58**
200 mm dia.	31.51	39.88	0.25	8.43	nr	**48.31**
250 mm dia.	36.92	46.73	0.25	8.43	nr	**55.16**
300 mm dia.	39.47	49.95	0.25	8.43	nr	**58.38**
350 mm dia.	43.66	55.26	0.25	8.43	nr	**63.69**
400 mm dia.	48.64	61.56	0.25	8.43	nr	**69.99**
Expansion bellows						
15 mm dia.	15.56	19.69	0.22	7.41	nr	**27.10**
20 mm dia.	15.98	20.23	0.22	7.41	nr	**27.64**
25 mm dia.	17.23	21.81	0.22	7.41	nr	**29.22**
32 mm dia.	18.34	23.21	0.22	7.41	nr	**30.62**
40 mm dia.	19.29	24.42	0.22	7.41	nr	**31.83**
50 mm dia.	21.83	27.63	0.22	7.41	nr	**35.04**
65 mm dia.	24.46	30.96	0.22	7.41	nr	**38.37**
80 mm dia.	26.66	33.75	0.29	9.77	nr	**43.52**
100 mm dia.	34.56	43.74	0.29	9.77	nr	**53.51**
125 mm dla.	39.07	49.45	0.29	9.77	nr	**59.22**
150 mm dia.	45.60	57.71	0.29	9.77	nr	**67.48**
200 mm dia.	63.02	79.76	0.36	12.13	nr	**91.89**
250 mm dia.	73.85	93.46	0.36	12.13	nr	**105.59**
300 mm dia.	78.96	99.93	0.36	12.13	nr	**112.06**
350 mm dia.	43.66	55.26	0.36	12.13	nr	**67.39**
400 mm dia.	97.31	123.16	0.36	12.13	nr	**135.29**
50 mm thick						
15 mm dia.	10.80	13.66	0.15	5.05	m	**18.71**
20 mm dia.	11.35	14.37	0.15	5.05	m	**19.42**
25 mm dia.	12.07	15.28	0.15	5.05	m	**20.33**
32 mm dia.	12.61	15.96	0.15	5.05	m	**21.01**
40 mm dia.	13.25	16.77	0.15	5.05	m	**21.82**
50 mm dia.	14.88	18.83	0.15	5.05	m	**23.88**
65 mm dia.	16.26	20.57	0.15	5.05	m	**25.62**

38 MECHANICAL/COOLING/HEATING SYSTEMS

Item	Net Price £	Material £	Labour hours	Labour £	Unit	Total rate £
THERMAL INSULATION – cont						
50 mm thick – cont						
80 mm dia.	17.39	22.01	0.22	7.41	m	29.42
100 mm dia.	22.26	28.17	0.22	7.41	m	35.58
125 mm dia.	24.97	31.61	0.22	7.41	m	39.02
150 mm dia.	28.82	36.48	0.22	7.41	m	43.89
200 mm dia.	39.35	49.81	0.25	8.43	m	58.24
250 mm dia.	45.44	57.51	0.25	8.43	m	65.94
300 mm dia.	48.15	60.94	0.25	8.43	m	69.37
350 mm dia.	53.15	67.27	0.25	8.43	m	75.70
400 mm dia.	58.95	74.60	0.25	8.43	m	83.03
Extra over for fittings concealed insulation						
Flange/union						
15 mm dia.	5.41	6.84	0.13	4.38	nr	11.22
20 mm dia.	5.69	7.20	0.13	4.38	nr	11.58
25 mm dia.	6.05	7.66	0.13	4.38	nr	12.04
32 mm dia.	6.32	8.00	0.13	4.38	nr	12.38
40 mm dia.	6.64	8.40	0.13	4.38	nr	12.78
50 mm dia.	7.42	9.39	0.13	4.38	nr	13.77
65 mm dia.	8.13	10.29	0.13	4.38	nr	14.67
80 mm dia.	8.70	11.01	0.18	6.06	nr	17.07
100 mm dia.	11.14	14.10	0.18	6.06	nr	20.16
125 mm dia.	12.49	15.80	0.18	6.06	nr	21.86
150 mm dia.	14.43	18.27	0.18	6.06	nr	24.33
200 mm dia.	19.66	24.89	0.22	7.41	nr	32.30
250 mm dia.	22.72	28.75	0.22	7.41	nr	36.16
300 mm dia.	24.08	30.48	0.22	7.41	nr	37.89
350 mm dia.	26.60	33.67	0.22	7.41	nr	41.08
400 mm dia.	29.48	37.31	0.22	7.41	nr	44.72
Valves						
15 mm dia.	10.80	13.66	0.15	5.05	nr	18.71
20 mm dia.	11.35	14.37	0.15	5.05	nr	19.42
25 mm dia.	12.07	15.28	0.15	5.05	nr	20.33
32 mm dia.	12.61	15.96	0.15	5.05	nr	21.01
40 mm dia.	13.25	16.77	0.15	5.05	nr	21.82
50 mm dia.	14.88	18.83	0.15	5.05	nr	23.88
65 mm dia.	16.26	20.57	0.15	5.05	nr	25.62
80 mm dia.	17.39	22.01	0.20	6.73	nr	28.74
100 mm dia.	22.26	28.17	0.20	6.73	nr	34.90
125 mm dia.	24.97	31.61	0.20	6.73	nr	38.34
150 mm dia.	28.82	36.48	0.20	6.73	nr	43.21
200 mm dia.	39.35	49.81	0.25	8.43	nr	58.24
250 mm dia.	45.44	57.51	0.25	8.43	nr	65.94
300 mm dia.	48.15	60.94	0.25	8.43	nr	69.37
350 mm dia.	53.15	67.27	0.25	8.43	nr	75.70
400 mm dia.	58.95	74.60	0.25	8.43	nr	83.03
Expansion bellows						
15 mm dia.	21.55	27.27	0.22	7.41	nr	34.68
20 mm dia.	22.69	28.72	0.22	7.41	nr	36.13
25 mm dia.	24.13	30.54	0.22	7.41	nr	37.95

38 MECHANICAL/COOLING/HEATING SYSTEMS

Item	Net Price £	Material £	Labour hours	Labour £	Unit	Total rate £
32 mm dia.	25.23	31.93	0.22	7.41	nr	39.34
40 mm dia.	26.54	33.59	0.22	7.41	nr	41.00
50 mm dia.	29.74	37.64	0.22	7.41	nr	45.05
65 mm dia.	32.50	41.13	0.22	7.41	nr	48.54
80 mm dia.	34.82	44.07	0.29	9.77	nr	53.84
100 mm dia.	44.47	56.28	0.29	9.77	nr	66.05
125 mm dia.	49.93	63.19	0.29	9.77	nr	72.96
150 mm dia.	57.64	72.95	0.29	9.77	nr	82.72
200 mm dia.	78.72	99.62	0.36	12.13	nr	111.75
250 mm dia.	90.87	115.00	0.36	12.13	nr	127.13
300 mm dia.	96.33	121.91	0.36	12.13	nr	134.04
350 mm dia.	106.31	134.55	0.36	12.13	nr	146.68
400 mm dia.	117.88	149.18	0.36	12.13	nr	161.31
Mineral fibre sectional insulation; bright class O foil faced; bright class O foil taped joints; 22 swg plain/embossed aluminium cladding; pop riveted						
Plantroom pipework						
20 mm thick						
15 mm dia.	6.92	8.76	0.44	14.82	m	23.58
20 mm dia.	7.31	9.25	0.44	14.82	m	24.07
25 mm dia.	7.83	9.91	0.44	14.82	m	24.73
32 mm dia.	8.48	10.73	0.44	14.82	m	25.55
40 mm dia.	8.94	11.31	0.44	14.82	m	26.13
50 mm dia.	9.95	12.59	0.44	14.82	m	27.41
Extra over for fittings plantroom insulation						
Flange/union						
15 mm dia.	7.73	9.78	0.58	19.54	nr	29.32
20 mm dia.	8.21	10.39	0.58	19.54	nr	29.93
25 mm dia.	8.79	11.12	0.58	19.54	nr	30.66
32 mm dia.	9.60	12.15	0.58	19.54	nr	31.69
40 mm dia.	10.14	12.84	0.58	19.54	nr	32.38
50 mm dia.	11.38	14.40	0.58	19.54	nr	33.94
Bends						
15 mm dia.	3.78	4.78	0.44	14.82	nr	19.60
20 mm dia.	4.04	5.12	0.44	14.82	nr	19.94
25 mm dia.	4.31	5.45	0.44	14.82	nr	20.27
32 mm dia.	4.66	5.90	0.44	14.82	nr	20.72
40 mm dia.	4.89	6.19	0.44	14.82	nr	21.01
50 mm dia.	5.48	6.93	0.44	14.82	nr	21.75
Tees						
15 mm dia.	2.27	2.88	0.44	14.82	nr	17.70
20 mm dia.	2.43	3.08	0.44	14.82	nr	17.90
25 mm dia.	2.58	3.27	0.44	14.82	nr	18.09
32 mm dia.	2.80	3.54	0.44	14.82	nr	18.36
40 mm dia.	2.96	3.74	0.44	14.82	nr	18.56
50 mm dia.	3.29	4.17	0.44	14.82	nr	18.99

38 MECHANICAL/COOLING/HEATING SYSTEMS

Item	Net Price £	Material £	Labour hours	Labour £	Unit	Total rate £
THERMAL INSULATION – cont						
Extra over for fittings plantroom						
insulation – cont						
Valves						
15 mm dia.	4.24	5.36	0.78	26.28	nr	**31.64**
20 mm dia.	4.84	6.13	0.78	26.28	nr	**32.41**
25 mm dia.	4.84	6.13	0.78	26.28	nr	**32.41**
32 mm dia.	5.42	6.85	0.78	26.28	nr	**33.13**
40 mm dia.	5.80	7.34	0.78	26.28	nr	**33.62**
50 mm dia.	6.63	8.39	0.78	26.28	nr	**34.67**
Pumps						
15 mm dia.	22.72	28.75	2.34	78.84	nr	**107.59**
20 mm dia.	24.09	30.49	2.34	78.84	nr	**109.33**
25 mm dia.	25.87	32.74	2.34	78.84	nr	**111.58**
32 mm dia.	28.23	35.73	2.34	78.84	nr	**114.57**
40 mm dia.	29.82	37.74	2.34	78.84	nr	**116.58**
50 mm dia.	33.53	42.44	2.34	78.84	nr	**121.28**
Expansion bellows						
15 mm dia.	18.17	22.99	1.05	35.37	nr	**58.36**
20 mm dia.	19.28	24.40	1.05	35.37	nr	**59.77**
25 mm dia.	20.71	26.21	1.05	35.37	nr	**61.58**
32 mm dia.	22.58	28.58	1.05	35.37	nr	**63.95**
40 mm dia.	23.83	30.16	1.05	35.37	nr	**65.53**
50 mm dia.	26.82	33.95	1.05	35.37	nr	**69.32**
25 mm thick						
15 mm dia.	7.59	9.61	0.44	14.82	m	**24.43**
20 mm dia.	8.21	10.39	0.44	14.82	m	**25.21**
25 mm dia.	8.99	11.38	0.44	14.82	m	**26.20**
32 mm dia.	9.57	12.11	0.44	14.82	m	**26.93**
40 mm dia.	10.30	13.04	0.44	14.82	m	**27.86**
50 mm dia.	11.54	14.60	0.44	14.82	m	**29.42**
65 mm dia.	13.04	16.51	0.44	14.82	m	**31.33**
80 mm dia.	14.13	17.89	0.52	17.52	m	**35.41**
100 mm dia.	17.54	22.20	0.52	17.52	m	**39.72**
125 mm dia.	20.37	25.78	0.52	17.52	m	**43.30**
150 mm dia.	23.71	30.00	0.52	17.52	m	**47.52**
200 mm dia.	32.39	40.99	0.60	20.22	m	**61.21**
250 mm dia.	38.49	48.71	0.60	20.22	m	**68.93**
300 mm dia.	43.09	54.53	0.60	20.22	m	**74.75**
Extra over for fittings plantroom insulation						
Flange/union						
15 mm dia.	8.48	10.73	0.58	19.54	nr	**30.27**
20 mm dia.	9.17	11.60	0.58	19.54	nr	**31.14**
25 mm dia.	10.12	12.81	0.58	19.54	nr	**32.35**

38 MECHANICAL/COOLING/HEATING SYSTEMS

Item	Net Price £	Material £	Labour hours	Labour £	Unit	Total rate £
32 mm dia.	10.88	13.76	0.58	19.54	nr	**33.30**
40 mm dia.	11.66	14.76	0.58	19.54	nr	**34.30**
50 mm dia.	13.14	16.63	0.58	19.54	nr	**36.17**
65 mm dia.	14.91	18.87	0.58	19.54	nr	**38.41**
80 mm dia.	16.21	20.52	0.67	22.58	nr	**43.10**
100 mm dia.	20.52	25.97	0.67	22.58	nr	**48.55**
125 mm dia.	23.82	30.15	0.67	22.58	nr	**52.73**
150 mm dia.	27.92	35.34	0.67	22.58	nr	**57.92**
200 mm dia.	38.58	48.83	0.87	29.31	nr	**78.14**
250 mm dia.	45.95	58.15	0.87	29.31	nr	**87.46**
300 mm dia.	50.80	64.29	0.87	29.31	nr	**93.60**
Bends						
15 mm dia.	4.17	5.28	0.44	14.82	nr	**20.10**
20 mm dia.	4.52	5.72	0.44	14.82	nr	**20.54**
25 mm dia.	4.94	6.25	0.44	14.82	nr	**21.07**
32 mm dia.	5.27	6.68	0.44	14.82	nr	**21.50**
40 mm dia.	5.69	7.20	0.44	14.82	nr	**22.02**
50 mm dia.	6.35	8.04	0.44	14.82	nr	**22.86**
65 mm dia.	7.16	9.06	0.44	14.82	nr	**23.88**
80 mm dia.	7.76	9.82	0.52	17.52	nr	**27.34**
100 mm dia.	9.66	12.23	0.52	17.52	nr	**29.75**
125 mm dia.	11.23	14.21	0.52	17.52	nr	**31.73**
150 mm dia.	13.04	16.51	0.52	17.52	nr	**34.03**
200 mm dia.	17.81	22.55	0.60	20.22	nr	**42.77**
250 mm dia.	21.17	26.79	0.60	20.22	nr	**47.01**
300 mm dia.	23.70	29.99	0.60	20.22	nr	**50.21**
Tees						
15 mm dia.	2.50	3.17	0.44	14.82	nr	**17.99**
20 mm dia.	2.69	3.40	0.44	14.82	nr	**18.22**
25 mm dia.	2.98	3.77	0.44	14.82	nr	**18.59**
32 mm dia.	3.16	4.00	0.44	14.82	nr	**18.82**
40 mm dia.	3.39	4.29	0.44	14.82	nr	**19.11**
50 mm dia.	3.78	4.78	0.44	14.82	nr	**19.60**
65 mm dia.	4.29	5.43	0.44	14.82	nr	**20.25**
80 mm dia.	4.66	5.90	0.52	17.52	nr	**23.42**
100 mm dia.	5.79	7.32	0.52	17.52	nr	**24.84**
125 mm dia.	6.74	8.53	0.52	17.52	nr	**26.05**
150 mm dia.	7.83	9.91	0.52	17.52	nr	**27.43**
200 mm dia.	10.67	13.51	0.60	20.22	nr	**33.73**
250 mm dia.	12.72	16.09	0.60	20.22	nr	**36.31**
300 mm dia.	14.24	18.02	0.60	20.22	nr	**38.24**
Valves						
15 mm dia.	13.48	17.06	0.78	26.28	nr	**43.34**
20 mm dia.	14.57	18.44	0.78	26.28	nr	**44.72**
25 mm dia.	16.09	20.36	0.78	26.28	nr	**46.64**
32 mm dia.	17.26	21.84	0.78	26.28	nr	**48.12**
40 mm dia.	18.52	23.44	0.78	26.28	nr	**49.72**
50 mm dia.	20.84	26.38	0.78	26.28	nr	**52.66**
65 mm dia.	23.68	29.97	0.78	26.28	nr	**56.25**

38 MECHANICAL/COOLING/HEATING SYSTEMS

Item	Net Price £	Material £	Labour hours	Labour £	Unit	Total rate £
THERMAL INSULATION – cont						
Extra over for fittings plantroom insulation – cont						
Valves – cont						
80 mm dia.	25.76	32.60	0.92	30.99	nr	**63.59**
100 mm dia.	32.56	41.20	0.92	30.99	nr	**72.19**
125 mm dia.	37.86	47.91	0.92	30.99	nr	**78.90**
150 mm dia.	44.35	56.13	0.92	30.99	nr	**87.12**
200 mm dia.	61.28	77.56	1.12	37.73	nr	**115.29**
250 mm dia.	73.00	92.39	1.12	37.73	nr	**130.12**
300 mm dia.	80.66	102.09	1.12	37.73	nr	**139.82**
Pumps						
15 mm dia.	24.96	31.58	2.34	78.84	nr	**110.42**
20 mm dia.	26.97	34.14	2.34	78.84	nr	**112.98**
25 mm dia.	29.80	37.71	2.34	78.84	nr	**116.55**
32 mm dia.	31.93	40.41	2.34	78.84	nr	**119.25**
40 mm dia.	34.28	43.39	2.34	78.84	nr	**122.23**
50 mm dia.	38.64	48.90	2.34	78.84	nr	**127.74**
65 mm dia.	43.83	55.47	2.34	78.84	nr	**134.31**
80 mm dia.	47.69	60.36	2.76	92.98	nr	**153.34**
100 mm dia.	60.34	76.36	2.76	92.98	nr	**169.34**
125 mm dia.	70.08	88.69	2.76	92.98	nr	**181.67**
150 mm dia.	82.12	103.94	2.76	92.98	nr	**196.92**
200 mm dia.	113.48	143.62	3.36	113.20	nr	**256.82**
250 mm dia.	135.18	171.08	3.36	113.20	nr	**284.28**
300 mm dia.	149.34	189.00	3.36	113.20	nr	**302.20**
Expansion bellows						
15 mm dia.	19.96	25.26	1.05	35.37	nr	**60.63**
20 mm dia.	21.56	27.28	1.05	35.37	nr	**62.65**
25 mm dia.	23.82	30.15	1.05	35.37	nr	**65.52**
32 mm dia.	25.57	32.36	1.05	35.37	nr	**67.73**
40 mm dia.	27.44	34.73	1.05	35.37	nr	**70.10**
50 mm dia.	30.92	39.13	1.05	35.37	nr	**74.50**
65 mm dia.	35.08	44.40	1.05	35.37	nr	**79.77**
80 mm dia.	38.17	48.31	1.26	42.45	nr	**90.76**
100 mm dia.	48.26	61.07	1.26	42.45	nr	**103.52**
125 mm dia.	56.03	70.91	1.26	42.45	nr	**113.36**
150 mm dia.	65.68	83.13	1.26	42.45	nr	**125.58**
200 mm dia.	90.79	114.90	1.53	51.54	nr	**166.44**
250 mm dia.	108.16	136.89	1.53	51.54	nr	**188.43**
300 mm dia.	119.46	151.19	1.53	51.54	nr	**202.73**
30 mm thick						
15 mm dia.	9.36	11.85	0.44	14.82	m	**26.67**
20 mm dia.	9.89	12.52	0.44	14.82	m	**27.34**
25 mm dia.	10.48	13.26	0.44	14.82	m	**28.08**
32 mm dia.	11.49	14.54	0.44	14.82	m	**29.36**
40 mm dia.	11.95	15.12	0.44	14.82	m	**29.94**
50 mm dia.	13.25	16.77	0.44	14.82	m	**31.59**
65 mm dia.	14.97	18.95	0.44	14.82	m	**33.77**

38 MECHANICAL/COOLING/HEATING SYSTEMS

Item	Net Price £	Material £	Labour hours	Labour £	Unit	Total rate £
80 mm dia.	16.36	20.71	0.52	17.52	m	38.23
100 mm dia.	20.10	25.44	0.52	17.52	m	42.96
125 mm dia.	22.95	29.04	0.52	17.52	m	46.56
150 mm dia.	26.74	33.85	0.52	17.52	m	51.37
200 mm dia.	35.98	45.54	0.60	20.22	m	65.76
250 mm dia.	42.54	53.84	0.60	20.22	m	74.06
300 mm dia.	46.83	59.27	0.60	20.22	m	79.49
350 mm dia.	52.14	65.99	0.60	20.22	m	86.21
Extra over for fittings plantroom insulation						
Flange/union						
15 mm dia.	10.63	13.45	0.58	19.54	nr	32.99
20 mm dia.	11.29	14.29	0.58	19.54	nr	33.83
25 mm dia.	11.97	15.15	0.58	19.54	nr	34.69
32 mm dia.	13.11	16.59	0.58	19.54	nr	36.13
40 mm dia.	13.72	17.36	0.58	19.54	nr	36.90
50 mm dia.	15.35	19.43	0.58	19.54	nr	38.97
65 mm dia.	17.35	21.96	0.58	19.54	nr	41.50
80 mm dia.	18.96	23.99	0.67	22.58	nr	46.57
100 mm dia.	23.68	29.97	0.67	22.58	nr	52.55
125 mm dia.	27.10	34.29	0.67	22.58	nr	56.87
150 mm dia.	31.70	40.12	0.67	22.58	nr	62.70
200 mm dia.	43.17	54.63	0.87	29.31	nr	83.94
250 mm dia.	51.18	64.77	0.87	29.31	nr	94.08
300 mm dia.	55.68	70.47	0.87	29.31	nr	99.78
350 mm dia.	61.78	78.19	0.87	29.31	nr	107.50
Bends						
15 mm dia.	5.14	6.51	0.44	14.82	nr	21.33
20 mm dia.	5.43	6.88	0.44	14.82	nr	21.70
25 mm dia.	5.77	7.30	0.44	14.82	nr	22.12
32 mm dia.	6.30	7.97	0.44	14.82	nr	22.79
40 mm dia.	6.56	8.30	0.44	14.82	nr	23.12
50 mm dia.	7.31	9.25	0.44	14.82	nr	24.07
65 mm dia.	8.24	10.43	0.44	14.82	nr	25.25
80 mm dia.	8.98	11.37	0.52	17.52	nr	28.89
100 mm dia.	11.05	13.99	0.52	17.52	nr	31.51
125 mm dia.	12.62	15.97	0.52	17.52	nr	33.49
150 mm dia.	14.72	18.63	0.52	17.52	nr	36.15
200 mm dia.	19.78	25.03	0.60	20.22	nr	45.25
250 mm dia.	23.42	29.64	0.60	20.22	nr	49.86
300 mm dia.	25.76	32.60	0.60	20.22	nr	52.82
350 mm dia.	28.70	36.32	0.60	20.22	nr	56.54
Tees						
15 mm dia.	3.07	3.89	0.44	14.82	nr	18.71
20 mm dia.	3.26	4.12	0.44	14.82	nr	18.94
25 mm dia.	3.44	4.36	0.44	14.82	nr	19.18
32 mm dia.	3.77	4.77	0.44	14.82	nr	19.59
40 mm dia.	3.94	4.98	0.44	14.82	nr	19.80
50 mm dia.	4.37	5.53	0.44	14.82	nr	20.35
65 mm dia.	4.93	6.24	0.44	14.82	nr	21.06

38 MECHANICAL/COOLING/HEATING SYSTEMS

Item	Net Price £	Material £	Labour hours	Labour £	Unit	Total rate £
THERMAL INSULATION – cont						
Extra over for fittings plantroom insulation – cont						
Tees – cont						
80 mm dia.	5.41	6.84	0.52	17.52	nr	24.36
100 mm dia.	6.64	8.40	0.52	17.52	nr	25.92
125 mm dia.	7.59	9.61	0.52	17.52	nr	27.13
150 mm dia.	8.85	11.20	0.52	17.52	nr	28.72
200 mm dia.	11.87	15.02	0.60	20.22	nr	35.24
250 mm dia.	14.06	17.80	0.60	20.22	nr	38.02
300 mm dia.	15.45	19.56	0.60	20.22	nr	39.78
350 mm dia.	17.23	21.81	0.60	20.22	nr	42.03
Valves						
15 mm dia.	16.90	21.39	0.78	26.28	nr	47.67
20 mm dia.	17.92	22.68	0.78	26.28	nr	48.96
25 mm dia.	19.01	24.06	0.78	26.28	nr	50.34
32 mm dia.	20.83	26.36	0.78	26.28	nr	52.64
40 mm dia.	21.78	27.56	0.78	26.28	nr	53.84
50 mm dia.	24.39	30.87	0.78	26.28	nr	57.15
65 mm dia.	27.57	34.89	0.78	26.28	nr	61.17
80 mm dia.	30.10	38.09	0.92	30.99	nr	69.08
100 mm dia.	37.60	47.59	0.92	30.99	nr	78.58
125 mm dia.	43.06	54.50	0.92	30.99	nr	85.49
150 mm dia.	50.34	63.71	0.92	30.99	nr	94.70
200 mm dia.	68.55	86.76	1.12	37.73	nr	124.49
250 mm dia.	81.28	102.87	1.12	37.73	nr	140.60
300 mm dia.	88.44	111.93	1.12	37.73	nr	149.66
350 mm dia.	98.11	124.16	1.12	37.73	nr	161.89
Pumps						
15 mm dia.	31.31	39.63	2.34	78.84	nr	118.47
20 mm dia.	33.22	42.04	2.34	78.84	nr	120.88
25 mm dia.	35.19	44.53	2.34	78.84	nr	123.37
32 mm dia.	38.56	48.80	2.34	78.84	nr	127.64
40 mm dia.	40.34	51.05	2.34	78.84	nr	129.89
50 mm dia.	45.19	57.19	2.34	78.84	nr	136.03
65 mm dia.	51.03	64.58	2.34	78.84	nr	143.42
80 mm dia.	55.72	70.52	2.76	92.98	nr	163.50
100 mm dia.	69.61	88.10	2.76	92.98	nr	181.08
125 mm dia.	79.73	100.90	2.76	92.98	nr	193.88
150 mm dia.	93.24	118.00	2.76	92.98	nr	210.98
200 mm dia.	126.96	160.68	3.36	113.20	nr	273.88
250 mm dia.	150.55	190.53	3.36	113.20	nr	303.73
300 mm dia.	163.78	207.28	3.36	113.20	nr	320.48
350 mm dia.	181.68	229.94	3.36	113.20	nr	343.14
Expansion bellows						
15 mm dia.	25.05	31.71	1.05	35.37	nr	67.08
20 mm dia.	26.58	33.64	1.05	35.37	nr	69.01
25 mm dia.	28.14	35.62	1.05	35.37	nr	70.99

38 MECHANICAL/COOLING/HEATING SYSTEMS

Item	Net Price £	Material £	Labour hours	Labour £	Unit	Total rate £
32 mm dia.	30.85	39.04	1.05	35.37	nr	**74.41**
40 mm dia.	32.27	40.85	1.05	35.37	nr	**76.22**
50 mm dia.	36.12	45.72	1.05	35.37	nr	**81.09**
65 mm dia.	40.83	51.68	1.05	35.37	nr	**87.05**
80 mm dia.	44.62	56.47	1.26	42.45	nr	**98.92**
100 mm dia.	55.70	70.49	1.26	42.45	nr	**112.94**
125 mm dia.	63.81	80.76	1.26	42.45	nr	**123.21**
150 mm dia.	74.60	94.42	1.26	42.45	nr	**136.87**
200 mm dia.	101.58	128.56	1.53	51.54	nr	**180.10**
250 mm dia.	120.44	152.43	1.53	51.54	nr	**203.97**
300 mm dia.	131.02	165.82	1.53	51.54	nr	**217.36**
350 mm dia.	145.34	183.94	1.53	51.54	nr	**235.48**
40 mm thick						
15 mm dia.	11.44	14.48	0.44	14.82	m	**29.30**
20 mm dia.	12.01	15.20	0.44	14.82	m	**30.02**
25 mm dia.	12.79	16.18	0.44	14.82	m	**31.00**
32 mm dia.	13.41	16.97	0.44	14.82	m	**31.79**
40 mm dia.	14.25	18.03	0.44	14.82	m	**32.85**
50 mm dia.	15.78	19.97	0.44	14.82	m	**34.79**
65 mm dia.	17.57	22.23	0.44	14.82	m	**37.05**
80 mm dia.	18.91	23.93	0.52	17.52	m	**41.45**
100 mm dia.	27.70	35.06	0.52	17.52	m	**52.58**
125 mm dia.	25.95	32.84	0.52	17.52	m	**50.36**
150 mm dia.	30.26	38.29	0.52	17.52	m	**55.81**
200 mm dia.	40.03	50.66	0.60	20.22	m	**70.88**
250 mm dia.	46.88	59.33	0.60	20.22	m	**79.55**
300 mm dia.	51.82	65.59	0.60	20.22	m	**85.81**
350 mm dia.	57.92	73.30	0.60	20.22	m	**93.52**
400 mm dia.	65.15	82.45	0.60	20.22	m	**102.67**
Extra over for fittings plantroom insulation						
Flange/union						
15 mm dia.	13.21	16.72	0.58	19.54	nr	**36.26**
20 mm dia.	13.79	17.45	0.58	19.54	nr	**36.99**
25 mm dia.	14.73	18.64	0.58	19.54	nr	**38.18**
32 mm dia.	15.53	19.66	0.58	19.54	nr	**39.20**
40 mm dia.	16.47	20.84	0.58	19.54	nr	**40.38**
50 mm dia.	18.34	23.21	0.58	19.54	nr	**42.75**
65 mm dia.	20.50	25.95	0.58	19.54	nr	**45.49**
80 mm dia.	22.11	27.98	0.67	22.58	nr	**50.56**
100 mm dia.	27.70	35.06	0.67	22.58	nr	**57.64**
125 mm dia.	31.02	39.26	0.67	22.58	nr	**61.84**
150 mm dia.	36.18	45.79	0.67	22.58	nr	**68.37**
200 mm dia.	48.55	61.44	0.87	29.31	nr	**90.75**
250 mm dia.	56.91	72.03	0.87	29.31	nr	**101.34**
300 mm dia.	62.26	78.79	0.87	29.31	nr	**108.10**
350 mm dia.	69.42	87.85	0.87	29.31	nr	**117.16**
400 mm dia.	77.83	98.50	0.87	29.31	nr	**127.81**

38 MECHANICAL/COOLING/HEATING SYSTEMS

Item	Net Price £	Material £	Labour hours	Labour £	Unit	Total rate £
THERMAL INSULATION – cont						
Extra over for fittings plantroom insulation – cont						
Bends						
15 mm dia.	6.29	7.96	0.44	14.82	nr	**22.78**
20 mm dia.	6.59	8.34	0.44	14.82	nr	**23.16**
25 mm dia.	7.04	8.92	0.44	14.82	nr	**23.74**
32 mm dia.	7.36	9.32	0.44	14.82	nr	**24.14**
40 mm dia.	7.84	9.92	0.44	14.82	nr	**24.74**
50 mm dia.	8.67	10.98	0.44	14.82	nr	**25.80**
65 mm dia.	9.69	12.26	0.44	14.82	nr	**27.08**
80 mm dia.	10.39	13.15	0.52	17.52	nr	**30.67**
100 mm dia.	12.83	16.24	0.52	17.52	nr	**33.76**
125 mm dia.	14.27	18.07	0.52	17.52	nr	**35.59**
150 mm dia.	16.64	21.06	0.52	17.52	nr	**38.58**
200 mm dia.	22.00	27.84	0.60	20.22	nr	**48.06**
250 mm dia.	25.61	32.41	0.60	20.22	nr	**52.63**
300 mm dia.	28.51	36.09	0.60	20.22	nr	**56.31**
350 mm dia.	31.88	40.34	0.60	20.22	nr	**60.56**
400 mm dia.	35.83	45.35	0.60	20.22	nr	**65.57**
Tees						
15 mm dia.	3.77	4.77	0.44	14.82	nr	**19.59**
20 mm dia.	3.94	4.98	0.44	14.82	nr	**19.80**
25 mm dia.	4.22	5.34	0.44	14.82	nr	**20.16**
32 mm dia.	4.42	5.59	0.44	14.82	nr	**20.41**
40 mm dia.	4.71	5.96	0.44	14.82	nr	**20.78**
50 mm dia.	5.19	6.56	0.44	14.82	nr	**21.38**
65 mm dia.	5.80	7.34	0.44	14.82	nr	**22.16**
80 mm dia.	6.24	7.90	0.52	17.52	nr	**25.42**
100 mm dia.	7.70	9.74	0.52	17.52	nr	**27.26**
125 mm dia.	8.58	10.86	0.52	17.52	nr	**28.38**
150 mm dia.	10.00	12.66	0.52	17.52	nr	**30.18**
200 mm dia.	13.21	16.72	0.60	20.22	nr	**36.94**
250 mm dia.	15.50	19.62	0.60	20.22	nr	**39.84**
300 mm dia.	17.09	21.63	0.60	20.22	nr	**41.85**
350 mm dia.	19.13	24.21	0.60	20.22	nr	**44.43**
400 mm dia.	21.49	27.19	0.60	20.22	nr	**47.41**
Valves						
15 mm dia.	20.99	26.57	0.78	26.28	nr	**52.85**
20 mm dia.	21.90	27.72	0.78	26.28	nr	**54.00**
25 mm dia.	23.42	29.64	0.78	26.28	nr	**55.92**
32 mm dia.	24.66	31.21	0.78	26.28	nr	**57.49**
40 mm dia.	26.13	33.07	0.78	26.28	nr	**59.35**
50 mm dia.	29.13	36.87	0.78	26.28	nr	**63.15**
65 mm dia.	32.53	41.17	0.78	26.28	nr	**67.45**
80 mm dia.	35.13	44.46	0.92	30.99	nr	**75.45**
100 mm dia.	44.02	55.71	0.92	30.99	nr	**86.70**
125 mm dia.	49.23	62.31	0.92	30.99	nr	**93.30**
150 mm dia.	57.48	72.74	0.92	30.99	nr	**103.73**

38 MECHANICAL/COOLING/HEATING SYSTEMS

Item	Net Price £	Material £	Labour hours	Labour £	Unit	Total rate £
200 mm dia.	77.10	97.57	1.12	37.73	nr	**135.30**
250 mm dia.	90.38	114.39	1.12	37.73	nr	**152.12**
300 mm dia.	98.91	125.18	1.12	37.73	nr	**162.91**
350 mm dia.	110.25	139.53	1.12	37.73	nr	**177.26**
400 mm dia.	123.63	156.46	1.12	37.73	nr	**194.19**
Pumps						
15 mm dia.	38.84	49.16	2.34	78.84	nr	**128.00**
20 mm dia.	40.53	51.30	2.34	78.84	nr	**130.14**
25 mm dia.	43.36	54.88	2.34	78.84	nr	**133.72**
32 mm dia.	45.61	57.72	2.34	78.84	nr	**136.56**
40 mm dia.	48.41	61.26	2.34	78.84	nr	**140.10**
50 mm dia.	53.96	68.29	2.34	78.84	nr	**147.13**
65 mm dia.	60.27	76.28	2.34	78.84	nr	**155.12**
80 mm dia.	65.06	82.34	2.76	92.98	nr	**175.32**
100 mm dia.	81.48	103.12	2.76	92.98	nr	**196.10**
125 mm dia.	91.16	115.37	2.76	92.98	nr	**208.35**
150 mm dia.	106.43	134.70	2.76	92.98	nr	**227.68**
200 mm dia.	142.78	180.70	3.36	113.20	nr	**293.90**
250 mm dia.	167.38	211.84	3.36	113.20	nr	**325.04**
300 mm dia.	239.49	303.09	3.36	113.20	nr	**416.29**
350 mm dia.	204.15	258.37	3.36	113.20	nr	**371.57**
400 mm dia.	228.93	289.73	3.36	113.20	nr	**402.93**
Expansion bellows						
15 mm dia.	31.06	39.31	1.05	35.37	nr	**74.68**
20 mm dia.	32.46	41.08	1.05	35.37	nr	**76.45**
25 mm dia.	34.68	43.89	1.05	35.37	nr	**79.26**
32 mm dia.	36.51	46.21	1.05	35.37	nr	**81.58**
40 mm dia.	38.73	49.01	1.05	35.37	nr	**84.38**
50 mm dia.	43.17	54.63	1.05	35.37	nr	**90.00**
65 mm dia.	48.23	61.04	1.05	35.37	nr	**96.41**
80 mm dia.	52.04	65.87	1.26	42.45	nr	**108.32**
100 mm dia.	65.19	82.50	1.26	42.45	nr	**124.95**
125 mm dia.	72.91	92.28	1.26	42.45	nr	**134.73**
150 mm dia.	85.14	107.76	1.26	42.45	nr	**150.21**
200 mm dia.	114.20	144.54	1.53	51.54	nr	**196.08**
250 mm dia.	133.93	169.50	1.53	51.54	nr	**221.04**
300 mm dia.	146.51	185.43	1.53	51.54	nr	**236.97**
350 mm dia.	163.31	206.68	1.53	51.54	nr	**258.22**
400 mm dia.	183.14	231.78	1.53	51.54	nr	**283.32**
50 mm thick						
15 mm dia.	14.92	18.88	0.44	14.82	m	**33.70**
20 mm dia.	15.72	19.89	0.44	14.82	m	**34.71**
25 mm dia.	16.66	21.09	0.44	14.82	m	**35.91**
32 mm dia.	17.52	22.18	0.44	14.82	m	**37.00**
40 mm dia.	18.32	23.18	0.44	14.82	m	**38.00**
50 mm dia.	20.22	25.59	0.44	14.82	m	**40.41**
65 mm dia.	21.72	27.48	0.44	14.82	m	**42.30**

38 MECHANICAL/COOLING/HEATING SYSTEMS

Item	Net Price £	Material £	Labour hours	Labour £	Unit	Total rate £
THERMAL INSULATION – cont						
50 mm thick – cont						
80 mm dia.	23.33	29.52	0.52	17.52	m	47.04
100 mm dia.	28.82	36.48	0.52	17.52	m	54.00
125 mm dia.	31.91	40.39	0.52	17.52	m	57.91
150 mm dia.	36.65	46.38	0.52	17.52	m	63.90
200 mm dia.	48.04	60.80	0.60	20.22	m	81.02
250 mm dia.	55.53	70.28	0.60	20.22	m	90.50
300 mm dia.	60.45	76.51	0.60	20.22	m	96.73
350 mm dia.	67.37	85.27	0.60	20.22	m	105.49
400 mm dia.	75.34	95.35	0.60	20.22	m	115.57
Extra over for fittings plantroom insulation						
Flange/union						
15 mm dia.	17.57	22.23	0.58	19.54	nr	41.77
20 mm dia.	18.55	23.48	0.58	19.54	nr	43.02
25 mm dia.	19.65	24.86	0.58	19.54	nr	44.40
32 mm dia.	20.62	26.10	0.58	19.54	nr	45.64
40 mm dia.	21.61	27.35	0.58	19.54	nr	46.89
50 mm dia.	23.99	30.36	0.58	19.54	nr	49.90
65 mm dia.	25.92	32.80	0.58	19.54	nr	52.34
80 mm dia.	27.77	35.15	0.67	22.58	nr	57.73
100 mm dia.	34.74	43.97	0.67	22.58	nr	66.55
125 mm dia.	38.63	48.89	0.67	22.58	nr	71.47
150 mm dia.	44.46	56.27	0.67	22.58	nr	78.85
200 mm dia.	59.08	74.77	0.87	29.31	nr	104.08
250 mm dia.	68.30	86.44	0.87	29.31	nr	115.75
300 mm dia.	73.67	93.24	0.87	29.31	nr	122.55
350 mm dia.	81.93	103.69	0.87	29.31	nr	133.00
400 mm dia.	91.39	115.66	0.87	29.31	nr	144.97
Bend						
15 mm dia.	8.23	10.42	0.44	14.82	nr	25.24
20 mm dia.	8.65	10.94	0.44	14.82	nr	25.76
25 mm dia.	9.17	11.60	0.44	14.82	nr	26.42
32 mm dia.	9.66	12.23	0.44	14.82	nr	27.05
40 mm dia.	10.07	12.75	0.44	14.82	nr	27.57
50 mm dia.	11.13	14.09	0.44	14.82	nr	28.91
65 mm dia.	11.95	15.12	0.44	14.82	nr	29.94
80 mm dia.	12.83	16.24	0.52	17.52	nr	33.76
100 mm dia.	15.85	20.06	0.52	17.52	nr	37.58
125 mm dia.	17.54	22.20	0.52	17.52	nr	39.72
150 mm dia.	20.17	25.52	0.52	17.52	nr	43.04
200 mm dia.	26.42	33.43	0.60	20.22	nr	53.65
250 mm dia.	30.54	38.65	0.60	20.22	nr	58.87
300 mm dia.	33.24	42.07	0.60	20.22	nr	62.29
350 mm dia.	37.05	46.89	0.60	20.22	nr	67.11
400 mm dia.	41.44	52.45	0.60	20.22	nr	72.67

38 MECHANICAL/COOLING/HEATING SYSTEMS

Item	Net Price £	Material £	Labour hours	Labour £	Unit	Total rate £
Tee						
15 mm dia.	4.93	6.24	0.44	14.82	nr	**21.06**
20 mm dia.	5.18	6.55	0.44	14.82	nr	**21.37**
25 mm dia.	5.49	6.94	0.44	14.82	nr	**21.76**
32 mm dia.	5.79	7.32	0.44	14.82	nr	**22.14**
40 mm dia.	6.05	7.66	0.44	14.82	nr	**22.48**
50 mm dia.	6.67	8.44	0.44	14.82	nr	**23.26**
65 mm dia.	7.16	9.06	0.44	14.82	nr	**23.88**
80 mm dia.	7.70	9.74	0.52	17.52	nr	**27.26**
100 mm dia.	9.51	12.04	0.52	17.52	nr	**29.56**
125 mm dia.	10.53	13.33	0.52	17.52	nr	**30.85**
150 mm dia.	12.08	15.29	0.52	17.52	nr	**32.81**
200 mm dia.	15.85	20.06	0.60	20.22	nr	**40.28**
250 mm dia.	18.33	23.20	0.60	20.22	nr	**43.42**
300 mm dia.	19.96	25.26	0.60	20.22	nr	**45.48**
350 mm dia.	22.24	28.15	0.60	20.22	nr	**48.37**
400 mm dia.	24.87	31.47	0.60	20.22	nr	**51.69**
Valves						
15 mm dia.	27.94	35.36	0.78	26.28	nr	**61.64**
20 mm dia.	29.44	37.26	0.78	26.28	nr	**63.54**
25 mm dia.	31.21	39.50	0.78	26.28	nr	**65.78**
32 mm dia.	32.76	41.46	0.78	26.28	nr	**67.74**
40 mm dia.	34.31	43.42	0.78	26.28	nr	**69.70**
50 mm dia.	38.08	48.19	0.78	26.28	nr	**74.47**
65 mm dia.	41.14	52.07	0.78	26.28	nr	**78.35**
80 mm dia.	44.13	55.85	0.92	30.99	nr	**86.84**
100 mm dia.	55.18	69.83	0.92	30.99	nr	**100.82**
125 mm dia.	61.35	77.65	0.92	30.99	nr	**108.64**
150 mm dia.	70.61	89.36	0.92	30.99	nr	**120.35**
200 mm dia.	93.88	118.81	1.12	37.73	nr	**156.54**
250 mm dia.	108.47	137.28	1.12	37.73	nr	**175.01**
300 mm dia.	117.50	148.71	1.12	37.73	nr	**186.44**
350 mm dia.	130.12	164.68	1.12	37.73	nr	**202.41**
400 mm dia.	145.17	183.72	1.12	37.73	nr	**221.45**
Pumps						
15 mm dia.	51.74	65.49	2.34	78.84	nr	**144.33**
20 mm dia.	54.54	69.03	2.34	78.84	nr	**147.87**
25 mm dia.	57.77	73.11	2.34	78.84	nr	**151.95**
32 mm dia.	60.67	76.79	2.34	78.84	nr	**155.63**
40 mm dia.	63.57	80.45	2.34	78.84	nr	**159.29**
50 mm dia.	70.49	89.21	2.34	78.84	nr	**168.05**
65 mm dia.	100.80	127.57	2.34	78.84	nr	**206.41**
80 mm dia.	81.75	103.47	2.76	92.98	nr	**196.45**
100 mm dia.	102.16	129.29	2.76	92.98	nr	**222.27**
125 mm dia.	113.62	143.80	2.76	92.98	nr	**236.78**
150 mm dia.	130.74	165.47	2.76	92.98	nr	**258.45**
200 mm dia.	173.83	220.00	3.36	113.20	nr	**333.20**
250 mm dia.	200.86	254.21	3.36	113.20	nr	**367.41**
300 mm dia.	216.74	274.31	3.36	113.20	nr	**387.51**
350 mm dia.	240.93	304.92	3.36	113.20	nr	**418.12**
400 mm dia.	268.84	340.24	3.36	113.20	nr	**453.44**

38 MECHANICAL/COOLING/HEATING SYSTEMS

Item	Net Price £	Material £	Labour hours	Labour £	Unit	Total rate £
THERMAL INSULATION – cont						
Extra over for fittings plantroom insulation – cont						
Expansion bellows						
15 mm dia.	41.37	52.36	1.05	35.37	nr	**87.73**
20 mm dia.	43.61	55.19	1.05	35.37	nr	**90.56**
25 mm dia.	46.21	58.49	1.05	35.37	nr	**93.86**
32 mm dia.	48.52	61.41	1.05	35.37	nr	**96.78**
40 mm dia.	50.84	64.34	1.05	35.37	nr	**99.71**
50 mm dia.	56.40	71.38	1.05	35.37	nr	**106.75**
65 mm dia.	60.94	77.12	1.05	35.37	nr	**112.49**
80 mm dia.	65.41	82.78	1.26	42.45	nr	**125.23**
100 mm dia.	81.74	103.45	1.26	42.45	nr	**145.90**
125 mm dia.	90.89	115.04	1.26	42.45	nr	**157.49**
150 mm dia.	104.61	132.40	1.26	42.45	nr	**174.85**
200 mm dia.	139.07	176.01	1.53	51.54	nr	**227.55**
250 mm dia.	160.69	203.37	1.53	51.54	nr	**254.91**
300 mm dia.	173.43	219.50	1.53	51.54	nr	**271.04**
350 mm dia.	192.75	243.95	1.53	51.54	nr	**295.49**
400 mm dia.	215.05	272.17	1.53	51.54	nr	**323.71**
Mineral fibre sectional insulation; bright class O foil faced; bright class O foil taped joints; 0.8 mm polyisobutylene sheeting; welded joints						
External pipework						
20 mm thick						
15 mm dia.	6.70	8.48	0.30	10.11	m	**18.59**
20 mm dia.	7.13	9.03	0.30	10.11	m	**19.14**
25 mm dia.	7.70	9.74	0.30	10.11	m	**19.85**
32 mm dia.	8.45	10.70	0.30	10.11	m	**20.81**
40 mm dia.	9.02	11.41	0.30	10.11	m	**21.52**
50 mm dia.	10.19	12.89	0.30	10.11	m	**23.00**
Extra over for fittings external insulation						
Flange/union						
15 mm dia.	10.26	12.98	0.75	25.26	nr	**38.24**
20 mm dia.	10.90	13.80	0.75	25.26	nr	**39.06**
25 mm dia.	11.70	14.81	0.75	25.26	nr	**40.07**
32 mm dia.	12.76	16.15	0.75	25.26	nr	**41.41**
40 mm dia.	13.47	17.05	0.75	25.26	nr	**42.31**
50 mm dia.	15.09	19.10	0.75	25.26	nr	**44.36**
Bends						
15 mm dia.	1.70	2.15	0.30	10.11	nr	**12.26**
20 mm dia.	1.79	2.26	0.30	10.11	nr	**12.37**
25 mm dia.	1.93	2.44	0.30	10.11	nr	**12.55**
32 mm dia.	2.12	2.69	0.30	10.11	nr	**12.80**
40 mm dia.	2.26	2.86	0.30	10.11	nr	**12.97**
50 mm dia.	2.54	3.21	0.30	10.11	nr	**13.32**

38 MECHANICAL/COOLING/HEATING SYSTEMS

Item	Net Price £	Material £	Labour hours	Labour £	Unit	Total rate £
Tees						
15 mm dia.	1.70	2.15	0.30	10.11	nr	**12.26**
20 mm dia.	1.79	2.26	0.30	10.11	nr	**12.37**
25 mm dia.	1.93	2.44	0.30	10.11	nr	**12.55**
32 mm dia.	2.12	2.69	0.30	10.11	nr	**12.80**
40 mm dia.	2.26	2.86	0.30	10.11	nr	**12.97**
50 mm dia.	2.54	3.21	0.30	10.11	nr	**13.32**
Valves						
15 mm dia.	16.31	20.64	1.03	34.70	nr	**55.34**
20 mm dia.	17.32	21.92	1.03	34.70	nr	**56.62**
25 mm dia.	18.59	23.53	1.03	34.70	nr	**58.23**
32 mm dia.	20.24	25.61	1.03	34.70	nr	**60.31**
40 mm dia.	21.37	27.05	1.03	34.70	nr	**61.75**
50 mm dia.	23.95	30.31	1.03	34.70	nr	**65.01**
Expansion bellows						
15 mm dia.	24.14	30.55	1.42	47.84	nr	**78.39**
20 mm dia.	25.67	32.49	1.42	47.84	nr	**80.33**
25 mm dia.	27.58	34.91	1.42	47.84	nr	**82.75**
32 mm dia.	30.00	37.97	1.42	47.84	nr	**85.81**
40 mm dia.	31.66	40.07	1.42	47.84	nr	**87.91**
50 mm dia.	35.49	44.91	1.42	47.84	nr	**92.75**
25 mm thick						
15 mm dia.	7.39	9.35	0.30	10.11	m	**19.46**
20 mm dia.	7.93	10.04	0.30	10.11	m	**20.15**
25 mm dia.	8.72	11.03	0.30	10.11	m	**21.14**
32 mm dia.	9.44	11.95	0.30	10.11	m	**22.06**
40 mm dia.	10.05	12.72	0.30	10.11	m	**22.83**
50 mm dia.	11.35	14.37	0.30	10.11	m	**24.48**
65 mm dia.	12.87	16.28	0.30	10.11	m	**26.39**
80 mm dia.	14.09	17.83	0.40	13.47	m	**31.30**
100 mm dia.	17.78	22.50	0.40	13.47	m	**35.97**
125 mm dia.	20.48	25.92	0.40	13.47	m	**39.39**
150 mm dia.	24.02	30.40	0.40	13.47	m	**43.87**
200 mm dia.	32.56	41.20	0.50	16.84	m	**58.04**
250 mm dia.	38.89	49.22	0.50	16.84	m	**66.06**
300 mm dia.	42.29	53.52	0.50	16.84	m	**70.36**
Extra over for fittings external insulation						
Flange/union						
15 mm dia.	11.29	14.29	0.75	25.26	nr	**39.55**
20 mm dia.	12.18	15.41	0.75	25.26	nr	**40.67**
25 mm dia.	13.37	16.92	0.75	25.26	nr	**42.18**
32 mm dia.	14.29	18.09	0.75	25.26	nr	**43.35**
40 mm dia.	15.30	19.36	0.75	25.26	nr	**44.62**
50 mm dia.	17.15	21.71	0.75	25.26	nr	**46.97**
65 mm dia.	19.39	24.54	0.75	25.26	nr	**49.80**
80 mm dia.	21.09	26.69	0.89	29.98	nr	**56.67**
100 mm dia.	26.21	33.17	0.89	29.98	nr	**63.15**
125 mm dia.	30.35	38.42	0.89	29.98	nr	**68.40**
150 mm dia.	35.33	44.71	0.89	29.98	nr	**74.69**
200 mm dia.	47.80	60.49	1.15	38.74	nr	**99.23**

38 MECHANICAL/COOLING/HEATING SYSTEMS

Item	Net Price £	Material £	Labour hours	Labour £	Unit	Total rate £
THERMAL INSULATION – cont						
Extra over for fittings external insulation – cont						
Flange/union – cont						
250 mm dia.	56.91	72.03	1.15	38.74	nr	**110.77**
300 mm dia.	63.22	80.01	1.15	38.74	nr	**118.75**
Bends						
15 mm dia.	1.84	2.33	0.30	10.11	nr	**12.44**
20 mm dia.	1.99	2.52	0.30	10.11	nr	**12.63**
25 mm dia.	2.16	2.73	0.30	10.11	nr	**12.84**
32 mm dia.	2.34	2.96	0.30	10.11	nr	**13.07**
40 mm dia.	2.51	3.18	0.30	10.11	nr	**13.29**
50 mm dia.	2.84	3.60	0.30	10.11	nr	**13.71**
65 mm dia.	3.23	4.09	0.30	10.11	nr	**14.20**
80 mm dia.	3.52	4.46	0.40	13.47	nr	**17.93**
100 mm dia.	4.44	5.62	0.40	13.47	nr	**19.09**
125 mm dia.	5.11	6.46	0.40	13.47	nr	**19.93**
150 mm dia.	6.02	7.62	0.40	13.47	nr	**21.09**
200 mm dia.	8.13	10.29	0.50	16.84	nr	**27.13**
250 mm dia.	9.73	12.31	0.50	16.84	nr	**29.15**
300 mm dia.	10.57	13.37	0.50	16.84	nr	**30.21**
Tees						
15 mm dia.	1.84	2.33	0.30	10.11	nr	**12.44**
20 mm dia.	1.99	2.52	0.30	10.11	nr	**12.63**
25 mm dia.	2.16	2.73	0.30	10.11	nr	**12.84**
32 mm dia.	2.34	2.96	0.30	10.11	nr	**13.07**
40 mm dia.	2.51	3.18	0.30	10.11	nr	**13.29**
50 mm dia.	2.84	3.60	0.30	10.11	nr	**13.71**
65 mm dia.	3.23	4.09	0.30	10.11	nr	**14.20**
80 mm dia.	3.52	4.46	0.40	13.47	nr	**17.93**
100 mm dia.	4.44	5.62	0.40	13.47	nr	**19.09**
125 mm dia.	5.11	6.46	0.40	13.47	nr	**19.93**
150 mm dia.	6.02	7.62	0.40	13.47	nr	**21.09**
200 mm dia.	8.13	10.29	0.50	16.84	nr	**27.13**
250 mm dia.	9.73	12.31	0.50	16.84	nr	**29.15**
300 mm dia.	10.57	13.37	0.50	16.84	nr	**30.21**
Valves						
15 mm dia.	17.95	22.71	1.03	34.70	nr	**57.41**
20 mm dia.	19.34	24.47	1.03	34.70	nr	**59.17**
25 mm dia.	21.21	26.85	1.03	34.70	nr	**61.55**
32 mm dia.	22.73	28.76	1.03	34.70	nr	**63.46**
40 mm dia.	24.28	30.73	1.03	34.70	nr	**65.43**
50 mm dia.	27.27	34.52	1.03	34.70	nr	**69.22**
65 mm dia.	30.84	39.03	1.03	34.70	nr	**73.73**
80 mm dia.	33.51	42.41	1.25	42.11	nr	**84.52**
100 mm dia.	41.66	52.73	1.25	42.11	nr	**94.84**
125 mm dia.	48.21	61.02	1.25	42.11	nr	**103.13**
150 mm dia.	56.13	71.04	1.25	42.11	nr	**113.15**
200 mm dia.	75.95	96.12	1.55	52.21	nr	**148.33**
250 mm dia.	90.38	114.39	1.55	52.21	nr	**166.60**
300 mm dia.	100.40	127.06	1.55	52.21	nr	**179.27**

38 MECHANICAL/COOLING/HEATING SYSTEMS

Item	Net Price £	Material £	Labour hours	Labour £	Unit	Total rate £
Expansion bellows						
15 mm dia.	26.60	33.67	1.42	47.84	nr	**81.51**
20 mm dia.	28.66	36.28	1.42	47.84	nr	**84.12**
25 mm dia.	31.43	39.78	1.42	47.84	nr	**87.62**
32 mm dia.	33.71	42.66	1.42	47.84	nr	**90.50**
40 mm dia.	36.01	45.57	1.42	47.84	nr	**93.41**
50 mm dia.	40.36	51.08	1.42	47.84	nr	**98.92**
65 mm dia.	45.66	57.79	1.42	47.84	nr	**105.63**
80 mm dia.	49.64	62.82	1.75	58.97	nr	**121.79**
100 mm dia.	61.71	78.10	1.75	58.97	nr	**137.07**
125 mm dia.	71.42	90.38	1.75	58.97	nr	**149.35**
150 mm dia.	83.14	105.22	1.75	58.97	nr	**164.19**
200 mm dia.	112.50	142.39	2.17	73.11	nr	**215.50**
250 mm dia.	133.89	169.46	2.17	73.11	nr	**242.57**
300 mm dia.	148.74	188.25	3.17	106.80	nr	**295.05**
30 mm thick						
15 mm dia.	10.71	13.55	0.30	10.11	m	**23.66**
20 mm dia.	9.69	12.26	0.30	10.11	m	**22.37**
25 mm dia.	10.26	12.98	0.30	10.11	m	**23.09**
32 mm dia.	11.14	14.10	0.30	10.11	m	**24.21**
40 mm dia.	11.73	14.84	0.30	10.11	m	**24.95**
50 mm dia.	13.21	16.72	0.30	10.11	m	**26.83**
65 mm dia.	14.85	18.79	0.30	10.11	m	**28.90**
80 mm dia.	16.14	20.43	0.40	13.47	m	**33.90**
100 mm dia.	20.19	25.55	0.40	13.47	m	**39.02**
125 mm dia.	23.12	29.27	0.40	13.47	m	**42.74**
150 mm dia.	26.85	33.98	0.40	13.47	m	**47.45**
200 mm dia.	36.16	45.76	0.50	16.84	m	**62.60**
250 mm dia.	42.91	54.31	0.50	16.84	m	**71.15**
300 mm dia.	46.28	58.58	0.50	16.84	m	**75.42**
350 mm dia.	50.60	64.04	0.50	16.84	m	**80.88**
Extra over for fittings external insulation						
Flange/union						
15 mm dia.	13.82	17.49	0.75	25.26	nr	**42.75**
20 mm dia.	14.63	18.51	0.75	25.26	nr	**43.77**
25 mm dia.	15.53	19.66	0.75	25.26	nr	**44.92**
32 mm dia.	16.94	21.44	0.75	25.26	nr	**46.70**
40 mm dia.	17.69	22.39	0.75	25.26	nr	**47.65**
50 mm dia.	19.74	24.99	0.75	25.26	nr	**50.25**
65 mm dia.	22.23	28.13	0.75	25.26	nr	**53.39**
80 mm dia.	24.22	30.65	0.89	29.98	nr	**60.63**
100 mm dia.	29.82	37.74	0.89	29.98	nr	**67.72**
125 mm dia.	34.08	43.13	0.89	29.98	nr	**73.11**
150 mm dia.	39.58	50.10	0.89	29.98	nr	**80.08**
200 mm dia.	52.87	66.91	1.15	38.74	nr	**105.65**
250 mm dia.	62.61	79.24	1.15	38.74	nr	**117.98**
300 mm dia.	68.62	86.84	1.15	38.74	nr	**125.58**
350 mm dia.	75.79	95.92	1.15	38.74	nr	**134.66**

38 MECHANICAL/COOLING/HEATING SYSTEMS

Item	Net Price £	Material £	Labour hours	Labour £	Unit	Total rate £
THERMAL INSULATION – cont						
Extra over for fittings external insulation – cont						
Bends						
15 mm dia.	2.27	2.88	0.30	10.11	nr	**12.99**
20 mm dia.	2.42	3.06	0.30	10.11	nr	**13.17**
25 mm dia.	2.58	3.27	0.30	10.11	nr	**13.38**
32 mm dia.	2.78	3.52	0.30	10.11	nr	**13.63**
40 mm dia.	2.93	3.71	0.30	10.11	nr	**13.82**
50 mm dia.	3.31	4.19	0.30	10.11	nr	**14.30**
65 mm dia.	3.72	4.70	0.30	10.11	nr	**14.81**
80 mm dia.	4.05	5.13	0.40	13.47	nr	**18.60**
100 mm dia.	5.07	6.42	0.40	13.47	nr	**19.89**
125 mm dia.	5.79	7.32	0.40	13.47	nr	**20.79**
150 mm dia.	6.71	8.49	0.40	13.47	nr	**21.96**
200 mm dia.	9.05	11.46	0.50	16.84	nr	**28.30**
250 mm dia.	10.75	13.61	0.50	16.84	nr	**30.45**
300 mm dia.	11.56	14.63	0.50	16.84	nr	**31.47**
350 mm dia.	12.65	16.00	0.50	16.84	nr	**32.84**
Tees						
15 mm dia.	2.27	2.88	0.30	10.11	nr	**12.99**
20 mm dia.	2.42	3.06	0.30	10.11	nr	**13.17**
25 mm dia.	2.58	3.27	0.30	10.11	nr	**13.38**
32 mm dia.	2.78	3.52	0.30	10.11	nr	**13.63**
40 mm dia.	2.93	3.71	0.30	10.11	nr	**13.82**
50 mm dia.	3.31	4.19	0.30	10.11	nr	**14.30**
65 mm dia.	3.72	4.70	0.30	10.11	nr	**14.81**
80 mm dia.	4.05	5.13	0.40	13.47	nr	**18.60**
100 mm dia.	5.07	6.42	0.40	13.47	nr	**19.89**
125 mm dia.	5.79	7.32	0.40	13.47	nr	**20.79**
150 mm dia.	6.71	8.49	0.40	13.47	nr	**21.96**
200 mm dia.	9.05	11.46	0.50	16.84	nr	**28.30**
250 mm dia.	10.75	13.61	0.50	16.84	nr	**30.45**
300 mm dia.	11.56	14.63	0.50	16.84	nr	**31.47**
350 mm dia.	12.65	16.00	0.50	16.84	nr	**32.84**
Valves						
15 mm dia.	21.95	27.78	1.03	34.70	nr	**62.48**
20 mm dia.	23.26	29.43	1.03	34.70	nr	**64.13**
25 mm dia.	24.67	31.23	1.03	34.70	nr	**65.93**
32 mm dia.	26.86	33.99	1.03	34.70	nr	**68.69**
40 mm dia.	28.11	35.57	1.03	34.70	nr	**70.27**
50 mm dia.	31.35	39.68	1.03	34.70	nr	**74.38**
65 mm dia.	35.30	44.68	1.03	34.70	nr	**79.38**
80 mm dia.	38.46	48.68	1.25	42.11	nr	**90.79**
100 mm dia.	47.32	59.89	1.25	42.11	nr	**102.00**
125 mm dia.	54.09	68.45	1.25	42.11	nr	**110.56**
150 mm dia.	62.84	79.53	1.25	42.11	nr	**121.64**
200 mm dia.	84.00	106.31	1.55	52.21	nr	**158.52**
250 mm dia.	99.46	125.88	1.55	52.21	nr	**178.09**
300 mm dia.	108.97	137.92	1.55	52.21	nr	**190.13**
350 mm dia.	120.38	152.35	1.55	52.21	nr	**204.56**

38 MECHANICAL/COOLING/HEATING SYSTEMS

Item	Net Price £	Material £	Labour hours	Labour £	Unit	Total rate £
Expansion bellows						
15 mm dia.	32.50	41.13	1.42	47.84	nr	**88.97**
20 mm dia.	34.47	43.62	1.42	47.84	nr	**91.46**
25 mm dia.	36.52	46.22	1.42	47.84	nr	**94.06**
32 mm dia.	39.81	50.39	1.42	47.84	nr	**98.23**
40 mm dia.	41.66	52.73	1.42	47.84	nr	**100.57**
50 mm dia.	46.44	58.78	1.42	47.84	nr	**106.62**
65 mm dia.	52.30	66.19	1.42	47.84	nr	**114.03**
80 mm dia.	57.01	72.15	1.75	58.97	nr	**131.12**
100 mm dia.	70.12	88.75	1.75	58.97	nr	**147.72**
125 mm dia.	80.14	101.43	1.75	58.97	nr	**160.40**
150 mm dia.	93.09	117.81	1.75	58.97	nr	**176.78**
200 mm dia.	124.43	157.48	2.17	73.11	nr	**230.59**
250 mm dia.	147.35	186.49	2.17	73.11	nr	**259.60**
300 mm dia.	161.44	204.32	2.17	73.11	nr	**277.43**
350 mm dia.	178.36	225.74	2.17	73.11	nr	**298.85**
40 mm thick						
15 mm dia.	11.38	14.40	0.30	10.11	m	**24.51**
20 mm dia.	11.78	14.91	0.30	10.11	m	**25.02**
25 mm dia.	12.58	15.93	0.30	10.11	m	**26.04**
32 mm dia.	13.40	16.96	0.30	10.11	m	**27.07**
40 mm dia.	14.00	17.72	0.30	10.11	m	**27.83**
50 mm dia.	15.61	19.76	0.30	10.11	m	**29.87**
65 mm dia.	17.43	22.06	0.30	10.11	m	**32.17**
80 mm dia.	18.89	23.91	0.40	13.47	m	**37.38**
100 mm dia.	23.57	29.83	0.40	13.47	m	**43.30**
125 mm dia.	26.56	33.61	0.40	13.47	m	**47.08**
150 mm dia.	30.67	38.82	0.40	13.47	m	**52.29**
200 mm dia.	40.91	51.78	0.50	16.84	m	**68.62**
250 mm dia.	47.91	60.64	0.50	16.84	m	**77.48**
300 mm dia.	51.86	65.63	0.50	16.84	m	**82.47**
350 mm dia.	57.01	72.15	0.50	16.84	m	**88.99**
400 mm dia.	63.44	80.29	0.50	16.84	m	**97.13**
Extra over for fittings external insulation						
Flange/union						
15 mm dia.	17.01	21.53	0.75	25.26	nr	**46.79**
20 mm dia.	17.77	22.49	0.75	25.26	nr	**47.75**
25 mm dia.	18.94	23.97	0.75	25.26	nr	**49.23**
32 mm dia.	19.96	25.26	0.75	25.26	nr	**50.52**
40 mm dia.	21.07	26.67	0.75	25.26	nr	**51.93**
50 mm dia.	23.34	29.53	0.75	25.26	nr	**54.79**
65 mm dia.	26.01	32.92	0.75	25.26	nr	**58.18**
80 mm dia.	28.04	35.49	0.89	29.98	nr	**65.47**
100 mm dia.	34.53	43.70	0.89	29.98	nr	**73.68**
125 mm dia.	38.63	48.89	0.89	29.98	nr	**78.87**
150 mm dia.	44.76	56.65	0.89	29.98	nr	**86.63**
200 mm dia.	59.02	74.69	1.15	38.74	nr	**113.43**
250 mm dia.	69.19	87.56	1.15	38.74	nr	**126.30**
300 mm dia.	76.02	96.21	1.15	38.74	nr	**134.95**
350 mm dia.	84.32	106.71	1.15	38.74	nr	**145.45**
400 mm dia.	94.42	119.49	1.15	38.74	nr	**158.23**

38 MECHANICAL/COOLING/HEATING SYSTEMS

Item	Net Price £	Material £	Labour hours	Labour £	Unit	Total rate £
THERMAL INSULATION – cont						
Extra over for fittings external insulation – cont						
Bends						
15 mm dia.	2.84	3.60	0.30	10.11	nr	**13.71**
20 mm dia.	2.95	3.73	0.30	10.11	nr	**13.84**
25 mm dia.	3.15	3.99	0.30	10.11	nr	**14.10**
32 mm dia.	3.34	4.22	0.30	10.11	nr	**14.33**
40 mm dia.	3.52	4.46	0.30	10.11	nr	**14.57**
50 mm dia.	3.93	4.97	0.30	10.11	nr	**15.08**
65 mm dia.	4.37	5.53	0.30	10.11	nr	**15.64**
80 mm dia.	4.74	6.00	0.40	13.47	nr	**19.47**
100 mm dia.	5.89	7.46	0.40	13.47	nr	**20.93**
125 mm dia.	6.64	8.40	0.40	13.47	nr	**21.87**
150 mm dia.	7.68	9.72	0.40	13.47	nr	**23.19**
200 mm dia.	10.23	12.95	0.50	16.84	nr	**29.79**
250 mm dia.	11.97	15.15	0.50	16.84	nr	**31.99**
300 mm dia.	12.96	16.40	0.50	16.84	nr	**33.24**
350 mm dia.	14.24	18.02	0.50	16.84	nr	**34.86**
400 mm dia.	15.87	20.08	0.50	16.84	nr	**36.92**
Tees						
15 mm dia.	2.84	3.60	0.30	10.11	nr	**13.71**
20 mm dia.	2.95	3.73	0.30	10.11	nr	**13.84**
25 mm dia.	3.15	3.99	0.30	10.11	nr	**14.10**
32 mm dia.	3.34	4.22	0.30	10.11	nr	**14.33**
40 mm dia.	3.52	4.46	0.30	10.11	nr	**14.57**
50 mm dia.	3.93	4.97	0.30	10.11	nr	**15.08**
65 mm dia.	4.37	5.53	0.30	10.11	nr	**15.64**
80 mm dia.	4.74	6.00	0.40	13.47	nr	**19.47**
100 mm dia.	5.89	7.46	0.40	13.47	nr	**20.93**
125 mm dia.	6.64	8.40	0.40	13.47	nr	**21.87**
150 mm dia.	7.68	9.72	0.40	13.47	nr	**23.19**
200 mm dia.	10.23	12.95	0.50	16.84	nr	**29.79**
250 mm dia.	11.97	15.15	0.50	16.84	nr	**31.99**
300 mm dia.	12.96	16.40	0.50	16.84	nr	**33.24**
350 mm dia.	14.24	18.02	0.50	16.84	nr	**34.86**
400 mm dia.	15.87	20.08	0.50	16.84	nr	**36.92**
Valves						
15 mm dia.	27.00	34.17	1.03	34.70	nr	**68.87**
20 mm dia.	28.22	35.72	1.03	34.70	nr	**70.42**
25 mm dia.	30.07	38.06	1.03	34.70	nr	**72.76**
32 mm dia.	31.67	40.08	1.03	34.70	nr	**74.78**
40 mm dia.	33.49	42.38	1.03	34.70	nr	**77.08**
50 mm dia.	37.11	46.96	1.03	34.70	nr	**81.66**
65 mm dia.	41.32	52.29	1.03	34.70	nr	**86.99**
80 mm dia.	44.53	56.36	1.25	42.11	nr	**98.47**
100 mm dia.	54.85	69.42	1.25	42.11	nr	**111.53**
125 mm dia.	61.35	77.65	1.25	42.11	nr	**119.76**
150 mm dia.	71.12	90.01	1.25	42.11	nr	**132.12**

38 MECHANICAL/COOLING/HEATING SYSTEMS

Item	Net Price £	Material £	Labour hours	Labour £	Unit	Total rate £
200 mm dia.	93.75	118.65	1.55	52.21	nr	**170.86**
250 mm dia.	109.84	139.01	1.55	52.21	nr	**191.22**
300 mm dia.	120.76	152.84	1.55	52.21	nr	**205.05**
350 mm dia.	133.93	169.50	1.55	52.21	nr	**221.71**
400 mm dia.	149.95	189.77	1.55	52.21	nr	**241.98**
Expansion bellows						
15 mm dia.	39.99	50.61	1.42	47.84	nr	**98.45**
20 mm dia.	41.81	52.92	1.42	47.84	nr	**100.76**
25 mm dia.	44.54	56.37	1.42	47.84	nr	**104.21**
32 mm dia.	46.93	59.39	1.42	47.84	nr	**107.23**
40 mm dia.	49.61	62.79	1.42	47.84	nr	**110.63**
50 mm dia.	54.99	69.60	1.42	47.84	nr	**117.44**
65 mm dia.	61.20	77.46	1.42	47.84	nr	**125.30**
80 mm dia.	65.97	83.50	1.75	58.97	nr	**142.47**
100 mm dia.	81.27	102.86	1.75	58.97	nr	**161.83**
125 mm dia.	90.89	115.04	1.75	58.97	nr	**174.01**
150 mm dia.	105.37	133.36	1.75	58.97	nr	**192.33**
200 mm dia.	138.89	175.78	2.17	73.11	nr	**248.89**
250 mm dia.	162.72	205.93	2.17	73.11	nr	**279.04**
300 mm dia.	178.90	226.42	2.17	73.11	nr	**299.53**
350 mm dia.	198.42	251.12	2.17	73.11	nr	**324.23**
400 mm dia.	222.17	281.18	2.17	73.11	nr	**354.29**
50 mm thick						
15 mm dia.	15.00	18.98	0.30	10.11	m	**29.09**
20 mm dia.	15.74	19.92	0.30	10.11	m	**30.03**
25 mm dia.	16.64	21.06	0.30	10.11	m	**31.17**
32 mm dia.	17.96	22.72	0.30	10.11	m	**32.83**
40 mm dia.	18.25	23.09	0.30	10.11	m	**33.20**
50 mm dia.	20.22	25.59	0.30	10.11	m	**35.70**
65 mm dia.	22.04	27.90	0.30	10.11	m	**38.01**
80 mm dia.	23.57	29.83	0.40	13.47	m	**43.30**
100 mm dia.	29.19	36.94	0.40	13.47	m	**50.41**
125 mm dia.	32.66	41.34	0.40	13.47	m	**54.81**
150 mm dia.	37.34	47.25	0.40	13.47	m	**60.72**
200 mm dia.	49.44	62.57	0.50	16.84	m	**79.41**
250 mm dia.	57.11	72.27	0.50	16.84	m	**89.11**
300 mm dia.	61.25	77.52	0.50	16.84	m	**94.36**
350 mm dia.	67.23	85.09	0.50	16.84	m	**101.93**
400 mm dia.	74.47	94.25	0.50	16.84	m	**111.09**
Extra over for fittings external insulation						
Flange/union						
15 mm dia.	22.11	27.98	0.75	25.26	nr	**53.24**
20 mm dia.	23.26	29.43	0.75	25.26	nr	**54.69**
25 mm dia.	24.58	31.11	0.75	25.26	nr	**56.37**
32 mm dia.	25.80	32.65	0.75	25.26	nr	**57.91**
40 mm dia.	27.00	34.17	0.75	25.26	nr	**59.43**
50 mm dia.	29.76	37.67	0.75	25.26	nr	**62.93**
65 mm dia.	32.18	40.72	0.75	25.26	nr	**65.98**

38 MECHANICAL/COOLING/HEATING SYSTEMS

Item	Net Price £	Material £	Labour hours	Labour £	Unit	Total rate £
THERMAL INSULATION – cont						
Extra over for fittings external insulation – cont						
Flange/union – cont						
80 mm dia.	34.50	43.67	0.89	29.98	nr	**73.65**
100 mm dia.	42.42	53.68	0.89	29.98	nr	**83.66**
125 mm dia.	47.15	59.67	0.89	29.98	nr	**89.65**
150 mm dia.	54.00	68.34	0.89	29.98	nr	**98.32**
200 mm dia.	70.63	89.39	1.15	38.74	nr	**128.13**
250 mm dia.	81.61	103.29	1.15	38.74	nr	**142.03**
300 mm dia.	88.55	112.07	1.15	38.74	nr	**150.81**
350 mm dia.	97.99	124.02	1.15	38.74	nr	**162.76**
400 mm dia.	109.19	138.19	1.15	38.74	nr	**176.93**
Bend						
15 mm dia.	3.75	4.75	0.30	10.11	nr	**14.86**
20 mm dia.	3.94	4.98	0.30	10.11	nr	**15.09**
25 mm dia.	4.16	5.26	0.30	10.11	nr	**15.37**
32 mm dia.	4.37	5.53	0.30	10.11	nr	**15.64**
40 mm dia.	4.58	5.80	0.30	10.11	nr	**15.91**
50 mm dia.	5.07	6.42	0.30	10.11	nr	**16.53**
65 mm dia.	5.50	6.96	0.30	10.11	nr	**17.07**
80 mm dia.	5.87	7.43	0.40	13.47	nr	**20.90**
100 mm dia.	7.30	9.24	0.40	13.47	nr	**22.71**
125 mm dia.	8.14	10.30	0.40	13.47	nr	**23.77**
150 mm dia.	9.35	11.84	0.40	13.47	nr	**25.31**
200 mm dia.	12.36	15.65	0.50	16.84	nr	**32.49**
250 mm dia.	14.28	18.08	0.50	16.84	nr	**34.92**
300 mm dia.	15.33	19.40	0.50	16.84	nr	**36.24**
350 mm dia.	16.81	21.28	0.50	16.84	nr	**38.12**
400 mm dia.	18.60	23.54	0.50	16.84	nr	**40.38**
Tee						
15 mm dia.	3.75	4.75	0.30	10.11	nr	**14.86**
20 mm dia.	3.94	4.98	0.30	10.11	nr	**15.09**
25 mm dia.	4.16	5.26	0.30	10.11	nr	**15.37**
32 mm dia.	4.37	5.53	0.30	10.11	nr	**15.64**
40 mm dia.	4.58	5.80	0.30	10.11	nr	**15.91**
50 mm dia.	5.07	6.42	0.30	10.11	nr	**16.53**
65 mm dia.	5.50	6.96	0.30	10.11	nr	**17.07**
80 mm dia.	5.87	7.43	0.40	13.47	nr	**20.90**
100 mm dia.	7.30	9.24	0.40	13.47	nr	**22.71**
125 mm dia.	8.14	10.30	0.40	13.47	nr	**23.77**
150 mm dia.	9.35	11.84	0.40	13.47	nr	**25.31**
200 mm dia.	12.36	15.65	0.50	16.84	nr	**32.49**
250 mm dia.	14.28	18.08	0.50	16.84	nr	**34.92**
300 mm dia.	15.33	19.40	0.50	16.84	nr	**36.24**
350 mm dia.	16.81	21.28	0.50	16.84	nr	**38.12**
400 mm dia.	18.60	23.54	0.50	16.84	nr	**40.38**
Valves						
15 mm dia.	35.11	44.43	1.03	34.70	nr	**79.13**
20 mm dia.	36.93	46.74	1.03	34.70	nr	**81.44**
25 mm dia.	39.04	49.41	1.03	34.70	nr	**84.11**

38 MECHANICAL/COOLING/HEATING SYSTEMS

Item	Net Price £	Material £	Labour hours	Labour £	Unit	Total rate £
32 mm dia.	40.99	51.88	1.03	34.70	nr	**86.58**
40 mm dia.	42.89	54.29	1.03	34.70	nr	**88.99**
50 mm dia.	47.30	59.86	1.03	34.70	nr	**94.56**
65 mm dia.	51.15	64.74	1.03	34.70	nr	**99.44**
80 mm dia.	54.80	69.35	1.25	42.11	nr	**111.46**
100 mm dia.	67.37	85.27	1.25	42.11	nr	**127.38**
125 mm dia.	74.88	94.76	1.25	42.11	nr	**136.87**
150 mm dia.	85.72	108.48	1.25	42.11	nr	**150.59**
200 mm dia.	112.16	141.95	1.55	52.21	nr	**194.16**
250 mm dia.	129.61	164.04	1.55	52.21	nr	**216.25**
300 mm dia.	140.63	177.98	1.55	52.21	nr	**230.19**
350 mm dia.	155.63	196.96	1.55	52.21	nr	**249.17**
400 mm dia.	173.39	219.44	1.55	52.21	nr	**271.65**
Expansion bellows						
15 mm dia.	52.01	65.82	1.42	47.84	nr	**113.66**
20 mm dia.	54.72	69.25	1.42	47.84	nr	**117.09**
25 mm dia.	57.84	73.20	1.42	47.84	nr	**121.04**
32 mm dia.	60.76	76.90	1.42	47.84	nr	**124.74**
40 mm dia.	63.53	80.40	1.42	47.84	nr	**128.24**
50 mm dia.	70.03	88.63	1.42	47.84	nr	**136.47**
65 mm dia.	75.76	95.88	1.42	47.84	nr	**143.72**
80 mm dia.	81.20	102.77	1.75	58.97	nr	**161.74**
100 mm dia.	99.81	126.32	1.75	58.97	nr	**185.29**
125 mm dia.	110.94	140.40	1.75	58.97	nr	**199.37**
150 mm dia.	127.01	160.74	1.75	58.97	nr	**219.71**
200 mm dia.	166.18	210.31	2.17	73.11	nr	**283.42**
250 mm dia.	192.03	243.03	2.17	73.11	nr	**316.14**
300 mm dia.	208.34	263.67	2.17	73.11	nr	**336.78**
350 mm dia.	230.56	291.79	2.17	73.11	nr	**364.90**
400 mm dia.	256.90	325.14	2.17	73.11	nr	**398.25**

38 MECHANICAL/COOLING/HEATING SYSTEMS

Item	Net Price £	Material £	Labour hours	Labour £	Unit	Total rate £
STEAM HEATING						
PIPELINES						
For pipework prices refer to section – Low Temperature Hot Water Heating						
PIPELINE ANCILLARIES						
Steam traps and accessories						
Cast iron; inverted bucket type; steam trap pressure range up to 17 bar at 210°C; screwed ends						
½" dia.	114.58	145.02	0.85	28.64	nr	**173.66**
¾" dia.	114.58	145.02	1.13	38.07	nr	**183.09**
1" dia.	193.06	244.34	1.35	45.48	nr	**289.82**
1½" dia.	596.05	754.36	1.80	60.65	nr	**815.01**
2" dia.	667.18	844.38	2.18	73.44	nr	**917.82**
Cast iron; inverted bucket type; steam trap pressure range up to 17 bar at 210°C; flanged ends to BS 4504 PN16; bolted connections						
15 mm dia.	186.32	235.80	1.15	38.74	nr	**274.54**
20 mm dia.	231.91	293.51	1.25	42.11	nr	**335.62**
25 mm dia.	287.90	364.37	1.33	44.80	nr	**409.17**
40 mm dia.	610.75	772.97	1.46	49.19	nr	**822.16**
50 mm dia.	681.88	862.98	1.60	53.91	nr	**916.89**
Steam traps and strainers						
Stainless steel; thermodynamic trap with pressure range up to 42 bar; temperature range to 400°C; screwed ends to steel						
15 mm dia.	74.48	94.26	0.84	28.31	nr	**122.57**
20 mm dia.	84.30	106.69	1.14	38.42	nr	**145.11**
Stainless steel; thermodynamic trap with pressure range up to 24 bar; temperature range to 288°C; flanged ends to DIN 2456 PN64; bolted connections						
15 mm dia.	155.94	197.36	1.24	41.80	nr	**239.16**
20 mm dia.	170.93	216.33	1.34	45.16	nr	**261.49**
25 mm dia.	197.60	250.08	1.40	47.19	nr	**297.27**
Malleable iron pipeline strainer; max steam working pressure 14 bar and temperature range to 230°C; screwed ends to steel						
½" dia.	22.34	28.27	0.84	28.31	nr	**56.58**
¾" dia.	27.92	35.34	1.14	38.42	nr	**73.76**
1" dia.	43.57	55.14	1.30	43.81	nr	**98.95**
1½" dia.	89.48	113.24	1.50	50.59	nr	**163.83**
2" dia.	125.63	159.00	1.74	58.70	nr	**217.70**

38 MECHANICAL/COOLING/HEATING SYSTEMS

Item	Net Price £	Material £	Labour hours	Labour £	Unit	Total rate £
Bronze pipeline strainer; max steam working pressure 25 bar; flanged ends to BS 4504 PN25; bolted connections						
15 mm dia.	191.92	242.89	1.24	41.80	nr	**284.69**
20 mm dia.	234.10	296.27	1.34	45.16	nr	**341.43**
25 mm dia.	268.40	339.68	1.40	47.19	nr	**386.87**
32 mm dia.	416.66	527.33	1.46	49.26	nr	**576.59**
40 mm dia.	472.84	598.43	1.54	51.91	nr	**650.34**
50 mm dia.	727.22	920.37	1.64	55.32	nr	**975.69**
65 mm dia.	805.22	1019.09	2.50	84.22	nr	**1103.31**
80 mm dia.	1001.85	1267.94	2.91	97.93	nr	**1365.87**
100 mm dia.	1735.30	2196.20	3.51	118.20	nr	**2314.40**
Balanced pressure thermostatic steam trap and strainer; max working pressure up to 13 bar; screwed ends to steel						
½" dia.	154.37	195.37	1.26	42.48	nr	**237.85**
¾" dia.	154.77	195.88	1.71	57.69	nr	**253.57**
Bimetallic thermostatic steam trap and strainer; max working pressure up to 21 bar; flanged ends						
15 mm	200.81	254.15	1.24	41.80	nr	**295.95**
20 mm	220.62	279.22	1.34	45.16	nr	**324.38**
Sight glasses						
Pressed brass; straight; single window; screwed ends to steel						
15 mm dia.	48.37	61.22	0.84	28.31	nr	**89.53**
20 mm dia.	53.68	67.94	1.14	38.42	nr	**106.36**
25 mm dia.	67.09	84.91	1.30	43.81	nr	**128.72**
Gunmetal; straight; double window; screwed ends to steel						
15 mm dia.	78.00	98.72	0.84	28.31	nr	**127.03**
20 mm dia.	85.84	108.64	1.14	38.42	nr	**147.06**
25 mm dia.	106.10	134.28	1.30	43.81	nr	**178.09**
32 mm dia.	174.79	221.21	1.35	45.53	nr	**266.74**
40 mm dia.	174.79	221.21	1.74	58.70	nr	**279.91**
50 mm dia.	212.24	268.61	2.08	70.19	nr	**338.80**
SG Iron flanged; BS 4504, PN 25						
15 mm dia.	159.17	201.44	1.00	33.69	nr	**235.13**
20 mm dia.	187.27	237.01	1.25	42.11	nr	**279.12**
25 mm dia.	238.78	302.20	1.50	50.59	nr	**352.79**
32 mm dia.	263.71	333.75	1.70	57.30	nr	**391.05**
40 mm dia.	344.87	436.46	2.00	67.38	nr	**503.84**
50 mm dia.	416.66	527.33	2.30	77.63	nr	**604.96**
Check valve and sight glass; gun metal; screwed						
15 mm dia.	78.78	99.70	0.84	28.31	nr	**128.01**
20 mm dia.	83.48	105.65	1.14	38.42	nr	**144.07**
25 mm dia.	140.44	177.74	1.30	43.81	nr	**221.55**

38 MECHANICAL/COOLING/HEATING SYSTEMS

Item	Net Price £	Material £	Labour hours	Labour £	Unit	Total rate £
STEAM HEATING – cont						
Pressure reducing valves						
Pressure reducing valve for steam; maximum range of 17 bar and 232°C; screwed ends to steel						
15 mm dia.	589.88	746.55	0.87	29.31	nr	**775.86**
20 mm dia.	638.24	807.76	0.91	30.65	nr	**838.41**
25 mm dia.	688.18	870.96	1.35	45.48	nr	**916.44**
Pressure reducing valve for steam; maximum range of 17 bar and 232°C; flanged ends to BS 4504 PN 25						
25 mm dia.	828.65	1048.73	1.70	57.27	nr	**1106.00**
32 mm dia.	941.00	1190.93	1.87	63.00	nr	**1253.93**
40 mm dia.	1123.60	1422.03	2.12	71.42	nr	**1493.45**
50 mm dia.	1296.81	1641.25	2.57	86.58	nr	**1727.83**
Safety and relief valves						
Bronze safety valve; 'pop' type; side outlet; including easing lever; working pressure saturated steam up to 20.7 bar; screwed ends to steel						
15 mm dia.	200.38	253.60	0.32	10.79	nr	**264.39**
20 mm dia.	226.51	286.68	0.40	13.47	nr	**300.15**
Bronze safety valve; 'pop' type; side outlet; including easing lever; working pressure saturated steam up to 17.2 bar; screwed ends to steel						
25 mm dia.	322.34	407.95	0.47	15.84	nr	**423.79**
32 mm dia.	385.10	487.38	0.56	18.87	nr	**506.25**
Bronze safety valve; 'pop' type; side outlet; including easing lever; working pressure saturated steam up to 13.8 bar; screwed ends to steel						
40 mm dia.	498.34	630.69	0.64	21.57	nr	**652.26**
50 mm dia.	693.51	877.71	0.76	25.60	nr	**903.31**
65 mm dia.	986.28	1248.24	0.94	31.66	nr	**1279.90**
80 mm dia.	1279.01	1618.71	1.10	37.06	nr	**1655.77**

38 MECHANICAL/COOLING/HEATING SYSTEMS

Item	Net Price £	Material £	Labour hours	Labour £	Unit	Total rate £
EQUIPMENT						
CALORIFIERS						
Non-storage calorifiers; mild steel shell construction with indirect steam heating for secondary LPHW at 82°C flow and 71°C return to BS 853; maximum test on shell 11 bar, tubes 26 bar						
Horizontal/vertical, for steam at 3 bar–5.5 bar						
88 kW capacity	789.21	998.83	8.00	269.52	nr	**1268.35**
176 kW capacity	1148.88	1454.02	12.05	405.90	nr	**1859.92**
293 kW capacity	1251.66	1584.11	14.08	474.51	nr	**2058.62**
586 kW capacity	1827.44	2312.81	37.04	1247.78	nr	**3560.59**
879 kW capacity	2275.39	2879.73	40.00	1347.61	nr	**4227.34**
1465 kW capacity	2774.73	3511.69	45.45	1531.38	nr	**5043.07**

38 MECHANICAL/COOLING/HEATING SYSTEMS

Item	Net Price £	Material £	Labour hours	Labour £	Unit	Total rate £
LOCAL HEATING UNITS						
Warm air unit heater for connection to LTHW or steam supplies; suitable for heights up to 3 m; recirculating type; mild steel casing; heating coil; adjustable discharge louvre; axial fan; horizontal or vertical discharge; normal speed; entering air temperature 15°C; complete with enclosures; includes fixing in position; includes connections to primary heating supply; electrical work elsewhere						
Low pressure hot water						
7.5 kW, 265 l/sec	832.85	1054.05	6.53	220.00	nr	**1274.05**
15.4 kW, 575 l/sec	1008.17	1275.94	7.54	254.02	nr	**1529.96**
26.9 kW, 1040 l/sec	1366.12	1728.97	8.65	291.41	nr	**2020.38**
48.0 kW, 1620 l/sec	1800.79	2279.08	9.35	315.01	nr	**2594.09**
Steam, 2 bar						
9.2 kW, 265 l/sec	1185.60	1500.50	6.53	220.00	nr	**1720.50**
18.8 kW, 575 l/sec	1283.80	1624.77	6.82	229.77	nr	**1854.54**
34.8 kW, 1040 l/sec	1476.59	1868.78	6.82	229.77	nr	**2098.55**
51.6 kW, 1625 l/sec	2000.33	2531.61	7.10	239.20	nr	**2770.81**

38 MECHANICAL/COOLING/HEATING SYSTEMS

Item	Net Price £	Material £	Labour hours	Labour £	Unit	Total rate £
CENTRAL REFRIGERATION PLANT						
CHILLERS; Air cooled						
Selection of air cooled chillers based on chilled water flow and return temperatures 6°C and 12°C, and an outdoor temperature of 35°C						
Air cooled liquid chiller; refrigerant 410 A; scroll compressors; twin circuit; integral controls; includes placing in position; electrical work elsewhere						
Cooling load						
100 kW	18717.83	23689.29	8.00	269.52	nr	**23958.81**
150 kW	23589.17	29854.45	8.00	269.52	nr	**30123.97**
200 kW	29198.17	36953.20	8.00	269.52	nr	**37222.72**
Air cooled liquid chiller; refrigerant 410 A; reciprocating compressors; twin circuit; integral controls; includes placing in position; electrical work elsewhere						
Cooling load						
400 kW	52087.00	65921.31	8.00	269.52	nr	**66190.83**
550 kW	74274.22	94001.45	8.00	269.52	nr	**94270.97**
700 kW	86287.12	109204.98	9.00	303.22	nr	**109508.20**
Air cooled liquid chiller; refrigerant R134a; screw compressors; twin circuit; integral controls; includes placing in position; electrical work elsewhere						
Cooling load						
250 kW	52519.89	66469.18	8.00	269.52	nr	**66738.70**
400 kW	62919.37	79630.76	8.00	269.52	nr	**79900.28**
600 kW	84464.73	106898.56	8.00	269.52	nr	**107168.08**
800 kW	101519.38	128482.93	9.00	303.22	nr	**128786.15**
1000 kW	119034.31	150649.82	9.00	303.22	nr	**150953.04**
1200 kW	151200.56	191359.43	10.00	336.91	nr	**191696.34**
Air cooled liquid chiller; ductable for indoor installation; refrigerant 410 A; scroll compressors; integral controls; includes placing in position; electrical work elsewhere						
Cooling load						
40 kW	11132.18	14088.88	6.00	202.14	nr	**14291.02**
80 kW	17432.83	22062.99	6.00	202.14	nr	**22265.13**

38 MECHANICAL/COOLING/HEATING SYSTEMS

Item	Net Price £	Material £	Labour hours	Labour £	Unit	Total rate £
CENTRAL REFRIGERATION PLANT – cont						
CHILLERS; Higher efficiency air cooled						
Selection of air cooled chillers based on chilled water flow and return temperatures of 6°C and 12°C and an outdoor temperature of 25°C						
These machines have significantly higher part load operating efficiencies than conventional air cooled machines						
Air cooled liquid chiller, refrigerant R410 A; scroll compressors; complete with free cooling facility; integral controls; including placing in position; electrical work elsewhere						
Cooling load						
250 kW	47532.85	60157.57	8.00	269.52	nr	**60427.09**
300 kW	52445.25	66374.71	8.00	269.52	nr	**66644.23**
350 kW	60604.38	76700.90	8.00	269.52	nr	**76970.42**
400 kW	68397.76	86564.21	8.00	269.52	nr	**86833.73**
450 kW	71071.03	89947.49	8.00	269.52	nr	**90217.01**
500 kW	77260.96	97781.47	8.00	269.52	nr	**98050.99**
600 kW	79286.13	100344.53	8.00	269.52	nr	**100614.05**
650 kW	92603.94	117199.54	9.00	303.22	nr	**117502.76**
700 kW	98079.84	124129.85	9.00	303.22	nr	**124433.07**
CHILLERS; Water cooled						
Selection of water cooled chillers based on chilled water flow and return temperatures of 6°C and 12°C, and condenser entering and leaving temperatures of 27°C and 33°C						
Water cooled liquid chiller; refrigerant 410 A; reciprocating compressors; twin circuit; integral controls; includes placing in position; electrical work elsewhere						
Cooling load						
200 kw	24060.63	30451.13	8.00	269.52	nr	**30720.65**
350 kW	42579.42	53888.51	8.00	269.52	nr	**54158.03**
500 kW	48312.82	61144.71	8.00	269.52	nr	**61414.23**
650 kW	55996.75	70869.49	9.00	303.22	nr	**71172.71**
750 kW	68207.43	86323.33	9.00	303.22	nr	**86626.55**
Water cooled condenserless liquid chiller; refrigerant 410 A; reciprocating compressors; twin circuit; integral controls; includes placing in position; electrical work elsewhere						
Cooling load						
200 kW	20765.96	26281.39	8.00	269.52	nr	**26550.91**
350 kW	36453.37	46135.39	8.00	269.52	nr	**46404.91**
500 kW	46072.41	58309.24	8.00	269.52	nr	**58578.76**
650 kW	56347.64	71313.57	9.00	303.22	nr	**71616.79**
750 kW	64117.68	81147.34	9.00	303.22	nr	**81450.56**

38 MECHANICAL/COOLING/HEATING SYSTEMS

Item	Net Price £	Material £	Labour hours	Labour £	Unit	Total rate £
Water cooled liquid chiller; refrigerant R134a; screw compressors; twin circuit; integral controls; includes placing in position; electrical work elsewhere						
Cooling load						
300 kW	40121.36	50777.60	8.00	269.52	nr	**51047.12**
500 kW	51136.61	64718.49	8.00	269.52	nr	**64988.01**
700 kW	64483.03	81609.72	9.00	303.22	nr	**81912.94**
900 kW	90385.96	114392.47	9.00	303.22	nr	**114695.69**
1100 kW	99845.02	126363.85	10.00	336.91	nr	**126700.76**
1300 kW	115441.75	146103.08	10.00	336.91	nr	**146439.99**
Water cooled liquid chiller; refrigerant R134a; centrifugal compressors; twin circuit; integral controls; includes placing in position; electrical work elsewhere						
Cooling load						
1600 kW	162859.09	206114.46	11.00	370.60	nr	**206485.06**
1900 kW	198347.04	251028.02	11.00	370.60	nr	**251398.62**
2200 kW	236310.78	299074.92	13.00	437.98	nr	**299512.90**
2500 kW	244151.18	308997.73	13.00	437.98	nr	**309435.71**
3000 kW	297107.86	376019.71	15.00	505.36	nr	**376525.07**
3500 kW	404396.80	511804.59	15.00	505.36	nr	**512309.95**
4000 kW	429505.14	543581.71	20.00	673.80	nr	**544255.51**
4500 kW	458236.59	579944.23	20.00	673.80	nr	**580618.03**
5000 kW	558861.24	707294.78	25.00	842.26	nr	**708137.04**
CHILLERS; Absorption						
Absorption chiller, for operation using low pressure steam; selection based on chilled water flow and return temperatures of 6°C and 12°C, steam at 1 bar gauge and condenser entering and leaving temperatures of 27°C and 33°C; integral controls; includes placing in position; electrical work elsewhere						
Cooling load						
400 kW	70166.99	88803.34	8.00	269.52	nr	**89072.86**
700 kW	89095.23	112758.92	9.00	303.22	nr	**113062.14**
1000 kW	101645.28	128642.27	10.00	336.91	nr	**128979.18**
1300 kW	121302.56	153520.52	12.00	404.29	nr	**153924.81**
1600 kW	143560.90	181690.68	14.00	471.67	nr	**182162.35**
2000 kW	170146.57	215337.49	15.00	505.36	nr	**215842.85**
Absorption chiller, for operation using low pressure hot water; selection based on chilled water flow and return temperatures of 6°C and 12°C, cooling water temperatures of 27°C and 33°C and hot water at 90°C; integral controls; includes placing in position; electrical work elsewhere						
Cooling load						
700 kW	101645.28	128642.27	9.00	303.22	nr	**128945.49**
1000 kW	127060.34	160807.56	10.00	336.91	nr	**161144.47**
1300 kW	147779.50	187029.73	12.00	404.29	nr	**187434.02**
1600 kW	170146.57	215337.49	14.00	471.67	nr	**215809.16**

38 MECHANICAL/COOLING/HEATING SYSTEMS

Item	Net Price £	Material £	Labour hours	Labour £	Unit	Total rate £
CENTRAL REFRIGERATION PLANT – cont						
HEAT REJECTION						
Dry air liquid coolers						
Flat coil configuration						
500 kW	30986.00	39215.88	15.00	505.36	nr	**39721.24**
Extra for inverter panels (factory wired and mounted on units)	2382.00	3014.66	15.00	505.36	nr	**3520.02**
800 kW	52208.00	66074.44	15.00	505.36	nr	**66579.80**
Extra for inverter panels (factory wired and mounted on units)	30986.00	39215.88	15.00	505.36	nr	**39721.24**
1100 kW	68128.00	86222.80	15.00	505.36	nr	**86728.16**
Extra for inverter panels (factory wired and mounted on units)	4764.00	6029.32	15.00	505.36	nr	**6534.68**
1400 kW	102366.00	129554.41	15.00	505.36	nr	**130059.77**
Extra for inverter panels (factory wired and mounted on units)	6600.00	8352.96	15.00	505.36	nr	**8858.32**
1700 kW	102366.00	129554.41	15.00	505.36	nr	**130059.77**
Extra for inverter panels (factory wired and mounted on units)	7146.00	9043.98	15.00	505.36	nr	**9549.34**
2000 kW	124599.00	157692.49	15.00	505.36	nr	**158197.85**
Extra for inverter panels (factory wired and mounted on units)	8238.00	10426.01	15.00	505.36	nr	**10931.37**
Note: heat rejection capacities above 500 kW require multiple units. Prices are therefore for total number of units.						
Vee type coil configuration						
500 kW	32675.00	41353.48	15.00	505.36	nr	**41858.84**
Extra for inverter panels (factory wired and mounted on units)	2200.00	2784.32	15.00	505.36	nr	**3289.68**
800 kW	53053.00	67143.88	15.00	505.36	nr	**67649.24**
Extra for inverter panels (factory wired and mounted on units)	2928.00	3705.68	15.00	505.36	nr	**4211.04**
1100 kW	69737.00	88259.15	15.00	505.36	nr	**88764.51**
Extra for inverter panels (factory wired and mounted on units)	3292.00	4166.36	15.00	505.36	nr	**4671.72**
1400 kW	94880.00	120080.13	15.00	505.36	nr	**120585.49**
Extra for inverter panels (factory wired and mounted on units)	5128.00	6490.00	15.00	505.36	nr	**6995.36**
1700 kW	107554.00	136120.34	15.00	505.36	nr	**136625.70**
Extra for inverter panels (factory wired and mounted on units)	6220.00	7872.03	15.00	505.36	nr	**8377.39**
2000 kW	127996.00	161991.74	15.00	505.36	nr	**162497.10**
Extra for inverter panels (factory wired and mounted on units)	6584.00	8332.71	15.00	505.36	nr	**8838.07**
Note: Heat rejection capacities above 1100 kW require multiple units. Prices are for total number of units.						

38 MECHANICAL/COOLING/HEATING SYSTEMS

Item	Net Price £	Material £	Labour hours	Labour £	Unit	Total rate £
Air cooled condensers						
Air cooled condenser; refrigerant 410 A; selection based on condensing temperature of 45°C at 32°C dry bulb ambient; includes placing in position; electrical work elsewhere						
Flat coil configuration						
500 kW	18276.10	23130.23	15.00	505.36	nr	**23635.59**
Extra for inverter panels (factory wired and mounted on units)	8362.63	10583.74	15.00	505.36	nr	**11089.10**
800 kW	29983.77	37947.46	15.00	505.36	nr	**38452.82**
Extra for inverter panels (factory wired and mounted on units)	12802.41	16202.73	15.00	505.36	nr	**16708.09**
1100 kW	40140.58	50801.92	15.00	505.36	nr	**51307.28**
Extra for inverter panels (factory wired and mounted on units)	18200.10	23034.04	15.00	505.36	nr	**23539.40**
1400 kW	50114.90	63425.42	15.00	505.36	nr	**63930.78**
Extra for inverter panels (factory wired and mounted on units)	22381.42	28325.92	15.00	505.36	nr	**28831.28**
1700 kW	65258.84	82591.59	15.00	505.36	nr	**83096.95**
Extra for inverter panels (factory wired and mounted on units)	33572.12	42488.88	15.00	505.36	nr	**42994.24**
2000 kW	75172.36	95138.14	15.00	505.36	nr	**95643.50**
Extra for inverter panels (factory wired and mounted on units)	33572.12	42488.88	15.00	505.36	nr	**42994.24**
Note: Heat rejection capacities above 500 kW require multiple units. Prices are for total number of units						
Vee type coil configuration						
500 kW	17941.61	22706.90	15.00	505.36	nr	**23212.26**
Extra for inverter panels (factory wired and mounted on units)	6401.22	8101.39	15.00	505.36	nr	**8606.75**
800 kW	26988.44	34156.57	15.00	505.36	nr	**34661.93**
Extra for inverter panels (factory wired and mounted on units)	13106.51	16587.60	15.00	505.36	nr	**17092.96**
1100 kW	39608.42	50128.41	15.00	505.36	nr	**50633.77**
Extra for inverter panels (factory wired and mounted on units)	22290.19	28210.46	15.00	505.36	nr	**28715.82**
1400 kW	46116.05	58364.48	15.00	505.36	nr	**58869.84**
Extra for inverter panels (factory wired and mounted on units)	26213.01	33175.18	15.00	505.36	nr	**33680.54**
1700 kW	59412.61	75192.60	15.00	505.36	nr	**75697.96**
Extra for inverter panels (factory wired and mounted on units)	33450.48	42334.92	15.00	505.36	nr	**42840.28**
2000 kW	69174.09	87546.73	15.00	505.36	nr	**88052.09**
Extra for inverter panels (factory wired and mounted on units)	39319.52	49762.79	15.00	505.36	nr	**50268.15**
Note: Heat rejection capacities above 1100 kW require multiple units. Prices are for total number of units.						

38 MECHANICAL/COOLING/HEATING SYSTEMS

Item	Net Price £	Material £	Labour hours	Labour £	Unit	Total rate £
CENTRAL REFRIGERATION PLANT – cont						
Air cooled condensers – cont						
Cooling towers						
Cooling towers; forced draught, centrifugal fan, conterflow design; based on water temperatures of 35°C on and 29°C off at 21°C wet bulb ambient temperature; includes placing in position; electrical work elsewhere						
Cooling towers: Open circuit type						
900 kW	18525.22	23445.52	20.00	673.80	nr	**24119.32**
Extra for stainless steel construction	8706.75	11019.27	–	–	nr	**11019.27**
Extra for intake and discharge sound attenuation	8767.65	11096.33	–	–	nr	**11096.33**
1500 kW	23278.28	29461.00	20.00	673.80	nr	**30134.80**
Extra for stainless steel construction	10940.66	13846.50	–	–	nr	**13846.50**
Extra for intake and discharge sound attenuation	13530.49	17124.18	–	–	nr	**17124.18**
2100 kW	29918.57	37864.94	20.00	673.80	nr	**38538.74**
Extra for stainless steel construction	14061.72	17796.51	–	–	nr	**17796.51**
Extra for intake and discharge sound attenuation	19558.65	24753.42	–	–	nr	**24753.42**
2700 kW	38485.11	48706.75	20.00	673.80	nr	**49380.55**
Extra for stainless steel construction	18087.95	22892.11	–	–	nr	**22892.11**
Extra for intake and discharge sound attenuation	18447.04	23346.58	–	–	nr	**23346.58**
3300 kW	46944.63	59413.12	20.00	673.80	nr	**60086.92**
Extra for stainless steel construction	22063.68	27923.80	–	–	nr	**27923.80**
Extra for intake and discharge sound attenuation	21827.20	27624.51	–	–	nr	**27624.51**
3900 kW	57377.98	72617.57	20.00	673.80	nr	**73291.37**
Extra for stainless steel construction	26967.19	34129.67	–	–	nr	**34129.67**
Extra for intake and discharge sound attenuation	25924.21	32809.68	–	–	nr	**32809.68**
4500 kW	59093.45	74788.67	23.00	774.87	nr	**75563.54**
Extra for stainless steel construction	27773.71	35150.40	–	–	nr	**35150.40**
Extra for intake and discharge sound attenuation	24724.71	31291.59	–	–	nr	**31291.59**
5100 kW	74867.06	94751.75	23.00	774.87	nr	**95526.62**
Extra for stainless steel construction	35187.49	44533.28	–	–	nr	**44533.28**
Extra for intake and discharge sound attenuation	34289.18	43396.38	–	–	nr	**43396.38**
5700 kW	83510.05	105690.32	30.00	1010.71	nr	**106701.03**
Extra for stainless steel construction	39249.70	49674.42	–	–	nr	**49674.42**
Extra for intake and discharge sound attenuation	44385.86	56174.74	–	–	nr	**56174.74**

38 MECHANICAL/COOLING/HEATING SYSTEMS

Item	Net Price £	Material £	Labour hours	Labour £	Unit	Total rate £
6300 kW	90255.69	114227.60	30.00	1010.71	nr	**115238.31**
Extra for stainless steel construction	42419.85	53686.56	–	–	nr	**53686.56**
Extra for intake and discharge sound attenuation	44385.55	56174.35	–	–	nr	**56174.35**
Cooling towers: Closed circuit type (includes 20% ethylene glycol)						
900 kW	46033.23	58259.66	20.00	673.80	nr	**58933.46**
Extra for stainless steel construction	28770.62	36412.10	–	–	nr	**36412.10**
Extra for intake and discharge sound attenuation	12903.56	16330.74	–	–	nr	**16330.74**
1500 kW	74039.48	93704.36	20.00	673.80	nr	**94378.16**
Extra for stainless steel construction	46274.37	58564.84	–	–	nr	**58564.84**
Extra for intake and discharge sound attenuation	24981.02	31615.98	–	–	nr	**31615.98**
2100 kW	86667.33	109686.17	20.00	673.80	nr	**110359.97**
Extra for stainless steel construction	54166.87	68553.59	–	–	nr	**68553.59**
Extra for intake and discharge sound attenuation	24981.34	31616.38	–	–	nr	**31616.38**
2700 kW	125777.96	159184.58	25.00	842.26	nr	**160026.84**
Extra for stainless steel construction	78610.63	99489.61	–	–	nr	**99489.61**
Extra for intake and discharge sound attenuation	33017.67	41787.17	–	–	nr	**41787.17**
3300 kW	157287.90	199063.57	25.00	842.26	nr	**199905.83**
Extra for stainless steel construction	98304.37	124414.01	–	–	nr	**124414.01**
Extra for intake and discharge sound attenuation	49774.31	62994.37	–	–	nr	**62994.37**
3900 kW	172123.83	217839.92	25.00	842.26	nr	**218682.18**
Extra for stainless steel construction	107576.87	136149.28	–	–	nr	**136149.28**
Extra for intake and discharge sound attenuation	50474.85	63880.97	–	–	nr	**63880.97**
4500 kW	205435.17	259998.75	40.00	1347.61	nr	**261346.36**
Extra for stainless steel construction	128396.45	162498.55	–	–	nr	**162498.55**
Extra for intake and discharge sound attenuation	74534.47	94330.82	–	–	nr	**94330.82**
5100 kW	232841.37	294684.04	40.00	1347.61	nr	**296031.65**
Extra for stainless steel construction	145525.62	184177.22	–	–	nr	**184177.22**
Extra for intake and discharge sound attenuation	82869.77	104879.98	–	–	nr	**104879.98**
5700 kW	281499.31	356265.53	40.00	1347.61	nr	**357613.14**
Extra for stainless steel construction	175936.04	222664.66	–	–	nr	**222664.66**
Extra for intake and discharge sound attenuation	82869.89	104880.14	–	–	nr	**104880.14**
6300 kW	314575.46	398126.70	40.00	1347.61	nr	**399474.31**
Extra for stainless steel construction	196609.37	248828.82	–	–	nr	**248828.82**
Extra for intake and discharge sound attenuation	99548.12	125988.11	–	–	nr	**125988.11**

38 MECHANICAL/COOLING/HEATING SYSTEMS

Item	Net Price £	Material £	Labour hours	Labour £	Unit	Total rate £
CHILLED WATER						
SCREWED STEEL						
PIPELINES						
For pipework prices refer to section – Low Temperature Hot Water Heating, with the exception of chilled water blocks within brackets as detailed hereafter. For minimum fixing distances, refer to the Tables and Memoranda to the rear of the book						
Fixings						
For steel pipes; black malleable iron						
Oversized pipe clip, to contain 30 mm insulation block for vapour barrier						
15 mm dia.	4.39	5.56	0.10	3.37	nr	8.93
20 mm dia.	4.43	5.61	0.11	3.71	nr	9.32
25 mm dia.	4.65	5.88	0.12	4.03	nr	9.91
32 mm dia.	4.79	6.06	0.14	4.72	nr	10.78
40 mm dia.	5.05	6.40	0.15	5.05	nr	11.45
50 mm dia.	7.22	9.14	0.16	5.39	nr	14.53
65 mm dia.	7.79	9.86	0.30	10.11	nr	19.97
80 mm dia.	8.26	10.45	0.35	11.79	nr	22.24
100 mm dia.	13.10	16.58	0.40	13.47	nr	30.05
125 mm dia.	15.20	19.24	0.60	20.23	nr	39.47
150 mm dia.	23.78	30.09	0.77	25.95	nr	56.04
200 mm dia.	30.06	38.05	0.90	30.32	nr	68.37
250 mm dia.	39.04	49.41	1.10	37.06	nr	86.47
300 mm dia.	46.46	58.80	1.25	42.11	nr	100.91
350 mm dia.	53.91	68.23	1.50	50.53	nr	118.76
400 mm dia.	60.37	76.41	1.75	58.97	nr	135.38
Screw on backplate (Male), black malleable iron; plugged and screwed						
M12	0.60	0.76	0.10	3.37	nr	4.13
Screw on backplate (Female), black malleable iron; plugged and screwed						
M12	0.60	1.50	0.10	3.37	nr	4.87
Extra over channel sections for fabricated hangers and brackets						
Galvanized steel; including inserts, bolts, nuts, washers; fixed to backgrounds						
41 × 21 mm	6.79	8.59	0.29	9.77	m	18.36
41 × 41 mm	8.14	10.30	0.29	9.77	m	20.07
Threaded rods; metric thread; including nuts, washers etc.						
12 mm dia. × 600 mm long	3.43	4.35	0.18	6.06	nr	10.41

38 MECHANICAL/COOLING/HEATING SYSTEMS

Item	Net Price £	Material £	Labour hours	Labour £	Unit	Total rate £
For plastic pipework suitable for chilled water systems, refer to ABS pipework details in Section – Cold Water, with the exception of chilled water blocks within brackets as detailed for the aforementioned steel pipe. For minimum fixing distances refer to the Tables and Memoranda to the rear of the book						
For copper pipework, refer to section – Cold Water with the exception of chilled water blocks within brackets as detailed hereafter. For minimum fixing distances refer to the Tables and Memoranda to the rear of the book						
Fixing						
For copper pipework						
Oversized pipe clip, to contain 30 mm insulation block for vapour barrier						
15 mm dia.	4.39	5.56	0.10	3.37	nr	**8.93**
22 mm dia.	4.43	5.61	0.11	3.71	nr	**9.32**
28 mm dia.	4.65	5.88	0.12	4.03	nr	**9.91**
35 mm dia.	4.79	6.06	0.14	4.72	nr	**10.78**
42 mm dia.	5.05	6.40	0.15	5.05	nr	**11.45**
54 mm dia.	7.22	9.14	0.16	5.39	nr	**14.53**
67 mm dia.	7.79	9.86	0.30	10.11	nr	**19.97**
76 mm dia.	8.26	10.45	0.35	11.79	nr	**22.24**
108 mm dia.	9.13	28.13	0.40	13.47	nr	**41.60**
133 mm dia.	15.20	19.24	0.60	20.23	nr	**39.47**
159 mm dia.	23.78	30.09	0.77	25.95	nr	**56.04**
Screw on backplate, female						
All sizes 15 mm to 54 mm × 10 mm	1.73	2.18	0.10	3.37	nr	**5.55**
Screw on backplate, male						
All sizes 15 mm to 54 mm × 10 mm	2.32	2.93	0.10	3.37	nr	**6.30**
Extra over channel sections for fabricated hangers and brackets						
Galvanized steel; including inserts, bolts, nuts, washers; fixed to backgrounds						
41 × 21 mm	6.79	8.59	0.29	9.77	m	**18.36**
41 × 41 mm	8.14	10.30	0.29	9.77	m	**20.07**
Threaded rods; metric thread; including nuts, washers etc.						
10 mm dia. × 600 mm long for ring clips up to 54 mm	2.19	2.77	0.18	6.06	nr	**8.83**
12 mm dia. × 600 mm long for ring clips from 54 mm	3.43	4.35	0.18	6.06	nr	**10.41**

38 MECHANICAL/COOLING/HEATING SYSTEMS

Item	Net Price £	Material £	Labour hours	Labour £	Unit	Total rate £
CHILLED WATER – cont						
PIPELINE ANCILLARIES						
For prices for ancillaries refer to section: Low Temperature Hot Water Heating						
HEAT EXCHANGERS						
Plate heat exchanger; for use in CHW systems; painted carbon steel frame, stainless steel plates, nitrile rubber gaskets, design pressure of 10 bar and operating temperature of 110/135°C						
Primary side; 13°C in, 8°C out; secondary side; 6°C in, 11°C out						
264 kW, 12.60 l/s	6238.94	7896.00	10.00	336.91	nr	**8232.91**
290 kW, 13.85 l/s	6576.42	8323.11	12.00	404.29	nr	**8727.40**
316 kW, 15.11 l/s	6846.10	8664.42	12.00	404.29	nr	**9068.71**
350 kW, 16.69 l/s	7362.41	9317.86	16.00	539.04	nr	**9856.90**
395 kW, 18.88 l/s	7878.72	9971.30	16.00	539.04	nr	**10510.34**
454 kW, 21.69 l/s	8557.99	10830.99	16.00	539.04	nr	**11370.03**
475 kW, 22.66 l/s	8723.85	11040.90	16.00	539.04	nr	**11579.94**
527 kW, 25.17 l/s	9453.60	11964.48	16.00	539.04	nr	**12503.52**
554 kW, 26.43 l/s	9704.54	12282.07	16.00	539.04	nr	**12821.11**
580 kW, 27.68 l/s	10037.06	12702.91	16.00	539.04	nr	**13241.95**
633 kW, 30.19 l/s	10640.53	13466.66	16.00	539.04	nr	**14005.70**
661 kW, 31.52 l/s	10897.25	13791.56	16.00	539.04	nr	**14330.60**
713 kW, 34.04 l/s	11632.74	14722.40	18.00	606.42	nr	**15328.82**
740 kW, 35.28 l/s	11886.58	15043.66	18.00	606.42	nr	**15650.08**
804 kW, 38.33 l/s	12656.72	16018.34	18.00	606.42	nr	**16624.76**
1925 kW, 91.82 l/s	21246.46	26889.52	20.00	673.80	nr	**27563.32**
2710 kW, 129.26 l/s	28220.91	35716.39	20.00	673.80	nr	**36390.19**
3100 kW, 147.87 l/s	31439.88	39790.31	20.00	673.80	nr	**40464.11**
Note: For temperature conditions different to those above, the cost of the units can vary significantly, and therefore the manufacturers advice should be sought.						
TRACE HEATING						
Trace heating; for freeze protection or temperature maintenance of pipework; to BS 6351; including fixing to parent structures by plastic pull ties						
Straight laid F-S-C Wintergard						
15 mm	23.66	29.95	0.27	9.09	m	**39.04**
25 mm	23.66	29.95	0.27	9.09	m	**39.04**
28 mm	23.66	29.95	0.27	9.09	m	**39.04**

38 MECHANICAL/COOLING/HEATING SYSTEMS

Item	Net Price £	Material £	Labour hours	Labour £	Unit	Total rate £
32 mm	23.66	29.95	0.30	10.11	m	**40.06**
35 mm	23.66	29.95	0.31	10.44	m	**40.39**
50 mm	23.66	29.95	0.34	11.46	m	**41.41**
100 mm	23.66	29.95	0.40	13.47	m	**43.42**
150 mm F-S-B Wintergard (to larger sizes)	23.66	29.95	0.40	13.47	m	**43.42**
Helically wound F-S-C Wintergard						
15 mm	30.16	38.17	1.00	33.69	m	**71.86**
25 mm	30.16	38.17	1.00	33.69	m	**71.86**
28 mm	30.16	38.17	1.00	33.69	m	**71.86**
32 mm	30.16	38.17	1.00	33.69	m	**71.86**
35 mm	30.16	38.17	1.00	33.69	m	**71.86**
50 mm	30.16	38.17	1.00	33.69	m	**71.86**
100 mm	30.16	38.17	1.00	33.69	m	**71.86**
150 mm F-S-B Wintergard (to larger sizes)	30.16	38.17	1.00	33.69	m	**71.86**
Accessories for trace heating; weatherproof; polycarbonate enclosure to IP standards; fully installed						
Connection junction box						
100 × 100 × 75 mm junction box incl power connecting kit	70.87	89.69	1.40	47.17	nr	**136.86**
Single air thermostat						
150 × 150 × 75 mm AT-TS–13	155.25	196.48	1.42	47.84	nr	**244.32**
Single capillary thermostat						
150 × 150 × 75 mm AT-TS–13	223.37	282.70	1.46	49.19	nr	**331.89**
Twin capillary thermostat						
150 × 150 × 75 mm	400.73	507.16	1.46	49.19	nr	**556.35**
PRESSURIZATION UNITS						
Chilled water packaged pressurization unit complete with expansion vessel(s), interconnecting pipework and necessary isolating and drain valves; includes placing in position; electrical work elsewhere. Selection based on a final working pressure of 4 bar, a 3 m static head and system operating temperatures of 6°/12°C						
System volume						
1800 litres	1266.30	1602.63	8.00	269.52	nr	**1872.15**
4500 litres	1378.65	1744.81	8.00	269.52	nr	**2014.33**
7200 litres	1451.92	1837.55	10.00	336.91	nr	**2174.46**
9900 litres	1946.48	2463.46	10.00	336.91	nr	**2800.37**
15300 litres	2035.63	2576.29	13.00	437.98	nr	**3014.27**
22500 litres	2127.21	2692.20	20.00	673.80	nr	**3366.00**
27000 litres	2127.21	2692.20	20.00	673.80	nr	**3366.00**

38 MECHANICAL/COOLING/HEATING SYSTEMS

Item	Net Price £	Material £	Labour hours	Labour £	Unit	Total rate £
CHILLED WATER – cont						
CHILLED BEAMS						
Static (passive) beams; cooling data based on 9 K waterside cooling dt (e.g. water 14°C flow and 16°C return, room temperature 24°C)						
Static passive beam in perforated metal casing for exposed mounting from solid ceiling or roof soffit, 2 pipe cooling only						
Length/Cooling Output						
1200 mm/300 W	402.90	509.91	1.31	44.13	m	554.04
1800 mm/450 W	525.30	664.82	1.53	51.54	m	716.36
2400 mm/600 W	627.30	793.91	1.75	58.97	m	852.88
3000 mm/750 W	744.60	942.37	2.01	67.73	m	1010.10
3600 mm/900 W	918.00	1161.82	2.18	73.44	m	1235.26
Static passive beam, chassis type for mounting above open grid or perforated metal ceiling or roof soffit, 2 pipe cooling only						
Length/Cooling Output						
1200 mm/300 W	285.60	361.46	1.31	44.13	m	405.59
1800 mm/450 W	392.70	497.00	1.53	51.54	m	548.54
2400 mm/600 W	484.50	613.18	1.75	58.97	m	672.15
3000 mm/750 W	586.50	742.28	2.01	67.73	m	810.01
3600 mm/900 W	693.60	877.82	2.18	73.44	m	951.26
Multi-service beams; for exposed mounting on solid ceiling slabs, with Integrated lighting and lighting control, provision of fascia openings for other services (electrical work and other services elsewhere). Performance criteria equivalent to standard beams as listed above						
2 way discharge exposed linear multi-service climate beam for exposed mounting on solid ceiling; closed type with integrated secondary air circulation, 2 pipe cooling only, with primary air						
Length/Cooling Output						
1200 mm/550 W	816.00	1032.73	2.26	76.14	m	1108.87
1800 mm/900 W	1122.00	1420.00	2.49	83.89	m	1503.89
2400 mm/1200 W	1428.00	1807.28	2.67	89.96	m	1897.24
3000 mm/1550 W	1734.00	2194.55	2.84	95.68	m	2290.23
3600 mm/1850 W	2091.00	2646.37	3.08	103.77	m	2750.14
2 way discharge exposed linear multi-service climate beam for exposed mounting on solid ceiling; closed type with integrated secondary air circulation, 4 pipe cooling and heating, with primary air						

38 MECHANICAL/COOLING/HEATING SYSTEMS

Item	Net Price £	Material £	Labour hours	Labour £	Unit	Total rate £
Length/Cooling Output/Heating Output						
1200 mm/550 W/495 W	846.60	1071.46	2.59	87.26	m	**1158.72**
1800 mm/900 W/810 W	1157.70	1465.18	2.82	95.01	m	**1560.19**
2400 mm/1200 W/1080 W	1468.80	1858.91	3.00	101.07	m	**1959.98**
3000 mm/1550 W/1395 W	1779.90	2252.64	3.18	107.13	m	**2359.77**
3600 mm/1850 W/1665 W	2142.00	2710.92	3.41	114.88	m	**2825.80**
Static passive multi-service beam for exposed mounting on solid ceiling, 2 pipe cooling only						
Length/Cooling Output						
1200 mm/300 W	831.30	1052.09	1.81	60.97	m	**1113.06**
1800 mm/450 W	1137.30	1439.37	2.03	68.40	m	**1507.77**
2400 mm/600 W	1443.30	1826.64	2.25	75.81	m	**1902.45**
3000 mm/750 W	1744.20	2207.46	2.51	84.57	m	**2292.03**
3600 mm/900 W	2101.20	2659.28	2.68	90.28	m	**2749.56**
Ventilated (active) beams; cooling data based on 9 K airside and 9 K waterside cooling dt (e.g. water 14°C flow and 16°C return, primary air 15°C, room temperature 24°C), heating data based on −4 K airside and 20 K waterside heating dt (e.g. water at 45°C flow and 35°C return, primary air 16°C, room temperature 20°C)						
1 way discharge recessed linear active climate beam mounted within a false ceiling; closed type with integrated secondary air circulation; 300 mm wide, 2 pipe cooling only, 5 l/s/m primary air						
Length/Cooling Output						
1200 mm/340 W	312.12	395.02	1.76	59.29	m	**454.31**
1800 mm/530 W	385.56	487.96	1.99	67.04	m	**555.00**
2400 mm/730 W	453.90	574.46	2.17	73.11	m	**647.57**
3000 mm/900 W	523.26	662.23	2.34	78.84	m	**741.07**
3600 mm/1100 W	565.08	715.16	2.58	86.92	m	**802.08**
1 way discharge recessed linear active climate beam mounted within a false ceiling; closed type with integrated secondary air circulation; 300 mm wide, 4 pipe cooling and heating, with 5 l/s/m primary air						
Length/Cooling Output/Heating Output						
1200 mm/340 W/180 W	326.40	413.09	2.09	70.41	m	**483.50**
1800 mm/530 W/310 W	403.92	511.20	2.32	78.16	m	**589.36**
2400 mm/730 W/450 W	477.36	604.15	2.50	84.22	m	**688.37**
3000 mm/900 W/600 W	541.62	685.47	2.68	90.28	m	**775.75**
3600 mm/1100 W/750 W	584.46	739.69	2.91	98.03	m	**837.72**
2 way discharge recessed linear active climate beam mounted within a false ceiling; closed type with integrated secondary air circulation, 600 mm wide, 2 pipe cooling only, with 10 l/s/m primary air						

38 MECHANICAL/COOLING/HEATING SYSTEMS

Item	Net Price £	Material £	Labour hours	Labour £	Unit	Total rate £
CHILLED WATER – cont						
Ventilated (active) beams – cont						
Length/Cooling Output						
1200 mm/550 W	316.20	400.19	1.76	59.29	m	**459.48**
1800 mm/900 W	433.50	548.63	1.99	67.04	m	**615.67**
2400 mm/1200 W	540.60	684.19	2.17	73.11	m	**757.30**
3000 mm/1550 W	632.40	800.36	2.34	78.84	m	**879.20**
3600 mm/1850 W	724.20	916.55	2.58	86.92	m	**1003.47**
2 way discharge recessed linear active climate beam mounted within a false ceiling; closed type with integrated secondary air circulation; 600 mm wide, 4 pipe cooling and heating, with 10 l/s/m primary air						
Length/Cooling Output/Heating Output						
1200 mm/550 W/495 W	346.80	438.91	2.09	70.41	m	**509.32**
1800 mm/900 W/810 W	469.20	593.82	2.32	78.16	m	**671.98**
2400 mm/1200 W/1080 W	581.40	735.82	2.50	84.22	m	**820.04**
3000 mm/1550 W/1395 W	678.30	858.46	2.68	90.28	m	**948.74**
3600 mm/1850 W/1665 W	775.20	981.10	2.91	98.03	m	**1079.13**
4 way discharge recessed modular active climate beam mounted within a false ceiling; closed type with integrated secondary air circulation; 600 mm wide, 2 pipe cooling only, with primary air at 10 l/s/m of beam discharge (600 × 600 unit, 12 l/s, 1200 × 600 unit 18 l/s)						
Length/Cooling Output						
600x600/500 W	331.50	419.54	1.56	52.56	m	**472.10**
1200x600/850 W	382.50	484.10	1.76	59.29	m	**543.39**
4 way discharge recessed modular active climate beam mounted within a false ceiling; closed type with integrated secondary air circulation; 600 mm wide, 4 pipe cooling and heating, with primary air at 10 l/s/m of beam discharge (600 × 600 unit 12 l/s, 1200 × 600 unit 18 l/s)						
Length/Cooling Output/Heating Output						
600x600/500 W/450 W	357.00	451.82	1.89	63.67	m	**515.49**
1200x600/850 W/765 W	413.10	522.82	2.09	70.41	m	**593.23**
1 way sidewall discharge active climate beam mounted within a ceiling bulkhead; closed type with integrated secondary air cisculation, 2 pipe cooling only, with 20 l/s/m primary air						
Length/Cooling Output						
900 mm/700 W	367.20	464.73	1.89	63.67	m	**528.40**
1100 mm/900 W	397.80	503.45	1.89	63.67	m	**567.12**
1300 mm/1100 W	469.20	593.82	2.09	70.41	m	**664.23**
1500 mm/1200 W	499.80	632.54	2.09	70.41	m	**702.95**
1 way sidewall discharge active climate beam mounted within a ceiling bulkhead; closed type with integrated secondary air circulation, 4 pipe only, with 20 l/s/m primary air						

38 MECHANICAL/COOLING/HEATING SYSTEMS

Item	Net Price £	Material £	Labour hours	Labour £	Unit	Total rate £
Length/Cooling Output/Heating Output						
900 mm/700 W/400 W	397.80	503.45	1.89	63.67	m	**567.12**
1100 mm/900 W/600 W	428.40	542.18	1.89	63.67	m	**605.85**
1300 mm/1100 W/800 W	499.80	632.54	2.09	70.41	m	**702.95**
1500 mm/1200 W/900 W	530.40	671.27	2.09	70.41	m	**741.68**
CHILLED CEILINGS						
Traditional (60% Radiant Absorption, 40% Convection) with bonded elements in perforated metal ceiling tiles (typically 30% free area and 2.5 mm perforated hole sizes)/ tartan grid arrangement. Maximum waterside cooling effect approx. 65 W/m² at 9 dTK.						
Radiant elements and metal ceilings combined	142.80	180.72	1.75	58.97	m	**239.69**
Hybrid (40% Radiant Absorption, 60% Convection) with radiant cooling chilled beam elements positioned above a perforated metal ceiling tiles (typically 30% free area and 2.5 mm perforated hole sizes)/tartan grid arrangement. Maximum waterside cooling effect approx. 100 W/m² at 9 dTK.						
Radiant chilled beams	61.20	77.46	0.60	20.22	m	**97.68**
Metal ceilings	35.70	45.18	1.15	38.74	m	**83.92**
Convective beam (5% Radiant Absorption, 95% Convection) with convective cooling fin coil batteries positioned above a perforated metal ceiling tiles (typically 50% free area and 5.0 mm perforated hole sizes)/tartan grid arrangement. Maximum waterside cooling effect approx. 100 W/m² at 9 dTK.						
Convective fin coil batteries	66.30	83.91	0.60	20.22	m	**104.13**
Metal ceilings	35.70	45.18	1.15	38.74	m	**83.92**
LEAK DETECTION						
Leak detection system consisting of a central control module connected by a leader cable to water sensing cables						
Control modules						
Alarm Only (1 zone)	553.00	699.88	4.00	154.56	nr	**854.44**
Alarm Only (8 zone)	1083.00	1370.64	4.00	154.56	nr	**1525.20**
Alarm and location	2852.00	3609.49	8.00	309.11	nr	**3918.60**
Cables						
Sensing – 3 m length	118.00	149.34	4.00	154.56	nr	**303.90**
Sensing – 5 m length	145.00	183.51	4.00	154.56	nr	**338.07**
Sensing – 7.5 m length	206.00	260.71	4.00	154.56	nr	**415.27**
Sensing – 15 m length	288.00	364.49	8.00	309.11	nr	**673.60**
Sensing – 25 m length	400.00	506.24	12.00	404.29	nr	**910.53**
Leader – 3.5 m length	50.00	63.28	2.00	77.28	nr	**140.56**
End terminal						
End terminal	25.00	31.64	0.05	1.94	nr	**33.58**

38 MECHANICAL/COOLING/HEATING SYSTEMS

Item	Net Price £	Material £	Labour hours	Labour £	Unit	Total rate £
CHILLED WATER – cont						
ENERGY METERS						
Ultrasonic						
Energy meter for measuring energy use in chilled water systems; includes ultrasonic flow meter (with sensor and signal converter), energy calculator, pair of temperature sensors with brass pockets, and 3 m of interconnecting cable; includes fixing in position; electrical work elsewhere						
Pipe size (flanged connections to PN16); maximum flow rate						
50 mm, 36 m³/hr	1389.38	1758.40	1.80	60.65	nr	**1819.05**
65 mm, 60 m³/hr	1531.50	1938.26	2.32	78.16	nr	**2016.42**
80 mm, 100 m³/hr	1716.70	2172.65	2.56	86.25	nr	**2258.90**
125 mm, 250 m³/hr	1988.96	2517.22	3.60	121.28	nr	**2638.50**
150 mm, 360 m³/hr	2159.14	2732.61	4.80	161.72	nr	**2894.33**
200 mm, 600 m³/hr	2412.38	3053.11	6.24	210.22	nr	**3263.33**
250 mm, 1000 m³/hr	2783.74	3523.11	9.60	323.42	nr	**3846.53**
300 mm, 1500 m³/hr	3272.23	4141.33	10.80	363.87	nr	**4505.20**
350 mm, 2000 m³/hr	3942.89	4990.13	13.20	444.71	nr	**5434.84**
400 mm, 2500 m³/hr	4513.46	5712.24	15.60	525.57	nr	**6237.81**
500 mm, 3000 m³/hr	5122.05	6482.47	24.00	808.56	nr	**7291.03**
600 mm, 3500 m³/hr	5754.67	7283.11	28.00	943.33	nr	**8226.44**
Electromagnetic						
Energy meter for measuring energy use in chilled water systems; includes electromagnetic flow meter (with sensor and signal converter), energy calculator, pair of temperature sensors with brass pockets, and 3 m of interconnecting cable; includes fixing in position; electrical work elsewhere						
Pipe size (flanged connections to PN40); maximum flow rate						
25 mm, 17.7 m³/hr	996.08	1260.64	1.48	49.86	nr	**1310.50**
40 mm, 45 m³/hr	1005.60	1272.69	1.55	52.21	nr	**1324.90**
Pipe size (flanged connections to PN16); maximum flow rate						
50 mm, 70 m³/hr	1016.87	1286.95	1.80	60.65	nr	**1347.60**
65 mm, 120 m³/hr	1021.20	1292.44	2.32	78.16	nr	**1370.60**
80 mm, 180 m³/hr	1026.41	1299.02	2.56	86.25	nr	**1385.27**
125 mm, 450 m³/hr	1117.35	1414.12	3.60	121.28	nr	**1535.40**
150 mm, 625 m³/hr	1185.78	1500.72	4.80	161.72	nr	**1662.44**
200 mm, 1100 m³/hr	1275.86	1614.73	6.24	210.22	nr	**1824.95**
250 mm, 1750 m³/hr	1431.77	1812.05	9.60	323.42	nr	**2135.47**
300 mm, 2550 m³/hr	1825.01	2309.73	10.80	363.87	nr	**2673.60**
350 mm, 3450 m³/hr	2374.15	3004.72	13.20	444.71	nr	**3449.43**
400 mm, 4500 m³/hr	2708.48	3427.85	15.60	525.57	nr	**3953.42**

38 MECHANICAL/COOLING/HEATING SYSTEMS

Item	Net Price £	Material £	Labour hours	Labour £	Unit	Total rate £
LOCAL COOLING UNITS						
Split system with ceiling void evaporator unit and external condensing unit						
Ceiling mounted 4 way blow cassette heat pump unit with remote fan speed and load control; refrigerant 410 A; includes outdoor unit						
Cooling 3.6 kW, heating 4.1 kW	1823.63	2307.98	35.00	1179.16	nr	**3487.14**
Cooling 4.9 kW, heating 5.5 kW	2009.61	2543.36	35.00	1179.16	nr	**3722.52**
Cooling 7.1 kW, heating 8.2 kW	2432.45	3078.51	35.00	1179.16	nr	**4257.67**
Cooling 10 kW, heating 11.2 kW	2877.10	3641.25	35.00	1179.16	nr	**4820.41**
Cooling 12.20 kW, heating 14.60 kW	3147.37	3983.31	35.00	1179.16	nr	**5162.47**
Ceiling mounted 4 way blow cooling only unit with remote fan speed and load control; refrigerant 410 A; includes outdoor unit						
Cooling 3.80 kW	1650.70	2089.12	35.00	1179.16	nr	**3268.28**
Cooling 5.20 kW	1843.95	2333.70	35.00	1179.16	nr	**3512.86**
Cooling 7.10 kW	2274.07	2878.06	35.00	1179.16	nr	**4057.22**
Cooling 10 kW	3641.70	4608.93	35.00	1179.16	nr	**5788.09**
Cooling 12.2 kW	4198.34	5313.42	35.00	1179.16	nr	**6492.58**
In ceiling, ducted heat pump unit with remote fan speed and load control; refrigerant 410 A; includes outdoor unit						
Cooling 3.60 kW, heating 4.10 kW	1336.84	1691.91	35.00	1179.16	nr	**2871.07**
Cooling 4.90 kW, heating 5.50 kW	1537.36	1945.69	35.00	1179.16	nr	**3124.85**
Cooling 7.10 kW, heating 8.20 kW	1537.36	1945.69	35.00	1179.16	nr	**3124.85**
Cooling 10 kW, heating 11.20 kW	1537.36	1945.69	35.00	1179.16	nr	**3124.85**
Cooling 12.20 kW, heating 14.50 kW	3099.43	3922.64	35.00	1179.16	nr	**5101.80**
In ceiling, ducted cooling only unit with remote fan speed and load control; refrigerant 410 A; includes outdoor unit						
Cooling 3.70 kW	1206.06	1526.39	35.00	1179.16	nr	**2705.55**
Cooling 4.90 kW	1434.20	1815.13	35.00	1179.16	nr	**2994.29**
Cooling 7.10 kW	2165.10	2740.15	35.00	1179.16	nr	**3919.31**
Cooling 10 kW	2468.79	3124.50	35.00	1179.16	nr	**4303.66**
Cooling 12.3 kW	2752.13	3483.10	35.00	1179.16	nr	**4662.26**
Room units						
Ceiling mounted 4 way blow cassette heat pump unit with remote fan speed and load control; refrigerant 410 A; excludes outdoor unit						
Cooling 3.6 kW, heating 4.1 kW	1150.84	1456.50	17.00	572.73	nr	**2029.23**
Cooling 4.9 kW, heating 5.5 kW	1178.45	1491.45	17.00	572.73	nr	**2064.18**
Cooling 7.1 kW, heating 8.2 kW	1275.81	1614.67	17.00	572.73	nr	**2187.40**
Cooling 10 kW, heating 11.2 kW	1381.89	1748.92	17.00	572.73	nr	**2321.65**
Cooling 12.20 kW, heating 14.60 kW	1498.14	1896.05	17.00	572.73	nr	**2468.78**

38 MECHANICAL/COOLING/HEATING SYSTEMS

Item	Net Price £	Material £	Labour hours	Labour £	Unit	Total rate £
LOCAL COOLING UNITS – cont						
Room units – cont						
Ceiling mounted 4 way blow cooling unit with remote fan speed and load control; refrigerant 410 A; excludes outdoor unit						
Cooling 3.80 kW	1086.89	1375.57	17.00	572.73	nr	**1948.30**
Cooling 5.20 kW	1095.62	1386.62	17.00	572.73	nr	**1959.35**
Cooling 7.10 kW	1275.81	1614.67	17.00	572.73	nr	**2187.40**
Cooling 10 kW	1381.89	1748.92	17.00	572.73	nr	**2321.65**
Cooling 12.2 kW	1498.14	1896.05	17.00	572.73	nr	**2468.78**
In ceiling, ducted heat pump unit with remote fan speed and load control; refrigerant 410 A; excludes outdoor unit						
Cooling 3.60 kW, heating 4.10 kW	664.06	840.44	17.00	572.73	nr	**1413.17**
Cooling 4.90 kW, heating 5.50 kW	706.20	893.77	17.00	572.73	nr	**1466.50**
Cooling 7.10 kW, heating 8.20 kW	1166.83	1476.74	17.00	572.73	nr	**2049.47**
Cooling 10 kW, heating 11.20 kW	1208.97	1530.08	17.00	572.73	nr	**2102.81**
Cooling 12.20 kW, heating 14.50 kW	1450.14	1835.30	17.00	572.73	nr	**2408.03**
In ceiling, ducted cooling unit only with remote fan speed and load control; refrigerant 410 A; excludes outdoor unit						
Cooling 3.70 kW	642.26	812.84	17.00	572.73	nr	**1385.57**
Cooling 4.90 kW	685.86	868.02	17.00	572.73	nr	**1440.75**
Cooling 7.10 kW	1166.83	1476.74	17.00	572.73	nr	**2049.47**
Cooling 10 kW	1208.97	1530.08	17.00	572.73	nr	**2102.81**
Cooling 12.3 kW	1450.19	1835.36	17.00	572.73	nr	**2408.09**
External condensing units suitable for connection to multiple indoor units; inverter driven; refrigerant 410 A						
Cooling only						
9 kW	2512.38	3179.67	17.00	572.73	nr	**3752.40**
Heat pump						
Cooling 5.20 kW, heating 6.10 kW	1790.22	2265.70	17.00	572.73	nr	**2838.43**
Cooling 6.80 kW, heating 2.50 kW	2282.79	2889.10	17.00	572.73	nr	**3461.83**
Cooling 8 kW, heating 9.60 kW	2643.17	3345.19	17.00	572.73	nr	**3917.92**
Cooling 14.50 kW, heating 16.50 kW	4386.86	5552.01	21.00	707.49	nr	**6259.50**

38 VENTILATION/AIR CONDITIONING SYSTEMS

Item	Net Price £	Material £	Labour hours	Labour £	Unit	Total rate £
DUCTWORK: CIRCULAR						
AIR DUCTLINES						
Galvanized sheet metal DW144 class B spirally wound circular section ductwork; including all necessary stiffeners, joints, couplers in the running length and duct supports						
Straight duct						
80 mm dia.	6.20	7.85	0.87	28.07	m	35.92
100 mm dia.	6.42	8.12	0.87	28.07	m	36.19
160 mm dia.	8.80	11.13	0.87	28.07	m	39.20
200 mm dia.	11.18	14.15	0.87	28.07	m	42.22
250 mm dia.	13.28	16.81	1.21	39.05	m	55.86
315 mm dia.	16.19	20.48	1.21	39.05	m	59.53
355 mm dia.	23.35	29.56	1.21	39.05	m	68.61
400 mm dia.	23.89	30.24	1.21	39.05	m	69.29
450 mm dia.	28.58	36.18	1.21	39.05	m	75.23
500 mm dia.	31.14	39.41	1.21	39.05	m	78.46
630 mm dia.	59.96	75.88	1.39	44.87	m	120.75
710 mm dia.	66.96	84.74	1.39	44.87	m	129.61
800 mm dia.	71.17	90.07	1.44	46.47	m	136.54
900 mm dia.	84.41	106.83	1.46	47.12	m	153.95
1000 mm dia.	108.46	137.27	1.65	53.26	m	190.53
1120 mm dia.	129.46	163.84	2.43	78.43	m	242.27
1250 mm dia.	141.10	178.57	2.43	78.43	m	257.00
1400 mm dia.	158.50	200.60	2.77	89.40	m	290.00
1600 mm dia.	177.39	224.50	3.06	98.75	m	323.25
Extra over fittings; circular duct class B						
End cap						
80 mm dia.	2.43	3.08	0.15	4.85	nr	7.93
100 mm dia.	2.64	3.34	0.15	4.85	nr	8.19
160 mm dia.	3.98	5.04	0.15	4.85	nr	9.89
200 mm dia.	5.04	6.38	0.20	6.45	nr	12.83
250 mm dia.	7.87	9.96	0.29	9.35	nr	19.31
315 mm dia.	9.19	11.63	0.29	9.35	nr	20.98
355 mm dia.	13.76	17.42	0.44	14.20	nr	31.62
400 mm dia.	15.09	19.10	0.44	14.20	nr	33.30
450 mm dia.	17.00	21.52	0.44	14.20	nr	35.72
500 mm dia.	17.91	22.67	0.44	14.20	nr	36.87
630 mm dia.	65.01	82.28	0.58	18.72	nr	101.00
710 mm dia.	73.81	93.42	0.69	22.27	nr	115.69
800 mm dia.	79.26	100.31	0.81	26.15	nr	126.46
900 mm dia.	101.14	128.00	0.92	29.69	nr	157.69
1000 mm dia.	149.49	189.19	1.04	33.57	nr	222.76
1120 mm dia.	178.11	225.41	1.16	37.44	nr	262.85
1250 mm dia.	201.43	254.93	1.16	37.44	nr	292.37
1400 mm dia.	258.98	327.77	1.16	37.44	nr	365.21
1600 mm dia.	287.97	364.46	1.16	37.44	nr	401.90

38 VENTILATION/AIR CONDITIONING SYSTEMS

Item	Net Price £	Material £	Labour hours	Labour £	Unit	Total rate £
DUCTWORK: CIRCULAR – cont						
Extra over fittings – cont						
Reducer						
80 mm dia.	12.17	15.40	0.29	9.35	nr	24.75
100 mm dia.	12.56	15.89	0.29	9.35	nr	25.24
160 mm dia.	16.79	21.25	0.29	9.35	nr	30.60
200 mm dia.	20.45	25.88	0.44	14.20	nr	40.08
250 mm dia.	22.95	29.04	0.58	18.72	nr	47.76
315 mm dia.	23.60	29.87	0.58	18.72	nr	48.59
355 mm dia.	25.80	32.65	0.87	28.07	nr	60.72
400 mm dia.	30.09	38.08	0.87	28.07	nr	66.15
450 mm dia.	32.52	41.16	0.87	28.07	nr	69.23
500 mm dia.	34.86	44.12	0.87	28.07	nr	72.19
630 mm dia.	96.98	122.74	0.87	28.07	nr	150.81
710 mm dia.	119.72	151.51	0.96	30.98	nr	182.49
800 mm dia.	130.51	165.18	1.06	34.20	nr	199.38
900 mm dia.	158.76	200.93	1.16	37.44	nr	238.37
1000 mm dia.	230.80	292.10	1.25	40.34	nr	332.44
1120 mm dia.	239.76	303.44	3.47	111.98	nr	415.42
1250 mm dia.	306.78	388.26	3.47	111.98	nr	500.24
1400 mm dia.	329.18	416.61	4.05	130.72	nr	547.33
1600 mm dia.	434.86	550.36	4.62	149.11	nr	699.47
90° segmented radius bend						
80 mm dia.	5.22	6.61	0.29	9.35	nr	15.96
100 mm dia.	5.56	7.03	0.29	9.35	nr	16.38
160 mm dia.	9.41	11.91	0.29	9.35	nr	21.26
200 mm dia.	11.68	14.78	0.44	14.20	nr	28.98
250 mm dia.	16.25	20.56	0.58	18.72	nr	39.28
315 mm dia.	15.39	19.48	0.58	18.72	nr	38.20
355 mm dia.	15.51	19.63	0.87	28.07	nr	47.70
400 mm dia.	19.10	24.17	0.87	28.07	nr	52.24
450 mm dia.	21.79	27.57	0.87	28.07	nr	55.64
500 mm dia.	27.54	34.85	0.87	28.07	nr	62.92
630 mm dia.	81.04	102.57	0.87	28.07	nr	130.64
710 mm dia.	106.17	134.37	0.96	30.98	nr	165.35
800 mm dia.	111.90	141.62	1.06	34.20	nr	175.82
900 mm dia.	156.62	198.22	1.16	37.44	nr	235.66
1000 mm dia.	244.10	308.93	1.25	40.34	nr	349.27
1120 mm dia.	291.76	369.25	3.47	111.98	nr	481.23
1250 mm dia.	314.45	397.97	3.47	111.98	nr	509.95
1400 mm dia.	642.87	813.61	4.05	130.72	nr	944.33
1600 mm dia.	718.74	909.64	4.62	149.11	nr	1058.75
45° radius bend						
80 mm dia.	4.49	5.68	0.29	9.35	nr	15.03
100 mm dia.	4.87	6.16	0.29	9.35	nr	15.51
160 mm dia.	8.14	10.30	0.29	9.35	nr	19.65
200 mm dia.	10.11	12.79	0.40	12.91	nr	25.70
250 mm dia.	13.99	17.71	0.58	18.72	nr	36.43
315 mm dia.	14.97	18.95	0.58	18.72	nr	37.67
355 mm dia.	17.34	21.94	0.87	28.07	nr	50.01

38 VENTILATION/AIR CONDITIONING SYSTEMS

Item	Net Price £	Material £	Labour hours	Labour £	Unit	Total rate £
400 mm dia.	18.79	23.78	0.87	28.07	nr	51.85
450 mm dia.	20.17	25.52	0.87	28.07	nr	53.59
500 mm dia.	23.60	29.87	0.87	28.07	nr	57.94
630 mm dia.	80.07	101.34	0.87	28.07	nr	129.41
710 mm dia.	93.18	117.92	0.96	30.98	nr	148.90
800 mm dia.	107.90	136.56	1.06	34.20	nr	170.76
900 mm dia.	131.80	166.80	1.16	37.44	nr	204.24
1000 mm dia.	220.77	279.41	1.25	40.34	nr	319.75
1120 mm dia.	268.64	339.99	3.47	111.98	nr	451.97
1250 mm dia.	292.07	369.64	3.47	111.98	nr	481.62
1400 mm dia.	435.80	551.54	4.05	130.72	nr	682.26
1600 mm dia.	468.56	593.01	4.62	149.11	nr	742.12
90° equal twin bend						
80 mm dia.	28.23	35.73	0.58	18.72	nr	54.45
100 mm dia.	28.71	36.33	0.58	18.72	nr	55.05
160 mm dia.	37.78	47.81	0.58	18.72	nr	66.53
200 mm dia.	49.12	62.17	0.87	28.07	nr	90.24
250 mm dia.	54.61	69.12	1.16	37.44	nr	106.56
315 mm dia.	84.06	106.39	1.16	37.44	nr	143.83
355 mm dia.	99.36	125.75	1.73	55.84	nr	181.59
400 mm dia.	115.34	145.97	1.73	55.84	nr	201.81
450 mm dia.	135.19	171.09	1.73	55.84	nr	226.93
500 mm dia.	142.94	180.90	1.73	55.84	nr	236.74
630 mm dia.	252.82	319.97	1.73	55.84	nr	375.81
710 mm dia.	312.41	395.38	1.82	58.73	nr	454.11
800 mm dia.	429.54	543.63	1.93	62.29	nr	605.92
900 mm dia.	563.36	712.99	2.02	65.20	nr	778.19
1000 mm dia.	867.39	1097.77	2.11	68.08	nr	1165.85
1120 mm dia.	1255.38	1588.81	4.62	149.11	nr	1737.92
1250 mm dia.	1266.53	1602.92	4.62	149.11	nr	1752.03
1400 mm dia.	1715.44	2171.06	4.62	149.11	nr	2320.17
1600 mm dia.	1872.02	2369.23	4.62	149.11	nr	2518.34
Conical branch						
80 mm dia.	21.20	26.84	0.58	18.72	nr	45.56
100 mm dia.	21.93	27.75	0.58	18.72	nr	46.47
160 mm dia.	25.06	31.72	0.58	18.72	nr	50.44
200 mm dia.	29.17	36.92	0.87	28.07	nr	64.99
250 mm dia.	32.31	40.89	1.16	37.44	nr	78.33
315 mm dia.	50.81	64.31	1.16	37.44	nr	101.75
355 mm dia.	59.97	75.90	1.73	55.84	nr	131.74
400 mm dia.	70.77	89.57	1.73	55.84	nr	145.41
450 mm dia.	79.04	100.04	1.73	55.84	nr	155.88
500 mm dia.	87.58	110.85	1.73	55.84	nr	166.69
630 mm dia.	128.34	162.42	1.73	55.84	nr	218.26
710 mm dia.	137.48	173.99	1.82	58.73	nr	232.72
800 mm dia.	147.72	186.95	1.93	62.29	nr	249.24
900 mm dia.	160.91	203.65	2.02	65.20	nr	268.85
1000 mm dia.	224.71	284.39	2.11	68.08	nr	352.47
1120 mm dia.	273.57	346.23	4.62	149.11	nr	495.34
1250 mm dia.	353.46	447.34	5.20	167.82	nr	615.16
1400 mm dia. .	392.12	496.27	5.20	167.82	nr	664.09
1600 mm dia.	449.06	568.33	5.20	167.82	nr	736.15

38 VENTILATION/AIR CONDITIONING SYSTEMS

Item	Net Price £	Material £	Labour hours	Labour £	Unit	Total rate £
DUCTWORK: CIRCULAR – cont						
Extra over fittings – cont						
45° branch						
80 mm dia.	23.57	29.83	0.58	18.72	nr	**48.55**
100 mm dia.	23.78	30.09	0.58	18.72	nr	**48.81**
160 mm dia.	25.27	31.99	0.58	18.72	nr	**50.71**
200 mm dia.	27.27	34.52	0.87	28.07	nr	**62.59**
250 mm dia.	30.53	38.64	1.16	37.44	nr	**76.08**
315 mm dia.	35.29	44.67	1.16	37.44	nr	**82.11**
355 mm dia.	38.47	48.69	1.73	55.84	nr	**104.53**
400 mm dia.	43.33	54.84	1.73	55.84	nr	**110.68**
450 mm dia.	48.75	61.70	1.73	55.84	nr	**117.54**
500 mm dia.	54.16	68.54	1.73	55.84	nr	**124.38**
630 mm dia.	95.85	121.31	1.73	55.84	nr	**177.15**
710 mm dia.	112.10	141.87	1.82	58.73	nr	**200.60**
800 mm dia.	122.81	155.43	2.13	68.73	nr	**224.16**
900 mm dia.	136.15	172.31	2.31	74.56	nr	**246.87**
1000 mm dia.	202.17	255.86	2.31	74.56	nr	**330.42**
1120 mm dia.	249.18	315.36	4.62	149.11	nr	**464.47**
1250 mm dia.	332.77	421.15	4.62	149.11	nr	**570.26**
1400 mm dia.	372.88	471.91	4.62	149.11	nr	**621.02**
1600 mm dia.	427.15	540.60	4.62	149.11	nr	**689.71**

For galvanized sheet metal DW144 class C rates, refer to Galvanized sheet metal DW144 class B

38 VENTILATION/AIR CONDITIONING SYSTEMS

Item	Net Price £	Material £	Labour hours	Labour £	Unit	Total rate £
DUCTWORK: FLAT OVAL						
AIR DUCTLINES						
Galvanized sheet metal DW144 class B spirally wound flat oval section ductwork; including all necessary stiffeners, joints, couplers in the running length and duct supports						
Straight duct						
345 × 102 mm	78.34	99.14	2.71	87.46	m	186.60
427 × 102 mm	83.11	105.18	2.99	96.50	m	201.68
508 × 102 mm	87.91	111.26	3.14	101.34	m	212.60
559 × 152 mm	93.08	117.80	3.43	110.70	m	228.50
531 × 203 mm	93.08	117.80	3.43	110.70	m	228.50
851 × 203 mm	139.28	176.28	5.72	184.60	m	360.88
582 × 254 mm	98.61	124.80	3.62	116.83	m	241.63
823 × 254 mm	139.39	176.41	5.80	187.19	m	363.60
1303 × 254 mm	256.42	324.52	8.13	262.37	m	586.89
632 × 305 mm	103.42	130.88	3.93	126.84	m	257.72
1275 × 305 mm	250.08	316.50	8.13	262.37	m	578.87
765 × 356 mm	118.01	149.35	5.72	184.60	m	333.95
1247 × 356 mm	258.14	326.70	8.13	262.37	m	589.07
1727 × 356 mm	307.95	389.74	10.41	335.96	m	725.70
737 × 406 mm	113.33	143.43	5.72	184.60	m	328.03
818 × 406 mm	137.68	174.25	6.21	200.41	m	374.66
978 × 406 mm	180.94	229.00	6.92	223.33	m	452.33
1379 × 406 mm	270.86	342.80	8.75	282.40	m	625.20
1699 × 406 mm	310.11	392.47	10.41	335.96	m	728.43
709 × 457 mm	122.73	155.32	5.72	184.60	m	339.92
1189 × 457 mm	236.21	298.95	8.80	284.00	m	582.95
1671 × 457 mm	307.02	388.56	10.31	332.73	m	721.29
678 × 508 mm	122.46	154.99	5.72	184.60	m	339.59
919 × 508 mm	154.32	195.31	7.30	235.59	m	430.90
1321 × 508 mm	270.17	341.92	8.75	282.40	m	624.32
Extra over fittings; flat oval duct class B						
End cap						
345 × 102 mm	38.02	48.12	0.20	6.45	nr	54.57
427 × 102 mm	47.68	60.35	0.20	6.45	nr	66.80
508 × 102 mm	51.33	64.96	0.20	6.45	nr	71.41
559 × 152 mm	53.36	67.54	0.29	9.35	nr	76.89
531 × 203 mm	53.38	67.56	0.29	9.35	nr	76.91
851 × 203 mm	65.93	83.44	0.44	14.20	nr	97.64
582 × 254 mm	56.84	71.94	0.44	14.20	nr	86.14
823 × 254 mm	65.95	83.46	0.44	14.20	nr	97.66
1303 × 254 mm	195.35	247.24	0.69	22.27	nr	269.51
632 × 305 mm	59.33	75.08	0.69	22.27	nr	97.35
1275 × 305 mm	191.65	242.55	0.69	22.27	nr	264.82

38 VENTILATION/AIR CONDITIONING SYSTEMS

Item	Net Price £	Material £	Labour hours	Labour £	Unit	Total rate £
DUCTWORK: FLAT OVAL – cont						
Extra over fittings – cont						
765 × 356 mm	65.95	83.46	0.69	22.27	nr	**105.73**
1247 × 356 mm	192.00	243.00	0.69	22.27	nr	**265.27**
1727 × 356 mm	303.62	384.26	0.69	22.27	nr	**406.53**
737 × 406 mm	66.01	83.54	1.04	33.57	nr	**117.11**
818 × 406 mm	72.05	91.19	0.69	22.27	nr	**113.46**
978 × 406 mm	97.83	123.82	0.69	22.27	nr	**146.09**
1379 × 406 mm	228.24	288.86	1.04	33.57	nr	**322.43**
1699 × 406 mm	303.93	384.65	1.04	33.57	nr	**418.22**
709 × 457 mm	65.96	83.47	1.04	33.57	nr	**117.04**
1189 × 457 mm	188.55	238.63	1.04	33.57	nr	**272.20**
1671 × 457 mm	298.25	377.46	1.04	33.57	nr	**411.03**
678 × 508 mm	65.95	83.46	1.04	33.57	nr	**117.03**
919 × 508 mm	97.82	123.80	1.04	33.57	nr	**157.37**
1321 × 508 mm	227.71	288.19	1.04	33.57	nr	**321.76**
Reducer						
345 × 102 mm	80.27	101.60	0.95	30.67	nr	**132.27**
427 × 102 mm	87.05	110.17	1.06	34.20	nr	**144.37**
508 × 102 mm	100.41	127.08	1.13	36.48	nr	**163.56**
559 × 152 mm	118.02	149.36	1.26	40.67	nr	**190.03**
531 × 203 mm	119.35	151.05	1.26	40.67	nr	**191.72**
851 × 203 mm	131.20	166.05	1.34	43.24	nr	**209.29**
582 × 254 mm	123.11	155.80	1.34	43.24	nr	**199.04**
823 × 254 mm	133.15	168.52	1.34	43.24	nr	**211.76**
1303 × 254 mm	354.38	448.50	1.34	43.24	nr	**491.74**
632 × 305 mm	125.71	159.10	0.70	22.59	nr	**181.69**
1275 × 305 mm	354.48	448.63	1.16	37.44	nr	**486.07**
765 × 356 mm	145.74	184.45	1.16	37.44	nr	**221.89**
1247 × 356 mm	353.45	447.33	1.16	37.44	nr	**484.77**
1727 × 356 mm	405.65	513.39	1.25	40.34	nr	**553.73**
737 × 406 mm	149.29	188.94	1.16	37.44	nr	**226.38**
818 × 406 mm	162.14	205.21	1.27	40.98	nr	**246.19**
978 × 406 mm	177.33	224.43	1.44	46.47	nr	**270.90**
1379 × 406 mm	354.13	448.19	1.44	46.47	nr	**494.66**
1699 × 406 mm	408.52	517.03	1.44	46.47	nr	**563.50**
709 × 457 mm	147.34	186.47	1.16	37.44	nr	**223.91**
1189 × 457 mm	367.75	465.43	1.34	43.24	nr	**508.67**
1671 × 457 mm	403.88	511.15	1.44	46.47	nr	**557.62**
678 × 508 mm	149.25	188.89	1.16	37.44	nr	**226.33**
919 × 508 mm	195.94	247.98	1.26	40.67	nr	**288.65**
1321 × 508 mm	361.64	457.69	1.44	46.47	nr	**504.16**
90° radius bend						
345 × 102 mm	205.91	260.60	0.29	9.35	nr	**269.95**
427 × 102 mm	228.48	289.16	0.58	18.72	nr	**307.88**
508 × 102 mm	281.77	356.61	0.58	18.72	nr	**375.33**
559 × 152 mm	327.30	414.23	0.58	18.72	nr	**432.95**
531 × 203 mm	331.43	419.46	0.87	28.07	nr	**447.53**
851 × 203 mm	337.06	426.59	0.87	28.07	nr	**454.66**
582 × 254 mm	347.80	440.17	0.87	28.07	nr	**468.24**

38 VENTILATION/AIR CONDITIONING SYSTEMS

Item	Net Price £	Material £	Labour hours	Labour £	Unit	Total rate £
823 × 254 mm	343.16	434.30	0.87	28.07	nr	462.37
1303 × 254 mm	360.75	456.57	0.96	30.98	nr	487.55
632 × 305 mm	364.42	461.20	0.87	28.07	nr	489.27
1275 × 305 mm	377.70	478.02	0.96	30.98	nr	509.00
765 × 356 mm	382.76	484.42	0.87	28.07	nr	512.49
1247 × 356 mm	372.87	471.90	0.96	30.98	nr	502.88
1727 × 356 mm	723.19	915.26	1.25	40.34	nr	955.60
737 × 406 mm	393.86	498.47	0.96	30.98	nr	529.45
818 × 406 mm	375.93	475.78	0.87	28.07	nr	503.85
978 × 406 mm	310.53	393.01	0.96	30.98	nr	423.99
1379 × 406 mm	604.39	764.92	1.16	37.44	nr	802.36
1699 × 406 mm	730.82	924.93	1.25	40.34	nr	965.27
709 × 457 mm	387.67	490.64	0.87	28.07	nr	518.71
1189 × 457 mm	433.32	548.41	0.96	30.98	nr	579.39
1671 × 457 mm	1052.23	1331.70	1.25	40.34	nr	1372.04
678 × 508 mm	393.73	498.30	0.87	28.07	nr	526.37
919 × 508 mm	368.98	466.98	0.96	30.98	nr	497.96
1321 × 508 mm	703.44	890.28	1.16	37.44	nr	927.72
45° radius bend						
345 × 102 mm	101.83	128.88	0.79	25.49	nr	154.37
427 × 102 mm	133.64	169.13	0.85	27.44	nr	196.57
508 × 102 mm	164.67	208.41	0.95	30.67	nr	239.08
559 × 152 mm	191.22	242.01	0.79	25.49	nr	267.50
531 × 203 mm	193.29	244.63	0.85	27.44	nr	272.07
851 × 203 mm	204.95	259.38	0.98	31.63	nr	291.01
582 × 254 mm	203.71	257.81	0.76	24.53	nr	282.34
823 × 254 mm	208.02	263.27	0.95	30.67	nr	293.94
1303 × 254 mm	388.56	491.76	1.16	37.44	nr	529.20
632 × 305 mm	214.19	271.07	0.58	18.72	nr	289.79
1275 × 305 mm	393.15	497.57	1.16	37.44	nr	535.01
765 × 356 mm	227.83	288.34	0.87	28.07	nr	316.41
1247 × 356 mm	391.13	495.02	1.16	37.44	nr	532.46
1727 × 356 mm	691.56	875.24	1.26	40.67	nr	915.91
737 × 406 mm	233.38	295.37	0.69	22.27	nr	317.64
818 × 406 mm	226.63	286.82	0.78	25.18	nr	312.00
978 × 406 mm	197.35	249.77	0.87	28.07	nr	277.84
1379 × 406 mm	560.26	709.06	1.16	37.44	nr	746.50
1699 × 406 mm	695.70	880.48	1.27	40.98	nr	921.46
709 × 457 mm	229.96	291.03	0.81	26.15	nr	317.18
1189 × 457 mm	417.72	528.66	0.95	30.67	nr	559.33
1671 × 457 mm	736.91	932.64	1.26	40.67	nr	973.31
678 × 508 mm	233.31	295.28	0.92	29.69	nr	324.97
919 × 508 mm	226.57	286.74	1.10	35.50	nr	322.24
1321 × 508 mm	572.68	724.79	1.25	40.34	nr	765.13

38 VENTILATION/AIR CONDITIONING SYSTEMS

Item	Net Price £	Material £	Labour hours	Labour £	Unit	Total rate £
DUCTWORK: FLAT OVAL – cont						
Extra over fittings – cont						
90° hard bend with turning vanes						
345 × 102 mm	87.35	110.56	0.55	17.75	nr	**128.31**
427 × 102 mm	150.34	190.27	1.16	37.44	nr	**227.71**
508 × 102 mm	204.05	258.25	1.16	37.44	nr	**295.69**
559 × 152 mm	229.40	290.33	1.16	37.44	nr	**327.77**
531 × 203 mm	229.84	290.89	1.73	55.84	nr	**346.73**
851 × 203 mm	335.81	425.01	1.73	55.84	nr	**480.85**
582 × 254 mm	246.19	311.57	1.73	55.84	nr	**367.41**
823 × 254 mm	335.81	425.01	1.73	55.84	nr	**480.85**
1303 × 254 mm	595.57	753.75	1.82	58.73	nr	**812.48**
632 × 305 mm	262.45	332.16	1.73	55.84	nr	**388.00**
1275 × 305 mm	591.78	748.96	1.82	58.73	nr	**807.69**
765 × 356 mm	320.08	405.09	1.73	55.84	nr	**460.93**
1247 × 356 mm	592.14	749.41	1.82	58.73	nr	**808.14**
1727 × 356 mm	782.04	989.76	1.82	58.73	nr	**1048.49**
737 × 406 mm	320.08	405.09	1.73	55.84	nr	**460.93**
818 × 406 mm	363.36	459.87	1.73	55.84	nr	**515.71**
978 × 406 mm	419.46	530.87	1.73	55.84	nr	**586.71**
1379 × 406 mm	652.64	825.98	1.82	58.73	nr	**884.71**
1699 × 406 mm	782.34	990.12	2.11	68.08	nr	**1058.20**
709 × 457 mm	320.09	405.10	1.73	55.84	nr	**460.94**
1189 × 457 mm	588.60	744.93	1.82	58.73	nr	**803.66**
1671 × 457 mm	776.53	982.78	2.11	68.08	nr	**1050.86**
678 × 508 mm	320.08	405.09	1.82	58.73	nr	**463.82**
919 × 508 mm	419.46	530.87	1.82	58.73	nr	**589.60**
1321 × 508 mm	652.09	825.28	2.11	68.08	nr	**893.36**
90° branch						
345 × 102 mm	164.70	208.44	0.58	18.72	nr	**227.16**
427 × 102 mm	167.02	211.38	0.58	18.72	nr	**230.10**
508 × 102 mm	176.23	223.04	1.16	37.44	nr	**260.48**
559 × 152 mm	205.53	260.12	1.16	37.44	nr	**297.56**
531 × 203 mm	208.16	263.45	1.16	37.44	nr	**300.89**
851 × 203 mm	304.91	385.90	1.73	55.84	nr	**441.74**
582 × 254 mm	218.66	276.74	1.73	55.84	nr	**332.58**
823 × 254 mm	308.78	390.79	1.73	55.84	nr	**446.63**
1303 × 254 mm	423.65	536.17	1.82	58.73	nr	**594.90**
632 × 305 mm	229.39	290.32	1.73	55.84	nr	**346.16**
1275 × 305 mm	432.26	547.06	1.82	58.73	nr	**605.79**
765 × 356 mm	336.22	425.52	1.73	55.84	nr	**481.36**
1247 × 356 mm	428.60	542.44	1.82	58.73	nr	**601.17**
1727 × 356 mm	642.66	813.36	2.11	68.08	nr	**881.44**
737 × 406 mm	343.80	435.11	1.73	55.84	nr	**490.95**
818 × 406 mm	383.99	485.98	1.73	55.84	nr	**541.82**
978 × 406 mm	338.17	427.99	1.82	58.73	nr	**486.72**
1379 × 406 mm	432.68	547.60	1.93	62.29	nr	**609.89**
1699 × 406 mm	647.51	819.49	2.11	68.08	nr	**887.57**

38 VENTILATION/AIR CONDITIONING SYSTEMS

Item	Net Price £	Material £	Labour hours	Labour £	Unit	Total rate £
709 × 457 mm	338.87	428.87	1.73	55.84	nr	484.71
1189 × 457 mm	466.77	590.74	1.82	58.73	nr	649.47
1671 × 457 mm	900.87	1140.14	2.11	68.08	nr	1208.22
678 × 508 mm	342.70	433.72	1.73	55.84	nr	489.56
919 × 508 mm	378.22	478.68	1.82	58.73	nr	537.41
1321 × 508 mm	547.65	693.10	2.11	68.08	nr	761.18
45° branch						
345 × 102 mm	99.26	125.62	0.58	18.72	nr	144.34
427 × 102 mm	171.14	216.60	0.58	18.72	nr	235.32
508 × 102 mm	232.43	294.17	1.16	37.44	nr	331.61
559 × 152 mm	261.80	331.33	1.73	55.84	nr	387.17
531 × 203 mm	261.81	331.35	1.73	55.84	nr	387.19
851 × 203 mm	364.44	461.24	1.73	55.84	nr	517.08
582 × 254 mm	280.35	354.82	1.73	55.84	nr	410.66
823 × 254 mm	364.46	461.26	1.82	58.73	nr	519.99
1303 × 254 mm	635.75	804.61	1.92	61.96	nr	866.57
632 × 305 mm	298.83	378.20	1.73	55.84	nr	434.04
1275 × 305 mm	632.14	800.04	1.82	58.73	nr	858.77
765 × 356 mm	364.46	461.26	1.73	55.84	nr	517.10
1247 × 356 mm	632.49	800.48	1.82	58.73	nr	859.21
1727 × 356 mm	836.27	1058.39	1.82	58.73	nr	1117.12
737 × 406 mm	364.46	461.26	1.73	55.84	nr	517.10
818 × 406 mm	394.35	499.09	1.73	55.84	nr	554.93
978 × 406 mm	455.24	576.15	1.73	55.84	nr	631.99
1379 × 406 mm	697.75	883.08	1.93	62.29	nr	945.37
1699 × 406 mm	836.58	1058.78	2.19	70.67	nr	1129.45
709 × 457 mm	364.47	461.27	1.73	55.84	nr	517.11
1189 × 457 mm	629.12	796.22	1.82	58.73	nr	854.95
1671 × 457 mm	831.03	1051.75	2.11	68.08	nr	1119.83
678 × 508 mm	364.46	461.26	1.73	55.84	nr	517.10
919 × 508 mm	455.23	576.14	1.82	58.73	nr	634.87
1321 × 508 mm	697.22	882.40	1.93	62.29	nr	944.69

For rates for access doors refer to ancillaries in Ductwork Ancillaries: Access Hatches

38 VENTILATION/AIR CONDITIONING SYSTEMS

Item	Net Price £	Material £	Labour hours	Labour £	Unit	Total rate £
DUCTWORK: FLEXIBLE						
AIR DUCTLINES						
Aluminium foil flexible ductwork, DW 144 class B; multiple aluminium polyester laminate fabric, with high tensile steel wire helix						
Duct						
102 mm dia.	1.62	2.05	0.33	10.64	m	**12.69**
152 mm dia.	2.38	3.01	0.33	10.64	m	**13.65**
203 mm dia.	3.12	3.95	0.33	10.64	m	**14.59**
254 mm dia.	3.92	4.96	0.33	10.64	m	**15.60**
304 mm dia.	4.80	6.07	0.33	10.64	m	**16.71**
355 mm dia.	6.53	8.27	0.33	10.64	m	**18.91**
406 mm dia.	7.27	9.21	0.33	10.64	m	**19.85**
Insulated aluminium foil flexible ductwork, DW144 class B; laminate construction of aluminium and polyester multiple inner core with 25 mm insulation; outer layer of multiple aluminium polyester laminate, with high tensile steel wire helix						
Duct						
102 mm dia.	3.63	4.59	0.50	16.14	m	**20.73**
152 mm dia.	4.74	6.00	0.50	16.14	m	**22.14**
203 mm dia.	5.74	7.27	0.50	16.14	m	**23.41**
254 mm dia.	7.06	8.94	0.50	16.14	m	**25.08**
304 mm dia.	9.08	11.49	0.50	16.14	m	**27.63**
355 mm dia.	11.27	14.27	0.50	16.14	m	**30.41**
406 mm dia.	12.48	15.79	0.50	16.14	m	**31.93**

38 VENTILATION/AIR CONDITIONING SYSTEMS

Item	Net Price £	Material £	Labour hours	Labour £	Unit	Total rate £
DUCTWORK: PLASTIC						
AIR DUCTLINES						
Rigid grey PVC DW 154 circular section ductwork; solvent welded or filler rod welded joints; excludes couplers and supports (these are detailed separately); ductwork to conform to curent HSE regulations						
Straight duct (standard length 6 m)						
110 mm	7.52	9.52	1.71	55.18	m	**64.70**
160 mm	11.50	14.56	1.74	56.16	m	**70.72**
200 mm	14.54	18.40	1.79	57.77	m	**76.17**
225 mm	17.60	22.28	1.96	63.26	m	**85.54**
250 mm	19.46	24.63	1.98	63.90	m	**88.53**
315 mm	24.95	31.57	2.23	71.97	m	**103.54**
355 mm	30.45	38.54	2.25	72.62	m	**111.16**
400 mm	35.70	45.18	2.39	77.13	m	**122.31**
450 mm	46.05	58.28	2.95	95.20	m	**153.48**
500 mm	61.32	77.60	2.98	96.17	m	**173.77**
600 mm	118.51	149.99	3.12	100.69	m	**250.68**
Extra for supports (BZP finish); Horizontal – Maximum 2.4 m centres; Vertical – Maximum 4.0 m centres						
Duct Size						
110 mm	28.59	36.19	0.59	19.04	m	**55.23**
160 mm	28.69	36.31	0.59	19.04	m	**55.35**
200 mm	28.78	36.42	0.59	19.04	m	**55.46**
225 mm	30.56	38.67	0.63	20.34	m	**59.01**
250 mm	30.57	38.68	0.63	20.34	m	**59.02**
315 mm	30.79	38.96	0.63	20.34	m	**59.30**
355 mm	29.45	37.27	0.81	26.15	m	**63.42**
400 mm	39.98	50.60	0.78	25.18	m	**75.78**
450 mm	41.10	52.01	0.80	25.82	m	**77.83**
500 mm	41.29	52.26	0.80	25.82	m	**78.08**
600 mm	43.85	55.50	0.85	27.44	m	**82.94**
Note: These are maximum figures and may be reduced subject to local conditions (i.e. a high number of changes of direction)						
Extra over fittings; Rigid grey PVC						
90° bend						
110 mm	22.96	29.05	1.11	35.82	m	**64.87**
160 mm	25.34	32.07	1.34	43.24	m	**75.31**
200 mm	21.21	26.85	2.00	64.55	m	**91.40**
225 mm	32.55	41.19	1.79	57.77	m	**98.96**
250 mm	34.65	43.85	2.01	64.87	m	**108.72**
315 mm	59.83	75.72	2.56	82.62	m	**158.34**
355 mm	71.24	90.16	2.71	87.46	m	**177.62**

38 VENTILATION/AIR CONDITIONING SYSTEMS

Item	Net Price £	Material £	Labour hours	Labour £	Unit	Total rate £
DUCTWORK: PLASTIC – cont						
Extra over fittings – cont						
90° bend – cont						
400 mm	90.49	114.52	3.58	115.54	m	**230.06**
450 mm	273.52	346.17	4.55	146.84	m	**493.01**
500 mm	317.46	401.78	5.01	161.68	m	**563.46**
600 mm	543.42	687.75	5.78	186.54	m	**874.29**
45° bend						
110 mm	12.82	16.23	0.71	22.92	m	**39.15**
160 mm	16.98	21.49	0.93	30.00	m	**51.49**
200 mm	18.99	24.04	1.14	36.79	m	**60.83**
225 mm	23.23	29.40	1.41	45.49	m	**74.89**
250 mm	26.51	33.56	1.62	52.28	m	**85.84**
315 mm	42.19	53.39	1.93	62.29	m	**115.68**
355 mm	54.51	68.99	2.22	71.65	m	**140.64**
400 mm	71.18	90.08	2.85	91.97	m	**182.05**
450 mm	160.20	202.75	3.77	121.68	m	**324.43**
500 mm	184.98	234.11	4.16	134.25	m	**368.36**
600 mm	277.57	351.29	4.81	155.24	m	**506.53**
Tee						
110 mm	42.10	53.28	1.04	33.57	m	**86.85**
160 mm	60.31	76.33	1.38	44.53	m	**120.86**
200 mm	82.61	104.55	1.79	57.77	m	**162.32**
225 mm	98.79	125.03	2.25	72.62	m	**197.65**
250 mm	118.98	150.58	2.62	84.56	m	**235.14**
315 mm	179.64	227.35	3.17	102.30	m	**329.65**
355 mm	246.70	312.22	3.73	120.37	m	**432.59**
400 mm	256.74	324.93	4.71	152.01	m	**476.94**
450 mm	284.65	360.25	5.72	184.60	m	**544.85**
500 mm	356.86	451.64	6.33	204.28	m	**655.92**
Coupler						
110 mm	8.32	10.53	0.70	22.59	m	**33.12**
160 mm	12.04	15.24	0.91	29.37	m	**44.61**
200 mm	15.77	19.96	1.12	36.14	m	**56.10**
225 mm	20.41	25.83	1.41	45.49	m	**71.32**
250 mm	23.60	29.87	1.60	51.63	m	**81.50**
315 mm	28.69	36.31	1.86	60.03	m	**96.34**
355 mm	31.52	39.89	2.43	78.43	m	**118.32**
400 mm	35.90	45.44	2.66	85.85	m	**131.29**
450 mm	78.31	99.11	3.20	103.28	m	**202.39**
500 mm	88.59	112.12	3.52	113.60	m	**225.72**
Damper						
110 mm	52.34	66.24	0.96	30.98	m	**97.22**
160 mm	55.16	69.81	1.26	40.67	m	**110.48**
200 mm	67.74	85.74	1.52	49.06	m	**134.80**
225 mm	72.65	91.94	1.88	60.67	m	**152.61**
250 mm	73.43	92.94	2.16	69.71	m	**162.65**
315 mm	86.73	109.76	2.94	94.89	m	**204.65**
355 mm	96.48	122.10	3.42	110.38	m	**232.48**
400 mm	101.51	128.48	3.79	122.30	m	**250.78**

38 VENTILATION/AIR CONDITIONING SYSTEMS

Item	Net Price £	Material £	Labour hours	Labour £	Unit	Total rate £
Reducer						
160 × 110 mm	9.56	12.10	0.85	27.44	m	**39.54**
200 × 110 mm	12.26	15.51	0.98	31.63	m	**47.14**
200 × 160 mm	23.60	29.87	1.07	34.54	m	**64.41**
225 × 200 mm	32.90	41.64	1.38	44.53	m	**86.17**
250 × 160 mm	30.46	38.55	1.38	44.53	m	**83.08**
250 × 200 mm	35.81	45.33	1.48	47.77	m	**93.10**
250 × 225 mm	31.09	39.35	1.58	50.99	m	**90.34**
315 × 200 mm	63.60	80.49	1.63	52.60	m	**133.09**
315 × 250 mm	39.34	49.78	1.84	59.38	m	**109.16**
355 × 200 mm	63.60	80.49	1.99	64.22	m	**144.71**
355 × 250 mm	59.05	74.74	1.97	63.57	m	**138.31**
355 × 315 mm	83.39	105.54	2.16	69.71	m	**175.25**
400 × 315 mm	69.39	87.82	2.75	88.76	m	**176.58**
400 × 355 mm	69.39	87.82	2.73	88.10	m	**175.92**
Flange						
110 mm	12.04	15.24	0.70	22.59	m	**37.83**
160 mm	14.64	18.52	0.91	29.37	m	**47.89**
200 mm	16.26	20.57	1.12	36.14	m	**56.71**
225 mm	16.72	21.16	1.35	43.56	m	**64.72**
250 mm	17.76	22.48	1.55	50.02	m	**72.50**
315 mm	25.69	32.51	1.85	59.70	m	**92.21**
355 mm	27.77	35.15	2.06	66.48	m	**101.63**
400 mm	30.12	38.12	2.62	84.56	m	**122.68**
Polypropylene (PPS) DW154 circular section ductwork; filler rod welded joints; excludes couplers and supports (these are detailed separately); ductwork to conform to current HSE regulations						
Straight duct (standard length 5 m)						
110 mm	9.91	12.54	0.65	20.97	m	**33.51**
160 mm	14.88	18.83	0.75	24.21	m	**43.04**
200 mm	19.28	24.40	0.85	27.44	m	**51.84**
225 mm	25.22	31.92	1.07	34.54	m	**66.46**
250 mm	27.98	35.41	1.18	38.08	m	**73.49**
315 mm	53.66	67.92	1.41	45.49	m	**113.41**
355 mm	60.92	77.10	1.53	49.38	m	**126.48**
400 mm	81.99	103.77	1.73	55.84	m	**159.61**
Extra for supports (BZP finish); Horizontal – Maximum 2.4 m centre; Vertical – Maximum 4.0 m centre						
Duct size						
110 mm	39.62	50.14	0.42	13.55	m	**63.69**
160 mm	39.62	50.14	0.42	13.55	m	**63.69**
200 mm	39.62	50.14	0.42	13.55	m	**63.69**
225 mm	39.62	50.14	0.42	13.55	m	**63.69**
250 mm	39.62	50.14	0.42	13.55	m	**63.69**
315 mm	39.62	50.14	0.42	13.55	m	**63.69**
355 mm	39.62	50.14	0.55	17.75	m	**67.89**
400 mm	56.08	70.97	0.55	17.75	m	**88.72**

38 VENTILATION/AIR CONDITIONING SYSTEMS

Item	Net Price £	Material £	Labour hours	Labour £	Unit	Total rate £
DUCTWORK: PLASTIC – cont						
Extra for supports (BZP finish) – cont						
Duct size – cont						
Note: These are maximum figures and may be reduced subject to local conditions (i.e. a high number of changes of direction)						
Extra over fittings; Polypropylene (DW 154)						
90° bend						
110 mm	17.52	22.18	1.01	32.59	m	**54.77**
160 mm	24.69	31.25	1.30	41.96	m	**73.21**
200 mm	28.96	36.65	1.58	50.99	m	**87.64**
225 mm	39.15	49.55	2.10	67.77	m	**117.32**
250 mm	42.57	53.87	2.36	76.16	m	**130.03**
315 mm	104.43	132.17	3.14	101.34	m	**233.51**
355 mm	126.03	159.50	3.52	113.60	m	**273.10**
400 mm	134.93	170.77	4.39	141.67	m	**312.44**
45° bend						
110 mm	12.90	16.33	0.84	27.10	m	**43.43**
160 mm	20.80	26.32	1.13	36.48	m	**62.80**
200 mm	23.84	30.17	1.39	44.87	m	**75.04**
225 mm	29.38	37.18	1.73	55.84	m	**93.02**
250 mm	33.52	42.43	1.99	64.22	m	**106.65**
315 mm	84.97	107.54	2.57	82.94	m	**190.48**
355 mm	94.73	119.88	2.93	94.56	m	**214.44**
400 mm	100.15	126.75	3.60	116.18	m	**242.93**
Tee						
110 mm	65.50	82.90	1.55	50.02	m	**132.92**
160 mm	91.25	115.48	2.17	70.03	m	**185.51**
200 mm	115.38	146.03	2.68	86.49	m	**232.52**
225 mm	130.56	165.23	3.38	109.08	m	**274.31**
250 mm	171.00	216.42	3.95	127.47	m	**343.89**
315 mm	225.27	285.11	4.69	151.36	m	**436.47**
355 mm	274.80	347.78	5.44	175.56	m	**523.34**
400 mm	310.20	392.59	6.09	196.54	m	**589.13**
Coupler						
110 mm	13.71	17.35	0.84	27.10	m	**44.45**
160 mm	16.67	21.10	1.10	35.50	m	**56.60**
200 mm	20.07	25.40	1.36	43.89	m	**69.29**
225 mm	21.58	27.32	1.67	53.91	m	**81.23**
250 mm	23.03	29.14	1.91	61.63	m	**90.77**
315 mm	30.92	39.13	2.24	72.30	m	**111.43**
355 mm	39.68	50.22	2.58	83.26	m	**133.48**
400 mm	45.22	57.23	3.30	106.50	m	**163.73**
Damper						
110 mm	67.42	85.32	0.79	25.49	m	**110.81**
160 mm	74.42	94.18	1.13	36.48	m	**130.66**
200 mm	81.72	103.42	1.44	46.47	m	**149.89**

38 VENTILATION/AIR CONDITIONING SYSTEMS

Item	Net Price £	Material £	Labour hours	Labour £	Unit	Total rate £
225 mm	87.44	110.67	1.81	58.42	m	**169.09**
250 mm	91.32	115.57	2.08	67.13	m	**182.70**
315 mm	103.72	131.26	2.40	77.46	m	**208.72**
355 mm	114.20	144.54	2.75	88.76	m	**233.30**
400 mm	123.36	156.13	3.57	115.20	m	**271.33**
Reducer						
160 × 110 mm	12.44	15.75	0.87	28.07	m	**43.82**
200 × 160 mm	11.97	15.15	1.16	37.44	m	**52.59**
225 × 200 mm	15.61	19.76	1.46	47.12	m	**66.88**
250 × 200 mm	17.52	22.18	1.61	51.96	m	**74.14**
250 × 225 mm	28.18	35.66	1.70	54.87	m	**90.53**
315 × 200 mm	38.83	49.15	1.66	53.57	m	**102.72**
315 × 250 mm	24.48	30.98	2.03	65.52	m	**96.50**
355 × 250 mm	32.01	40.51	2.05	66.16	m	**106.67**
355 × 315 mm	50.31	63.67	2.22	71.65	m	**135.32**
400 × 315 mm	49.14	62.19	2.88	92.95	m	**155.14**
400 × 355 mm	46.06	58.30	2.92	94.24	m	**152.54**
Flange						
110 mm	13.27	16.80	0.76	24.53	m	**41.33**
160 mm	17.33	21.93	1.02	32.92	m	**54.85**
200 mm	19.77	25.02	1.27	40.98	m	**66.00**
225 mm	20.91	26.47	1.55	50.02	m	**76.49**
250 mm	23.50	29.75	1.79	57.77	m	**87.52**
315 mm	28.08	35.54	2.08	67.13	m	**102.67**
355 mm	32.09	40.61	2.36	76.16	m	**116.77**
400 mm	35.40	44.80	3.07	99.08	m	**143.88**

38 VENTILATION/AIR CONDITIONING SYSTEMS

Item	Net Price £	Material £	Labour hours	Labour £	Unit	Total rate £
DUCTWORK: RECTANGULAR – CLASS B						
AIR DUCTLINES						
Galvanized sheet metal DW144 class B rectangular section ductwork; including all necessary stiffeners, joints, couplers in the running length and duct supports						
Ductwork up to 400 mm longest side						
Sum of two sides 200 mm	25.37	32.11	1.16	37.44	m	**69.55**
Sum of two sides 300 mm	26.80	33.91	1.16	37.44	m	**71.35**
Sum of two sides 400 mm	23.12	29.27	1.19	38.39	m	**67.66**
Sum of two sides 500 mm	25.11	31.77	1.19	38.39	m	**70.16**
Sum of two sides 600 mm	26.84	33.97	1.27	40.98	m	**74.95**
Sum of two sides 700 mm	28.56	36.14	1.27	40.98	m	**77.12**
Sum of two sides 800 mm	30.42	38.49	1.27	40.98	m	**79.47**
Extra over fittings; Rectangular ductwork class B; up to 400 mm longest side						
End cap						
Sum of two sides 200 mm	15.46	19.57	0.38	12.26	nr	**31.83**
Sum of two sides 300 mm	17.23	21.81	0.38	12.26	nr	**34.07**
Sum of two sides 400 mm	18.99	24.04	0.38	12.26	nr	**36.30**
Sum of two sides 500 mm	20.76	26.28	0.38	12.26	nr	**38.54**
Sum of two sides 600 mm	22.53	28.52	0.38	12.26	nr	**40.78**
Sum of two sides 700 mm	24.29	30.74	0.38	12.26	nr	**43.00**
Sum of two sides 800 mm	26.07	33.00	0.38	12.26	nr	**45.26**
Reducer						
Sum of two sides 200 mm	25.25	31.95	1.40	45.18	nr	**77.13**
Sum of two sides 300 mm	28.51	36.09	1.40	45.18	nr	**81.27**
Sum of two sides 400 mm	46.69	59.09	1.42	45.83	nr	**104.92**
Sum of two sides 500 mm	50.54	63.96	1.42	45.83	nr	**109.79**
Sum of two sides 600 mm	54.39	68.84	1.69	54.53	nr	**123.37**
Sum of two sides 700 mm	58.24	73.71	1.69	54.53	nr	**128.24**
Sum of two sides 800 mm	62.05	78.53	1.92	61.96	nr	**140.49**
Offset						
Sum of two sides 200 mm	37.27	47.17	1.63	52.60	nr	**99.77**
Sum of two sides 300 mm	42.12	53.31	1.63	52.60	nr	**105.91**
Sum of two sides 400 mm	62.30	78.85	1.65	53.26	nr	**132.11**
Sum of two sides 500 mm	67.73	85.71	1.65	53.26	nr	**138.97**
Sum of two sides 600 mm	72.17	91.34	1.92	61.96	nr	**153.30**
Sum of two sides 700 mm	77.12	97.61	1.92	61.96	nr	**159.57**
Sum of two sides 800 mm	81.42	103.04	1.92	61.96	nr	**165.00**
Square to round						
Sum of two sides 200 mm	32.76	41.46	1.63	52.60	nr	**94.06**
Sum of two sides 300 mm	36.82	46.60	1.63	52.60	nr	**99.20**
Sum of two sides 400 mm	49.03	62.05	1.65	53.26	nr	**115.31**
Sum of two sides 500 mm	53.35	67.52	1.65	53.26	nr	**120.78**
Sum of two sides 600 mm	57.68	73.00	1.92	61.96	nr	**134.96**
Sum of two sides 700 mm	62.01	78.48	1.92	61.96	nr	**140.44**
Sum of two sides 800 mm	66.29	83.90	1.92	61.96	nr	**145.86**

38 VENTILATION/AIR CONDITIONING SYSTEMS

Item	Net Price £	Material £	Labour hours	Labour £	Unit	Total rate £
90° radius bend						
Sum of two sides 200 mm	24.66	31.21	1.22	39.37	nr	**70.58**
Sum of two sides 300 mm	26.60	33.67	1.22	39.37	nr	**73.04**
Sum of two sides 400 mm	44.83	56.74	1.25	40.34	nr	**97.08**
Sum of two sides 500 mm	47.70	60.37	1.25	40.34	nr	**100.71**
Sum of two sides 600 mm	51.50	65.18	1.33	42.92	nr	**108.10**
Sum of two sides 700 mm	54.82	69.38	1.33	42.92	nr	**112.30**
Sum of two sides 800 mm	58.56	74.11	1.40	45.18	nr	**119.29**
45° radius bend						
Sum of two sides 200 mm	26.63	33.70	0.89	28.73	nr	**62.43**
Sum of two sides 300 mm	29.23	36.99	1.12	36.14	nr	**73.13**
Sum of two sides 400 mm	46.98	59.46	1.10	35.50	nr	**94.96**
Sum of two sides 500 mm	50.35	63.73	1.10	35.50	nr	**99.23**
Sum of two sides 600 mm	54.19	68.58	1.16	37.44	nr	**106.02**
Sum of two sides 700 mm	57.78	73.12	1.16	37.44	nr	**110.56**
Sum of two sides 800 mm	61.57	77.92	1.22	39.37	nr	**117.29**
90° mitre bend						
Sum of two sides 200 mm	41.21	52.16	1.29	41.64	nr	**93.80**
Sum of two sides 300 mm	45.28	57.31	1.29	41.64	nr	**98.95**
Sum of two sides 400 mm	65.84	83.33	1.29	41.64	nr	**124.97**
Sum of two sides 500 mm	71.10	89.98	1.29	41.64	nr	**131.62**
Sum of two sides 600 mm	77.86	98.54	1.39	44.87	nr	**143.41**
Sum of two sides 700 mm	84.13	106.48	1.39	44.87	nr	**151.35**
Sum of two sides 800 mm	90.98	115.15	1.46	47.12	nr	**162.27**
Branch						
Sum of two sides 200 mm	40.82	51.67	0.92	29.69	nr	**81.36**
Sum of two sides 300 mm	45.36	57.41	0.92	29.69	nr	**87.10**
Sum of two sides 400 mm	56.90	72.02	0.95	30.67	nr	**102.69**
Sum of two sides 500 mm	61.98	78.44	0.95	30.67	nr	**109.11**
Sum of two sides 600 mm	66.91	84.68	1.03	33.25	nr	**117.93**
Sum of two sides 700 mm	71.85	90.93	1.03	33.25	nr	**124.18**
Sum of two sides 800 mm	76.79	97.18	1.03	33.25	nr	**130.43**
Grille neck						
Sum of two sides 200 mm	46.81	59.25	1.10	35.50	nr	**94.75**
Sum of two sides 300 mm	52.42	66.34	1.10	35.50	nr	**101.84**
Sum of two sides 400 mm	58.02	73.43	1.16	37.44	nr	**110.87**
Sum of two sides 500 mm	63.62	80.52	1.16	37.44	nr	**117.96**
Sum of two sides 600 mm	69.24	87.63	1.18	38.08	nr	**125.71**
Sum of two sides 700 mm	74.83	94.71	1.18	38.08	nr	**132.79**
Sum of two sides 800 mm	80.43	101.80	1.18	38.08	nr	**139.88**
Ductwork 401 to 600 mm longest side						
Sum of two sides 600 mm	31.30	39.61	1.27	40.98	m	**80.59**
Sum of two sides 700 mm	33.80	42.77	1.27	40.98	m	**83.75**
Sum of two sides 800 mm	36.26	45.89	1.27	40.98	m	**86.87**
Sum of two sides 900 mm	38.57	48.81	1.27	40.98	m	**89.79**
Sum of two sides 1000 mm	40.91	51.78	1.37	44.22	m	**96.00**
Sum of two sides 1100 mm	43.53	55.09	1.37	44.22	m	**99.31**
Sum of two sides 1200 mm	45.85	58.03	1.37	44.22	m	**102.25**

38 VENTILATION/AIR CONDITIONING SYSTEMS

Item	Net Price £	Material £	Labour hours	Labour £	Unit	Total rate £
DUCTWORK: RECTANGULAR – CLASS B – cont						
Extra over fittings; Ductwork 401 to 600 mm longest side						
End cap						
Sum of two sides 600 mm	23.16	29.31	0.38	12.26	nr	**41.57**
Sum of two sides 700 mm	25.01	31.65	0.38	12.26	nr	**43.91**
Sum of two sides 800 mm	26.84	33.97	0.38	12.26	nr	**46.23**
Sum of two sides 900 mm	28.69	36.31	0.58	18.72	nr	**55.03**
Sum of two sides 1000 mm	30.53	38.64	0.58	18.72	nr	**57.36**
Sum of two sides 1100 mm	32.37	40.97	0.58	18.72	nr	**59.69**
Sum of two sides 1200 mm	34.21	43.30	0.58	18.72	nr	**62.02**
Reducer						
Sum of two sides 600 mm	53.51	67.73	1.69	54.53	nr	**122.26**
Sum of two sides 700 mm	57.37	72.61	1.69	54.53	nr	**127.14**
Sum of two sides 800 mm	61.17	77.41	1.92	61.96	nr	**139.37**
Sum of two sides 900 mm	65.03	82.30	1.92	61.96	nr	**144.26**
Sum of two sides 1000 mm	68.90	87.20	2.18	70.36	nr	**157.56**
Sum of two sides 1100 mm	73.08	92.49	2.18	70.36	nr	**162.85**
Sum of two sides 1200 mm	76.94	97.37	2.18	70.36	nr	**167.73**
Offset						
Sum of two sides 600 mm	74.41	94.17	1.92	61.96	nr	**156.13**
Sum of two sides 700 mm	79.89	101.11	1.92	61.96	nr	**163.07**
Sum of two sides 800 mm	84.23	106.60	1.92	61.96	nr	**168.56**
Sum of two sides 900 mm	88.63	112.17	1.92	61.96	nr	**174.13**
Sum of two sides 1000 mm	93.53	118.37	2.18	70.36	nr	**188.73**
Sum of two sides 1100 mm	98.09	124.14	2.18	70.36	nr	**194.50**
Sum of two sides 1200 mm	102.37	129.56	2.18	70.36	nr	**199.92**
Square to round						
Sum of two sides 600 mm	56.62	71.66	1.33	42.92	nr	**114.58**
Sum of two sides 700 mm	60.97	77.17	1.33	42.92	nr	**120.09**
Sum of two sides 800 mm	65.25	82.58	1.40	45.18	nr	**127.76**
Sum of two sides 900 mm	69.62	88.11	1.40	45.18	nr	**133.29**
Sum of two sides 1000 mm	73.96	93.60	1.82	58.73	nr	**152.33**
Sum of two sides 1100 mm	78.39	99.21	1.82	58.73	nr	**157.94**
Sum of two sides 1200 mm	82.75	104.73	1.82	58.73	nr	**163.46**
90° radius bend						
Sum of two sides 600 mm	52.13	65.98	1.16	37.44	nr	**103.42**
Sum of two sides 700 mm	55.01	69.62	1.16	37.44	nr	**107.06**
Sum of two sides 800 mm	59.29	75.04	1.22	39.37	nr	**114.41**
Sum of two sides 900 mm	63.66	80.57	1.22	39.37	nr	**119.94**
Sum of two sides 1000 mm	67.31	85.19	1.40	45.18	nr	**130.37**
Sum of two sides 1100 mm	71.92	91.02	1.40	45.18	nr	**136.20**
Sum of two sides 1200 mm	76.32	96.59	1.40	45.18	nr	**141.77**
45° bend						
Sum of two sides 600 mm	55.20	69.87	1.16	37.44	nr	**107.31**
Sum of two sides 700 mm	58.61	74.18	1.39	44.87	nr	**119.05**
Sum of two sides 800 mm	62.74	79.41	1.46	47.12	nr	**126.53**

38 VENTILATION/AIR CONDITIONING SYSTEMS

Item	Net Price £	Material £	Labour hours	Labour £	Unit	Total rate £
Sum of two sides 900 mm	66.95	84.73	1.46	47.12	nr	**131.85**
Sum of two sides 1000 mm	70.77	89.57	1.88	60.67	nr	**150.24**
Sum of two sides 1100 mm	75.29	95.29	1.88	60.67	nr	**155.96**
Sum of two sides 1200 mm	79.51	100.63	1.88	60.67	nr	**161.30**
90° mitre bend						
Sum of two sides 600 mm	87.67	110.96	1.39	44.87	nr	**155.83**
Sum of two sides 700 mm	92.89	117.57	2.16	69.71	nr	**187.28**
Sum of two sides 800 mm	100.40	127.06	2.26	72.93	nr	**199.99**
Sum of two sides 900 mm	108.03	136.72	2.26	72.93	nr	**209.65**
Sum of two sides 1000 mm	114.89	145.41	3.01	97.14	nr	**242.55**
Sum of two sides 1100 mm	122.88	155.51	3.01	97.14	nr	**252.65**
Sum of two sides 1200 mm	130.70	165.41	3.01	97.14	nr	**262.55**
Branch						
Sum of two sides 600 mm	69.10	87.45	1.03	33.25	nr	**120.70**
Sum of two sides 700 mm	74.26	93.98	1.03	33.25	nr	**127.23**
Sum of two sides 800 mm	79.43	100.53	1.03	33.25	nr	**133.78**
Sum of two sides 900 mm	84.61	107.08	1.03	33.25	nr	**140.33**
Sum of two sides 1000 mm	89.79	113.64	1.29	41.64	nr	**155.28**
Sum of two sides 1100 mm	95.18	120.46	1.29	41.64	nr	**162.10**
Sum of two sides 1200 mm	100.35	127.01	1.29	41.64	nr	**168.65**
Grille neck						
Sum of two sides 600 mm	71.74	90.80	1.18	38.08	nr	**128.88**
Sum of two sides 700 mm	77.67	98.30	1.18	38.08	nr	**136.38**
Sum of two sides 800 mm	83.61	105.82	1.18	38.08	nr	**143.90**
Sum of two sides 900 mm	89.54	113.32	1.18	38.08	nr	**151.40**
Sum of two sides 1000 mm	95.48	120.84	1.44	46.47	nr	**167.31**
Sum of two sides 1100 mm	101.41	128.34	1.44	46.47	nr	**174.81**
Sum of two sides 1200 mm	107.35	135.87	1.44	46.47	nr	**182.34**
Ductwork 601 to 800 mm longest side						
Sum of two sides 900 mm	43.06	54.50	1.27	40.98	m	**95.48**
Sum of two sides 1000 mm	45.38	57.43	1.37	44.22	m	**101.65**
Sum of two sides 1100 mm	47.70	60.37	1.37	44.22	m	**104.59**
Sum of two sides 1200 mm	50.34	63.71	1.37	44.22	m	**107.93**
Sum of two sides 1300 mm	52.65	66.63	1.40	45.18	m	**111.81**
Sum of two sides 1400 mm	54.98	69.59	1.40	45.18	m	**114.77**
Sum of two sides 1500 mm	57.31	72.53	1.48	47.77	m	**120.30**
Sum of two sides 1600 mm	59.64	75.48	1.55	50.02	m	**125.50**
Extra over fittings: Ductwork 601 to 800 mm longest side						
End cap						
Sum of two sides 900 mm	28.69	36.31	0.58	18.72	nr	**55.03**
Sum of two sides 1000 mm	30.53	38.64	0.58	18.72	nr	**57.36**
Sum of two sides 1100 mm	32.37	40.97	0.58	18.72	nr	**59.69**
Sum of two sides 1200 mm	34.21	43.30	0.58	18.72	nr	**62.02**
Sum of two sides 1300 mm	36.06	45.64	0.58	18.72	nr	**64.36**
Sum of two sides 1400 mm	37.64	47.63	0.58	18.72	nr	**66.35**
Sum of two sides 1500 mm	43.72	55.33	0.58	18.72	nr	**74.05**
Sum of two sides 1600 mm	49.82	63.06	0.58	18.72	nr	**81.78**

38 VENTILATION/AIR CONDITIONING SYSTEMS

Item	Net Price £	Material £	Labour hours	Labour £	Unit	Total rate £
DUCTWORK: RECTANGULAR – CLASS B – cont						
Extra over fittings – cont						
Reducer						
Sum of two sides 900 mm	65.65	83.08	1.92	61.96	nr	**145.04**
Sum of two sides 1000 mm	69.51	87.98	2.18	70.36	nr	**158.34**
Sum of two sides 1100 mm	73.39	92.88	2.18	70.36	nr	**163.24**
Sum of two sides 1200 mm	77.55	98.15	2.18	70.36	nr	**168.51**
Sum of two sides 1300 mm	81.41	103.03	2.30	74.22	nr	**177.25**
Sum of two sides 1400 mm	84.75	107.26	2.30	74.22	nr	**181.48**
Sum of two sides 1500 mm	97.13	122.93	2.47	79.71	nr	**202.64**
Sum of two sides 1600 mm	109.49	138.57	2.47	79.71	nr	**218.28**
Offset						
Sum of two sides 900 mm	91.12	115.33	1.92	61.96	nr	**177.29**
Sum of two sides 1000 mm	95.20	120.49	2.18	70.36	nr	**190.85**
Sum of two sides 1100 mm	99.27	125.64	2.18	70.36	nr	**196.00**
Sum of two sides 1200 mm	103.50	131.00	2.18	70.36	nr	**201.36**
Sum of two sides 1300 mm	107.48	136.02	2.30	74.22	nr	**210.24**
Sum of two sides 1400 mm	111.39	140.97	2.30	74.22	nr	**215.19**
Sum of two sides 1500 mm	128.08	162.10	2.47	79.71	nr	**241.81**
Sum of two sides 1600 mm	144.72	183.15	2.47	79.71	nr	**262.86**
Square to round						
Sum of two sides 900 mm	69.04	87.38	1.40	45.18	nr	**132.56**
Sum of two sides 1000 mm	73.40	92.89	1.82	58.73	nr	**151.62**
Sum of two sides 1100 mm	77.75	98.40	1.82	58.73	nr	**157.13**
Sum of two sides 1200 mm	82.18	104.00	1.82	58.73	nr	**162.73**
Sum of two sides 1300 mm	86.54	109.52	2.15	69.40	nr	**178.92**
Sum of two sides 1400 mm	90.24	114.21	2.15	69.40	nr	**183.61**
Sum of two sides 1500 mm	105.21	133.16	2.38	76.81	nr	**209.97**
Sum of two sides 1600 mm	120.19	152.11	2.38	76.81	nr	**228.92**
90° radius bend						
Sum of two sides 900 mm	59.62	75.45	1.22	39.37	nr	**114.82**
Sum of two sides 1000 mm	64.08	81.10	1.40	45.18	nr	**126.28**
Sum of two sides 1100 mm	68.52	86.72	1.40	45.18	nr	**131.90**
Sum of two sides 1200 mm	73.15	92.58	1.40	45.18	nr	**137.76**
Sum of two sides 1300 mm	77.61	98.22	1.91	61.63	nr	**159.85**
Sum of two sides 1400 mm	80.62	102.03	1.91	61.63	nr	**163.66**
Sum of two sides 1500 mm	93.60	118.46	2.11	68.08	nr	**186.54**
Sum of two sides 1600 mm	106.57	134.87	2.11	68.08	nr	**202.95**
45° bend						
Sum of two sides 900 mm	65.81	83.29	1.22	39.37	nr	**122.66**
Sum of two sides 1000 mm	70.05	88.66	1.40	45.18	nr	**133.84**
Sum of two sides 1100 mm	74.28	94.01	1.88	60.67	nr	**154.68**
Sum of two sides 1200 mm	78.84	99.78	1.88	60.67	nr	**160.45**
Sum of two sides 1300 mm	20.03	25.35	2.26	72.93	nr	**98.28**
Sum of two sides 1400 mm	86.31	109.23	2.26	72.93	nr	**182.16**
Sum of two sides 1500 mm	99.07	125.38	2.49	80.35	nr	**205.73**
Sum of two sides 1600 mm	105.51	133.54	2.49	80.35	nr	**213.89**

38 VENTILATION/AIR CONDITIONING SYSTEMS

Item	Net Price £	Material £	Labour hours	Labour £	Unit	Total rate £
90° mitre bend						
Sum of two sides 900 mm	101.73	128.74	1.22	39.37	nr	**168.11**
Sum of two sides 1000 mm	110.09	139.33	1.40	45.18	nr	**184.51**
Sum of two sides 1100 mm	118.44	149.90	3.01	97.14	nr	**247.04**
Sum of two sides 1200 mm	126.95	160.66	3.01	97.14	nr	**257.80**
Sum of two sides 1300 mm	135.31	171.25	3.67	118.44	nr	**289.69**
Sum of two sides 1400 mm	141.86	179.54	3.67	118.44	nr	**297.98**
Sum of two sides 1500 mm	163.14	206.47	4.07	131.34	nr	**337.81**
Sum of two sides 1600 mm	184.43	233.42	4.07	131.34	nr	**364.76**
Branch						
Sum of two sides 900 mm	87.11	110.24	1.22	39.37	nr	**149.61**
Sum of two sides 1000 mm	92.42	116.96	1.40	45.18	nr	**162.14**
Sum of two sides 1100 mm	97.73	123.68	1.29	41.64	nr	**165.32**
Sum of two sides 1200 mm	103.28	130.72	1.29	41.64	nr	**172.36**
Sum of two sides 1300 mm	108.60	137.45	1.39	44.87	nr	**182.32**
Sum of two sides 1400 mm	113.22	143.29	1.39	44.87	nr	**188.16**
Sum of two sides 1500 mm	129.85	164.34	1.64	52.93	nr	**217.27**
Sum of two sides 1600 mm	146.50	185.42	1.64	52.93	nr	**238.35**
Grille neck						
Sum of two sides 900 mm	89.54	113.32	1.22	39.37	nr	**152.69**
Sum of two sides 1000 mm	95.48	120.84	1.40	45.18	nr	**166.02**
Sum of two sides 1100 mm	101.41	128.34	1.44	46.47	nr	**174.81**
Sum of two sides 1200 mm	107.35	135.87	1.44	46.47	nr	**182.34**
Sum of two sides 1300 mm	113.29	143.38	1.69	54.53	nr	**197.91**
Sum of two sides 1400 mm	119.22	150.89	1.69	54.53	nr	**205.42**
Sum of two sides 1500 mm	125.16	158.40	1.79	57.77	nr	**216.17**
Sum of two sides 1600 mm	131.10	165.92	1.79	57.77	nr	**223.69**
Ductwork 801 to 1000 mm longest side						
Sum of two sides 1100 mm	65.30	82.64	1.37	44.22	m	**126.86**
Sum of two sides 1200 mm	69.15	87.52	1.37	44.22	m	**131.74**
Sum of two sides 1300 mm	73.01	92.40	1.40	45.18	m	**137.58**
Sum of two sides 1400 mm	77.17	97.66	1.40	45.18	m	**142.84**
Sum of two sides 1500 mm	81.01	102.52	1.48	47.77	m	**150.29**
Sum of two sides 1600 mm	84.87	107.41	1.55	50.02	m	**157.43**
Sum of two sides 1700 mm	88.71	112.27	1.55	50.02	m	**162.29**
Sum of two sides 1800 mm	92.88	117.54	1.61	51.96	m	**169.50**
Sum of two sides 1900 mm	96.71	122.39	1.61	51.96	m	**174.35**
Sum of two sides 2000 mm	100.56	127.27	1.61	51.96	m	**179.23**
Extra over fittings; Ductwork 801 to 1000 mm longest side						
End cap						
Sum of two sides 1100 mm	32.37	40.97	1.44	46.47	nr	**87.44**
Sum of two sides 1200 mm	34.21	43.30	1.44	46.47	nr	**89.77**
Sum of two sides 1300 mm	36.06	45.64	1.44	46.47	nr	**92.11**
Sum of two sides 1400 mm	37.64	47.63	1.44	46.47	nr	**94.10**
Sum of two sides 1500 mm	43.72	55.33	1.44	46.47	nr	**101.80**
Sum of two sides 1600 mm	49.82	63.06	1.44	46.47	nr	**109.53**
Sum of two sides 1700 mm	55.91	70.76	1.44	46.47	nr	**117.23**

38 VENTILATION/AIR CONDITIONING SYSTEMS

Item	Net Price £	Material £	Labour hours	Labour £	Unit	Total rate £
DUCTWORK: RECTANGULAR – CLASS B – cont						
Extra over fittings – cont						
End cap – cont						
Sum of two sides 1800 mm	62.01	78.48	1.44	46.47	nr	**124.95**
Sum of two sides 1900 mm	68.10	86.18	1.44	46.47	nr	**132.65**
Sum of two sides 2000 mm	74.19	93.89	1.44	46.47	nr	**140.36**
Reducer						
Sum of two sides 1100 mm	61.04	77.26	1.44	46.47	nr	**123.73**
Sum of two sides 1200 mm	63.97	80.96	1.44	46.47	nr	**127.43**
Sum of two sides 1300 mm	66.91	84.68	1.69	54.53	nr	**139.21**
Sum of two sides 1400 mm	69.59	88.08	1.69	54.53	nr	**142.61**
Sum of two sides 1500 mm	81.02	102.54	2.47	79.71	nr	**182.25**
Sum of two sides 1600 mm	92.46	117.02	2.47	79.71	nr	**196.73**
Sum of two sides 1700 mm	103.90	131.50	2.47	79.71	nr	**211.21**
Sum of two sides 1800 mm	115.60	146.31	2.59	83.60	nr	**229.91**
Sum of two sides 1900 mm	127.03	160.76	2.71	87.46	nr	**248.22**
Sum of two sides 2000 mm	138.48	175.26	2.71	87.46	nr	**262.72**
Offset						
Sum of two sides 1100 mm	93.73	118.62	1.44	46.47	nr	**165.09**
Sum of two sides 1200 mm	95.57	120.95	1.44	46.47	nr	**167.42**
Sum of two sides 1300 mm	97.14	122.94	1.69	54.53	nr	**177.47**
Sum of two sides 1400 mm	97.86	123.85	1.69	54.53	nr	**178.38**
Sum of two sides 1500 mm	111.64	141.29	2.47	79.71	nr	**221.00**
Sum of two sides 1600 mm	125.12	158.36	2.47	79.71	nr	**238.07**
Sum of two sides 1700 mm	138.37	175.12	2.59	83.60	nr	**258.72**
Sum of two sides 1800 mm	151.45	191.68	2.61	84.24	nr	**275.92**
Sum of two sides 1900 mm	166.06	210.17	2.71	87.46	nr	**297.63**
Sum of two sides 2000 mm	178.54	225.96	2.71	87.46	nr	**313.42**
Square to round						
Sum of two sides 1100 mm	65.18	82.49	1.44	46.47	nr	**128.96**
Sum of two sides 1200 mm	68.60	86.82	1.44	46.47	nr	**133.29**
Sum of two sides 1300 mm	72.03	91.16	1.69	54.53	nr	**145.69**
Sum of two sides 1400 mm	74.83	94.71	1.69	54.53	nr	**149.24**
Sum of two sides 1500 mm	88.89	112.50	2.38	76.81	nr	**189.31**
Sum of two sides 1600 mm	102.94	130.28	2.38	76.81	nr	**207.09**
Sum of two sides 1700 mm	117.00	148.08	2.55	82.30	nr	**230.38**
Sum of two sides 1800 mm	131.07	165.88	2.55	82.30	nr	**248.18**
Sum of two sides 1900 mm	145.12	183.67	2.83	91.34	nr	**275.01**
Sum of two sides 2000 mm	159.17	201.44	2.83	91.34	nr	**292.78**
90° radius bend						
Sum of two sides 1100 mm	49.44	62.57	1.44	46.47	nr	**109.04**
Sum of two sides 1200 mm	53.36	67.54	1.44	46.47	nr	**114.01**
Sum of two sides 1300 mm	57.28	72.50	1.69	54.53	nr	**127.03**
Sum of two sides 1400 mm	60.81	76.97	1.69	54.53	nr	**131.50**
Sum of two sides 1500 mm	73.22	92.67	2.11	68.08	nr	**160.75**
Sum of two sides 1600 mm	85.64	108.38	2.11	68.08	nr	**176.46**
Sum of two sides 1700 mm	98.06	124.11	2.26	72.93	nr	**197.04**
Sum of two sides 1800 mm	110.61	139.99	2.26	72.93	nr	**212.92**
Sum of two sides 1900 mm	120.81	152.90	2.48	80.04	nr	**232.94**
Sum of two sides 2000 mm	133.23	168.62	2.48	80.04	nr	**248.66**

38 VENTILATION/AIR CONDITIONING SYSTEMS

Item	Net Price £	Material £	Labour hours	Labour £	Unit	Total rate £
45° bend						
Sum of two sides 1100 mm	64.12	81.16	1.44	46.47	nr	**127.63**
Sum of two sides 1200 mm	68.07	86.15	1.44	46.47	nr	**132.62**
Sum of two sides 1300 mm	72.01	91.13	1.69	54.53	nr	**145.66**
Sum of two sides 1400 mm	75.72	95.83	1.69	54.53	nr	**150.36**
Sum of two sides 1500 mm	88.16	111.57	2.49	80.35	nr	**191.92**
Sum of two sides 1600 mm	100.62	127.34	2.49	80.35	nr	**207.69**
Sum of two sides 1700 mm	113.06	143.09	2.67	86.16	nr	**229.25**
Sum of two sides 1800 mm	125.78	159.19	2.67	86.16	nr	**245.35**
Sum of two sides 1900 mm	137.07	173.48	3.06	98.75	nr	**272.23**
Sum of two sides 2000 mm	149.53	189.25	3.06	98.75	nr	**288.00**
90° mitre bend						
Sum of two sides 1100 mm	102.31	129.48	1.44	46.47	nr	**175.95**
Sum of two sides 1200 mm	109.74	138.89	1.44	46.47	nr	**185.36**
Sum of two sides 1300 mm	117.16	148.28	1.69	54.53	nr	**202.81**
Sum of two sides 1400 mm	123.89	156.80	1.69	54.53	nr	**211.33**
Sum of two sides 1500 mm	144.06	182.32	2.07	66.80	nr	**249.12**
Sum of two sides 1600 mm	164.23	207.85	2.07	66.80	nr	**274.65**
Sum of two sides 1700 mm	184.41	233.39	2.80	90.36	nr	**323.75**
Sum of two sides 1800 mm	204.67	259.03	2.67	86.16	nr	**345.19**
Sum of two sides 1900 mm	222.49	281.58	2.95	95.20	nr	**376.78**
Sum of two sides 2000 mm	242.77	307.25	2.95	95.20	nr	**402.45**
Branch						
Sum of two sides 1100 mm	97.95	123.96	1.44	46.47	nr	**170.43**
Sum of two sides 1200 mm	103.28	130.72	1.44	46.47	nr	**177.19**
Sum of two sides 1300 mm	108.60	137.45	1.64	52.93	nr	**190.38**
Sum of two sides 1400 mm	113.44	143.57	1.64	52.93	nr	**196.50**
Sum of two sides 1500 mm	130.09	164.64	1.64	52.93	nr	**217.57**
Sum of two sides 1600 mm	146.10	184.90	1.64	52.93	nr	**237.83**
Sum of two sides 1700 mm	163.37	206.76	1.69	54.53	nr	**261.29**
Sum of two sides 1800 mm	180.24	228.11	1.69	54.53	nr	**282.64**
Sum of two sides 1900 mm	196.89	249.19	1.85	59.70	nr	**308.89**
Sum of two sides 2000 mm	213.53	270.24	1.85	59.70	nr	**329.94**
Grille neck						
Sum of two sides 1100 mm	101.41	128.34	1.44	46.47	nr	**174.81**
Sum of two sides 1200 mm	107.35	135.87	1.44	46.47	nr	**182.34**
Sum of two sides 1300 mm	113.29	143.38	1.69	54.53	nr	**197.91**
Sum of two sides 1400 mm	118.43	149.89	1.69	54.53	nr	**204.42**
Sum of two sides 1500 mm	137.12	173.54	1.79	57.77	nr	**231.31**
Sum of two sides 1600 mm	155.81	197.20	1.79	57.77	nr	**254.97**
Sum of two sides 1700 mm	174.49	220.83	1.86	60.03	nr	**280.86**
Sum of two sides 1800 mm	193.18	244.48	2.02	65.20	nr	**309.68**
Sum of two sides 1900 mm	211.86	268.13	2.02	65.20	nr	**333.33**
Sum of two sides 2000 mm	230.54	291.77	2.02	65.20	nr	**356.97**
Ductwork 1001 to 1250 mm longest side						
Sum of two sides 1300 mm	88.26	111.70	1.40	45.18	m	**156.88**
Sum of two sides 1400 mm	93.08	117.80	1.40	45.18	m	**162.98**
Sum of two sides 1500 mm	97.56	123.47	1.48	47.77	m	**171.24**

38 VENTILATION/AIR CONDITIONING SYSTEMS

Item	Net Price £	Material £	Labour hours	Labour £	Unit	Total rate £
DUCTWORK: RECTANGULAR – CLASS B – cont						
Ductwork 1001 to 1250 mm longest side – cont						
Sum of two sides 1600 mm	102.36	129.55	1.55	50.02	m	**179.57**
Sum of two sides 1700 mm	106.84	135.22	1.55	50.02	m	**185.24**
Sum of two sides 1800 mm	111.33	140.90	1.61	51.96	m	**192.86**
Sum of two sides 1900 mm	115.82	146.59	1.61	51.96	m	**198.55**
Sum of two sides 2000 mm	120.62	152.66	1.61	51.96	m	**204.62**
Sum of two sides 2100 mm	125.10	158.32	2.17	70.03	m	**228.35**
Sum of two sides 2200 mm	129.60	164.02	2.19	70.67	m	**234.69**
Sum of two sides 2300 mm	134.72	170.50	2.19	70.67	m	**241.17**
Sum of two sides 2400 mm	139.20	176.18	2.38	76.81	m	**252.99**
Sum of two sides 2500 mm	144.00	182.25	2.38	76.81	m	**259.06**
Extra over fittings; Ductwork 1001 to 1250 mm longest side						
End cap						
Sum of two sides 1300 mm	36.79	46.56	1.69	54.53	nr	**101.09**
Sum of two sides 1400 mm	38.43	48.64	1.69	54.53	nr	**103.17**
Sum of two sides 1500 mm	44.58	56.43	1.69	54.53	nr	**110.96**
Sum of two sides 1600 mm	50.74	64.22	1.69	54.53	nr	**118.75**
Sum of two sides 1700 mm	56.88	71.98	1.69	54.53	nr	**126.51**
Sum of two sides 1800 mm	63.03	79.77	1.69	54.53	nr	**134.30**
Sum of two sides 1900 mm	69.19	87.56	1.69	54.53	nr	**142.09**
Sum of two sides 2000 mm	75.32	95.32	1.69	54.53	nr	**149.85**
Sum of two sides 2100 mm	81.47	103.11	1.69	54.53	nr	**157.64**
Sum of two sides 2200 mm	87.63	110.90	1.69	54.53	nr	**165.43**
Sum of two sides 2300 mm	93.78	118.69	1.69	54.53	nr	**173.22**
Sum of two sides 2400 mm	99.92	126.46	1.69	54.53	nr	**180.99**
Sum of two sides 2500 mm	106.08	134.25	1.69	54.53	nr	**188.78**
Reducer						
Sum of two sides 1300 mm	64.86	82.08	1.69	54.53	nr	**136.61**
Sum of two sides 1400 mm	67.27	85.14	1.69	54.53	nr	**139.67**
Sum of two sides 1500 mm	78.68	99.58	2.47	79.71	nr	**179.29**
Sum of two sides 1600 mm	90.35	114.35	2.47	79.71	nr	**194.06**
Sum of two sides 1700 mm	101.76	128.79	2.47	79.71	nr	**208.50**
Sum of two sides 1800 mm	113.17	143.23	2.59	83.60	nr	**226.83**
Sum of two sides 1900 mm	124.58	157.67	2.71	87.46	nr	**245.13**
Sum of two sides 2000 mm	136.26	172.45	2.59	83.60	nr	**256.05**
Sum of two sides 2100 mm	147.67	186.89	2.92	94.24	nr	**281.13**
Sum of two sides 2200 mm	159.09	201.34	2.92	94.24	nr	**295.58**
Sum of two sides 2300 mm	170.56	215.86	2.92	94.24	nr	**310.10**
Sum of two sides 2400 mm	182.18	230.56	3.12	100.69	nr	**331.25**
Sum of two sides 2500 mm	193.85	245.34	3.12	100.69	nr	**346.03**
Offset						
Sum of two sides 1300 mm	126.88	160.57	1.69	54.53	nr	**215.10**
Sum of two sides 1400 mm	131.23	166.08	1.69	54.53	nr	**220.61**
Sum of two sides 1500 mm	146.94	185.96	2.47	79.71	nr	**265.67**

38 VENTILATION/AIR CONDITIONING SYSTEMS

Item	Net Price £	Material £	Labour hours	Labour £	Unit	Total rate £
Sum of two sides 1600 mm	162.56	205.73	2.47	79.71	nr	**285.44**
Sum of two sides 1700 mm	177.70	224.90	2.59	83.60	nr	**308.50**
Sum of two sides 1800 mm	192.58	243.73	2.61	84.24	nr	**327.97**
Sum of two sides 1900 mm	207.15	262.17	2.71	87.46	nr	**349.63**
Sum of two sides 2000 mm	221.60	280.46	2.71	87.46	nr	**367.92**
Sum of two sides 2100 mm	235.63	298.21	2.92	94.24	nr	**392.45**
Sum of two sides 2200 mm	251.72	318.57	3.26	105.21	nr	**423.78**
Sum of two sides 2300 mm	270.17	341.92	3.26	105.21	nr	**447.13**
Sum of two sides 2400 mm	288.55	365.19	3.48	112.31	nr	**477.50**
Sum of two sides 2500 mm	307.00	388.54	3.48	112.31	nr	**500.85**
Square to round						
Sum of two sides 1300 mm	69.39	87.82	1.69	54.53	nr	**142.35**
Sum of two sides 1400 mm	72.14	91.30	1.69	54.53	nr	**145.83**
Sum of two sides 1500 mm	86.17	109.05	2.38	76.81	nr	**185.86**
Sum of two sides 1600 mm	100.24	126.86	2.38	76.81	nr	**203.67**
Sum of two sides 1700 mm	114.27	144.63	2.55	82.30	nr	**226.93**
Sum of two sides 1800 mm	128.28	162.36	2.55	82.30	nr	**244.66**
Sum of two sides 1900 mm	142.30	180.10	2.83	91.34	nr	**271.44**
Sum of two sides 2000 mm	156.37	197.90	2.83	91.34	nr	**289.24**
Sum of two sides 2100 mm	170.39	215.64	3.85	124.24	nr	**339.88**
Sum of two sides 2200 mm	184.42	233.40	4.18	134.90	nr	**368.30**
Sum of two sides 2300 mm	198.48	251.19	4.22	136.19	nr	**387.38**
Sum of two sides 2400 mm	212.51	268.96	4.68	151.03	nr	**419.99**
Sum of two sides 2500 mm	226.56	286.73	4.70	151.68	nr	**438.41**
90° radius bend						
Sum of two sides 1300 mm	46.82	59.26	1.69	54.53	nr	**113.79**
Sum of two sides 1400 mm	47.92	60.65	1.69	54.53	nr	**115.18**
Sum of two sides 1500 mm	60.80	76.94	2.11	68.08	nr	**145.02**
Sum of two sides 1600 mm	73.74	93.33	2.11	68.08	nr	**161.41**
Sum of two sides 1700 mm	86.62	109.63	2.19	70.67	nr	**180.30**
Sum of two sides 1800 mm	99.48	125.90	2.19	70.67	nr	**196.57**
Sum of two sides 1900 mm	112.36	142.21	2.48	80.04	nr	**222.25**
Sum of two sides 2000 mm	125.33	158.61	2.26	72.93	nr	**231.54**
Sum of two sides 2100 mm	138.18	174.88	2.48	80.04	nr	**254.92**
Sum of two sides 2200 mm	151.07	191.20	2.48	80.04	nr	**271.24**
Sum of two sides 2300 mm	158.34	200.39	2.48	80.04	nr	**280.43**
Sum of two sides 2400 mm	171.25	216.73	3.90	125.87	nr	**342.60**
Sum of two sides 2500 mm	184.18	233.09	3.90	125.87	nr	**358.96**
45° bend						
Sum of two sides 1300 mm	67.01	84.81	1.69	54.53	nr	**139.34**
Sum of two sides 1400 mm	69.26	87.65	1.69	54.53	nr	**142.18**
Sum of two sides 1500 mm	81.95	103.71	2.49	80.35	nr	**184.06**
Sum of two sides 1600 mm	94.91	120.12	2.49	80.35	nr	**200.47**
Sum of two sides 1700 mm	107.61	136.19	2.67	86.16	nr	**222.35**
Sum of two sides 1800 mm	120.30	152.25	2.67	86.16	nr	**238.41**
Sum of two sides 1900 mm	133.01	168.34	3.06	98.75	nr	**267.09**
Sum of two sides 2000 mm	145.96	184.72	3.06	98.75	nr	**283.47**
Sum of two sides 2100 mm	158.67	200.82	4.05	130.72	nr	**331.54**
Sum of two sides 2200 mm	171.37	216.89	4.05	130.72	nr	**347.61**
Sum of two sides 2300 mm	181.36	229.53	4.39	141.67	nr	**371.20**
Sum of two sides 2400 mm	194.07	245.62	4.85	156.52	nr	**402.14**
Sum of two sides 2500 mm	207.03	262.01	4.85	156.52	nr	**418.53**

38 VENTILATION/AIR CONDITIONING SYSTEMS

Item	Net Price £	Material £	Labour hours	Labour £	Unit	Total rate £
DUCTWORK: RECTANGULAR – CLASS B – cont						
Extra over fittings – cont						
90° mitre bend						
Sum of two sides 1300 mm	138.70	175.54	1.69	54.53	nr	**230.07**
Sum of two sides 1400 mm	144.74	183.19	1.69	54.53	nr	**237.72**
Sum of two sides 1500 mm	168.12	212.78	2.80	90.36	nr	**303.14**
Sum of two sides 1600 mm	191.53	242.40	2.80	90.36	nr	**332.76**
Sum of two sides 1700 mm	214.90	271.98	2.95	95.20	nr	**367.18**
Sum of two sides 1800 mm	238.30	301.59	2.95	95.20	nr	**396.79**
Sum of two sides 1900 mm	261.68	331.18	4.05	130.72	nr	**461.90**
Sum of two sides 2000 mm	285.09	360.81	4.05	130.72	nr	**491.53**
Sum of two sides 2100 mm	308.47	390.40	4.07	131.34	nr	**521.74**
Sum of two sides 2200 mm	331.86	420.00	4.07	131.34	nr	**551.34**
Sum of two sides 2300 mm	349.08	441.80	4.39	141.67	nr	**583.47**
Sum of two sides 2400 mm	372.66	471.64	4.85	156.52	nr	**628.16**
Sum of two sides 2500 mm	396.24	501.48	4.85	156.52	nr	**658.00**
Branch						
Sum of two sides 1300 mm	111.39	140.97	1.44	46.47	nr	**187.44**
Sum of two sides 1400 mm	116.17	147.02	1.44	46.47	nr	**193.49**
Sum of two sides 1500 mm	132.99	168.31	1.64	52.93	nr	**221.24**
Sum of two sides 1600 mm	150.04	189.90	1.64	52.93	nr	**242.83**
Sum of two sides 1700 mm	166.85	211.16	1.64	52.93	nr	**264.09**
Sum of two sides 1800 mm	183.67	232.46	1.64	52.93	nr	**285.39**
Sum of two sides 1900 mm	200.48	253.72	1.69	54.53	nr	**308.25**
Sum of two sides 2000 mm	217.52	275.30	1.69	54.53	nr	**329.83**
Sum of two sides 2100 mm	234.34	296.58	1.85	59.70	nr	**356.28**
Sum of two sides 2200 mm	251.14	317.84	1.85	59.70	nr	**377.54**
Sum of two sides 2300 mm	268.19	339.42	2.61	84.24	nr	**423.66**
Sum of two sides 2400 mm	285.00	360.70	2.61	84.24	nr	**444.94**
Sum of two sides 2500 mm	302.05	382.28	2.61	84.24	nr	**466.52**
Grille neck						
Sum of two sides 1300 mm	116.97	148.04	1.79	57.77	nr	**205.81**
Sum of two sides 1400 mm	122.39	154.90	1.79	57.77	nr	**212.67**
Sum of two sides 1500 mm	141.34	178.88	1.79	57.77	nr	**236.65**
Sum of two sides 1600 mm	160.33	202.91	1.79	57.77	nr	**260.68**
Sum of two sides 1700 mm	179.29	226.91	1.86	60.03	nr	**286.94**
Sum of two sides 1800 mm	198.26	250.91	2.02	65.20	nr	**316.11**
Sum of two sides 1900 mm	217.25	274.95	2.02	65.20	nr	**340.15**
Sum of two sides 2000 mm	236.20	298.94	2.02	65.20	nr	**364.14**
Sum of two sides 2100 mm	255.17	322.94	2.61	84.24	nr	**407.18**
Sum of two sides 2200 mm	274.13	346.94	2.61	84.24	nr	**431.18**
Sum of two sides 2300 mm	293.11	370.96	2.61	84.24	nr	**455.20**
Sum of two sides 2400 mm	312.07	394.96	2.88	92.95	nr	**487.91**
Sum of two sides 2500 mm	331.04	418.97	2.88	92.95	nr	**511.92**
Ductwork 1251 to 1600 mm longest side						
Sum of two sides 1700 mm	135.16	171.06	1.55	50.02	m	**221.08**
Sum of two sides 1800 mm	140.47	177.78	1.61	51.96	m	**229.74**
Sum of two sides 1900 mm	145.85	184.59	1.61	51.96	m	**236.55**

38 VENTILATION/AIR CONDITIONING SYSTEMS

Item	Net Price £	Material £	Labour hours	Labour £	Unit	Total rate £
Sum of two sides 2000 mm	150.91	190.99	1.61	51.96	m	242.95
Sum of two sides 2100 mm	155.98	197.41	2.17	70.03	m	267.44
Sum of two sides 2200 mm	161.05	203.83	2.19	70.67	m	274.50
Sum of two sides 2300 mm	166.42	210.62	2.19	70.67	m	281.29
Sum of two sides 2400 mm	171.73	217.34	2.38	76.81	m	294.15
Sum of two sides 2500 mm	176.81	223.78	2.38	76.81	m	300.59
Sum of two sides 2600 mm	181.87	230.17	2.64	85.20	m	315.37
Sum of two sides 2700 mm	186.93	236.58	2.66	85.85	m	322.43
Sum of two sides 2800 mm	192.25	243.31	2.95	95.20	m	338.51
Sum of two sides 2900 mm	213.64	270.38	2.96	95.52	m	365.90
Sum of two sides 3000 mm	218.71	276.80	3.15	101.66	m	378.46
Sum of two sides 3100 mm	224.03	283.53	3.15	101.66	m	385.19
Sum of two sides 3200 mm	229.09	289.93	3.18	102.63	m	392.56
Extra over fittings; Ductwork 1251 to 1600 mm longest side						
End cap						
Sum of two sides 1700 mm	56.88	71.98	0.58	18.72	nr	90.70
Sum of two sides 1800 mm	63.03	79.77	0.58	18.72	nr	98.49
Sum of two sides 1900 mm	69.19	87.56	0.58	18.72	nr	106.28
Sum of two sides 2000 mm	75.32	95.32	0.58	18.72	nr	114.04
Sum of two sides 2100 mm	81.47	103.11	0.87	28.07	nr	131.18
Sum of two sides 2200 mm	87.63	110.90	0.87	28.07	nr	138.97
Sum of two sides 2300 mm	93.78	118.69	0.87	28.07	nr	146.76
Sum of two sides 2400 mm	99.92	126.46	0.87	28.07	nr	154.53
Sum of two sides 2500 mm	106.08	134.25	0.87	28.07	nr	162.32
Sum of two sides 2600 mm	112.23	142.04	0.87	28.07	nr	170.11
Sum of two sides 2700 mm	118.37	149.81	0.87	28.07	nr	177.88
Sum of two sides 2800 mm	124.53	157.61	1.16	37.44	nr	195.05
Sum of two sides 2900 mm	130.68	165.39	1.16	37.44	nr	202.83
Sum of two sides 3000 mm	136.83	173.17	1.30	41.96	nr	215.13
Sum of two sides 3100 mm	141.72	179.36	1.80	58.09	nr	237.45
Sum of two sides 3200 mm	146.15	184.97	1.80	58.09	nr	243.06
Reducer						
Sum of two sides 1700 mm	69.46	87.91	2.47	79.71	nr	167.62
Sum of two sides 1800 mm	80.13	101.42	2.59	83.60	nr	185.02
Sum of two sides 1900 mm	90.92	115.07	2.71	87.46	nr	202.53
Sum of two sides 2000 mm	101.51	128.48	2.71	87.46	nr	215.94
Sum of two sides 2100 mm	112.11	141.88	2.92	94.24	nr	236.12
Sum of two sides 2200 mm	122.71	155.30	2.92	94.24	nr	249.54
Sum of two sides 2300 mm	133.49	168.94	2.92	94.24	nr	263.18
Sum of two sides 2400 mm	144.15	182.44	3.12	100.69	nr	283.13
Sum of two sides 2500 mm	154.75	195.85	3.12	100.69	nr	296.54
Sum of two sides 2600 mm	165.35	209.27	3.12	100.69	nr	309.96
Sum of two sides 2700 mm	175.93	222.66	3.12	100.69	nr	323.35
Sum of two sides 2800 mm	186.59	236.15	3.95	127.47	nr	363.62
Sum of two sides 2900 mm	184.81	233.90	3.97	128.12	nr	362.02
Sum of two sides 3000 mm	195.41	247.31	4.52	145.87	nr	393.18
Sum of two sides 3100 mm	203.25	257.23	4.52	145.87	nr	403.10
Sum of two sides 3200 mm	210.43	266.32	4.52	145.87	nr	412.19

38 VENTILATION/AIR CONDITIONING SYSTEMS

Item	Net Price £	Material £	Labour hours	Labour £	Unit	Total rate £
DUCTWORK: RECTANGULAR – CLASS B – cont						
Extra over fittings – cont						
Offset						
Sum of two sides 1700 mm	175.75	222.43	2.59	83.60	nr	**306.03**
Sum of two sides 1800 mm	195.27	247.14	2.61	84.24	nr	**331.38**
Sum of two sides 1900 mm	208.41	263.76	2.71	87.46	nr	**351.22**
Sum of two sides 2000 mm	220.98	279.68	2.71	87.46	nr	**367.14**
Sum of two sides 2100 mm	233.17	295.10	2.92	94.24	nr	**389.34**
Sum of two sides 2200 mm	245.00	310.07	3.26	105.21	nr	**415.28**
Sum of two sides 2300 mm	256.35	324.44	3.26	105.21	nr	**429.65**
Sum of two sides 2400 mm	275.22	348.32	3.47	111.98	nr	**460.30**
Sum of two sides 2500 mm	286.13	362.13	3.48	112.31	nr	**474.44**
Sum of two sides 2600 mm	304.51	385.39	3.49	112.64	nr	**498.03**
Sum of two sides 2700 mm	322.88	408.63	3.50	112.95	nr	**521.58**
Sum of two sides 2800 mm	341.36	432.03	4.34	140.07	nr	**572.10**
Sum of two sides 2900 mm	341.07	431.66	4.76	153.62	nr	**585.28**
Sum of two sides 3000 mm	359.45	454.92	5.32	171.70	nr	**626.62**
Sum of two sides 3100 mm	373.70	472.95	5.35	172.67	nr	**645.62**
Sum of two sides 3200 mm	386.94	489.71	5.35	172.67	nr	**662.38**
Square to round						
Sum of two sides 1700 mm	81.97	103.75	2.55	82.30	nr	**186.05**
Sum of two sides 1800 mm	95.24	120.53	2.55	82.30	nr	**202.83**
Sum of two sides 1900 mm	108.41	137.20	2.83	91.34	nr	**228.54**
Sum of two sides 2000 mm	121.62	153.92	2.83	91.34	nr	**245.26**
Sum of two sides 2100 mm	134.84	170.65	3.85	124.24	nr	**294.89**
Sum of two sides 2200 mm	148.04	187.36	4.18	134.90	nr	**322.26**
Sum of two sides 2300 mm	161.21	204.03	4.22	136.19	nr	**340.22**
Sum of two sides 2400 mm	174.48	220.82	4.68	151.03	nr	**371.85**
Sum of two sides 2500 mm	187.68	237.53	4.70	151.68	nr	**389.21**
Sum of two sides 2600 mm	200.89	254.25	4.70	151.68	nr	**405.93**
Sum of two sides 2700 mm	214.11	270.97	4.71	152.01	nr	**422.98**
Sum of two sides 2800 mm	227.38	287.77	8.19	264.31	nr	**552.08**
Sum of two sides 2900 mm	227.98	288.53	8.62	278.20	nr	**566.73**
Sum of two sides 3000 mm	241.18	305.23	8.75	282.40	nr	**587.63**
Sum of two sides 3100 mm	251.01	317.68	8.75	282.40	nr	**600.08**
Sum of two sides 3200 mm	259.93	328.97	8.75	282.40	nr	**611.37**
90° radius bend						
Sum of two sides 1700 mm	174.81	221.24	2.19	70.67	nr	**291.91**
Sum of two sides 1800 mm	191.13	241.90	2.19	70.67	nr	**312.57**
Sum of two sides 1900 mm	216.15	273.56	2.26	72.93	nr	**346.49**
Sum of two sides 2000 mm	240.70	304.63	2.26	72.93	nr	**377.56**
Sum of two sides 2100 mm	265.26	335.71	2.48	80.04	nr	**415.75**
Sum of two sides 2200 mm	289.82	366.80	2.48	80.04	nr	**446.84**
Sum of two sides 2300 mm	314.83	398.45	2.48	80.04	nr	**478.49**
Sum of two sides 2400 mm	330.60	418.41	3.90	125.87	nr	**544.28**
Sum of two sides 2500 mm	355.13	449.46	3.90	125.87	nr	**575.33**
Sum of two sides 2600 mm	379.67	480.51	4.26	137.48	nr	**617.99**
Sum of two sides 2700 mm	404.19	511.54	4.55	146.84	nr	**658.38**

38 VENTILATION/AIR CONDITIONING SYSTEMS

Item	Net Price £	Material £	Labour hours	Labour £	Unit	Total rate £
Sum of two sides 2800 mm	419.54	530.97	4.55	146.84	nr	**677.81**
Sum of two sides 2900 mm	420.66	532.39	6.87	221.72	nr	**754.11**
Sum of two sides 3000 mm	445.15	563.38	7.00	225.90	nr	**789.28**
Sum of two sides 3100 mm	464.17	587.45	7.00	225.90	nr	**813.35**
Sum of two sides 3200 mm	481.81	609.78	7.00	225.90	nr	**835.68**
45° bend						
Sum of two sides 1700 mm	84.00	106.31	2.67	86.16	nr	**192.47**
Sum of two sides 1800 mm	92.05	116.50	2.67	86.16	nr	**202.66**
Sum of two sides 1900 mm	104.55	132.32	3.06	98.75	nr	**231.07**
Sum of two sides 2000 mm	116.84	147.87	3.06	98.75	nr	**246.62**
Sum of two sides 2100 mm	129.14	163.44	4.05	130.72	nr	**294.16**
Sum of two sides 2200 mm	141.43	179.00	4.05	130.72	nr	**309.72**
Sum of two sides 2300 mm	153.94	194.82	4.39	141.67	nr	**336.49**
Sum of two sides 2400 mm	161.71	204.66	4.85	156.52	nr	**361.18**
Sum of two sides 2500 mm	173.99	220.20	4.85	156.52	nr	**376.72**
Sum of two sides 2600 mm	186.27	235.75	4.87	157.17	nr	**392.92**
Sum of two sides 2700 mm	198.55	251.28	4.87	157.17	nr	**408.45**
Sum of two sides 2800 mm	206.10	260.84	8.81	284.33	nr	**545.17**
Sum of two sides 2900 mm	205.88	260.56	8.81	284.33	nr	**544.89**
Sum of two sides 3000 mm	218.14	276.08	9.31	300.46	nr	**576.54**
Sum of two sides 3100 mm	227.65	288.11	9.31	300.46	nr	**588.57**
Sum of two sides 3200 mm	236.49	299.30	9.39	303.04	nr	**602.34**
90° mitre bend						
Sum of two sides 1700 mm	188.48	238.54	2.67	86.16	nr	**324.70**
Sum of two sides 1800 mm	201.22	254.67	2.80	90.36	nr	**345.03**
Sum of two sides 1900 mm	225.14	284.94	2.95	95.20	nr	**380.14**
Sum of two sides 2000 mm	249.11	315.27	2.95	95.20	nr	**410.47**
Sum of two sides 2100 mm	273.09	345.62	4.05	130.72	nr	**476.34**
Sum of two sides 2200 mm	297.07	375.97	4.05	130.72	nr	**506.69**
Sum of two sides 2300 mm	215.92	273.27	4.39	141.67	nr	**414.94**
Sum of two sides 2400 mm	333.98	422.69	4.85	156.52	nr	**579.21**
Sum of two sides 2500 mm	358.11	453.22	4.85	156.52	nr	**609.74**
Sum of two sides 2600 mm	382.23	483.75	4.87	157.17	nr	**640.92**
Sum of two sides 2700 mm	406.37	514.30	4.87	157.17	nr	**671.47**
Sum of two sides 2800 mm	419.35	530.73	8.81	284.33	nr	**815.06**
Sum of two sides 2900 mm	423.13	535.52	14.81	477.95	nr	**1013.47**
Sum of two sides 3000 mm	447.42	566.25	15.20	490.55	nr	**1056.80**
Sum of two sides 3100 mm	467.40	591.54	15.60	503.45	nr	**1094.99**
Sum of two sides 3200 mm	486.54	615.76	15.60	503.45	nr	**1119.21**
Branch						
Sum of two sides 1700 mm	166.85	211.16	1.69	54.53	nr	**265.69**
Sum of two sides 1800 mm	183.67	232.46	1.69	54.53	nr	**286.99**
Sum of two sides 1900 mm	200.71	254.02	1.85	59.70	nr	**313.72**
Sum of two sides 2000 mm	217.52	275.30	1.85	59.70	nr	**335.00**
Sum of two sides 2100 mm	234.34	296.58	2.61	84.24	nr	**380.82**
Sum of two sides 2200 mm	251.14	317.84	2.61	84.24	nr	**402.08**
Sum of two sides 2300 mm	268.19	339.42	2.61	84.24	nr	**423.66**
Sum of two sides 2400 mm	285.00	360.70	2.88	92.95	nr	**453.65**
Sum of two sides 2500 mm	301.82	381.99	2.88	92.95	nr	**474.94**
Sum of two sides 2600 mm	318.63	403.26	2.88	92.95	nr	**496.21**
Sum of two sides 2700 mm	335.44	424.54	2.88	92.95	nr	**517.49**

38 VENTILATION/AIR CONDITIONING SYSTEMS

Item	Net Price £	Material £	Labour hours	Labour £	Unit	Total rate £
DUCTWORK: RECTANGULAR – CLASS B – cont						
Extra over fittings – cont						
Branch – cont						
Sum of two sides 2800 mm	352.26	445.82	3.94	127.15	nr	**572.97**
Sum of two sides 2900 mm	369.30	467.39	3.94	127.15	nr	**594.54**
Sum of two sides 3000 mm	386.11	488.66	4.83	155.88	nr	**644.54**
Sum of two sides 3100 mm	399.55	505.67	4.83	155.88	nr	**661.55**
Sum of two sides 3200 mm	411.79	521.16	4.83	155.88	nr	**677.04**
Grille neck						
Sum of two sides 1700 mm	179.29	226.91	1.86	60.03	nr	**286.94**
Sum of two sides 1800 mm	198.26	250.91	2.02	65.20	nr	**316.11**
Sum of two sides 1900 mm	217.25	274.95	2.02	65.20	nr	**340.15**
Sum of two sides 2000 mm	236.20	298.94	2.02	65.20	nr	**364.14**
Sum of two sides 2100 mm	255.17	322.94	2.61	84.24	nr	**407.18**
Sum of two sides 2200 mm	274.13	346.94	2.61	84.24	nr	**431.18**
Sum of two sides 2300 mm	293.11	370.96	2.61	84.24	nr	**455.20**
Sum of two sides 2400 mm	312.07	394.96	2.88	92.95	nr	**487.91**
Sum of two sides 2500 mm	331.04	418.97	2.88	92.95	nr	**511.92**
Sum of two sides 2600 mm	350.01	442.97	2.88	92.95	nr	**535.92**
Sum of two sides 2700 mm	368.98	466.98	2.88	92.95	nr	**559.93**
Sum of two sides 2800 mm	387.95	490.99	3.94	127.15	nr	**618.14**
Sum of two sides 2900 mm	406.91	514.99	4.12	132.97	nr	**647.96**
Sum of two sides 3000 mm	425.88	538.99	5.00	161.36	nr	**700.35**
Sum of two sides 3100 mm	441.05	558.20	5.00	161.36	nr	**719.56**
Sum of two sides 3200 mm	454.87	575.68	5.00	161.36	nr	**737.04**
Ductwork 1601 to 2000 mm longest side						
Sum of two sides 2100 mm	173.57	219.67	2.17	70.03	m	**289.70**
Sum of two sides 2200 mm	179.10	226.67	2.17	70.03	m	**296.70**
Sum of two sides 2300 mm	184.68	233.73	2.19	70.67	m	**304.40**
Sum of two sides 2400 mm	190.16	240.67	2.38	76.81	m	**317.48**
Sum of two sides 2500 mm	195.73	247.71	2.38	76.81	m	**324.52**
Sum of two sides 2600 mm	201.01	254.40	2.64	85.20	m	**339.60**
Sum of two sides 2700 mm	206.29	261.08	2.66	85.85	m	**346.93**
Sum of two sides 2800 mm	211.56	267.75	2.95	95.20	m	**362.95**
Sum of two sides 2900 mm	217.13	274.80	2.96	95.52	m	**370.32**
Sum of two sides 3000 mm	222.42	281.49	2.96	95.52	m	**377.01**
Sum of two sides 3100 mm	239.91	303.63	2.96	95.52	m	**399.15**
Sum of two sides 3200 mm	245.19	310.31	3.15	101.66	m	**411.97**
Sum of two sides 3300 mm	267.10	338.04	3.15	101.66	m	**439.70**
Sum of two sides 3400 mm	272.39	344.74	3.15	101.66	m	**446.40**
Sum of two sides 3500 mm	277.67	351.42	3.15	101.66	m	**453.08**
Sum of two sides 3600 mm	282.94	358.09	3.18	102.63	m	**460.72**
Sum of two sides 3700 mm	288.52	365.15	3.18	102.63	m	**467.78**
Sum of two sides 3800 mm	293.80	371.83	3.18	102.63	m	**474.46**
Sum of two sides 3900 mm	299.07	378.50	3.18	102.63	m	**481.13**
Sum of two sides 4000 mm	304.35	385.19	3.18	102.63	m	**487.82**

38 VENTILATION/AIR CONDITIONING SYSTEMS

Item	Net Price £	Material £	Labour hours	Labour £	Unit	Total rate £
Extra over fittings; Ductwork 1601 to 2000 mm longest side						
End cap						
Sum of two sides 2100 mm	81.47	103.11	0.87	28.07	nr	**131.18**
Sum of two sides 2200 mm	87.63	110.90	0.87	28.07	nr	**138.97**
Sum of two sides 2300 mm	93.78	118.69	0.87	28.07	nr	**146.76**
Sum of two sides 2400 mm	99.92	126.46	0.87	28.07	nr	**154.53**
Sum of two sides 2500 mm	106.08	134.25	0.87	28.07	nr	**162.32**
Sum of two sides 2600 mm	112.23	142.04	0.87	28.07	nr	**170.11**
Sum of two sides 2700 mm	118.37	149.81	0.87	28.07	nr	**177.88**
Sum of two sides 2800 mm	124.53	157.61	1.16	37.44	nr	**195.05**
Sum of two sides 2900 mm	130.68	165.39	1.16	37.44	nr	**202.83**
Sum of two sides 3000 mm	136.83	173.17	1.73	55.84	nr	**229.01**
Sum of two sides 3100 mm	141.72	179.36	1.80	58.09	nr	**237.45**
Sum of two sides 3200 mm	146.15	184.97	1.80	58.09	nr	**243.06**
Sum of two sides 3300 mm	150.59	190.59	1.80	58.09	nr	**248.68**
Sum of two sides 3400 mm	155.01	196.18	1.80	58.09	nr	**254.27**
Sum of two sides 3500 mm	159.45	201.80	1.80	58.09	nr	**259.89**
Sum of two sides 3600 mm	163.88	207.40	1.80	58.09	nr	**265.49**
Sum of two sides 3700 mm	168.31	213.01	1.80	58.09	nr	**271.10**
Sum of two sides 3800 mm	172.75	218.64	1.80	58.09	nr	**276.73**
Sum of two sides 3900 mm	177.18	224.24	1.80	58.09	nr	**282.33**
Sum of two sides 4000 mm	181.63	229.87	1.80	58.09	nr	**287.96**
Reducer						
Sum of two sides 2100 mm	111.46	141.06	2.61	84.24	nr	**225.30**
Sum of two sides 2200 mm	121.94	154.32	2.61	84.24	nr	**238.56**
Sum of two sides 2300 mm	132.61	167.83	2.61	84.24	nr	**252.07**
Sum of two sides 2400 mm	146.06	184.86	2.88	92.95	nr	**277.81**
Sum of two sides 2500 mm	153.57	194.35	3.12	100.69	nr	**295.04**
Sum of two sides 2600 mm	164.05	207.63	3.12	100.69	nr	**308.32**
Sum of two sides 2700 mm	174.52	220.88	3.12	100.69	nr	**321.57**
Sum of two sides 2800 mm	184.99	234.12	3.95	127.47	nr	**361.59**
Sum of two sides 2900 mm	195.66	247.63	3.97	128.12	nr	**375.75**
Sum of two sides 3000 mm	206.14	260.89	4.52	145.87	nr	**406.76**
Sum of two sides 3100 mm	204.57	258.90	4.52	145.87	nr	**404.77**
Sum of two sides 3200 mm	211.61	267.81	4.52	145.87	nr	**413.68**
Sum of two sides 3300 mm	206.00	260.71	4.52	145.87	nr	**406.58**
Sum of two sides 3400 mm	213.06	269.65	4.52	145.87	nr	**415.52**
Sum of two sides 3500 mm	220.09	278.54	4.52	145.87	nr	**424.41**
Sum of two sides 3600 mm	227.14	287.47	4.52	145.87	nr	**433.34**
Sum of two sides 3700 mm	245.61	310.84	4.52	145.87	nr	**456.71**
Sum of two sides 3800 mm	252.66	319.77	4.52	145.87	nr	**465.64**
Sum of two sides 3900 mm	259.69	328.66	4.52	145.87	nr	**474.53**
Sum of two sides 4000 mm	266.75	337.60	4.52	145.87	nr	**483.47**

38 VENTILATION/AIR CONDITIONING SYSTEMS

Item	Net Price £	Material £	Labour hours	Labour £	Unit	Total rate £
DUCTWORK: RECTANGULAR – CLASS B – cont						
Extra over fittings – cont						
Offset						
Sum of two sides 2100 mm	261.33	330.74	2.61	84.24	nr	**414.98**
Sum of two sides 2200 mm	281.01	355.64	2.61	84.24	nr	**439.88**
Sum of two sides 2300 mm	292.29	369.92	2.61	84.24	nr	**454.16**
Sum of two sides 2400 mm	320.41	405.51	2.88	92.95	nr	**498.46**
Sum of two sides 2500 mm	331.30	419.29	3.48	112.31	nr	**531.60**
Sum of two sides 2600 mm	341.57	432.29	3.49	112.64	nr	**544.93**
Sum of two sides 2700 mm	351.46	444.81	3.50	112.95	nr	**557.76**
Sum of two sides 2800 mm	360.92	456.78	4.34	140.07	nr	**596.85**
Sum of two sides 2900 mm	370.10	468.40	4.76	153.62	nr	**622.02**
Sum of two sides 3000 mm	378.74	479.34	5.32	171.70	nr	**651.04**
Sum of two sides 3100 mm	379.26	479.99	5.35	172.67	nr	**652.66**
Sum of two sides 3200 mm	392.59	496.87	5.35	172.67	nr	**669.54**
Sum of two sides 3300 mm	386.86	489.61	5.35	172.67	nr	**662.28**
Sum of two sides 3400 mm	400.21	506.51	5.35	172.67	nr	**679.18**
Sum of two sides 3500 mm	413.56	523.40	5.35	172.67	nr	**696.07**
Sum of two sides 3600 mm	426.91	540.30	5.35	172.67	nr	**712.97**
Sum of two sides 3700 mm	450.52	570.18	5.35	172.67	nr	**742.85**
Sum of two sides 3800 mm	464.92	588.40	5.35	172.67	nr	**761.07**
Sum of two sides 3900 mm	478.26	605.28	5.35	172.67	nr	**777.95**
Sum of two sides 4000 mm	491.62	622.19	5.35	172.67	nr	**794.86**
Square to round						
Sum of two sides 2100 mm	165.86	209.91	2.61	84.24	nr	**294.15**
Sum of two sides 2200 mm	181.89	230.20	2.61	84.24	nr	**314.44**
Sum of two sides 2300 mm	197.87	250.42	2.61	84.24	nr	**334.66**
Sum of two sides 2400 mm	213.97	270.80	2.88	92.95	nr	**363.75**
Sum of two sides 2500 mm	229.95	291.02	4.70	151.68	nr	**442.70**
Sum of two sides 2600 mm	245.94	311.26	4.70	151.68	nr	**462.94**
Sum of two sides 2700 mm	261.95	331.52	4.71	152.01	nr	**483.53**
Sum of two sides 2800 mm	277.95	351.77	8.19	264.31	nr	**616.08**
Sum of two sides 2900 mm	293.92	371.99	8.19	264.31	nr	**636.30**
Sum of two sides 3000 mm	309.92	392.24	8.19	264.31	nr	**656.55**
Sum of two sides 3100 mm	312.61	395.64	8.19	264.31	nr	**659.95**
Sum of two sides 3200 mm	323.46	409.37	8.19	264.31	nr	**673.68**
Sum of two sides 3300 mm	321.44	406.82	8.19	264.31	nr	**671.13**
Sum of two sides 3400 mm	332.30	420.56	8.62	278.20	nr	**698.76**
Sum of two sides 3500 mm	343.14	434.28	8.62	278.20	nr	**712.48**
Sum of two sides 3600 mm	354.00	448.02	8.62	278.20	nr	**726.22**
Sum of two sides 3700 mm	370.45	468.84	8.62	278.20	nr	**747.04**
Sum of two sides 3800 mm	381.31	482.59	8.75	282.40	nr	**764.99**
Sum of two sides 3900 mm	392.16	496.32	8.75	282.40	nr	**778.72**
Sum of two sides 4000 mm	403.02	510.06	8.75	282.40	nr	**792.46**
90° radius bend						
Sum of two sides 2100 mm	241.00	305.01	2.61	84.24	nr	**389.25**
Sum of two sides 2200 mm	461.90	584.58	2.61	84.24	nr	**668.82**
Sum of two sides 2300 mm	500.52	633.46	2.61	84.24	nr	**717.70**

38 VENTILATION/AIR CONDITIONING SYSTEMS

Item	Net Price £	Material £	Labour hours	Labour £	Unit	Total rate £
Sum of two sides 2400 mm	512.35	648.44	2.88	92.95	nr	741.39
Sum of two sides 2500 mm	550.92	697.24	3.90	125.87	nr	823.11
Sum of two sides 2600 mm	588.70	745.06	4.26	137.48	nr	882.54
Sum of two sides 2700 mm	626.48	792.87	4.55	146.84	nr	939.71
Sum of two sides 2800 mm	664.26	840.68	4.55	146.84	nr	987.52
Sum of two sides 2900 mm	702.80	889.46	6.87	221.72	nr	1111.18
Sum of two sides 3000 mm	740.58	937.28	6.87	221.72	nr	1159.00
Sum of two sides 3100 mm	749.03	947.97	6.87	221.72	nr	1169.69
Sum of two sides 3200 mm	777.35	983.82	6.87	221.72	nr	1205.54
Sum of two sides 3300 mm	775.41	981.36	6.87	221.72	nr	1203.08
Sum of two sides 3400 mm	802.90	1016.15	7.00	225.90	nr	1242.05
Sum of two sides 3500 mm	830.41	1050.96	7.00	225.90	nr	1276.86
Sum of two sides 3600 mm	857.89	1085.75	7.00	225.90	nr	1311.65
Sum of two sides 3700 mm	919.88	1164.20	7.00	225.90	nr	1390.10
Sum of two sides 3800 mm	947.36	1198.98	7.00	225.90	nr	1424.88
Sum of two sides 3900 mm	974.86	1233.78	7.00	225.90	nr	1459.68
Sum of two sides 4000 mm	1002.34	1268.56	7.00	225.90	nr	1494.46
45° bend						
Sum of two sides 2100 mm	116.05	146.88	2.61	84.24	nr	231.12
Sum of two sides 2200 mm	330.17	417.86	2.61	84.24	nr	502.10
Sum of two sides 2300 mm	356.00	450.55	2.61	84.24	nr	534.79
Sum of two sides 2400 mm	367.60	465.24	2.88	92.95	nr	558.19
Sum of two sides 2500 mm	393.39	497.87	4.85	156.52	nr	654.39
Sum of two sides 2600 mm	418.61	529.79	4.87	157.17	nr	686.96
Sum of two sides 2700 mm	443.82	561.70	4.87	157.17	nr	718.87
Sum of two sides 2800 mm	469.03	593.60	8.81	284.33	nr	877.93
Sum of two sides 2900 mm	494.84	626.27	8.81	284.33	nr	910.60
Sum of two sides 3000 mm	520.05	658.18	9.31	300.46	nr	958.64
Sum of two sides 3100 mm	528.73	669.16	9.31	300.46	nr	969.62
Sum of two sides 3200 mm	547.06	692.36	9.31	300.46	nr	992.82
Sum of two sides 3300 mm	550.53	696.75	9.31	300.46	nr	997.21
Sum of two sides 3400 mm	568.89	719.99	9.31	300.46	nr	1020.45
Sum of two sides 3500 mm	587.24	743.21	9.39	303.04	nr	1046.25
Sum of two sides 3600 mm	605.59	766.44	9.39	303.04	nr	1069.48
Sum of two sides 3700 mm	647.04	818.90	9.39	303.04	nr	1121.94
Sum of two sides 3800 mm	665.38	842.11	9.39	303.04	nr	1145.15
Sum of two sides 3900 mm	683.73	865.32	9.39	303.04	nr	1168.36
Sum of two sides 4000 mm	702.09	888.56	9.39	303.04	nr	1191.60
90° mitre bend						
Sum of two sides 2100 mm	536.77	679.34	2.61	84.24	nr	763.58
Sum of two sides 2200 mm	567.11	717.73	2.61	84.24	nr	801.97
Sum of two sides 2300 mm	612.71	775.44	2.61	84.24	nr	859.68
Sum of two sides 2400 mm	626.29	792.64	2.88	92.95	nr	885.59
Sum of two sides 2500 mm	672.09	850.60	4.85	156.52	nr	1007.12
Sum of two sides 2600 mm	718.16	908.90	4.87	157.17	nr	1066.07
Sum of two sides 2700 mm	764.23	967.21	4.87	157.17	nr	1124.38
Sum of two sides 2800 mm	810.30	1025.52	8.81	284.33	nr	1309.85
Sum of two sides 2900 mm	856.09	1083.47	14.81	477.95	nr	1561.42
Sum of two sides 3000 mm	902.16	1141.77	15.20	490.55	nr	1632.32
Sum of two sides 3100 mm	914.45	1157.33	15.20	490.55	nr	1647.88

38 VENTILATION/AIR CONDITIONING SYSTEMS

Item	Net Price £	Material £	Labour hours	Labour £	Unit	Total rate £
DUCTWORK: RECTANGULAR – CLASS B – cont						
Extra over fittings – cont						
90° mitre bend – cont						
Sum of two sides 3200 mm	950.23	1202.61	15.20	490.55	nr	1693.16
Sum of two sides 3300 mm	960.27	1215.32	15.20	490.55	nr	1705.87
Sum of two sides 3400 mm	996.04	1260.59	15.20	490.55	nr	1751.14
Sum of two sides 3500 mm	1031.81	1305.86	15.60	503.45	nr	1809.31
Sum of two sides 3600 mm	1067.58	1351.13	15.60	503.45	nr	1854.58
Sum of two sides 3700 mm	1114.35	1410.33	15.60	503.45	nr	1913.78
Sum of two sides 3800 mm	1150.12	1455.60	15.60	503.45	nr	1959.05
Sum of two sides 3900 mm	1185.90	1500.88	15.60	503.45	nr	2004.33
Sum of two sides 4000 mm	1221.67	1546.15	15.60	503.45	nr	2049.60
Branch						
Sum of two sides 2100 mm	240.64	304.55	2.61	84.24	nr	388.79
Sum of two sides 2200 mm	257.46	325.84	2.61	84.24	nr	410.08
Sum of two sides 2300 mm	274.53	347.45	2.61	84.24	nr	431.69
Sum of two sides 2400 mm	291.08	368.39	2.88	92.95	nr	461.34
Sum of two sides 2500 mm	308.16	390.01	2.88	92.95	nr	482.96
Sum of two sides 2600 mm	325.01	411.33	2.88	92.95	nr	504.28
Sum of two sides 2700 mm	341.85	432.64	2.88	92.95	nr	525.59
Sum of two sides 2800 mm	358.69	453.96	3.94	127.15	nr	581.11
Sum of two sides 2900 mm	375.76	475.56	3.94	127.15	nr	602.71
Sum of two sides 3000 mm	392.59	496.87	3.94	127.15	nr	624.02
Sum of two sides 3100 mm	406.08	513.93	3.94	127.15	nr	641.08
Sum of two sides 3200 mm	418.34	529.45	3.94	127.15	nr	656.60
Sum of two sides 3300 mm	430.84	545.27	3.94	127.15	nr	672.42
Sum of two sides 3400 mm	443.12	560.82	4.83	155.88	nr	716.70
Sum of two sides 3500 mm	455.38	576.33	4.83	155.88	nr	732.21
Sum of two sides 3600 mm	467.66	591.88	4.83	155.88	nr	747.76
Sum of two sides 3700 mm	485.78	614.80	4.83	155.88	nr	770.68
Sum of two sides 3800 mm	498.06	630.35	4.83	155.88	nr	786.23
Sum of two sides 3900 mm	510.32	645.86	4.83	155.88	nr	801.74
Sum of two sides 4000 mm	522.60	661.40	4.83	155.88	nr	817.28
Grille neck						
Sum of two sides 2100 mm	255.17	322.94	2.61	84.24	nr	407.18
Sum of two sides 2200 mm	274.13	346.94	2.61	84.24	nr	431.18
Sum of two sides 2300 mm	293.11	370.96	2.61	84.24	nr	455.20
Sum of two sides 2400 mm	312.07	394.96	2.88	92.95	nr	487.91
Sum of two sides 2500 mm	331.04	418.97	2.88	92.95	nr	511.92
Sum of two sides 2600 mm	350.01	442.97	2.88	92.95	nr	535.92
Sum of two sides 2700 mm	368.98	466.98	2.88	92.95	nr	559.93
Sum of two sides 2800 mm	387.95	490.99	3.94	127.15	nr	618.14
Sum of two sides 2900 mm	406.91	514.99	4.12	132.97	nr	647.96
Sum of two sides 3000 mm	425.88	538.99	4.12	132.97	nr	671.96
Sum of two sides 3100 mm	441.05	558.20	4.12	132.97	nr	691.17
Sum of two sides 3200 mm	454.87	575.68	4.12	132.97	nr	708.65
Sum of two sides 3300 mm	468.67	593.15	4.12	132.97	nr	726.12
Sum of two sides 3400 mm	482.52	610.68	5.00	161.36	nr	772.04
Sum of two sides 3500 mm	496.35	628.19	5.00	161.36	nr	789.55

38 VENTILATION/AIR CONDITIONING SYSTEMS

Item	Net Price £	Material £	Labour hours	Labour £	Unit	Total rate £
Sum of two sides 3600 mm	510.16	645.66	5.00	161.36	nr	**807.02**
Sum of two sides 3700 mm	523.99	663.16	5.00	161.36	nr	**824.52**
Sum of two sides 3800 mm	537.80	680.64	5.00	161.36	nr	**842.00**
Sum of two sides 3900 mm	551.63	698.14	5.00	161.36	nr	**859.50**
Sum of two sides 4000 mm	565.46	715.65	5.00	161.36	nr	**877.01**
Ductwork 2001 to 2500 mm longest side						
Sum of two sides 2500 mm	251.45	318.24	2.38	76.81	m	**395.05**
Sum of two sides 2600 mm	258.05	326.59	2.64	85.20	m	**411.79**
Sum of two sides 2700 mm	263.60	333.61	2.66	85.85	m	**419.46**
Sum of two sides 2800 mm	270.76	342.68	2.95	95.20	m	**437.88**
Sum of two sides 2900 mm	276.32	349.71	2.96	95.52	m	**445.23**
Sum of two sides 3000 mm	281.58	356.37	3.15	101.66	m	**458.03**
Sum of two sides 3100 mm	287.10	363.35	3.15	101.66	m	**465.01**
Sum of two sides 3200 mm	292.37	370.03	3.15	101.66	m	**471.69**
Sum of two sides 3300 mm	297.92	377.05	3.15	101.66	m	**478.71**
Sum of two sides 3400 mm	303.18	383.70	2.66	85.85	m	**469.55**
Sum of two sides 3500 mm	323.21	409.06	3.15	101.66	m	**510.72**
Sum of two sides 3600 mm	328.49	415.73	3.18	102.63	m	**518.36**
Sum of two sides 3700 mm	350.73	443.88	3.18	102.63	m	**546.51**
Sum of two sides 3800 mm	355.98	450.53	3.18	102.63	m	**553.16**
Sum of two sides 3900 mm	361.24	457.18	3.18	102.63	m	**559.81**
Sum of two sides 4000 mm	368.20	466.00	3.18	102.63	m	**568.63**
Extra over fittings; Ductwork 2001 to 2500 mm longest side						
End cap						
Sum of two sides 2500 mm	106.08	134.25	0.87	28.07	nr	**162.32**
Sum of two sides 2600 mm	112.23	142.04	0.87	28.07	nr	**170.11**
Sum of two sides 2700 mm	118.37	149.81	0.87	28.07	nr	**177.88**
Sum of two sides 2800 mm	124.53	157.61	1.16	37.44	nr	**195.05**
Sum of two sides 2900 mm	130.68	165.39	1.16	37.44	nr	**202.83**
Sum of two sides 3000 mm	136.83	173.17	1.73	55.84	nr	**229.01**
Sum of two sides 3100 mm	141.72	179.36	1.73	55.84	nr	**235.20**
Sum of two sides 3200 mm	146.15	184.97	1.73	55.84	nr	**240.81**
Sum of two sides 3300 mm	150.59	190.59	1.73	55.84	nr	**246.43**
Sum of two sides 3400 mm	155.01	196.18	1.80	58.09	nr	**254.27**
Sum of two sides 3500 mm	159.45	201.80	1.80	58.09	nr	**259.89**
Sum of two sides 3600 mm	163.88	207.40	1.80	58.09	nr	**265.49**
Sum of two sides 3700 mm	168.31	213.01	1.80	58.09	nr	**271.10**
Sum of two sides 3800 mm	172.75	218.64	1.80	58.09	nr	**276.73**
Sum of two sides 3900 mm	177.18	224.24	1.80	58.09	nr	**282.33**
Sum of two sides 4000 mm	181.63	229.87	1.80	58.09	nr	**287.96**
Reducer						
Sum of two sides 2500 mm	125.16	158.40	3.12	100.69	nr	**259.09**
Sum of two sides 2600 mm	135.58	171.60	3.12	100.69	nr	**272.29**
Sum of two sides 2700 mm	146.24	185.08	3.12	100.69	nr	**285.77**
Sum of two sides 2800 mm	156.44	197.99	3.95	127.47	nr	**325.46**
Sum of two sides 2900 mm	167.09	211.47	3.97	128.12	nr	**339.59**
Sum of two sides 3000 mm	177.54	224.69	3.97	128.12	nr	**352.81**
Sum of two sides 3100 mm	185.26	234.46	3.97	128.12	nr	**362.58**

38 VENTILATION/AIR CONDITIONING SYSTEMS

Item	Net Price £	Material £	Labour hours	Labour £	Unit	Total rate £
DUCTWORK: RECTANGULAR – CLASS B – cont						
Extra over fittings – cont						
Reducer – cont						
Sum of two sides 3200 mm	192.26	243.32	3.97	128.12	nr	**371.44**
Sum of two sides 3300 mm	199.51	252.50	3.97	128.12	nr	**380.62**
Sum of two sides 3400 mm	206.51	261.36	4.52	145.87	nr	**407.23**
Sum of two sides 3500 mm	203.25	257.23	4.52	145.87	nr	**403.10**
Sum of two sides 3600 mm	210.27	266.12	4.52	145.87	nr	**411.99**
Sum of two sides 3700 mm	204.37	258.65	4.52	145.87	nr	**404.52**
Sum of two sides 3800 mm	211.40	267.55	4.52	145.87	nr	**413.42**
Sum of two sides 3900 mm	218.43	276.45	4.52	145.87	nr	**422.32**
Sum of two sides 4000 mm	225.56	285.47	4.52	145.87	nr	**431.34**
Offset						
Sum of two sides 2500 mm	314.80	398.41	3.48	112.31	nr	**510.72**
Sum of two sides 2600 mm	334.45	423.28	3.49	112.64	nr	**535.92**
Sum of two sides 2700 mm	342.00	432.84	3.50	112.95	nr	**545.79**
Sum of two sides 2800 mm	373.71	472.96	4.34	140.07	nr	**613.03**
Sum of two sides 2900 mm	380.88	482.04	4.76	153.62	nr	**635.66**
Sum of two sides 3000 mm	387.42	490.31	5.32	171.70	nr	**662.01**
Sum of two sides 3100 mm	389.48	492.92	5.35	172.67	nr	**665.59**
Sum of two sides 3200 mm	390.04	493.64	5.32	171.70	nr	**665.34**
Sum of two sides 3300 mm	390.32	493.99	5.32	171.70	nr	**665.69**
Sum of two sides 3400 mm	390.04	493.64	5.35	172.67	nr	**666.31**
Sum of two sides 3500 mm	387.43	490.34	5.32	171.70	nr	**662.04**
Sum of two sides 3600 mm	400.74	507.18	5.35	172.67	nr	**679.85**
Sum of two sides 3700 mm	394.57	499.36	5.35	172.67	nr	**672.03**
Sum of two sides 3800 mm	407.88	516.21	5.35	172.67	nr	**688.88**
Sum of two sides 3900 mm	421.22	533.10	5.35	172.67	nr	**705.77**
Sum of two sides 4000 mm	434.51	549.92	5.35	172.67	nr	**722.59**
Square to round						
Sum of two sides 2500 mm	200.97	254.35	4.70	151.68	nr	**406.03**
Sum of two sides 2600 mm	216.95	274.57	4.70	151.68	nr	**426.25**
Sum of two sides 2700 mm	232.91	294.77	4.71	152.01	nr	**446.78**
Sum of two sides 2800 mm	248.92	315.03	8.19	264.31	nr	**579.34**
Sum of two sides 2900 mm	264.87	335.22	8.19	264.31	nr	**599.53**
Sum of two sides 3000 mm	280.86	355.45	8.19	264.31	nr	**619.76**
Sum of two sides 3100 mm	292.81	370.59	8.19	264.31	nr	**634.90**
Sum of two sides 3200 mm	303.64	384.28	8.62	278.20	nr	**662.48**
Sum of two sides 3300 mm	314.48	398.00	8.62	278.20	nr	**676.20**
Sum of two sides 3400 mm	325.29	411.69	8.62	278.20	nr	**689.89**
Sum of two sides 3500 mm	325.25	411.63	8.62	278.20	nr	**689.83**
Sum of two sides 3600 mm	336.09	425.35	8.75	282.40	nr	**707.75**
Sum of two sides 3700 mm	333.78	422.43	8.75	282.40	nr	**704.83**
Sum of two sides 3800 mm	344.61	436.14	8.75	282.40	nr	**718.54**
Sum of two sides 3900 mm	355.45	449.86	8.75	282.40	nr	**732.26**
Sum of two sides 4000 mm	366.20	463.47	8.75	282.40	nr	**745.87**
90° radius bend						
Sum of two sides 2500 mm	470.73	595.75	3.90	125.87	nr	**721.62**
Sum of two sides 2600 mm	491.92	622.57	4.26	137.48	nr	**760.05**
Sum of two sides 2700 mm	530.40	671.27	4.55	146.84	nr	**818.11**

38 VENTILATION/AIR CONDITIONING SYSTEMS

Item	Net Price £	Material £	Labour hours	Labour £	Unit	Total rate £
Sum of two sides 2800 mm	533.10	674.69	4.55	146.84	nr	821.53
Sum of two sides 2900 mm	571.51	723.31	6.87	221.72	nr	945.03
Sum of two sides 3000 mm	609.21	771.02	6.87	221.72	nr	992.74
Sum of two sides 3100 mm	638.73	808.37	6.87	221.72	nr	1030.09
Sum of two sides 3200 mm	666.14	843.07	6.87	221.72	nr	1064.79
Sum of two sides 3300 mm	694.27	878.67	6.87	221.72	nr	1100.39
Sum of two sides 3400 mm	721.70	913.38	6.87	221.72	nr	1135.10
Sum of two sides 3500 mm	721.91	913.65	7.00	225.90	nr	1139.55
Sum of two sides 3600 mm	749.34	948.36	7.00	225.90	nr	1174.26
Sum of two sides 3700 mm	742.31	939.47	7.00	225.90	nr	1165.37
Sum of two sides 3800 mm	769.72	974.15	7.00	225.90	nr	1200.05
Sum of two sides 3900 mm	797.14	1008.86	7.00	225.90	nr	1234.76
Sum of two sides 4000 mm	804.76	1018.51	7.00	225.90	nr	1244.41
45° bend						
Sum of two sides 2500 mm	351.82	445.27	4.85	156.52	nr	601.79
Sum of two sides 2600 mm	368.51	466.39	4.87	157.17	nr	623.56
Sum of two sides 2700 mm	394.29	499.02	4.87	157.17	nr	656.19
Sum of two sides 2800 mm	401.29	507.88	8.81	284.33	nr	792.21
Sum of two sides 2900 mm	427.03	540.44	8.81	284.33	nr	824.77
Sum of two sides 3000 mm	452.20	572.31	9.31	300.46	nr	872.77
Sum of two sides 3100 mm	472.01	597.37	9.31	300.46	nr	897.83
Sum of two sides 3200 mm	490.30	620.52	9.31	300.46	nr	920.98
Sum of two sides 3300 mm	509.19	644.43	9.31	300.46	nr	944.89
Sum of two sides 3400 mm	527.50	667.61	9.31	300.46	nr	968.07
Sum of two sides 3500 mm	532.10	673.42	9.31	300.46	nr	973.88
Sum of two sides 3600 mm	550.42	696.61	9.39	303.04	nr	999.65
Sum of two sides 3700 mm	550.91	697.23	9.39	303.04	nr	1000.27
Sum of two sides 3800 mm	569.21	720.40	9.39	303.04	nr	1023.44
Sum of two sides 3900 mm	587.51	743.56	9.39	303.04	nr	1046.60
Sum of two sides 4000 mm	595.91	754.19	9.39	303.04	nr	1057.23
90° mitre bend						
Sum of two sides 2500 mm	570.94	722.58	4.85	156.52	nr	879.10
Sum of two sides 2600 mm	605.65	766.51	4.87	157.17	nr	923.68
Sum of two sides 2700 mm	651.70	824.79	4.87	157.17	nr	981.96
Sum of two sides 2800 mm	653.05	826.50	8.81	284.33	nr	1110.83
Sum of two sides 2900 mm	699.29	885.02	14.81	477.95	nr	1362.97
Sum of two sides 3000 mm	744.92	942.77	14.81	477.95	nr	1420.72
Sum of two sides 3100 mm	784.13	992.40	15.20	490.55	nr	1482.95
Sum of two sides 3200 mm	820.41	1038.31	15.20	490.55	nr	1528.86
Sum of two sides 3300 mm	856.35	1083.80	15.20	490.55	nr	1574.35
Sum of two sides 3400 mm	892.65	1129.73	15.20	490.55	nr	1620.28
Sum of two sides 3500 mm	892.82	1129.96	15.20	490.55	nr	1620.51
Sum of two sides 3600 mm	929.10	1175.87	15.60	503.45	nr	1679.32
Sum of two sides 3700 mm	934.22	1182.35	15.60	503.45	nr	1685.80
Sum of two sides 3800 mm	970.51	1228.28	15.60	503.45	nr	1731.73
Sum of two sides 3900 mm	1006.79	1274.19	15.60	503.45	nr	1777.64
Sum of two sides 4000 mm	1017.54	1287.80	15.60	503.45	nr	1791.25
Branch						
Sum of two sides 2500 mm	308.72	390.71	2.88	92.95	nr	483.66
Sum of two sides 2600 mm	325.53	411.99	2.88	92.95	nr	504.94
Sum of two sides 2700 mm	342.60	433.60	2.88	92.95	nr	526.55

38 VENTILATION/AIR CONDITIONING SYSTEMS

Item	Net Price £	Material £	Labour hours	Labour £	Unit	Total rate £
DUCTWORK: RECTANGULAR – CLASS B – cont						
Extra over fittings – cont						
Branch – cont						
Sum of two sides 2800 mm	359.16	454.55	3.94	127.15	nr	**581.70**
Sum of two sides 2900 mm	376.23	476.16	3.94	127.15	nr	**603.31**
Sum of two sides 3000 mm	393.07	497.47	3.94	127.15	nr	**624.62**
Sum of two sides 3100 mm	406.53	514.51	3.94	127.15	nr	**641.66**
Sum of two sides 3200 mm	418.81	530.05	3.94	127.15	nr	**657.20**
Sum of two sides 3300 mm	431.31	545.87	3.94	127.15	nr	**673.02**
Sum of two sides 3400 mm	443.58	561.40	3.94	127.15	nr	**688.55**
Sum of two sides 3500 mm	456.45	577.68	4.83	155.88	nr	**733.56**
Sum of two sides 3600 mm	468.72	593.21	4.83	155.88	nr	**749.09**
Sum of two sides 3700 mm	481.23	609.04	4.83	155.88	nr	**764.92**
Sum of two sides 3800 mm	493.49	624.56	4.83	155.88	nr	**780.44**
Sum of two sides 3900 mm	505.77	640.10	4.83	155.88	nr	**795.98**
Sum of two sides 4000 mm	518.23	655.87	4.83	155.88	nr	**811.75**
Grille neck						
Sum of two sides 2500 mm	331.04	418.97	2.88	92.95	nr	**511.92**
Sum of two sides 2600 mm	350.01	442.97	2.88	92.95	nr	**535.92**
Sum of two sides 2700 mm	368.98	466.98	2.88	92.95	nr	**559.93**
Sum of two sides 2800 mm	387.95	490.99	3.94	127.15	nr	**618.14**
Sum of two sides 2900 mm	406.80	514.84	3.94	127.15	nr	**641.99**
Sum of two sides 3000 mm	425.88	538.99	3.94	127.15	nr	**666.14**
Sum of two sides 3100 mm	441.05	558.20	4.12	132.97	nr	**691.17**
Sum of two sides 3200 mm	454.87	575.68	4.12	132.97	nr	**708.65**
Sum of two sides 3300 mm	468.68	593.16	4.12	132.97	nr	**726.13**
Sum of two sides 3400 mm	482.52	610.68	4.12	132.97	nr	**743.65**
Sum of two sides 3500 mm	496.35	628.19	5.00	161.36	nr	**789.55**
Sum of two sides 3600 mm	510.16	645.66	5.00	161.36	nr	**807.02**
Sum of two sides 3700 mm	523.99	663.16	5.00	161.36	nr	**824.52**
Sum of two sides 3800 mm	537.80	680.64	5.00	161.36	nr	**842.00**
Sum of two sides 3900 mm	551.63	698.14	5.00	161.36	nr	**859.50**
Sum of two sides 4000 mm	565.46	715.65	5.00	161.36	nr	**877.01**
Ductwork 2501 to 4000 mm longest side						
Sum of two sides 3000 mm	398.57	504.43	2.38	76.81	m	**581.24**
Sum of two sides 3100 mm	408.10	516.49	2.38	76.81	m	**593.30**
Sum of two sides 3200 mm	417.10	527.88	2.38	76.81	m	**604.69**
Sum of two sides 3300 mm	426.09	539.26	2.38	76.81	m	**616.07**
Sum of two sides 3400 mm	437.52	553.73	2.64	85.20	m	**638.93**
Sum of two sides 3500 mm	446.52	565.12	2.66	85.85	m	**650.97**
Sum of two sides 3600 mm	455.53	576.52	2.95	95.20	m	**671.72**
Sum of two sides 3700 mm	467.23	591.33	2.96	95.52	m	**686.85**
Sum of two sides 3800 mm	476.24	602.73	3.15	101.66	m	**704.39**
Sum of two sides 3900 mm	501.53	634.74	3.15	101.66	m	**736.40**
Sum of two sides 4000 mm	510.53	646.13	3.15	101.66	m	**747.79**
Sum of two sides 4100 mm	519.82	657.89	3.35	108.10	m	**765.99**
Sum of two sides 4200 mm	528.82	669.28	3.35	108.10	m	**777.38**
Sum of two sides 4300 mm	537.81	680.66	3.60	116.18	m	**796.84**
Sum of two sides 4400 mm	550.69	696.95	3.60	116.18	m	**813.13**

38 VENTILATION/AIR CONDITIONING SYSTEMS

Item	Net Price £	Material £	Labour hours	Labour £	Unit	Total rate £
Sum of two sides 4500 mm	559.70	708.36	3.60	116.18	m	**824.54**

Extra over fittings; Ductwork 2501 to 4000 mm longest side
End cap

Item	Net Price £	Material £	Labour hours	Labour £	Unit	Total rate £
Sum of two sides 3000 mm	138.33	175.07	1.73	55.84	nr	**230.91**
Sum of two sides 3100 mm	143.26	181.31	1.73	55.84	nr	**237.15**
Sum of two sides 3200 mm	147.74	186.98	1.73	55.84	nr	**242.82**
Sum of two sides 3300 mm	152.14	192.55	1.73	55.84	nr	**248.39**
Sum of two sides 3400 mm	156.71	198.33	1.73	55.84	nr	**254.17**
Sum of two sides 3500 mm	161.21	204.03	1.73	55.84	nr	**259.87**
Sum of two sides 3600 mm	165.69	209.70	1.73	55.84	nr	**265.54**
Sum of two sides 3700 mm	170.18	215.38	1.73	55.84	nr	**271.22**
Sum of two sides 3800 mm	174.66	221.05	1.73	55.84	nr	**276.89**
Sum of two sides 3900 mm	179.15	226.73	1.80	58.09	nr	**284.82**
Sum of two sides 4000 mm	183.63	232.40	1.80	58.09	nr	**290.49**
Sum of two sides 4100 mm	188.12	238.09	1.80	58.09	nr	**296.18**
Sum of two sides 4200 mm	192.60	243.76	1.80	58.09	nr	**301.85**
Sum of two sides 4300 mm	197.08	249.42	1.88	60.67	nr	**310.09**
Sum of two sides 4400 mm	201.57	255.10	1.88	60.67	nr	**315.77**
Sum of two sides 4500 mm	206.05	260.78	1.88	60.67	nr	**321.45**
Reducer						
Sum of two sides 3000 mm	140.72	178.09	3.12	100.69	nr	**278.78**
Sum of two sides 3100 mm	147.35	186.49	3.12	100.69	nr	**287.18**
Sum of two sides 3200 mm	153.07	193.73	3.12	100.69	nr	**294.42**
Sum of two sides 3300 mm	158.82	201.01	3.12	100.69	nr	**301.70**
Sum of two sides 3400 mm	164.48	208.16	3.12	100.69	nr	**308.85**
Sum of two sides 3500 mm	170.22	215.43	3.12	100.69	nr	**316.12**
Sum of two sides 3600 mm	175.95	222.68	3.95	127.47	nr	**350.15**
Sum of two sides 3700 mm	181.85	230.15	3.97	128.12	nr	**358.27**
Sum of two sides 3800 mm	187.57	237.38	4.52	145.87	nr	**383.25**
Sum of two sides 3900 mm	182.96	231.55	4.52	145.87	nr	**377.42**
Sum of two sides 4000 mm	188.69	238.81	4.52	145.87	nr	**384.68**
Sum of two sides 4100 mm	194.63	246.32	4.52	145.87	nr	**392.19**
Sum of two sides 4200 mm	200.37	253.59	4.52	145.87	nr	**399.46**
Sum of two sides 4300 mm	206.09	260.83	4.92	158.78	nr	**419.61**
Sum of two sides 4400 mm	211.95	268.24	4.92	158.78	nr	**427.02**
Sum of two sides 4500 mm	217.68	275.50	5.12	165.23	nr	**440.73**
Offset						
Sum of two sides 3000 mm	391.04	494.91	3.48	112.31	nr	**607.22**
Sum of two sides 3100 mm	384.75	486.94	3.48	112.31	nr	**599.25**
Sum of two sides 3200 mm	376.17	476.08	3.48	112.31	nr	**588.39**
Sum of two sides 3300 mm	366.67	464.06	3.48	112.31	nr	**576.37**
Sum of two sides 3400 mm	401.71	508.40	3.50	112.95	nr	**621.35**
Sum of two sides 3500 mm	391.30	495.23	3.49	112.64	nr	**607.87**
Sum of two sides 3600 mm	379.96	480.87	4.25	137.17	nr	**618.04**
Sum of two sides 3700 mm	416.09	526.60	4.76	153.62	nr	**680.22**
Sum of two sides 3800 mm	403.82	511.08	5.32	171.70	nr	**682.78**
Sum of two sides 3900 mm	428.18	541.90	5.35	172.67	nr	**714.57**
Sum of two sides 4000 mm	413.86	523.78	5.35	172.67	nr	**696.45**

38 VENTILATION/AIR CONDITIONING SYSTEMS

Item	Net Price £	Material £	Labour hours	Labour £	Unit	Total rate £
DUCTWORK: RECTANGULAR – CLASS B – cont						
Extra over fittings – cont						
Offset – cont						
Sum of two sides 4100 mm	398.74	504.65	5.85	188.80	nr	**693.45**
Sum of two sides 4200 mm	382.57	484.18	5.85	188.80	nr	**672.98**
Sum of two sides 4300 mm	393.53	498.05	6.15	198.48	nr	**696.53**
Sum of two sides 4400 mm	433.07	548.09	6.30	203.31	nr	**751.40**
Sum of two sides 4500 mm	415.51	525.87	6.15	198.48	nr	**724.35**
Square to round						
Sum of two sides 3000 mm	286.52	362.62	4.70	151.68	nr	**514.30**
Sum of two sides 3100 mm	298.82	378.19	4.70	151.68	nr	**529.87**
Sum of two sides 3200 mm	309.78	392.06	4.70	151.68	nr	**543.74**
Sum of two sides 3300 mm	320.75	405.94	4.70	151.68	nr	**557.62**
Sum of two sides 3400 mm	331.72	419.82	4.71	152.01	nr	**571.83**
Sum of two sides 3500 mm	342.68	433.70	4.71	152.01	nr	**585.71**
Sum of two sides 3600 mm	353.64	447.56	8.19	264.31	nr	**711.87**
Sum of two sides 3700 mm	364.60	461.44	8.62	278.20	nr	**739.64**
Sum of two sides 3800 mm	375.57	475.32	8.62	278.20	nr	**753.52**
Sum of two sides 3900 mm	459.15	581.10	8.62	278.20	nr	**859.30**
Sum of two sides 4000 mm	472.21	597.63	8.75	282.40	nr	**880.03**
Sum of two sides 4100 mm	428.51	542.33	11.23	362.43	nr	**904.76**
Sum of two sides 4200 mm	498.30	630.65	11.23	362.43	nr	**993.08**
Sum of two sides 4300 mm	511.34	647.15	11.25	363.07	nr	**1010.22**
Sum of two sides 4400 mm	524.39	663.67	11.25	363.07	nr	**1026.74**
Sum of two sides 4500 mm	537.45	680.20	11.26	363.40	nr	**1043.60**
90° radius bend						
Sum of two sides 3000 mm	831.29	1052.08	3.90	125.87	nr	**1177.95**
Sum of two sides 3100 mm	870.81	1102.10	3.90	125.87	nr	**1227.97**
Sum of two sides 3200 mm	906.27	1146.98	4.26	137.48	nr	**1284.46**
Sum of two sides 3300 mm	941.73	1191.85	4.26	137.48	nr	**1329.33**
Sum of two sides 3400 mm	919.66	1163.93	4.26	137.48	nr	**1301.41**
Sum of two sides 3500 mm	954.65	1208.20	4.55	146.84	nr	**1355.04**
Sum of two sides 3600 mm	989.65	1252.50	4.55	146.84	nr	**1399.34**
Sum of two sides 3700 mm	964.96	1221.25	6.87	221.72	nr	**1442.97**
Sum of two sides 3800 mm	999.50	1264.97	6.87	221.72	nr	**1486.69**
Sum of two sides 3900 mm	935.52	1184.00	6.87	221.72	nr	**1405.72**
Sum of two sides 4000 mm	969.62	1227.15	7.00	225.90	nr	**1453.05**
Sum of two sides 4100 mm	1004.69	1271.54	7.20	232.37	nr	**1503.91**
Sum of two sides 4200 mm	1038.78	1314.68	7.20	232.37	nr	**1547.05**
Sum of two sides 4300 mm	1072.87	1357.82	7.41	239.14	nr	**1596.96**
Sum of two sides 4400 mm	1002.31	1268.52	7.41	239.14	nr	**1507.66**
Sum of two sides 4500 mm	1035.74	1310.84	7.55	243.67	nr	**1554.51**
45° bend						
Sum of two sides 3000 mm	403.29	510.41	4.85	156.52	nr	**666.93**
Sum of two sides 3100 mm	422.96	535.29	4.85	156.52	nr	**691.81**
Sum of two sides 3200 mm	440.62	557.65	4.85	156.52	nr	**714.17**

38 VENTILATION/AIR CONDITIONING SYSTEMS

Item	Net Price £	Material £	Labour hours	Labour £	Unit	Total rate £
Sum of two sides 3300 mm	451.98	572.03	4.87	157.17	nr	729.20
Sum of two sides 3400 mm	446.73	565.38	4.87	157.17	nr	722.55
Sum of two sides 3500 mm	464.18	587.46	4.87	157.17	nr	744.63
Sum of two sides 3600 mm	481.62	609.54	8.81	284.33	nr	893.87
Sum of two sides 3700 mm	468.73	593.22	8.81	284.33	nr	877.55
Sum of two sides 3800 mm	485.94	615.00	9.31	300.46	nr	915.46
Sum of two sides 3900 mm	452.72	572.96	9.31	300.46	nr	873.42
Sum of two sides 4000 mm	469.70	594.45	9.31	300.46	nr	894.91
Sum of two sides 4100 mm	487.16	616.55	10.01	323.04	nr	939.59
Sum of two sides 4200 mm	504.15	638.05	9.31	300.46	nr	938.51
Sum of two sides 4300 mm	521.14	659.56	10.01	323.04	nr	982.60
Sum of two sides 4400 mm	485.08	613.92	10.52	339.51	nr	953.43
Sum of two sides 4500 mm	501.74	635.01	10.52	339.51	nr	974.52
90° mitre bend						
Sum of two sides 3000 mm	977.86	1237.58	4.85	156.52	nr	1394.10
Sum of two sides 3100 mm	1026.06	1298.58	4.85	156.52	nr	1455.10
Sum of two sides 3200 mm	1071.39	1355.95	4.87	157.17	nr	1513.12
Sum of two sides 3300 mm	1116.73	1413.33	4.87	157.17	nr	1570.50
Sum of two sides 3400 mm	1089.10	1378.36	4.87	157.17	nr	1535.53
Sum of two sides 3500 mm	1134.03	1435.22	8.81	284.33	nr	1719.55
Sum of two sides 3600 mm	1178.96	1492.09	8.81	284.33	nr	1776.42
Sum of two sides 3700 mm	1156.65	1463.85	14.81	477.95	nr	1941.80
Sum of two sides 3800 mm	1201.24	1520.29	14.81	477.95	nr	1998.24
Sum of two sides 3900 mm	1117.02	1413.70	14.81	477.95	nr	1891.65
Sum of two sides 4000 mm	1161.22	1469.64	15.20	490.55	nr	1960.19
Sum of two sides 4100 mm	1204.82	1524.82	16.30	526.04	nr	2050.86
Sum of two sides 4200 mm	1249.01	1580.75	16.50	532.50	nr	2113.25
Sum of two sides 4300 mm	1293.21	1636.69	17.01	548.96	nr	2185.65
Sum of two sides 4400 mm	1212.37	1534.38	17.01	548.96	nr	2083.34
Sum of two sides 4500 mm	1256.01	1589.60	17.01	548.96	nr	2138.56
Branch						
Sum of two sides 3000 mm	419.60	531.05	2.88	92.95	nr	624.00
Sum of two sides 3100 mm	434.11	549.40	2.88	92.95	nr	642.35
Sum of two sides 3200 mm	447.21	565.99	2.88	92.95	nr	658.94
Sum of two sides 3300 mm	460.29	582.55	2.88	92.95	nr	675.50
Sum of two sides 3400 mm	473.33	599.04	2.88	92.95	nr	691.99
Sum of two sides 3500 mm	486.42	615.61	3.94	127.15	nr	742.76
Sum of two sides 3600 mm	499.52	632.20	3.94	127.15	nr	759.35
Sum of two sides 3700 mm	512.76	648.95	3.94	127.15	nr	776.10
Sum of two sides 3800 mm	525.86	665.53	4.83	155.88	nr	821.41
Sum of two sides 3900 mm	539.47	682.75	4.83	155.88	nr	838.63
Sum of two sides 4000 mm	552.57	699.33	4.83	155.88	nr	855.21
Sum of two sides 4100 mm	565.88	716.17	5.44	175.56	nr	891.73
Sum of two sides 4200 mm	578.97	732.75	5.44	175.56	nr	908.31
Sum of two sides 4300 mm	592.05	749.30	5.85	188.80	nr	938.10
Sum of two sides 4400 mm	595.83	754.08	5.85	188.80	nr	942.88
Sum of two sides 4500 mm	618.38	782.62	5.85	188.80	nr	971.42
Grille neck						
Sum of two sides 3000 mm	433.41	548.52	2.88	92.95	nr	641.47
Sum of two sides 3100 mm	448.83	568.04	2.88	92.95	nr	660.99
Sum of two sides 3200 mm	462.90	585.85	2.88	92.95	nr	678.80

38 VENTILATION/AIR CONDITIONING SYSTEMS

Item	Net Price £	Material £	Labour hours	Labour £	Unit	Total rate £
DUCTWORK: RECTANGULAR – CLASS B – cont						
Extra over fittings – cont						
Grille neck – cont						
Sum of two sides 3300 mm	476.98	603.67	2.88	92.95	nr	**696.62**
Sum of two sides 3400 mm	491.06	621.49	3.94	127.15	nr	**748.64**
Sum of two sides 3500 mm	505.14	639.31	3.94	127.15	nr	**766.46**
Sum of two sides 3600 mm	435.14	550.72	3.94	127.15	nr	**677.87**
Sum of two sides 3700 mm	533.27	674.91	4.12	132.97	nr	**807.88**
Sum of two sides 3800 mm	547.35	692.73	4.12	132.97	nr	**825.70**
Sum of two sides 3900 mm	561.43	710.55	4.12	132.97	nr	**843.52**
Sum of two sides 4000 mm	575.50	728.35	5.00	161.36	nr	**889.71**
Sum of two sides 4100 mm	589.57	746.16	5.00	161.36	nr	**907.52**
Sum of two sides 4200 mm	603.65	763.97	5.00	161.36	nr	**925.33**
Sum of two sides 4300 mm	617.73	781.79	5.23	168.80	nr	**950.59**
Sum of two sides 4400 mm	631.80	799.60	5.23	168.80	nr	**968.40**
Sum of two sides 4500 mm	645.88	817.42	5.39	173.95	nr	**991.37**

38 VENTILATION/AIR CONDITIONING SYSTEMS

Item	Net Price £	Material £	Labour hours	Labour £	Unit	Total rate £
DUCTWORK: RECTANGULAR – CLASS C						
AIR DUCTLINES						
Galvanized sheet metal DW144 class C rectangular section ductwork; including all necessary stiffeners, joints, couplers in the running length and duct supports						
Ductwork up to 400 mm longest side						
Sum of two sides 200 mm	26.16	33.11	1.17	37.77	m	**70.88**
Sum of two sides 300 mm	28.14	35.62	1.17	37.77	m	**73.39**
Sum of two sides 400 mm	25.74	32.58	1.19	38.39	m	**70.97**
Sum of two sides 500 mm	28.32	35.84	1.16	37.44	m	**73.28**
Sum of two sides 600 mm	30.64	38.77	1.19	38.39	m	**77.16**
Sum of two sides 700 mm	32.97	41.73	1.19	38.39	m	**80.12**
Sum of two sides 800 mm	35.41	44.81	1.19	38.39	m	**83.20**
Extra over fittings; Ductwork up to 400 mm longest side						
End cap						
Sum of two sides 200 mm	15.78	19.97	0.38	12.26	nr	**32.23**
Sum of two sides 300 mm	17.63	22.31	0.38	12.26	nr	**34.57**
Sum of two sides 400 mm	19.47	24.64	0.38	12.26	nr	**36.90**
Sum of two sides 500 mm	21.31	26.97	0.38	12.26	nr	**39.23**
Sum of two sides 600 mm	23.16	29.31	0.38	12.26	nr	**41.57**
Sum of two sides 700 mm	25.01	31.65	0.38	12.26	nr	**43.91**
Sum of two sides 800 mm	26.84	33.97	0.38	12.26	nr	**46.23**
Reducer						
Sum of two sides 200 mm	25.63	32.44	1.40	45.18	nr	**77.62**
Sum of two sides 300 mm	28.92	36.60	1.40	45.18	nr	**81.78**
Sum of two sides 400 mm	47.34	59.91	1.42	45.83	nr	**105.74**
Sum of two sides 500 mm	51.23	64.84	1.42	45.83	nr	**110.67**
Sum of two sides 600 mm	55.13	69.78	1.69	54.53	nr	**124.31**
Sum of two sides 700 mm	59.01	74.68	1.69	54.53	nr	**129.21**
Sum of two sides 800 mm	62.86	79.55	1.92	61.96	nr	**141.51**
Offset						
Sum of two sides 200 mm	37.80	47.84	1.63	52.60	nr	**100.44**
Sum of two sides 300 mm	42.73	54.07	1.63	52.60	nr	**106.67**
Sum of two sides 400 mm	63.13	79.90	1.65	53.26	nr	**133.16**
Sum of two sides 500 mm	68.63	86.86	1.65	53.26	nr	**140.12**
Sum of two sides 600 mm	73.10	92.51	1.92	61.96	nr	**154.47**
Sum of two sides 700 mm	78.07	98.81	1.92	61.96	nr	**160.77**
Sum of two sides 800 mm	82.41	104.29	1.92	61.96	nr	**166.25**
Square to round						
Sum of two sides 200 mm	33.25	42.08	1.22	39.37	nr	**81.45**
Sum of two sides 300 mm	37.36	47.29	1.22	39.37	nr	**86.66**
Sum of two sides 400 mm	49.71	62.91	1.25	40.34	nr	**103.25**
Sum of two sides 500 mm	54.09	68.45	1.25	40.34	nr	**108.79**
Sum of two sides 600 mm	58.45	73.98	1.33	42.92	nr	**116.90**
Sum of two sides 700 mm	62.85	79.54	1.33	42.92	nr	**122.46**
Sum of two sides 800 mm	67.18	85.02	1.40	45.18	nr	**130.20**

38 VENTILATION/AIR CONDITIONING SYSTEMS

Item	Net Price £	Material £	Labour hours	Labour £	Unit	Total rate £
DUCTWORK: RECTANGULAR – CLASS C – cont						
Extra over fittings – cont						
90° radius bend						
Sum of two sides 200 mm	25.02	31.66	1.10	35.50	nr	67.16
Sum of two sides 300 mm	26.99	34.16	1.10	35.50	nr	69.66
Sum of two sides 400 mm	45.42	57.48	1.12	36.14	nr	93.62
Sum of two sides 500 mm	48.30	61.13	1.12	36.14	nr	97.27
Sum of two sides 600 mm	52.15	66.00	1.16	37.44	nr	103.44
Sum of two sides 700 mm	55.49	70.22	1.16	37.44	nr	107.66
Sum of two sides 800 mm	59.16	74.87	1.22	39.37	nr	114.24
45° radius bend						
Sum of two sides 200 mm	27.01	34.18	1.29	41.64	nr	75.82
Sum of two sides 300 mm	29.65	37.52	1.29	41.64	nr	79.16
Sum of two sides 400 mm	47.73	60.40	1.29	41.64	nr	102.04
Sum of two sides 500 mm	51.04	64.60	1.29	41.64	nr	106.24
Sum of two sides 600 mm	54.93	69.52	1.39	44.87	nr	114.39
Sum of two sides 700 mm	58.56	74.11	1.39	44.87	nr	118.98
Sum of two sides 800 mm	62.40	78.97	1.46	47.12	nr	126.09
90° mitre bend						
Sum of two sides 200 mm	41.82	52.93	2.04	65.83	nr	118.76
Sum of two sides 300 mm	45.94	58.14	2.04	65.83	nr	123.97
Sum of two sides 400 mm	66.74	84.47	2.09	67.46	nr	151.93
Sum of two sides 500 mm	72.05	91.19	2.09	67.46	nr	158.65
Sum of two sides 600 mm	78.88	99.83	2.15	69.40	nr	169.23
Sum of two sides 700 mm	85.21	107.84	2.15	69.40	nr	177.24
Sum of two sides 800 mm	92.14	116.61	2.26	72.93	nr	189.54
Branch						
Sum of two sides 200 mm	41.56	52.60	0.92	29.69	nr	82.29
Sum of two sides 300 mm	46.26	58.54	0.92	29.69	nr	88.23
Sum of two sides 400 mm	57.81	73.17	0.95	30.67	nr	103.84
Sum of two sides 500 mm	63.38	80.21	0.95	30.67	nr	110.88
Sum of two sides 600 mm	68.48	86.67	1.03	33.25	nr	119.92
Sum of two sides 700 mm	73.60	93.15	1.03	33.25	nr	126.40
Sum of two sides 800 mm	78.71	99.61	1.03	33.25	nr	132.86
Grille neck						
Sum of two sides 200 mm	47.99	60.74	1.10	35.50	nr	96.24
Sum of two sides 300 mm	53.93	68.25	1.10	35.50	nr	103.75
Sum of two sides 400 mm	59.86	75.76	1.16	37.44	nr	113.20
Sum of two sides 500 mm	65.80	83.27	1.16	37.44	nr	120.71
Sum of two sides 600 mm	71.74	90.80	1.18	38.08	nr	128.88
Sum of two sides 700 mm	77.67	98.30	1.18	38.08	nr	136.38
Sum of two sides 800 mm	83.61	105.82	1.18	38.08	nr	143.90
Ductwork 401 to 600 mm longest side						
Sum of two sides 600 mm	40.46	51.21	1.17	37.77	m	88.98
Sum of two sides 700 mm	44.50	56.32	1.17	37.77	m	94.09
Sum of two sides 800 mm	48.46	61.33	1.17	37.77	m	99.10
Sum of two sides 900 mm	52.32	66.21	1.27	40.98	m	107.19
Sum of two sides 1000 mm	56.17	71.09	1.48	47.77	m	118.86
Sum of two sides 1100 mm	60.33	76.35	1.49	48.08	m	124.43
Sum of two sides 1200 mm	64.18	81.22	1.49	48.08	m	129.30

38 VENTILATION/AIR CONDITIONING SYSTEMS

Item	Net Price £	Material £	Labour hours	Labour £	Unit	Total rate £
Extra over fittings: Ductwork 401 to 600 mm longest side						
End cap						
Sum of two sides 600 mm	23.16	29.31	0.38	12.26	nr	**41.57**
Sum of two sides 700 mm	25.01	31.65	0.38	12.26	nr	**43.91**
Sum of two sides 800 mm	26.84	33.97	0.38	12.26	nr	**46.23**
Sum of two sides 900 mm	28.69	36.31	0.38	12.26	nr	**48.57**
Sum of two sides 1000 mm	30.53	38.64	0.38	12.26	nr	**50.90**
Sum of two sides 1100 mm	32.37	40.97	0.38	12.26	nr	**53.23**
Sum of two sides 1200 mm	34.21	43.30	0.38	12.26	nr	**55.56**
Reducer						
Sum of two sides 600 mm	49.46	62.60	1.69	54.53	nr	**117.13**
Sum of two sides 700 mm	52.64	66.62	1.69	54.53	nr	**121.15**
Sum of two sides 800 mm	55.76	70.57	1.92	61.96	nr	**132.53**
Sum of two sides 900 mm	58.97	74.64	1.92	61.96	nr	**136.60**
Sum of two sides 1000 mm	62.15	78.66	2.18	70.36	nr	**149.02**
Sum of two sides 1100 mm	65.65	83.08	2.18	70.36	nr	**153.44**
Sum of two sides 1200 mm	68.83	87.11	2.18	70.36	nr	**157.47**
Offset						
Sum of two sides 600 mm	69.68	88.19	1.92	61.96	nr	**150.15**
Sum of two sides 700 mm	74.39	94.15	1.92	61.96	nr	**156.11**
Sum of two sides 800 mm	77.01	97.46	1.92	61.96	nr	**159.42**
Sum of two sides 900 mm	79.42	100.51	1.92	61.96	nr	**162.47**
Sum of two sides 1000 mm	82.74	104.72	2.18	70.36	nr	**175.08**
Sum of two sides 1100 mm	84.89	107.44	2.18	70.36	nr	**177.80**
Sum of two sides 1200 mm	86.56	109.55	2.18	70.36	nr	**179.91**
Square to round						
Sum of two sides 600 mm	52.57	66.53	1.33	42.92	nr	**109.45**
Sum of two sides 700 mm	56.24	71.18	1.33	42.92	nr	**114.10**
Sum of two sides 800 mm	59.86	75.76	1.40	45.18	nr	**120.94**
Sum of two sides 900 mm	63.53	80.40	1.40	45.18	nr	**125.58**
Sum of two sides 1000 mm	67.22	85.08	1.82	58.73	nr	**143.81**
Sum of two sides 1100 mm	70.97	89.82	1.82	58.73	nr	**148.55**
Sum of two sides 1200 mm	74.65	94.47	1.82	58.73	nr	**153.20**
90° radius bend						
Sum of two sides 600 mm	46.65	59.04	1.16	37.44	nr	**96.48**
Sum of two sides 700 mm	47.77	60.46	1.16	37.44	nr	**97.90**
Sum of two sides 800 mm	51.02	64.57	1.22	39.37	nr	**103.94**
Sum of two sides 900 mm	54.36	68.80	1.22	39.37	nr	**108.17**
Sum of two sides 1000 mm	56.39	71.37	1.40	45.18	nr	**116.55**
Sum of two sides 1100 mm	59.92	75.84	1.40	45.18	nr	**121.02**
Sum of two sides 1200 mm	63.21	80.00	1.40	45.18	nr	**125.18**
45° bend						
Sum of two sides 600 mm	52.00	65.81	1.39	44.87	nr	**110.68**
Sum of two sides 700 mm	54.47	68.94	1.39	44.87	nr	**113.81**
Sum of two sides 800 mm	58.01	73.42	1.46	47.12	nr	**120.54**
Sum of two sides 900 mm	61.61	77.97	1.46	47.12	nr	**125.09**
Sum of two sides 1000 mm	64.55	81.69	1.88	60.67	nr	**142.36**
Sum of two sides 1100 mm	68.46	86.64	1.88	60.67	nr	**147.31**
Sum of two sides 1200 mm	72.06	91.20	1.88	60.67	nr	**151.87**

38 VENTILATION/AIR CONDITIONING SYSTEMS

Item	Net Price £	Material £	Labour hours	Labour £	Unit	Total rate £
DUCTWORK: RECTANGULAR – CLASS C – cont						
Extra over fittings – cont						
90° mitre bend						
Sum of two sides 600 mm	81.82	103.56	2.15	69.40	nr	**172.96**
Sum of two sides 700 mm	85.02	107.60	2.15	69.40	nr	**177.00**
Sum of two sides 800 mm	91.40	115.67	2.26	72.93	nr	**188.60**
Sum of two sides 900 mm	97.91	123.92	2.26	72.93	nr	**196.85**
Sum of two sides 1000 mm	102.89	130.22	3.03	97.80	nr	**228.02**
Sum of two sides 1100 mm	109.68	138.81	3.03	97.80	nr	**236.61**
Sum of two sides 1200 mm	116.30	147.19	3.03	97.80	nr	**244.99**
Branch						
Sum of two sides 600 mm	69.10	87.45	1.03	33.25	nr	**120.70**
Sum of two sides 700 mm	74.26	93.98	1.03	33.25	nr	**127.23**
Sum of two sides 800 mm	79.43	100.53	1.03	33.25	nr	**133.78**
Sum of two sides 900 mm	84.61	107.08	1.03	33.25	nr	**140.33**
Sum of two sides 1000 mm	89.79	113.64	1.29	41.64	nr	**155.28**
Sum of two sides 1100 mm	95.18	120.46	1.29	41.64	nr	**162.10**
Sum of two sides 1200 mm	100.35	127.01	1.29	41.64	nr	**168.65**
Grille neck						
Sum of two sides 600 mm	71.74	90.80	1.18	38.08	nr	**128.88**
Sum of two sides 700 mm	77.67	98.30	1.18	38.08	nr	**136.38**
Sum of two sides 800 mm	83.61	105.82	1.18	38.08	nr	**143.90**
Sum of two sides 900 mm	89.54	113.32	1.18	38.08	nr	**151.40**
Sum of two sides 1000 mm	95.48	120.84	1.44	46.47	nr	**167.31**
Sum of two sides 1100 mm	101.41	128.34	1.44	46.47	nr	**174.81**
Sum of two sides 1200 mm	107.35	135.87	1.44	46.47	nr	**182.34**
Ductwork 601 to 800 mm longest side						
Sum of two sides 900 mm	56.79	71.87	1.27	40.98	m	**112.85**
Sum of two sides 1000 mm	60.65	76.75	1.48	47.77	m	**124.52**
Sum of two sides 1100 mm	64.51	81.65	1.49	48.08	m	**129.73**
Sum of two sides 1200 mm	68.66	86.90	1.49	48.08	m	**134.98**
Sum of two sides 1300 mm	72.51	91.77	1.51	48.74	m	**140.51**
Sum of two sides 1400 mm	75.73	95.84	1.55	50.02	m	**145.86**
Sum of two sides 1500 mm	80.22	101.53	1.61	51.96	m	**153.49**
Sum of two sides 1600 mm	84.07	106.40	1.62	52.28	m	**158.68**
Extra over fittings; Ductwork 601 to 800 mm longest side						
End cap						
Sum of two sides 900 mm	28.69	36.31	0.38	12.26	nr	**48.57**
Sum of two sides 1000 mm	30.53	38.64	0.38	12.26	nr	**50.90**
Sum of two sides 1100 mm	32.37	40.97	0.38	12.26	nr	**53.23**
Sum of two sides 1200 mm	34.21	43.30	0.38	12.26	nr	**55.56**
Sum of two sides 1300 mm	36.06	45.64	0.38	12.26	nr	**57.90**
Sum of two sides 1400 mm	37.64	47.63	0.38	12.26	nr	**59.89**
Sum of two sides 1500 mm	43.70	55.31	0.38	12.26	nr	**67.57**
Sum of two sides 1600 mm	49.82	63.06	0.38	12.26	nr	**75.32**

38 VENTILATION/AIR CONDITIONING SYSTEMS

Item	Net Price £	Material £	Labour hours	Labour £	Unit	Total rate £
Reducer						
Sum of two sides 900 mm	59.59	75.42	1.92	61.96	nr	**137.38**
Sum of two sides 1000 mm	62.78	79.45	2.18	70.36	nr	**149.81**
Sum of two sides 1100 mm	65.96	83.47	2.18	70.36	nr	**153.83**
Sum of two sides 1200 mm	69.45	87.90	2.18	70.36	nr	**158.26**
Sum of two sides 1300 mm	72.64	91.93	2.30	74.22	nr	**166.15**
Sum of two sides 1400 mm	75.31	95.31	2.30	74.22	nr	**169.53**
Sum of two sides 1500 mm	87.01	110.12	2.47	79.71	nr	**189.83**
Sum of two sides 1600 mm	98.70	124.91	2.47	79.71	nr	**204.62**
Offset						
Sum of two sides 900 mm	84.04	106.37	1.92	61.96	nr	**168.33**
Sum of two sides 1000 mm	86.16	109.04	2.18	70.36	nr	**179.40**
Sum of two sides 1100 mm	88.03	111.41	2.18	70.36	nr	**181.77**
Sum of two sides 1200 mm	89.82	113.68	2.18	70.36	nr	**184.04**
Sum of two sides 1300 mm	91.12	115.33	2.47	79.71	nr	**195.04**
Sum of two sides 1400 mm	92.96	117.64	2.47	79.71	nr	**197.35**
Sum of two sides 1500 mm	106.55	134.85	2.47	79.71	nr	**214.56**
Sum of two sides 1600 mm	119.87	151.70	2.47	79.71	nr	**231.41**
Square to round						
Sum of two sides 900 mm	62.97	79.70	1.40	45.18	nr	**124.88**
Sum of two sides 1000 mm	66.65	84.35	1.82	58.73	nr	**143.08**
Sum of two sides 1100 mm	70.33	89.01	1.82	58.73	nr	**147.74**
Sum of two sides 1200 mm	74.09	93.77	1.82	58.73	nr	**152.50**
Sum of two sides 1300 mm	77.77	98.43	2.32	74.87	nr	**173.30**
Sum of two sides 1400 mm	80.79	102.24	2.32	74.87	nr	**177.11**
Sum of two sides 1500 mm	95.09	120.34	2.56	82.62	nr	**202.96**
Sum of two sides 1600 mm	109.40	138.45	2.58	83.26	nr	**221.71**
90° radius bend						
Sum of two sides 900 mm	48.20	61.01	1.22	39.37	nr	**100.38**
Sum of two sides 1000 mm	51.39	65.04	1.40	45.18	nr	**110.22**
Sum of two sides 1100 mm	54.57	69.06	1.40	45.18	nr	**114.24**
Sum of two sides 1200 mm	57.94	73.33	1.40	45.18	nr	**118.51**
Sum of two sides 1300 mm	61.12	77.36	1.91	61.63	nr	**138.99**
Sum of two sides 1400 mm	62.05	78.53	1.91	61.63	nr	**140.16**
Sum of two sides 1500 mm	73.69	93.26	2.11	68.08	nr	**161.34**
Sum of two sides 1600 mm	85.33	107.99	2.11	68.08	nr	**176.07**
45° bend						
Sum of two sides 900 mm	59.43	75.22	1.46	47.12	nr	**122.34**
Sum of two sides 1000 mm	62.96	79.68	1.88	60.67	nr	**140.35**
Sum of two sides 1100 mm	66.50	84.16	1.88	60.67	nr	**144.83**
Sum of two sides 1200 mm	70.33	89.01	1.88	60.67	nr	**149.68**
Sum of two sides 1300 mm	73.86	93.48	2.26	72.93	nr	**166.41**
Sum of two sides 1400 mm	75.98	96.16	2.44	78.75	nr	**174.91**
Sum of two sides 1500 mm	87.97	111.34	2.44	78.75	nr	**190.09**
Sum of two sides 1600 mm	99.99	126.55	2.68	86.49	nr	**213.04**
90° mitre bend						
Sum of two sides 900 mm	88.91	112.53	2.26	72.93	nr	**185.46**
Sum of two sides 1000 mm	95.83	121.28	3.03	97.80	nr	**219.08**
Sum of two sides 1100 mm	102.78	130.08	3.03	97.80	nr	**227.88**

38 VENTILATION/AIR CONDITIONING SYSTEMS

Item	Net Price £	Material £	Labour hours	Labour £	Unit	Total rate £
DUCTWORK: RECTANGULAR – CLASS C – cont						
Extra over fittings – cont						
90° mitre bend – cont						
Sum of two sides 1200 mm	109.85	139.03	3.03	97.80	nr	**236.83**
Sum of two sides 1300 mm	116.78	147.80	3.85	124.24	nr	**272.04**
Sum of two sides 1400 mm	120.86	152.96	3.85	124.24	nr	**277.20**
Sum of two sides 1500 mm	140.66	178.02	4.25	137.17	nr	**315.19**
Sum of two sides 1600 mm	160.44	203.06	4.26	137.48	nr	**340.54**
Branch						
Sum of two sides 900 mm	87.11	110.24	1.03	33.25	nr	**143.49**
Sum of two sides 1000 mm	92.42	116.96	1.29	41.64	nr	**158.60**
Sum of two sides 1100 mm	97.73	123.68	1.29	41.64	nr	**165.32**
Sum of two sides 1200 mm	103.28	130.72	1.29	41.64	nr	**172.36**
Sum of two sides 1300 mm	108.60	137.45	1.39	44.87	nr	**182.32**
Sum of two sides 1400 mm	113.22	143.29	1.39	44.87	nr	**188.16**
Sum of two sides 1500 mm	129.85	164.34	1.64	52.93	nr	**217.27**
Sum of two sides 1600 mm	146.50	185.42	1.64	52.93	nr	**238.35**
Grille neck						
Sum of two sides 900 mm	89.54	113.32	1.18	38.08	nr	**151.40**
Sum of two sides 1000 mm	95.48	120.84	1.44	46.47	nr	**167.31**
Sum of two sides 1100 mm	101.41	128.34	1.44	46.47	nr	**174.81**
Sum of two sides 1200 mm	107.35	135.87	1.44	46.47	nr	**182.34**
Sum of two sides 1300 mm	113.29	143.38	1.69	54.53	nr	**197.91**
Sum of two sides 1400 mm	118.43	149.89	1.69	54.53	nr	**204.42**
Sum of two sides 1500 mm	137.12	173.54	1.79	57.77	nr	**231.31**
Sum of two sides 1600 mm	208.34	263.67	1.79	57.77	nr	**321.44**
Ductwork 801 to 1000 mm longest side						
Sum of two sides 1100 mm	65.30	82.64	1.49	48.08	m	**130.72**
Sum of two sides 1200 mm	69.15	87.52	1.49	48.08	m	**135.60**
Sum of two sides 1300 mm	73.01	92.40	1.51	48.74	m	**141.14**
Sum of two sides 1400 mm	77.16	97.65	1.55	50.02	m	**147.67**
Sum of two sides 1500 mm	81.02	102.54	1.61	51.96	m	**154.50**
Sum of two sides 1600 mm	84.87	107.41	1.62	52.28	m	**159.69**
Sum of two sides 1700 mm	88.72	112.28	1.74	56.16	m	**168.44**
Sum of two sides 1800 mm	92.88	117.54	1.76	56.80	m	**174.34**
Sum of two sides 1900 mm	96.71	122.39	1.81	58.42	m	**180.81**
Sum of two sides 2000 mm	100.56	127.27	1.82	58.73	m	**186.00**
Extra over fittings; Ductwork 801 to 1000 mm longest side						
End cap						
Sum of two sides 1100 mm	32.36	40.96	0.38	12.26	nr	**53.22**
Sum of two sides 1200 mm	34.21	43.30	0.38	12.26	nr	**55.56**
Sum of two sides 1300 mm	36.06	45.64	0.38	12.26	nr	**57.90**
Sum of two sides 1400 mm	37.65	47.64	0.38	12.26	nr	**59.90**
Sum of two sides 1500 mm	43.72	55.33	0.38	12.26	nr	**67.59**
Sum of two sides 1600 mm	49.82	63.06	0.38	12.26	nr	**75.32**
Sum of two sides 1700 mm	55.91	70.76	0.58	18.72	nr	**89.48**

38 VENTILATION/AIR CONDITIONING SYSTEMS

Item	Net Price £	Material £	Labour hours	Labour £	Unit	Total rate £
Sum of two sides 1800 mm	62.01	78.48	0.58	18.72	nr	97.20
Sum of two sides 1900 mm	68.10	86.18	0.58	18.72	nr	104.90
Sum of two sides 2000 mm	74.19	93.89	0.58	18.72	nr	112.61
Reducer						
Sum of two sides 1100 mm	61.03	77.24	2.18	70.36	nr	147.60
Sum of two sides 1200 mm	63.97	80.96	2.18	70.36	nr	151.32
Sum of two sides 1300 mm	66.91	84.68	2.30	74.22	nr	158.90
Sum of two sides 1400 mm	69.59	88.08	2.30	74.22	nr	162.30
Sum of two sides 1500 mm	81.03	102.55	2.47	79.71	nr	182.26
Sum of two sides 1600 mm	92.45	117.01	2.47	79.71	nr	196.72
Sum of two sides 1700 mm	103.90	131.50	2.59	83.60	nr	215.10
Sum of two sides 1800 mm	115.60	146.31	2.59	83.60	nr	229.91
Sum of two sides 1900 mm	127.03	160.76	2.71	87.46	nr	248.22
Sum of two sides 2000 mm	138.48	175.26	2.71	87.46	nr	262.72
Offset						
Sum of two sides 1100 mm	93.73	118.62	2.18	70.36	nr	188.98
Sum of two sides 1200 mm	95.57	120.95	2.18	70.36	nr	191.31
Sum of two sides 1300 mm	97.14	122.94	2.47	79.71	nr	202.65
Sum of two sides 1400 mm	97.86	123.85	2.47	79.71	nr	203.56
Sum of two sides 1500 mm	111.64	141.29	2.47	79.71	nr	221.00
Sum of two sides 1600 mm	125.12	158.36	2.47	79.71	nr	238.07
Sum of two sides 1700 mm	138.37	175.12	2.61	84.24	nr	259.36
Sum of two sides 1800 mm	151.44	191.67	2.61	84.24	nr	275.91
Sum of two sides 1900 mm	166.06	210.17	2.71	87.46	nr	297.63
Sum of two sides 2000 mm	178.54	225.96	2.71	87.46	nr	313.42
Square to round						
Sum of two sides 1100 mm	65.19	82.50	1.82	58.73	nr	141.23
Sum of two sides 1200 mm	68.60	86.82	1.82	58.73	nr	145.55
Sum of two sides 1300 mm	72.03	91.16	2.32	74.87	nr	166.03
Sum of two sides 1400 mm	74.83	94.71	2.32	74.87	nr	169.58
Sum of two sides 1500 mm	88.88	112.48	2.56	82.62	nr	195.10
Sum of two sides 1600 mm	102.94	130.28	2.58	83.26	nr	213.54
Sum of two sides 1700 mm	116.99	148.06	2.84	91.65	nr	239.71
Sum of two sides 1800 mm	131.07	165.88	2.84	91.65	nr	257.53
Sum of two sides 1900 mm	145.12	183.67	3.13	101.01	nr	284.68
Sum of two sides 2000 mm	159.18	201.45	3.13	101.01	nr	302.46
90° radius bend						
Sum of two sides 1100 mm	49.44	62.57	1.40	45.18	nr	107.75
Sum of two sides 1200 mm	53.36	67.54	1.40	45.18	nr	112.72
Sum of two sides 1300 mm	57.29	72.51	1.91	61.63	nr	134.14
Sum of two sides 1400 mm	60.81	76.97	1.91	61.63	nr	138.60
Sum of two sides 1500 mm	73.22	92.67	2.11	68.08	nr	160.75
Sum of two sides 1600 mm	85.64	108.38	2.11	68.08	nr	176.46
Sum of two sides 1700 mm	98.06	124.11	2.55	82.30	nr	206.41
Sum of two sides 1800 mm	110.61	139.99	2.55	82.30	nr	222.29
Sum of two sides 1900 mm	120.80	152.88	2.80	90.36	nr	243.24
Sum of two sides 2000 mm	133.21	168.59	2.80	90.36	nr	258.95
45° bend						
Sum of two sides 1100 mm	64.12	81.16	1.88	60.67	nr	141.83
Sum of two sides 1200 mm	68.06	86.14	1.88	60.67	nr	146.81
Sum of two sides 1300 mm	72.02	91.15	2.26	72.93	nr	164.08

38 VENTILATION/AIR CONDITIONING SYSTEMS

Item	Net Price £	Material £	Labour hours	Labour £	Unit	Total rate £
DUCTWORK: RECTANGULAR – CLASS C – cont						
Extra over fittings – cont						
45° bend – cont						
Sum of two sides 1400 mm	75.72	95.83	2.44	78.75	nr	174.58
Sum of two sides 1500 mm	88.17	111.59	2.68	86.49	nr	198.08
Sum of two sides 1600 mm	100.61	127.33	2.69	86.81	nr	214.14
Sum of two sides 1700 mm	113.06	143.09	2.96	95.52	nr	238.61
Sum of two sides 1800 mm	125.77	159.17	2.96	95.52	nr	254.69
Sum of two sides 1900 mm	137.07	173.48	3.26	105.21	nr	278.69
Sum of two sides 2000 mm	149.53	189.25	3.26	105.21	nr	294.46
90° mitre bend						
Sum of two sides 1100 mm	102.32	129.49	3.03	97.80	nr	227.29
Sum of two sides 1200 mm	109.74	138.89	3.03	97.80	nr	236.69
Sum of two sides 1300 mm	117.16	148.28	3.85	124.24	nr	272.52
Sum of two sides 1400 mm	123.89	156.80	3.85	124.24	nr	281.04
Sum of two sides 1500 mm	144.06	182.32	4.25	137.17	nr	319.49
Sum of two sides 1600 mm	164.23	207.85	4.26	137.48	nr	345.33
Sum of two sides 1700 mm	184.41	233.39	4.68	151.03	nr	384.42
Sum of two sides 1800 mm	204.67	259.03	4.68	151.03	nr	410.06
Sum of two sides 1900 mm	222.48	281.57	4.87	157.17	nr	438.74
Sum of two sides 2000 mm	242.77	307.25	4.87	157.17	nr	464.42
Branch						
Sum of two sides 1100 mm	97.96	123.97	1.29	41.64	nr	165.61
Sum of two sides 1200 mm	103.28	130.72	1.29	41.64	nr	172.36
Sum of two sides 1300 mm	108.60	137.45	1.39	44.87	nr	182.32
Sum of two sides 1400 mm	113.43	143.56	1.39	44.87	nr	188.43
Sum of two sides 1500 mm	130.09	164.64	1.64	52.93	nr	217.57
Sum of two sides 1600 mm	146.72	185.68	1.64	52.93	nr	238.61
Sum of two sides 1700 mm	163.37	206.76	1.69	54.53	nr	261.29
Sum of two sides 1800 mm	180.25	228.12	1.69	54.53	nr	282.65
Sum of two sides 1900 mm	196.89	249.19	1.85	59.70	nr	308.89
Sum of two sides 2000 mm	213.53	270.24	1.85	59.70	nr	329.94
Grille neck						
Sum of two sides 1100 mm	101.41	128.34	1.44	46.47	nr	174.81
Sum of two sides 1200 mm	107.35	135.87	1.44	46.47	nr	182.34
Sum of two sides 1300 mm	113.29	143.38	1.69	54.53	nr	197.91
Sum of two sides 1400 mm	118.43	149.89	1.69	54.53	nr	204.42
Sum of two sides 1500 mm	137.12	173.54	1.79	57.77	nr	231.31
Sum of two sides 1600 mm	155.80	197.18	1.79	57.77	nr	254.95
Sum of two sides 1700 mm	174.49	220.83	1.86	60.03	nr	280.86
Sum of two sides 1800 mm	193.17	244.47	1.86	60.03	nr	304.50
Sum of two sides 1900 mm	211.86	268.13	2.02	65.20	nr	333.33
Sum of two sides 2000 mm	230.54	291.77	2.02	65.20	nr	356.97
Ductwork 1001 to 1250 mm longest side						
Sum of two sides 1300 mm	109.60	138.71	1.51	48.74	m	187.45
Sum of two sides 1400 mm	114.97	145.51	1.55	50.02	m	195.53
Sum of two sides 1500 mm	120.03	151.91	1.61	51.96	m	203.87

38 VENTILATION/AIR CONDITIONING SYSTEMS

Item	Net Price £	Material £	Labour hours	Labour £	Unit	Total rate £
Sum of two sides 1600 mm	125.40	158.70	1.62	52.28	m	**210.98**
Sum of two sides 1700 mm	130.47	165.12	1.74	56.16	m	**221.28**
Sum of two sides 1800 mm	135.54	171.54	1.76	56.80	m	**228.34**
Sum of two sides 1900 mm	140.61	177.96	1.81	58.42	m	**236.38**
Sum of two sides 2000 mm	145.97	184.74	1.82	58.73	m	**243.47**
Sum of two sides 2100 mm	151.05	191.17	2.53	81.66	m	**272.83**
Sum of two sides 2200 mm	171.77	217.39	2.55	82.30	m	**299.69**
Sum of two sides 2300 mm	177.47	224.60	2.56	82.62	m	**307.22**
Sum of two sides 2400 mm	182.53	231.01	2.76	89.07	m	**320.08**
Sum of two sides 2500 mm	187.90	237.81	2.77	89.40	m	**327.21**
Extra over fittings; Ductwork 1001 to 1250 mm longest side						
End cap						
Sum of two sides 1300 mm	36.79	46.56	0.38	12.26	nr	**58.82**
Sum of two sides 1400 mm	38.44	48.65	0.38	12.26	nr	**60.91**
Sum of two sides 1500 mm	44.58	56.43	0.38	12.26	nr	**68.69**
Sum of two sides 1600 mm	50.74	64.22	0.38	12.26	nr	**76.48**
Sum of two sides 1700 mm	56.88	71.98	0.58	18.72	nr	**90.70**
Sum of two sides 1800 mm	63.03	79.77	0.58	18.72	nr	**98.49**
Sum of two sides 1900 mm	69.17	87.54	0.58	18.72	nr	**106.26**
Sum of two sides 2000 mm	75.33	95.33	0.58	18.72	nr	**114.05**
Sum of two sides 2100 mm	81.47	103.11	0.87	28.07	nr	**131.18**
Sum of two sides 2200 mm	87.63	110.90	0.87	28.07	nr	**138.97**
Sum of two sides 2300 mm	93.78	118.69	0.87	28.07	nr	**146.76**
Sum of two sides 2400 mm	99.92	126.46	0.87	28.07	nr	**154.53**
Sum of two sides 2500 mm	106.07	134.24	0.87	28.07	nr	**162.31**
Reducer						
Sum of two sides 1300 mm	52.20	66.07	2.30	74.22	nr	**140.29**
Sum of two sides 1400 mm	54.28	68.70	2.30	74.22	nr	**142.92**
Sum of two sides 1500 mm	65.32	82.67	2.47	79.71	nr	**162.38**
Sum of two sides 1600 mm	76.64	96.99	2.47	79.71	nr	**176.70**
Sum of two sides 1700 mm	87.70	110.99	2.59	83.60	nr	**194.59**
Sum of two sides 1800 mm	98.77	125.00	2.59	83.60	nr	**208.60**
Sum of two sides 1900 mm	109.82	138.99	2.71	87.46	nr	**226.45**
Sum of two sides 2000 mm	121.13	153.31	2.71	87.46	nr	**240.77**
Sum of two sides 2100 mm	132.20	167.32	2.92	94.24	nr	**261.56**
Sum of two sides 2200 mm	134.05	169.66	2.92	94.24	nr	**263.90**
Sum of two sides 2300 mm	145.37	183.98	2.92	94.24	nr	**278.22**
Sum of two sides 2400 mm	156.44	197.99	3.12	100.69	nr	**298.68**
Sum of two sides 2500 mm	167.74	212.30	3.12	100.69	nr	**312.99**
Offset						
Sum of two sides 1300 mm	124.35	157.38	2.47	79.71	nr	**237.09**
Sum of two sides 1400 mm	128.05	162.06	2.47	79.71	nr	**241.77**
Sum of two sides 1500 mm	142.35	180.16	2.47	79.71	nr	**259.87**
Sum of two sides 1600 mm	156.50	198.06	2.47	79.71	nr	**277.77**
Sum of two sides 1700 mm	170.03	215.19	2.61	84.24	nr	**299.43**
Sum of two sides 1800 mm	183.22	231.88	2.61	84.24	nr	**316.12**
Sum of two sides 1900 mm	196.04	248.11	2.71	87.46	nr	**335.57**
Sum of two sides 2000 mm	208.61	264.02	2.71	87.46	nr	**351.48**

38 VENTILATION/AIR CONDITIONING SYSTEMS

Item	Net Price £	Material £	Labour hours	Labour £	Unit	Total rate £
DUCTWORK: RECTANGULAR – CLASS C – cont						
Extra over fittings – cont						
Offset – cont						
Sum of two sides 2100 mm	220.67	279.28	2.92	94.24	nr	**373.52**
Sum of two sides 2200 mm	217.48	275.24	3.26	105.21	nr	**380.45**
Sum of two sides 2300 mm	235.93	298.59	3.26	105.21	nr	**403.80**
Sum of two sides 2400 mm	254.31	321.85	3.47	111.98	nr	**433.83**
Sum of two sides 2500 mm	272.75	345.20	3.48	112.31	nr	**457.51**
Square to round						
Sum of two sides 1300 mm	56.74	71.81	2.32	74.87	nr	**146.68**
Sum of two sides 1400 mm	59.17	74.88	2.32	74.87	nr	**149.75**
Sum of two sides 1500 mm	72.83	92.18	2.56	82.62	nr	**174.80**
Sum of two sides 1600 mm	86.53	109.51	2.58	83.26	nr	**192.77**
Sum of two sides 1700 mm	100.21	126.83	2.84	91.65	nr	**218.48**
Sum of two sides 1800 mm	113.87	144.11	2.84	91.65	nr	**235.76**
Sum of two sides 1900 mm	127.54	161.41	3.13	101.01	nr	**262.42**
Sum of two sides 2000 mm	141.24	178.75	3.13	101.01	nr	**279.76**
Sum of two sides 2100 mm	154.90	196.04	4.26	137.48	nr	**333.52**
Sum of two sides 2200 mm	159.37	201.70	4.27	137.82	nr	**339.52**
Sum of two sides 2300 mm	173.08	219.05	4.30	138.77	nr	**357.82**
Sum of two sides 2400 mm	186.76	236.36	4.77	153.94	nr	**390.30**
Sum of two sides 2500 mm	200.46	253.70	4.79	154.58	nr	**408.28**
90° radius bend						
Sum of two sides 1300 mm	23.19	29.34	1.91	61.63	nr	**90.97**
Sum of two sides 1400 mm	22.78	28.83	1.91	61.63	nr	**90.46**
Sum of two sides 1500 mm	34.99	44.28	2.11	68.08	nr	**112.36**
Sum of two sides 1600 mm	47.25	59.80	2.11	68.08	nr	**127.88**
Sum of two sides 1700 mm	59.43	75.22	2.55	82.30	nr	**157.52**
Sum of two sides 1800 mm	71.61	90.63	2.55	82.30	nr	**172.93**
Sum of two sides 1900 mm	83.78	106.03	2.80	90.36	nr	**196.39**
Sum of two sides 2000 mm	96.07	121.59	2.80	90.36	nr	**211.95**
Sum of two sides 2100 mm	108.23	136.98	2.80	90.36	nr	**227.34**
Sum of two sides 2200 mm	102.65	129.91	2.80	90.36	nr	**220.27**
Sum of two sides 2300 mm	105.91	134.04	4.35	140.38	nr	**274.42**
Sum of two sides 2400 mm	118.07	149.43	4.35	140.38	nr	**289.81**
Sum of two sides 2500 mm	130.26	164.85	4.35	140.38	nr	**305.23**
45° bend						
Sum of two sides 1300 mm	54.14	68.52	2.26	72.93	nr	**141.45**
Sum of two sides 1400 mm	55.61	70.38	2.44	78.75	nr	**149.13**
Sum of two sides 1500 mm	67.93	85.97	2.68	86.49	nr	**172.46**
Sum of two sides 1600 mm	80.52	101.91	2.69	86.81	nr	**188.72**
Sum of two sides 1700 mm	92.84	117.50	2.96	95.52	nr	**213.02**
Sum of two sides 1800 mm	105.16	133.09	2.96	95.52	nr	**228.61**
Sum of two sides 1900 mm	117.50	148.71	3.26	105.21	nr	**253.92**
Sum of two sides 2000 mm	130.08	164.63	3.26	105.21	nr	**269.84**
Sum of two sides 2100 mm	142.40	180.22	7.50	242.04	nr	**422.26**
Sum of two sides 2200 mm	145.08	183.61	7.50	242.04	nr	**425.65**
Sum of two sides 2300 mm	153.02	193.66	7.55	243.67	nr	**437.33**
Sum of two sides 2400 mm	165.35	209.27	8.13	262.37	nr	**471.64**
Sum of two sides 2500 mm	177.89	225.14	8.30	267.86	nr	**493.00**

38 VENTILATION/AIR CONDITIONING SYSTEMS

Item	Net Price £	Material £	Labour hours	Labour £	Unit	Total rate £
90° mitre bend						
Sum of two sides 1300 mm	123.65	156.49	3.85	124.24	nr	**280.73**
Sum of two sides 1400 mm	128.83	163.05	3.85	124.24	nr	**287.29**
Sum of two sides 1500 mm	152.38	192.85	4.25	137.17	nr	**330.02**
Sum of two sides 1600 mm	175.92	222.64	4.26	137.48	nr	**360.12**
Sum of two sides 1700 mm	199.46	252.44	4.68	151.03	nr	**403.47**
Sum of two sides 1800 mm	223.00	282.23	4.68	151.03	nr	**433.26**
Sum of two sides 1900 mm	246.53	312.01	4.87	157.17	nr	**469.18**
Sum of two sides 2000 mm	270.09	341.82	4.87	157.17	nr	**498.99**
Sum of two sides 2100 mm	293.62	371.60	7.50	242.04	nr	**613.64**
Sum of two sides 2200 mm	296.46	375.20	7.50	242.04	nr	**617.24**
Sum of two sides 2300 mm	309.60	391.83	7.55	243.67	nr	**635.50**
Sum of two sides 2400 mm	333.29	421.81	8.13	262.37	nr	**684.18**
Sum of two sides 2500 mm	356.98	451.80	8.30	267.86	nr	**719.66**
Branch						
Sum of two sides 1300 mm	111.39	140.97	1.39	44.87	nr	**185.84**
Sum of two sides 1400 mm	116.17	147.02	1.39	44.87	nr	**191.89**
Sum of two sides 1500 mm	132.99	168.31	1.64	52.93	nr	**221.24**
Sum of two sides 1600 mm	150.04	189.90	1.64	52.93	nr	**242.83**
Sum of two sides 1700 mm	166.85	211.16	1.69	54.53	nr	**265.69**
Sum of two sides 1800 mm	183.66	232.44	1.69	54.53	nr	**286.97**
Sum of two sides 1900 mm	200.48	253.72	1.85	59.70	nr	**313.42**
Sum of two sides 2000 mm	217.52	275.30	1.85	59.70	nr	**335.00**
Sum of two sides 2100 mm	234.34	296.58	2.61	84.24	nr	**380.82**
Sum of two sides 2200 mm	251.13	317.83	2.61	84.24	nr	**402.07**
Sum of two sides 2300 mm	268.19	339.42	2.61	84.24	nr	**423.66**
Sum of two sides 2400 mm	285.00	360.70	2.88	92.95	nr	**453.65**
Sum of two sides 2500 mm	302.05	382.28	2.88	92.95	nr	**475.23**
Grille neck						
Sum of two sides 1300 mm	116.97	148.04	1.69	54.53	nr	**202.57**
Sum of two sides 1400 mm	122.39	154.90	1.69	54.53	nr	**209.43**
Sum of two sides 1500 mm	141.37	178.92	1.79	57.77	nr	**236.69**
Sum of two sides 1600 mm	160.33	202.91	1.79	57.77	nr	**260.68**
Sum of two sides 1700 mm	179.29	226.91	1.86	60.03	nr	**286.94**
Sum of two sides 1800 mm	198.26	250.91	1.86	60.03	nr	**310.94**
Sum of two sides 1900 mm	217.24	274.94	2.02	65.20	nr	**340.14**
Sum of two sides 2000 mm	236.20	298.94	2.02	65.20	nr	**364.14**
Sum of two sides 2100 mm	255.17	322.94	2.61	84.24	nr	**407.18**
Sum of two sides 2200 mm	274.14	346.95	2.80	90.36	nr	**437.31**
Sum of two sides 2300 mm	293.11	370.96	2.80	90.36	nr	**461.32**
Sum of two sides 2400 mm	312.11	395.00	3.06	98.75	nr	**493.75**
Sum of two sides 2500 mm	331.04	418.97	3.06	98.75	nr	**517.72**
Ductwork 1251 to 1600 mm longest side						
Sum of two sides 1700 mm	138.72	175.56	1.74	56.16	m	**231.72**
Sum of two sides 1800 mm	144.25	182.56	1.76	56.80	m	**239.36**
Sum of two sides 1900 mm	149.83	189.63	1.81	58.42	m	**248.05**
Sum of two sides 2000 mm	155.10	196.29	1.82	58.73	m	**255.02**
Sum of two sides 2100 mm	160.37	202.97	2.53	81.66	m	**284.63**
Sum of two sides 2200 mm	165.65	209.64	2.55	82.30	m	**291.94**
Sum of two sides 2300 mm	171.24	216.72	2.56	82.62	m	**299.34**

38 VENTILATION/AIR CONDITIONING SYSTEMS

Item	Net Price £	Material £	Labour hours	Labour £	Unit	Total rate £
DUCTWORK: RECTANGULAR – CLASS C – cont						
Ductwork 1251 to 1600 mm longest side – cont						
Grille neck – cont						
Sum of two sides 2400 mm	176.75	223.70	2.76	89.07	m	**312.77**
Sum of two sides 2500 mm	182.04	230.40	2.77	89.40	m	**319.80**
Sum of two sides 2600 mm	203.26	257.24	2.97	95.84	m	**353.08**
Sum of two sides 2700 mm	208.53	263.92	2.99	96.50	m	**360.42**
Sum of two sides 2800 mm	214.16	271.04	3.30	106.50	m	**377.54**
Sum of two sides 2900 mm	219.72	278.07	3.31	106.83	m	**384.90**
Sum of two sides 3000 mm	225.00	284.76	3.53	113.93	m	**398.69**
Sum of two sides 3100 mm	230.53	291.76	3.55	114.58	m	**406.34**
Sum of two sides 3200 mm	235.81	298.45	3.56	114.89	m	**413.34**
Extra over fittings; Ductwork 1251 to 1600 mm longest side						
End cap						
Sum of two sides 1700 mm	56.88	71.98	0.58	18.72	nr	**90.70**
Sum of two sides 1800 mm	63.21	80.00	0.58	18.72	nr	**98.72**
Sum of two sides 1900 mm	69.17	87.54	0.58	18.72	nr	**106.26**
Sum of two sides 2000 mm	81.47	103.11	0.58	18.72	nr	**121.83**
Sum of two sides 2100 mm	87.63	110.90	0.87	28.07	nr	**138.97**
Sum of two sides 2200 mm	93.78	118.69	0.87	28.07	nr	**146.76**
Sum of two sides 2300 mm	99.92	126.46	0.87	28.07	nr	**154.53**
Sum of two sides 2400 mm	106.07	134.24	0.87	28.07	nr	**162.31**
Sum of two sides 2500 mm	112.24	142.05	0.87	28.07	nr	**170.12**
Sum of two sides 2600 mm	118.37	149.81	0.87	28.07	nr	**177.88**
Sum of two sides 2700 mm	124.53	157.61	0.87	28.07	nr	**185.68**
Sum of two sides 2800 mm	130.68	165.39	1.16	37.44	nr	**202.83**
Sum of two sides 2900 mm	136.83	173.17	1.16	37.44	nr	**210.61**
Sum of two sides 3000 mm	211.86	268.13	1.73	55.84	nr	**323.97**
Sum of two sides 3100 mm	141.72	179.36	1.73	55.84	nr	**235.20**
Sum of two sides 3200 mm	146.15	184.97	1.73	55.84	nr	**240.81**
Reducer						
Sum of two sides 1700 mm	76.67	97.04	2.59	83.60	nr	**180.64**
Sum of two sides 1800 mm	87.15	110.30	2.59	83.60	nr	**193.90**
Sum of two sides 1900 mm	97.81	123.79	2.71	87.46	nr	**211.25**
Sum of two sides 2000 mm	75.33	95.33	2.71	87.46	nr	**182.79**
Sum of two sides 2100 mm	108.27	137.03	2.92	94.24	nr	**231.27**
Sum of two sides 2200 mm	118.76	150.30	2.92	94.24	nr	**244.54**
Sum of two sides 2300 mm	129.23	163.55	2.92	94.24	nr	**257.79**
Sum of two sides 2400 mm	139.89	177.05	3.12	100.69	nr	**277.74**
Sum of two sides 2500 mm	150.39	190.33	3.12	100.69	nr	**291.02**
Sum of two sides 2600 mm	160.86	203.58	3.16	101.99	nr	**305.57**
Sum of two sides 2700 mm	158.82	201.01	3.16	101.99	nr	**303.00**
Sum of two sides 2800 mm	169.30	214.27	4.00	129.09	nr	**343.36**
Sum of two sides 2900 mm	179.72	227.45	4.01	129.40	nr	**356.85**
Sum of two sides 3000 mm	201.38	254.87	4.56	147.17	nr	**402.04**
Sum of two sides 3100 mm	219.58	277.91	4.56	147.17	nr	**425.08**
Sum of two sides 3200 mm	226.63	286.82	4.56	147.17	nr	**433.99**

38 VENTILATION/AIR CONDITIONING SYSTEMS

Item	Net Price £	Material £	Labour hours	Labour £	Unit	Total rate £
Offset						
Sum of two sides 1700 mm	188.72	238.84	2.61	84.24	nr	**323.08**
Sum of two sides 1800 mm	208.41	263.76	2.61	84.24	nr	**348.00**
Sum of two sides 1900 mm	221.44	280.26	2.71	87.46	nr	**367.72**
Sum of two sides 2000 mm	233.83	295.94	2.71	87.46	nr	**383.40**
Sum of two sides 2100 mm	245.84	311.14	2.92	94.24	nr	**405.38**
Sum of two sides 2200 mm	257.46	325.84	3.26	105.21	nr	**431.05**
Sum of two sides 2300 mm	268.76	340.14	3.26	105.21	nr	**445.35**
Sum of two sides 2400 mm	287.47	363.82	3.47	111.98	nr	**475.80**
Sum of two sides 2500 mm	298.06	377.23	3.48	112.31	nr	**489.54**
Sum of two sides 2600 mm	297.94	377.07	3.49	112.64	nr	**489.71**
Sum of two sides 2700 mm	316.43	400.48	3.50	112.95	nr	**513.43**
Sum of two sides 2800 mm	334.85	423.79	4.33	139.74	nr	**563.53**
Sum of two sides 2900 mm	364.41	461.19	4.74	152.97	nr	**614.16**
Sum of two sides 3000 mm	382.91	484.61	5.31	171.37	nr	**655.98**
Sum of two sides 3100 mm	397.27	502.79	5.34	172.33	nr	**675.12**
Sum of two sides 3200 mm	410.63	519.69	5.35	172.67	nr	**692.36**
Square to round						
Sum of two sides 1700 mm	84.04	106.37	2.84	91.65	nr	**198.02**
Sum of two sides 1800 mm	97.13	122.93	2.84	91.65	nr	**214.58**
Sum of two sides 1900 mm	110.15	139.41	3.13	101.01	nr	**240.42**
Sum of two sides 2000 mm	123.21	155.94	3.13	101.01	nr	**256.95**
Sum of two sides 2100 mm	136.25	172.44	4.26	137.48	nr	**309.92**
Sum of two sides 2200 mm	149.31	188.97	4.27	137.82	nr	**326.79**
Sum of two sides 2300 mm	162.33	205.44	4.30	138.77	nr	**344.21**
Sum of two sides 2400 mm	175.42	222.01	4.77	153.94	nr	**375.95**
Sum of two sides 2500 mm	188.48	238.54	4.79	154.58	nr	**393.12**
Sum of two sides 2600 mm	189.04	239.25	4.95	159.75	nr	**399.00**
Sum of two sides 2700 mm	202.11	255.79	4.95	159.75	nr	**415.54**
Sum of two sides 2800 mm	215.13	272.27	8.49	274.00	nr	**546.27**
Sum of two sides 2900 mm	233.63	295.68	8.88	286.59	nr	**582.27**
Sum of two sides 3000 mm	246.71	312.23	9.02	291.10	nr	**603.33**
Sum of two sides 3100 mm	256.37	324.46	9.02	291.10	nr	**615.56**
Sum of two sides 3200 mm	265.14	335.56	9.09	293.35	nr	**628.91**
90° radius bend						
Sum of two sides 1700 mm	190.43	241.01	2.55	82.30	nr	**323.31**
Sum of two sides 1800 mm	206.13	260.88	2.55	82.30	nr	**343.18**
Sum of two sides 1900 mm	230.99	292.34	2.80	90.36	nr	**382.70**
Sum of two sides 2000 mm	255.35	323.18	2.80	90.36	nr	**413.54**
Sum of two sides 2100 mm	279.72	354.01	2.61	84.24	nr	**438.25**
Sum of two sides 2200 mm	304.09	384.85	2.62	84.56	nr	**469.41**
Sum of two sides 2300 mm	328.92	416.28	2.63	84.87	nr	**501.15**
Sum of two sides 2400 mm	343.97	435.33	4.34	140.07	nr	**575.40**
Sum of two sides 2500 mm	368.29	466.11	4.35	140.38	nr	**606.49**
Sum of two sides 2600 mm	370.19	468.51	4.53	146.20	nr	**614.71**
Sum of two sides 2700 mm	394.51	499.30	4.53	146.20	nr	**645.50**
Sum of two sides 2800 mm	407.66	515.94	7.13	230.10	nr	**746.04**
Sum of two sides 2900 mm	454.35	575.03	7.17	231.40	nr	**806.43**
Sum of two sides 3000 mm	478.63	605.75	7.26	234.30	nr	**840.05**
Sum of two sides 3100 mm	497.43	629.55	7.26	234.30	nr	**863.85**
Sum of two sides 3200 mm	514.86	651.60	7.31	235.92	nr	**887.52**

38 VENTILATION/AIR CONDITIONING SYSTEMS

Item	Net Price £	Material £	Labour hours	Labour £	Unit	Total rate £
DUCTWORK: RECTANGULAR – CLASS C – cont						
Extra over fittings – cont						
45° bend						
Sum of two sides 1700 mm	91.61	115.94	2.96	95.52	nr	**211.46**
Sum of two sides 1800 mm	99.34	125.72	2.96	95.52	nr	**221.24**
Sum of two sides 1900 mm	111.75	141.43	3.26	105.21	nr	**246.64**
Sum of two sides 2000 mm	123.95	156.87	3.26	105.21	nr	**262.08**
Sum of two sides 2100 mm	136.12	172.28	7.50	242.04	nr	**414.32**
Sum of two sides 2200 mm	148.31	187.70	7.50	242.04	nr	**429.74**
Sum of two sides 2300 mm	160.73	203.41	7.55	243.67	nr	**447.08**
Sum of two sides 2400 mm	168.12	212.78	8.13	262.37	nr	**475.15**
Sum of two sides 2500 mm	180.28	228.17	8.30	267.86	nr	**496.03**
Sum of two sides 2600 mm	147.69	186.92	8.56	276.26	nr	**463.18**
Sum of two sides 2700 mm	192.63	243.79	8.62	278.20	nr	**521.99**
Sum of two sides 2800 mm	199.07	251.94	9.09	293.35	nr	**545.29**
Sum of two sides 2900 mm	222.40	281.47	9.09	293.35	nr	**574.82**
Sum of two sides 3000 mm	234.56	296.86	9.62	310.46	nr	**607.32**
Sum of two sides 3100 mm	243.93	308.72	9.62	310.46	nr	**619.18**
Sum of two sides 3200 mm	252.65	319.75	9.62	310.46	nr	**630.21**
90° mitre bend						
Sum of two sides 1700 mm	200.39	253.61	4.68	151.03	nr	**404.64**
Sum of two sides 1800 mm	212.81	269.34	4.68	151.03	nr	**420.37**
Sum of two sides 1900 mm	236.86	299.77	4.87	157.17	nr	**456.94**
Sum of two sides 2000 mm	260.97	330.29	4.87	157.17	nr	**487.46**
Sum of two sides 2100 mm	285.09	360.81	7.50	242.04	nr	**602.85**
Sum of two sides 2200 mm	309.21	391.34	7.50	242.04	nr	**633.38**
Sum of two sides 2300 mm	333.25	421.76	7.55	243.67	nr	**665.43**
Sum of two sides 2400 mm	345.77	437.61	8.13	262.37	nr	**699.98**
Sum of two sides 2500 mm	370.02	468.29	8.30	267.86	nr	**736.15**
Sum of two sides 2600 mm	367.70	465.36	8.56	276.26	nr	**741.62**
Sum of two sides 2700 mm	391.94	496.04	8.62	278.20	nr	**774.24**
Sum of two sides 2800 mm	402.63	509.57	15.20	490.55	nr	**1000.12**
Sum of two sides 2900 mm	445.77	564.17	15.20	490.55	nr	**1054.72**
Sum of two sides 3000 mm	470.14	595.01	15.60	503.45	nr	**1098.46**
Sum of two sides 3100 mm	490.22	620.42	16.04	517.65	nr	**1138.07**
Sum of two sides 3200 mm	509.45	644.76	16.04	517.65	nr	**1162.41**
Branch						
Sum of two sides 1700 mm	171.97	217.65	1.69	54.53	nr	**272.18**
Sum of two sides 1800 mm	188.79	238.93	1.69	54.53	nr	**293.46**
Sum of two sides 1900 mm	205.87	260.55	1.85	59.70	nr	**320.25**
Sum of two sides 2000 mm	222.70	281.85	1.85	59.70	nr	**341.55**
Sum of two sides 2100 mm	239.55	303.17	2.61	84.24	nr	**387.41**
Sum of two sides 2200 mm	256.41	324.51	2.61	84.24	nr	**408.75**
Sum of two sides 2300 mm	273.48	346.11	2.61	84.24	nr	**430.35**
Sum of two sides 2400 mm	290.28	367.38	2.88	92.95	nr	**460.33**
Sum of two sides 2500 mm	307.14	388.72	2.88	92.95	nr	**481.67**
Sum of two sides 2600 mm	323.98	410.03	2.88	92.95	nr	**502.98**
Sum of two sides 2700 mm	340.83	431.36	2.88	92.95	nr	**524.31**

38 VENTILATION/AIR CONDITIONING SYSTEMS

Item	Net Price £	Material £	Labour hours	Labour £	Unit	Total rate £
Sum of two sides 2800 mm	357.63	452.61	3.94	127.15	nr	**579.76**
Sum of two sides 2900 mm	380.20	481.19	3.94	127.15	nr	**608.34**
Sum of two sides 3000 mm	397.05	502.51	4.83	155.88	nr	**658.39**
Sum of two sides 3100 mm	410.52	519.56	4.83	155.88	nr	**675.44**
Sum of two sides 3200 mm	422.79	535.08	4.83	155.88	nr	**690.96**
Grille neck						
Sum of two sides 1700 mm	179.29	226.91	1.86	60.03	nr	**286.94**
Sum of two sides 1800 mm	198.26	250.91	1.86	60.03	nr	**310.94**
Sum of two sides 1900 mm	217.24	274.94	2.02	65.20	nr	**340.14**
Sum of two sides 2000 mm	236.20	298.94	2.02	65.20	nr	**364.14**
Sum of two sides 2100 mm	255.17	322.94	2.80	90.36	nr	**413.30**
Sum of two sides 2200 mm	274.14	346.95	2.80	90.36	nr	**437.31**
Sum of two sides 2300 mm	293.11	370.96	2.80	90.36	nr	**461.32**
Sum of two sides 2400 mm	312.07	394.96	3.06	98.75	nr	**493.71**
Sum of two sides 2500 mm	331.04	418.97	3.06	98.75	nr	**517.72**
Sum of two sides 2600 mm	350.00	442.96	3.08	99.40	nr	**542.36**
Sum of two sides 2700 mm	368.98	466.98	3.08	99.40	nr	**566.38**
Sum of two sides 2800 mm	387.93	490.96	4.13	133.28	nr	**624.24**
Sum of two sides 2900 mm	406.91	514.99	4.13	133.28	nr	**648.27**
Sum of two sides 3000 mm	425.88	538.99	5.02	162.01	nr	**701.00**
Sum of two sides 3100 mm	441.06	558.21	5.02	162.01	nr	**720.22**
Sum of two sides 3200 mm	454.87	575.68	5.02	162.01	nr	**737.69**
Ductwork 1601 to 2000 mm longest side						
Sum of two sides 2100 mm	185.75	235.09	2.53	81.66	m	**316.75**
Sum of two sides 2200 mm	191.85	242.80	2.55	82.30	m	**325.10**
Sum of two sides 2300 mm	198.01	250.60	2.55	82.30	m	**332.90**
Sum of two sides 2400 mm	204.05	258.25	2.56	82.62	m	**340.87**
Sum of two sides 2500 mm	210.21	266.04	2.76	89.07	m	**355.11**
Sum of two sides 2600 mm	216.07	273.46	2.77	89.40	m	**362.86**
Sum of two sides 2700 mm	221.91	280.85	2.97	95.84	m	**376.69**
Sum of two sides 2800 mm	227.79	288.29	2.99	96.50	m	**384.79**
Sum of two sides 2900 mm	233.93	296.06	3.30	106.50	m	**402.56**
Sum of two sides 3000 mm	239.81	303.51	3.31	106.83	m	**410.34**
Sum of two sides 3100 mm	274.21	347.04	3.53	113.93	m	**460.97**
Sum of two sides 3200 mm	280.07	354.46	3.53	113.93	m	**468.39**
Sum of two sides 3300 mm	286.23	362.25	3.53	113.93	m	**476.18**
Sum of two sides 3400 mm	292.09	369.67	3.55	114.58	m	**484.25**
Sum of two sides 3500 mm	297.95	377.08	3.55	114.58	m	**491.66**
Sum of two sides 3600 mm	303.80	384.48	3.55	114.58	m	**499.06**
Sum of two sides 3700 mm	309.96	392.28	3.56	114.89	m	**507.17**
Sum of two sides 3800 mm	315.81	399.69	3.56	114.89	m	**514.58**
Sum of two sides 3900 mm	321.68	407.12	3.56	114.89	m	**522.01**
Sum of two sides 4000 mm	327.54	414.53	3.56	114.89	m	**529.42**
Extra over fittings; Ductwork 1601 to 2000 mm longest side						
End cap						
Sum of two sides 2100 mm	82.53	104.45	0.87	28.07	nr	**132.52**
Sum of two sides 2200 mm	88.73	112.29	0.87	28.07	nr	**140.36**
Sum of two sides 2300 mm	94.94	120.15	0.87	28.07	nr	**148.22**

38 VENTILATION/AIR CONDITIONING SYSTEMS

Item	Net Price £	Material £	Labour hours	Labour £	Unit	Total rate £
DUCTWORK: RECTANGULAR – CLASS C – cont						
Extra over fittings – cont						
End cap – cont						
Sum of two sides 2400 mm	101.14	128.00	0.87	28.07	nr	**156.07**
Sum of two sides 2500 mm	107.34	135.84	0.87	28.07	nr	**163.91**
Sum of two sides 2600 mm	113.54	143.70	0.87	28.07	nr	**171.77**
Sum of two sides 2700 mm	119.73	151.52	0.87	28.07	nr	**179.59**
Sum of two sides 2800 mm	125.93	159.38	1.16	37.44	nr	**196.82**
Sum of two sides 2900 mm	132.14	167.24	1.16	37.44	nr	**204.68**
Sum of two sides 3000 mm	138.33	175.07	1.73	55.84	nr	**230.91**
Sum of two sides 3100 mm	143.26	181.31	1.73	55.84	nr	**237.15**
Sum of two sides 3200 mm	147.74	186.98	1.73	55.84	nr	**242.82**
Sum of two sides 3300 mm	152.23	192.66	1.73	55.84	nr	**248.50**
Sum of two sides 3400 mm	156.72	198.34	1.73	55.84	nr	**254.18**
Sum of two sides 3500 mm	161.21	204.03	1.73	55.84	nr	**259.87**
Sum of two sides 3600 mm	165.69	209.70	1.73	55.84	nr	**265.54**
Sum of two sides 3700 mm	170.18	215.38	1.73	55.84	nr	**271.22**
Sum of two sides 3800 mm	174.65	221.03	1.73	55.84	nr	**276.87**
Sum of two sides 3900 mm	179.15	226.73	1.73	55.84	nr	**282.57**
Sum of two sides 4000 mm	183.63	232.40	1.73	55.84	nr	**288.24**
Reducer						
Sum of two sides 2100 mm	110.86	140.30	2.92	94.24	nr	**234.54**
Sum of two sides 2200 mm	121.31	153.53	2.92	94.24	nr	**247.77**
Sum of two sides 2300 mm	131.94	166.98	2.92	94.24	nr	**261.22**
Sum of two sides 2400 mm	142.23	180.01	3.12	100.69	nr	**280.70**
Sum of two sides 2500 mm	152.86	193.46	3.12	100.69	nr	**294.15**
Sum of two sides 2600 mm	163.31	206.68	3.16	101.99	nr	**308.67**
Sum of two sides 2700 mm	173.75	219.90	3.16	101.99	nr	**321.89**
Sum of two sides 2800 mm	184.19	233.11	4.00	129.09	nr	**362.20**
Sum of two sides 2900 mm	194.82	246.57	4.01	129.40	nr	**375.97**
Sum of two sides 3000 mm	205.29	259.82	4.01	129.40	nr	**389.22**
Sum of two sides 3100 mm	190.84	241.53	4.01	129.40	nr	**370.93**
Sum of two sides 3200 mm	197.84	250.39	4.56	147.17	nr	**397.56**
Sum of two sides 3300 mm	216.30	273.75	4.56	147.17	nr	**420.92**
Sum of two sides 3400 mm	223.31	282.62	4.56	147.17	nr	**429.79**
Sum of two sides 3500 mm	230.32	291.49	4.56	147.17	nr	**438.66**
Sum of two sides 3600 mm	237.34	300.37	4.56	147.17	nr	**447.54**
Sum of two sides 3700 mm	244.55	309.50	4.56	147.17	nr	**456.67**
Sum of two sides 3800 mm	251.57	318.38	4.56	147.17	nr	**465.55**
Sum of two sides 3900 mm	258.58	327.26	4.56	147.17	nr	**474.43**
Sum of two sides 4000 mm	265.60	336.15	4.56	147.17	nr	**483.32**
Offset						
Sum of two sides 2100 mm	260.79	330.05	2.92	94.24	nr	**424.29**
Sum of two sides 2200 mm	280.45	354.94	3.26	105.21	nr	**460.15**
Sum of two sides 2300 mm	291.64	369.10	3.26	105.21	nr	**474.31**
Sum of two sides 2400 mm	319.79	404.72	3.47	111.98	nr	**516.70**
Sum of two sides 2500 mm	330.58	418.39	3.48	112.31	nr	**530.70**
Sum of two sides 2600 mm	340.75	431.26	3.49	112.64	nr	**543.90**
Sum of two sides 2700 mm	350.52	443.62	3.50	112.95	nr	**556.57**

38 VENTILATION/AIR CONDITIONING SYSTEMS

Item	Net Price £	Material £	Labour hours	Labour £	Unit	Total rate £
Sum of two sides 2800 mm	359.89	455.48	4.33	139.74	nr	**595.22**
Sum of two sides 2900 mm	368.94	466.93	4.74	152.97	nr	**619.90**
Sum of two sides 3000 mm	377.46	477.71	5.34	172.33	nr	**650.04**
Sum of two sides 3100 mm	358.77	454.06	5.35	172.67	nr	**626.73**
Sum of two sides 3200 mm	372.08	470.90	5.31	171.37	nr	**642.27**
Sum of two sides 3300 mm	396.69	502.05	5.35	172.67	nr	**674.72**
Sum of two sides 3400 mm	410.00	518.90	5.35	172.67	nr	**691.57**
Sum of two sides 3500 mm	423.32	535.75	5.35	172.67	nr	**708.42**
Sum of two sides 3600 mm	436.61	552.57	5.35	172.67	nr	**725.24**
Sum of two sides 3700 mm	449.99	569.51	5.35	172.67	nr	**742.18**
Sum of two sides 3800 mm	463.29	586.34	5.35	172.67	nr	**759.01**
Sum of two sides 3900 mm	476.59	603.18	5.35	172.67	nr	**775.85**
Sum of two sides 4000 mm	489.90	620.02	5.35	172.67	nr	**792.69**
Square to round						
Sum of two sides 2100 mm	165.25	209.14	4.26	137.48	nr	**346.62**
Sum of two sides 2200 mm	181.27	229.42	4.27	137.82	nr	**367.24**
Sum of two sides 2300 mm	197.20	249.58	4.30	138.77	nr	**388.35**
Sum of two sides 2400 mm	213.29	269.94	4.77	153.94	nr	**423.88**
Sum of two sides 2500 mm	229.23	290.11	4.79	154.58	nr	**444.69**
Sum of two sides 2600 mm	245.20	310.33	4.95	159.75	nr	**470.08**
Sum of two sides 2700 mm	261.17	330.53	4.95	159.75	nr	**490.28**
Sum of two sides 2800 mm	277.15	350.76	8.49	274.00	nr	**624.76**
Sum of two sides 2900 mm	293.09	370.93	8.88	286.59	nr	**657.52**
Sum of two sides 3000 mm	309.05	391.14	9.02	291.10	nr	**682.24**
Sum of two sides 3100 mm	298.85	378.22	9.02	291.10	nr	**669.32**
Sum of two sides 3200 mm	309.69	391.94	9.09	293.35	nr	**685.29**
Sum of two sides 3300 mm	326.11	412.72	9.09	293.35	nr	**706.07**
Sum of two sides 3400 mm	336.92	426.41	9.09	293.35	nr	**719.76**
Sum of two sides 3500 mm	347.76	440.13	9.09	293.35	nr	**733.48**
Sum of two sides 3600 mm	358.59	453.84	9.09	293.35	nr	**747.19**
Sum of two sides 3700 mm	369.38	467.49	9.09	293.35	nr	**760.84**
Sum of two sides 3800 mm	380.21	481.20	9.09	293.35	nr	**774.55**
Sum of two sides 3900 mm	391.04	494.91	9.09	293.35	nr	**788.26**
Sum of two sides 4000 mm	401.86	508.59	9.09	293.35	nr	**801.94**
90° radius bend						
Sum of two sides 2100 mm	243.14	307.72	2.61	84.24	nr	**391.96**
Sum of two sides 2200 mm	464.08	587.34	2.62	84.56	nr	**671.90**
Sum of two sides 2300 mm	503.52	637.26	2.63	84.87	nr	**722.13**
Sum of two sides 2400 mm	514.59	651.27	4.34	140.07	nr	**791.34**
Sum of two sides 2500 mm	554.03	701.18	4.35	140.38	nr	**841.56**
Sum of two sides 2600 mm	592.69	750.11	4.53	146.20	nr	**896.31**
Sum of two sides 2700 mm	631.35	799.04	4.53	146.20	nr	**945.24**
Sum of two sides 2800 mm	670.01	847.96	7.13	230.10	nr	**1078.06**
Sum of two sides 2900 mm	709.45	897.88	7.17	231.40	nr	**1129.28**
Sum of two sides 3000 mm	748.12	946.83	7.26	234.30	nr	**1181.13**
Sum of two sides 3100 mm	728.08	921.46	7.26	234.30	nr	**1155.76**
Sum of two sides 3200 mm	756.45	957.36	7.31	235.92	nr	**1193.28**
Sum of two sides 3300 mm	819.32	1036.93	7.31	235.92	nr	**1272.85**
Sum of two sides 3400 mm	847.70	1072.85	7.31	235.92	nr	**1308.77**
Sum of two sides 3500 mm	876.08	1108.77	7.31	235.92	nr	**1344.69**

38 VENTILATION/AIR CONDITIONING SYSTEMS

Item	Net Price £	Material £	Labour hours	Labour £	Unit	Total rate £
DUCTWORK: RECTANGULAR – CLASS C – cont						
Extra over fittings – cont						
90° radius bend – cont						
Sum of two sides 3600 mm	904.46	1144.68	7.31	235.92	nr	**1380.60**
Sum of two sides 3700 mm	933.60	1181.57	7.31	235.92	nr	**1417.49**
Sum of two sides 3800 mm	961.98	1217.48	7.31	235.92	nr	**1453.40**
Sum of two sides 3900 mm	990.36	1253.40	7.31	235.92	nr	**1489.32**
Sum of two sides 4000 mm	1018.72	1289.29	7.31	235.92	nr	**1525.21**
45° bend						
Sum of two sides 2100 mm	117.22	148.36	7.50	242.04	nr	**390.40**
Sum of two sides 2200 mm	331.36	419.37	7.50	242.04	nr	**661.41**
Sum of two sides 2300 mm	357.65	452.64	7.55	243.67	nr	**696.31**
Sum of two sides 2400 mm	368.81	466.77	8.13	262.37	nr	**729.14**
Sum of two sides 2500 mm	395.08	500.01	8.30	267.86	nr	**767.87**
Sum of two sides 2600 mm	420.78	532.54	8.56	276.26	nr	**808.80**
Sum of two sides 2700 mm	446.46	565.04	8.62	278.20	nr	**843.24**
Sum of two sides 2800 mm	472.17	597.58	9.09	293.35	nr	**890.93**
Sum of two sides 2900 mm	498.45	630.84	9.09	293.35	nr	**924.19**
Sum of two sides 3000 mm	524.15	663.36	9.62	310.46	nr	**973.82**
Sum of two sides 3100 mm	517.81	655.35	9.62	310.46	nr	**965.81**
Sum of two sides 3200 mm	536.65	679.18	9.62	310.46	nr	**989.64**
Sum of two sides 3300 mm	578.54	732.20	9.62	310.46	nr	**1042.66**
Sum of two sides 3400 mm	597.38	756.04	9.62	310.46	nr	**1066.50**
Sum of two sides 3500 mm	616.22	779.89	9.62	310.46	nr	**1090.35**
Sum of two sides 3600 mm	635.04	803.71	9.62	310.46	nr	**1114.17**
Sum of two sides 3700 mm	654.47	828.30	9.62	310.46	nr	**1138.76**
Sum of two sides 3800 mm	673.30	852.13	9.62	310.46	nr	**1162.59**
Sum of two sides 3900 mm	692.13	875.96	9.62	310.46	nr	**1186.42**
Sum of two sides 4000 mm	710.95	899.77	9.62	310.46	nr	**1210.23**
90° mitre bend						
Sum of two sides 2100 mm	535.33	677.51	7.50	242.04	nr	**919.55**
Sum of two sides 2200 mm	565.52	715.72	7.50	242.04	nr	**957.76**
Sum of two sides 2300 mm	611.06	773.36	7.55	243.67	nr	**1017.03**
Sum of two sides 2400 mm	624.41	790.25	8.13	262.37	nr	**1052.62**
Sum of two sides 2500 mm	670.13	848.12	8.30	267.86	nr	**1115.98**
Sum of two sides 2600 mm	716.12	906.33	8.56	276.26	nr	**1182.59**
Sum of two sides 2700 mm	762.10	964.51	8.62	278.20	nr	**1242.71**
Sum of two sides 2800 mm	808.07	1022.69	15.20	490.55	nr	**1513.24**
Sum of two sides 2900 mm	853.80	1080.56	15.20	490.55	nr	**1571.11**
Sum of two sides 3000 mm	899.79	1138.77	16.04	517.65	nr	**1656.42**
Sum of two sides 3100 mm	876.67	1109.52	15.60	503.45	nr	**1612.97**
Sum of two sides 3200 mm	912.37	1154.70	16.04	517.65	nr	**1672.35**
Sum of two sides 3300 mm	968.90	1226.24	16.04	517.65	nr	**1743.89**
Sum of two sides 3400 mm	1004.60	1271.42	16.04	517.65	nr	**1789.07**
Sum of two sides 3500 mm	1040.29	1316.59	16.04	517.65	nr	**1834.24**
Sum of two sides 3600 mm	1075.99	1361.77	16.04	517.65	nr	**1879.42**
Sum of two sides 3700 mm	1111.43	1406.63	16.04	517.65	nr	**1924.28**
Sum of two sides 3800 mm	1147.13	1451.81	16.04	517.65	nr	**1969.46**
Sum of two sides 3900 mm	1182.82	1496.98	16.04	517.65	nr	**2014.63**
Sum of two sides 4000 mm	1218.51	1542.15	16.04	517.65	nr	**2059.80**

38 VENTILATION/AIR CONDITIONING SYSTEMS

Item	Net Price £	Material £	Labour hours	Labour £	Unit	Total rate £
Branch						
Sum of two sides 2100 mm	243.81	308.57	2.61	84.24	nr	**392.81**
Sum of two sides 2200 mm	260.78	330.04	2.61	84.24	nr	**414.28**
Sum of two sides 2300 mm	277.99	351.83	2.61	84.24	nr	**436.07**
Sum of two sides 2400 mm	294.70	372.97	2.88	92.95	nr	**465.92**
Sum of two sides 2500 mm	311.94	394.79	2.88	92.95	nr	**487.74**
Sum of two sides 2600 mm	328.92	416.28	2.88	92.95	nr	**509.23**
Sum of two sides 2700 mm	345.90	437.77	2.88	92.95	nr	**530.72**
Sum of two sides 2800 mm	362.92	459.31	3.94	127.15	nr	**586.46**
Sum of two sides 2900 mm	380.13	481.10	3.94	127.15	nr	**608.25**
Sum of two sides 3000 mm	397.14	502.62	4.83	155.88	nr	**658.50**
Sum of two sides 3100 mm	410.75	519.85	4.83	155.88	nr	**675.73**
Sum of two sides 3200 mm	423.17	535.56	4.83	155.88	nr	**691.44**
Sum of two sides 3300 mm	441.44	558.69	4.83	155.88	nr	**714.57**
Sum of two sides 3400 mm	453.87	574.41	4.83	155.88	nr	**730.29**
Sum of two sides 3500 mm	466.29	590.14	4.83	155.88	nr	**746.02**
Sum of two sides 3600 mm	478.71	605.85	4.83	155.88	nr	**761.73**
Sum of two sides 3700 mm	491.36	621.87	4.83	155.88	nr	**777.75**
Sum of two sides 3800 mm	503.77	637.57	4.83	155.88	nr	**793.45**
Sum of two sides 3900 mm	516.22	653.33	4.83	155.88	nr	**809.21**
Sum of two sides 4000 mm	528.63	669.03	4.83	155.88	nr	**824.91**
Grille neck						
Sum of two sides 2100 mm	260.45	329.63	2.80	90.36	nr	**419.99**
Sum of two sides 2200 mm	279.66	353.94	2.80	90.36	nr	**444.30**
Sum of two sides 2300 mm	298.89	378.28	2.80	90.36	nr	**468.64**
Sum of two sides 2400 mm	318.12	402.62	3.06	98.75	nr	**501.37**
Sum of two sides 2500 mm	337.31	426.90	3.06	98.75	nr	**525.65**
Sum of two sides 2600 mm	356.54	451.24	3.08	99.40	nr	**550.64**
Sum of two sides 2700 mm	375.75	475.55	3.08	99.40	nr	**574.95**
Sum of two sides 2800 mm	394.97	499.88	4.13	133.28	nr	**633.16**
Sum of two sides 2900 mm	414.19	524.19	4.13	133.28	nr	**657.47**
Sum of two sides 3000 mm	433.41	548.52	5.02	162.01	nr	**710.53**
Sum of two sides 3100 mm	448.82	568.03	5.02	162.01	nr	**730.04**
Sum of two sides 3200 mm	462.91	585.86	5.02	162.01	nr	**747.87**
Sum of two sides 3300 mm	476.98	603.67	5.02	162.01	nr	**765.68**
Sum of two sides 3400 mm	491.06	621.49	5.02	162.01	nr	**783.50**
Sum of two sides 3500 mm	505.13	639.30	5.02	162.01	nr	**801.31**
Sum of two sides 3600 mm	519.20	657.10	5.02	162.01	nr	**819.11**
Sum of two sides 3700 mm	533.27	674.91	5.02	162.01	nr	**836.92**
Sum of two sides 3800 mm	547.35	692.73	5.02	162.01	nr	**854.74**
Sum of two sides 3900 mm	561.42	710.53	5.02	162.01	nr	**872.54**
Sum of two sides 4000 mm	575.50	728.35	5.02	162.01	nr	**890.36**
Ductwork 2001 to 2500 mm longest side						
Sum of two sides 2500 mm	347.34	439.59	2.77	89.40	m	**528.99**
Sum of two sides 2600 mm	357.67	452.67	2.97	95.84	m	**548.51**
Sum of two sides 2700 mm	366.95	464.41	2.99	96.50	m	**560.91**
Sum of two sides 2800 mm	377.85	478.21	3.30	106.50	m	**584.71**
Sum of two sides 2900 mm	387.14	489.97	3.31	106.83	m	**596.80**
Sum of two sides 3000 mm	396.14	501.36	3.53	113.93	m	**615.29**
Sum of two sides 3100 mm	405.40	513.07	3.55	114.58	m	**627.65**

38 VENTILATION/AIR CONDITIONING SYSTEMS

Item	Net Price £	Material £	Labour hours	Labour £	Unit	Total rate £
DUCTWORK: RECTANGULAR – CLASS C – cont						
Ductwork 2001 to 2500 mm longest side – cont						
Sum of two sides 3200 mm	414.40	524.46	3.56	114.89	m	**639.35**
Sum of two sides 3300 mm	423.68	536.21	3.56	114.89	m	**651.10**
Sum of two sides 3400 mm	432.69	547.61	3.56	114.89	m	**662.50**
Sum of two sides 3500 mm	473.11	598.76	3.56	114.89	m	**713.65**
Sum of two sides 3600 mm	482.11	610.15	3.56	114.89	m	**725.04**
Sum of two sides 3700 mm	491.39	621.90	3.56	114.89	m	**736.79**
Sum of two sides 3800 mm	500.40	633.30	3.56	114.89	m	**748.19**
Sum of two sides 3900 mm	509.41	644.71	3.56	114.89	m	**759.60**
Sum of two sides 4000 mm	520.10	658.24	3.56	114.89	m	**773.13**
Extra over fittings; Ductwork 2001 to 2500 mm longest side						
End cap						
Sum of two sides 2500 mm	107.34	135.84	0.87	28.07	nr	**163.91**
Sum of two sides 2600 mm	113.54	143.70	0.87	28.07	nr	**171.77**
Sum of two sides 2700 mm	119.73	151.52	0.87	28.07	nr	**179.59**
Sum of two sides 2800 mm	125.93	159.38	1.16	37.44	nr	**196.82**
Sum of two sides 2900 mm	132.14	167.24	1.16	37.44	nr	**204.68**
Sum of two sides 3000 mm	138.33	175.07	1.73	55.84	nr	**230.91**
Sum of two sides 3100 mm	143.26	181.31	1.73	55.84	nr	**237.15**
Sum of two sides 3200 mm	147.74	186.98	1.73	55.84	nr	**242.82**
Sum of two sides 3300 mm	152.23	192.66	1.73	55.84	nr	**248.50**
Sum of two sides 3400 mm	156.72	198.34	1.73	55.84	nr	**254.18**
Sum of two sides 3500 mm	161.21	204.03	1.73	55.84	nr	**259.87**
Sum of two sides 3600 mm	165.69	209.70	1.73	55.84	nr	**265.54**
Sum of two sides 3700 mm	170.18	215.38	1.73	55.84	nr	**271.22**
Sum of two sides 3800 mm	174.65	221.03	1.73	55.84	nr	**276.87**
Sum of two sides 3900 mm	179.15	226.73	1.73	55.84	nr	**282.57**
Sum of two sides 4000 mm	183.63	232.40	1.73	55.84	nr	**288.24**
Reducer						
Sum of two sides 2500 mm	94.85	120.04	3.12	100.69	nr	**220.73**
Sum of two sides 2600 mm	103.99	131.61	3.16	101.99	nr	**233.60**
Sum of two sides 2700 mm	113.36	143.47	3.16	101.99	nr	**245.46**
Sum of two sides 2800 mm	122.25	154.72	4.00	129.09	nr	**283.81**
Sum of two sides 2900 mm	131.62	166.58	4.01	129.40	nr	**295.98**
Sum of two sides 3000 mm	140.78	178.17	4.56	147.17	nr	**325.34**
Sum of two sides 3100 mm	147.19	186.28	4.56	147.17	nr	**333.45**
Sum of two sides 3200 mm	152.93	193.55	4.56	147.17	nr	**340.72**
Sum of two sides 3300 mm	158.87	201.06	4.56	147.17	nr	**348.23**
Sum of two sides 3400 mm	164.60	208.32	4.56	147.17	nr	**355.49**
Sum of two sides 3500 mm	146.92	185.94	4.56	147.17	nr	**333.11**
Sum of two sides 3600 mm	152.65	193.19	4.56	147.17	nr	**340.36**
Sum of two sides 3700 mm	170.07	215.24	4.56	147.17	nr	**362.41**
Sum of two sides 3800 mm	175.82	222.52	4.56	147.17	nr	**369.69**
Sum of two sides 3900 mm	181.55	229.77	4.56	147.17	nr	**376.94**
Sum of two sides 4000 mm	187.45	237.24	4.56	147.17	nr	**384.41**

38 VENTILATION/AIR CONDITIONING SYSTEMS

Item	Net Price £	Material £	Labour hours	Labour £	Unit	Total rate £
Offset						
Sum of two sides 2500 mm	296.67	375.47	3.48	112.31	nr	**487.78**
Sum of two sides 2600 mm	315.50	399.30	3.49	112.64	nr	**511.94**
Sum of two sides 2700 mm	315.08	398.76	3.50	112.95	nr	**511.71**
Sum of two sides 2800 mm	353.15	446.95	4.33	139.74	nr	**586.69**
Sum of two sides 2900 mm	351.82	445.27	4.74	152.97	nr	**598.24**
Sum of two sides 3000 mm	349.33	442.11	5.31	171.37	nr	**613.48**
Sum of two sides 3100 mm	341.88	432.68	5.34	172.33	nr	**605.01**
Sum of two sides 3200 mm	332.39	420.67	5.35	172.67	nr	**593.34**
Sum of two sides 3300 mm	322.09	407.64	5.35	172.67	nr	**580.31**
Sum of two sides 3400 mm	310.75	393.29	5.35	172.67	nr	**565.96**
Sum of two sides 3500 mm	286.22	362.24	5.35	172.67	nr	**534.91**
Sum of two sides 3600 mm	297.18	376.11	5.35	172.67	nr	**548.78**
Sum of two sides 3700 mm	319.70	404.61	5.35	172.67	nr	**577.28**
Sum of two sides 3800 mm	330.68	418.51	5.35	172.67	nr	**591.18**
Sum of two sides 3900 mm	341.63	432.36	5.35	172.67	nr	**605.03**
Sum of two sides 4000 mm	352.62	446.28	5.35	172.67	nr	**618.95**
Square to round						
Sum of two sides 2500 mm	151.90	192.25	4.79	154.58	nr	**346.83**
Sum of two sides 2600 mm	165.93	210.00	4.95	159.75	nr	**369.75**
Sum of two sides 2700 mm	179.96	227.75	4.95	159.75	nr	**387.50**
Sum of two sides 2800 mm	193.99	245.52	8.49	274.00	nr	**519.52**
Sum of two sides 2900 mm	207.99	263.23	8.88	286.59	nr	**549.82**
Sum of two sides 3000 mm	222.03	281.00	9.02	291.10	nr	**572.10**
Sum of two sides 3100 mm	232.02	293.64	9.02	291.10	nr	**584.74**
Sum of two sides 3200 mm	240.91	304.90	9.09	293.35	nr	**598.25**
Sum of two sides 3300 mm	249.75	316.09	9.09	293.35	nr	**609.44**
Sum of two sides 3400 mm	258.64	327.33	9.09	293.35	nr	**620.68**
Sum of two sides 3500 mm	243.50	308.18	9.09	293.35	nr	**601.53**
Sum of two sides 3600 mm	252.38	319.41	9.09	293.35	nr	**612.76**
Sum of two sides 3700 mm	266.99	337.90	9.09	293.35	nr	**631.25**
Sum of two sides 3800 mm	275.86	349.13	9.09	293.35	nr	**642.48**
Sum of two sides 3900 mm	284.74	360.37	9.09	293.35	nr	**653.72**
Sum of two sides 4000 mm	293.57	371.54	9.09	293.35	nr	**664.89**
90° radius bend						
Sum of two sides 2500 mm	425.54	538.56	4.35	140.38	nr	**678.94**
Sum of two sides 2600 mm	437.71	553.96	4.53	146.20	nr	**700.16**
Sum of two sides 2700 mm	474.71	600.79	4.53	146.20	nr	**746.99**
Sum of two sides 2800 mm	459.30	581.29	7.13	230.10	nr	**811.39**
Sum of two sides 2900 mm	495.81	627.50	7.17	231.40	nr	**858.90**
Sum of two sides 3000 mm	531.59	672.78	7.26	234.30	nr	**907.08**
Sum of two sides 3100 mm	559.18	707.69	7.26	234.30	nr	**941.99**
Sum of two sides 3200 mm	584.68	739.97	7.31	235.92	nr	**975.89**
Sum of two sides 3300 mm	610.89	773.15	7.31	235.92	nr	**1009.07**
Sum of two sides 3400 mm	636.39	805.41	7.31	235.92	nr	**1041.33**
Sum of two sides 3500 mm	599.53	758.77	7.31	235.92	nr	**994.69**
Sum of two sides 3600 mm	625.03	791.03	7.31	235.92	nr	**1026.95**
Sum of two sides 3700 mm	685.69	867.81	7.31	235.92	nr	**1103.73**
Sum of two sides 3800 mm	711.19	900.08	7.31	235.92	nr	**1136.00**
Sum of two sides 3900 mm	736.68	932.34	7.31	235.92	nr	**1168.26**
Sum of two sides 4000 mm	731.52	925.81	7.31	235.92	nr	**1161.73**

38 VENTILATION/AIR CONDITIONING SYSTEMS

Item	Net Price £	Material £	Labour hours	Labour £	Unit	Total rate £
DUCTWORK: RECTANGULAR – CLASS C – cont						
Extra over fittings – cont						
45° bend						
Sum of two sides 2500 mm	343.89	435.23	8.30	267.86	nr	**703.09**
Sum of two sides 2600 mm	356.57	451.27	8.56	276.26	nr	**727.53**
Sum of two sides 2700 mm	382.13	483.63	8.62	278.20	nr	**761.83**
Sum of two sides 2800 mm	380.55	481.62	9.09	293.35	nr	**774.97**
Sum of two sides 2900 mm	405.88	513.68	9.09	293.35	nr	**807.03**
Sum of two sides 3000 mm	430.64	545.01	9.62	310.46	nr	**855.47**
Sum of two sides 3100 mm	450.01	569.53	9.62	310.46	nr	**879.99**
Sum of two sides 3200 mm	467.92	592.20	9.62	310.46	nr	**902.66**
Sum of two sides 3300 mm	486.38	615.56	9.62	310.46	nr	**926.02**
Sum of two sides 3400 mm	504.27	638.21	9.62	310.46	nr	**948.67**
Sum of two sides 3500 mm	490.03	620.18	9.62	310.46	nr	**930.64**
Sum of two sides 3600 mm	507.91	642.81	9.62	310.46	nr	**953.27**
Sum of two sides 3700 mm	549.36	695.27	9.62	310.46	nr	**1005.73**
Sum of two sides 3800 mm	567.26	717.92	9.62	310.46	nr	**1028.38**
Sum of two sides 3900 mm	585.14	740.56	9.62	310.46	nr	**1051.02**
Sum of two sides 4000 mm	587.68	743.77	9.62	310.46	nr	**1054.23**
90° mitre bend						
Sum of two sides 2500 mm	427.01	540.42	8.30	267.86	nr	**808.28**
Sum of two sides 2600 mm	447.24	566.03	8.56	276.26	nr	**842.29**
Sum of two sides 2700 mm	487.30	616.73	8.62	278.20	nr	**894.93**
Sum of two sides 2800 mm	463.75	586.92	15.20	490.55	nr	**1077.47**
Sum of two sides 2900 mm	503.33	637.01	15.20	490.55	nr	**1127.56**
Sum of two sides 3000 mm	543.24	687.52	15.20	490.55	nr	**1178.07**
Sum of two sides 3100 mm	574.86	727.54	16.04	517.65	nr	**1245.19**
Sum of two sides 3200 mm	604.49	765.04	16.04	517.65	nr	**1282.69**
Sum of two sides 3300 mm	633.79	802.12	16.04	517.65	nr	**1319.77**
Sum of two sides 3400 mm	663.42	839.62	16.04	517.65	nr	**1357.27**
Sum of two sides 3500 mm	614.27	777.43	16.04	517.65	nr	**1295.08**
Sum of two sides 3600 mm	643.90	814.92	16.04	517.65	nr	**1332.57**
Sum of two sides 3700 mm	696.52	881.52	16.04	517.65	nr	**1399.17**
Sum of two sides 3800 mm	726.15	919.02	16.04	517.65	nr	**1436.67**
Sum of two sides 3900 mm	755.78	956.51	16.04	517.65	nr	**1474.16**
Sum of two sides 4000 mm	746.53	944.81	16.04	517.65	nr	**1462.46**
Branch						
Sum of two sides 2500 mm	331.23	419.20	2.88	92.95	nr	**512.15**
Sum of two sides 2600 mm	348.85	441.50	2.88	92.95	nr	**534.45**
Sum of two sides 2700 mm	366.74	464.15	2.88	92.95	nr	**557.10**
Sum of two sides 2800 mm	384.11	486.12	3.94	127.15	nr	**613.27**
Sum of two sides 2900 mm	402.00	508.77	3.94	127.15	nr	**635.92**
Sum of two sides 3000 mm	419.66	531.13	4.83	155.88	nr	**687.01**
Sum of two sides 3100 mm	433.95	549.20	4.83	155.88	nr	**705.08**
Sum of two sides 3200 mm	447.04	565.78	4.83	155.88	nr	**721.66**
Sum of two sides 3300 mm	460.36	582.64	4.83	155.88	nr	**738.52**
Sum of two sides 3400 mm	473.44	599.19	4.83	155.88	nr	**755.07**
Sum of two sides 3500 mm	487.13	616.52	4.83	155.88	nr	**772.40**

38 VENTILATION/AIR CONDITIONING SYSTEMS

Item	Net Price £	Material £	Labour hours	Labour £	Unit	Total rate £
Sum of two sides 3600 mm	500.21	633.07	4.83	155.88	nr	**788.95**
Sum of two sides 3700 mm	519.28	657.20	4.83	155.88	nr	**813.08**
Sum of two sides 3800 mm	532.38	673.78	4.83	155.88	nr	**829.66**
Sum of two sides 3900 mm	545.46	690.33	4.83	155.88	nr	**846.21**
Sum of two sides 4000 mm	558.80	707.21	4.83	155.88	nr	**863.09**
Grille neck						
Sum of two sides 2500 mm	337.31	426.90	3.06	98.75	nr	**525.65**
Sum of two sides 2600 mm	356.54	451.24	3.08	99.40	nr	**550.64**
Sum of two sides 2700 mm	375.75	475.55	3.08	99.40	nr	**574.95**
Sum of two sides 2800 mm	394.97	499.88	4.13	133.28	nr	**633.16**
Sum of two sides 2900 mm	414.19	524.19	4.13	133.28	nr	**657.47**
Sum of two sides 3000 mm	433.41	548.52	5.02	162.01	nr	**710.53**
Sum of two sides 3100 mm	448.82	568.03	5.02	162.01	nr	**730.04**
Sum of two sides 3200 mm	462.91	585.86	5.02	162.01	nr	**747.87**
Sum of two sides 3300 mm	476.98	603.67	5.02	162.01	nr	**765.68**
Sum of two sides 3400 mm	491.06	621.49	5.02	162.01	nr	**783.50**
Sum of two sides 3500 mm	505.13	639.30	5.02	162.01	nr	**801.31**
Sum of two sides 3600 mm	519.20	657.10	5.02	162.01	nr	**819.11**
Sum of two sides 3700 mm	533.27	674.91	5.02	162.01	nr	**836.92**
Sum of two sides 3800 mm	547.35	692.73	5.02	162.01	nr	**854.74**
Sum of two sides 3900 mm	561.42	710.53	5.02	162.01	nr	**872.54**
Sum of two sides 4000 mm	575.50	728.35	5.02	162.01	nr	**890.36**
Ductwork 2501 to 4000 mm longest side						
Sum of two sides 3000 mm	398.57	504.43	2.38	76.81	m	**581.24**
Sum of two sides 3100 mm	408.10	516.49	2.38	76.81	m	**593.30**
Sum of two sides 3200 mm	417.10	527.88	2.38	76.81	m	**604.69**
Sum of two sides 3300 mm	426.09	539.26	2.38	76.81	m	**616.07**
Sum of two sides 3400 mm	437.52	553.73	2.64	85.20	m	**638.93**
Sum of two sides 3500 mm	446.52	565.12	2.66	85.85	m	**650.97**
Sum of two sides 3600 mm	455.53	576.52	2.95	95.20	m	**671.72**
Sum of two sides 3700 mm	467.23	591.33	2.96	95.52	m	**686.85**
Sum of two sides 3800 mm	476.24	602.73	3.15	101.66	m	**704.39**
Sum of two sides 3900 mm	501.53	634.74	3.15	101.66	m	**736.40**
Sum of two sides 4000 mm	510.53	646.13	3.15	101.66	m	**747.79**
Sum of two sides 4100 mm	519.82	657.89	3.35	108.10	m	**765.99**
Sum of two sides 4200 mm	528.82	669.28	3.35	108.10	m	**777.38**
Sum of two sides 4300 mm	537.81	680.66	3.60	116.18	m	**796.84**
Sum of two sides 4400 mm	550.69	696.95	3.60	116.18	m	**813.13**
Sum of two sides 4500 mm	559.70	708.36	3.60	116.18	m	**824.54**
Extra over fittings; Ductwork 2501 to 4000 mm longest side						
End cap						
Sum of two sides 3000 mm	138.33	175.07	1.73	55.84	nr	**230.91**
Sum of two sides 3100 mm	143.26	181.31	1.73	55.84	nr	**237.15**
Sum of two sides 3200 mm	147.74	186.98	1.73	55.84	nr	**242.82**
Sum of two sides 3300 mm	152.14	192.55	1.73	55.84	nr	**248.39**
Sum of two sides 3400 mm	156.71	198.33	1.73	55.84	nr	**254.17**
Sum of two sides 3500 mm	161.21	204.03	1.73	55.84	nr	**259.87**
Sum of two sides 3600 mm	165.69	209.70	1.73	55.84	nr	**265.54**

38 VENTILATION/AIR CONDITIONING SYSTEMS

Item	Net Price £	Material £	Labour hours	Labour £	Unit	Total rate £
DUCTWORK: RECTANGULAR – CLASS C – cont						
Extra over fittings – cont						
End cap – cont						
Sum of two sides 3700 mm	170.18	215.38	1.73	55.84	nr	**271.22**
Sum of two sides 3800 mm	174.66	221.05	1.73	55.84	nr	**276.89**
Sum of two sides 3900 mm	179.15	226.73	1.80	58.09	nr	**284.82**
Sum of two sides 4000 mm	183.63	232.40	1.80	58.09	nr	**290.49**
Sum of two sides 4100 mm	188.12	238.09	1.80	58.09	nr	**296.18**
Sum of two sides 4200 mm	192.60	243.76	1.80	58.09	nr	**301.85**
Sum of two sides 4300 mm	197.08	249.42	1.88	60.67	nr	**310.09**
Sum of two sides 4400 mm	201.57	255.10	1.88	60.67	nr	**315.77**
Sum of two sides 4500 mm	206.05	260.78	1.88	60.67	nr	**321.45**
Reducer						
Sum of two sides 3000 mm	140.72	178.09	3.12	100.69	nr	**278.78**
Sum of two sides 3100 mm	147.35	186.49	3.12	100.69	nr	**287.18**
Sum of two sides 3200 mm	153.07	193.73	3.12	100.69	nr	**294.42**
Sum of two sides 3300 mm	158.82	201.01	3.12	100.69	nr	**301.70**
Sum of two sides 3400 mm	164.48	208.16	3.12	100.69	nr	**308.85**
Sum of two sides 3500 mm	170.22	215.43	3.12	100.69	nr	**316.12**
Sum of two sides 3600 mm	175.95	222.68	3.95	127.47	nr	**350.15**
Sum of two sides 3700 mm	181.85	230.15	3.97	128.12	nr	**358.27**
Sum of two sides 3800 mm	187.57	237.38	4.52	145.87	nr	**383.25**
Sum of two sides 3900 mm	182.96	231.55	4.52	145.87	nr	**377.42**
Sum of two sides 4000 mm	188.69	238.81	4.52	145.87	nr	**384.68**
Sum of two sides 4100 mm	194.63	246.32	4.52	145.87	nr	**392.19**
Sum of two sides 4200 mm	200.37	253.59	4.52	145.87	nr	**399.46**
Sum of two sides 4300 mm	206.09	260.83	4.92	158.78	nr	**419.61**
Sum of two sides 4400 mm	211.95	268.24	4.92	158.78	nr	**427.02**
Sum of two sides 4500 mm	217.68	275.50	5.12	165.23	nr	**440.73**
Offset						
Sum of two sides 3000 mm	391.04	494.91	3.48	112.31	nr	**607.22**
Sum of two sides 3100 mm	384.75	486.94	3.48	112.31	nr	**599.25**
Sum of two sides 3200 mm	376.17	476.08	3.48	112.31	nr	**588.39**
Sum of two sides 3300 mm	366.67	464.06	3.48	112.31	nr	**576.37**
Sum of two sides 3400 mm	401.71	508.40	3.50	112.95	nr	**621.35**
Sum of two sides 3500 mm	391.30	495.23	3.49	112.64	nr	**607.87**
Sum of two sides 3600 mm	379.96	480.87	4.25	137.17	nr	**618.04**
Sum of two sides 3700 mm	416.09	526.60	4.76	153.62	nr	**680.22**
Sum of two sides 3800 mm	403.82	511.08	5.32	171.70	nr	**682.78**
Sum of two sides 3900 mm	428.18	541.90	5.35	172.67	nr	**714.57**
Sum of two sides 4000 mm	413.86	523.78	5.35	172.67	nr	**696.45**
Sum of two sides 4100 mm	398.74	504.65	5.85	188.80	nr	**693.45**
Sum of two sides 4200 mm	382.57	484.18	5.85	188.80	nr	**672.98**
Sum of two sides 4300 mm	393.53	498.05	6.15	198.48	nr	**696.53**
Sum of two sides 4400 mm	433.07	548.09	6.30	203.31	nr	**751.40**
Sum of two sides 4500 mm	415.51	525.87	6.15	198.48	nr	**724.35**

38 VENTILATION/AIR CONDITIONING SYSTEMS

Item	Net Price £	Material £	Labour hours	Labour £	Unit	Total rate £
Square to round						
Sum of two sides 3000 mm	286.52	362.62	4.70	151.68	nr	**514.30**
Sum of two sides 3100 mm	298.82	378.19	4.70	151.68	nr	**529.87**
Sum of two sides 3200 mm	309.78	392.06	4.70	151.68	nr	**543.74**
Sum of two sides 3300 mm	320.75	405.94	4.70	151.68	nr	**557.62**
Sum of two sides 3400 mm	331.72	419.82	4.71	152.01	nr	**571.83**
Sum of two sides 3500 mm	342.68	433.70	4.71	152.01	nr	**585.71**
Sum of two sides 3600 mm	353.64	447.56	8.19	264.31	nr	**711.87**
Sum of two sides 3700 mm	364.60	461.44	8.62	278.20	nr	**739.64**
Sum of two sides 3800 mm	375.57	475.32	8.62	278.20	nr	**753.52**
Sum of two sides 3900 mm	459.15	581.10	8.62	278.20	nr	**859.30**
Sum of two sides 4000 mm	472.21	597.63	8.75	282.40	nr	**880.03**
Sum of two sides 4100 mm	428.51	542.33	11.23	362.43	nr	**904.76**
Sum of two sides 4200 mm	498.30	630.65	11.23	362.43	nr	**993.08**
Sum of two sides 4300 mm	511.34	647.15	11.25	363.07	nr	**1010.22**
Sum of two sides 4400 mm	524.39	663.67	11.25	363.07	nr	**1026.74**
Sum of two sides 4500 mm	537.45	680.20	11.26	363.40	nr	**1043.60**
90° radius bend						
Sum of two sides 3000 mm	831.29	1052.08	3.90	125.87	nr	**1177.95**
Sum of two sides 3100 mm	870.81	1102.10	3.90	125.87	nr	**1227.97**
Sum of two sides 3200 mm	906.27	1146.98	4.26	137.48	nr	**1284.46**
Sum of two sides 3300 mm	941.73	1191.85	4.26	137.48	nr	**1329.33**
Sum of two sides 3400 mm	919.66	1163.93	4.26	137.48	nr	**1301.41**
Sum of two sides 3500 mm	954.65	1208.20	4.55	146.84	nr	**1355.04**
Sum of two sides 3600 mm	989.65	1252.50	4.55	146.84	nr	**1399.34**
Sum of two sides 3700 mm	964.96	1221.25	6.87	221.72	nr	**1442.97**
Sum of two sides 3800 mm	999.50	1264.97	6.87	221.72	nr	**1486.69**
Sum of two sides 3900 mm	935.52	1184.00	6.87	221.72	nr	**1405.72**
Sum of two sides 4000 mm	969.62	1227.15	7.00	225.90	nr	**1453.05**
Sum of two sides 4100 mm	1004.69	1271.54	7.20	232.37	nr	**1503.91**
Sum of two sides 4200 mm	1038.78	1314.68	7.20	232.37	nr	**1547.05**
Sum of two sides 4300 mm	1072.87	1357.82	7.41	239.14	nr	**1596.96**
Sum of two sides 4400 mm	1002.31	1268.52	7.41	239.14	nr	**1507.66**
Sum of two sides 4500 mm	1035.74	1310.84	7.55	243.67	nr	**1554.51**
45° bend						
Sum of two sides 3000 mm	403.29	510.41	4.85	156.52	nr	**666.93**
Sum of two sides 3100 mm	422.96	535.29	4.85	156.52	nr	**691.81**
Sum of two sides 3200 mm	440.62	557.65	4.85	156.52	nr	**714.17**
Sum of two sides 3300 mm	451.98	572.03	4.87	157.17	nr	**729.20**
Sum of two sides 3400 mm	446.73	565.38	4.87	157.17	nr	**722.55**
Sum of two sides 3500 mm	464.18	587.46	4.87	157.17	nr	**744.63**
Sum of two sides 3600 mm	481.62	609.54	8.81	284.33	nr	**893.87**
Sum of two sides 3700 mm	468.73	593.22	8.81	284.33	nr	**877.55**
Sum of two sides 3800 mm	485.94	615.00	9.31	300.46	nr	**915.46**
Sum of two sides 3900 mm	452.72	572.96	9.31	300.46	nr	**873.42**
Sum of two sides 4000 mm	469.70	594.45	9.31	300.46	nr	**894.91**
Sum of two sides 4100 mm	487.16	616.55	10.01	323.04	nr	**939.59**
Sum of two sides 4200 mm	504.15	638.05	9.31	300.46	nr	**938.51**
Sum of two sides 4300 mm	521.14	659.56	10.01	323.04	nr	**982.60**
Sum of two sides 4400 mm	485.08	613.92	10.52	339.51	nr	**953.43**
Sum of two sides 4500 mm	501.74	635.01	10.52	339.51	nr	**974.52**

38 VENTILATION/AIR CONDITIONING SYSTEMS

Item	Net Price £	Material £	Labour hours	Labour £	Unit	Total rate £
DUCTWORK: RECTANGULAR – CLASS C – cont						
Extra over fittings – cont						
90° mitre bend						
Sum of two sides 3000 mm	977.86	1237.58	4.85	156.52	nr	**1394.10**
Sum of two sides 3100 mm	1026.06	1298.58	4.85	156.52	nr	**1455.10**
Sum of two sides 3200 mm	1071.39	1355.95	4.87	157.17	nr	**1513.12**
Sum of two sides 3300 mm	1116.73	1413.33	4.87	157.17	nr	**1570.50**
Sum of two sides 3400 mm	1089.10	1378.36	4.87	157.17	nr	**1535.53**
Sum of two sides 3500 mm	1134.03	1435.22	8.81	284.33	nr	**1719.55**
Sum of two sides 3600 mm	1178.96	1492.09	8.81	284.33	nr	**1776.42**
Sum of two sides 3700 mm	1156.65	1463.85	14.81	477.95	nr	**1941.80**
Sum of two sides 3800 mm	1201.24	1520.29	14.81	477.95	nr	**1998.24**
Sum of two sides 3900 mm	1117.02	1413.70	14.81	477.95	nr	**1891.65**
Sum of two sides 4000 mm	1161.22	1469.64	15.20	490.55	nr	**1960.19**
Sum of two sides 4100 mm	1204.82	1524.82	16.30	526.04	nr	**2050.86**
Sum of two sides 4200 mm	1249.01	1580.75	16.50	532.50	nr	**2113.25**
Sum of two sides 4300 mm	1293.21	1636.69	17.01	548.96	nr	**2185.65**
Sum of two sides 4400 mm	1212.37	1534.38	17.01	548.96	nr	**2083.34**
Sum of two sides 4500 mm	1256.01	1589.60	17.01	548.96	nr	**2138.56**
Branch						
Sum of two sides 3000 mm	419.60	531.05	2.88	92.95	nr	**624.00**
Sum of two sides 3100 mm	434.11	549.40	2.88	92.95	nr	**642.35**
Sum of two sides 3200 mm	447.21	565.99	2.88	92.95	nr	**658.94**
Sum of two sides 3300 mm	460.29	582.55	2.88	92.95	nr	**675.50**
Sum of two sides 3400 mm	473.33	599.04	2.88	92.95	nr	**691.99**
Sum of two sides 3500 mm	486.42	615.61	3.94	127.15	nr	**742.76**
Sum of two sides 3600 mm	499.52	632.20	3.94	127.15	nr	**759.35**
Sum of two sides 3700 mm	512.76	648.95	3.94	127.15	nr	**776.10**
Sum of two sides 3800 mm	525.86	665.53	4.83	155.88	nr	**821.41**
Sum of two sides 3900 mm	539.47	682.75	4.83	155.88	nr	**838.63**
Sum of two sides 4000 mm	552.57	699.33	4.83	155.88	nr	**855.21**
Sum of two sides 4100 mm	565.88	716.17	5.44	175.56	nr	**891.73**
Sum of two sides 4200 mm	578.97	732.75	5.44	175.56	nr	**908.31**
Sum of two sides 4300 mm	592.05	749.30	5.85	188.80	nr	**938.10**
Sum of two sides 4400 mm	595.83	754.08	5.85	188.80	nr	**942.88**
Sum of two sides 4500 mm	618.38	782.62	5.85	188.80	nr	**971.42**
Grille neck						
Sum of two sides 3000 mm	433.41	548.52	2.88	92.95	nr	**641.47**
Sum of two sides 3100 mm	448.83	568.04	2.88	92.95	nr	**660.99**
Sum of two sides 3200 mm	462.90	585.85	2.88	92.95	nr	**678.80**
Sum of two sides 3300 mm	476.98	603.67	2.88	92.95	nr	**696.62**
Sum of two sides 3400 mm	491.06	621.49	3.94	127.15	nr	**748.64**
Sum of two sides 3500 mm	505.14	639.31	3.94	127.15	nr	**766.46**
Sum of two sides 3600 mm	435.14	550.72	3.94	127.15	nr	**677.87**
Sum of two sides 3700 mm	533.27	674.91	4.12	132.97	nr	**807.88**
Sum of two sides 3800 mm	547.35	692.73	4.12	132.97	nr	**825.70**
Sum of two sides 3900 mm	561.43	710.55	4.12	132.97	nr	**843.52**
Sum of two sides 4000 mm	575.50	728.35	5.00	161.36	nr	**889.71**

38 VENTILATION/AIR CONDITIONING SYSTEMS

Item	Net Price £	Material £	Labour hours	Labour £	Unit	Total rate £
Sum of two sides 4100 mm	589.57	746.16	5.00	161.36	nr	**907.52**
Sum of two sides 4200 mm	603.65	763.97	5.00	161.36	nr	**925.33**
Sum of two sides 4300 mm	617.73	781.79	5.23	168.80	nr	**950.59**
Sum of two sides 4400 mm	631.80	799.60	5.23	168.80	nr	**968.40**
Sum of two sides 4500 mm	645.88	817.42	5.39	173.95	nr	**991.37**

DUCTWORK ANCILLARIES: ACCESS DOORS

Refer to ancillaries in Ductwork Ancillaries: Access Hatches

38 VENTILATION/AIR CONDITIONING SYSTEMS

Item	Net Price £	Material £	Labour hours	Labour £	Unit	Total rate £
DUCTWORK ANCILLARIES: VOLUME CONTROL AND FIRE DAMPERS						
Volume control damper; opposed blade; galvanized steel casing; aluminium aerofoil blades; manually operated						
Rectangular						
Sum of two sides 200 mm	23.60	29.87	1.60	51.63	nr	**81.50**
Sum of two sides 300 mm	25.24	31.94	1.60	51.63	nr	**83.57**
Sum of two sides 400 mm	27.60	34.93	1.60	51.63	nr	**86.56**
Sum of two sides 500 mm	30.21	38.24	1.60	51.63	nr	**89.87**
Sum of two sides 600 mm	33.40	42.27	1.70	54.89	nr	**97.16**
Sum of two sides 700 mm	36.71	46.46	2.10	67.79	nr	**114.25**
Sum of two sides 800 mm	40.10	50.75	2.15	69.40	nr	**120.15**
Sum of two sides 900 mm	44.01	55.70	2.30	74.22	nr	**129.92**
Sum of two sides 1000 mm	47.79	60.48	2.40	77.46	nr	**137.94**
Sum of two sides 1100 mm	52.06	65.89	2.60	83.91	nr	**149.80**
Sum of two sides 1200 mm	58.88	74.51	2.80	90.36	nr	**164.87**
Sum of two sides 1300 mm	63.49	80.35	3.10	100.05	nr	**180.40**
Sum of two sides 1400 mm	68.45	86.63	3.25	104.89	nr	**191.52**
Sum of two sides 1500 mm	74.10	93.78	3.40	109.73	nr	**203.51**
Sum of two sides 1600 mm	79.66	100.82	3.45	111.34	nr	**212.16**
Sum of two sides 1700 mm	84.97	107.54	3.60	116.18	nr	**223.72**
Sum of two sides 1800 mm	91.33	115.58	3.90	125.87	nr	**241.45**
Sum of two sides 1900 mm	97.24	123.07	4.20	135.60	nr	**258.67**
Sum of two sides 2000 mm	104.31	132.01	4.33	139.74	nr	**271.75**
Circular						
100 mm dia.	31.51	39.88	0.80	25.82	nr	**65.70**
150 mm dia.	35.99	45.55	0.80	25.82	nr	**71.37**
160 mm dia.	37.52	47.49	0.90	29.04	nr	**76.53**
200 mm dia.	40.71	51.52	1.05	33.90	nr	**85.42**
250 mm dia.	45.32	57.36	1.20	38.74	nr	**96.10**
300 mm dia.	50.17	63.49	1.35	43.56	nr	**107.05**
315 mm dia.	52.40	66.32	1.35	43.61	nr	**109.93**
350 mm dia.	55.12	69.76	1.65	53.26	nr	**123.02**
400 mm dia.	60.18	76.16	1.90	61.35	nr	**137.51**
450 mm dia.	65.01	82.28	2.10	67.79	nr	**150.07**
500 mm dia.	70.82	89.63	2.95	95.20	nr	**184.83**
550 mm dia.	76.48	96.79	2.94	94.89	nr	**191.68**
600 mm dia.	82.84	104.84	2.94	94.89	nr	**199.73**
650 mm dia.	89.45	113.21	4.55	146.84	nr	**260.05**
700 mm dia.	96.65	122.32	5.20	167.82	nr	**290.14**
750 mm dia.	104.31	132.01	5.20	167.82	nr	**299.83**
800 mm dia.	112.24	142.05	5.80	187.19	nr	**329.24**
850 mm dia.	120.49	152.49	5.80	187.19	nr	**339.68**
900 mm dia.	129.10	163.39	6.40	206.55	nr	**369.94**
950 mm dia.	137.96	174.60	6.40	206.55	nr	**381.15**
1000 mm dia.	146.92	185.94	7.00	225.90	nr	**411.84**

38 VENTILATION/AIR CONDITIONING SYSTEMS

Item	Net Price £	Material £	Labour hours	Labour £	Unit	Total rate £
Flat oval						
345 × 102 mm	53.23	67.37	1.20	38.73	nr	**106.10**
508 × 102 mm	59.96	75.88	1.60	51.63	nr	**127.51**
559 × 152 mm	67.75	85.75	1.90	61.35	nr	**147.10**
531 × 203 mm	74.01	93.67	1.90	61.35	nr	**155.02**
851 × 203 mm	78.24	99.02	4.55	146.84	nr	**245.86**
582 × 254 mm	82.15	103.97	2.10	67.79	nr	**171.76**
823 × 254 mm	89.94	113.83	4.10	132.32	nr	**246.15**
632 × 305 mm	90.52	114.56	2.95	95.20	nr	**209.76**
765 × 356 mm	101.25	128.14	4.55	146.84	nr	**274.98**
737 × 406 mm	107.86	136.51	4.55	146.84	nr	**283.35**
818 × 406 mm	110.12	139.37	5.20	167.82	nr	**307.19**
978 × 406 mm	119.08	150.71	5.50	177.50	nr	**328.21**
709 × 457 mm	112.24	142.05	4.50	145.36	nr	**287.41**
678 × 508 mm	118.73	150.26	4.55	146.84	nr	**297.10**
919 × 508 mm	133.24	168.63	6.00	193.64	nr	**362.27**
Fire damper; galvanized steel casing; stainless steel folding shutter; fusible link with manual reset; BS 476 4 hour fire-rated						
Rectangular						
Sum of two sides 200 mm	43.24	54.72	1.60	51.63	nr	**106.35**
Sum of two sides 300 mm	44.62	56.47	1.60	51.63	nr	**108.10**
Sum of two sides 400 mm	46.53	58.89	1.60	51.63	nr	**110.52**
Sum of two sides 500 mm	50.90	64.42	1.60	51.63	nr	**116.05**
Sum of two sides 600 mm	55.76	70.57	1.70	54.89	nr	**125.46**
Sum of two sides 700 mm	60.76	76.90	2.10	67.79	nr	**144.69**
Sum of two sides 800 mm	65.95	83.46	2.15	69.41	nr	**152.87**
Sum of two sides 900 mm	71.16	90.06	2.30	74.36	nr	**164.42**
Sum of two sides 1000 mm	76.48	96.79	2.40	77.58	nr	**174.37**
Sum of two sides 1100 mm	81.22	102.79	2.60	84.04	nr	**186.83**
Sum of two sides 1200 mm	86.88	109.95	2.80	90.41	nr	**200.36**
Sum of two sides 1300 mm	92.67	117.29	3.10	100.05	nr	**217.34**
Sum of two sides 1400 mm	98.91	125.18	3.25	104.89	nr	**230.07**
Sum of two sides 1500 mm	104.34	132.05	3.40	109.78	nr	**241.83**
Sum of two sides 1600 mm	110.33	139.63	3.45	111.34	nr	**250.97**
Sum of two sides 1700 mm	116.36	147.27	3.60	116.18	nr	**263.45**
Sum of two sides 1800 mm	122.70	155.29	3.90	125.87	nr	**281.16**
Sum of two sides 1900 mm	128.71	162.89	4.20	135.60	nr	**298.49**
Sum of two sides 2000 mm	134.37	170.06	4.33	139.74	nr	**309.80**
Sum of two sides 2100 mm	142.99	180.97	4.43	142.98	nr	**323.95**
Sum of two sides 2200 mm	151.28	191.46	4.55	146.84	nr	**338.30**
Circular						
100 mm dia.	52.68	66.67	0.80	25.82	nr	**92.49**
150 mm dia.	55.43	70.16	0.80	25.82	nr	**95.98**
160 mm dia.	55.76	70.57	0.90	29.04	nr	**99.61**
200 mm dia.	58.52	74.07	1.05	33.88	nr	**107.95**
250 mm dia.	65.22	82.54	1.20	38.74	nr	**121.28**
300 mm dia.	72.49	91.74	1.35	43.56	nr	**135.30**
315 mm dia.	75.64	95.73	1.35	43.61	nr	**139.34**

38 VENTILATION/AIR CONDITIONING SYSTEMS

Item	Net Price £	Material £	Labour hours	Labour £	Unit	Total rate £
DUCTWORK ANCILLARIES: VOLUME CONTROL AND FIRE DAMPERS – cont						
Fire damper – cont						
Circular – cont						
350 mm dia.	80.23	101.54	1.65	53.26	nr	**154.80**
400 mm dia.	88.64	112.18	1.90	61.35	nr	**173.53**
450 mm dia.	96.93	122.67	2.10	67.79	nr	**190.46**
500 mm dia.	106.03	134.19	2.95	95.20	nr	**229.39**
550 mm dia.	115.15	145.73	2.94	94.89	nr	**240.62**
600 mm dia.	124.81	157.96	4.55	146.84	nr	**304.80**
650 mm dia.	135.05	170.92	4.55	146.84	nr	**317.76**
700 mm dia.	145.55	184.21	5.20	167.82	nr	**352.03**
750 mm dia.	156.58	198.17	5.20	167.82	nr	**365.99**
800 mm dia.	167.86	212.44	5.80	187.19	nr	**399.63**
850 mm dia.	179.72	227.45	5.80	187.19	nr	**414.64**
900 mm dia.	191.92	242.89	6.40	206.55	nr	**449.44**
950 mm dia.	204.44	258.74	6.40	206.55	nr	**465.29**
1000 mm dia.	217.58	275.37	7.00	225.90	nr	**501.27**
Flat oval						
345 × 102 mm	72.45	91.69	1.20	38.74	nr	**130.43**
427 × 102 mm	78.67	99.57	1.35	43.61	nr	**143.18**
508 × 102 mm	81.33	102.93	1.60	51.63	nr	**154.56**
559 × 152 mm	88.66	112.21	1.90	61.35	nr	**173.56**
531 × 203 mm	93.26	118.03	1.90	61.35	nr	**179.38**
851 × 203 mm	117.19	148.31	4.55	146.84	nr	**295.15**
582 × 254 mm	123.66	156.51	2.10	67.79	nr	**224.30**
632 × 305 mm	129.76	164.23	2.95	95.20	nr	**259.43**
765 × 356 mm	149.10	188.70	4.55	146.84	nr	**335.54**
737 × 406 mm	155.54	196.85	4.55	146.84	nr	**343.69**
818 × 406 mm	165.10	208.95	5.20	167.82	nr	**376.77**
978 × 406 mm	179.49	227.16	5.50	177.50	nr	**404.66**
709 × 457 mm	156.72	198.34	4.50	145.36	nr	**343.70**
678 × 508 mm	163.04	206.35	4.55	146.84	nr	**353.19**
Smoke/fire damper; galvanized steel casing; stainless steel folding shutter; fusible link and 24 V DC electromagnetic shutter release mechanism; spring operated; BS 476 4 hour fire-rating						
Rectangular						
Sum of two sides 200 mm	331.21	419.18	1.60	51.63	nr	**470.81**
Sum of two sides 300 mm	332.16	420.38	1.60	51.63	nr	**472.01**
Sum of two sides 400 mm	333.12	421.60	1.60	51.63	nr	**473.23**
Sum of two sides 500 mm	341.10	431.69	1.60	51.63	nr	**483.32**
Sum of two sides 600 mm	349.45	442.27	1.70	54.89	nr	**497.16**
Sum of two sides 700 mm	358.17	453.30	2.10	67.79	nr	**521.09**
Sum of two sides 800 mm	367.10	464.60	2.15	69.41	nr	**534.01**
Sum of two sides 900 mm	376.42	476.39	2.30	74.36	nr	**550.75**
Sum of two sides 1000 mm	386.06	488.60	2.40	77.58	nr	**566.18**
Sum of two sides 1100 mm	395.96	501.12	2.60	84.04	nr	**585.16**
Sum of two sides 1200 mm	406.21	514.10	2.80	90.41	nr	**604.51**

38 VENTILATION/AIR CONDITIONING SYSTEMS

Item	Net Price £	Material £	Labour hours	Labour £	Unit	Total rate £
Sum of two sides 1300 mm	416.83	527.54	3.10	100.05	nr	**627.59**
Sum of two sides 1400 mm	427.67	541.26	3.25	104.89	nr	**646.15**
Sum of two sides 1500 mm	438.86	555.42	3.40	109.78	nr	**665.20**
Sum of two sides 1600 mm	450.43	570.07	3.45	111.34	nr	**681.41**
Sum of two sides 1700 mm	462.24	585.01	3.60	116.18	nr	**701.19**
Sum of two sides 1800 mm	474.40	600.40	3.90	125.87	nr	**726.27**
Sum of two sides 1900 mm	486.78	616.07	4.20	135.60	nr	**751.67**
Sum of two sides 2000 mm	499.53	632.21	4.33	139.74	nr	**771.95**
Circular						
100 mm dia.	161.55	204.46	0.80	25.82	nr	**230.28**
150 mm dia.	161.55	204.46	0.90	29.04	nr	**233.50**
200 mm dia.	161.55	204.46	1.05	33.88	nr	**238.34**
250 mm dia.	170.33	215.57	1.20	38.74	nr	**254.31**
300 mm dia.	179.46	227.12	1.35	43.61	nr	**270.73**
350 mm dia.	204.24	258.48	1.65	53.26	nr	**311.74**
400 mm dia.	224.61	284.27	1.90	61.35	nr	**345.62**
450 mm dia.	234.22	296.43	2.10	67.79	nr	**364.22**
500 mm dia.	255.08	322.83	2.95	95.20	nr	**418.03**
550 mm dia.	276.63	350.10	2.95	95.20	nr	**445.30**
600 mm dia.	289.66	366.60	4.55	146.84	nr	**513.44**
650 mm dia.	313.42	396.66	4.55	146.84	nr	**543.50**
700 mm dia.	339.27	429.39	5.20	167.82	nr	**597.21**
750 mm dia.	355.33	449.70	5.20	167.82	nr	**617.52**
800 mm dia.	386.24	488.82	5.80	187.19	nr	**676.01**
850 mm dia.	414.08	524.06	5.80	187.19	nr	**711.25**
900 mm dia.	429.67	543.79	6.40	206.55	nr	**750.34**
950 mm dia.	458.54	580.33	6.40	206.55	nr	**786.88**
1000 mm dia.	479.22	606.50	7.00	225.90	nr	**832.40**
Flat oval						
531 × 203 mm	286.62	362.75	1.90	61.35	nr	**424.10**
851 × 203 mm	319.97	404.96	4.55	146.84	nr	**551.80**
582 × 254 mm	312.01	394.88	2.10	67.79	nr	**462.67**
632 × 305 mm	332.61	420.95	2.95	95.20	nr	**516.15**
765 × 356 mm	358.83	454.14	4.55	146.84	nr	**600.98**
737 × 406 mm	375.02	474.62	4.55	146.84	nr	**621.46**
818 × 406 mm	380.28	481.29	5.20	167.82	nr	**649.11**
978 × 406 mm	401.71	508.40	5.50	177.50	nr	**685.90**
709 × 457 mm	385.79	488.25	4.50	145.36	nr	**633.61**
678 × 508 mm	401.61	508.28	4.55	146.84	nr	**655.12**

38 VENTILATION/AIR CONDITIONING SYSTEMS

Item	Net Price £	Material £	Labour hours	Labour £	Unit	Total rate £
DUCTWORK ANCILLARIES: ACCESS HATCHES						
Access doors, hollow steel construction; 25 mm mineral wool insulation; removeable or hinged; fixed with cams; including subframe and integral sealing gaskets						
Rectangular duct						
150 × 150 mm	18.75	23.73	1.25	40.34	nr	**64.07**
200 × 200 mm	20.63	26.11	1.25	40.34	nr	**66.45**
300 × 150 mm	21.21	26.85	1.25	40.34	nr	**67.19**
300 × 300 mm	23.86	30.20	1.25	40.34	nr	**70.54**
400 × 400 mm	27.22	34.45	1.35	43.61	nr	**78.06**
450 × 300 mm	27.22	34.45	1.50	48.46	nr	**82.91**
450 × 450 mm	29.99	37.96	1.50	48.46	nr	**86.42**
Access doors, hollow steel construction; 25 mm mineral wool insulation; removeable or hinged; fixed with cams; including subframe and integral sealing gaskets						
Flat oval duct						
235 × 90 mm	38.10	48.22	1.25	40.34	nr	**88.56**
235 × 140 mm	40.60	51.39	1.35	43.61	nr	**95.00**
335 × 235 mm	46.39	58.71	1.50	48.46	nr	**107.17**
535 × 235 mm	52.19	66.05	1.50	48.46	nr	**114.51**

38 VENTILATION/AIR CONDITIONING SYSTEMS

Item	Net Price £	Material £	Labour hours	Labour £	Unit	Total rate £
GRILLES/DIFFUSERS/LOUVRES						
Supply grilles; single deflection; extruded aluminium alloy frame and adjustable horizontal vanes; silver grey polyester powder coated; screw fixed						
Rectangular; for duct, ceiling and sidewall applications						
150 × 100 mm	13.13	16.62	0.60	19.36	nr	**35.98**
150 × 150 mm	16.05	20.32	0.60	19.36	nr	**39.68**
200 × 150 mm	19.08	24.15	0.65	20.97	nr	**45.12**
200 × 200 mm	16.51	20.90	0.72	23.24	nr	**44.14**
300 × 100 mm	14.16	17.92	0.72	23.24	nr	**41.16**
300 × 150 mm	16.43	20.80	0.80	25.82	nr	**46.62**
300 × 200 mm	18.68	23.64	0.88	28.40	nr	**52.04**
300 × 300 mm	23.17	29.32	1.04	33.57	nr	**62.89**
400 × 100 mm	15.87	20.08	0.88	28.40	nr	**48.48**
400 × 150 mm	18.34	23.21	0.94	30.34	nr	**53.55**
400 × 200 mm	20.84	26.38	1.04	33.57	nr	**59.95**
400 × 300 mm	25.84	32.70	1.12	36.14	nr	**68.84**
600 × 200 mm	31.79	40.23	1.26	40.67	nr	**80.90**
600 × 300 mm	51.45	65.12	1.40	45.18	nr	**110.30**
600 × 400 mm	61.27	77.55	1.61	51.96	nr	**129.51**
600 × 500 mm	71.13	90.03	1.76	56.80	nr	**146.83**
600 × 600 mm	80.97	102.48	2.17	70.03	nr	**172.51**
800 × 300 mm	60.21	76.20	1.76	56.80	nr	**133.00**
800 × 400 mm	71.68	90.72	2.17	70.03	nr	**160.75**
800 × 600 mm	94.61	119.74	3.00	96.82	nr	**216.56**
1000 × 300 mm	69.00	87.33	2.60	83.91	nr	**171.24**
1000 × 400 mm	82.08	103.88	3.00	96.82	nr	**200.70**
1000 × 600 mm	108.25	137.00	3.80	122.64	nr	**259.64**
1000 × 800 mm	134.41	170.11	3.80	122.64	nr	**292.75**
1200 × 600 mm	121.87	154.24	4.61	148.78	nr	**303.02**
1200 × 800 mm	151.28	191.46	4.61	148.78	nr	**340.24**
1200 × 1000 mm	180.68	228.67	4.61	148.78	nr	**377.45**
Rectangular; for duct, ceiling and sidewall applications; including opposed blade damper volume regulator						
150 × 100 mm	24.48	30.98	0.72	23.25	nr	**54.23**
150 × 150 mm	28.35	35.88	0.72	23.25	nr	**59.13**
200 × 150 mm	30.43	38.52	0.83	26.80	nr	**65.32**
200 × 200 mm	32.90	41.64	0.90	29.04	nr	**70.68**
300 × 100 mm	34.94	44.22	0.90	29.04	nr	**73.26**
300 × 150 mm	37.24	47.13	0.98	31.63	nr	**78.76**
300 × 200 mm	40.13	50.79	1.06	34.23	nr	**85.02**
300 × 300 mm	45.87	58.05	1.20	38.74	nr	**96.79**
400 × 100 mm	45.41	57.47	1.06	34.23	nr	**91.70**
400 × 150 mm	47.95	60.68	1.13	36.47	nr	**97.15**
400 × 200 mm	48.18	60.97	1.20	38.74	nr	**99.71**
400 × 300 mm	54.26	68.67	1.34	43.25	nr	**111.92**
600 × 200 mm	66.12	83.69	1.50	48.46	nr	**132.15**
600 × 300 mm	87.84	111.17	1.66	53.61	nr	**164.78**

38 VENTILATION/AIR CONDITIONING SYSTEMS

Item	Net Price £	Material £	Labour hours	Labour £	Unit	Total rate £
GRILLES/DIFFUSERS/LOUVRES – cont						
Supply grilles – cont						
Rectangular – cont						
600 × 400 mm	99.86	126.38	1.80	58.15	nr	**184.53**
600 × 500 mm	113.08	143.11	2.00	64.55	nr	**207.66**
600 × 600 mm	125.10	158.32	2.60	84.04	nr	**242.36**
800 × 300 mm	120.46	152.45	2.00	64.55	nr	**217.00**
800 × 400 mm	135.07	170.95	2.60	84.04	nr	**254.99**
800 × 600 mm	165.93	210.00	3.61	116.51	nr	**326.51**
1000 × 300 mm	140.16	177.39	3.00	96.92	nr	**274.31**
1000 × 400 mm	157.05	198.77	3.61	116.51	nr	**315.28**
1000 × 600 mm	192.84	244.06	4.61	148.72	nr	**392.78**
1000 × 800 mm	285.19	360.93	4.61	148.72	nr	**509.65**
1200 × 600 mm	214.22	271.12	5.62	181.31	nr	**452.43**
1200 × 800 mm	314.34	397.82	6.10	196.86	nr	**594.68**
1200 × 1000 mm	357.42	452.35	6.50	209.78	nr	**662.13**
Supply grilles; double deflection; extruded aluminium alloy frame and adjustable horizontal and vertical vanes; white polyester powder coated; screw fixed						
Rectangular; for duct, ceiling and sidewall applications						
150 × 100 mm	13.13	16.62	0.88	28.40	nr	**45.02**
150 × 150 mm	16.95	21.45	0.88	28.40	nr	**49.85**
200 × 150 mm	19.08	24.15	1.08	34.85	nr	**59.00**
200 × 200 mm	21.26	26.90	1.25	40.34	nr	**67.24**
300 × 100 mm	20.71	26.21	1.25	40.34	nr	**66.55**
300 × 150 mm	23.38	29.59	1.50	48.41	nr	**78.00**
300 × 200 mm	26.06	32.98	1.75	56.48	nr	**89.46**
300 × 300 mm	31.45	39.80	2.15	69.40	nr	**109.20**
400 × 100 mm	24.50	31.01	1.75	56.48	nr	**87.49**
400 × 150 mm	27.69	35.04	1.95	62.93	nr	**97.97**
400 × 200 mm	30.89	39.10	2.15	69.40	nr	**108.50**
400 × 300 mm	37.27	47.17	2.55	82.30	nr	**129.47**
600 × 200 mm	68.86	87.15	3.01	97.14	nr	**184.29**
600 × 300 mm	83.21	105.31	3.36	108.44	nr	**213.75**
600 × 400 mm	97.58	123.50	3.80	122.64	nr	**246.14**
600 × 500 mm	111.91	141.64	4.20	135.54	nr	**277.18**
600 × 600 mm	126.26	159.79	4.51	145.54	nr	**305.33**
800 × 300 mm	103.06	130.44	4.20	135.54	nr	**265.98**
800 × 400 mm	120.89	153.00	4.51	145.54	nr	**298.54**
800 × 600 mm	156.56	198.14	5.10	164.60	nr	**362.74**
1000 × 300 mm	122.89	155.53	4.80	154.91	nr	**310.44**
1000 × 400 mm	144.21	182.52	5.10	164.60	nr	**347.12**
1000 × 600 mm	186.83	236.45	5.72	184.60	nr	**421.05**
1000 × 800 mm	229.48	290.43	6.33	204.28	nr	**494.71**
1200 × 600 mm	217.11	274.77	5.72	184.60	nr	**459.37**
1200 × 800 mm	266.72	337.56	6.33	204.28	nr	**541.84**
1200 × 1000 mm	316.31	400.32	6.33	204.28	nr	**604.60**

38 VENTILATION/AIR CONDITIONING SYSTEMS

Item	Net Price £	Material £	Labour hours	Labour £	Unit	Total rate £
Rectangular; for duct, ceiling and sidewall applications; including opposed blade damper volume regulator						
150 × 100 mm	26.76	33.87	1.00	32.27	nr	**66.14**
150 × 150 mm	31.75	40.19	1.00	32.27	nr	**72.46**
200 × 150 mm	35.02	44.32	1.26	40.67	nr	**84.99**
200 × 200 mm	37.63	47.62	1.43	46.16	nr	**93.78**
300 × 100 mm	41.49	52.51	1.43	46.16	nr	**98.67**
300 × 150 mm	44.22	55.97	1.68	54.22	nr	**110.19**
300 × 200 mm	47.51	60.13	1.93	62.29	nr	**122.42**
300 × 300 mm	54.15	68.53	2.31	74.56	nr	**143.09**
400 × 100 mm	54.05	68.41	1.93	62.29	nr	**130.70**
400 × 150 mm	57.29	72.51	2.14	69.06	nr	**141.57**
400 × 200 mm	58.20	73.66	2.31	74.56	nr	**148.22**
400 × 300 mm	66.20	83.79	2.77	89.40	nr	**173.19**
600 × 200 mm	103.79	131.35	3.25	104.89	nr	**236.24**
600 × 300 mm	120.41	152.39	3.62	116.83	nr	**269.22**
600 × 400 mm	137.06	173.47	3.99	128.78	nr	**302.25**
600 × 500 mm	154.91	196.06	4.44	143.29	nr	**339.35**
600 × 600 mm	171.54	217.10	4.94	159.43	nr	**376.53**
800 × 300 mm	164.41	208.07	4.44	143.29	nr	**351.36**
800 × 400 mm	185.50	234.76	4.94	159.43	nr	**394.19**
800 × 600 mm	229.38	290.30	5.71	184.27	nr	**474.57**
1000 × 300 mm	195.34	247.22	5.20	167.82	nr	**415.04**
1000 × 400 mm	220.60	279.19	5.71	184.27	nr	**463.46**
1000 × 600 mm	273.18	345.73	6.53	210.74	nr	**556.47**
1000 × 800 mm	382.86	484.55	6.53	210.74	nr	**695.29**
1200 × 600 mm	311.40	394.11	7.34	236.88	nr	**630.99**
1200 × 800 mm	432.63	547.53	8.80	284.00	nr	**831.53**
1200 × 1000 mm	496.30	628.12	8.80	284.00	nr	**912.12**
Floor grille suitable for mounting in raised access floors; heavy duty; extruded aluminium; standard mill finish; complete with opposed blade volume control damper						
Diffuser						
600 mm × 600 mm	154.34	195.33	0.70	22.59	nr	**217.92**
Extra for nylon coated black finish	23.62	29.89	–	–	nr	**29.89**
Exhaust grilles; aluminium						
0° fixed blade core						
150 × 150 mm	15.98	20.23	0.60	19.44	nr	**39.67**
200 × 200 mm	18.80	23.79	0.72	23.25	nr	**47.04**
250 × 250 mm	21.88	27.69	0.80	25.82	nr	**53.51**
300 × 300 mm	25.23	31.93	1.00	32.27	nr	**64.20**
350 × 350 mm	28.84	36.50	1.20	38.73	nr	**75.23**

38 VENTILATION/AIR CONDITIONING SYSTEMS

Item	Net Price £	Material £	Labour hours	Labour £	Unit	Total rate £
GRILLES/DIFFUSERS/LOUVRES – cont						
Exhaust grilles – cont						
0° fixed blade core; including opposed blade damper volume regulator						
150 × 150 mm	30.09	38.08	0.62	20.03	nr	**58.11**
200 × 200 mm	34.39	43.52	0.72	23.25	nr	**66.77**
250 × 250 mm	39.15	49.55	0.80	25.82	nr	**75.37**
300 × 300 mm	46.84	59.28	1.00	32.27	nr	**91.55**
350 × 350 mm	52.44	66.37	1.20	38.73	nr	**105.10**
45° fixed blade core						
150 × 150 mm	15.98	20.23	0.62	20.03	nr	**40.26**
200 × 200 mm	18.80	23.79	0.72	23.25	nr	**47.04**
250 × 250 mm	21.88	27.69	0.80	25.82	nr	**53.51**
300 × 300 mm	25.23	31.93	1.00	32.27	nr	**64.20**
350 × 350 mm	28.84	36.50	1.20	38.73	nr	**75.23**
45° fixed blade core; including opposed blade damper volume regulator						
150 × 150 mm	30.09	38.08	0.62	20.03	nr	**58.11**
200 × 200 mm	34.39	43.52	0.72	23.25	nr	**66.77**
250 × 250 mm	39.15	49.55	0.80	25.82	nr	**75.37**
300 × 300 mm	46.84	59.28	1.00	32.27	nr	**91.55**
350 × 350 mm	52.44	66.37	1.20	38.73	nr	**105.10**
Eggcrate core						
150 × 150 mm	10.72	13.56	0.62	20.03	nr	**33.59**
200 × 200 mm	13.25	16.77	1.00	32.27	nr	**49.04**
250 × 250 mm	16.14	20.43	0.80	25.82	nr	**46.25**
300 × 300 mm	20.54	26.00	1.00	32.27	nr	**58.27**
350 × 350 mm	22.95	29.04	1.20	38.73	nr	**67.77**
Eggcrate core; including opposed blade damper volume regulator						
150 × 150 mm	24.82	31.42	0.62	20.03	nr	**51.45**
200 × 200 mm	28.84	36.50	0.72	23.25	nr	**59.75**
250 × 250 mm	33.38	42.25	0.80	25.82	nr	**68.07**
300 × 300 mm	42.16	53.36	1.00	32.27	nr	**85.63**
350 × 350 mm	46.57	58.93	1.20	38.73	nr	**97.66**
Mesh/perforated plate core						
150 × 150 mm	11.79	14.92	0.62	20.01	nr	**34.93**
200 × 200 mm	14.57	18.44	0.72	23.24	nr	**41.68**
250 × 250 mm	17.76	22.48	0.80	25.82	nr	**48.30**
300 × 300 mm	22.61	28.62	1.00	32.27	nr	**60.89**
350 × 350 mm	25.27	31.99	1.20	38.73	nr	**70.72**
Mesh/perforated plate core; including opposed blade damper volume regulator						
150 × 150 mm	25.90	32.78	0.62	20.01	nr	**52.79**
200 × 200 mm	30.17	38.18	0.72	23.24	nr	**61.42**
250 × 250 mm	35.01	44.31	0.80	25.82	nr	**70.13**
300 × 300 mm	44.23	55.98	0.80	25.82	nr	**81.80**
350 × 350 mm	48.85	61.82	1.20	38.73	nr	**100.55**

38 VENTILATION/AIR CONDITIONING SYSTEMS

Item	Net Price £	Material £	Labour hours	Labour £	Unit	Total rate £
Plastic air diffusion system						
Eggcrate grilles						
150 × 150 mm	4.81	6.09	0.62	20.90	nr	26.99
200 × 200 mm	8.42	10.65	0.72	24.27	nr	34.92
250 × 250 mm	9.61	12.16	0.80	26.96	nr	39.12
300 × 300 mm	9.61	12.16	1.00	33.69	nr	45.85
Single deflection grilles						
150 × 150 mm	7.57	9.58	0.62	20.90	nr	30.48
200 × 200 mm	9.38	11.87	0.72	24.27	nr	36.14
250 × 250 mm	10.32	13.06	0.80	26.96	nr	40.02
300 × 300 mm	12.98	16.43	1.00	33.69	nr	50.12
Double deflection grilles						
150 × 150 mm	9.43	11.94	0.62	20.90	nr	32.84
200 × 200 mm	12.51	15.84	0.72	24.27	nr	40.11
250 × 250 mm	14.30	18.10	0.80	26.96	nr	45.06
300 × 300 mm	20.19	25.55	1.00	33.69	nr	59.24
Door transfer grilles						
150 × 150 mm	14.17	17.93	0.62	20.90	nr	38.83
200 × 200 mm	16.46	20.83	0.72	24.27	nr	45.10
250 × 250 mm	17.05	21.58	0.80	26.96	nr	48.54
300 × 300 mm	18.50	23.41	1.00	33.69	nr	57.10
Opposed blade dampers						
150 × 150 mm	8.58	10.86	0.62	20.90	nr	31.76
200 × 200 mm	10.09	12.77	0.72	24.27	nr	37.04
250 × 250 mm	15.13	19.15	0.80	26.96	nr	46.11
300 × 300 mm	16.46	20.83	1.00	33.69	nr	54.52
Ceiling mounted diffusers; circular aluminium multi-cone diffuser						
Circular; for ceiling mounting						
141 mm dia. neck	54.79	69.34	0.80	25.82	nr	95.16
197 mm dia. neck	80.21	101.52	1.10	35.50	nr	137.02
309 mm dia. neck	108.64	137.49	1.40	45.20	nr	182.69
365 mm dia. neck	92.95	117.63	1.50	48.46	nr	166.09
457 mm dia. neck	146.54	185.46	2.00	64.55	nr	250.01
Circular; for ceiling mounting; including louvre damper volume control						
141 mm dia. neck	81.89	103.64	1.00	32.27	nr	135.91
197 mm dia. neck	107.31	135.81	1.20	38.74	nr	174.55
309 mm dia. neck	141.58	179.19	1.60	51.63	nr	230.82
365 mm dia. neck	128.36	162.46	1.90	61.35	nr	223.81
457 mm dia. neck	188.20	238.19	2.40	77.58	nr	315.77
Ceiling mounted diffusers; rectangular aluminium multi-cone diffuser; four way flow						
Rectangular; for ceiling mounting						
150 × 150 mm neck	22.55	28.54	1.80	58.15	nr	86.69
300 × 150 mm neck	38.46	48.68	2.30	74.22	nr	122.90
300 × 300 mm neck	34.35	43.48	2.80	90.36	nr	133.84

38 VENTILATION/AIR CONDITIONING SYSTEMS

Item	Net Price £	Material £	Labour hours	Labour £	Unit	Total rate £
GRILLES/DIFFUSERS/LOUVRES – cont						
Ceiling mounted diffusers – cont						
Rectangular – cont						
450 × 150 mm neck	49.01	62.03	2.80	90.36	nr	**152.39**
450 × 300 mm neck	47.97	60.72	3.20	103.28	nr	**164.00**
450 × 450 mm neck	47.90	60.63	3.40	109.78	nr	**170.41**
600 × 150 mm neck	60.03	75.97	3.20	103.28	nr	**179.25**
600 × 300 mm neck	59.70	75.56	3.50	112.95	nr	**188.51**
600 × 600 mm neck	66.61	84.30	4.00	129.09	nr	**213.39**
Rectangular; for ceiling mounting; including opposed blade damper volume regulator						
150 × 150 mm neck	28.94	36.62	1.80	58.15	nr	**94.77**
300 × 150 mm neck	58.30	73.79	2.30	74.22	nr	**148.01**
300 × 300 mm neck	43.18	54.64	2.80	90.41	nr	**145.05**
450 × 150 mm neck	78.16	98.92	2.80	90.36	nr	**189.28**
450 × 300 mm neck	76.73	97.10	3.30	106.50	nr	**203.60**
450 × 450 mm neck	60.71	76.83	3.51	113.23	nr	**190.06**
600 × 150 mm neck	95.84	121.30	3.30	106.50	nr	**227.80**
600 × 300 mm neck	95.13	120.40	4.00	129.09	nr	**249.49**
600 × 600 mm neck	98.06	124.11	5.62	181.31	nr	**305.42**
Slot diffusers; continuous aluminium slot diffuser with flanged frame (1500 mm sections)						
Diffuser						
1 slot	37.06	46.91	3.76	121.32	m	**168.23**
2 slot	45.11	57.09	3.76	121.30	m	**178.39**
3 slot	55.52	70.27	3.76	121.30	m	**191.57**
4 slot	62.49	79.08	4.50	145.36	m	**224.44**
6 slot	82.33	104.19	4.50	145.36	m	**249.55**
Diffuser; including equalizing deflector						
1 slot	76.71	97.08	5.26	169.86	m	**266.94**
2 slot	95.17	120.44	5.26	169.86	m	**290.30**
3 slot	116.40	147.31	5.26	169.86	m	**317.17**
4 slot	130.71	165.42	6.33	204.25	m	**369.67**
6 slot	171.94	217.60	6.33	204.25	m	**421.85**
Extra over for ends						
1 slot	4.14	5.24	1.00	32.27	nr	**37.51**
2 slot	4.36	5.52	1.00	32.27	nr	**37.79**
3 slot	4.62	5.85	1.00	32.27	nr	**38.12**
4 slot	4.85	6.14	1.30	41.96	nr	**48.10**
6 slot	5.54	7.01	1.40	45.18	nr	**52.19**
Plenum boxes; 1.0 m long; circular spigot; including cord operated flap damper						
1 slot	52.81	66.84	2.75	88.91	nr	**155.75**
2 slot	54.19	68.58	2.75	88.91	nr	**157.49**
3 slot	54.89	69.47	2.75	88.91	nr	**158.38**
4 slot	59.88	75.78	3.51	113.23	nr	**189.01**
6 slot	62.83	79.52	3.51	113.23	nr	**192.75**

38 VENTILATION/AIR CONDITIONING SYSTEMS

Item	Net Price £	Material £	Labour hours	Labour £	Unit	Total rate £
Plenum boxes; 2.0 m long; circular spigot; including cord operated flap damper						
1 slot	58.47	74.00	3.26	105.12	nr	**179.12**
2 slot	61.31	77.59	3.26	105.12	nr	**182.71**
3 slot	62.18	78.69	3.26	105.12	nr	**183.81**
4 slot	69.25	87.64	3.76	121.32	nr	**208.96**
6 slot	73.84	93.45	3.76	121.32	nr	**214.77**
Perforated diffusers; rectangular face aluminium perforated diffuser; quick release face plate; for integration with rectangular ceiling tiles						
Circular spigot; rectangular diffuser						
150 mm dia. spigot; 300 × 300 mm diffuser	53.98	68.32	1.00	32.27	nr	**100.59**
300 mm dia. spigot; 600 × 600 mm diffuser	80.55	101.94	1.40	45.20	nr	**147.14**
Circular spigot; rectangular diffuser; including louvre damper volume regulator						
150 mm dia. spigot; 300 × 300 mm diffuser	81.08	102.61	1.00	32.27	nr	**134.88**
300 mm dia. spigot; 600 × 600 mm diffuser	113.48	143.62	1.60	51.63	nr	**195.25**
Rectangular spigot; rectangular diffuser						
150 × 150 mm dia. spigot; 300 × 300 mm diffuser	53.98	68.32	1.00	33.69	nr	**102.01**
300 × 150 mm dia. spigot; 600 × 300 mm diffuser	72.66	91.96	1.20	40.42	nr	**132.38**
300 × 300 mm dia. spigot; 600 × 600 mm diffuser	80.55	101.94	1.40	47.17	nr	**149.11**
600 × 300 mm dia. spigot; 1200 × 600 mm diffuser	84.13	106.48	1.60	53.91	nr	**160.39**
Rectangular spigot; rectangular diffuser; including opposed blade damper volume regulator						
150 × 150 mm dia. spigot; 300 × 300 mm diffuser	81.08	102.61	1.20	40.42	nr	**143.03**
300 × 150 mm dia. spigot; 600 × 300 mm diffuser	108.09	136.80	1.40	47.19	nr	**183.99**
300 × 300 mm dia. spigot; 600 × 600 mm diffuser	113.48	143.62	1.60	53.91	nr	**197.53**
600 × 300 mm dia. spigot; 1200 × 600 mm diffuser	132.54	167.74	1.80	60.65	nr	**228.39**
Floor swirl diffuser; manual adjustment of air discharge direction; complete with damper and dirt trap						
Plastic diffuser						
150 mm dia.	29.91	37.86	0.50	16.14	nr	**54.00**
200 mm dia.	29.91	37.86	0.50	16.14	nr	**54.00**
Aluminium diffuser						
150 mm dia.	32.02	40.52	0.50	16.14	nr	**56.66**
200 mm dia.	46.80	59.23	0.50	16.14	nr	**75.37**

38 VENTILATION/AIR CONDITIONING SYSTEMS

Item	Net Price £	Material £	Labour hours	Labour £	Unit	Total rate £
GRILLES/DIFFUSERS/LOUVRES – cont						
Ceiling mounted diffusers – cont						
Plastic air diffusion system						
Cellular diffusers						
300 × 300 mm	13.75	17.40	2.80	94.37	nr	**111.77**
600 × 600 mm	31.83	40.29	4.00	134.76	nr	**175.05**
Multi-cone diffusers						
300 × 300 mm	11.10	14.04	2.80	94.37	nr	**108.41**
450 × 450 mm	18.26	23.11	3.40	114.60	nr	**137.71**
500 × 500 mm	19.30	24.43	3.80	128.11	nr	**152.54**
600 × 600 mm	26.71	33.80	4.00	134.76	nr	**168.56**
625 × 625 mm	45.95	58.15	4.26	143.37	nr	**201.52**
Opposed blade dampers						
300 × 300 mm	14.80	18.73	1.20	40.44	nr	**59.17**
450 × 450 mm	16.27	20.60	1.50	50.59	nr	**71.19**
600 × 600 mm	33.68	42.63	2.60	87.73	nr	**130.36**
Plenum boxes						
300 mm	9.38	11.87	2.80	94.37	nr	**106.24**
450 mm	14.56	18.42	3.40	114.60	nr	**133.02**
600 mm	23.32	29.51	4.00	134.76	nr	**164.27**
Plenum spigot reducer						
600 mm	18.50	23.41	1.00	33.69	nr	**57.10**
Blanking kits for cellular diffusers						
300 mm	6.55	8.29	0.88	29.66	nr	**37.95**
600 mm	7.78	9.84	1.10	37.06	nr	**46.90**
Blanking kits for multi-cone diffusers						
300 mm	3.46	4.38	0.88	29.66	nr	**34.04**
450 mm	10.36	13.12	0.90	30.32	nr	**43.44**
600 mm	12.09	15.30	1.10	37.06	nr	**52.36**
Acoustic louvres; opening mounted; 300 mm deep steel louvres with blades packed with acoustic infill; 12 mm galvanized mesh birdscreen; screw fixing in opening						
Louvre units; self finished galvanized steel						
900 mm high × 600 mm wide	135.30	171.24	3.00	101.07	nr	**272.31**
900 mm high × 900 mm wide	167.86	212.44	3.00	101.07	nr	**313.51**
900 mm high × 1200 mm wide	198.74	251.53	3.34	112.53	nr	**364.06**
900 mm high × 1500 mm wide	259.35	328.24	3.34	112.53	nr	**440.77**
900 mm high × 1800 mm wide	290.80	368.03	3.34	112.53	nr	**480.56**
900 mm high × 2100 mm wide	322.25	407.84	3.34	112.53	nr	**520.37**
900 mm high × 2400 mm wide	353.11	446.89	3.68	123.97	nr	**570.86**
900 mm high × 2700 mm wide	399.15	505.16	3.68	123.97	nr	**629.13**
900 mm high × 3000 mm wide	427.21	540.68	3.68	123.97	nr	**664.65**
1200 mm high × 600 mm wide	177.40	224.52	3.00	101.07	nr	**325.59**
1200 mm high × 900 mm wide	218.38	276.38	3.34	112.53	nr	**388.91**

38 VENTILATION/AIR CONDITIONING SYSTEMS

Item	Net Price £	Material £	Labour hours	Labour £	Unit	Total rate £
1200 mm high × 1200 mm wide	258.80	327.53	3.34	112.53	nr	**440.06**
1200 mm high × 1500 mm wide	343.01	434.11	3.34	112.53	nr	**546.64**
1200 mm high × 1800 mm wide	383.42	485.25	3.68	123.97	nr	**609.22**
1200 mm high × 2100 mm wide	424.97	537.85	3.68	123.97	nr	**661.82**
1200 mm high × 2400 mm wide	465.40	589.01	3.68	123.97	nr	**712.98**
1500 mm high × 600 mm wide	220.06	278.51	3.00	101.07	nr	**379.58**
1500 mm high × 900 mm wide	270.05	341.78	3.34	112.53	nr	**454.31**
1500 mm high × 1200 mm wide	319.99	404.98	3.34	112.53	nr	**517.51**
1500 mm high × 1500 mm wide	426.67	540.00	3.68	123.97	nr	**663.97**
1500 mm high × 1800 mm wide	476.63	603.22	3.68	123.97	nr	**727.19**
1500 mm high × 2100 mm wide	527.16	667.17	4.00	134.76	nr	**801.93**
1800 mm high × 600 mm wide	261.60	331.08	3.34	112.53	nr	**443.61**
1800 mm high × 900 mm wide	321.11	406.39	3.34	112.53	nr	**518.92**
1800 mm high × 1200 mm wide	381.17	482.41	3.68	123.97	nr	**606.38**
1800 mm high × 1500 mm wide	510.87	646.55	3.68	123.97	nr	**770.52**
Louvre units; polyester powder coated steel						
900 mm high × 600 mm wide	196.50	248.70	3.00	101.07	nr	**349.77**
900 mm high × 900 mm wide	258.80	327.53	3.00	101.07	nr	**428.60**
900 mm high × 1200 mm wide	319.99	404.98	3.34	112.53	nr	**517.51**
900 mm high × 1500 mm wide	410.94	520.08	3.34	112.53	nr	**632.61**
900 mm high × 1800 mm wide	472.68	598.23	3.34	112.53	nr	**710.76**
900 mm high × 2100 mm wide	535.01	677.11	3.34	112.53	nr	**789.64**
900 mm high × 2400 mm wide	596.18	754.52	3.68	123.97	nr	**878.49**
900 mm high × 2700 mm wide	671.98	850.46	3.68	123.97	nr	**974.43**
900 mm high × 3000 mm wide	730.36	924.35	3.68	123.97	nr	**1048.32**
1200 mm high × 600 mm wide	258.23	326.82	3.00	101.07	nr	**427.89**
1200 mm high × 900 mm wide	339.63	429.83	3.34	112.53	nr	**542.36**
1200 mm high × 1200 mm wide	421.04	532.87	3.34	112.53	nr	**645.40**
1200 mm high × 1500 mm wide	545.10	689.88	3.34	112.53	nr	**802.41**
1200 mm high × 1800 mm wide	625.94	792.19	3.68	123.97	nr	**916.16**
1200 mm high × 2100 mm wide	707.92	895.94	3.68	123.97	nr	**1019.91**
1200 mm high × 2400 mm wide	788.74	998.23	3.68	123.97	nr	**1122.20**
1500 mm high × 600 mm wide	321.11	406.39	3.00	101.07	nr	**507.46**
1500 mm high × 900 mm wide	421.59	533.57	3.34	112.53	nr	**646.10**
1500 mm high × 1200 mm wide	522.09	660.76	3.34	112.53	nr	**773.29**
1500 mm high × 1500 mm wide	679.28	859.70	3.68	123.97	nr	**983.67**
1500 mm high × 1800 mm wide	780.33	987.58	3.68	123.97	nr	**1111.55**
1500 mm high × 2100 mm wide	880.83	1114.78	4.00	134.76	nr	**1249.54**
1800 mm high × 600 mm wide	382.84	484.52	3.34	112.53	nr	**597.05**
1800 mm high × 900 mm wide	503.00	636.60	3.34	112.53	nr	**749.13**
1800 mm high × 1200 mm wide	623.71	789.36	3.68	123.97	nr	**913.33**
1800 mm high × 1500 mm wide	814.02	1030.22	3.68	123.97	nr	**1154.19**
Weather louvres; opening mounted; 300 mm deep galvanized steel louvres; screw fixing in position						
Louvre units; including 12 mm galvanized mesh birdscreen						
900 × 600 mm	123.69	156.54	2.25	75.88	nr	**232.42**
900 × 900 mm	178.72	226.18	2.25	75.88	nr	**302.06**
900 × 1200 mm	215.04	272.16	2.50	84.22	nr	**356.38**

38 VENTILATION/AIR CONDITIONING SYSTEMS

Item	Net Price £	Material £	Labour hours	Labour £	Unit	Total rate £
GRILLES/DIFFUSERS/LOUVRES – cont						
Weather louvres – cont						
Louvre units – cont						
900 × 1500 mm	261.70	331.21	2.50	84.22	nr	**415.43**
900 × 1800 mm	308.37	390.28	2.50	84.22	nr	**474.50**
900 × 2100 mm	386.54	489.20	2.50	84.22	nr	**573.42**
900 × 2400 mm	415.23	525.52	2.76	93.07	nr	**618.59**
900 × 2700 mm	468.79	593.30	2.76	93.07	nr	**686.37**
900 × 3000 mm	511.96	647.93	2.76	93.07	nr	**741.00**
1200 × 600 mm	167.53	212.03	2.25	75.88	nr	**287.91**
1200 × 900 mm	234.08	296.25	2.50	84.22	nr	**380.47**
1200 × 1200 mm	301.16	381.15	2.50	84.22	nr	**465.37**
1200 × 1500 mm	351.53	444.90	2.50	84.22	nr	**529.12**
1200 × 1800 mm	459.73	581.83	2.76	93.07	nr	**674.90**
1200 × 2100 mm	522.84	661.71	2.76	93.07	nr	**754.78**
1200 × 2400 mm	546.01	691.03	2.76	93.07	nr	**784.10**
1500 × 600 mm	198.00	250.59	2.25	75.88	nr	**326.47**
1500 × 900 mm	272.85	345.32	2.50	84.22	nr	**429.54**
1500 × 1200 mm	359.39	454.84	2.50	84.22	nr	**539.06**
1500 × 1500 mm	421.36	533.28	2.76	93.07	nr	**626.35**
1500 × 1800 mm	496.84	628.80	2.76	93.07	nr	**721.87**
1500 × 2100 mm	621.39	786.43	3.00	101.17	nr	**887.60**
1800 × 600 mm	221.48	280.30	2.50	84.22	nr	**364.52**
1800 × 900 mm	306.18	387.50	2.50	84.22	nr	**471.72**
1800 × 1200 mm	401.48	508.11	2.76	93.07	nr	**601.18**
1800 × 1500 mm	474.91	601.05	3.00	101.17	nr	**702.22**

38 VENTILATION/AIR CONDITIONING SYSTEMS

Item	Net Price £	Material £	Labour hours	Labour £	Unit	Total rate £
PLANT/EQUIPMENT: FANS						
Axial flow fan; including ancillaries, anti-vibration mountings, mounting feet, matching flanges, flexible connectors and clips; 415 V, 3 phase, 50 Hz motor; includes fixing in position; electrical work elsewhere						
Aerofoil blade fan unit; short duct case						
315 mm dia.; 0.47 m³/s duty; 147 Pa	819.72	1037.43	4.50	151.60	nr	**1189.03**
500 mm dia.; 1.89 m³/s duty; 500 Pa	1359.51	1720.60	5.00	168.45	nr	**1889.05**
560 mm dia.; 2.36 m³/s duty; 147 Pa	1300.81	1646.31	5.50	185.29	nr	**1831.60**
710 mm dia.; 5.67 m³/s duty; 245 Pa	1746.57	2210.45	6.00	202.14	nr	**2412.59**
Aerofoil blade fan unit; long duct case						
315 mm dia.; 0.47 m³/s duty; 147 Pa	819.72	1037.43	4.50	151.60	nr	**1189.03**
500 mm dia.; 1.89 m³/s duty; 500 Pa	1359.51	1720.60	5.00	168.45	nr	**1889.05**
560 mm dia.; 2.36 m³/s duty; 147 Pa	1300.81	1646.31	5.50	185.29	nr	**1831.60**
710 mm dia.; 5.67 m³/s duty; 245 Pa	1746.57	2210.45	6.00	202.14	nr	**2412.59**
Aerofoil blade fan unit; two stage parallel fan arrangement; long duct case						
315 mm; 0.47 m³/s @ 500 Pa	819.72	1037.43	4.50	151.60	nr	**1189.03**
355 mm; 0.83 m³/s @ 147 Pa	1359.51	1720.60	4.75	160.03	nr	**1880.63**
710 mm; 3.77 m³/s @ 431 Pa	1300.81	1646.31	6.00	202.14	nr	**1848.45**
710 mm; 6.61 m³/s @ 500 Pa	1746.57	2210.45	6.00	202.14	nr	**2412.59**
Axial flow fan; suitable for operation at 300°C for 90 minutes; including ancillaries, anti-vibration mountings, mounting feet, matching flanges, flexible connectors and clips; 415 V, 3 phase, 50 Hz motor; includes fixing in position; price includes air operated damper; electrical work elsewhere						
450 mm; 2.0 m³/s @ 300 Pa	1751.13	2216.23	5.00	168.45	nr	**2384.68**
630 mm; 4.6 m³/s @ 200 Pa	2462.90	3117.05	5.50	185.29	nr	**3302.34**
900 mm; 9.0 m³/s @ 300 Pa	3920.11	4961.29	6.50	218.98	nr	**5180.27**
1000 mm; 15.0 m³/s @ 400 Pa	6041.98	7646.73	7.50	252.67	nr	**7899.40**
Bifurcated fan; suitable for temperature up to 200°C with motor protection to IP55; including ancillaries, anti-vibration mountings, mounting feet, matching flanges, flexible connectors and clips; 415 V, 3 phase, 50 Hz motor; includes fixing in position; electrical work elsewhere						
300 mm; 0.50 m³/s @ 100 Pa	1597.37	2021.63	4.50	151.60	nr	**2173.23**
400 mm; 1.97 m³/s @ 200 Pa	1987.60	2515.51	5.00	168.45	nr	**2683.96**
800 mm; 3.86 m³/s @ 200 Pa	3183.81	4029.44	5.50	185.29	nr	**4214.73**
1000 mm; 6.10 m³/s @ 400 Pa	4481.97	5672.39	6.50	218.98	nr	**5891.37**

38 VENTILATION/AIR CONDITIONING SYSTEMS

Item	Net Price £	Material £	Labour hours	Labour £	Unit	Total rate £
PLANT/EQUIPMENT: FANS – cont						
Duct mounted in line fan with backward curved centrifugal impellor; including ancillaries, matching flanges, flexible connectors and clips; 415 V, 3 phase, 50 Hz motor; includes fixing in position; electrical work elsewhere (NB: inverter included)						
0.5 m³/s @ 200 Pa	3288.65	4162.11	4.50	151.60	nr	**4313.71**
1.0 m³/s @ 300 Pa	3288.65	4162.11	5.00	168.45	nr	**4330.56**
3.0 m³/s @ 500 Pa	3170.73	4012.87	5.50	185.29	nr	**4198.16**
5.0 m³/s @ 750 Pa	6085.97	7702.40	6.50	218.98	nr	**7921.38**
7.0 m³/s @ 1000 Pa	7651.01	9683.12	7.00	235.83	nr	**9918.95**
Twin fan extract unit; belt driven; located internally; complete with anti-vibration mounts and non-return shutter; including ancillaries, matching flanges, flexible connectors and clips; 3 phase, 50 Hz motor; includes fixing in position; electrical work elsewhere (NB: inverter and auto changeover included)						
0.25 m³/s @ 150 Pa	4968.12	6287.66	4.50	151.60	nr	**6439.26**
1.00 m³/s @ 200 Pa	4968.12	6287.66	5.00	168.45	nr	**6456.11**
2.00 m³/s @ 250 Pa	5194.31	6573.92	6.50	218.98	nr	**6792.90**
Roof mounted extract fan; including ancillaries, fibreglass cowling, fitted shutters and bird guard; 415 V, 3 phase, 50 Hz motor; includes fixing in position; electrical work elsewhere (NB: inverter included)						
Flat roof installation, fixed to curb						
355 mm	1948.14	2465.57	4.50	151.60	nr	**2617.17**
400 mm	2134.41	2701.31	5.50	185.29	nr	**2886.60**
450 mm	2134.41	2701.31	7.00	235.83	nr	**2937.14**
500 mm	2517.30	3185.90	8.00	269.52	nr	**3455.42**
Centrifugal fan; single speed for internal domestic kitchens/utility rooms; fitted with standard overload protection; complete with housing; includes placing in position; electrical work elsewhere						
Window mounted						
245 m³/hr	318.73	403.38	0.50	16.84	nr	**420.22**
500 m³/hr	616.53	780.28	0.50	16.84	nr	**797.12**
Wall mounted						
245 m³/hr	403.05	510.10	0.83	28.07	nr	**538.17**
500 m³/hr	780.66	988.01	0.83	28.07	nr	**1016.08**

38 VENTILATION/AIR CONDITIONING SYSTEMS

Item	Net Price £	Material £	Labour hours	Labour £	Unit	Total rate £
Centrifugal fan; various speeds, simultaneous ventilation from separate areas fitted with standard overload protection; complete with housing; includes placing in position; ducting and electrical work elsewhere						
Fan unit						
147–300 m³/hr	635.92	804.82	1.00	33.69	nr	**838.51**
175–411 m³/hr	635.92	804.82	1.00	33.69	nr	**838.51**
Toilet extract units; centrifugal fan; various speeds for internal domestic bathrooms/ WCs, complete with housing; includes placing in position; electrical work elsewhere						
Fan unit; fixed to wall; including shutter						
Single speed 85 m³/hr	115.31	145.94	0.75	25.26	nr	**171.20**
Two speed 60–85 m³/hr	151.56	191.81	0.83	28.07	nr	**219.88**
Humidity controlled; autospeed; fixed to wall; including shutter						
30–60–85 m³/hr	267.97	339.15	1.00	33.69	nr	**372.84**

38 VENTILATION/AIR CONDITIONING SYSTEMS

Item	Net Price £	Material £	Labour hours	Labour £	Unit	Total rate £
PLANT/EQUIPMENT: AIR FILTRATION						
High efficiency duct mounted filters; 99.997% H13 (EU13); tested to BS 3928 Standard; 1700 m³/hr air volume; continuous rating up to 80°C; sealed wood case, aluminium spacers, neoprene gaskets; water repellant filter media; includes placing in position						
610 × 610 × 292 mm	289.60	366.52	1.00	32.27	nr	**398.79**
Side withdrawal frame	109.71	138.85	2.50	80.68	nr	**219.53**
High capacity; 3400 m³/hr air volume; continuous rating up to 80°C; anti-corrosion coated mild steel frame, polyurethane sealant and neoprene gaskets; water repellant filter media; includes placing in position						
610 × 610 × 292 mm	314.88	398.51	1.00	32.27	nr	**430.78**
Side withdrawal frame	109.71	138.85	2.50	80.68	nr	**219.53**
Bag filters; 40/60% F5 (EU5); tested to BSEN 779 Duct mounted bag filter; continuous rating up to 60°C; rigid filter assembly; sealed into one piece coated mild steel header with sealed pocket separators; includes placing in position						
6 pocket, 592 × 592 × 25 mm header; pockets 380 mm long; 1690 m³/hr	20.46	25.89	1.00	32.27	nr	**58.16**
Side withdrawal frame	18.99	24.04	2.00	64.55	nr	**88.59**
6 pocket, 592 × 592 × 25 mm header; pockets 500 mm long; 2550 m³/hr	20.79	26.31	1.50	48.41	nr	**74.72**
Side withdrawal frame	18.99	24.04	2.50	80.68	nr	**104.72**
6 pocket, 592 × 592 × 25 mm header; pockets 600 mm long; 3380 m³/hr	24.53	31.05	1.50	48.41	nr	**79.46**
Side withdrawal frame	18.99	24.04	3.00	96.82	nr	**120.86**
Bag filters; 80/90% F7, (EU7); tested to BSEN 779 Duct mounted bag filter; continuous rating up to 60°C; rigid filter assembly; sealed into one piece coated mild steel header with sealed pocket separators; includes placing in position						
6 pocket, 592 × 592 × 25 mm header; pockets 500 mm long; 1688 m³/hr	20.79	26.31	1.00	32.27	nr	**58.58**
Side withdrawal frame	18.99	24.04	2.00	64.55	nr	**88.59**
6 pocket, 592 × 592 × 25 mm header; pockets 635 mm long; 2047 m³/hr	24.53	31.05	1.50	48.41	nr	**79.46**
Side withdrawal frame	18.99	24.04	2.50	80.68	nr	**104.72**
6 pocket, 592 × 592 × 25 mm header; pockets 762 mm long; 2729 m³/hr	28.28	35.80	1.50	48.41	nr	**84.21**
Side withdrawal frame	18.99	24.04	3.00	96.82	nr	**120.86**

38 VENTILATION/AIR CONDITIONING SYSTEMS

Item	Net Price £	Material £	Labour hours	Labour £	Unit	Total rate £
Grease filters, washable; minimum 65% Double sided extract unit; lightweight stainless steel construction; demountable composite filter media of woven metal mat and expanded metal mesh supports; for mounting on hood and extract systems (hood not included); includes placing in position						
500 × 686 × 565 mm, 4080 m³/hr	426.83	540.20	2.00	64.55	nr	**604.75**
1000 × 686 × 565 mm, 8160 m³/hr	650.90	823.78	3.00	96.82	nr	**920.60**
1500 × 686 × 565 mm, 12240 m³/hr	892.80	1129.92	3.50	112.95	nr	**1242.87**
Panel filters; 82% G3 (EU3); tested to BS EN779 Modular duct mounted filter panels; continuous rating up to 100°C; graduated density media; rigid cardboard frame; includes placing in position						
596 × 596 × 47 mm, 2360 m³/hr	5.17	6.54	1.00	32.27	nr	**38.81**
Side withdrawal frame	37.08	46.93	2.50	80.68	nr	**127.61**
596 × 287 × 47 mm, 1140 m³/hr	3.57	4.51	1.00	32.27	nr	**36.78**
Side withdrawal frame	37.08	46.93	2.50	80.68	nr	**127.61**
Panel filters; 90% G4 (EU4); tested to BS EN779 Modular duct mounted filter panels; continuous rating up to 100°C; pleated media with wire support; rigid cardboard frame; includes placing in position						
596 × 596 × 47 mm, 2560 m³/hr	7.82	9.90	1.00	32.27	nr	**42.17**
Side withdrawal frame	46.80	59.23	3.00	96.82	nr	**156.05**
596 × 287 × 47 mm, 1230 m³/hr	5.40	6.83	1.00	32.27	nr	**39.10**
Side withdrawal frame	37.08	46.93	3.00	96.82	nr	**143.75**
Carbon filters; standard duty disposable carbon filters; steel frame with bonded carbon panels; for fixing to ductwork; including placing in position 12 panels						
597 × 597 × 298 mm, 1460 m³/hr	363.77	460.39	0.33	10.64	nr	**471.03**
597 × 597 × 451 mm, 2200 m³/hr	410.08	519.00	0.33	10.64	nr	**529.64**
597 × 597 × 597 mm, 2930 m³/hr	457.32	578.78	0.33	10.64	nr	**589.42**
8 panels						
451 × 451 × 298 mm, 740 m³/hr	274.24	347.08	0.29	9.35	nr	**356.43**
451 × 451 × 451 mm, 1105 m³/hr	303.53	384.15	0.29	9.35	nr	**393.50**
451 × 451 × 597 mm, 1460 m³/hr	331.47	419.51	0.29	9.35	nr	**428.86**
6 panels						
298 × 298 × 298 mm, 365 m³/hr	194.33	245.94	0.25	8.08	nr	**254.02**
298 × 298 × 451 mm, 550 m³/hr	208.06	263.32	0.25	8.08	nr	**271.40**
298 × 298 × 597 mm, 780 m³/hr	221.20	279.96	0.25	8.08	nr	**288.04**

38 VENTILATION/AIR CONDITIONING SYSTEMS

Item	Net Price £	Material £	Labour hours	Labour £	Unit	Total rate £
PLANT/EQUIPMENT: AIR CURTAINS						
The selection of air curtains requires consideration of the particular conditions involved; climatic conditions, wind influence, construction and position all influence selection; consultation with a specialist manufacturer is therefore advisable.						
Commercial grade air curtains; recessed or exposed units with rigid sheet steel casing; aluminium grilles; high quality motor/centrifugal fan assembly; includes fixing in position; electrical work elsewhere						
Ambient temperature; 240 V single phase supply; mounting height 2.40 m						
1000 × 590 × 270 mm	2749.53	3479.81	12.05	405.90	nr	**3885.71**
1500 × 590 × 270 mm	3519.21	4453.92	12.05	405.90	nr	**4859.82**
2000 × 590 × 270 mm	4257.44	5388.22	12.05	405.97	nr	**5794.19**
2500 × 590 × 270 mm	4732.74	5989.76	13.00	437.98	nr	**6427.74**
Ambient temperature; 240 V single phase supply; mounting height 2.80 m						
1000 × 590 × 270 mm	3176.50	4020.17	16.13	543.39	nr	**4563.56**
1500 × 590 × 270 mm	4055.18	5132.23	16.13	543.39	nr	**5675.62**
2000 × 590 × 270 mm	5055.22	6397.89	16.13	543.39	nr	**6941.28**
2500 × 590 × 270 mm	5845.13	7397.60	17.10	576.11	nr	**7973.71**
Ambient temperature 240 V single phase supply; mounting height 3.30 m						
1000 × 774 × 370 mm	4149.57	5251.69	17.24	580.88	nr	**5832.57**
1500 × 774 × 370 mm	5611.42	7101.81	17.24	580.88	nr	**7682.69**
2000 × 774 × 370 mm	6940.67	8784.12	17.24	580.88	nr	**9365.00**
2500 × 774 × 370 mm	8773.32	11103.51	18.30	616.54	nr	**11720.05**
Ambient temperature; 240 V single phase supply; mounting height 4.00 m						
1000 × 774 × 370 mm	4583.30	5800.63	19.10	643.48	nr	**6444.11**
1500 × 774 × 370 mm	6012.55	7609.48	19.10	643.48	nr	**8252.96**
2000 × 774 × 370 mm	7485.64	9473.82	19.10	643.48	nr	**10117.30**
2500 × 774 × 370 mm	8773.32	11103.51	19.90	670.44	nr	**11773.95**
Water heated; 240 V single phase supply; mounting height 2.40 m						
1000 × 590 × 270 mm; 2.30–9.40 kW output	3234.94	4094.14	12.05	405.90	nr	**4500.04**
1500 × 590 × 270 mm; 3.50–14.20 kW output	4140.59	5240.33	12.05	405.90	nr	**5646.23**
2000 × 590 × 270 mm; 4.70–19.00 kW output	5009.15	6339.58	12.05	405.90	nr	**6745.48**
2500 × 590 × 270 mm; 5.90–23.70 kW output	5568.72	7047.77	13.00	437.98	nr	**7485.75**

38 VENTILATION/AIR CONDITIONING SYSTEMS

Item	Net Price £	Material £	Labour hours	Labour £	Unit	Total rate £
Water heated; 240 V single phase supply; mounting height 2.80 m						
1000 × 590 × 270 mm; 3.30–11.90 kW output	3738.32	4731.22	16.13	543.39	nr	**5274.61**
1500 × 590 × 270 mm; 5.00–17.90 kW output	4772.07	6039.53	16.13	543.39	nr	**6582.92**
2000 × 590 × 270 mm; 6.70–23.90 kW output	5948.51	7528.44	16.13	543.39	nr	**8071.83**
2500 × 590 × 270 mm; 8.30–29.80 kW output	6877.75	8704.48	17.10	576.11	nr	**9280.59**
Water heated; 240 V single phase supply; mounting height 3.30 m						
1000 × 774 × 370 mm; 6.10–21.80 kW output	4882.18	6178.88	17.24	580.88	nr	**6759.76**
1500 × 774 × 370 mm; 9.20–32.80 kW output	6602.46	8356.07	17.24	580.88	nr	**8936.95**
2000 × 774 × 370 mm; 12.30–43.70 kW output	8166.56	10335.60	17.24	580.88	nr	**10916.48**
2500 × 774 × 370 mm; 15.30–54.60 kW output	9666.60	12234.05	18.30	616.54	nr	**12850.59**
Water heated; 240 V single phase supply; mounting height 4.00 m						
1000 × 774 × 370 mm; 7.20–24.20 kW output	5392.31	6824.51	19.10	643.48	nr	**7467.99**
1500 × 774 × 370 mm; 10.90–36.30 kW output	7074.38	8953.34	19.10	643.48	nr	**9596.82**
2000 × 774 × 370 mm; 14.50–48.40 kW output	8807.03	11146.17	19.10	643.48	nr	**11789.65**
2500 × 774 × 370 mm; 18.10–60.60 kW output	10321.69	13063.13	19.90	670.44	nr	**13733.57**
Electrically heated; 415 V three phase supply; mounting height 2.40 m						
1000 × 590 × 270 mm; 2.30–9.40 kW output	3934.96	4980.08	12.05	405.90	nr	**5385.98**
1500 × 590 × 270 mm; 3.50–14.20 kW output	4895.66	6195.95	12.05	405.90	nr	**6601.85**
2000 × 590 × 270 mm; 4.70–19.00 kW output	5835.02	7384.80	12.05	465.53	nr	**7850.33**
2500 × 590 × 270 mm; 5.90–23.70 kW output	6633.93	8395.90	13.00	502.31	nr	**8898.21**
Electrically heated; 415 V three phase supply; mounting height 2.80 m						
1000 × 590 × 270 mm; 3.30–11.90 kW output	4632.74	5863.20	16.13	543.39	nr	**6406.59**
1500 × 590 × 270 mm; 5.00–17.90 kW output	5696.81	7209.89	16.13	543.39	nr	**7753.28**
2000 × 590 × 270 mm; 6.70–23.90 kW output	7193.49	9104.08	16.13	543.39	nr	**9647.47**
2500 × 590 × 270 mm; 8.30–29.80 kW output	8039.59	10174.91	17.10	576.11	nr	**10751.02**

38 VENTILATION/AIR CONDITIONING SYSTEMS

Item	Net Price £	Material £	Labour hours	Labour £	Unit	Total rate £
PLANT/EQUIPMENT: AIR CURTAINS – cont						
Commercial grade air curtains – cont						
Electrically heated; 415 V three phase supply; mounting height 3.30 m						
1000 × 774 × 370 mm; 6.10–21.80 kW output	7039.55	8909.25	17.24	580.82	nr	**9490.07**
1500 × 774 × 370 mm; 9.20–32.80 kW output	9431.77	11936.85	17.24	580.82	nr	**12517.67**
2000 × 774 × 370 mm; 12.30–43.70 kW output	11383.52	14406.99	17.24	580.82	nr	**14987.81**
2500 × 774 × 370 mm; 15.30–54.60 kW output	13475.72	17054.87	18.30	616.54	nr	**17671.41**
Electrically heated; 415 V three phase supply; mounting height 4.00 m						
1000 × 774 × 370 mm; 7.20–24.20 kW output	7768.79	9832.18	19.10	643.48	nr	**10475.66**
1500 × 774 × 370 mm; 10.90–36.30 kW output	10078.98	12755.96	19.10	643.48	nr	**13399.44**
2000 × 774 × 370 mm; 14.50–48.40 kW output	12190.29	15428.03	19.10	643.48	nr	**16071.51**
2500 × 774 × 370 mm; 18.10–60.60 kW output	14439.79	18275.00	19.90	670.44	nr	**18945.44**
Industrial grade air curtains; recessed or exposed units with rigid sheet steel casing; aluminium grilles; high quality motor/ centrifugal fan assembly; includes fixing in position; electrical work elsewhere						
Ambient temperature; 230 V single phase supply; including wiring between multiple units; horizontally or vertically mounted; opening maximum 6.00 m						
1500 × 585 × 853 mm; 3.0 A supply	2571.79	3254.85	17.24	580.88	nr	**3835.73**
2000 × 585 × 853 mm; 4.0 A supply	3116.58	3944.35	17.24	580.88	nr	**4525.23**
Water heated; 230 V single phase supply; including wiring between multiple units; horizontally or vertically mounted; opening maximum 6.00 m						
1500 × 585 × 956 mm; 3.0 A supply; 34.80 kW output	2947.85	3730.80	17.24	580.88	nr	**4311.68**
2000 × 585 × 956 mm; 4.0 A supply; 50.70 kW output	3601.82	4558.47	17.24	580.88	nr	**5139.35**
Water heated; 230 V single phase supply; including wiring between multiple units; vertically mounted in single bank for openings maximum 6.00 m wide or opposing twin banks for openings maximum 10.00 m wide						
1500 × 585 mm; 3.0 A supply; 41.1 kW output	2947.85	3730.80	17.24	580.88	nr	**4311.68**
2000 × 585 mm; 4.0 A supply; 57.7 kW output	3601.82	4558.47	17.24	580.88	nr	**5139.35**
Remote mounted electronic controller unit; excluding wiring to units						
0–10V control	45.22	57.23	5.00	168.45	nr	**225.68**

38 VENTILATION/AIR CONDITIONING SYSTEMS

Item	Net Price £	Material £	Labour hours	Labour £	Unit	Total rate £
SILENCERS/ACOUSTIC TREATMENT						
Attenuators; DW144 galvanized construction c/w splitters; self securing; fitted to ductwork						
To suit rectangular ducts; unit length 600 mm						
100 × 100 mm	109.40	138.45	0.75	24.21	nr	162.66
150 × 150 mm	114.19	144.51	0.75	24.21	nr	168.72
200 × 200 mm	118.98	150.58	0.75	24.21	nr	174.79
300 × 300 mm	133.11	168.46	0.75	24.21	nr	192.67
400 × 400 mm	149.43	189.12	1.00	32.27	nr	221.39
500 × 500 mm	174.40	220.72	1.25	40.34	nr	261.06
600 × 300 mm	167.99	212.61	1.25	40.34	nr	252.95
600 × 600 mm	235.11	297.55	1.25	40.34	nr	337.89
700 × 300 mm	178.12	225.43	1.50	48.41	nr	273.84
700 × 700 mm	265.22	335.66	1.50	48.41	nr	384.07
800 × 300 mm	184.51	233.52	2.00	64.55	nr	298.07
800 × 800 mm	305.99	387.26	2.00	64.55	nr	451.81
1000 × 1000 mm	391.68	495.71	3.00	96.82	nr	592.53
To suit rectangular ducts; unit length 1200 mm						
200 × 200 mm	131.21	166.06	1.00	32.27	nr	198.33
300 × 300 mm	162.21	205.30	1.00	32.27	nr	237.57
400 × 400 mm	195.03	246.83	1.33	42.92	nr	289.75
500 × 500 mm	235.40	297.92	1.66	53.57	nr	351.49
600 × 300 mm	224.45	284.07	1.66	53.57	nr	337.64
600 × 600 mm	328.40	415.62	1.66	53.57	nr	469.19
700 × 300 mm	240.60	304.51	2.00	64.55	nr	369.06
700 × 700 mm	375.80	475.61	2.00	64.55	nr	540.16
800 × 300 mm	251.54	318.35	2.66	85.85	nr	404.20
800 × 800 mm	433.63	548.80	2.66	85.85	nr	634.65
1000 × 1000 mm	587.01	742.92	4.00	129.09	nr	872.01
1300 × 1300 mm	1007.62	1275.24	8.00	258.18	nr	1533.42
1500 × 1500 mm	1153.34	1459.66	8.00	258.18	nr	1717.84
1800 × 1800 mm	1578.04	1997.17	10.66	344.03	nr	2341.20
2000 × 2000 mm	1770.74	2241.05	13.33	430.20	nr	2671.25
To suit rectangular ducts; unit length 1800 mm						
200 × 200 mm	165.59	209.57	1.00	32.27	nr	241.84
300 × 300 mm	220.03	278.47	1.00	32.27	nr	310.74
400 × 400 mm	263.02	332.88	1.33	42.92	nr	375.80
500 × 500 mm	306.52	387.93	1.66	53.57	nr	441.50
600 × 300 mm	320.58	405.73	1.66	53.57	nr	459.30
600 × 600 mm	447.97	566.96	1.66	53.57	nr	620.53
700 × 300 mm	339.86	430.12	2.00	64.55	nr	494.67
700 × 700 mm	535.22	677.38	2.00	64.55	nr	741.93
800 × 300 mm	359.38	454.83	2.66	85.85	nr	540.68
800 × 800 mm	622.74	788.14	2.66	85.85	nr	873.99
1000 × 1000 mm	837.08	1059.41	4.00	129.09	nr	1188.50
1300 × 1300 mm	1398.36	1769.77	8.00	258.18	nr	2027.95
1500 × 1500 mm	1584.72	2005.62	8.00	258.18	nr	2263.80
1800 × 1800 mm	2200.09	2784.43	13.33	430.20	nr	3214.63
2000 × 2000 mm	2491.29	3152.98	10.66	344.03	nr	3497.01
2300 × 2300 mm	3114.06	3941.16	16.00	516.36	nr	4457.52
2500 × 2500 mm	4343.74	5497.44	18.66	602.21	nr	6099.65

38 VENTILATION/AIR CONDITIONING SYSTEMS

Item	Net Price £	Material £	Labour hours	Labour £	Unit	Total rate £
SILENCERS/ACOUSTIC TREATMENT – cont						
Attenuators – cont						
To suit rectangular ducts; unit length 2400 mm						
500 × 500 mm	396.38	501.66	2.08	67.23	nr	**568.89**
600 × 300 mm	378.42	478.92	2.08	67.23	nr	**546.15**
600 × 600 mm	563.62	713.32	2.08	67.23	nr	**780.55**
700 × 300 mm	410.20	519.15	2.50	80.68	nr	**599.83**
700 × 700 mm	665.75	842.58	2.50	80.68	nr	**923.26**
800 × 300 mm	433.63	548.80	3.33	107.48	nr	**656.28**
800 × 800 mm	770.44	975.07	3.33	107.48	nr	**1082.55**
1000 × 1000 mm	1039.47	1315.55	5.00	161.36	nr	**1476.91**
1300 × 1300 mm	1683.61	2130.78	10.00	322.73	nr	**2453.51**
1500 × 1500 mm	1921.54	2431.90	10.00	322.73	nr	**2754.63**
1800 × 1800 mm	2741.67	3469.86	13.33	430.12	nr	**3899.98**
2000 × 2000 mm	3076.61	3893.76	16.66	537.67	nr	**4431.43**
2300 × 2300 mm	3806.45	4817.44	20.00	645.46	nr	**5462.90**
2500 × 2500 mm	5344.82	6764.41	23.32	752.73	nr	**7517.14**
To suit circular ducts; unit length 600 mm						
100 mm dia.	95.00	120.23	0.75	24.21	nr	**144.44**
200 mm dia.	111.78	141.47	0.75	24.21	nr	**165.68**
250 mm dia.	126.71	160.36	0.75	24.21	nr	**184.57**
315 mm dia.	147.49	186.66	1.00	32.27	nr	**218.93**
355 mm dia.	170.67	216.00	1.00	32.27	nr	**248.27**
400 mm dia.	183.46	232.19	1.00	32.27	nr	**264.46**
450 mm dia.	221.54	280.38	1.00	32.27	nr	**312.65**
500 mm dia.	243.93	308.72	1.26	40.60	nr	**349.32**
630 mm dia.	287.59	363.98	1.50	48.41	nr	**412.39**
710 mm dia.	313.18	396.36	1.50	48.41	nr	**444.77**
800 mm dia.	348.34	440.85	2.00	64.55	nr	**505.40**
1000 mm dia.	467.59	591.79	3.00	96.82	nr	**688.61**
To suit circular ducts; unit length 1200 mm						
100 mm dia.	139.77	176.89	1.00	32.27	nr	**209.16**
200 mm dia.	153.34	194.06	1.00	32.27	nr	**226.33**
250 mm dia.	172.27	218.03	1.00	32.27	nr	**250.30**
315 mm dia.	197.04	249.38	1.33	42.92	nr	**292.30**
355 mm dia.	236.20	298.94	1.33	42.92	nr	**341.86**
400 mm dia.	250.59	317.15	1.33	42.92	nr	**360.07**
450 mm dia.	290.26	367.35	1.33	42.92	nr	**410.27**
500 mm dia.	297.48	376.49	1.66	53.57	nr	**430.06**
630 mm dia.	393.90	498.52	2.00	64.55	nr	**563.07**
710 mm dia.	435.45	551.11	2.00	64.55	nr	**615.66**
800 mm dia.	434.65	550.09	2.66	85.85	nr	**635.94**
1000 mm dia.	583.49	738.46	4.00	129.09	nr	**867.55**
1250 mm dia.	1099.16	1391.10	8.00	258.18	nr	**1649.28**
1400 mm dia.	1195.04	1512.45	8.00	258.18	nr	**1770.63**
1600 mm dia.	1324.87	1676.75	10.66	344.03	nr	**2020.78**
To suit circular ducts; unit length 1800 mm						
200 mm dia.	214.88	271.95	1.25	40.34	nr	**312.29**
250 mm dia.	240.46	304.33	1.25	40.34	nr	**344.67**
315 mm dia.	285.76	361.66	1.66	53.57	nr	**415.23**

38 VENTILATION/AIR CONDITIONING SYSTEMS

Item	Net Price £	Material £	Labour hours	Labour £	Unit	Total rate £
355 mm dia.	336.90	426.38	1.66	53.57	nr	**479.95**
400 mm dia.	348.61	441.20	1.66	53.57	nr	**494.77**
450 mm dia.	356.09	450.67	1.66	53.57	nr	**504.24**
500 mm dia.	391.46	495.43	2.08	67.13	nr	**562.56**
630 mm dia.	433.05	548.07	2.50	80.68	nr	**628.75**
710 mm dia.	506.57	641.11	2.50	80.68	nr	**721.79**
800 mm dia.	524.15	663.36	3.33	107.48	nr	**770.84**
1000 mm dia.	1368.47	1731.93	6.66	214.94	nr	**1946.87**
1250 mm dia.	1390.86	1760.27	6.66	214.94	nr	**1975.21**
1400 mm dia.	1536.65	1944.78	5.00	161.36	nr	**2106.14**
1600 mm dia.	1537.58	1945.97	9.90	319.50	nr	**2265.47**

38 VENTILATION/AIR CONDITIONING SYSTEMS

Item	Net Price £	Material £	Labour hours	Labour £	Unit	Total rate £
THERMAL INSULATION						
Concealed ductwork						
Flexible wrap; 20 kg–45 kg Bright Class O aluminium foil faced; Bright Class O foil taped joints; 62 mm metal pins and washers; aluminium bands						
40 mm thick insulation	16.24	20.55	0.40	13.47	m²	**34.02**
Semi-rigid slab; 45 kg Bright Class O aluminium foil faced mineral fibre; Bright Class O foil taped joints; 62 mm metal pins and washers; aluminium bands						
40 mm thick insulation	21.24	26.88	0.65	21.90	m²	**48.78**
Plantroom ductwork						
Semi-rigid slab; 45 kg Bright Class O aluminium foil faced mineral fibre; Bright Class O foil taped joints; 62 mm metal pins and washers; 22 swg plain/embossed aluminium cladding; pop rivited						
50 mm thick insulation	53.59	67.83	1.50	50.53	m²	**118.36**
External ductwork						
Semi-rigid slab; 45 kg Bright Class O aluminium foil faced mineral fibre; Bright Class O foil taped joints; 62 mm metal pins and washers; 0.8 mm polyisobutylene sheeting; welded joints						
50 mm thick insulation	39.09	49.47	1.25	42.11	m²	**91.58**

38 VENTILATION/AIR CONDITIONING SYSTEMS

Item	Net Price £	Material £	Labour hours	Labour £	Unit	Total rate £
DUCTWORK: FIRE RATED						
DUCTLINES						
The relevant BS requires that the fire-rating of ductwork meets 3 criteria; stability (hours), integrity (hours) and insulation (hours). The least of the 3 periods defines the fire-rating. The BS does however allow stability and integrity to be considered in isolation. Rates are therefore provided for both types of system.						
Care should to be taken when using the rates within this section to ensure thta the requiremements for stability, integrity and insulation are known and the appropriate rates are used.						
High density single layer mineral wool fire-rated ductwork slab, in accordance with BSEN1366, ducts 'Type A' and 'Type B'; 165 kg class O foil faced mineral fibre; 100 mm wide bright class O foil taped joints; welded pins; includes protection to all supports.						
½ hour stability, integrity and insulation						
25 mm thick, vertical and horizontal ductwork	95.40	120.74	1.25	42.11	m²	**162.85**
1 hour stability, integrity and insulation						
30 mm thick, vertical ductwork	111.99	141.74	1.50	50.53	m²	**192.27**
40 mm thick, horizontal ductwork	131.72	166.70	1.50	50.53	m²	**217.23**
1½ hour stability, integrity and insulation						
50 mm thick, vertical ductwork	157.35	199.15	1.75	58.97	m²	**258.12**
70 mm thick, horizontal ductwork	196.82	249.10	1.75	58.97	m²	**308.07**
2 hour stability, integrity and insulation						
70 mm, vertical ductwork	202.79	256.65	2.00	67.38	m²	**324.03**
90 mm horizontal ductwork	242.26	306.60	2.00	67.38	m²	**373.98**
Kitchen extract, 1 hour stability, integrity and insulation						
90 mm, vertical and horizontal	242.26	306.60	2.00	67.38	m²	**373.98**
Galvanized sheet metal rectangular section ductwork to BSEN1366, ducts 'Type A' and 'Type B'; provides 2 hours stability and 2 hours integrity at 1100°C (no rating for insulation); including all necessary stiffeners, joints and supports in the running length						
Ductwork up to 600 mm longest side						
Sum of two sides 200 mm	121.61	153.91	2.91	93.91	m	**247.82**
Sum of two sides 300 mm	132.03	167.09	2.99	96.50	m	**263.59**
Sum of two sides 400 mm	142.42	180.24	3.17	102.30	m	**282.54**

38 VENTILATION/AIR CONDITIONING SYSTEMS

Item	Net Price £	Material £	Labour hours	Labour £	Unit	Total rate £
DUCTWORK: FIRE RATED – cont						
Ductwork up to 600 mm longest side – cont						
Sum of two sides 500 mm	152.84	193.44	3.37	108.76	m	**302.20**
Sum of two sides 600 mm	163.23	206.58	3.54	114.25	m	**320.83**
Sum of two sides 700 mm	173.66	219.79	3.72	120.05	m	**339.84**
Sum of two sides 800 mm	184.07	232.96	3.90	125.87	m	**358.83**
Sum of two sides 900 mm	194.46	246.11	5.04	162.66	m	**408.77**
Sum of two sides 1000 mm	204.89	259.31	5.58	180.08	m	**439.39**
Sum of two sides 1100 mm	215.28	272.46	5.84	188.47	m	**460.93**
Sum of two sides 1200 mm	225.67	285.61	6.11	197.20	m	**482.81**
Extra over fittings; Ductwork up to 600 mm longest side						
End cap						
Sum of two sides 200 mm	34.54	43.71	0.81	26.15	nr	**69.86**
Sum of two sides 300 mm	36.64	46.37	0.84	27.10	nr	**73.47**
Sum of two sides 400 mm	38.73	49.01	0.87	28.07	nr	**77.08**
Sum of two sides 500 mm	40.82	51.67	0.90	29.04	nr	**80.71**
Sum of two sides 600 mm	42.91	54.31	0.93	30.00	nr	**84.31**
Sum of two sides 700 mm	45.03	56.99	0.96	30.98	nr	**87.97**
Sum of two sides 800 mm	47.17	59.70	0.98	31.63	nr	**91.33**
Sum of two sides 900 mm	49.28	62.37	1.17	37.77	nr	**100.14**
Sum of two sides 1000 mm	51.34	64.97	1.22	39.37	nr	**104.34**
Sum of two sides 1100 mm	53.47	67.67	1.25	40.34	nr	**108.01**
Sum of two sides 1200 mm	55.55	70.30	1.28	41.31	nr	**111.61**
Reducer						
Sum of two sides 200 mm	136.95	173.32	2.23	71.97	nr	**245.29**
Sum of two sides 300 mm	143.25	181.29	2.37	76.50	nr	**257.79**
Sum of two sides 400 mm	149.51	189.22	2.51	81.00	nr	**270.22**
Sum of two sides 500 mm	155.82	197.21	2.65	85.53	nr	**282.74**
Sum of two sides 600 mm	162.10	205.15	2.79	90.04	nr	**295.19**
Sum of two sides 700 mm	168.39	213.11	2.87	92.62	nr	**305.73**
Sum of two sides 800 mm	174.70	221.10	3.01	97.14	nr	**318.24**
Sum of two sides 900 mm	180.99	229.06	3.06	98.75	nr	**327.81**
Sum of two sides 1000 mm	187.26	236.99	3.17	102.30	nr	**339.29**
Sum of two sides 1100 mm	193.53	244.93	3.22	103.91	nr	**348.84**
Sum of two sides 1200 mm	199.86	252.94	3.28	105.85	nr	**358.79**
Offset						
Sum of two sides 200 mm	330.53	418.32	2.95	95.20	nr	**513.52**
Sum of two sides 300 mm	338.72	428.68	3.11	100.37	nr	**529.05**
Sum of two sides 400 mm	346.91	439.05	3.26	105.21	nr	**544.26**
Sum of two sides 500 mm	355.13	449.46	3.42	110.38	nr	**559.84**
Sum of two sides 600 mm	363.33	459.83	3.57	115.20	nr	**575.03**
Sum of two sides 700 mm	379.74	480.60	3.23	104.24	nr	**584.84**
Sum of two sides 800 mm	371.54	470.22	3.73	120.37	nr	**590.59**
Sum of two sides 900 mm	387.97	491.02	3.45	111.34	nr	**602.36**
Sum of two sides 1000 mm	396.15	501.37	3.67	118.44	nr	**619.81**
Sum of two sides 1100 mm	404.33	511.72	3.78	121.99	nr	**633.71**
Sum of two sides 1200 mm	412.58	522.17	3.89	125.53	nr	**647.70**

38 VENTILATION/AIR CONDITIONING SYSTEMS

Item	Net Price £	Material £	Labour hours	Labour £	Unit	Total rate £
90° radius bend						
Sum of two sides 200 mm	134.01	169.60	2.06	66.48	nr	**236.08**
Sum of two sides 300 mm	144.17	182.46	2.21	71.32	nr	**253.78**
Sum of two sides 400 mm	154.30	195.28	2.36	76.16	nr	**271.44**
Sum of two sides 500 mm	164.49	208.17	2.51	81.00	nr	**289.17**
Sum of two sides 600 mm	174.65	221.03	2.66	85.85	nr	**306.88**
Sum of two sides 700 mm	184.75	233.82	2.81	90.70	nr	**324.52**
Sum of two sides 800 mm	194.94	246.71	2.97	95.84	nr	**342.55**
Sum of two sides 900 mm	205.04	259.50	2.99	96.50	nr	**356.00**
Sum of two sides 1000 mm	215.26	272.43	3.04	98.11	nr	**370.54**
Sum of two sides 1100 mm	225.36	285.22	3.07	99.08	nr	**384.30**
Sum of two sides 1200 mm	235.49	298.03	3.09	99.72	nr	**397.75**
45° radius bend						
Sum of two sides 200 mm	165.24	209.13	1.52	49.06	nr	**258.19**
Sum of two sides 300 mm	169.35	214.33	1.59	51.32	nr	**265.65**
Sum of two sides 400 mm	173.49	219.56	1.66	53.57	nr	**273.13**
Sum of two sides 500 mm	177.58	224.75	1.73	55.84	nr	**280.59**
Sum of two sides 600 mm	181.66	229.91	1.80	58.09	nr	**288.00**
Sum of two sides 700 mm	185.80	235.14	1.87	60.36	nr	**295.50**
Sum of two sides 800 mm	189.86	240.28	1.93	62.29	nr	**302.57**
Sum of two sides 900 mm	193.99	245.52	1.99	64.22	nr	**309.74**
Sum of two sides 1000 mm	198.08	250.69	2.05	66.16	nr	**316.85**
Sum of two sides 1100 mm	202.19	255.89	2.11	68.08	nr	**323.97**
Sum of two sides 1200 mm	206.28	261.07	2.17	70.03	nr	**331.10**
90° mitre bend						
Sum of two sides 200 mm	138.94	175.84	2.47	79.71	nr	**255.55**
Sum of two sides 300 mm	161.03	203.80	2.65	85.53	nr	**289.33**
Sum of two sides 400 mm	183.10	231.73	2.84	91.65	nr	**323.38**
Sum of two sides 500 mm	205.23	259.74	3.02	97.46	nr	**357.20**
Sum of two sides 600 mm	227.30	287.67	3.20	103.28	nr	**390.95**
Sum of two sides 700 mm	249.41	315.65	3.38	109.08	nr	**424.73**
Sum of two sides 800 mm	271.52	343.64	3.56	114.89	nr	**458.53**
Sum of two sides 900 mm	293.60	371.58	3.59	115.86	nr	**487.44**
Sum of two sides 1000 mm	315.74	399.60	3.66	118.12	nr	**517.72**
Sum of two sides 1100 mm	337.78	427.49	3.69	119.08	nr	**546.57**
Sum of two sides 1200 mm	359.94	455.54	3.72	120.05	nr	**575.59**
Branch (Side-on Shoe)						
Sum of two sides 200 mm	51.58	65.28	0.98	31.63	nr	**96.91**
Sum of two sides 300 mm	51.86	65.63	1.06	34.20	nr	**99.83**
Sum of two sides 400 mm	52.15	66.00	1.14	36.79	nr	**102.79**
Sum of two sides 500 mm	52.44	66.37	1.22	39.37	nr	**105.74**
Sum of two sides 600 mm	52.71	66.71	1.30	41.96	nr	**108.67**
Sum of two sides 700 mm	52.98	67.05	1.38	44.53	nr	**111.58**
Sum of two sides 800 mm	53.66	67.92	1.37	44.22	nr	**112.14**
Sum of two sides 900 mm	53.25	67.39	1.46	47.12	nr	**114.51**
Sum of two sides 1000 mm	53.80	68.08	1.46	47.12	nr	**115.20**
Sum of two sides 1100 mm	54.09	68.45	1.50	48.41	nr	**116.86**
Sum of two sides 1200 mm	54.37	68.81	1.55	50.02	nr	**118.83**

38 VENTILATION/AIR CONDITIONING SYSTEMS

Item	Net Price £	Material £	Labour hours	Labour £	Unit	Total rate £
DUCTWORK: FIRE RATED – cont						
Ductwork 601 to 800 mm longest side						
Sum of two sides 900 mm	194.43	246.08	5.04	162.66	m	**408.74**
Sum of two sides 1000 mm	204.87	259.28	5.58	180.08	m	**439.36**
Sum of two sides 1100 mm	215.28	272.46	5.84	188.47	m	**460.93**
Sum of two sides 1200 mm	225.67	285.61	6.11	197.20	m	**482.81**
Sum of two sides 1300 mm	236.15	298.87	6.38	205.90	m	**504.77**
Sum of two sides 1400 mm	246.54	312.02	6.65	214.60	m	**526.62**
Sum of two sides 1500 mm	257.00	325.26	6.91	223.01	m	**548.27**
Sum of two sides 1600 mm	267.41	338.43	7.18	231.72	m	**570.15**
Extra over fittings; Ductwork 601 to 800 mm longest side						
End cap						
Sum of two sides 900 mm	45.76	57.92	1.17	37.77	nr	**95.69**
Sum of two sides 1000 mm	49.04	62.07	1.22	39.37	nr	**101.44**
Sum of two sides 1100 mm	52.33	66.23	1.25	40.34	nr	**106.57**
Sum of two sides 1200 mm	55.59	70.36	1.28	41.31	nr	**111.67**
Sum of two sides 1300 mm	58.85	74.48	1.31	42.28	nr	**116.76**
Sum of two sides 1400 mm	62.13	78.64	1.34	43.24	nr	**121.88**
Sum of two sides 1500 mm	65.39	82.76	1.36	43.89	nr	**126.65**
Sum of two sides 1600 mm	68.67	86.91	1.39	44.87	nr	**131.78**
Reducer						
Sum of two sides 900 mm	171.30	216.80	3.06	98.75	nr	**315.55**
Sum of two sides 1000 mm	180.82	228.85	3.17	102.30	nr	**331.15**
Sum of two sides 1100 mm	190.32	240.87	3.22	103.91	nr	**344.78**
Sum of two sides 1200 mm	199.88	252.96	3.28	105.85	nr	**358.81**
Sum of two sides 1300 mm	209.42	265.04	3.33	107.48	nr	**372.52**
Sum of two sides 1400 mm	218.95	277.10	3.38	109.08	nr	**386.18**
Sum of two sides 1500 mm	228.46	289.14	3.44	111.01	nr	**400.15**
Sum of two sides 1600 mm	238.01	301.22	3.49	112.64	nr	**413.86**
Offset						
Sum of two sides 900 mm	364.29	461.05	3.45	111.34	nr	**572.39**
Sum of two sides 1000 mm	381.63	482.99	3.67	118.44	nr	**601.43**
Sum of two sides 1100 mm	398.99	504.96	3.78	121.99	nr	**626.95**
Sum of two sides 1200 mm	416.33	526.90	3.89	125.53	nr	**652.43**
Sum of two sides 1300 mm	433.64	548.81	4.00	129.09	nr	**677.90**
Sum of two sides 1400 mm	451.00	570.79	4.11	132.65	nr	**703.44**
Sum of two sides 1500 mm	468.34	592.73	4.23	136.51	nr	**729.24**
Sum of two sides 1600 mm	486.22	615.36	4.34	140.07	nr	**755.43**
90° radius bend						
Sum of two sides 900 mm	182.13	230.51	2.99	96.50	nr	**327.01**
Sum of two sides 1000 mm	200.04	253.18	3.04	98.11	nr	**351.29**
Sum of two sides 1100 mm	217.88	275.74	3.07	99.08	nr	**374.82**
Sum of two sides 1200 mm	235.78	298.40	3.09	99.72	nr	**398.12**
Sum of two sides 1300 mm	253.64	321.00	3.12	100.69	nr	**421.69**
Sum of two sides 1400 mm	271.52	343.64	3.15	101.66	nr	**445.30**
Sum of two sides 1500 mm	289.40	366.26	3.17	102.30	nr	**468.56**
Sum of two sides 1600 mm	307.29	388.91	3.20	103.28	nr	**492.19**

38 VENTILATION/AIR CONDITIONING SYSTEMS

Item	Net Price £	Material £	Labour hours	Labour £	Unit	Total rate £
45° bend						
Sum of two sides 900 mm	180.85	228.88	1.74	56.16	nr	285.04
Sum of two sides 1000 mm	189.33	239.61	1.85	59.70	nr	299.31
Sum of two sides 1100 mm	197.84	250.39	1.91	61.63	nr	312.02
Sum of two sides 1200 mm	206.30	261.09	1.96	63.26	nr	324.35
Sum of two sides 1300 mm	214.78	271.82	2.02	65.20	nr	337.02
Sum of two sides 1400 mm	223.27	282.58	2.07	66.80	nr	349.38
Sum of two sides 1500 mm	231.74	293.29	2.13	68.73	nr	362.02
Sum of two sides 1600 mm	240.24	304.05	2.18	70.36	nr	374.41
90° mitre bend						
Sum of two sides 900 mm	272.69	345.12	3.59	115.86	nr	460.98
Sum of two sides 1000 mm	301.78	381.93	3.66	118.12	nr	500.05
Sum of two sides 1100 mm	330.84	418.71	3.69	119.08	nr	537.79
Sum of two sides 1200 mm	359.94	455.54	3.72	120.05	nr	575.59
Sum of two sides 1300 mm	388.96	492.26	3.75	121.03	nr	613.29
Sum of two sides 1400 mm	418.05	529.09	3.78	121.99	nr	651.08
Sum of two sides 1500 mm	447.10	565.85	3.81	122.96	nr	688.81
Sum of two sides 1600 mm	476.19	602.66	3.84	123.93	nr	726.59
Branch (Side-on Shoe)						
Sum of two sides 900 mm	50.61	64.05	1.37	44.22	nr	108.27
Sum of two sides 1000 mm	51.86	65.63	1.46	47.12	nr	112.75
Sum of two sides 1100 mm	53.14	67.26	1.50	48.41	nr	115.67
Sum of two sides 1200 mm	54.39	68.84	1.55	50.02	nr	118.86
Sum of two sides 1300 mm	55.66	70.45	1.59	51.32	nr	121.77
Sum of two sides 1400 mm	54.68	69.20	1.68	54.22	nr	123.42
Sum of two sides 1500 mm	56.98	72.12	1.63	52.60	nr	124.72
Sum of two sides 1600 mm	59.48	75.28	1.72	55.51	nr	130.79
Ductwork 801 to 1000 mm longest side						
Sum of two sides 1100 mm	215.28	272.46	5.84	188.47	m	460.93
Sum of two sides 1200 mm	225.67	285.61	6.11	197.20	m	482.81
Sum of two sides 1300 mm	236.10	298.80	6.38	205.90	m	504.70
Sum of two sides 1400 mm	246.52	312.00	6.65	214.60	m	526.60
Sum of two sides 1500 mm	256.93	325.17	6.91	223.01	m	548.18
Sum of two sides 1600 mm	267.35	338.36	7.18	231.72	m	570.08
Sum of two sides 1700 mm	277.79	351.57	7.45	240.43	m	592.00
Sum of two sides 1800 mm	288.21	364.76	7.71	248.82	m	613.58
Sum of two sides 1900 mm	298.63	377.94	7.98	257.53	m	635.47
Sum of two sides 2000 mm	309.06	391.15	8.25	266.26	m	657.41
Extra over fittings; Ductwork 801 to 1000 mm longest side						
End cap						
Sum of two sides 1100 mm	52.33	66.23	1.25	40.34	nr	106.57
Sum of two sides 1200 mm	55.59	70.36	1.28	41.31	nr	111.67
Sum of two sides 1300 mm	58.85	74.48	1.31	42.28	nr	116.76
Sum of two sides 1400 mm	62.13	78.64	1.34	43.24	nr	121.88
Sum of two sides 1500 mm	65.39	82.76	1.36	43.89	nr	126.65
Sum of two sides 1600 mm	68.64	86.87	1.39	44.87	nr	131.74
Sum of two sides 1700 mm	71.91	91.01	1.42	45.83	nr	136.84

38 VENTILATION/AIR CONDITIONING SYSTEMS

Item	Net Price £	Material £	Labour hours	Labour £	Unit	Total rate £
DUCTWORK: FIRE RATED – cont						
Extra over fittings – cont						
Sum of two sides 1800 mm	75.19	95.16	1.45	46.80	nr	**141.96**
Sum of two sides 1900 mm	78.44	99.28	1.48	47.77	nr	**147.05**
Sum of two sides 2000 mm	81.70	103.40	1.51	48.74	nr	**152.14**
Reducer						
Sum of two sides 1100 mm	190.32	240.87	3.22	103.91	nr	**344.78**
Sum of two sides 1200 mm	199.92	253.02	3.28	105.85	nr	**358.87**
Sum of two sides 1300 mm	209.49	265.13	3.33	107.48	nr	**372.61**
Sum of two sides 1400 mm	219.10	277.29	3.38	109.08	nr	**386.37**
Sum of two sides 1500 mm	228.64	289.36	3.44	111.01	nr	**400.37**
Sum of two sides 1600 mm	238.26	301.54	3.49	112.64	nr	**414.18**
Sum of two sides 1700 mm	247.82	313.64	3.54	114.25	nr	**427.89**
Sum of two sides 1800 mm	257.40	325.76	3.60	116.18	nr	**441.94**
Sum of two sides 1900 mm	267.02	337.94	3.65	117.80	nr	**455.74**
Sum of two sides 2000 mm	281.35	356.08	3.70	119.41	nr	**475.49**
Offset						
Sum of two sides 1100 mm	395.84	500.98	3.78	121.99	nr	**622.97**
Sum of two sides 1200 mm	412.58	522.17	3.89	125.53	nr	**647.70**
Sum of two sides 1300 mm	429.29	543.31	4.00	129.09	nr	**672.40**
Sum of two sides 1400 mm	446.03	564.49	4.11	132.65	nr	**697.14**
Sum of two sides 1500 mm	462.77	585.68	4.23	136.51	nr	**722.19**
Sum of two sides 1600 mm	479.53	606.89	4.34	140.07	nr	**746.96**
Sum of two sides 1700 mm	496.24	628.04	4.45	143.61	nr	**771.65**
Sum of two sides 1800 mm	512.98	649.23	4.56	147.17	nr	**796.40**
Sum of two sides 1900 mm	529.68	670.36	4.67	150.71	nr	**821.07**
Sum of two sides 2000 mm	554.79	702.14	5.53	178.46	nr	**880.60**
90° radius bend						
Sum of two sides 1100 mm	217.55	275.33	3.07	99.08	nr	**374.41**
Sum of two sides 1200 mm	235.46	298.00	3.09	99.72	nr	**397.72**
Sum of two sides 1300 mm	253.40	320.70	3.12	100.69	nr	**421.39**
Sum of two sides 1400 mm	271.32	343.38	3.15	101.66	nr	**445.04**
Sum of two sides 1500 mm	289.24	366.06	3.17	102.30	nr	**468.36**
Sum of two sides 1600 mm	307.15	388.73	3.20	103.28	nr	**492.01**
Sum of two sides 1700 mm	325.06	411.40	3.22	103.91	nr	**515.31**
Sum of two sides 1800 mm	342.98	434.08	3.25	104.89	nr	**538.97**
Sum of two sides 1900 mm	360.90	456.76	3.28	105.85	nr	**562.61**
Sum of two sides 2000 mm	378.28	478.76	3.30	106.50	nr	**585.26**
45° bend						
Sum of two sides 1100 mm	199.13	252.02	1.91	61.63	nr	**313.65**
Sum of two sides 1200 mm	207.45	262.55	1.96	63.26	nr	**325.81**
Sum of two sides 1300 mm	215.78	273.09	2.02	65.20	nr	**338.29**
Sum of two sides 1400 mm	224.13	283.66	2.07	66.80	nr	**350.46**
Sum of two sides 1500 mm	232.46	294.20	2.13	68.73	nr	**362.93**
Sum of two sides 1600 mm	240.78	304.73	2.18	70.36	nr	**375.09**
Sum of two sides 1700 mm	249.10	315.26	2.24	72.30	nr	**387.56**
Sum of two sides 1800 mm	257.43	325.81	2.30	74.22	nr	**400.03**
Sum of two sides 1900 mm	265.73	336.30	2.35	75.84	nr	**412.14**
Sum of two sides 2000 mm	274.11	346.91	2.76	89.07	nr	**435.98**

38 VENTILATION/AIR CONDITIONING SYSTEMS

Item	Net Price £	Material £	Labour hours	Labour £	Unit	Total rate £
90° mitre bend						
Sum of two sides 1100 mm	330.84	418.71	3.69	119.08	nr	**537.79**
Sum of two sides 1200 mm	359.94	455.54	3.72	120.05	nr	**575.59**
Sum of two sides 1300 mm	388.96	492.26	3.75	121.03	nr	**613.29**
Sum of two sides 1400 mm	418.05	529.09	3.78	121.99	nr	**651.08**
Sum of two sides 1500 mm	447.16	565.92	3.81	122.96	nr	**688.88**
Sum of two sides 1600 mm	476.19	602.66	3.84	123.93	nr	**726.59**
Sum of two sides 1700 mm	505.28	639.49	3.87	124.90	nr	**764.39**
Sum of two sides 1800 mm	534.36	676.29	3.90	125.87	nr	**802.16**
Sum of two sides 1900 mm	563.42	713.06	3.93	126.84	nr	**839.90**
Sum of two sides 2000 mm	592.50	749.87	3.96	127.80	nr	**877.67**
Branch (Side-on Shoe)						
Sum of two sides 1100 mm	53.10	67.20	1.50	48.41	nr	**115.61**
Sum of two sides 1200 mm	54.32	68.75	1.55	50.02	nr	**118.77**
Sum of two sides 1300 mm	55.64	70.41	1.59	51.32	nr	**121.73**
Sum of two sides 1400 mm	56.94	72.06	1.63	52.60	nr	**124.66**
Sum of two sides 1500 mm	58.22	73.68	1.68	54.22	nr	**127.90**
Sum of two sides 1600 mm	59.48	75.28	1.72	55.51	nr	**130.79**
Sum of two sides 1700 mm	60.79	76.93	1.77	57.12	nr	**134.05**
Sum of two sides 1800 mm	62.08	78.57	1.81	58.42	nr	**136.99**
Sum of two sides 1900 mm	63.36	80.19	1.86	60.03	nr	**140.22**
Sum of two sides 2000 mm	65.27	82.61	1.90	61.32	nr	**143.93**
Ductwork 1001 to 1250 mm longest side						
Sum of two sides 1300 mm	321.66	407.10	6.38	205.90	m	**613.00**
Sum of two sides 1400 mm	335.52	424.64	6.65	214.60	m	**639.24**
Sum of two sides 1500 mm	349.41	442.21	6.91	223.01	m	**665.22**
Sum of two sides 1600 mm	359.79	455.35	6.91	223.01	m	**678.36**
Sum of two sides 1700 mm	370.17	468.48	7.45	240.43	m	**708.91**
Sum of two sides 1800 mm	383.98	485.97	7.71	248.82	m	**734.79**
Sum of two sides 1900 mm	397.83	503.50	7.98	257.53	m	**761.03**
Sum of two sides 2000 mm	411.70	521.05	8.25	266.26	m	**787.31**
Sum of two sides 2100 mm	425.54	538.56	9.62	310.46	m	**849.02**
Sum of two sides 2200 mm	435.94	551.72	9.62	310.46	m	**862.18**
Sum of two sides 2300 mm	446.31	564.85	10.07	324.99	m	**889.84**
Sum of two sides 2400 mm	460.16	582.38	10.47	337.90	m	**920.28**
Sum of two sides 2500 mm	474.01	599.91	10.87	350.81	m	**950.72**
Extra over fittings; Ductwork 1001 to 1250 mm longest side						
End cap						
Sum of two sides 1300 mm	58.87	74.50	1.31	42.28	nr	**116.78**
Sum of two sides 1400 mm	62.13	78.64	1.34	43.24	nr	**121.88**
Sum of two sides 1500 mm	65.35	82.71	1.36	43.89	nr	**126.60**
Sum of two sides 1600 mm	68.64	86.87	1.39	44.87	nr	**131.74**
Sum of two sides 1700 mm	71.91	91.01	1.42	45.83	nr	**136.84**
Sum of two sides 1800 mm	75.14	95.10	1.45	46.80	nr	**141.90**
Sum of two sides 1900 mm	78.44	99.28	1.48	47.77	nr	**147.05**

38 VENTILATION/AIR CONDITIONING SYSTEMS

Item	Net Price £	Material £	Labour hours	Labour £	Unit	Total rate £
DUCTWORK: FIRE RATED – cont						
Extra over fittings – cont						
End cap – cont						
Sum of two sides 2000 mm	81.70	103.40	1.51	48.74	nr	**152.14**
Sum of two sides 2100 mm	84.97	107.54	2.66	85.85	nr	**193.39**
Sum of two sides 2200 mm	88.23	111.66	2.80	90.36	nr	**202.02**
Sum of two sides 2300 mm	91.51	115.82	2.95	95.20	nr	**211.02**
Sum of two sides 2400 mm	94.75	119.92	3.10	100.05	nr	**219.97**
Sum of two sides 2500 mm	98.01	124.04	3.24	104.56	nr	**228.60**
Reducer						
Sum of two sides 1300 mm	209.42	265.04	3.33	107.48	nr	**372.52**
Sum of two sides 1400 mm	218.84	276.96	3.38	109.08	nr	**386.04**
Sum of two sides 1500 mm	228.27	288.90	3.44	111.01	nr	**399.91**
Sum of two sides 1600 mm	237.74	300.89	3.49	112.64	nr	**413.53**
Sum of two sides 1700 mm	247.24	312.91	3.54	114.25	nr	**427.16**
Sum of two sides 1800 mm	256.73	324.91	3.60	116.18	nr	**441.09**
Sum of two sides 1900 mm	266.17	336.86	3.65	117.80	nr	**454.66**
Sum of two sides 2000 mm	275.65	348.86	3.70	119.41	nr	**468.27**
Sum of two sides 2100 mm	285.11	360.83	3.75	121.03	nr	**481.86**
Sum of two sides 2200 mm	294.56	372.79	3.80	122.64	nr	**495.43**
Sum of two sides 2300 mm	304.04	384.80	3.85	124.24	nr	**509.04**
Sum of two sides 2400 mm	313.51	396.78	3.90	125.87	nr	**522.65**
Sum of two sides 2500 mm	323.01	408.80	3.95	127.47	nr	**536.27**
Offset						
Sum of two sides 1300 mm	427.86	541.50	4.00	129.09	nr	**670.59**
Sum of two sides 1400 mm	444.85	563.00	4.11	132.65	nr	**695.65**
Sum of two sides 1500 mm	461.81	584.47	4.23	136.51	nr	**720.98**
Sum of two sides 1600 mm	478.79	605.95	4.34	140.07	nr	**746.02**
Sum of two sides 1700 mm	495.75	627.42	4.45	143.61	nr	**771.03**
Sum of two sides 1800 mm	512.71	648.88	4.56	147.17	nr	**796.05**
Sum of two sides 1900 mm	529.66	670.34	4.67	150.71	nr	**821.05**
Sum of two sides 2000 mm	546.62	691.80	5.53	178.46	nr	**870.26**
Sum of two sides 2100 mm	563.57	713.25	5.76	185.89	nr	**899.14**
Sum of two sides 2200 mm	580.25	734.36	5.99	193.32	nr	**927.68**
Sum of two sides 2300 mm	597.50	756.19	6.22	200.74	nr	**956.93**
Sum of two sides 2400 mm	614.44	777.64	6.45	208.15	nr	**985.79**
Sum of two sides 2500 mm	631.44	799.15	6.68	215.58	nr	**1014.73**
90° radius bend						
Sum of two sides 1300 mm	253.06	320.28	3.12	100.69	nr	**420.97**
Sum of two sides 1400 mm	281.60	356.40	3.15	101.66	nr	**458.06**
Sum of two sides 1500 mm	289.03	365.79	3.17	102.30	nr	**468.09**
Sum of two sides 1600 mm	306.96	388.48	3.20	103.28	nr	**491.76**
Sum of two sides 1700 mm	324.94	411.24	3.22	103.91	nr	**515.15**
Sum of two sides 1800 mm	342.90	433.98	3.25	104.89	nr	**538.87**
Sum of two sides 1900 mm	360.87	456.71	3.28	105.85	nr	**562.56**
Sum of two sides 2000 mm	378.85	479.47	3.30	106.50	nr	**585.97**
Sum of two sides 2100 mm	396.82	502.22	3.32	107.15	nr	**609.37**
Sum of two sides 2200 mm	414.77	524.93	3.34	107.79	nr	**632.72**
Sum of two sides 2300 mm	432.76	547.70	3.36	108.44	nr	**656.14**
Sum of two sides 2400 mm	450.74	570.46	3.38	109.08	nr	**679.54**
Sum of two sides 2500 mm	468.70	593.19	3.40	109.73	nr	**702.92**

38 VENTILATION/AIR CONDITIONING SYSTEMS

Item	Net Price £	Material £	Labour hours	Labour £	Unit	Total rate £
45° bend						
Sum of two sides 1300 mm	204.98	259.43	2.02	65.20	nr	**324.63**
Sum of two sides 1400 mm	214.21	271.11	2.07	66.80	nr	**337.91**
Sum of two sides 1500 mm	223.42	282.76	2.13	68.73	nr	**351.49**
Sum of two sides 1600 mm	232.66	294.46	2.18	70.36	nr	**364.82**
Sum of two sides 1700 mm	241.87	306.11	2.24	72.30	nr	**378.41**
Sum of two sides 1800 mm	251.12	317.82	2.30	74.22	nr	**392.04**
Sum of two sides 1900 mm	260.33	329.47	2.35	75.84	nr	**405.31**
Sum of two sides 2000 mm	269.58	341.19	2.76	89.07	nr	**430.26**
Sum of two sides 2100 mm	278.82	352.88	2.89	93.26	nr	**446.14**
Sum of two sides 2200 mm	288.06	364.57	3.01	97.14	nr	**461.71**
Sum of two sides 2300 mm	297.27	376.23	3.13	101.01	nr	**477.24**
Sum of two sides 2400 mm	306.53	387.95	3.26	105.21	nr	**493.16**
Sum of two sides 2500 mm	315.74	399.60	3.37	108.76	nr	**508.36**
90° mitre bend						
Sum of two sides 1300 mm	388.08	491.15	3.75	121.03	nr	**612.18**
Sum of two sides 1400 mm	416.89	527.62	3.78	121.99	nr	**649.61**
Sum of two sides 1500 mm	446.84	565.52	3.81	122.96	nr	**688.48**
Sum of two sides 1600 mm	476.17	602.64	3.84	123.93	nr	**726.57**
Sum of two sides 1700 mm	505.49	639.74	3.87	124.90	nr	**764.64**
Sum of two sides 1800 mm	534.81	676.86	3.90	125.87	nr	**802.73**
Sum of two sides 1900 mm	564.16	714.00	3.93	126.84	nr	**840.84**
Sum of two sides 2000 mm	593.53	751.17	3.96	127.80	nr	**878.97**
Sum of two sides 2100 mm	622.85	788.28	3.99	128.78	nr	**917.06**
Sum of two sides 2200 mm	652.20	825.43	4.02	129.74	nr	**955.17**
Sum of two sides 2300 mm	681.55	862.57	4.05	130.72	nr	**993.29**
Sum of two sides 2400 mm	710.83	899.63	4.08	131.68	nr	**1031.31**
Sum of two sides 2500 mm	740.17	936.76	4.11	132.65	nr	**1069.41**
Branch (Side-on Shoe)						
Sum of two sides 1300 mm	55.73	70.53	1.59	51.32	nr	**121.85**
Sum of two sides 1400 mm	57.02	72.16	1.63	52.60	nr	**124.76**
Sum of two sides 1500 mm	58.28	73.76	1.68	54.22	nr	**127.98**
Sum of two sides 1600 mm	59.53	75.34	1.72	55.51	nr	**130.85**
Sum of two sides 1700 mm	60.83	76.99	1.77	57.12	nr	**134.11**
Sum of two sides 1800 mm	62.13	78.64	1.81	58.42	nr	**137.06**
Sum of two sides 1900 mm	63.39	80.23	1.86	60.03	nr	**140.26**
Sum of two sides 2000 mm	64.69	81.87	1.90	61.32	nr	**143.19**
Sum of two sides 2100 mm	65.95	83.46	2.58	83.26	nr	**166.72**
Sum of two sides 2200 mm	67.20	85.05	2.61	84.24	nr	**169.29**
Sum of two sides 2300 mm	68.52	86.72	2.64	85.20	nr	**171.92**
Sum of two sides 2400 mm	69.80	88.33	2.88	92.95	nr	**181.28**
Sum of two sides 2500 mm	71.05	89.92	2.91	93.91	nr	**183.83**
Ductwork 1251 to 2000 mm longest side						
Sum of two sides 1700 mm	393.81	498.41	7.45	240.43	m	**738.84**
Sum of two sides 1800 mm	419.06	530.36	7.71	248.82	m	**779.18**
Sum of two sides 1900 mm	444.29	562.30	7.98	257.53	m	**819.83**
Sum of two sides 2000 mm	469.50	594.19	8.25	266.26	m	**860.45**
Sum of two sides 2100 mm	494.78	626.19	9.62	310.46	m	**936.65**
Sum of two sides 2200 mm	519.98	658.09	9.66	311.75	m	**969.84**
Sum of two sides 2300 mm	545.18	689.98	10.07	324.99	m	**1014.97**

38 VENTILATION/AIR CONDITIONING SYSTEMS

Item	Net Price £	Material £	Labour hours	Labour £	Unit	Total rate £
DUCTWORK: FIRE RATED – cont						
Ductwork 1251 to 2000 mm longest side – cont						
Sum of two sides 2400 mm	570.41	721.91	10.47	337.90	m	1059.81
Sum of two sides 2500 mm	595.64	753.84	10.87	350.81	m	1104.65
Sum of two sides 2600 mm	620.93	785.85	11.27	363.71	m	1149.56
Sum of two sides 2700 mm	646.13	817.75	11.67	376.62	m	1194.37
Sum of two sides 2800 mm	671.37	849.69	12.08	389.86	m	1239.55
Sum of two sides 2900 mm	696.57	881.57	12.48	402.76	m	1284.33
Sum of two sides 3000 mm	721.80	913.51	12.88	415.68	m	1329.19
Sum of two sides 3100 mm	747.05	945.47	13.26	427.94	m	1373.41
Sum of two sides 3200 mm	772.28	977.40	13.69	441.82	m	1419.22
Sum of two sides 3300 mm	475.14	601.34	14.09	454.73	m	1056.07
Sum of two sides 3400 mm	490.14	620.32	14.49	467.63	m	1087.95
Sum of two sides 3500 mm	505.16	639.33	14.89	480.54	m	1119.87
Sum of two sides 3600 mm	520.17	658.32	15.29	493.45	m	1151.77
Sum of two sides 3700 mm	535.25	677.41	15.69	506.36	m	1183.77
Sum of two sides 3800 mm	550.25	696.39	16.09	519.28	m	1215.67
Sum of two sides 3900 mm	565.29	715.43	16.49	532.18	m	1247.61
Sum of two sides 4000 mm	580.31	734.44	16.89	545.08	m	1279.52
Extra over fittings; Ductwork 1251 to 2000 mm longest side						
End cap						
Sum of two sides 1700 mm	106.82	135.20	1.42	45.83	nr	181.03
Sum of two sides 1800 mm	113.70	143.90	1.45	46.80	nr	190.70
Sum of two sides 1900 mm	120.56	152.58	1.48	47.77	nr	200.35
Sum of two sides 2000 mm	127.45	161.30	1.51	48.74	nr	210.04
Sum of two sides 2100 mm	134.29	169.96	2.66	85.85	nr	255.81
Sum of two sides 2200 mm	141.19	178.68	2.80	90.36	nr	269.04
Sum of two sides 2300 mm	148.02	187.33	2.95	95.20	nr	282.53
Sum of two sides 2400 mm	154.90	196.04	3.10	100.05	nr	296.09
Sum of two sides 2500 mm	161.77	204.74	3.24	104.56	nr	309.30
Sum of two sides 2600 mm	168.66	213.46	3.39	109.40	nr	322.86
Sum of two sides 2700 mm	175.48	222.08	3.54	114.25	nr	336.33
Sum of two sides 2800 mm	182.37	230.81	3.68	118.76	nr	349.57
Sum of two sides 2900 mm	189.29	239.57	3.83	123.61	nr	363.18
Sum of two sides 3000 mm	196.13	248.23	3.98	128.44	nr	376.67
Sum of two sides 3100 mm	203.00	256.92	4.12	132.97	nr	389.89
Sum of two sides 3200 mm	209.89	265.64	4.27	137.82	nr	403.46
Sum of two sides 3300 mm	153.10	193.76	4.42	142.64	nr	336.40
Sum of two sides 3400 mm	157.97	199.93	4.57	147.48	nr	347.41
Sum of two sides 3500 mm	161.96	204.97	4.72	152.33	nr	357.30
Sum of two sides 3600 mm	165.21	209.09	4.87	157.17	nr	366.26
Sum of two sides 3700 mm	170.08	215.25	5.02	162.01	nr	377.26
Sum of two sides 3800 mm	255.36	323.19	5.17	166.86	nr	490.05
Sum of two sides 3900 mm	182.22	230.62	5.32	171.70	nr	402.32
Sum of two sides 4000 mm	187.09	236.78	5.47	176.53	nr	413.31

38 VENTILATION/AIR CONDITIONING SYSTEMS

Item	Net Price £	Material £	Labour hours	Labour £	Unit	Total rate £
Reducer						
Sum of two sides 1700 mm	230.69	291.96	3.54	114.25	nr	**406.21**
Sum of two sides 1800 mm	258.79	327.52	3.60	116.18	nr	**443.70**
Sum of two sides 1900 mm	286.86	363.05	3.65	117.80	nr	**480.85**
Sum of two sides 2000 mm	343.04	434.16	2.92	94.24	nr	**528.40**
Sum of two sides 2100 mm	314.97	398.63	3.70	119.41	nr	**518.04**
Sum of two sides 2200 mm	371.14	469.72	3.10	100.05	nr	**569.77**
Sum of two sides 2300 mm	399.22	505.25	3.29	106.18	nr	**611.43**
Sum of two sides 2400 mm	427.32	540.81	3.48	112.31	nr	**653.12**
Sum of two sides 2500 mm	455.37	576.32	3.66	118.12	nr	**694.44**
Sum of two sides 2600 mm	483.49	611.90	3.85	124.24	nr	**736.14**
Sum of two sides 2700 mm	511.58	647.46	4.03	130.07	nr	**777.53**
Sum of two sides 2800 mm	539.68	683.02	4.22	136.19	nr	**819.21**
Sum of two sides 2900 mm	567.77	718.57	4.40	142.00	nr	**860.57**
Sum of two sides 3000 mm	609.84	771.81	4.59	148.14	nr	**919.95**
Sum of two sides 3100 mm	609.84	771.81	4.78	154.27	nr	**926.08**
Sum of two sides 3200 mm	637.96	807.40	4.96	160.07	nr	**967.47**
Sum of two sides 3300 mm	480.46	608.07	5.15	166.20	nr	**774.27**
Sum of two sides 3400 mm	500.30	633.18	5.34	172.33	nr	**805.51**
Sum of two sides 3500 mm	520.15	658.30	5.53	178.46	nr	**836.76**
Sum of two sides 3600 mm	539.97	683.39	5.72	184.60	nr	**867.99**
Sum of two sides 3700 mm	559.83	708.52	5.91	190.74	nr	**899.26**
Sum of two sides 3800 mm	579.66	733.62	6.10	196.86	nr	**930.48**
Sum of two sides 3900 mm	599.53	758.77	6.29	203.00	nr	**961.77**
Sum of two sides 4000 mm	619.37	783.88	6.48	209.13	nr	**993.01**
Offset						
Sum of two sides 1700 mm	230.69	291.96	4.45	143.61	nr	**435.57**
Sum of two sides 1800 mm	249.70	316.02	4.56	147.17	nr	**463.19**
Sum of two sides 1900 mm	268.75	340.13	4.67	150.71	nr	**490.84**
Sum of two sides 2000 mm	287.67	364.08	5.53	178.46	nr	**542.54**
Sum of two sides 2100 mm	306.69	388.15	5.76	185.89	nr	**574.04**
Sum of two sides 2200 mm	325.70	412.20	5.99	193.32	nr	**605.52**
Sum of two sides 2300 mm	344.74	436.31	6.22	200.74	nr	**637.05**
Sum of two sides 2400 mm	363.72	460.32	6.45	208.15	nr	**668.47**
Sum of two sides 2500 mm	382.73	484.38	6.68	215.58	nr	**699.96**
Sum of two sides 2600 mm	401.76	508.47	6.91	223.01	nr	**731.48**
Sum of two sides 2700 mm	420.77	532.53	7.14	230.43	nr	**762.96**
Sum of two sides 2800 mm	439.80	556.61	7.37	237.85	nr	**794.46**
Sum of two sides 2900 mm	458.75	580.60	7.60	245.27	nr	**825.87**
Sum of two sides 3000 mm	477.80	604.70	7.83	252.69	nr	**857.39**
Sum of two sides 3100 mm	496.82	628.78	8.06	260.12	nr	**888.90**
Sum of two sides 3200 mm	515.82	652.83	8.29	267.55	nr	**920.38**
Sum of two sides 3300 mm	377.82	478.17	8.52	274.96	nr	**753.13**
Sum of two sides 3400 mm	391.24	495.15	8.75	282.40	nr	**777.55**
Sum of two sides 3500 mm	404.68	512.16	8.98	289.81	nr	**801.97**
Sum of two sides 3600 mm	418.10	529.14	9.21	297.23	nr	**826.37**
Sum of two sides 3700 mm	431.54	546.16	9.44	304.65	nr	**850.81**
Sum of two sides 3800 mm	444.96	563.14	9.67	312.09	nr	**875.23**
Sum of two sides 3900 mm	458.39	580.14	9.90	319.50	nr	**899.64**
Sum of two sides 4000 mm	471.84	597.16	10.13	326.92	nr	**924.08**

38 VENTILATION/AIR CONDITIONING SYSTEMS

Item	Net Price £	Material £	Labour hours	Labour £	Unit	Total rate £
DUCTWORK: FIRE RATED – cont						
Extra over fittings – cont						
90° radius bend						
Sum of two sides 1700 mm	321.05	406.32	3.22	103.91	nr	**510.23**
Sum of two sides 1800 mm	350.06	443.04	3.25	104.89	nr	**547.93**
Sum of two sides 1900 mm	369.46	467.59	3.32	107.15	nr	**574.74**
Sum of two sides 2000 mm	379.09	479.77	3.28	105.85	nr	**585.62**
Sum of two sides 2100 mm	408.10	516.49	3.30	106.50	nr	**622.99**
Sum of two sides 2200 mm	466.92	590.93	3.32	107.15	nr	**698.08**
Sum of two sides 2300 mm	495.15	626.66	3.36	108.44	nr	**735.10**
Sum of two sides 2400 mm	524.11	663.31	3.38	109.08	nr	**772.39**
Sum of two sides 2500 mm	553.16	700.08	3.40	109.73	nr	**809.81**
Sum of two sides 2600 mm	582.19	736.81	3.42	110.38	nr	**847.19**
Sum of two sides 2700 mm	611.22	773.56	3.44	111.01	nr	**884.57**
Sum of two sides 2800 mm	640.24	810.29	3.46	111.66	nr	**921.95**
Sum of two sides 2900 mm	669.19	846.92	3.48	112.31	nr	**959.23**
Sum of two sides 3000 mm	698.23	883.68	3.50	112.95	nr	**996.63**
Sum of two sides 3100 mm	727.23	920.38	3.52	113.60	nr	**1033.98**
Sum of two sides 3200 mm	731.27	925.50	3.54	114.25	nr	**1039.75**
Sum of two sides 3300 mm	518.11	655.72	3.56	114.89	nr	**770.61**
Sum of two sides 3400 mm	763.21	965.92	3.58	115.54	nr	**1081.46**
Sum of two sides 3500 mm	536.20	678.62	3.60	116.18	nr	**794.80**
Sum of two sides 3600 mm	554.63	701.94	3.62	116.83	nr	**818.77**
Sum of two sides 3700 mm	573.06	725.27	3.64	117.48	nr	**842.75**
Sum of two sides 3800 mm	591.53	748.64	3.66	118.12	nr	**866.76**
Sum of two sides 3900 mm	609.99	772.00	3.68	118.76	nr	**890.76**
Sum of two sides 4000 mm	628.43	795.35	3.70	119.41	nr	**914.76**
45° bend						
Sum of two sides 1700 mm	238.01	301.22	2.24	72.30	nr	**373.52**
Sum of two sides 1800 mm	264.32	334.52	2.30	74.22	nr	**408.74**
Sum of two sides 1900 mm	290.54	367.71	2.35	75.84	nr	**443.55**
Sum of two sides 2000 mm	316.88	401.04	2.76	89.07	nr	**490.11**
Sum of two sides 2100 mm	408.10	516.49	3.01	97.14	nr	**613.63**
Sum of two sides 2200 mm	395.79	500.91	3.13	101.01	nr	**601.92**
Sum of two sides 2300 mm	437.10	553.19	2.89	93.26	nr	**646.45**
Sum of two sides 2400 mm	422.04	534.14	3.26	105.21	nr	**639.35**
Sum of two sides 2500 mm	448.34	567.41	3.37	108.76	nr	**676.17**
Sum of two sides 2600 mm	474.59	600.64	3.49	112.64	nr	**713.28**
Sum of two sides 2700 mm	500.94	633.99	3.62	116.83	nr	**750.82**
Sum of two sides 2800 mm	527.22	667.25	3.74	120.70	nr	**787.95**
Sum of two sides 2900 mm	553.50	700.50	3.86	124.58	nr	**825.08**
Sum of two sides 3000 mm	579.79	733.78	3.98	128.44	nr	**862.22**
Sum of two sides 3100 mm	606.09	767.07	4.10	132.32	nr	**899.39**
Sum of two sides 3200 mm	632.35	800.31	4.22	136.19	nr	**936.50**
Sum of two sides 3300 mm	465.33	588.92	4.34	140.07	nr	**728.99**
Sum of two sides 3400 mm	483.90	612.43	4.46	143.93	nr	**756.36**
Sum of two sides 3500 mm	502.48	635.94	4.58	147.81	nr	**783.75**

38 VENTILATION/AIR CONDITIONING SYSTEMS

Item	Net Price £	Material £	Labour hours	Labour £	Unit	Total rate £
Sum of two sides 3600 mm	521.07	659.47	4.70	151.68	nr	**811.15**
Sum of two sides 3700 mm	539.60	682.92	4.82	155.56	nr	**838.48**
Sum of two sides 3800 mm	558.19	706.44	4.94	159.43	nr	**865.87**
Sum of two sides 3900 mm	576.74	729.93	5.06	163.30	nr	**893.23**
Sum of two sides 4000 mm	595.35	753.48	5.18	167.17	nr	**920.65**
90° mitre bend						
Sum of two sides 1700 mm	480.85	608.56	3.87	124.90	nr	**733.46**
Sum of two sides 1800 mm	563.95	713.73	3.90	125.87	nr	**839.60**
Sum of two sides 1900 mm	647.07	818.93	3.93	126.84	nr	**945.77**
Sum of two sides 2000 mm	730.11	924.02	3.96	127.80	nr	**1051.82**
Sum of two sides 2100 mm	813.24	1029.24	3.82	123.28	nr	**1152.52**
Sum of two sides 2200 mm	896.33	1134.39	3.83	123.61	nr	**1258.00**
Sum of two sides 2300 mm	979.45	1239.59	3.83	123.61	nr	**1363.20**
Sum of two sides 2400 mm	1062.55	1344.76	3.84	123.93	nr	**1468.69**
Sum of two sides 2500 mm	1145.64	1449.92	3.85	124.24	nr	**1574.16**
Sum of two sides 2600 mm	1228.69	1555.03	3.85	124.24	nr	**1679.27**
Sum of two sides 2700 mm	1311.80	1660.21	3.86	124.58	nr	**1784.79**
Sum of two sides 2800 mm	1394.90	1765.39	3.86	124.58	nr	**1889.97**
Sum of two sides 2900 mm	1478.00	1870.56	3.87	124.90	nr	**1995.46**
Sum of two sides 3000 mm	1561.09	1975.71	3.87	124.90	nr	**2100.61**
Sum of two sides 3100 mm	1644.22	2080.93	3.88	125.22	nr	**2206.15**
Sum of two sides 3200 mm	1727.29	2186.06	3.88	125.22	nr	**2311.28**
Sum of two sides 3300 mm	1278.99	1618.69	3.89	125.53	nr	**1744.22**
Sum of two sides 3400 mm	1337.69	1692.98	3.90	125.87	nr	**1818.85**
Sum of two sides 3500 mm	1396.40	1767.28	3.91	126.18	nr	**1893.46**
Sum of two sides 3600 mm	1455.11	1841.58	3.92	126.50	nr	**1968.08**
Sum of two sides 3700 mm	1513.82	1915.89	3.93	126.84	nr	**2042.73**
Sum of two sides 3800 mm	1572.52	1990.18	3.94	127.15	nr	**2117.33**
Sum of two sides 3900 mm	1631.24	2064.50	3.95	127.47	nr	**2191.97**
Sum of two sides 4000 mm	1689.94	2138.79	3.96	127.80	nr	**2266.59**
Branch (Side-on Shoe)						
Sum of two sides 1700 mm	54.52	69.00	1.77	57.12	nr	**126.12**
Sum of two sides 1800 mm	59.55	75.36	1.81	58.42	nr	**133.78**
Sum of two sides 1900 mm	64.66	81.84	1.86	60.03	nr	**141.87**
Sum of two sides 2000 mm	69.73	88.24	1.90	61.32	nr	**149.56**
Sum of two sides 2100 mm	74.80	94.66	2.58	83.26	nr	**177.92**
Sum of two sides 2200 mm	79.85	101.06	2.61	84.24	nr	**185.30**
Sum of two sides 2300 mm	84.95	107.51	2.65	85.53	nr	**193.04**
Sum of two sides 2400 mm	90.00	113.90	2.68	86.49	nr	**200.39**
Sum of two sides 2500 mm	95.08	120.33	2.71	87.46	nr	**207.79**
Sum of two sides 2600 mm	100.17	126.77	2.75	88.76	nr	**215.53**
Sum of two sides 2700 mm	105.19	133.12	2.78	89.72	nr	**222.84**
Sum of two sides 2800 mm	110.30	139.60	2.81	90.70	nr	**230.30**
Sum of two sides 2900 mm	115.37	146.01	2.84	91.65	nr	**237.66**
Sum of two sides 3000 mm	120.43	152.42	2.87	92.62	nr	**245.04**
Sum of two sides 3100 mm	125.50	158.84	2.90	93.59	nr	**252.43**
Sum of two sides 3200 mm	130.59	165.28	2.93	94.56	nr	**259.84**
Sum of two sides 3300 mm	95.83	121.28	2.93	94.56	nr	**215.84**
Sum of two sides 3400 mm	99.41	125.81	3.00	96.82	nr	**222.63**
Sum of two sides 3500 mm	103.00	130.36	3.03	97.80	nr	**228.16**

38 VENTILATION/AIR CONDITIONING SYSTEMS

Item	Net Price £	Material £	Labour hours	Labour £	Unit	Total rate £
DUCTWORK: FIRE RATED – cont						
Extra over fittings – cont						
Branch (Side-on Shoe) – cont						
Sum of two sides 3600 mm	106.55	134.85	3.06	98.75	nr	**233.60**
Sum of two sides 3700 mm	110.16	139.42	3.09	99.72	nr	**239.14**
Sum of two sides 3800 mm	113.76	143.98	3.12	100.69	nr	**244.67**
Sum of two sides 3900 mm	117.33	148.49	3.15	101.66	nr	**250.15**
Sum of two sides 4000 mm	120.89	153.00	3.18	102.63	nr	**255.63**
Rectangular section ductwork to BSEN1366 (Duraduct LT), ducts 'Type A' and 'Type B'; manufactured from 6 mm thick laminate fire board consisting of steel circular hole punched facings pressed to a fibre cement core; provides up to 4 hours stability, 4 hours integrity and 32 minutes insulation; including all necessary stiffeners, joints and supports in the running length						
Ductwork up to 600 mm longest side						
Sum of two sides 200 mm	504.46	638.44	4.00	129.09	m	**767.53**
Sum of two sides 400 mm	516.77	654.02	4.00	129.09	m	**783.11**
Sum of two sides 600 mm	540.02	683.45	4.00	129.09	m	**812.54**
Sum of two sides 800 mm	712.61	901.88	5.50	177.50	m	**1079.38**
Sum of two sides 1000 mm	960.02	1215.00	5.50	177.50	m	**1392.50**
Sum of two sides 1200 mm	992.31	1255.87	6.00	193.64	m	**1449.51**
Extra over fittings; Ductwork up to 600 mm longest side						
End cap						
Sum of two sides 200 mm	119.72	151.51	0.81	26.15	m	**177.66**
Sum of two sides 400 mm	119.72	151.51	0.87	28.07	m	**179.58**
Sum of two sides 600 mm	119.72	151.51	0.93	30.00	m	**181.51**
Sum of two sides 800 mm	193.38	244.74	0.98	31.63	m	**276.37**
Sum of two sides 1000 mm	193.38	244.74	1.22	39.37	m	**284.11**
Sum of two sides 1200 mm	294.64	372.89	1.28	41.31	m	**414.20**
Reducer						
Sum of two sides 200 mm	105.87	133.99	2.23	71.97	m	**205.96**
Sum of two sides 400 mm	105.87	133.99	2.51	81.00	m	**214.99**
Sum of two sides 600 mm	105.87	133.99	2.79	90.04	m	**224.03**
Sum of two sides 800 mm	207.18	262.20	3.01	97.14	m	**359.34**
Sum of two sides 1000 mm	207.18	262.20	3.17	102.30	m	**364.50**
Sum of two sides 1200 mm	262.44	332.15	3.28	105.85	m	**438.00**
Offset						
Sum of two sides 200 mm	105.87	133.99	2.95	95.20	m	**229.19**
Sum of two sides 400 mm	105.87	133.99	3.26	105.21	m	**239.20**
Sum of two sides 600 mm	105.87	133.99	3.57	115.20	m	**249.19**
Sum of two sides 800 mm	207.18	262.20	3.23	104.24	m	**366.44**
Sum of two sides 1000 mm	207.18	262.20	3.67	118.44	m	**380.64**
Sum of two sides 1200 mm	262.44	332.15	3.89	125.53	m	**457.68**

38 VENTILATION/AIR CONDITIONING SYSTEMS

Item	Net Price £	Material £	Labour hours	Labour £	Unit	Total rate £
90° radius bend						
Sum of two sides 200 mm	354.49	448.64	2.06	66.48	m	**515.12**
Sum of two sides 400 mm	354.49	448.64	2.36	76.16	m	**524.80**
Sum of two sides 600 mm	354.49	448.64	2.66	85.85	m	**534.49**
Sum of two sides 800 mm	414.36	524.42	2.97	95.84	m	**620.26**
Sum of two sides 1000 mm	414.36	524.42	3.04	98.11	m	**622.53**
Sum of two sides 1200 mm	469.62	594.35	3.09	99.72	m	**694.07**
45° radius bend						
Sum of two sides 200 mm	354.49	448.64	1.52	49.06	m	**497.70**
Sum of two sides 400 mm	354.49	448.64	1.66	53.57	m	**502.21**
Sum of two sides 600 mm	354.49	448.64	1.80	58.09	m	**506.73**
Sum of two sides 800 mm	414.36	524.42	1.93	62.29	m	**586.71**
Sum of two sides 1000 mm	414.36	524.42	2.05	66.16	m	**590.58**
Sum of two sides 1200 mm	469.62	594.35	2.17	70.03	m	**664.38**
90° mitre bend						
Sum of two sides 200 mm	354.49	448.64	2.47	79.71	m	**528.35**
Sum of two sides 400 mm	354.49	448.64	2.84	91.65	m	**540.29**
Sum of two sides 600 mm	354.49	448.64	3.20	103.28	m	**551.92**
Sum of two sides 800 mm	414.36	524.42	3.56	114.89	m	**639.31**
Sum of two sides 1000 mm	414.36	524.42	3.66	118.12	m	**642.54**
Sum of two sides 1200 mm	469.62	594.35	3.72	120.05	m	**714.40**
Branch						
Sum of two sides 200 mm	128.92	163.16	0.98	31.63	m	**194.79**
Sum of two sides 400 mm	128.92	163.16	1.14	36.79	m	**199.95**
Sum of two sides 600 mm	128.92	163.16	1.30	41.96	m	**205.12**
Sum of two sides 800 mm	161.13	203.93	1.46	47.12	m	**251.05**
Sum of two sides 1000 mm	175.69	222.35	1.46	47.12	m	**269.47**
Sum of two sides 1200 mm	216.35	273.82	1.55	50.02	m	**323.84**
Ductwork 601 to 1000 mm longest side						
Sum of two sides 1000 mm	712.61	901.88	6.00	193.64	m	**1095.52**
Sum of two sides 1100 mm	992.31	1255.87	6.00	193.64	m	**1449.51**
Sum of two sides 1300 mm	1008.87	1276.82	6.00	193.64	m	**1470.46**
Sum of two sides 1500 mm	1032.39	1306.59	6.00	193.64	m	**1500.23**
Sum of two sides 1700 mm	1247.72	1579.11	6.00	193.64	m	**1772.75**
Sum of two sides 1900 mm	1429.86	1809.63	6.00	193.64	m	**2003.27**
Extra over fittings; Ductwork 601 to 1000 mm longest side						
End cap						
Sum of two sides 1000 mm	193.38	244.74	1.25	40.34	m	**285.08**
Sum of two sides 1100 mm	294.64	372.89	1.25	40.34	m	**413.23**
Sum of two sides 1300 mm	294.64	372.89	1.31	42.28	m	**415.17**
Sum of two sides 1500 mm	294.85	373.16	1.36	43.89	m	**417.05**
Sum of two sides 1700 mm	520.25	658.43	1.42	45.83	m	**704.26**
Sum of two sides 1900 mm	520.25	658.43	1.48	47.77	m	**706.20**
Reducer						
Sum of two sides 1000 mm	207.18	262.20	3.22	103.91	m	**366.11**
Sum of two sides 1100 mm	262.44	332.15	3.22	103.91	m	**436.06**
Sum of two sides 1300 mm	262.44	332.15	3.33	107.48	m	**439.63**

38 VENTILATION/AIR CONDITIONING SYSTEMS

Item	Net Price £	Material £	Labour hours	Labour £	Unit	Total rate £
DUCTWORK: FIRE RATED – cont						
Extra over fittings – cont						
Sum of two sides 1500 mm	262.44	332.15	3.44	111.01	m	**443.16**
Sum of two sides 1700 mm	322.30	407.90	3.54	114.25	m	**522.15**
Sum of two sides 1900 mm	322.30	407.90	3.65	117.80	m	**525.70**
Offset						
Sum of two sides 1000 mm	207.18	262.20	3.78	121.99	m	**384.19**
Sum of two sides 1100 mm	262.44	332.15	3.78	121.99	m	**454.14**
Sum of two sides 1300 mm	262.44	332.15	4.00	129.09	m	**461.24**
Sum of two sides 1500 mm	262.44	332.15	4.23	136.51	m	**468.66**
Sum of two sides 1700 mm	322.30	407.90	4.45	143.61	m	**551.51**
Sum of two sides 1900 mm	322.30	407.90	4.67	150.71	m	**558.61**
90° radius bend						
Sum of two sides 1000 mm	414.36	524.42	3.78	121.99	m	**646.41**
Sum of two sides 1100 mm	469.62	594.35	3.07	99.08	m	**693.43**
Sum of two sides 1300 mm	469.62	594.35	3.12	100.69	m	**695.04**
Sum of two sides 1500 mm	469.62	594.35	3.17	102.30	m	**696.65**
Sum of two sides 1700 mm	616.94	780.80	3.22	103.91	m	**884.71**
Sum of two sides 1900 mm	616.94	780.80	3.28	105.85	m	**886.65**
45° bend						
Sum of two sides 1000 mm	469.62	594.35	1.91	61.63	m	**655.98**
Sum of two sides 1100 mm	469.62	594.35	2.00	64.55	m	**658.90**
Sum of two sides 1300 mm	469.62	594.35	2.02	65.20	m	**659.55**
Sum of two sides 1500 mm	469.62	594.35	2.13	68.73	m	**663.08**
Sum of two sides 1700 mm	616.94	780.80	2.24	72.30	m	**853.10**
Sum of two sides 1900 mm	616.94	780.80	2.35	75.84	m	**856.64**
90° mitre bend						
Sum of two sides 1000 mm	414.36	524.42	3.78	121.99	m	**646.41**
Sum of two sides 1100 mm	469.62	594.35	3.69	119.08	m	**713.43**
Sum of two sides 1300 mm	469.62	594.35	3.75	121.03	m	**715.38**
Sum of two sides 1500 mm	469.62	594.35	3.81	122.96	m	**717.31**
Sum of two sides 1700 mm	616.94	780.80	3.87	124.90	m	**905.70**
Sum of two sides 1900 mm	616.94	780.80	3.93	126.84	m	**907.64**
Branch						
Sum of two sides 1000 mm	167.53	212.03	2.38	76.81	m	**288.84**
Sum of two sides 1100 mm	216.35	273.82	1.50	48.41	m	**322.23**
Sum of two sides 1300 mm	216.35	273.82	1.59	51.32	m	**325.14**
Sum of two sides 1500 mm	216.35	273.82	1.68	54.22	m	**328.04**
Sum of two sides 1700 mm	276.24	349.61	1.77	57.12	m	**406.73**
Sum of two sides 1900 mm	276.24	349.61	1.86	60.03	m	**409.64**
Ductwork 1001 to 1250 mm longest side						
Sum of two sides 1300 mm	992.31	1255.87	6.00	193.64	m	**1449.51**
Sum of two sides 1500 mm	1054.71	1334.84	6.00	193.64	m	**1528.48**
Sum of two sides 1700 mm	1247.72	1579.11	6.00	193.64	m	**1772.75**
Sum of two sides 1900 mm	1268.51	1605.43	6.00	193.64	m	**1799.07**
Sum of two sides 2100 mm	1662.32	2103.83	6.50	209.78	m	**2313.61**
Sum of two sides 2300 mm	1752.63	2218.13	6.50	209.78	m	**2427.91**
Sum of two sides 2500 mm	2086.43	2640.59	6.50	209.78	m	**2850.37**

38 VENTILATION/AIR CONDITIONING SYSTEMS

Item	Net Price £	Material £	Labour hours	Labour £	Unit	Total rate £
Extra over fittings; Ductwork 1001 to 1250 mm longest side						
End cap						
Sum of two sides 1300 mm	294.64	372.89	1.31	42.28	m	**415.17**
Sum of two sides 1500 mm	294.64	372.89	1.36	43.89	m	**416.78**
Sum of two sides 1700 mm	520.25	658.43	1.42	45.83	m	**704.26**
Sum of two sides 1900 mm	520.25	658.43	1.48	47.77	m	**706.20**
Sum of two sides 2100 mm	709.00	897.31	2.66	85.85	m	**983.16**
Sum of two sides 2300 mm	709.00	897.31	2.95	95.20	m	**992.51**
Sum of two sides 2500 mm	709.00	897.31	3.24	104.56	m	**1001.87**
Reducer						
Sum of two sides 1300 mm	262.44	332.15	3.33	107.48	m	**439.63**
Sum of two sides 1500 mm	262.44	332.15	3.44	111.01	m	**443.16**
Sum of two sides 1700 mm	322.30	407.90	3.54	114.25	m	**522.15**
Sum of two sides 1900 mm	322.30	407.90	3.65	117.80	m	**525.70**
Sum of two sides 2100 mm	382.11	483.59	3.75	121.03	m	**604.62**
Sum of two sides 2300 mm	382.11	483.59	3.85	124.24	m	**607.83**
Sum of two sides 2500 mm	382.11	483.59	3.95	127.47	m	**611.06**
Offset						
Sum of two sides 1300 mm	262.44	332.15	4.00	129.09	m	**461.24**
Sum of two sides 1500 mm	262.44	332.15	4.23	136.51	m	**468.66**
Sum of two sides 1700 mm	322.30	407.90	4.45	143.61	m	**551.51**
Sum of two sides 1900 mm	322.30	407.90	4.67	150.71	m	**558.61**
Sum of two sides 2100 mm	382.11	483.59	5.76	185.89	m	**669.48**
Sum of two sides 2300 mm	382.11	483.59	6.22	200.74	m	**684.33**
Sum of two sides 2500 mm	382.11	483.59	6.68	215.58	m	**699.17**
90° radius bend						
Sum of two sides 1300 mm	469.62	594.35	3.12	100.69	m	**695.04**
Sum of two sides 1500 mm	469.62	594.35	3.17	102.30	m	**696.65**
Sum of two sides 1700 mm	616.94	780.80	3.22	103.91	m	**884.71**
Sum of two sides 1900 mm	616.94	780.80	3.28	105.85	m	**886.65**
Sum of two sides 2100 mm	630.74	798.27	3.32	107.15	m	**905.42**
Sum of two sides 2300 mm	630.74	798.27	3.36	108.44	m	**906.71**
Sum of two sides 2500 mm	630.74	798.27	3.40	109.73	m	**908.00**
45° bend						
Sum of two sides 1300 mm	469.62	594.35	2.02	65.20	m	**659.55**
Sum of two sides 1500 mm	469.62	594.35	2.13	68.73	m	**663.08**
Sum of two sides 1700 mm	616.94	780.80	2.24	72.30	m	**853.10**
Sum of two sides 1900 mm	616.94	780.80	2.35	75.84	m	**856.64**
Sum of two sides 2100 mm	630.74	798.27	2.89	93.26	m	**891.53**
Sum of two sides 2300 mm	630.74	798.27	3.13	101.01	m	**899.28**
Sum of two sides 2500 mm	630.74	798.27	3.37	108.76	m	**907.03**

38 VENTILATION/AIR CONDITIONING SYSTEMS

Item	Net Price £	Material £	Labour hours	Labour £	Unit	Total rate £
DUCTWORK: FIRE RATED – cont						
Extra over fittings – cont						
90° mitre bend						
Sum of two sides 1300 mm	469.62	594.35	3.75	121.03	m	**715.38**
Sum of two sides 1500 mm	469.62	594.35	3.81	122.96	m	**717.31**
Sum of two sides 1700 mm	616.94	780.80	3.87	124.90	m	**905.70**
Sum of two sides 1900 mm	616.94	780.80	3.93	126.84	m	**907.64**
Sum of two sides 2100 mm	630.74	798.27	3.99	128.78	m	**927.05**
Sum of two sides 2300 mm	630.74	798.27	4.05	130.72	m	**928.99**
Sum of two sides 2500 mm	630.74	798.27	4.11	132.65	m	**930.92**
Branch						
Sum of two sides 1300 mm	216.35	273.82	1.59	51.32	m	**325.14**
Sum of two sides 1500 mm	216.35	273.82	1.68	54.22	m	**328.04**
Sum of two sides 1700 mm	276.24	349.61	1.77	57.12	m	**406.73**
Sum of two sides 1900 mm	276.24	349.61	1.86	60.03	m	**409.64**
Sum of two sides 2100 mm	290.01	367.04	2.58	83.26	m	**450.30**
Sum of two sides 2300 mm	290.01	367.04	2.64	85.20	m	**452.24**
Sum of two sides 2500 mm	290.01	367.04	2.91	93.91	m	**460.95**
Ductwork 1251 to 2000 mm longest side						
Sum of two sides 1800 mm	1247.72	1579.11	6.00	193.64	m	**1772.75**
Sum of two sides 2000 mm	1455.64	1842.25	6.00	193.64	m	**2035.89**
Sum of two sides 2200 mm	1662.32	2103.83	6.50	209.78	m	**2313.61**
Sum of two sides 2400 mm	1869.36	2365.87	6.50	209.78	m	**2575.65**
Sum of two sides 2600 mm	1969.41	2492.48	6.66	214.94	m	**2707.42**
Sum of two sides 2800 mm	2160.24	2734.00	6.66	214.94	m	**2948.94**
Sum of two sides 3000 mm	2349.90	2974.04	6.66	214.94	m	**3188.98**
Sum of two sides 3200 mm	2367.47	2996.27	9.00	290.46	m	**3286.73**
Sum of two sides 3400 mm	2604.21	3295.89	9.00	290.46	m	**3586.35**
Sum of two sides 3600 mm	2651.23	3355.40	11.70	377.60	m	**3733.00**
Sum of two sides 3800 mm	2887.13	3653.96	11.70	377.60	m	**4031.56**
Sum of two sides 4000 mm	3339.97	4227.07	11.70	377.60	m	**4604.67**
Extra over fittings; Ductwork 1251 to 2000 mm longest sides						
End cap						
Sum of two sides 1800 mm	520.25	658.43	1.45	46.80	m	**705.23**
Sum of two sides 2000 mm	520.25	658.43	1.51	48.74	m	**707.17**
Sum of two sides 2200 mm	709.00	897.31	2.80	90.36	m	**987.67**
Sum of two sides 2400 mm	709.00	897.31	3.10	100.05	m	**997.36**
Sum of two sides 2600 mm	971.38	1229.38	3.39	109.40	m	**1338.78**
Sum of two sides 2800 mm	971.38	1229.38	3.68	118.76	m	**1348.14**
Sum of two sides 3000 mm	971.38	1229.38	3.98	128.44	m	**1357.82**
Sum of two sides 3200 mm	1312.07	1660.56	4.27	137.82	m	**1798.38**
Sum of two sides 3400 mm	1312.07	1660.56	4.57	147.48	m	**1808.04**
Sum of two sides 3600 mm	1615.90	2045.09	4.87	157.17	m	**2202.26**
Sum of two sides 3800 mm	1615.90	2045.09	5.17	166.86	m	**2211.95**
Sum of two sides 4000 mm	1615.90	2045.09	5.47	176.53	m	**2221.62**

38 VENTILATION/AIR CONDITIONING SYSTEMS

Item	Net Price £	Material £	Labour hours	Labour £	Unit	Total rate £
Reducer						
Sum of two sides 1800 mm	322.30	407.90	3.60	116.18	m	524.08
Sum of two sides 2000 mm	322.30	407.90	3.70	119.41	m	527.31
Sum of two sides 2200 mm	382.11	483.59	3.10	100.05	m	583.64
Sum of two sides 2400 mm	382.11	483.59	3.48	112.31	m	595.90
Sum of two sides 2600 mm	529.42	670.03	3.85	124.24	m	794.27
Sum of two sides 2800 mm	529.42	670.03	4.22	136.19	m	806.22
Sum of two sides 3000 mm	529.42	670.03	4.59	148.14	m	818.17
Sum of two sides 3200 mm	589.27	745.79	4.96	160.07	m	905.86
Sum of two sides 3400 mm	589.27	745.79	5.34	172.33	m	918.12
Sum of two sides 3600 mm	644.54	815.73	5.72	184.60	m	1000.33
Sum of two sides 3800 mm	644.54	815.73	6.10	196.86	m	1012.59
Sum of two sides 4000 mm	644.54	815.73	6.48	209.13	m	1024.86
Offset						
Sum of two sides 1800 mm	322.30	407.90	4.56	147.17	m	555.07
Sum of two sides 2000 mm	322.30	407.90	5.53	178.46	m	586.36
Sum of two sides 2200 mm	382.11	483.59	5.99	193.32	m	676.91
Sum of two sides 2400 mm	382.11	483.59	6.45	208.15	m	691.74
Sum of two sides 2600 mm	529.42	670.03	6.91	223.01	m	893.04
Sum of two sides 2800 mm	529.42	670.03	7.37	237.85	m	907.88
Sum of two sides 3000 mm	529.42	670.03	7.83	252.69	m	922.72
Sum of two sides 3200 mm	589.27	745.79	8.29	267.55	m	1013.34
Sum of two sides 3400 mm	589.27	745.79	8.75	282.40	m	1028.19
Sum of two sides 3600 mm	644.54	815.73	9.21	297.23	m	1112.96
Sum of two sides 3800 mm	644.54	815.73	9.67	312.09	m	1127.82
Sum of two sides 4000 mm	644.54	815.73	10.13	326.92	m	1142.65
90° radius bend						
Sum of two sides 1800 mm	616.94	780.80	3.25	104.89	m	885.69
Sum of two sides 2000 mm	616.94	780.80	3.30	106.50	m	887.30
Sum of two sides 2200 mm	630.74	798.27	3.34	107.79	m	906.06
Sum of two sides 2400 mm	630.74	798.27	3.38	109.08	m	907.35
Sum of two sides 2600 mm	966.80	1223.58	3.42	110.38	m	1333.96
Sum of two sides 2800 mm	966.80	1223.58	3.46	111.66	m	1335.24
Sum of two sides 3000 mm	966.80	1223.58	3.50	112.95	m	1336.53
Sum of two sides 3200 mm	1160.16	1468.30	3.54	114.25	m	1582.55
Sum of two sides 3400 mm	1160.16	1468.30	3.58	115.54	m	1583.84
Sum of two sides 3600 mm	1358.09	1718.80	3.62	116.83	m	1835.63
Sum of two sides 3800 mm	1358.09	1718.80	3.66	118.12	m	1836.92
Sum of two sides 4000 mm	1358.09	1718.80	3.70	119.41	m	1838.21
45° bend						
Sum of two sides 1800 mm	616.94	780.80	2.30	74.22	m	855.02
Sum of two sides 2000 mm	616.94	780.80	2.76	89.07	m	869.87
Sum of two sides 2200 mm	630.74	798.27	3.01	97.14	m	895.41
Sum of two sides 2400 mm	630.74	798.27	3.26	105.21	m	903.48
Sum of two sides 2600 mm	966.80	1223.58	3.49	112.64	m	1336.22
Sum of two sides 2800 mm	966.80	1223.58	3.74	120.70	m	1344.28
Sum of two sides 3000 mm	966.80	1223.58	3.98	128.44	m	1352.02
Sum of two sides 3200 mm	1160.16	1468.30	4.22	136.19	m	1604.49
Sum of two sides 3400 mm	1160.16	1468.30	4.46	143.93	m	1612.23
Sum of two sides 3600 mm	1358.09	1718.80	4.70	151.68	m	1870.48
Sum of two sides 3800 mm	1358.09	1718.80	4.94	159.43	m	1878.23
Sum of two sides 4000 mm	1358.09	1718.80	5.18	167.17	m	1885.97

38 VENTILATION/AIR CONDITIONING SYSTEMS

Item	Net Price £	Material £	Labour hours	Labour £	Unit	Total rate £
DUCTWORK: FIRE RATED – cont						
Extra over fittings – cont						
90° mitre bend						
Sum of two sides 1800 mm	616.94	780.80	3.90	125.87	m	**906.67**
Sum of two sides 2000 mm	616.94	780.80	3.96	127.80	m	**908.60**
Sum of two sides 2200 mm	630.74	798.27	3.83	123.61	m	**921.88**
Sum of two sides 2400 mm	838.66	1061.41	3.84	123.93	m	**1185.34**
Sum of two sides 2600 mm	966.80	1223.58	3.85	124.24	m	**1347.82**
Sum of two sides 2800 mm	966.80	1223.58	3.86	124.58	m	**1348.16**
Sum of two sides 3000 mm	966.80	1223.58	3.87	124.90	m	**1348.48**
Sum of two sides 3200 mm	1160.16	1468.30	3.88	125.22	m	**1593.52**
Sum of two sides 3400 mm	1160.16	1468.30	3.90	125.87	m	**1594.17**
Sum of two sides 3600 mm	1358.09	1718.80	3.92	126.50	m	**1845.30**
Sum of two sides 3800 mm	1358.09	1718.80	3.94	127.15	m	**1845.95**
Sum of two sides 4000 mm	1358.09	1718.80	3.96	127.80	m	**1846.60**
Branch						
Sum of two sides 1800 mm	276.24	349.61	1.81	58.42	m	**408.03**
Sum of two sides 2000 mm	276.24	349.61	1.90	61.32	m	**410.93**
Sum of two sides 2200 mm	290.01	367.04	2.61	84.24	m	**451.28**
Sum of two sides 2400 mm	290.01	367.04	2.68	86.49	m	**453.53**
Sum of two sides 2600 mm	345.27	436.98	2.75	88.76	m	**525.74**
Sum of two sides 2800 mm	345.27	436.98	2.81	90.70	m	**527.68**
Sum of two sides 3000 mm	345.27	436.98	2.87	92.62	m	**529.60**
Sum of two sides 3200 mm	359.12	454.51	2.93	94.56	m	**549.07**
Sum of two sides 3400 mm	359.12	454.51	3.00	96.82	m	**551.33**
Sum of two sides 3600 mm	460.40	582.68	3.06	98.75	m	**681.43**
Sum of two sides 3800 mm	460.40	582.68	3.12	100.69	m	**683.37**
Sum of two sides 4000 mm	460.40	582.68	3.18	102.63	m	**685.31**

38 VENTILATION/AIR CONDITIONING SYSTEMS

Item	Net Price £	Material £	Labour hours	Labour £	Unit	Total rate £
LOW VELOCITY AIR CONDITIONING: AIR HANDLING UNITS						
Supply air handling unit; inlet with motorized damper, LTHW frost coil (at −5°C to +5°C), panel filter (EU4), bag filter (EU6), cooling coil (at 28°C db/20°C wb to 12°C db/11.5°C wb), LTHW heating coil (at 5°C to 21°C), supply fan, outlet plenum; includes access sections; all units located internally; Includes placing in position and fitting of sections together; electrical work elsewhere.						
Volume, external pressure						
2 m³/s @ 350 Pa	8748.79	11072.47	40.00	1347.61	nr	**12420.08**
2 m³/s @ 700 Pa	9346.84	11829.36	40.00	1347.61	nr	**13176.97**
5 m³/s @ 350 Pa	14752.21	18670.40	65.00	2189.87	nr	**20860.27**
5 m³/s @ 700 Pa	15268.27	19323.53	65.00	2189.87	nr	**21513.40**
8 m³/s @ 350 Pa	22095.68	27964.29	77.00	2594.16	nr	**30558.45**
8 m³/s @ 700 Pa	22440.13	28400.23	77.00	2594.16	nr	**30994.39**
10 m³/s @ 350 Pa	24331.60	30794.08	100.00	3369.03	nr	**34163.11**
10 m³/s @ 700 Pa	25493.25	32264.25	100.00	3369.03	nr	**35633.28**
13 m³/s @ 350 Pa	31216.77	39507.94	108.00	3638.54	nr	**43146.48**
13 m³/s @ 700 Pa	32180.90	40728.15	108.00	3638.54	nr	**44366.69**
15 m³/s @ 350 Pa	35750.08	45245.30	120.00	4042.83	nr	**49288.13**
15 m³/s @ 700 Pa	36400.57	46068.56	120.00	4042.83	nr	**50111.39**
18 m³/s @ 350 Pa	40567.16	51341.80	133.00	4480.81	nr	**55822.61**
18 m³/s @ 700 Pa	41312.91	52285.62	133.00	4480.81	nr	**56766.43**
20 m³/s @ 350 Pa	47010.45	59496.43	142.00	4784.02	nr	**64280.45**
20 m³/s @ 700 Pa	47686.51	60352.05	142.00	4784.02	nr	**65136.07**
Extra for inlet and discharge attenuators at 900 mm long						
2 m³/s @ 350 Pa	2703.04	3420.97	5.00	168.45	nr	**3589.42**
5 m³/s @ 350 Pa	5673.05	7179.82	10.00	336.91	nr	**7516.73**
10 m³/s @ 700 Pa	8071.55	10215.35	13.00	437.98	nr	**10653.33**
15 m³/s @ 700 Pa	12239.56	15490.38	16.00	539.04	nr	**16029.42**
20 m³/s @ 700 Pa	15795.23	19990.44	20.00	673.80	nr	**20664.24**
Extra for locating units externally						
2 m³/s @ 350 Pa	2002.82	2534.77	–	–	nr	**2534.77**
5 m³/s @ 350 Pa	2541.19	3216.12	–	–	nr	**3216.12**
10 m³/s @ 700 Pa	4065.42	5145.19	–	–	nr	**5145.19**
15 m³/s @ 700 Pa	5875.13	7435.57	–	–	nr	**7435.57**
20 m³/s @ 700 Pa	10183.50	12888.24	–	–	nr	**12888.24**

38 VENTILATION/AIR CONDITIONING SYSTEMS

Item	Net Price £	Material £	Labour hours	Labour £	Unit	Total rate £
LOW VELOCITY AIR CONDITIONING: AIR HANDLING UNITS – cont						
Modular air handling unit with supply and extract sections.Supply side; inlet with motorized damper, LTHW frost coil (at −5°C to 5°C), panel filter (EU4), bag filter (EU6), cooling coil at 28°Cdb/20°Cwb to 12°Cdb/ 11.5°C wb), LTHW heating coil (at 5°C to 21°C), supply fan, outlet plenum. Extract side; inlet with motorized damper, extract fan; includes access sections; placing in position and fitting of sections together; electrical work elsewhere						
Volume, external pressure						
2 m³/s @ 350 Pa	12221.73	15467.82	50.00	1684.51	nr	**17152.33**
2 m³/s @ 700 Pa	12813.10	16216.26	50.00	1684.51	nr	**17900.77**
5 m³/s @ 350 Pa	21063.55	26658.03	86.00	2897.36	nr	**29555.39**
5 m³/s @ 700 Pa	21458.91	27158.40	86.00	2897.36	nr	**30055.76**
8 m³/s @ 350 Pa	29249.77	37018.51	105.00	3537.48	nr	**40555.99**
8 5 m³/s @ 700 Pa	29951.89	37907.12	105.00	3537.48	nr	**41444.60**
10 5 m³/s @ 350 Pa	33774.76	42745.34	120.00	4042.83	nr	**46788.17**
10 m³/s @ 700 Pa	34502.37	43666.20	120.00	4042.83	nr	**47709.03**
13 m³/s @ 350 Pa	43463.12	55006.93	130.00	4379.74	nr	**59386.67**
13 m³/s @ 700 Pa	44826.36	56732.24	130.00	4379.74	nr	**61111.98**
15 m³/s @ 350 Pa	48566.98	61466.37	145.00	4885.09	nr	**66351.46**
15 m³/s @ 700 Pa	50864.60	64374.24	145.00	4885.09	nr	**69259.33**
18 m³/s @ 350 Pa	55517.90	70263.46	160.00	5390.45	nr	**75653.91**
18 m³/s @ 700 Pa	56817.71	71908.49	160.00	5390.45	nr	**77298.94**
20 m³/s @ 350 Pa	60678.81	76795.11	175.00	5895.80	nr	**82690.91**
20 m³/s @ 700 Pa	62788.27	79464.84	175.00	5895.80	nr	**85360.64**
Extra for inlet and discharge attenuators at 900 mm long						
2 m³/s @ 350 Pa	4927.79	6236.61	8.00	269.52	nr	**6506.13**
5 m³/s @ 350 Pa	9730.98	12315.53	10.00	336.91	nr	**12652.44**
10 m³/s @ 700 Pa	13924.22	17622.49	13.00	437.98	nr	**18060.47**
15 m³/s @ 700 Pa	23818.00	30144.06	16.00	539.04	nr	**30683.10**
20 m³/s @ 700 Pa	27803.68	35188.34	20.00	673.80	nr	**35862.14**
Extra for locating units externally						
2 m³/s @ 350 Pa	3944.48	4992.13	–	–	nr	**4992.13**
5 m³/s @ 350 Pa	5601.13	7088.79	–	–	nr	**7088.79**
10 m³/s @ 700 Pa	7995.23	10118.76	–	–	nr	**10118.76**
15 m³/s @ 700 Pa	15180.85	19212.88	–	–	nr	**19212.88**
20 m³/s @ 700 Pa	23337.62	29536.09	–	–	nr	**29536.09**
Extra for humidifier, self generating type						
2 m³/s @ 350 Pa (10 kg/hr)	3437.34	4350.29	5.00	168.45	nr	**4518.74**
5 m³/s @ 350 Pa (18 kg/hr)	4253.77	5383.57	5.00	168.45	nr	**5552.02**
10 m³/s @ 700 Pa (30 kg/hr)	4840.58	6126.24	6.00	202.14	nr	**6328.38**
15 m³/s @ 700 Pa (60 kg/hr)	8964.45	11345.41	8.00	269.52	nr	**11614.93**
20 m³/s @ 700 Pa (90 kg/hr)	13478.00	17057.76	10.00	336.91	nr	**17394.67**

38 VENTILATION/AIR CONDITIONING SYSTEMS

Item	Net Price £	Material £	Labour hours	Labour £	Unit	Total rate £
Extra for mixing box						
2 m³/s @ 350 Pa	1650.54	2088.92	4.00	134.76	nr	**2223.68**
5 m³/s @ 350 Pa	2326.52	2944.45	4.00	134.76	nr	**3079.21**
10 m³/s @ 700 Pa	3259.44	4125.15	5.00	168.45	nr	**4293.60**
15 m³/s @ 700 Pa	4350.14	5505.54	6.00	202.14	nr	**5707.68**
20 m³/s @ 700 Pa	6631.67	8393.04	6.00	202.14	nr	**8595.18**
Extra for runaround coil, including pump and associated pipework; typical outputs in brackets (based on minimal distance between the supply and extract units)						
2 m³/s @350 Pa (26 kW)	4941.19	6253.56	30.00	1010.71	nr	**7264.27**
5 m³/s @ 350 Pa (37 kW)	8760.83	11087.71	30.00	1010.71	nr	**12098.42**
10 m³/s @ 700 Pa (85 kW)	15168.04	19196.68	40.00	1347.61	nr	**20544.29**
15 m³/s @ 700 Pa (151 kW)	21341.98	27010.41	50.00	1684.51	nr	**28694.92**
20 m³/s @ 700 Pa (158 kW)	29092.14	36819.01	60.00	2021.42	nr	**38840.43**
Extra for thermal wheel (typical outputs in brackets)						
2 m³/s @350 Pa (37 kW)	10220.19	12934.67	12.00	404.29	nr	**13338.96**
5 m³/s @ 350 Pa (65 kW)	14678.67	18577.33	12.00	404.29	nr	**18981.62**
10 m³/s @ 700 Pa (127 kW)	19575.08	24774.22	15.00	505.36	nr	**25279.58**
15 m³/s @ 700 Pa (160 kW)	36675.39	46416.37	17.00	572.73	nr	**46989.10**
20 m³/s @ 700 Pa (262 kW)	42841.12	54219.73	19.00	640.11	nr	**54859.84**
Extra for plate heat exchanger, including additional filtration in extract leg (typical outputs in brackets)						
2 m³/s @350 Pa (25 kW)	6477.36	8197.75	12.00	404.29	nr	**8602.04**
5 m³/s @ 350 Pa (51 kW)	12662.43	16025.58	12.00	404.29	nr	**16429.87**
10 m³/s @ 700 Pa (98 kW)	18579.30	23513.96	15.00	505.36	nr	**24019.32**
15 m³/s @ 700 Pa (160 kW)	32084.91	40606.66	17.00	572.73	nr	**41179.39**
20 m³/s @ 700 Pa (190 kW)	41434.98	52440.11	19.00	640.11	nr	**53080.22**
Extra for electric heating in lieu of LTHW						
2 m³/s @350 Pa	2244.98	2841.25	–	–	nr	**2841.25**
5 m³/s @ 350 Pa	4206.21	5323.38	–	–	nr	**5323.38**
10 m³/s @ 700 Pa	5679.05	7187.41	–	–	nr	**7187.41**
15 m³/s @ 700 Pa	5955.02	7536.67	–	–	nr	**7536.67**
20 m³/s @ 700 Pa	7027.77	8894.35	–	–	nr	**8894.35**

38 VENTILATION/AIR CONDITIONING SYSTEMS

Item	Net Price £	Material £	Labour hours	Labour £	Unit	Total rate £
VAV AIR CONDITIONING						
VAV TERMINAL BOXES						
VAV terminal box; integral acoustic silencer; factory installed and prewired control components (excluding electronic controller); selected at 200 Pa at entry to unit; includes fixing in position; electrical work elsewhere						
80 l/s–110 l/s	545.17	689.96	2.00	64.55	nr	**754.51**
Extra for secondary silencer	132.58	167.80	0.50	16.14	nr	**183.94**
Extra for 2 row LTHW heating coil	327.54	414.53	–	–	nr	**414.53**
150 l/s–190 l/s	573.67	726.04	2.00	64.55	nr	**790.59**
Extra for secondary silencer	145.58	184.25	0.50	16.14	nr	**200.39**
Extra for 2 row LTHW heating coil	350.92	444.12	–	–	nr	**444.12**
250 l/s–310 l/s	639.96	809.93	2.00	64.55	nr	**874.48**
Extra for secondary silencer	192.68	243.86	0.50	16.14	nr	**260.00**
Extra for 2 row LTHW heating coil	397.72	503.35	–	–	nr	**503.35**
420 l/s–520 l/s	693.24	877.36	2.00	64.55	nr	**941.91**
Extra for secondary silencer	219.30	277.55	0.50	16.14	nr	**293.69**
Extra for 2 row LTHW heating coil	484.28	612.91	–	–	nr	**612.91**
650 l/s–790 l/s	839.44	1062.40	2.00	64.55	nr	**1126.95**
Extra for secondary silencer	280.64	355.17	0.50	16.14	nr	**371.31**
Extra for 2 row LTHW heating coil	507.68	642.52	–	–	nr	**642.52**
1130 l/s–1370 l/s	975.12	1234.12	2.00	64.55	nr	**1298.67**
Extra for secondary silencer	392.15	496.31	0.50	16.14	nr	**512.45**
Extra for 2 row LTHW heating coil	531.06	672.11	–	–	nr	**672.11**
Extra for electric heater & thyristor controls, 3 kW/1ph (per box)	432.80	547.75	–	–	nr	**547.75**
Extra for fitting free issue box controller	–	–	2.13	68.73	nr	**68.73**
Fan assisted VAV terminal box; factory installed and prewired control components (excluding electronic controller); selected at 40 Pa external static pressure; includes fixing in position, electrical work elsewhere						
100 l/s–175 l/s	1206.83	1527.37	3.00	96.82	nr	**1624.19**
Extra for secondary silencer	143.72	181.89	0.50	16.14	nr	**198.03**
Extra for 1 row LTHW heating coil	386.02	488.54	–	–	nr	**488.54**
170 l/s–360 l/s	1316.47	1666.12	3.00	96.82	nr	**1762.94**
Extra for secondary silencer	182.14	230.52	0.50	16.14	nr	**246.66**
Extra for 1 row LTHW heating coil	409.42	518.16	–	–	nr	**518.16**
300 l/s–640 l/s	1509.14	1909.97	3.00	96.82	nr	**2006.79**
Extra for secondary silencer	278.79	352.83	0.50	16.14	nr	**368.97**
Extra for 1 row LTHW heating coil	444.51	562.58	–	–	nr	**562.58**
620 l/s–850 l/s	1509.14	1909.97	3.00	96.82	nr	**2006.79**
Extra for secondary silencer	278.79	352.83	0.50	16.14	nr	**368.97**
Extra for 1 row LTHW heating coil	479.60	606.98	–	–	nr	**606.98**
Extra for electric heater plus thyristor controls, 3 kW/1ph (per box)	444.51	562.58	–	–	nr	**562.58**
Extra for fitting free issue controller	–	–	2.13	68.73	nr	**68.73**

38 VENTILATION/AIR CONDITIONING SYSTEMS

Item	Net Price £	Material £	Labour hours	Labour £	Unit	Total rate £
FAN COIL AIR CONDITIONING						
All selections based on summer return air condition of 23°C @ 50% RH, CHW @ 6°/ 12°C, LTHW @ 82°/71°C (where applicable), medium speed, external resistance of 30 Pa.						
All selections are based on heating and cooling units. For waterside control units there is no significant reduction in cost between 4 pipe heating and cooling and 2 pipe cooling only units (excluding controls). For airside control units, there is a marginal reduction (less than 5%) between 4 pipe heating and cooling units and 2 pipe cooling only units (excluding controls).						
Ceiling void mounted horizontal waterside control fan coil unit; cooling coil ; LTHW heating coil; multi-tapped speed transformer; fine wire mesh filter; includes fixing in position; electrical work elsewhere						
Total cooling load, heating load						
2800 W, 1000 W	529.73	670.42	4.00	134.76	nr	**805.18**
4000 W, 1700 W	504.17	638.08	4.00	134.76	nr	**772.84**
4500 W, 1900 W	781.98	989.68	4.00	134.76	nr	**1124.44**
6000 W, 2600 W	1022.77	1294.42	4.00	134.76	nr	**1429.18**
Ceiling void mounted horizontal waterside control fan coil unit; cooling coil; electric heating coil; multi-tapped speed transformer; fine wire mesh filter; includes fixing in position; electrical work elsewhere + thyristor and 2 No. HTCOs						
Total cooling load, heating load						
2800 W, 1500 W.	734.76	929.91	4.00	134.76	nr	**1064.67**
4000 W, 2000 W	760.22	962.14	4.00	134.76	nr	**1096.90**
4500 W, 2000 W	967.98	1225.08	4.00	134.76	nr	**1359.84**
6000 W, 3000 W	1195.88	1513.50	4.00	134.76	nr	**1648.26**
Ceiling void mounted horizontal airside control fan coil unit; cooling coil ; LHTW heating coil; multi-tapped speed transformer; fine wire mesh filter, damper actuator & fixing kit; includes fixing in position; electrical work elsewhere						
Total cooling load, heating load						
2600 W, 2200 W.	832.43	1053.53	4.00	134.76	nr	**1188.29**
3600 W, 3200 W	856.51	1084.00	4.00	134.76	nr	**1218.76**
4000 W, 3600 W	958.56	1213.15	4.00	134.76	nr	**1347.91**
5400 W, 5000 W	1144.31	1448.24	4.00	134.76	nr	**1583.00**

38 VENTILATION/AIR CONDITIONING SYSTEMS

Item	Net Price £	Material £	Labour hours	Labour £	Unit	Total rate £
FAN COIL AIR CONDITIONING – cont						
Ceiling void mounted horizontal airside control fan coil unit; cooling coil ; electric heating coil; multi-tapped speed transformer; fine wire mesh filter, damper actuator & fixing kit; includes fixing in position; electrical work elsewhere + thyristor and 2 No. HTCOs						
Total cooling load, heating load						
2600 W, 1500 W	886.32	1121.72	4.00	134.76	nr	**1256.48**
3600 W, 2000 W	909.25	1150.74	4.00	134.76	nr	**1285.50**
4000 W, 2000 W	1005.57	1272.64	4.00	134.76	nr	**1407.40**
5400 W, 3000 W	1194.75	1512.08	4.00	134.76	nr	**1646.84**
Ceiling void mounted slimline horizontal waterside control fan coil unit, 170 mm deep; cooilng coil; LTHW heating coil; multi-tapped speed transformer; fine wire mesh filter; includes fixing in position; electrical work elsewhere						
Total cooling load, heating load						
1100 W, 1500 W	653.56	827.14	3.50	117.91	nr	**945.05**
3200 W, 3700 W	1248.64	1580.28	4.00	134.76	nr	**1715.04**
Ceiling void mounted slimline horizontal waterside control fan coil unit, 170 mm deep; cooilng coil; electric heating coil; multi-tapped speed transformer; fine wire mesh filter; includes fixing in position; electrical work elsewhere						
Total cooling load, heating load						
1100 W, 1000 W	482.72	610.93	3.50	117.91	nr	**728.84**
3400 W, 2000 W	1359.87	1721.05	4.00	134.76	nr	**1855.81**
Ceiling void mounted slimline horizontal airside control fan coil unit, 170 mm deep; cooling coil; multi-tapped speed transformer; fine wire mesh filter, damper actuator & fixing kit; includes fixing in position; electrical work elsewhere						
Total cooling load						
1000 W	848.48	1073.83	3.50	117.91	nr	**1191.74**
3000 W	1131.70	1432.28	4.00	134.76	nr	**1567.04**
Ceiling void mounted slimline horizontal airside control fan coil unit, 170 mm deep; cooilng coil; electric heating coil; multi-tapped speed transformer; fine wire mesh filter, damper actuator & fixing kit; includes fixing in position; electrical work elsewhere						
Total cooling load, heating load						
1000 W, 1000 W	628.34	795.22	3.50	117.91	nr	**913.13**
3000 W, 2000 W	1242.91	1573.03	4.00	134.76	nr	**1707.79**

38 VENTILATION/AIR CONDITIONING SYSTEMS

Item	Net Price £	Material £	Labour hours	Labour £	Unit	Total rate £
Low level perimeter waterside control fan coil unit; cooling coil; LTHW heating coil; multi-tapped speed transformer; fine wire mesh filter; includes fixing in position; electrical work elsewhere						
Total cooling load, heating load						
1700 W, 1400 W	575.60	728.48	3.50	117.91	nr	846.39
Extra over for standard cabinet	227.84	288.36	1.00	33.69	nr	322.05
2200 W, 1900 W	744.14	941.79	3.50	117.91	nr	1059.70
Extra over for standard cabinet	277.53	351.24	1.00	33.69	nr	384.93
2600 W, 2200 W	744.14	941.79	3.50	117.91	nr	1059.70
Extra over for standard cabinet	284.38	359.91	1.00	33.69	nr	393.60
3900 W, 3200 W	948.24	1200.09	3.50	117.91	nr	1318.00
Extra over for standard cabinet	310.08	392.44	1.00	33.69	nr	426.13
Low level perimeter waterside control fan coil unit; cooling coil; electric heating coil; multi-tapped speed transformer; fine wire mesh filter; includes fixing in position; electrical work elsewhere						
Total cooling load, heating load						
1700 W, 1500 W	684.52	866.33	3.50	117.91	nr	984.24
Extra over for standard cabinet	227.84	288.36	1.00	33.69	nr	322.05
2200 W, 2000 W	855.36	1082.55	3.50	117.91	nr	1200.46
Extra over for standard cabinet	277.53	351.24	1.00	33.69	nr	384.93
2600 W, 2000 W	855.36	1082.55	3.50	117.91	nr	1200.46
Extra over for standard cabinet	284.38	359.91	1.00	33.69	nr	393.60
3800 W, 3000 W	1061.75	1343.75	3.50	117.91	nr	1461.66
Extra over for standard cabinet	310.08	392.44	1.00	33.69	nr	426.13

38 VENTILATION/AIR CONDITIONING SYSTEMS

Item	Net Price £	Material £	Labour hours	Labour £	Unit	Total rate £
MECHANICAL VENTILATION WITH HEAT RECOVERY (MVHR) UNITS						
Mechanical ventilation with heat recovery (MVHR) units						
A ventilation unit with high efficiency (up to 130%) rotary heat exchanger (rotary heat exchangers do not require a condense drain or heater battery installed). The unit recovers hot, cold and humid air. Includes main unit, built-in controller and delivery						
Air flow rate						
40–200 m³/h, 55 l/s	1346.92	1704.66	3.00	101.07	nr	**1805.73**
50–250 m³/h, 69 l/s	1484.26	1878.48	3.00	101.07	nr	**1979.55**
60–300 m³/h, 83 l/s	1623.92	2055.23	3.00	101.07	nr	**2156.30**
60–350 m³/h, 97 l/s	1703.56	2156.02	3.00	101.07	nr	**2257.09**
70–450 m³/h, 125 l/s	1857.06	2350.30	3.00	101.07	nr	**2451.37**
High performance ventilation unit with enthalpy exchanger, class leading efficiency, automatic modulating true summer by-pass, integrated humidity sensor, four variable speed flow rate set points, dial-a-duty motor						
Air flow rate						
70–600 m³/h	4046.31	5121.01	3.00	101.07	nr	**5222.08**
Extra over for additional cooling module with 1.5 kW cooling	4030.01	5100.38	–	–	nr	**5100.38**
Extra over for post-treatment air to water exchanger						
300 m³/h	637.25	806.50	3.00	101.07	nr	**907.57**
400 m³/h	824.63	1043.65	3.00	101.07	nr	**1144.72**
500 m³/h	732.83	927.47	3.00	101.07	nr	**1028.54**
600 m³/h	967.65	1224.65	3.00	101.07	nr	**1325.72**
Extra over for subsoil heat exchanger	4227.97	5350.92	3.00	101.07	nr	**5451.99**

PART 4

Material Costs/Measured Work Prices – Electrical Installations

Contractual Procedures in the Construction Industry, 7th edition

Allan Ashworth and Srinath Perera

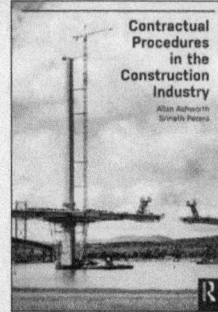

Contractual Procedures in the Construction Industry 7th edition aims to provide students with a comprehensive understanding of the subject, and reinforces the changes that are taking place within the construction industry. The book looks at contract law within the context of construction contracts, it examines the different procurement routes that have evolved over time and the particular aspects relating to design and construction, lean methods of construction and the advantages and disadvantages of PFI/PPP and its variants. It covers the development of partnering, supply chain management, design and build and the way that the clients and professions have adapted to change in the procurement of buildings and engineering projects.

Key features of the new edition include:

- A revised chapter covering the concept of value for money in line with the greater emphasis on added value throughout the industry today.
- A new chapter covering developments in information technology applications (building information modelling, blockchains, data analytics, smart contracts and others) and construction procurement.
- Deeper coverage of the strategies that need to be considered in respect of contract selection.
- Improved discussion of sustainability and the increasing importance of resilience in the built environment.
- Concise descriptions of some the more important construction case laws.

March 2018: 246 x 174 mm: 458 pp
Pb: 978-1-138-69393-7: £50.99

To Order: Tel: +44 (0) 1235 400524 Fax: +44 (0) 1235 400525
or Post: Taylor and Francis Customer Services,
Bookpoint Ltd, Unit T1, 200 Milton Park, Abingdon, Oxon, OX14 4TA UK
Email: book.orders@tandf.co.uk

For a complete listing of all our titles visit:
www.tandf.co.uk

Taylor & Francis
Taylor & Francis Group

Material Costs/Measured Work Prices

DIRECTIONS

The following explanations are given for each of the column headings and letter codes.

Unit	Prices for each unit are given as singular (i.e. 1 metre, 1 nr) unless stated otherwise.
Net price	Industry tender prices, plus nominal allowance for fixings, waste and applicable trade discounts.
Material cost	Net price plus percentage allowance for overheads (7%), profit (5%) and preliminaries (13%).
Labour norms	In man-hours for each operation.
Labour cost	Labour constant multiplied by the appropriate all-in man-hour cost based on gang rate (See also relevant Rates of Wages Section) plus percentage allowance for overheads, profit and preliminaries.
Measured work Price (total rate)	Material cost plus Labour cost.

MATERIAL COSTS

The Material Costs given are based at Second Quarter 2020 but exclude any charges in respect of VAT. The average rate of copper during this quarter is US$5,179/UK £4,096 per tonne. Users of the book are advised to register on the Spon's website www.pricebooks.co.uk/updates to receive the free quarterly updates – alerts will then be provided by e-mail as changes arise.

MEASURED WORK PRICES

These prices are intended to apply to new work in the London area. The prices are for reasonable quantities of work and the user should make suitable adjustments if the quantities are especially small or especially large. Adjustments may also be required for locality (e.g. outside London – refer to cost indices in Approximate Estimating section for details of adjustment factors) and for the market conditions (e.g. volume of work secured or being tendered) at the time of use.

ELECTRICAL INSTALLATIONS

The labour rate has been based on average gang rates per man-hour effective from 6th January 2020 including allowances for all other emoluments and expenses. Future changes will be published in the free Spon's quarterly update by registering on their website. To this rate, has been added 13% and 7% to cover site and head office overheads and preliminary items together with a further 5% for profit, resulting in an inclusive rate of £38.64 per man hour. The rate has been calculated on a working year of 2,016 hours; a detailed build-up of the rate is given at the end of these directions.

In calculating the 'Measured Work Prices' the following assumptions have been made:

DIRECTIONS

(a)   That the work is carried out as a subcontract under the Standard Form of Building Contract.

(b)   That, unless otherwise stated, the work is being carried out in open areas at a height which would not require more than simple scaffolding.

(c)   That the building in which the work is being carried out is no more than six storeys high.

Where these assumptions are not valid, as for example where work is carried out in ducts and similar confined spaces or in multi-storey structures when additional time is needed to get to and from upper floors, then an appropriate adjustment must be made to the prices. Such adjustment will normally be to the labour element only.

DIRECTIONS

LABOUR RATE – ELECTRICAL

The annual cost of a notional eleven man gang

	TECHNICIAN 1 NR	APPROVED ELECTRICIANS 4 NR	ELECTRICIANS 4 NR	LABOURERS 2 NR	SUB-TOTALS
Hourly Rate from 6th January 2020	21.77	19.34	17.84	14.32	
Working hours per annum per man	1,680.00	1,680.00	1,680.00	1,680.00	
x Hourly rate × nr of men = £ per annum	36,573.60	129,964.80	119,884.80	48,115.20	334,538.40
Overtime Rate	32.66	29.01	26.76	21.48	
Overtime hours per annum per man	336.00	336.00	336.00	336.00	
x Hourly rate × nr of men = £ per annum	10,972.08	38,989.44	35,965.44	14,434.56	100,361.52
Total	47,545.68	168,954.24	155,850.24	62,549.76	434,899.92
Incentive schemes (insert percentage) 0.00%	0.00	0.00	0.00	0.00	0.00
Daily Travel Time Allowance (15–20 miles each way) effective from 2 January 2020	7.04	7.04	7.04	7.04	
Days per annum per man	224.00	224.00	224.00	224.00	
x nr of men = £ per annum	1,576.96	6,307.84	6,307.84	3,153.92	17,346.56
Daily Travel Fare Allowance (15–20 miles each way) effective from 7 January 2019	8.80	8.80	8.80	8.80	
Days per annum per man	224.00	224.00	224.00	224.00	
x nr of men = £ per annum	1,971.20	7,884.80	7,884.80	3,942.40	21,683.20
JIB Pension Scheme @ 3.0%	1,647.49	6,052.92	5,605.40	2,277.60	15,630.41
JIB combined benefits scheme (nr of weeks per man)	52.00	52.00	52.00	52.00	
Benefit Credit effective from 6 January 2020	74.61	67.71	63.27	53.31	
x nr of men = £ per annum	3,879.72	14,083.68	13,160.16	5,544.24	36,667.80
Holiday Top-up Funding	66.93	59.70	55.43	44.93	
x nr of men @ 7.5 hrs per day = £ per annum	3,480.26	12,418.38	11,528.40	4,672.20	32,099.24
National Insurance Contributions:					
Annual gross pay (subject to NI) each	58,453.82	209,648.94	194,731.44	79,862.52	
% of NI Contributions	13.80	13.80	13.80	13.80	
£ Contributions/annum	6,687.87	23,416.52	21,357.91	8,263.51	59,725.81

SUB-TOTAL		618,052.94
TRAINING (INCLUDING ANY TRADE REGISTRATIONS) – SAY	1.00%	6,180.53
SEVERANCE PAY AND SUNDRY COSTS – SAY	1.50%	9,363.50
EMPLOYER'S LIABILITY AND THIRD PARTY INSURANCE – SAY	2.00%	12,671.94
ANNUAL COST OF NOTIONAL GANG		646,268.91
MEN ACTUALLY WORKING = 10.5 THEREFORE ANNUAL COST PER PRODUCTIVE MAN		61,549.42
AVERAGE NR OF HOURS WORKED PER MAN = 2016		
THEREFORE ALL-IN MAN HOUR		30.53
PRELIMINARY ITEMS – SAY	13%	3.97
SITE AND HEAD OFFICE OVERHEADS AND PROFIT (7% & 5% RESPECTIVELY) – SAY	12%	4.14
THEREFORE INCLUSIVE MAN HOUR		38.64

DIRECTIONS

Notes:

(1)   Hourly wage rates are those effective from 6th January 2020.
(2)   The following assumptions have been made in the above calculations:
        (a)   Hourly rates are based on London rate and job reporting own transport.
        (b)   The working week of 37.5 hours is made up of 7.5 hours Monday to Friday.
        (c)   Five days in the year are lost through sickness or similar reason.
        (d)   A working year of 2,016 hours.
(3)   The incentive scheme addition of 0% is intended to reflect bonus schemes typically in use.
(4)   National insurance contributions are those effective from 1st April 2020.
(5)   Weekly JIB Combined Benefit Credit Scheme are those effective from 6 January 2020 – Contact JOB for ECI (0330 221 0240).
(6)   Overtime is paid after 37.5 hours.

ELECTRICAL SUPPLY/POWER/LIGHTING

Item	Net Price £	Material £	Labour hours	Labour £	Unit	Total rate £
ELECTRICAL GENERATION PLANT						
STANDBY GENERATORS						
Standby diesel generating sets; supply and installation; fixing to base; all supports and fixings; all necessary connections to equipment						
Three phase, 400 V, four wire 50 Hz packaged standby diesel generating set, complete with radio and television suppressors, daily service fuel tank and associated piping, 4 metres of exhaust pipe and primary exhaust silencer, control panel, mains failure relay, starting battery with charger, all internal wiring, interconnections, earthing and labels. Rated for standby duty; including UK delivery, installation and commissioning						
60 kVA	19069.53	24134.40	100.00	3863.88	nr	27998.28
100 kVA	26733.54	33833.97	100.00	3863.88	nr	37697.85
150 kVA	30574.13	38694.62	100.00	3863.88	nr	42558.50
315 kVA	55123.36	69764.13	120.00	4636.65	nr	74400.78
500 kVA	64973.45	82230.40	120.00	4636.65	nr	86867.05
750 kVA	110568.26	139935.19	140.00	5409.43	nr	145344.62
1000 kVA	146954.68	185985.84	140.00	5409.43	nr	191395.27
1500 kVA	172951.46	218887.37	170.00	6568.59	nr	225455.96
2000 kVA	266409.93	337168.41	170.00	6568.59	nr	343737.00
2500 kVA	364751.17	461629.08	210.00	8114.14	nr	469743.22
Extra for residential silencer; peformance 75 dBA at 1 m; including connection to exhaust pipe						
60 kVA	1023.85	1295.78	10.00	386.39	nr	1682.17
100 kVA	1223.40	1548.33	10.00	386.39	nr	1934.72
150 kVA	1402.99	1775.63	10.00	386.39	nr	2162.02
315 kVA	2729.23	3454.11	15.00	579.58	nr	4033.69
500 kVA	3490.60	4417.71	15.00	579.58	nr	4997.29
750 kVA	5200.56	6581.83	20.00	772.78	nr	7354.61
1000 kVA	7004.21	8864.53	20.00	772.78	nr	9637.31
1500 kVA	11157.90	14121.44	20.00	772.78	nr	14894.22
2000 kVA	12085.05	15294.84	30.00	1159.17	nr	16454.01
2500 kVA	13818.06	17488.14	30.00	1159.17	nr	18647.31
Synchronization panel for paralleling generators – not generators to mains; including interconnecting cables; commissioning and testing; fixing to backgrounds						
2 × 60 kVA	8528.45	10793.61	80.00	3091.10	nr	13884.71
2 × 100 kVA	9168.55	11603.72	80.00	3091.10	nr	14694.82
2 × 150 kVA	11533.97	14597.40	80.00	3091.10	nr	17688.50

ELECTRICAL SUPPLY/POWER/LIGHTING

Item	Net Price £	Material £	Labour hours	Labour £	Unit	Total rate £
ELECTRICAL GENERATION PLANT – cont						
Standby diesel generating sets – cont						
2 × 315 kVA	19608.08	24815.99	80.00	3091.10	nr	**27907.09**
2 × 500 kVA	28308.46	35827.19	80.00	3091.10	nr	**38918.29**
2 × 750 kVA	31944.86	40429.41	80.00	3091.10	nr	**43520.51**
2 × 1000 kVA	34428.49	43572.69	120.00	4636.65	nr	**48209.34**
2 × 1500 kVA	40757.32	51582.46	120.00	4636.65	nr	**56219.11**
2 × 2000 kVA	44160.40	55889.40	120.00	4636.65	nr	**60526.05**
2 × 2500 kVA	51483.89	65158.02	120.00	4636.65	nr	**69794.67**
Prefabricated drop-over acoustic housing; performance 85 dBA at 1 m over the range from 60 kVA to 315 kVA, 75 dBA from 500 kVA to 2500 kVA						
100 kVA	3811.39	4823.69	7.00	270.47	nr	**5094.16**
150 kVA	6189.12	7832.96	15.00	579.58	nr	**8412.54**
315 kVA	13205.58	16712.99	25.00	965.97	nr	**17678.96**
500 kVA	21810.80	27603.74	40.00	1545.56	nr	**29149.30**
750 kVA	34371.71	43500.83	40.00	1545.56	nr	**45046.39**
1000 kVA	43485.00	55034.62	40.00	1545.56	nr	**56580.18**
1500 kVA	43485.00	55034.62	40.00	1545.56	nr	**56580.18**
2000 kVA	86256.21	109165.86	60.00	2318.32	nr	**111484.18**
2500 kVA	103678.45	131215.45	70.00	2704.71	nr	**133920.16**
COMBINED HEAT AND POWER (CHP) UNITS						
Gas fired engine; skid-mounted with acoustic enclosure complete with exhaust fan and attenuators; exhaust gas attenuation; includes 6 m long pipe connections; dry air cooler for secondary water circuit to reject excess heat; controls and panel; commissioning						
Electrical output; Heat output						
12 kW; 30 kW	46002.00	58220.13	–	–	nr	**58220.13**
15 kW; 34 kW	47684.00	60348.87	–	–	nr	**60348.87**
20 kW; 43 kW	49477.00	62618.09	–	–	nr	**62618.09**
43 kW; 65 kW	90094.00	114022.97	–	–	nr	**114022.97**
50 kW; 81 kW	92414.00	116959.16	–	–	nr	**116959.16**
70 kW; 114 kW	101861.00	128915.28	–	–	nr	**128915.28**
104 kW, 142 kW	133905.00	169470.17	–	–	nr	**169470.17**
133 kW, 193 kW	134678.00	170448.48	–	–	nr	**170448.48**
210 kW, 253 kW	184842.00	233936.04	–	–	nr	**233936.04**
263 kW, 375 kW	211968.00	268266.70	–	–	nr	**268266.70**
356 kW, 426 kW	251857.00	318750.22	–	–	nr	**318750.22**
434 kW; 516 kW	278356.00	352287.35	–	–	nr	**352287.35**
532 kW, 665 kW	296165.00	374826.42	–	–	nr	**374826.42**
Heat dump; emergency cooling						
12 kW; 30 kW	3717.00	4704.24	–	–	nr	**4704.24**
15 kW; 34 kW	3713.00	4699.17	–	–	nr	**4699.17**
20 kW; 43 kW	3717.00	4704.24	–	–	nr	**4704.24**

ELECTRICAL SUPPLY/POWER/LIGHTING

Item	Net Price £	Material £	Labour hours	Labour £	Unit	Total rate £
43 kW; 65 kW	3868.00	4895.34	–	–	nr	**4895.34**
50 kW; 81 kW	4162.00	5267.43	–	–	nr	**5267.43**
70 kW; 114 kW	5258.00	6654.52	–	–	nr	**6654.52**
104 kW, 142 kW	5809.00	7351.87	–	–	nr	**7351.87**
133 kW, 193 kW	6903.00	8736.44	–	–	nr	**8736.44**
210 kW, 253 kW	9894.00	12521.85	–	–	nr	**12521.85**
263 kW, 375 kW	14660.00	18553.70	–	–	nr	**18553.70**
356 kW, 426 kW	16299.00	20628.01	–	–	nr	**20628.01**
434 kW; 516 kW	18599.00	23538.89	–	–	nr	**23538.89**
532 kW, 665 kW	20579.00	26044.78	–	–	nr	**26044.78**
Note: The costs detailed are based on a specialist subcontract package, as part of the M&E contract works, and include installation						
Upgraded catalytic converter						
43 kW; 65 kW	1194.00	1511.13	–	–	nr	**1511.13**
50 kW; 81 kW	1194.00	1511.13	–	–	nr	**1511.13**
70 kW; 114 kW	2614.00	3308.28	–	–	nr	**3308.28**
133 kW, 193 kW	3544.00	4485.29	–	–	nr	**4485.29**
263 kW, 375 kW	6314.00	7991.00	–	–	nr	**7991.00**
Flexible connections						
12 kW; 30 kW	503.00	636.60	–	–	nr	**636.60**
15 kW; 34 kW	503.00	636.60	–	–	nr	**636.60**
20 kW; 43 kW	503.00	636.60	–	–	nr	**636.60**
43 kW; 65 kW	677.00	856.81	–	–	nr	**856.81**
50 kW; 81 kW	677.00	856.81	–	–	nr	**856.81**
70 kW; 114 kW	868.00	1098.54	–	–	nr	**1098.54**
104 kW, 142 kW	1169.00	1479.49	–	–	nr	**1479.49**
133 kW, 193 kW	1229.00	1555.42	–	–	nr	**1555.42**
210 kW, 253 kW	1706.00	2159.11	–	–	nr	**2159.11**
263 kW, 375 kW	1655.00	2094.57	–	–	nr	**2094.57**
356 kW, 426 kW	1867.00	2362.88	–	–	nr	**2362.88**
434 kW; 516 kW	2260.00	2860.26	–	–	nr	**2860.26**
532 kW, 665 kW	2407.00	3046.30	–	–	nr	**3046.30**

ELECTRICAL SUPPLY/POWER/LIGHTING

Item	Net Price £	Material £	Labour hours	Labour £	Unit	Total rate £
HV SUPPLY						
Cable; 6350/11000 volts, 3 core, XLPE;						
stranded copper conductors; steel wire						
armoured; LSOH to BS 7835						
Laid in trench/duct including marker tape						
(cable tiles measured elsewhere)						
95 mm²	26.83	33.96	0.23	8.88	m	42.84
120 mm²	35.50	44.93	0.23	8.88	m	53.81
150 mm²	40.00	50.62	0.25	9.65	m	60.27
185 mm²	47.19	59.72	0.25	9.65	m	69.37
240 mm²	65.16	82.47	0.27	10.43	m	92.90
300 mm²	77.99	98.71	0.29	11.20	m	109.91
Pile tape; ES 1–12–23; 1 m	38.70	48.98	0.01	0.39	m	49.37
Clipped direct to backgrounds including cleats						
95 mm²	27.29	34.54	0.47	18.17	m	52.71
120 mm²	36.12	45.72	0.50	19.33	m	65.05
150 mm²	40.68	51.49	0.53	20.47	m	71.96
185 mm²	48.00	60.75	0.55	21.25	m	82.00
240 mm²	66.28	83.89	0.60	23.18	m	107.07
300 mm²	79.33	100.40	0.68	26.28	m	126.68
Terminations for above cables, including						
heat-shrink kit and glanding off						
95 × 3 Core XLPE SWA LSF 11 kV						
Termination	738.58	934.75	4.75	183.53	Nr	1118.28
120 × 3 Core XLPE SWA LSF 11 kV						
Termination	771.20	976.04	5.00	193.19	Nr	1169.23
150 × 3 Core XLPE SWA LSF 11 kV						
Termination	805.23	1019.10	6.00	231.83	Nr	1250.93
185 × 3 Core XLPE SWA LSF 11 kV						
Termination	817.29	1034.36	6.90	266.62	Nr	1300.98
240 × 3 Core XLPE SWA LSF 11 kV						
Termination	852.16	1078.49	7.50	289.79	Nr	1368.28
300 × 3 Core XLPE SWA LSF 11 kV						
Termination	911.20	1153.22	8.75	338.09	Nr	1491.31
Cable tiles; single width; laid in trench above						
cables on prepared sand bed (cost of						
excavation excluded); reinforced concrete						
covers; concave/convex ends						
914 × 152 × 63/38 mm	38.70	48.98	0.11	4.26	m	53.24
914 × 229 × 63/38 mm	41.08	51.99	0.11	4.26	m	56.25
914 × 305 × 63/38 mm	43.46	55.00	0.11	4.26	m	59.26

ELECTRICAL SUPPLY/POWER/LIGHTING

Item	Net Price £	Material £	Labour hours	Labour £	Unit	Total rate £
HV SWITCHGEAR AND TRANSFORMERS						
HV circuit breakers; installed on prepared foundations including all supports, fixings and interpanel connections where relevant. Excludes main and multicore cabling and heat shrink cable termination kits. Three phase 11 kV, 630 amp, Air or SF6 insulated, with fixed pattern vacuum or SF6 circuit breaker panels; hand charged spring closing operation; prospective fault level up to 25 kA for 3 seconds. Feeders include ammeter with selector switch, VIP relays, overcurrent and earth fault relays with necessary current relays with necessary current transformers; incomers include 3 phase VT, voltmeter and phase selector switch; Includes IDMT overcurrent and earth fault relays/CTs.						
Single panel with cable chamber	26820.00	33943.39	31.70	1224.84	nr	**35168.23**
Three panel with one incomer and two feeders; with cable chambers	49555.00	62716.81	67.83	2620.87	nr	**65337.68**
Five panel with two incoming, two feeders and a bus section; with cable chambers	88775.00	112353.64	99.17	3831.81	nr	**116185.45**
Tripping Batteries						
Battery chargers; switchgear tripping and closing; double wound transfomer and earth screen; including fixing to background, commissioning and testing Valve regulated lead acid battery NGTS 3.12.2; BS 6290 Part 2; TPS 9/3; IEEE485						
30 volt; 19 Ah; 3 A	3172.52	4015.14	6.50	251.15	nr	**4266.29**
40 volt; 29 Ah; 3 A	5082.62	6432.56	8.50	328.42	nr	**6760.98**
110 volt; 19 Ah; 3 A	5184.78	6561.86	6.50	251.15	nr	**6813.01**
110 volt; 29 Ah; 3 A	5657.20	7159.76	8.50	328.42	nr	**7488.18**
100 volt; 38 Ah; 3 A	6045.24	7650.85	10.00	386.39	nr	**8037.24**
Step down transformers; 11/0.415 kV, Dyn 11, 50 Hz. Complete with lifting lugs, mounting skids, provisions for wheels, undrilled gland plates to air-filled cable boxes, off load tapping facility, including UK delivery Oil-filled in free breathing ventilated steel tank						
500 kVA	11720.26	14833.16	30.00	1159.17	nr	**15992.33**
800 kVA	13347.53	16892.64	30.00	1159.17	nr	**18051.81**
1000 kVA	15127.69	19145.60	30.00	1159.17	nr	**20304.77**
1250 kVA	18382.17	23264.47	35.00	1352.36	nr	**24616.83**
1500 kVA	21783.01	27568.58	35.00	1352.36	nr	**28920.94**
2000 kVA	28291.97	35806.32	35.00	1352.36	nr	**37158.68**

ELECTRICAL SUPPLY/POWER/LIGHTING

Item	Net Price £	Material £	Labour hours	Labour £	Unit	Total rate £
HV SUPPLY – cont						
Step down transformers – cont						
MIDEL – filled in gasket-sealed steel tank						
500 kVA	15523.65	19646.73	30.00	1159.17	nr	**20805.90**
800 kVA	17693.32	22392.66	30.00	1159.17	nr	**23551.83**
1000 kVA	20144.99	25495.50	30.00	1159.17	nr	**26654.67**
1250 kVA	24348.67	30815.68	35.00	1352.36	nr	**32168.04**
1500 kVA	28969.98	36664.41	35.00	1352.36	nr	**38016.77**
2000 kVA	37512.97	47476.42	40.00	1545.56	nr	**49021.98**
Extra for						
Fluid temperature indicator with 2 N/O contacts	435.48	551.14	2.00	77.28	nr	**628.42**
Winding temperature indicator with 2 N/O contacts	1066.94	1350.32	2.00	77.28	nr	**1427.60**
Dehydrating breather	108.87	137.78	2.00	77.28	nr	**215.06**
Plain rollers	326.61	413.36	2.00	77.28	nr	**490.64**
Pressure relief device with 1 N/O contact	653.21	826.71	2.00	77.28	nr	**903.99**
Step down transformers; 11/0.415 kV, Dyn 11, 50 Hz. Complete with lifting lugs, mounting skids, provisions for wheels, undrilled gland plates to air-filled cable boxes, off load tapping facility, including UK delivery						
Cast resin type in ventilated steel encloure, AN – Air Natural including winding temperture indicator with 2 N/O contacts						
500 kVA	16608.47	21019.68	40.00	1545.56	nr	**22565.24**
800 kVA	19320.53	24452.06	40.00	1545.56	nr	**25997.62**
1000 kVA	23372.34	29580.03	40.00	1545.56	nr	**31125.59**
1250 kVA	25297.89	32017.01	45.00	1738.74	nr	**33755.75**
1600 kVA	28698.79	36321.19	45.00	1738.74	nr	**38059.93**
2000 kVA	32495.70	41126.56	50.00	1931.93	nr	**43058.49**
Cast Resin type in ventilated steel enclosure with temperature controlled fans to achieve 40% increase to AN/AF rating. Includes winding temperature indicator with 2 N/O contacts						
500/700 kVA	18235.69	23079.09	42.00	1622.82	nr	**24701.91**
800/1120 kVA	21218.98	26854.74	42.00	1622.82	nr	**28477.56**
1000/1400 kVA	26253.61	33226.57	42.00	1622.82	nr	**34849.39**
12501750 kVA	27338.42	34599.50	47.00	1816.02	nr	**36415.52**
1600/2240 kVA	30885.74	39089.00	47.00	1816.02	nr	**40905.02**
2000/2800 kVA	35225.03	44580.79	52.00	2009.21	nr	**46590.00**

ELECTRICAL SUPPLY/POWER/LIGHTING

Item	Net Price £	Material £	Labour hours	Labour £	Unit	Total rate £
LV DISTRIBUTION: CONDUIT AND CABLE TRUNKING						
Heavy gauge, screwed drawn steel; surface fixed on saddles to backgrounds, with standard pattern boxes and fittings including all fixings and supports (forming holes, conduit entry, draw wires etc. and components for earth continuity are included)						
Black enamelled						
20 mm dia.	3.73	4.72	0.49	18.93	m	**23.65**
25 mm dia.	5.07	6.42	0.56	21.64	m	**28.06**
32 mm dia.	10.88	13.76	0.64	24.73	m	**38.49**
38 mm dia.	14.10	17.84	0.73	28.21	m	**46.05**
50 mm dia.	26.73	33.82	1.04	40.19	m	**74.01**
Galvanized						
20 mm dia.	3.45	4.37	0.49	18.93	m	**23.30**
25 mm dia.	4.70	5.95	0.56	21.64	m	**27.59**
32 mm dia.	10.07	12.75	0.64	24.73	m	**37.48**
38 mm dia.	13.04	16.51	0.73	28.21	m	**44.72**
50 mm dia.	24.75	31.33	1.04	40.19	m	**71.52**
High impact PVC; surface fixed on saddles to backgrounds; with standard pattern boxes and fittings; including all fixings and supports						
Light gauge						
16 mm dia.	0.72	0.91	0.27	10.43	m	**11.34**
20 mm dia.	0.81	1.03	0.28	10.82	m	**11.85**
25 mm dia.	1.11	1.40	0.33	12.75	m	**14.15**
32 mm dia.	1.77	2.24	0.38	14.68	m	**16.92**
38 mm dia.	2.35	2.98	0.44	17.00	m	**19.98**
50 mm dia.	3.88	4.91	0.48	18.54	m	**23.45**
Heavy gauge						
16 mm dia.	1.03	1.30	0.27	10.43	m	**11.73**
20 mm dia.	1.17	1.48	0.28	10.82	m	**12.30**
25 mm dia.	1.60	2.03	0.33	12.75	m	**14.78**
32 mm dia.	2.54	3.21	0.38	14.68	m	**17.89**
38 mm dia.	3.36	4.26	0.44	17.00	m	**21.26**
50 mm dia.	5.56	7.03	0.48	18.54	m	**25.57**
Flexible conduits; including adaptors and locknuts (for connections to equipment)						
Metallic, PVC covered conduit; not exceeding 1 m long; including zinc plated mild steel adaptors, lock nuts and earth conductor						
16 mm dia.	14.24	18.02	0.42	16.23	nr	**34.25**
20 mm dia.	15.10	19.11	0.46	17.77	nr	**36.88**
25 mm dia.	22.39	28.34	0.43	16.62	nr	**44.96**
32 mm dia.	34.67	43.88	0.51	19.70	nr	**63.58**
38 mm dia.	43.79	55.42	0.56	21.64	nr	**77.06**
50 mm dia.	151.94	192.29	0.82	31.67	nr	**223.96**

ELECTRICAL SUPPLY/POWER/LIGHTING

Item	Net Price £	Material £	Labour hours	Labour £	Unit	Total rate £
LV DISTRIBUTION: CONDUIT AND CABLE TRUNKING – cont						
Flexible conduits – cont						
PVC conduit; not exceeding 1 m long; including nylon adaptors, lock nuts						
16 mm dia.	5.55	7.02	0.46	17.77	nr	**24.79**
20 mm dia.	5.55	7.02	0.48	18.54	nr	**25.56**
25 mm dia.	6.97	8.83	0.50	19.33	nr	**28.16**
32 mm.dia.	10.11	12.79	0.58	22.41	nr	**35.20**
PVC adaptable boxes; fixed to backgrounds; including all supports and fixings (cutting and connecting conduit to boxes is included)						
Square pattern						
75 × 75 × 53 mm	2.98	3.77	0.69	26.67	nr	**30.44**
100 × 100 × 75 mm	5.06	6.41	0.71	27.44	nr	**33.85**
150 × 150 × 75 mm	6.47	8.19	0.80	30.90	nr	**39.09**
Terminal strips to be fixed in metal or polythene adaptable boxes)						
20 amp high density polythene						
2 way	1.25	1.58	0.23	8.88	nr	**10.46**
3 way	1.56	1.97	0.23	8.88	nr	**10.85**
4 way	2.09	2.64	0.23	8.88	nr	**11.52**
5 way	2.60	3.29	0.23	8.88	nr	**12.17**
6 way	3.12	3.95	0.25	9.65	nr	**13.60**
7 way	3.65	4.61	0.25	9.65	nr	**14.26**
8 way	4.15	5.25	0.29	11.20	nr	**16.45**
9 way	4.68	5.92	0.30	11.59	nr	**17.51**
10 way	5.21	6.60	0.34	13.14	nr	**19.74**
11 way	5.72	7.24	0.34	13.14	nr	**20.38**
13 way	6.75	8.55	0.37	14.30	nr	**22.85**
14 way	7.27	9.21	0.37	14.30	nr	**23.51**
15 way	7.80	9.87	0.39	15.08	nr	**24.95**
16 way	8.31	10.52	0.45	17.39	nr	**27.91**
18 way	9.36	11.85	0.45	17.39	nr	**29.24**

ELECTRICAL SUPPLY/POWER/LIGHTING

Item	Net Price £	Material £	Labour hours	Labour £	Unit	Total rate £
LV DISTRIBUTION: TRUNKING						
Galvanized steel trunking; fixed to backgrounds; jointed with standard connectors (including plates for air gap between trunking and background); earth continuity straps included						
Single compartment						
50 × 50 mm	3.98	5.04	0.39	15.08	m	20.12
75 × 50 mm	5.47	6.92	0.44	17.00	m	23.92
75 × 75 mm	6.35	8.04	0.47	18.17	m	26.21
100 × 50 mm	6.79	8.59	0.50	19.33	m	27.92
100 × 75 mm	7.34	9.28	0.57	22.02	m	31.30
100 × 100 mm	7.45	9.43	0.62	23.96	m	33.39
150 × 50 mm	10.99	13.91	0.78	30.14	m	44.05
150 × 100 mm	12.52	15.85	0.78	30.14	m	45.99
150 × 150 mm	13.37	16.92	0.86	33.23	m	50.15
225 × 75 mm	17.68	22.38	0.88	34.00	m	56.38
225 × 150 mm	22.10	27.97	0.84	32.46	m	60.43
225 × 225 mm	22.60	28.60	0.99	38.25	m	66.85
300 × 75 mm	18.18	23.00	0.96	37.09	m	60.09
300 × 100 mm	20.93	26.49	0.99	38.25	m	64.74
300 × 150 mm	16.02	20.27	0.99	38.25	m	58.52
300 × 225 mm	13.63	17.25	1.09	42.12	m	59.37
300 × 300 mm	31.90	40.38	1.16	44.81	m	85.19
Double compartment						
100 × 50 mm	5.21	6.60	0.54	20.87	m	27.47
100 × 75 mm	5.62	7.11	0.62	23.96	m	31.07
100 × 100 mm	10.66	13.50	0.66	25.50	m	39.00
150 × 50 mm	7.92	10.02	0.70	27.05	m	37.07
150 × 100 mm	9.81	12.42	0.83	32.07	m	44.49
150 × 150 mm	18.25	23.09	0.92	35.55	m	58.64
Triple compartment						
150 × 50 mm	9.12	11.55	0.79	30.53	m	42.08
150 × 100 mm	9.36	11.85	0.78	30.14	m	41.99
150 × 150 mm	19.35	24.49	1.01	39.03	m	63.52
Galvanized steel trunking fittings; cutting and jointing trunking to fittings is included						
Stop end						
50 × 50 mm	1.38	1.75	0.19	7.34	nr	9.09
75 × 50 mm	1.46	1.85	0.20	7.73	nr	9.58
75 × 75 mm	1.46	1.85	0.21	8.11	nr	9.96
100 × 50 mm	1.52	1.93	0.31	11.97	nr	13.90
100 × 75 mm	1.64	2.07	0.27	10.43	nr	12.50
100 × 100 mm	1.61	2.04	0.27	10.43	nr	12.47
150 × 50 mm	1.79	2.26	0.28	10.82	nr	13.08
150 × 100 mm	1.95	2.46	0.30	11.59	nr	14.05
150 × 150 mm	2.01	2.54	0.32	12.36	nr	14.90
225 × 75 mm	2.07	2.62	0.35	13.53	nr	16.15
225 × 150 mm	2.30	2.91	0.37	14.30	nr	17.21
225 × 225 mm	3.58	4.54	0.38	14.68	nr	19.22

ELECTRICAL SUPPLY/POWER/LIGHTING

Item	Net Price £	Material £	Labour hours	Labour £	Unit	Total rate £
LV DISTRIBUTION: TRUNKING – cont						
Galvanized steel trunking fittings – cont						
Stop end – cont						
300 × 75 mm	3.60	4.56	0.42	16.23	nr	20.79
300 × 100 mm	3.08	3.90	0.42	16.23	nr	20.13
300 × 150 mm	3.60	4.56	0.43	16.62	nr	21.18
300 × 225 mm	4.32	5.47	0.45	17.39	nr	22.86
300 × 300 mm	4.62	5.85	0.48	18.54	nr	24.39
Flanged connector						
50 × 50 mm	1.46	1.85	0.19	7.34	nr	9.19
75 × 50 mm	2.24	2.83	0.20	7.73	nr	10.56
75 × 75 mm	1.81	2.30	0.21	8.11	nr	10.41
100 × 50 mm	2.33	2.95	0.26	10.05	nr	13.00
100 × 75 mm	2.40	3.04	0.27	10.43	nr	13.47
100 × 100 mm	2.33	2.95	0.27	10.43	nr	13.38
150 × 50 mm	2.35	2.98	0.28	10.82	nr	13.80
150 × 100 mm	2.51	3.18	0.30	11.59	nr	14.77
150 × 150 mm	2.55	3.23	0.32	12.36	nr	15.59
225 × 75 mm	2.01	2.54	0.35	13.53	nr	16.07
225 × 150 mm	2.46	3.11	0.37	14.30	nr	17.41
225 × 225 mm	3.61	4.57	0.38	14.68	nr	19.25
300 × 75 mm	3.02	3.82	0.42	16.23	nr	20.05
300 × 100 mm	3.49	4.41	0.42	16.23	nr	20.64
300 × 150 mm	3.95	5.00	0.43	16.62	nr	21.62
300 × 225 mm	3.95	5.00	0.45	17.39	nr	22.39
300 × 300 mm	4.53	5.73	0.48	18.54	nr	24.27
Bends 90°; single compartment						
50 × 50 mm	6.06	7.67	0.42	16.23	nr	23.90
75 × 50 mm	7.83	9.91	0.45	17.39	nr	27.30
75 × 75 mm	7.33	9.27	0.48	18.54	nr	27.81
100 × 50 mm	8.00	10.12	0.53	20.47	nr	30.59
100 × 75 mm	8.20	10.38	0.56	21.64	nr	32.02
100 × 100 mm	8.20	10.38	0.58	22.41	nr	32.79
150 × 50 mm	10.15	12.85	0.64	24.73	nr	37.58
150 × 100 mm	12.95	16.39	0.91	35.16	nr	51.55
150 × 150 mm	12.35	15.64	0.89	34.38	nr	50.02
225 × 75 mm	18.87	23.88	0.76	29.37	nr	53.25
225 × 150 mm	23.45	29.68	0.82	31.67	nr	61.35
225 × 225 mm	27.61	34.94	0.83	32.07	nr	67.01
300 × 75 mm	25.19	31.88	0.85	32.84	nr	64.72
300 × 100 mm	25.49	32.26	0.90	34.78	nr	67.04
300 × 150 mm	28.04	35.49	0.96	37.09	nr	72.58
300 × 225 mm	30.84	39.03	0.98	37.87	nr	76.90
300 × 300 mm	31.53	39.91	1.06	40.96	nr	80.87
Bends 90°; double compartment						
100 × 50 mm	7.28	9.22	0.53	20.47	nr	29.69
100 × 75 mm	7.67	9.71	0.56	21.64	nr	31.35
100 × 100 mm	12.52	15.85	0.58	22.41	nr	38.26
150 × 50 mm	8.00	10.12	0.65	25.11	nr	35.23
150 × 100 mm	8.43	10.67	0.69	26.67	nr	37.34
150 × 150 mm	21.05	26.64	0.73	28.21	nr	54.85

ELECTRICAL SUPPLY/POWER/LIGHTING

Item	Net Price £	Material £	Labour hours	Labour £	Unit	Total rate £
Bends 90°; triple compartment						
150 × 50 mm	10.52	13.32	0.68	26.28	nr	39.60
150 × 100 mm	14.71	18.61	0.73	28.21	nr	46.82
150 × 150 mm	25.66	32.48	0.77	29.76	nr	62.24
Tees; single compartment						
50 × 50 mm	7.36	9.32	0.56	21.64	nr	30.96
75 × 50 mm	8.76	11.09	0.57	22.02	nr	33.11
75 × 75 mm	8.32	10.53	0.60	23.18	nr	33.71
100 × 50 mm	10.26	12.98	0.65	25.11	nr	38.09
100 × 75 mm	10.32	13.06	0.72	27.82	nr	40.88
100 × 100 mm	9.90	12.53	0.71	27.44	nr	39.97
150 × 50 mm	11.74	14.86	0.82	31.67	nr	46.53
150 × 100 mm	14.58	18.46	0.84	32.46	nr	50.92
150 × 150 mm	14.14	17.90	0.91	35.16	nr	53.06
225 × 75 mm	24.63	31.17	0.94	36.32	nr	67.49
225 × 150 mm	33.27	42.11	1.01	39.03	nr	81.14
225 × 225 mm	38.93	49.27	1.02	39.41	nr	88.68
300 × 75 mm	35.56	45.00	1.07	41.35	nr	86.35
300 × 100 mm	36.84	46.63	1.07	41.35	nr	87.98
300 × 150 mm	39.69	50.23	1.14	44.04	nr	94.27
300 × 225 mm	42.98	54.40	1.19	45.98	nr	100.38
300 × 300 mm	46.49	58.83	1.26	48.69	nr	107.52
Tees; double compartment						
100 × 50 mm	7.07	8.95	0.65	25.11	nr	34.06
100 × 75 mm	16.65	21.07	0.71	27.44	nr	48.51
100 × 100 mm	16.98	21.49	0.72	27.82	nr	49.31
150 × 50 mm	9.29	11.76	0.82	31.67	nr	43.43
150 × 100 mm	9.73	12.31	0.85	32.84	nr	45.15
150 × 150 mm	28.76	36.40	0.91	35.16	nr	71.56
Tees; triple compartment						
150 × 50 mm	10.63	13.45	0.87	33.61	nr	47.06
150 × 100 mm	11.68	14.78	0.89	34.38	nr	49.16
150 × 150 mm	34.07	43.12	0.96	37.09	nr	80.21
Crossovers; single compartment						
50 × 50 mm	9.34	11.82	0.65	25.11	nr	36.93
75 × 50 mm	12.66	16.03	0.66	25.50	nr	41.53
75 × 75 mm	12.95	16.39	0.69	26.67	nr	43.06
100 × 50 mm	16.00	20.25	0.74	28.59	nr	48.84
100 × 75 mm	11.79	14.92	0.80	30.90	nr	45.82
100 × 100 mm	16.13	20.42	0.81	31.29	nr	51.71
150 × 50 mm	19.10	24.17	0.91	35.16	nr	59.33
150 × 100 mm	19.38	24.53	0.94	36.32	nr	60.85
150 × 150 mm	18.57	23.50	0.99	38.25	nr	61.75
225 × 75 mm	34.01	43.04	1.01	39.03	nr	82.07
225 × 150 mm	37.12	46.98	1.08	41.73	nr	88.71
225 × 225 mm	50.50	63.92	1.09	42.12	nr	106.04
300 × 75 mm	47.55	60.18	1.14	44.04	nr	104.22
300 × 100 mm	48.98	61.99	1.16	44.81	nr	106.80
300 × 150 mm	52.88	66.92	1.19	45.98	nr	112.90
300 × 225 mm	55.95	70.81	1.21	46.75	nr	117.56
300 × 300 mm	59.65	75.49	1.29	49.84	nr	125.33

ELECTRICAL SUPPLY/POWER/LIGHTING

Item	Net Price £	Material £	Labour hours	Labour £	Unit	Total rate £
LV DISTRIBUTION: TRUNKING – cont						
Galvanized steel trunking fittings – cont						
Stop end – cont						
Crossovers; double compartment						
100 × 50 mm	9.39	11.88	0.74	28.59	nr	**40.47**
100 × 75 mm	9.85	12.47	0.80	30.90	nr	**43.37**
100 × 100 mm	19.11	24.18	0.81	31.29	nr	**55.47**
150 × 50 mm	11.12	14.08	0.86	33.23	nr	**47.31**
150 × 100 mm	11.16	14.12	0.94	36.32	nr	**50.44**
150 × 150 mm	32.47	41.09	1.00	38.64	nr	**79.73**
Crossovers; triple compartment						
150 × 50 mm	12.17	15.40	0.97	37.48	nr	**52.88**
150 × 100 mm	13.39	16.95	0.99	38.25	nr	**55.20**
150 × 150 mm	38.74	49.03	1.06	40.96	nr	**89.99**
Galvanized steel flush floor trunking; fixed to backgrounds; supports and fixings; standard coupling joints; earth continuity straps included						
Triple compartment						
350 × 60 mm	46.42	58.74	1.32	51.00	m	**109.74**
Four compartment						
350 × 60 mm	48.33	61.16	1.32	51.00	m	**112.16**
Galvanized steel flush floor trunking; fittings (cutting and jointing trunking to fittings is included)						
Stop end; triple compartment						
350 × 60 mm	4.87	6.16	0.53	20.47	nr	**26.63**
Stop end; four compartment						
350 × 60 mm	11.29	14.29	0.53	20.47	nr	**34.76**
Rising bend; standard; triple compartment						
350 × 60 mm	40.15	50.81	1.30	50.23	nr	**101.04**
Rising bend; standard; four compartment						
350 × 60 mm	42.07	53.24	1.30	50.23	nr	**103.47**
Rising bend; skirting; triple compartment						
350 × 60 mm	78.77	99.69	1.33	51.39	nr	**151.08**
Rising bend; skirting; four compartment						
350 × 60 mm	89.40	113.14	1.33	51.39	nr	**164.53**
Junction box; triple compartment						
350 × 60 mm	52.35	66.26	1.16	44.81	nr	**111.07**
Junction box; four compartment						
350 × 60 mm	54.45	68.91	1.16	44.81	nr	**113.72**
Body coupler (pair)						
3 and 4 compartment	2.70	3.42	0.16	6.17	nr	**9.59**
Service outlet module comprising flat lid with flanged carpet trim; twin 13 A outlet and drilled plate for mounting 2 telephone outlets; one blank plate; triple compartment						
3 compartment	64.94	82.19	0.47	18.17	nr	**100.36**

ELECTRICAL SUPPLY/POWER/LIGHTING

Item	Net Price £	Material £	Labour hours	Labour £	Unit	Total rate £
Service outlet module comprising flat lid with flanged carpet trim; twin 13 A outlet and drilled plate for mounting 2 telephone outlets; two blank plates; four compartment						
4 compartment	72.33	91.54	0.47	18.17	nr	**109.71**
Single compartment PVC trunking; grey finish; clip on lid; fixed to backgrounds; including supports and fixings (standard coupling joints)						
Dimensions						
50 × 50 mm	11.18	14.15	0.27	10.43	m	**24.58**
75 × 50 mm	12.12	15.34	0.28	10.82	m	**26.16**
75 × 75 mm	13.73	17.37	0.29	11.20	m	**28.57**
100 × 50 mm	15.55	19.68	0.34	13.14	m	**32.82**
100 × 75 mm	17.06	21.59	0.37	14.30	m	**35.89**
100 × 100 mm	18.01	22.79	0.37	14.30	m	**37.09**
150 × 50 mm	15.63	19.78	0.41	15.85	m	**35.63**
150 × 75 mm	27.86	35.26	0.44	17.00	m	**52.26**
150 × 100 mm	33.53	42.44	0.44	17.00	m	**59.44**
150 × 150 mm	34.30	43.41	0.48	18.54	m	**61.95**
Single compartment PVC trunking; fittings (cutting and jointing trunking to fittings is included)						
Crossover						
50 × 50 mm	18.32	23.18	0.29	11.20	nr	**34.38**
75 × 50 mm	20.31	25.70	0.30	11.59	nr	**37.29**
75 × 75 mm	22.04	27.90	0.31	11.97	nr	**39.87**
100 × 50 mm	29.38	37.18	0.35	13.53	nr	**50.71**
100 × 75 mm	37.06	46.91	0.36	13.91	nr	**60.82**
100 × 100 mm	32.78	41.48	0.40	15.46	nr	**56.94**
150 × 75 mm	41.99	53.14	0.45	17.39	nr	**70.53**
150 × 100 mm	50.37	63.75	0.46	17.77	nr	**81.52**
150 × 150 mm	78.19	98.95	0.47	18.17	nr	**117.12**
Stop end						
50 × 50 mm	0.76	0.96	0.12	4.64	nr	**5.60**
75 × 50 mm	1.43	1.81	0.12	4.64	nr	**6.45**
75 × 75 mm	1.09	1.38	0.13	5.03	nr	**6.41**
100 × 50 mm	1.95	2.46	0.16	6.17	nr	**8.63**
100 × 75 mm	3.05	3.86	0.16	6.17	nr	**10.03**
100 × 100 mm	3.06	3.88	0.18	6.96	nr	**10.84**
150 × 75 mm	7.97	10.09	0.20	7.73	nr	**17.82**
150 × 100 mm	9.87	12.49	0.21	8.11	nr	**20.60**
150 × 150 mm	10.07	12.75	0.22	8.50	nr	**21.25**
Flanged coupling						
50 × 50 mm	4.59	5.81	0.32	12.36	nr	**18.17**
75 × 50 mm	5.25	6.64	0.33	12.75	nr	**19.39**
75 × 75 mm	6.30	7.97	0.34	13.14	nr	**21.11**

ELECTRICAL SUPPLY/POWER/LIGHTING

Item	Net Price £	Material £	Labour hours	Labour £	Unit	Total rate £
LV DISTRIBUTION: TRUNKING – cont						
Single compartment PVC trunking – cont						
Flanged coupling – cont						
100 × 50 mm	7.09	8.97	0.44	17.00	nr	**25.97**
100 × 75 mm	8.03	10.16	0.45	17.39	nr	**27.55**
100 × 100 mm	8.55	10.82	0.46	17.77	nr	**28.59**
150 × 75 mm	9.06	11.47	0.57	22.02	nr	**33.49**
150 × 100 mm	9.55	12.08	0.57	22.02	nr	**34.10**
150 × 150 mm	10.02	12.68	0.59	22.79	nr	**35.47**
Internal coupling						
50 × 50 mm	1.62	2.05	0.07	2.71	nr	**4.76**
75 × 50 mm	1.92	2.43	0.07	2.71	nr	**5.14**
75 × 75 mm	1.93	2.44	0.07	2.71	nr	**5.15**
100 × 50 mm	2.58	3.27	0.08	3.09	nr	**6.36**
100 × 75 mm	2.89	3.66	0.08	3.09	nr	**6.75**
100 × 100 mm	3.19	4.03	0.08	3.09	nr	**7.12**
External coupling						
50 × 50 mm	1.77	2.24	0.09	3.48	nr	**5.72**
75 × 50 mm	2.12	2.69	0.09	3.48	nr	**6.17**
75 × 75 mm	2.10	2.65	0.09	3.48	nr	**6.13**
100 × 50 mm	2.82	3.57	0.10	3.86	nr	**7.43**
100 × 75 mm	3.21	4.07	0.10	3.86	nr	**7.93**
100 × 100 mm	3.50	4.44	0.10	3.86	nr	**8.30**
150 × 75 mm	4.01	5.07	0.11	4.26	nr	**9.33**
150 × 100 mm	4.18	5.29	0.11	4.26	nr	**9.55**
150 × 150 mm	4.32	5.47	0.11	4.26	nr	**9.73**
Angle; flat cover						
50 × 50 mm	6.50	8.22	0.18	6.96	nr	**15.18**
75 × 50 mm	8.66	10.96	0.19	7.34	nr	**18.30**
75 × 75 mm	9.96	12.60	0.20	7.73	nr	**20.33**
100 × 50 mm	14.88	18.83	0.23	8.88	nr	**27.71**
100 × 75 mm	22.66	28.68	0.26	10.05	nr	**38.73**
100 × 100 mm	20.67	26.16	0.26	10.05	nr	**36.21**
150 × 75 mm	28.17	35.65	0.30	11.59	nr	**47.24**
150 × 100 mm	33.51	42.41	0.33	12.75	nr	**55.16**
150 × 150 mm	50.78	64.27	0.34	13.14	nr	**77.41**
Angle; internal or external cover						
50 × 50 mm	7.91	10.01	0.18	6.96	nr	**16.97**
75 × 50 mm	10.90	13.80	0.19	7.34	nr	**21.14**
75 × 75 mm	14.00	17.72	0.20	7.73	nr	**25.45**
100 × 50 mm	15.08	19.08	0.23	8.88	nr	**27.96**
100 × 75 mm	24.25	30.69	0.26	10.05	nr	**40.74**
100 × 100 mm	24.36	30.83	0.26	10.05	nr	**40.88**
150 × 75 mm	29.78	37.69	0.30	11.59	nr	**49.28**
150 × 100 mm	35.10	44.42	0.33	12.75	nr	**57.17**
150 × 150 mm	49.30	62.40	0.34	13.14	nr	**75.54**
Tee; flat cover						
50 × 50 mm	5.44	6.89	0.24	9.27	nr	**16.16**
75 × 50 mm	8.30	10.51	0.25	9.65	nr	**20.16**
75 × 75 mm	8.53	10.80	0.26	10.05	nr	**20.85**

ELECTRICAL SUPPLY/POWER/LIGHTING

Item	Net Price £	Material £	Labour hours	Labour £	Unit	Total rate £
100 × 50 mm	18.43	23.33	0.32	12.36	nr	35.69
100 × 75 mm	19.58	24.79	0.33	12.75	nr	37.54
100 × 100 mm	25.55	32.33	0.34	13.14	nr	45.47
150 × 75 mm	33.79	42.76	0.41	15.85	nr	58.61
150 × 100 mm	43.34	54.85	0.42	16.23	nr	71.08
150 × 150 mm	58.96	74.61	0.44	17.00	nr	91.61
Tee; internal or external cover						
50 × 50 mm	15.42	19.51	0.24	9.27	nr	28.78
75 × 50 mm	16.94	21.44	0.25	9.65	nr	31.09
75 × 75 mm	18.84	23.84	0.26	10.05	nr	33.89
100 × 50 mm	24.18	30.60	0.32	12.36	nr	42.96
100 × 75 mm	27.62	34.96	0.33	12.75	nr	47.71
100 × 100 mm	30.99	39.22	0.34	13.14	nr	52.36
150 × 75 mm	40.15	50.81	0.41	15.85	nr	66.66
150 × 100 mm	48.39	61.24	0.42	16.23	nr	77.47
150 × 150 mm	64.61	81.77	0.44	17.00	nr	98.77
Division strip (1.8 m long)						
50 mm	8.36	10.58	0.07	2.71	nr	13.29
75 mm	10.71	13.55	0.07	2.71	nr	16.26
100 mm	13.61	17.23	0.08	3.09	nr	20.32
PVC miniature trunking; white finish; fixed to backgrounds; including supports and fixing; standard coupling joints						
Single compartment						
16 × 16 mm	1.51	1.92	0.20	7.73	m	9.65
25 × 16 mm	1.84	2.33	0.21	8.11	m	10.44
38 × 16 mm	2.30	2.91	0.24	9.27	m	12.18
38 × 25 mm	2.75	3.48	0.25	9.65	m	13.13
Compartmented						
38 × 16 mm	2.69	3.40	0.24	9.27	m	12.67
38 × 25 mm	3.21	4.07	0.25	9.65	m	13.72
PVC miniature trunking fittings; single compartment; white finish; cutting and jointing trunking to fittings is included						
Coupling						
16 × 16 mm	0.47	0.59	0.10	3.86	nr	4.45
25 × 16 mm	0.47	0.59	0.12	4.45	nr	5.04
38 × 16 mm	0.47	0.59	0.12	4.64	nr	5.23
38 × 25 mm	1.10	1.39	0.14	5.41	nr	6.80
Stop end						
16 × 16 mm	0.47	0.59	0.12	4.64	nr	5.23
25 × 16 mm	0.47	0.59	0.13	4.91	nr	5.50
38 × 16 mm	0.47	0.59	0.15	5.80	nr	6.39
38 × 25 mm	1.10	1.39	0.17	6.56	nr	7.95
Bend; flat, internal or external						
16 × 16 mm	0.47	0.59	0.18	6.96	nr	7.55
25 × 16 mm	0.47	0.59	0.20	7.65	nr	8.24
38 × 16 mm	0.47	0.59	0.21	8.11	nr	8.70
38 × 25 mm	1.10	1.39	0.23	8.88	nr	10.27

ELECTRICAL SUPPLY/POWER/LIGHTING

Item	Net Price £	Material £	Labour hours	Labour £	Unit	Total rate £
LV DISTRIBUTION: TRUNKING – cont						
PVC miniature trunking fittings – cont						
Tee						
16 × 16 mm	0.78	0.99	0.19	7.34	nr	8.33
25 × 16 mm	0.78	0.99	0.23	8.88	nr	9.87
38 × 16 mm	0.78	0.99	0.26	10.05	nr	11.04
38 × 25 mm	1.09	1.38	0.29	11.20	nr	12.58
PVC bench trunking; white or grey finish; fixed to backgrounds; including supports and fixings; standard coupling joints						
Trunking						
90 × 90 mm	24.36	30.83	0.33	12.75	m	43.58
PVC bench trunking fittings; white or grey finish; cutting and jointing trunking to fittings is included						
Stop end						
90 × 90 mm	5.36	6.79	0.09	3.48	nr	10.27
Coupling						
90 × 90 mm	3.32	4.20	0.09	3.48	nr	7.68
Internal or external bend						
90 × 90 mm	18.21	23.05	0.28	10.82	nr	33.87
Socket plate						
90 × 90 mm – 1 gang	1.03	1.30	0.10	3.86	nr	5.16
90 × 90 mm – 2 gang	1.23	1.56	0.10	3.86	nr	5.42
PVC underfloor trunking; single compartment; fitted in floor screed; standard coupling joints						
Trunking						
60 × 25 mm	12.42	15.71	0.22	8.50	m	24.21
90 × 35 mm	17.89	22.65	0.27	10.43	m	33.08
PVC underfloor trunking fittings; single compartment; fitted in floor screed; (cutting and jointing trunking to fittings is included)						
Jointing sleeve						
60 × 25 mm	0.83	1.05	0.08	3.09	nr	4.14
90 × 35 mm	1.38	1.75	0.10	3.86	nr	5.61
Duct connector						
90 × 35 mm	0.64	0.81	0.17	6.56	nr	7.37
Socket reducer						
90 × 35 mm	1.14	1.44	0.12	4.64	nr	6.08
Vertical access box; 2 compartment						
Shallow	64.25	81.31	0.37	14.30	nr	95.61

ELECTRICAL SUPPLY/POWER/LIGHTING

Item	Net Price £	Material £	Labour hours	Labour £	Unit	Total rate £
Duct bend; vertical						
60 × 25 mm	12.53	15.86	0.27	10.43	nr	**26.29**
90 × 35 mm	14.14	17.90	0.35	13.53	nr	**31.43**
Duct bend; horizontal						
60 × 25 mm	14.78	18.70	0.30	11.59	nr	**30.29**
90 × 35 mm	15.00	18.98	0.37	14.30	nr	**33.28**
Zinc coated steel underfloor ducting; fixed to backgrounds; standard coupling joints; earth continuity straps; (Including supports and fixing, packing shims where required)						
Double compartment						
150 × 25 mm	10.11	12.79	0.57	22.02	m	**34.81**
Triple compartment						
225 × 25 mm	17.91	22.67	0.93	35.93	m	**58.60**
Zinc coated steel underfloor ducting fittings; (cutting and jointing to fittings is included)						
Stop end; double compartment						
150 × 25 mm	3.18	4.02	0.31	11.97	nr	**15.99**
Stop end; triple compartment						
225 × 25 mm	3.64	4.60	0.37	14.30	nr	**18.90**
Rising bend; double compartment; standard trunking						
150 × 25 mm	20.47	25.91	0.71	27.44	nr	**53.35**
Rising bend; triple compartment; standard trunking						
225 × 25 mm	36.78	46.55	0.85	32.84	nr	**79.39**
Rising bend; double compartment; to skirting						
150 × 25 mm	45.56	57.66	0.90	34.78	nr	**92.44**
Rising bend; triple compartment; to skirting						
225 × 25 mm	53.72	67.98	0.95	36.70	nr	**104.68**
Horizontal bend; double compartment						
150 × 25 mm	31.12	39.39	0.64	24.73	nr	**64.12**
Horizontal bend; triple compartment						
225 × 25 mm	35.89	45.43	0.77	29.76	nr	**75.19**
Junction or service outlet boxes; terminal; double compartment						
150 mm	32.20	40.76	0.91	35.16	nr	**75.92**
Junction or service outlet boxes; terminal; triple compartment						
225 mm	36.80	46.57	1.11	42.90	nr	**89.47**
Junction or service outlet boxes; through or angle; double compartment						
150 mm	42.76	54.12	0.97	37.48	nr	**91.60**
Junction or service outlet boxes; through or angle; triple compartment						
225 mm	47.50	60.11	1.17	45.20	nr	**105.31**

ELECTRICAL SUPPLY/POWER/LIGHTING

Item	Net Price £	Material £	Labour hours	Labour £	Unit	Total rate £
LV DISTRIBUTION: TRUNKING – cont						
Zinc coated steel underfloor ducting fittings – cont						
Junction or service outlet boxes; tee; double compartment						
150 mm	42.76	54.12	1.02	39.41	nr	**93.53**
Junction or service outlet boxes; tee; triple compartment						
225 mm	47.50	60.11	1.22	47.14	nr	**107.25**
Junction or service outlet boxes; cross; double compartment						
up to 150 mm	42.76	54.12	1.03	39.80	nr	**93.92**
Junction or service outlet boxes; cross; triple compartment						
225 mm	47.50	60.11	1.23	47.52	nr	**107.63**
Plates for junction/inspection boxes; double and triple compartment						
Blank plate	9.34	11.82	0.92	35.55	nr	**47.37**
Conduit entry plate	11.72	14.83	0.86	33.23	nr	**48.06**
Trunking entry plate	11.72	14.83	0.86	33.23	nr	**48.06**
Service outlet box comprising flat lid with flanged carpet trim; twin 13 A outlet and drilled plate for mounting 2 telephone outlets and terminal blocks; terminal outlet box; double compartment						
150 × 25 mm trunking	66.68	84.39	1.68	64.92	nr	**149.31**
Service outlet box comprising flat lid with flanged carpet trim; twin 13 A outlet and drilled plate for mounting 2 telephone outlets and terminal blocks; terminal outlet box; triple compartment						
225 × 25 mm trunking	74.22	93.93	1.93	74.57	nr	**168.50**
PVC skirting/dado modular trunking; white (cutting and jointing trunking to fittings and backplates for fixing to walls is included)						
Main carrier/backplate						
50 × 170 mm	19.85	25.12	0.22	8.50	m	**33.62**
62 × 190 mm	22.04	27.90	0.22	8.50	m	**36.40**
Extension carrier/backplate						
50 × 42 mm	12.06	15.27	0.58	22.41	m	**37.68**
Carrier/backplate						
Including cover seal	7.04	8.92	0.53	20.47	m	**29.39**
Chamfered covers for fixing to backplates						
50 × 42 mm	4.09	5.17	0.33	12.75	m	**17.92**
Square covers for fixing to backplates						
50 × 42 mm	8.20	10.38	0.33	12.75	m	**23.13**

ELECTRICAL SUPPLY/POWER/LIGHTING

Item	Net Price £	Material £	Labour hours	Labour £	Unit	Total rate £
Plain covers for fixing to backplates						
85 mm	4.09	5.17	0.34	13.14	m	**18.31**
Retainers-clip to backplates to hold cables						
For chamfered covers	1.13	1.43	0.07	2.71	m	**4.14**
For square-recessed covers	0.98	1.24	0.07	2.71	m	**3.95**
For plain covers	3.77	4.77	0.07	2.71	m	**7.48**
Pre-packaged corner assemblies						
Internal; for 170 × 50 Assy	7.93	10.04	0.51	19.70	nr	**29.74**
Internal; for 190 × 62 Assy	8.81	11.16	0.51	19.70	nr	**30.86**
Internal; for 215 × 50 Assy	9.86	12.48	0.53	20.47	nr	**32.95**
Internal; for 254 × 50 Assy	11.74	14.86	0.53	20.47	nr	**35.33**
External; for 170 × 50 Assy	7.93	10.04	0.56	21.64	nr	**31.68**
External; for 190 × 62 Assy	8.81	11.16	0.56	21.64	nr	**32.80**
External; for 215 × 50 Assy	9.86	12.48	0.58	22.41	nr	**34.89**
External; for 254 × 50 Assy	11.74	14.86	0.58	22.41	nr	**37.27**
Clip on end caps						
170 × 50 Assy	4.75	6.01	0.11	4.26	nr	**10.27**
215 × 50 Assy	5.62	7.11	0.11	4.26	nr	**11.37**
254 × 50 Assy	6.59	8.34	0.11	4.26	nr	**12.60**
190 × 62 Assy	6.64	8.40	0.11	4.26	nr	**12.66**
Outlet box						
1 Gang; in horizontal trunking; clip in	4.24	5.36	0.34	13.14	nr	**18.50**
2 Gang; in horizontal trunking; clip in	5.29	6.70	0.34	13.14	nr	**19.84**
1 Gang; in vertical trunking; clip in	4.24	5.36	0.34	13.14	nr	**18.50**
Sheet steel adaptable boxes; with plain or knockout sides; fixed to backgrounds; including supports and fixings (cutting and connecting conduit to boxes is included)						
Square pattern – black						
75 × 75 × 37 mm	2.21	2.80	0.69	26.67	nr	**29.47**
75 × 75 × 50 mm	2.67	3.38	0.69	26.67	nr	**30.05**
75 × 75 × 75 mm	2.40	3.04	0.69	26.67	nr	**29.71**
100 × 100 × 50 mm	4.52	5.72	0.71	27.44	nr	**33.16**
150 × 150 × 50 mm	3.35	4.24	0.79	30.53	nr	**34.77**
150 × 150 × 75 mm	5.79	7.32	0.80	30.90	nr	**38.22**
150 × 150 × 100 mm	6.55	8.29	0.80	30.90	nr	**39.19**
200 × 200 × 50 mm	5.25	6.64	0.80	30.90	nr	**37.54**
225 × 225 × 50 mm	6.71	8.49	0.93	35.93	nr	**44.42**
225 × 225 × 100 mm	8.90	11.27	0.94	36.32	nr	**47.59**
300 × 300 × 100 mm	9.55	12.08	0.99	38.25	nr	**50.33**
Square pattern – galvanized						
75 × 75 × 37 mm	2.03	2.56	0.69	26.67	nr	**29.23**
75 × 75 × 50 mm	2.37	3.00	0.70	27.05	nr	**30.05**
75 × 75 × 75 mm	2.76	3.49	0.69	26.67	nr	**30.16**
100 × 100 × 50 mm	2.51	3.18	0.71	27.44	nr	**30.62**
150 × 150 × 50 mm	2.93	3.71	0.84	32.46	nr	**36.17**
150 × 150 × 75 mm	3.53	4.47	0.80	30.90	nr	**35.37**
150 × 150 × 100 mm	4.27	5.41	0.80	30.90	nr	**36.31**

ELECTRICAL SUPPLY/POWER/LIGHTING

Item	Net Price £	Material £	Labour hours	Labour £	Unit	Total rate £
LV DISTRIBUTION: TRUNKING – cont						
Sheet steel adaptable boxes – cont						
Square pattern – galvanized – cont						
225 × 225 × 50 mm	5.52	6.99	0.93	35.93	nr	**42.92**
225 × 225 × 100 mm	6.71	8.49	0.94	36.32	nr	**44.81**
300 × 300 × 100 mm	11.41	–	–	–	nr	**-**
Rectangular pattern – black						
100 × 75 × 50 mm	3.74	4.74	0.69	26.67	nr	**31.41**
150 × 75 × 50 mm	3.92	4.96	0.70	27.05	nr	**32.01**
150 × 75 × 75 mm	4.25	5.38	0.71	27.44	nr	**32.82**
150 × 100 × 75 mm	9.47	11.98	0.71	27.44	nr	**39.42**
225 × 75 × 50 mm	8.18	10.35	0.78	30.14	nr	**40.49**
225 × 150 × 75 mm	13.07	16.54	0.81	31.29	nr	**47.83**
225 × 150 × 100 mm	24.53	31.05	0.81	31.29	nr	**62.34**
300 × 150 × 50 mm	24.53	31.05	0.93	35.93	nr	**66.98**
300 × 150 × 75 mm	24.53	31.05	0.94	36.32	nr	**67.37**
300 × 150 × 100 mm	24.53	31.05	0.96	37.09	nr	**68.14**
Rectangular pattern – galvanized						
100 × 75 × 50 mm	3.55	4.49	0.69	26.67	nr	**31.16**
150 × 75 × 50 mm	3.72	4.70	0.70	27.05	nr	**31.75**
150 × 75 × 75 mm	4.04	5.12	0.71	27.44	nr	**32.56**
150 × 100 × 75 mm	8.98	11.37	0.71	27.44	nr	**38.81**
225 × 75 × 50 mm	7.78	9.84	0.89	34.38	nr	**44.22**
225 × 150 × 75 mm	12.41	15.70	0.81	31.29	nr	**46.99**
225 × 150 × 100 mm	23.31	29.50	0.81	31.29	nr	**60.79**
300 × 150 × 50 mm	23.31	29.50	0.93	35.93	nr	**65.43**
300 × 150 × 75 mm	23.31	29.50	0.94	36.32	nr	**65.82**
300 × 150 × 100 mm	23.31	29.50	0.96	37.09	nr	**66.59**

ELECTRICAL SUPPLY/POWER/LIGHTING

Item	Net Price £	Material £	Labour hours	Labour £	Unit	Total rate £
LV DISTRIBUTION: CABLES AND WIRING						
ARMOURED CABLE						
Cable; XLPE insulated; PVC sheathed; copper stranded conductors to BS 5467; laid in trench/duct including marker tape; (cable tiles measured elsewhere)						
600/1000 volt grade; single core (aluminium wire armour)						
25 mm²	0.45	0.57	0.15	5.80	m	6.37
35 mm²	0.50	0.64	0.15	5.80	m	6.44
50 mm²	0.73	0.92	0.17	6.56	m	7.48
70 mm²	6.14	7.77	0.18	6.96	m	14.73
95 mm²	7.97	10.09	0.20	7.73	m	17.82
120 mm²	9.93	12.57	0.22	8.50	m	21.07
150 mm²	12.04	15.24	0.24	9.27	m	24.51
185 mm²	14.70	18.60	0.26	10.05	m	28.65
240 mm²	18.18	23.00	0.30	11.59	m	34.59
300 mm²	23.70	29.99	0.31	11.97	m	41.96
400 mm²	30.06	38.05	0.38	14.68	m	52.73
500 mm²	38.19	48.33	0.44	17.00	m	65.33
630 mm²	48.87	61.85	0.52	20.09	m	81.94
800 mm²	31.67	40.08	0.62	23.96	m	64.04
1000 mm²	36.68	46.42	0.65	25.11	m	71.53
600/1000 volt grade; two core (galvanized steel wire armour)						
1.5 mm²	1.07	1.36	0.06	2.32	m	3.68
2.5 mm²	1.27	1.61	0.06	2.32	m	3.93
4 mm²	1.55	1.96	0.08	3.09	m	5.05
6 mm²	1.96	2.48	0.08	3.09	m	5.57
10 mm²	2.60	3.29	0.10	3.86	m	7.15
16 mm²	3.71	4.69	0.10	3.86	m	8.55
25 mm²	4.89	6.19	0.15	5.80	m	11.99
35 mm²	6.71	8.49	0.15	5.80	m	14.29
50 mm²	8.40	10.63	0.17	6.56	m	17.19
70 mm²	11.47	14.52	0.18	6.96	m	21.48
95 mm²	15.58	19.72	0.20	7.73	m	27.45
120 mm²	19.64	24.85	0.22	8.50	m	33.35
150 mm²	23.66	29.95	0.24	9.27	m	39.22
185 mm²	29.86	37.79	0.26	10.05	m	47.84
240 mm²	35.48	44.90	0.30	11.59	m	56.49
300 mm²	46.89	59.35	0.31	11.97	m	71.32
400 mm²	49.51	62.66	0.35	13.53	m	76.19
600/1000 volt grade; three core (galvanized steel wire armour)						
1.5 mm²	1.15	1.46	0.07	2.71	m	4.17
2.5 mm²	1.42	1.79	0.07	2.71	m	4.50
4 mm²	1.78	2.25	0.09	3.48	m	5.73

ELECTRICAL SUPPLY/POWER/LIGHTING

Item	Net Price £	Material £	Labour hours	Labour £	Unit	Total rate £
LV DISTRIBUTION: CABLES AND WIRING – cont						
Cable – cont						
600/1000 volt grade – cont						
6 mm²	2.01	2.54	0.10	3.86	m	**6.40**
10 mm²	3.00	3.80	0.11	4.26	m	**8.06**
16 mm²	4.29	5.43	0.11	4.26	m	**9.69**
25 mm²	6.36	8.05	0.16	6.17	m	**14.22**
35 mm²	8.41	10.64	0.16	6.17	m	**16.81**
50 mm²	10.84	13.72	0.19	7.34	m	**21.06**
70 mm²	15.17	19.20	0.21	8.11	m	**27.31**
95 mm²	20.91	26.47	0.23	8.88	m	**35.35**
120 mm²	25.80	32.65	0.24	9.27	m	**41.92**
150 mm²	31.85	40.31	0.27	10.43	m	**50.74**
185 mm²	39.70	50.24	0.30	11.59	m	**61.83**
240 mm²	51.12	64.70	0.33	12.75	m	**77.45**
300 mm²	64.61	81.77	0.35	13.53	m	**95.30**
400 mm²	64.61	81.77	0.41	15.85	m	**97.62**
600/1000 volt grade; four core (galvanized steel wire armour)						
1.5 mm²	1.15	1.46	0.08	3.09	m	**4.55**
2.5 mm²	1.51	1.92	0.09	3.48	m	**5.40**
4 mm²	1.99	2.52	0.10	3.86	m	**6.38**
6 mm²	2.82	3.57	0.10	3.86	m	**7.43**
10 mm²	4.03	5.10	0.12	4.64	m	**9.74**
16 mm²	5.86	7.41	0.12	4.64	m	**12.05**
25 mm²	8.54	10.81	0.18	6.96	m	**17.77**
35 mm²	11.46	14.50	0.19	7.34	m	**21.84**
50 mm²	15.03	19.02	0.21	8.11	m	**27.13**
70 mm²	21.33	26.99	0.23	8.88	m	**35.87**
95 mm²	28.81	36.47	0.26	10.05	m	**46.52**
120 mm²	36.05	45.63	0.28	10.82	m	**56.45**
150 mm²	44.52	56.35	0.32	12.36	m	**68.71**
185 mm²	55.01	69.62	0.35	13.53	m	**83.15**
240 mm²	71.31	90.25	0.36	13.91	m	**104.16**
300 mm²	89.75	113.59	0.40	15.46	m	**129.05**
400 mm²	99.33	125.71	0.45	17.39	m	**143.10**
600/1000 volt grade; seven core (galvanized steel wire armour)						
1.5 mm²	1.89	2.40	0.10	3.86	m	**6.26**
2.5 mm²	2.46	3.11	0.10	3.86	m	**6.97**
4 mm²	2.50	3.17	0.11	4.26	m	**7.43**
600/1000 volt grade; twelve core (galvanized steel wire armour)						
1.5 mm²	2.86	3.62	0.11	4.26	m	**7.88**
2.5 mm²	4.03	5.10	0.11	4.26	m	**9.36**
600/1000 volt grade; nineteen core (galvanized steel wire armour)						
1.5 mm²	4.05	5.13	0.13	5.03	m	**10.16**
2.5 mm²	5.71	7.22	0.14	5.41	m	**12.63**

ELECTRICAL SUPPLY/POWER/LIGHTING

Item	Net Price £	Material £	Labour hours	Labour £	Unit	Total rate £
600/1000 volt grade; twenty seven core (galvanized steel wire armour)						
1.5 mm²	5.51	6.98	0.14	5.41	m	12.39
2.5 mm²	7.75	9.81	0.16	6.17	m	15.98
600/1000 volt grade; thirty seven core (galvanized steel wire armour)						
1.5 mm²	7.38	9.34	0.15	5.80	m	15.14
2.5 mm²	10.25	12.97	0.17	6.56	m	19.53
Cable; XLPE insulated; PVC sheathed copper stranded conductors to BS 5467; clipped direct to backgrounds including cleat						
600/1000 volt grade; single core (aluminium wire armour)						
25 mm²	2.03	2.56	0.35	13.53	m	16.09
35 mm²	2.23	2.82	0.36	13.91	m	16.73
50 mm²	2.45	3.10	0.37	14.30	m	17.40
70 mm²	8.43	10.67	0.39	15.08	m	25.75
95 mm²	11.60	14.68	0.42	16.23	m	30.91
120 mm²	13.85	17.53	0.47	18.17	m	35.70
150 mm²	15.95	20.18	0.51	19.70	m	39.88
185 mm²	18.62	23.56	0.59	22.79	m	46.35
240 mm²	22.28	28.20	0.68	26.28	m	54.48
300 mm²	27.81	35.20	0.74	28.59	m	63.79
400 mm²	33.90	42.91	0.88	34.00	m	76.91
500 mm²	43.76	55.38	0.88	34.00	m	89.38
630 mm²	55.03	69.64	1.05	40.58	m	110.22
800 mm²	38.43	48.64	1.33	51.39	m	100.03
1000 mm²	42.97	54.39	1.40	54.10	m	108.49
600/1000 volt grade; two core (galvanized steel wire armour)						
1.5 mm²	1.45	1.84	0.20	7.73	m	9.57
2.5 mm²	1.65	2.08	0.20	7.73	m	9.81
4.0 mm²	1.93	2.44	0.21	8.11	m	10.55
6.0 mm²	2.38	3.01	0.22	8.50	m	11.51
10.0 mm²	3.25	4.11	0.24	9.27	m	13.38
16.0 mm²	4.38	5.54	0.25	9.65	m	15.19
25 mm²	5.94	7.52	0.35	13.53	m	21.05
35 mm²	7.56	9.56	0.36	13.91	m	23.47
50 mm²	9.33	11.80	0.37	14.30	m	26.10
70 mm²	11.93	15.10	0.39	15.08	m	30.18
95 mm²	16.09	20.36	0.42	16.23	m	36.59
120 mm²	20.14	25.49	0.47	18.17	m	43.66
150 mm²	24.47	30.97	0.51	19.70	m	50.67
185 mm²	30.83	39.02	0.59	22.79	m	61.81
240 mm²	40.03	50.66	0.68	26.28	m	76.94
300 mm²	52.20	66.07	0.74	28.59	m	94.66
400 mm²	39.53	50.03	0.88	34.00	m	84.03

ELECTRICAL SUPPLY/POWER/LIGHTING

Item	Net Price £	Material £	Labour hours	Labour £	Unit	Total rate £
LV DISTRIBUTION: CABLES AND WIRING – cont						
Cable – cont						
600/1000 volt grade; three core (galvanized steel wire armour)						
1.5 mm²	1.57	1.98	0.20	7.73	m	9.71
2.5 mm²	1.87	2.36	0.21	8.11	m	10.47
4.0 mm²	2.25	2.84	0.22	8.50	m	11.34
6.0 mm²	2.79	3.53	0.22	8.50	m	12.03
10.0 mm²	4.12	5.22	0.25	9.65	m	14.87
16.0 mm²	5.51	6.98	0.26	10.05	m	17.03
25 mm²	7.92	10.02	0.37	14.30	m	24.32
35 mm²	10.24	12.96	0.39	15.08	m	28.04
50 mm²	12.85	16.26	0.40	15.46	m	31.72
70 mm²	16.92	21.41	0.42	16.23	m	37.64
95 mm²	23.44	29.67	0.45	17.39	m	47.06
120 mm²	28.77	36.41	0.52	20.09	m	56.50
150 mm²	35.85	45.37	0.55	21.25	m	66.62
185 mm²	45.00	56.95	0.63	24.34	m	81.29
240 mm²	57.94	73.33	0.71	27.44	m	100.77
300 mm²	71.96	91.07	0.78	30.14	m	121.21
400 mm²	77.59	98.20	0.87	33.61	m	131.81
600/1000 volt grade; four core (galvanized steel wire armour)						
1.5 mm²	1.75	2.22	0.21	8.11	m	10.33
2.5 mm²	2.11	2.67	0.22	8.50	m	11.17
4.0 mm²	2.64	3.34	0.22	8.50	m	11.84
6.0 mm²	3.64	4.60	0.23	8.88	m	13.48
10.0 mm²	5.10	6.45	0.26	10.05	m	16.50
16.0 mm²	6.94	8.78	0.26	10.05	m	18.83
25 mm²	9.71	12.29	0.39	15.08	m	27.37
35 mm²	12.65	16.00	0.40	15.46	m	31.46
50 mm²	16.33	20.66	0.41	15.85	m	36.51
70 mm²	22.18	28.07	0.45	17.39	m	45.46
95 mm²	29.72	37.61	0.50	19.33	m	56.94
120 mm²	37.43	47.38	0.54	20.87	m	68.25
150 mm²	46.63	59.01	0.60	23.18	m	82.19
185 mm²	57.42	72.67	0.67	25.89	m	98.56
240 mm²	74.02	93.68	0.75	28.99	m	122.67
300 mm²	92.92	117.60	0.83	32.07	m	149.67
400 mm²	91.72	116.08	0.91	35.16	m	151.24
600/1000 volt grade; seven core (galvanized steel wire armour)						
1.5 mm²	2.57	3.25	0.20	7.73	m	10.98
2.5 mm²	3.14	3.98	0.20	7.73	m	11.71
4.0 mm²	4.08	5.16	0.23	8.88	m	14.04
600/1000 volt grade; twelve core (galvanized steel wire armour)						
1.5 mm²	4.54	12.29	0.23	8.88	m	21.17
2.5 mm²	5.14	6.51	0.24	9.27	m	15.78

ELECTRICAL SUPPLY/POWER/LIGHTING

Item	Net Price £	Material £	Labour hours	Labour £	Unit	Total rate £
600/1000 volt grade; nineteen core (galvanized steel wire armour)						
1.5 mm²	5.17	6.54	0.26	10.05	m	16.59
2.5 mm²	6.93	8.77	0.28	10.82	m	19.59
600/1000 volt grade; twenty seven core (galvanized steel wire armour)						
1.5 mm²	6.71	8.49	0.29	11.20	m	19.69
2.5 mm²	9.06	11.47	0.30	11.59	m	23.06
600/1000 volt grade; thirty seven core (galvanized steel wire armour)						
1.5 mm²	8.13	10.29	0.32	12.36	m	22.65
2.5 mm²	11.52	14.58	0.33	12.75	m	27.33
Cable termination; brass weatherproof gland with inner and outer seal, shroud, brass locknut and earth ring (including drilling and cutting mild steel gland plate)						
600/1000 volt grade; single core (aluminium wire armour)						
25 mm²	7.33	9.27	1.70	65.69	nr	74.96
35 mm²	7.33	9.27	1.79	69.16	nr	78.43
50 mm²	8.02	10.15	2.06	79.60	nr	89.75
70 mm²	8.95	11.32	2.12	81.91	nr	93.23
95 mm²	9.00	11.39	2.39	92.36	nr	103.75
120 mm²	9.04	11.45	2.47	95.44	nr	106.89
150 mm²	18.33	23.20	2.73	105.49	nr	128.69
185 mm²	18.61	23.55	3.05	117.86	nr	141.41
240 mm²	18.52	23.44	3.45	133.30	nr	156.74
300 mm²	18.93	23.96	3.84	148.38	nr	172.34
400 mm²	27.08	34.27	4.21	162.67	nr	196.94
500 mm²	28.89	36.57	5.70	220.24	m	256.81
630 mm²	35.25	44.61	6.20	239.57	m	284.18
800 mm²	71.89	90.99	7.50	289.79	m	380.78
1000 mm²	86.77	109.82	10.00	386.39	m	496.21
600/1000 volt grade; two core (galvanized steel wire armour)						
1.5 mm²	6.94	8.78	0.58	22.41	nr	31.19
2.5 mm²	6.94	8.78	0.58	22.46	nr	31.24
4 mm²	6.94	8.78	0.58	22.41	nr	31.19
6 mm²	6.96	8.80	0.67	25.89	nr	34.69
10 mm²	7.74	9.80	1.00	38.64	nr	48.44
16 mm²	7.93	10.04	1.11	42.90	nr	52.94
25 mm²	9.12	11.55	1.70	65.69	nr	77.24
35 mm²	9.05	11.46	1.79	69.16	nr	80.62
50 mm²	17.45	22.09	2.06	79.60	nr	101.69
70 mm²	17.60	22.28	2.12	81.91	nr	104.19
95 mm²	16.75	21.20	2.39	92.36	nr	113.56
120 mm²	16.83	21.30	2.47	95.44	nr	116.74
150 mm²	26.31	33.30	2.73	105.49	nr	138.79
185 mm²	26.91	34.06	3.05	117.86	nr	151.92
240 mm²	28.61	36.21	3.45	133.30	nr	169.51

ELECTRICAL SUPPLY/POWER/LIGHTING

Item	Net Price £	Material £	Labour hours	Labour £	Unit	Total rate £
LV DISTRIBUTION: CABLES AND WIRING – cont						
Cable termination – cont						
600/1000 volt grade – cont						
300 mm²	41.70	52.77	3.84	148.38	nr	201.15
400 mm²	53.90	68.22	4.21	162.67	nr	230.89
600/1000 volt grade; three core (galvanized steel wire armour)						
1.5 mm²	7.09	8.97	0.61	23.76	nr	32.73
2.5 mm²	7.09	8.97	0.62	23.96	nr	32.93
4 mm²	7.09	8.97	0.62	23.96	nr	32.93
6 mm²	7.91	10.01	0.71	27.44	nr	37.45
10 mm²	9.16	11.59	1.06	40.96	nr	52.55
16 mm²	9.44	11.95	1.19	45.98	nr	57.93
25 mm²	17.65	22.33	1.81	69.93	nr	92.26
35 mm²	17.68	22.38	1.99	76.89	nr	99.27
50 mm²	17.71	22.41	2.23	86.16	nr	108.57
70 mm²	16.97	21.48	2.40	92.74	nr	114.22
95 mm²	17.13	21.68	2.63	101.62	nr	123.30
120 mm²	17.22	21.80	2.83	109.35	nr	131.15
150 mm²	27.78	35.16	3.22	124.42	nr	159.58
185 mm²	28.64	36.24	3.44	132.91	nr	169.15
240 mm²	43.52	55.08	3.83	147.99	nr	203.07
300 mm²	44.75	56.64	4.28	165.38	nr	222.02
400 mm²	47.81	60.51	5.00	193.19	nr	253.70
600/1000 volt grade; four core (galvanized steel wire armour)						
1.5 mm²	7.20	9.12	0.67	25.89	nr	35.01
2.5 mm²	7.20	9.12	0.67	25.89	nr	35.01
4 mm²	7.20	9.12	0.71	27.44	nr	36.56
6 mm²	9.27	11.74	0.76	29.37	nr	41.11
10 mm²	9.27	11.74	1.14	44.04	nr	55.78
16 mm²	9.64	12.20	1.29	49.84	nr	62.04
25 mm²	17.79	22.51	1.99	76.89	nr	99.40
35 mm²	17.83	22.57	2.16	83.45	nr	106.02
50 mm²	16.92	21.41	2.49	96.21	nr	117.62
70 mm²	17.20	21.77	2.65	102.39	nr	124.16
95 mm²	24.82	31.42	2.98	115.15	nr	146.57
120 mm²	24.94	31.56	3.15	121.71	nr	153.27
150 mm²	29.07	36.79	3.50	135.24	nr	172.03
185 mm²	42.45	53.73	3.72	143.73	nr	197.46
240 mm²	45.88	58.06	4.33	167.29	nr	225.35
300 mm²	66.32	83.93	4.86	187.79	nr	271.72
400 mm²	50.23	63.57	5.46	210.96	nr	274.53
600/1000 volt grade; seven core (galvanized steel wire armour)						
1.5 mm²	8.05	10.19	0.81	31.29	nr	41.48
2.5 mm²	8.05	10.19	0.85	32.84	nr	43.03
4 mm²	9.00	11.39	0.93	35.93	nr	47.32

ELECTRICAL SUPPLY/POWER/LIGHTING

Item	Net Price £	Material £	Labour hours	Labour £	Unit	Total rate £
600/1000 volt grade; twelve core (galvanized steel wire armour)						
1.5 mm²	9.63	12.19	1.14	44.04	nr	56.23
2.5 mm²	10.95	13.85	1.13	43.67	nr	57.52
600/1000 volt grade; nineteen core (galvanized steel wire armour)						
1.5 mm²	12.01	15.20	1.54	59.51	nr	74.71
2.5 mm²	12.01	15.20	1.54	59.51	nr	74.71
600/1000 volt grade; twenty seven core (galvanized steel wire armour)						
1.5 mm²	13.22	16.73	1.94	74.96	nr	91.69
2.5 mm²	21.95	27.78	2.31	89.25	nr	117.03
600/1000 volt grade; thirty seven core (galvanized steel wire armour)						
1.5 mm²	23.46	29.69	2.53	97.75	nr	127.44
2.5 mm²	23.46	29.69	2.87	110.89	nr	140.58
Cable; XLPE insulated; LSOH sheathed (LSF); copper stranded conductors to BS 6724; laid in trench/duct including marker tape (cable tiles measured elsewhere)						
600/1000 volt grade; single core (aluminium wire armour)						
50 mm²	5.08	6.43	0.17	6.56	m	12.99
70 mm²	7.00	8.86	0.18	6.96	m	15.82
95 mm²	9.20	11.65	0.20	7.73	m	19.38
120 mm²	11.03	13.96	0.22	8.50	m	22.46
150 mm²	13.17	16.67	0.24	9.27	m	25.94
185 mm²	15.86	20.07	0.26	10.05	m	30.12
240 mm²	21.19	26.81	0.30	11.59	m	38.40
300 mm²	25.74	32.58	0.31	11.97	m	44.55
400 mm²	33.64	42.57	0.35	13.53	m	56.10
500 mm²	42.56	53.86	0.44	17.00	m	70.86
630 mm²	56.54	71.56	0.52	20.09	m	91.65
800 mm²	71.11	89.99	0.62	23.96	m	113.95
1000 mm²	90.76	114.87	0.65	25.11	m	139.98
600/1000 volt grade; two core (galvanized steel wire armour)						
1.5 mm²	1.04	1.32	0.06	2.32	m	3.64
2.5 mm²	1.24	1.57	0.06	2.32	m	3.89
4 mm²	1.63	2.06	0.08	3.09	m	5.15
6 mm²	2.24	2.83	0.08	3.09	m	5.92
10 mm²	3.32	4.20	0.10	3.86	m	8.06
16 mm²	4.54	5.75	0.10	3.86	m	9.61
25 mm²	6.09	7.71	0.15	5.80	m	13.51
35 mm²	7.69	9.73	0.17	6.56	m	16.29
50 mm²	9.65	12.21	0.15	5.80	m	18.01
70 mm²	13.16	16.65	0.18	6.96	m	23.61
95 mm²	17.25	21.83	0.20	7.73	m	29.56

ELECTRICAL SUPPLY/POWER/LIGHTING

Item	Net Price £	Material £	Labour hours	Labour £	Unit	Total rate £
LV DISTRIBUTION: CABLES AND WIRING – cont						
Cable – cont						
600/1000 volt grade – cont						
120 mm²	21.91	27.73	0.22	8.50	m	**36.23**
150 mm²	26.49	33.52	0.24	9.27	m	**42.79**
185 mm²	33.11	41.90	0.26	10.05	m	**51.95**
240 mm²	42.13	53.32	0.30	11.59	m	**64.91**
300 mm²	39.35	49.81	0.31	11.97	m	**61.78**
400 mm²	60.78	76.92	0.35	13.53	m	**90.45**
600/1000 volt grade; three core (galvanized steel wire armour)						
1.5 mm²	1.18	1.49	0.07	2.71	m	**4.20**
2.5 mm²	1.54	1.95	0.07	2.71	m	**4.66**
4 mm²	2.01	2.54	0.09	3.48	m	**6.02**
6 mm²	2.72	3.44	0.10	3.86	m	**7.30**
10 mm²	4.18	5.29	0.11	4.26	m	**9.55**
16 mm²	5.72	7.24	0.11	4.26	m	**11.50**
25 mm²	7.97	10.09	0.16	6.17	m	**16.26**
35 mm²	10.28	13.01	0.16	6.17	m	**19.18**
50 mm²	13.28	16.81	0.19	7.34	m	**24.15**
70 mm²	17.90	22.66	0.21	8.11	m	**30.77**
95 mm²	24.85	31.45	0.23	8.88	m	**40.33**
120 mm²	30.55	38.66	0.24	9.27	m	**47.93**
150 mm²	37.74	47.77	0.27	10.43	m	**58.20**
185 mm²	46.53	58.89	0.30	11.59	m	**70.48**
240 mm²	61.05	77.27	0.33	12.75	m	**90.02**
300 mm²	74.78	94.64	0.35	13.53	m	**108.17**
400 mm²	105.17	133.10	0.41	15.85	m	**148.95**
600/1000 volt grade; four core (galvanized steel wire armour)						
1.5 mm²	1.34	1.69	0.08	3.09	m	**4.78**
2.5 mm²	1.80	2.27	0.09	3.48	m	**5.75**
4 mm²	2.56	3.24	0.10	3.86	m	**7.10**
6 mm²	3.64	4.60	0.10	3.86	m	**8.46**
10 mm²	5.07	6.42	0.12	4.64	m	**11.06**
16 mm²	7.15	9.05	0.12	4.64	m	**13.69**
25 mm²	10.41	13.17	0.18	6.96	m	**20.13**
35 mm²	13.75	17.40	0.19	7.34	m	**24.74**
50 mm²	18.00	22.78	0.21	8.11	m	**30.89**
70 mm²	23.61	29.88	0.23	8.88	m	**38.76**
95 mm²	31.63	40.03	0.26	10.05	m	**50.08**
120 mm²	39.54	50.04	0.28	10.82	m	**60.86**
150 mm²	53.69	67.95	0.32	12.36	m	**80.31**
185 mm²	66.34	83.96	0.35	13.53	m	**97.49**
240 mm²	85.81	108.61	0.36	13.91	m	**122.52**
300 mm²	106.66	134.99	0.40	15.46	m	**150.45**
400 mm²	142.41	180.23	0.45	17.39	m	**197.62**

ELECTRICAL SUPPLY/POWER/LIGHTING

Item	Net Price £	Material £	Labour hours	Labour £	Unit	Total rate £
600/1000 volt grade; seven core (galvanized steel wire armour)						
1.5 mm²	1.94	2.45	0.10	3.86	m	6.31
2.5 mm²	2.62	3.32	0.10	3.86	m	7.18
4 mm²	4.84	6.13	0.11	4.26	m	10.39
600/1000 volt grade; twelve core (galvanized steel wire armour)						
1.5 mm²	3.22	4.08	0.11	4.26	m	8.34
2.5 mm²	4.60	5.82	0.11	4.26	m	10.08
600/1000 volt grade; nineteen core (galvanized steel wire armour)						
1.5 mm²	4.68	5.92	0.13	5.03	m	10.95
2.5 mm²	6.73	8.51	0.14	5.41	m	13.92
600/1000 volt grade; twenty seven core (galvanized steel wire armour)						
1.5 mm²	6.43	8.14	0.14	5.41	m	13.55
2.5 mm²	9.12	11.55	0.16	6.17	m	17.72
600/1000 volt grade; thirty seven core (galvanized steel wire armour)						
1.5 mm²	8.16	10.33	0.15	5.80	m	16.13
2.5 mm²	11.60	14.68	0.17	6.56	m	21.24
Cable; XLPE insulated; LSOH sheathed (LSF) copper stranded conductors to BS 6724; clipped direct to backgrounds including cleat						
600/1000 volt grade; single core (aluminium wire armour)						
50 mm²	4.03	5.10	0.37	14.30	m	19.40
70 mm²	6.12	7.75	0.39	15.08	m	22.83
95 mm²	8.23	10.42	0.42	16.23	m	26.65
120 mm²	9.81	12.42	0.47	18.17	m	30.59
150 mm²	12.12	15.34	0.51	19.70	m	35.04
185 mm²	14.56	18.42	0.59	22.79	m	41.21
240 mm²	18.77	23.76	0.68	26.28	m	50.04
300 mm²	22.87	28.94	0.74	28.59	m	57.53
400 mm²	29.72	37.61	0.81	31.29	m	68.90
500 mm²	38.13	48.26	0.88	34.00	m	82.26
630 mm²	47.82	60.52	1.05	40.58	m	101.10
800 mm²	61.62	77.99	1.33	51.39	m	129.38
1000 mm²	71.35	90.31	1.40	54.10	m	144.41
600/1000 volt grade; two core (galvanized steel wire armour)						
1.5 mm²	0.88	1.11	0.20	7.73	m	8.84
2.5 mm²	1.01	1.28	0.20	7.73	m	9.01
4.0 mm²	1.31	1.66	0.21	8.11	m	9.77
6.0 mm²	1.59	2.02	0.22	8.50	m	10.52
10.0 mm²	2.13	2.70	0.24	9.27	m	11.97
16.0 mm²	3.21	4.07	0.25	9.65	m	13.72
25 mm²	3.21	4.07	0.35	13.53	m	17.60

ELECTRICAL SUPPLY/POWER/LIGHTING

Item	Net Price £	Material £	Labour hours	Labour £	Unit	Total rate £
LV DISTRIBUTION: CABLES AND WIRING – cont						
Cable – cont						
600/1000 volt grade – cont						
35 mm²	6.48	8.20	0.36	13.91	m	22.11
50 mm²	8.03	10.16	0.37	14.30	m	24.46
70 mm²	11.93	15.10	0.39	15.08	m	30.18
95 mm²	16.03	20.28	0.42	16.23	m	36.51
120 mm²	19.63	24.84	0.47	18.17	m	43.01
150 mm²	24.35	30.82	0.51	19.70	m	50.52
185 mm²	31.02	39.26	0.59	22.79	m	62.05
240 mm²	39.72	50.27	0.68	26.28	m	76.55
300 mm²	53.34	67.50	0.74	28.59	m	96.09
400 mm²	70.57	89.31	0.81	31.29	m	120.60
600/1000 volt grade; three core (galvanized steel wire armour)						
1.5 mm²	0.96	1.21	0.20	7.73	m	8.94
2.5 mm²	1.22	1.55	0.21	8.11	m	9.66
4.0 mm²	1.50	1.90	0.22	8.50	m	10.40
6.0 mm²	1.94	2.45	0.22	8.50	m	10.95
10.0 mm²	2.96	3.74	0.25	9.65	m	13.39
16.0 mm²	4.15	5.25	0.26	10.05	m	15.30
25 mm²	5.60	7.09	0.37	14.30	m	21.39
35 mm²	7.79	9.86	0.39	15.08	m	24.94
50 mm²	11.12	14.08	0.40	15.46	m	29.54
70 mm²	16.51	20.90	0.42	16.23	m	37.13
95 mm²	22.73	28.76	0.45	17.39	m	46.15
120 mm²	28.26	35.76	0.52	20.09	m	55.85
150 mm²	35.11	44.43	0.55	21.25	m	65.68
185 mm²	43.26	54.75	0.63	24.34	m	79.09
240 mm²	55.35	70.06	0.71	27.44	m	97.50
300 mm²	70.15	88.78	0.78	30.14	m	118.92
400 mm²	83.31	105.44	0.87	33.61	m	139.05
600/1000 volt grade; four core (galvanized steel wire armour)						
1.5 mm²	1.13	1.43	0.21	8.11	m	9.54
2.5 mm²	1.41	1.78	0.22	8.50	m	10.28
4.0 mm²	1.75	2.22	0.22	8.50	m	10.72
6.0 mm²	2.57	3.25	0.23	8.88	m	12.13
10.0 mm²	3.61	4.57	0.26	10.05	m	14.62
16.0 mm²	5.21	6.60	0.26	10.05	m	16.65
25 mm²	7.52	9.52	0.39	15.08	m	24.60
35 mm²	9.81	12.42	0.40	15.46	m	27.88
50 mm²	14.20	17.98	0.41	15.85	m	33.83
70 mm²	21.90	27.72	0.45	17.39	m	45.11
95 mm²	29.34	37.13	0.50	19.33	m	56.46
120 mm²	36.85	46.64	0.54	20.87	m	67.51
150 mm²	44.61	56.46	0.60	23.18	m	79.64
185 mm²	54.74	69.28	0.67	25.89	m	95.17
240 mm²	70.76	89.56	0.75	28.99	m	118.55

ELECTRICAL SUPPLY/POWER/LIGHTING

Item	Net Price £	Material £	Labour hours	Labour £	Unit	Total rate £
300 mm²	88.78	112.36	0.83	32.07	m	**144.43**
400 mm²	92.30	116.82	0.91	35.16	m	**151.98**
600/1000 volt grade; seven core (galvanized steel wire armour)						
1.5 mm²	1.62	2.05	0.20	7.73	m	**9.78**
2.5 mm²	2.26	2.86	0.20	7.73	m	**10.59**
4.0 mm²	4.10	5.19	0.23	8.88	m	**14.07**
600/1000 volt grade; twelve core (galvanized steel wire armour)						
1.5 mm²	2.57	3.25	0.23	8.88	m	**12.13**
2.5 mm²	3.86	4.88	0.24	9.27	m	**14.15**
600/1000 volt grade; nineteen core (galvanized steel wire armour)						
1.5 mm²	3.89	4.93	0.26	10.05	m	**14.98**
2.5 mm²	5.83	7.38	0.28	10.82	m	**18.20**
600/1000 volt grade; twenty seven core (galvanized steel wire armour)						
1.5 mm²	5.34	6.75	0.29	11.20	m	**17.95**
2.5 mm²	7.45	9.43	0.30	11.59	m	**21.02**
600/1000 volt grade; thirty seven core (galvanized steel wire armour)						
1.5 mm²	6.67	8.44	0.32	12.36	m	**20.80**
2.5 mm²	9.44	11.95	0.33	12.75	m	**24.70**
Cable termination; brass weatherproof gland with inner and outer seal, shroud, brass locknut and earth ring (including drilling and cutting mild steel gland plate)						
600/1000 volt grade; single core (aluminium wire armour)						
25 mm²	3.71	4.69	1.70	65.69	nr	**70.38**
35 mm²	4.04	5.12	1.79	69.16	nr	**74.28**
50 mm²	4.48	5.67	2.06	79.60	nr	**85.27**
70 mm²	6.41	8.11	2.12	81.91	nr	**90.02**
95 mm²	8.56	10.83	2.39	92.36	nr	**103.19**
120 mm²	10.14	12.84	2.47	95.44	nr	**108.28**
150 mm²	12.47	15.78	2.73	105.49	nr	**121.27**
185 mm²	14.85	18.79	3.05	117.86	nr	**136.65**
240 mm²	19.46	24.63	3.45	133.30	nr	**157.93**
300 mm²	23.52	29.77	3.84	148.38	nr	**178.15**
400 mm²	34.34	43.46	4.21	162.67	nr	**206.13**
500 mm²	43.27	54.77	5.70	220.24	m	**275.01**
630 mm²	56.28	71.23	6.20	239.57	m	**310.80**
800 mm²	69.98	88.57	7.50	289.79	m	**378.36**
1000 mm²	78.26	99.04	10.00	386.39	m	**485.43**
600/1000 volt grade; two core (galvanized steel wire armour)						
1.5 mm²	1.09	1.38	0.52	19.97	nr	**21.35**
2.5 mm²	1.26	1.59	0.58	22.41	nr	**24.00**
4 mm²	1.59	2.02	0.58	22.41	nr	**24.43**

ELECTRICAL SUPPLY/POWER/LIGHTING

Item	Net Price £	Material £	Labour hours	Labour £	Unit	Total rate £
LV DISTRIBUTION: CABLES AND WIRING – cont						
Cable termination – cont						
600/1000 volt grade – cont						
6 mm²	1.85	2.34	0.67	25.89	nr	28.23
10 mm²	2.40	3.04	1.00	38.64	nr	41.68
16 mm²	3.48	4.40	1.11	42.90	nr	47.30
25 mm²	3.03	3.83	1.70	65.69	nr	69.52
35 mm²	6.76	8.56	1.79	69.16	nr	77.72
50 mm²	8.32	10.53	2.06	79.60	nr	90.13
70 mm²	12.37	15.66	2.12	81.91	nr	97.57
95 mm²	16.11	20.38	2.39	92.36	nr	112.74
120 mm²	19.82	25.09	2.47	95.44	nr	120.53
150 mm²	24.54	31.06	2.73	105.49	nr	136.55
185 mm²	31.24	39.54	3.05	117.86	nr	157.40
240 mm²	38.88	49.20	3.45	133.30	nr	182.50
300 mm²	54.44	68.90	3.84	148.38	nr	217.28
400 mm²	67.78	85.78	4.21	162.67	nr	248.45
600/1000 volt grade; three core (galvanized steel wire armour)						
1.5 mm²	1.17	1.48	0.62	23.96	nr	25.44
2.5 mm²	1.40	1.77	0.62	23.96	nr	25.73
4 mm²	1.78	2.25	0.62	23.96	nr	26.21
6 mm²	2.20	2.79	0.71	27.44	nr	30.23
10 mm²	3.22	4.08	1.06	40.96	nr	45.04
16 mm²	4.43	5.61	1.19	45.98	nr	51.59
25 mm²	6.21	7.86	1.81	69.93	nr	77.79
35 mm²	8.07	10.21	1.99	76.89	nr	87.10
50 mm²	11.49	14.54	2.23	86.16	nr	100.70
70 mm²	16.68	21.11	2.40	92.74	nr	113.85
95 mm²	22.91	29.00	2.63	101.62	nr	130.62
120 mm²	28.44	36.00	2.83	109.35	nr	145.35
150 mm²	35.40	44.80	3.22	124.42	nr	169.22
185 mm²	43.53	55.09	3.44	132.91	nr	188.00
240 mm²	57.79	73.14	3.83	147.99	nr	221.13
300 mm²	72.75	92.08	4.28	165.38	nr	257.46
400 mm²	86.36	109.30	5.00	193.19	nr	302.49
600/1000 volt grade; four core (galvanized steel wire armour)						
1.5 mm²	1.35	1.71	0.67	25.89	nr	27.60
2.5 mm²	1.67	2.12	0.67	25.89	nr	28.01
4 mm²	2.04	2.59	0.71	27.44	nr	30.03
6 mm²	2.85	3.61	0.76	29.37	nr	32.98
10 mm²	3.91	4.95	1.14	44.04	nr	48.99
16 mm²	5.48	6.93	1.29	49.84	nr	56.77
25 mm²	7.81	9.89	1.99	76.89	nr	86.78
35 mm²	10.19	12.89	2.16	83.45	nr	96.34
50 mm²	14.58	18.46	2.49	96.21	nr	114.67
70 mm²	22.08	27.94	2.65	102.39	nr	130.33
95 mm²	29.60	37.46	2.98	115.15	nr	152.61

ELECTRICAL SUPPLY/POWER/LIGHTING

Item	Net Price £	Material £	Labour hours	Labour £	Unit	Total rate £
120 mm²	37.09	46.94	3.15	121.71	nr	**168.65**
150 mm²	46.80	59.23	3.50	135.24	nr	**194.47**
185 mm²	57.19	72.37	3.72	143.73	nr	**216.10**
240 mm²	74.04	93.71	4.33	167.29	nr	**261.00**
300 mm²	92.92	117.60	4.86	187.79	nr	**305.39**
400 mm²	96.39	121.99	5.46	210.96	nr	**332.95**
600/1000 volt grade; seven core (galvanized steel wire armour)						
1.5 mm²	1.88	2.37	0.81	31.29	nr	**33.66**
2.5 mm²	2.58	3.27	0.85	32.84	nr	**36.11**
4 mm²	4.68	5.92	0.93	35.93	nr	**41.85**
600/1000 volt grade; twelve core (galvanized steel wire armour)						
1.5 mm²	2.75	3.48	1.14	44.04	nr	**47.52**
2.5 mm²	4.13	5.23	1.13	43.67	nr	**48.90**
600/1000 volt grade; nineteen core (galvanized steel wire armour)						
1.5 mm²	4.16	5.26	1.54	59.51	nr	**64.77**
2.5 mm²	6.11	7.73	1.54	59.51	nr	**67.24**
600/1000 volt grade; twenty seven core (galvanized steel wire armour)						
1.5 mm²	5.61	7.10	1.94	74.96	nr	**82.06**
2.5 mm²	7.81	9.89	2.31	89.25	nr	**99.14**
600/1000 volt grade; thirty seven core (galvanized steel wire armour)						
1.5 mm²	7.01	8.87	2.53	97.75	nr	**106.62**
2.5 mm²	9.80	12.40	2.87	110.89	nr	**123.29**
UNARMOURED CABLE						
Cable: XLPE insulated; PVC sheathed 90c copper to CMA Code 6181e; for internal wiring; clipped to backgrounds; (Supports and fixings included)						
300/500 volt grade; single core						
6.0 mm²	0.76	0.96	0.09	3.48	m	**4.44**
10 mm²	1.13	1.43	0.10	3.86	m	**5.29**
16 mm²	1.90	2.41	0.12	4.64	m	**7.05**
Cable; LSF insulated to CMA Code 6491B; non-sheathed copper; laid/drawn in trunking/conduit						
450/750 volt grade; single core						
1.5 mm²	0.13	0.17	0.03	1.16	m	**1.33**
2.5 mm²	0.22	0.28	0.03	1.16	m	**1.44**
4.0 mm²	0.33	0.41	0.03	1.16	m	**1.57**
6.0 mm²	0.48	0.60	0.04	1.55	m	**2.15**
10.0 mm²	0.83	1.05	0.04	1.55	m	**2.60**
16.0 mm²	1.30	1.65	0.05	1.94	m	**3.59**
25.0 mm²	2.00	2.53	0.06	2.32	m	**4.85**

ELECTRICAL SUPPLY/POWER/LIGHTING

Item	Net Price £	Material £	Labour hours	Labour £	Unit	Total rate £
LV DISTRIBUTION: CABLES AND WIRING – cont						
Cable – cont						
450/750 volt grade – cont						
35.0 mm²	2.74	3.47	0.06	2.32	m	**5.79**
50.0 mm²	3.82	4.84	0.07	2.71	m	**7.55**
70.0 mm²	5.39	6.82	0.08	3.09	m	**9.91**
95.0 mm²	7.38	9.34	0.08	3.09	m	**12.43**
120.0 mm²	9.31	11.78	0.10	3.86	m	**15.64**
150.0 mm²	11.53	14.59	0.13	5.03	m	**19.62**
185.0 mm²	14.34	18.14	0.13	5.03	m	**23.17**
240.0 mm²	18.75	23.73	0.13	5.03	m	**28.76**
300.0 mm²	23.38	29.59	0.13	5.03	m	**34.62**
Cable; twin & earth to CMA code 6242Y; clipped to backgrounds						
300/500 volt grade; PVC/PVC						
1.5 mm² 2C+E	0.31	0.39	0.01	0.39	m	**0.78**
1.5 mm² 3C+E	0.47	0.59	0.02	0.77	m	**1.36**
2.5 mm² 2C+E	0.46	0.58	0.02	0.77	m	**1.35**
4.0 mm² 2C+E	0.72	0.91	0.02	0.77	m	**1.68**
6.0 mm² 2C+E	1.04	1.32	0.02	0.77	m	**2.09**
10.0 mm² 2C+E	1.68	2.13	0.03	1.16	m	**3.29**
16.0 mm² 2C+E	2.52	3.19	0.03	1.16	m	**4.35**
300/500 volt grade; LSF/LSF						
1.5 mm² 2C+E	0.52	0.66	0.01	0.39	m	**1.05**
1.5 mm² 3C+E	0.74	0.94	0.02	0.77	m	**1.71**
2.5 mm² 2C+E	0.92	1.16	0.02	0.77	m	**1.93**
4.0 mm² 2C+E	1.10	1.39	0.02	0.77	m	**2.16**
6.0 mm² 2C+E	1.60	2.03	0.02	0.77	m	**2.80**
10.0 mm² 2C+E	3.66	4.64	0.03	1.16	m	**5.80**
16.0 mm² 2C+E	5.87	7.43	0.03	1.16	m	**8.59**
EARTH CABLE						
Cable; LSF insulated to CMA Code 6491B; non-sheathed copper; laid/drawn in trunking/conduit						
450/750 volt grade; single core						
1.5 mm²	0.14	0.18	0.03	1.16	m	**1.34**
2.5 mm²	0.22	0.28	0.03	1.16	m	**1.44**
4.0 mm²	0.34	0.43	0.03	1.16	m	**1.59**
6.0 mm²	0.48	0.60	0.04	1.55	m	**2.15**
10.0 mm²	0.84	1.06	0.04	1.55	m	**2.61**
16.0 mm²	1.31	1.66	0.05	1.94	m	**3.60**
25.0 mm²	2.03	2.56	0.06	2.32	m	**4.88**
35.0 mm²	2.77	3.51	0.06	2.32	m	**5.83**
50.0 mm²	3.86	4.88	0.07	2.71	m	**7.59**
70.0 mm²	5.46	6.91	0.08	3.09	m	**10.00**
95.0 mm²	7.47	9.45	0.08	3.09	m	**12.54**

ELECTRICAL SUPPLY/POWER/LIGHTING

Item	Net Price £	Material £	Labour hours	Labour £	Unit	Total rate £
120.0 mm²	9.42	11.92	0.10	3.86	m	**15.78**
150.0 mm²	11.67	14.77	0.13	5.03	m	**19.80**
185.0 mm²	14.51	18.37	0.16	6.17	m	**24.54**
240.0 mm²	18.97	24.01	0.20	7.73	m	**31.74**
300.0 mm²	23.67	29.96	0.13	5.03	m	**34.99**
FLEXIBLE CABLE						
Flexible cord; PVC insulated; PVC sheathed; copper stranded to CMA Code 218Y (laid loose)						
300 volt grade; two core						
0.50 mm²	0.17	0.21	0.07	2.71	m	**2.92**
0.75 mm²	0.22	0.28	0.07	2.71	m	**2.99**
300 volt grade; three core						
0.50 mm²	0.25	0.31	0.07	2.71	m	**3.02**
0.75 mm²	0.34	0.43	0.07	2.71	m	**3.14**
1.0 mm²	0.36	0.46	0.07	2.71	m	**3.17**
1.5 mm²	0.57	0.72	0.07	2.71	m	**3.43**
2.5 mm²	0.77	0.97	0.08	3.09	m	**4.06**
Flexible cord; PVC insulated; PVC sheathed; copper stranded to CMA Code 318Y (laid loose)						
300/500 volt grade; two core						
1.0 mm²	0.68	0.86	0.07	2.71	m	**3.57**
1.5 mm²	0.98	1.24	0.07	2.71	m	**3.95**
2.5 mm²	2.06	2.61	0.07	2.71	m	**5.32**
300/500 volt grade; three core						
0.75 mm²	0.42	0.53	0.07	2.71	m	**3.24**
1.5 mm²	0.69	0.87	0.07	2.71	m	**3.58**
2.5 mm²	1.80	2.27	0.08	3.09	m	**5.36**
300/500 volt grade; four core						
0.75 mm²	1.42	1.79	0.08	3.09	m	**4.88**
1.0 mm²	1.66	2.11	0.08	3.09	m	**5.20**
1.5 mm²	2.36	2.99	0.08	3.09	m	**6.08**
2.5 mm²	3.66	4.64	0.09	3.48	m	**8.12**
Flexible cord; PVC insulated; PVC sheathed for use in high temperature zones; copper stranded to CMA Code 309Y (laid loose)						
300/500 volt grade; two core						
0.50 mm²	0.80	1.01	0.07	2.71	m	**3.72**
0.75 mm²	0.88	1.11	0.07	2.71	m	**3.82**
1.0 mm²	0.36	0.46	0.07	2.71	m	**3.17**
1.5 mm²	0.49	0.62	0.07	2.71	m	**3.33**
2.5 mm²	0.73	0.92	0.07	2.71	m	**3.63**

ELECTRICAL SUPPLY/POWER/LIGHTING

Item	Net Price £	Material £	Labour hours	Labour £	Unit	Total rate £
LV DISTRIBUTION: CABLES AND WIRING – cont						
Flexible cord – cont						
300/500 volt grade; three core						
0.50 mm²	0.36	0.46	0.07	2.71	m	**3.17**
0.75 mm²	1.06	1.34	0.07	2.71	m	**4.05**
1.0 mm²	1.54	1.95	0.07	2.71	m	**4.66**
1.5 mm²	2.14	2.71	0.07	2.71	m	**5.42**
2.5 mm²	3.21	4.07	0.07	2.71	m	**6.78**
Flexible cord; rubber insulated; rubber sheathed; copper stranded to CMA Code 318 (laid loose)						
300/500 volt grade; two core						
0.50 mm²	0.43	0.55	0.07	2.71	m	**3.26**
0.75 mm²	0.53	0.67	0.07	2.71	m	**3.38**
1.0 mm²	0.64	0.81	0.07	2.71	m	**3.52**
1.5 mm²	0.81	1.03	0.07	2.71	m	**3.74**
2.5 mm²	1.21	1.53	0.07	2.71	m	**4.24**
300/500 volt grade; three core						
0.50 mm²	0.53	0.67	0.07	2.71	m	**3.38**
0.75 mm²	1.07	1.36	0.07	2.71	m	**4.07**
1.0 mm²	1.35	1.71	0.07	2.71	m	**4.42**
1.5 mm²	1.71	2.16	0.07	2.71	m	**4.87**
2.5 mm²	2.12	2.69	0.07	2.71	m	**5.40**
300/500 volt grade; four core						
0.50 mm²	0.60	0.76	0.08	3.09	m	**3.85**
0.75 mm²	1.42	1.79	0.08	3.09	m	**4.88**
1.0 mm²	1.65	2.08	0.08	3.09	m	**5.17**
1.5 mm²	2.19	2.77	0.08	3.09	m	**5.86**
2.5 mm²	2.48	3.14	0.08	3.09	m	**6.23**
Flexible cord; rubber insulated; rubber sheathed; for 90C operation; copper stranded to CMA Code 318 (laid loose)						
450/750 volt grade; two core						
0.50 mm²	0.38	0.48	0.07	2.71	m	**3.19**
0.75 mm²	0.59	0.75	0.07	2.71	m	**3.46**
1.0 mm²	0.60	0.76	0.07	2.71	m	**3.47**
1.5 mm²	0.76	0.96	0.07	2.71	m	**3.67**
2.5 mm²	1.10	1.39	0.07	2.71	m	**4.10**
450/750 volt grade; three core						
0.50 mm²	0.61	0.77	0.07	2.71	m	**3.48**
0.75 mm²	0.69	0.87	0.07	2.71	m	**3.58**
1.0 mm²	0.68	0.86	0.07	2.71	m	**3.57**
1.5 mm²	0.93	1.18	0.07	2.71	m	**3.89**
2.5 mm²	1.42	1.79	0.07	2.71	m	**4.50**

ELECTRICAL SUPPLY/POWER/LIGHTING

Item	Net Price £	Material £	Labour hours	Labour £	Unit	Total rate £
450/750 volt grade; four core						
0.75 mm²	0.66	0.84	0.07	2.71	m	3.55
1.0 mm²	0.86	1.09	0.08	3.09	m	4.18
1.5 mm²	1.31	1.66	0.08	3.09	m	4.75
2.5 mm²	1.84	2.33	0.08	3.09	m	5.42
Heavy flexible cable; rubber insulated; rubber sheathed; copper stranded to CMA Code 638P (laid loose)						
450/750 volt grade; two core						
1.0 mm²	0.59	0.75	0.08	3.09	m	3.84
1.5 mm²	0.70	0.88	0.08	3.09	m	3.97
2.5 mm²	0.96	1.21	0.08	3.09	m	4.30
450/750 volt grade; three core						
1.0 mm²	0.70	0.88	0.08	3.09	m	3.97
1.5 mm²	0.82	1.04	0.08	3.09	m	4.13
2.5 mm²	0.94	1.19	0.08	3.09	m	4.28
450/750 volt grade; four core						
1.0 mm²	0.96	1.21	0.08	3.09	m	4.30
1.5 mm²	1.05	1.33	0.08	3.09	m	4.42
2.5 mm²	1.44	1.83	0.08	3.09	m	4.92
FIRE-RATED CABLE						
Cable, mineral insulated; copper sheathed with copper conductors; fixed with clips to backgrounds. BASEC approval to BS 6207 Part 1 1995; complies with BS 6387 Category CWZ						
Light duty 500 volt grade; bare						
2L 1.0	5.18	6.55	0.23	8.88	m	15.43
2L 1.5	5.91	7.48	0.23	8.88	m	16.36
2L 2.5	7.12	9.02	0.25	9.65	m	18.67
2L 4.0	10.06	12.73	0.25	9.65	m	22.38
3L 1.0	6.12	7.75	0.24	9.27	m	17.02
3L 1.5	7.28	9.22	0.25	9.65	m	18.87
3L 2.5	10.40	13.16	0.25	9.65	m	22.81
4L 1.0	6.99	8.85	0.25	9.65	m	18.50
4L 1.5	7.84	9.92	0.25	9.65	m	19.57
4L 2.5	12.32	15.59	0.26	10.05	m	25.64
7L 1.5	12.42	15.71	0.28	10.82	m	26.53
7L 2.5	15.74	19.92	0.27	10.43	m	30.35
Light duty 500 volt grade; LSF sheathed						
2L 1.0	4.13	5.23	0.23	8.88	m	14.11
2L 1.5	4.69	5.94	0.23	8.88	m	14.82
2L 2.5	5.16	6.53	0.25	9.65	m	16.18
2L 4.0	7.74	9.80	0.25	9.65	m	19.45
3L 1.0	4.85	6.14	0.24	9.27	m	15.41
3L 1.5	5.80	7.34	0.25	9.65	m	16.99
3L 2.5	7.49	9.48	0.25	9.65	m	19.13

ELECTRICAL SUPPLY/POWER/LIGHTING

Item	Net Price £	Material £	Labour hours	Labour £	Unit	Total rate £
LV DISTRIBUTION: CABLES AND WIRING – cont						
Cable, mineral insulated – cont						
Light duty 500 volt grade – cont						
4L 1.0	5.47	6.92	0.25	9.65	m	**16.57**
4L 1.5	6.67	8.44	0.25	9.65	m	**18.09**
4L 2.5	9.35	11.84	0.26	10.05	m	**21.89**
7L 1.5	9.65	12.21	0.28	10.82	m	**23.03**
7L 2.5	12.08	15.29	0.27	10.43	m	**25.72**
7L 1.0	7.88	9.97	0.27	10.43	m	**20.40**
Heavy duty 750 volt grade; bare						
1H 10	10.27	13.00	0.25	9.65	m	**22.65**
1H 16	13.90	17.60	0.26	10.05	m	**27.65**
1H 25	19.32	24.45	0.27	10.43	m	**34.88**
1H 35	27.89	35.30	0.32	12.36	m	**47.66**
1H 50	30.53	38.64	0.35	13.53	m	**52.17**
1H 70	40.75	51.58	0.38	14.68	m	**66.26**
1H 95	60.42	76.46	0.41	15.85	m	**92.31**
1H 120	63.56	80.44	0.46	17.77	m	**98.21**
1H 150	77.96	98.66	0.50	19.33	m	**117.99**
1H 185	95.07	120.32	0.56	21.64	m	**141.96**
1H 240	123.69	156.54	0.69	26.67	m	**183.21**
2H 1.5	9.48	12.00	0.25	9.65	m	**21.65**
2H 2.5	11.44	14.48	0.26	10.05	m	**24.53**
2H 4.0	14.00	17.72	0.26	10.05	m	**27.77**
2H 6.0	18.01	22.79	0.29	11.20	m	**33.99**
2H 10.0	23.01	29.12	0.34	13.14	m	**42.26**
2H 16.0	32.92	41.66	0.40	15.46	m	**57.12**
2H 25.0	45.39	57.44	0.44	17.00	m	**74.44**
3H 1.5	10.53	13.33	0.25	9.65	m	**22.98**
3H 2.5	13.04	16.51	0.25	9.65	m	**26.16**
3H 4.0	15.89	20.12	0.27	10.43	m	**30.55**
3H 6.0	19.62	24.83	0.30	11.59	m	**36.42**
3H 10.0	29.10	36.83	0.35	13.53	m	**50.36**
3H 16.0	39.20	49.62	0.41	15.85	m	**65.47**
3H 25.0	57.51	72.79	0.47	18.17	m	**90.96**
4H 1.5	12.78	16.17	0.24	9.27	m	**25.44**
4H 2.5	15.50	19.62	0.26	10.05	m	**29.67**
4H 4.0	19.14	24.23	0.29	11.20	m	**35.43**
4H 6.0	24.21	30.64	0.31	11.97	unit	**42.61**
4H 10.0	34.39	43.52	0.37	14.30	m	**57.82**
4H 16.0	49.28	62.37	0.44	17.00	m	**79.37**
4H 25.0	70.36	89.05	0.52	20.09	m	**109.14**
7H 1.5	17.24	21.82	0.30	11.59	m	**33.41**
7H 2.5	22.98	29.09	0.32	12.36	m	**41.45**
12H 2.5	39.20	49.62	0.39	15.08	m	**64.70**
19H 1.5	57.46	72.72	0.42	16.23	m	**88.95**

ELECTRICAL SUPPLY/POWER/LIGHTING

Item	Net Price £	Material £	Labour hours	Labour £	Unit	Total rate £
Heavy duty 750 volt grade; LSF sheathed						
1H 10	8.11	10.26	0.25	9.65	m	**19.91**
1H 16	11.03	13.96	0.26	10.05	m	**24.01**
1H 25	15.20	19.24	0.27	10.43	m	**29.67**
1H 35	20.59	26.06	0.32	12.36	m	**38.42**
1H 50	25.03	31.67	0.35	13.53	m	**45.20**
1H 70	31.78	40.22	0.38	14.68	m	**54.90**
1H 95	44.58	56.43	0.41	15.85	m	**72.28**
1H 120	49.67	62.87	0.46	17.77	m	**80.64**
1H 150	60.42	76.46	0.50	19.33	m	**95.79**
1H 185	73.34	92.81	0.56	21.64	m	**114.45**
1H 240	95.05	120.30	0.68	26.28	m	**146.58**
2H 1.5	7.57	9.58	0.25	9.65	m	**19.23**
2H 2.5	9.07	11.48	0.26	10.05	m	**21.53**
2H 4.0	11.16	14.12	0.26	10.05	m	**24.17**
2H 6.0	14.26	18.04	0.29	11.20	m	**29.24**
2H 10.0	18.87	23.88	0.34	13.14	m	**37.02**
2H 16.0	25.89	32.77	0.40	15.46	m	**48.23**
2H 25.0	36.04	45.62	0.44	17.00	m	**62.62**
3H 1.5	8.25	10.44	0.25	9.65	m	**20.09**
3H 2.5	10.31	13.05	0.25	9.65	m	**22.70**
3H 4.0	12.67	16.04	0.27	10.43	m	**26.47**
3H 6.0	16.29	20.62	0.30	11.59	m	**32.21**
3H 10.0	21.58	27.32	0.35	13.53	m	**40.85**
3H 16.0	30.22	38.25	0.41	15.85	m	**54.10**
3H 25.0	43.45	54.99	0.47	18.17	m	**73.16**
4H 1.5	10.14	12.84	0.24	9.27	m	**22.11**
4H 2.5	12.32	15.59	0.26	10.05	m	**25.64**
4H 4.0	15.68	19.85	0.29	11.20	m	**31.05**
4H 6.0	19.91	25.20	0.31	11.97	m	**37.17**
4H 10.0	26.21	33.17	0.37	14.30	m	**47.47**
4H 16.0	37.45	47.40	0.44	17.00	m	**64.40**
4H 25.0	52.96	67.02	0.52	20.09	m	**87.11**
7H 1.5	13.06	16.53	0.30	11.59	m	**28.12**
7H 2.5	17.89	22.65	0.32	12.36	m	**35.01**
12H 1.5	22.46	28.43	0.39	15.08	m	**43.51**
12H 2.5	29.91	37.86	0.39	15.08	m	**52.94**
19H 1.5	43.39	54.91	0.42	16.23	m	**71.14**
Cable terminations for MI Cable; Polymeric one piece moulding; containing grey sealing compound; testing; phase marking and connection						
Light duty 500 volt grade; brass gland; polymeric one moulding containing grey sealing compound; coloured conductor sleeving; Earth tag; plastic gland shroud						
2L 1.5	13.01	16.46	0.27	10.43	m	**26.89**
2L 2.5	13.01	16.46	0.27	10.43	m	**26.89**
3L 1.5	13.05	16.52	0.27	10.43	m	**26.95**
4L 1.5	13.05	16.52	0.27	10.43	m	**26.95**

ELECTRICAL SUPPLY/POWER/LIGHTING

Item	Net Price £	Material £	Labour hours	Labour £	Unit	Total rate £
LV DISTRIBUTION: CABLES AND WIRING – cont						
Cable Terminations; for MI copper sheathed cable. Certified for installation in potentially explosive atmospheres; testing; phase marking and connection; BS 6207 Part 2 1995						
Light duty 500 volt grade; brass gland; brass pot with earth tail; pot closure; sealing compound; conductor sleeving; plastic gland shroud; identification markers						
2L 1.0	12.74	16.13	0.39	15.08	nr	**31.21**
2L 1.5	12.74	16.13	0.41	15.85	nr	**31.98**
2L 2.5	12.74	16.13	0.44	16.84	nr	**32.97**
2L 4.0	12.74	16.13	0.46	17.77	nr	**33.90**
3L 1.0	12.76	16.15	0.43	16.62	nr	**32.77**
3L 1.5	12.76	16.15	0.44	16.91	nr	**33.06**
3L 2.5	12.76	16.15	0.44	17.00	nr	**33.15**
4L 1.0	12.76	16.15	0.47	18.17	nr	**34.32**
4L 1.5	12.76	16.15	0.47	18.33	nr	**34.48**
4L 2.5	12.76	16.15	0.50	19.33	nr	**35.48**
7L 1.0	27.28	34.53	0.69	26.67	nr	**61.20**
7L 1.5	27.82	35.21	0.70	27.05	nr	**62.26**
7L 2.5	27.82	35.21	0.74	28.59	nr	**63.80**
Heavy duty 750 volt grade; brass gland; brass pot with earth tail; pot closure; sealing compound; conductor sleeving; plastic gland shroud; identification markers						
1H 10	12.95	16.39	0.37	14.30	nr	**30.69**
1H 16	12.95	16.39	0.39	15.08	nr	**31.47**
1H 25	12.95	16.39	0.56	21.64	nr	**38.03**
1H 35	12.95	16.39	0.57	22.02	nr	**38.41**
1H 50	27.68	35.03	0.60	23.18	nr	**58.21**
1H 70	27.68	35.03	0.67	25.89	nr	**60.92**
1H 95	27.68	35.03	0.75	28.99	nr	**64.02**
1H 120	48.01	60.76	0.94	36.32	nr	**97.08**
1H 150	48.01	60.76	0.99	38.25	nr	**99.01**
1H 185	48.01	60.76	1.26	48.69	nr	**109.45**
1H 240	75.74	95.86	1.37	52.94	nr	**148.80**
2H 1.5	12.95	16.39	0.42	16.23	nr	**32.62**
2H 2.5	12.95	16.39	0.44	17.11	nr	**33.50**
2H 4	12.95	16.39	0.47	18.17	nr	**34.56**
2H 6	12.95	16.39	0.54	20.87	nr	**37.26**
2H 10	27.68	35.03	0.58	22.41	nr	**57.44**
2H 16	27.68	35.03	0.69	26.67	nr	**61.70**
2H 25	43.03	54.45	0.77	29.76	nr	**84.21**
3H 1.5	12.96	16.40	0.44	17.00	nr	**33.40**
3H 2.5	12.96	16.40	0.47	18.28	nr	**34.68**
3H 4	12.96	16.40	0.57	22.02	nr	**38.42**
3H 6	27.75	35.12	0.61	23.56	nr	**58.68**
3H 10	27.75	35.12	0.65	25.11	nr	**60.23**

ELECTRICAL SUPPLY/POWER/LIGHTING

Item	Net Price £	Material £	Labour hours	Labour £	Unit	Total rate £
3H 16	27.75	35.12	0.78	30.14	nr	**65.26**
3H 25	66.30	83.91	0.85	32.84	nr	**116.75**
4H 1.5	12.96	16.40	0.52	20.09	nr	**36.49**
4H 2.5	12.96	16.40	0.53	20.47	nr	**36.87**
4H 4	27.75	35.12	0.60	23.18	nr	**58.30**
4H 6	27.75	35.12	0.65	25.11	nr	**60.23**
4H 10	27.75	35.12	0.69	26.67	nr	**61.79**
4H 16	43.03	54.45	0.88	34.00	nr	**88.45**
4H 25	66.30	83.91	0.93	35.93	nr	**119.84**
7H 1.5	27.68	35.03	0.71	27.44	nr	**62.47**
7H 2.5	27.68	35.03	0.74	28.59	nr	**63.62**
12H 1.5	43.03	54.45	0.85	32.84	nr	**87.29**
12H 2.5	43.03	54.45	1.00	38.64	nr	**93.09**
19H 2.5	66.30	83.91	1.11	42.90	nr	**126.81**
Cable; FP100; LOSH insulated; non-sheathed fire-resistant to LPCB Approved to BS 6387 Category CWZ; in conduit or trunking including terminations						
450/750 volt grade; single core						
1.0 mm²	0.80	1.01	0.13	5.03	m	**6.04**
1.5 mm²	0.35	0.45	0.13	5.03	m	**5.48**
2.5 mm²	0.49	0.62	0.13	5.03	m	**5.65**
4.0 mm²	0.72	0.91	0.14	5.29	m	**6.20**
6.0 mm²	0.89	1.13	0.13	5.03	m	**6.16**
10 mm²	1.63	2.06	0.16	6.17	m	**8.23**
16 mm²	2.49	3.15	0.16	6.17	m	**9.32**
Cable; FP200; Insudite insulated; LSOH sheathed screened fire-resistant; BASEC Approved to BS 7629; fixed with clips to backgrounds						
300/500 volt grade; two core						
1.0 mm²	2.82	3.57	0.20	7.73	m	**11.30**
1.5 mm²	3.47	4.39	0.20	7.73	m	**12.12**
2.5 mm²	4.66	5.90	0.20	7.73	m	**13.63**
4.0 mm²	7.31	9.25	0.23	8.88	m	**18.13**
300/500 volt grade; three core						
1.0 mm²	3.94	4.98	0.23	8.88	m	**13.86**
1.5 mm²	4.74	6.00	0.23	8.88	m	**14.88**
2.5 mm²	5.72	7.24	0.23	8.88	m	**16.12**
4.0 mm²	9.38	11.87	0.25	9.65	m	**21.52**
300/500 volt grade; four core						
1.0 mm²	4.61	5.84	0.25	9.65	m	**15.49**
1.5 mm²	5.58	7.07	0.25	9.65	m	**16.72**
2.5 mm²	7.68	9.72	0.25	9.65	m	**19.37**
4.0 mm	11.75	14.87	0.28	10.82	m	**25.69**

ELECTRICAL SUPPLY/POWER/LIGHTING

Item	Net Price £	Material £	Labour hours	Labour £	Unit	Total rate £
LV DISTRIBUTION: CABLES AND WIRING – cont						
Terminations; including glanding-off, connection to equipment						
Two core						
1.0 mm²	0.71	0.90	0.35	13.53	nr	**14.43**
1.5 mm²	0.71	0.90	0.35	13.53	nr	**14.43**
2.5 mm²	0.71	0.90	0.35	13.53	nr	**14.43**
4.0 mm²	0.71	0.90	0.35	13.53	nr	**14.43**
Three core						
1.0 mm²	0.71	0.90	0.35	13.53	nr	**14.43**
1.5 mm²	0.71	0.90	0.35	13.53	nr	**14.43**
2.5 mm²	0.71	0.90	0.35	13.65	nr	**14.55**
4.0 mm²	0.71	0.90	0.35	13.53	nr	**14.43**
Four core						
1.0 mm²	0.71	0.90	0.35	13.53	nr	**14.43**
1.5 mm²	0.71	0.90	0.35	13.53	nr	**14.43**
2.5 mm²	0.71	0.90	0.35	13.53	nr	**14.43**
4.0 mm²	0.71	0.90	0.35	13.53	nr	**14.43**
Cable; FP400; polymeric insulated; LSOH sheathed fire-resistant; armoured; with copper stranded copper conductors; BASEC Approved to BS 7846; fixed with clips to backgrounds						
600/1000 volt grade; two core						
1.5 mm²	3.22	4.08	0.20	7.73	m	**11.81**
2.5 mm²	3.73	4.72	0.20	7.73	m	**12.45**
4.0 mm²	4.13	5.23	0.21	8.11	m	**13.34**
6.0 mm²	5.09	6.44	0.22	8.50	m	**14.94**
10 mm²	5.49	6.94	0.24	9.27	m	**16.21**
16 mm²	8.35	10.57	0.25	9.65	m	**20.22**
25 mm²	11.54	14.60	0.35	13.53	m	**28.13**
600/1000 volt grade; three core						
1.5 mm²	3.57	4.51	0.20	7.73	m	**12.24**
2.5 mm²	4.16	5.26	0.21	8.11	m	**13.37**
4.0 mm²	4.91	6.22	0.22	8.50	m	**14.72**
6.0 mm²	5.20	6.59	0.22	8.50	m	**15.09**
10 mm²	6.31	7.99	0.25	9.65	m	**17.64**
16 mm²	10.34	13.08	0.26	10.05	m	**23.13**
25 mm²	12.91	16.34	0.37	14.30	m	**30.64**
600/1000 volt grade; four core						
1.5 mm²	4.76	6.03	0.21	8.11	m	**14.14**
2.5 mm²	3.48	4.40	0.22	8.50	m	**12.90**
4.0 mm²	6.53	8.27	0.22	8.50	m	**16.77**
6.0 mm²	6.92	8.76	0.23	8.88	m	**17.64**
10 mm²	7.04	8.92	0.26	10.05	m	**18.97**
16 mm²	10.01	12.67	0.26	10.05	m	**22.72**
25 mm²	11.57	14.64	0.39	15.08	m	**29.72**

ELECTRICAL SUPPLY/POWER/LIGHTING

Item	Net Price £	Material £	Labour hours	Labour £	Unit	Total rate £
Terminations; including glanding-off, connection to equipment						
Two core						
1.5 mm²	2.13	2.70	0.58	22.41	nr	**25.11**
2.5 mm²	2.26	2.86	0.58	22.41	nr	**25.27**
4.0 mm²	2.47	3.12	0.61	23.68	nr	**26.80**
6.0 mm²	2.73	3.45	0.67	25.89	nr	**29.34**
10 mm²	3.00	3.80	1.00	38.64	nr	**42.44**
16 mm²	3.30	4.18	1.11	42.90	nr	**47.08**
25 mm²	3.62	4.58	1.70	65.69	nr	**70.27**
Three core						
1.5 mm²	3.21	4.07	0.62	23.96	nr	**28.03**
2.5 mm²	3.37	4.27	0.62	23.96	nr	**28.23**
4.0 mm²	3.72	4.70	0.66	25.66	nr	**30.36**
6.0 mm²	4.08	5.16	0.71	27.44	nr	**32.60**
10 mm²	4.48	5.67	1.06	40.96	nr	**46.63**
16 mm²	4.95	6.26	1.19	45.98	nr	**52.24**
25 mm²	5.43	6.88	1.81	69.93	nr	**76.81**
Four core						
1.5 mm²	4.27	5.41	0.67	25.89	nr	**31.30**
2.5 mm²	6.37	8.06	0.69	26.47	nr	**34.53**
4.0 mm²	4.72	5.97	0.71	27.44	nr	**33.41**
6.0 mm²	5.01	6.34	0.76	29.37	nr	**35.71**
10 mm²	7.28	9.22	1.14	44.04	nr	**53.26**
16 mm²	7.65	9.68	1.29	49.84	nr	**59.52**
25 mm²	13.45	17.02	1.35	52.17	nr	**69.19**
Cable; FP600; polymeric insulated; LSOH sheathed fire-resistant; armoured; with copper stranded copper conductors; BASEC Approved to BS 7846; fixed with clips to backgrounds						
600/1000 volt grade; two core						
4.0 mm²	5.66	7.17	0.21	8.11	m	**15.28**
10 mm²	6.58	8.33	0.24	9.27	m	**17.60**
16 mm²	7.78	9.84	0.25	9.65	m	**19.49**
600/1000 volt grade; three core						
4.0 mm²	6.58	8.33	0.22	8.50	m	**16.83**
6.0 mm²	6.49	8.21	0.22	8.50	m	**16.71**
10 mm²	8.24	10.43	0.25	9.65	m	**20.08**
16 mm²	9.90	12.53	0.26	10.05	m	**22.58**
25 mm²	13.18	16.68	0.37	14.30	m	**30.98**
35 mm²	15.89	20.12	0.38	14.68	m	**34.80**
50 mm²	21.12	26.73	0.40	15.46	m	**42.19**
70 mm²	25.83	32.69	0.42	16.23	m	**48.92**
95 mm²	38.17	48.31	0.45	17.39	m	**65.70**
600/1000 volt grade; four core						
4.0 mm²	6.46	8.18	0.22	8.50	nr	**16.68**
6.0 mm²	7.12	9.02	0.23	8.88	nr	**17.90**
10 mm²	7.95	10.06	0.26	10.05	nr	**20.11**

ELECTRICAL SUPPLY/POWER/LIGHTING

Item	Net Price £	Material £	Labour hours	Labour £	Unit	Total rate £
LV DISTRIBUTION: CABLES AND WIRING – cont						
Cable – cont						
600/1000 volt grade – cont						
16 mm²	9.13	11.56	0.26	10.05	nr	**21.61**
25 mm²	11.14	14.10	0.39	15.08	nr	**29.18**
35 mm²	14.26	18.04	0.40	15.46	m	**33.50**
50 mm²	17.71	22.41	0.42	16.23	m	**38.64**
70 mm²	22.88	28.95	0.45	17.39	m	**46.34**
95 mm²	30.28	38.33	0.47	18.17	m	**56.50**
120 mm²	37.60	47.59	0.48	18.54	m	**66.13**
150 mm²	45.19	57.19	0.51	19.70	m	**76.89**
185 mm²	55.57	70.32	0.52	20.09	m	**90.41**
240 mm²	71.44	90.42	0.55	21.25	m	**111.67**
300 mm²	87.53	110.78	0.59	22.79	m	**133.57**
400 mm²	109.64	138.76	0.63	24.34	m	**163.10**
Terminations; including glanding-off, connection to equipment						
Two core						
4.0 mm²	3.51	4.45	0.35	13.53	nr	**17.98**
10 mm²	4.08	5.16	0.35	13.53	nr	**18.69**
16 mm²	4.82	6.10	0.35	13.53	nr	**19.63**
Three core						
4.0 mm²	4.08	5.16	0.35	13.53	nr	**18.69**
6.0 mm²	4.02	5.08	0.35	13.53	nr	**18.61**
10 mm²	5.11	6.46	0.35	13.53	nr	**19.99**
16 mm²	6.14	7.77	0.35	13.53	nr	**21.30**
25 mm²	8.17	10.34	0.35	13.53	nr	**23.87**
35 mm²	9.85	12.47	0.35	13.53	m	**26.00**
50 mm²	13.10	16.58	0.35	13.53	m	**30.11**
70 mm²	16.01	20.26	0.35	13.53	m	**33.79**
95 mm²	23.66	29.95	0.35	13.53	m	**43.48**
Four core						
4.0 mm²	4.00	5.06	0.35	13.53	nr	**18.59**
6.0 mm²	4.42	5.59	0.35	13.53	nr	**19.12**
10 mm²	4.93	6.24	0.35	13.53	nr	**19.77**
16 mm²	5.66	7.17	0.35	13.53	nr	**20.70**
25 mm²	6.91	8.75	0.35	13.53	nr	**22.28**
35 mm²	8.84	11.19	0.35	13.53	Unit	**24.72**
50 mm²	10.98	13.90	0.35	13.53	m	**27.43**
70 mm²	14.19	17.95	0.35	13.53	m	**31.48**
95 mm²	18.78	23.77	0.35	13.53	m	**37.30**
120 mm²	23.31	29.50	0.35	13.53	m	**43.03**
150 mm²	28.02	35.46	0.35	13.53	m	**48.99**
185 mm²	34.46	43.61	0.35	13.53	m	**57.14**
240 mm²	44.29	56.06	0.35	13.53	m	**69.59**
300 mm²	54.26	68.67	0.35	13.53	m	**82.20**
400 mm²	67.97	86.03	0.35	13.53	m	**99.56**

ELECTRICAL SUPPLY/POWER/LIGHTING

Item	Net Price £	Material £	Labour hours	Labour £	Unit	Total rate £
Cable; Firetuff fire-resistant to BS 6387; fixed with clips to backgrounds						
Two core						
1.5 mm²	2.44	3.09	0.20	7.73	m	**10.82**
2.5 mm²	2.95	3.73	0.20	7.73	m	**11.46**
4.0 mm²	3.83	4.85	0.21	8.11	m	**12.96**
Three core						
1.5 mm²	2.98	3.77	0.20	7.73	m	**11.50**
2.5 mm²	3.37	4.27	0.21	8.11	m	**12.38**
4.0 mm²	4.87	6.16	0.22	8.50	m	**14.66**
Four core						
1.5 mm²	3.33	4.21	0.21	8.11	m	**12.32**
2.5 mm²	4.08	5.16	0.22	8.50	m	**13.66**
4.0 mm²	5.56	7.03	0.22	8.50	m	**15.53**

ELECTRICAL SUPPLY/POWER/LIGHTING

Item	Net Price £	Material £	Labour hours	Labour £	Unit	Total rate £
LV DISTRIBUTION: MODULAR WIRING						
Modular wiring systems; including commissioning						
Master distribution box; steel; fixed to backgrounds; 6 Port						
4.0 mm 18 core armoured home run cable	176.85	223.82	0.90	34.78	nr	**258.60**
4.0 mm 24 core armoured home run cable	176.85	223.82	0.95	36.70	nr	**260.52**
4.0 mm 18 core armoured home run cable & data cable	187.88	237.78	0.95	36.70	nr	**274.48**
6.0 mm 18 core armoured home run cable	176.85	223.82	1.00	38.64	nr	**262.46**
6.0 mm 24 core armoured home run cable	176.85	223.82	1.10	42.50	nr	**266.32**
6.0 mm 18 core armoured home run cable & data cable	187.88	237.78	1.10	42.50	nr	**280.28**
Master distribution box; steel; fixed to backgrounds; 9 Port						
4.0 mm 27 core armoured home run cable	221.04	279.75	1.30	50.23	nr	**329.98**
4.0 mm 27 core armoured home run cable & data cable	221.04	279.75	1.45	56.03	nr	**335.78**
6.0 mm 27 core armoured home run cable	232.10	293.74	1.45	56.03	nr	**349.77**
6.0 mm 27 core armoured home run cable & data cable	232.10	293.74	1.55	59.89	nr	**353.63**
Metal clad cable; BSEN 60439 Part 2 1993; BASEC approved						
4.0 mm 18 core	15.43	19.53	0.30	11.59	m	**31.12**
4.0 mm 24 core	23.75	30.06	0.32	12.36	m	**42.42**
4.0 mm 27 core	23.94	30.30	0.35	13.53	m	**43.83**
6.0 mm 18 core	20.56	26.02	0.32	12.36	m	**38.38**
6.0 mm 27 core	28.64	36.24	0.35	13.53	m	**49.77**
Metal clad data cable						
Single twisted pair	3.01	3.81	0.18	6.96	m	**10.77**
Twin twisted pair	4.85	6.14	0.18	6.96	m	**13.10**
Distribution cables; armoured; BSEN 60439 Part 2 1993; BASEC approved						
3 wire; 6.1 metre long	70.98	89.84	0.92	35.55	nr	**125.39**
4 wire; 6.1 metre long	80.05	101.32	0.96	37.09	nr	**138.41**
Extender cables; armoured; BSEN 60439 Part 2 1993; BASEC approved						
3 wire						
0.9 metre long	44.55	56.38	0.13	5.03	nr	**61.41**
1.5 metre long	49.05	62.08	0.23	8.88	nr	**70.96**
2.1 metre long	53.54	67.76	0.31	11.97	nr	**79.73**
2.7 metre long	58.04	73.46	0.40	15.46	nr	**88.92**
3.4 metre long	63.29	80.10	0.51	19.70	nr	**99.80**
4.6 metre long	72.29	91.49	0.69	26.67	nr	**118.16**
6.1 metre long	83.53	105.72	0.92	35.55	nr	**141.27**
7.6 metre long	94.79	119.96	1.14	44.04	nr	**164.00**
9.1 metre long	106.02	134.18	1.37	52.94	nr	**187.12**
10.7 metre long	118.02	149.36	1.61	62.20	nr	**211.56**

ELECTRICAL SUPPLY/POWER/LIGHTING

Item	Net Price £	Material £	Labour hours	Labour £	Unit	Total rate £
4 wire						
0.9 metre long	47.68	60.35	0.14	5.41	nr	65.76
1.5 metre long	53.06	67.16	0.24	9.27	nr	76.43
2.1 metre long	58.43	73.95	0.32	12.36	nr	86.31
2.7 metre long	63.81	80.76	0.43	16.62	nr	97.38
3.4 metre long	70.09	88.70	0.51	19.70	nr	108.40
4.6 metre long	80.85	102.32	0.67	25.89	nr	128.21
6.1 metre long	94.58	119.71	0.92	35.55	nr	155.26
7.6 metre long	107.73	136.34	1.22	47.14	nr	183.48
9.1 metre long	121.18	153.36	1.46	56.40	nr	209.76
10.7 metre long	135.52	171.52	1.71	66.08	nr	237.60
3 wire; including twisted pair						
0.9 metre long	61.06	77.28	0.13	5.03	nr	82.31
1.5 metre long	68.61	86.83	0.23	8.88	nr	95.71
2.1 metre long	76.14	96.36	0.31	11.97	nr	108.33
2.7 metre long	83.67	105.90	0.40	15.46	nr	121.36
3.4 metre long	92.47	117.03	0.51	19.70	nr	136.73
4.6 metre long	107.54	136.10	0.69	26.67	nr	162.77
6.1 metre long	125.26	158.52	0.92	35.55	nr	194.07
7.6 metre long	145.20	183.77	1.14	44.04	nr	227.81
9.1 metre long	164.06	207.64	1.37	52.94	nr	260.58
10.7 metre long	184.14	233.05	1.61	62.20	nr	295.25
Extender whip ended cables; armoured; BSEN 60439 Part 2 1993; BASEC approved						
3 wire; 3.0 metre long	42.65	53.97	0.30	11.59	nr	65.56
4 wire; 3.0 metre long	47.27	59.83	0.30	11.59	nr	71.42
T connectors						
3 wire						
6 pole 20a Tee	52.77	66.79	0.10	3.86	nr	70.65
Snap fix	26.89	34.04	0.10	3.86	nr	37.90
1.5 metre flexible cable	31.19	39.47	0.10	3.86	nr	43.33
1.5 metre armoured cable	35.21	44.56	0.15	5.80	nr	50.36
1.5 metre armoured cable with twisted pair	44.88	56.80	0.15	5.80	nr	62.60
4 Wire						
Snap fix	27.66	35.01	0.10	3.86	nr	38.87
1.5 metre flexible cable	32.47	41.09	0.10	3.86	nr	44.95
1.5 metre armoured cable	36.51	46.21	0.18	6.96	nr	53.17
Splitters						
5 wire	24.53	31.05	0.20	7.73	nr	38.78
5 wire converter	31.62	40.02	0.20	7.73	nr	47.75
Switch modules						
3 wire; 6.1 metre long armoured cable	57.97	73.37	0.75	28.99	nr	102.36
4 wire; 6.1 metre long armoured cable	65.23	82.56	0.80	30.90	nr	113.46
Distribution cables; unarmoured; IEC 998 DIN/VDE 0628						
3 wire; 6.1 metre long	27.71	35.07	0.70	27.05	nr	62.12
4 wire; 6.1 metre long	34.66	43.87	0.75	28.99	nr	72.86

ELECTRICAL SUPPLY/POWER/LIGHTING

Item	Net Price £	Material £	Labour hours	Labour £	Unit	Total rate £
LV DISTRIBUTION: MODULAR WIRING – cont						
Extender cables; unarmoured; IEC 998 DIN/VDE 0628						
3 wire						
0.9 metre long	16.92	21.41	0.07	2.71	nr	**24.12**
1.5 metre long	18.52	23.44	0.12	4.64	nr	**28.08**
2.1 metre long	20.14	25.49	0.17	6.56	nr	**32.05**
2.7 metre long	26.65	33.72	0.27	10.43	nr	**44.15**
3.4 metre long	23.62	29.89	0.22	8.50	nr	**38.39**
4.6 metre long	26.86	33.99	0.37	14.30	nr	**48.29**
6.1 metre long	30.88	39.08	0.49	18.93	nr	**58.01**
7.6 metre long	34.89	44.16	0.61	23.56	nr	**67.72**
9.1 metre long	38.91	49.25	0.73	28.21	nr	**77.46**
10.7 metre long	43.21	54.69	0.86	33.23	nr	**87.92**
4 wire						
0.9 metre long	20.37	25.78	0.08	3.09	nr	**28.87**
1.5 metre long	22.47	28.44	0.14	5.41	nr	**33.85**
2.1 metre long	24.57	31.09	0.19	7.34	nr	**38.43**
2.7 metre long	26.65	33.72	0.24	9.27	nr	**42.99**
3.4 metre long	28.53	36.11	0.31	11.97	nr	**48.08**
4.6 metre long	33.29	42.13	0.41	15.85	nr	**57.98**
6.1 metre long	38.52	48.75	0.55	21.25	nr	**70.00**
7.6 metre long	43.74	55.36	0.68	26.28	nr	**81.64**
9.1 metre long	48.98	61.99	0.82	31.67	nr	**93.66**
10.7 metre long	54.56	69.05	0.96	37.09	nr	**106.14**
5 wire						
0.9 metre long	30.90	39.11	0.09	3.48	nr	**42.59**
1.5 metre long	33.61	42.54	0.15	5.80	nr	**48.34**
2.1 metre long	36.30	45.94	0.21	8.11	nr	**54.05**
2.7 metre long	39.02	49.38	0.27	10.43	nr	**59.81**
3.4 metre long	42.17	53.37	0.34	13.14	nr	**66.51**
4.6 metre long	47.58	60.22	0.46	17.77	nr	**77.99**
6.1 metre long	54.35	68.79	0.61	23.56	nr	**92.35**
7.6 metre long	61.12	77.36	0.76	29.37	nr	**106.73**
9.1 metre long	67.87	85.89	0.91	35.16	nr	**121.05**
10.7 metre long	75.09	95.03	1.07	41.35	nr	**136.38**
Extender whip ended cables; armoured; IEC 998 DIN/VDE 0628						
3 wire; 2.5 mm; 3.0 metre long	16.34	20.68	0.30	11.59	nr	**32.27**
4 wire; 2.5 mm; 3.0 metre long	19.23	24.34	0.30	11.59	nr	**35.93**
T connectors						
3 wire						
5 pin; direct fix	15.08	19.08	0.10	3.86	nr	**22.94**
5 pin; 1.5 mm flexible cable; 0.3 metre long	21.12	26.73	0.15	5.80	nr	**32.53**

ELECTRICAL SUPPLY/POWER/LIGHTING

Item	Net Price £	Material £	Labour hours	Labour £	Unit	Total rate £
4 wire						
5 pin; direct fix	19.23	24.34	0.20	7.73	nr	**32.07**
5 pin; 1.5 mm flexible cable; 0.3 metre long	23.93	30.28	0.20	7.73	nr	**38.01**
5 wire						
5 pin; direct fix	21.88	27.69	0.20	7.73	nr	**35.42**
Splitters						
3 way; 5 pin	12.60	15.95	0.25	9.65	nr	**25.60**
Switch modules						
3 wire	35.93	45.47	0.20	7.73	nr	**53.20**
4 wire	36.92	46.73	0.22	8.50	nr	**55.23**

ELECTRICAL SUPPLY/POWER/LIGHTING

Item	Net Price £	Material £	Labour hours	Labour £	Unit	Total rate £
LV DISTRIBUTION: BUSBAR TRUNKING						
MAINS BUSBAR						
Low impedance busbar trunking; fixed to backgrounds including supports, fixings and connections/jointing to equipment						
Straight copper busbar						
1000 amp TP&N	440.51	557.51	3.41	131.76	m	**689.27**
1350 amp TP&N	541.45	685.26	3.58	138.33	m	**823.59**
2000 amp TP&N	679.11	859.48	5.00	193.19	m	**1052.67**
2500 amp TP&N	1026.03	1298.54	5.90	227.98	m	**1526.52**
Extra for fittings mains busbar						
IP54 protection						
1000 amp TP&N	30.04	38.02	2.16	83.45	m	**121.47**
1350 amp TP&N	32.28	40.86	2.61	100.84	m	**141.70**
2000 amp TP&N	44.78	56.67	3.51	135.62	m	**192.29**
2500 amp TP&N	53.17	67.29	3.96	153.01	m	**220.30**
End cover						
1000 amp TP&N	34.55	43.72	0.56	21.64	nr	**65.36**
1350 amp TP&N	36.03	45.60	0.56	21.64	nr	**67.24**
2000 amp TP&N	57.82	73.18	0.66	25.50	nr	**98.68**
2500 amp TP&N	58.57	74.12	0.66	25.50	nr	**99.62**
Edge elbow						
1000 amp TP&N	494.86	626.29	2.01	77.67	nr	**703.96**
1350 amp TP&N	556.93	704.85	2.01	77.67	nr	**782.52**
2000 amp TP&N	863.80	1093.22	2.40	92.74	nr	**1185.96**
2500 amp TP&N	1134.54	1435.87	2.40	92.74	nr	**1528.61**
Flat elbow						
1000 amp TP&N	429.34	543.37	2.01	77.67	nr	**621.04**
1350 amp TP&N	465.54	589.19	2.01	77.67	nr	**666.86**
2000 amp TP&N	660.37	835.77	2.40	92.74	nr	**928.51**
2500 amp TP&N	822.45	1040.89	2.40	92.74	nr	**1133.63**
Offset						
1000 amp TP&N	862.11	1091.08	3.00	115.92	nr	**1207.00**
1350 amp TP&N	1060.38	1342.02	3.00	115.92	nr	**1457.94**
2000 amp TP&N	1670.77	2114.53	3.50	135.24	nr	**2249.77**
2500 amp TP&N	1901.78	2406.89	3.50	135.24	nr	**2542.13**
Edge Z unit						
1000 amp TP&N	1291.43	1634.44	3.00	115.92	nr	**1750.36**
1350 amp TP&N	1653.52	2092.70	3.00	115.92	nr	**2208.62**
2000 amp TP&N	2477.70	3135.78	3.50	135.24	nr	**3271.02**
2500 amp TP&N	2829.43	3580.93	3.50	135.24	nr	**3716.17**
Flat Z unit						
1000 amp TP&N	1396.30	1767.16	3.00	115.92	nr	**1883.08**
1350 amp TP&N	1396.62	1767.56	3.00	115.92	nr	**1883.48**
2000 amp TP&N	2079.40	2631.69	3.50	135.24	nr	**2766.93**
2500 amp TP&N	2500.10	3164.12	3.50	135.24	nr	**3299.36**

ELECTRICAL SUPPLY/POWER/LIGHTING

Item	Net Price £	Material £	Labour hours	Labour £	Unit	Total rate £
Edge tee						
1000 amp TP&N	1291.43	1634.44	2.20	85.01	nr	**1719.45**
1350 amp TP&N	1653.52	2092.70	2.20	85.01	nr	**2177.71**
2000 amp TP&N	2477.70	3135.78	2.60	100.46	nr	**3236.24**
2500 amp TP&N	2830.27	3582.00	2.60	100.46	nr	**3682.46**
Tap off; TP&N integral contactor/breaker						
18 amp	210.06	265.85	0.82	31.67	nr	**297.52**
Tap off; TP&N fusable with on-load switch; excludes fuses						
32 amp	594.34	752.19	0.82	31.67	nr	**783.86**
63 amp	607.27	768.57	0.88	34.00	nr	**802.57**
100 amp	742.66	939.92	1.18	45.60	nr	**985.52**
160 amp	844.21	1068.44	1.41	54.49	nr	**1122.93**
250 amp	1090.10	1379.63	1.76	68.00	nr	**1447.63**
315 amp	1287.21	1629.10	2.06	79.60	nr	**1708.70**
Tap off; TP&N MCCB						
63 amp	751.62	951.25	0.88	34.00	nr	**985.25**
125 amp	900.95	1140.24	1.18	45.60	nr	**1185.84**
160 amp	983.58	1244.82	1.41	54.49	nr	**1299.31**
250 amp	1265.31	1601.38	1.76	68.00	nr	**1669.38**
400 amp	1612.75	2041.10	2.06	79.60	nr	**2120.70**
RISING MAINS BUSBAR						
Rising mains busbar; insulated supports, earth continuity bar; including couplers; fixed to backgrounds						
Straight aluminium bar						
200 amp TP&N	191.95	242.93	2.13	82.30	m	**325.23**
315 amp TP&N	215.05	272.17	2.15	83.07	m	**355.24**
400 amp TP&N	249.05	315.20	2.15	83.07	m	**398.27**
630 amp TP&N	307.38	389.02	2.47	95.44	m	**484.46**
800 amp TP&N	461.66	584.28	2.88	111.28	m	**695.56**
Extra for fittings rising busbar						
End feed unit						
200 amp TP&N	383.92	485.89	2.57	99.30	nr	**585.19**
315 amp TP&N	385.12	487.41	2.76	106.64	nr	**594.05**
400 amp TP&N	430.08	544.31	2.76	106.64	nr	**650.95**
630 amp TP&N	431.29	545.84	3.64	140.65	nr	**686.49**
800 amp TP&N	478.68	605.82	4.54	175.43	nr	**781.25**
Top feeder unit						
200 amp TP&N	383.92	485.89	2.57	99.30	nr	**585.19**
315 amp TP&N	385.12	487.41	2.76	106.64	nr	**594.05**
400 amp TP&N	430.08	544.31	2.76	106.64	nr	**650.95**
630 amp TP&N	431.29	545.84	3.64	140.65	nr	**686.49**
800 amp TP&N	478.68	605.82	4.54	175.43	nr	**781.25**
End cap						
200 amp TP&N	32.80	41.51	0.18	6.96	nr	**48.47**
315 amp TP&N	32.80	41.51	0.27	10.43	nr	**51.94**
400 amp TP&N	36.44	46.12	0.27	10.43	nr	**56.55**

ELECTRICAL SUPPLY/POWER/LIGHTING

Item	Net Price £	Material £	Labour hours	Labour £	Unit	Total rate £
LV DISTRIBUTION: BUSBAR TRUNKING – cont						
Extra for fittings rising busbar – cont						
End cap – cont						
630 amp TP&N	36.44	46.12	0.41	15.85	nr	**61.97**
800 amp TP&N	105.69	133.76	0.41	15.85	nr	**149.61**
Edge elbow						
200 amp TP&N	44.95	56.88	0.55	21.25	nr	**78.13**
315 amp TP&N	46.17	58.43	0.94	36.32	nr	**94.75**
400 amp TP&N	343.82	435.14	0.94	36.32	nr	**471.46**
630 amp TP&N	345.04	436.69	1.45	56.03	nr	**492.72**
800 amp TP&N	326.81	413.62	1.45	56.03	nr	**469.65**
Flat elbow						
200 amp TP&N	148.22	187.59	0.55	21.25	nr	**208.84**
315 amp TP&N	148.22	187.59	0.94	36.32	nr	**223.91**
400 amp TP&N	200.46	253.70	0.94	36.32	nr	**290.02**
630 amp TP&N	201.67	255.24	1.45	56.03	nr	**311.27**
800 amp TP&N	278.22	352.12	1.45	56.03	nr	**408.15**
Edge tee						
200 amp TP&N	207.74	262.92	0.61	23.56	nr	**286.48**
315 amp TP&N	208.97	264.48	1.02	39.41	nr	**303.89**
400 amp TP&N	292.80	370.56	1.02	39.41	nr	**409.97**
630 amp TP&N	292.80	370.56	1.57	60.66	nr	**431.22**
800 amp TP&N	478.68	605.82	1.57	60.66	nr	**666.48**
Flat tee						
200 amp TP&N	267.28	338.27	0.61	23.56	nr	**361.83**
315 amp TP&N	208.97	264.48	1.02	39.41	nr	**303.89**
400 amp TP&N	420.36	532.01	1.02	39.41	nr	**571.42**
630 amp TP&N	421.58	533.56	1.57	60.66	nr	**594.22**
800 amp TP&N	589.23	745.73	1.57	60.66	nr	**806.39**
Tap off units						
TP&N fusable with on-load switch; excludes fuses						
32 amp	195.61	247.56	0.82	31.67	nr	**279.23**
63 amp	258.77	327.50	0.88	34.00	nr	**361.50**
100 amp	347.46	439.75	1.18	45.60	nr	**485.35**
250 amp	521.20	659.64	1.41	54.49	nr	**714.13**
400 amp	759.32	960.99	2.06	79.60	nr	**1040.59**
TP&N MCCB						
32 amp	199.25	252.17	0.82	31.67	nr	**283.84**
63 amp	275.79	349.04	0.88	34.00	nr	**383.04**
100 amp	442.22	559.68	1.18	45.60	nr	**605.28**
250 amp	749.60	948.70	1.41	54.49	nr	**1003.19**
400 amp	1316.97	1666.76	2.06	79.60	nr	**1746.36**

ELECTRICAL SUPPLY/POWER/LIGHTING

Item	Net Price £	Material £	Labour hours	Labour £	Unit	Total rate £
LIGHTING BUSBAR						
Prewired busbar, plug-in trunking for lighting; galvanized sheet steel housing (PE); tin-plated copper conductors with tap off units at 1 m intervals						
Straight lengths – 25 amp						
2 pole & PE	31.91	40.39	0.16	6.17	m	**46.56**
4 pole & PE	35.45	44.87	0.16	6.17	m	**51.04**
Straight lengths – 40 amp						
2 pole & PE	31.91	40.39	0.16	6.17	m	**46.56**
4 pole & PE	42.53	53.83	0.16	6.17	m	**60.00**
Components for pre-wired busbars, plug-in trunking for lighting						
Plug-in tap off units						
10 amp 4 pole & PE; 3 m of cable	27.17	34.38	0.10	3.86	nr	**38.24**
16 amp 4 pole & PE; 3 m of cable	29.54	37.39	0.10	3.86	nr	**41.25**
16 amp with phase selection, 2 pole & PE; no cable	22.45	28.41	0.10	3.86	nr	**32.27**
Trunking components						
End feed unit & cover; 4 pole & PE	34.26	43.36	0.23	8.88	nr	**52.24**
Centre feed unit	171.34	216.84	0.29	11.20	nr	**228.04**
Right hand, intermediate terminal box feed unit	36.63	46.36	0.23	8.88	nr	**55.24**
End cover (for R/hand feed)	9.46	11.97	0.06	2.32	nr	**14.29**
Flexible elbow unit	81.54	103.20	0.12	4.64	nr	**107.84**
Fixing bracket – universal	5.92	7.49	0.10	3.86	nr	**11.35**
Suspension bracket – flat	4.72	5.97	0.10	3.86	nr	**9.83**
UNDERFLOOR BUSBAR						
Prewired busbar, plug-in trunking for underfloor power distribution; galvanized sheet steel housing (PE); copper conductors with tap off units at 300 mm intervals						
Straight lengths – 63 amp						
2 pole & PE	20.08	25.41	0.28	10.82	m	**36.23**
3 pole & PE; Clean Earth System	26.00	32.91	0.28	10.82	m	**43.73**
Components for prewired busbars, plug-in trunking for underfloor power distribution						
Plug-in tap off units						
32 amp 2 pole & PE; 3 m metal flexible prewired conduit	33.09	41.88	0.25	9.65	nr	**51.53**
32 amp 3 pole & PE; clean earth; 3 m metal flexible prewired conduit	40.18	50.85	0.28	10.82	nr	**61.67**

ELECTRICAL SUPPLY/POWER/LIGHTING

Item	Net Price £	Material £	Labour hours	Labour £	Unit	Total rate £
LV DISTRIBUTION: BUSBAR TRUNKING – cont						
Components for prewired busbars, plug-in trunking for underfloor power distribution – cont						
Trunking components						
End feed unit & cover; 2 pole & PE	35.45	44.87	0.35	13.53	nr	**58.40**
End feed unit & cover; 3 pole & PE; clean earth	38.99	49.35	0.38	14.68	nr	**64.03**
End cover; 2 pole & PE	10.64	13.46	0.11	4.26	nr	**17.72**
End cover; 3 pole & PE	11.81	14.95	0.11	4.26	nr	**19.21**
Flexible interlink/corner; 2 pole & PE; 1 m long	61.44	77.76	0.34	13.14	nr	**90.90**
Flexible interlink/corner; 3 pole & PE; 1 m long	69.72	88.23	0.35	13.53	nr	**101.76**
Flexible interlink/corner; 2 pole & PE; 2 m long	75.62	95.70	0.37	14.30	nr	**110.00**
Flexible interlink/corner; 3 pole & PE; 2 m long	82.71	104.68	0.37	14.30	nr	**118.98**

ELECTRICAL SUPPLY/POWER/LIGHTING

Item	Net Price £	Material £	Labour hours	Labour £	Unit	Total rate £
LV DISTRIBUTION: CABLE SUPPORTS						
LADDER RACK						
Light duty galvanized steel ladder rack; fixed to backgrounds; including supports, fixings and brackets; earth continuity straps						
Straight lengths						
150 mm wide ladder	31.40	39.74	0.69	26.67	m	**66.41**
300 mm wide ladder	34.02	43.05	0.88	34.00	m	**77.05**
450 mm wide ladder	21.06	26.66	1.26	48.69	m	**75.35**
600 mm wide ladder	38.57	48.81	1.51	58.34	m	**107.15**
750 mm wide ladder	29.08	36.80	1.69	65.31	m	**102.11**
900 mm wide ladder	30.76	38.93	1.75	67.63	m	**106.56**
Extra over (cutting and jointing racking to fittings is included)						
Inside riser bend						
150 mm wide ladder	112.65	142.56	0.33	12.75	nr	**155.31**
300 mm wide ladder	114.91	145.43	0.56	21.64	nr	**167.07**
450 mm wide ladder	96.67	122.35	0.85	32.84	nr	**155.19**
600 mm wide ladder	155.95	197.37	0.99	38.25	nr	**235.62**
750 mm wide ladder	83.85	106.12	1.07	41.35	nr	**147.47**
900 mm wide ladder	87.56	110.81	1.12	43.27	nr	**154.08**
Outside riser bend						
150 mm wide ladder	112.65	142.56	0.43	16.62	nr	**159.18**
300 mm wide ladder	114.91	145.43	0.43	16.62	nr	**162.05**
450 mm wide ladder	76.37	96.66	0.73	28.21	nr	**124.87**
600 mm wide ladder	155.95	197.37	0.86	33.23	nr	**230.60**
750 mm wide ladder	83.85	106.12	0.97	37.48	nr	**143.60**
900 mm wide ladder	87.56	110.81	1.15	44.43	nr	**155.24**
Equal tee						
150 mm wide ladder	93.57	118.42	0.62	23.96	nr	**142.38**
300 mm wide ladder	131.26	166.12	0.62	23.96	nr	**190.08**
450 mm wide ladder	120.65	152.69	1.09	42.12	nr	**194.81**
600 mm wide ladder	107.84	136.48	1.12	43.27	nr	**179.75**
750 mm wide ladder	176.69	223.62	1.16	44.81	nr	**268.43**
900 mm wide ladder	182.75	231.29	1.21	46.75	nr	**278.04**
Unequal tee						
150 mm wide ladder	82.80	104.79	0.57	22.02	nr	**126.81**
300 mm wide ladder	76.82	97.23	0.57	22.02	nr	**119.25**
450 mm wide ladder	97.68	123.63	1.17	45.20	nr	**168.83**
600 mm wide ladder	83.63	105.84	1.17	45.20	nr	**151.04**
750 mm wide ladder	100.77	127.53	1.37	52.94	nr	**180.47**
900 mm wide ladder	101.61	128.60	1.37	52.94	nr	**181.54**
4 way crossovers						
150 mm wide ladder	239.64	303.28	0.72	27.82	nr	**331.10**
300 mm wide ladder	246.60	312.10	0.72	27.82	nr	**339.92**
450 mm wide ladder	176.17	222.96	1.13	43.67	nr	**266.63**

ELECTRICAL SUPPLY/POWER/LIGHTING

Item	Net Price £	Material £	Labour hours	Labour £	Unit	Total rate £
LV DISTRIBUTION: CABLE SUPPORTS – cont						
Extra over (cutting and jointing racking to fittings is included) – cont						
4 way crossovers – cont						
600 mm wide ladder	274.05	346.84	1.29	49.84	nr	**396.68**
750 mm wide ladder	170.01	215.16	1.41	54.49	nr	**269.65**
900 mm wide ladder	174.60	220.98	1.64	63.37	nr	**284.35**
Flat bend (light duty)						
150 mm wide ladder	102.59	129.84	0.36	13.91	nr	**143.75**
300 mm wide ladder	106.94	135.34	0.40	15.46	nr	**150.80**
450 mm wide ladder	82.47	104.37	0.42	16.23	nr	**120.60**
600 mm wide ladder	133.91	169.48	0.59	22.79	nr	**192.27**
750 mm wide ladder	109.72	138.86	0.78	30.14	nr	**169.00**
900 mm wide ladder	119.79	151.60	0.86	33.23	nr	**184.83**
Heavy duty galvanized steel ladder rack; fixed to backgrounds; including supports, fixings and brackets; earth continuity straps						
Straight lengths						
150 mm wide ladder	41.73	52.81	0.68	26.28	m	**79.09**
300 mm wide ladder	45.00	56.95	0.79	30.53	m	**87.48**
450 mm wide ladder	48.12	60.91	1.07	41.35	m	**102.26**
600 mm wide ladder	52.06	65.89	1.24	47.91	m	**113.80**
750 mm wide ladder	60.33	76.35	1.49	57.57	m	**133.92**
900 mm wide ladder	52.20	66.07	1.67	64.53	m	**130.60**
Extra over (cutting and jointing racking to fittings is included)						
Flat bend						
150 mm wide ladder	112.73	142.67	0.34	13.14	nr	**155.81**
300 mm wide ladder	118.99	150.60	0.39	15.08	nr	**165.68**
450 mm wide ladder	132.33	167.47	0.43	16.62	nr	**184.09**
600 mm wide ladder	147.45	186.61	0.61	23.56	nr	**210.17**
750 mm wide ladder	166.05	210.16	0.82	31.67	nr	**241.83**
900 mm wide ladder	178.34	225.70	0.97	37.48	nr	**263.18**
Inside riser bend						
150 mm wide ladder	144.47	182.84	0.27	10.43	nr	**193.27**
300 mm wide ladder	146.47	185.37	0.45	17.39	nr	**202.76**
450 mm wide ladder	156.88	198.54	0.65	25.11	nr	**223.65**
600 mm wide ladder	167.43	211.90	0.81	31.29	nr	**243.19**
750 mm wide ladder	173.97	220.18	0.92	35.55	nr	**255.73**
900 mm wide ladder	187.99	237.92	1.06	40.96	nr	**278.88**
Outside riser bend						
150 mm wide ladder	144.47	182.84	0.27	10.43	nr	**193.27**
300 mm wide ladder	146.47	185.37	0.33	12.75	nr	**198.12**
450 mm wide ladder	156.88	198.54	0.61	23.56	nr	**222.10**
600 mm wide ladder	167.43	211.90	0.76	29.37	nr	**241.27**
750 mm wide ladder	173.97	220.18	0.94	36.32	nr	**256.50**
900 mm wide ladder	191.75	242.68	1.05	40.58	nr	**283.26**

ELECTRICAL SUPPLY/POWER/LIGHTING

Item	Net Price £	Material £	Labour hours	Labour £	Unit	Total rate £
Equal tee						
150 mm wide ladder	163.85	207.37	0.37	14.30	nr	**221.67**
300 mm wide ladder	187.57	237.38	0.57	22.02	nr	**259.40**
450 mm wide ladder	202.25	255.96	0.83	32.07	nr	**288.03**
600 mm wide ladder	223.06	282.31	0.92	35.55	nr	**317.86**
750 mm wide ladder	279.06	353.18	1.13	43.67	nr	**396.85**
900 mm wide ladder	293.74	371.76	1.20	46.37	nr	**418.13**
Unequal tee						
300 mm wide ladder	176.83	223.80	0.57	22.02	nr	**245.82**
450 mm wide ladder	188.87	239.03	1.17	45.20	nr	**284.23**
600 mm wide ladder	199.54	252.54	1.17	45.20	nr	**297.74**
750 mm wide ladder	247.74	313.54	1.25	48.29	nr	**361.83**
900 mm wide ladder	251.92	318.83	1.33	51.39	nr	**370.22**
4 way crossovers						
150 mm wide ladder	255.25	323.04	0.50	19.33	nr	**342.37**
300 mm wide ladder	267.92	339.08	0.67	25.89	nr	**364.97**
450 mm wide ladder	330.93	418.82	0.92	35.55	nr	**454.37**
600 mm wide ladder	352.11	445.63	1.07	41.35	nr	**486.98**
750 mm wide ladder	368.64	466.55	1.25	48.29	nr	**514.84**
900 mm wide ladder	394.73	499.56	1.36	52.55	nr	**552.11**
Double set						
150 mm wide ladder	210.42	266.30	0.51	19.70	nr	**286.00**
300 mm wide ladder	222.10	281.09	0.63	24.34	nr	**305.43**
450 mm wide ladder	247.00	312.60	0.86	33.23	nr	**345.83**
600 mm wide ladder	275.20	348.30	0.99	38.25	nr	**386.55**
750 mm wide ladder	309.94	392.26	1.16	44.81	nr	**437.07**
900 mm wide ladder	139.54	176.60	1.31	50.61	nr	**227.21**
Extra heavy duty galvanized steel ladder rack; fixed to backgrounds; including supports, fixings and brackets; earth continuity straps						
Straight lengths						
150 mm wide ladder	52.70	66.70	0.63	24.34	m	**91.04**
300 mm wide ladder	55.31	70.00	0.70	27.05	m	**97.05**
450 mm wide ladder	58.05	73.47	0.83	32.07	m	**105.54**
600 mm wide ladder	58.83	74.46	0.89	34.38	m	**108.84**
750 mm wide ladder	58.99	74.66	1.22	47.14	m	**121.80**
900 mm wide ladder	61.56	77.91	1.44	55.63	m	**133.54**
Extra over (cutting and jointing racking to fittings is included)						
Flat bend						
150 mm wide ladder	127.14	160.91	0.36	13.91	nr	**174.82**
300 mm wide ladder	132.30	167.44	0.39	15.08	nr	**182.52**
450 mm wide ladder	142.86	180.80	0.43	16.62	nr	**197.42**
600 mm wide ladder	161.61	204.53	0.61	23.56	nr	**228.09**
750 mm wide ladder	175.20	221.74	0.82	31.67	nr	**253.41**
900 mm wide ladder	189.75	240.15	0.97	37.48	nr	**277.63**

ELECTRICAL SUPPLY/POWER/LIGHTING

Item	Net Price £	Material £	Labour hours	Labour £	Unit	Total rate £
LV DISTRIBUTION: CABLE SUPPORTS – cont						
Extra over (cutting and jointing racking to fittings is included) – cont						
Inside riser bend						
150 mm wide ladder	139.38	176.40	0.36	13.91	nr	**190.31**
300 mm wide ladder	141.64	179.26	0.39	15.08	nr	**194.34**
450 mm wide ladder	149.72	189.48	0.43	16.62	nr	**206.10**
600 mm wide ladder	165.13	208.99	0.61	23.56	nr	**232.55**
750 mm wide ladder	169.41	214.40	0.82	31.67	nr	**246.07**
900 mm wide ladder	182.67	231.19	0.97	37.48	nr	**268.67**
Outside riser bend						
150 mm wide ladder	139.75	176.87	0.36	13.91	nr	**190.78**
300 mm wide ladder	141.64	179.26	0.39	15.08	nr	**194.34**
450 mm wide ladder	149.72	189.48	0.41	15.85	nr	**205.33**
600 mm wide ladder	165.13	208.99	0.57	22.02	nr	**231.01**
750 mm wide ladder	169.41	214.40	0.82	31.67	nr	**246.07**
900 mm wide ladder	182.67	231.19	0.93	35.93	nr	**267.12**
Equal tee						
150 mm wide ladder	180.48	228.41	0.37	14.30	nr	**242.71**
300 mm wide ladder	199.26	252.18	0.57	22.02	nr	**274.20**
450 mm wide ladder	211.05	267.11	0.83	32.07	nr	**299.18**
600 mm wide ladder	237.87	301.04	0.92	35.55	nr	**336.59**
750 mm wide ladder	277.94	351.76	1.13	43.67	nr	**395.43**
900 mm wide ladder	288.00	364.49	1.20	46.37	nr	**410.86**
Unequal tee						
150 mm wide ladder	179.63	227.34	0.37	14.30	nr	**241.64**
300 mm wide ladder	197.82	250.36	0.57	22.02	nr	**272.38**
450 mm wide ladder	209.25	264.82	1.17	45.20	nr	**310.02**
600 mm wide ladder	235.41	297.93	1.17	45.20	nr	**343.13**
750 mm wide ladder	287.52	363.89	1.25	48.29	nr	**412.18**
900 mm wide ladder	289.38	366.24	1.33	51.39	nr	**417.63**
4 way crossovers						
150 mm wide ladder	234.23	296.44	0.50	19.33	nr	**315.77**
300 mm wide ladder	243.42	308.07	0.67	25.89	nr	**333.96**
450 mm wide ladder	284.76	360.39	0.92	35.55	nr	**395.94**
600 mm wide ladder	316.57	400.65	1.07	41.35	nr	**442.00**
750 mm wide ladder	329.08	416.48	1.25	48.29	nr	**464.77**
900 mm wide ladder	343.85	435.18	1.36	52.55	nr	**487.73**

ELECTRICAL SUPPLY/POWER/LIGHTING

Item	Net Price £	Material £	Labour hours	Labour £	Unit	Total rate £
LV DISTRIBUTION: CABLE TRAYS						
Galvanized steel cable tray to BS 729; including standard coupling joints, fixings and earth continuity straps (supports and hangers are excluded)						
Light duty tray						
Straight lengths						
50 mm wide	4.00	5.06	0.19	7.34	m	12.40
75 mm wide	5.04	6.38	0.23	8.88	m	15.26
100 mm wide	6.21	7.86	0.31	11.97	m	19.83
150 mm wide	8.13	10.29	0.33	12.75	m	23.04
225 mm wide	15.53	19.66	0.39	15.08	m	34.74
300 mm wide	21.84	27.64	0.49	18.93	m	46.57
450 mm wide	12.30	15.57	0.60	23.18	m	38.75
600 mm wide	31.08	39.33	0.79	30.53	m	69.86
750 mm wide	39.48	49.96	1.04	40.19	m	90.15
900 mm wide	49.10	62.14	1.26	48.69	m	110.83
Extra over (cutting and jointing tray to fittings is included)						
Straight reducer						
75 mm wide	16.28	20.61	0.22	8.50	nr	29.11
100 mm wide	16.60	21.01	0.25	9.65	nr	30.66
150 mm wide	21.70	27.46	0.27	10.43	nr	37.89
225 mm wide	28.49	36.05	0.34	13.14	nr	49.19
300 mm wide	36.27	45.91	0.39	15.08	nr	60.99
450 mm wide	28.41	35.95	0.49	18.93	nr	54.88
600 mm wide	45.86	58.04	0.54	20.87	nr	78.91
750 mm wide	60.25	76.25	0.61	23.56	nr	99.81
900 mm wide	70.00	88.59	0.69	26.67	nr	115.26
Flat bend; 90°						
50 mm wide	9.77	12.36	0.19	7.34	nr	19.70
75 mm wide	10.17	12.87	0.24	9.27	nr	22.14
100 mm wide	11.44	14.48	0.28	10.82	nr	25.30
150 mm wide	12.37	15.66	0.30	11.59	nr	27.25
225 mm wide	16.06	20.33	0.36	13.91	nr	34.24
300 mm wide	23.49	29.72	0.44	17.00	nr	46.72
450 mm wide	17.13	21.68	0.57	22.02	nr	43.70
600 mm wide	37.86	47.91	0.69	26.67	nr	74.58
750 mm wide	53.62	67.86	0.81	31.29	nr	99.15
900 mm wide	78.86	99.80	0.94	36.32	nr	136.12
Adjustable riser						
50 mm wide	19.63	24.84	0.26	10.05	nr	34.89
75 mm wide	21.24	26.88	0.29	11.20	nr	38.08
100 mm wide	20.56	26.02	0.32	12.36	nr	38.38
150 mm wide	23.80	30.12	0.36	13.91	nr	44.03
225 mm wide	31.34	39.66	0.44	17.00	nr	56.66
300 mm wide	33.58	42.50	0.52	20.09	nr	62.59
450 mm wide	29.01	36.71	0.66	25.50	nr	62.21

ELECTRICAL SUPPLY/POWER/LIGHTING

Item	Net Price £	Material £	Labour hours	Labour £	Unit	Total rate £
LV DISTRIBUTION: CABLE TRAYS – cont						
Extra over (cutting and jointing tray to fittings is included) – cont						
Adjustable riser – cont						
600 mm wide	44.85	56.76	0.79	30.53	nr	**87.29**
750 mm wide	57.72	73.05	1.03	39.80	nr	**112.85**
900 mm wide	67.93	85.97	1.10	42.50	nr	**128.47**
Inside riser; 90°						
50 mm wide	6.17	7.81	0.28	10.82	nr	**18.63**
75 mm wide	6.45	8.16	0.31	11.97	nr	**20.13**
100 mm wide	7.09	17.12	0.33	12.75	nr	**29.87**
150 mm wide	8.42	10.65	0.37	14.30	nr	**24.95**
225 mm wide	10.35	13.10	0.44	17.00	nr	**30.10**
300 mm wide	10.95	13.85	0.53	20.47	nr	**34.32**
450 mm wide	17.52	22.18	0.67	25.89	nr	**48.07**
600 mm wide	44.74	56.63	0.79	30.53	nr	**87.16**
750 mm wide	55.19	69.84	0.95	36.70	nr	**106.54**
900 mm wide	65.57	82.98	1.11	42.90	nr	**125.88**
Outside riser; 90°						
50 mm wide	6.17	7.81	0.28	10.82	nr	**18.63**
75 mm wide	6.45	8.16	0.31	11.97	nr	**20.13**
100 mm wide	7.09	8.97	0.33	12.75	nr	**21.72**
150 mm wide	8.42	10.65	0.37	14.30	nr	**24.95**
225 mm wide	10.35	13.10	0.44	17.00	nr	**30.10**
300 mm wide	10.95	13.85	0.53	20.47	nr	**34.32**
450 mm wide	17.52	22.18	0.67	25.89	nr	**48.07**
600 mm wide	24.88	31.48	0.79	30.53	nr	**62.01**
750 mm wide	33.78	42.75	0.95	36.70	nr	**79.45**
900 mm wide	33.53	42.44	1.11	42.90	nr	**85.34**
Equal tee						
50 mm wide	14.76	18.68	0.30	11.59	nr	**30.27**
75 mm wide	15.34	19.41	0.31	11.97	nr	**31.38**
100 mm wide	15.80	19.99	0.35	13.53	nr	**33.52**
150 mm wide	19.18	24.27	0.36	13.91	nr	**38.18**
225 mm wide	25.39	32.13	0.44	17.00	nr	**49.13**
300 mm wide	35.27	44.64	0.54	20.87	nr	**65.51**
450 mm wide	25.10	31.76	0.71	27.44	nr	**59.20**
600 mm wide	49.95	63.21	0.92	35.55	nr	**98.76**
750 mm wide	74.88	94.76	1.19	45.98	nr	**140.74**
900 mm wide	106.50	134.79	1.44	55.63	nr	**190.42**
Unequal tee						
75 mm wide	9.51	12.04	0.38	14.68	nr	**26.72**
100 mm wide	9.51	12.04	0.39	15.08	nr	**27.12**
150 mm wide	9.79	12.39	0.43	16.62	nr	**29.01**
225 mm wide	11.36	14.38	0.50	19.33	nr	**33.71**
300 mm wide	14.31	18.11	0.63	24.34	nr	**42.45**
450 mm wide	18.98	24.02	0.80	30.90	nr	**54.92**
600 mm wide	29.77	37.68	1.02	39.41	nr	**77.09**
750 mm wide	85.04	107.63	1.12	43.27	nr	**150.90**
900 mm wide	112.10	141.87	1.35	52.17	nr	**194.04**

ELECTRICAL SUPPLY/POWER/LIGHTING

Item	Net Price £	Material £	Labour hours	Labour £	Unit	Total rate £
4 way crossovers						
50 mm wide	21.66	27.42	0.38	14.68	nr	42.10
75 mm wide	22.07	27.93	0.40	15.46	nr	43.39
100 mm wide	23.22	29.39	0.40	15.46	nr	44.85
150 mm wide	27.51	34.82	0.44	17.00	nr	51.82
225 mm wide	37.06	46.91	0.53	20.47	nr	67.38
300 mm wide	51.01	64.56	0.64	24.73	nr	89.29
450 mm wide	44.36	56.15	0.84	32.46	nr	88.61
600 mm wide	49.95	63.21	1.03	39.80	nr	103.01
750 mm wide	74.88	94.76	1.13	43.67	nr	138.43
900 mm wide	106.50	134.79	1.36	52.55	nr	187.34
Medium duty tray with return flange						
Straight lengths						
50 mm wide	8.90	11.27	0.33	12.75	m	24.02
75 mm wide	8.73	11.04	0.33	12.75	m	23.79
100 mm wide	9.46	11.97	0.35	13.53	m	25.50
150 mm wide	11.47	14.52	0.39	15.08	m	29.60
225 mm wide	14.83	18.77	0.45	17.39	m	36.16
300 mm wide	22.90	28.99	0.57	22.02	m	51.01
450 mm wide	32.00	40.50	0.69	26.67	m	67.17
600 mm wide	42.85	54.23	0.91	35.16	m	89.39
Extra over (cutting and jointing tray to fittings is included)						
Straight reducer						
100 mm wide	31.33	39.65	0.25	9.65	nr	49.30
150 mm wide	34.23	43.32	0.27	10.43	nr	53.75
225 mm wide	39.43	49.91	0.34	13.14	nr	63.05
300 mm wide	45.66	57.79	0.39	15.08	nr	72.87
450 mm wide	59.01	74.68	0.49	18.93	nr	93.61
600 mm wide	73.11	92.52	0.54	20.87	nr	113.39
Flat bend; 90°						
75 mm wide	43.54	55.10	0.24	9.27	nr	64.37
100 mm wide	45.88	58.06	0.28	10.82	nr	68.88
150 mm wide	49.09	62.13	0.30	11.59	nr	73.72
225 mm wide	56.48	71.48	0.36	13.91	nr	85.39
300 mm wide	67.76	85.76	0.44	17.00	nr	102.76
450 mm wide	72.68	91.99	0.57	22.02	nr	114.01
600 mm wide	116.13	146.98	0.69	26.67	nr	173.65
Adjustable bend						
75 mm wide	47.77	60.46	0.29	11.20	nr	71.66
100 mm wide	51.99	65.80	0.32	12.36	nr	78.16
150 mm wide	59.46	75.25	0.36	13.91	nr	89.16
225 mm wide	66.63	84.32	0.44	17.00	nr	101.32
300 mm wide	75.30	95.30	0.52	20.09	nr	115.39
Adjustable riser						
75 mm wide	42.61	53.93	0.29	11.20	nr	65.13
100 mm wide	43.31	54.81	0.32	12.36	nr	67.17
150 mm wide	47.74	60.42	0.36	13.91	nr	74.33

ELECTRICAL SUPPLY/POWER/LIGHTING

Item	Net Price £	Material £	Labour hours	Labour £	Unit	Total rate £
LV DISTRIBUTION: CABLE TRAYS – cont						
Extra over (cutting and jointing tray to fittings is included) – cont						
Adjustable riser – cont						
225 mm wide ·	50.64	64.09	0.44	17.00	nr	**81.09**
300 mm wide	53.99	68.33	0.52	20.09	nr	**88.42**
450 mm wide	70.82	89.63	0.66	25.50	nr	**115.13**
600 mm wide	86.86	109.93	0.79	30.53	nr	**140.46**
Inside riser; 90°						
75 mm wide	25.60	32.40	0.31	11.97	nr	**44.37**
100 mm wide	25.85	32.72	0.33	12.75	nr	**45.47**
150 mm wide	29.44	37.26	0.37	14.30	nr	**51.56**
225 mm wide	35.97	45.53	0.44	17.00	nr	**62.53**
300 mm wide	43.69	55.29	0.53	20.47	nr	**75.76**
450 mm wide	63.06	79.81	0.67	25.89	nr	**105.70**
600 mm wide	101.20	128.08	0.79	30.53	nr	**158.61**
Outside riser; 90°						
75 mm wide	25.60	32.40	0.31	11.97	nr	**44.37**
100 mm wide	25.85	32.72	0.33	12.75	nr	**45.47**
150 mm wide	29.44	37.26	0.37	14.30	nr	**51.56**
225 mm wide	35.97	45.53	0.44	17.00	nr	**62.53**
300 mm wide	43.69	55.29	0.53	20.47	nr	**75.76**
450 mm wide	63.06	79.81	0.67	25.89	nr	**105.70**
600 mm wide	101.20	128.08	0.79	30.53	nr	**158.61**
Equal tee						
75 mm wide	61.95	78.40	0.31	11.97	nr	**90.37**
100 mm wide	64.46	81.58	0.35	13.53	nr	**95.11**
150 mm wide	68.25	86.37	0.36	13.91	nr	**100.28**
225 mm wide	73.69	93.26	0.74	28.59	nr	**121.85**
300 mm wide	90.49	114.52	0.54	20.87	nr	**135.39**
450 mm wide	117.07	148.16	0.71	27.44	nr	**175.60**
600 mm wide	167.32	211.76	0.92	35.55	nr	**247.31**
Unequal tee						
100 mm wide	60.67	76.79	0.39	15.08	nr	**91.87**
150 mm wide	60.67	76.79	0.43	16.62	nr	**93.41**
225 mm wide	70.96	89.80	0.50	19.33	nr	**109.13**
300 mm wide	86.14	109.02	0.63	24.34	nr	**133.36**
450 mm wide	113.47	143.61	0.80	30.90	nr	**174.51**
600 mm wide	163.01	206.30	1.02	39.41	nr	**245.71**
4 way crossovers						
75 mm wide	80.68	102.11	0.40	15.46	nr	**117.57**
100 mm wide	86.95	110.04	0.40	15.46	nr	**125.50**
150 mm wide	75.57	95.64	0.44	17.00	nr	**112.64**
225 mm wide	109.26	138.28	0.53	20.47	nr	**158.75**
300 mm wide	127.13	160.90	0.64	24.73	nr	**185.63**
450 mm wide	161.80	204.77	0.84	32.46	nr	**237.23**
600 mm wide	235.86	298.50	1.03	39.80	nr	**338.30**

ELECTRICAL SUPPLY/POWER/LIGHTING

Item	Net Price £	Material £	Labour hours	Labour £	Unit	Total rate £
Heavy duty tray with return flange						
Straight lengths						
75 mm	15.85	20.06	0.34	13.14	m	33.20
100 mm	16.08	20.35	0.36	13.91	m	34.26
150 mm	17.77	22.49	0.40	15.46	m	37.95
225 mm	21.56	27.28	0.46	17.77	m	45.05
300 mm	24.47	30.97	0.58	22.41	m	53.38
450 mm	43.05	54.49	0.70	27.05	m	81.54
600 mm	59.99	75.92	0.92	35.55	m	111.47
750 mm	82.97	105.01	1.01	39.03	m	144.04
900 mm	86.85	109.92	1.14	44.04	m	153.96
Extra over (cutting and jointing tray to fittings is included)						
Straight reducer						
100 mm wide	45.91	58.11	0.25	9.65	nr	67.76
150 mm wide	47.67	60.33	0.27	10.43	nr	70.76
225 mm wide	54.14	68.52	0.34	13.14	nr	81.66
300 mm wide	60.35	76.38	0.39	15.08	nr	91.46
450 mm wide	86.32	109.24	0.49	18.93	nr	128.17
600 mm wide	95.97	121.46	0.54	20.87	nr	142.33
750 mm wide	121.71	154.03	0.60	23.18	nr	177.21
900 mm wide	133.51	168.97	0.66	25.50	nr	194.47
Flat bend; 90°						
75 mm wide	57.69	73.01	0.24	9.27	nr	82.28
100 mm wide	62.50	79.11	0.28	10.82	nr	89.93
150 mm wide	68.07	86.15	0.30	11.59	nr	97.74
225 mm wide	69.10	87.45	0.36	13.91	nr	101.36
300 mm wide	78.46	99.30	0.44	17.00	nr	116.30
450 mm wide	126.01	159.48	0.57	22.02	nr	181.50
600 mm wide	171.52	217.08	0.69	26.67	nr	243.75
750 mm wide	218.88	277.01	0.83	32.07	nr	309.08
900 mm wide	228.80	289.56	1.01	39.03	nr	328.59
Adjustable bend						
75 mm wide	53.41	67.59	0.29	11.20	nr	78.79
100 mm wide	59.05	74.74	0.32	12.36	nr	87.10
150 mm wide	62.39	78.96	0.36	13.91	nr	92.87
225 mm wide	68.67	86.91	0.44	17.00	nr	103.91
300 mm wide	81.47	103.11	0.52	20.09	nr	123.20
Adjustable riser						
75 mm wide	47.07	59.57	0.29	11.20	nr	70.77
100 mm wide	45.47	57.55	0.32	12.36	nr	69.91
150 mm wide	54.19	68.58	0.36	13.91	nr	82.49
225 mm wide	58.03	73.44	0.44	17.00	nr	90.44
300 mm wide	62.85	79.54	0.52	20.09	nr	99.63
450 mm wide	76.50	96.82	0.66	25.50	nr	122.32
600 mm wide	91.14	115.35	0.79	30.53	nr	145.88
750 mm wide	110.37	139.69	1.03	39.80	nr	179.49
900 mm wide	128.17	162.21	1.10	42.50	nr	204.71

ELECTRICAL SUPPLY/POWER/LIGHTING

Item	Net Price £	Material £	Labour hours	Labour £	Unit	Total rate £
LV DISTRIBUTION: CABLE TRAYS – cont						
Extra over (cutting and jointing tray to fittings is included) – cont						
Inside riser; 90°						
75 mm wide	41.98	53.13	0.31	11.97	nr	65.10
100 mm wide	42.53	53.83	0.33	12.75	nr	66.58
150 mm wide	46.07	58.31	0.37	14.30	nr	72.61
225 mm wide	48.16	60.95	0.44	17.00	nr	77.95
300 mm wide	49.69	62.89	0.53	20.47	nr	83.36
450 mm wide	85.24	107.88	0.67	25.89	nr	133.77
600 mm wide	104.58	132.36	0.79	30.53	nr	162.89
750 mm wide	129.44	163.82	0.95	36.70	nr	200.52
900 mm wide	152.60	193.13	1.11	42.90	nr	236.03
Outside riser; 90°						
75 mm wide	41.98	53.13	0.31	11.97	nr	65.10
100 mm wide	42.53	53.83	0.33	12.75	nr	66.58
150 mm wide	46.07	58.31	0.37	14.30	nr	72.61
225 mm wide	48.16	60.95	0.44	17.00	nr	77.95
300 mm wide	49.69	62.89	0.53	20.47	nr	83.36
450 mm wide	85.24	107.88	0.67	25.89	nr	133.77
600 mm wide	104.58	132.36	0.79	30.53	nr	162.89
750 mm wide	129.44	163.82	0.95	36.70	nr	200.52
900 mm wide	152.60	193.13	1.11	42.90	nr	236.03
Equal tee						
75 mm wide	76.44	96.75	0.31	11.97	nr	108.72
100 mm wide	81.50	103.15	0.35	13.53	nr	116.68
150 mm wide	88.49	111.99	0.36	13.91	nr	125.90
225 mm wide	98.21	124.30	0.44	17.00	nr	141.30
300 mm wide	111.41	141.00	0.54	20.87	nr	161.87
450 mm wide	161.76	204.72	0.71	27.44	nr	232.16
600 mm wide	210.23	266.07	0.92	35.55	nr	301.62
750 mm wide	259.34	328.22	1.19	45.98	nr	374.20
900 mm wide	316.49	400.55	1.45	56.03	nr	456.58
Unequal tee						
75 mm wide	73.13	92.56	0.38	14.68	nr	107.24
100 mm wide	81.68	103.38	0.39	15.08	nr	118.46
150 mm wide	89.54	113.32	0.43	16.62	nr	129.94
225 mm wide	103.04	130.41	0.50	19.33	nr	149.74
300 mm wide	111.41	141.00	0.63	24.34	nr	165.34
450 mm wide	161.76	204.72	0.80	30.90	nr	235.62
600 mm wide	213.57	270.29	1.02	39.41	nr	309.70
750 mm wide	281.68	356.50	1.12	43.27	nr	399.77
900 mm wide	340.74	431.24	1.35	52.17	nr	483.41
4 way crossovers						
75 mm wide	74.22	93.93	0.40	15.46	nr	109.39
100 mm wide	106.39	134.65	0.40	15.46	nr	150.11
150 mm wide	106.39	134.65	0.44	17.00	nr	151.65
225 mm wide	147.92	187.21	0.53	20.47	nr	207.68
300 mm wide	163.11	206.43	0.64	24.73	nr	231.16
450 mm wide	234.63	296.95	0.84	32.46	nr	329.41
600 mm wide	318.89	403.59	1.03	39.80	nr	443.39

ELECTRICAL SUPPLY/POWER/LIGHTING

Item	Net Price £	Material £	Labour hours	Labour £	Unit	Total rate £
750 mm wide	388.94	492.24	1.13	43.67	nr	**535.91**
900 mm wide	469.17	593.78	1.36	52.55	nr	**646.33**
GRP cable tray including standard coupling joints and fixings (supports and hangers excluded)						
Tray						
100 mm wide	44.76	56.65	0.34	13.14	m	**69.79**
200 mm wide	57.32	72.54	0.39	15.08	m	**87.62**
400 mm wide	89.88	113.75	0.53	20.47	m	**134.22**
Cover						
100 mm wide	25.22	31.92	0.10	3.86	m	**35.78**
200 mm wide	33.41	42.28	0.11	4.26	m	**46.54**
400 mm wide	57.05	72.21	0.14	5.41	m	**77.62**
Extra for (cutting and jointing to fittings included)						
Reducer						
200 mm wide	118.94	150.53	0.23	8.88	nr	**159.41**
400 mm wide	155.39	196.66	0.30	11.59	nr	**208.25**
Reducer cover						
200 mm wide	74.44	94.21	0.25	9.65	nr	**103.86**
400 mm wide	108.80	137.69	0.28	10.82	nr	**148.51**
Bend						
100 mm wide	99.46	125.88	0.34	13.14	nr	**139.02**
200 mm wide	115.45	146.12	0.40	15.46	nr	**161.58**
400 mm wide	148.38	187.79	0.32	12.36	nr	**200.15**
Bend cover						
100 mm wide	49.42	62.54	0.10	3.86	nr	**66.40**
200 mm wide	66.01	83.54	0.10	3.86	nr	**87.40**
400 mm wide	91.19	115.40	0.13	5.03	nr	**120.43**
Tee						
100 mm wide	126.59	160.22	0.37	14.30	nr	**174.52**
200 mm wide	139.58	176.66	0.43	16.62	nr	**193.28**
400 mm wide	173.13	219.12	0.56	21.64	nr	**240.76**
Tee cover						
100 mm wide	63.78	80.72	0.27	10.43	nr	**91.15**
200 mm wide	76.77	97.16	0.31	11.97	nr	**109.13**
400 mm wide	108.29	137.05	0.37	14.30	nr	**151.35**

ELECTRICAL SUPPLY/POWER/LIGHTING

Item	Net Price £	Material £	Labour hours	Labour £	Unit	Total rate £
LV DISTRIBUTION: BASKET TRAY						
Mild steel cable basket; zinc plated including standard coupling joints, fixings and earth continuity straps (supports and hangers are excluded)						
Basket 54 mm deep						
100 mm wide	6.25	7.91	0.22	8.50	m	**16.41**
150 mm wide	6.97	8.83	0.25	9.65	m	**18.48**
200 mm wide	7.62	9.64	0.28	10.82	m	**20.46**
300 mm wide	8.94	11.31	0.34	13.14	m	**24.45**
450 mm wide	10.84	13.72	0.44	17.00	m	**30.72**
600 mm wide	13.26	16.78	0.70	27.05	m	**43.83**
Extra for (cutting and jointing to fittings is included)						
Reducer						
150 mm wide	20.46	25.89	0.25	9.65	nr	**35.54**
200 mm wide	24.01	30.39	0.28	10.82	nr	**41.21**
300 mm wide	24.40	30.88	0.38	14.68	nr	**45.56**
450 mm wide	26.78	33.89	0.48	18.54	nr	**52.43**
600 mm wide	33.36	42.22	0.48	18.54	nr	**60.76**
Bend						
100 mm wide	19.54	24.73	0.23	8.88	nr	**33.61**
150 mm wide	22.89	28.97	0.26	10.05	nr	**39.02**
200 mm wide	23.28	29.47	0.30	11.59	nr	**41.06**
300 mm wide	25.49	32.26	0.35	13.53	nr	**45.79**
450 mm wide	31.76	40.20	0.50	19.33	nr	**59.53**
600 mm wide	37.36	47.29	0.58	22.41	nr	**69.70**
Tee						
100 mm wide	24.56	31.08	0.28	10.82	nr	**41.90**
150 mm wide	26.05	32.97	0.30	11.59	nr	**44.56**
200 mm wide	26.60	33.67	0.33	12.75	nr	**46.42**
300 mm wide	33.45	42.34	0.39	15.08	nr	**57.42**
450 mm wide	43.81	55.45	0.56	21.64	nr	**77.09**
600 mm wide	45.29	57.32	0.65	25.11	nr	**82.43**
Cross over						
100 mm wide	33.51	42.41	0.40	15.46	nr	**57.87**
150 mm wide	34.07	43.12	0.42	16.23	nr	**59.35**
200 mm wide	37.01	46.84	0.46	17.77	nr	**64.61**
300 mm wide	42.04	53.21	0.51	19.70	nr	**72.91**
450 mm wide	49.30	62.40	0.74	28.59	nr	**90.99**
600 mm wide	50.89	64.41	0.82	31.67	nr	**96.08**

ELECTRICAL SUPPLY/POWER/LIGHTING

Item	Net Price £	Material £	Labour hours	Labour £	Unit	Total rate £
Mild steel cable basket; expoxy coated including standard coupling joints, fixings and earth continuity straps (supports and hangers are excluded)						
Basket 54 mm deep						
100 mm wide	14.00	17.72	0.22	8.50	m	**26.22**
150 mm wide	15.83	20.04	0.25	9.65	m	**29.69**
200 mm wide	17.63	22.31	0.28	10.82	m	**33.13**
300 mm wide	20.02	25.33	0.34	13.14	m	**38.47**
450 mm wide	24.71	31.27	0.44	17.00	m	**48.27**
600 mm wide	28.14	35.62	0.70	27.05	m	**62.67**
Extra for (cutting and jointing to fittings is included)						
Reducer						
150 mm wide	29.28	37.06	0.28	10.82	nr	**47.88**
200 mm wide	32.81	41.53	0.28	10.82	nr	**52.35**
300 mm wide	35.00	44.30	0.38	14.68	nr	**58.98**
450 mm wide	40.91	51.78	0.48	18.54	nr	**70.32**
600 mm wide	50.99	64.53	0.48	18.54	nr	**83.07**
Bend						
100 mm wide	28.39	35.93	0.23	8.88	nr	**44.81**
150 mm wide	31.72	40.14	0.26	10.05	nr	**50.19**
200 mm wide	32.12	40.66	0.30	11.59	nr	**52.25**
300 mm wide	36.14	45.74	0.35	13.53	nr	**59.27**
450 mm wide	45.86	58.04	0.50	19.33	nr	**77.37**
600 mm wide	55.07	69.70	0.58	22.41	nr	**92.11**
Tee						
100 mm wide	33.37	42.24	0.28	10.82	nr	**53.06**
150 mm wide	34.91	44.18	0.30	11.59	nr	**55.77**
200 mm wide	35.41	44.81	0.33	12.75	nr	**57.56**
300 mm wide	44.04	55.74	0.39	15.08	nr	**70.82**
450 mm wide	56.12	71.03	0.56	21.64	nr	**92.67**
600 mm wide	62.95	79.67	0.65	25.11	nr	**104.78**
Cross over						
100 mm wide	42.33	53.57	0.40	15.46	nr	**69.03**
150 mm wide	42.81	54.19	0.42	16.23	nr	**70.42**
200 mm wide	45.86	58.04	0.46	17.77	nr	**75.81**
300 mm wide	52.64	66.62	0.51	19.70	nr	**86.32**
450 mm wide	63.40	80.24	0.74	28.59	nr	**108.83**
600 mm wide	68.50	86.69	0.82	31.67	nr	**118.36**

ELECTRICAL SUPPLY/POWER/LIGHTING

Item	Net Price £	Material £	Labour hours	Labour £	Unit	Total rate £
LV DISTRIBUTION: SWITCHGEAR AND DISTRIBUTION BOARDS						
LV switchboard components, factory-assembled modular construction to IP41; form 4, type 5; 2400 mm high, with front and rear access; top cable entry/exit; includes delivery, offloading, positioning and commissioning (hence separate labour costs are not detailed below); excludes cabling and cable terminations						
Air circuit breakers (ACBs) to BSEN 60947–2, withdrawable type, fitted with adjustable instantaneous and overload protection. Includes enclosure and copper links, assembled into LV switchboard.						
ACB–100 kA fault rated.						
4 pole, 6300 A (1600 mm wide)	41803.96	52907.09	–	–	nr	**52907.09**
4 pole, 5000 A (1600 mm wide)	33177.75	41989.76	–	–	nr	**41989.76**
4 pole, 4000 A (1600 mm wide)	19011.18	24060.55	–	–	nr	**24060.55**
4 pole, 3200 A (1600 mm wide)	15208.94	19248.43	–	–	nr	**19248.43**
4 pole, 2500 A (1600 mm wide)	11051.29	13986.52	–	–	nr	**13986.52**
4 pole, 2000 A (1600 mm wide) 85ka	8841.03	11189.20	–	–	nr	**11189.20**
4 pole, 1600 A (1600 mm wide) 85ka	8034.64	10168.64	–	–	nr	**10168.64**
4 pole, 1250 A (1600 mm wide)	6277.07	7944.26	–	–	nr	**7944.26**
4 pole, 1000 A (1600 mm wide)	6159.03	7794.86	–	–	nr	**9295.89**
4 pole, 800 A (1600 mm wide) 85ka	5876.03	7436.70	–	–	nr	**7436.70**
3 pole, 6300 A (1600 mm wide)	34601.14	43791.20	–	–	nr	**43791.20**
3 pole, 5000 A (1600 mm wide)	27461.22	34754.92	–	–	nr	**34754.92**
3 pole, 4000 A (1600 mm wide)	14611.75	18492.63	–	–	nr	**18492.63**
3 pole, 3200 A (1600 mm wide)	11689.40	14794.10	–	–	nr	**14794.10**
3 pole, 2500 A (1600 mm wide)	9132.33	11557.87	–	–	nr	**11557.87**
3 pole, 2000 A (1600 mm wide)	8092.48	10241.84	–	–	nr	**10241.84**
3 pole, 1600 A (1600 mm wide)	6473.98	8193.47	–	–	nr	**8193.47**
3 pole, 1250 A (1600 mm wide)	5057.81	6401.17	–	–	nr	**6401.17**
3 pole, 1000 A (1600 mm wide)	4983.26	6306.81	–	–	nr	**6306.81**
3 pole, 800 A (1600 mm wide)	3986.61	5045.45	–	–	nr	**5045.45**
ACB–65 kA fault rated						
4 pole, 4000 A (1600 mm wide)	41803.96	52907.09	–	–	nr	**52907.09**
4 pole, 3200 A (1600 mm wide)	14030.39	17756.86	–	–	nr	**17756.86**
4 pole, 2500 A (1600 mm wide)	10508.52	13299.59	–	–	nr	**13299.59**
4 pole, 2000 A (1600 mm wide)	8406.82	10639.68	–	–	nr	**10639.68**
4 pole, 1600 A (1600 mm wide)	7600.44	9619.12	–	–	nr	**9619.12**
4 pole, 1250 A (1600 mm wide)	7168.13	9071.99	–	–	nr	**9071.99**
4 pole, 1000 A (1600 mm wide)	6965.12	8815.06	–	–	nr	**8815.06**
4 pole, 800 A (1600 mm wide)	5572.09	7052.04	–	–	nr	**7052.04**
3 pole, 6300 A (1600 mm wide)	34601.14	43791.20	–	–	nr	**43791.20**
3 pole, 4000 A (1600 mm wide)	21968.97	27803.93	–	–	nr	**27803.93**
3 pole, 3200 A (1600 mm wide)	10820.98	13695.04	–	–	nr	**13695.04**

ELECTRICAL SUPPLY/POWER/LIGHTING

Item	Net Price £	Material £	Labour hours	Labour £	Unit	Total rate £
3 pole, 2500 A (1600 mm wide)	8453.89	10699.25	–	–	nr	10699.25
3 pole, 2000 A (1600 mm wide)	7642.79	9672.71	–	–	nr	9672.71
3 pole, 1600 A (1600 mm wide)	6114.22	7738.16	–	–	nr	7738.16
3 pole, 1250 A (1600 mm wide)	4776.73	6045.42	–	–	nr	6045.42
3 pole, 1000 A (1600 mm wide)	4758.42	6022.25	–	–	nr	6022.25
3 pole, 800 A (1600 mm wide)	3806.73	4817.79	–	–	nr	4817.79
Extra for						
Cable box (one per ACB for form 4, types 6 & 7)	366.83	464.26	–	–	nr	464.26
Opening coil (shunt trip)	166.86	211.18	–	–	nr	211.18
Closing coil	110.54	139.90	–	–	nr	139.90
Undervoltage release	197.87	250.42	–	–	nr	250.42
Motor operator	665.80	842.63	–	–	nr	842.63
Mechnical Interlock (per ACB)	460.26	582.50	–	–	nr	582.50
ACB Fortress/Castell adaptor kit (one per ACB)	99.25	125.61	–	–	nr	125.61
Fortress/Castell ACB lock (one per ACB)	312.80	395.88	–	–	nr	395.88
Fortress/Castell key	77.89	98.58	–	–	nr	98.58
Moulded case circuit breakers (MCCBs) to BS EN 60947–2; plug-in type, fitted with electronic trip unit. Includes metalwork section and copper links, assembled into LV switchboard						
MCCB–150 kA fault rated						
4 pole, 630 A (800 mm wide, 600 mm high)	2818.40	3566.96	–	–	nr	3566.96
4 pole, 400 A (800 mm wide, 400 mm high)	1887.24	2388.49	–	–	nr	2388.49
4 pole, 250 A (800 mm wide, 400 mm high)	1742.64	2205.48	–	–	nr	2205.48
4 pole, 160 A (800 mm wide, 300 mm high)	1193.70	1510.75	–	–	nr	1510.75
4 pole, 100 A (800 mm wide, 200 mm high)	1881.68	2381.46	–	–	nr	2381.46
3 pole, 630 A (800 mm wide, 600 mm high)	2227.04	2818.55	–	–	nr	2818.55
3 pole, 400 A (800 mm wide, 400 mm high)	1515.84	1918.45	–	–	nr	1918.45
3 pole, 250 A (800 mm wide, 400 mm high)	991.32	1254.61	–	–	nr	1254.61
3 pole, 160 A (800 mm wide, 300 mm high)	712.28	901.47	–	–	nr	901.47
3 pole, 100 A (800 mm wide, 200 mm high)	579.61	733.56	–	–	nr	733.56
MCCB–70 kA fault rated						
4 pole, 630 A (800 mm wide, 600 mm high)	2353.80	2978.96	–	–	nr	2978.96
4 pole, 400 A (800 mm wide, 400 mm high)	1617.99	2047.73	–	–	nr	2047.73
4 pole, 250 A (800 mm wide, 400 mm high)	1178.28	1491.24	–	–	nr	1491.24
4 pole, 160 A (800 mm wide, 300 mm high)	888.07	1123.94	–	–	nr	1123.94
4 pole, 100 A (800 mm wide, 200 mm high)	731.81	926.18	–	–	nr	926.18
3 pole, 630 A (800 mm wide, 600 mm high)	1725.08	2183.26	–	–	nr	2183.26
3 pole, 400 A (800 mm wide, 400 mm high)	1164.31	1473.55	–	–	nr	1473.55
3 pole, 250 A (800 mm wide, 400 mm high)	811.97	1027.63	–	–	nr	1027.63
3 pole, 160 A (800 mm wide, 300 mm high)	602.74	762.83	–	–	nr	762.83
3 pole, 100 A (800 mm wide, 200 mm high)	477.36	604.15	–	–	nr	604.15

ELECTRICAL SUPPLY/POWER/LIGHTING

Item	Net Price £	Material £	Labour hours	Labour £	Unit	Total rate £
LV DISTRIBUTION: SWITCHGEAR AND DISTRIBUTION BOARDS – cont						
Moulded case circuit breakers (MCCBs) to BS EN 60947–2 – cont						
MCCB–45 kA fault rated						
4 pole, 630 A (800 mm wide, 600 mm, LI, SSKA high)	2167.80	2743.56	–	–	nr	2743.56
4 pole, 400 A (800 mm wide, 400 mm, LI, SSKA high)	1504.90	1904.60	–	–	nr	1904.60
3 pole, 630 A (800 mm wide, 600 mm, LI, SSKA high)	1616.06	2045.29	–	–	nr	2045.29
3 pole, 400 A (800 mm wide, 400 mm, LI, SSKA high)	1109.68	1404.41	–	–	nr	1404.41
MCCB–36kA fault rated						
4 pole, 250 A (800 mm wide, 400 mm high)	1804.71	2284.04	–	–	nr	2284.04
4 pole, 160 A (800 mm wide, 300 mm high)	1367.11	1730.21	–	–	nr	1730.21
3 pole, 250 A (800 mm wide, 400 mm high)	1025.33	1297.65	–	–	nr	1297.65
3 pole, 160 A (800 mm wide, 300 mm high)	902.98	1142.81	–	–	nr	1142.81
Extra for						
Cable box (one per MCCB for form 4, types 6 & 7)	155.14	196.35	–	–	nr	196.35
Shunt trip (for ratings 100 A to 630 A)	47.82	60.52	–	–	nr	60.52
Undervoltage release (for ratings 100 A to 630 A)	62.03	78.50	–	–	nr	78.50
Motor operator for 630 A MCCB	114.14	144.46	–	–	nr	144.46
Motor operator for 400 A MCCB	114.14	144.46	–	–	nr	144.46
Motor operator for 250 A MCCB	55.40	70.11	–	–	nr	70.11
Motor operator for 160 A/100 A MCCB	49.25	62.33	–	–	nr	62.33
Door handle for 630/400 A MCCB	97.10	122.89	–	–	nr	122.89
Door handle for 250/160/100 A MCCB	76.94	97.37	–	–	nr	97.37
MCCB earth fault protection	585.18	740.60	–	–	nr	740.60
LV switchboard busbar						
Copper busbar assembled into LV switchboard, ASTA type tested to appropriate fault level. Busbar length may be estimated by adding the widths of the ACB sections to the width of the MCCB sections. ACBs up to 2000 A rating may be stacked two high; larger ratings are one per section. To determine the number of MCCB sections, add together all the MCCB heights and divide by 1800 mm, rounding up as necessary						
6000 A (6 × 10 mm × 100 mm)	3044.97	3853.72	–	–	nr	3853.72
5000 A (4 × 10 mm × 100 mm)	2502.78	3167.52	–	–	nr	3167.52
4000 A (4 × 10 mm × 100 mm)	2502.78	3167.52	–	–	nr	3167.52
3200 A (3 × 10 mm × 100 mm)	1738.35	2200.06	–	–	nr	2200.06
2500 A (2 × 10 mm × 100 mm)	1466.63	1856.16	–	–	nr	1856.16
2000 A (2 × 10 mm × 80 mm)	1078.05	1364.38	–	–	nr	1364.38
1600 A (2 × 10 mm × 50 mm)	782.19	989.93	–	–	nr	989.93

ELECTRICAL SUPPLY/POWER/LIGHTING

Item	Net Price £	Material £	Labour hours	Labour £	Unit	Total rate £
1250 A (2 × 10 mm × 40 mm)	613.31	776.20	–	–	nr	**776.20**
1000 A (2 × 10 mm × 30 mm)	613.31	776.20	–	–	nr	**776.20**
800 A (2 × 10 mm × 20 mm)	477.43	604.24	–	–	nr	**604.24**
630 A (2 × 10 mm × 20 mm)	477.43	604.24	–	–	nr	**604.24**
400 A (2 × 10 mm × 10 mm)	408.88	517.47	–	–	nr	**517.47**
Automatic power factor correction (PFC); floor standing steel enclosure to IP 42, complete with microprocessor based relay and status indication; includes delivery, offloading, positioning and commissioning; excludes cabling and cable terminations						
Standard PFC (no detuning)						
100 kVAr	5970.60	7556.39	–	–	nr	**7556.39**
200 kVAr	8350.19	10568.00	–	–	nr	**10568.00**
400 kVAr	14254.74	18040.80	–	–	nr	**18040.80**
600 kVAr	19525.67	24711.69	–	–	nr	**24711.69**
PFC with detuning reactors						
100 kVAr	9713.95	12293.97	–	–	nr	**12293.97**
200 kVAr	13575.40	17181.02	–	–	nr	**17181.02**
400 kVAr	23397.28	29611.60	–	–	nr	**29611.60**
600 kVAr	34408.96	43547.97	–	–	nr	**43547.97**

ELECTRICAL SUPPLY/POWER/LIGHTING

Item	Net Price £	Material £	Labour hours	Labour £	Unit	Total rate £
LV DISTRIBUTION: AUTOMATIC TRANSFER SWITCHES						
Automatic transfer switches; steel enclosure; solenoid operating; programmable controller, keypad and LCD display; fixed to backgrounds; including commissioning and testing						
Panel mounting type 4 pole M6 s; non BMS connection						
40 amp	2161.75	2735.91	2.40	92.74	nr	**2828.65**
63 amp	2294.12	2903.44	2.50	96.60	nr	**3000.04**
80 amp	2346.83	2970.15	2.60	100.46	nr	**3070.61**
100 amp	2381.22	3013.67	2.60	100.46	nr	**3114.13**
125 amp	2445.39	3094.88	2.70	104.33	nr	**3199.21**
160 amp	2571.85	3254.93	2.80	108.18	nr	**3363.11**
Panel mounting type 4 pole M6e; BMS connection						
40 amp	2983.06	3775.36	2.50	96.60	nr	**3871.96**
63 amp	3159.59	3998.78	2.60	100.46	nr	**4099.24**
80 amp	3229.85	4087.70	2.70	104.33	nr	**4192.03**
100 amp	3275.70	4145.72	2.80	108.18	nr	**4253.90**
125 amp	3361.27	4254.03	2.90	112.06	nr	**4366.09**
160 amp	3529.85	4467.38	3.00	115.92	nr	**4583.30**
Enclosed type 3 pole or 4 pole						
125 amp	4434.15	5611.86	2.60	100.46	nr	**5712.32**
160 amp	4650.45	5885.61	2.90	112.06	nr	**5997.67**
200amp	4866.75	6159.36	3.30	127.51	nr	**6286.87**
250amp	5072.24	6419.43	4.30	166.15	nr	**6585.58**
300amp	5245.28	6638.43	4.84	187.02	nr	**6825.45**
400amp	5818.47	7363.85	5.12	197.83	nr	**7561.68**
500amp	6110.48	7733.42	5.50	212.51	nr	**7945.93**
630amp	7029.75	8896.85	6.00	231.83	nr	**9128.68**
800amp	7970.66	10087.67	6.20	239.57	nr	**10327.24**
1000amp	8543.85	10813.10	6.90	266.62	nr	**11079.72**
1250amp	8814.23	11155.29	7.70	297.52	nr	**11452.81**
1600amp	9463.13	11976.54	8.50	328.42	nr	**12304.96**
2000amp	11896.50	15056.22	–	–	nr	**15056.22**
Enclosed type 3 pole or 4 pole; with single by-pass						
40 amp	10284.57	13016.15	2.60	100.46	nr	**13116.61**
63 amp	10484.22	13268.83	2.60	100.46	nr	**13369.29**
80 amp	10616.99	13436.86	2.60	100.46	nr	**13537.32**
100 amp	10762.53	13621.06	2.60	100.46	nr	**13721.52**
125 amp	11319.23	14325.62	2.90	112.06	nr	**14437.68**
160 amp	12247.05	15499.87	2.90	112.06	nr	**15611.93**
250 amp	15939.64	20173.20	3.30	127.51	nr	**20300.71**
400 amp	16748.51	21196.92	4.30	166.15	nr	**21363.07**
630 amp	22973.10	29074.75	4.84	187.02	nr	**29261.77**
800 amp	24680.08	31235.11	5.12	197.83	nr	**31432.94**

ELECTRICAL SUPPLY/POWER/LIGHTING

Item	Net Price £	Material £	Labour hours	Labour £	Unit	Total rate £
1000 amp	34889.54	44156.20	5.50	212.51	nr	44368.71
1250 amp	42557.31	53860.53	6.00	231.83	nr	54092.36
1600 amp	51740.03	65482.18	6.20	239.57	nr	65721.75
2000 amp	63085.52	79841.04	6.90	266.62	nr	80107.66
2500 amp	73306.68	92776.94	7.70	297.52	nr	93074.46
3200 amp	86942.30	110034.18	8.50	328.42	nr	110362.60
Enclosed type 3 pole or 4 pole; with dual by-pass						
40 amp	13466.05	17042.64	2.60	100.46	nr	17143.10
63 amp	13978.95	17691.76	2.60	100.46	nr	17792.22
80 amp	14151.85	17910.58	2.60	100.46	nr	18011.04
100 amp	14350.08	18161.46	2.60	100.46	nr	18261.92
125 amp	15092.31	19100.83	2.90	112.06	nr	19212.89
160 amp	16329.38	20666.46	2.90	112.06	nr	20778.52
250 amp	21252.86	26897.62	3.30	127.51	nr	27025.13
400 amp	22331.37	28262.58	4.30	166.15	nr	28428.73
630 amp	30630.76	38766.29	4.84	187.02	nr	38953.31
800 amp	32906.77	41646.81	5.12	197.83	nr	41844.64
1000 amp	46519.40	58874.95	5.50	212.51	nr	59087.46
1250 amp	54533.48	69017.57	6.00	231.83	nr	69249.40
1600 amp	63869.19	80832.84	6.20	239.57	nr	81072.41
2000 amp	76160.36	96388.56	6.90	266.62	nr	96655.18
2500 amp	86814.92	109872.96	7.70	297.52	nr	110170.48
3200 amp	102599.45	129849.87	8.50	328.42	nr	130178.29

ELECTRICAL SUPPLY/POWER/LIGHTING

Item	Net Price £	Material £	Labour hours	Labour £	Unit	Total rate £
LV DISTRIBUTION: BREAKERS AND FUSES						
MCCB panelboards; IP4X construction, 50 kA busbars and fully rated neutral; fitted with doorlock, removable glandplate; form 3b Type 2; BSEN 60439–1; including fixing to backgrounds						
Panelboards cubicle with MCCB incomer						
Up to 250 A						
4 way TPN	1194.60	1511.89	1.00	38.64	nr	**1550.53**
Extra over for integral incomer metering	1223.51	1548.48	1.50	57.96	nr	**1606.44**
Up to 630 A						
6 way TPN	2148.91	2719.66	2.00	77.28	nr	**2796.94**
12 way TPN	2426.98	3071.59	2.50	96.60	nr	**3168.19**
18 way TPN	2825.17	3575.53	3.00	115.92	nr	**3691.45**
Extra over for integral incomer metering	1445.95	1829.99	1.50	57.96	nr	**1887.95**
Up to 800 A						
6 way TPN	3677.18	4653.84	2.00	77.28	nr	**4731.12**
12 way TPN	4411.31	5582.95	2.50	96.60	nr	**5679.55**
18 way TPN	4696.04	5943.31	3.00	115.92	nr	**6059.23**
Extra over for integral incomer metering	1445.95	1829.99	1.50	57.96	nr	**1887.95**
Up to 1200 A						
20 way TPN	9839.22	12452.52	3.50	135.24	nr	**12587.76**
Up to 1600 A						
28 way TPN	12742.23	16126.57	3.50	135.24	nr	**16261.81**
Up to 2000 A						
28 way TPN	13870.08	17553.97	4.00	154.56	nr	**17708.53**
Feeder MCCBs						
Single pole						
32 A	100.90	127.70	0.75	28.99	nr	**156.69**
63 A	103.20	130.61	0.75	28.99	nr	**159.60**
100 A	105.48	133.49	0.75	28.99	nr	**162.48**
160 A	112.37	142.22	1.00	38.64	nr	**180.86**
Double pole						
32 A	151.36	191.56	0.75	28.99	nr	**220.55**
63 A	153.64	194.44	0.75	28.99	nr	**223.43**
100 A	224.74	284.44	0.75	28.99	nr	**313.43**
160 A	279.77	354.08	1.00	38.64	nr	**392.72**
Triple pole						
32 A	201.82	255.43	0.75	28.99	nr	**284.42**
63 A	206.38	261.20	0.75	28.99	nr	**290.19**
100 A	268.30	339.56	0.75	28.99	nr	**368.55**
160 A	346.30	438.28	1.00	38.64	nr	**476.92**
250 A	520.58	658.85	1.00	38.64	nr	**697.49**
400 A	704.05	891.05	1.25	48.29	nr	**939.34**
630 A	1155.88	1462.88	1.50	57.96	nr	**1520.84**

ELECTRICAL SUPPLY/POWER/LIGHTING

Item	Net Price £	Material £	Labour hours	Labour £	Unit	Total rate £
MCB distribution boards; IP3X external protection enclosure; removable earth and neutral bars and DIN rail; 125/250 amp incomers; including fixing to backgrounds						
SP&N						
6 way	98.09	124.14	2.00	77.28	nr	**201.42**
8 way	116.33	147.22	2.50	96.60	nr	**243.82**
12 way	133.52	168.99	3.00	115.92	nr	**284.91**
16 way	158.70	200.85	4.00	154.56	nr	**355.41**
24 way	334.32	423.11	5.00	193.19	nr	**616.30**
TP&N						
4 way	653.60	827.20	3.00	115.92	nr	**943.12**
6 way	676.57	856.26	3.50	135.24	nr	**991.50**
8 way	736.72	932.39	4.00	154.56	nr	**1086.95**
12 way	785.52	994.16	4.00	154.56	nr	**1148.72**
16 way	898.86	1137.60	5.00	193.19	nr	**1330.79**
24 way	1131.96	1432.60	6.40	247.28	nr	**1679.88**
Miniature circuit breakers for distribution boards; BS EN 60 898; DIN rail mounting; including connecting to circuit						
SP&N; including connecting of wiring						
6 amp	11.84	14.99	0.10	3.86	nr	**18.85**
10–40 amp	12.32	15.59	0.10	3.86	nr	**19.45**
50–63 amp	12.90	16.33	0.14	5.41	nr	**21.74**
TP&N; including connecting of wiring						
6 amp	50.21	63.55	0.30	11.59	nr	**75.14**
10–40 amp	52.19	66.05	0.45	17.39	nr	**83.44**
50–63 amp	54.67	69.19	0.45	17.39	nr	**86.58**
Residual current circuit breakers for distribution boards; DIN rail mounting; including connecting to circuit						
SP&N						
10 mA						
6 amp	77.75	98.40	0.21	8.11	nr	**106.51**
10–32 amp	76.27	96.53	0.26	10.05	nr	**106.58**
45 amp	77.52	98.11	0.26	10.05	nr	**108.16**
30 mA						
6 amp	77.75	98.40	0.21	8.11	nr	**106.51**
10–40 amp	76.27	96.53	0.21	8.11	nr	**104.64**
50 –63 amp	78.77	99.69	0.26	10.05	nr	**109.74**
100 mA						
6 amp	143.76	181.94	0.21	8.11	nr	**190.05**
10–40 amp	143.76	181.94	0.23	8.99	nr	**190.93**
50 –63 amp	143.76	181.94	0.26	10.05	nr	**191.99**

ELECTRICAL SUPPLY/POWER/LIGHTING

Item	Net Price £	Material £	Labour hours	Labour £	Unit	Total rate £
LV DISTRIBUTION: BREAKERS AND FUSES – cont						
HRC fused distribution boards; IP4X external protection enclosure; including earth and neutral bars; fixing to backgrounds						
SP&N						
20 amp incomer						
4 way	185.29	234.51	1.00	38.64	nr	273.15
6 way	223.70	283.11	1.20	46.37	nr	329.48
8 way	262.24	331.89	1.40	54.10	nr	385.99
12 way	339.35	429.49	1.80	69.54	nr	499.03
32 amp incomer						
4 way	223.09	282.34	1.00	38.64	nr	320.98
6 way	293.38	371.30	1.20	46.37	nr	417.67
8 way	345.17	436.84	1.40	54.10	nr	490.94
12 way	444.39	562.42	1.80	69.54	nr	631.96
TP&N						
20 amp incomer						
4 way	349.61	442.47	1.50	57.96	nr	500.43
6 way	442.17	559.61	2.10	81.13	nr	640.74
8 way	522.79	661.64	2.70	104.33	nr	765.97
12 way	733.46	928.27	3.90	150.70	nr	1078.97
32 amp incomer						
4 way	418.24	529.32	1.50	57.96	nr	587.28
6 way	561.51	710.65	2.10	81.13	nr	791.78
8 way	685.91	868.09	2.70	104.33	nr	972.42
12 way	951.35	1204.03	3.90	150.70	nr	1354.73
63 amp incomer						
4 way	889.04	1125.17	2.17	83.84	nr	1209.01
6 way	1140.19	1443.02	2.83	109.35	nr	1552.37
8 way	1373.05	1737.74	2.57	99.30	nr	1837.04
100 amp incomer						
4 way	1406.01	1779.44	2.40	92.74	nr	1872.18
6 way	1837.78	2325.89	2.73	105.49	nr	2431.38
8 way	2247.16	2844.00	3.87	149.53	nr	2993.53
200 amp incomer						
4 way	3482.55	4407.51	5.36	207.10	nr	4614.61
6 way	4602.94	5825.48	6.17	238.40	nr	6063.88
HRC fuse; includes fixing to fuse holder						
2–30 amp	3.78	4.78	0.10	3.86	nr	8.64
35–63 amp	8.15	10.32	0.12	4.64	nr	14.96
80 amp	12.04	15.24	0.15	5.80	nr	21.04
100 amp	14.49	18.33	0.15	5.80	nr	24.13
125 amp	21.88	27.69	0.15	5.80	nr	33.49
160 amp	22.95	29.04	0.15	5.80	nr	34.84
200 amp	23.77	30.08	0.15	5.80	nr	35.88

ELECTRICAL SUPPLY/POWER/LIGHTING

Item	Net Price £	Material £	Labour hours	Labour £	Unit	Total rate £
Consumer units; fixed to backgrounds; including supports, fixings, connections/ jointing to equipment.						
Switched and insulated; moulded plastic case, 63 amp 230 volt SP&N; earth and neutral bars; 30 mA RCCB protection; fitted MCBs						
2 way	142.51	180.36	1.67	64.53	nr	**244.89**
4 way	159.90	202.37	1.59	61.43	nr	**263.80**
6 way	173.82	219.99	2.50	96.60	nr	**316.59**
8 way	187.57	237.38	3.00	115.92	nr	**353.30**
12 way	217.89	275.77	4.00	154.56	nr	**430.33**
16 way	263.27	333.20	5.50	212.51	nr	**545.71**
Switched and insulated; moulded plastic case, 100 amp 230 volt SP&N; earth and neutral bars; 30 mA RCCB protection; fitted MCBs						
2 way	142.51	180.36	1.67	64.53	nr	**244.89**
4 way	159.90	202.37	1.59	61.43	nr	**263.80**
6 way	173.82	219.99	2.50	96.60	nr	**316.59**
8 way	187.57	237.38	3.00	115.92	nr	**353.30**
12 way	217.89	275.77	4.00	154.56	nr	**430.33**
16 way	263.27	333.20	5.50	212.51	nr	**545.71**
Extra for						
Residual current device; double pole; 230 volt/30 mA tripping current						
16 amp	77.91	98.60	0.22	8.50	nr	**107.10**
30 amp	79.15	100.17	0.22	8.50	nr	**108.67**
40 amp	80.38	101.73	0.22	8.50	nr	**110.23**
63 amp	99.52	125.96	0.22	8.50	nr	**134.46**
80 amp	110.70	140.10	0.22	8.50	nr	**148.60**
100 amp	136.25	172.44	0.25	9.65	nr	**182.09**
Residual current device; double pole; 230 volt/100 mA tripping current						
63 amp	90.98	115.15	0.22	8.50	nr	**123.65**
80 amp	105.25	133.20	0.22	8.50	nr	**141.70**
100 amp	136.27	172.47	0.25	9.65	nr	**182.12**
Heavy duty fuse switches; with HRC fuses BS 5419; short circuit rating 65 kA, 500 volt; including retractable operating switches						
SP&N						
63 amp	358.62	453.87	1.30	50.23	nr	**504.10**
100 amp	524.23	663.47	1.95	75.34	nr	**738.81**
TP&N						
63 amp	451.81	571.82	1.83	70.71	nr	**642.53**
100 amp	634.88	803.50	2.48	95.82	nr	**899.32**
200 amp	979.34	1239.45	3.13	120.94	nr	**1360.39**
300 amp	1701.39	2153.28	4.45	171.94	nr	**2325.22**
400 amp	1867.81	2363.91	4.45	171.94	nr	**2535.85**
600 amp	2819.37	3568.20	5.72	221.01	nr	**3789.21**
800 amp	4404.19	5573.94	7.88	304.48	nr	**5878.42**

ELECTRICAL SUPPLY/POWER/LIGHTING

Item	Net Price £	Material £	Labour hours	Labour £	Unit	Total rate £
LV DISTRIBUTION: BREAKERS AND FUSES – cont						
Switch disconnectors to BSEN 60947–3; in sheet steel case; IP41 with door interlock fixed to backgrounds						
Double pole						
20 amp	72.57	91.84	1.02	39.41	nr	**131.25**
32 amp	87.47	110.70	1.02	39.41	nr	**150.11**
63 amp	319.18	403.95	1.21	46.75	nr	**450.70**
100 amp	292.89	370.69	1.86	71.87	nr	**442.56**
TP&N						
20 amp	91.10	115.29	1.29	49.84	nr	**165.13**
32 amp	106.01	134.16	1.83	70.71	nr	**204.87**
63 amp	359.73	455.27	2.48	95.82	nr	**551.09**
100 amp	362.71	459.04	2.48	95.82	nr	**554.86**
125 amp	378.31	478.79	2.48	95.82	nr	**574.61**
160 amp	870.43	1101.62	2.48	95.82	nr	**1197.44**
Enclosed switch disconnector to BSEN 60947–3; enclosure minimum IP55 rating; complete with earth connection bar; fixed to backgrounds.						
TP						
20 amp	72.57	91.84	1.02	39.41	nr	**131.25**
32 amp	87.47	110.70	1.02	39.41	nr	**150.11**
63 amp	319.18	403.95	1.21	46.75	nr	**450.70**
TP&N						
20 amp	91.10	115.29	1.29	49.84	nr	**165.13**
32 amp	106.01	134.16	1.83	70.71	nr	**204.87**
63 amp	359.73	455.27	2.48	95.82	nr	**551.09**
Busbar chambers; fixed to background including all supports, fixings, connections/jointing to equipment.						
Sheet steel case enclosing 4 pole 550 volt copper bars, detachable metal end plates						
600 mm long						
200 amp	661.09	836.67	2.62	101.24	nr	**937.91**
300 amp	849.52	1075.16	3.03	117.08	nr	**1192.24**
500 amp	1455.63	1842.24	4.48	173.10	nr	**2015.34**
900 mm long						
200 amp	952.25	1205.16	3.04	117.47	nr	**1322.63**
300 amp	1122.26	1420.33	3.59	138.71	nr	**1559.04**
500 amp	1660.02	2100.92	4.42	170.78	nr	**2271.70**
1350 mm long						
200 amp	1300.70	1646.16	3.38	130.59	nr	**1776.75**
300 amp	1530.90	1937.51	3.94	152.24	nr	**2089.75**
500 amp	2450.53	3101.39	4.82	186.23	nr	**3287.62**

ELECTRICAL SUPPLY/POWER/LIGHTING

Item	Net Price £	Material £	Labour hours	Labour £	Unit	Total rate £
Contactor relays; pressed steel enclosure; fixed to backgrounds including supports, fixings, connections/jointing to equipment						
Relays						
6 Amp, 415/240 Volt, 4 pole N/O	76.56	96.89	0.52	20.09	nr	**116.98**
6 Amp, 415/240 Volt, 8 pole N/O	93.64	118.51	0.85	32.84	nr	**151.35**
Push button stations; heavy gauge pressed steel enclosure; polycarbonate cover; IP65; fixed to backgrounds including supports, fixings, connections/ joining to equipment						
Standard units						
One button (start or stop)	96.91	122.65	0.39	15.08	nr	**137.73**
Two button (start or stop)	102.85	130.17	0.47	18.17	nr	**148.34**
Three button (forward-reverse-stop)	145.77	184.49	0.57	22.02	nr	**206.51**
Weatherproof junction boxes; enclosures with rail mounted terminal blocks; side hung door to receive padlock; fixed to backgrounds, including all supports and fixings (suitable for cable up to 2.5 mm²; including glandplates and gaskets.)						
Sheet steel with zinc spray finish enclosure						
Overall size 229 × 152; suitable to receive						
3 × 20(A) glands per gland plate	98.13	124.20	1.43	55.26	nr	**179.46**
Overall size 306 × 306; suitable to receive						
14 × 20(A) glands per gland plate	131.65	166.61	2.17	83.84	nr	**250.45**
Overall size 458 × 382; suitable to receive						
18 × 20(A) glands per gland plate	191.70	242.61	3.51	135.62	nr	**378.23**
Overall size 762 x508; suitable to receive						
26 × 20(A) glands per gland plate	202.75	256.60	4.85	187.40	nr	**444.00**
Overall size 914 × 610; suitable to receive						
45 × 20(A) glands per gland plate	224.10	283.62	7.01	270.86	nr	**554.48**
Weatherproof junction boxes; enclosures with rail mounted terminal blocks; screw fixed lid; fixed to backgrounds, including all supports and fixings (suitable for cable up to 2.5 mm²; including glandplates and gaskets)						
Glassfibre reinforced polycarbonate enclosure						
Overall size 190 × 190 × 130 mm	129.05	163.33	1.43	55.26	nr	**218.59**
Overall size 190 × 190 × 180 mm	188.96	239.14	1.53	59.11	nr	**298.25**
Overall size 280 × 190 × 130 mm	213.43	270.12	2.17	83.84	nr	**353.96**
Overall size 280 × 190 × 180 mm	239.10	302.60	2.37	91.58	nr	**394.18**
Overall size 380 × 190 × 130 mm	266.78	337.64	3.33	128.67	nr	**466.31**
Overall size 380 × 190 × 180 mm	286.51	362.61	3.30	127.51	nr	**490.12**
Overall size 380 × 280 × 130 mm	306.29	387.64	4.66	180.06	nr	**567.70**
Overall size 380 × 280 × 180 mm	329.99	417.64	5.36	207.10	nr	**624.74**
Overall size 560 × 280 × 130 mm	397.19	502.68	7.01	270.86	nr	**773.54**
Overall size 560 × 380 × 180 mm	409.04	517.69	7.67	296.36	nr	**814.05**

ELECTRICAL SUPPLY/POWER/LIGHTING

Item	Net Price £	Material £	Labour hours	Labour £	Unit	Total rate £
GENERAL LIGHTING						
LED tube luminaires; surface fixed to backgrounds						
Batten type; surface mounted						
600 mm Single – 10 W	30.19	38.20	0.58	22.41	nr	**60.61**
600 mm Twin – 20 W	41.02	51.91	0.59	22.79	nr	**74.70**
1200 mm Single – 20 W	40.30	51.00	0.76	29.37	nr	**80.37**
1200 mm Twin – 40 W	64.82	82.04	0.84	32.46	nr	**114.50**
1500 mm Single – 26 W	49.61	62.79	0.77	29.76	nr	**92.55**
1500 mm Twin – 50 W	78.74	99.66	0.85	32.84	nr	**132.50**
1800 mm Single – 30 W	56.54	71.56	1.05	40.58	nr	**112.14**
1800 mm Twin – 60 W	101.88	128.93	1.06	40.96	nr	**169.89**
Extra over cost for emergency (remote)	83.00	105.04	–	–	nr	**105.04**
Extra over cost for DALI dimming (remote)	30.00	37.97	–	–	nr	**37.97**
Surface mounted LED batten, non-dimmable, opal diffuser						
600 mm Single – 10 W	60.00	75.94	0.62	23.96	nr	**99.90**
600 mm Twin – 20 W	87.23	110.40	0.62	23.96	nr	**134.36**
1200 mm Single – 20 W	69.12	87.48	0.79	30.53	nr	**118.01**
1200 mm Twin – 40 W	90.12	114.06	0.80	30.90	nr	**144.96**
1500 mm Single – 26 W	79.00	99.98	0.88	34.00	nr	**133.98**
1500 mm Twin – 50 W	95.83	121.28	0.90	34.78	nr	**156.06**
1800 mm Single – 30 W	82.36	104.24	1.09	42.12	nr	**146.36**
1800 mm Twin – 60 W	110.32	139.62	1.10	42.50	nr	**182.12**
Extra over cost for emergency (remote)	83.00	105.04	1.30	50.23	nr	**155.27**
Extra over cost for DALI dimming (remote)	30.00	37.97	1.31	50.61	nr	**88.58**
Surface mounted linear LED batten, Non-dimmable, MPO diffuser						
600 mm Single – 10 W	63.00	79.73	1.09	42.12	nr	**121.85**
600 mm Twin – 20 W	92.35	116.88	1.09	42.12	nr	**159.00**
1200 mm Single – 20 W	73.17	92.60	0.79	30.53	nr	**123.13**
1200 mm Twin – 40 W	95.12	120.39	1.25	48.29	nr	**168.68**
1500 mm Single – 26 W	82.00	103.78	0.90	34.78	nr	**138.56**
1500 mm Twin – 50 W	99.97	126.53	0.90	34.78	nr	**161.31**
1800 mm Single – 30 W	87.36	110.57	0.90	34.78	nr	**145.35**
1800 mm Twin – 60 W	117.45	148.65	0.90	34.78	nr	**183.43**
Extra Over cost for emergency (remote)	83.00	105.04	–	–	nr	**105.04**
Extra Over cost for DALI dimming (remote)	30.00	37.97	–	–	nr	**37.97**
Modular recessed LED; Non-dimmable, low brightness; MPO diffuser fitted to exposed T grid ceiling						
300 × 300 mm, 2 W	35.00	44.30	0.84	32.46	nr	**76.76**
Over cost for emergency	80.00	101.25	–	–	nr	**101.25**
Over cost for DALI dimming	42.00	53.16	–	–	nr	**53.16**
600 × 600 mm, 40 W	32.00	40.50	0.87	33.61	nr	**74.11**
Over cost for emergency	80.00	101.25	–	–	nr	**101.25**
Over cost for DALI dimming	30.00	37.97	–	–	nr	**37.97**
300 × 1200 mm, 70 W	80.50	101.88	0.87	33.61	nr	**135.49**
Over cost for emergency (remote)	110.00	139.22	–	–	nr	**139.22**
Over cost for DALI dimming (remote)	45.00	56.95	–	–	nr	**56.95**

ELECTRICAL SUPPLY/POWER/LIGHTING

Item	Net Price £	Material £	Labour hours	Labour £	Unit	Total rate £
Ceiling recessed round asymetric LED downlighter; for wall washing with non-dimmable gear						
1 × 12 W	72.00	91.12	0.75	28.99	nr	**120.11**
1 × 18 W	98.00	124.03	0.75	28.99	nr	**153.02**
1 × 26 W	123.00	155.67	0.75	28.99	nr	**184.66**
1 × 38 W	115.50	146.18	0.75	28.99	nr	**175.17**
1 × 50 W	145.50	184.14	0.75	28.99	nr	**213.13**
Extra Over cost for emergency (remote)	145.00	183.51	–	–	nr	**183.51**
Extra Over cost for DALI dimming (remote)	30.00	37.97	–	–	nr	**37.97**
Ceiling recessed asymetric LED wall washer; non-dimmable control gear; linear 80 mm × 600 mm						
1 × 35 W	220.00	278.43	0.75	28.99	nr	**307.42**
Extra Over cost for emergency (integral)	145.00	183.51	–	–	nr	**183.51**
Extra Over cost for DALI dimming (integral)	30.00	37.97	–	–	nr	**37.97**
Wall mounted LED uplighter; Integral driver; non-dimmable						
1 × 15 W	110.00	139.22	0.84	32.46	nr	**171.68**
1 × 30 W	145.00	183.51	0.84	32.46	nr	**215.97**
1 × 45 W	175.00	221.48	0.84	32.46	nr	**253.94**
Extra Over cost for emergency (remote)	145.00	183.51	–	–	nr	**183.51**
Extra Over cost for DALI dimming (remote)	30.00	37.97	–	–	nr	**37.97**
Suspended linear LED up/downlight 1.5 m Integral gear; MPO diffuser cut-off; 30% uplight, 70% downlight						
1 × 45 W (downlight), 1 × 30 W (uplight)	240.00	303.74	0.75	28.99	nr	**332.73**
Extra Over cost for emergency (remote)	145.00	183.51	–	–	nr	**183.51**
Extra Over cost for DALI dimming (remote)	30.00	37.97	–	–	nr	**37.97**
Recessed 'architectural' LED, Integral gear, Bat wing, low brightness, delivers direct, ceiling and graduated wall washing illumination						
600 × 600 mm, 40 W	220.00	278.43	0.87	33.61	nr	**312.04**
600 × 600 mm, 50 W	280.00	354.37	0.87	33.61	nr	**387.98**
Extra Over cost for emergency (remote)	145.00	183.51	–	–	nr	**183.51**
Extra Over cost for DALI dimming (remote)	40.00	50.62	–	–	nr	**50.62**
500 × 500 mm, 70 W	200.00	253.12	0.87	33.61	nr	**286.73**
500 × 500 mm, 85 W	220.00	278.43	0.87	33.61	nr	**312.04**
Extra Over cost for emergency (remote)	180.00	227.81	–	–	nr	**227.81**
Extra Over cost for DALI dimming (remote)	50.00	63.28	–	–	nr	**63.28**
Ceiling recessed round fixed LED downlighter with non-dimmable gear						
1 × 12 W	62.00	78.47	0.75	28.99	nr	**107.46**
1 × 18 W	88.00	111.37	0.75	28.99	nr	**140.36**
1 × 26 W	112.00	141.75	0.75	28.99	nr	**170.74**
1 × 38 W	105.00	132.89	0.75	28.99	nr	**161.88**
1 × 50 W	135.00	170.86	0.75	28.99	nr	**199.85**
Extra Over cost for emergency (remote)	145.00	183.51	–	–	nr	**183.51**
Extra Over cost for DALI dimming (remote)	30.00	37.97	–	–	nr	**37.97**

ELECTRICAL SUPPLY/POWER/LIGHTING

Item	Net Price £	Material £	Labour hours	Labour £	Unit	Total rate £
GENERAL LIGHTING – cont						
Ceiling recessed round adjustable LED downlighter with non dimmable gear						
1 × 12 W	75.00	94.92	0.75	28.99	nr	**123.91**
1 × 18 W	92.00	116.44	0.75	28.99	nr	**145.43**
1 × 26 W	115.70	146.43	0.75	28.99	nr	**175.42**
1 × 38 W	131.00	165.79	0.75	28.99	nr	**194.78**
1 × 50 W	165.00	208.82	0.75	28.99	nr	**237.81**
Extra Over cost for emergency (remote)	145.00	183.51	–	–	nr	**183.51**
Extra Over cost for DALI dimming (remote)	30.00	37.97	–	–	nr	**37.97**
LUMINAIRES						
High/low bay, LED non-dimmable; aluminium reflector						
35 W	210.00	265.78	1.50	57.96	nr	**323.74**
45 W	245.00	310.07	1.50	57.96	nr	**368.03**
55 W	265.00	335.38	1.50	57.96	nr	**393.34**
LED batten fitting						
Corrosion resistant GRP body; gasket sealed; acrylic diffuser						
600 mm – 35 W	82.00	103.78	0.49	18.93	nr	**122.71**
1200 mm – 40 W	110.00	139.22	0.64	24.73	nr	**163.95**
1500 mm – 45 W	138.00	174.65	0.72	27.82	nr	**202.47**
1800 mm – 50 W	169.00	213.89	0.94	36.32	nr	**250.21**
Extra Over cost for emergency (Integral)	145.00	183.51	–	–	nr	**183.51**
Flameproof to IIA/IIB,I.P. 64; Aluminium Body; BS 229 and 899						
600 mm – 35 W	391.00	494.85	1.04	40.19	nr	**535.04**
1200 mm – 40 W	468.00	592.30	1.31	50.61	nr	**642.91**
1500 mm – 45 W	525.00	664.44	1.64	63.37	nr	**727.81**
1800 mm – 50 W	582.00	736.58	1.97	76.12	nr	**812.70**
Extra Over cost for emergency (Integral)	145.00	183.51	–	–	nr	**183.51**
LED downlight/spotlight; track mounted (3ct); adjustable fixing; including driver, polycarbonate optical lenses/diffuser, heat sink and 3000 K colour temperature 100 mm dia.						
20 W	199.00	251.85	0.75	28.99	nr	**280.84**
30 W	220.00	278.43	0.75	28.99	nr	**307.42**
Extra for emergency pack (track mounted remote)	150.00	189.84	0.25	9.65	nr	**199.49**
Extra for DALI lighting control (with DALI track)	45.00	56.95	0.25	9.65	nr	**66.60**
LED square modular spotlight; recessed into ceiling; including driver, polycarbonate optical lenses/diffuser, heat sink and 3000 K colour temperature						
1 × 15 W	155.00	196.17	0.84	32.46	nr	**228.63**
Extra for emergency pack 1 lamp (remote)	150.00	189.84	0.25	9.65	nr	**199.49**
Extra for DALI lighting control	30.00	37.97	0.25	9.65	nr	**47.62**

ELECTRICAL SUPPLY/POWER/LIGHTING

Item	Net Price £	Material £	Labour hours	Labour £	Unit	Total rate £
2 × 15 W	210.00	265.78	0.89	34.38	nr	300.16
Extra for emergency pack 1 lamp (remote)	150.00	189.84	0.25	9.65	nr	199.49
Extra for DALI lighting control	55.00	69.61	0.25	9.65	nr	79.26
3 × 15 W	285.00	360.70	0.89	34.38	nr	395.08
Extra for emergency pack 1 lamp (remote)	150.00	189.84	0.25	9.65	nr	199.49
Extra for DALI lighting control	95.00	120.23	0.25	9.65	nr	129.88
LED downlight/spotlight; track mounted; adjustable fixing; including driver, polycarbonate optical lenses/diffuser, heat sink, RGB colour						
100 mm dia.	282.00	356.90	0.75	28.99	nr	385.89
Internal flexible linear LED, IP20, including driver, extrusion and opal cover white 3000 K colour temperature, per/m						
7.2 W	55.60	70.37	0.25	9.65	nr	80.02
14.4 W	72.60	91.88	0.25	9.65	nr	101.53
17.3 W	85.60	108.34	0.25	9.65	nr	117.99
Extra for DALI lighting control (remote)	50.00	63.28	0.25	9.65	nr	72.93
Internal flexible linear LED, IP20, including DMX driver and opal cover RGB, per/m						
6 W	75.00	94.92	0.25	9.65	nr	104.57
9 W	83.00	105.04	0.25	9.65	nr	114.69
12 W	95.00	120.23	0.25	9.65	nr	129.88
15 W	110.00	139.22	0.25	9.65	nr	148.87
LED in-ground adjustable performance uplighters, IP68, including driver, polycarbonate optical lenses/diffuser, heat sink, 4000 K colour temperature						
15 W	320.00	404.99	0.25	9.65	nr	414.64
30 W	355.00	449.29	0.25	9.65	nr	458.94
45 W	395.00	499.91	0.25	9.65	nr	509.56
Extra for DALI lighting control	50.00	63.28	0.25	9.65	nr	72.93
LED in-ground adjustable performance uplighters, IP68, including driver, polycarbonate optical lenses/diffuser, heat sink, RGB						
15 W	335.00	423.98	0.25	9.65	nr	433.63
30 W	370.00	468.27	0.25	9.65	nr	477.92
45 W	420.00	531.55	0.25	9.65	nr	541.20
LED in-ground fixed decorative uplighters, IP68, including driver, polycarbonate optical lenses/diffuser, heat sink, 4000 K colour, 50 mm dia.						
3 W	185.00	234.14	0.25	9.65	nr	243.79
6 W	200.00	253.12	0.25	9.65	nr	262.77
9 W	220.00	278.43	0.25	9.65	nr	288.08
Extra for DALI lighting control	55.00	69.61	0.25	9.65	nr	79.26

ELECTRICAL SUPPLY/POWER/LIGHTING

Item	Net Price £	Material £	Labour hours	Labour £	Unit	Total rate £
GENERAL LIGHTING – cont						
LUMINAIRES – cont						
LED in-ground fixed decorative uplighters, IP68, including driver, polycarbonate optical lenses/diffuser, heat sink, RGB 50 mm dia.						
3 W	155.00	196.17	0.25	9.65	nr	205.82
6 W	165.00	208.82	0.25	9.65	nr	218.47
9 W	175.00	221.48	0.25	9.65	nr	231.13
External flexible linear LED, IP66, including driver, extrusion and opal cover, 3000 K colour temperature, per/m						
6 W	85.00	107.58	0.25	9.65	nr	117.23
9 W	97.00	122.76	0.25	9.65	nr	132.41
12 W	110.00	139.22	0.25	9.65	nr	148.87
15 W	125.00	158.20	0.25	9.65	nr	167.85
Extra for DALI lighting control	50.00	63.28	0.25	9.65	nr	72.93
External flexible linear LED, IP66, including driver, extrusion and opal cover, RGB, per/m						
6 W	95.00	120.23	0.25	9.65	nr	129.88
9 W	109.00	137.95	0.25	9.65	nr	147.60
12 W	121.00	153.14	0.25	9.65	nr	162.79
15 W	139.00	175.92	0.25	9.65	nr	185.57
Extra for DALI lighting control	50.00	63.28	0.25	9.65	nr	72.93
Handrail puck, LED, IP66, including driver and controller, optical lenses/diffuser 3000 K colour temperature						
9w Asymetric	145.00	183.51	0.25	9.65	nr	193.16
9w Down	165.00	208.82	0.25	9.65	nr	218.47
Extra for emergency pack (per unit)	150.00	189.84	0.25	9.65	nr	199.49
Extra for DALI lighting control	50.00	63.28	0.25	9.65	nr	72.93
Handrail puck, LED, IP66, including driver and controller, optical lenses/diffuser, RGB						
9w Asymetric	160.00	202.50	0.25	9.65	nr	212.15
9w Down	178.00	225.28	0.25	9.65	nr	234.93
Extra for emergency pack (per unit)	150.00	189.84	0.25	9.65	nr	199.49
Extra for DALI lighting control	50.00	63.28	0.25	9.65	nr	72.93
External lighting						
Bulkhead; aluminium body and polycarbonate bowl; vandal-resistant; IP65						
15 W	190.00	240.46	0.75	28.99	nr	269.45
25 W	210.00	265.78	0.75	28.99	nr	294.77
40 W	235.00	297.42	0.75	28.99	nr	326.41
Extra for emergency pack (Integral)	150.00	189.84	0.25	9.65	nr	199.49
Photocell	25.00	31.64	0.75	28.99	nr	60.63
1500 mm high LED circular bollard; polycarbonate visor; vandal-resistant; IP54						
15 W	310.00	392.34	1.75	67.63	nr	459.97
25 W	335.00	423.98	1.75	67.63	nr	491.61
40 W	355.00	449.29	1.75	67.63	nr	516.92

ELECTRICAL SUPPLY/POWER/LIGHTING

Item	Net Price £	Material £	Labour hours	Labour £	Unit	Total rate £
Floodlight; enclosed high performance LED; integeral control gear; reflector; toughened glass; IP65						
20 W	185.00	234.14	1.25	48.29	nr	**282.43**
35 W	210.00	265.78	1.25	48.29	nr	**314.07**
50 W	245.00	310.07	1.25	48.29	nr	**358.36**
65 W	275.00	348.04	1.25	48.29	nr	**396.33**
80 W	305.00	386.01	1.25	48.29	nr	**434.30**
Extra for Photocell	25.00	31.64	0.75	28.99	nr	**60.63**
Lighting Track						
Single circuit; 25 A 2 P&E steel trunking; low voltage with copper conductors; including couplers and supports; suspended						
Straight track/m	22.00	27.84	0.50	19.33	m	**47.17**
Live end feed unit complete with end stop	6.50	8.22	0.33	12.75	nr	**20.97**
Flexible couplers 0.5 m	8.52	10.79	0.33	12.75	nr	**23.54**
Tap off complete with 0.8 m of cable	9.35	11.84	0.25	9.65	nr	**21.49**
Three circuit; 25 A 2 P&E steel trunking; low voltage with copper conductors; including couplers and supports incorporating integral twisted pair comms bus bracket; fixed to backgrounds						
Straight track	25.35	32.09	0.75	28.99	nr	**61.08**
Live end feed unit complete with end stop	7.50	9.49	0.50	19.33	nr	**28.82**
Flexible couplers 0.5 m	9.89	12.52	0.45	17.39	nr	**29.91**
Tap off complete with 0.8 m of cable	10.35	13.10	0.25	9.65	nr	**22.75**
LIGHTING ACCESSORIES						
Switches						
6 amp metal clad surface mounted switch, gridswitch; one way						
1 gang	9.38	11.87	0.43	16.62	nr	**28.49**
2 gang	12.80	16.20	0.55	21.25	nr	**37.45**
3 gang	22.47	28.44	0.77	29.76	nr	**58.20**
4 gang	23.82	30.15	0.88	34.00	nr	**64.15**
6 gang	41.04	51.95	1.00	38.64	nr	**90.59**
8 gang	56.51	71.52	1.28	49.46	nr	**120.98**
10 gang	89.14	112.82	1.67	64.53	nr	**177.35**
Extra for						
10 amp – Two way switch	3.35	4.24	0.03	1.16	nr	**5.40**
20 amp – Two way switch	4.36	5.52	0.04	1.55	nr	**7.07**
20 amp – Intermediate	8.19	10.36	0.08	3.09	nr	**13.45**
20 amp – One way SP switch	3.25	4.11	0.08	3.09	nr	**7.20**
Steel blank plate; 1 gang	1.98	2.51	0.07	2.71	nr	**5.22**
Steel blank plate; 2 gang	3.36	4.26	0.08	3.09	nr	**7.35**

ELECTRICAL SUPPLY/POWER/LIGHTING

Item	Net Price £	Material £	Labour hours	Labour £	Unit	Total rate £
GENERAL LIGHTING – cont						
Switches – cont						
6 amp modular type switch; galvanized steel box, bronze or satin chrome coverplate; metal clad switches; flush mounting; one way						
1 gang	25.84	32.70	0.43	16.62	nr	**49.32**
2 gang	29.27	37.05	0.55	21.25	nr	**58.30**
3 gang	42.28	53.51	0.77	29.76	nr	**83.27**
4 gang	60.90	77.08	0.88	34.00	nr	**111.08**
6 gang	102.67	129.94	1.18	45.60	nr	**175.54**
8 gang	122.48	155.01	1.63	62.98	nr	**217.99**
9 gang	153.62	194.42	1.83	70.71	nr	**265.13**
12 gang	182.97	231.57	2.29	88.48	nr	**320.05**
6 amp modular type swtich; galvanized steel box; bronze or satin chrome coverplate; flush mounting; two way						
1 gang	26.87	34.00	0.43	16.62	nr	**50.62**
2 gang	36.82	46.60	0.55	21.25	nr	**67.85**
3 gang	53.19	67.31	0.77	29.76	nr	**97.07**
4 gang	63.33	80.15	0.88	34.00	nr	**114.15**
6 gang	106.77	135.13	1.18	45.60	nr	**180.73**
8 gang	127.37	161.20	1.63	62.98	nr	**224.18**
9 gang	159.78	202.22	1.83	70.71	nr	**272.93**
12 gang	190.29	240.83	2.22	85.78	nr	**326.61**
Plate switches; 10 amp flush mounted, white plastic fronted; 16 mm metal box; fitted brass earth terminal						
1 gang 1 way, single pole	10.24	12.96	0.28	10.82	nr	**23.78**
1 gang 2 way, single pole	11.47	14.52	0.33	12.75	nr	**27.27**
2 gang 2 way, single pole	14.33	18.13	0.44	17.00	nr	**35.13**
3 gang 2 way, single pole	16.82	21.29	0.56	21.64	nr	**42.93**
1 gang intermediate	22.68	28.71	0.43	16.62	nr	**45.33**
1 gang 1 way, double pole	19.61	24.82	0.33	12.75	nr	**37.57**
1 gang single pole with bell symbol	15.69	19.86	0.23	8.88	nr	**28.74**
1 gang single pole marked PRESS	12.55	15.88	0.23	8.88	nr	**24.76**
Time delay switch, suppressed	69.47	87.92	0.49	18.93	nr	**106.85**
Plate switches; 6 amp flush mounted white plastic fronted; 25 mm metal box; fitted brass earth terminal						
4 gang 2 way, single pole	30.62	38.75	0.42	16.23	nr	**54.98**
6 gang 2 way, single pole	48.63	61.54	0.47	18.17	nr	**79.71**
Architrave plate switches; 6 amp flush mounted, white plastic fronted; 27 mm metal box; brass earth terminal						
1 gang 2 way, single pole	4.10	5.19	0.30	11.59	nr	**16.78**
2 gang 2 way, single pole	8.31	10.52	0.36	13.91	nr	**24.43**

ELECTRICAL SUPPLY/POWER/LIGHTING

Item	Net Price £	Material £	Labour hours	Labour £	Unit	Total rate £
Ceiling switches, white moulded plastic, pull cord; standard unit						
6 amp, 1 way, single pole	4.50	5.69	0.32	12.36	nr	**18.05**
6 amp, 2 way, single pole	5.25	6.64	0.34	13.14	nr	**19.78**
16 amp, 1 way, double pole	7.82	9.90	0.37	14.30	nr	**24.20**
45 amp, 1 way, double pole with neon indicator	14.40	18.22	0.47	18.17	nr	**36.39**
10 amp splash proof moulded switch with plain, threaded or PVC entry						
1 gang, 2 way single pole	25.21	31.91	0.34	13.14	nr	**45.05**
2 gang, 1 way single pole	28.49	36.05	0.36	13.91	nr	**49.96**
2 gang, 2 way single pole	36.09	45.67	0.40	15.46	nr	**61.13**
6 amp watertight switch; metalclad; BS 3676; ingress protected to IP65 surface mounted						
1 gang, 2 way; terminal entry	21.49	27.19	0.41	15.85	nr	**43.04**
1 gang, 2 way; through entry	21.49	27.19	0.42	16.23	nr	**43.42**
2 gang, 2 way; terminal entry	26.20	33.16	0.54	20.87	nr	**54.03**
2 gang, 2 way; through entry	26.20	33.16	0.53	20.47	nr	**53.63**
2 way replacement switch	19.32	24.45	0.10	3.86	nr	**28.31**
15 amp watertight switch; metalclad; BS 3676; ingress protected to IP65; surface mounted						
1 gang 2 way, terminal entry	32.58	41.24	0.42	16.23	nr	**57.47**
1 gang 2 way, through entry	32.58	41.24	0.43	16.62	nr	**57.86**
2 gang 2 way, terminal entry	76.47	96.78	0.55	21.25	nr	**118.03**
2 gang 2 way, through entry	76.47	96.78	0.54	20.87	nr	**117.65**
Intermediate interior only	19.32	24.45	0.11	4.26	nr	**28.71**
2 way interior only	19.32	24.45	0.11	4.26	nr	**28.71**
Double pole interior only	19.32	24.45	0.11	4.26	nr	**28.71**
Electrical accessories; fixed to backgrounds (Including fixings)						
Dimmer switches; rotary action; for individual lights; moulded plastic case; metal backbox; flush mounted						
1 gang, 1 way; 250 W	20.07	25.40	0.28	10.82	nr	**36.22**
1 gang, 1 way; 400 W	26.44	33.47	0.28	10.82	nr	**44.29**
Dimmer switches; push on/off action; for individual lights; moulded plastic case; metal backbox; flush mounted						
1 gang, 2 way; 250 W	30.68	38.83	0.34	13.14	nr	**51.97**
3 gang, 2 way; 250 W	45.22	57.23	0.48	18.54	nr	**75.77**
4 gang, 2 way; 250 W	59.50	75.31	0.57	22.02	nr	**97.33**
Dimmer switches; rotary action; metal cald; metal backbox; BS 5518 and BS 800; flush mounted						
1 gang, 1 way; 400 W	55.73	70.53	0.33	12.75	nr	**83.28**

ELECTRICAL SUPPLY/POWER/LIGHTING

Item	Net Price £	Material £	Labour hours	Labour £	Unit	Total rate £
GENERAL LIGHTING – cont						
Ceiling roses						
Ceiling rose: white moulded plastic; flush fixed to conduit box						
Plug in type; ceiling socket with 2 terminals, loop-in and ceiling plug with 3 terminals and cover	11.22	14.20	0.34	13.14	nr	**27.34**
BC lampholder; white moulded plastic; heat resistent PVC insulated and sheathed cable; flush fixed						
2 core; 0.75 mm²	3.55	4.49	0.33	12.75	nr	**17.24**
Batten holder: white moulded plastic; 3 terminals; BS 5042; fixed to conduit						
Straight pattern; 2 terminals with loop-in and Earth	5.20	6.59	0.29	11.20	nr	**17.79**
Angled pattern; looped in terminal	–	–	0.29	11.20	nr	**11.20**
LIGHTING CONTROL MODULES						
SPV.92 Lighting control module; plug in; 9 output, 9 channel – switching						
Base and lid assembly	189.48	239.80	2.05	79.22	nr	**319.02**
SPV.92 Lighting control module; plug in; 9 output, 9 channel – dimming (DALI, DSI, 1–10 V)						
Base and lid assembly	231.72	293.26	2.05	79.22	nr	**372.48**
SPH.12+ Lighting control module; hard wired; 4 circuit switching						
Base and lid assembly	204.47	258.78	1.85	71.48	nr	**330.26**
SPH.27r Lighting control module; hard wired; 2 circuit 40 ballast drive DALI (with relays)						
Base and lid assembly	293.08	370.92	1.85	71.48	nr	**442.40**
SPH.27 Lighting control module; hard wired; 2 circuit 40 ballast drive DALI (without relays)						
Base and lid assembly	279.45	353.67	1.85	71.48	nr	**425.15**
SPV.56 Compact lighting control module; 3 output 18 ballast drive; dimmable (DALI)						
Base and lid assembly	258.98	327.77	1.85	71.48	nr	**399.25**
SPB.82 Blind control module; 8 outputs						
Base and lid assembly	204.47	258.78	1.85	71.48	nr	**330.26**
Interfaces						
SPO.20 DALI to DMX converter	272.61	345.02	1.85	71.48	nr	**416.50**
Dual bus presence detectors (DALI & E-Bus)						
SPU.7-S Dual-bus presence detector and infra-red sensor	68.16	86.26	0.60	23.18	nr	**109.44**
SPU.6-S Dual-bus universal sensor	74.98	94.90	0.60	23.18	nr	**118.08**
SPU.6-C Compact Dual-bus universal sensor	76.19	191.32	0.60	23.18	nr	**214.50**
SPU.6-C1 Compact Dual-bus universal sensor conduit/surface mount	80.73	102.17	0.60	23.18	nr	**125.35**

ELECTRICAL SUPPLY/POWER/LIGHTING

Item	Net Price £	Material £	Labour hours	Labour £	Unit	Total rate £
Scene switch plate; anodized aluminium finish						
SPK.9–4+4 Eight button intelligent switch plate	136.30	172.50	2.00	77.28	nr	**249.78**
SPK.9–4-S Four button intelligent switch plate	81.79	103.51	1.60	61.82	nr	**165.33**
SPK.9–2-S Two button intelligent switch plate	68.16	86.26	1.20	46.37	nr	**132.63**
Input devices/Interfaces						
SPF.4d Compact DALI interface for 4 gang retractive switch plate	34.08	43.13	1.20	46.37	nr	**89.50**
SPL.20 External photocell 20,000 Lux range	163.56	207.00	1.50	57.96	nr	**264.96**
SPF.6-T AV interface; RS232 interface	177.21	224.28	1.85	71.48	nr	**295.76**
Sub-addressing for DALI strings						
Sub-addressing charge per SPH.27R (40 Sub-addresses)	133.21	168.59	3.00	115.92	nr	**284.51**
Sub-addressing charge per SPV.56 (20 Sub-addresses)	66.60	252.88	3.00	115.92	nr	**368.80**
Commissioning						
Not included – worked out based on complexity of project						

ELECTRICAL SUPPLY/POWER/LIGHTING

Item	Net Price £	Material £	Labour hours	Labour £	Unit	Total rate £
GENERAL LV POWER						
ACCESSORIES						
Outlets						
Socket outlet: unswitched; 13 amp metal clad; BS 1363; galvanized steel box and coverplate with white plastic inserts; fixed surface mounted						
1 gang	10.81	13.69	0.41	15.85	nr	**29.54**
2 gang	18.01	22.79	0.41	15.85	nr	**38.64**
Socket outlet: switched; 13 amp metal clad; BS 1363; galvanized steel box and coverplate with white plastic inserts; fixed surface mounted						
1 gang	12.63	15.98	0.43	16.62	nr	**32.60**
2 gang	22.99	29.10	0.45	17.39	nr	**46.49**
Socket outlet: switched with neon indicator; 13 amp metal clad; BS 1363; galvanized steel box and coverplate withwhite plastic inserts; fixed surface mounted						
1 gang	26.73	33.82	0.43	16.62	nr	**50.44**
2 gang	48.60	61.51	0.45	17.39	nr	**78.90**
Socket outlet: unswitched; 13 Amp; BS 1363; white moulded plastic box and coverplate; fixed surface mounted						
1 gang	6.66	8.43	0.41	15.85	nr	**24.28**
2 gang	13.14	16.63	0.41	15.85	nr	**32.48**
Socket outlet; switched; 13 Amp; BS 1363; white moulded plastic box and coverplate; fixed surface mounted						
1 gang	4.86	6.15	0.43	16.62	nr	**22.77**
2 gang	7.80	9.87	0.45	17.39	nr	**27.26**
Socket outlet: switched with neon indicator; 13 Amp; BS 1363; white moulded plastic box and coverplate; fixed surface mounted						
1 gang	22.19	28.08	0.43	16.62	nr	**44.70**
2 gang	29.92	37.87	0.45	17.39	nr	**55.26**
Socket outlet: switched; 13 Amp; BS 1363; galvanized steel box, white moulded coverplate; flush fitted						
1 gang	6.05	7.66	0.43	16.62	nr	**24.28**
2 gang	16.70	21.13	0.45	17.39	nr	**38.52**
Socket outlet: switched with neon indicator; 13 Amp; BS 1363; galvanized steel box, white moulded coverplate; flush fixed						
1 gang	22.19	28.08	0.43	16.62	nr	**44.70**
2 gang	38.37	48.56	0.45	17.39	nr	**65.95**

ELECTRICAL SUPPLY/POWER/LIGHTING

Item	Net Price £	Material £	Labour hours	Labour £	Unit	Total rate £
Socket outlet: switched; 13 amp; BS 1363; galvanized steel box, satin chrome coverplate; BS 4662; flush fixed						
1 gang	24.05	30.44	0.43	16.62	nr	**47.06**
2 gang	30.42	38.49	0.45	17.39	nr	**55.88**
Socket outlet: switched with neon indicator; 13 amp; BS 1363; steel backbox, satin chrome coverplate; BS 4662; flush fixed						
1 gang	30.29	38.34	0.43	16.62	nr	**54.96**
2 gang	54.62	69.13	0.45	17.39	nr	**86.52**
RCD protected socket outlets, 13 amp, to BS 1363; galvanized steel box, white moulded cover plate; flush fitted						
2 gang, 10 mA tripping (active control)	136.02	172.14	0.45	17.39	nr	**189.53**
2 gang, 30 mA tripping (active control)	121.17	153.35	0.45	17.39	nr	**170.74**
2 gang, 30 mA tripping (passive control)	121.17	153.35	0.45	17.39	nr	**170.74**
Filtered socket outlets, 13 amp, to BS 1363, with separate 'clean earth' terminal; galvanized steel box, white moulded cover plate; flush fitted						
2 gang (spike protected)	102.42	129.62	0.50	19.33	nr	**148.95**
2 gang (spike and RFI protected)	129.64	164.07	0.55	21.25	nr	**185.32**
Replacement filter cassette	35.93	45.47	0.15	5.80	nr	**51.27**
Non-standard socket outlets, 13 amp, to BS 1363, with separate 'clean earth' terminal; for plugs with T-shaped earth pin; galvanized steel box, white moulded cover plate; flush fitted						
1 gang	19.01	24.06	0.43	16.62	nr	**40.68**
2 gang	34.21	43.30	0.43	16.62	nr	**59.92**
2 gang coloured RED	47.51	60.13	0.43	16.62	nr	**76.75**
Weatherproof socket outlet: 40 amp; switched; single gang; RCD protected; water and dust protected to I.P.66; surface mounted						
40 A 30 mA tripping current protecting 1 socket	150.56	190.55	0.52	20.09	nr	**210.64**
40 A 30 mA tripping current protecting 2 sockets	66.92	84.69	0.64	24.73	nr	**109.42**
Plug for weatherproof socket outlet: protected to I.P.66						
13 amp plug	6.59	8.34	0.21	8.11	nr	**16.45**
Floor service outlet box; comprising flat lid with flanged carpet trim; twin 13 A switched socket outlets; punched plate for mounting 2 telephone outlets; one blank plate; triple compartment						
3 compartment	67.77	85.77	0.88	34.00	nr	**119.77**
Floor service outlet box; comprising flat lid with flanged carpet trim; 2 twin 13 A switched socket outlets; punched plate for mounting 1 telephone outlet; one blank plate; triple compartment						
3 compartment	60.48	76.54	0.88	34.00	nr	**110.54**

ELECTRICAL SUPPLY/POWER/LIGHTING

Item	Net Price £	Material £	Labour hours	Labour £	Unit	Total rate £
GENERAL LV POWER – cont						
Outlets – cont						
Floor service outlet box; comprising flat lid with flanged carpet trim; twin 13 A switched socket outlets; punched plate for mounting 2 telephone outlets; two blank plates; four compartment						
4 compartment	79.54	100.67	0.88	34.00	nr	**134.67**
Floor service outlet box; comprising flat lid with flanged carpet trim; single 13 A unswitched socket outlet; single compartment; circular						
1 compartment	84.85	107.39	0.79	30.53	nr	**137.92**
Floor service grommet, comprising flat lid with flanged carpet trim; circular						
Floor Grommet	38.38	48.57	0.49	18.93	nr	**67.50**
Power Posts/Poles/Pillars						
Power Post						
Power post; aluminium painted body; PVC-u cover; 5 nr outlets	423.70	536.23	4.00	154.56	nr	**690.79**
Power Pole						
Power pole; 3.6 m high; aluminium painted body; PVC-u cover; 6 nr outlets	543.31	687.61	4.00	154.56	nr	**842.17**
Extra for						
Power pole extension bar; 900 mm long	59.37	75.14	1.50	57.96	nr	**133.10**
Vertical multi-compartment pillar; PVC-u; BS 4678 Part 4 EN60529; excludes accessories						
Single						
630 mm long	189.24	239.50	2.00	77.28	nr	**316.78**
3000 mm long	545.35	690.20	2.00	77.28	nr	**767.48**
Double						
630 mm long	189.24	239.50	3.00	115.92	nr	**355.42**
3000 mm long	578.80	732.52	3.00	115.92	nr	**848.44**
Connection Units						
Connection units: moulded pattern; BS 5733; moulded plastic box; white coverplate; knockout for flex outlet; surface mounted – standard fused						
DP switched	11.00	13.92	0.49	18.93	nr	**32.85**
Unswitched	10.11	12.79	0.49	18.93	nr	**31.72**
DP switched with neon indicator	13.96	17.66	0.49	18.93	nr	**36.59**
Connection units: moulded pattern; BS 5733; galvanized steel box; white coverplate; knockout for flex outlet; surface mounted						
DP switched	13.86	31.46	0.49	18.93	nr	**50.39**
DP unswitched	12.97	16.42	0.49	18.93	nr	**35.35**
DP switched with neon indicator	16.83	21.30	0.49	18.93	nr	**40.23**

ELECTRICAL SUPPLY/POWER/LIGHTING

Item	Net Price £	Material £	Labour hours	Labour £	Unit	Total rate £
Connection units: galvanized pressed steel pattern; galvanized steel box; satin chrome or satin brass finish; white moulded plastic inserts; flush mounted – standard fused						
DP switched	20.32	25.72	0.49	18.93	nr	**44.65**
Unswitched	19.10	24.17	0.49	18.93	nr	**43.10**
DP switched with neon indicator	27.42	34.70	0.49	18.93	nr	**53.63**
Connection units: galvanized steel box; satin chrome or satin brass finish; white moulded plastic inserts; flex outlet; flush mounted – standard fused						
Switched	19.49	24.66	0.49	18.93	nr	**43.59**
Unswitched	18.49	23.40	0.49	18.93	nr	**42.33**
Switched with neon indicator	25.08	31.74	0.49	18.93	nr	**50.67**
Shaver Sockets						
Shaver unit: self setting overload device; 200/250 voltage supply; white moulded plastic faceplate; unswitched						
Surface type with moulded plastic box	38.19	48.33	0.55	21.25	nr	**69.58**
Flush type with galvanized steel box	40.09	50.74	0.57	22.02	nr	**72.76**
Shaver unit: dual voltage supply unit; white moulded plastic faceplate; unswitched						
Surface type with moulded plastic box	46.05	58.28	0.62	23.96	nr	**82.24**
Flush type with galvanized steel box	47.87	60.58	0.64	24.73	nr	**85.31**
Cooker Control Units						
Cooker control unit: BS 4177; 45 amp D.P. main switch; 13 amp switched socket outlet; metal coverplate; plastic inserts; neon indicators						
Surface mounted with mounting box	54.07	68.43	0.61	23.56	nr	**91.99**
Flush mounted with galvanized steel box	51.68	65.41	0.61	23.56	nr	**88.97**
Cooker control unit: BS 4177; 45 amp D.P. main switch; 13 amp switched socket outlet; moulded plastic box and coverplate; surface mounted						
Standard	37.52	47.49	0.61	23.56	nr	**71.05**
With neon indicators	44.02	55.71	0.61	23.56	nr	**79.27**
Control Components						
Connector unit: moulded white plastic cover and block; galvanized steel back box; to immersion heaters						
3 kW up to 915 mm long; fitted to thermostat	40.20	50.88	0.75	28.99	nr	**79.87**
Water heater switch: 20 amp; switched with neon indicator						
DP switched with neon indicator	19.97	25.28	0.45	17.39	nr	**42.67**

ELECTRICAL SUPPLY/POWER/LIGHTING

Item	Net Price £	Material £	Labour hours	Labour £	Unit	Total rate £
GENERAL LV POWER – cont						
Switch Disconnectors						
Switch disconnectors; moulded plastic						
enclosure; fixed to backgrounds						
3 pole; IP54; Grey						
16 amp	37.23	47.12	0.80	30.90	nr	78.02
25 amp	44.08	55.79	0.80	30.90	nr	86.69
40 amp	71.78	90.84	0.80	30.90	nr	121.74
63 amp	111.73	141.40	1.00	38.64	nr	180.04
80 amp	193.70	245.15	1.25	48.29	nr	293.44
6 pole; IP54; Grey						
25 amp	61.93	78.38	1.00	38.64	nr	117.02
63 amp	104.58	132.36	1.25	48.29	nr	180.65
80 amp	198.24	250.89	1.80	69.54	nr	320.43
3 pole; IP54; Yellow						
16 amp	40.92	51.79	0.80	30.90	nr	82.69
25 amp	48.38	61.23	0.80	30.90	nr	92.13
40 amp	78.49	99.33	0.80	30.90	nr	130.23
63 amp	122.11	154.54	1.00	38.64	nr	193.18
6 pole; IP54; Yellow						
25 amp	60.75	76.89	1.00	38.64	nr	115.53
Industrial Sockets/Plugs						
Plugs; Splashproof; 100–130 volts, 50–60 Hz;						
IP 44 (Yellow)						
2 pole and earth						
16 amp	8.90	11.27	0.55	21.25	nr	32.52
32 amp	15.46	19.57	0.60	23.18	nr	42.75
3 pole and earth						
16 amp	16.75	21.20	0.65	25.11	nr	46.31
32 amp	17.65	22.33	0.72	27.82	nr	50.15
3 pole; neutral and earth						
16 amp	23.47	29.70	0.72	27.82	nr	57.52
32 amp	20.90	26.45	0.78	30.14	nr	56.59
Connectors; Splashproof; 100–130 volts,						
50–60 Hz; IP 44 (Yellow)						
2 pole and earth						
16 amp	8.90	11.27	0.42	16.23	nr	27.50
32 amp	15.46	19.57	0.50	19.33	nr	38.90
3 pole and earth						
16 amp	16.75	21.20	0.48	18.54	nr	39.74
32 amp	18.07	22.87	0.58	22.41	nr	45.28
3 pole; neutral and earth						
16 amp	23.47	29.70	0.52	20.09	nr	49.79
32 amp	23.79	30.11	0.73	28.21	nr	58.32

ELECTRICAL SUPPLY/POWER/LIGHTING

Item	Net Price £	Material £	Labour hours	Labour £	Unit	Total rate £
Angled sockets; surface mounted; Splashproof; 100–130 volts, 50–60 Hz; IP 44 (Yellow)						
2 pole and earth						
16 amp	5.44	6.89	0.55	21.25	nr	**28.14**
32 amp	19.80	25.05	0.60	23.18	nr	**48.23**
3 pole and earth						
16 amp	18.86	23.87	0.65	25.11	nr	**48.98**
32 amp	35.41	44.81	0.72	27.82	nr	**72.63**
3 pole; neutral and earth						
16 amp	26.18	33.13	0.72	27.82	nr	**60.95**
32 amp	34.23	43.32	0.78	30.14	nr	**73.46**
Plugs; Watertight; 100–130 volts, 50–60 Hz; IP67 (Yellow)						
2 pole and earth						
16 amp	14.06	17.80	0.55	21.25	nr	**39.05**
32 amp	24.50	31.01	0.60	23.18	nr	**54.19**
63 amp	67.18	85.02	0.75	28.99	nr	**114.01**
Connectors; Watertight; 100–130 volts, 50–60 Hz; IP 67 (Yellow)						
2 pole and earth						
16 amp	28.72	36.34	0.42	16.23	nr	**52.57**
32 amp	48.67	61.60	0.50	19.33	nr	**80.93**
63 amp	117.80	149.08	0.67	25.89	nr	**174.97**
Angled sockets; surface mounted; Watertight; 100–130 volts, 50–60 Hz; IP 67 (Yellow)						
2 pole and earth						
16 amp	23.73	30.03	0.55	21.25	nr	**51.28**
32 amp	46.41	58.73	0.60	23.18	nr	**81.91**
Plugs; Splashproof; 200–250 volts, 50–60 Hz; IP 44 (Blue)						
2 pole and earth						
16 amp	3.39	4.29	0.55	21.25	nr	**25.54**
32 amp	12.16	15.39	0.60	23.18	nr	**38.57**
63 amp	58.59	74.16	0.75	28.99	nr	**103.15**
3 pole and earth						
16 amp	13.46	17.04	0.65	25.11	nr	**42.15**
32 amp	18.08	22.88	0.72	27.82	nr	**50.70**
63 amp	58.80	74.41	0.83	32.07	nr	**106.48**
3 pole; neutral and earth						
16 amp	14.35	18.17	0.72	27.82	nr	**45.99**
32 amp	21.77	27.55	0.78	30.14	nr	**57.69**
Connectors; Splashproof; 200–250 volts, 50–60 Hz; IP 44 (Blue)						
2 pole and earth						
16 amp	8.37	10.60	0.42	16.23	nr	**26.83**
32 amp	20.79	26.31	0.50	19.33	nr	**45.64**
63 amp	73.42	92.92	0.67	25.89	nr	**118.81**

ELECTRICAL SUPPLY/POWER/LIGHTING

Item	Net Price £	Material £	Labour hours	Labour £	Unit	Total rate £
GENERAL LV POWER – cont						
Industrial Sockets/Plugs – cont						
Connectors – cont						
3 pole and earth						
16 amp	22.16	28.04	0.48	18.54	nr	**46.58**
32 amp	29.28	37.06	0.58	22.41	nr	**59.47**
63 amp	60.40	76.44	0.75	28.99	nr	**105.43**
3 pole; neutral and earth						
16 amp	25.75	32.59	0.52	20.09	nr	**52.68**
32 amp	88.20	111.63	0.73	28.21	nr	**139.84**
Angled sockets; surface mounted; Splashproof; 200–250 volts, 50–60 Hz; IP 44 (Blue)						
2 pole and earth						
16 amp	11.00	13.92	0.55	21.25	nr	**35.17**
32 amp	19.04	24.10	0.60	23.18	nr	**47.28**
63 amp	83.70	105.93	0.75	28.99	nr	**134.92**
3 pole and earth						
16 amp	21.40	27.08	0.65	25.11	nr	**52.19**
32 amp	37.03	46.86	0.72	27.82	nr	**74.68**
63 amp	81.52	103.17	0.83	32.07	nr	**135.24**
3 pole; neutral and earth						
16 amp	18.62	23.56	0.72	27.82	nr	**51.38**
32 amp	32.58	41.24	0.78	30.14	nr	**71.38**
Plugs; Watertight; 200–250 volts, 50–60 Hz; IP67 (Blue)						
2 pole and earth						
16 amp	14.35	18.17	0.41	15.85	nr	**34.02**
32 amp	26.35	33.35	0.50	19.33	nr	**52.68**
63 amp	72.31	91.52	0.66	25.50	nr	**117.02**
125 amp	185.28	234.49	0.86	33.23	nr	**267.72**
Connectors; Watertight; 200–250 volts, 50–60 Hz; IP 67 (Blue)						
2 pole and earth						
16 amp	14.35	18.17	0.41	15.85	nr	**34.02**
32 amp	24.46	30.96	0.50	19.33	nr	**50.29**
63 amp	62.80	79.48	0.67	25.89	nr	**105.37**
125 amp	249.65	315.95	0.87	33.61	nr	**349.56**
Angled sockets; surface mounted; Watertight; 200–250 volts, 50–60 Hz; IP 67 (Blue)						
2 pole and earth						
16 amp	23.73	30.03	0.55	21.25	nr	**51.28**
32 amp	46.40	58.72	0.60	23.18	nr	**81.90**
125 amp	337.96	427.72	1.00	38.64	nr	**466.36**

ELECTRICAL SUPPLY/POWER/LIGHTING

Item	Net Price £	Material £	Labour hours	Labour £	Unit	Total rate £
UNINTERRUPTIBLE POWER SUPPLY						
Uninterruptible power supply; sheet steel enclosure; self contained battery pack; including installation, testing and commissioning						
Single phase input and output; 5 year battery life; standard 13 A socket outlet connection						
1.0 kVA (10 minute supply)	1141.85	1445.12	0.30	11.59	nr	**1456.71**
1.0 kVA (30 minute supply)	1627.31	2059.52	0.50	19.33	nr	**2078.85**
2.0 kVA (10 minute supply)	2037.16	2578.23	0.50	19.33	nr	**2597.56**
2.0 kVA (60 minute supply)	3418.94	4327.01	0.50	19.33	nr	**4346.34**
3.0 kVA (10 minute supply)	2691.92	3406.89	0.50	19.33	nr	**3426.22**
3.0 kVA (40 minute supply)	3475.92	4399.12	1.00	38.64	nr	**4437.76**
5.0 kVA (30 minute supply)	4406.18	5576.46	1.00	38.64	nr	**5615.10**
8.0 kVA (10 minute supply)	5300.13	6707.85	2.00	77.28	nr	**6785.13**
8.0 kVA (30 minute supply)	7167.00	9070.56	2.00	77.28	nr	**9147.84**
Uninterruptible power supply; including final connections and testing and commissioning						
Medium size static; single phase input and output; 10 year battery life; in cubicle						
10.0 kVA (10 minutes supply)	5299.84	6707.48	10.00	386.39	nr	**7093.87**
10.0 kVA (30 minutes supply)	7166.68	9070.15	15.00	579.58	nr	**9649.73**
15.0 kVA (10 minutes supply)	8650.64	10948.25	10.00	386.39	nr	**11334.64**
15.0 kVA (30 minutes supply)	11957.09	15132.89	15.00	579.58	nr	**15712.47**
20.0 kVA (10 minutes supply)	8650.64	10948.25	10.00	386.39	nr	**11334.64**
20.0 kVA (30 minutes supply)	15391.17	19479.06	15.00	579.58	nr	**20058.64**
Medium size static; three phase input and output; 10 year battery life; in cubicle						
10.0 kVA (10 minutes supply)	7153.50	9053.48	10.00	386.39	nr	**9439.87**
10.0 kVA (30 minutes supply)	8961.31	11341.43	15.00	579.58	nr	**11921.01**
15.0 kVA (10 minutes supply)	8158.48	10325.37	15.00	579.58	nr	**10904.95**
15.0 kVA (30 minutes supply)	10081.88	12759.62	20.00	772.78	nr	**13532.40**
20.0 kVA (10 minutes supply)	8236.89	10424.61	20.00	772.78	nr	**11197.39**
20.0 kVA (30 minutes supply)	12462.78	15772.89	25.00	965.97	nr	**16738.86**
30.0 kVA (10 minutes supply)	12326.39	15600.28	25.00	965.97	nr	**16566.25**
30.0 kVA (30 minutes supply)	15495.57	19611.19	30.00	1159.17	nr	**20770.36**
Large size static; three phase input and output; 10 year battery life; in cubicle						
40 kVA (10 minutes supply)	12326.39	15600.28	30.00	1159.17	nr	**16759.45**
40 kVA (30 minutes supply)	18618.72	23563.85	30.00	1159.17	nr	**24723.02**
60 kVA (10 minutes supply)	19827.31	25093.44	35.00	1352.36	nr	**26445.80**
60 kVA (30 minutes supply)	23093.45	29227.07	35.00	1352.36	nr	**30579.43**
100 kVA (10 minutes supply)	29138.69	36877.93	40.00	1545.56	nr	**38423.49**
200 kVA (10 minutes supply)	48088.66	60861.01	40.00	1545.56	nr	**62406.57**
300 kVA (10 minutes supply)	78260.58	99046.60	40.00	1545.56	nr	**100592.16**
400 kVA (10 minutes supply)	91825.18	116213.94	50.00	1931.93	nr	**118145.87**
500 kVA (10 minutes supply)	107763.80	136385.86	60.00	2318.32	nr	**138704.18**
600 kVA (10 minutes supply)	144980.07	183486.78	70.00	2704.71	nr	**186191.49**
800 kVA (10 minutes supply)	181062.47	229152.66	80.00	3091.10	nr	**232243.76**

ELECTRICAL SUPPLY/POWER/LIGHTING

Item	Net Price £	Material £	Labour hours	Labour £	Unit	Total rate £
UNINTERRUPTIBLE POWER SUPPLY – cont						
Uninterruptible power supply – cont						
Integral diesel rotary; 400 V three phase input and output; no break supply; including ventilation and acoustic attenuation oil day tank and interconnecting pipework						
300 kVA	529309.58	669894.20	120.00	4636.65	nr	**1760267.55**
500 kVA	545124.07	689909.02	120.00	4636.65	nr	**694545.67**
800 kVA	595203.56	753289.62	140.00	5409.43	nr	**758699.05**
1000 kVA	660339.73	835725.96	140.00	5409.43	nr	**841135.39**
1670 kVA	786568.35	995480.91	170.00	6568.59	nr	**1002049.50**
2000 kVA	958230.71	1212736.78	200.00	7727.75	nr	**1220464.53**
2500 kVA	1107398.22	1401523.19	230.00	8886.92	nr	**1410410.11**
Uninterruptible power supply (UPS) consisting of a single cabinet, sheet steel enclosure with single 3 phase input and output. Internaly consists of multiple power modules to provide the rated output, with an additional power module for redundancy; 10 or 30 minute back-up, 10 year design life VRLA battery either internal or external (sizes dependent), including delivery, testing and commissioning						
20 kVA N+1 (10 mins)	12807.00	16208.54	10.00	386.39	nr	**16594.93**
20 kVA N+1 (30 mins)	14123.00	17874.07	10.00	386.39	nr	**18260.46**
40 kVA N+1 (10 mins)	17278.00	21867.04	30.00	1159.17	nr	**23026.21**
40 kVA N+1 (30 mins)	21850.00	27653.36	30.00	1159.17	nr	**28812.53**
60 kVA N+1 (10 mins)	23608.00	29878.28	35.00	1352.36	nr	**31230.64**
60 kVA N+1 (30 mins)	29032.00	36742.90	35.00	1352.36	nr	**38095.26**
100 kVA N+1 (10 mins)	34138.00	43205.05	40.00	1545.56	nr	**44750.61**
160 kVA N+1 (10 mins)	49679.00	62873.74	40.00	1545.56	nr	**64419.30**
200 kVA N+1 (10 mins)	63483.00	80344.08	40.00	1545.56	nr	**81889.64**

ELECTRICAL SUPPLY/POWER/LIGHTING

Item	Net Price £	Material £	Labour hours	Labour £	Unit	Total rate £
COMMERCIAL BATTERY STORAGE SYSTEMS						
Commercial Battery Storage Systems						
Battery storage system; self-contained compact battery pack; flexible layout; rated for outdoor use; includes labour, materials, civils, design and project management						
100 kW nominal/169 kWh batteries (1.6hr capacity)	108131.05	136850.66	48.00	1854.66	nr	**138705.32**
500 kW nominal/845 kWh batteries (1.6hr capacity)	394054.90	498715.88	58.00	2241.05	nr	**500956.93**
1000 kW nominal/1690 kWh batteries (1.6hr capacity)	745717.75	943780.39	76.00	2936.55	nr	**946716.94**

ELECTRICAL SUPPLY/POWER/LIGHTING

Item	Net Price £	Material £	Labour hours	Labour £	Unit	Total rate £
EMERGENCY LIGHTING						
24 volt/50 volt/110 volt fluorescent slave luminaires						
For use with DC central battery systems						
Indoor, 8 W	46.04	58.27	0.80	30.90	nr	**89.17**
Indoor, exit sign box	57.35	72.59	0.80	30.90	nr	**103.49**
Outdoor, 8 W weatherproof	51.31	64.94	0.80	30.90	nr	**95.84**
Conversion module AC/DC	57.35	72.59	0.25	9.65	nr	**82.24**
Self contained; polycarbonate base and diffuser; LED charging light to European sign directive; 3 hour standby						
Non maintained						
Indoor, 8 W	50.83	64.33	1.00	38.64	nr	**102.97**
Outdoor, 8 W weatherproof, vandal-resistant IP65	72.99	92.38	1.00	38.64	nr	**131.02**
Maintained						
Indoor, 8 W	36.39	46.05	1.00	38.64	nr	**84.69**
Outdoor, 8 W weatherproof, vandal-resistant IP65	107.66	136.26	1.00	38.64	nr	**174.90**
Exit signage						
Exit sign; gold effect, pendular including brackets						
Non-maintained, 8 W	138.24	174.96	1.00	38.64	nr	**213.60**
Maintained, 8 W	148.79	188.31	1.00	38.64	nr	**226.95**
Modification kit						
Module & battery for 58 W fluorescent modification from mains fitting to emergency; 3 hour standby	47.65	60.30	0.50	19.33	nr	**79.63**
Extra for remote box (when fitting is too small for modification)	23.66	29.95	0.50	19.33	nr	**49.28**
12 volt low voltage lighting; non maintained; 3 hour standby						
2 × 20 W lamp load	164.18	207.78	1.20	46.37	nr	**254.15**
1 × 50 W lamp load	179.23	226.83	1.00	38.64	nr	**265.47**
Maintained 3 hour standby						
2 × 20 W lamp load	155.42	196.69	1.20	46.37	nr	**243.06**
1 × 50 W lamp load	170.46	215.73	1.00	38.64	nr	**254.37**
DC central battery systems BS5266 compliant 24/50/110 volt						
DC supply to luminaires on mains failure; metal cubicle with battery charger, changeover device and battery as integral unit; ICEL 1001 compliant; 10 year design life valve regulated lead acid battery; 24 hour recharge; LCD display & LED indication; ICEL alarm pack; Includes on-site commissioning on 110 volt systems only						

ELECTRICAL SUPPLY/POWER/LIGHTING

Item	Net Price £	Material £	Labour hours	Labour £	Unit	Total rate £
24 Volt, wall mounted						
300 W maintained, 1 hour	2618.05	3313.41	4.00	154.56	nr	3467.97
635 W maintained, 3 hour	3176.18	4019.77	6.00	231.83	nr	4251.60
470 W non maintained, 1 hour	2334.09	2954.02	4.00	154.56	nr	3108.58
780 W non maintained, 3 hour	2757.58	3490.00	6.00	231.83	nr	3721.83
50 volt						
935 W maintained, 3 hour	2507.91	3174.01	8.00	309.11	nr	3483.12
1965 W maintained, 3 hour	3490.73	4417.86	8.00	309.11	nr	4726.97
1311 W non maintained, 3 hour	3417.30	4324.94	8.00	309.11	nr	4634.05
2510 W non maintained 3, hour	5218.97	6605.13	8.00	309.11	nr	6914.24
110 volt						
1603 W maintained, 3 hour	4919.10	6225.61	8.00	309.11	nr	6534.72
4446 W maintained, 3 hour	5617.99	7110.13	10.00	386.39	nr	7496.52
2492 W non maintained, 3 hour	6240.98	7898.59	10.00	386.39	nr	8284.98
5429 W non maintained, 3 hour	8518.77	10781.36	12.00	463.67	nr	11245.03
DC central battery systems; BS EN 50171 compliant; 24/50/110 volt						
Central power systems						
DC supply to luminaires on mains failure; metal cubicle with battery charger, changeover device and battery as integral unit; 10 year design life valve regulated lead acid battery; 12 hour recharge to 80% of specified duty; low volts discount; LCD display & LED indication; includes on-site commissioning for CPS systems only; battery sized for 'end of life' @ 20°C test pushbutton						
24 Volt, floor standing						
400 W non maintained, 1 hour	2458.94	3112.03	4.00	154.56	nr	3266.59
600 W maintained, 3 hour	3161.50	4001.20	6.00	231.83	nr	4233.03
50 volt						
2133 W non maintained, 3 hour	5527.41	6995.49	8.00	309.11	nr	7304.60
1900 W maintained, 3 hour	5282.61	6685.67	8.00	309.11	nr	6994.78
110 volt						
2200 W non maintained, 3 hour	5112.48	6470.35	8.00	309.11	nr	6779.46
4000 W maintained, 3 hour	5706.11	7221.65	12.00	463.67	nr	7685.32
Low power systems						
DC supply to luminaires on mains failure; metal cubicle with battery charger, changeover device and battery as integral unit; 5 year design life valve regulated lead acid battery; low volts discount; LED display & LED indication; battery sized for 'end of life' @ 20°C test pushbutton						
24 Volt, floor standing						
300 W non maintained, 1 hour	2696.39	3412.55	4.00	154.56	nr	3567.11
600 W maintained, 3 hour	2985.24	3778.12	6.00	231.83	nr	4009.95

ELECTRICAL SUPPLY/POWER/LIGHTING

Item	Net Price £	Material £	Labour hours	Labour £	Unit	Total rate £
EMERGENCY LIGHTING – cont						
AC static inverter system; BS5266 compliant; one hour standby						
Central system supplying AC power on mains failure to mains luminaires; ICEL 1001 compliant metal cubicle(s) with changeover device, battery charger, battery & static inverter; 10 year design life valve regulated lead acid battery; 24 hour recharge; LED indication and LCD display; pure sinewave output						
One hour						
750 VA, 600 W single phase I/P & O/P	3185.98	4032.18	6.00	231.83	nr	**4264.01**
3 kVA, 2.55 kW single phase I/P & O/P	5226.32	6614.43	8.00	309.11	nr	**6923.54**
5 kVA, 4.25 kW single phase I/P & O/P	6901.93	8735.08	10.00	386.39	nr	**9121.47**
8 kVA, 6.80 kW single phase I/P & O/P	8130.77	10290.30	12.00	463.67	nr	**10753.97**
10 kVA, 8.5 kW single phase I/P & O/P	10496.70	13284.62	14.00	540.94	nr	**13825.56**
13 kVA, 11.05 kW single phase I/P & O/P	14042.51	17772.20	16.00	618.22	nr	**18390.42**
15 kVA, 12.75 kW single phase I/P & O/P	14554.13	18419.71	30.00	1159.17	nr	**19578.88**
20 kVA, 17.0 kW 3 phase I/P & single phase O/P	20101.13	25439.99	40.00	1545.56	nr	**26985.55**
30 kVA, 25.5 kW 3 phase I/P & O/P	27164.61	34379.53	60.00	2318.32	nr	**36697.85**
40 kVA, 34.0 kW 3 phase I/P & O/P	33865.79	42860.54	80.00	3091.10	nr	**45951.64**
50 kVA, 42.5 kW 3 phase I/P & O/P	44169.13	55900.45	90.00	3477.49	nr	**59377.94**
65 kVA, 55.25 kW 3 phase I/P & O/P	54183.59	68574.76	100.00	3863.88	nr	**72438.64**
90 kVA, 68.85 kW 3 phase I/P & O/P	71601.80	90619.23	120.00	4636.65	nr	**95255.88**
120 kVA, 102 kW 3 phase I/P & O/P	79924.72	101152.72	150.00	5795.82	nr	**106948.54**
Three hour						
750 VA, 600 W single phase I/P & O/P	3728.18	4718.38	6.00	231.83	nr	**4950.21**
3 kVA, 2.55 kW single phase I/P & O/P	6274.02	7940.40	8.00	309.11	nr	**8249.51**
5 kVA, 4.25 kW single phase I/P & O/P	9945.92	12587.56	10.00	386.39	nr	**12973.95**
8 kVA, 6.80 kW single phase I/P & O/P	12374.27	15660.88	12.00	463.67	nr	**16124.55**
10 kVA, 8.5 kW single phase I/P & O/P	16322.77	20658.10	14.00	540.94	nr	**21199.04**
13 kVA, 11.05 kW single phase I/P & O/P	20126.84	25472.53	16.00	618.22	nr	**26090.75**
15 kVA, 12.75 kW single phase I/P & O/P	20457.31	25890.77	30.00	1159.17	nr	**27049.94**
20 kVA, 17.0 kW 3 phase I/P & single phase O/P	29624.78	37493.12	40.00	1545.56	nr	**39038.68**
30 kVA, 25.5 kW 3 phase I/P & O/P	44516.72	56340.36	60.00	2318.32	nr	**58658.68**
40 kVA, 34.0 kW 3 phase I/P & O/P	53222.78	67358.75	80.00	3091.10	nr	**70449.85**
50 kVA, 42.5 kW 3 phase I/P & O/P	71785.38	90851.58	90.00	3477.49	nr	**94329.07**
65 kVA, 55.25 kW 3 phase I/P & O/P	86796.05	109849.08	100.00	3863.88	nr	**113712.96**
90 kVA, 68.85 kW 3 phase I/P & O/P	102723.47	130006.82	120.00	4636.65	nr	**134643.47**
120 kVA, 102 kW 3 phase I/P & O/P	126609.10	160236.47	150.00	5795.82	nr	**166032.29**

ELECTRICAL SUPPLY/POWER/LIGHTING

Item	Net Price £	Material £	Labour hours	Labour £	Unit	Total rate £
AC static inverter system; BS EN 50171 compliant; one hour standby; low power system (typically wall mounted) Central system supplying AC power on mains failure to mains luminaires; metal cubicle(s) with changeover device, battery charger, battery & static inverter; 5 year design life valve regulated lead acid battery; LED indication and LCD display; 12 hour recharge to 80% duty; inverter rated for 120% of load for 100% of duty; battery sized for 'end of life' @ 20°C test pushbutton						
One hour						
300 VA, 240 W single phase I/P & O/P	1424.70	1803.10	3.00	115.92	nr	**1919.02**
600 VA, 480 W single phase I/P & O/P	1717.22	2173.32	4.00	154.56	nr	**2327.88**
750 VA, 600 W single phase I/P & O/P	3168.84	4010.48	6.00	231.83	nr	**4242.31**
Three hour						
150 VA, 120 W single phase I/P & O/P	1570.34	1987.42	3.00	115.92	nr	**2103.34**
450 VA, 360 W single phase I/P & O/P	1862.87	2357.64	4.00	154.56	nr	**2512.20**
750 VA, 600 W single phase I/P & O/P	3708.59	4693.60	6.00	231.83	nr	**4925.43**
AC static inverter system central power system; CPS BS EN 50171 compliant; one hour standby Central system supplying AC power on mains failure to mains luminaires; metal cubicle(s) with changeover device, battery charger, battery & static inverter; LED indication and LCD display; pure sinewave output; 10 year design life valve regulated lead acid battery; 12 hour recharge to 80% duty specified; inverter rated for 120% of load for 100% of duty; battery sized for 'end of life' @ 20°C test push button; includes on-site commissioning						
One hour						
750 VA, 600 W single phase I/P & O/P	3545.81	4487.58	6.00	231.83	nr	**4719.41**
3 kVA, 2.55 kW single phase I/P & O/P	5649.80	7150.38	8.00	309.11	nr	**7459.49**
5 kVA, 4.25 kW single phase I/P & O/P	7315.62	9258.65	10.00	386.39	nr	**9645.04**
8 kVA, 6.80 kW single phase I/P & O/P	8538.36	10806.15	12.00	463.67	nr	**11269.82**
10 kVA, 8.5 kW single phase I/P & O/P	10964.25	13876.35	14.00	540.94	nr	**14417.29**
13 kVA, 11.05 kW single phase I/P & O/P	14491.71	18340.71	16.00	618.22	nr	**18958.93**
15 kVA, 12.75 kW single phase I/P & O/P	14999.66	18983.57	30.00	1159.17	nr	**20142.74**
20 kVA, 17.0 kW 3 phase I/P & single phase O/P	20662.92	26150.99	40.00	1545.56	nr	**27696.55**
30 kVA, 25.5 kW 3 phase I/P & O/P	27834.12	35226.87	60.00	2318.32	nr	**37545.19**
40 kVA, 34.0 kW 3 phase I/P & O/P	34498.60	43661.43	80.00	3091.10	nr	**46752.53**
50 kVA, 42.5 kW 3 phase I/P & O/P	44929.20	56862.40	90.00	3477.49	nr	**60339.89**
65 kVA, 55.25 kW 3 phase I/P & O/P	55493.21	70232.21	100.00	3863.88	nr	**74096.09**
90 kVA, 68.85 kW 3 phase I/P & O/P	72818.40	92158.96	120.00	4636.65	nr	**96795.61**
120 kVA, 102 kW 3 phase I/P & O/P	81097.28	102636.72	150.00	5795.82	nr	**108432.54**

ELECTRICAL SUPPLY/POWER/LIGHTING

Item	Net Price £	Material £	Labour hours	Labour £	Unit	Total rate £
EMERGENCY LIGHTING – cont						
AC static inverter system central power system – cont						
Three hour						
750 VA, 600 W single phase I/P & O/P	4102.73	5192.41	6.00	231.83	nr	5424.24
3 kVA, 2.55 kW single phase I/P & O/P	6691.40	8468.63	8.00	309.11	nr	8777.74
5 kVA, 4.25 kW single phase I/P & O/P	10368.18	13121.96	10.00	386.39	nr	13508.35
8 kVA, 6.80 kW single phase I/P & O/P	12759.80	16148.80	12.00	463.67	nr	16612.47
10 kVA, 8.5 kW single phase I/P & O/P	16760.94	21212.64	14.00	540.94	nr	21753.58
13 kVA, 11.05 kW single phase I/P & O/P	20544.20	26000.74	16.00	618.22	nr	26618.96
15 kVA, 12.75 kW single phase I/P & O/P	20870.99	26414.33	30.00	1159.17	nr	27573.50
20 kVA, 17.0 kW 3 phase I/P & single phase O/P	30414.24	38492.26	40.00	1545.56	nr	40037.82
30 kVA, 25.5 kW 3 phase I/P & O/P	45091.98	57068.41	60.00	2318.32	nr	59386.73
40 kVA, 34.0 kW 3 phase I/P & O/P	53752.74	68029.47	80.00	3091.10	nr	71120.57
50 kVA, 42.5 kW 3 phase I/P & O/P	72397.36	91626.10	90.00	3477.49	nr	95103.59
65 kVA, 55.25 kW 3 phase I/P & O/P	87931.89	111286.60	100.00	3863.88	nr	115150.48
90 kVA, 68.85 kW 3 phase I/P & O/P	109319.39	138354.62	120.00	4636.65	nr	142991.27
120 kVA, 102 kW 3 phase I/P & O/P	127529.51	161401.35	150.00	5795.82	nr	167197.17

COMMUNICATIONS/SECURITY/CONTROL

Item	Net Price £	Material £	Labour hours	Labour £	Unit	Total rate £
TELECOMMUNICATIONS						
CABLES						
Multipair internal telephone cable; BS 6746; loose laid on tray/basket 0.5 millimetre dia. conductor LSZH insulated and sheathed multipair cables; BT specification CW 1308						
3 pair	0.21	0.27	0.03	1.16	m	1.43
4 pair	0.24	0.30	0.03	1.16	m	1.46
6 pair	0.37	0.47	0.03	1.16	m	1.63
10 pair	0.63	0.80	0.05	1.94	m	2.74
15 pair	0.95	1.20	0.06	2.32	m	3.52
20 pair + 1 wire	1.35	1.71	0.06	2.32	m	4.03
25 pair	1.71	2.16	0.08	3.09	m	5.25
40 pair + earth	2.42	3.06	0.08	3.09	m	6.15
50 pair + earth	2.91	3.68	0.10	3.86	m	7.54
80 pair + earth	3.89	4.93	0.10	3.86	m	8.79
100 pair + earth	6.20	7.85	0.12	4.64	m	12.49
Multipair internal telephone cable; BS 6746; installed in conduit/trunking 0.5 millimetre dia. conductor LSZH insulated and sheathed multipair cables; BT specification CW 1308						
3 pair	0.21	0.27	0.05	1.94	m	2.21
4 pair	0.24	0.30	0.06	2.32	m	2.62
6 pair	0.37	0.47	0.06	2.32	m	2.79
10 pair	0.63	0.80	0.07	2.71	m	3.51
15 pair	0.95	1.20	0.07	2.71	m	3.91
20 pair + 1 wire	1.35	1.71	0.09	3.48	m	5.19
25 pair	1.71	2.16	0.10	3.86	m	6.02
40 pair + earth	2.42	3.06	0.12	4.64	m	7.70
50 pair + earth	2.91	3.68	0.14	5.41	m	9.09
80 pair + earth	3.89	4.93	0.14	5.41	m	10.34
100 pair + earth	6.20	7.85	0.15	5.80	m	13.65
Low speed data; unshielded twisted pair; solid copper conductors; LSOH sheath; nominal impedance 100 ohm; Category 3 to ISO IS 1801/EIA/TIA 568B and EN50173/ 50174 standards to current revisions Installed in riser						
25 pair 24 AWG	1.88	2.37	0.03	1.16	m	3.53
50 pair 24 AWG	3.16	4.00	0.06	2.32	m	6.32
100 pair 24 AWG	7.05	8.93	0.10	3.86	m	12.79
Installed below floor						
25 pair 24 AWG	1.88	2.37	0.02	0.77	m	3.14
50 pair 24 AWG	3.16	4.00	0.05	1.94	m	5.94
100 pair 24 AWG	7.05	8.93	0.08	3.09	m	12.02

COMMUNICATIONS/SECURITY/CONTROL

Item	Net Price £	Material £	Labour hours	Labour £	Unit	Total rate £
TELECOMMUNICATIONS – cont						
ACCESSORIES						
Telephone outlet: moulded plastic plate with box; fitted and connected; flush or surface mounted						
Single master outlet	8.01	10.14	0.35	13.53	nr	**23.67**
Single secondary outlet	5.92	7.49	0.35	13.53	nr	**21.02**
Telephone outlet: bronze or satin chromeplate; with box; fitted and connected; flush or surface mounted						
Single master outlet	13.03	16.49	0.35	13.53	nr	**30.02**
Single secondary outlet	14.23	18.01	0.35	13.53	nr	**31.54**
Frames and box connections						
Provision and installation of a dual vertical Krone 108 A voice distribution frame that can accommodate a total of 138 × Krone 237 A Strips	270.78	342.70	1.50	57.96	nr	**400.66**
Label frame (Traffolyte style)	1.71	2.16	0.27	10.43	nr	**12.59**
Provision and installation of a box connection 301 A voice termination unit that can accommodate a total of 10 × Krone 237 A Strips	23.48	29.71	0.25	9.65	nr	**39.36**
Label frame (Traffolyte style)	0.57	0.72	0.08	3.20	nr	**3.92**
Provision and installation of a box connection 201 voice termination unit that can accommodate 20 pairs	12.49	15.80	0.17	6.42	nr	**22.22**
Label frame (Traffolyte style)	0.57	0.72	0.08	3.20	nr	**3.92**
Terminate, Test and Label Voice Multicore System						
Patch panels						
Voice; 19' wide fully loaded, finished in black including termination and forming of cables (assuming 2 pairs per port)						
25 port – RJ45 UTP – Krone	58.75	74.36	2.60	100.46	nr	**174.82**
50 port – RJ45 UTP – Krone	80.09	101.36	4.65	179.66	nr	**281.02**
900 pair fully loaded Systimax style of frame including forming and termination of 9 × 100 pair cables	661.80	837.57	25.00	965.97	nr	**1803.54**
Installation and termination of Krone Strip (10 pair block – 237 A) including designation label strip	3.95	5.00	0.50	19.33	nr	**24.33**
Patch panel and Outlet labelling per port (Traffolyte style)	0.20	0.26	0.02	0.77	nr	**1.03**
Provision and installation of a voice jumper, for cross termination on Krone Termination Strips	0.06	0.08	0.06	2.32	nr	**2.40**
CW1308/Cat 3 cable circuit test per pair	–	–	0.05	1.94	nr	**1.94**

COMMUNICATIONS/SECURITY/CONTROL

Item	Net Price £	Material £	Labour hours	Labour £	Unit	Total rate £
RADIO AND TELEVISION						
RADIO						
Cables						
Radio frequency cable; BS 2316; PVC sheathed; laid loose						
7/0.41 mm tinned copper inner conductor; solid polyethylene dielectric insulation; bare copper wire braid; PVC sheath; 75 ohm impedance						
Cable	1.16	1.47	0.05	1.94	m	**3.41**
Twin 1/0.58 mm copper covered steel solid core wire conductor; solid polyethylene dielectric insulation; barecopper wire braid; PVC sheath; 75 ohm impedance						
Cable	0.27	0.35	0.05	1.94	m	**2.29**
TELEVISION						
Cables						
Television aerial cable; coaxial; PVC sheathed; fixed to backgrounds						
General purpose TV aerial downlead; copper stranded inner conductor; cellular polythene insulation; copper braid outer conductor; 75 ohm impedance						
7/0.25 mm	0.26	0.32	0.06	2.32	m	**2.64**
Low loss TV aerial downlead; solid copper inner conductor; cellular polythene insulation; copper braid outer; conductor; 75 ohm impedance						
1/1.12 mm	0.27	0.35	0.06	2.32	m	**2.67**
Low loss air spaced; solid copper inner conductor; air spaced polythene insulation; copper braid outer conductor; 75 ohm impedance						
1/1.00 mm	0.36	0.46	0.06	2.32	m	**2.78**
Satelite aerial downlead; solid copper inner conductor; air spaced polythene insulation; copper tape and braid outer conductor; 75 ohm impedance						
1/1.00 mm	0.31	0.39	0.06	2.32	m	**2.71**
Satelite TV coaxial; solid copper inner conductor; semi air spaced polyethylene dielectric insulation; plain annealed copper foil and copper braid screen in outer conductor; PVC sheath; 75 ohm impedance						
1/1.25 mm	0.51	0.65	0.08	3.09	m	**3.74**

COMMUNICATIONS/SECURITY/CONTROL

Item	Net Price £	Material £	Labour hours	Labour £	Unit	Total rate £
RADIO AND TELEVISION – cont						
Television aerial cable – cont						
Satelite TV coaxial; solid copper inner conductor; air spaced polyethylene dielectric insulation; plain annealed copper foil and copper braid screen in outer conductor; PVC sheath; 75 ohm impedance						
1/1.67 mm	0.84	1.06	0.09	3.48	m	**4.54**
Video cable; PVC flame retardant sheath; laid loose						
7/0.1 mm silver coated copper covered annealed steel wire conductor; polyethylene dielectric insulation with tin coated copper wire braid; 75 ohm impedance						
Cable	0.20	0.26	0.05	1.94	m	**2.20**
ACCESSORIES						
TV co-axial socket outlet: moulded plastic box; flush or surface mounted						
One way Direct Connection	8.96	11.33	0.35	13.53	nr	**24.86**
Two way Direct Connection	12.48	15.79	0.35	13.53	nr	**29.32**
One way Isolated UHF/VHF	15.77	19.96	0.35	13.53	nr	**33.49**
Two way Isolated UHF/VHF	21.47	27.17	0.35	13.53	nr	**40.70**

COMMUNICATIONS/SECURITY/CONTROL

Item	Net Price £	Material £	Labour hours	Labour £	Unit	Total rate £
CLOCKS						
Master clock						
Master clock module 230 V with programming circuits, NTP client and server for synchronization over Ethernet network, sounders Harmonys or Melodys, control of DHF and NTP clock systems						
Rack mounted	1045.00	1322.55	3.00	115.92	nr	**1438.47**
Wall mounted	940.00	1189.66	5.00	193.19	nr	**1382.85**
Master clock module 230 V with programming circuits for control of wired and DHF clock systems, relays or sounders, NTP client and server for synchronization over Ethernet network						
Rack mounted	1411.00	1785.76	3.00	115.92	nr	**1901.68**
Wall mounted	1146.00	1450.38	5.00	193.19	nr	**1643.57**
Clocks and bells, master clock module 230 V with 3 programming circuits for control of wired and DHF clock systems, relays and sounders						
Rack mounted	911.00	1152.96	3.00	115.92	nr	**1268.88**
Wall mounted	720.00	911.23	5.00	193.19	nr	**1104.42**
Clocks only, master clock module 230 V for control of wired and DHF clock systems						
Rack mounted	689.00	872.00	3.00	115.92	nr	**987.92**
Wall mounted	535.00	677.10	5.00	193.19	nr	**870.29**
Extra over for						
GPS antenna with 20 m cable	210.00	265.78	2.50	96.60	nr	**362.38**
DHF transmitter	309.00	391.07	1.50	57.96	nr	**449.03**
DHF repeater, signal booster up to 200 m	354.00	448.02	3.00	115.92	nr	**563.94**
UPS	915.00	1158.02	5.00	193.19	nr	**1351.21**
Rack mounted external AC/DC power supply	526.00	665.71	3.00	115.92	nr	**781.63**
Indoor clock, 30 cm dia.						
Stand-alone						
HMS display, quartz movement, battery operated	50.00	63.28	0.80	30.90	nr	**94.18**
DHF wireless						
Hour and minute display, battery operated	114.00	144.28	0.80	30.90	nr	**175.18**
Hour, minute and second display, battery operated	125.00	158.20	0.80	30.90	nr	**189.10**
HM display, TBT	118.00	149.34	0.80	30.90	nr	**180.24**
HMS display, TBT	127.00	160.73	0.80	30.90	nr	**191.63**
Slave wired						
Hour and minute display, 24 V	68.00	86.06	1.00	38.64	nr	**124.70**
Hour, minute and second display, 24 V	82.00	103.78	1.00	38.64	nr	**142.42**
Hour, minute display, TBT, AFNOR	113.00	143.01	1.00	38.64	nr	**181.65**
Hour, minute and second display, TBT AFNOR	125.00	158.20	1.00	38.64	nr	**196.84**
Hour, minute display, POE, NTP	158.00	199.96	1.00	38.64	nr	**238.60**
Hour, minute and second display, POE, NTP	168.00	212.62	1.00	38.64	nr	**251.26**

COMMUNICATIONS/SECURITY/CONTROL

Item	Net Price £	Material £	Labour hours	Labour £	Unit	Total rate £
CLOCKS – cont						
Indoor clock, 40 cm dia.						
Stand-alone						
HMS display, quartz movement, battery operated	95.00	120.23	0.80	30.90	nr	**151.13**
DHF wireless						
Hour and minute display, battery operated	167.00	211.36	0.80	30.90	nr	**242.26**
Hour, minute and second display, battery operated	172.00	217.68	0.80	30.90	nr	**248.58**
HM display, TBT	167.00	211.36	0.80	30.90	nr	**242.26**
HMS display, TBT	179.00	226.54	0.80	30.90	nr	**257.44**
Slave wired						
Hour and minute display, 24 V	126.00	159.47	1.00	38.64	nr	**198.11**
Hour, minute and second display, 24 V	116.00	146.81	1.00	38.64	nr	**185.45**
Hour, minute display, TBT, AFNOR	177.00	224.01	1.00	38.64	nr	**262.65**
Hour, minute and second display, TBT AFNOR	183.00	231.60	1.00	38.64	nr	**270.24**
Hour, minute display, POE, NTP	209.00	264.51	1.00	38.64	nr	**303.15**
Hour, minute and second display, POE, NTP	215.00	272.10	1.00	38.64	nr	**310.74**
Digital clock, 7 cm digit height, 230 V						
Independent quartz movement						
Red display	210.00	265.78	0.80	30.90	nr	**296.68**
Green display	210.00	265.78	0.80	30.90	nr	**296.68**
Yellow display	210.00	265.78	0.80	30.90	nr	**296.68**
Blue display	285.00	360.70	0.80	30.90	nr	**391.60**
White display	285.00	360.70	0.80	30.90	nr	**391.60**
AFNOR, 230 V						
Red display	215.00	272.10	0.80	30.90	nr	**303.00**
Green display	215.00	272.10	0.80	30.90	nr	**303.00**
Yellow display	215.00	272.10	0.80	30.90	nr	**303.00**
Blue display	290.00	367.02	0.80	30.90	nr	**397.92**
White display	290.00	367.02	0.80	30.90	nr	**397.92**
DHF wireless, 230 V						
Red display	230.00	291.09	0.80	30.90	nr	**321.99**
Green display	230.00	291.09	0.80	30.90	nr	**321.99**
Yellow display	230.00	291.09	0.80	30.90	nr	**321.99**
Blue display	305.00	386.01	0.80	30.90	nr	**416.91**
White display	305.00	386.01	0.80	30.90	nr	**416.91**
NTP POE						
Red display	250.00	316.40	0.80	30.90	nr	**347.30**
Green display	250.00	316.40	0.80	30.90	nr	**347.30**
Yellow display	250.00	316.40	0.80	30.90	nr	**347.30**
Blue display	325.00	411.32	0.80	30.90	nr	**442.22**
White display	325.00	411.32	0.80	30.90	nr	**442.22**

COMMUNICATIONS/SECURITY/CONTROL

Item	Net Price £	Material £	Labour hours	Labour £	Unit	Total rate £
Digital clock, 5 cm digit height, 230 V						
Independent quartz movement						
Red display	175.00	221.48	0.80	30.90	nr	252.38
Green display	175.00	221.48	0.80	30.90	nr	252.38
Yellow display	175.00	221.48	0.80	30.90	nr	252.38
Blue display	230.00	291.09	0.80	30.90	nr	321.99
White display	230.00	291.09	0.80	30.90	nr	321.99
AFNOR, 230 V						
Red display	175.00	221.48	0.80	30.90	nr	252.38
Green display	175.00	221.48	0.80	30.90	nr	252.38
Yellow display	175.00	221.48	0.80	30.90	nr	252.38
Blue display	230.00	291.09	0.80	30.90	nr	321.99
White display	230.00	291.09	0.80	30.90	nr	321.99
DHF wireless, 230 V						
Red display	205.00	259.45	0.80	30.90	nr	290.35
Green display	205.00	259.45	0.80	30.90	nr	290.35
Yellow display	205.00	259.45	0.80	30.90	nr	290.35
Blue display	260.00	329.06	0.80	30.90	nr	359.96
White display	260.00	329.06	0.80	30.90	nr	359.96
NTP POE						
Red display	220.00	278.43	0.80	30.90	nr	309.33
Green display	220.00	278.43	0.80	30.90	nr	309.33
Yellow display	220.00	278.43	0.80	30.90	nr	309.33
Blue display	275.00	348.04	0.80	30.90	nr	378.94
White display	275.00	348.04	0.80	30.90	nr	378.94
Digital clock, 10 cm digit height, 230 V						
Independent quartz movement						
Red display	375.00	474.60	0.80	30.90	nr	505.50
Green display	375.00	474.60	0.80	30.90	nr	505.50
Yellow display	375.00	474.60	0.80	30.90	nr	505.50
Blue display	465.00	588.50	0.80	30.90	nr	619.40
White display	465.00	588.50	0.80	30.90	nr	619.40
AFNOR, 230 V						
Red display	375.00	474.60	0.80	30.90	nr	505.50
Green display	375.00	474.60	0.80	30.90	nr	505.50
Yellow display	375.00	474.60	0.80	30.90	nr	505.50
Blue display	465.00	588.50	0.80	30.90	nr	619.40
White display	465.00	588.50	0.80	30.90	nr	619.40
DHF wireless, 230 V						
Red display	390.00	493.58	0.80	30.90	nr	524.48
Green display	390.00	493.58	0.80	30.90	nr	524.48
Yellow display	390.00	493.58	0.80	30.90	nr	524.48
Blue display	480.00	607.49	0.80	30.90	nr	638.39
White display	480.00	607.49	0.80	30.90	nr	638.39
NTP POE						
Red display	405.00	512.57	0.80	30.90	nr	543.47
Green display	405.00	512.57	0.80	30.90	nr	543.47
Yellow display	405.00	512.57	0.80	30.90	nr	543.47
Blue display	495.00	626.47	0.80	30.90	nr	657.37
White display	495.00	626.47	0.80	30.90	nr	657.37

COMMUNICATIONS/SECURITY/CONTROL

Item	Net Price £	Material £	Labour hours	Labour £	Unit	Total rate £
DATA TRANSMISSION						
Cabinets						
Floor standing; suitable for 19' patch panels with glass lockable doors, metal rear doors, side panels, vertical cable management, 2 × 4 way PDUs, 4 way fan, earth bonding kit; installed on raised floor						
600 mm wide × 800 mm deep – 18U	795.48	1006.76	3.00	115.92	nr	**1122.68**
600 mm wide × 800 mm deep – 24U	828.61	1048.69	3.00	115.92	nr	**1164.61**
600 mm wide × 800 mm deep – 33U	889.93	1126.29	4.00	154.56	nr	**1280.85**
600 mm wide × 800 mm deep – 42U	954.57	1208.10	4.00	154.56	nr	**1362.66**
600 mm wide × 800 mm deep – 47U	1007.70	1275.34	4.00	154.56	nr	**1429.90**
800 mm wide × 800 mm deep – 42U	1080.54	1367.53	4.00	154.56	nr	**1522.09**
800 mm wide × 800 mm deep – 47U	1126.98	1426.31	4.00	154.56	nr	**1580.87**
Label cabinet	2.45	3.10	0.25	9.65	nr	**12.75**
Wall mounted; suitable for 19' patch panels with glass lockable doors, side panels, vertical cable management, 2 × 4 way PDUs, 4 way fan, earth bonding kit; fixed to wall						
19 mm wide × 500 mm deep – 9U	530.56	671.47	3.00	115.92	nr	**787.39**
19 mm wide × 500 mm deep – 12U	543.61	687.99	3.00	115.92	nr	**803.91**
19 mm wide × 500 mm deep – 15U	563.34	712.96	3.00	115.92	nr	**828.88**
19 mm wide × 500 mm deep – 18U	662.90	838.97	3.00	115.92	nr	**954.89**
19 mm wide × 500 mm deep – 21U	689.60	872.76	3.00	115.92	nr	**988.68**
Label cabinet	2.55	3.23	0.25	9.65	nr	**12.88**
Frames						
Floor standing; suitable for 19' patch panels with supports, vertical cable management, earth bonding kit; installed on raised floor						
19 mm wide × 500 mm deep – 25U	596.73	755.22	2.50	96.60	nr	**851.82**
19 mm wide × 500 mm deep – 39U	729.06	922.70	2.50	96.60	nr	**1019.30**
19 mm wide × 500 mm deep – 42U	795.53	1006.82	2.50	96.60	nr	**1103.42**
19 mm wide × 500 mm deep – 47U	862.00	1090.95	2.50	96.60	nr	**1187.55**
Label frame	2.55	3.23	0.25	9.65	nr	**12.88**
Copper data cabling						
Unshielded twisted pair; solid copper conductors; PVC insulation; nominal impedance 100 ohm; Cat 5e to ISO 11801, EIA/TIA 568B and EN 50173/50174 standards to the current revisions						
4 pair 24 AWG; nominal outside dia. 5.6 mm; installed above ceiling	0.22	0.28	0.02	0.77	m	**1.05**
4 pair 24 AWG; nominal outside dia. 5.6 mm; installed in riser	0.22	0.28	0.02	0.77	m	**1.05**
4 pair 24 AWG; nominal outside dia. 5.6 mm; installed below floor	0.22	0.28	0.01	0.39	m	**0.67**
4 pair 24 AWG; nominal outside dia. 5.6 mm; installed in trunking	0.22	0.28	0.02	0.77	m	**1.05**

COMMUNICATIONS/SECURITY/CONTROL

Item	Net Price £	Material £	Labour hours	Labour £	Unit	Total rate £
Unshielded twisted pair; solid copper conductors; LSOH sheathed; nominal impedance 100 ohm; Cat 5e to ISO 11801, EIA/TIA 568B and EN 50173/50174 standards to the current revisions						
4 pair 24 AWG; nominal outside dia. 5.6 mm; installed above ceiling	0.26	0.32	0.02	0.77	m	1.09
4 pair 24 AWG; nominal outside dia. 5.6 mm; installed in riser	0.26	0.32	0.02	0.77	m	1.09
4 pair 24 AWG; nominal outside dia. 5.6 mm; installed below floor	0.26	0.32	0.01	0.39	m	0.71
4 pair 24 AWG; nominal outside dia. 5.6 mm; installed in trunking	0.26	0.32	0.02	0.77	m	1.09
Unshielded twisted pair; solid copper conductors; PVC insulation; nominal impedance 100 ohm; Cat 6 to ISO 11801, EIA/TIA 568B and EN 50173/50174 standards to the current revisions						
4 pair 24 AWG; nominal outside dia. 5.6 mm; installed above ceiling	0.35	0.45	0.02	0.58	m	1.03
4 pair 24 AWG; nominal outside dia. 5.6 mm; installed in riser	0.35	0.45	0.02	0.58	m	1.03
4 pair 24 AWG; nominal outside dia. 5.6 mm; installed below floor	0.35	0.45	0.01	0.30	m	0.75
4 pair 24 AWG; nominal outside dia. 5.6 mm; installed in trunking	0.35	0.45	0.02	0.77	m	1.22
Unshielded twisted pair; solid copper conductors; LSOH sheathed; nominal impedance 100 ohm; Cat 6 to ISO 11801, EIA/TIA 568B and EN 50173/50174 standards to the current revisions						
4 pair 24 AWG; nominal outside dia. 5.6 mm; installed above ceiling	0.36	0.46	0.02	0.85	m	1.31
4 pair 24 AWG; nominal outside dia. 5.6 mm; installed in riser	0.36	0.46	0.02	0.85	m	1.31
4 pair 24 AWG; nominal outside dia. 5.6 mm; installed below floor	0.36	0.46	0.01	0.43	m	0.89
4 pair 24 AWG; nominal outside dia. 5.6 mm; installed in trunking	0.36	0.46	0.02	0.85	m	1.31
Patch panels						
Category 5e; 19' wide fully loaded, finished in black including termination and forming of cables						
24 port – RJ45 UTP – Krone/110	59.26	75.00	4.75	183.53	nr	258.53
48 port – RJ45 UTP – Krone/110	111.15	140.67	9.35	361.28	nr	501.95
Patch panel labelling per port	0.27	0.35	0.02	0.77	nr	1.12

COMMUNICATIONS/SECURITY/CONTROL

Item	Net Price £	Material £	Labour hours	Labour £	Unit	Total rate £
DATA TRANSMISSION – cont						
Patch panels – cont						
Category 6; 19' wide fully loaded, finished in black including termination and forming of cables						
24 port – RJ45 UTP – Krone/110	113.18	143.24	5.00	193.19	nr	**336.43**
48 port – RJ45 UTP – Krone/110	206.43	261.26	9.80	378.65	nr	**639.91**
Patch panel labelling per port	0.27	0.35	0.02	0.77	nr	**1.12**
Work station						
Category 5e RJ45 data outlet plate and multiway outlet boxes for wall, ceiling and below floor installations including label to ISO 11801 standards						
Wall mounted; fully loaded						
One gang LSOH PVC plate	5.42	6.85	0.15	5.80	nr	**12.65**
Two gang LSOH PVC plate	7.76	9.82	0.20	7.73	nr	**17.55**
Four gang LSOH PVC plate	13.29	16.82	0.40	15.46	nr	**32.28**
One gang satin brass plate	17.59	22.27	0.25	9.65	nr	**31.92**
Two gang satin brass plate	19.85	25.12	0.33	12.75	nr	**37.87**
Ceiling mounted; fully loaded						
One gang metal clad plate	9.34	11.82	0.33	12.75	nr	**24.57**
Two gang metal clad plate	12.69	16.06	0.45	17.39	nr	**33.45**
Below floor; fully loaded						
Four way outlet box, 5 m length 20 mm flexible conduit with glands and strain relief bracket	31.59	39.98	0.80	30.90	nr	**70.88**
Six way outlet box, 5 m length 25 mm flexible conduit with glands and strain relief bracket	40.84	51.69	1.20	46.37	nr	**98.06**
Eight way outlet box, 5 m length 25 mm flexible conduit with glands and strain relief bracket	51.42	65.07	1.40	54.10	nr	**119.17**
Installation of outlet boxes to desks	–	–	0.40	15.46	nr	**15.46**
Category 6 RJ45 data outlet plate and multiway outlet boxes for wall, ceiling and below floor installations including label to ISO 11801 standards						
Wall mounted; fully loaded						
One gang LSOH PVC plate	6.92	8.76	0.17	6.38	nr	**15.14**
Two gang LSOH PVC plate	10.16	12.86	0.22	8.50	nr	**21.36**
Four gang LSOH PVC plate	18.34	23.21	0.44	17.00	nr	**40.21**
One gang satin brass plate	16.91	21.40	0.28	10.63	nr	**32.03**
Two gang satin brass plate	21.16	26.78	0.36	14.02	nr	**40.80**
Ceiling mounted; fully loaded						
One gang metal clad plate	11.21	14.19	0.36	14.02	nr	**28.21**
Two gang metal clad plate	15.83	20.04	0.50	19.12	nr	**39.16**

COMMUNICATIONS/SECURITY/CONTROL

Item	Net Price £	Material £	Labour hours	Labour £	Unit	Total rate £
Below floor; fully loaded						
Four way outlet box, 5 m length 20 mm flexible conduit with glands and strain relief bracket	38.72	49.00	0.88	34.00	nr	**83.00**
Six way outlet box, 5 m length 25 mm flexible conduit with glands and strain relief bracket	51.50	65.18	1.32	51.00	nr	**116.18**
Eight way outlet box, 5 m length 25 mm flexible conduit with glands and strain relief bracket	65.59	83.01	1.54	59.51	nr	**142.52**
Installation of outlet boxes to desks	–	–	0.40	15.46	nr	**15.46**
Category 5e cable test	–	–	0.10	3.86	nr	**3.86**
Category 6 cable test	–	–	0.11	4.26	nr	**4.26**
Copper patch leads						
Category 5e; straight through booted RJ45 UTP – RJ45 UTP						
Patch lead 1 m length	1.94	2.45	0.09	3.48	nr	**5.93**
Patch lead 3 m length	2.31	2.92	0.09	3.48	nr	**6.40**
Patch lead 5 m length	3.27	4.14	0.10	3.86	nr	**8.00**
Patch lead 7 m length	4.80	6.07	0.10	3.86	nr	**9.93**
Category 6; straight through booted RJ45 UTP – RJ45 UTP						
Patch lead 1 m length	5.15	6.52	0.09	3.48	nr	**10.00**
Patch lead 3 m length	7.77	9.83	0.09	3.48	nr	**13.31**
Patch lead 5 m length	9.11	11.52	0.10	3.86	nr	**15.38**
Patch lead 7 m length	11.55	14.62	0.10	3.86	nr	**18.48**
Note 1 – With an intelligent Patching System, add a 25% uplift to the Cat 5e and Cat 6 System price. This would be dependent on the System and IMS solution chosen, i.e. iPatch, RIT or iTRACs.						
Note 2 – For a Cat 6 augmented solution (10 Gbps capable), add 25% to a Cat 6 System price.						
Fibre infrastructure						
Fibre optic cable, tight buffered, internal/external application, single mode, LSOH sheathed						
4 core fibre optic cable	1.54	1.95	0.10	3.86	m	**5.81**
8 core fibre optic cable	2.57	3.25	0.10	3.86	m	**7.11**
12 core fibre optic cable	3.18	4.02	0.10	3.86	m	**7.88**
16 core fibre optic cable	3.87	4.89	0.10	3.86	m	**8.75**
24 core fibre optic cable	4.67	5.91	0.10	3.86	m	**9.77**
Fibre optic cable OM1 and OM2, tight buffered, internal/external application, 62.5/125 multimode fibre, LSOH sheathed						
4 core fibre optic cable	1.64	2.07	0.10	3.86	m	**5.93**
8 core fibre optic cable	2.72	3.44	0.10	3.86	m	**7.30**
12 core fibre optic cable	3.48	4.40	0.10	3.86	m	**8.26**

COMMUNICATIONS/SECURITY/CONTROL

Item	Net Price £	Material £	Labour hours	Labour £	Unit	Total rate £
DATA TRANSMISSION – cont						
Fibre infrastructure – cont						
16 core fibre optic cable	4.31	5.45	0.10	3.86	m	**9.31**
24 core fibre optic cable	5.00	6.33	0.10	3.86	m	**10.19**
Fibre optic cable OM3, tight buffered, internal only application, 50/125 multimode fibre, LSOH sheathed						
4 core fibre optic cable	2.04	2.59	0.10	3.86	nr	**6.45**
8 core fibre optic cable	3.18	4.02	0.10	3.86	nr	**7.88**
12 core fibre optic cable	4.61	5.84	0.10	3.86	nr	**9.70**
24 core fibre cable	8.81	11.16	0.10	3.86	nr	**15.02**
Fibre optic single and multimode connectors and couplers including termination						
ST singlemode booted connector	7.76	9.82	0.25	9.65	nr	**19.47**
ST multimode booted connector	3.75	4.75	0.25	9.65	nr	**14.40**
SC simplex singlemode booted connector	11.00	13.92	0.25	9.65	nr	**23.57**
SC simplex multimode booted connector	4.01	5.07	0.25	9.65	nr	**14.72**
SC duplex multimode booted connector	8.00	10.12	0.25	9.65	nr	**19.77**
ST – SC duplex adaptor	19.44	24.61	0.01	0.39	nr	**25.00**
ST inline bulkhead coupler	4.17	5.28	0.01	0.39	nr	**5.67**
SC duplex coupler	7.78	9.84	0.01	0.39	nr	**10.23**
MTRJ small form factor duplex connector	7.75	9.81	0.25	9.65	nr	**19.46**
LC simplex multimode booted connector	–	–	0.25	9.65	nr	**9.65**
Singlemode core test per core	–	–	0.20	7.73	nr	**7.73**
Multimode core test per core	–	–	0.20	7.73	nr	**7.73**
Fibre; 19' wide fully loaded, labelled, alluminium alloy c/w couplers, fibre management and glands (excludes termination of fibre cores)						
8 way ST; fixed drawer	83.51	105.69	0.50	19.33	nr	**125.02**
16 way ST; fixed drawer	125.24	158.50	0.50	19.33	nr	**177.83**
24 way ST; fixed drawer	174.00	220.21	0.50	19.33	nr	**239.54**
8 way ST; sliding drawer	106.24	134.46	0.50	19.33	nr	**153.79**
16 way ST; sliding drawer	145.98	184.76	0.50	19.33	nr	**204.09**
24 way ST; sliding drawer	192.12	243.15	0.50	19.33	nr	**262.48**
8 way (4 duplex) SC; fixed drawer	99.55	125.99	0.50	19.33	nr	**145.32**
16 way (8 duplex) SC; fixed drawer	145.69	184.39	0.50	19.33	nr	**203.72**
24 way (12 duplex) SC; fixed drawer	179.08	226.64	0.50	19.33	nr	**245.97**
8 way (4 duplex) SC; sliding drawer	125.96	159.41	0.50	19.33	nr	**178.74**
16 way (8 duplex) SC; sliding drawer	172.40	218.19	0.50	19.33	nr	**237.52**
24 way (12 duplex) SC; sliding drawer	205.48	260.05	0.50	19.33	nr	**279.38**
8 way (4 duplex) MTRJ; fixed drawer	112.91	142.90	0.50	19.33	nr	**162.23**
16 way (8 duplex) MTRJ; fixed drawer	172.40	218.19	0.50	19.33	nr	**237.52**
24 way (12 duplex) MTRJ; fixed drawer	205.48	260.05	0.50	19.33	nr	**279.38**
8 way (4 duplex) FC/PC; fixed drawer	118.37	149.81	0.50	19.33	nr	**169.14**
16 way (8 duplex) FC/PC; fixed drawer	180.90	228.95	0.50	19.33	nr	**248.28**
24 way (12 duplex) FC/PC; fixed drawer	215.81	273.13	0.50	19.33	nr	**292.46**
Patch panel label per way	0.27	0.35	0.02	0.77	nr	**1.12**

COMMUNICATIONS/SECURITY/CONTROL

Item	Net Price £	Material £	Labour hours	Labour £	Unit	Total rate £
Fibre patch leads						
Single Mode						
Duplex OS1 LC – LC						
Fibre patch lead 1 m length	25.61	32.41	0.08	3.09	nr	**35.50**
Fibre patch lead 3 m length	23.96	30.32	0.08	3.09	nr	**33.41**
Fibre patch lead 5 m length	25.48	32.24	0.10	3.86	nr	**36.10**
Multimode						
Duplex 50/125 OM3 MTRJ – MTRJ						
Fibre patch lead 1 m length	22.48	28.45	0.08	3.09	nr	**31.54**
Fibre patch lead 3 m length	24.43	30.92	0.08	3.09	nr	**34.01**
Fibre patch lead 5 m length	25.66	32.48	0.10	3.86	nr	**36.34**
Duplex 50/125 OM3 ST – ST						
Fibre patch lead 1 m length	28.03	35.47	0.08	3.09	nr	**38.56**
Fibre patch lead 3 m length	30.49	38.58	0.08	3.09	nr	**41.67**
Fibre patch lead 5 m length	32.02	40.52	0.10	3.86	nr	**44.38**
Duplex 50/125 OM3 SC – SC						
Fibre patch lead 1 m length	25.15	31.83	0.08	3.09	nr	**34.92**
Firbe patch lead 3 m length	27.37	34.64	0.08	3.09	nr	**37.73**
Fibre patch lead 5 m length	28.73	36.36	0.10	3.86	nr	**40.22**
Duplex 50/125 OM3 LC – LC						
Fibre patch lead 1 m length	26.95	34.10	0.08	3.09	nr	**37.19**
Fibre patch lead 3 m length	29.32	37.11	0.08	3.09	nr	**40.20**
Fibre patch lead 5 m length	30.79	38.96	0.10	3.86	nr	**42.82**

COMMUNICATIONS/SECURITY/CONTROL

Item	Net Price £	Material £	Labour hours	Labour £	Unit	Total rate £
ACCESS CONTROL EQUIPMENT						
Equipment to control the movement of personnel into defined spaces; includes fixing to backgrounds, termination of power and data cables; excludes cable containment and cable installation						
Access control						
Proximity reader	114.44	144.84	1.50	57.96	nr	**202.80**
Exit button	20.81	26.34	1.50	57.96	nr	**84.30**
Exit PIR	137.33	173.80	2.00	77.28	nr	**251.08**
Emergency break glass double pole	31.21	39.50	1.00	38.64	nr	**78.14**
Alarm contact flush	5.20	6.59	1.00	38.64	nr	**45.23**
Alarm contact surface	5.20	6.59	1.00	38.64	nr	**45.23**
Reader controller 16 door	4463.32	5648.78	4.00	154.56	nr	**5803.34**
Reader controller 8 door	2962.02	3748.73	4.00	154.56	nr	**3903.29**
Reader controller 2 door	1037.28	1312.79	4.00	154.56	nr	**1467.35**
Reader interface	420.69	532.43	1.50	57.96	nr	**590.39**
Lock power supply 12 volt 3 amp	250.74	317.34	2.00	77.28	nr	**394.62**
Lock power supply 24 volt 3 amp	438.01	554.34	2.00	77.28	nr	**631.62**
Rechargeable battery 12 volt 7AHr	12.48	15.79	0.25	9.65	nr	**25.44**
Lock equipment						
Single slimline magnetic lock, monitored	108.20	136.94	2.00	77.28	nr	**214.22**
Single slimline magnetic lock, unmonitored	93.64	118.51	1.75	67.63	nr	**186.14**
Double slimline magnetic lock, monitored	214.32	271.24	4.00	154.56	nr	**425.80**
Double slimline magnetic lock, unmonitored	184.15	233.06	4.50	173.87	nr	**406.93**
Standard single magnetic lock, monitored	137.33	173.80	1.25	48.29	nr	**222.09**
Standard single magnetic lock, unmonitored	110.28	139.57	1.00	38.64	nr	**178.21**
Standard single magnetic lock, double monitored	184.15	233.06	1.50	57.96	nr	**291.02**
Standard double magnetic lock, monitored	274.67	347.63	1.50	57.96	nr	**405.59**
Standard double magnetic lock, unmonitored	215.36	272.56	1.25	48.29	nr	**320.85**
12 V electric release fail safe, monitored	143.58	181.72	1.25	48.29	nr	**230.01**
12 V electric release fail secure, monitored	143.58	181.72	1.00	38.64	nr	**220.36**
Solenoid bolt	271.54	343.66	1.50	57.96	nr	**401.62**
Electric mortice lock	101.96	129.04	1.00	38.64	nr	**167.68**

COMMUNICATIONS/SECURITY/CONTROL

Item	Net Price £	Material £	Labour hours	Labour £	Unit	Total rate £
SECURITY DETECTION AND ALARM						
Detection and alarm systems for the protection of property and persons; includes fixing of equipment to backgrounds and termination of power and data cabling; excludes cable containment and cable installation						
Detection, alarm equipment						
Alarm contact flush	6.17	7.81	1.00	38.64	nr	**46.45**
Alarm contact surface	6.17	7.81	1.00	38.64	nr	**46.45**
Roller shutter contact	19.76	25.01	1.00	38.64	nr	**63.65**
Personal attack button	20.99	26.57	1.00	38.64	nr	**65.21**
Acoustic break glass detectors	28.40	35.94	2.00	77.28	nr	**113.22**
Vibration detectors	37.04	46.88	2.00	77.28	nr	**124.16**
12 metre PIR detector	43.22	54.70	1.50	57.96	nr	**112.66**
15 metre dual detector	71.62	90.64	1.50	57.96	nr	**148.60**
8 zone alarm panel	380.32	481.33	3.00	115.92	nr	**597.25**
8–24 zone end station	373.73	472.99	4.00	154.56	nr	**627.55**
Remote keypad	125.95	159.40	2.00	77.28	nr	**236.68**
8 zone expansion	96.32	121.90	1.50	57.96	nr	**179.86**
Final exit set button	10.27	13.00	1.00	38.64	nr	**51.64**
Self contained external sounder	92.61	117.21	2.00	77.28	nr	**194.49**
Internal loudspeaker	18.52	23.44	2.00	77.28	nr	**100.72**
Rechargeable battery 12 volt 7 AHr (ampere hours)	14.82	18.76	0.25	9.65	nr	**28.41**
Surveillance equipment						
Vandal-resistant camera, colour	198.45	251.16	3.00	115.92	nr	**367.08**
External camera, colour	396.90	502.32	5.00	193.19	nr	**695.51**
Auto dome external, colour	992.25	1255.79	5.00	193.19	nr	**1448.98**
Auto dome external, colour/monochrome	992.25	1255.79	5.00	193.19	nr	**1448.98**
Auto dome internal, colour	396.90	502.32	4.00	154.56	nr	**656.88**
Auto dome internal colour/monochrome	396.90	502.32	4.00	154.56	nr	**656.88**
Mini internal domes	659.07	834.12	3.00	115.92	nr	**950.04**
Camera switcher, 32 inputs	1430.34	1810.23	4.00	154.56	nr	**1964.79**
Full function keyboard	269.01	340.46	1.00	38.64	nr	**379.10**
16 CH multiplexors, duplex	792.72	1003.26	2.00	77.28	nr	**1080.54**
16 way DVR, 250 GB harddrive	992.25	1255.79	3.00	115.92	nr	**1371.71**
Additional 250 GB harddrive	264.60	334.88	3.00	115.92	nr	**450.80**
16 way DVR, 500 GB harddrive	1323.00	2930.18	3.00	115.92	nr	**3046.10**
16 way DVR, 750 GB harddrive	1323.00	5023.17	1.00	38.64	nr	**5061.81**
10" colour monitor	171.99	217.67	2.00	77.28	nr	**294.95**
15" colour monitor	198.45	251.16	2.00	77.28	nr	**328.44**
17" colour monitor, high resolution	238.14	301.39	2.00	77.28	nr	**378.67**
21" colour monitor, high resolution	264.60	334.88	2.00	77.28	nr	**412.16**

COMMUNICATIONS/SECURITY/CONTROL

Item	Net Price £	Material £	Labour hours	Labour £	Unit	Total rate £
SECURITY DETECTION AND ALARM – cont						
Detection and alarm systems for the protection of property and persons – cont IP CCTV						
Internal fixed dome IP camera h.264	441.00	558.13	–	–	nr	**558.13**
Internal fixed dome IP dome camera 1MP	441.00	558.13	–	–	nr	**558.13**
External fixed dome IP	496.13	627.91	–	–	nr	**627.91**
External PTZ IP	1764.00	2232.52	–	–	nr	**2232.52**
24 Port PoE switch	1929.38	2441.82	–	–	nr	**2441.82**
16 Base channel network video recorder with 4TB storage	2205.00	2790.65	–	–	nr	**2790.65**
IP camera channel licence	88.20	111.63	–	–	nr	**111.63**

COMMUNICATIONS/SECURITY/CONTROL

Item	Net Price £	Material £	Labour hours	Labour £	Unit	Total rate £
FIRE DETECTION AND ALARM						
STANDARD FIRE DETECTION						
Control panel						
Zone control panel; 16 zone, mild steel case						
Single loop	448.59	567.74	3.00	115.92	nr	**683.66**
Single loop (flush mounted)	472.66	598.20	0.25	9.65	nr	**607.85**
Loop extension card (1 loop)	234.58	296.89	3.51	135.58	nr	**432.47**
Two loop	670.07	848.04	4.00	154.56	nr	**1002.60**
Two loop (flush mounted)	694.14	878.51	5.00	193.19	nr	**1071.70**
Repeater panels						
Standard network card	389.20	492.58	4.00	154.56	nr	**647.14**
FR – focus repeater	389.20	492.58	4.00	154.56	nr	**647.14**
Focus repeater with controls	429.32	543.35	0.25	9.65	nr	**553.00**
Equipment						
Manual call point units: plastic covered						
Surface mounted						
Call point	42.84	54.22	0.50	19.33	nr	**73.55**
Call point; Weatherproof	155.16	196.37	0.80	30.90	nr	**227.27**
Flush mounted						
Call point	42.84	54.22	0.56	21.64	nr	**75.86**
Call point; Weatherproof	155.16	196.37	0.86	33.25	nr	**229.62**
Detectors						
Smoke, ionization type with mounting base	17.49	22.13	0.75	28.99	nr	**51.12**
Smoke, optical type with mounting base	16.85	21.32	0.75	28.99	nr	**50.31**
Fixed temperature heat detector with mounting base (60°C)	20.07	25.40	0.75	28.99	nr	**54.39**
Rate of Rise heat detector with mounting base (90°C)	22.85	28.92	0.75	28.99	nr	**57.91**
Duct detector including optical smoke detector and base	241.57	305.73	2.00	77.28	nr	**383.01**
Remote smoke detector LED indicator with base	5.75	7.28	0.50	19.33	nr	**26.61**
Sounders						
Intelligent wall sounder beacon	90.80	114.91	0.75	28.99	nr	**143.90**
Waterproof intelligent sounder beacon	153.93	194.81	0.75	28.99	nr	**223.80**
Siren; 230 V	75.43	95.47	1.25	48.29	nr	**143.76**
Magnetic Door Holder; 230 V; surface fixed	71.14	90.04	1.50	58.02	nr	**148.06**

COMMUNICATIONS/SECURITY/CONTROL

Item	Net Price £	Material £	Labour hours	Labour £	Unit	Total rate £
FIRE DETECTION AND ALARM – cont						
ADDRESSABLE FIRE DETECTION						
Control panel						
Analogue addressable panel; BS EN54 Part 2 and 4 1998; incorporating 120 addresses per loop (maximum 1–2 km length); sounders wired on loop; sealed lead acid integral battery standby providing 48 hour standby; 24 volt DC; mild steel case; surface fixed						
1 loop; 4 × 12 volt batteries	1501.48	1900.27	6.00	231.83	nr	**2132.10**
Extra for 1 loop panel						
Loop expander card	84.89	107.44	1.00	38.64	nr	**146.08**
Repeater panel	1042.39	1319.25	6.00	231.83	nr	**1551.08**
Network nodes	1783.76	2257.53	6.00	231.83	nr	**2489.36**
Interface unit; for other systems						
Mains powered	534.18	676.05	1.50	57.96	nr	**734.01**
Loop powered	234.47	296.74	1.00	38.64	nr	**335.38**
Single channel I/O	121.85	154.21	1.00	38.64	nr	**192.85**
Zone module	151.55	191.80	1.50	57.96	nr	**249.76**
4 loop; 4 × 12 volt batteries; 24 hour standby; 30 minute alarm	2517.20	3185.77	8.00	309.11	nr	**3494.88**
Extra for 4 loop panel						
Loop card	527.85	668.05	1.00	38.64	nr	**706.69**
Repeater panel	1494.72	1891.71	6.00	231.83	nr	**2123.54**
Mimic panel	3669.71	4644.38	5.00	193.19	nr	**4837.57**
Network nodes	1782.38	2255.78	6.00	231.83	nr	**2487.61**
Interface unit; for other systems						
Mains powered	534.18	676.05	1.50	57.96	nr	**734.01**
Loop powered	234.47	296.74	1.00	38.64	nr	**335.38**
Single channel I/O	121.85	154.21	1.00	38.64	nr	**192.85**
Zone module	151.55	191.80	1.50	57.96	nr	**249.76**
Line modules	27.94	35.36	1.00	38.64	nr	**74.00**
8 loop; 4 × 12 volt batteries; 24 hour standby; 30 minute alarm	5131.43	6494.34	12.00	463.67	nr	**6958.01**
Extra for 8 loop panel						
Loop card	527.85	668.05	1.00	38.64	nr	**706.69**
Repeater panel	1494.72	1891.71	6.00	231.83	nr	**2123.54**
Mimic panel	3669.71	4644.38	5.00	193.19	nr	**4837.57**
Network nodes	1782.38	2255.78	6.00	231.83	nr	**2487.61**
Interface unit; for other systems						
Mains powered	534.18	676.05	1.50	57.96	nr	**734.01**
Loop powered	234.50	296.78	1.00	38.64	nr	**335.42**
Single channel I/O	51.15	64.74	1.00	38.64	nr	**103.38**
Zone module	151.55	191.80	1.50	57.96	nr	**249.76**
Line modules	27.94	35.36	1.00	38.64	nr	**74.00**

COMMUNICATIONS/SECURITY/CONTROL

Item	Net Price £	Material £	Labour hours	Labour £	Unit	Total rate £
Equipment						
Manual Call Point						
Surface mounted						
Call point	70.38	89.07	1.00	38.64	nr	**127.71**
Call point; Weatherproof	257.90	326.40	1.25	48.29	nr	**374.69**
Flush mounted						
Call point	70.38	89.07	1.00	38.64	nr	**127.71**
Call point; Weatherproof	257.90	326.40	1.25	48.29	nr	**374.69**
Detectors						
Smoke, ionization type with mounting base	71.67	90.71	0.75	28.99	nr	**119.70**
Smoke, optical type with mounting base	70.95	89.79	0.75	28.99	nr	**118.78**
Fixed temperature heat detector with mounting base (60°C)	71.31	90.25	0.75	28.99	nr	**119.24**
Rate of Rise heat detector with mounting base (90°C)	71.31	90.25	0.75	28.99	nr	**119.24**
Duct detector including optical smoke detector and addressable base	454.44	575.14	2.00	77.28	nr	**652.42**
Beam smoke detector with transmitter and receiver unit	700.59	886.67	2.00	77.28	nr	**963.95**
Zone short circuit isolator	45.06	57.03	0.75	28.99	nr	**86.02**
Plant interface unit	33.20	42.02	0.50	19.33	nr	**61.35**
Sounders						
Xenon flasher, 24 volt, conduit box	75.67	95.77	0.50	19.33	nr	**115.10**
Xenon flasher, 24 volt, conduit box; weatherproof	112.64	142.55	0.50	19.33	nr	**161.88**
6" bell, conduit box	90.80	307.53	0.75	28.99	nr	**336.52**
6" bell, conduit box; weatherproof	61.39	77.69	0.75	28.99	nr	**106.68**
Siren; 24 V polarised	76.64	96.99	1.00	38.64	nr	**135.63**
Siren; 240 V	75.43	95.47	1.25	48.29	nr	**143.76**
Magnetic Door Holder; 240 V; surface fixed	71.14	90.04	1.50	58.02	nr	**148.06**
Addressable wireless fire detection system, BS 5839 and EN54 Part 25 compliant, 868 Mhz operating frequency						
Control panel						
Wireless fire alarm panel, complete with back-up battery						
8 zone	502.52	635.99	4.00	154.56	nr	**790.55**
20 zone	677.82	857.85	5.00	193.19	nr	**1051.04**

COMMUNICATIONS/SECURITY/CONTROL

Item	Net Price £	Material £	Labour hours	Labour £	Unit	Total rate £
FIRE DETECTION AND ALARM – cont						
Equipment						
Smoke detector	134.40	170.09	0.50	19.33	nr	**189.42**
Combined smoke detector and sounder	257.10	325.38	0.50	19.33	nr	**344.71**
Combined smoke detector, sounder and beacon	303.85	384.55	0.50	19.33	nr	**403.88**
Heat detector	128.55	162.69	0.50	19.33	nr	**182.02**
Combined heat detector and sounder	251.27	318.01	0.50	19.33	nr	**337.34**
Combined heat detector, sounder and beacon	292.16	369.76	0.50	19.33	nr	**389.09**
Manual call point	134.40	170.09	0.50	19.33	nr	**189.42**
Manual call point (weatherproof)	315.54	399.35	0.75	28.99	nr	**428.34**
Sounder	181.15	229.26	0.50	19.33	nr	**248.59**
Beacon	222.04	281.02	0.50	19.33	nr	**300.35**
Combined sounder and beacon	227.88	288.40	0.50	19.33	nr	**307.73**
Remote silence button	128.55	162.69	0.50	19.33	nr	**182.02**
Output unit	239.58	303.22	0.50	19.33	nr	**322.55**
Signal booster panel	444.09	562.04	0.50	19.33	nr	**581.37**
Extra over for additional wired antenna	233.73	295.80	1.00	38.64	nr	**334.44**

COMMUNICATIONS/SECURITY/CONTROL

Item	Net Price £	Material £	Labour hours	Labour £	Unit	Total rate £
EARTHING AND BONDING						
Earth bar; polymer insulators and base mounting; including connections						
Non-disconnect link						
6 way	190.05	240.53	0.81	31.29	nr	**271.82**
8 way	210.44	266.34	0.81	31.29	nr	**297.63**
10 way	244.26	309.13	0.81	31.29	nr	**340.42**
Disconnect link						
6 way	214.10	270.96	1.01	39.03	nr	**309.99**
8 way	236.71	299.58	1.01	39.03	nr	**338.61**
10 way	262.50	332.23	1.01	39.03	nr	**371.26**
Soild earth bar; including connections						
150 × 50 × 6 mm	57.92	73.30	1.01	39.03	nr	**112.33**
Extra for earthing						
Disconnecting link						
300 × 50 × 6 mm	65.65	83.08	1.16	44.81	nr	**127.89**
500 × 50 × 6 mm	72.23	91.41	1.16	44.81	nr	**136.22**
Crimp lugs; including screws and connections to cable						
25 mm	0.54	0.68	0.31	11.97	nr	**12.65**
35 mm	0.73	0.92	0.31	11.97	nr	**12.89**
50 mm	0.84	1.06	0.32	12.36	nr	**13.42**
70 mm	1.51	1.92	0.32	12.36	nr	**14.28**
95 mm	2.52	3.19	0.46	17.77	nr	**20.96**
120 mm	1.78	2.25	1.25	48.29	nr	**50.54**
Earth clamps; connection to pipework						
15 mm to 32 mm dia.	1.48	1.87	0.15	5.80	nr	**7.67**
32 mm to 50 mm dia.	1.85	2.34	0.18	6.96	nr	**9.30**
50 mm to 75 mm dia.	2.16	2.73	0.20	7.73	nr	**10.46**

COMMUNICATIONS/SECURITY/CONTROL

Item	Net Price £	Material £	Labour hours	Labour £	Unit	Total rate £
LIGHTNING PROTECTION						
Conductor tape						
PVC sheathed copper tape						
25 × 3 mm	23.57	29.83	0.30	11.59	m	41.42
25 × 6 mm	44.56	56.39	0.30	11.59	m	67.98
50 × 6 mm	91.08	115.27	0.30	11.59	m	126.86
PVC sheathed copper solid circular conductor						
8 mm	13.75	17.40	0.50	19.33	m	36.73
Bare copper tape						
20 × 3 mm	19.33	24.46	0.30	11.59	m	36.05
25 × 3 mm	22.14	28.02	0.30	11.59	m	39.61
25 × 6 mm	40.61	51.40	0.40	15.46	m	66.86
50 × 6 mm	78.56	99.42	0.50	19.33	m	118.75
Bare copper solid circular conductor						
8 mm	12.50	15.83	0.50	19.33	m	35.16
Tape fixings; flat; metallic						
PVC sheathed copper						
25 × 3 mm	23.57	29.83	0.33	12.75	nr	42.58
25 × 6 mm	44.56	56.39	0.33	12.75	nr	69.14
50 × 6 mm	91.08	115.27	0.33	12.75	nr	128.02
8 mm	7.90	10.00	0.50	19.33	nr	29.33
Bare copper						
20 × 3 mm	20.30	25.69	0.30	11.59	nr	37.28
25 × 3 mm	22.14	28.02	0.30	11.59	nr	39.61
25 × 6 mm	40.61	51.40	0.40	15.46	nr	66.86
50 × 6 mm	78.56	99.42	0.50	19.33	nr	118.75
8 mm	13.13	16.62	0.50	19.33	nr	35.95
Tape fixings; flat; non-metallic; PVC sheathed copper						
25 × 3 mm	2.75	3.48	0.30	11.59	nr	15.07
Tape fixings; flat; non-metallic; bare copper						
20 × 3 mm	1.08	1.37	0.30	11.59	nr	12.96
25 × 3 mm	1.08	1.37	0.30	11.59	nr	12.96
50 × 6 mm	5.73	7.25	0.30	11.59	nr	18.84
Puddle flanges; copper						
600 mm long	125.56	158.91	0.93	35.93	nr	194.84
Air rods						
Pointed air rod fixed to structure; copper 10 mm dia.						
500 mm long	20.90	26.45	1.00	38.64	nr	65.09
1000 mm long	35.23	44.59	1.50	57.96	nr	102.55
Extra for						
Air terminal base	26.09	33.02	0.35	13.53	nr	46.55
Strike pad	27.71	35.07	0.35	13.53	nr	48.60

COMMUNICATIONS/SECURITY/CONTROL

Item	Net Price £	Material £	Labour hours	Labour £	Unit	Total rate £
16 mm dia.						
500 mm long	33.13	41.93	0.91	35.16	nr	77.09
1000 mm long	60.53	76.61	1.75	67.63	nr	144.24
2000 mm long	110.52	139.88	2.50	96.60	nr	236.48
Extra for						
Multiple point	55.65	70.43	0.35	13.53	nr	83.96
Air terminal base	30.58	38.71	0.35	13.53	nr	52.24
Ridge saddle	57.42	72.67	0.35	13.53	nr	86.20
Side mounting bracket	57.14	72.32	0.50	19.33	nr	91.65
Rod to tape coupling	23.05	29.18	0.50	19.33	nr	48.51
Strike Pad	27.71	35.07	0.35	13.53	nr	48.60
Air terminals						
16 mm dia.						
500 mm long	33.13	41.93	0.65	25.11	nr	67.04
1000 mm long	60.53	76.61	0.78	30.14	nr	106.75
2000 mm long	110.52	139.88	1.50	57.96	nr	197.84
Extra for						
Multiple point	55.65	70.43	0.35	13.53	nr	83.96
Flat saddle	57.14	72.32	0.35	13.53	nr	85.85
Side bracket	28.56	36.14	0.50	19.33	nr	55.47
Rod to cable coupling	23.71	30.00	0.50	19.33	nr	49.33
Bonds and clamps						
Bond to flat surface; copper						
26 mm	4.97	6.29	0.45	17.39	nr	23.68
8 mm dia.	20.48	25.92	0.33	12.75	nr	38.67
Pipe bond						
26 mm	4.97	6.29	0.45	17.39	nr	23.68
8 mm dia.	50.92	64.44	0.33	12.75	nr	77.19
Rod to tape clamp						
26 mm	9.19	11.63	0.45	17.39	nr	29.02
Square clamp; copper						
25 × 3 mm	10.43	13.20	0.33	12.75	nr	25.95
50 × 6 mm	53.41	67.59	0.50	19.33	nr	86.92
8 mm dia.	11.15	14.11	0.33	12.75	nr	26.86
Test clamp; copper						
26 × 8 mm; oblong	15.55	19.68	0.50	19.33	nr	39.01
26 × 8 mm; plate type	47.33	59.90	0.50	19.33	nr	79.23
26 × 8 mm; screw down	42.54	53.84	0.50	19.33	nr	73.17
Cast in earth points						
2 hole	34.57	43.75	0.75	28.99	nr	72.74
4 hole	51.30	64.93	1.00	38.64	nr	103.57
Extra for cast in earth points						
Cover plate; 25 × 3 mm	38.80	49.10	0.25	9.65	nr	58.75
Cover plate; 8 mm	38.80	49.10	0.25	9.65	nr	58.75
Rebar clamp; 8 mm	26.72	33.81	0.25	9.65	nr	43.46
Static earth receptacle	181.79	230.07	0.50	19.33	nr	249.40

COMMUNICATIONS/SECURITY/CONTROL

Item	Net Price £	Material £	Labour hours	Labour £	Unit	Total rate £
LIGHTNING PROTECTION – cont						
Bonds and clamps – cont						
Copper braided bonds						
25 × 3 mm						
200 mm hole centres	17.95	22.71	0.33	12.75	nr	**35.46**
400 mm holes centres	27.51	34.82	0.40	15.46	nr	**50.28**
U bolt clamps						
16 mm	11.85	15.00	0.33	12.75	nr	**27.75**
20 mm	13.44	17.01	0.33	12.75	nr	**29.76**
25 mm	16.47	20.84	0.33	12.75	nr	**33.59**
Earth pits/mats						
Earth inspection pit; hand to others for fixing						
Concrete	53.31	67.47	1.00	38.64	nr	**106.11**
Polypropylene	52.10	65.93	1.00	38.64	nr	**104.57**
Extra for						
5 hole copper earth bar; concrete pit	51.69	130.84	0.35	13.53	nr	**144.37**
5 hole earth bar; polypropylene	43.30	54.80	0.35	13.53	nr	**68.33**
Water proof electrode seal						
Single flange	324.90	411.20	0.93	35.93	nr	**447.13**
Double flange	546.70	691.90	0.93	35.93	nr	**727.83**
Earth electrode mat; laid in ground and connected						
Copper tape lattice						
600 × 600 × 3 mm	146.64	185.58	0.93	35.93	nr	**221.51**
900 × 900 × 3 mm	262.96	332.80	0.93	35.93	nr	**368.73**
Copper tape plate						
600 × 600 × 1.5 mm	127.04	160.79	0.93	35.93	nr	**196.72**
600 × 600 × 3 mm	251.93	318.84	0.93	35.93	nr	**354.77**
900 × 900 × 1.5 mm	282.47	357.49	0.93	35.93	nr	**393.42**
900 × 900 × 3 mm	547.84	693.35	0.93	35.93	nr	**729.28**
Earth rods						
Solid cored copper earth electrodes driven into ground and connected						
15 mm dia.						
1200 mm long	51.61	65.32	0.93	35.93	nr	**101.25**
Extra for						
Coupling	1.56	1.97	0.06	2.32	nr	**4.29**
Driving stud	1.92	2.43	0.06	2.32	nr	**4.75**
Spike	1.77	2.24	0.06	2.32	nr	**4.56**
Rod clamp; flat tape	9.19	11.63	0.25	9.65	nr	**21.28**
Rod clamp; solid conductor	4.11	5.20	0.25	9.65	nr	**14.85**
20 mm dia.						
1200 mm long	100.11	126.69	0.98	37.87	nr	**164.56**
Extra for						
Coupling	1.56	1.97	0.06	2.32	nr	**4.29**
Driving stud	3.14	3.98	0.06	2.32	nr	**6.30**
Spike	2.89	3.66	0.06	2.32	nr	**5.98**
Rod clamp; flat tape	9.19	11.63	0.25	9.65	nr	**21.28**
Rod clamp; solid conductor	4.55	5.76	0.25	9.65	nr	**15.41**

COMMUNICATIONS/SECURITY/CONTROL

Item	Net Price £	Material £	Labour hours	Labour £	Unit	Total rate £
Stainless steel earth electrodes driven into ground and connected						
16 mm dia.						
1200 mm long	65.93	83.44	0.93	35.93	nr	**119.37**
Extra for						
Coupling	2.01	2.54	0.06	2.32	nr	**4.86**
Driving head	1.92	2.43	0.06	2.32	nr	**4.75**
Spike	1.77	2.24	0.06	2.32	nr	**4.56**
Rod clamp; flat tape	9.19	11.63	0.25	9.65	nr	**21.28**
Rod clamp; solid conductor	4.11	5.20	0.25	9.65	nr	**14.85**
Surge protection						
Single phase; including connection to equipment						
90–150 V	376.82	476.91	5.00	193.19	nr	**670.10**
200–280 V	357.23	452.11	5.00	193.19	nr	**645.30**
Three phase; including connection to equipment						
156–260 V	760.55	962.55	10.00	386.39	nr	**1348.94**
346–484 V	708.69	896.92	10.00	386.39	nr	**1283.31**
349–484 V; remote display	788.21	997.56	10.00	386.39	nr	**1383.95**
346–484 V; 60 kA	1309.07	1656.76	10.00	386.39	nr	**2043.15**
346–484 V; 120 kA	2504.05	3169.13	10.00	386.39	nr	**3555.52**

COMMUNICATIONS/SECURITY/CONTROL

Item	Net Price £	Material £	Labour hours	Labour £	Unit	Total rate £
CENTRAL CONTROL/BUILDING MANAGEMENT						
Equipment						
Switches/sensors; includes fixing in position; electrical work elsewhere. Note: These are normally free issued to the mechanical contractor for fitting. The labour times applied assume the installation has been prepared for the fitting of the component.						
Pressure devices						
Liquid differential pressure sensor	169.78	214.87	0.50	19.33	nr	234.20
Liquid differential pressure switch	93.04	117.76	0.50	19.33	nr	137.09
Air differential pressure transmitter	122.79	155.40	0.50	19.33	nr	174.73
Air differential pressure switch	16.78	21.24	0.50	19.33	nr	40.57
Liquid level switch	84.72	107.22	0.50	19.33	nr	126.55
Static pressure sensor	95.94	121.42	0.50	19.33	nr	140.75
High pressure switch	84.72	107.22	0.50	19.33	nr	126.55
Low pressure switch	84.72	107.22	0.50	19.33	nr	126.55
Water pressure switch	129.86	164.35	0.50	19.33	nr	183.68
Duct averaging temperature sensor	56.49	71.49	1.00	38.64	nr	110.13
Temperature devices						
Return air sensor (fan coils)	5.53	7.00	1.00	38.64	nr	45.64
Frost thermostat	36.23	45.85	0.50	19.33	nr	65.18
Immersion thermostat	58.63	74.20	0.50	19.33	nr	93.53
Temperature high limit	58.63	74.20	0.50	19.33	nr	93.53
Temperature sensor with averaging element	56.49	71.49	0.50	19.33	nr	90.82
Immersion temperature sensor	23.99	30.36	0.50	19.33	nr	49.69
Space temperature sensor	6.95	8.79	1.00	38.64	nr	47.43
Combined space temperature & humidity sensor	95.94	121.42	1.00	38.64	nr	160.06
Outside air temperature sensor	13.43	17.00	2.00	77.28	nr	94.28
Outside air temperature & humidity sensor	314.64	398.20	2.00	77.28	nr	475.48
Duct humidity sensor	123.76	156.63	0.50	19.33	nr	175.96
Space humidity sensor	95.94	121.42	1.00	38.64	nr	160.06
Immersion water flow sensor	23.99	30.36	0.50	19.33	nr	49.69
Rain sensor	224.48	284.10	2.00	77.28	nr	361.38
Wind speed and direction sensor	709.89	898.44	2.00	77.28	nr	975.72
Controllers; includes fixing in position; electrical work elsewhere						
Zone						
Fan coil controller	181.31	229.47	2.00	77.28	nr	306.75
VAV controller	237.09	300.06	2.00	77.28	nr	377.34

COMMUNICATIONS/SECURITY/CONTROL

Item	Net Price £	Material £	Labour hours	Labour £	Unit	Total rate £
Plant						
Controller, 96 I/O points (exact configuration is dependent upon the number of I/O boards added)	3327.78	4211.64	0.50	19.33	nr	**4230.97**
Controller, 48 I/O points (exact configuration is dependent upon the number of I/O boards added)	2778.12	3515.99	0.50	19.33	nr	**3535.32**
Controller, 32 I/O points (exact configuration is dependent upon the number of I/O boards added)	2307.03	2919.77	0.50	19.33	nr	**2939.10**
Additional Digital Input Boards (12 DI)	549.67	695.67	0.20	7.73	nr	**703.40**
Additional Digital Output Boards (6 DO)	470.06	594.91	0.20	7.73	nr	**602.64**
Additional Analogue Input Boards (8 AI)	470.06	594.91	0.20	7.73	nr	**602.64**
Additional Analogue Output Boards (8 AO)	470.06	594.91	0.20	7.73	nr	**602.64**
Outstation Enclosure (fitted in riser with space allowance for controller and network device)	470.06	594.91	5.00	193.19	nr	**788.10**
Damper actuator; electrical work elsewhere						
Damper actuator 0–10 v	99.77	126.27	1.00	38.64	nr	**164.91**
Damper actuator with auxiliary switches	66.19	83.76	1.00	38.64	nr	**122.40**
Frequency inverters: not mounted within MCC; includes fixing in position; electrical work elsewhere						
2.2 kW	759.18	960.81	2.00	77.28	nr	**1038.09**
3 kW	839.45	1062.41	2.00	77.28	nr	**1139.69**
7.5 kW	1242.07	1571.96	2.00	77.28	nr	**1649.24**
11 kW	1518.34	1921.61	2.00	77.28	nr	**1998.89**
15 kW	1834.76	2322.07	2.00	77.28	nr	**2399.35**
18.5 kW	2144.10	2713.57	2.50	96.60	nr	**2810.17**
20 kW	2420.38	3063.23	2.50	96.60	nr	**3159.83**
30 kW	2932.79	3711.74	2.50	96.60	nr	**3808.34**
55 kW	5296.49	6703.23	3.00	115.92	nr	**6819.15**
Miscellaneous; includes fixing in position; electrical work elsewhere						
1 kW Thyristor	67.60	85.56	2.00	77.28	nr	**162.84**
10 kW Thyristor	130.16	164.73	2.00	77.28	nr	**242.01**
Front end and networking; electrical work elsewhere						
PC/monitor	1003.57	1270.11	2.00	77.28	nr	**1347.39**
Printer	295.17	373.56	2.00	77.28	nr	**450.84**
PC software	–	–	–	–	nr	**-**
Network server software	2413.29	3054.26	–	–	nr	**3054.26**
Router (allows connection to a network)	121.61	153.91	–	–	nr	**153.91**

COMMUNICATIONS/SECURITY/CONTROL

Item	Net Price £	Material £	Labour hours	Labour £	Unit	Total rate £
WI-FI						
Wi-Fi						
Zone director controller, 1000 Mbps, RJ 45, 2 ports, auto MDX and auto sensing						
Up to 2,000 clients	895.74	1133.65	4.00	154.56	nr	**1288.21**
Up to 5,000 clients, 500 access points	6199.59	7846.20	4.00	154.56	nr	**8000.76**
Up to 20,000 clients, 1,000 access points	7355.30	9308.87	4.00	154.56	nr	**9463.43**
Wi-Fi access points, suitable for large developments (commercial buildings, stadiums, hotels, education)						
Indoor access points, 802.11ac Wi-Fi, dual band, 2.4 GHz and 5 GHz						
Up to 100 connected devices	298.95	378.35	1.20	46.37	nr	**424.72**
Up to 100 connected devices, 867 Mbps	404.85	512.38	1.20	46.37	nr	**558.75**
Up to 300 connected devices, 867 Mbps	527.30	667.35	1.20	46.37	nr	**713.72**
Up to 400 connected devices, 1300 Mbps	649.74	822.32	1.20	46.37	nr	**868.69**
Up to 500 connected devices, 1300 Mbps	813.01	1028.94	1.20	46.37	nr	**1075.31**
Up to 500 connected devices, 1733 Mbps	1070.04	1354.25	1.20	46.37	nr	**1400.62**
Outdoor access points, 802.11ac Wi-Fi, dual band, 2.4 GHz and 5 GHz, IP67, plastic enclosure, with flexible wall or pole mounting						
Standard	1057.90	1338.88	1.80	69.54	nr	**1408.42**
Up to 500 Mbps	1952.55	2471.15	1.80	69.54	nr	**2540.69**
Wi-Fi access points, suitable for small/ medium sized and residential						
Indoor access points, 802.11ac Wi-Fi, dual band, 2.4 GHz and 5 GHz						
Up to 100 connected devices, 867 Mbps	404.85	512.38	1.20	46.37	nr	**558.75**
Up to 300 connected devices, 567 Mbps	527.30	667.35	1.20	46.37	nr	**713.72**
Up to 400 connected devices, 1300 Mbps	649.74	822.32	1.20	46.37	nr	**868.69**
Outdoor access points, 802.11ac Wi-Fi, dual band, 2.4 GHz and 5 GHz, IP67, plastic enclosure, with flexible wall or pole mounting						
Standard	1057.90	1338.88	1.80	69.54	nr	**1408.42**

PART 5

Rates of Wages

Quantity Surveyor's Pocket Book, 3rd Edition

D. Cartlidge

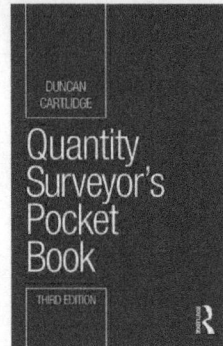

The third edition of the Quantity Surveyor's Pocket Book has been updated in line with NRM1, NRM2 and NRM3, and remains a must-have guide for students and qualified practitioners. Its focused coverage of the data, techniques and skills essential to the quantity surveying role makes it an invaluable companion for everything from initial cost advice to the final account stage.

Key features and updates included in this new edition:
- An up-to-date analysis of NRM1, 2 and 3;
- Measurement and estimating examples in NRM2 format;
- Changes in procurement practice;
- Changes in professional development, guidance notes and schemes of work;
- The increased use of NEC3 form of contract;
- The impact of BIM.

This text includes recommended formats for cost plans, developer's budgets, financial reports, financial statements and final accounts. This is the ideal concise reference for quantity surveyors, project and commercial managers, and students of any of the above.

March 2017; 186 × 123 mm, 466 pp
Pbk: 978-1-138-69836-9; £23.99

To Order: Tel: +44 (0) 1235 400524 Fax: +44 (0) 1235 400525
or Post: Taylor and Francis Customer Services,
Bookpoint Ltd, Unit T1, 200 Milton Park, Abingdon, Oxon, OX14 4TA UK
Email: book.orders@tandf.co.uk

For a complete listing of all our titles visit:
www.tandf.co.uk

Taylor & Francis
Taylor & Francis Group

Mechanical Installations

Rates of Wages

HEATING, VENTILATING, AIR CONDITIONING, PIPING AND DOMESTIC ENGINEERING INDUSTRY

For full details of the wage agreement and the Heating, Ventilating, Air Conditioning, Piping and Domestic Engineering Industry's National Working Rule Agreement, contact:

Building & Engineering Services Association
Lincoln House,
137–143 Hammersmith Road,
London W14 0QL
Telephone: 020 7313 4900
Internet: www.b-es.org

WAGE RATES, ALLOWANCES AND OTHER PROVISIONS

Hourly rates of wages

All districts of the United Kingdom

Main grades	From 7 October 2019 p/hr
Foreman	17.70
Senior Craftsman (+2nd welding skill)	15.79
Senior Craftsman	14.63
Craftsman (+2nd welding skill)	14.63
Craftsman	13.47
Operative	12.16
Adult Trainee	10.25
Mate (18 and over)	10.25
Mate (16–17)	4.76
Modern Apprentices	
Junior Apprentice	6.65
Intermediate Apprentice	9.43
Senior Apprentice	12.16

Note: Ductwork Erection Operatives are entitled to the same rates and allowances as the parallel Fitter grades shown.

HEATING, VENTILATING, AIR CONDITIONING, PIPING AND DOMESTIC ENGINEERING INDUSTRY

Trainee *Rates of Pay*

Junior Ductwork Trainees (Probationary)

Age at entry	From 7 October 2019 *p/hr*
17	6.03
18	6.03
19	6.03
20	6.03

Junior Ductwork Erectors (Year of Training)

	From 7 October 2019		
	1 yr	2 yr	3 yr
Age at entry	p/h	p/hr	p/hr
17	7.51	9.35	10.60
18	7.51	9.35	10.60
19	7.51	9.35	10.60
20	7.51	9.35	10.60

Responsibility Allowance (Craftsmen)	*From 7 October 2019 p/hr*
Second welding skill or supervisory responsibility (one unit)	0.58
Second welding skill and supervisory responsibility (two units)	0.58

Responsibility Allowance (Senior Craftsmen)	*From 7 October 2019 p/hr*
Second welding skill	0.58
Supervising responsibility	1.16
Second welding skill and supervisory responsibility	0.58

Daily travelling allowance – Scale 2

C: Craftsmen including Installers

M&A: Mates, Apprentices and Adult Trainees

Direct distance from centre to job in miles

		From 7 October 2019	
Over	*Not exceeding*	*C p/hr*	*p/hr M&A*
15	20	2.82	2.41
20	30	7.23	6.26
30	40	10.41	9.00
40	50	13.72	11.72

HEATING, VENTILATING, AIR CONDITIONING, PIPING AND DOMESTIC ENGINEERING INDUSTRY

Daily travelling allowance – Scale 1

C: Craftsmen including Installers

M&A: Mates, Apprentices and Adult Trainees

Direct distance from centre to job in miles

		From 7 October 2019	
Over	*Not exceeding*	*C p/hr*	*M&A p/hr*
0	15	7.69	7.69
15	20	10.51	10.11
20	30	14.93	13.95
30	40	18.10	16.71
40	50	21.39	19.44

Weekly Holiday Credit and Welfare Contributions

	From 7 October 2019						
	£	£	£	£	£	£	£
	a	b	c	d	e	f	g
Weekly Holiday Credit	82.43	76.25	73.55	70.84	68.12	65.45	62.78
Combined Weekly/Welfare Holiday Credit and Contribution	92.08	85.90	83.20	80.49	77.77	75.10	72.43

	From 7 October 2019			
	£	£	£	£
	h	i	j	K
Weekly Holiday Credit	56.63	47.74	43.92	30.99
Combined Weekly/Welfare Holiday Credit and Contribution	66.28	57.39	53.57	40.64

HEATING, VENTILATING, AIR CONDITIONING, PIPING AND DOMESTIC ENGINEERING INDUSTRY

The grades of H&V Operatives entitled to the different rates of Weekly Holiday Credit and Welfare Contribution are as follows:

a	b	c	d
Foreman	Senior Craftsman (RAS & RAW)	Senior Craftsman (RAS)	Senior Craftsman (RAW)
e	f	g	h
Senior Craftsman Craftsman (+2 RA)	Craftsman (+ 1RA)	Craftsman	Installer Senior Modern Apprentice
i	j	k	l
Adult Trainee Mate (over 18)	Intermediate Modern Apprentice	Junior Modern Apprentice	No grade allocated to this Credit Value Category

Daily abnormal conditions money	*From 7 October 2019*
Per day	3.45

Lodging allowance	*From 1 October 2018*
Per night	40.40

Explanatory Notes

1.   Working Hours

     The normal working week (Monday to Friday) shall be 37.5 hours.

2.   Overtime

     Time worked in excess of 37.5 hours during the normal working week shall be paid at time and a half until 12 hours have been worked since the actual starting time. Thereafter double time shall be paid until normal starting time the following morning. Weekend overtime shall be paid at time and a half for the first 5 hours worked on a Saturday and at double time thereafter until normal starting time on Monday morning.

PLUMBING MECHANICAL ENGINEERING SERVICES INDUSTRY

The Joint Industry Board for Plumbing Mechanical Engineering Services has agreed a three year wage agreement for 2018, 2019 and 2020 with effect from 1st January 2018 and subsequently from 7th January 2019, and 6th January 2020.

For full details of this wage agreement and the JIB PMES National Working Rules, contact:

The Joint Industry Board for Plumbing Mechanical Engineering Services in England and Wales
JIB-PMES
PO Box 267
PE19 9DN
Telephone: 01480 476925
E-mail: info@jib-pmes.org.uk

WAGE RATES, ALLOWANCES AND OTHER PROVISIONS
EFFECTIVE FROM 6 January 2020

Basic Rates of Hourly Pay

Applicable in England and Wales

	Hourly rate £
Operatives	
Technical plumber and gas service technician	17.19
Advanced plumber and gas service engineer	15.48
Trained plumber and gas service fitter	13.29
Apprentices	
4th year of training with NVQ level 3	12.85
4th year of training with NVQ level 2	11.64
4th year of training	10.24
3rd year of training with NVQ level 2	10.12
3rd year of training	8.33
2nd year of training	7.39
1st year of training	6.43
Adult Trainees	
3rd 6 months of employment	11.57
2nd 6 months of employment	11.12
1st 6 months of employment	10.37

PLUMBING MECHANICAL ENGINEERING SERVICES INDUSTRY

Major Projects Agreement

Where a job is designated as being a Major Project then the following Major Project Performance Payment hourly rate supplement shall be payable:

Employee Category

	National Payment £	London* Payment £
Technical Plumber and Gas Service Technician	2.20	3.57
Advanced Plumber and Gas Service Engineer	2.20	3.57
Trained Plumber and Gas Service Fitter	2.20	3.57
All 4th year apprentices	1.76	2.86
All 3rd year apprentices	1.32	2.68
2nd year apprentice	1.21	1.96
1st year apprentice	0.88	1.43
All adult trainees	1.76	2.86

* The London Payment Supplement applies only to designated Major Projects that are within the M25 London orbital motorway and are effective from 1 February 2007.

* National payment hourly rates are unchanged for 2007 and will continue to be at the rates shown in Promulgation 138 A issued 14 October 2003.

PLUMBING MECHANICAL ENGINEERING SERVICES INDUSTRY

Allowances

Daily travel time allowance plus return fares

All daily travel allowances are to be paid at the daily rate as follows:

Over	Not exceeding	All Operatives	3rd & 4th Year Apprentices	1st & 2nd Year Apprentices
20	30	£4.86	3.13	1.95
30	40	11.34	7.30	4.68
40	50	12.96	7.74	4.86

Responsibility/Incentive Pay Allowance

As from 3 September 2003, Employers may, in consultation with the employees concerned, enhance the basic graded rates of pay by the payment of an additional amount, as per the bands shown below, where it is agreed that their work involves extra responsibility, productivity or flexibility.

Band 1   an additional rate of up to £ 0.31 per hour
Band 2   an additional rate of up to £ 0.52 per hour
Band 3   an additional rate of up to £ 0.78 per hour
Band 4   an additional rate of up to £ 1.02 per hour

This allowance forms part of an operative's basic rate of pay and shall be used to calculate premium payments.

Mileage allowance	£0.45 per mile
Lodging allowance	£38.60 per night
Subsistence Allowance (London Only)	£5.58 per night

Plumbers welding supplement

Possession of Gas or Arc Certificate	£0.34 per hour
Possession of Gas and Arc Certificate	£0.54 per hour

PLUMBING MECHANICAL ENGINEERING SERVICES INDUSTRY

JIB-PMES Additional Holiday Pay (AHP) (From 7 January 2019)

All PHMES Operatives, Apprentices etc. who are in current membership of 'Unite the Union' at the time a holiday is taken, shall also be entitled to receive an additional payment of AHP from the JIB-PMES – to be paid via their employer – for each Credit funding their holiday pay. 62 Credits is the maximum number payable in a year.

The amount of AHP payable per credit for all holidays is as follows:

From January 2019

Operatives etc. £2.49 per credit (Max – £154.38)

All Apprentices £1.25 per credit (Max – £77.50)

Explanatory Notes

1. Working Hours

    The normal working week (Monday to Friday) shall be 37½ hours, with 45 hours to be worked in the same period before overtime rates become applicable.

2. Overtime

    Overtime shall be paid at time and a half up to 8.00pm (Monday to Friday) and up to 1.00pm (Saturday). Overtime worked after these times shall be paid at double time.

3. Major Projects Agreement

    Under the Major Projects Agreement the normal working week shall be 38 hours (Monday to Friday) with overtime rates payable for all hours worked in excess of 38 hours in accordance with 2 above. However, it should be noted that the hourly rate supplement shall be paid for each hour worked but does not attract premium time enhancement.

4. Pension

    In addition to their hourly rates of pay, plumbing employees are entitled to inclusion within the Industry Pension Scheme (or one providing equivalent benefits). The current levels of industry scheme contributions are 6½% (employers) and 3¼% (employees).

5. Additional Holiday Pay (AHP) Contributions

    As last year, JIB-PMES Holiday Pay Schemes shall apply for the Holiday Credit Period 2019–2020, with 62 credits being the maximum number payable in a year.

    For full details please refer to the Additional Holiday Pay (AHP) section as published by the JIB for Plumbing Mechanical Engineering Services in England and Wales.

Electrical Installations

Rates of Wages

ELECTRICAL CONTRACTING INDUSTRY

For full details of this wage agreement and the Joint Industry Board for the Electrical Contracting Industry's National Working Rules, contact:

Joint Industry Board
PO Box 127
Swanley
Kent
BR8 9BH
Telephone: 0333 321 230
Internet: www.jib.org.uk

WAGES (Graded Operatives)

Rates

Since 7 January 2002 two different wage rates have applied to JIB Graded Operatives working on site, depending on whether the Employer transports them to site or whether they provide their own transport. The two categories are:

Job Employed (Transport Provided)

Payable to an Operative who is transported to and from the job by his Employer. The Operative shall also be entitled to payment for Travel Time, when travelling in his own time, as detailed in the appropriate scale.

Job Employed (Own Transport)

Payable to an Operative who travels by his own means to and from the job. The Operative shall be entitled to payment for Travel Allowance and also Travel Time, when travelling in his own time, as detailed in the appropriate scale.

ELECTRICAL CONTRACTING INDUSTRY

The JIB rates of wages are set out below:

From and including 6 January 2020, the JIB hourly rates of wages shall be as set out below:

(i)   National Standard Rate

Grade	Transport Provided	Own Transport
Technician (or equivalent specialist grade)	£ 18.51	£ 19.45
Approved Electrician (or equivalent specialist grade)	£ 16.37	£ 17.27
Electrician (or equivalent specialist grade)	£ 14.99	£ 15.92
Senior Graded Electrical Trainee	£ 14.26	£ 15.13
Electrical Improver	£ 13.47	£ 14.35
Labourer	£ 11.90	£ 12.78
Adult Trainee	£ 11.90	£ 12.78

(ii)   London Rate

Grade	Transport Provided	Own Transport
Technician (or equivalent specialist grade)	£ 20.75	£ 21.77
Approved Electrician (or equivalent specialist grade)	£ 18.32	£ 19.34
Electrician (or equivalent specialist grade)	£ 16.78	£ 17.84
Senior Graded Electrical Trainee	£ 15.93	£ 16.96
Electrical Improver	£ 15.09	£ 16.07
Labourer	£ 13.35	£ 14.32
Adult Trainee	£ 13.35	£ 14.32

From and including 6 January 2020, the JIB hourly rates for Job Employed apprentices shall be:

(i)   National Standard Rates

	Transport Provided	Own Transport
Stage 1	£ 5.24	£ 6.13
Stage 2	£ 7.71	£ 8.63
Stage 3	£ 11.17	£ 12.09
Stage 4	£ 11.81	£ 12.75

(ii)   London Rate

	Transport Provided	Own Transport
Stage 1	£ 5.86	£ 6.87
Stage 2	£ 8.65	£ 9.64
Stage 3	£ 12.52	£ 13.55
Stage 4	£ 13.25	£ 14.29

ELECTRICAL CONTRACTING INDUSTRY

Travelling Time and Travel Allowances

From and including 6 January 2020

With effect from 2 January 2017, the existing Travel Allowance and Travelling Time tables will be replaced by:

(i)   A Mileage Allowance for operatives and apprentices using their own vehicular transport from the shop to their place of work. This allowance will not be taxable because it is within HMRC Approved Mileage Rates.

(ii) A taxable Mileage Rate for those operatives and apprentices for whom transport has been provided.

These will be introduced on jobs over 15 actual miles travelled from the shop to the job and vice versa using the shortest route. The shortest route will be calculated using the RAC Route Planner and will be paid both ways.

The non-taxable Mileage Allowance payments for operatives and apprentices using their own transport will be:

21p per mile with effect from Monday 2 January 2017

22p per mile with effect from 7 January 2019

Operatives and apprentices making their way to site in transport provided by their employer will receive taxable Mileage Rates of:

11p per mile with effect from Monday 2 January 2017

12p per mile with effect from 7 January 2019

For the avoidance of doubt, jobs less than 15 miles from the shop to the job and vice versa will continue to receive no payment.

ELECTRICAL CONTRACTING INDUSTRY

Lodging Allowances

£41.70 from and including 6 January 2020

Lodgings weekend retention fee, maximum reimbursement

£41.70 from and including 6 January 2020

Annual Holiday Lodging Allowance Retention

Maximum £13.72 per night (£96.04 per week) from and including 6 January 2020

Responsibility money

From and including 30 March 1998 the minimum payment increased to 10p per hour and the maximum to £1.00 per hour (no change)

From and including 4 January 1992 responsibility payments are enhanced by overtime and shift premiums where appropriate (no change)

Combined JIB Benefits Stamp Value (from week commencing 6 January 2020)

JIB grade	Weekly JIB combined £	Holiday value £
Technician	£ 66.93	£ 59.61
Approved Electrician	£ 59.70	£ 52.71
Electrician	£ 55.43	£ 48.27
Labourer & Adult Trainee	£ 44.93	£ 38.31

Explanatory Notes

1. Working Hours

    The normal working week (Monday to Friday) shall be 37½ hours, with 38 hours to be worked in the same period before overtime rates become applicable.

2. Overtime

    Overtime shall be paid at time and a half for all weekday overtime. Saturday overtime shall be paid at time and a half for the first 6 hours, or up to 3.00pm (whichever comes first). Thereafter double time shall be paid until normal starting time on Monday.

PART 6

Daywork

When work is carried out in connection with a contract that cannot be valued in any other way, it is usual to assess the value on a cost basis with suitable allowances to cover overheads and profit. The basis of costing is a matter for agreement between the parties concerned but definitions of prime cost for the Heating and Ventilating and Electrical Industries have been published jointly by the Royal Institution of Chartered Surveyors and the appropriate bodies of the industries concerned, for those who wish to use them.

These, together with a schedule of basic plant hire charges are reproduced on the following pages, with the kind permission of the Royal Institution of Chartered Surveyors, who own the copyright.

Quality Auditing in Construction Projects

Abdul Razzak Rumane

This book provides construction professionals, designers, contractors and quality auditors involved in construction projects with the auditing skills and processes required to improve construction quality and make their projects more competitive and economical.

The processes within the book focus on auditing compliance to ISO, corporate quality management systems, project specific quality management systems, contract management, regulatory authorities' requirements, safety, and environmental considerations. The book is divided into seven chapters and each chapter is divided into numbered sections covering auditing-related topics that have importance or relevance for understanding quality auditing concepts for construction projects.

No other book covers construction quality auditing in such detail and with this level of practical application. It is an essential guide for construction and quality professionals, but also for students and academics interested in learning about quality auditing in construction projects.

June 2019: 234 × 156 mm: 600pp
Hb: 978-0-8153-8531-8 : £130.00

To Order: Tel: +44 (0) 1235 400524 Fax: +44 (0) 1235 400525
or Post: Taylor and Francis Customer Services,
Bookpoint Ltd, Unit T1, 200 Milton Park, Abingdon, Oxon, OX14 4TA UK
Email: book.orders@tandf.co.uk

For a complete listing of all our titles visit:
www.tandf.co.uk

Taylor & Francis
Taylor & Francis Group

HEATING AND VENTILATING INDUSTRY

DEFINITION OF PRIME COST OF DAYWORK CARRIED OUT UNDER A HEATING, VENTILATING, AIR CONDITIONING, REFRIGERATION, PIPEWORK AND/OR DOMESTIC ENGINEERING CONTRACT (JULY 1980 EDITION)

This Definition of Prime Cost is published by the Royal Institution of Chartered Surveyors and the Heating and Ventilating Contractors Association for convenience, and for use by people who choose to use it. Members of the Heating and Ventilating Contractors Association are not in any way debarred from defining Prime Cost and rendering accounts for work carried out on that basis in any way they choose. Building owners are advised to reach agreement with contractors on the Definition of Prime Cost to be used prior to entering into a contract or subcontract.

SECTION 1: APPLICATION

1.1 This Definition provides a basis for the valuation of daywork executed under such heating, ventilating, air conditioning, refrigeration, pipework and or domestic engineering contracts as provide for its use.

1.2 It is not applicable in any other circumstances, such as jobbing or other work carried out as a separate or main contract nor in the case of daywork executed after a date of practical completion.

1.3 The terms 'contract' and 'contractor' herein shall be read as 'subcontract' and 'subcontractor' as applicable.

SECTION 2: COMPOSITION OF TOTAL CHARGES

2.1 The Prime Cost of daywork comprises the sum of the following costs:
   (a) Labour as defined in Section 3.
   (b) Materials and goods as defined in Section 4.
   (c) Plant as defined in Section 5.

2.2 Incidental costs, overheads and profit as defined in Section 6, as provided in the contract and expressed therein as percentage adjustments, are applicable to each of 2.1 (a)–(c).

SECTION 3: LABOUR

3.1 The standard wage rates, emoluments and expenses referred to below and the standard working hours referred to in 3.2 are those laid down for the time being in the rules or decisions or agreements of the Joint Conciliation Committee of the Heating, Ventilating and Domestic Engineering Industry applicable to the works (or those of such other body as may be appropriate) and to the grade of operative concerned at the time when and the area where the daywork is executed.

3.2 Hourly base rates for labour are computed by dividing the annual prime cost of labour, based upon the standard working hours and as defined in 3.4, by the number of standard working hours per annum. See example.

3.3 The hourly rates computed in accordance with 3.2 shall be applied in respect of the time spent by operatives directly engaged on daywork, including those operating mechanical plant and transport and erecting and dismantling other plant (unless otherwise expressly provided in the contract) and handling and distributing the materials and goods used in the daywork.

3.4 The annual prime cost of labour comprises the following:
   (a) Standard weekly earnings (i.e. the standard working week as determined at the appropriate rate for the operative concerned).
   (b) Any supplemental payments.
   (c) Any guaranteed minimum payments (unless included in Section 6.1(a)–(p)).
   (d) Merit money.
   (e) Differentials or extra payments in respect of skill, responsibility, discomfort, inconvenience or risk (excluding those in respect of supervisory responsibility – see 3.5)
   (f) Payments in respect of public holidays.
   (g) Any amounts which may become payable by the contractor to or in respect of operatives arising from the rules etc. referred to in 3.1 which are not provided for in 3.4 (a)–(f) nor in Section 6.1 (a)–(p).
   (h) Employers contributions to the WELPLAN, the HVACR Welfare and Holiday Scheme or payments in lieu thereof.

HEATING AND VENTILATING INDUSTRY

    (i)   Employers National Insurance contributions as applicable to 3.4 (a)–(h).

    (j)   Any contribution, levy or tax imposed by Statute, payable by the contractor in his capacity as an employer.

3.5   Differentials or extra payments in respect of supervisory responsibility are excluded from the annual prime cost (see Section 6). The time of principals, staff, foremen, chargehands and the like when working manually is admissible under this Section at the rates for the appropriate grades.

SECTION 4: MATERIALS AND GOODS

4.1   The prime cost of materials and goods obtained specifically for the daywork is the invoice cost after deducting all trade discounts and any portion of cash discounts in excess of 5%.

4.2   The prime cost of all other materials and goods used in the daywork is based upon the current market prices plus any appropriate handling charges.

4.3   The prime cost referred to in 4.1 and 4.2 includes the cost of delivery to site.

4.4   Any Value Added Tax which is treated, or is capable of being treated, as input tax (as defined by the Finance Act 1972, or any re-enactment or amendment thereof or substitution therefore) by the contractor is excluded.

SECTION 5: PLANT

5.1   Unless otherwise stated in the contract, the prime cost of plant comprises the cost of the following:

    (a)   use or hire of mechanically-operated plant and transport for the time employed on and/or provided or retained for the daywork;

    (b)   use of non-mechanical plant (excluding non-mechanical hand tools) for the time employed on and/or provided or retained for the daywork;

    (c)   transport to and from the site and erection and dismantling where applicable.

5.2   The use of non-mechanical hand tools and of erected scaffolding, staging, trestles or the like is excluded (see Section 6), unless specifically retained for the daywork.

SECTION 6: INCIDENTAL COSTS, OVERHEADS AND PROFIT

6.1   The percentage adjustments provided in the contract which are applicable to each of the totals of Sections 3, 4 and 5 comprise the following:

    (a)   Head office charges.

    (b)   Site staff including site supervision.

    (c)   The additional cost of overtime (other than that referred to in 6.2).

    (d)   Time lost due to inclement weather.

    (e)   The additional cost of bonuses and all other incentive payments in excess of any included in 3.4.

    (f)   Apprentices' study time.

    (g)   Fares and travelling allowances.

    (h)   Country, lodging and periodic allowances.

    (i)   Sick pay or insurances in respect thereof, other than as included in 3.4.

    (j)   Third party and employers' liability insurance.

    (k)   Liability in respect of redundancy payments to employees.

    (l)   Employer's National Insurance contributions not included in 3.4.

    (m)   Use and maintenance of non-mechanical hand tools.

    (n)   Use of erected scaffolding, staging, trestles or the like (but see 5.2).

    (o)   Use of tarpaulins, protective clothing, artificial lighting, safety and welfare facilities, storage and the like that may be available on site.

    (p)   Any variation to basic rates required by the contractor in cases where the contract provides for the use of a specified schedule of basic plant charges (to the extent that no other provision is made for such variation – see 5.1).

    (q)   In the case of a sub-contract which provides that the sub-contractor shall allow a cash discount, such provision as is necessary for the allowance of the prescribed rate of discount.

HEATING AND VENTILATING INDUSTRY

(r)   All other liabilities and obligations whatsoever not specifically referred to in this Section nor chargeable under any other Section.

(s)   Profit.

6.2   The additional cost of overtime where specifically ordered by the Architect/Supervising Officer shall only be chargeable in the terms of a prior written agreement between the parties.

MECHANICAL INSTALLATIONS

Calculation of Hourly Base Rate of Labour for Typical Main Grades applicable from 7th October 2019, refer to notes within Section Four – Rates of Wages.

	Foreman	Senior Craftsman (+ 2nd Welding Skill)	Senior Craftsman	Craftsman	Installer	Mate over 18
Hourly Rate from 7th October 2019	17.70	15.79	14.63	13.47	12.16	10.25
Annual standard earnings excluding all holidays, 45.6 weeks x 38 hours	30,399.75	27,119.33	25,127.03	23,134.73	20,884.80	17,604.38
Employers national insurance contributions for this year	3,641.10	3,146.50	2,814.97	2,599.47	2,150.29	1,636.31
Weekly holiday credit and welfare contributions (52 weeks) for this year	4,788.16	4,466.80	4,044.04	4,466.80	3,446.56	2,984.28
Annual prime cost of labour	38,657.90	34,579.20	31,845.27	30,068.21	26,364.09	22,125.61
Hourly base rate	22.51	20.13	18.54	17.51	15.35	12.88

Notes:

(1)   Annual industry holiday (4.6 weeks x 37.5 hours) and public holidays (1.6 weeks x 37.5 hours) are paid through weekly holiday credit and welfare stamp scheme.

(2)   Where applicable, Merit money and other variables (e.g. daily abnormal conditions money), which attract Employer's National Insurance contribution, should be included.

(3)   Contractors in Northern Ireland should add the appropriate amount of CITB Levy to the annual prime cost of labour prior to calculating the hourly base rate.

(4)   Hourly rate based on 1,717.50 hours per annum and calculated as follows

52 Weeks @ 37.5 hrs/wk	=		1,950.00
Less			
Public Holiday = 8/5 = 1.6 weeks @ 3.75 hrs/wk	=	60	
Annual holidays = 4.6 weeks @ 37.5 hrs/wk	=	172.50	232.50
Hours	=		1,717.50

(5)   For calculation of Holiday Credits and ENI refer to detailed labour rate evaluation

(6)   National Insurance contributions are those effective from 1 April 2020.

(7)   Weekly holiday credit/welfare stamp values are those assumed and effective from 7 October 2019

(8)   Hourly rates of wages are those assumed and effective from 7 October 2019.

ELECTRICAL INDUSTRY

DEFINITION OF PRIME COST OF DAYWORK CARRIED OUT UNDER AN ELECTRICAL CONTRACT (MARCH 1981 EDITION)

This Definition of Prime Cost is published by The Royal Institution of Chartered Surveyors and The Electrical Contractors' Associations for convenience and for use by people who choose to use it. Members of The Electrical Contractors' Association are not in any way debarred from defining Prime Cost and rendering accounts for work carried out on that basis in any way they choose. Building owners are advised to reach agreement with contractors on the Definition of Prime Cost to be used prior to entering into a contract or subcontract.

SECTION 1: APPLICATION

1.1   This Definition provides a basis for the valuation of daywork executed under such electrical contracts as provide for its use.
1.2   It is not applicable in any other circumstances, such as jobbing, or other work carried out as a separate or main contract, nor in the case of daywork executed after the date of practical completion.
1.3   The terms 'contract' and 'contractor' herein shall be read as 'subcontract' and 'subcontractor' as the context may require.

SECTION 2: COMPOSITION OF TOTAL CHARGES

2.1   The Prime Cost of daywork comprises the sum of the following costs:
    (a)   Labour as defined in Section 3.
    (b)   Materials and goods as defined in Section 4.
    (c)   Plant as defined in Section 5.
2.2   Incidental costs, overheads and profit as defined in Section 6, as provided in the contract and expressed therein as percentage adjustments, are applicable to each of 2.1 (a)-(c).

SECTION 3: LABOUR

3.1   The standard wage rates, emoluments and expenses referred to below and the standard working hours referred to in 3.2 are those laid down for the time being in the rules and determinations or decisions of the Joint Industry Board or the Scottish Joint Industry Board for the Electrical Contracting Industry (or those of such other body as may be appropriate) applicable to the works and relating to the grade of operative concerned at the time when and in the area where daywork is executed.
3.2   Hourly base rates for labour are computed by dividing the annual prime cost of labour, based upon the standard working hours and as defined in 3.4 by the number of standard working hours per annum. See examples.
3.3   The hourly rates computed in accordance with 3.2 shall be applied in respect of the time spent by operatives directly engaged on daywork, including those operating mechanical plant and transport and erecting and dismantling other plant (unless otherwise expressly provided in the contract) and handling and distributing the materials and goods used in the daywork.
3.4   The annual prime cost of labour comprises the following:
    (a)   Standard weekly earnings (i.e. the standard working week as determined at the appropriate rate for the operative concerned).
    (b)   Payments in respect of public holidays.
    (c)   Any amounts which may become payable by the Contractor to or in respect of operatives arising from operation of the rules etc. referred to in 3.1 which are not provided for in 3.4(a) and (b) nor in Section 6.
    (d)   Employer's National Insurance Contributions as applicable to 3.4 (a)-(c).
    (e)   Employer's contributions to the Joint Industry Board Combined Benefits Scheme or Scottish Joint Industry Board Holiday and Welfare Stamp Scheme, and holiday payments made to apprentices in compliance with the Joint Industry Board National Working Rules and Industrial Determinations as an employer.
    (f)   Any contribution, levy or tax imposed by Statute, payable by the Contractor in his capacity as an employer.

ELECTRICAL INDUSTRY

3.5   Differentials or extra payments in respect of supervisory responsibility are excluded from the annual prime cost (see Section 6). The time of principals and similar categories, when working manually, is admissible under this Section at the rates for the appropriate grades.

SECTION 4: MATERIALS AND GOODS

4.1   The prime cost of materials and goods obtained specifically for the daywork is the invoice cost after deducting all trade discounts and any portion of cash discounts in excess of 5%.

4.2   The prime cost of all other materials and goods used in the daywork is based upon the current market prices plus any appropriate handling charges.

4.3   The prime cost referred to in 4.1 and 4.2 includes the cost of delivery to site.

4.4   Any Value Added Tax which is treated, or is capable of being treated, as input tax (as defined by the Finance Act 1972, or any re-enactment or amendment thereof or substitution therefore) by the Contractor is excluded.

SECTION 5: PLANT

5.1   Unless otherwise stated in the contract, the prime cost of plant comprises the cost of the following:
- (a)   Use or hire of mechanically-operated plant and transport for the time employed on and/or provided or retained for the daywork;
- (b)   Use of non-mechanical plant (excluding non-mechanical hand tools) for the time employed on and/or provided or retained for the daywork;
- (c)   Transport to and from the site and erection and dismantling where applicable.

5.2   The use of non-mechanical hand tools and of erected scaffolding, staging, trestles or the likes is excluded (see Section 6), unless specifically retained for daywork.

5.3   Note: Where hired or other plant is operated by the Electrical Contractor's operatives, such time is to be included under Section 3 unless otherwise provided in the contract.

SECTION 6: INCIDENTAL COSTS, OVERHEADS AND PROFIT

6.1   The percentage adjustments provided in the contract which are applicable to each of the totals of Sections 3, 4 and 5, compromise the following:
- (a)   Head Office charges.
- (b)   Site staff including site supervision.
- (c)   The additional cost of overtime (other than that referred to in 6.2).
- (d)   Time lost due to inclement weather.
- (e)   The additional cost of bonuses and other incentive payments.
- (f)   Apprentices' study time.
- (g)   Travelling time and fares.
- (h)   Country and lodging allowances.
- (i)   Sick pay or insurance in lieu thereof, in respect of apprentices.
- (j)   Third party and employers' liability insurance.
- (k)   Liability in respect of redundancy payments to employees.
- (l)   Employers' National Insurance Contributions not included in 3.4.
- (m)   Use and maintenance of non-mechanical hand tools.
- (n)   Use of erected scaffolding, staging, trestles or the like (but see 5.2.).
- (o)   Use of tarpaulins, protective clothing, artificial lighting, safety and welfare facilities, storage and the like that may be available on site.
- (p)   Any variation to basic rates required by the Contractor in cases where the contract provides for the use of a specified schedule of basic plant charges (to the extent that no other provision is made for such variation - see 5.1).
- (q)   All other liabilities and obligations whatsoever not specifically referred to in this Section nor chargeable under any other Section.
- (r)   Profit.

ELECTRICAL INDUSTRY

(s)   In the case of a subcontract which provides that the subcontractor shall allow a cash discount, such provision as is necessary for the allowance of the prescribed rate of discount.

6.2   The additional cost of overtime where specifically ordered by the Architect/Supervising Officer shall only be chargeable in the terms of a prior written agreement between the parties.

ELECTRICAL INSTALLATIONS

Calculation of Hourly Base Rate of Labour for Typical Main Grades applicable from 6January 2020.

	Technician	Approved Electrician	Electrician	Labourer
Hourly Rate from 6 January 2020 (London Rates)	21.77	19.34	17.84	14.32
Annual standard earnings excluding all holidays, 45.8 weeks x 37.5 hours	37,389.98	33,216.45	30,640.20	24,594.60
Employers national insurance contributions from 6 April 2018	4,774.73	4,120.82	3,717.17	2,769.94
JIB Combined benefits from 1 January 2018	3,879.72	3,520.92	3,290.04	2,772.12
Holiday top up funding	1,961.81	1,755.63	1,637.76	1,337.28
Annual prime cost of labour	48,006.23	42,613.82	39,285.17	31,473.94
Hourly base rate	27.95	24.81	22.87	18.33

Notes:

(1)   Annual industry holiday (4.4 weeks x 37.5 hours) and public holidays (1.6 weeks x 37.5 hours)
(2)   It should be noted that all labour costs incurred by the Contractor in his capacity as an Employer, other than those contained in the hourly rate above, must be taken into account under Section 6.
(3)   Public Holidays are paid through weekly holiday credit and welfare stamp scheme.
(4)   Contractors in Northern Ireland should add the appropriate amount of CITB Levy to the annual prime cost of labour prior to calculating the hourly base rate.
(5)   Hourly rate based on 1,710 .00 hours per annum and calculated as follows:

52 Weeks @ 37.5 hrs/wk	=		1,950.00
Less			
Public Holiday = 8/5 = 1.6 weeks @ 37.5 hrs/wk	=	60	
Annual holidays = 4.8 weeks @ 37.5 hrs/wk	=	180	240
Hours	=		1,710.00

(6)   For calculation of holiday credits and ENI refer to detailed labour rate evaluation.
(7)   Hourly wage rates are those effective from 6January 2020.
(8)   National Insurance contributions are those effective from 1 April 2020.

BUILDING INDUSTRY PLANT HIRE COSTS

SCHEDULE OF BASIC PLANT CHARGES (JULY 2010)

This Schedule is published by the Royal Institution of Chartered Surveyors and is for use in connection with Dayworks under a Building Contract.

EXPLANATORY NOTES

1. The rates in the Schedule are intended to apply solely to daywork carried out under and incidental to a Building Contract. They are NOT intended to apply to:
   (i) jobbing or any other work carried out as a main or separate contract; or
   (ii) work carried out after the date of commencement of the Defects Liability Period.
2. The rates apply to plant and machinery already on site, whether hired or owned by the Contractor.
3. The rates, unless otherwise stated, include the cost of fuel and power of every description, lubricating oils, grease, maintenance, sharpening of tools, replacement of spare parts, all consumable stores and for licences and insurances applicable to items of plant.
4. The rates, unless otherwise stated, do not include the costs of drivers and attendants (unless otherwise stated).
5. The rates in the Schedule are base costs and may be subject to an overall adjustment for price movement, overheads and profit, quoted by the Contractor prior to the placing of the Contract.
6. The rates should be applied to the time during which the plant is actually engaged in daywork.
7. Whether or not plant is chargeable on daywork depends on the daywork agreement in use and the inclusion of an item of plant in this schedule does not necessarily indicate that item is chargeable.
8. Rates for plant not included in the Schedule or which is not already on site and is specifically provided or hired for daywork shall be settled at prices which are reasonably related to the rates in the Schedule having regard to any overall adjustment quoted by the Contractor in the Conditions of Contract.

BUILDING INDUSTRY PLANT HIRE COSTS

MECHANICAL PLANT AND TOOLS

Item of plant	Size/Rating	Unit	Rate per Hour (£)
PUMPS			
Mobile Pumps			
Including pump hoses, values and strainers, etc.			
Diaphragm	50 mm diameter	Each	1.17
Diaphragm	76 mm diameter	Each	1.89
Diaphragm	102 mm diameter	Each	3.54
Submersible	50 mm diameter	Each	0.76
Submersible	76 mm diameter	Each	0.86
Submersible	102 mm diameter	Each	1.03
Induced Flow	50 mm diameter	Each	0.77
Induced Flow	76 mm diameter	Each	1.67
Centrifugal, self priming	25 mm diameter	Each	1.30
Centrifugal, self priming	50 mm diameter	Each	1.92
Centrifugal, self priming	75 mm diameter	Each	2.74
Centrifugal, self priming	102 mm diameter	Each	3.35
Centrifugal, self priming	152 mm diameter	Each	4.27
SCAFFOLDING, SHORING, FENCING			
Complete Scaffolding			
Mobile working towers, single width	2.0 m x 0.72 m base x 7.45 m high	Each	3.36
Mobile working towers, single width	2.0 m x 0.72 m base x 8.84 m high	Each	3.79
Mobile working towers, double width	2.0 m x 1.35 m × 7.45 m high	Each	3.79
Mobile /working towers, double width	2.0 m x 1.35 m × 15.8 m high	Each	7.13
Chimney scaffold, single unit		Each	1.92
Chimney scaffold, twin unit		Each	3.59
Push along access platform	1.63 – 3.1 m	Each	5.00
Push along access platform	1.80 m x 0.70 m	Each	1.79
Trestles			
Trestle, adjustable	Any height	Pair	0.41
Trestle, painters	1.8 m high	Pair	0.31
Trestle, painters	2.4 m high	Pair	0.36
Shoring, Planking and Strutting			
'Acrow' adjustable prop	Sizes up to 4.9 m (open)	Each	0.06
'Strong boy' support attachment		Each	0.22
Adjustable trench strut	Sizes up to 1.67m (open)	Each	0.16

BUILDING INDUSTRY PLANT HIRE COSTS

Item of plant	Size/Rating	Unit	Rate per Hour (£)
Trench sheet		Metre	0.03
Backhoe trench box	Base unit	Each	1.23
Backhoe trench box	Top unit	Each	0.87
Temporary Fencing			
Including block and coupler			
Site fencing steel grid panel	3.5 m x 2.0 m	Each	0.05
Anti-climb site steel grid fence panel	3.5 m x 2.0 m	Each	0.08
Solid panel Heras	2.0 m x 2.0 m	Each	0.09
Pedestrian gate		Each	0.36
Roadway gate		Each	0.60

LIFTING APPLIANCES AND CONVEYORS

Cranes

Mobile Cranes

Rates are inclusive of drivers

Lorry mounted, telescopic jib

Two wheel drive	5 tonnes	Each	19.00
Two wheel drive	8 tonnes	Each	42.00
Two wheel drive	10 tonnes	Each	50.00
Two wheel drive	12 tonnes	Each	77.00
Two wheel drive	20 tonnes	Each	89.69
Four wheel drive	18 tonnes	Each	46.51
Four wheel drive	25 tonnes	Each	35.90
Four wheel drive	30 tonnes	Each	38.46
Four wheel drive	45 tonnes	Each	46.15
Four wheel drive	50 tonnes	Each	53.85
Four wheel drive	60 tonnes	Each	61.54
Four wheel drive	70 tonnes	Each	71.79

Static tower crane

Rates inclusive of driver

Note: Capacity equals maximum lift in tonnes times maximum radius at which it can be lifted

	Capacity (Metre/tonnes) Up to	Height under hook above ground (m) Up to		
Tower crane	30	22	Each	22.23
Tower crane	40	22	Each	26.62
Tower crane	40	30	Each	33.33

BUILDING INDUSTRY PLANT HIRE COSTS

Item of plant	Size/Rating		Unit	Rate per Hour (£)
Tower crane	50	22	Each	29.16
Tower crane	60	22	Each	35.90
Tower crane	60	36	Each	35.90
Tower crane	70	22	Each	41.03
Tower crane	80	22	Each	39.12
Tower crane	90	42	Each	37.18
Tower crane	110	36	Each	47.62
Tower crane	140	36	Each	55.77
Tower crane	170	36	Each	64.11
Tower crane	200	36	Each	71.95
Tower crane	250	36	Each	84.77
Tower crane with luffing jig	30	25	Each	22.23
Tower crane with luffing jig	40	30	Each	26.62
Tower crane with luffing jig	50	30	Each	29.16
Tower crane with luffing jig	60	36	Each	41.03
Tower crane with luffing jig	65	30	Each	33.13
Tower crane with luffing jig	80	22	Each	48.72
Tower crane with luffing jig	100	45	Each	48.72
Tower crane with luffing jig	125	30	Each	53.85
Tower crane with luffing jig	160	50	Each	53.85
Tower crane with luffing jig	200	50	Each	74.36
Tower crane with luffing jig	300	60	Each	100.00
Crane Equipment				
Muck tipping skip	Up to 200 litres		Each	0.67
Muck tipping skip	500 litres		Each	0.82
Muck tipping skip	750 litres		Each	1.08
Muck tipping skip	1000 litres		Each	1.28
Muck tipping skip	1500 litres		Each	1.41
Muck tipping skip	2000 litres		Each	1.67
Mortar skip	250 litres, plastic		Each	0.41
Mortar skip	350 litres, steel		Each	0.77
Boat skip	250 litres		Each	0.92
Boat skip	500 litres		Each	1.08
Boat skip	750 litres		Each	1.23
Boat skip	1000 litres		Each	1.38
Boat skip	1500 litres		Each	1.64
Boat skip	2000 litres		Each	1.90

BUILDING INDUSTRY PLANT HIRE COSTS

Item of plant	Size/Rating	Unit	Rate per Hour (£)
Boat skip	3000 litres	Each	2.82
Boat skip	4000 litres	Each	3.23
Master flow skip	250 litres	Each	0.77
Master flow skip	500 litres	Each	1.03
Master flow skip	750 litres	Each	1.28
Master flow skip	1000 litres	Each	1.44
Master flow skip	1500 litres	Each	1.69
Master flow skip	2000 litres	Each	1.85
Grand master flow skip	500 litres	Each	1.28
Grand master flow skip	750 litres	Each	1.64
Grand master flow skip	1000 litres	Each	1.69
Grand master flow skip	1500 litres	Each	1.95
Grand master flow skip	2000 litres	Each	2.21
Cone flow skip	500 litres	Each	1.33
Cone flow skip	1000 litres	Each	1.69
Geared rollover skip	500 litres	Each	1.28
Geared rollover skip	750 litres	Each	1.64
Geared rollover skip	1000 litres	Each	1.69
Geared rollover skip	1500 litres	Each	1.95
Geared rollover skip	2000 litres	Each	2.21
Multi skip, rope operated	200 mm outlet size, 500 litres	Each	1.49
Multi skip, rope operated	200 mm outlet size, 750 litres	Each	1.64
Multi skip, rope operated	200 mm outlet size, 1000 litres	Each	1.74
Multi skip, rope operated	200 mm outlet size, 1500 litres	Each	2.00
Multi skip, rope operated	200 mm outlet size, 2000 litres	Each	2.26
Multi skip, man riding	200 mm outlet size, 1000 litres	Each	2.00
Multi skip	4 point lifting frame	Each	0.90
Multi skip	Chain brothers	Set	0.87
Crane Accessories			
Multi-purpose crane forks	1.5 and 2 tonnes S.W.L.	Each	1.13
Self levelling crane forks		Each	1.28
Man cage	1 man, 230 kg S.W.L.	Each	1.90
Man cage	2 man, 500 kg S.W.L.	Each	1.95
Man cage	4 man, 750 kg S.W.L.	Each	2.15
Man cage	8 man, 1000 kg S.W.L.	Each	3.33
Stretcher cage	500 kg, S.W.L.	Each	2.69
Goods carrying cage	1500 kg, S.W.L.	Each	1.33
Goods carrying cage	3000 kg, S.W.L.	Each	1.85

BUILDING INDUSTRY PLANT HIRE COSTS

Item of plant	Size/Rating		Unit	Rate per Hour (£)
Builders' skip lifting cradle	12 tonnes, S.W.L.		Each	2.31
Board/pallet fork	1600 kg, S.W.L.		Each	1.90
Gas bottle carrier	500 kg, S.W.L.		Each	0.92
Hoists				
Scaffold hoist	200 kg		Each	2.46
Rack and pinion (goods only)	500 kg		Each	4.56
Rack and pinion (goods only)	1100 kg		Each	5.90
Rack and pinion (goods and passenger)	8 person, 80 kg		Each	7.44
Rack and pinion (goods and passenger)	14 person, 1400 kg		Each	8.72
Wheelbarrow chain sling			Each	1.67
Conveyors				
<u>Belt conveyors</u>				
Conveyor	8 m long x 450 mm wide		Each	5.90
Miniveyor, control box and loading hopper	3 m unit		Each	4.49
<u>Other Conveying Equipment</u>				
Wheelbarrow			Each	0.62
Hydraulic superlift			Each	4.56
Pavac slab lifter (tile hoist)			Each	4.49
High lift pallet truck			Each	3.08
Lifting Trucks	Payload	Maximum Lift		
Fork lift, two wheel drive	1100 kg	up to 3.0 m	Each	5.64
Fork lift, two wheel drive	2540 kg	up to 3.7 m	Each	5.64
Fork lift, four wheel drive	1524 kg	up to 6.0 m	Each	5.64
Fork lift, four wheel drive	2600 kg	up to 5.4 m	Each	7.44
Fork life, four wheel drive	4000 kg	up to 17 m	Each	10.77
Lifting Platforms				
Hydraulic platform (Cherry picker)	9 m		Each	4.62
Hydraulic platform (Cherry picker)	12 m		Each	7.56
Hydraulic platform (Cherry picker)	15 m		Each	10.13
Hydraulic platform (Cherry picker)	17 m		Each	15.63
Hydraulic platform (Cherry picker)	20 m		Each	18.13
Hydraulic platform (Cherry picker)	25.6 m		Each	32.38
Scissor lift	7.6 m, electric		Each	3.85
Scissor lift	7.8 m, electric		Each	5.13
Scissor lift	9.7 m, electric		Each	4.23
Scissor lift	10 m, diesel		Each	6.41
Telescopic handler	7 m, 2 tonnes		Each	5.13

BUILDING INDUSTRY PLANT HIRE COSTS

Item of plant	Size/Rating	Unit	Rate per Hour (£)
Telescopic handler	13 m, 3 tonnes	Each	7.18
Lifting and Jacking Gear			
Pipe winch including gantry	1 tonne	Set	1.92
Pipe winch including gantry	3 tonnes	Set	3.21
Chain block	1 tonne	Each	0.35
Chain block	2 tonnes	Each	0.58
Chain block	5 tonnes	Each	1.14
Pull lift (Tirfor winch)	1 tonne	Each	0.64
Pull lift (Tirfor winch)	1.6 tonnes	Each	0.90
Pull lift (Tirfor winch)	3.2 tonnes	Each	1.15
Brother or chain slings, two legs	not exceeding 3.1 tonnes	Set	0.21
Brother or chain slings, two legs	not exceeding 4.25 tonnes	Set	0.31
Brother or chain slings, four legs	not exceeding 11.2 tonnes	Set	1.09
CONSTRUCTION VEHICLES			
Lorries			
Plated lorries (Rates are inclusive of driver)			
Platform lorry	7.5 tonnes	Each	16.21
Platform lorry	17 tonnes	Each	22.90
Platform lorry	24 tonnes	Each	30.68
Extra for lorry with crane attachment	up to 2.5 tonnes	Each	3.25
Extra for lorry with crane attachment	up to 5 tonnes	Each	6.00
Extra for lorry with crane attachment	up to 7.5 tonnes	Each	9.10
Tipper Lorries			
(Rates are inclusive of driver)			
Tipper lorry	up to 11 tonnes	Each	15.78
Tipper lorry	up to 17 tonnes	Each	23.95
Tipper lorry	up to 25 tonnes	Each	31.35
Tipper lorry	up to 31 tonnes	Each	37.79
Dumpers			
Site use only (excl. tax, insurance and extra cost of DERV etc. when operating on highway)	*Makers Capacity*		
Two wheel drive	1 tonnes	Each	1.71
Four wheel drive	2 tonnes	Each	2.43
Four wheel drive	3 tonnes	Each	2.44

BUILDING INDUSTRY PLANT HIRE COSTS

Item of plant	Size/Rating	Unit	Rate per Hour (£)
Four wheel drive	5 tonnes	Each	3.08
Four wheel drive	6 tonnes	Each	3.85
Four wheel drive	9 tonnes	Each	5.65
Tracked	0.5 tonnes	Each	3.33
Tracked	1.5 tonnes	Each	4.23
Tracked	3.0 tonnes	Each	8.33
Tracked	6.0 tonnes	Each	16.03
Dumper Trucks (*Rates are inclusive of drivers*)			
Dumper truck	up to 15 tonnes	Each	28.56
Dumper truck	up to 17 tonnes	Each	32.82
Dumper truck	up to 23 tonnes	Each	54.64
Dumper truck	up to 30 tonnes	Each	63.50
Dumper truck	up to 35 tonnes	Each	73.02
Dumper truck	up to 40 tonnes	Each	87.84
Dumper truck	up to 50 tonnes	Each	133.44
Tractors			
Agricultural Type			
Wheeled, rubber-clad tyred	up to 40 kW	Each	8.63
Wheeled, rubber-clad tyred	up to 90 kW	Each	25.31
Wheeled, rubber-clad tyred	up to 140 kW	Each	36.49
Crawler Tractors			
With bull or angle dozer	up to 70 kW	Each	29.38
With bull or angle dozer	up to 85 kW	Each	38.63
With bull or angle dozer	up to 100 kW	Each	52.59
With bull or angle dozer	up to 115 kW	Each	55.85
With bull or angle dozer	up to 135 kW	Each	60.43
With bull or angle dozer	up to 185 kW	Each	76.44
With bull or angle dozer	up to 200 kW	Each	96.43
With bull or angle dozer	up to 250 kW	Each	117.68
With bull or angle dozer	up to 350 kW	Each	160.03
With bull or angle dozer	up to 450 kW	Each	219.86
With loading shovel	0.8 m³	Each	26.92
With loading shovel	1.0 m³	Each	32.59
With loading shovel	1.2 m³	Each	37.53
With loading shovel	1.4 m³	Each	42.89
With loading shovel	1.8 m³	Each	52.22
With loading shovel	2.0 m³	Each	57.22

BUILDING INDUSTRY PLANT HIRE COSTS

Item of plant	Size/Rating	Unit	Rate per Hour (£)
With loading shovel	2.1 m³	Each	60.12
With loading shovel	3.5 m³	Each	87.26
Light vans			
VW Caddivan or the like		Each	5.26
VW Transport transit or the like	1.0 tonnes	Each	6.03
Luton Box Van or the like	1.8 tonnes	Each	9.87
Water/Fuel Storage			
Mobile water container	110 litres	Each	0.62
Water bowser	1100 litres	Each	0.72
Water bowser	3000 litres	Each	0.87
Mobile fuel container	110 litres	Each	0.62
Fuel bowser	1100 litres	Each	1.23
Fuel bowser	3000 litres	Each	1.87

EXCAVATIONS AND LOADERS

Excavators			
Wheeled, hydraulic	up to 11 tonnes	Each	25.86
Wheeled, hydraulic	up to 14 tonnes	Each	30.82
Wheeled, hydraulic	up to 16 tonnes	Each	34.50
Wheeled, hydraulic	up to 21 tonnes	Each	39.10
Wheeled, hydraulic	up to 25 tonnes	Each	43.81
Wheeled, hydraulic	up to 30 tonnes	Each	55.30
Crawler, hydraulic	up to 11 tonnes	Each	25.86
Crawler, hydraulic	up to 14 tonnes	Each	30.82
Crawler, hydraulic	up to 17 tonnes	Each	34.50
Crawler, hydraulic	up to 23 tonnes	Each	39.10
Crawler, hydraulic	up to 30 tonnes	Each	43.81
Crawler, hydraulic	up to 35 tonnes	Each	55.30
Crawler, hydraulic	up to 38 tonnes	Each	71.73
Crawler, hydraulic	up to 55 tonnes	Each	95.63
Mini excavator	1000/1500 kg	Each	4.87
Mini excavator	2150/2400 kg	Each	6.67
Mini excavator	2700/3500 kg	Each	7.31
Mini excavator	3500/4500 kg	Each	8.21
Mini excavator	4500/6000 kg	Each	9.23
Mini excavator	7000 kg	Each	14.10
Micro excavator	725mm wide	Each	5.13

BUILDING INDUSTRY PLANT HIRE COSTS

Item of plant	Size/Rating	Unit	Rate per Hour (£)
Loaders			
Shovel loader	0.4 m³	Each	7.69
Shovel loader	1.57 m³	Each	8.97
Shovel loader, four wheel drive	1.7 m³	Each	4.83
Shovel loader, four wheel drive	2.3 m³	Each	4.38
Shovel loader, four wheel drive	3.3 m³	Each	5.06
Skid steer loader wheeled	300/400 kg payload	Each	7.31
Skid steer loader wheeled	625 kg payload	Each	7.67
Tracked skip loader	650 kg	Each	4.42
Excavator Loaders			
Wheeled tractor type with back-hoe excavator			
Four wheel drive			
Four wheel drive, 2 wheel steer	6 tonnes	Each	6.41
Four wheel drive, 2 wheel steer	8 tonnes	Each	8.59
Attachments			
Breakers for excavator		Each	8.72
Breakers for mini excavator		Each	1.75
Breakers for back-hoe excavator/loader		Each	5.13
COMPACTION EQUIPMENT			
Rollers			
Vibrating roller	368–420 kg	Each	1.43
Single roller	533 kg	Each	1.94
Single roller	750 kg	Each	3.43
Twin roller	up to 650 kg	Each	6.03
Twin roller	up to 950 kg	Each	6.62
Twin roller with seat end steering wheel	up to 1400 kg	Each	7.68
Twin roller with seat end steering wheel	up to 2500 kg	Each	10.61
Pavement roller	3–4 tonnes dead weight	Each	6.00
Pavement roller	4–6 tonnes	Each	6.86
Pavement roller	6–10 tonnes	Each	7.17
Pavement roller	10–13 tonnes	Each	19.86
Rammers			
Tamper rammer 2 stroke – petrol	225 mm – 275 mm	Each	1.52

BUILDING INDUSTRY PLANT HIRE COSTS

Item of plant	Size/Rating	Unit	Rate per Hour (£)
Soil Compactors			
Plate compactor	75 mm – 400 mm	Each	1.53
Plate compactor rubber pad	375 mm – 1400 mm	Each	1.53
Plate compactor reversible plate-petrol	400 mm	Each	2.44
CONCRETE EQUIPEMENT			
Concrete/Mortar Mixers			
Open drum without hopper	0.09/0.06 m³	Each	0.61
Open drum without hopper	0.12/0.09 m³	Each	1.22
Open drum without hopper	0.15/0.10 m³	Each	0.72
Concrete/Mortar Transport Equipment			
Concrete pump incl. hose, valve and couplers			
Lorry mounted concrete pump	24 m max. distance	Each	50.00
Lorry mounted concrete pump	34 m max. distance	Each	66.00
Lorry mounted concrete pump	42 m max. distance	Each	91.50
Concrete Equipment			
Vibrator, poker, petrol type	up to 75 mm dia.	Each	0.69
Air vibrator (*excluding compressor and hose*)	up to 75 mm dia.	Each	0.64
Extra poker heads	25/36/60 mm diameter	Each	0.76
Vibrating screed unit with beam	5.00 m	Each	2.48
Vibrating screed unit with adjustable beam	3.00–5.00 m	Each	3.54
Power float	725 mm–900 mm	Each	2.56
Power float finishing pan		Each	0.62
Floor grinder	660 × 1016 mm, 110 V electric	Each	4.31
Floor plane	450 × 1100 mm	Each	4.31
TESTING EQUIPMENT			
Pipe Testing Equipment			
Pressure testing pump, electric		Set	2.19
Pressure test pump		Set	0.80
SITE ACCOMODATION AND TEMPORARY SERVICES			Rate per Hour (£)
Heating equipment			
Space heater – propane	80,000 Btu/hr	Each	1.03

BUILDING INDUSTRY PLANT HIRE COSTS

Item of plant	Size/Rating	Unit	Rate per Hour (£)
Space heater – propane/electric	125,000 Btu/hr	Each	2.09
Space heater – propane/electric	250,000 Btu/hr	Each	2.33
Space heater – propane	125,000 Btu/hr	Each	1.54
Space heater – propane	260,000 Btu/hr	Each	1.88
Cabinet heater		Each	0.82
Cabinet heater, catalytic		Each	0.57
Electric halogen heater		Each	1.27
Ceramic heater	3 kW	Each	0.99
Fan heater	3 kW	Each	0.66
Cooling fan		Each	1.92
Mobile cooling unit, small		Each	3.60
Mobile cooling unit, large		Each	4.98
Air conditioning unit		Each	2.81
Site Lighting and Equipment			
Tripod floodlight	500 W	Each	0.48
Tripod floodlight	1000 W	Each	0.62
Towable floodlight	4 × 100 W	Each	3.85
Hand held floodlight	500 W	Each	0.51
Rechargeable light		Each	0.41
Inspection light		Each	0.37
Plasterer's light		Each	0.65
Lighting mast		Each	2.87
Festoon light string	25 m	Each	0.55
Site Electrical Equipment			
Extension leads	240 V/14 m	Each	0.26
Extension leads	110 V/14 m	Each	0.36
Cable reel	25 m 110 V/240 V	Each	0.46
Cable reel	50 m 110 V/240 V	Each	0.88
4 way junction box	110 V	Each	0.56
Power Generating Units			
Generator – petrol	2 kVA	Each	1.23
Generator – silenced petrol	2 kVA	Each	2.87
Generator – petrol	3 kVA	Each	1.47
Generator – diesel	5 kVA	Each	2.44
Generator – silenced diesel	10 kVA	Each	1.90
Generator – silenced diesel	15 kVA	Each	2.26
Generator – silenced diesel	30 kVA	Each	3.33
Generator – silenced diesel	50 kVA	Each	4.10

BUILDING INDUSTRY PLANT HIRE COSTS

Item of plant	Size/Rating	Unit	Rate per Hour (£)
Generator – silenced diesel	75kVA	Each	4.62
Generator – silenced diesel	100kVA	Each	5.64
Generator – silenced diesel	150kVA	Each	7.18
Generator – silenced diesel	200kVA	Each	9.74
Generator – silenced diesel	250kVA	Each	11.28
Generator – silenced diesel	350kVA	Each	14.36
Generator – silenced diesel	500kVA	Each	15.38
Tail adaptor	240V	Each	0.10
Transformers			
Transformer	3kVA	Each	0.32
Transformer	5kVA	Each	1.23
Transformer	7.5kVA	Each	0.59
Transformer	10kVA	Each	2.00
Rubbish Collection and Disposal Equipment			
Rubbish Chutes			
Standard plastic module	1 m section	Each	0.15
Steel liner insert		Each	0.30
Steel top hopper		Each	0.22
Plastic side entry hopper		Each	0.22
Plastic side entry hopper liner		Each	0.22
Dust Extraction Plant			
Dust extraction unit, light duty		Each	2.97
Dust extraction unit, heavy duty		Each	2.97
SITE EQUIPMENT - Welding Equipment			
Arc-(Electric) Complete With Leads			
Welder generator – petrol	200 amp	Each	3.53
Welder generator – diesel	300/350 amp	Each	3.78
Welder generator – diesel	4000 amp	Each	7.92
Extra welding lead sets		Each	0.69
Gas-Oxy Welder			
Welding and cutting set (including oxygen and acetylene, excluding underwater equipment and thermic boring)			
Small		Each	2.24
Large		Each	3.75

BUILDING INDUSTRY PLANT HIRE COSTS

Item of plant	Size/Rating	Unit	Rate per Hour (£)
Lead burning gun		Each	0.50
Mig welder		Each	1.38
Fume extractor		Each	2.46
Road Works Equipment			
Traffic lights, mains/generator	2-way	Set	10.94
Traffic lights, mains/generator	3-way	Set	11.56
Traffic lights, mains/generator	4-way	Set	12.19
Flashing light		Each	0.10
Road safety cone	450 mm	Each	0.08
Safety cone	750 mm	Each	0.10
Safety barrier plank	1.25 m	Each	0.13
Safety barrier plank	2 m	Each	0.15
Safety barrier plank post		Each	0.13
Safety barrier plank post base		Each	0.10
Safety four gate barrier	1 m each gate	Set	0.77
Guard barrier	2 m	Each	0.19
Road sign	750 mm	Each	0.23
Road sign	900 mm	Each	0.31
Road sign	1200 mm	Each	0.42
Speed ramp/cable protection	500 mm section	Each	0.14
Hose ramp open top	3m section	Each	0.07
DPC Equipment			
Damp-proofing injection machine		Each	2.56
Cleaning Equipment			
Vacuum cleaner (industrial wet) single motor		Each	1.08
Vacuum cleaner (industrial wet) twin motor	30 litre capacity	Each	1.79
Vacuum cleaner (industrial wet) twin motor	70 litre capacity	Each	2.21
Steam cleaner	Diesel/electric 1 phase	Each	3.33
Steam cleaner	Diesel/electric 3 phase	Each	3.85
Pressure washer, light duty electric	1450 PSI	Each	0.72
Pressure washer, heavy duty, diesel	2500 PSI	Each	1.33
Pressure washer, heavy duty, diesel	4000 PSI	Each	2.18
Cold pressure washer, electric		Each	2.39
Hot pressure washer, petrol		Each	4.19
Hot pressure washer, electric		Each	5.13

BUILDING INDUSTRY PLANT HIRE COSTS

Item of plant	Size/Rating	Unit	Rate per Hour (£)
Cold pressure washer, petrol		Each	2.92
Sandblast attachment to last washer		Each	1.23
Drain cleaning attachment to last washer		Each	1.03
Surface Preparation Equipment			
Rotavator	5 h.p.	Each	2.46
Rotavator	9 h.p.	Each	5.00
Scabbler, up to three heads		Each	1.53
Scabbler, pole		Each	2.68
Scrabbler, multi-headed floor		Each	3.89
Floor preparation machine		Each	1.05
Compressors and Equipment			
Portable Compressors			
Compressor – electric	4 cfm	Each	1.36
Compressor – electric	8 cfm lightweight	Each	1.31
Compressor – electric	8 cfm	Each	1.36
Compressor – electric	14 cfm	Each	1.56
Compressor – petrol	24 cfm	Each	2.15
Compressor – electric	25 cfm	Each	2.10
Compressor – electric	30 cfm	Each	2.36
Compressor – diesel	100 cfm	Each	2.56
Compressor – diesel	250 cfm	Each	5.54
Compressor – diesel	400 cfm	Each	8.72
Mobile Compressors			
Lorry mounted compressor			
(machine plus lorry only)	up to 3 m³	Each	41.47
(machine plus lorry only)	up to 5 m³	Each	48.94
Tractor mounted compressor			
(machine plus rubber tyred tractor)	Up to 4 m³	Each	21.03
Accessories (Pneumatic Tools) *(with and including up to 15 m of air hose)*			
Demolition pick, medium duty		Each	0.90
Demolition pick, heavy duty		Each	1.03
Breakers (with six steels) light	up to 150 kg	Each	1.19
Breakers (with six steels) medium	295 kg	Each	1.24
Breakers (with six steels) heavy	386 kg	Each	1.44
Rock drill (for use with compressor) hand held		Each	1.18
Additional hoses	15 m	Each	0.09

BUILDING INDUSTRY PLANT HIRE COSTS

Item of plant	Size/Rating	Unit	Rate per Hour (£)
Breakers			
Demolition hammer drill, heavy duty, electric		Each	1.54
Road breaker, electric		Each	2.41
Road breaker, 2 stroke, petrol		Each	4.06
Hydraulic breaker unit, light duty, petrol		Each	3.06
Hydraulic breaker unit, heavy duty, petrol		Each	3.46
Hydraulic breaker unit, heavy duty, diesel		Each	4.62
Quarrying and Tooling Equipment			
Block and stone splitter, hydraulic	600 mm × 600 mm	Each	1.90
Block and stone splitter, manual		Each	1.64
Steel Reinforcement Equipment			
Bar bending machine – manual	up to 13 mm dia. rods	Each	1.03
Bar bending machine – manual	up to 20 mm dia. rods	Each	1.41
Bar shearing machine – electric	up to 38 mm dia. rods	Each	3.08
Bar shearing machine – electric	up to 40 mm dia. rods	Each	4.62
Bar cropper machine – electric	up to 13 mm dia. rods	Each	2.05
Bar cropper machine – electric	up to 20 mm dia. rods	Each	2.56
Bar cropper machine – electric	up to 40 mm dia. rods	Each	4.62
Bar cropper machine – 3 phase	up to 40 mm dia. rods	Each	4.62
Dehumidifiers			
110/240v Water	68 litres extraction per 24 hours	Each	2.46
110/240v Water	90 litres extraction per 24 hours	Each	3.38
SMALL TOOLS			
Saws			
Masonry bench saw	350 mm–500 mm dia.	Each	1.13
Floor saw	125 mm max. cut	Each	1.15
Floor saw	150 mm max. cut	Each	3.83
Floor saw, reversible	350 mm max. cut	Each	3.32
Wall saw, electric		Each	2.05
Chop/cut off saw, electric	350 mm dia.	Each	1.79
Circular saw, electric	230 mm dia.	Each	0.72
Tyrannosaw		Each	1.74
Reciprocating saw		Each	0.79
Door trimmer		Each	1.17
Stone saw	300 mm	Each	1.44

BUILDING INDUSTRY PLANT HIRE COSTS

Item of plant	Size/Rating	Unit	Rate per Hour (£)
Chainsaw, petrol	500 mm	Each	3.92
Full chainsaw safety kit		Each	0.41
Worktop jig		Each	1.08
Pipe Work Equipment			
Pipe bender	15 mm–22 mm	Each	0.92
Pipe bender, hydraulic	50 mm	Each	1.76
Pipe bender, electric	50 mm–150 mm dia.	Each	2.19
Pipe cutter, hydraulic		Each	0.46
Tripod pipe vice		Set	0.75
Ratchet threader	12 mm–32 mm	Each	0.93
Pipe threading machine, electric	12 mm–75 mm	Each	3.07
Pipe threading machine, electric	12 mm–100 mm	Each	4.93
Impact wrench, electric		Each	1.33
Hand-held Drills and Equipment			
Impact or hammer drill	up to 25 mm dia.	Each	1.03
Impact or hammer drill	35 mm diameter	Each	1.29
Dry diamond core cutter		Each	0.99
Angle head drill		Each	0.90
Stirrer, mixer drill		Each	1.13
Paint, Insulation Application Equipment			
Airless spray unit		Each	4.13
Portaspray unit		Each	1.16
HPVL turbine spray unit		Each	2.23
Compressor and spray gun		Each	1.91
Other Handtools			
Staple gun		Each	0.96
Air nail gun	110 V	Each	1.01
Cartridge hammer		Each	1.08
Tongue and groove nailer complete with mallet		Each	1.59
Diamond wall chasing machine		Each	2.63
Masonry chain saw	300 mm	Each	5.49
Floor grinder		Each	3.99
Floor plane		Each	1.79
Diamond concrete planer		Each	1.93
Autofeed screwdriver, electric		Each	1.38
Laminate trimmer		Each	0.91

BUILDING INDUSTRY PLANT HIRE COSTS

Item of plant	Size/Rating	Unit	Rate per Hour (£)
Biscuit jointer		Each	1.49
Random orbital sander		Each	0.97
Floor sander		Each	1.54
Palm, delta, flap or belt sander		Each	0.75
Disc cutter, electric	300 mm	Each	1.49
Disc cutter, 2 stroke petrol	300 mm	Each	1.24
Dust suppressor for petrol disc cutter		Each	0.51
Cutter cart for petrol disc cutter		Each	1.21
Grinder, angle or cutter	up to 225 mm	Each	0.50
Grinder, angle or cutter	300 mm	Each	1.41
Mortar raking tool attachment		Each	0.19
Floor/polisher scrubber	325 mm	Each	1.76
Floor tile stripper		Each	2.44
Wallpaper stripper, electric		Each	0.81
Hot air paint stripper		Each	0.50
Electric diamond tile cutter	all sizes	Each	2.42
Hand tile cutter		Each	0.82
Electric needle gun		Each	1.29
Needle chipping gun		Each	1.85
Pedestrian floor sweeper	250 mm dia.	Each	0.82
Pedestrian floor sweeper	Petrol	Each	2.20
Diamond tile saw		Each	1.84
Blow lamp equipment and glass		Set	0.50

PART 7

Tables and Memoranda

This part contains the following sections:

Building Services Design for Energy Efficient Buildings, 2nd edition

Paul Tymkow *et al.*

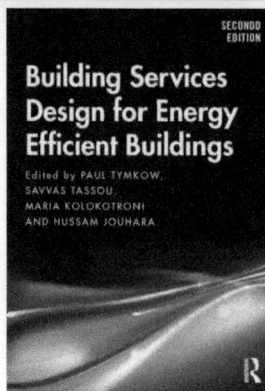

SECOND EDITION

Building Services Design for Energy Efficient Buildings

Edited by PAUL TYMKOW, SAVVAS TASSOU, MARIA KOLOKOTRONI AND HUSSAM JOUHARA

The role and influence of building services engineers are undergoing rapid change and are pivotal to achieving low-carbon buildings.

The essential conceptual design issues for planning the principal building services systems that influence energy efficiency are examined in detail, namely HVAC and electrical systems. In addition, the following issues are addressed:

- background issues on climate change, whole-life performance and design collaboration
- generic strategies for energy efficient, low-carbon design
- health and wellbeing and post occupancy evaluation
- building ventilation
- air conditioning and HVAC system selection
- thermal energy generation and distribution systems
- low-energy approaches for thermal control
- electrical systems, data collection, controls and monitoring
- building thermal load assessment
- building electric power load assessment
- space planning and design integration with other disciplines.

In order to deliver buildings that help mitigate climate change impacts, a new perspective is required for building services engineers, from the initial conceptual design and throughout the design collaboration with other disciplines. This book provides a contemporary introduction and guide to this new approach, for students and practitioners alike.

July 2020: 376 pp
Pb: 978-0-8153-6561-7 : £42.99

To Order: Tel: +44 (0) 1235 400524 Fax: +44 (0) 1235 400525
or Post: Taylor and Francis Customer Services,
Bookpoint Ltd, Unit T1, 200 Milton Park, Abingdon, Oxon, OX14 4TA UK
Email: book.orders@tandf.co.uk

For a complete listing of all our titles visit:
www.tandf.co.uk

Taylor & Francis
Taylor & Francis Group

CONVERSION TABLES

Length	Unit	Conversion factors			
Millimetre	mm	1 in	= 25.4 mm	1 mm	= 0.0394 in
Centimetre	cm	1 in	= 2.54 cm	1 cm	= 0.3937 in
Metre	m	1 ft	= 0.3048 m	1 m	= 3.2808 ft
		1 yd	= 0.9144 m		= 1.0936 yd
Kilometre	km	1 mile	= 1.6093 km	1 km	= 0.6214 mile

Note:
1 cm	= 10 mm	1 ft	= 12 in
1 m	= 1 000 mm	1 yd	= 3 ft
1 km	= 1 000 m	1 mile	= 1 760 yd

Area	Unit	Conversion factors			
Square Millimetre	mm^2	$1\ in^2$	$= 645.2\ mm^2$	$1\ mm^2$	$= 0.0016\ in^2$
Square Centimetre	cm^2	$1\ in^2$	$= 6.4516\ cm^2$	$1\ cm^2$	$= 1.1550\ in^2$
Square Metre	m^2	$1\ ft^2$	$= 0.0929\ m^2$	$1\ m^2$	$= 10.764\ ft^2$
		$1\ yd^2$	$= 0.8361\ m^2$	$1\ m^2$	$= 1.1960\ yd^2$
Square Kilometre	km^2	$1\ mile^2$	$= 2.590\ km^2$	$1\ km^2$	$= 0.3861\ mile^2$

Note:
$1\ cm^2$	$= 100\ mm^2$	$1\ ft^2$	$= 144\ in^2$
$1\ m^2$	$= 10\ 000\ cm^2$	$1\ yd^2$	$= 9\ ft^2$
$1\ km^2$	$= 100$ hectares	1 acre	$= 4\ 840\ yd^2$
		$1\ mile^2$	$= 640$ acres

Volume	Unit	Conversion factors			
Cubic Centimetre	cm^3	$1\ cm^3$	$= 0.0610\ in^3$	$1\ in^3$	$= 16.387\ cm^3$
Cubic Decimetre	dm^3	$1\ dm^3$	$= 0.0353\ ft^3$	$1\ ft^3$	$= 28.329\ dm^3$
Cubic Metre	m^3	$1\ m^3$	$= 35.3147\ ft^3$	$1\ ft^3$	$= 0.0283\ m^3$
		$1\ m^3$	$= 1.3080\ yd^3$	$1\ yd^3$	$= 0.7646\ m^3$
Litre	l	1 l	= 1.76 pint	1 pint	= 0.5683 l
			= 2.113 US pt		= 0.4733 US l

Note:
$1\ dm^3$	$= 1\ 000\ cm^3$	$1\ ft^3$	$= 1\ 728\ in^3$	1 pint	= 20 fl oz
$1\ m^3$	$= 1\ 000\ dm^3$	$1\ yd^3$	$= 27\ ft^3$	1 gal	= 8 pints
1 l	$= 1\ dm^3$				

Neither the Centimetre nor Decimetre are SI units, and as such their use, particularly that of the Decimetre, is not widespread outside educational circles.

Mass	Unit	Conversion factors			
Milligram	mg	1 mg	= 0.0154 grain	1 grain	= 64.935 mg
Gram	g	1 g	= 0.0353 oz	1 oz	= 28.35 g
Kilogram	kg	1 kg	= 2.2046 lb	1 lb	= 0.4536 kg
Tonne	t	1 t	= 0.9842 ton	1 ton	= 1.016 t

Note:
1 g	= 1000 mg	1 oz	= 437.5 grains	1 cwt	= 112 lb
1 kg	= 1000 g	1 lb	= 16 oz	1 ton	= 20 cwt
1 t	= 1000 kg	1 stone	= 14 lb		

Force	Unit	Conversion factors			
Newton	N	1 lbf	= 4.448 N	1 kgf	= 9.807 N
Kilonewton	kN	1 lbf	= 0.004448 kN	1 ton f	= 9.964 kN
Meganewton	MN	100 tonf	= 0.9964 MN		

CONVERSION TABLES

Pressure and stress	Unit	Conversion factors
Kilonewton per square metre	kN/m^2	1 lbf/in^2 = 6.895 kN/m^2
		1 bar = 100 kN/m^2
Meganewton per square metre	MN/m^2	1 tonf/ft^2 = 107.3 kN/m^2 = 0.1073 MN/m^2
		1 kgf/cm^2 = 98.07 kN/m^2
		1 lbf/ft^2 = 0.04788 kN/m^2

Coefficient of consolidation (Cv) or swelling	Unit	Conversion factors
Square metre per year	m^2/year	1 cm^2/s = 3 154 m^2/year
		1 ft^2/year = 0.0929 m^2/year

Coefficient of permeability	Unit	Conversion factors
Metre per second	m/s	1 cm/s = 0.01 m/s
Metre per year	m/year	1 ft/year = 0.3048 m/year
		= 0.9651 × (10)^{8}m/s

Temperature	Unit	Conversion factors	
Degree Celsius	°C	°C = 5/9 × (°F − 32)	°F = (9 × °C)/ 5 + 32

SPEED CONVERSION

km/h	m/min	mph	fpm
1	16.7	0.6	54.7
2	33.3	1.2	109.4
3	50.0	1.9	164.0
4	66.7	2.5	218.7
5	83.3	3.1	273.4
6	100.0	3.7	328.1
7	116.7	4.3	382.8
8	133.3	5.0	437.4
9	150.0	5.6	492.1
10	166.7	6.2	546.8
11	183.3	6.8	601.5
12	200.0	7.5	656.2
13	216.7	8.1	710.8
14	233.3	8.7	765.5
15	250.0	9.3	820.2
16	266.7	9.9	874.9
17	283.3	10.6	929.6
18	300.0	11.2	984.3
19	316.7	11.8	1038.9
20	333.3	12.4	1093.6
21	350.0	13.0	1148.3
22	366.7	13.7	1203.0
23	383.3	14.3	1257.7
24	400.0	14.9	1312.3
25	416.7	15.5	1367.0
26	433.3	16.2	1421.7
27	450.0	16.8	1476.4
28	466.7	17.4	1531.1
29	483.3	18.0	1585.7
30	500.0	18.6	1640.4
31	516.7	19.3	1695.1
32	533.3	19.9	1749.8
33	550.0	20.5	1804.5
34	566.7	21.1	1859.1
35	583.3	21.7	1913.8
36	600.0	22.4	1968.5
37	616.7	23.0	2023.2
38	633.3	23.6	2077.9
39	650.0	24.2	2132.5
40	666.7	24.9	2187.2
41	683.3	25.5	2241.9
42	700.0	26.1	2296.6
43	716.7	26.7	2351.3
44	733.3	27.3	2405.9
45	750.0	28.0	2460.6

CONVERSION TABLES

km/h	m/min	mph	fpm
46	766.7	28.6	2515.3
47	783.3	29.2	2570.0
48	800.0	29.8	2624.7
49	816.7	30.4	2679.4
50	833.3	31.1	2734.0

GEOMETRY

Two dimensional figures

Figure	Diagram of figure	Surface area	Perimeter
Square		a^2	$4a$
Rectangle		ab	$2(a+b)$
Triangle		$\frac{1}{2}ch$	$a+b+c$
Circle		πr^2 $\frac{1}{4}\pi d^2$ where $2r=d$	$2\pi r$ πd
Parallelogram		ah	$2(a+b)$
Trapezium		$\frac{1}{2}h(a+b)$	$a+b+c+d$
Ellipse		Approximately πab	$\pi(a+b)$
Hexagon		$2.6 \times a^2$	

GEOMETRY

Figure	Diagram of figure	Surface area	Perimeter
Octagon		$4.83 \times a^2$	$6a$
Sector of a circle		$\frac{1}{2}rb$ or $\frac{q}{360}\pi r^2$ note b = angle $\frac{q}{360} \times \pi 2r$	
Segment of a circle		$S - T$ where S = area of sector, T = area of triangle	
Bellmouth		$\frac{3}{14} \times r^2$	

GEOMETRY

Three dimensional figures

Figure	Diagram of figure	Surface area	Volume
Cube		$6a^2$	a^3
Cuboid/ rectangular block		$2(ab + ac + bc)$	abc
Prism/ triangular block		$bd + hc + dc + ad$	$\frac{1}{2}hcd$
Cylinder		$2\pi r^2 + 2\pi h$	$\pi r^2 h$
Sphere		$4\pi r^2$	$\frac{4}{3}\pi r^3$
Segment of sphere		$2\pi Rh$	$\frac{1}{6}\pi h(3r^2 + h^2)$ $\frac{1}{3}\pi h^2(3R - H)$
Pyramid		$(a + b)l + ab$	$\frac{1}{3}abh$

GEOMETRY

Figure	Diagram of figure	Surface area	Volume
Frustum of a pyramid		$l(a + b + c + d) + \sqrt{(ab + cd)}$ [rectangular figure only]	$\frac{h}{3}(ab + cd + \sqrt{abcd})$
Cone		πrl (excluding base) $\pi rl + \pi r^2$ (including base)	$\frac{1}{3}\pi r^2 h$ $\frac{1}{12}\pi d^2 h$
Frustum of a cone		$\pi r^2 + \pi R^2 + \pi l(R + r)$	$\frac{1}{3}\pi(R^2 + Rr + r^2)$

FORMULAE

Formulae

Formula	Description
Pythagoras Theorem	$A^2 = B^2 + C^2$ where A is the hypotenuse of a right-angled triangle and B and C are the two adjacent sides

Simpsons Rule	The Area is divided into an even number of strips of equal width, and therefore has an odd number of ordinates at the division points
	
	$$\text{area} = \frac{S(A + 2B + 4C)}{3}$$
	
	where S = common interval (strip width)
	              A = sum of first and last ordinates
	              B = sum of remaining odd ordinates
	              C = sum of the even ordinates
	
	The Volume can be calculated by the same formula, but by substituting the area of each coordinate rather than its length

Trapezoidal Rule	A given trench is divided into two equal sections, giving three ordinates, the first, the middle and the last
	
	$$\text{volume} = \frac{S \times (A + B + 2C)}{2}$$
	
	where S = width of the strips
	              A = area of the first section
	              B = area of the last section
	              C = area of the rest of the sections

Prismoidal Rule	A given trench is divided into two equal sections, giving three ordinates, the first, the middle and the last
	
	$$\text{volume} = \frac{L \times (A + 4B + C)}{6}$$
	
	where L = total length of trench
	              A = area of the first section
	              B = area of the middle section
	              C = area of the last section

TYPICAL THERMAL CONDUCTIVITY OF BUILDING MATERIALS

(Always check manufacturer's details – variation will occur depending on product and nature of materials)

	Thermal conductivity (W/mK)		Thermal conductivity (W/mK)
Acoustic plasterboard	0.25	Oriented strand board	0.13
Aerated concrete slab (500 kg/m^3)	0.16	Outer leaf brick	0.77
Aluminium	237	Plasterboard	0.22
Asphalt (1700 kg/m^3)	0.5	Plaster dense (1300 kg/m^3)	0.5
Bitumen-impregnated fibreboard	0.05	Plaster lightweight (600 kg/m^3)	0.16
Blocks (standard grade 600 kg/m^3)	0.15	Plywood (950 kg/m^3)	0.16
Blocks (solar grade 460 kg/m^3)	0.11	Prefabricated timber wall panels (check manufacturer)	0.12
Brickwork (outer leaf 1700 kg/m^3)	0.84	Screed (1200 kg/m^3)	0.41
Brickwork (inner leaf 1700 kg/m^3)	0.62	Stone chippings (1800 kg/m^3)	0.96
Dense aggregate concrete block 1800 kg/m^3 (exposed)	1.21	Tile hanging (1900 kg/m^3)	0.84
Dense aggregate concrete block 1800 kg/m^3 (protected)	1.13	Timber (650 kg/m^3)	0.14
Calcium silicate board (600 kg/m^3)	0.17	Timber flooring (650 kg/m^3)	0.14
Concrete general	1.28	Timber rafters	0.13
Concrete (heavyweight 2300 kg/m^3)	1.63	Timber roof or floor joists	0.13
Concrete (dense 2100 kg/m^3 typical floor)	1.4	Roof tile (1900 kg/m^3)	0.84
Concrete (dense 2000 kg/m^3 typical floor)	1.13	Timber blocks (650 kg/m^3)	0.14
Concrete (medium 1400 kg/m^3)	0.51	Cellular glass	0.045
Concrete (lightweight 1200 kg/m^3)	0.38	Expanded polystyrene	0.034
Concrete (lightweight 600 kg/m^3)	0.19	Expanded polystyrene slab (25 kg/m^3)	0.035
Concrete slab (aerated 500 kg/m^3)	0.16	Extruded polystyrene	0.035
Copper	390	Glass mineral wool	0.04
External render sand/cement finish	1	Mineral quilt (12 kg/m^3)	0.04
External render (1300 kg/m^3)	0.5	Mineral wool slab (25 kg/m^3)	0.035
Felt – Bitumen layers (1700 kg/m^3)	0.5	Phenolic foam	0.022
Fibreboard (300 kg/m^3)	0.06	Polyisocyanurate	0.025
Glass	0.93	Polyurethane	0.025
Marble	3	Rigid polyurethane	0.025
Metal tray used in wriggly tin concrete floors (7800 kg/m^3)	50	Rock mineral wool	0.038
Mortar (1750 kg/m^3)	0.8		

EARTHWORK

Weights of Typical Materials Handled by Excavators

The weight of the material is that of the state in its natural bed and includes moisture.
Adjustments should be made to allow for loose or compacted states

Material	Mass (kg/m³)	Mass (lb/cu yd)
Ashes, dry	610	1028
Ashes, wet	810	1365
Basalt, broken	1954	3293
Basalt, solid	2933	4943
Bauxite, crushed	1281	2159
Borax, fine	849	1431
Caliche	1440	2427
Cement, clinker	1415	2385
Chalk, fine	1221	2058
Chalk, solid	2406	4055
Cinders, coal, ash	641	1080
Cinders, furnace	913	1538
Clay, compacted	1746	2942
Clay, dry	1073	1808
Clay, wet	1602	2700
Coal, anthracite, solid	1506	2538
Coal, bituminous	1351	2277
Coke	610	1028
Dolomite, lumpy	1522	2565
Dolomite, solid	2886	4864
Earth, dense	2002	3374
Earth, dry, loam	1249	2105
Earth, Fullers, raw	673	1134
Earth, moist	1442	2430
Earth, wet	1602	2700
Felsite	2495	4205
Fieldspar, solid	2613	4404
Fluorite	3093	5213
Gabbro	3093	5213
Gneiss	2696	4544
Granite	2690	4534
Gravel, dry ¼ to 2 inch	1682	2835
Gravel, dry, loose	1522	2565
Gravel, wet ¼ to 2 inch	2002	3374
Gypsum, broken	1450	2444
Gypsum, solid	2787	4697
Hardcore (consolidated)	1928	3249
Lignite, dry	801	1350
Limestone, broken	1554	2619
Limestone, solid	2596	4375
Magnesite, magnesium ore	2993	5044
Marble	2679	4515
Marl, wet	2216	3735
Mica, broken	1602	2700
Mica, solid	2883	4859
Peat, dry	400	674
Peat, moist	700	1179

EARTHWORK

Material	Mass (kg/m³)	Mass (lb/cu yd)
Peat, wet	1121	1889
Potash	1281	2159
Pumice, stone	640	1078
Quarry waste	1438	2423
Quartz sand	1201	2024
Quartz, solid	2584	4355
Rhyolite	2400	4045
Sand and gravel, dry	1650	2781
Sand and gravel, wet	2020	3404
Sand, dry	1602	2700
Sand, wet	1831	3086
Sandstone, solid	2412	4065
Shale, solid	2637	4444
Slag, broken	2114	3563
Slag, furnace granulated	961	1619
Slate, broken	1370	2309
Slate, solid	2667	4495
Snow, compacted	481	810
Snow, freshly fallen	160	269
Taconite	2803	4724
Trachyte	2400	4045
Trap rock, solid	2791	4704
Turf	400	674
Water	1000	1685

Transport Capacities

Type of vehicle	Capacity of vehicle	
	Payload	Heaped capacity
Wheelbarrow	150	0.10
1 tonne dumper	1250	1.00
2.5 tonne dumper	4000	2.50
Articulated dump truck (Volvo A20 6 × 4)	18500	11.00
Articulated dump truck (Volvo A35 6 × 6)	32000	19.00
Large capacity rear dumper (Euclid R35)	35000	22.00
Large capacity rear dumper (Euclid R85)	85000	50.00

EARTHWORK

Machine Volumes for Excavating and Filling

Machine type	Cycles per minute	Volume per minute (m^3)
1.5 tonne excavator	1	0.04
	2	0.08
	3	0.12
3 tonne excavator	1	0.13
	2	0.26
	3	0.39
5 tonne excavator	1	0.28
	2	0.56
	3	0.84
7 tonne excavator	1	0.28
	2	0.56
	3	0.84
21 tonne excavator	1	1.21
	2	2.42
	3	3.63
Backhoe loader JCB3CX excavator	1	0.28
Rear bucket capacity 0.28 m^3	2	0.56
	3	0.84
Backhoe loader JCB3CX loading	1	1.00
Front bucket capacity 1.00 m^3	2	2.00

Machine Volumes for Excavating and Filling

Machine type	Loads per hour	Volume per hour (m^3)
1 tonne high tip skip loader	5	2.43
Volume 0.485 m^3	7	3.40
	10	4.85
3 tonne dumper	4	7.60
Max volume 2.40 m^3	5	9.50
Available volume 1.9 m^3	7	13.30
	10	19.00
6 tonne dumper	4	15.08
Max volume 3.40 m^3	5	18.85
Available volume 3.77 m^3	7	26.39
	10	37.70

EARTHWORK

Bulkage of Soils (after excavation)

Type of soil	Approximate bulking of 1 m³ after excavation
Vegetable soil and loam	25–30%
Soft clay	30–40%
Stiff clay	10–20%
Gravel	20–25%
Sand	40–50%
Chalk	40–50%
Rock, weathered	30–40%
Rock, unweathered	50–60%

Shrinkage of Materials (on being deposited)

Type of soil	Approximate bulking of 1 m³ after excavation
Clay	10%
Gravel	8%
Gravel and sand	9%
Loam and light sandy soils	12%
Loose vegetable soils	15%

Voids in Material Used as Subbases or Beddings

Material	m³ of voids/m³
Alluvium	0.37
River grit	0.29
Quarry sand	0.24
Shingle	0.37
Gravel	0.39
Broken stone	0.45
Broken bricks	0.42

Angles of Repose

Type of soil		Degrees
Clay	– dry	30
	– damp, well drained	45
	– wet	15–20
Earth	– dry	30
	– damp	45
Gravel	– moist	48
Sand	– dry or moist	35
	– wet	25
Loam		40

EARTHWORK

Slopes and Angles

Ratio of base to height	Angle in degrees
5:1	11
4:1	14
3:1	18
2:1	27
1½:1	34
1:1	45
1:1½	56
1:2	63
1:3	72
1:4	76
1:5	79

Grades (in Degrees and Percents)

Degrees	Percent	Degrees	Percent
1	1.8	24	44.5
2	3.5	25	46.6
3	5.2	26	48.8
4	7.0	27	51.0
5	8.8	28	53.2
6	10.5	29	55.4
7	12.3	30	57.7
8	14.0	31	60.0
9	15.8	32	62.5
10	17.6	33	64.9
11	19.4	34	67.4
12	21.3	35	70.0
13	23.1	36	72.7
14	24.9	37	75.4
15	26.8	38	78.1
16	28.7	39	81.0
17	30.6	40	83.9
18	32.5	41	86.9
19	34.4	42	90.0
20	36.4	43	93.3
21	38.4	44	96.6
22	40.4	45	100.0
23	42.4		

EARTHWORK

Bearing Powers

Ground conditions		Bearing power		
		kg/m²	lb/in²	Metric t/m²
Rock,	broken	483	70	50
	solid	2415	350	240
Clay,	dry or hard	380	55	40
	medium dry	190	27	20
	soft or wet	100	14	10
Gravel,	cemented	760	110	80
Sand,	compacted	380	55	40
	clean dry	190	27	20
Swamp and alluvial soils		48	7	5

Earthwork Support

Maximum depth of excavation in various soils without the use of earthwork support

Ground conditions	Feet (ft)	Metres (m)
Compact soil	12	3.66
Drained loam	6	1.83
Dry sand	1	0.3
Gravelly earth	2	0.61
Ordinary earth	3	0.91
Stiff clay	10	3.05

It is important to note that the above table should only be used as a guide. Each case must be taken on its merits and, as the limited distances given above are approached, careful watch must be kept for the slightest signs of caving in

CONCRETE WORK

Weights of Concrete and Concrete Elements

Type of material		kg/m³	lb/cu ft
Ordinary concrete (dense aggregates)			
Non-reinforced plain or mass concrete			
Nominal weight		2305	144
Aggregate	– limestone	2162 to 2407	135 to 150
	– gravel	2244 to 2407	140 to 150
	– broken brick	2000 (av)	125 (av)
	– other crushed stone	2326 to 2489	145 to 155
Reinforced concrete			
Nominal weight		2407	150
Reinforcement	– 1%	2305 to 2468	144 to 154
	– 2%	2356 to 2519	147 to 157
	– 4%	2448 to 2703	153 to 163
Special concretes			
Heavy concrete			
Aggregates	– barytes, magnetite	3210 (min)	200 (min)
	– steel shot, punchings	5280	330
Lean mixes			
Dry-lean (gravel aggregate)		2244	140
Soil-cement (normal mix)		1601	100

CONCRETE WORK

Type of material		kg/m² per mm thick	lb/sq ft per inch thick
Ordinary concrete (dense aggregates)			
Solid slabs (floors, walls etc.)			
Thickness:	75 mm or 3 in	184	37.5
	100 mm or 4 in	245	50
	150 mm or 6 in	378	75
	250 mm or 10 in	612	125
	300 mm or 12 in	734	150
Ribbed slabs			
Thickness:	125 mm or 5 in	204	42
	150 mm or 6 in	219	45
	225 mm or 9 in	281	57
	300 mm or 12 in	342	70
Special concretes			
Finishes etc.			
	Rendering, screed etc.		
	Granolithic, terrazzo	1928 to 2401	10 to 12.5
	Glass-block (hollow) concrete	1734 (approx)	9 (approx)
Prestressed concrete		Weights as for reinforced concrete (upper limits)	
Air-entrained concrete		Weights as for plain or reinforced concrete	

CONCRETE WORK

Average Weight of Aggregates

Materials	Voids %	Weight kg/m³
Sand	39	1660
Gravel 10–20 mm	45	1440
Gravel 35–75 mm	42	1555
Crushed stone	50	1330
Crushed granite (over 15 mm)	50	1345
(n.e. 15 mm)	47	1440
'All-in' ballast	32	1800–2000

Material	kg/m³	lb/cu yd
Vermiculite (aggregate)	64–80	108–135
All-in aggregate	1999	125

Applications and Mix Design

Site mixed concrete

Recommended mix	Class of work suitable for	Cement (kg)	Sand (kg)	Coarse aggregate (kg)	Nr 25 kg bags cement per m³ of combined aggregate
1:3:6	Roughest type of mass concrete such as footings, road haunching over 300 mm thick	208	905	1509	8.30
1:2.5:5	Mass concrete of better class than 1:3:6 such as bases for machinery, walls below ground etc.	249	881	1474	10.00
1:2:4	Most ordinary uses of concrete, such as mass walls above ground, road slabs etc. and general reinforced concrete work	304	889	1431	12.20
1:1.5:3	Watertight floors, pavements and walls, tanks, pits, steps, paths, surface of 2 course roads, reinforced concrete where extra strength is required	371	801	1336	14.90
1:1:2	Works of thin section such as fence posts and small precast work	511	720	1206	20.40

CONCRETE WORK

Ready mixed concrete

Application	Designated concrete	Standardized prescribed concrete	Recommended consistence (nominal slump class)
Foundations			
Mass concrete fill or blinding	GEN 1	ST2	S3
Strip footings	GEN 1	ST2	S3
Mass concrete foundations			
Single storey buildings	GEN 1	ST2	S3
Double storey buildings	GEN 3	ST4	S3
Trench fill foundations			
Single storey buildings	GEN 1	ST2	S4
Double storey buildings	GEN 3	ST4	S4
General applications			
Kerb bedding and haunching	GEN 0	ST1	S1
Drainage works – immediate support	GEN 1	ST2	S1
Other drainage works	GEN 1	ST2	S3
Oversite below suspended slabs	GEN 1	ST2	S3
Floors			
Garage and house floors with no embedded steel	GEN 3	ST4	S2
Wearing surface: Light foot and trolley traffic	RC30	ST4	S2
Wearing surface: General industrial	RC40	N/A	S2
Wearing surface: Heavy industrial	RC50	N/A	S2
Paving			
House drives, domestic parking and external parking	PAV 1	N/A	S2
Heavy-duty external paving	PAV 2	N/A	S2

CONCRETE WORK

Prescribed Mixes for Ordinary Structural Concrete

Weights of cement and total dry aggregates in kg to produce approximately one cubic metre of fully compacted concrete together with the percentages by weight of fine aggregate in total dry aggregates

Conc. grade	Nominal max size of aggregate (mm)	40		20		14		10	
	Workability	Med.	High	Med.	High	Med.	High	Med.	High
	Limits to slump that may be expected (mm)	50–100	100–150	25–75	75–125	10–50	50–100	10–25	25–50
7	Cement (kg)	180	200	210	230	–	–	–	–
	Total aggregate (kg)	1950	1850	1900	1800	–	–	–	–
	Fine aggregate (%)	30–45	30–45	35–50	35–50	–	–	–	–
10	Cement (kg)	210	230	240	260	–	–	–	–
	Total aggregate (kg)	1900	1850	1850	1800	–	–	–	–
	Fine aggregate (%)	30–45	30–45	35–50	35–50	–	–	–	–
15	Cement (kg)	250	270	280	310	–	–	–	–
	Total aggregate (kg)	1850	1800	1800	1750	–	–	–	–
	Fine aggregate (%)	30–45	30–45	35–50	35–50	–	–	–	–
20	Cement (kg)	300	320	320	350	340	380	360	410
	Total aggregate (kg)	1850	1750	1800	1750	1750	1700	1750	1650
	Sand								
	Zone 1 (%)	35	40	40	45	45	50	50	55
	Zone 2 (%)	30	35	35	40	40	45	45	50
	Zone 3 (%)	30	30	30	35	35	40	40	45
25	Cement (kg)	340	360	360	390	380	420	400	450
	Total aggregate (kg)	1800	1750	1750	1700	1700	1650	1700	1600
	Sand								
	Zone 1 (%)	35	40	40	45	45	50	50	55
	Zone 2 (%)	30	35	35	40	40	45	45	50
	Zone 3 (%)	30	30	30	35	35	40	40	45
30	Cement (kg)	370	390	400	430	430	470	460	510
	Total aggregate (kg)	1750	1700	1700	1650	1700	1600	1650	1550
	Sand								
	Zone 1 (%)	35	40	40	45	45	50	50	55
	Zone 2 (%)	30	35	35	40	40	45	45	50
	Zone 3 (%)	30	30	30	35	35	40	40	45

REINFORCEMENT

Weights of Bar Reinforcement

Nominal sizes (mm)	Cross-sectional area (mm²)	Mass (kg/m)	Length of bar (m/tonne)
6	28.27	0.222	4505
8	50.27	0.395	2534
10	78.54	0.617	1622
12	113.10	0.888	1126
16	201.06	1.578	634
20	314.16	2.466	405
25	490.87	3.853	260
32	804.25	6.313	158
40	1265.64	9.865	101
50	1963.50	15.413	65

Weights of Bars (at specific spacings)

Weights of metric bars in kilogrammes per square metre

Size (mm)	Spacing of bars in millimetres									
	75	100	125	150	175	200	225	250	275	300
6	2.96	2.220	1.776	1.480	1.27	1.110	0.99	0.89	0.81	0.74
8	5.26	3.95	3.16	2.63	2.26	1.97	1.75	1.58	1.44	1.32
10	8.22	6.17	4.93	4.11	3.52	3.08	2.74	2.47	2.24	2.06
12	11.84	8.88	7.10	5.92	5.07	4.44	3.95	3.55	3.23	2.96
16	21.04	15.78	12.63	10.52	9.02	7.89	7.02	6.31	5.74	5.26
20	32.88	24.66	19.73	16.44	14.09	12.33	10.96	9.87	8.97	8.22
25	51.38	38.53	30.83	25.69	22.02	19.27	17.13	15.41	14.01	12.84
32	84.18	63.13	50.51	42.09	36.08	31.57	28.06	25.25	22.96	21.04
40	131.53	98.65	78.92	65.76	56.37	49.32	43.84	39.46	35.87	32.88
50	205.51	154.13	123.31	102.76	88.08	77.07	68.50	61.65	56.05	51.38

Basic weight of steelwork taken as 7850 kg/m³
Basic weight of bar reinforcement per metre run = 0.00785 kg/mm²
The value of π has been taken as 3.141592654

REINFORCEMENT

Fabric Reinforcement

Preferred range of designated fabric types and stock sheet sizes

Fabric reference	Longitudinal wires			Cross wires			
	Nominal wire size (mm)	Pitch (mm)	Area (mm²/m)	Nominal wire size (mm)	Pitch (mm)	Area (mm²/m)	Mass (kg/m²)
Square mesh							
A393	10	200	393	10	200	393	6.16
A252	8	200	252	8	200	252	3.95
A193	7	200	193	7	200	193	3.02
A142	6	200	142	6	200	142	2.22
A98	5	200	98	5	200	98	1.54
Structural mesh							
B1131	12	100	1131	8	200	252	10.90
B785	10	100	785	8	200	252	8.14
B503	8	100	503	8	200	252	5.93
B385	7	100	385	7	200	193	4.53
B283	6	100	283	7	200	193	3.73
B196	5	100	196	7	200	193	3.05
Long mesh							
C785	10	100	785	6	400	70.8	6.72
C636	9	100	636	6	400	70.8	5.55
C503	8	100	503	5	400	49.0	4.34
C385	7	100	385	5	400	49.0	3.41
C283	6	100	283	5	400	49.0	2.61
Wrapping mesh							
D98	5	200	98	5	200	98	1.54
D49	2.5	100	49	2.5	100	49	0.77

Stock sheet size 4.8 m × 2.4 m, Area 11.52 m²

Average weight kg/m³ of steelwork reinforcement in concrete for various building elements

Substructure	kg/m³ concrete	Substructure	kg/m³ concrete
Pile caps	110–150	Plate slab	150–220
Tie beams	130–170	Cant slab	145–210
Ground beams	230–330	Ribbed floors	130–200
Bases	125–180	Topping to block floor	30–40
Footings	100–150	Columns	210–310
Retaining walls	150–210	Beams	250–350
Raft	60–70	Stairs	130–170
Slabs – one way	120–200	Walls – normal	40–100
Slabs – two way	110–220	Walls – wind	70–125

Note: For exposed elements add the following %:
Walls 50%, Beams 100%, Columns 15%

FORMWORK

Formwork Stripping Times – Normal Curing Periods

Conditions under which concrete is maturing	Minimum periods of protection for different types of cement					
	Number of days (where the average surface temperature of the concrete exceeds 10°C during the whole period)			Equivalent maturity (degree hours) calculated as the age of the concrete in hours multiplied by the number of degrees Celsius by which the average surface temperature of the concrete exceeds 10°C		
	Other	SRPC	OPC or RHPC	Other	SRPC	OPC or RHPC
1. Hot weather or drying winds	7	4	3	3500	2000	1500
2. Conditions not covered by 1	4	3	2	2000	1500	1000

KEY
OPC – Ordinary Portland Cement
RHPC – Rapid-hardening Portland Cement
SRPC – Sulphate-resisting Portland Cement

Minimum Period before Striking Formwork

	Minimum period before striking		
	Surface temperature of concrete		
	16°C	17°C	t°C (0–25)
Vertical formwork to columns, walls and large beams	12 hours	18 hours	300 hours t+10
Soffit formwork to slabs	4 days	6 days	100 days t+10
Props to slabs	10 days	15 days	250 days t+10
Soffit formwork to beams	9 days	14 days	230 days t+10
Props to beams	14 days	21 days	360 days t+10

MASONRY

Number of Bricks Required for Various Types of Work per m² of Walling

Description	Brick size	
	215 × 102.5 × 50 mm	215 × 102.5 × 65 mm
Half brick thick		
Stretcher bond	74	59
English bond	108	86
English garden wall bond	90	72
Flemish bond	96	79
Flemish garden wall bond	83	66
One brick thick and cavity wall of two half brick skins		
Stretcher bond	148	119

Quantities of Bricks and Mortar Required per m² of Walling

	Unit	No of bricks required	Mortar required (cubic metres)		
Standard bricks			No frogs	Single frogs	Double frogs
Brick size 215 × 102.5 × 50 mm					
half brick wall (103 mm)	m²	72	0.022	0.027	0.032
2 × half brick cavity wall (270 mm)	m²	144	0.044	0.054	0.064
one brick wall (215 mm)	m²	144	0.052	0.064	0.076
one and a half brick wall (322 mm)	m²	216	0.073	0.091	0.108
Mass brickwork	m³	576	0.347	0.413	0.480
Brick size 215 × 102.5 × 65 mm					
half brick wall (103 mm)	m²	58	0.019	0.022	0.026
2 × half brick cavity wall (270 mm)	m²	116	0.038	0.045	0.055
one brick wall (215 mm)	m²	116	0.046	0.055	0.064
one and a half brick wall (322 mm)	m²	174	0.063	0.074	0.088
Mass brickwork	m³	464	0.307	0.360	0.413
Metric modular bricks			Perforated		
Brick size 200 × 100 × 75 mm					
90 mm thick	m²	67	0.016	0.019	
190 mm thick	m²	133	0.042	0.048	
290 mm thick	m²	200	0.068	0.078	
Brick size 200 × 100 × 100 mm					
90 mm thick	m²	50	0.013	0.016	
190 mm thick	m²	100	0.036	0.041	
290 mm thick	m²	150	0.059	0.067	
Brick size 300 × 100 × 75 mm					
90 mm thick	m²	33	–	0.015	
Brick size 300 × 100 × 100 mm					
90 mm thick	m²	44	0.015	0.018	

Note: Assuming 10 mm thick joints

MASONRY

Mortar Required per m² Blockwork (9.88 blocks/m²)

Wall thickness	75	90	100	125	140	190	215
Mortar m³/m²	0.005	0.006	0.007	0.008	0.009	0.013	0.014

Mortar Group	Cement: lime: sand	Masonry cement: sand	Cement: sand with plasticizer
1	1:0–0.25:3		
2	1:0.5:4–4.5	1:2.5-3.5	1:3–4
3	1:1:5–6	1:4–5	1:5–6
4	1:2:8–9	1:5.5–6.5	1:7–8
5	1:3:10–12	1:6.5–7	1:8

Group 1: strong inflexible mortar
Group 5: weak but flexible

All mixes within a group are of approximately similar strength
Frost resistance increases with the use of plasticizers
Cement: lime: sand mixes give the strongest bond and greatest resistance to rain penetration
Masonry cement equals ordinary Portland cement plus a fine neutral mineral filler and an air entraining agent

Calcium Silicate Bricks

Type	Strength	Location
Class 2 crushing strength	14.0 N/mm²	not suitable for walls
Class 3	20.5 N/mm²	walls above dpc
Class 4	27.5 N/mm²	cappings and copings
Class 5	34.5 N/mm²	retaining walls
Class 6	41.5 N/mm²	walls below ground
Class 7	48.5 N/mm²	walls below ground

The Class 7 calcium silicate bricks are therefore equal in strength to Class B bricks
Calcium silicate bricks are not suitable for DPCs

Durability of Bricks	
FL	Frost resistant with low salt content
FN	Frost resistant with normal salt content
ML	Moderately frost resistant with low salt content
MN	Moderately frost resistant with normal salt content

MASONRY

Brickwork Dimensions

No. of horizontal bricks	Dimensions (mm)	No. of vertical courses	Height of vertical courses (mm)
½	112.5	1	75
1	225.0	2	150
1½	337.5	3	225
2	450.0	4	300
2½	562.5	5	375
3	675.0	6	450
3½	787.5	7	525
4	900.0	8	600
4½	1012.5	9	675
5	1125.0	10	750
5½	1237.5	11	825
6	1350.0	12	900
6½	1462.5	13	975
7	1575.0	14	1050
7½	1687.5	15	1125
8	1800.0	16	1200
8½	1912.5	17	1275
9	2025.0	18	1350
9½	2137.5	19	1425
10	2250.0	20	1500
20	4500.0	24	1575
40	9000.0	28	2100
50	11250.0	32	2400
60	13500.0	36	2700
75	16875.0	40	3000

TIMBER

Weights of Timber

Material	kg/m³	lb/cu ft
General	806 (avg)	50 (avg)
Douglas fir	479	30
Yellow pine, spruce	479	30
Pitch pine	673	42
Larch, elm	561	35
Oak (English)	724 to 959	45 to 60
Teak	643 to 877	40 to 55
Jarrah	959	60
Greenheart	1040 to 1204	65 to 75
Quebracho	1285	80
Material	kg/m² per mm thickness	lb/sq ft per inch thickness
Wooden boarding and blocks		
Softwood	0.48	2.5
Hardwood	0.76	4
Hardboard	1.06	5.5
Chipboard	0.76	4
Plywood	0.62	3.25
Blockboard	0.48	2.5
Fibreboard	0.29	1.5
Wood-wool	0.58	3
Plasterboard	0.96	5
Weather boarding	0.35	1.8

TIMBER

Conversion Tables (for timber only)

Inches	Millimetres	Feet	Metres
1	25	1	0.300
2	50	2	0.600
3	75	3	0.900
4	100	4	1.200
5	125	5	1.500
6	150	6	1.800
7	175	7	2.100
8	200	8	2.400
9	225	9	2.700
10	250	10	3.000
11	275	11	3.300
12	300	12	3.600
13	325	13	3.900
14	350	14	4.200
15	375	15	4.500
16	400	16	4.800
17	425	17	5.100
18	450	18	5.400
19	475	19	5.700
20	500	20	6.000
21	525	21	6.300
22	550	22	6.600
23	575	23	6.900
24	600	24	7.200

Planed Softwood
The finished end section size of planed timber is usually 3/16" less than the original size from which it is produced. This however varies slightly depending upon availability of material and origin of the species used.

Standards (timber) to cubic metres and cubic metres to standards (timber)

Cubic metres	Cubic metres standards	Standards
4.672	1	0.214
9.344	2	0.428
14.017	3	0.642
18.689	4	0.856
23.361	5	1.070
28.033	6	1.284
32.706	7	1.498
37.378	8	1.712
42.050	9	1.926
46.722	10	2.140
93.445	20	4.281
140.167	30	6.421
186.890	40	8.561
233.612	50	10.702
280.335	60	12.842
327.057	70	14.982
373.779	80	17.122

TIMBER

1 cu metre = 35.3148 cu ft = 0.21403 std

1 cu ft = 0.028317 cu metres

1 std = 4.67227 cu metres

Basic sizes of sawn softwood available (cross-sectional areas)

Thickness (mm)	Width (mm)								
	75	100	125	150	175	200	225	250	300
16	X	X	X	X					
19	X	X	X	X					
22	X	X	X	X					
25	X	X	X	X	X	X	X	X	X
32	X	X	X	X	X	X	X	X	X
36	X	X	X	X					
38	X	X	X	X	X	X	X		
44	X	X	X	X	X	X	X	X	X
47*	X	X	X	X	X	X	X	X	X
50	X	X	X	X	X	X	X	X	X
63	X	X	X	X	X	X	X		
75	X	X	X	X	X	X	X	X	
100		X		X		X		X	X
150				X		X			X
200						X			
250								X	
300									X

* This range of widths for 47 mm thickness will usually be found to be available in construction quality only

Note: The smaller sizes below 100 mm thick and 250 mm width are normally but not exclusively of European origin. Sizes beyond this are usually of North and South American origin

Basic lengths of sawn softwood available (metres)

1.80	2.10	3.00	4.20	5.10	6.00	7.20
	2.40	3.30	4.50	5.40	6.30	
	2.70	3.60	4.80	5.70	6.60	
		3.90			6.90	

Note: Lengths of 6.00 m and over will generally only be available from North American species and may have to be recut from larger sizes

TIMBER

Reductions from basic size to finished size by planning of two opposed faces

Purpose	Reductions from basic sizes for timber			
	15–35 mm	**36–100 mm**	**101–150 mm**	**over 150 mm**
a) Constructional timber	3 mm	3 mm	5 mm	6 mm
b) Matching interlocking boards	4 mm	4 mm	6 mm	6 mm
c) Wood trim not specified in BS 584	5 mm	7 mm	7 mm	9 mm
d) Joinery and cabinet work	7 mm	9 mm	11 mm	13 mm

Note: The reduction of width or depth is overall the extreme size and is exclusive of any reduction of the face by the machining of a tongue or lap joints

Maximum Spans for Various Roof Trusses

Maximum permissible spans for rafters for Fink trussed rafters

Basic size (mm)	Actual size (mm)	Pitch (degrees)								
		15 (m)	**17.5 (m)**	**20 (m)**	**22.5 (m)**	**25 (m)**	**27.5 (m)**	**30 (m)**	**32.5 (m)**	**35 (m)**
38 × 75	35 × 72	6.03	6.16	6.29	6.41	6.51	6.60	6.70	6.80	6.90
38 × 100	35 × 97	7.48	7.67	7.83	7.97	8.10	8.22	8.34	8.47	8.61
38 × 125	35 × 120	8.80	9.00	9.20	9.37	9.54	9.68	9.82	9.98	10.16
44 × 75	41 × 72	6.45	6.59	6.71	6.83	6.93	7.03	7.14	7.24	7.35
44 × 100	41 × 97	8.05	8.23	8.40	8.55	8.68	8.81	8.93	9.09	9.22
44 × 125	41 × 120	9.38	9.60	9.81	9.99	10.15	10.31	10.45	10.64	10.81
50 × 75	47 × 72	6.87	7.01	7.13	7.25	7.35	7.45	7.53	7.67	7.78
50 × 100	47 × 97	8.62	8.80	8.97	9.12	9.25	9.38	9.50	9.66	9.80
50 × 125	47 × 120	10.01	10.24	10.44	10.62	10.77	10.94	11.00	11.00	11.00

TIMBER

Sizes of Internal and External Doorsets

Description	Internal size (mm)	Permissible deviation	External size (mm)	Permissible deviation
Coordinating dimension: height of door leaf height sets	2100		2100	
Coordinating dimension: height of ceiling height set	2300 2350 2400 2700 3000		2300 2350 2400 2700 3000	
Coordinating dimension: width of all doorsets S = Single leaf set D = Double leaf set	600 S 700 S 800 S&D 900 S&D 1000 S&D 1200 D 1500 D 1800 D 2100 D		900 S 1000 S 1200 D 1800 D 2100 D	
Work size: height of door leaf height set	2090	± 2.0	2095	± 2.0
Work size: height of ceiling height set	2285 2335 2385 2685 2985	± 2.0	2295 2345 2395 2695 2995	± 2.0
Work size: width of all doorsets S = Single leaf set D = Double leaf set	590 S 690 S 790 S&D 890 S&D 990 S&D 1190 D 1490 D 1790 D 2090 D	± 2.0	895 S 995 S 1195 D 1495 D 1795 D 2095 D	± 2.0
Width of door leaf in single leaf sets F = Flush leaf P = Panel leaf	526 F 626 F 726 F&P 826 F&P 926 F&P	± 1.5	806 F&P 906 F&P	± 1.5
Width of door leaf in double leaf sets F = Flush leaf P = Panel leaf	362 F 412 F 426 F 562 F&P 712 F&P 826 F&P 1012 F&P	± 1.5	552 F&P 702 F&P 852 F&P 1002 F&P	± 1.5
Door leaf height for all doorsets	2040	± 1.5	1994	± 1.5

ROOFING

Total Roof Loadings for Various Types of Tiles/Slates

	Roof load (slope) kg/m^2		
	Slate/Tile	Roofing underlay and battens^2	Total dead load kg/m
Asbestos cement slate (600 × 300)	21.50	3.14	24.64
Clay tile interlocking	67.00	5.50	72.50
plain	43.50	2.87	46.37
Concrete tile interlocking	47.20	2.69	49.89
plain	78.20	5.50	83.70
Natural slate (18" × 10")	35.40	3.40	38.80
	Roof load (plan) kg/m^2		
Asbestos cement slate (600 × 300)	28.45	76.50	104.95
Clay tile interlocking	53.54	76.50	130.04
plain	83.71	76.50	60.21
Concrete tile interlocking	57.60	76.50	134.10
plain	96.64	76.50	173.14

ROOFING

Tiling Data

Product		Lap (mm)	Gauge of battens	No. slates per m²	Battens (m/m²)	Weight as laid (kg/m²)
CEMENT SLATES						
Eternit slates	600 × 300 mm	100	250	13.4	4.00	19.50
(Duracem)		90	255	13.1	3.92	19.20
		80	260	12.9	3.85	19.00
		70	265	12.7	3.77	18.60
	600 × 350 mm	100	250	11.5	4.00	19.50
		90	255	11.2	3.92	19.20
	500 × 250 mm	100	200	20.0	5.00	20.00
		90	205	19.5	4.88	19.50
		80	210	19.1	4.76	19.00
		70	215	18.6	4.65	18.60
	400 × 200 mm	90	155	32.3	6.45	20.80
		80	160	31.3	6.25	20.20
		70	165	30.3	6.06	19.60
CONCRETE TILES/SLATES						
Redland Roofing						
Stonewold slate	430 × 380 mm	75	355	8.2	2.82	51.20
Double Roman tile	418 × 330 mm	75	355	8.2	2.91	45.50
Grovebury pantile	418 × 332 mm	75	343	9.7	2.91	47.90
Norfolk pantile	381 × 227 mm	75	306	16.3	3.26	44.01
		100	281	17.8	3.56	48.06
Renown interlocking tile	418 × 330 mm	75	343	9.7	2.91	46.40
'49' tile	381 × 227 mm	75	306	16.3	3.26	44.80
		100	281	17.8	3.56	48.95
Plain, vertical tiling	265 × 165 mm	35	115	52.7	8.70	62.20
Marley Roofing						
Bold roll tile	420 × 330 mm	75	344	9.7	2.90	47.00
		100	–	10.5	3.20	51.00
Modern roof tile	420 × 330 mm	75	338	10.2	3.00	54.00
		100	–	11.0	3.20	58.00
Ludlow major	420 × 330 mm	75	338	10.2	3.00	45.00
		100	–	11.0	3.20	49.00
Ludlow plus	387 × 229 mm	75	305	16.1	3.30	47.00
		100	–	17.5	3.60	51.00
Mendip tile	420 × 330 mm	75	338	10.2	3.00	47.00
		100	–	11.0	3.20	51.00
Wessex	413 × 330 mm	75	338	10.2	3.00	54.00
		100	–	11.0	3.20	58.00
Plain tile	267 × 165 mm	65	100	60.0	10.00	76.00
		75	95	64.0	10.50	81.00
		85	90	68.0	11.30	86.00
Plain vertical tiles (feature)	267 × 165 mm	35	110	53.0	8.70	67.00
		34	115	56.0	9.10	71.00

ROOFING

Slate Nails, Quantity per Kilogram

Length	Type			
	Plain wire	Galvanized wire	Copper nail	Zinc nail
28.5 mm	325	305	325	415
34.4 mm	286	256	254	292
50.8 mm	242	224	194	200

Metal Sheet Coverings

Thicknesses and weights of sheet metal coverings								
Lead to BS 1178								
BS Code No	3	4	5	6	7	8		
Colour code	Green	Blue	Red	Black	White	Orange		
Thickness (mm)	1.25	1.80	2.24	2.50	3.15	3.55		
Density kg/m^2	14.18	20.41	25.40	30.05	35.72	40.26		
Copper to BS 2870								
Thickness (mm)		0.60	0.70					
Bay width								
Roll (mm)		500	650					
Seam (mm)		525	600					
Standard width to form bay	600	750						
Normal length of sheet	1.80	1.80						
Zinc to BS 849								
Zinc Gauge (Nr)	9	10	11	12	13	14	15	16
Thickness (mm)	0.43	0.48	0.56	0.64	0.71	0.79	0.91	1.04
Density (kg/m^2)	3.1	3.2	3.8	4.3	4.8	5.3	6.2	7.0
Aluminium to BS 4868								
Thickness (mm)	0.5	0.6	0.7	0.8	0.9	1.0	1.2	
Density (kg/m^2)	12.8	15.4	17.9	20.5	23.0	25.6	30.7	

ROOFING

Type of felt	Nominal mass per unit area (kg/10 m)	Nominal mass per unit area of fibre base (g/m^2)	Nominal length of roll (m)
Class 1			
1B fine granule	14	220	10 or 20
surfaced bitumen	18	330	10 or 20
	25	470	10
1E mineral surfaced bitumen	38	470	10
1F reinforced bitumen	15	160 (fibre) 110 (hessian)	15
1F reinforced bitumen, aluminium faced	13	160 (fibre) 110 (hessian)	15
Class 2			
2B fine granule surfaced bitumen asbestos	18	500	10 or 20
2E mineral surfaced bitumen asbestos	38	600	10
Class 3			
3B fine granule surfaced bitumen glass fibre	18	60	20
3E mineral surfaced bitumen glass fibre	28	60	10
3E venting base layer bitumen glass fibre	32	60*	10
3H venting base layer bitumen glass fibre	17	60*	20

* Excluding effect of perforations

GLAZING

GLAZING

Nominal thickness (mm)	Tolerance on thickness (mm)	Approximate weight (kg/m²)	Normal maximum size (mm)
Float and polished plate glass			
3	+ 0.2	7.50	2140 × 1220
4	+ 0.2	10.00	2760 × 1220
5	+ 0.2	12.50	3180 × 2100
6	+ 0.2	15.00	4600 × 3180
10	+ 0.3	25.00)	6000 × 3300
12	+ 0.3	30.00)	
15	+ 0.5	37.50	3050 × 3000
19	+ 1.0	47.50)	3000 × 2900
25	+ 1.0	63.50)	
Clear sheet glass			
2 *	+ 0.2	5.00	1920 × 1220
3	+ 0.3	7.50	2130 × 1320
4	+ 0.3	10.00	2760 × 1220
5 *	+ 0.3	12.50)	2130 × 2400
6 *	+ 0.3	15.00)	
Cast glass			
3	+ 0.4		
	− 0.2	6.00)	2140 × 1280
4	+ 0.5	7.50)	
5	+ 0.5	9.50	2140 × 1320
6	+ 0.5	11.50)	3700 × 1280
10	+ 0.8	21.50)	
Wired glass			
(Cast wired glass)			
6	+ 0.3	−)	3700 × 1840
	− 0.7	)	
7	+ 0.7	−)	
(Polished wire glass)			
6	+ 1.0	−	330 × 1830

* The 5 mm and 6 mm thickness are known as *thick drawn sheet*. Although 2 mm sheet glass is available it is not recommended for general glazing purposes

METAL

Weights of Metals

Material	kg/m³	lb/cu ft
Metals, steel construction, etc.		
Iron		
– cast	7207	450
– wrought	7687	480
– ore – general	2407	150
– (crushed) Swedish	3682	230
Steel	7854	490
Copper		
– cast	8731	545
– wrought	8945	558
Brass	8497	530
Bronze	8945	558
Aluminium	2774	173
Lead	11322	707
Zinc (rolled)	7140	446
	g/mm² per metre	**lb/sq ft per foot**
Steel bars	7.85	3.4
Structural steelwork	Net weight of member @ 7854 kg/m³	
riveted	+ 10% for cleats, rivets, bolts, etc.	
welded	+ 1.25% to 2.5% for welds, etc.	
Rolled sections		
beams	+ 2.5%	
stanchions	+ 5% (extra for caps and bases)	
Plate		
web girders	+ 10% for rivets or welds, stiffeners, etc.	
	kg/m	**lb/ft**
Steel stairs: industrial type		
1 m or 3 ft wide	84	56
Steel tubes		
50 mm or 2 in bore	5 to 6	3 to 4
Gas piping		
20 mm or ¾ in	2	1¼

METAL

Universal Beams BS 4: Part 1: 2005

Designation	Mass (kg/m)	Depth of section (mm)	Width of section (mm)	Thickness		Surface area (m²/m)
				Web (mm)	Flange (mm)	
1016 × 305 × 487	487.0	1036.1	308.5	30.0	54.1	3.20
1016 × 305 × 438	438.0	1025.9	305.4	26.9	49.0	3.17
1016 × 305 × 393	393.0	1016.0	303.0	24.4	43.9	3.15
1016 × 305 × 349	349.0	1008.1	302.0	21.1	40.0	3.13
1016 × 305 × 314	314.0	1000.0	300.0	19.1	35.9	3.11
1016 × 305 × 272	272.0	990.1	300.0	16.5	31.0	3.10
1016 × 305 × 249	249.0	980.2	300.0	16.5	26.0	3.08
1016 × 305 × 222	222.0	970.3	300.0	16.0	21.1	3.06
914 × 419 × 388	388.0	921.0	420.5	21.4	36.6	3.44
914 × 419 × 343	343.3	911.8	418.5	19.4	32.0	3.42
914 × 305 × 289	289.1	926.6	307.7	19.5	32.0	3.01
914 × 305 × 253	253.4	918.4	305.5	17.3	27.9	2.99
914 × 305 × 224	224.2	910.4	304.1	15.9	23.9	2.97
914 × 305 × 201	200.9	903.0	303.3	15.1	20.2	2.96
838 × 292 × 226	226.5	850.9	293.8	16.1	26.8	2.81
838 × 292 × 194	193.8	840.7	292.4	14.7	21.7	2.79
838 × 292 × 176	175.9	834.9	291.7	14.0	18.8	2.78
762 × 267 × 197	196.8	769.8	268.0	15.6	25.4	2.55
762 × 267 × 173	173.0	762.2	266.7	14.3	21.6	2.53
762 × 267 × 147	146.9	754.0	265.2	12.8	17.5	2.51
762 × 267 × 134	133.9	750.0	264.4	12.0	15.5	2.51
686 × 254 × 170	170.2	692.9	255.8	14.5	23.7	2.35
686 × 254 × 152	152.4	687.5	254.5	13.2	21.0	2.34
686 × 254 × 140	140.1	383.5	253.7	12.4	19.0	2.33
686 × 254 × 125	125.2	677.9	253.0	11.7	16.2	2.32
610 × 305 × 238	238.1	635.8	311.4	18.4	31.4	2.45
610 × 305 × 179	179.0	620.2	307.1	14.1	23.6	2.41
610 × 305 × 149	149.1	612.4	304.8	11.8	19.7	2.39
610 × 229 × 140	139.9	617.2	230.2	13.1	22.1	2.11
610 × 229 × 125	125.1	612.2	229.0	11.9	19.6	2.09
610 × 229 × 113	113.0	607.6	228.2	11.1	17.3	2.08
610 × 229 × 101	101.2	602.6	227.6	10.5	14.8	2.07
533 × 210 × 122	122.0	544.5	211.9	12.7	21.3	1.89
533 × 210 × 109	109.0	539.5	210.8	11.6	18.8	1.88
533 × 210 × 101	101.0	536.7	210.0	10.8	17.4	1.87
533 × 210 × 92	92.1	533.1	209.3	10.1	15.6	1.86
533 × 210 × 82	82.2	528.3	208.8	9.6	13.2	1.85
457 × 191 × 98	98.3	467.2	192.8	11.4	19.6	1.67
457 × 191 × 89	89.3	463.4	191.9	10.5	17.7	1.66
457 × 191 × 82	82.0	460.0	191.3	9.9	16.0	1.65
457 × 191 × 74	74.3	457.0	190.4	9.0	14.5	1.64
457 × 191 × 67	67.1	453.4	189.9	8.5	12.7	1.63
457 × 152 × 82	82.1	465.8	155.3	10.5	18.9	1.51
457 × 152 × 74	74.2	462.0	154.4	9.6	17.0	1.50
457 × 152 × 67	67.2	458.0	153.8	9.0	15.0	1.50
457 × 152 × 60	59.8	454.6	152.9	8.1	13.3	1.50
457 × 152 × 52	52.3	449.8	152.4	7.6	10.9	1.48
406 × 178 × 74	74.2	412.8	179.5	9.5	16.0	1.51
406 × 178 × 67	67.1	409.4	178.8	8.8	14.3	1.50
406 × 178 × 60	60.1	406.4	177.9	7.9	12.8	1.49

METAL

Designation	Mass (kg/m)	Depth of section (mm)	Width of section (mm)	Thickness		Surface area (m²/m)
				Web (mm)	Flange (mm)	
406 × 178 × 50	54.1	402.6	177.7	7.7	10.9	1.48
406 × 140 × 46	46.0	403.2	142.2	6.8	11.2	1.34
406 × 140 × 39	39.0	398.0	141.8	6.4	8.6	1.33
356 × 171 × 67	67.1	363.4	173.2	9.1	15.7	1.38
356 × 171 × 57	57.0	358.0	172.2	8.1	13.0	1.37
356 × 171 × 51	51.0	355.0	171.5	7.4	11.5	1.36
356 × 171 × 45	45.0	351.4	171.1	7.0	9.7	1.36
356 × 127 × 39	39.1	353.4	126.0	6.6	10.7	1.18
356 × 127 × 33	33.1	349.0	125.4	6.0	8.5	1.17
305 × 165 × 54	54.0	310.4	166.9	7.9	13.7	1.26
305 × 165 × 46	46.1	306.6	165.7	6.7	11.8	1.25
305 × 165 × 40	40.3	303.4	165.0	6.0	10.2	1.24
305 × 127 × 48	48.1	311.0	125.3	9.0	14.0	1.09
305 × 127 × 42	41.9	307.2	124.3	8.0	12.1	1.08
305 × 127 × 37	37.0	304.4	123.3	7.1	10.7	1.07
305 × 102 × 33	32.8	312.7	102.4	6.6	10.8	1.01
305 × 102 × 28	28.2	308.7	101.8	6.0	8.8	1.00
305 × 102 × 25	24.8	305.1	101.6	5.8	7.0	0.992
254 × 146 × 43	43.0	259.6	147.3	7.2	12.7	1.08
254 × 146 × 37	37.0	256.0	146.4	6.3	10.9	1.07
254 × 146 × 31	31.1	251.4	146.1	6.0	8.6	1.06
254 × 102 × 28	28.3	260.4	102.2	6.3	10.0	0.904
254 × 102 × 25	25.2	257.2	101.9	6.0	8.4	0.897
254 × 102 × 22	22.0	254.0	101.6	5.7	6.8	0.890
203 × 133 × 30	30.0	206.8	133.9	6.4	9.6	0.923
203 × 133 × 25	25.1	203.2	133.2	5.7	7.8	0.915
203 × 102 × 23	23.1	203.2	101.8	5.4	9.3	0.790
178 × 102 × 19	19.0	177.8	101.2	4.8	7.9	0.738
152 × 89 × 16	16.0	152.4	88.7	4.5	7.7	0.638
127 × 76 × 13	13.0	127.0	76.0	4.0	7.6	0.537

METAL

Universal Columns BS 4: Part 1: 2005

Designation	Mass (kg/m)	Depth of section (mm)	Width of section (mm)	Thickness		Surface area (m²/m)
				Web (mm)	Flange (mm)	
356 × 406 × 634	633.9	474.7	424.0	47.6	77.0	2.52
356 × 406 × 551	551.0	455.6	418.5	42.1	67.5	2.47
356 × 406 × 467	467.0	436.6	412.2	35.8	58.0	2.42
356 × 406 × 393	393.0	419.0	407.0	30.6	49.2	2.38
356 × 406 × 340	339.9	406.4	403.0	26.6	42.9	2.35
356 × 406 × 287	287.1	393.6	399.0	22.6	36.5	2.31
356 × 406 × 235	235.1	381.0	384.8	18.4	30.2	2.28
356 × 368 × 202	201.9	374.6	374.7	16.5	27.0	2.19
356 × 368 × 177	177.0	368.2	372.6	14.4	23.8	2.17
356 × 368 × 153	152.9	362.0	370.5	12.3	20.7	2.16
356 × 368 × 129	129.0	355.6	368.6	10.4	17.5	2.14
305 × 305 × 283	282.9	365.3	322.2	26.8	44.1	1.94
305 × 305 × 240	240.0	352.5	318.4	23.0	37.7	1.91
305 × 305 × 198	198.1	339.9	314.5	19.1	31.4	1.87
305 × 305 × 158	158.1	327.1	311.2	15.8	25.0	1.84
305 × 305 × 137	136.9	320.5	309.2	13.8	21.7	1.82
305 × 305 × 118	117.9	314.5	307.4	12.0	18.7	1.81
305 × 305 × 97	96.9	307.9	305.3	9.9	15.4	1.79
254 × 254 × 167	167.1	289.1	265.2	19.2	31.7	1.58
254 × 254 × 132	132.0	276.3	261.3	15.3	25.3	1.55
254 × 254 × 107	107.1	266.7	258.8	12.8	20.5	1.52
254 × 254 × 89	88.9	260.3	256.3	10.3	17.3	1.50
254 × 254 × 73	73.1	254.1	254.6	8.6	14.2	1.49
203 × 203 × 86	86.1	222.2	209.1	12.7	20.5	1.24
203 × 203 × 71	71.0	215.8	206.4	10.0	17.3	1.22
203 × 203 × 60	60.0	209.6	205.8	9.4	14.2	1.21
203 × 203 × 52	52.0	206.2	204.3	7.9	12.5	1.20
203 × 203 × 46	46.1	203.2	203.6	7.2	11.0	1.19
152 × 152 × 37	37.0	161.8	154.4	8.0	11.5	0.912
152 × 152 × 30	30.0	157.6	152.9	6.5	9.4	0.901
152 × 152 × 23	23.0	152.4	152.2	5.8	6.8	0.889

METAL

Joists BS 4: Part 1: 2005 (retained for reference, Corus have ceased manufacture in UK)

Designation	Mass (kg/m)	Depth of section (mm)	Width of section (mm)	Thickness		Surface area (m²/m)
				Web (mm)	Flange (mm)	
254 × 203 × 82	82.0	254.0	203.2	10.2	19.9	1.210
203 × 152 × 52	52.3	203.2	152.4	8.9	16.5	0.932
152 × 127 × 37	37.3	152.4	127.0	10.4	13.2	0.737
127 × 114 × 29	29.3	127.0	114.3	10.2	11.5	0.646
127 × 114 × 27	26.9	127.0	114.3	7.4	11.4	0.650
102 × 102 × 23	23.0	101.6	101.6	9.5	10.3	0.549
102 × 44 × 7	7.5	101.6	44.5	4.3	6.1	0.350
89 × 89 × 19	19.5	88.9	88.9	9.5	9.9	0.476
76 × 76 × 13	12.8	76.2	76.2	5.1	8.4	0.411

Parallel Flange Channels

Designation	Mass (kg/m)	Depth of section (mm)	Width of section (mm)	Thickness		Surface area (m²/m)
				Web (mm)	Flange (mm)	
430 × 100 × 64	64.4	430	100	11.0	19.0	1.23
380 × 100 × 54	54.0	380	100	9.5	17.5	1.13
300 × 100 × 46	45.5	300	100	9.0	16.5	0.969
300 × 90 × 41	41.4	300	90	9.0	15.5	0.932
260 × 90 × 35	34.8	260	90	8.0	14.0	0.854
260 × 75 × 28	27.6	260	75	7.0	12.0	0.79
230 × 90 × 32	32.2	230	90	7.5	14.0	0.795
230 × 75 × 26	25.7	230	75	6.5	12.5	0.737
200 × 90 × 30	29.7	200	90	7.0	14.0	0.736
200 × 75 × 23	23.4	200	75	6.0	12.5	0.678
180 × 90 × 26	26.1	180	90	6.5	12.5	0.697
180 × 75 × 20	20.3	180	75	6.0	10.5	0.638
150 × 90 × 24	23.9	150	90	6.5	12.0	0.637
150 × 75 × 18	17.9	150	75	5.5	10.0	0.579
125 × 65 × 15	14.8	125	65	5.5	9.5	0.489
100 × 50 × 10	10.2	100	50	5.0	8.5	0.382

METAL

Equal Angles BS EN 10056-1

Designation	Mass (kg/m)	Surface area (m²/m)
200 × 200 × 24	71.1	0.790
200 × 200 × 20	59.9	0.790
200 × 200 × 18	54.2	0.790
200 × 200 × 16	48.5	0.790
150 × 150 × 18	40.1	0.59
150 × 150 × 15	33.8	0.59
150 × 150 × 12	27.3	0.59
150 × 150 × 10	23.0	0.59
120 × 120 × 15	26.6	0.47
120 × 120 × 12	21.6	0.47
120 × 120 × 10	18.2	0.47
120 × 120 × 8	14.7	0.47
100 × 100 × 15	21.9	0.39
100 × 100 × 12	17.8	0.39
100 × 100 × 10	15.0	0.39
100 × 100 × 8	12.2	0.39
90 × 90 × 12	15.9	0.35
90 × 90 × 10	13.4	0.35
90 × 90 × 8	10.9	0.35
90 × 90 × 7	9.61	0.35
90 × 90 × 6	8.30	0.35

Unequal Angles BS EN 10056-1

Designation	Mass (kg/m)	Surface area (m²/m)
200 × 150 × 18	47.1	0.69
200 × 150 × 15	39.6	0.69
200 × 150 × 12	32.0	0.69
200 × 100 × 15	33.7	0.59
200 × 100 × 12	27.3	0.59
200 × 100 × 10	23.0	0.59
150 × 90 × 15	26.6	0.47
150 × 90 × 12	21.6	0.47
150 × 90 × 10	18.2	0.47
150 × 75 × 15	24.8	0.44
150 × 75 × 12	20.2	0.44
150 × 75 × 10	17.0	0.44
125 × 75 × 12	17.8	0.40
125 × 75 × 10	15.0	0.40
125 × 75 × 8	12.2	0.40
100 × 75 × 12	15.4	0.34
100 × 75 × 10	13.0	0.34
100 × 75 × 8	10.6	0.34
100 × 65 × 10	12.3	0.32
100 × 65 × 8	9.94	0.32
100 × 65 × 7	8.77	0.32

METAL

Structural Tees Split from Universal Beams BS 4: Part 1: 2005

Designation	Mass (kg/m)	Surface area (m²/m)
305 × 305 × 90	89.5	1.22
305 × 305 × 75	74.6	1.22
254 × 343 × 63	62.6	1.19
229 × 305 × 70	69.9	1.07
229 × 305 × 63	62.5	1.07
229 × 305 × 57	56.5	1.07
229 × 305 × 51	50.6	1.07
210 × 267 × 61	61.0	0.95
210 × 267 × 55	54.5	0.95
210 × 267 × 51	50.5	0.95
210 × 267 × 46	46.1	0.95
210 × 267 × 41	41.1	0.95
191 × 229 × 49	49.2	0.84
191 × 229 × 45	44.6	0.84
191 × 229 × 41	41.0	0.84
191 × 229 × 37	37.1	0.84
191 × 229 × 34	33.6	0.84
152 × 229 × 41	41.0	0.76
152 × 229 × 37	37.1	0.76
152 × 229 × 34	33.6	0.76
152 × 229 × 30	29.9	0.76
152 × 229 × 26	26.2	0.76

Universal Bearing Piles BS 4: Part 1: 2005

Designation	Mass (kg/m)	Depth of Section (mm)	Width of section (mm)	Thickness	
				Web (mm)	Flange (mm)
356 × 368 × 174	173.9	361.4	378.5	20.3	20.4
356 × 368 × 152	152.0	356.4	376.0	17.8	17.9
356 × 368 × 133	133.0	352.0	373.8	15.6	15.7
356 × 368 × 109	108.9	346.4	371.0	12.8	12.9
305 × 305 × 223	222.9	337.9	325.7	30.3	30.4
305 × 305 × 186	186.0	328.3	320.9	25.5	25.6
305 × 305 × 149	149.1	318.5	316.0	20.6	20.7
305 × 305 × 126	126.1	312.3	312.9	17.5	17.6
305 × 305 × 110	110.0	307.9	310.7	15.3	15.4
305 × 305 × 95	94.9	303.7	308.7	13.3	13.3
305 × 305 × 88	88.0	301.7	307.8	12.4	12.3
305 × 305 × 79	78.9	299.3	306.4	11.0	11.1
254 × 254 × 85	85.1	254.3	260.4	14.4	14.3
254 × 254 × 71	71.0	249.7	258.0	12.0	12.0
254 × 254 × 63	63.0	247.1	256.6	10.6	10.7
203 × 203 × 54	53.9	204.0	207.7	11.3	11.4
203 × 203 × 45	44.9	200.2	205.9	9.5	9.5

METAL

Hot Formed Square Hollow Sections EN 10210 S275J2H & S355J2H

Size (mm)	Wall thickness (mm)	Mass (kg/m)	Superficial area (m²/m)
40 × 40	2.5	2.89	0.154
	3.0	3.41	0.152
	3.2	3.61	0.152
	3.6	4.01	0.151
	4.0	4.39	0.150
	5.0	5.28	0.147
50 × 50	2.5	3.68	0.194
	3.0	4.35	0.192
	3.2	4.62	0.192
	3.6	5.14	0.191
	4.0	5.64	0.190
	5.0	6.85	0.187
	6.0	7.99	0.185
	6.3	8.31	0.184
60 × 60	3.0	5.29	0.232
	3.2	5.62	0.232
	3.6	6.27	0.231
	4.0	6.90	0.230
	5.0	8.42	0.227
	6.0	9.87	0.225
	6.3	10.30	0.224
	8.0	12.50	0.219
70 × 70	3.0	6.24	0.272
	3.2	6.63	0.272
	3.6	7.40	0.271
	4.0	8.15	0.270
	5.0	9.99	0.267
	6.0	11.80	0.265
	6.3	12.30	0.264
	8.0	15.00	0.259
80 × 80	3.2	7.63	0.312
	3.6	8.53	0.311
	4.0	9.41	0.310
	5.0	11.60	0.307
	6.0	13.60	0.305
	6.3	14.20	0.304
	8.0	17.50	0.299
90 × 90	3.6	9.66	0.351
	4.0	10.70	0.350
	5.0	13.10	0.347
	6.0	15.50	0.345
	6.3	16.20	0.344
	8.0	20.10	0.339
100 × 100	3.6	10.80	0.391
	4.0	11.90	0.390
	5.0	14.70	0.387
	6.0	17.40	0.385
	6.3	18.20	0.384
	8.0	22.60	0.379
	10.0	27.40	0.374
120 × 120	4.0	14.40	0.470
	5.0	17.80	0.467
	6.0	21.20	0.465

METAL

Size (mm)	Wall thickness (mm)	Mass (kg/m)	Superficial area (m²/m)
	6.3	22.20	0.464
	8.0	27.60	0.459
	10.0	33.70	0.454
	12.0	39.50	0.449
	12.5	40.90	0.448
140 × 140	5.0	21.00	0.547
	6.0	24.90	0.545
	6.3	26.10	0.544
	8.0	32.60	0.539
	10.0	40.00	0.534
	12.0	47.00	0.529
	12.5	48.70	0.528
150 × 150	5.0	22.60	0.587
	6.0	26.80	0.585
	6.3	28.10	0.584
	8.0	35.10	0.579
	10.0	43.10	0.574
	12.0	50.80	0.569
	12.5	52.70	0.568
Hot formed from seamless hollow	16.0	65.2	0.559
160 × 160	5.0	24.10	0.627
	6.0	28.70	0.625
	6.3	30.10	0.624
	8.0	37.60	0.619
	10.0	46.30	0.614
	12.0	54.60	0.609
	12.5	56.60	0.608
	16.0	70.20	0.599
180 × 180	5.0	27.30	0.707
	6.0	32.50	0.705
	6.3	34.00	0.704
	8.0	42.70	0.699
	10.0	52.50	0.694
	12.0	62.10	0.689
	12.5	64.40	0.688
	16.0	80.20	0.679
200 × 200	5.0	30.40	0.787
	6.0	36.20	0.785
	6.3	38.00	0.784
	8.0	47.70	0.779
	10.0	58.80	0.774
	12.0	69.60	0.769
	12.5	72.30	0.768
	16.0	90.30	0.759
250 × 250	5.0	38.30	0.987
	6.0	45.70	0.985
	6.3	47.90	0.984
	8.0	60.30	0.979
	10.0	74.50	0.974
	12.0	88.50	0.969
	12.5	91.90	0.968
	16.0	115.00	0.959

METAL

Size (mm)	Wall thickness (mm)	Mass (kg/m)	Superficial area (m²/m)
300 × 300	6.0	55.10	1.18
	6.3	57.80	1.18
	8.0	72.80	1.18
	10.0	90.20	1.17
	12.0	107.00	1.17
	12.5	112.00	1.17
	16.0	141.00	1.16
350 × 350	8.0	85.40	1.38
	10.0	106.00	1.37
	12.0	126.00	1.37
	12.5	131.00	1.37
	16.0	166.00	1.36
400 × 400	8.0	97.90	1.58
	10.0	122.00	1.57
	12.0	145.00	1.57
	12.5	151.00	1.57
	16.0	191.00	1.56
(Grade S355J2H only)	20.00*	235.00	1.55

Note: * SAW process

METAL

Hot Formed Square Hollow Sections JUMBO RHS: JIS G3136

Size (mm)	Wall thickness (mm)	Mass (kg/m)	Superficial area (m²/m)
350 × 350	19.0	190.00	1.33
	22.0	217.00	1.32
	25.0	242.00	1.31
400 × 400	22.0	251.00	1.52
	25.0	282.00	1.51
450 × 450	12.0	162.00	1.76
	16.0	213.00	1.75
	19.0	250.00	1.73
	22.0	286.00	1.72
	25.0	321.00	1.71
	28.0 *	355.00	1.70
	32.0 *	399.00	1.69
500 × 500	12.0	181.00	1.96
	16.0	238.00	1.95
	19.0	280.00	1.93
	22.0	320.00	1.92
	25.0	360.00	1.91
	28.0 *	399.00	1.90
	32.0 *	450.00	1.89
	36.0 *	498.00	1.88
550 × 550	16.0	263.00	2.15
	19.0	309.00	2.13
	22.0	355.00	2.12
	25.0	399.00	2.11
	28.0 *	443.00	2.10
	32.0 *	500.00	2.09
	36.0 *	555.00	2.08
	40.0 *	608.00	2.06
600 × 600	25.0 *	439.00	2.31
	28.0 *	487.00	2.30
	32.0 *	550.00	2.29
	36.0 *	611.00	2.28
	40.0 *	671.00	2.26
700 × 700	25.0 *	517.00	2.71
	28.0 *	575.00	2.70
	32.0 *	651.00	2.69
	36.0 *	724.00	2.68
	40.0 *	797.00	2.68

Note: * SAW process

METAL

Hot Formed Rectangular Hollow Sections: EN10210 S275J2h & S355J2H

Size (mm)	Wall thickness (mm)	Mass (kg/m)	Superficial area (m²/m)
50 × 30	2.5	2.89	0.154
	3.0	3.41	0.152
	3.2	3.61	0.152
	3.6	4.01	0.151
	4.0	4.39	0.150
	5.0	5.28	0.147
60 × 40	2.5	3.68	0.194
	3.0	4.35	0.192
	3.2	4.62	0.192
	3.6	5.14	0.191
	4.0	5.64	0.190
	5.0	6.85	0.187
	6.0	7.99	0.185
	6.3	8.31	0.184
80 × 40	3.0	5.29	0.232
	3.2	5.62	0.232
	3.6	6.27	0.231
	4.0	6.90	0.230
	5.0	8.42	0.227
	6.0	9.87	0.225
	6.3	10.30	0.224
	8.0	12.50	0.219
76.2 × 50.8	3.0	5.62	0.246
	3.2	5.97	0.246
	3.6	6.66	0.245
	4.0	7.34	0.244
	5.0	8.97	0.241
	6.0	10.50	0.239
	6.3	11.00	0.238
	8.0	13.40	0.233
90 × 50	3.0	6.24	0.272
	3.2	6.63	0.272
	3.6	7.40	0.271
	4.0	8.15	0.270
	5.0	9.99	0.267
	6.0	11.80	0.265
	6.3	12.30	0.264
	8.0	15.00	0.259
100 × 50	3.0	6.71	0.292
	3.2	7.13	0.292
	3.6	7.96	0.291
	4.0	8.78	0.290
	5.0	10.80	0.287
	6.0	12.70	0.285
	6.3	13.30	0.284
	8.0	16.30	0.279

METAL

Size (mm)	Wall thickness (mm)	Mass (kg/m)	Superficial area (m²/m)
100 × 60	3.0	7.18	0.312
	3.2	7.63	0.312
	3.6	8.53	0.311
	4.0	9.41	0.310
	5.0	11.60	0.307
	6.0	13.60	0.305
	6.3	14.20	0.304
	8.0	17.50	0.299
120 × 60	3.6	9.70	0.351
	4.0	10.70	0.350
	5.0	13.10	0.347
	6.0	15.50	0.345
	6.3	16.20	0.344
	8.0	20.10	0.339
120 × 80	3.6	10.80	0.391
	4.0	11.90	0.390
	5.0	14.70	0.387
	6.0	17.40	0.385
	6.3	18.20	0.384
	8.0	22.60	0.379
	10.0	27.40	0.374
150 × 100	4.0	15.10	0.490
	5.0	18.60	0.487
	6.0	22.10	0.485
	6.3	23.10	0.484
	8.0	28.90	0.479
	10.0	35.30	0.474
	12.0	41.40	0.469
	12.5	42.80	0.468
160 × 80	4.0	14.40	0.470
	5.0	17.80	0.467
	6.0	21.20	0.465
	6.3	22.20	0.464
	8.0	27.60	0.459
	10.0	33.70	0.454
	12.0	39.50	0.449
	12.5	40.90	0.448
200 × 100	5.0	22.60	0.587
	6.0	26.80	0.585
	6.3	28.10	0.584
	8.0	35.10	0.579
	10.0	43.10	0.574
	12.0	50.80	0.569
	12.5	52.70	0.568
	16.0	65.20	0.559
250 × 150	5.0	30.40	0.787
	6.0	36.20	0.785
	6.3	38.00	0.784
	8.0	47.70	0.779
	10.0	58.80	0.774
	12.0	69.60	0.769
	12.5	72.30	0.768
	16.0	90.30	0.759

METAL

Size (mm)	Wall thickness (mm)	Mass (kg/m)	Superficial area (m²/m)
300 × 200	5.0	38.30	0.987
	6.0	45.70	0.985
	6.3	47.90	0.984
	8.0	60.30	0.979
	10.0	74.50	0.974
	12.0	88.50	0.969
	12.5	91.90	0.968
	16.0	115.00	0.959
400 × 200	6.0	55.10	1.18
	6.3	57.80	1.18
	8.0	72.80	1.18
	10.0	90.20	1.17
	12.0	107.00	1.17
	12.5	112.00	1.17
	16.0	141.00	1.16
450 × 250	8.0	85.40	1.38
	10.0	106.00	1.37
	12.0	126.00	1.37
	12.5	131.00	1.37
	16.0	166.00	1.36
500 × 300	8.0	98.00	1.58
	10.0	122.00	1.57
	12.0	145.00	1.57
	12.5	151.00	1.57
	16.0	191.00	1.56
	20.0	235.00	1.55

METAL

Hot Formed Circular Hollow Sections EN 10210 S275J2H & S355J2H

Outside diameter (mm)	Wall thickness (mm)	Mass (kg/m)	Superficial area (m²/m)
21.3	3.2	1.43	0.067
26.9	3.2	1.87	0.085
33.7	3.0	2.27	0.106
	3.2	2.41	0.106
	3.6	2.67	0.106
	4.0	2.93	0.106
42.4	3.0	2.91	0.133
	3.2	3.09	0.133
	3.6	3.44	0.133
	4.0	3.79	0.133
48.3	2.5	2.82	0.152
	3.0	3.35	0.152
	3.2	3.56	0.152
	3.6	3.97	0.152
	4.0	4.37	0.152
	5.0	5.34	0.152
60.3	2.5	3.56	0.189
	3.0	4.24	0.189
	3.2	4.51	0.189
	3.6	5.03	0.189
	4.0	5.55	0.189
	5.0	6.82	0.189
76.1	2.5	4.54	0.239
	3.0	5.41	0.239
	3.2	5.75	0.239
	3.6	6.44	0.239
	4.0	7.11	0.239
	5.0	8.77	0.239
	6.0	10.40	0.239
	6.3	10.80	0.239
88.9	2.5	5.33	0.279
	3.0	6.36	0.279
	3.2	6.76	0.27
	3.6	7.57	0.279
	4.0	8.38	0.279
	5.0	10.30	0.279
	6.0	12.30	0.279
	6.3	12.80	0.279
114.3	3.0	8.23	0.359
	3.2	8.77	0.359
	3.6	9.83	0.359
	4.0	10.09	0.359
	5.0	13.50	0.359
	6.0	16.00	0.359
	6.3	16.80	0.359

METAL

Outside diameter (mm)	Wall thickness (mm)	Mass (kg/m)	Superficial area (m²/m)
139.7	3.2	10.80	0.439
	3.6	12.10	0.439
	4.0	13.40	0.439
	5.0	16.60	0.439
	6.0	19.80	0.439
	6.3	20.70	0.439
	8.0	26.00	0.439
	10.0	32.00	0.439
168.3	3.2	13.00	0.529
	3.6	14.60	0.529
	4.0	16.20	0.529
	5.0	20.10	0.529
	6.0	24.00	0.529
	6.3	25.20	0.529
	8.0	31.60	0.529
	10.0	39.00	0.529
	12.0	46.30	0.529
	12.5	48.00	0.529
193.7	5.0	23.30	0.609
	6.0	27.80	0.609
	6.3	29.10	0.609
	8.0	36.60	0.609
	10.0	45.30	0.609
	12.0	53.80	0.609
	12.5	55.90	0.609
219.1	5.0	26.40	0.688
	6.0	31.50	0.688
	6.3	33.10	0.688
	8.0	41.60	0.688
	10.0	51.60	0.688
	12.0	61.30	0.688
	12.5	63.70	0.688
	16.0	80.10	0.688
244.5	5.0	29.50	0.768
	6.0	35.30	0.768
	6.3	37.00	0.768
	8.0	46.70	0.768
	10.0	57.80	0.768
	12.0	68.80	0.768
	12.5	71.50	0.768
	16.0	90.20	0.768
273.0	5.0	33.00	0.858
	6.0	39.50	0.858
	6.3	41.40	0.858
	8.0	52.30	0.858
	10.0	64.90	0.858
	12.0	77.20	0.858
	12.5	80.30	0.858
	16.0	101.00	0.858
323.9	5.0	39.30	1.02
	6.0	47.00	1.02
	6.3	49.30	1.02
	8.0	62.30	1.02
	10.0	77.40	1.02

METAL

Outside diameter (mm)	Wall thickness (mm)	Mass (kg/m)	Superficial area (m²/m)
	12.0	92.30	1.02
	12.5	96.00	1.02
	16.0	121.00	1.02
355.6	6.3	54.30	1.12
	8.0	68.60	1.12
	10.0	85.30	1.12
	12.0	102.00	1.12
	12.5	106.00	1.12
	16.0	134.00	1.12
406.4	6.3	62.20	1.28
	8.0	79.60	1.28
	10.0	97.80	1.28
	12.0	117.00	1.28
	12.5	121.00	1.28
	16.0	154.00	1.28
457.0	6.3	70.00	1.44
	8.0	88.60	1.44
	10.0	110.00	1.44
	12.0	132.00	1.44
	12.5	137.00	1.44
	16.0	174.00	1.44
508.0	6.3	77.90	1.60
	8.0	98.60	1.60
	10.0	123.00	1.60
	12.0	147.00	1.60
	12.5	153.00	1.60
	16.0	194.00	1.60

METAL

Spacing of Holes in Angles

Nominal leg length (mm)	Spacing of holes						Maximum diameter of bolt or rivet		
	A	B	C	D	E	F	A	B and C	D, E and F
200		75	75	55	55	55		30	20
150		55	55					20	
125		45	60					20	
120									
100	55						24		
90	50						24		
80	45						20		
75	45						20		
70	40						20		
65	35						20		
60	35						16		
50	28						12		
45	25								
40	23								
30	20								
25	15								

KERBS, PAVING, ETC.

KERBS/EDGINGS/CHANNELS

Precast Concrete Kerbs to BS 7263

Straight kerb units: length from 450 to 915 mm

150 mm high × 125 mm thick		
bullnosed	type BN	
half battered	type HB3	
255 mm high × 125 mm thick		
45° splayed	type SP	
half battered	type HB2	
305 mm high × 150 mm thick		
half battered	type HB1	
Quadrant kerb units		
150 mm high × 305 and 455 mm radius to match	type BN	type QBN
150 mm high × 305 and 455 mm radius to match	type HB2, HB3	type QHB
150 mm high × 305 and 455 mm radius to match	type SP	type QSP
255 mm high × 305 and 455 mm radius to match	type BN	type QBN
255 mm high × 305 and 455 mm radius to match	type HB2, HB3	type QHB
225 mm high × 305 and 455 mm radius to match	type SP	type QSP
Angle kerb units		
305 × 305 × 225 mm high × 125 mm thick		
bullnosed external angle	type XA	
splayed external angle to match type SP	type XA	
bullnosed internal angle	type IA	
splayed internal angle to match type SP	type IA	
Channels		
255 mm wide × 125 mm high flat	type CS1	
150 mm wide × 125 mm high flat type	CS2	
255 mm wide × 125 mm high dished	type CD	

KERBS, PAVING, ETC.

Transition kerb units			
from kerb type SP to HB	left handed	type TL	
	right handed	type TR	
from kerb type BN to HB	left handed	type DL1	
	right handed	type DR1	
from kerb type BN to SP	left handed	type DL2	
	right handed	type DR2	

Number of kerbs required per quarter circle (780 mm kerb lengths)

Radius (m)	Number in quarter circle
12	24
10	20
8	16
6	12
5	10
4	8
3	6
2	4
1	2

Precast Concrete Edgings

Round top type ER	Flat top type EF	Bullnosed top type EBN
150 × 50 mm	150 × 50 mm	150 × 50 mm
200 × 50 mm	200 × 50 mm	200 × 50 mm
250 × 50 mm	250 × 50 mm	250 × 50 mm

KERBS, PAVING, ETC.

BASES

Cement Bound Material for Bases and Subbases

CBM1:	very carefully graded aggregate from 37.5–75 mm, with a 7-day strength of 4.5 N/mm^2
CBM2:	same range of aggregate as CBM1 but with more tolerance in each size of aggregate with a 7-day strength of 7.0 N/mm^2
CBM3:	crushed natural aggregate or blast furnace slag, graded from 37.5–150 mm for 40 mm aggregate, and from 20–75 mm for 20 mm aggregate, with a 7-day strength of 10 N/mm^2
CBM4:	crushed natural aggregate or blast furnace slag, graded from 37.5–150 mm for 40 mm aggregate, and from 20–75 mm for 20 mm aggregate, with a 7-day strength of 15 N/mm^2

INTERLOCKING BRICK/BLOCK ROADS/PAVINGS

Sizes of Precast Concrete Paving Blocks

Type R blocks
200 × 100 × 60 mm
200 × 100 × 65 mm
200 × 100 × 80 mm
200 × 100 × 100 mm

Type S
Any shape within a 295 mm space

Sizes of clay brick pavers
200 × 100 × 50 mm
200 × 100 × 65 mm
210 × 105 × 50 mm
210 × 105 × 65 mm
215 × 102.5 × 50 mm
215 × 102.5 × 65 mm

Type PA: 3 kN
Footpaths and pedestrian areas, private driveways, car parks, light vehicle traffic and over-run

Type PB: 7 kN
Residential roads, lorry parks, factory yards, docks, petrol station forecourts, hardstandings, bus stations

KERBS, PAVING, ETC.

PAVING AND SURFACING

Weights and Sizes of Paving and Surfacing

Description of item	Size	Quantity per tonne
Paving 50 mm thick	900 × 600 mm	15
Paving 50 mm thick	750 × 600 mm	18
Paving 50 mm thick	600 × 600 mm	23
Paving 50 mm thick	450 × 600 mm	30
Paving 38 mm thick	600 × 600 mm	30
Path edging	914 × 50 × 150 mm	60
Kerb (including radius and tapers)	125 × 254 × 914 mm	15
Kerb (including radius and tapers)	125 × 150 × 914 mm	25
Square channel	125 × 254 × 914 mm	15
Dished channel	125 × 254 × 914 mm	15
Quadrants	300 × 300 × 254 mm	19
Quadrants	450 × 450 × 254 mm	12
Quadrants	300 × 300 × 150 mm	30
Internal angles	300 × 300 × 254 mm	30
Fluted pavement channel	255 × 75 × 914 mm	25
Corner stones	300 × 300 mm	80
Corner stones	360 × 360 mm	60
Cable covers	914 × 175 mm	55
Gulley kerbs	220 × 220 × 150 mm	60
Gulley kerbs	220 × 200 × 75 mm	120

KERBS, PAVING, ETC.

Weights and Sizes of Paving and Surfacing

Material	kg/m³	lb/cu yd
Tarmacadam	2306	3891
Macadam (waterbound)	2563	4325
Vermiculite (aggregate)	64–80	108–135
Terracotta	2114	3568
Cork – compressed	388	24
	kg/m²	lb/sq ft
Clay floor tiles, 12.7 mm	27.3	5.6
Pavement lights	122	25
Damp-proof course	5	1
	kg/m² per mm thickness	lb/sq ft per inch thickness
Paving slabs (stone)	2.3	12
Granite setts	2.88	15
Asphalt	2.30	12
Rubber flooring	1.68	9
Polyvinyl chloride	1.94 (avg)	10 (avg)

Coverage (m²) Per Cubic Metre of Materials Used as Subbases or Capping Layers

Consolidated thickness laid in (mm)	Square metre coverage		
	Gravel	Sand	Hardcore
50	15.80	16.50	–
75	10.50	11.00	–
100	7.92	8.20	7.42
125	6.34	6.60	5.90
150	5.28	5.50	4.95
175	–	–	4.23
200	–	–	3.71
225	–	–	3.30
300	–	–	2.47

KERBS, PAVING, ETC.

Approximate Rate of Spreads

Average thickness of course (mm)	Description	Approximate rate of spread			
		Open Textured		Dense, Medium & Fine Textured	
		(kg/m²)	(m²/t)	(kg/m²)	(m²/t)
35	14 mm open textured or dense wearing course	60–75	13–17	70–85	12–14
40	20 mm open textured or dense base course	70–85	12–14	80–100	10–12
45	20 mm open textured or dense base course	80–100	10–12	95–100	9–10
50	20 mm open textured or dense, or 28 mm dense base course	85–110	9–12	110–120	8–9
60	28 mm dense base course, 40 mm open textured of dense base course or 40 mm single course as base course		8–10	130–150	7–8
65	28 mm dense base course, 40 mm open textured or dense base course or 40 mm single course	100–135	7–10	140–160	6–7
75	40 mm single course, 40 mm open textured or dense base course, 40 mm dense roadbase	120–150	7–8	165–185	5–6
100	40 mm dense base course or roadbase	–	–	220–240	4–4.5

KERBS, PAVING, ETC.

Surface Dressing Roads: Coverage (m²) per Tonne of Material

Size in mm	Sand	Granite chips	Gravel	Limestone chips
Sand	168	–	–	–
3	–	148	152	165
6	–	130	133	144
9	–	111	114	123
13	–	85	87	95
19	–	68	71	78

Sizes of Flags

Reference	Nominal size (mm)	Thickness (mm)
A	600 × 450	50 and 63
B	600 × 600	50 and 63
C	600 × 750	50 and 63
D	600 × 900	50 and 63
E	450 × 450	50 and 70 chamfered top surface
F	400 × 400	50 and 65 chamfered top surface
G	300 × 300	50 and 60 chamfered top surface

Sizes of Natural Stone Setts

Width (mm)		Length (mm)		Depth (mm)
100	×	100	×	100
75	×	150 to 250	×	125
75	×	150 to 250	×	150
100	×	150 to 250	×	100
100	×	150 to 250	×	150

SEEDING/TURFING AND PLANTING

Topsoil Quality

Topsoil grade	Properties
Premium	Natural topsoil, high fertility, loamy texture, good soil structure, suitable for intensive cultivation.
General purpose	Natural or manufactured topsoil of lesser quality than Premium, suitable for agriculture or amenity landscape, may need fertilizer or soil structure improvement.
Economy	Selected subsoil, natural mineral deposit such as river silt or greensand. The grade comprises two subgrades; 'Low clay' and 'High clay' which is more liable to compaction in handling. This grade is suitable for low-production agricultural land and amenity woodland or conservation planting areas.

Forms of Trees

Standards:	Shall be clear with substantially straight stems. Grafted and budded trees shall have no more than a slight bend at the union. Standards shall be designated as Half, Extra light, Light, Standard, Selected standard, Heavy, and Extra heavy.
Sizes of Standards	
Heavy standard	12–14 cm girth × 3.50 to 5.00 m high
Extra Heavy standard	14–16 cm girth × 4.25 to 5.00 m high
Extra Heavy standard	16–18 cm girth × 4.25 to 6.00 m high
Extra Heavy standard	18–20 cm girth × 5.00 to 6.00 m high
Semi-mature trees:	Between 6.0 m and 12.0 m tall with a girth of 20 to 75 cm at 1.0 m above ground.
Feathered trees:	Shall have a defined upright central leader, with stem furnished with evenly spread and balanced lateral shoots down to or near the ground.
Whips:	Shall be without significant feather growth as determined by visual inspection.
Multi-stemmed trees:	Shall have two or more main stems at, near, above or below ground.

Seedlings grown from seed and not transplanted shall be specified when ordered for sale as:

1+0	one year old seedling
2+0	two year old seedling
1+1	one year seed bed, one year transplanted = two year old seedling
1+2	one year seed bed, two years transplanted = three year old seedling
2+1	two years seed bed, one year transplanted = three year old seedling
1u1	two years seed bed, undercut after 1 year = two year old seedling
2u2	four years seed bed, undercut after 2 years = four year old seedling

SEEDING/TURFING AND PLANTING

Cuttings

The age of cuttings (plants grown from shoots, stems, or roots of the mother plant) shall be specified when ordered for sale. The height of transplants and undercut seedlings/cuttings (which have been transplanted or undercut at least once) shall be stated in centimetres. The number of growing seasons before and after transplanting or undercutting shall be stated.

0 + 1	one year cutting
0 + 2	two year cutting
0 + 1 + 1	one year cutting bed, one year transplanted = two year old seedling
0 + 1 + 2	one year cutting bed, two years transplanted = three year old seedling

Grass Cutting Capacities in m² per hour

Speed mph	Width of cut in metres												
	0.5	0.7	1.0	1.2	1.5	1.7	2.0	2.0	2.1	2.5	2.8	3.0	3.4
1.0	724	1127	1529	1931	2334	2736	3138	3219	3380	4023	4506	4828	5472
1.5	1086	1690	2293	2897	3500	4104	4707	4828	5069	6035	6759	7242	8208
2.0	1448	2253	3058	3862	4667	5472	6276	6437	6759	8047	9012	9656	10944
2.5	1811	2816	3822	4828	5834	6840	7846	8047	8449	10058	11265	12070	13679
3.0	2173	3380	4587	5794	7001	8208	9415	9656	10139	12070	13518	14484	16415
3.5	2535	3943	5351	6759	8167	9576	10984	11265	11829	14082	15772	16898	19151
4.0	2897	4506	6115	7725	9334	10944	12553	12875	13518	16093	18025	19312	21887
4.5	3259	5069	6880	8690	10501	12311	14122	14484	15208	18105	20278	21726	24623
5.0	3621	5633	7644	9656	11668	13679	15691	16093	16898	20117	22531	24140	27359
5.5	3983	6196	8409	10622	12834	15047	17260	17703	18588	22128	24784	26554	30095
6.0	4345	6759	9173	11587	14001	16415	18829	19312	20278	24140	27037	28968	32831
6.5	4707	7322	9938	12553	15168	17783	20398	20921	21967	26152	29290	31382	35566
7.0	5069	7886	10702	13518	16335	19151	21967	22531	23657	28163	31543	33796	38302

Number of Plants per m² at the following offset spacings

(All plants equidistant horizontally and vertically)

Distance mm	Nr of Plants
100	114.43
200	28.22
250	18.17
300	12.35
400	6.72
500	4.29
600	3.04
700	2.16
750	1.88
900	1.26
1000	1.05
1200	0.68
1500	0.42
2000	0.23

SEEDING/TURFING AND PLANTING

Grass Clippings Wet: Based on 3.5 m³/tonne

Annual kg/100 m²		Average 20 cuts kg/100 m²		m²/tonne	m²/m³
32.0		1.6		61162.1	214067.3

Nr of cuts	22	20	18	16	12	4
kg/cut	1.45	1.60	1.78	2.00	2.67	8.00
		Area capacity of 3 tonne vehicle per load				
m²	206250	187500	168750	150000	112500	37500
Load m³		100 m² units/m³ of vehicle space				
1	196.4	178.6	160.7	142.9	107.1	35.7
2	392.9	357.1	321.4	285.7	214.3	71.4
3	589.3	535.7	482.1	428.6	321.4	107.1
4	785.7	714.3	642.9	571.4	428.6	142.9
5	982.1	892.9	803.6	714.3	535.7	178.6

Transportation of Trees

To unload large trees a machine with the necessary lifting strength is required. The weight of the trees must therefore be known in advance. The following table gives a rough overview. The additional columns with root ball dimensions and the number of plants per trailer provide additional information, for example about preparing planting holes and calculating unloading times.

Girth in cm	Rootball diameter in cm	Ball height in cm	Weight in kg	Numbers of trees per trailer
16–18	50–60	40	150	100–120
18–20	60–70	40–50	200	80–100
20–25	60–70	40–50	270	50–70
25–30	80	50–60	350	50
30–35	90–100	60–70	500	12–18
35–40	100–110	60–70	650	10–15
40–45	110–120	60–70	850	8–12
45–50	110–120	60–70	1100	5–7
50–60	130–140	60–70	1600	1–3
60–70	150–160	60–70	2500	1
70–80	180–200	70	4000	1
80–90	200–220	70–80	5500	1
90–100	230–250	80–90	7500	1
100–120	250–270	80–90	9500	1

Data supplied by Lorenz von Ehren GmbH
The information in the table is approximate; deviations depend on soil type, genus and weather

FENCING AND GATES

Types of Preservative

Creosote (tar oil) can be 'factory' applied	by pressure to BS 144: pts 1&2 by immersion to BS 144: pt 1 by hot and cold open tank to BS 144: pts 1&2
Copper/chromium/arsenic (CCA)	by full cell process to BS 4072 pts 1&2
Organic solvent (OS)	by double vacuum (vacvac) to BS 5707 pts 1&3 by immersion to BS 5057 pts 1&3
Pentachlorophenol (PCP)	by heavy oil double vacuum to BS 5705 pts 2&3

Boron diffusion process (treated with disodium octaborate to BWPA Manual 1986)

Note: Boron is used on green timber at source and the timber is supplied dry

Cleft Chestnut Pale Fences

Pales	Pale spacing	Wire lines	
900 mm	75 mm	2	temporary protection
1050 mm	75 or 100 mm	2	light protective fences
1200 mm	75 mm	3	perimeter fences
1350 mm	75 mm	3	perimeter fences
1500 mm	50 mm	3	narrow perimeter fences
1800 mm	50 mm	3	light security fences

Close-Boarded Fences

Close-boarded fences 1.05 to 1.8 m high
Type BCR (recessed) or BCM (morticed) with concrete posts 140 × 115 mm tapered and Type BW with timber posts

Palisade Fences

Wooden palisade fences
Type WPC with concrete posts 140 × 115 mm tapered and Type WPW with timber posts

For both types of fence:
Height of fence 1050 mm: two rails
Height of fence 1200 mm: two rails
Height of fence 1500 mm: three rails
Height of fence 1650 mm: three rails
Height of fence 1800 mm: three rails

FENCING AND GATES

Post and Rail Fences

Wooden post and rail fences
Type MPR 11/3 morticed rails and Type SPR 11/3 nailed rails
Height to top of rail 1100 mm
Rails: three rails 87 mm, 38 mm

Type MPR 11/4 morticed rails and Type SPR 11/4 nailed rails
Height to top of rail 1100 mm
Rails: four rails 87 mm, 38 mm

Type MPR 13/4 morticed rails and Type SPR 13/4 nailed rails
Height to top of rail 1300 mm
Rail spacing 250 mm, 250 mm, and 225 mm from top
Rails: four rails 87 mm, 38 mm

Steel Posts

Rolled steel angle iron posts for chain link fencing

Posts	Fence height	Strut	Straining post
1500 × 40 × 40 × 5 mm	900 mm	1500 × 40 × 40 × 5 mm	1500 × 50 × 50 × 6 mm
1800 × 40 × 40 × 5 mm	1200 mm	1800 × 40 × 40 × 5 mm	1800 × 50 × 50 × 6 mm
2000 × 45 × 45 × 5 mm	1400 mm	2000 × 45 × 45 × 5 mm	2000 × 60 × 60 × 6 mm
2600 × 45 × 45 × 5 mm	1800 mm	2600 × 45 × 45 × 5 mm	2600 × 60 × 60 × 6 mm
3000 × 50 × 50 × 6 mm with arms	1800 mm	2600 × 45 × 45 × 5 mm	3000 × 60 × 60 × 6 mm

Concrete Posts

Concrete posts for chain link fencing

Posts and straining posts	Fence height	Strut
1570 mm 100 × 100 mm	900 mm	1500 mm × 75 × 75 mm
1870 mm 125 × 125 mm	1200 mm	1830 mm × 100 × 75 mm
2070 mm 125 × 125 mm	1400 mm	1980 mm × 100 × 75 mm
2620 mm 125 × 125 mm	1800 mm	2590 mm × 100 × 85 mm
3040 mm 125 × 125 mm	1800 mm	2590 mm × 100 × 85 mm (with arms)

Tables and Memoranda

FENCING AND GATES

Rolled Steel Angle Posts

Rolled steel angle posts for rectangular wire mesh (field) fencing

Posts	Fence height	Strut	Straining post
1200 × 40 × 40 × 5 mm	600 mm	1200 × 75 × 75 mm	1350 × 100 × 100 mm
1400 × 40 × 40 × 5 mm	800 mm	1400 × 75 × 75 mm	1550 × 100 × 100 mm
1500 × 40 × 40 × 5 mm	900 mm	1500 × 75 × 75 mm	1650 × 100 × 100 mm
1600 × 40 × 40 × 5 mm	1000 mm	1600 × 75 × 75 mm	1750 × 100 × 100 mm
1750 × 40 × 40 × 5 mm	1150 mm	1750 × 75 × 100 mm	1900 × 125 × 125 mm

Concrete Posts

Concrete posts for rectangular wire mesh (field) fencing

Posts	Fence height	Strut	Straining post
1270 × 100 × 100 mm	600 mm	1200 × 75 × 75 mm	1420 × 100 × 100 mm
1470 × 100 × 100 mm	800 mm	1350 × 75 × 75 mm	1620 × 100 × 100 mm
1570 × 100 × 100 mm	900 mm	1500 × 75 × 75 mm	1720 × 100 × 100 mm
1670 × 100 × 100 mm	600 mm	1650 × 75 × 75 mm	1820 × 100 × 100 mm
1820 × 125 × 125 mm	1150 mm	1830 × 75 × 100 mm	1970 × 125 × 125 mm

Cleft Chestnut Pale Fences

Timber Posts

Timber posts for wire mesh and hexagonal wire netting fences

Round timber for general fences

Posts	Fence height	Strut	Straining post
1300 × 65 mm dia.	600 mm	1200 × 80 mm dia.	1450 × 100 mm dia.
1500 × 65 mm dia.	800 mm	1400 × 80 mm dia.	1650 × 100 mm dia.
1600 × 65 mm dia.	900 mm	1500 × 80 mm dia.	1750 × 100 mm dia.
1700 × 65 mm dia.	1050 mm	1600 × 80 mm dia.	1850 × 100 mm dia.
1800 × 65 mm dia.	1150 mm	1750 × 80 mm dia.	2000 × 120 mm dia.

Squared timber for general fences

Posts	Fence height	Strut	Straining post
1300 × 75 × 75 mm	600 mm	1200 × 75 × 75 mm	1450 × 100 × 100 mm
1500 × 75 × 75 mm	800 mm	1400 × 75 × 75 mm	1650 × 100 × 100 mm
1600 × 75 × 75 mm	900 mm	1500 × 75 × 75 mm	1750 × 100 × 100 mm
1700 × 75 × 75 mm	1050 mm	1600 × 75 × 75 mm	1850 × 100 × 100 mm
1800 × 75 × 75 mm	1150 mm	1750 × 75 × 75 mm	2000 × 125 × 100 mm

FENCING AND GATES

Steel Fences to BS 1722: Part 9: 1992

	Fence height	Top/bottom rails and flat posts	Vertical bars
Light	1000 mm	40 × 10 mm 450 mm in ground	12 mm dia. at 115 mm cs
	1200 mm	40 × 10 mm 550 mm in ground	12 mm dia. at 115 mm cs
	1400 mm	40 × 10 mm 550 mm in ground	12 mm dia. at 115 mm cs
Light	1000 mm	40 × 10 mm 450 mm in ground	16 mm dia. at 120 mm cs
	1200 mm	40 × 10 mm 550 mm in ground	16 mm dia. at 120 mm cs
	1400 mm	40 × 10 mm 550 mm in ground	16 mm dia. at 120 mm cs
Medium	1200 mm	50 × 10 mm 550 mm in ground	20 mm dia. at 125 mm cs
	1400 mm	50 × 10 mm 550 mm in ground	20 mm dia. at 125 mm cs
	1600 mm	50 × 10 mm 600 mm in ground	22 mm dia. at 145 mm cs
	1800 mm	50 × 10 mm 600 mm in ground	22 mm dia. at 145 mm cs
Heavy	1600 mm	50 × 10 mm 600 mm in ground	22 mm dia. at 145 mm cs
	1800 mm	50 × 10 mm 600 mm in ground	22 mm dia. at 145 mm cs
	2000 mm	50 × 10 mm 600 mm in ground	22 mm dia. at 145 mm cs
	2200 mm	50 × 10 mm 600 mm in ground	22 mm dia. at 145 mm cs

Notes: Mild steel fences: round or square verticals; flat standards and horizontals. Tops of vertical bars may be bow-top, blunt, or pointed. Round or square bar railings

Timber Field Gates to BS 3470: 1975

Gates made to this standard are designed to open one way only
All timber gates are 1100 mm high
Width over stiles 2400, 2700, 3000, 3300, 3600, and 4200 mm
Gates over 4200 mm should be made in two leaves

Steel Field Gates to BS 3470: 1975

All steel gates are 1100 mm high
Heavy duty: width over stiles 2400, 3000, 3600 and 4500 mm
Light duty: width over stiles 2400, 3000, and 3600 mm

FENCING AND GATES

Domestic Front Entrance Gates to BS 4092: Part 1: 1966

Metal gates:	Single gates are 900 mm high minimum, 900 mm, 1000 mm and 1100 mm wide

Domestic Front Entrance Gates to BS 4092: Part 2: 1966

Wooden gates:	All rails shall be tenoned into the stiles Single gates are 840 mm high minimum, 801 mm and 1020 mm wide Double gates are 840 mm high minimum, 2130, 2340 and 2640 mm wide

Timber Bridle Gates to BS 5709:1979 (Horse or Hunting Gates)

Gates open one way only Minimum width between posts Minimum height	1525 mm 1100 mm

Timber Kissing Gates to BS 5709:1979

Minimum width	700 mm
Minimum height	1000 mm
Minimum distance between shutting posts	600 mm
Minimum clearance at mid-point	600 mm

Metal Kissing Gates to BS 5709:1979

Sizes are the same as those for timber kissing gates Maximum gaps between rails 120 mm

Categories of Pedestrian Guard Rail to BS 3049:1976

Class A for normal use Class B where vandalism is expected Class C where crowd pressure is likely

DRAINAGE

Width Required for Trenches for Various Diameters of Pipes

Pipe diameter (mm)	Trench n.e. 1.50 m deep	Trench over 1.50 m deep
n.e. 100 mm	450 mm	600 mm
100–150 mm	500 mm	650 mm
150–225 mm	600 mm	750 mm
225–300 mm	650 mm	800 mm
300–400 mm	750 mm	900 mm
400–450 mm	900 mm	1050 mm
450–600 mm	1100 mm	1300 mm

Weights and Dimensions – Vitrified Clay Pipes

Product	Nominal diameter (mm)	Effective length (mm)	BS 65 limits of tolerance min (mm)	max (mm)	Crushing strength (kN/m)	Weight (kg/pipe)	(kg/m)
Supersleve	100	1600	96	105	35.00	14.71	9.19
	150	1750	146	158	35.00	29.24	16.71
Hepsleve	225	1850	221	236	28.00	84.03	45.42
	300	2500	295	313	34.00	193.05	77.22
	150	1500	146	158	22.00	37.04	24.69
Hepseal	225	1750	221	236	28.00	85.47	48.84
	300	2500	295	313	34.00	204.08	81.63
	400	2500	394	414	44.00	357.14	142.86
	450	2500	444	464	44.00	454.55	181.63
	500	2500	494	514	48.00	555.56	222.22
	600	2500	591	615	57.00	796.23	307.69
	700	3000	689	719	67.00	1111.11	370.45
	800	3000	788	822	72.00	1351.35	450.45
Hepline	100	1600	95	107	22.00	14.71	9.19
	150	1750	145	160	22.00	29.24	16.71
	225	1850	219	239	28.00	84.03	45.42
	300	1850	292	317	34.00	142.86	77.22
Hepduct (conduit)	90	1500	–	–	28.00	12.05	8.03
	100	1600	–	–	28.00	14.71	9.19
	125	1750	–	–	28.00	20.73	11.84
	150	1750	–	–	28.00	29.24	16.71
	225	1850	–	–	28.00	84.03	45.42
	300	1850	–	–	34.00	142.86	77.22

DRAINAGE

Weights and Dimensions – Vitrified Clay Pipes

Nominal internal diameter (mm)	Nominal wall thickness (mm)	Approximate weight (kg/m)
150	25	45
225	29	71
300	32	122
375	35	162
450	38	191
600	48	317
750	54	454
900	60	616
1200	76	912
1500	89	1458
1800	102	1884
2100	127	2619

Wall thickness, weights and pipe lengths vary, depending on type of pipe required

The particulars shown above represent a selection of available diameters and are applicable to strength class 1 pipes with flexible rubber ring joints

Tubes with Ogee joints are also available

Weights and Dimensions – PVC-u Pipes

	Nominal size	Mean outside diameter (mm)		Wall thickness	Weight
		min	max	(mm)	(kg/m)
Standard pipes	82.4	82.4	82.7	3.2	1.2
	110.0	110.0	110.4	3.2	1.6
	160.0	160.0	160.6	4.1	3.0
	200.0	200.0	200.6	4.9	4.6
	250.0	250.0	250.7	6.1	7.2
Perforated pipes heavy grade	As above	As above	As above	As above	As above
thin wall	82.4	82.4	82.7	1.7	–
	110.0	110.0	110.4	2.2	–
	160.0	160.0	160.6	3.2	–

Width of Trenches Required for Various Diameters of Pipes

Pipe diameter (mm)	Trench n.e. 1.5 m deep (mm)	Trench over 1.5 m deep (mm)
n.e. 100	450	600
100–150	500	650
150–225	600	750
225–300	650	800
300–400	750	900
400–450	900	1050
450–600	1100	1300

DRAINAGE

DRAINAGE BELOW GROUND AND LAND DRAINAGE

Flow of Water Which Can Be Carried by Various Sizes of Pipe

Clay or concrete pipes

	Gradient of pipeline							
	1:10	1:20	1:30	1:40	1:50	1:60	1:80	1:100
Pipe size	Flow in litres per second							
DN 100 15.0	8.5	6.8	5.8	5.2	4.7	4.0	3.5	
DN 150 28.0	19.0	16.0	14.0	12.0	11.0	9.1	8.0	
DN 225 140.0	95.0	76.0	66.0	58.0	53.0	46.0	40.0	

Plastic pipes

	Gradient of pipeline							
	1:10	1:20	1:30	1:40	1:50	1:60	1:80	1:100
Pipe size	Flow in litres per second							
82.4 mm i/dia.	12.0	8.5	6.8	5.8	5.2	4.7	4.0	3.5
110 mm i/dia.	28.0	19.0	16.0	14.0	12.0	11.0	9.1	8.0
160 mm i/dia.	76.0	53.0	43.0	37.0	33.0	29.0	25.0	22.0
200 mm i/dia.	140.0	95.0	76.0	66.0	58.0	53.0	46.0	40.0

Vitrified (Perforated) Clay Pipes and Fittings to BS En 295-5 1994

Length not specified		
75 mm bore	250 mm bore	600 mm bore
100	300	700
125	350	800
150	400	1000
200	450	1200
225	500	

Precast Concrete Pipes: Prestressed Non-pressure Pipes and Fittings: Flexible Joints to BS 5911: Pt. 103: 1994

Rationalized metric nominal sizes: 450, 500	
Length:	500–1000 by 100 increments
	1000–2200 by 200 increments
	2200–2800 by 300 increments
Angles: length:	450–600 angles 45, 22.5,11.25°
	600 or more angles 22.5, 11.25°

DRAINAGE

Precast Concrete Pipes: Unreinforced and Circular Manholes and Soakaways to BS 5911: Pt. 200: 1994

Nominal sizes:	
Shafts:	675, 900 mm
Chambers:	900, 1050, 1200, 1350, 1500, 1800, 2100, 2400, 2700, 3000 mm
Large chambers:	To have either tapered reducing rings or a flat reducing slab in order to accept the standard cover
Ring depths:	1. 300–1200 mm by 300 mm increments except for bottom slab and rings below cover slab, these are by 150 mm increments
	2. 250–1000 mm by 250 mm increments except for bottom slab and rings below cover slab, these are by 125 mm increments
Access hole:	750 × 750 mm for DN 1050 chamber 1200 × 675 mm for DN 1350 chamber

Calculation of Soakaway Depth

The following formula determines the depth of concrete ring soakaway that would be required for draining given amounts of water.

$$h = \frac{4ar}{3\pi D^2}$$

h = depth of the chamber below the invert pipe
a = the area to be drained
r = the hourly rate of rainfall (50 mm per hour)
π = pi
D = internal diameter of the soakaway

This table shows the depth of chambers in each ring size which would be required to contain the volume of water specified. These allow a recommended storage capacity of ⅓ (one third of the hourly rainfall figure).

Table Showing Required Depth of Concrete Ring Chambers in Metres

Area m²	50	100	150	200	300	400	500
Ring size							
0.9	1.31	2.62	3.93	5.24	7.86	10.48	13.10
1.1	0.96	1.92	2.89	3.85	5.77	7.70	9.62
1.2	0.74	1.47	2.21	2.95	4.42	5.89	7.37
1.4	0.58	1.16	1.75	2.33	3.49	4.66	5.82
1.5	0.47	0.94	1.41	1.89	2.83	3.77	4.72
1.8	0.33	0.65	0.98	1.31	1.96	2.62	3.27
2.1	0.24	0.48	0.72	0.96	1.44	1.92	2.41
2.4	0.18	0.37	0.55	0.74	1.11	1.47	1.84
2.7	0.15	0.29	0.44	0.58	0.87	1.16	1.46
3.0	0.12	0.24	0.35	0.47	0.71	0.94	1.18

Precast Concrete Inspection Chambers and Gullies to BS 5911: Part 230: 1994

Nominal sizes:	375 diameter, 750, 900 mm deep
	450 diameter, 750, 900, 1050, 1200 mm deep
Depths:	from the top for trapped or untrapped units:
	centre of outlet 300 mm
	invert (bottom) of the outlet pipe 400 mm
Depth of water seal for trapped gullies:	
	85 mm, rodding eye int. dia. 100 mm
Cover slab:	65 mm min

Bedding Flexible Pipes: PVC-u Or Ductile Iron

Type 1 =	100 mm fill below pipe, 300 mm above pipe: single size material
Type 2 =	100 mm fill below pipe, 300 mm above pipe: single size or graded material
Type 3 =	100 mm fill below pipe, 75 mm above pipe with concrete protective slab over
Type 4 =	100 mm fill below pipe, fill laid level with top of pipe
Type 5 =	200 mm fill below pipe, fill laid level with top of pipe
Concrete =	25 mm sand blinding to bottom of trench, pipe supported on chocks, 100 mm concrete under the pipe, 150 mm concrete over the pipe

DRAINAGE

Bedding Rigid Pipes: Clay or Concrete
(for vitrified clay pipes the manufacturer should be consulted)

Class D:	Pipe laid on natural ground with cut-outs for joints, soil screened to remove stones over 40 mm and returned over pipe to 150 m min depth. Suitable for firm ground with trenches trimmed by hand.
Class N:	Pipe laid on 50 mm granular material of graded aggregate to Table 4 of BS 882, or 10 mm aggregate to Table 6 of BS 882, or as dug light soil (not clay) screened to remove stones over 10 mm. Suitable for machine dug trenches.
Class B:	As Class N, but with granular bedding extending half way up the pipe diameter.
Class F:	Pipe laid on 100 mm granular fill to BS 882 below pipe, minimum 150 mm granular fill above pipe: single size material. Suitable for machine dug trenches.
Class A:	Concrete 100 mm thick under the pipe extending half way up the pipe, backfilled with the appropriate class of fill. Used where there is only a very shallow fall to the drain. Class A bedding allows the pipes to be laid to an exact gradient.
Concrete surround:	25 mm sand blinding to bottom of trench, pipe supported on chocks, 100 mm concrete under the pipe, 150 mm concrete over the pipe. It is preferable to bed pipes under slabs or wall in granular material.

PIPED SUPPLY SYSTEMS

Identification of Service Tubes From Utility to Dwellings

Utility	Colour	Size	Depth
British Telecom	grey	54 mm od	450 mm
Electricity	black	38 mm od	450 mm
Gas	yellow	42 mm od rigid 60 mm od convoluted	450 mm
Water	may be blue	(normally untubed)	750 mm

ELECTRICAL SUPPLY/POWER/LIGHTING SYSTEMS

Electrical Insulation Class En 60.598 BS 4533

Class 1:	luminaires comply with class 1 (I) earthed electrical requirements
Class 2:	luminaires comply with class 2 (II) double insulated electrical requirements
Class 3:	luminaires comply with class 3 (III) electrical requirements

Protection to Light Fittings

BS EN 60529:1992 Classification for degrees of protection provided by enclosures.
(IP Code – International or ingress Protection)

1st characteristic: against ingress of solid foreign objects

The figure	2	indicates that fingers cannot enter
	3	that a 2.5 mm diameter probe cannot enter
	4	that a 1.0 mm diameter probe cannot enter
	5	the fitting is dust proof (no dust around live parts)
	6	the fitting is dust tight (no dust entry)

2nd characteristic: ingress of water with harmful effects

The figure	0	indicates unprotected
	1	vertically dripping water cannot enter
	2	water dripping 15° (tilt) cannot enter
	3	spraying water cannot enter
	4	splashing water cannot enter
	5	jetting water cannot enter
	6	powerful jetting water cannot enter
	7	proof against temporary immersion
	8	proof against continuous immersion

Optional additional codes:		A–D protects against access to hazardous parts
	H	high voltage apparatus
	M	fitting was in motion during water test
	S	fitting was static during water test
	W	protects against weather

Marking code arrangement:	(example) IPX5S = IP (International or Ingress Protection)
	X (denotes omission of first characteristic)
	5 = jetting
	S = static during water test

RAIL TRACKS

	kg/m of track	lb/ft of track
Standard gauge		
Bull-head rails, chairs, transverse timber (softwood) sleepers etc.	245	165
Main lines		
Flat-bottom rails, transverse prestressed concrete sleepers, etc.	418	280
Add for electric third rail	51	35
Add for crushed stone ballast	2600	1750
	kg/m²	**lb/sq ft**
Overall average weight – rails connections, sleepers, ballast, etc.	733	150
	kg/m of track	**lb/ft of track**
Bridge rails, longitudinal timber sleepers, etc.	112	75

Tables and Memoranda

RAIL TRACKS

Heavy Rails

British Standard Section No.	Rail height (mm)	Foot width (mm)	Head width (mm)	Min web thickness (mm)	Section weight (kg/m)
Flat Bottom Rails					
60 A	114.30	109.54	57.15	11.11	30.62
70 A	123.82	111.12	60.32	12.30	34.81
75 A	128.59	114.30	61.91	12.70	37.45
80 A	133.35	117.47	63.50	13.10	39.76
90 A	142.88	127.00	66.67	13.89	45.10
95 A	147.64	130.17	69.85	14.68	47.31
100 A	152.40	133.35	69.85	15.08	50.18
110 A	158.75	139.70	69.85	15.87	54.52
113 A	158.75	139.70	69.85	20.00	56.22
50 'O'	100.01	100.01	52.39	10.32	24.82
80 'O'	127.00	127.00	63.50	13.89	39.74
60R	114.30	109.54	57.15	11.11	29.85
75R	128.59	122.24	61.91	13.10	37.09
80R	133.35	127.00	63.50	13.49	39.72
90R	142.88	136.53	66.67	13.89	44.58
95R	147.64	141.29	68.26	14.29	47.21
100R	152.40	146.05	69.85	14.29	49.60
95N	147.64	139.70	69.85	13.89	47.27
Bull Head Rails					
95R BH	145.26	69.85	69.85	19.05	47.07

Light Rails

British Standard Section No.	Rail height (mm)	Foot width (mm)	Head width (mm)	Min web thickness (mm)	Section weight (kg/m)
Flat Bottom Rails					
20M	65.09	55.56	30.96	6.75	9.88
30M	75.41	69.85	38.10	9.13	14.79
35M	80.96	76.20	42.86	9.13	17.39
35R	85.73	82.55	44.45	8.33	17.40
40	88.11	80.57	45.64	12.3	19.89
Bridge Rails					
13	48.00	92	36.00	18.0	13.31
16	54.00	108	44.50	16.0	16.06
20	55.50	127	50.00	20.5	19.86
28	67.00	152	50.00	31.0	28.62
35	76.00	160	58.00	34.5	35.38
50	76.00	165	58.50	–	50.18
Crane Rails					
A65	75.00	175.00	65.00	38.0	43.10
A75	85.00	200.00	75.00	45.0	56.20
A100	95.00	200.00	100.00	60.0	74.30
A120	105.00	220.00	120.00	72.0	100.00
175CR	152.40	152.40	107.95	38.1	86.92

RAIL TRACKS

Fish Plates

British Standard Section No.	Overall plate length		Hole diameter	Finished weight per pair	
	4 Hole (mm)	6 Hole (mm)	(mm)	4 Hole (kg/pair)	6 Hole (kg/pair)
For British Standard Heavy Rails: Flat Bottom Rails					
60 A	406.40	609.60	20.64	9.87	14.76
70 A	406.40	609.60	22.22	11.15	16.65
75 A	406.40	–	23.81	11.82	17.73
80 A	406.40	609.60	23.81	13.15	19.72
90 A	457.20	685.80	25.40	17.49	26.23
100 A	508.00	–	pear	25.02	–
110 A (shallow)	507.00	–	27.00	30.11	54.64
113 A (heavy)	507.00	–	27.00	30.11	54.64
50 'O' (shallow)	406.40	–	–	6.68	10.14
80 'O' (shallow)	495.30	–	23.81	14.72	22.69
60R (shallow)	406.40	609.60	20.64	8.76	13.13
60R (angled)	406.40	609.60	20.64	11.27	16.90
75R (shallow)	406.40	–	23.81	10.94	16.42
75R (angled)	406.40	–	23.81	13.67	–
80R (shallow)	406.40	609.60	23.81	11.93	17.89
80R (angled)	406.40	609.60	23.81	14.90	22.33
For British Standard Heavy Rails: Bull head rails					
95R BH (shallow)	–	457.20	27.00	14.59	14.61
For British Standard Light Rails: Flat Bottom Rails					
30M	355.6	–	–	–	2.72
35M	355.6	–	–	–	2.83
40	355.6	–	–	3.76	–

FRACTIONS, DECIMALS AND MILLIMETRE EQUIVALENTS

FRACTIONS, DECIMALS AND MILLIMETRE EQUIVALENTS

Fractions	Decimals	(mm)		Fractions	Decimals	(mm)
1/64	0.015625	0.396875		33/64	0.515625	13.096875
1/32	0.03125	0.79375		17/32	0.53125	13.49375
3/64	0.046875	1.190625		35/64	0.546875	13.890625
1/16	0.0625	1.5875		9/16	0.5625	14.2875
5/64	0.078125	1.984375		37/64	0.578125	14.684375
3/32	0.09375	2.38125		19/32	0.59375	15.08125
7/64	0.109375	2.778125		39/64	0.609375	15.478125
1/8	0.125	3.175		5/8	0.625	15.875
9/64	0.140625	3.571875		41/64	0.640625	16.271875
5/32	0.15625	3.96875		21/32	0.65625	16.66875
11/64	0.171875	4.365625		43/64	0.671875	17.065625
3/16	0.1875	4.7625		11/16	0.6875	17.4625
13/64	0.203125	5.159375		45/64	0.703125	17.859375
7/32	0.21875	5.55625		23/32	0.71875	18.25625
15/64	0.234375	5.953125		47/64	0.734375	18.653125
1/4	0.25	6.35		3/4	0.75	19.05
17/64	0.265625	6.746875		49/64	0.765625	19.446875
9/32	0.28125	7.14375		25/32	0.78125	19.84375
19/64	0.296875	7.540625		51/64	0.796875	20.240625
5/16	0.3125	7.9375		13/16	0.8125	20.6375
21/64	0.328125	8.334375		53/64	0.828125	21.034375
11/32	0.34375	8.73125		27/32	0.84375	21.43125
23/64	0.359375	9.128125		55/64	0.859375	21.828125
3/8	0.375	9.525		7/8	0.875	22.225
25/64	0.390625	9.921875		57/64	0.890625	22.621875
13/32	0.40625	10.31875		29/32	0.90625	23.01875
27/64	0.421875	10.71563		59/64	0.921875	23.415625
7/16	0.4375	11.1125		15/16	0.9375	23.8125
29/64	0.453125	11.50938		61/64	0.953125	24.209375
15/32	0.46875	11.90625		31/32	0.96875	24.60625
31/64	0.484375	12.30313		63/64	0.984375	25.003125
1/2	0.5	12.7		1.0	1	25.4

IMPERIAL STANDARD WIRE GAUGE (SWG)

SWG No.	Diameter (inches)	Diameter (mm)	SWG No.	Diameter (inches)	Diameter (mm)
7/0	0.5	12.7	23	0.024	0.61
6/0	0.464	11.79	24	0.022	0.559
5/0	0.432	10.97	25	0.02	0.508
4/0	0.4	10.16	26	0.018	0.457
3/0	0.372	9.45	27	0.0164	0.417
2/0	0.348	8.84	28	0.0148	0.376
1/0	0.324	8.23	29	0.0136	0.345
1	0.3	7.62	30	0.0124	0.315
2	0.276	7.01	31	0.0116	0.295
3	0.252	6.4	32	0.0108	0.274
4	0.232	5.89	33	0.01	0.254
5	0.212	5.38	34	0.009	0.234
6	0.192	4.88	35	0.008	0.213
7	0.176	4.47	36	0.008	0.193
8	0.16	4.06	37	0.007	0.173
9	0.144	3.66	38	0.006	0.152
10	0.128	3.25	39	0.005	0.132
11	0.116	2.95	40	0.005	0.122
12	0.104	2.64	41	0.004	0.112
13	0.092	2.34	42	0.004	0.102
14	0.08	2.03	43	0.004	0.091
15	0.072	1.83	44	0.003	0.081
16	0.064	1.63	45	0.003	0.071
17	0.056	1.42	46	0.002	0.061
18	0.048	1.22	47	0.002	0.051
19	0.04	1.016	48	0.002	0.041
20	0.036	0.914	49	0.001	0.031
21	0.032	0.813	50	0.001	0.025
22	0.028	0.711			

PIPES, WATER, STORAGE, INSULATION

WATER PRESSURE DUE TO HEIGHT

Imperial

Head (Feet)	Pressure (lb/in^2)		Head (Feet)	Pressure (lb/in^2)
1	0.43		70	30.35
5	2.17		75	32.51
10	4.34		80	34.68
15	6.5		85	36.85
20	8.67		90	39.02
25	10.84		95	41.18
30	13.01		100	43.35
35	15.17		105	45.52
40	17.34		110	47.69
45	19.51		120	52.02
50	21.68		130	56.36
55	23.84		140	60.69
60	26.01		150	65.03
65	28.18			

Metric

Head (m)	Pressure (bar)		Head (m)	Pressure (bar)
0.5	0.049		18.0	1.766
1.0	0.098		19.0	1.864
1.5	0.147		20.0	1.962
2.0	0.196		21.0	2.06
3.0	0.294		22.0	2.158
4.0	0.392		23.0	2.256
5.0	0.491		24.0	2.354
6.0	0.589		25.0	2.453
7.0	0.687		26.0	2.551
8.0	0.785		27.0	2.649
9.0	0.883		28.0	2.747
10.0	0.981		29.0	2.845
11.0	1.079		30.0	2.943
12.0	1.177		32.5	3.188
13.0	1.275		35.0	3.434
14.0	1.373		37.5	3.679
15.0	1.472		40.0	3.924
16.0	1.57		42.5	4.169
17.0	1.668		45.0	4.415

1 bar	=	14.5038 lbf/in^2
1 lbf/in^2	=	0.06895 bar
1 metre	=	3.2808 ft or 39.3701 in
1 foot	=	0.3048 metres
1 in wg	=	2.5 mbar (249.1 N/m^2)

PIPES, WATER, STORAGE, INSULATION

Dimensions and Weights of Copper Pipes to BSEN 1057, BSEN 12499, BSEN 14251

Outside diameter (mm)	Internal diameter (mm)	Weight per metre (kg)	Internal diameter (mm)	Weight per metre (kg)	Internal siameter (mm)	Weight per metre (kg)
	Formerly Table X		Formerly Table Y		Formerly Table Z	
6	4.80	0.0911	4.40	0.1170	5.00	0.0774
8	6.80	0.1246	6.40	0.1617	7.00	0.1054
10	8.80	0.1580	8.40	0.2064	9.00	0.1334
12	10.80	0.1914	10.40	0.2511	11.00	0.1612
15	13.60	0.2796	13.00	0.3923	14.00	0.2031
18	16.40	0.3852	16.00	0.4760	16.80	0.2918
22	20.22	0.5308	19.62	0.6974	20.82	0.3589
28	26.22	0.6814	25.62	0.8985	26.82	0.4594
35	32.63	1.1334	32.03	1.4085	33.63	0.6701
42	39.63	1.3675	39.03	1.6996	40.43	0.9216
54	51.63	1.7691	50.03	2.9052	52.23	1.3343
76.1	73.22	3.1287	72.22	4.1437	73.82	2.5131
108	105.12	4.4666	103.12	7.3745	105.72	3.5834
133	130.38	5.5151	–	–	130.38	5.5151
159	155.38	8.7795	–	–	156.38	6.6056

Dimensions of Stainless Steel Pipes to BS 4127

Outside siameter (mm)	Maximum outside siameter (mm)	Minimum outside diameter (mm)	Wall thickness (mm)	Working pressure (bar)
6	6.045	5.940	0.6	330
8	8.045	7.940	0.6	260
10	10.045	9.940	0.6	210
12	12.045	11.940	0.6	170
15	15.045	14.940	0.6	140
18	18.045	17.940	0.7	135
22	22.055	21.950	0.7	110
28	28.055	27.950	0.8	121
35	35.070	34.965	1.0	100
42	42.070	41.965	1.1	91
54	54.090	53.940	1.2	77

PIPES, WATER, STORAGE, INSULATION

Dimensions of Steel Pipes to BS 1387

Nominal Size	Approx. Outside Diameter	Outside diameter				Thickness		
		Light		Medium & Heavy		Light	Medium	Heavy
		Max	Min	Max	Min			
(mm)	(mm)	(mm)	(mm)	(mm)	(mm)	(mm)	(mm)	(mm)
6	10.20	10.10	9.70	10.40	9.80	1.80	2.00	2.65
8	13.50	13.60	13.20	13.90	13.30	1.80	2.35	2.90
10	17.20	17.10	16.70	17.40	16.80	1.80	2.35	2.90
15	21.30	21.40	21.00	21.70	21.10	2.00	2.65	3.25
20	26.90	26.90	26.40	27.20	26.60	2.35	2.65	3.25
25	33.70	33.80	33.20	34.20	33.40	2.65	3.25	4.05
32	42.40	42.50	41.90	42.90	42.10	2.65	3.25	4.05
40	48.30	48.40	47.80	48.80	48.00	2.90	3.25	4.05
50	60.30	60.20	59.60	60.80	59.80	2.90	3.65	4.50
65	76.10	76.00	75.20	76.60	75.40	3.25	3.65	4.50
80	88.90	88.70	87.90	89.50	88.10	3.25	4.05	4.85
100	114.30	113.90	113.00	114.90	113.30	3.65	4.50	5.40
125	139.70	–	–	140.60	138.70	–	4.85	5.40
150	165.1*	–	–	166.10	164.10	–	4.85	5.40

* 165.1 mm (6.5in) outside diameter is not generally recommended except where screwing to BS 21 is necessary
All dimensions are in accordance with ISO R65 except approximate outside diameters which are in accordance with ISO R64
Light quality is equivalent to ISO R65 Light Series II

Approximate Metres Per Tonne of Tubes to BS 1387

Nom. size	BLACK						GALVANIZED					
	Plain/screwed ends			Screwed & socketed			Plain/screwed ends			Screwed & socketed		
	L	M	H	L	M	H	L	M	H	L	M	H
(mm)	(m)	(m)	(m)	(m)	(m)	(m)	(m)	(m)	(m)	(m)	(m)	(m)
6	2765	2461	2030	2743	2443	2018	2604	2333	1948	2584	2317	1937
8	1936	1538	1300	1920	1527	1292	1826	1467	1254	1811	1458	1247
10	1483	1173	979	1471	1165	974	1400	1120	944	1386	1113	939
15	1050	817	688	1040	811	684	996	785	665	987	779	661
20	712	634	529	704	628	525	679	609	512	673	603	508
25	498	410	336	494	407	334	478	396	327	474	394	325
32	388	319	260	384	316	259	373	308	254	369	305	252
40	307	277	226	303	273	223	296	268	220	292	264	217
50	244	196	162	239	194	160	235	191	158	231	188	157
65	172	153	127	169	151	125	167	149	124	163	146	122
80	147	118	99	143	116	98	142	115	97	139	113	96
100	101	82	69	98	81	68	98	81	68	95	79	67
125	–	62	56	–	60	55	–	60	55	–	59	54
150	–	52	47	–	50	46	–	51	46	–	49	45

The figures for 'plain or screwed ends' apply also to tubes to BS 1775 of equivalent size and thickness
Key:
L – Light
M – Medium
H – Heavy

PIPES, WATER, STORAGE, INSULATION

Flange Dimension Chart to BS 4504 & BS 10

Normal Pressure Rating (PN 6) 6 Bar

Nom. size	Flange outside dia.	Table 6/2 Forged Welding Neck	Table 6/3 Plate Slip on	Table 6/4 Forged Bossed Screwed	Table 6/5 Forged Bossed Slip on	Table 6/8 Plate Blank	Raised face Dia.	Raised face T'ness	Nr. bolt hole	Size of bolt
15	80	12	12	12	12	12	40	2	4	M10 × 40
20	90	14	14	14	14	14	50	2	4	M10 × 45
25	100	14	14	14	14	14	60	2	4	M10 × 45
32	120	14	16	14	14	14	70	2	4	M12 × 45
40	130	14	16	14	14	14	80	3	4	M12 × 45
50	140	14	16	14	14	14	90	3	4	M12 × 45
65	160	14	16	14	14	14	110	3	4	M12 × 45
80	190	16	18	16	16	16	128	3	4	M16 × 55
100	210	16	18	16	16	16	148	3	4	M16 × 55
125	240	18	20	18	18	18	178	3	8	M16 × 60
150	265	18	20	18	18	18	202	3	8	M16 × 60
200	320	20	22	–	20	20	258	3	8	M16 × 60
250	375	22	24	–	22	22	312	3	12	M16 × 65
300	440	22	24	–	22	22	365	4	12	M20 × 70

Normal Pressure Rating (PN 16) 16 Bar

Nom. size	Flange outside dia.	Table 6/2 Forged Welding Neck	Table 6/3 Plate Slip on	Table 6/4 Forged Bossed Screwed	Table 6/5 Forged Bossed Slip on	Table 6/8 Plate Blank	Raised face Dia.	Raised face T'ness	Nr. bolt hole	Size of bolt
15	95	14	14	14	14	14	45	2	4	M12 × 45
20	105	16	16	16	16	16	58	2	4	M12 × 50
25	115	16	16	16	16	16	68	2	4	M12 × 50
32	140	16	16	16	16	16	78	2	4	M16 × 55
40	150	16	16	16	16	16	88	3	4	M16 × 55
50	165	18	18	18	18	18	102	3	4	M16 × 60
65	185	18	18	18	18	18	122	3	4	M16 × 60
80	200	20	20	20	20	20	138	3	8	M16 × 60
100	220	20	20	20	20	20	158	3	8	M16 × 65
125	250	22	22	22	22	22	188	3	8	M16 × 70
150	285	22	22	22	22	22	212	3	8	M20 × 70
200	340	24	24	–	24	24	268	3	12	M20 × 75
250	405	26	26	–	26	26	320	3	12	M24 × 90
300	460	28	28	–	28	28	378	4	12	M24 × 90

PIPES, WATER, STORAGE, INSULATION

Minimum Distances Between Supports/Fixings

Material	BS Nominal pipe size		Pipes – Vertical	Pipes – Horizontal on to low gradients
	(inch)	(mm)	Support distance in metres	Support distance in metres
Copper	0.50	15.00	1.90	1.30
	0.75	22.00	2.50	1.90
	1.00	28.00	2.50	1.90
	1.25	35.00	2.80	2.50
	1.50	42.00	2.80	2.50
	2.00	54.00	3.90	2.50
	2.50	67.00	3.90	2.80
	3.00	76.10	3.90	2.80
	4.00	108.00	3.90	2.80
	5.00	133.00	3.90	2.80
	6.00	159.00	3.90	2.80
muPVC	1.25	32.00	1.20	0.50
	1.50	40.00	1.20	0.50
	2.00	50.00	1.20	0.60
Polypropylene	1.25	32.00	1.20	0.50
	1.50	40.00	1.20	0.50
uPVC	–	82.40	1.20	0.50
	–	110.00	1.80	0.90
	–	160.00	1.80	1.20
Steel	0.50	15.00	2.40	1.80
	0.75	20.00	3.00	2.40
	1.00	25.00	3.00	2.40
	1.25	32.00	3.00	2.40
	1.50	40.00	3.70	2.40
	2.00	50.00	3.70	2.40
	2.50	65.00	4.60	3.00
	3.00	80.40	4.60	3.00
	4.00	100.00	4.60	3.00
	5.00	125.00	5.50	3.70
	6.00	150.00	5.50	4.50
	8.00	200.00	8.50	6.00
	10.00	250.00	9.00	6.50
	12.00	300.00	10.00	7.00
	16.00	400.00	10.00	8.25

PIPES, WATER, STORAGE, INSULATION

Litres of Water Storage Required Per Person Per Building Type

Type of building	Storage (litres)
Houses and flats (up to 4 bedrooms)	120/bedroom
Houses and flats (more than 4 bedrooms)	100/bedroom
Hostels	90/bed
Hotels	200/bed
Nurses homes and medical quarters	120/bed
Offices with canteen	45/person
Offices without canteen	40/person
Restaurants	7/meal
Boarding schools	90/person
Day schools – Primary	15/person
Day schools – Secondary	20/person

Recommended Air Conditioning Design Loads

Building type	Design loading
Computer rooms	500 W/m² of floor area
Restaurants	150 W/m² of floor area
Banks (main area)	100 W/m² of floor area
Supermarkets	25 W/m² of floor area
Large office block (exterior zone)	100 W/m² of floor area
Large office block (interior zone)	80 W/m² of floor area
Small office block (interior zone)	80 W/m² of floor area

PIPES, WATER, STORAGE, INSULATION

Capacity and Dimensions of Galvanized Mild Steel Cisterns – BS 417

Capacity (litres)	BS type (SCM)	Dimensions		
		Length (mm)	Width (mm)	Depth (mm)
18	45	457	305	305
36	70	610	305	371
54	90	610	406	371
68	110	610	432	432
86	135	610	457	482
114	180	686	508	508
159	230	736	559	559
191	270	762	584	610
227	320	914	610	584
264	360	914	660	610
327	450/1	1220	610	610
336	450/2	965	686	686
423	570	965	762	787
491	680	1090	864	736
709	910	1070	889	889

Capacity of Cold Water Polypropylene Storage Cisterns – BS 4213

Capacity (litres)	BS type (PC)	Maximum height (mm)
18	4	310
36	8	380
68	15	430
91	20	510
114	25	530
182	40	610
227	50	660
273	60	660
318	70	660
455	100	760

PIPES, WATER, STORAGE, INSULATION

Minimum Insulation Thickness to Protect Against Freezing for Domestic Cold Water Systems (8 Hour Evaluation Period)

Pipe size (mm)	Insulation thickness (mm)					
	Condition 1			Condition 2		
	λ = 0.020	λ = 0.030	λ = 0.040	λ = 0.020	λ = 0.030	λ = 0.040
Copper pipes						
15	11	20	34	12	23	41
22	6	9	13	6	10	15
28	4	6	9	4	7	10
35	3	5	7	4	5	7
42	3	4	5	8	4	6
54	2	3	4	2	3	4
76	2	2	3	2	2	3
Steel pipes						
15	9	15	24	10	18	29
20	6	9	13	6	10	15
25	4	7	9	5	7	10
32	3	5	6	3	5	7
40	3	4	5	3	4	6
50	2	3	4	2	3	4
65	2	2	3	2	3	3

Condition 1: water temperature 7°C; ambient temperature –6°C; evaluation period 8 h; permitted ice formation 50%; normal installation, i.e. inside the building and inside the envelope of the structural insulation
Condition 2: water temperature 2°C; ambient temperature –6°C; evaluation period 8 h; permitted ice formation 50%; extreme installation, i.e. inside the building but outside the envelope of the structural insulation
λ = thermal conductivity [W/(mK)]

Insulation Thickness for Chilled And Cold Water Supplies to Prevent Condensation

On a Low Emissivity Outer Surface (0.05, i.e. Bright Reinforced Aluminium Foil) with an Ambient Temperature of +25°C and a Relative Humidity of 80%

Steel pipe size (mm)	t = +10			t = +5			t = 0		
	Insulation thickness (mm)			Insulation thickness (mm)			Insulation thickness (mm)		
	λ = 0.030	λ = 0.040	λ = 0.050	λ = 0.030	λ = 0.040	λ = 0.050	λ = 0.030	λ = 0.040	λ = 0.050
15	16	20	25	22	28	34	28	36	43
25	18	24	29	25	32	39	32	41	50
50	22	28	34	30	39	47	38	49	60
100	26	34	41	36	47	57	46	60	73
150	29	38	46	40	52	64	51	67	82
250	33	43	53	46	60	74	59	77	94
Flat surfaces	39	52	65	56	75	93	73	97	122

t = temperature of contents (°C)
λ = thermal conductivity at mean temperature of insulation [W/(mK)]

PIPES, WATER, STORAGE, INSULATION

Insulation Thickness for Non-domestic Heating Installations to Control Heat Loss

Steel pipe size (mm)	t = 75			t = 100			t = 150		
	Insulation thickness (mm)			Insulation thickness (mm)			Insulation thickness (mm)		
	$\lambda = 0.030$	$\lambda = 0.040$	$\lambda = 0.050$	$\lambda = 0.030$	$\lambda = 0.040$	$\lambda = 0.050$	$\lambda = 0.030$	$\lambda = 0.040$	$\lambda = 0.050$
10	18	32	55	20	36	62	23	44	77
15	19	34	56	21	38	64	26	47	80
20	21	36	57	23	40	65	28	50	83
25	23	38	58	26	43	68	31	53	85
32	24	39	59	28	45	69	33	55	87
40	25	40	60	29	47	70	35	57	88
50	27	42	61	31	49	72	37	59	90
65	29	43	62	33	51	74	40	63	92
80	30	44	62	35	52	75	42	65	94
100	31	46	63	37	54	76	45	68	96
150	33	48	64	40	57	77	50	73	100
200	35	49	65	42	59	79	53	76	103
250	36	50	66	43	61	80	55	78	105

t = hot face temperature (°C)
λ = thermal conductivity at mean temperature of insulation [W/(mK)]

Index

Ebook Single-User Licence Agreement

We welcome you as a user of this Spon Price Book ebook and hope that you find it a useful and valuable tool. Please read this document carefully. **This is a legal agreement** between you (hereinafter referred to as the "Licensee") and Taylor and Francis Books Ltd. (the "Publisher"), which defines the terms under which you may use the Product. **By accessing and retrieving the access code on the label inside the front cover of this book you agree to these terms and conditions outlined herein. If you do not agree to these terms you must return the Product to your supplier intact, with the seal on the label unbroken and with the access code not accessed**.

1. **Definition of the Product**
    The product which is the subject of this Agreement, (the "Product") consists of online and offline access to the VitalSource ebook edition of *Spon's Mechanical & Electrical Services Price Book 2021*.

2. **Commencement and licence**
    2.1   This Agreement commences upon the breaking open of the document containing the access code by the Licensee (the "Commencement Date").
    2.2   This is a licence agreement (the "Agreement") for the use of the Product by the Licensee, and not an agreement for sale.
    2.3   The Publisher licenses the Licensee on a non-exclusive and non-transferable basis to use the Product on condition that the Licensee complies with this Agreement. The Licensee acknowledges that it is only permitted to use the Product in accordance with this Agreement.

3. **Multiple use**
    Use of the Product is not provided or allowed for more than one user or for a wide area network or consortium.

4. **Installation and Use**
    4.1   The Licensee may provide access to the Product for individual study in the following manner: The Licensee may install the Product on a secure local area network on a single site for use by one user.
    4.2   The Licensee shall be responsible for installing the Product and for the effectiveness of such installation.
    4.3   Text from the Product may be incorporated in a coursepack. Such use is only permissible with the express permission of the Publisher in writing and requires the payment of the appropriate fee as specified by the Publisher and signature of a separate licence agreement.
    4.4   The Product is a free addition to the book and the Publisher is under no obligation to provide any technical support.

5. **Permitted Activities**
    5.1   The Licensee shall be entitled to use the Product for its own internal purposes;
    5.2   The Licensee acknowledges that its rights to use the Product are strictly set out in this Agreement, and all other uses (whether expressly mentioned in Clause 6 below or not) are prohibited.

6. **Prohibited Activities**
    The following are prohibited without the express permission of the Publisher:
    6.1   The commercial exploitation of any part of the Product.
    6.2   The rental, loan, (free or for money or money's worth) or hire purchase of this product, save with the express consent of the Publisher.
    6.3   Any activity which raises the reasonable prospect of impeding the Publisher's ability or opportunities to market the Product.
    6.4   Any networking, physical or electronic distribution or dissemination of the product save as expressly permitted by this Agreement.
    6.5   Any reverse engineering, decompilation, disassembly or other alteration of the Product save in accordance with applicable national laws.
    6.6   The right to create any derivative product or service from the Product save as expressly provided for in this Agreement.
    6.7   Any alteration, amendment, modification or deletion from the Product, whether for the purposes of error correction or otherwise.

7. **General Responsibilities of the License**
   7.1 The Licensee will take all reasonable steps to ensure that the Product is used in accordance with the terms and conditions of this Agreement.
   7.2 The Licensee acknowledges that damages may not be a sufficient remedy for the Publisher in the event of breach of this Agreement by the Licensee, and that an injunction may be appropriate.
   7.3 The Licensee undertakes to keep the Product safe and to use its best endeavours to ensure that the product does not fall into the hands of third parties, whether as a result of theft or otherwise.
   7.4 Where information of a confidential nature relating to the product of the business affairs of the Publisher comes into the possession of the Licensee pursuant to this Agreement (or otherwise), the Licensee agrees to use such information solely for the purposes of this Agreement, and under no circumstances to disclose any element of the information to any third party save strictly as permitted under this Agreement. For the avoidance of doubt, the Licensee's obligations under this sub-clause 7.4 shall survive the termination of this Agreement.

8. **Warrant and Liability**
   8.1 The Publisher warrants that it has the authority to enter into this agreement and that it has secured all rights and permissions necessary to enable the Licensee to use the Product in accordance with this Agreement.
   8.2 The Publisher warrants that the Product as supplied on the Commencement Date shall be free of defects in materials and workmanship, and undertakes to replace any defective Product within 28 days of notice of such defect being received provided such notice is received within 30 days of such supply. As an alternative to replacement, the Publisher agrees fully to refund the Licensee in such circumstances, if the Licensee so requests, provided that the Licensee returns this copy of *Spon's Mechanical & Electrical Services Price Book 2021* to the Publisher. The provisions of this sub-clause 8.2 do not apply where the defect results from an accident or from misuse of the product by the Licensee.
   8.3 Sub-clause 8.2 sets out the sole and exclusive remedy of the Licensee in relation to defects in the Product.
   8.4 The Publisher and the Licensee acknowledge that the Publisher supplies the Product on an "as is" basis. The Publisher gives no warranties:
   8.4.1 that the Product satisfies the individual requirements of the Licensee; or
   8.4.2 that the Product is otherwise fit for the Licensee's purpose; or
   8.4.3 that the Product is compatible with the Licensee's hardware equipment and software operating environment.
   8.5 The Publisher hereby disclaims all warranties and conditions, express or implied, which are not stated above.
   8.6 Nothing in this Clause 8 limits the Publisher's liability to the Licensee in the event of death or personal injury resulting from the Publisher's negligence.
   8.7 The Publisher hereby excludes liability for loss of revenue, reputation, business, profits, or for indirect or consequential losses, irrespective of whether the Publisher was advised by the Licensee of the potential of such losses.
   8.8 The Licensee acknowledges the merit of independently verifying the price book data prior to taking any decisions of material significance (commercial or otherwise) based on such data. It is agreed that the Publisher shall not be liable for any losses which result from the Licensee placing reliance on the data under any circumstances.
   8.9 Subject to sub-clause 8.6 above, the Publisher's liability under this Agreement shall be limited to the purchase price.

9. **Intellectual Property Rights**
   9.1 Nothing in this Agreement affects the ownership of copyright or other intellectual property rights in the Product.
   9.2 The Licensee agrees to display the Publishers' copyright notice in the manner described in the Product.
   9.3 The Licensee hereby agrees to abide by copyright and similar notice requirements required by the Publisher, details of which are as follows:
   "© 2021 Taylor & Francis. All rights reserved. All materials in *Spon's Mechanical & Electrical Services Price Book 2021* are copyright protected. All rights reserved. No such materials may be used, displayed, modified, adapted, distributed, transmitted, transferred, published or otherwise reproduced in any form or by any means now or hereafter developed other than strictly in accordance with the terms of the licence agreement enclosed with *Spon's Mechanical & Electrical Services Price Book 2021*. However, text and images may be printed and copied for research and private

study within the preset program limitations. Please note the copyright notice above, and that any text or images printed or copied must credit the source."

9.4   This Product contains material proprietary to and copyedited by the Publisher and others. Except for the licence granted herein, all rights, title and interest in the Product, in all languages, formats and media throughout the world, including copyrights therein, are and remain the property of the Publisher or other copyright holders identified in the Product.

10.   Non-assignment

This Agreement and the licence contained within it may not be assigned to any other person or entity without the written consent of the Publisher.

11.   Termination and Consequences of Termination.

11.1   The Publisher shall have the right to terminate this Agreement if:

11.1.1 the Licensee is in material breach of this Agreement and fails to remedy such breach (where capable of remedy) within 14 days of a written notice from the Publisher requiring it to do so; or

11.1.2 the Licensee becomes insolvent, becomes subject to receivership, liquidation or similar external administration; or

11.1.3 the Licensee ceases to operate in business.

11.2   The Licensee shall have the right to terminate this Agreement for any reason upon two month's written notice. The Licensee shall not be entitled to any refund for payments made under this Agreement prior to termination under this sub-clause 11.2.

11.3   Termination by either of the parties is without prejudice to any other rights or remedies under the general law to which they may be entitled, or which survive such termination (including rights of the Publisher under sub-clause 7.4 above).

11.4   Upon termination of this Agreement, or expiry of its terms, the Licensee must destroy all copies and any back up copies of the product or part thereof.

12.   General

12.1   *Compliance with export provisions*
The Publisher hereby agrees to comply fully with all relevant export laws and regulations of the United Kingdom to ensure that the Product is not exported, directly or indirectly, in violation of English law.

12.2   *Force majeure*
The parties accept no responsibility for breaches of this Agreement occurring as a result of circumstances beyond their control.

12.3   *No waiver*
Any failure or delay by either party to exercise or enforce any right conferred by this Agreement shall not be deemed to be a waiver of such right.

12.4   *Entire agreement*
This Agreement represents the entire agreement between the Publisher and the Licensee concerning the Product. The terms of this Agreement supersede all prior purchase orders, written terms and conditions, written or verbal representations, advertising or statements relating in any way to the Product.

12.5   *Severability*
If any provision of this Agreement is found to be invalid or unenforceable by a court of law of competent jurisdiction, such a finding shall not affect the other provisions of this Agreement and all provisions of this Agreement unaffected by such a finding shall remain in full force and effect.

12.6   *Variations*
This agreement may only be varied in writing by means of variation signed in writing by both parties.

12.7   *Notices*
All notices to be delivered to: Spon's Price Books, Taylor & Francis Books Ltd., 3 Park Square, Milton Park, Abingdon, Oxfordshire, OX14 4RN, UK.

12.8   *Governing law*
This Agreement is governed by English law and the parties hereby agree that any dispute arising under this Agreement shall be subject to the jurisdiction of the English courts.

If you have any queries about the terms of this licence, please contact:

Spon's Price Books
Taylor & Francis Books Ltd.

Spon Press
an imprint of Taylor & Francis

Ebook options

Your print copy comes with a free ebook on the VitalSource® Bookshelf platform. Further copies of the ebook are available from https://www.crcpress.com/search/results?kw=spon+2021
Pick your price book and select the ebook under the 'Select Format' option.

To buy Spon's ebooks for five or more users of in your organisation please contact:

Spon's Price Books
eBooks & Online Sales
Taylor & Francis Books Ltd.
3 Park Square, Milton Park, Oxfordshire, OX14 4RN
Tel: (020) 337 73480
onlinesales@informa.com